Ruminant physiology

INRA – Institut National de la Recherche Agronomique

XIth International Symposium on Ruminant Physiology
Clermont-Ferrand, France - September 6-9, 2009

Ruminant physiology

Digestion, metabolism, and effects of nutrition on reproduction and welfare

edited by:
Y. Chilliard
F. Glasser
Y. Faulconnier
F. Bocquier
I. Veissier
M. Doreau

Proceedings of the XIth International Symposium on Ruminant Physiology

Wageningen Academic Publishers

ISBN: 978-90-8686-119-4

First published, 2009

The XIth International Symposium on Ruminant Physiology (ISRP) was organised in Clermont-Ferrand, France, September 6-9, 2009, by:

INRA (Institut National de la Recherche Agronomique)
UR1213 (Herbivore Research Unit)
Scientific Division for Animal Physiology and Livestock Systems
Centre de Recherches de Clermont-Ferrand-Theix
F-63122 Saint Genès Champanelle, France

The scientific programme was organised and the communications reviewed by the National Scientific Committee:

Y. Chilliard (INRA Clermont-Ferrand-Theix) (Chair, 'Metabolism' coordinator)[1]
M. Doreau (INRA Clermont-Ferrand-Theix) ('Digestion' coordinator)
F. Bocquier (Montpellier SupAgro) ('Nutrition-Reproduction' coordinator)
I. Veissier (INRA Clermont-Ferrand-Theix) ('Nutrition-Welfare' coordinator)
H. Archimède (INRA Antilles-Guyanne)
R. Baumont (INRA Clermont-Ferrand-Theix)
L. Bernard (INRA Clermont-Ferrand-Theix)
F. Blanc (ENITA Clermont-Ferrand)
A. Boudon (INRA Rennes)
P. Chavatte-Palmer (INRA Jouy-en-Josas)
J. Dupont (INRA Tours)
C. Duvaux-Ponter (AgroParisTech)
A. Ferlay (INRA Clermont-Ferrand-Theix)
J. Flament (INRA Rennes)
B. Graulet (INRA Clermont-Ferrand-Theix)
D. Gruffat (INRA Clermont-Ferrand-Theix)
S. Lemosquet (INRA Rennes)
C. Martin (INRA Clermont-Ferrand-Theix)
L. Mounier (ENV Lyon)
C. Philippeau (ENESA Dijon)
B. Picard (INRA Clermont-Ferrand-Theix)
G. Rychen (ENSAIA Nancy)
P. Schmidely (AgroParisTech)
A. Troegeler-Meynadier (ENV Toulouse)

In interaction with the International Scientific Committee:

A.W. Bell (Australia)
H. Ben Salem (Tunisia)
G. Bertoni (Italy)
Y.R. Boisclair (USA)
G. Breves (Germany)
P.J. Buttery (UK)
M.D. Carro (Spain)
Y. Chilliard (France)

[1] Who wishes to dedicate his activity for ISRP to Dr. R. Jarrige, Head of the "Station d'Elevage" (INRA Theix) during the previous 1979 ISRP symposium in Clermont-Ferrand, for his learning and his visionary views on ruminant physiology and husbandry.

J.L. Firkins (USA)
G.A. Gagliostro (Argentina)
J.K. Ha (South Korea)
G. Huszenicza (Hungary)
C.H. Knight (Denmark)
G.B. Martin (Australia)
M.O. Nielsen (Denmark)
P. Norgaard (Denmark)
H.V. Petit (Canada)
W.Y. Zhu (China)

On the practical level, the Symposium was organised by the Local Organising Committee[2]:

F. Glasser (Secretary General), Y. Faulconnier (Scientific Secretary), M. Auguergouz, P. Béraud, O. Bernard, A. Cornu, C. Espinasse, M. Doreau, S. Rudel, M. Teyssedre, Y. Chilliard (Chair), with the help of Clermont-Ferrand Tourist Centre, and of colleagues in charge of organising technical visits to INRA and to Saint Nectaire cheese-making farms (Y. Faulconnier, coordination, I. Cassar-Malek, Y. Chilliard, D. Durand, A. Farrugia, A. Ferlay, M. Hébraud, S. Lerch, C. Leroux, B. Martin, C. Martin, D. Micol, D. Morgavi, P. Nozière, B. Picard, D. Pomiès, E. Pujos-Guillot).

The Managing Editorial Committee was in charge of the proceedings:

W. Brand-Williams (coordinator, language editor), P. Béraud, M.H. Farce, N. Miraux, D. Montagu, D. Poirel.

The scientific sessions were chaired by:

D. Pethick (Australia), V. Fievez (Belgium), H.V. Petit (Canada), M.O. Nielsen (Denmark), G. Breves (Germany), G. Huszenicza (Hungary), R.J. Dewhurst (Ireland), R.J. Bessa (Portugal), M.D. Carro and P. Frutos (Spain), J.K. Ha (South Korea), P.J. Buttery, N.D. Scollan and R.J. Scaramuzzi (United Kingdom), J.L. Firkins (United States), and R. Baumont, A. Boudon, P. Chavatte-Palmer, M. Doreau, F. Enjalbert, J. Dupont, S. Lemosquet, C. Leroux, C. Martin, B. Picard, G. Rychen, D. Sauvant, P. Schmidely, A. Troegeler-Meynadier, I. Veissier (France)

[2] We thank our colleagues, organizers of previous symposia, for their advice and for having given us access to their files: the organizers of ISEP 2007 (in particular Dr I. Ortigues-Marty and J.-P. Brun), and INRA-RRI 2008 (in particular Dr C. Martin and Dr E. Forano).

Foreword

The XIth International Symposium on Ruminant Physiology (ISRP2009) takes place in Clermont-Ferrand, France, September 6 to 9, 2009. The previous meetings were held in Nottingham, UK (1960), Ames, USA (1964), Cambridge, UK (1969), Sydney, Australia (1974), Clermont-Ferrand, France (1979), Banff, Canada (1984), Sendai, Japan (1989), Willingen, Germany (1994), Pretoria, South Africa (1999), and Copenhagen, Denmark (2004).

More than 500 scientists from 48 countries attended ISRP2009. The countryside around Clermont-Ferrand is delightful in late summer and certainly contributes to create a stimulating and fruitful atmosphere for exchanging scientific information and deepening mutual friendships. A joint symposium on "Modelling digestion and metabolism in farm animals" was organised in Paris, September 10 to 12.

The programme of ISRP 2009 included 21 invited conferences that address ruminant comparative physiology, the rumen ecosystem and metagenomics, nutrient digestion and absorption, methanogenesis, tissue metabolism and gene expression, pregnancy, lactation and growth, adaptation to heat-stress, nitrogen use, nutrition and reproduction, nutrition and welfare and nutrition for sustainable ruminant production. The abstracts of these conferences are published in the following pages. Full papers will be published in the spring 2010 in a special issue of the scientific journal *Animal* (http://www.editorialmanager.com/animal/). Several invited speakers have associated recognised scientists from other international teams as co-authors of their presentation so that this symposium brings forth a wide and diversified approach of the topics chosen.

The ISRP2009 programme also includes 373 short communications, presented either orally (43) or as posters (330), whose abstracts are published in the present volume[3]. These communications come from research units in 49 countries[4] of all continents, showing a world-wide interest in ruminant nutrition and physiology.

As in previous ISRP meetings, the programme of ISRP 2009 covered a wide range of topics. The ultimate goal of the congress is to provide a comprehensive view of current knowledge and to draw perspectives on ruminant physiology to improve animal production efficiency, meat and milk quality, animal welfare and health, and to limit environmental impact (greenhouse gases, nitrogen use), with all these isssues being in line with the current challenges for sustainable animal breeding. The approaches chosen to address these challenges are also very diverse, including the development of physiological approaches, molecular high throughput technologies and modelling, in order to unravel the mechanisms involved in digestion, metabolism and the interactions between nutrition, reproduction and welfare.

The scientific and organising committees of ISRP2009 hope that the proceedings will be a tool and an inspiration for future progress in the field of ruminant nutrition and physiology. We would like to express our warm thanks to the many contributors in the process, especially to the organising committee members and to all scientists and technicians of the INRA Clermont-Ferrand/Theix Research Center, who made the organisation of the symposium a pleasant task; to the editing committee and chairpersons, whose expertise was essential to the publication and discussion

[3] The oral communications are presented at the beginning of each of the 4 thematic sections of this volume.

[4] France, 46, Australia, 31, China, 25, Iran, 25, Germany, 24, Japan, 18, United States, 18, Brazil, 17, Canada, 16, The Netherlands, 13, Spain,12, Denmark, 11, Belgium, 10, United Kingdom, 9, Mexico, 8, Sweden, 8, Finland, 7, New Zealand, 6, South Africa, 6, Italy, 5, Poland, 5, Thailand, 5, India, 4, South Korea, 4, Switzerland, 4, Uruguay, 4, Tunisia, 3, Uganda, 3, Croatia, 2, Indonesia, 2, Israel, 2, Norway, 2, Portugal, 2, Argentina, 1, Czech Republic, 1, Egypt, 1, Estonia, 1, Ghana, 1, Greece, 1, Hungary, 1, Ireland, 1, Kosovo, 1, Morocco, 1, Russia, 1, Saudi Arabia, 1, Slovenia, 1, Sudan, 1, Turkey, 1, Venezuela, 1.

of the scientific contributions; and to our sponsors[5]. Finally this congress would not have taken place without the special support from the INRA Scientific Division for Animal Physiology and Livestock Systems (PHASE).
A discussion was initiated in 2007 and continued during spring 2009 between the International Scientific Committee of ISRP and the International Advisory Committee of the International Symposium on Nutrition of Herbivores (ISNH) in order to combine their efforts in the future and make these two events even more complementary and thus more efficient. We hope this symposium will be an opportunity to deepen this discussion and eventually decide about the future of these two symposia.

On behalf of the scientific and organising committees of ISRP2009,

Y. Chilliard
F. Glasser
Y. Faulconnier
F. Bocquier
I. Veissier
M. Doreau

[5] Conseil Régional d'Auvergne, INRA Division 'Animal Physiology and Livestock Systems', INRA Scientific Direction 'Animal and Animal Products', LALLEMAND Animal Nutrition, ADISSEO France S.A.S., CCPA Group, INZO, CNIEL, TIMAB, PHYTOSYNTHESE, MERCODIA, Conseil Général du Puy-de-Dôme, DSM Nutritional Products France, DANISCO France, LIMAGRAIN, VALOREX, McKEY Food France, DANONE, Association des fromages d'Auvergne-PDO cheeses, Volvic.

Table of contents

Invited contributions

Evolutionary adaptations of ruminants and their potential relevance for modern production systems

M. Clauss[1], I.D. Hume[2] and J. Hummel[3]
[1]Clinic for Zoo Animals, Exotic Pets and Wildlife, Vetsuisse Faculty, University of Zurich, Switzerland; [2]School of Biological Sciences, University of Sydney, Australia; [3]Institute of Animal Science, Animal Nutrition Group, University of Bonn, Germany

Vertebrate herbivores cannot digest plant fibre auto-enzymatically but rely on gut microflora for this purpose. The efficiency of energy uptake, and hence the level of metabolism at which the organism can operate, is a function of many factors – including food intake level, digestive efficiency (itself a function of diet quality/selection, time available for digestion, and particle size), and nitrogen (and other nutrient) balance. Herbivores face the challenge that food intake level is mostly negatively correlated to factors determining digestive efficiency.
The major digestion types among mammalian herbivores – hindgut and foregut fermenters – represent different solutions to this challenge. Amongst these is a sorting mechanism facilitating nitrogen retention and coprophagy in small hindgut fermenters; in larger hindgut fermenters, microbial nitrogen from fermentation is presumably lost. Foregut fermenters use this microbial nitrogen, but are intake-limited because high intakes would translate into fast throughput of ingesta through the foregut, with incomplete fibre digestion but yet rather complete fermentation of easily digestible components that could be used more efficiently by auto-enzymatic digestion. It is the sorting mechanism in ruminants, which allows a differentially faster outflow of digested material from the rumen, and submits retained material to a digestion-enhancing second step of particle size reduction, that allows comparatively high food intake, and metabolism, in foregut fermenters. Comparative investigations in wild ruminants have increased our understanding of the forestomach sorting mechanism. Wild ruminants vary distinctively in the degree to which their rumen contents 'stratify' (with associated morphological adaptations); in particular, browsing species have rather homogenous rumen contents without separate gas and fluid layers. Yet all ruminants uniformly achieve efficient selective particle retention, suggesting that functions other than particle retention also played an important role in the evolution of stratification-enhancing adaptations. One interesting emerging hypothesis is that the high fluid turnover observed in cattle-type ruminants is an adaptation that not only leads to a shift of the sorting mechanism from the reticulum to the whole reticulorumen, but also optimises the harvest of microbial nitrogen from the forestomach.
In modern production, the major factor by which humans influence the efficiency of energy uptake is diet quality. Selective breeding for conversion efficiency has resulted in notable differences between wild and domestic animals, e.g. an increased gut volume in domestic ruminants. With increased knowledge on the relevance of individual factors, e.g. indicators of fluid throughput (saliva production, fluid turnover, omasum size), more specific selection parameters for breeding could be defined to increase productivity of domestic ruminants by continuing certain evolutionary trajectories.

Recent advances in metagenomics applied to ruminant gastrointestinal ecosystem

K.E. Nelson
Director of Human Microbiology, The J. Craig Venter Institute, MD 20850, Rockville, USA

Over the past few years, significant technological advances in the field of genomics have allowed for the advancement of the field of microbial metagenomics whereby we can now sequence DNA from any environment of choice without the need for culturing the microbial species that may be present in this environment. These metagenomic approaches have been applied to a range of environments including the human gastrointestinal tract (GIT), soils and the oceans, and all of these initial studies have revealed significant and unanticipated levels of microbial diversity. Cost-reductions associated with sequencing via the so-called 'next generation' technologies make it such that it is possible to interrogate these environments at a greater depth of sequence coverage. The metagenomics approaches have allowed for the identification of novel biochemical pathways in the oceans, the likely agents of gaining weight and other metabolic diseases in humans, and have allowed for the identification of novel antimicrobials in soils.

Initial whole genome sequencing projects have been completed for the rumen isolates *Ruminococcus albus*, *Fibrobacter succinogenes*, and *Prevotella ruminicola*, and a number of suppressive subtractive hybridisation (SSH) papers that have studied the complete genomes of closely related isolates have been presented by groups in North America. The technological developments described above have spread to metagenomics investigations of animals used in food production, inclusive of chickens, sheep and cattle. The significance lies in the promise that metagenomics holds for increasing the understanding of the diversity of species present in the rumen of both dairy and beef production animals, how these species can be manipulated to improve feed efficiency and decrease the emission of green house gases, and what is their potential use as fuel cells. What are the species that are associated with fiber degradation, and can the enzyme systems that they encode be applied to other environments for creation of renewal energy. Although the ruminant GIT system has been examined extensively using culture-based and 16S rRNA phylogenetic analyses, metagenomics holds much promise for advancing the field.

White and colleagues at The University of Illinois have published a series of papers where they have used SSH, Sanger sequencing, as well as next generation pyrosequencing sequencing and data analysis to investigate the rumen metagenome. Other groups have explored the rumen metagenome for enzymatic activities to reveal novel cyclodextrinases, polyphenol oxidases, and hydrolases. The metagenomic data that is being generated from these analyses is revealing extensive diversity, and the potential for enzymatic systems could possibly be used in industrial settings. The current developments in the field of metagenomics and the promise that it holds will be described in this presentation.

Gene expression in the digestive tissues of ruminants and their relationships with feeding and digestive processes

E.E. Connor, R.W. Li, R.L. Baldwin, VI and C. Li
U.S. Department of Agriculture-Agricultural Research Service, Bovine Functional Genomics Laboratory, MD 20705, Beltsville, USA

The gastrointestinal (GI) tract has multiple functions including digestion, nutrient absorption, secretion of hormones, and excretion of wastes. In the ruminant animal, development of this organ system is more complex than that of the monogastric animal due to the necessity to establish a fully functional and differentiated rumen, in which a diverse microbial population of bacteria, fungi and protozoa support fermentation and digestion of dietary fiber. Central to the goal of animal scientists to enhance nutrient uptake and production efficiency of ruminants is the need for a comprehensive understanding of GI development, as well as conditions that alter the digestion process. The relatively recent availability of genome sequence information has permitted physiological investigations related to the process of digestion for many agriculturally-important species at the gene transcript level. For instance, numerous studies have evaluated the expression of ruminant GI tract genes to gain insight into mechanisms involved in normal function, physiology, and development, such as nutrient uptake and transport across the epithelial cell barrier throughout the alimentary canal, maintenance of rumen pH, and regulation of GI tract motility, and cell proliferation. Further, multiple studies have examined the effects of dietary modification, including feeding of supplemental fat, starch and protein, or a forage- versus concentrate-based diet on expression of critical gene pathways in the gut. In addition, the expression of genes in the GI tract in response to disease, such as infection with GI parasites, has been investigated. This review will summarize some of the recent scientific literature related to gene expression in the GI tract of ruminants, primarily cattle, sheep and goats, as it pertains to (1) normal physiology, (2) dietary effects, (3) developmental effects, and (4) disease effects to provide an overview of critical proteins participating in the overall digestive processes, and their physiological functions. Recent findings from our laboratory will be highlighted also related to expression of the glucagon-like peptide 2 hormone pathway in the GI tract of dairy cattle during various stages of development and lactation, alterations in gene pathways associated with rumen development and differentiation in the weaning calf, and genes of the GI tract responding to Ostertagia, a common nematode infection of cattle. Finally, prospective areas of investigation will be discussed.

The role of microbes in rumen lipolysis and fatty acid biohydrogenation

R.J. Wallace
University of Aberdeen Rowett Institute of Nutrition and Health, Bucksburn, AB21 9SB, Aberdeen, United Kingdom; john.wallace@abdn.ac.uk

Despite the fact that the ruminant diet is rich in polyunsaturated fatty acids (PUFA), ruminant products – meat, milk and dairy – contain mainly saturated fatty acids (SFA) due to bacterial lipolysis and subsequent biohydrogenation of ingested PUFA in the rumen. The link between SFA consumption by man and coronary heart disease is well established. On the other hand, ruminant products also contain fatty acids that are known to be beneficial to health, namely conjugated linoleic acids (CLA). The aims of research in this field have been to understand the microbial ecology of lipolysis and biohydrogenation. During the course of the research, a number of findings have emerged that seem anomalous and paradoxical. Some of these are discussed here. Intuitively, it may appear that inhibiting ruminal lipase would cause more dietary PUFA to reach the mammary gland. However, lipolysis releases the nonesterified fatty acids that form the substrates for biohydrogenation, but which can, if they accumulate, inhibit the whole process. Thus, increasing lipase activity could be beneficial if the increased release of non-esterified PUFA were inhibitory. Rumen ciliate protozoa do not carry out biohydrogenation, yet protozoal lipids are much more highly enriched in CLA than bacterial lipids. How could this happen if protozoa do not metabolize PUFA? The presence of the unmodified PUFA may be explained by the engulfment of chloroplasts, which protects their constituent PUFA from bacterial lipolysis and biohydrogenation. But why biohydrogenation intermediates, the result of bacterial not protozoal activity, also accumulate in protozoa is much less clear. Linoleate isomerase has proved elusive both to purify and to clone. Enzyme reactions carried out in deuterium oxide solution indicate that one reason for these failings may be that the enzyme is a radical enzyme that requires a cofactor for activity. Also puzzling is which bacteria produce *trans*-10,*cis*-12-CLA, an isomer produced occasionally that leads to metabolic dysfunction and milk fat depression in cows. Only *Megasphaera elsdenii* and *Propionibacterium acnes* have been isolated as *trans*-10,*cis*-12-CLA producers, yet their numbers do not seem compatible with the '*trans*-10 shift'. The only ruminal bacterium isolated thus far that is able to form stearic acid (18:0) is *Butyrivibrio proteoclasticus*, the re-named *Clostridium proteoclasticum*, which in turn is the same organism isolated decades ago and named *Fusocillus*. New data will be presented in which low correlations between rumen bacterial community structure, particularly numbers of *B. proteoclasticus*, correlate poorly with fatty acid composition in digesta flowing from the rumen to the duodenum and in milk. Do we misunderstand the role of different bacterial species in biohydrogenation? Or are there uncultivated species that we need to understand and include in the analysis?

Microbial ecosystem and methanogenesis in ruminants

D.P. Morgavi[1], E. Forano[2], C. Martin[1] and C.J. Newbold[3]
[1]INRA UR1213 Herbivores, Site de Theix, 63122, Saint-Genès-Champanelle, France; [2]INRA UR454 Microbiologie, Site de Theix, 63122, Saint-Genès-Champanelle, France; [3]Department of Rural Sciences, The University of Wales, Aberystwyth, SY23 2AX, Ceredigion, United Kingdom

Cattle production is under increased public pressure because ruminants are major producers of the greenhouse gas methane. Methanogenesis is performed by a specialised group of microbes present in several anaerobic environments: the methanogenic archaea. In the rumen, methanogens utilise predominantly hydrogen and carbon dioxide as substrates to produce methane, filling an important functional niche in the ecosystem. The ecology and genomics of rumen methanogens is an area of active research. However, in addition to methanogens, other microbes also have an influence on methane production either because they are involved in hydrogen metabolism or because they affect methanogen numbers or other members of the microbiota. This review explores the relationship existing between some of these microbes and methanogenesis and highlights some functional groups that could play a role in reducing methane emissions.
Hydrogen is a key element that drives methane production in the rumen. Among hydrogen producers, protozoa have a preponderant position, which is strengthened by their close physical association with methanogens that favours hydrogen transfer. There is a causal relationship between protozoa and methane emissions and because this group is not essential for the normal development of the animal, protozoa might be a target for methane mitigation. An important function that is associated with production of hydrogen is the degradation of fibrous plant material, which is performed by a specialised group of microbes. However, not all members of the rumen fibrolytic community produce hydrogen. Increasing the proportion of non-hydrogen producers might decrease methane without affecting forage degradability.
Alternative pathways of hydrogen utilisation also exist in the rumen. Reductive acetogenesis, the conversion of hydrogen and carbon dioxide to acetate, has received much attention as it is the predominant mechanism of hydrogen elimination in the gastrointestinal tract of many herbivorous animals. Other bacteria are also capable of using electron acceptors other than CO_2 to oxidise hydrogen. These bacteria normally occupy a distinct ecological niche and they are not dominant members of the microbiota but their numbers can increase if the right potential electron acceptor is present in the diet. Nitrate and sulphate are alternative electron sinks that can promote the growth of these particular bacteria, although the reduced end products of these compounds, if they accumulate, can rapidly attain toxic levels. Fumarate-reducing bacteria, which use hydrogen to reduce fumarate to succinate, are also an alternative H_2 scavenging pathway. Methanotrophy, i.e. the oxidation of methane, is another metabolic pathway that exists in the rumen that has yet to be quantified *in vivo*.
Methanogens in the rumen coexist with other microbes which have contrasting activities. A better understanding of these populations and the pathways that compete with methanogenesis may provide novel targets for emissions abatement in ruminant production.

Transport of cations and anions across forestomach epithelia

S. Leonhard-Marek[1], F. Stumpff[2] and H. Martens[2]
[1]Department of Physiology, School of Veterinary Medicine Hannover, Germany; [2]Department of Veterinary Physiology, Free University of Berlin, Germany;
sabine.leonhard-marek@tiho-hannover.de

An efficient digestion of plant material by ruminants requires a large community of microorganisms within the forestomachs, with ample secretion of saliva ensuring an optimal environment. Daily salivary secretion of sodium (Na^+) exceeds that found in plasma while secretion of bicarbonate (HCO_3^-) is even higher. This implies the provision of efficient absorptive mechanisms across forestomach epithelia to allow for an early recycling. While Na^+ is absorbed from all forestomachs, HCO_3^- is secreted by the rumen but absorbed by the omasum to prevent liberation of CO_2 in the abomasum. Fermentation provides additional cations and anions that have to be absorbed to meet nutrient requirements (short chain fatty acids, SCFA) and to supply essential minerals (Mg^{2+}). Others can be absorbed, if delivered in high amounts (Ca^{2+}, P_i, K^+, Cl^-, NH_4^+). Throughout, the requirements of ruminal homeostasis and systemic acid/base and potassium balance must be met. While the presence of transport mechanisms for these electrolytes has been described previously, our knowledge about their nature, regulation and crosstalk has increased greatly in the last years. In the rumen, electroneutral absorption of Na^+ is performed via Na^+/H^+ exchange (NHE) and can be stimulated at acidic pH, with intracellular protons being provided by dissociation of SCFA, NH_4^+ and conversion of CO_2. Immunohistological presence of NHE3 and H^+-ATPases in the upper layers of the stratified rumen epithelium confirm functional data showing that luminal H^+ extrusion prevents an overwhelming acidification of the epithelium. However, at alkaline pH or high ruminal NH_3, uptake via NHE is low and absorption of Na^+ must be maintained by an additional pathway. Electrogenic influx of Na^+ occurs via apical non-selective cation channels which show increased conductance at low concentrations of Mg^{2+} or Ca^{2+} or H^+ in the luminal microclima and at low intracellular Mg^{2+} (Mg_i). A decrease in Mg_i may also cause an increased Na^+ absorption at high ruminal K^+ concentrations. Intracellular cAMP (due to prostaglandins or ß-agonists) decrease the activity of apical NHE and enhance the electrogenic uptake of Na^+ via increased activity of basolateral Na^+/Mg^{2+} with reduction of Mg_i. Basolateral extrusion of Na^+ occurs via Na^+/K^+-ATPase.
Different anion transporters have been shown on the mRNA level in rumen and omasum that are linked to functional apical exchange of HCO_3^-, Cl^- and the anions of $SCFA^-$. Basolaterally, these anions permeate a large conductance anion channel. Larger anions leave via monocarboxylate transport (MCT1). NHE1 and Na^+-HCO_3^- cotransporters contribute to pH_i regulation in basal epithelial layers.
Na^+ transport in the omasum is similar and involves both NHE and a Ca^{2+} and Mg^{2+} sensitive Na^+ conductance. An additional apical Na^+-Cl^- cotransport creates the gradients necessary for the absorption of HCO_3^- via Cl^-/HCO_3^- exchange. High individual variations in basal and gradient induced HCO_3^- absorption may explain why ruminants on the same diet show different susceptibility to abomasal gas accumulation and displacement.

Carbohydrate quantitative digestion and absorption in ruminants: from feed starch and fiber to nutrients available for tissues

P. Nozière[1], I. Ortigues-Marty[1], C. Loncke[1] and D. Sauvant[2]
[1]INRA UR 1213 Herbivores, Site de Theix, 63122, Saint-Genès-Champanelle, France; [2]AgroParisTech-INRA, 16 rue Claude Bernard, 75231, Paris, Cedex 05, France

Carbohydrates are the main source of energy in ruminants. Their site, extent, and kinetics of digestion highly impact the amount and profile of nutrients delivered to peripheral tissues, and the responses of the animal, i.e. ingestion, efficiency of performances, N and methane excretion, quality of products, and welfare. Development of multi-objectives feed evaluation systems thus requires a more integrated quantitative knowledge on carbohydrates digestion and yield of terminal products, as well as on their metabolism by splanchnic tissues. The objective of this paper is to review (1) quantitative knowledge on fiber, starch and sugar digestion, volatile fatty acids and glucose production and splanchnic metabolism, and (2) modelling approaches which aim at representing and/or predicting nutrient fluxes in digestive tract, portal and hepatic drainage. The first part of this review describes mechanisms and variations factors involved in site, extent, and rate of carbohydrates digestion, as well as in production, absorption and splanchnic metabolism of VFAs and glucose. Digestive mechanisms are more briefly presented, since they have already been extensively reviewed. Portal drained-viscera (PDV) and liver metabolism have been less studied, but the amount of published results is substantial. The need of quantitative integration is enlightened. The second part of this review is focused on published models of digestion and splanchnic metabolism, including both mechanistic and empirical approaches. A qualitative comparison of published mechanistic models of rumen digestion is presented. It shows that the representation of carbohydrate digestion and VFA yield is relatively homogeneous among models. Although published quantitative comparisons of these models are scarce, they stress that prediction of fiber digestion and VFA yield is still not good enough for use in feed formulation, whereas prediction of microbial N yield and starch ruminal digestion seems to be more satisfactory. Response of microbial metabolism to digestive interactions, as well as VFA stoichiometric coefficients and absorption rates may partly explain the poor predictions. Hardly any mechanistic models have been developed on PDV metabolism whereas a few exist for liver metabolism. A qualitative comparison of these models is presented. Most are focused on dairy cows and their level of aggregation in the representation of nutrient fluxes and metabolism highly differs depending on their objectives. Quantitative comparison of these models is still lacking. However, recent advances have been achieved with the empirical prediction of VFA and glucose production and fluxes through PDV and liver based on the current INRA feed evaluation system. These advances are presented. They illustrate that empirical prediction of ruminal VFA and intestinal glucose production can be evaluated by comparison with measured portal net fluxes measurements. We also illustrate the potential synergy between empirical and mechanistic modelling. It is concluded that concomitant empirical and mechanistic approach may likely help to progress towards development of multi-objectives feed evaluation systems based on nutrient fluxes.

Nutritional regulation of foetal growth and implications for productive life in ruminants

M.E. Symonds
Centre for Reproduction and Early Life, Institute of Clinical Research, University Hospital, NG7 2UH, Nottingham, United Kingdom

The maternal nutritional and metabolic environment is critical in determining not only reproductive success but long term health and viability of the offspring. Changes in maternal diet at defined stages of gestation coincident with different stages of development can have pronounced effects on organ and tissue function in later life. These are dependent in part on the particular stage of organogenesis targeted although another important factor is the extent to which postnatal growth is reset.

One early, critical window of organ development in the ruminant relates to the period covering uterine attachment, or implantation, and rapid placental growth. During this period there is pronounced cell division within developing organelles leading to their structural development, in many fetal tissues. In sheep, we have shown that a 50% global reduction in caloric intake over this specific period profoundly effects placental growth and morphology, resulting in reduced placentome weight. This occurs in conjunction with a lower capacity to inactivate maternal cortisol through the enzyme 11β-hydroxysteroid dehydrogenase type 2 that occurs in response to a decrease in maternal plasma cortisol. Placental gene expression of both, the glucocorticoid receptor, and the mitochondrial protein, uncoupling protein 2, are enhanced and this may contribute, in part, to the accompanying reduction in placental cell proliferation following nutrient restriction. Birth weight of the offspring is, however, unaffected by this dietary manipulation and although they possess more fat this adaptation does not persist into adulthood even when made obese. Then after birth further changes in fat development occur that impact on both glucocorticoid action and inflammatory responses.

When previously, *in utero*, nutrient restricted offspring are exposed to an obeseogenic environment they exhibit an amplified insulin response that is accompanied by a range of amplified and thus, adverse, physiological or metabolic responses following obesity. These types of adaptations are in marked contrast to the effect of late gestational nutrient restriction that results in reduced fat mass at birth. As young adults, however, fat mass is increased and these offspring are insulin resistant although basal insulin is unaffected, adaptations that are dependent on maternal parity.

In conclusion, changes in nutrient supply to either the mother and/or her fetus can have profound effects on a range of metabolically important tissues. These have the potential to either exacerbate or protect the resulting offspring from the adverse effects of later obesity and accompanying complications.

Adipose tissue and muscle growth interactions in cattle

M. Bonnet, I. Cassar-Malek, Y. Chilliard and B. Picard
INRA, UR1213 Herbivores, Site de Theix, 63122, Saint-Genès Champanelle, France; m uriel. bonnet@clermont.inra.fr

Producing meat animals with adequate muscular and adipose masses (i.e. lean-to-fat ratio) is an economic challenge for the beef industry. The lean-to-fat ratio is the result of a dynamic balance between the number and size of muscular and adipose cells respectively. Understanding how rearing practices affect the mechanisms governing this balance has implications for the quality and value of the carcass and the meat from cattle.

Muscular and adipose masses grow by an increase in the number of cells (hyperplasia) mainly during foetal life in cattle. The total number of muscle fibres is set by the end of the second trimester of gestation. It results from the successive proliferation and differentiation of the primary and secondary generations of fibres. The molecular pattern of developing bovine muscle cells has been studied using genomic tools. This reveals that the total number of fibres is controlled by a balance between proliferation and apoptosis, and that complex modifications in proportions of protein isoforms govern contractile and metabolic differentiation during the last third of gestation. Conversely, the number of adipocytes is set by birth or by early adulthood, depending on the anatomical location of the adipose tissue. Hyperplasia concerns brown adipocytes during foetal life and white adipocytes from a few weeks after birth. Molecular and histological data first suggested a possible transformation of brown into white adipocytes. However, the recent characterisation of distinct precursor cells for brown and white adipocytes in non-ruminants suggests that both types of adipocytes may participate in the growth of the adipose tissue. The molecular events involved in fat cell proliferation remain to be unravelled. Recent investigations have established that muscle and brown fat cells originate from the same precursor cells, highlighting a balance in the fate of these cell lineages.

Increased nutrient storage in fully differentiated muscle fibres or adipocytes, resulting in cell enlargement (hypertrophy), is thought to be the main mechanism whereby muscular and fat masses increase in growing cattle. Competition or prioritisation between adipose and muscular cells for the uptake and metabolism of nutrients is suggested by the inhibited or delayed adipose tissue growth in bovine genotypes exhibiting strong muscular development. Such competition or prioritisation occurs through cellular signalling pathways and the secretion of proteins by adipose tissues (adipokines) and muscle (myokines), putatively regulating their hypertrophy in a reciprocal manner. These mechanisms may be altered by feeding strategies and rearing practices.

Despite the remarkable insights gained on the hyperplasia and hypertrophy of adipose and muscular cells, our understanding of how cattle rearing practices may affect lean-to-fat ratio is still limited. Studies of the mechanisms underlying the 'adipose-muscular' cross-talk are fertile areas for future investigations directed to proposing new rearing practices for the sustainable production of carcasses and meat with an optimised lean-to-fat ratio.

The relationship between energy intake and efficiency of energy utilisation in lactating ruminants

B.J. Tolkamp
SAC, Animal and Health Department, West Mains Road, Edinburgh EH9 3JG, United Kingdom; bert.tolkamp@sac.ac.uk

Variation in feed net energy (NE) intake is not only one of the determinants of domestic ruminant productivity but is, in ecology, recognised as one of the defining contributions to variation in long-term (i.e. lifetime) inclusive fitness. Current models that predict energy intake in lactating ruminants all assume that this long-term contribution can be captured by some short-term 'proxy' of long-term fitness. In this proxy, it is generally assumed that lactating ruminants are short-term energy intake rate maximisers (at least up to the level of perceived requirements) and that variation in energy intake is a direct result of one or more constraints that find their origin in characteristics of the animal and the feed. These models predict energy intake in lactating ruminants on the basis of descriptions of the feed and of the animal but ignore known effects of feed and animal characteristics on the efficiency with which animals utilise feed metabolisable energy (ME) for NE. The same feed and animal characteristics that contribute to variation in energetic efficiency, however, are also associated with variation in short-term energy intake. Models have been developed that predict energy intake of non-reproducing ruminants on the basis of variation in energetic efficiency caused by differences between feeds and between animals. The recent literature has been reviewed to extend this approach to lactating ruminants. The current paper highlights the evidence that is in favour of an approach that directly links energetic efficiency with voluntary energy intake in lactating ruminants. This will not only lead to considerable simplifications of the energy evaluation systems for ruminants that are currently in use but also offers the opportunity to predict variation in voluntary energy intake of lactating ruminants by models that form a direct link between the short-term 'proxy' and its long-term fitness consequences.

Genomics of metabolic adaptations in the peri-partum cow

J.J. Loor
Mammalian NutriPhysioGenomics, Department of Animal Sciences and Division of Nutritional Sciences, University of Illinois at Urbana-Champaign, Urbana, 61801 IL, USA

The *peripartum* period is characterised by dramatic alterations in metabolism and function of key tissues such as liver, adipose, and mammary. Metabolic regulation in complex organisms relies partly on transcriptional control of gene networks as a long-term mechanism affecting the level of expression of several key enzymes. A cellular gene network can be defined as a collection of DNA segments which interact either with a regulator such as a transcription factor or nuclear receptor, but also with each other through their RNA and protein products and with other molecules in the cell. These 'global' interactions in a tissue can govern the rates at which genes in the network are transcribed into mRNA. Therefore, elucidating the gene networks driving adaptations from the transition between the non-lactating period, parturition, and established lactation is of high priority. Development of cattle-specific high-throughput transcriptomics technologies coupled with genome sequencing and annotation has made this undertaking achievable. Use of bovine DNA microarrays on liver, adipose, and mammary tissue from *peripartum* dairy cows already has demonstrated the potential of transcriptomics for identifying clusters of genes involved in regulating and coordinating function and crosstalk among tissues. Work examining the role of *pre partum* level of dietary energy has revealed unique clusters encompassing functional categories including signal transduction, cell-to-cell signaling, molecular transport, insulin signaling, lipid metabolism (synthesis and beta-oxidation), immune or inflammatory processes, and cell death in subcutaneous adipose as well as liver. The adipose transcriptome at 2 wk *pre partum* from cows overconsuming dietary energy (ca. 150% of estimated requirements) during the dry period is characterised by upregulation of lipogenic transcription regulators (e.g. CEBPB, CEBPA, THRSP) and their putative target genes (e.g. DGAT2, SCD), adipokines (LEP, ADIPOQ), and also differential expression of immune-related proteins (e.g. IFNB1, CCL5). Analysis of the liver transcriptome under the same dietary management suggests an alteration of immune and inflammatory responses (e.g. GPX3, SAA1), which could have implications on the ability of animals to make a smooth transition into lactation. Major advances in understanding the metabolic adaptations of the *peripartum* cow will come from coupling existing knowledge of enzyme kinetics, biochemistry, and hormone action with transcriptomics, proteomics, and metabolomics. Using a systems biology approach to integrate data generated at the mRNA, protein, metabolite, and tissue level can allow the assembly of the important components needed to model the *peripartum* cow. Such models could be useful in determining how and why we can manipulate complex processes that could have significant long-term economic impact including lactation persistency, fertility, and efficiency of conversion of feed to milk. An important goal of the future will be to apply additional experimental tools (e.g. gene silencing, chromatin immunoprecipitation) and bioinformatics (e.g. transcription factor binding site identification) to studies focused on periparturient cows.

Trans fatty acids and mammary lipogenesis in ruminants

K.J. Shingfield[1], L. Bernard[2], C. Leroux[2] and Y. Chilliard[2]
[1]MTT Agrifood Research Finland, FIN-30100, Jokioinen, Finland; [2]INRA, UR1213 Herbivores, Site de Theix, 63122, Saint-Genès-Champanelle, France; kevin.shingfield@mtt.fi

Fat is an important constituent contributing to the organoleptic, processing and physical properties of ruminant milk. Understanding the regulation of milk fat synthesis in ruminants is important for enhancing the nutritional value of milk and development of nutritional strategies to reduce milk energy secretion and improve energy balance. Nutrition is the major environmental factor influencing both the concentration and composition of fat in ruminant milk. Feeding low fibre/high starch diets and/or lipid supplements rich in polyunsaturated fatty acids are known to induce milk fat depression (MFD) in the bovine, typically increase milk fat secretion in the caprine, whereas responses in the ovine tend to be intermediate.

The role of diet on the regulation of mammary lipogenesis has been extensively investigated, and a number of hypotheses have been proposed to explain diet-induced MFD. Early considerations attributed MFD to reductions in substrate supply and/or changes in insulin secretion. Following the observation that reductions in milk fat synthesis during diet-induced MFD are associated with increases in the concentration of specific trans fatty acids (TFA) in milk, the biohydrogenation theory of MFD was proposed which attributes the causal mechanism to changes in ruminal lipid metabolism leading to increased formation of specific biohydrogenation intermediates that exert anti-lipogenic effects. *Trans*-10,*cis*-12 conjugated linoleic acid (CLA) formed during ruminal 18:2n-6 metabolism is the only biohydrogenation intermediate shown unequivocally to inhibit milk fat synthesis in ruminant species. However, increases in ruminal *trans*-10,*cis*-12 CLA formation do not explain entirely diet-induced MFD, suggesting that other biohydrogenation intermediates and/or other mechanisms may also be involved. Post-ruminal infusion experiments have provided evidence that *cis*-10,*trans*-12 CLA, *trans*-9,*cis*-11 CLA and *trans*-10 18:1 in high amounts may also exert anti-lipogenic effects.

More recent investigations exploiting molecular-based approaches have shown that mammary abundance of transcripts encoding for key lipogenic genes are reduced during MFD in the bovine, changes that are accompanied by decreases in sterol response element binding protein 1 (SREBP-1, a known transcription factor regulating lipogenesis) and SREBP-1 activation proteins. Furthermore, there is emerging evidence indicating that the expression of key regulators of lipid synthesis is increased in adipose of cows during MFD. Feeding diets of comparable composition do not induce MFD or significantly alter mammary lipogenic gene expression in the goat. Available evidence suggests that variation in mammary lipid secretion and lipogenic responses in the cow and goat may reflect inherent inter-species differences in ruminal lipid metabolism and/or mammary specific regulation of cellular processes and the relative roles of key lipogenic enzymes in the synthesis of milk fat triacylglycerides.

Metabolic and hormonal adaptations to heat stress in ruminants

U. Bernabucci[1], N. Lacetera[1], L.H. Baumgard[2], R.P. Rhoads[2], B. Ronchi[1] and A. Nardone[1]
[1]Dipartimento di Produzioni Animali, Università degli Studi della Tuscia, 01100-Viterbo, Italy;
[2]Department of Animal Sciences, The University of Arizona, AZ 85721, Tucson, USA;
bernab@unitus.it

Environmentally-induced periods of heat stress decrease productivity with devastating economic consequences to global animal agriculture. Heat stress can be defined as a physiological condition when the core body temperature of a given species exceeds its range specified for normal activity resulting from a total heat load (internal production and environment) exceeding the capacity for heat dissipation and this prompts physiological and behavioural responses to reduce the strain. The ability of ruminants to maintain homeothermia is species and breed dependant. For a variety of reasons, dairy breeds are typically more sensitive to heat stress than beef, and higher producing animals are presumably more susceptible to experience heat stress because they generate more metabolic heat.

During heat stress, ruminants, like other homeothermic animals, increase avenues of heat loss and reduce heat production in an attempt to remain euthermic. The immediate response to a heat load is increased respiration rates, decreased feed intake, and increased water intake. Acclimatisation is a process by which animals adapt to environmental conditions and engage behavioural, hormonal and metabolic changes that are characteristic of homeorhetic mechanisms utilised by the animals to survive in a new physiological state. For example, alterations in the hormonal profile are mainly characterised by a decline and increase in anabolic and catabolic hormones, respectively. The immediate response to heat load and the heat-induced change in homeorhetic modifiers alters post-absorptive energy, lipid and protein metabolism, impairs liver function, causes oxidative stress, jeopardises the immune response and decreases reproductive performance. These physiological modifications alter nutrient partitioning and may prevent the heat-stressed lactating cow from enlisting glucose sparing mechanisms (despite the reduced nutrient intake) and this is in large part why decreased feed intake only accounts for a minor portion of the reduced milk yield from environmentally-induced hyperthermic cows. How these metabolic changes are initiated and regulated is not clear. It also remains unclear how these changes differ between short-term versus long-term heat acclimation to impact animal productivity and well-being. A better understanding of adaptations enlisted by ruminants during heat stress periods is necessary to enhance the likelihood of developing strategies to simultaneously improve heat tolerance and increase productivity.

Strategies for optimising nitrogen use by ruminants: digestive and metabolic mechanisms

S. Calsamiglia[1], A. Ferret[1], C.K. Reynolds[2], N.B. Kristensen[3] and A.M. van Vuuren[4]
[1]Dpt. Ciencia Animal i dels Aliments, Universitat Autonoma de Barcelona, 08193-Bellaterra, Spain; [2]Department of Agriculture, University of Reading, Earley Gate, RG6 6AR, Reading, United Kingdom; [3]Faculty of Agricultural Sciences, Aarhus University, 8830, Tjele, Denmark; [4]Wageningen UR Livestock Research, P.O. Box 65, 8200 AB, Lelystad, the Netherlands

The efficiency of N utilisation in ruminants is typically low and highly variable (13 to 32%) compared with the higher efficiency in other production animals. The low efficiency has implications for the production performance and for the emission of contaminants to the environment. In the last several decades, many efforts have been devoted to improving the efficiency of N utilisation in ruminants, and while major improvements in our understanding of N requirements have been achieved, the overall efficiency remains low. The ruminant can be regarded as one link within the N cycle, besides soil and plants. Reducing N emission from this cycle by reducing N input through fertiliser and feed supplements will also reduce feed quality and N supply to the animal. The reduced N supply will reduce animal performance, although the overall efficiency of N utilisation will increase. To stop this vicious circle, efforts directed towards improving the efficiency of N capture in the rumen (amount of N captured by bacteria as a percentage of rumen degradable N) should continue. Modification of rate and extent of protein degradation and energy fermentation in the rumen has been traditionally approached by modifying the feed (chemical or temperature treatments). Attempts to achieve this objective by modifying the rumen microflora involved in protein degradation and amino acid deamination needs to be addressed. The effect of some additives on specific bacterial groups involved in protein degradation and deamination processes as well as the potential use of specific polyclonal antibodies against these specific bacterial groups are alternatives currently being studied. In addition, approaches that improve the efficiency with which absorbed amino acids are retained in milk and meat, without compromising overall efficiency of production, should be pursued. Whilst there is scope for improving the postabsorptive efficiency of N utilisation through genetic selection and other technologies, the inability to predictably alter postabsorptive essential amino acid supply relative to requirements for production remains an impediment to the development of practical feeding approaches that reduce N excretion in ruminants. However, the ability of the ruminant to recycle metabolic N to the rumen has promise as a mechanism by which to improve the efficiency of N utilisation in ruminants.

Nutritional sub-fertility in the dairy cow: towards improved reproductive management through a better biological understanding

N.C. Friggens[1], C. Disenhaus[2] and H.V. Petit[3]
[1]University of Aarhus, Faculty of Agricultural Sciences, Research Centre Foulum, 8830 Tjele, Denmark; [2]Agrocampus Ouest UMR 1080, 65 Rue St Brieuc, 35042, Rennes, France and INRA UMR 1080, 35570, Saint-Gilles, France; [3]Agr & Agri-Food Canada, Dairy & Swine Res & Dev Ctr, QC J1M 1Z3, Sherbrooke, Canada; n.friggens@agrsci.dk

The reproductive performance of dairy cattle has declined in recent decades. Cows are taking longer to return to oestrus after calving, have poorer conception rates, and show fewer behavioural signs of oestrus. Achieving a good level of reproductive performance is a major, and increasing, challenge for the dairy producer. Recently, new tools for monitoring (e.g. in-line milk progesterone) and nutritional methods (e.g. lipid sources) for manipulating reproductive status have become available. However, the effective use of such reproductive management aids depends upon them being deployed in the appropriate biological context, i.e. working with the cow rather than against her. In this paper we focus on understanding the overall patterns of the phenomena relating nutrition and fertility rather than the underlying multiplicity of physiological interactions.These patterns are important because they represent the natural adaptations of the animal for dealing with variations in the environment. We show how they can be used to monitor and modulate reproductive performance on-farm.
Selection for milk production, one aspect of reproduction has decreased conception rates (another aspect of reproduction). There appears to be an underlying trade-off between two aspects of reproduction: investment in the viability of the current calf and investment in future offspring. As the viability of the current calf is related to maternal milk production, we can expect that level of milk production *per se* has effects on subsequent reproductive performance. Indeed, embryo quality is reduced in high- compared to medium-genetic merit cows. Moreover, milk production on the day of oestrus affects oestrus expression. Another important phenomenon relating nutrition to reproduction is the adaptive use of body reserves in support of reproduction. There are orchestrated endocrine changes in pregnancy and lactation that facilitate deposition of body lipid during pregnancy and mobilisation in early lactation. We argue that both rate of body fat mobilisation and level of body fatness act as signals for reproduction. Excessive body fat mobilisation indicates that current conditions are worse than expected. Body fatness indicates the cow ability to safeguard future reproductive function, with thinness indicating dependency on environmental availability of energy. Both conditions delay further reproductive commitment.
The relationship between – milk production as an index of maternal investment, body fatness as an index of ability to safeguard reproductive investment, and body fat mobilisation as an index of the current nutritional environment – and reproductive performance is examined. Examples of incorporating these relationships into (1) models to monitor reproductive status using in-line progesterone profiles (2) nutritional strategies modulating body mobilisation by the use of glucogenic and lipogenic diets are presented.

Nutrition and reproduction in the male ruminant in natural or artificial reproductive management

G.B. Martin
UWA Institute of Agriculture M082, University of Western Australia, 6009, Crawley, Australia

In mature males, changes in feed intake seem to have little effect on gonadal function in bulls, but induce profound changes on sperm production in rams and bucks. These outcomes are due to changes in the size of the seminiferous tubules and in spermatogenic efficiency. Except with severe under-feeding, there are only minor changes in the endocrine function of the testis (testosterone production) unless season-long treatments are imposed. These outcomes also differ from those for malnutrition – for example, in young rams, zinc deficiency dramatically impairs both spermatogenic and steroidogenic capacity by inducing specific biochemical lesions in the testis.

We are developing a clear picture of the metabolic signals, neuroendocrine processes and hormonal control systems that are involved, particularly for the mature male sheep. The energetic components of the diet, rather than protein, seem to be responsible, so we have envisaged a model of the relationship between energy balance and reproduction that has 4 'dimensions': genotype, structure (organs), communication (chemical and neural signals, nutrient sensing) and time (dynamics, metabolic memory, programming). We have linked these perspectives to 'resource allocation theory' and incorporated them into strategies for 'clean, green and ethical animal production'.

In contrast to the clear outcomes with respect to spermatogenesis, the effects of nutrition on sexual behaviour are more difficult to define, perhaps because the behaviour is affected by a complex mix of physiological factors and because of flawed methods for quantifying male behaviour. For example, sexual behaviour is compromised by severe feed restriction (>30% of body mass lost), but male sexual behaviour requires intensive motor activity so a decline in libido could be caused by general weakness rather than nutritional limitations. In fact, motor activity is greater in rams with smaller testes than in rams with larger testes, suggesting that the cost of reproduction is greatest in animals that produce least spermatozoa. The interaction between sexual activity and feeding behaviour also complicates the issue under field conditions – rams can show little loss of body mass, but significant loss of testicular mass, during the breeding season. At the other end of the scale, overweight males can show reduced sexual success because they have difficulty courting and mounting. For this reason, exercise can enhance the fertilising capacity of rams. This will be important in extensive mating systems where males need to assemble and guard a harem and then mate many times for several weeks. For AI centres, there seems to be very few data on the nutritional management of males, but problems with overfed animals appear to be common.

In conclusion, feed intake exerts powerful effects on reproduction in mature male sheep and goats. At gonadal level, the gametogenic tissue responds rapidly to changes in nutrition, but the endocrine functions are less affected. Overall, libido is more sensitive to severe under-nutrition than sperm production, but increases in the level of nutrition stimulate sperm production before affecting libido.

Effects of pollutants on the reproduction and welfare of ruminants

S.M. Rhind[1], M. Bellingham[2], R.M. Sharpe[3], C. Cotinot[4], N.P. Evans[2], K.D. Sinclair[5], E. van der Zalm[6], K. Hart[6], J.S. Schmidt[6], B. Fischer[6], B. Mandon-Pepin[4], P. Pocar[7], T. Amezaga[8], R.G. Lea[5] and P.A. Fowler[8]

[1]Macaulay Land Use Research Institute, Aberdeen, United Kingdom; [2]University of Glasgow Veterinary School, United Kingdom; [3]MRC Human Reproductive Sciences Unit, Edinburgh, United Kingdom; [4]INRA, Jouy-en-Josas, France; [5]University of Nottingham School of Veterinary Sciences, Loughborough, United Kingdom; [6]University of Halle, Germany; [7]University of Milan, Italy; [8]University of Aberdeen, United Kingdom.

Endocrine disrupting compounds (EDC), which comprise, primarily, organic compounds and heavy metals, are largely anthropogenic. They are ubiquitous and persistent and exposure of ruminants is unavoidable although levels of tissue exposure are usually very low. Effects are exerted through various mechanisms, depending on the class of chemical involved. These include alteration of oestrogen or androgen function, either modulating steroidogenesis or activating or blocking their receptors, while others act through receptors like the aryl hydrocarbon receptor and PPARs. Some can also act directly on enzyme systems and on gene expression and so a mixture of pollutants may act through multiple mechanisms and can perturb multiple physiological systems. Since there are multiple mechanisms of action, depending on the class of chemical involved, biological effects are diverse, although they are often subtle and do not result in acute loss of production. Early stages of development are particularly susceptible to EDC effects. Studies of sheep subject to prolonged, low level exposure to multiple EDC, through grazing pastures fertilised with sewage sludge, have demonstrated perturbation of: fetal testis structure (reduced testis weight and reduced numbers of Sertoli cells, Leydig cells and gonocytes); fetal ovary development (reduced oocyte densities, increased expression of pro-apoptotic BAX gene and altered expression of multiple proteins); fetal neuroendocrine function (reduced hypothalamic and pituitary expression of genes); fetal uterus function (altered expression of multiple proteins); offspring behaviour (male lambs exhibited female exploratory behaviour patterns); adult ovary structure (reduced number of healthy primordial follicles); adult bone structure (sex-dependent changes in bone density and stiffness); adult mammary structure (reduced alveolar/stromal ratio and altered expression of multiple proteins). EDC are also known to have the capacity to compromise thyroid and immune systems, potentially making animals more susceptible to disease and indirectly compromising their welfare and fitness for breeding. It is concluded that while observed physiological effects are not obviously related to reductions in animal performance at this time, anthropogenic pollutants pose a threat to ruminant performance and welfare in the longer term and that future research must focus on improved understanding of exposure, additive and synergistic effects, the effects of mixtures of EDC on gene and protein expression and identification of previously unrecognised effects and mechanisms of action.

Impact of nutritional factors on the welfare of ruminants

G. Bertoni, L. Calamari and E. Trevisi
Istituto di Zootecnica, Facoltà di Agraria, Università Cattolica del Sacro Cuore, Via Emilia Parmense 84, 29100, Piacenza, Italy

Animal welfare has been defined in several different ways. Brambell's definition depicts animal welfare as a combination of both physical and mental well-being. Therefore, in order to reach optimal welfare, physical and mental discomfort and suffering must be prevented. Proper nutrition is crucial for optimal performance and then would sustain optimal fitness as well. At the same time a good availability of feed and water would avoid any physical and psychological suffering from hunger and thirst. Malnutrition can on the contrary cause various health disorders leading to animal distress. Within this issue we can include many different situations:

- nutrient deficiency or excess, which can lead to metabolic disorders or cause impaired immune function, therefore heightening the risk of infection;
- poor quality feeds which have been subjected to bacteria or fungi spoilage. The spoilage implies undesirable substances that can cause toxicity or digestive disorders;
- inability to satisfy extraordinary nutritional needs of high genetic merit animals. This problem is a little more complicated than the issue of simple nutrient deficiency, but metabolic and infectious diseases are usual consequences.

The nutritional problems listed above are directly or indirectly responsible for pro-inflammatory (i.e. IL-1, IL-6 and TNFα) cytokine release. This exits in inflammatory phenomena which in turn leads to the 'sickness behaviour' of the animal; namely various symptoms including apathy, anorexia, fatigue and anxiety, as well as irritability and some mild cognitive disorders. All of them are causes of mental and body discomfort.

Thus, the most direct effect of feeding or nutrition occurs via the release of pro-inflammatory cytokines; this release can occur for the digestive disorders, rumen and/or intestine, that causes endotoxins and/or bacteria translocation through the mucosal barrier. The consequent inflammation worsens the welfare status as suggested above, but also interesting it is the negative effect on appetite and feed efficiency. Particularly in transition period of dairy cows, but also in late pregnancy for sheep and goats, both latter conditions cause a worsening of negative energy balance and an increased risk for metabolic diseases, liver lipidosis and infectious diseases.

Nevertheless, nutrition can even slow down the inflammatory response which is important but also dangerous and painful (e.g. some nutrients can reduce the NFkB activity on expression of pro-inflammatory genes: cytokines, cyclooxigenases, NO synthetases, etc.).

For the future, according to the knowledge in human epigenetic, it is likely that deficiencies in foetal nutrition could result in variations in the frequency of diseases later in life. This could enlarge the relationship between nutrition and welfare that anyhow seems more frequent in case of intensively reared animals (i.e. dairy cows) which seem more susceptible to many stress factors including malnutrition.

Links between ruminants' feeding behaviour and their welfare

J.J. Villalba[1], F.D. Provenza[1] and X. Manteca[2]
[1]Department of Wildland Resources, Utah State University, Logan, Utah, 84322-5230, USA; [2]School of Veterinary Science, Universitat Autónoma de Barcelona, 08193, Barcelona, Spain

Ruminants select diets that are higher in nutrients and lower in secondary compounds than the average available in the environment. Individuals can better meet their needs for nutrients and regulate their intake of secondary compounds when offered a variety of foods than when constrained to a single food, even if the food is nutritionally balanced. The concept of food variety is central because single flavours, nutrients and secondary compounds all cause animals to satiate, and satiety may be aversive which limits food intake. The satiety hypothesis attributes changes in palatability to transient food aversions due to flavours, nutrients, and toxins interacting along concentration gradients. Gustatory, olfactory, and visual neurons stop responding to the taste, odor, and sight of a particular food eaten to satiety, yet they continue to respond to other foods. If 'other foods' are not available, animals stop responding and intake will decrease. Moreover, if monotony is aversive, then animal welfare may be compromised, even if monotony implies consuming a balanced diet. A diverse diet may increase resistance to disease in ruminants, by allowing consumption of small amounts of compounds with antimicrobial/antiparasitic effects and immunity-enhancing properties. Animals that manifest less fear towards the unfamiliar will typically accept more readily new foods, leading to a more diverse diet. Thus, fear responses towards novelty, may interact with availability of diverse foods to influence diet selection. There are at least three lines of evidence suggesting such a relationship in ruminants. First, the neural substrate responsible for the fear response is also involved in diet selection. Neuronal networks in the amygdala and the hippocampus are involved in both fear and diet selection responses. Second, experimental work in some domestic species has shown that when animals are placed in a new environment likely to elicit a stress response, they show a greater reluctance to eat novel foods compared with the same animals being offered new food in a familiar environment. Finally, work with farm animals has shown that environmental enrichment that allows animals to show a more flexible foraging behaviour decreases chronic stress. Individuals also differ in their temperaments and this may explain part of their differences in diet selection. Such differences may originate from experiences with the food (nutrients, toxins, flavours) and non-food (location, predation risk) environments *in utero* and early in life. Experiences early in life cause changes – neurological, morphological, physiological – in animals that influence their subsequent behaviour. Experiences with the environment enable animals to adapt to local diets, habitats and stressors. This implies that what constitutes a low-quality diet or stressor will differ among individuals reared in different habitats and with different dietary and non-dietary experiences. In conclusion, food diversity and early experiences with food will influence how animals perceive the stress created by the food environment, which will affect animal welfare and performance and thus profitability of the people who manage animals.

Stress and microbial endocrinology: prospects for ruminant nutrition

P. Freestone[1] and M. Lyte[2]
[1]Department of Infection, Immunity and Inflammation, University of Leicester School of Medicine, University Road, LE1 9HN, Leicester, United Kingdom; [2]Department of Pharmacy Practice, School of Pharmacy, Texas Tech University Health Sciences Center, Lubbock, Texas 79430-8162, USA; ppef1@le.ac.uk

The feed efficiency of ruminant meat and dairy livestock can be significantly influenced by factors within their living environments. In particular, events perceived by the animals as stressful (such as parturition, transport or handling) have been found to affect susceptibility to infection. It has been well documented that even minor stress such as weighing can result in an increase in colonisation and faecal shedding of enteric pathogens such as *Salmonella enterica* and *Escherichia coli* O157:H7. Such infections affect both ruminant overall health and therefore performance, and are a particular problem for the meat production industries. Prior explanations for stress enhancing the likelihood of infection is that activation of the sympathetic nervous system under stress leads to the release of neuroendocrine mediators such as the catecholamine stress hormones noradrenaline and adrenaline, which may impair innate and adaptive immunity. More recently, however, another equally compelling explanation, viewed through the lens of the newly recognised microbiological discipline of microbial endocrinology, is that the myriad of bacteria within the ruminant digestive tract are as responsive to the hormonal output of stress as the cells of their host. Work from our laboratories has shown that enteric pathogens have evolved systems for directly sensing stress hormones. We have demonstrated that even brief exposure of enteric pathogens to physiological concentrations of stress hormones can result in massive increases in growth and marked changes in expression of virulence factors such as adhesins and toxins. Happy, less stressed ruminants may therefore be better nourished animals and safer sources of meat. This article reviews evidence that as well as affecting nutrition, stress in ruminants is correlated with increased risk of enteric bacterial infections, and examines the molecular mechanisms that may be at work in both processes.

Feeding practices for sustainable ruminant production facing environmental changes and human food crisis

F. Bocquier[1,2] *and E. González-García*[2]
[1]*Montpellier SupAgro, UMR868 ERRC, 2 Place Pierre Viala, 34060, Montpellier;*[2]*INRA, UMR868 ERRC, 34060, Montpellier, France*

Numerous recent studies highlight sustainability problems for the development of ruminant production systems in the face of increasing human food needs. The global challenge is to improve the efficiency of the greater world population of ruminants present in developing countries (DC), while in industrialised countries (IC), highly productive animals reared with high input levels and often cereal-fed may be judged unsustainable. Although the context is very complex, in our view the main objectives of ruminant physiology should converge, for IC and DC countries, in a common, global knowledge advancement strategy. In DC this means improving the efficiency of systems, given that livestock farming is often the only possible use of rangelands. For IC settings, systems should be revisited to promote autonomy and environment-friendly feeding and managing practices. Assuming that competition for feed/food use is still a crucial criterion, future ruminant feeding systems should preferably focus on lignocellulosic sources. According to natural biome distributions and the recent increases in volume of crop residues and by-products, the annually renewed volumes of these biomasses are considerable. Therefore, we need to redesign our strategies for efficient use at local levels. For this purpose, digestion processes and rumen functioning need to be better understood. The renewed vision of ruminal digestion through the reduction of greenhouse gas emissions is also a key factor since it is an environmental demand that cannot be ignored. Concerning other physiological functions of ruminants, accumulated knowledge can be mobilised into an integrative approach that emphasises the adaptive capacities of animals to face variability in quantity and quality of supplied feeds. Basically, the reduction of inputs traditionally used to ensure feeding systems will need more flexible animals. In that sense, concepts of homeostasis and teleophoresis need to be updated and adapted to domestic species and breeds that were until now largely excluded from the dominant productive systems.

In conclusion, a more holistic approach to research targets is required where physiological functions and farming practices must converge and respond to every specific situation in an integral, dynamic and flexible conceptual perspective. From a scientific angle, both for IC and DC a broader range of experimental scenarios should be explored in order to arrive at innovative practices and solutions that address environmental, ethical and economic issues. There is a clear challenge in evaluating sustainability of ruminant production systems. This includes, in our opinion, strong interactions with other disciplines (multi- and trans-disciplinary design) to build new relevant indicators for sustainability evaluation.

Short communications

Digestion and absorption

Prediction of starch digestion in the small intestine of lactating cows

A. Bannink[1], J.L. Ellis[2], J. France[2] and J. Dijkstra[3]
[1]Animal Sciences Group, Animal Production, Wageningen University and Research Centre, Lelystad, the Netherlands; [2]Centre for Nutrition Modelling, Department of Animal and Poultry Science, University of Guelph, Guelph, ON, Canada; [3]Animal Nutrition Group, Wageningen University, Wageningen, the Netherlands; andre.bannink@wur.nl

Introduction

Glucose is an important nutrient for high-yielding dairy cows and particularly during early lactation glucose requirements are high. The most important sources are propionic acid absorbed from the rumen and rumen bypass starch digested in the small intestine (SI). Recently, an attempt was made to improve the prediction of rumen production rate of propionic acid as a function of fermented substrates and rumen acidity (Bannink *et al.*, 2008). New relationships were derived from *in vivo* data from rumen digestion trials. Intestinal digestion of rumen bypass starch is based on a relationship derived in a review on starch digestion by Nocek and Tamminga (1991). This relationship relates the extent of digestion of rumen by-pass starch in the SI (SISD) to the extent of digestion of feed starch in the rumen (RSD). Application for predictive purposes (Dijkstra *et al.*, 1996) presumes intestinal digestion is fully determined by rumen digestion. However, factors affecting intestinal starch digestion may differ from factors affecting microbial starch degradation in the rumen. The aim of the present study was to compare SISD prediction by a mechanistic, dynamic approach or by an empirical relationship (Nocek and Tamminga, 1991).

Material and methods

A mechanistic, dynamic model was developed which represents the effect of various factors on the enzymatic digestion of starch in the SI of dairy cows. Three separate intestinal compartments were identified in the model: the duodenum, the jejunum and the ileum compartment. A total of nine state variables have been represented for degradable bypass starch, microbial starch and amylase in every compartment. Factors included are the intrinsic degradation characteristics of starch, digesta pH, passage rate of digesta, amylase secretion and amylase degradation. Mills *et al.* (1999) reviewed the impact of these factors on post-rumen starch digestion. As an empirical approach, the relationship derived by Nocek and Tamminga (1991) between SISD (as % of starch entering) and RSD (as % of starch ingested) was used: ISD = $-0.728\times(100 - \text{RSD}) + 87.9$. Both the mechanistic and the empirical model were evaluated against independent data (twenty-two treatments; seven trials) on RSD and SISD reviewed by Reynolds (2006). Observed starch inflow to the SI (kg/d) was 1.90 ± 1.17 kg starch /d and considered feed starch, observed SISD was 71.5 ± 10.1% of starch inflow. Prediction results were evaluated by the square root of the mean squared prediction error as a proportion of observed mean (RMSPE%), decomposed into overall prediction bias (ECT%), deviation of the regression slope from unity (ER%) and error due to disturbance (ED%).

Results and discussion

The mechanistic model showed the lowest RMSPE% for the amount of starch digested in the SI (kg/d) of 20.6%, compared to 30.0% with the empirical model. Also prediction bias and the deviation from the regression slope from unity were the lowest with the mechanistic model (ECT% = 1.0%, ER% = 0.0%, ED% = 90.0%) compared to the empirical model (ECT% = 32.9%, ER% = 35.5%, ED% = 31.6%). Figure 1 demonstrates that the mechanistic model overestimated SISD with about 20% at the lowest amounts of starch entering the SI, whereas no such bias occurred at the higher end. In contrast, the empirical model estimated accurately at the lower end, but strongly underestimated SISD with about 20% at the higher end. The mechanistic model overestimated the absolute amount of starch digested (Figure 2) with about 150 g/d at the lower end of starch inflow to the SI, whereas the empirical model underestimates with more than 700 g/d at the higher end.

Conclusion

The preliminary results of the present study indicate that empirical equations for the prediction of starch digestion in the SI can be improved by adopting a more mechanistic approach. The empirical equation tested here takes RDS as fully explanatory for starch digestibility in the SI. Although intrinsic starch degradability is an important factor indeed, it seems feasible to base predictions on other factors related to SI conditions and processes as well.

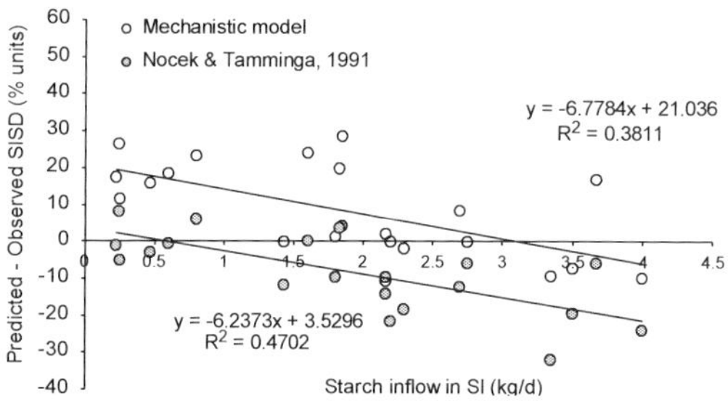

Figure 1. Prediction error of SI starch digestion (% of starch inflow) with a mechanistic (○) and an empirical model (●).

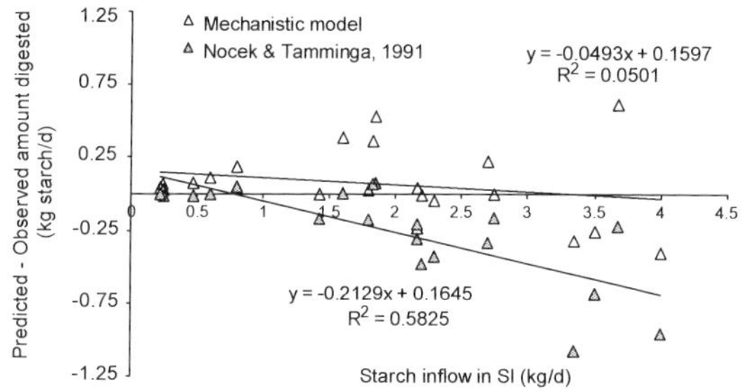

Figure 2. Prediction error of amount of starch digested in the SI (kg starch/d) with a mechanistic (Δ) and an empirical model (▲).

References

Bannink, A., J. France, S. Lopez, W.J.J. Gerrits, E. Kebreab, S. Tamminga and J. Dijkstra, 2008. Modelling the implications of feeding strategy on rumen fermentation and functioning of the rumen wall. Anim. Feed Sci. Techn. 143: 3-26.

Dijkstra, J., J. France, H.D. St. C. Neal, A.G. Assis, L.J.M. Aroeira and O.F. Campos, 1996. Simulation of digestion in cattle fed sugarcane: model development. J. Agric. Sci. 127: 231-246.

Mills, J.A.N., J. France and J. Dijkstra, 1999. A review of starch digestion in the lactating dairy cow and proposal for a mechanistic model: (2) Postrumen starch digestion and small intestinal glucose absorption. J. Anim. Feed Sci. 8: 451-481.

Nocek, J.E. and S. Tamminga, 1991. Site of digestion of starch in the gastrointestinal tract of dairy cows and its effect on milk yield and composition. J. Dairy Sci. 74: 3598-3629.

Reynolds, C.K., 2006. Production and metabolic effects of site of starch digestion in dairy cattle. Anim. Feed Sci. Techn. 130: 78-94.

Rate of propionate absorption influences intake in dairy cows fed ryegrass

A. Boudon[1,2], J. Juton[1,2], R. Delagarde[1,2], P. Faverdin[1,2] and J.-L. Peyraud[1,2]
[1]INRA, UMR 1080 Dairy Production, F-35590 St-Gilles, France; [2]Agrocampus Ouest, UMR 1080 Dairy Production, F-35000 Rennes, France; anne.boudon@rennes.inra.fr

Introduction

In Western European dairy systems, increasing the use of grazed herbage can reduce feed costs, limit the impact of livestock breeding on the environment and improve the quality of animal products. Highly digestible herbage, such as perennial ryegrass, generally has a high nutritive value, equivalent to that of a total mixed ration based on maize silage. However, its intake remains low and variable compared with a total mixed ration based on maize silage. A hypothesis that could explain this, at least partly, is that the satiating effect of energy nutrients absorbed from highly digestible herbage, i.e. mainly VFA provided by ruminal digestion, would be higher than that of glucose that represents a substantive part of the energy nutrient absorbed when maize silage is fed (Faverdin, 1999). The objective of this study was to test this hypothesis by comparing the satiating effect of propionate, the most satiating VFA (Faverdin, 1999), and of glucose, on dairy cows fed highly digestible ryegrass and to determine the impact of propionate rate of absorption on its satiating effect.

Material and methods

Two experiments were conducted successively in the spring, on 4 dairy cows each, according to a Latin square design with 4 treatments and 4 periods of 7 days at Rennes (France). In both experiments, dairy cows were fed fresh perennial ryegrass indoors. In exp 1, the satiating effects of propionate and glucose were compared with 4 treatments: ruminal infusions of 0.625 or 1.25 kg of propionate ('C3 Low' or 'C3 High'), duodenal infusion of 1.5 kg of glucose ('Gluc') and control. The infusion NE_L supply was supposed to be 8.3, 16.6 and 16.5 MJ for C3 Low, C3 High and Gluc respectively. All infusions lasted 18 hours per day and were adjusted with salts to be isotonic. In experiment 2, the impact of the rate of infusion of propionate on intake was tested with a factorial arrangement of 2 infusion types (infusion of 1 kg of propionate 'C3' vs. isotonic control) and 2 infusion rates (2.25 mol/h, i.e. high vs. 0.75 mol/h, i.e. low). The high rate infusions lasted 6 h per day and the low infusion rate lasted 18 h. In both experiments, intake was recorded daily. Data were analysed by ANOVA using the MIXED procedure of SAS®. For each experiment, the model included the effects treatment, period and cow (as random). In experiment 1, three orthogonal contrasts were used: 'Infusion' (Control vs. the 3 other treatments), 'Nutri' (C3 Low and High vs. Gluc) and 'Dose' (C3 Low vs. C3 High). In experiment 2, the treatment effect was divided in infusion 'Inf. Type' effect (C3 vs. Control), 'Inf. Rate' effect (6 h vs. 8 h infusions) and their interaction.

Results and discussion

In experiment 1, the glycemia was clearly higher with C3 or glucose infusion compared to the control, and with glucose infusions compared with propionate infusions (Table 1). The rumen concentration of propionate was higher with propionate infusions compared to the control or glucose infusions. The infusions of propionate and glucose only tended to reduce intake ($P<0.10$) compared with the control. Daily eating time and number of meals per day tended to decrease with C3 compared with glucose infusions ($P= 0.08$) but intake was unaffected by the nutrient infused ($P>0.10$). Intake significantly decreased between low dose and high doses of propionate (-1.2 kg, $P<0.05$).

Table 1. Effect of propionate or glucose on intake in dairy cows fed ryegrass (experiment 1).

	Control	C3 Low	C3 High	Gluc	SD	Infusion	Nutri	Dose
Intake, kg DM/d	16.7	16.6	15.4	16.0	0.58	0.08	0.97	0.03
Eating time, min/d	537	532	510	576	37.3	0.93	0.05	0.44
Number meals/d	10.3	9.3	10.0	10.8	0.88	0.64	0.08	0.27
Glycemia, mg/100ml	65.4	66.9	68.8	70.9	1.35	0.00	0.01	0.09
Rumen C3, mmoles/l	24.0	32.6	44.4	24.3	5.03	0.01	0.00	0.01

Decrease of intake with propionate infusions in experiment 1 was low compared to similar studies on dairy cows fed a total mixed ration based on maize silage (Oba and Allen, 2003). This difference may be explained by the low intake rates of dairy cows fed fresh herbage compared to that of cows fed a total mixed ration. Thus, the absorption fluxes of energy nutrients are likely more steady and low during the daytime on diet based on herbage. Exp 2 aimed to test this hypothesis.

In experiment 2, intake decreased with propionate infusions compared with control ($P<0.05$) and with high rate compared with low rate infusions ($P<0.01$) (Table 2). The decrease of intake with propionate infusion was largely higher when the infusion rate was high (-0.7 kg for high rate infusions and -2.2 kg for low rate infusions, $P<0.05$). The infusion rate only tended to decrease daily eating time and infusion type and rate did not affect the number of meals per day. Treatments clearly affected glycaemia and ruminal propionate concentration as expected.

Table 2. Effect of rate of propionate infusions on intake in dairy cows fed ryegrass (exp. 2).

	Control		C3		SD	Inf. type	Inf. rate	Rate × type
	High rate	Low rate	High rate	Low rate				
Intake, kg DM/d	17.9	18.0	15.7	17.3	0.76	0.00	0.04	0.05
Eating time, min/d	534	561	497	547	39.9	0.19	0.06	0.54
Number meals, /d	10.0	9.9	10.1	10.0	1.58	0.85	0.91	0.98
Glycemia, mg/100ml	65.3	64.5	68.4	68.0	1.65	0.00	0.41	0.78
Rumen C3, mmoles/l	23.6	24.3	42.2	33.6	3.50	0.00	0.04	0.02

Conclusion

These experiments suggested that the satiating effect of propionate in ruminants depends on its dynamics of absorption in the digestive tract. When the dynamics of absorption of propionate is very smooth, its satiating effect may not differ from that of glucose. Consequently, the nature of energy nutrients absorbed by grazing cows should not be the explanation of their low intake.

References

Faverdin, P., 1999. The effect of nutrients on feed intake in ruminants. Br. J. Nutr. 58: 523-531.

Oba, M. and M.S. Allen, 2003. Intraruminal infusion of propionate alters feeding behaviour and decreases energy intake of lactating dairy cows. J. Nutr. 133: 1094-1099.

Ruminal calcium (Ca) transport as affected by luminal Ca concentrations and Ca sources

G. Breves, M. Wilkens, G. Ricken and B. Schröder
Department of Physiology, School of Veterinary Medicine, Bischofsholer Damm 15/102, 30173 Hannover, Germany; gerhard.breves@tiho-hannover.de

Introduction

Although little is known about the underlying mechanisms, it is well accepted that the rumen wall is a considerable site for calcium (Ca) absorption. However, controversy still exists on the quantitative proportion of the forestomachs to overall gastrointestinal Ca absorption as well as the relevance of the Ca source. From a series of balance studies with multiple fistulated sheep and cows it was suggested that the importance of ruminal Ca transport increases as a function of Ca intake (Schröder and Breves, 2006). The relevance of ruminal Ca absorption is supported by the observation that oral administration of calcium salts to cows resulted in a significant increase in serum total calcium concentration within minutes after administration and Ca chloride seemed to increase plasma Ca more rapidly than Ca propionate (Goff and Horst, 1994), but a direct effect of the Ca source on its gastrointestinal absorption has never been further investigated. Therefore, the aim of the present study was to examine whether the Ca source, Ca concentration and luminal pH have any direct effects on Ca transport across the rumen wall.

Material and methods

The Ussing chamber technique was used to determine unidirectional Ca flux rates (mucosal-to-serosal flux rate J_{ms}, opposite direction J_{sm}) as described earlier (Schröder *et al.*, 1999). Tissues were obtained from adult cattle of both sexes slaughtered at the local abattoir. Within 20 minutes after exsanguination, pieces of about 20×25 cm were taken from the ventral rumen sac and rinsed in a modified Krebs-Henseleit-buffer (pH 7.4, 38 °C). The mucosa was stripped immediately from the underlying muscle layers and kept in buffer solution gassed with carbogen at 38 °C until it was mounted into the Ussing chambers 15 min later. By means of a radioisotope tracer technique with ^{45}Ca, unidirectional Ca flux rates were determined under varying conditions. In experiment 1, acute effects of either chloride, formate, acetate or propionate on J_{ms} Ca flux rates were tested in simple Krebs-Henseleit-buffer. In experiment 2, *in vivo* conditions were mimicked since the mucosal buffer was enriched with short chain fatty acids (SCFA) at physiological molar proportions and the pH was adjusted to 6.4. Ca flux rates were determined using different concentrations (1.2 and 3.7 mmol/L) of either Ca formate, Ca acetate or Ca propionate as Ca source in the mucosal compartment. In this experiment, the proportion of ionised calcium (Ca^{2+}) was analysed with a blood gas and electrolyte analyser (Rapidlab 248). In experiment 3, the aim was to simulate the oral administration of a Ca bolus. For this purpose a buffer enriched with a mixture of SCFA was employed and after a control period, the Ca concentration in the mucosal buffer solution was increased up to 11.2 mmol/l.

Results

Experiment 1: J_{ms} Ca flux rates across rumen tissues of cattle as measured in simple Krebs-Henseleit-buffer were in the order of 10 nmol/cm^2/h and did not differ significantly from those incubated in a mucosal buffer containing 40 mmol/l acetate. The presence of the same concentrations of chloride, formate and propionate induced slight increases of J_{ms} flux rates, but the differences were not statistically significant.

Experiment 2: In the presence of SCFA (acetate 36, propionate 15, butyrate 9 mmol/l) the increase in mucosal Ca concentrations from 1.2 to 3.4 mmol/l raised J_{ms} flux rates from approximately 16 to 70 nmol cm^2/h. Neither the basal nor the elevated Ca flux rates were affected by the Ca source used in these experiments (Ca chloride, Ca formate or Ca propionate). Irrespective of the Ca source, the fractions of ionised calcium measured in the mucosal buffer solutions ranged between 55% and 59%. *Experiment 3*: Raising the mucosal Ca concentration from 1.2 to 11.2 mmol/l resulted in a remarkable increase of the J_{ms} Ca flux rates from approx. 11 nmol/cm^2/h to almost 115 nmol/cm^2/h. The respective J_{sm} flux rates remained unaffected.

Discussion

In experiment 1 the mucosal buffer did not contain SCFA and the pH was adjusted to 7.4. These conditions were chosen in order to exclude any stimulating effect apart from a potential influence of the different anions investigated. Chloride, formate and propionate seemed to slightly increase the Ca flux rates from the mucosal to the serosal compartment. However, this was not statistically confirmed. Since there was obviously no difference between J_{ms} measured in the presence of control buffer or acetate, the anions chloride, formate and propionate were elected for further experiments. In experiment 2, the mucosal buffer contained SCFA and the pH was adjusted to 6.5. When the concentration of the Ca salts was increased from 1.2 to 3.7 mmol/l, J_{ms} was stimulated significantly, irrespective of the different anions. The 4.5-fold increment of J_{ms} cannot simply be explained by the 3.1-fold higher concentration of Ca in the mucosal buffer and the higher chemical gradient. In recent studies on ovine rumen, a stimulating effect of chloride could be demonstrated (Leonhard-Marek *et al.*, 2007) and the beneficial influence of SCFA on ruminal Ca transport was demonstrated in former studies (Schröder *et al.*, 1999). Thus, the increment could be due to a combination of the higher concentration of Ca and the respective anion. To exclude interferences due to different availabilities of Ca from the Ca salts, the concentrations of ionised Ca were determined in the buffer solutions. The proportion of free Ca ions, the form of Ca which can be absorbed by gastrointestinal epithelia, was neither affected by the Ca source nor the concentration. The results of the simulation of a Ca bolus *in vitro* in experiment 3 demonstrate that the capacity of the ruminal Ca transport mechanism is not saturable within a wide range of luminal Ca concentrations. This is in line with the rapid increase in serum Ca after oral administration reported in several *in vivo* studies (Goff and Horst, 1994).

References

Goff, J.P. and R.L. Horst, 1994. Calcium salts for treating hypocalcemia: carrier effects, acid-base balance, and oral versus rectal administration. J. Dairy Sci. 77:1451-1456.

Leonhard-Marek, S., G. Becker, B. Schröder and G. Breves, 2007. Chloride, gluconate, sulfate, and short-chain fatty acids affect calcium flux rates across the sheep forestomach epithelium. J. Dairy Sci. 90: 1516-1526.

Schröder, B. and G. Breves. 2006. Mechanisms and regulation of calcium absorption from the gastrointestinal tract in pigs and ruminants: comparative aspects with special emphasis on hypocalcemia in dairy cows. Anim. Health Res. Rev. 7:31-41.

Schröder, B., S. Vössing and G. Breves, 1999. *In vitro* studies on active calcium absorption from ovine rumen. J. Comp. Physiol. [B] 169: 487-494.

Estimation of microbial N flow from purine derivative urinary excretion in sheep and goats fed diets with different alfalfa hay:concentrate ratios

G. Cantalapiedra-Hijar[1], E. Molina-Alcaide[1], S. Ramos[2], M.L. Tejido[2], D.R. Yáñez-Ruiz[1] and M.D. Carro[2]
[1]Estación Experimental del Zaidín (CSIC), Profesor Albareda, 1, 18008 Granada, Spain; [2]Departamento de Producción Animal, Universidad de León, 24071 León, Spain; mdcart@unileon.es

Introduction

The urinary excretion of purine derivative (UEPD) analysis has been proposed as a non-invasive method of estimating microbial N flow (MNF) to the duodenum in ruminants and, equations have been developed in sheep (Cheng *et al.*, 1992) and goats (Belenguer *et al.*, 2002) assuming a constant purine base (PB):N ratio in rumen microbes. However, PB:N ratio may be affected by diet and sampling time (Carro, 2001) as well as by the type of bacterial pellet used (Firkins *et al.*, 987). The aim of this study was to estimate MNF from UEPD in sheep and goats fed two different diets by using the existing equations for both animal species, and to compare the values with those obtained by replacing the fixed PB:N ratio value proposed by the equations with PB:N ratios in solid- (SAB) and liquid- (LAB) associated bacteria isolated from the rumen in the present study.

Material and methods

Four Granadina goats (46.6±1.32 kg BW) and four Merino sheep (55.9±4.80 kg) fitted with ruminal cannulas were fed two diets with alfalfa hay:concentrate ratios of 70:30 (HA) and 30:70 (LA) following a cross-over design. Animals were fed at a daily rate of 56 g dry matter (DM)/kg $BW^{0.75}$ to minimise feed selection. After 10 d of adaptation in each experimental period, total faeces and urine were collected for 6 d to determine diet digestibility and UEPD (Balcells *et al.*, 1992), respectively. On day 17, rumen content from each animal was collected 2 h after feeding, and LAB and SAB were isolated (Ranilla and Carro, 2003) and analysed for N and PB (Balcells *et al.*, 1992). The MNF was estimated from the equations proposed by Cheng *et al.* (1992) for sheep and Belenguer *et al.* (2002) for goats. Alternatively, MNF was also estimated by replacing in the equations the fixed PB:N ratios by the values obtained in the present study. Data obtained were analysed independently by the GLM procedure of SPSS v.15.0 (SPSS Inc., 2007, Chicago, IL, USA). The effects of diet, species, diet × species interaction, and period were considered fixed, and animal within species was considered random.

Results

Daily intake of digestible organic matter (g/d) was higher (P=0.03) for sheep than for goats (Table 1), but no differences (P=0.16) were detected when the intake was expressed as g/kg $BW^{0.75}$ (values not shown). The fact that PB:N ratios in LAB were higher in goats compared to sheep only for HA diet (species x diet interaction, P=0.002) together with the lower (P=0.001) values for SAB found in goats vs. sheep, supports the hypothesis that the PB:N ratios may vary between animal species, diet received and even the type of pellet isolated. Estimations of PB duodenal flow were not affected by animal species (P=0.33) nor diet (P=0.36). However, using the PB:N ratios in LAB and SAB isolated in our conditions in sheep and goats, the MNF values were around 1.7 and 2.2 times greater, respectively, than those estimated using the PB:N ratio proposed in the equations. Although the PB:N ratio values were affected by animal species and diet, the MNF calculated using these values or those proposed in the equations was not influenced by animal species nor diet.

Table 1. Average values of digestible organic matter intake (DOMI), purine bases(PB):N ratio in liquid- (LAB) and solid- (SAB) associated bacteria, urinary excretion of purine derivatives (UEPD) and estimations of duodenal flow of PB (DFPB), and microbial N flow estimated using either the fixed PB:N ratio value existing in the equations (MNF-eq) or values obtained from LAB (MNF-LAB) and SAB (MNF-SAB) isolated from the rumen of sheep and goats fed diets with 70:30 (HA) and 30:70 (LA) alfalfa hay:concentrate ratio.

Item	Sheep		Goat		SEM	Significance of effects (*P*=)		
	HA	LA	HA	LA		Species	Diet	Species ×Diet
DOMI, g/d	720^{b}	747^{b}	541^{a}	641^{ab}	28.5	0.03	0.08	0.26
PB:N-LAB, μmol/mg N	0.93^{a}	0.92^{a}	1.11^{b}	0.92^{a}	0.011	0.004	0.001	0.002
PB:N-SAB, μmol/mg	0.92^{ab}	1.03^{b}	0.83^{a}	0.81^{a}	0.018	0.001	0.19	0.08
UEDP, mmol/d	12.1	13.9	12.7	14.0	0.82	0.84	0.37	0.91
DFPD, mmol/d	14.5	16.5	16.7	18.4	1.01	0.33	0.36	0.95
MNF-Eq, g/d	10.5	12.0	9.2	10.2	0.66	0.26	0.37	0.85
MNF-LAB, g/d	18.8	21.6	16.4	22.0	1.24	0.72	0.11	0.56
MNF-SAB, g/d	19.0	19.3	22.0	24.7	1.26	0.13	0.57	0.63

$^{a, b}$ Means within a row with unlike superscripts differ (P<0.05).

Conclusion

The MNF values in sheep and goats were highly overestimated since the fixed PB:N ratio value in the equations was replaced with those found in LAB and SAB isolated in the present study. However none of the estimates detected significant differences between animal species or diets, indicating that although values may be affected by the change, the interpretation of results would be unaltered.

Acknowledgement

This research was supported by the CICYT of Spain (projects AGL2004-04755-C02-01 and C02-02). Thanks are given to Dr. J. Balcells for his advice on calculations. G. Cantalapiedra-Hijar and S. Ramos gratefully acknowledge the receipt of scholarships from the MEC of Spain (AP2004-1908 and BES-2005-6842, respectively).

References

Chen, X.B., Y.K. Chen, M.F. Franklin, E.R. Ørskov and W.J. Shand, 1992. The effect of feed intake and body weight on purine derivative excretion and microbial protein supply in sheep. J. Anim. Sci., 70: 1534-1542.

Balcells, J., J.A. Guada, J.M. Peiró, and D.S. Parker, 1992. Simultaneous determination of allantoin and oxipurines in biological fluids by high-performance liquid chromatography. J. Chromat. 575: 153-157.

Belenguer, A., D. Yáñez, J. Balcells, N.H. Ozdemir Baber and M. González Ronquillo, 2002. Urinary excretion of purine derivatives and prediction of rumen microbial outflow in goats. Livest. Prod. Sci. 77:127-135.

Carro, M.D., 2001. La determinación de la síntesis de proteína microbiana en el rumen: comparación entre marcadores microbianos (Determination of microbial protein synthesis in the rumen: a comparison of microbial markers (Review)). Invest. Agr.: Prod. Sanid. Anim. 16 (1): 5-27.

Firkins, J.L., L.L. Berger, N.R. Merchen, G.C. Faey Jr. and R. L. Mulvaney, 1987. Ruminal nitrogen metabolism in steers as affected by feed intake and dietary urea concentration. J. Dairy Sci. 70: 2302-2311.

Ranilla, M.J. and M.D. Carro, 2003. Diet and procedures to detach particle-associated microbes from ruminal digesta influence chemical composition of microbes and estimation of microbial growth in Rusitec fermenters. J. Anim. Sci. 81: 537-544.

High sulfur content of dried distiller's grains: effects on ruminal fermentation

J.S. Drouillard[1], L.K. Thompson[1], S. Uwituze[1], K.K. Karges[2] and L.C. Hollis[1]
[1]Kansas State University, Manhattan, KS, USA; [2]Dakota Gold Research Assn., Sioux Falls, SD, USA; jdrouill@ksu.edu

Introduction

Rapid expansion of fuel ethanol production has resulted in widespread availability of distiller's grains (DG), which have become important staples of cattle diets. Sulfur content of DG varies due to the addition of sulfuric acid to fermenters for pH adjustment. Within the rumen, sulfur is reduced to hydrogen sulfide (H_2S) by ruminal microbes. The H_2S is eructated from the rumen and subsequently respired into the lungs, and when in excess can induce polioencephalomalacia (Gould, 1998). Elevated sulfur levels may exceed the maximum tolerable level of 0.4% of the diet (NRC, 1996) and may have detrimental effects on growth performance and carcass characteristics (Zinn *et al.*, 1997). The present studies compared ruminal fermentation and gaseous end-products when dietary sulfurs were varied in diets that combined DG with dry-rolled corn (DRC) or steam-flaked corn (SFC).

Material and methods

Trial 1 utilised 12 ruminally fistulated Angus steers in a randomised incomplete block experiment with a 2 x 2 factorial treatment arrangement. Factors consisted of dietary sulfur (0.42 or 0.65%; LS and HS respectively), and grain processing (SFC or DRC). Sulfuric acid was spiked into DG to achieve HS. All diets included 8.6% roughage and 30% DG (DM basis). Steers were randomly assigned to treatments and individual feeding pens, and allowed *ad libitum* access to feed and water. Two 15-d experimental periods were used, consisting of a 12-d adaptation and 3-d collection phase. Diets were mixed and delivered to cattle daily at 0800h. Before daily feeding, orts were weighed and dried to determine dry matter intake. Starting 7d before the collection phase, chromic oxide (10 g) in gelatin capsules (Torpac Inc., Fairfield, NJ, USA) was placed into the rumen prior to daily feeding to estimate faecal output. Ruminal digesta and faeces were collected at 2-h intervals after feeding during the collection phase, and ruminal pH was immediately measured. Concentrations of ruminal ammonia and VFA were determined, and faecal samples were used to estimate total tract digestibilities of DM, OM, NDF, CP, starch, and ether extract. Volatile fatty acid profiles, pH, and ammonia concentration were analysed as repeated measures using the MIXED procedure of SAS® (version 8.1; SAS Inst; Cary, NC, USA, 2002). Treatment means were calculated using least-squares means and separated using a protected *F*-test. Digestibility characteristics were analysed using the MIXED procedure, with treatment as a fixed effect, and animal and period as the random effects. A $P<0.05$ was considered significant. Trial 2 utilised 76 crossbred steers (410 kg initial weight) fed in individual pens. Steers were randomly allocated to the same dietary treatments used in trial 1. On d 69, 83, 90, 97, and 104, ruminal gas samples were aspirated through a needle at 0, 4, 8, and 12 h after feeding by puncturing the ruminal wall at the *paralumbar fossa,* and then were analysed for concentrations of H_2S and methane by gas chromatography. Performance data were analysed using the MIXED procedure, with fixed effects of grain processing and sulfur level. The model for H_2S and methane also included effects of sampling day and time after feeding.

Results

In trial 1, HS was associated with lower feed intake (P=0.08), but increases in apparent total tract digestibilities of DM and ether extract (P<0.05). Cattle fed HS also had greater ruminal ammonia concentrations (P<0.01), which was most pronounced in cattle fed DRC. Ruminal pH increased (P<0.05) in HS diets, which may be attributable, in part, to lower VFA concentrations (P=0.05) and higher ruminal ammonia when HS was fed.
Acetate was not affected (P=0.12) by sulfur level, but both butyrate (P<0.0001) and propionate were lower (P=0.002; Figure 1) in cattle fed HS compared to LS. Lactate also decreased (P=0.05) in cattle fed HS diets (Figure 1). It is conceivable that some free hydrogen ions that normally would be used to generate propionate were used to produce H_2S in cattle fed the HS diets. As in trial 1, steers in trial 2 fed HS had higher lower feed intakes compared to those fed LS (P<0.001). Additionally, steers fed HS had higher ruminal concentrations of H_2S compared to cattle fed LS diets, and H_2S was inversely related (P<0.01) to average daily gain (r=-0.42), feed intake (r=-0.43), and Gain:Feed (r=-0.20).

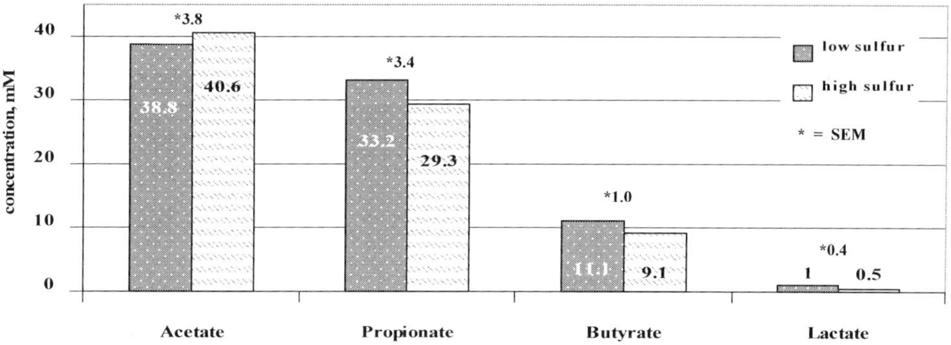

Figure 1. Influence of dietary sulfur on fermentative end products.

Conclusion

Cattle decrease feed intake in response to chronic exposure to high sulfur levels. High levels of dietary sulfur alter ruminal fermentation, fermentative end-products, and cattle performance.

References

Gould, D. H. 1998. Polioencephalomalacia. J. Anim. Sci. 76: 309-314.

NRC. 1996. Nutrient Requirements of Beef Cattle. 7th ed. Nat. Acad. Press, Washington, DC.

Zinn, R.A., E. Alvarez, M. Mendez, M. Montano, E. Ramirez and Y. Shen. 1997. Influence of dietary sulfur level on growth performance and digestive function in feedlot cattle. J. Anim. Sci. 75: 1723-1728.

The effect of accelerated diet step-up rate on performance of feedlot steers dosed with *Megasphaera elsdenii* NCIMB 41125

P.H. Henning[1], A.A. Campbell[2], F.H. Hagg[1], H.H. Meissner[1] and C.H. Horn[1]
[1]KK Animal Nutrition, The Greens Office Park, 26 Charles de Gaulle Street, Centurion 0046, South Africa; [2]Department of Animal and Wildlife Sciences, Faculty of Natural and Agricultural Sciences, University of Pretoria, Pretoria 0002, South Africa; phenning@kkan.com

Introduction

Newly-arrived feedlot cattle are usually adapted to the high-starch grower/finisher diet (Top Diet) by first feeding them a high-roughage starter diet and then gradually decreasing roughage content, over ca. 3 wk, by feeding a series of step-up diets. This practice was developed to prevent digestive disturbances, notably acidosis, that occur when cattle go directly onto the Top Diet. The step-up regime allows *M. elsdenii*, the primary lactic acid-utilising bacteria in the rumen, to increase from low levels associated with roughage feeding to that commensurate with ruminants adapted to high starch diets. *M. elsdenii* NCIMB 41125 was isolated from the rumen of cattle adapted to high starch diets (C.H. Horn *et al.*, personal communication). Earlier studies (unpublished) showed that giving *M. elsdenii* NCIMB 41125 to cattle as a single oral dose established a viable, lactate-utilising population of *M. elsdenii* in the rumen right away, with levels of ca. 10^9 live cells/ml being reached within 24 h. It is hypothesised that orally dosing cattle on arrival at the feedlot with *M. elsdenii* NCIMB 41125 will allow them to go onto the Top Diet sooner, without any adverse effects. This may allow a significant reduction in the number of step-up diets needed and the amount of roughage required in feedlots. It will also enable cattle to attain a higher energy intake earlier, when efficiency of energy utilisation is relatively higher. The objective of this study was to see how rapidly newly arrived feedlot cattle may be stepped-up when they are orally dosed with *M. elsdenii* NCIMB 41125.

Material and methods

The study followed a regression approach with six treatments, each having a different starting roughage level and then the roughage level decreasing, at a similar rate for all treatments, over subsequent days until the final Top Diet roughage level of 5% was reached (Table 1).
The decrease in roughage level for the respective treatments was accomplished by daily mixing Starter Diet (S) and Top Diet (T) in the desired ratios. Diets S and T, respectively, contained (g/kg DM) grass hay (175, 50), ground corn (564, 734), wheat bran (100, 60), protein mix (85, 80), molasses (50, 50), urea (10, 10), mineral-vitamin mix (16, 16) and starch (407, 520), NDF (302,

Table 1. Dietary treatments with different starting roughage levels (% of DM) and subsequent different number of days required to reach the Top Diet containing 5% (DM) roughage.

Treatment number	Starting roughage level on d 1	Roughage level of Top Diet	Day reaching Top Diet
1	17.5	5.0	21
2	15.0	5.0	17
3	12.5	5.0	13
4	10.0	5.0	9
5	7.5	5.0	5
6	5.0	5.0	1

198) and CP (147, 146). Once the Top Diet was reached within each treatment, steers continued on it until the end of the trial on d 35. Thirty-six Bonsmara type steers, with no previous exposure to concentrate diets, were used. Upon arrival at the feedlot they received only roughage and water *ad libitum*. Steers were blocked according to BW and allocated randomly to the 6 treatments. They were kept in single pens and fed *ad libitum*, twice per day, throughout the trial. Individual feed intake and BW was determined daily and weekly, respectively, from d 1 to 35. At 08.00 h on d 1, just prior to feeding the respective concentrate diets for the first time, each steer received a single 100 ml oral dose of a suspension containing *M. elsdenii* NCIMB 41125 (minimum 10^8 live cells/ml). Regression analysis was done, using GenStat® (VSN International Ltd, 2007, Hemel Hempstead, UK) to determine the relationship between performance variables and diet step-up rate. No non-linear relationships were found and data are given for linear analysis. Statistical significance was set at $P<0.05$ and trends are discussed at $P<0.10$.

Results

Linear regression between performance variables and diet step-up rate was non-significant ($P>0.05$) for DM intake and feed conversion ratio, but showed a positive trend ($P<0.10$) for BW gain (Table 2). Practical experience in the feedlot industry would suggest that increasing diet step-up rate, beyond the established norm (roughly corresponding to treatment 1 in the present study) will lead to a loss in performance and to health problems. The results of this study indicate that in cattle dosed with *M. elsdenii* NCIMB 41125 step-up rate could be increased, even to the point where cattle go directly onto the Top Diet (treatment 6), without any adverse effect. On the contrary, the higher energy intake, made possible by reaching the Top Diet earlier, can result in improved performance, as suggested by the increase in BW gain (Table 2).

Table 2. The effect of diet step-up rate (i.e. days taken to reach the Top Diet) on performance (d 1 to 35) of feedlot steers receiving a single oral dose of M. elsdenii NCIMB 41125 at the start of the feeding period.

	Day reaching Top Diet						Regression	
	21	17	13	9	5	1	SE	*P*-value
Starting BW, kg	228	228	228	230	234	228		
DM intake, kg/d	4.94	5.19	5.07	4.43	5.43	5.32	0.380	0.57
BW gain, kg/d	1.04	1.05	1.21	1.27	1.14	1.25	0.075	0.09
kg feed DM/kg gain[1]	5.25	5.23	4.31	3.54	4.91	4.28	0.634	0.28
No. of pulls[2]	1	0	1	1	1	1		

[1] Feed conversion ratio.

[2] Animals showing symptoms of digestive or respiratory health problems.

Conclusion

Dosing cattle with *M. elsdenii* NCIMB 41125 may reduce or eliminate the need for a series of step-up diets in the feedlot, resulting in savings on the cost associated with procurement, storage and handling of roughage. In the present study the accelerated step-up rate did not result in lower performance, on the contrary there was a tendency for increased BW gain. However, the trial covered only the critical feedlot adaptation period and follow-up work is required to evaluate the response over the entire fattening period.

The fate of glycerol entering the rumen of dairy cows

K. Holtenius, A. Werner Omazic and C. Kronqvist
Swedish University of Agriculture, Department of Animal Nutrition and Management, Kungsängens Research Centre, SE-753 23, Uppsala, Sweden; kjell.holtenius@huv.slu.se

Introduction

Glycerol, a by-product from bio-diesel production may be used in diets fed to cattle. The fate of glycerol entering the rumen is not fully understood. Glycerol may disappear from the rumen by microbial digestion, absorption and outflow through the omasal orifice. Absorbed glycerol is efficiently converted to glucose via gluconeogenesis. However it has generally been thought that glycerol is extensively metabolised in the rumen (Sudekum, 2007). However, the rate of fermentation of glycerol *in vitro* is only about 6%/h (Traube *et al.*, 2007). The aim of the present study was to quantify the rate of disappearance of glycerol through the different routes in dairy cows.

Material and methods

Three rumen fistulated dairy cows of the Swedish Red and White Breed were used. The cows were not pregnant or lactating. The cows were held in individual stalls. They were fed 4 kg DM hay, 1.25 kg DM concentrate and 2 kg DM straw.

In vivo-study: At 10:00 about 2 hours after the morning feeding the three cows each received a bolus dose of 500 g glycerol via the fistula and, as a marker of fluid outflow from the rumen, 8 g of CoLi-EDTA dissolved in 1000 ml water. Samples from the rumen fluid were collected before the glycerol load and 15, 30, 45, 60, 90, 120, 180, 240, 300 and 360 min after the load.

In vitro study: Microbial degradation of glycerol was studied in a macro scale *in vitro* system developed by P. Udén (Personal communication, 2009). Two polyethylene pipes, i.d. 250 mm and height 700 mm were used. The rumens of two of the cows were emptied manually. Each sixth handful taken from the rumen, 12 and 16 kg digesta respectively, was transferred to the pipes. Six litres of McDougal buffer and 140 g glycerol was added to each tube. The temperature was kept at 39 °C. The initial concentration of glycerol was similar to that in the rumen of the cows in the *in vivo*-study. The content was continuously gassed with carbon dioxide and thoroughly mixed for 20 sec every 5th minute during the first hr and thereafter every 10th minute. About 10 ml of fluid was collected by suction -5, 15, 30, 45, 60, 90, 120, 150, 180, 210 and 240 min after the glycerol was added to the pipes.

The washed rumen study: The rumen of the three cows was completely emptied, the digesta stored in containers kept in water baths with a temperature of approximately 40 °C. The rumens were then washed with body tempered saline solution twice and thereafter with a rinsing solution mimicking the rumen fluid composition. Thereafter 10 litres of one of the three experimental solutions was introduced to the rumen and left for 60 min. The experimental solution was basically the rinsing solution containing LiCo-EDTA and 15, 30 or 45 mmol/l of glycerol at the expense of NaCl. Samples of the solution, about 40 ml, were collected before it was introduced to the rumen and then every 10 min until 60 min. After 60 min the fluid was removed and the withdrawn volume was determined. The experimental set up was a 3×3 Latin square design. One wk was allowed between experiments. The absorption of glycerol during the 1 h measurement period was calculated by the following equation: $A=(V_0*C_0)-(V_{60}*C_{60})$ where A is the absorption (mmol/h), V_0, C_0, V_{60} and C_{60} are the volume of fluid in the rumen and the concentration of glycerol in the fluid at time 0 and 60 min respectively.

Results

In vivo study: The total glycerol disappearance from the rumen was best described by first order kinetics (R^2 ranging from 0.97-0.99). The fractional disappearance rate that ranged from 43 to 54%/h presumably reflected the sum of the three disappearance routes. The fractional outflow of CoLi-EDTA through the omasal orifice was calculated assuming that the steady state occurred for fluid in the rumen and that CoLi-EDTA was evenly distributed in rumen fluid. It was further assumed that the rate of Co-outflow via the omasal orifice was mirrored by the glycerol outflow. The fractional outflow of glycerol thus ranged from 7.2 to 8.4%/h.
In vitro study: The decrease in glycerol concentration with time was best described by first order kinetics. (R^2: 0.89 and 0.94 respectively). The fractional disappearance rates were 6.6% and 10.8% in the two pipes respectively. The fractional rate of disappearance is assumed to reflect microbial degradation since no absorption or outflow could occur from the tubes.
The washed rumen study: A linear equation was fitted to the Co concentration *versus* time relationship for each cow × treatment combination. These linear equations did always show a better fit than exponential equations as shown by higher r^2 values (range 0.84-0.99). This indicates that Co dilution followed 0-order kinetics. Thus it might be assumed that there was virtually no outflow of Co through the abomasal orifice and the dilution could be explained by fluid inflow via saliva and/or influx of fluid across the rumen epithelium. The relationship between the initial glycerol concentration in the artificial rumen fluid and the calculated rate of absorption of glycerol is shown (Figure 1).

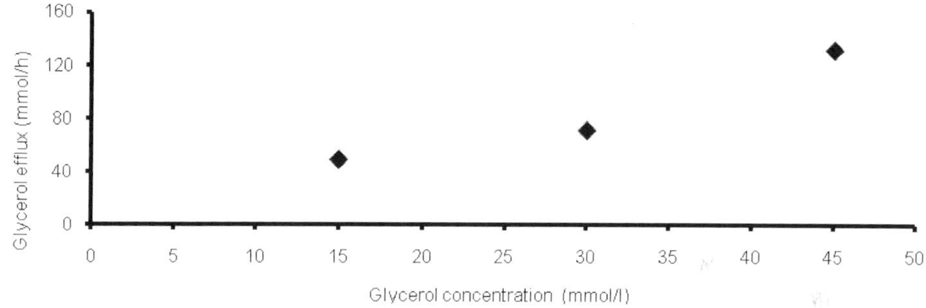

Figure 1. Relationship between the glycerol concentration in artificial rumen fluid and glycerol efflux. Points are means of data from three cows.

Conclusion

If the *in vitro* system used in the present study accurately reflected microbial degradation of glycerol, the rapid rate of glycerol disappearance from the rumen observed *in vivo* was mainly due to glycerol absorption and to a lesser extent due to outflow and microbial digestion in the rumen. The fractional absorption rate was not reduced when the glycerol concentration increased. This indicates that glycerol was absorbed by passive, non-carrier mediated, diffusion. The results suggest that glycerol might be used as an effective glucogenic precursor.

References

Sudekum, K.H., 2007. By-products from biodieselproduction: glycerine and rapeseed products in rations for farm animals. In: Doppenberg, J. and P. van der Aar (Eds.) Biofuels: implications for the feed industry. Wageningen Academic Publishers, Wageningen, the Netherlands.

Traube, S., K. Scoggin, S. Tjandrakusuma, M. Rasmussen and P. Reilly, 2007. Ruminal fermentation of propylene glycol and glycerol. J. Agric. Food Chem. 55: 7043-7051.

The effects of incremental fish oil supplementation on bacterial populations in the rumen

S. Huws[1], E.J. Kim[1], M.R.F. Lee[1], E. Pinloche[1], R.J. Wallace[2] and N.D. Scollan[1]
[1]IBERS, Aberystwyth University, Aberystwyth, SY23 3EB, United Kingdom; [2]Rowett Research Institute, Aberdeen University, Aberdeen, AB21 9SB, United Kingdom; nigel.scollan@aber.ac.uk

Introduction

Biohydrogenation of dietary polyunsaturated fatty acids (PUFA) in the rumen is believed to be carried out by ruminal bacteria belonging to the *Butyrivibrio* group (Paillard *et al.*, 2007), with the recently re-classified *Butyrivibrio proteoclasticus* (formerly *Clostridium proteoclasticum*; Moon *et al.*, 2008) thought to be a major contributor, particularly in the final step, the reduction of vaccenic acid (VA; *trans*-11 18:1) to stearic acid (18:0). Fish oil supplementation of the ruminant diet is beneficial due to the resultant enhanced concentrations of 20:5*n*-3 and 22:6*n*-3 in muscle and milk whilst fish oil also inhibits the complete metabolism of PUFA to 18:0, resulting in an increase in the flow of VA leaving the rumen. The aim of the present study was to evaluate how incremental fish oil inclusion in the diet influences bacterial diversity in the rumen using Terminal-Restriction Fragment Length Polymorphisms (T-RFLP) and quantitative PCR (qPCR) using both DNA and RNA as genomic and expression markers, respectively.

Material and methods

Eight Holstein × Friesian steers prepared with ruminal cannulae were offered grass silage and fish oil at 0, 1 and 3% on a dry matter intake basis in a Latin Square 3-period changeover design. Following 17 d adaptation to each diet, ruminal samples were collected for fatty acid analysis and fractionation into liquid- and solid-associated bacterial fractions (LAB and SAB, respectively). Both DNA and RNA were extracted from LAB and SAB samples, with the RNA reverse transcribed to cDNA, before T-RFLP was performed. Fatty acid methyl esters were prepared by alkaline hydrolysis followed by methylation with diazomethane and analysed on a CP-Sil 88, 100 m × 0.25 mm ID column (Chrompack, UK). The associations between concentration of 18:0 and the nucleic acids of 18:0-producing bacteria were conducted using GenStat (VSN International Ltd, Hemel Hempstead, UK).

Results

Increasing dietary fish oil altered ruminal fatty acid metabolism by increasing *trans*-11 18:1 and decreasing in 18:0 composition. A T-RFLP-derived unweighted pair-group method with arithmetic mean (UPGMA) dendrogram, using both DNA and RNA as markers and the endonucleases *Alu1*, *HaeIII*, and *MSP1*, showed that LAB (planktonic) and SAB (biofilm) populations generally clustered distinctly, but fish oil did not globally alter the bacterial populations. Nonetheless, when evaluating bacterial changes within individual steers, fish oil altered the total rumen eubacterial populations (Figure 1 shows a dendrogram generated using RNA as a marker and the endonuclease *Alu1*). Neither RNA and DNA concentration for *Butyrivibrio proteoclasticus* were correlated with 18:0 concentration (LAB r^2=0.0539 and SAB r^2=0.0132 for DNA; LAB r^2=0.033 and SAB r^2=0.017 for RNA).

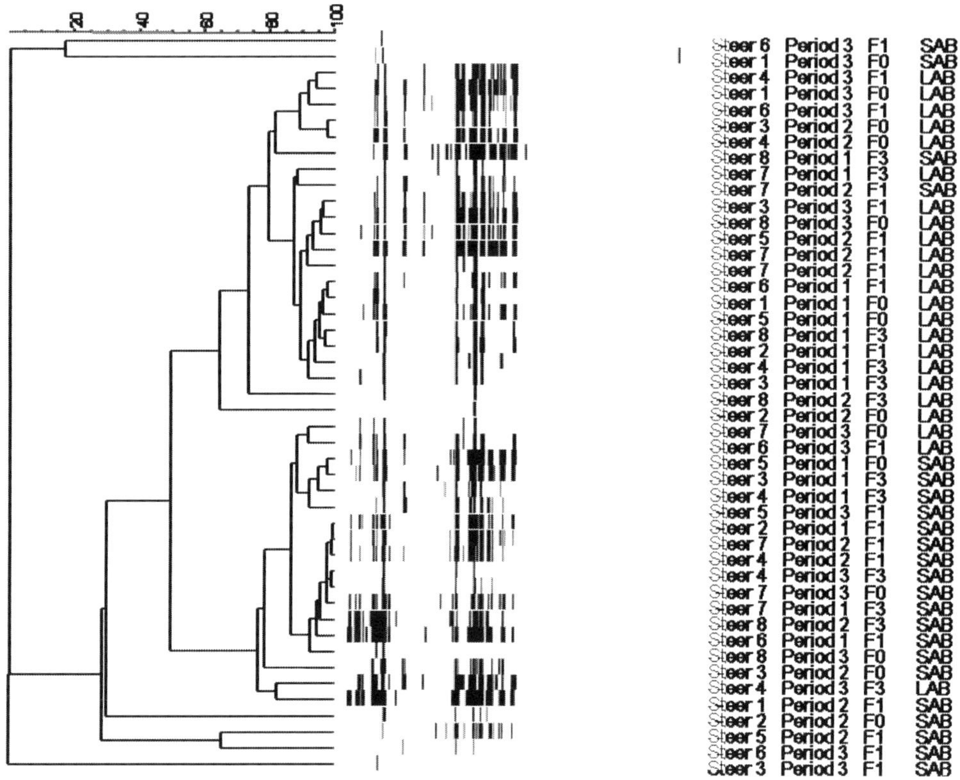

Figure 1. Total eubacterial 16S rRNA PCR-T-RFLP (Alu1)-derived unweighted pair group method with arithmetic mean (UPGMA) dendrogram showing the effect of fish oil on the total eubacterial LAB and SAB populations in the rumen. F0, F1 and F3 relate to fish levels used. Scale relates to percent similarity.

Conclusion

Use of both genomic level (DNA) and activity level (RNA) as markers did not reveal a strong correlation between *B. proteoclasticus* and 18:0 concentration in the rumen. This may suggest that other bacteria, perhaps unculturable, may play a role in biohydrogenation.

Acknowledgement

This work was supported by the Biotechnology and Biological Sciences Research Council (BBSRC), European ProSafeBeef project, and the Scottish Government Rural and Environmental Research Directorate (RERAD).

References

Moon, C.D., D.M. Pacheco, W.J. Kelly, S.C. Leahy, D. Li, J. Kopečný and G.T. Attwood, 2008. Reclassification of *Clostridium proteoclasticum* as *Butyrivibrio proteoclasticus* comb. nov., a butyrate-producing ruminal bacterium. Int. J. Syst. Evol. Microbiol. 58: 2041-2045.

Paillard, D., N. McKain, M.T. Rincon, K.J. Shingfield, D.I Givens and R.J. Wallace, 2007. Quantification of ruminal Clostridium proteoclasticum by real-time PCR using a molecular beacon approach. J. Appl. Microbiol. 103: 1251-1261.

Methane emissions and liveweight gain of cattle fed supplements of cottonseed and coconut oil

A.V. Klieve[1,2], *S.R. McLennan*[1], *D. Ouwerkerk*[1] *and R.S. Hegarty*[3]
[1]*DPI&F, Yeerongpilly, Queensland, Australia;* [2]*University of Queensland, Gatton, Queensland, Australia;* [3]*NSW DPI, Armidale, NSW, Australia; a.klieve@uq.edu.au*

Introduction

Methane is a by-product in the digestion of plant material. Effectively it is wasted feed material and energy that could otherwise be available for animal production. Methanogenesis is a microbiological mechanism that removes hydrogen from the rumen. Methane is a major greenhouse gas contributing heavily to global warming with a warming potential 21 times greater than carbon dioxide. Cattle and sheep contribute 53% of Australia's total methane emissions (3 million tonnes annually or the equivalent of 63 million tonnes of carbon dioxide) (NGGIC, 2007) and 14% of the nation's total greenhouse gas emissions.

Many lipid containing feeds are known to reduce methane generation and are attractive as nutritional supplements due to their high content of concentrated energy and/or protein. Coconut oil has been shown to significantly suppress methane production. Studies have shown that on average, methane suppression could be equated to an increase in capture of ingested energy of up to 27% (Machmuller and Kreuzer, 1999). The impact of a commonly used feed supplement in northern Australia, cottonseed (ca. 20% lipids) on methane generation had not been previously studied, although in temperate climes it has been shown to reduce emissions (Grainger *et al.*, 2008). In this study we evaluated the impact of supplements containing variable concentrations of coconut and cottonseed oils on rumen methanogenesis, methanogen populations, and on performance of *Bos indicus* cross cattle.

Material and Method

The experimental design was a randomised complete block design with 2 basal diets, 2 supplement types and 4 rates of supplement feeding. The basal feeds were pangola grass (*Digitaria eriantha*) hay fed *ad libitum* (H) or the same hay fed at 0.75% of liveweight (W; as fed basis)/d plus a molasses mix fed *ad libitum* (HM). The supplements were cottonseed meal plus cottonseed oil or copra meal plus coconut oil, both formulated to contain 20% oil (DM basis). For each basal feed type there was an unsupplemented control and the two supplements were fed at 0.2, 0.4 and 0.6%W/d. There were three heifers for each treatment making a total of 42 heifers. Liveweight and feed consumption were determined on a weekly basis.

The experiment consisted of a 7 d equilibration period followed by a 49 d growth study and a 10 d methane collection period. On day 38 of the growth study, rumen fluid was collected from all heifers and samples were taken to determine total methanogen numbers by real-time PCR, and methanogen community composition by denaturing gradient gel electrophoresis (DGGE) (Klieve *et al.*, 2006). Methane production was determined using the sulphur hexafluoride (SF_6) tracer approach as modified by Hegarty *et al.* (2007).

Results

The inclusion of oil fortified copra and cottonseed meals increased liveweight gain in cattle by up to 20 kg without the addition of molasses and up to 30 kg with molasses added, over a 7 wk period. Concomitant with the increased productivity there was no overall change in methane emissions per head but a substantial decrease in methane emitted per unit of production (1000 g to less that 100 g CH_4/kg ADG) (Figure 1).

In hay only (H) heifers, methanogen population density decreased with increasing coconut oil but not cottonseed oil in the diet. The unsupplemented HM heifers had fewer methanogens/ml of rumen fluid than did the unsupplemented H heifers. DGGE profiles indicated the presence of between 6 and 10 prevalent rumen methanogens but there were no consistent differences between heifers on different supplements.

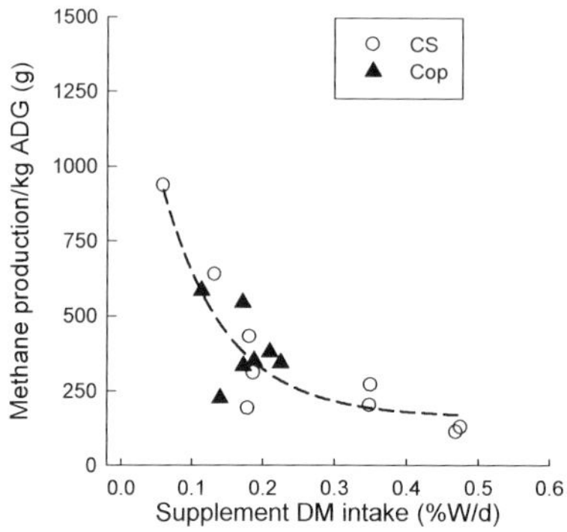

Figure 1. Effect of supplements on methane production per unit of average daily gain (ADG) for heifers given a basal diet of low quality pangola grass hay ad libitum. Symbols represent values for individual heifers. CS – cottonseed meal/oil; Cop – copra meal/oil.

Conclusion

Coconut and cottonseed supplements, with oil content up to 6% of DM, increase liveweight gain in cattle without increasing overall methane generation and they markedly decrease methane per kilogram of average daily gain.

Acknowledgement

This work was co-funded by the Department of Climate Change (formerly the Australian Greenhouse Office), Meat and Livestock Australia and the Department of Primary Industries and Fisheries (Queensland).

References

Grainger, C., T. Clarke, K.A. Beauchemin, S.M. McGinn and R.J. Eckard, 2008. Supplementation with whole cottonseed reduces methane emissions and can profitably increase milk production of dairy cows offered a forage and cereal grain diet. Aust. J. Expt. Agric. 48:73-76.

Klieve, A.V., M.N. O'Leary, A.J. Maguire, L. McMillen and D. Ouwerkerk, 2006. *In vitro* impact of lipids on ruminal methanogen populations and methane production. Book of Abstracts; 11 th International Symposium on Microbial Ecology, Pg. 250.

Machmuller, A. and M. Kreuzer, 1999. Methane suppression by coconut oil and associated effects on nutrient and energy balance in sheep. Can. J. Anim. Sci. 79:65-72.

NGGIC, 2007. National Greenhouse Gas Inventory 2005 with Methodology Supplement. Environment Australia: Canberra, Australia.

Hegarty R.S., J.P. Goopy, R.M. Herd and B. McCorkell, 2007. Cattle selected for lower residual feed intake have reduced daily methane production. J. Anim. Sci. 85:1479-1486.

Effects of a niacin supplementation to different diets on rumen fermentation, amounts of niacin at the duodenum and its concentration in blood and milk of dairy cows

P. Lebzien, I.-D. Niehoff, L. Hüther, W. Bigalke, S. Dänicke and G. Flachowsky
Institute of Animal Nutrition, Friedrich-Loeffler-Institute (FLI), Federal Research Institute for Animal Health, Bundesallee 50, 38116 Braunschweig, Germany; peter.lebzien@fli.bund.de

Introduction

Since niacin is incorporated in the coenzymes NAD(H) and NADPH(H), it is of great importance for the metabolism. Ruminants have different sources of niacin: feed, endogenous synthesis and synthesis by rumen microorganisms. It is believed that ruminal niacin synthesis covers the requirements of dairy cows. According to NRC (2001), the niacin requirement of a cow (650 kg liveweight and 35 kg of fat-corrected milk) is only 289 mg/d. But even so, beneficial effects of niacin supplementation on metabolism or performance have been observed (Niehoff *et al.*, 2009). However, experiments without an effect of supplemental niacin can also be found. Thus it was suggested that ruminal niacin degradation and/or synthesis is influenced by the ration fed. Hence, the present study was aimed at investigating the effect of niacin supplementation at different forage-to-concentrate ratios in the diet on rumen fermentation, amounts of niacin at the duodenum and its concentration in blood and milk of dairy cows.

Material and methods

A total of 7 midlactating and 3 dry Friesian cows (average weight: 599 kg), equipped with a large cannula in the dorsal sac of the rumen and a simple T-shaped cannula in the proximal duodenum were used. The concentrate (C) to forage (F) ratios in the rations, on dry matter (DM) basis, were either 1/3: 2/3 (LC), 1/2 : 1/2 (MC) or 2/3 : 1/3 (HC). Forage consisted of 60% maize silage and 40% grass silage on DM basis. The cows were fed in one period without and in the following period with a supplementation of 6 g nicotinic acid (NA) per cow and day. Each period was divided into 2 wks for adaptation, 1 wk for rumen, blood and milk sampling and 1 wk for duodenal sampling. The basal niacin content was 35, 35 and 34 mg/kg DM for the LC, MC and HC diet, respectively. Blood was sampled three times in the morning from a *vena jugularis externa*, milk on 2 d in the morning and in the evening and rumen fluid seven times around the first morning feeding. Spot samples of duodenal chyme were collected every 2 h for 5 consecutive days. Cr_2O_3 was used as a flow marker for duodenal flow and microbial portion in duodenal N was estimated by NIRS.

Results

Mean organic matter intake (12.3 kg) was comparable for all treatments. Effects of NA supplementation on some parameters of rumen metabolism are shown in Table 1. NA supplementation caused several significant changes in rumen metabolism: rumen ammonia and microbial protein synthesis increased, while the short chain fatty acid concentration decreased and pH-value was not affected. With regards to the given parameters, there were no significant interactions between NA supplementation and F:C ratio in the diet. The amounts of niacin reaching the duodenum were less in the LC diet than with higher concentrate levels and rose with NA supplementation (Table 2). In blood and milk only nicotinamide (NAM) could be detected. NAM concentration in the blood was increased by concentrate portion as well as by NA supplement. No interactions occurred between F:C ratio and NA supplementation. NAM in milk was increased only by increased concentrate levels.

Table 1. Effects of NA supplementation to different diets on some parameters of ruminal metabolism (LS means).

	LC (1/3C)		MC (1/2C)		HC (2/3C)		ANOVA, *P*-values		
	Control	+6g NA	Control	+6g NA	Control	+6g NA	C	NA	C × NA
pH-value	6.36	6.37	6.33	6.35	6.25	6.35	0.16	0.12	0.37
NH_3 (mmol/l)	6.5	8.0	7.1	9.5	6.3	8.5	0.04	<0.001	0.49
SCFA (mmol/l)	113.6	99.1	105.8	99.2	114.9	108.9	<0.01	<0.001	0.20
Microb.Prot. (g/d)	946	1,053	1,139	1,163	1,166	1,412	<0.001	<0.01	0.10

LC, MC, HC = Low, Medium and High concentrate (C) rations; NA = nicotinic acid; SCFA = short chain fatty acids.

Table 2. Niacin intake and duodenal flow (mg/d), as well as serum and milk concentrations (mg/l) (LS means).

	LC (1/3C)		MC (1/2C)		HC (2/3C)		ANOVA, *P*-values		
	Control	+6g NA	Control	+6g NA	Control	+6g NA	C	NA	C × NA
Niacin intake	553	6,449	325	6,337	476	6,370			
Duodenal niacin flow	1,602	2,021	1,886	2,221	1,895	2,630	<0.001	<0.001	0.25
Apparent niacin synth.[1]	1,057	-4,419	1,575	-4,089	1,421	-3,738	<0.01	<0.001	0.22
NAM in blood	0.32	0.36	0.35	0.46	0.41	0.52	<0.001	<0.001	0.11
NAM in milk	0.54	0.55	0.66	0.72	0.61	0.64	0.02	0.26	0.80

NAM = nicotinamide; [1] Apparent niacin synthesis = duodenal niacin flow – niacin intake.
There was a close relationship ($y = 1.96\,x - 172.3$; $r^2 = 0.85$) between microbial protein synthesis (x, g/d) and niacin flow at the duodenum (y, mg/d).

Conclusion

The niacin flow at the duodenum as well as blood niacin concentration was enhanced by supplemental niacin and also with increasing proportions of concentrate in the diet.
Niacin supplementation increased NH_3–concentration and microbial protein synthesis in the rumen, but decreased SCFA-concentration in the rumen.
The amount of niacin reaching the duodenum seems to cover the requirement of the cows.
The effects of supplemental niacin seem to be fairly independent from the ration fed.

References

NRC, 2001. Vitamins. In: Nutrient requirements of dairy cattle. 7th edition. NRC. National Academy Press, Washington D.C., 162-177.

Niehoff, I.-D., L. Hüther and P. Lebzien, 2009. Niacin for dairy cattle: a review. Br. J. Nutr. 101: 5-19.

Alpine vegetation essential oils and their effect on rumen lipid metabolism *in vitro*

M. Lourenço[1], L. Falchero[2], A. Tava[3] and V. Fievez[1]
[1]Laboratory for Animal Nutrition and Animal Product Quality, Department of Animal Production, Ghent University, Proefhoevestraat 10, 9090 Melle, Belgium; [2]Grazing-Land Management Unit, Department AGROSELVITER, University of Turin, Italy; [3]Research Centre for Fodder Crops and Dairy Production, Agricultural Research Council, Lodi, Italy; marta.lourenco@ugent.be

Introduction

Feed additives are strictly regulated in European animal production systems creating a need for natural alternatives, such as natural plant compounds that are both safe for the animals and for humans. Recently, some essential oils were reported to inhibit the growth of *Butyrivibrio fibrisolvens* JW11 and of *Clostridium proteoclasticum* P18, two important biohydrogenating bacteria (Durmic *et al.*, 2008). In addition, the botanical composition of pastures has been shown to affect the rumen lipid metabolism and this was suggested to be linked to the presence of plant secondary metabolites (Lourenço *et al.*, 2008). Thus, the objectives of this study were to assess the effect of plant secondary metabolites present in the essential oils prepared from 4 botanically different alpine pastures on *in vitro* rumen lipid metabolism.

Material and methods

Plant secondary metabolites were extracted from fresh vegetation samples of alpine pastures (Asiago: nutrient-rich red fescue pasture, FP; Asiago: nutrient-poor red fescue pasture, FM; Stura: red fescue pasture, FV and Stura: alpine clover pasture, TV) and essential oils (EO) prepared by steam-distillation (non-endogenous compounds represented 15% and 5% of the total EO composition for TV and FV pastures, and for FP and FM pastures, respectively). Alpine pastures were situated in the Italian Alps (Asiago: Malga Dosso di Sotto, Eastern Alpine region – Lat. 45°57'44" Long. 11°24'10"; Stura: Alpe Valcavera, Demonte – Western Alpine region – Lat. 44°22'54" Long. 7°6'43"). Essential oils (5.0 mg/incubation) were incubated *in vitro* with rumen fluid collected from 2 non-lactating Holstein dairy cows and screened for their effect on rumen lipid metabolism with or without addition of an external lipid source (20 mg of a 50:50 (v/v) mixture of linseed and sunflower oil). Glass gas tight flasks were incubated in triplicate on separate days, for 24 h at 39 °C with constant agitation. After 24 h, incubation fluid was prepared for long chain fatty acid (LCFA) analysis. In order to correct for eventual (trace) amounts of LCFA in EO extracts, the fatty acid metabolism of the external oil source (net LCFA) was assessed from the difference between the amounts of LCFA present in the incubation fluid with and without added lipid source. Apparent biohydrogenation was calculated from C18:2 n-6 and C18:3 n-3 proportions in dietary and incubation fluid C18 fatty acids. A general linear ANOVA model was used to evaluate the effect of the different essential oils on rumen lipid metabolism, according to $Y_{ij} = \mu + EO_{i=1,4} + R_{j=1,3} + \xi_{ij}$. Essential oils ($EO_{i=1,4}$; FP, FM, FV and TV) were introduced as fixed factors and runs ($R_{k=1,3}$; 1, 2 and 3) were introduced as random factor, μ = the overall mean and ξ_{ijk} = the residual error. Treatments were compared to the control using a 2-sided Dunnett post-hoc test. Differences were declared at $P<0.05$. When EO showed a significant effect, the different EO were compared using a Duncan *post-hoc* test.

Results

Proportions and apparent biohydrogenation of both C18:2 n-6 and C18:3 n-3 were not affected by any of the tested alpine vegetation EO (Table 1) compared with the control. However, the major biohydrogenation pathway of C18:2 n-6 in the presence of FP alpine vegetation EO seemed to be affected based on the trend for higher CLA *c9t11* proportions *vs* the control. This was further reflected in numerically lower C18:0 proportions for FP alpine vegetation EO, although not reaching a statistical significance (Table 1). This could suggest shifts in the rumen biohydrogenating microbial population in the presence of FP alpine vegetation EO. The latter consisted of 75% monoterpenes, of which limonene (35%) and carvone (31%) were the predominant compounds, suggesting that limonene and carvone could have antimicrobial effects on the biohydrogenating microbial population, impairing their growth and/or activity *in vitro*.

Table 1. Effect of the different alpine vegetation EO on net individual LCFA (expressed relative to the total amount of C18 fatty acids) and on apparent biohydrogenation (g C18:2 n-6 hydrogenated/100 g of C18:2 n-6 or input) after 24 h in vitro incubation (n=3).

	TV	FV	FP	FM	Control	SEM	Oil	Run
C18:3 n-3	14.9	14.0	14.4	13.3	14.5	1.44	0.945	0.202
C18:2 n-6	23.5	22.9	24.1	23.2	25.6	1.74	0.824	0.219
C18:2 *t11c15*	4.99	5.65	5.35	4.60	4.55	0.473	0.462	0.023
CLA *c9t11*	1.72^{c}	2.34^{bc}	3.88^{*a}	2.98^{ab}	2.26	0.382	0.030	0.128
C18:1 *t11*	9.02	9.84	9.29	8.28	7.93	1.69	0.926	0.354
C18:0	10.7	8.79	7.01	9.75	11.0	1.68	0.490	0.068
Apparent biohydrogenation								
C18:2 n-6	32.0	33.7	30.0	32.8	26.1	5.11	0.840	0.203
C18:3 n-3	47.4	50.4	49.0	53.0	48.8	5.14	0.947	0.197

* Trend (P<0.1) for significant difference from control.

^{a,b,c} Means with different superscripts in the same row differ significantly (P<0.05).

Conclusion

Based on our results, limonene and carvone are suggested to affect the ruminal biohydrogenating microbial population. Nevertheless, further research is needed to assess the true potential of plant secondary metabolites as manipulators of the ruminal lipid metabolism.

Acknowledgement

Post-doctoral grant of M. Lourenço from the Special Research Fund, Ghent University, Belgium.

References

Durmic, Z., C.S. McSweeney, G.W. Kemp, P. Hutton, R.J. Wallace and P.E. Vercoe, 2008. Australian plants with potential to inhibit bacteria and processes involved in ruminal biohydrogenation of fatty acids. Amin. Feed Sci. Technol. 145: 271-284.

Lourenço, M., B. Vlaeminck, G. Van Ranst, S. De Smet and V. Fievez, 2008. Influence of different dietary forages on the fatty acid composition of rumen digesta and ruminant meat and milk. Amin. Feed. Sci. Technol. 145: 418-437.

Level of intake and physiological state influences methane emissions from sheep fed fresh pasture

S. Muetzel, T.W. Knight, S.O. Hoskin, G. Molano, S. Maclean, D. Silva-Villacorta and H. Clark
AgResearch Grasslands, Private Bag 11-008, Palmerston North, New Zealand;
Stefan.Muetzel@agresearch.co.nz

Introduction

Hydrogen is formed in the rumen through the fermentation of substrates to short chain fatty acids. The principal route for disposal of hydrogen is methane formation. Methane yield (g CH_4/kg dry matter intake) however varies depending on the source of feed and feed composition (Blaxter and Clapperton, 1965). Evidence for the effect of level of feeding on methane yield is equivocal (Molano and Clark, 2008). Animal factors such as rumen pool size and fractional outflow rate can influence methane yield (Pinares-Patino *et al.*, 2003) and there is also animal to animal variation in methane yield for ruminants (Pinares-Patino, unpublished data). The aim of this experiment was to determine the effect of DMI (expressed as levels of maintenance) on methane yield (g CH_4/kg DMI).

Material and methods

The experiment was carried out between May and December 2008 at the AgResearch Grasslands Centre, Palmerston North Campus, using ewes in different physiological states (lactation, pregnancy). Lactating ewes were included in order to achieve high levels of intake of >2× their maintenance energy requirement (MEr). Twenty 18-month old ewes were used in the experiment and there were 6 measurement periods (Periods 1 to 6). Ten ewes were non-pregnant and non-lactating (Control group) and 10 were pregnant with a single lamb (Treatment group). The ewes in the Treatment group were in early, mid, late pregnancy, mid, late lactation and post weaning respectively in Periods 1 to 6. The Control group were fed to MEr in each of the 6 periods. The Treatment group was offered 1.0, 1.5, 2.0, 3.0, 2.0 and 1.0 × MEr during periods 1 to 6 respectively. During the 6 measurement periods, ewes were fed fresh cut grass indoors and were adapted to their feeding level for at least 8 days prior to methane measurements made on 2 consecutive days in open circuit calorimeters. During the lactation periods (Periods 4 and 5), when indoors, lambs were kept separated from the ewes and were trained to suckle twice a day before the ewes were fed. Between measurement periods the ewes grazed in pure ryegrass paddocks. Diets were offered twice a day 09:00 and 16:00 h. Samples of feed offered and refused were collected daily and analysed for DM content. Nutrient composition of fresh grass offered to the ewes was analysed by NIR spectroscopy. Simple and multiple regression analysis were done using Genstat v10.2 (2009).

Results and discussion

The composition of the diets varied considerably during the experiments (ash 64 - 102, crude protein 66 to 156, NDF 445 – 523 g/kg DM) with quality decreasing markedly in the summer (Period 6). Despite the differences in feed composition, the methane yield (yCH_4) from the control group, fed at maintenance level, was similar except for period 2 where one ewe showed very low methane emissions for which there was no obvious explanation.

Increasing the DMI (multiples of maintenance) decreased the yCH_4 by an average of 5.3 g/kg DMI including data for all 20 ewes. DMI accounted for 21.4% of the variance of yCH_4. Pregnancy status did not appear to affect yCH_4 (P=0.76). When DMI was increased during pregnancy from 1 to 2× MEr the yCH_4 decreased from 25.4 to 21.1 g CH_4/kg DMI. During lactation, when ewes were fed 2 and 3x MEr, yCH_4 decreased from 24.2 to 22.0 g CH_4/kg DMI. A multiple regression approach indicated that physiological status of the ewes (lactating, pregnant) accounted for 11.3% of the variation in yCH_4. Adding DMI increased this value to 52.7%. The slope and intercept values of the regressions of the non pregnant/non lactating and pregnant ewes were similar and therefore the datasets were combined into one group (non-lactating group; Figure 1). The full dataset is best described (adjusted $r^2 = 0.531$) by two parallel lines with a slope of -5.51 and significantly different intercepts ($P<0.001$) of 32.6 and 39.5 for non-lactating and lactating ewes respectively (Figure 1). The predicted methane yields (mean±se) at a common feed intake level of 2.2×MEr were 27.2±0.50 and 20.4±0.36 for lactating and non-lactating ewes respectively.

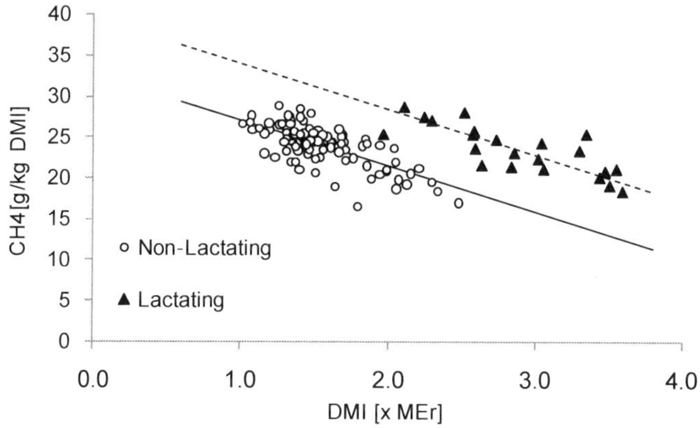

Figure 1. Relationship between methane yield and DMI (multiples of maintenance) for lactating and non-lactating ewes fed fresh ryegrass.

Conclusion

The methane yield at a maintenance level of feeding was generally consistent across the 6 periods for non-pregnant and non-lactating ewes. Pregnancy status did not appear to affect methane yield. An increase in DMI (multiples of maintenance) resulted in a decreased methane yield regardless of physiological state, however lactating ewes appeared to have a significantly higher methane yield than non-lactating ewes at a common level of intake (about 2.2× MEr). This experiment was not specifically designed to test the effect of lactation status on methane yield therefore further studies are required to confirm our observation and evaluate whether the effect is driven by changes in fatty acid production, turnover or substrate digestibility.

References

Blaxter, K.L. and J.L. Clapperton, 1965. Prediction of the amount of methane produced by ruminants. Br. J. Nutr. 19: 511-522.

Molano, G. and H. Clark, 2008. The effect of level of intake and forage quality on methane production by sheep. Austr. J. Exp. Agric. 48: 219-222.

Pinares-Patino, C.S., M.J. Ulyatt, K.R. Lassey, T.N. Barry and C.W. Holmes, 2003. Rumen function and digestion parameters associated with differences between sheep in methane emissions when fed chaffed lucerne hay. J. Agric. Sci. 140: 205-214.

Ruminal metabolism of soluble rapeseed meal protein *in vitro*

T. Stefanski and S. Ahvenjärvi
MTT Agrifood Research Finland, 31600, Jokioinen, Finland; tomasz.stefanski@mtt.fi

Introduction

The proportion of dietary protein metabolised in the rumen and escaping ruminal degradation are essential information for the optimisation of dairy cow rations. Several *in vitro* methods to assess rumen protein degradation have been introduced but thus far none has been widely adopted. The objective of the current study was to develop an *in vitro* method to determine the rate of soluble rapeseed meal protein (SRSM) degradation by rumen microbes using two intrinsically identical sources of SRSM of which one was labelled with 15N (L-SRSM) and another was unlabelled SRSM (U-SRSM). This approach was based on a theoretical assumption that the combined use of labelled and unlabelled protein substantially increases the amount of information available for model development and parameter estimation relative to the use of labelled or unlabelled protein separately.

Material and methods

To prepare SRSM 12 g of rapeseed meal was mixed with 240 ml of buffer solution in a centrifugation tube. The mixture was continuously stirred for 1 h at 39 °C and then centrifuged at 10,000 x *g* for 15 min at 4 °C. The supernatant was removed by aspiration and filtered through Whatman no. 1 filter paper to remove all insoluble protein. Subsamples of L-SRSM and U-SRSM approximately 12.5 mg of N were incubated in the presence of buffer, carbohydrate mixture and rumen microbes for 10 h. During incubation, samples were obtained at 0, 20 min, 40 min, 1, 1.5, 2, 2.5, 3, 4, 6, 8, and 10 h; i.e. 12 times in total. To provide a constant supply of energy for the microbes over the entire incubation period the carbohydrate mixture contained 60 mg of citrus fruit pectin, 80 mg of potato starch, and 360 mg of neutral detergent extracted grass silage. To prepare the microbial inoculum, rumen contents were collected 2 h after morning feeding from 2 ruminally cannulated Finnish Ayrshire cows in late lactation. The rumen contents were pooled and squeezed through two layers of cheesecloth. To enrich the inoculum with microbes attached to feed particles the retained solids were blended with warm buffer and squeezed through the cheesecloth. The rumen fluid was mixed with buffer (1:1) and filtered through eight layers of cheesecloth. The inoculum was kept under a continuous CO_2 stream for 30 min at 39 °C to allow small feed particles to flocculate on the top of inoculum and then those were removed by aspiration. The incubations were carried out at 39 °C in 30 stir bar equipped spinner flasks with a capacity of 120 ml. The vessels were divided into four groups. The first group contained 3 flasks filled with inoculum and buffer, the second group contained 3 flasks with buffer, inoculum, carbohydrate mixture, and L-SRSM and U-SRSM mixed in a 50:50 ratio. The third group contained 12 bottles filled with buffer, inoculum, carbohydrate mixture, ammonium chloride, and approximately 12.5 mg of L-SRSM. The fourth group was similar to the third one with the exception that ammonium chloride was labelled with 15N and U-SRSM instead of L-SRSM was used. The rate of energy supply for the rumen microbes was determined measuring the rate of gas production over 24 h from groups one and two. After incubation the flask contents were fractionated into ammonia N (AN), soluble non ammonia N (SNAN) and insoluble N (ISN) and pool size of 14N and 15N isotopes in these N pools was determined (AN^{14},AN^{15}, $SNAN^{14}$, $SNAN^{15}$, ISN^{14}, ISN^{15}). The model presented in Figure 1 was fitted to the observed pattern of exchange of each isotope between the different N pools to describe the kinetics of SRSM metabolism. In the model, SRSM was either incorporated directly into microbial N or was metabolised into ammonia N. A delay pool between SRSM and ammonia N pool was needed to describe the delay between the disappearance of N from the SRSM pool and the appearance in the ammonia N pool. Microbial N synthesis from the ammonia N pool was related to the rate of carbohydrate synthesis determined by the rate of gas production.

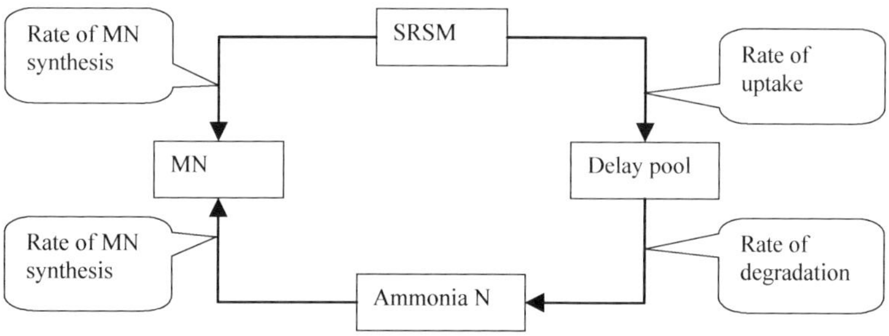

Figure 1. Model describing the kinetics of soluble rapeseed meal protein N degradation. SRSM = Soluble rapeseed meal protein; MN = Microbial nitrogen.

Results

The pool size of $SNAN^{14}$ and $SNAN^{15}$ for L-SRSM decreased in a linear manner during the 10 h incubation period from 13.23 to 5.05 mg and from 245 to 63 µg, respectively. In a similar fashion the $SNAN^{14}$ pool for U-SRSM decreased from 13.45 to 4.26 mg. The ISN^{14} pool for L-SRSM exhibited a rapid linear increase from 9.47 to 16.41 mg between 0 and 4 h and this was followed by a slower rate of increase up to 17.88 mg at 10 h. The ISN^{15} pool for L-SRSM increased from 79 to 163 µg between 0 and 4 h and from 163 to 175 µg between 6 and 10 h. The pattern of increases in ISN^{14} pool for U-SRSM was essentially similar to that observed for L-SRSM but the increases in pool size of ISN^{15} for U-SRSM exhibited a sigmoid pattern with a slow rate of increase up to 2 h (39 to 82 µg) followed by a steeper increase from 2 to 4 h (82 to 150 µg) and then slower rate of increase between 6 and 10 h (150 to 218 µg). The pattern of increases in ISN pools was closely associated with decreases observed in the ammonium pools with the exception of rapid initial increases in AN pools for L-SRSM. A delay pool between SRSM and AN pools in the model was needed to describe this phenomenon. The AN^{14} pools for L-RSM and U-LRSM first increased up to 1 h and then decreased up to 4 h and then increased up to 10 h. The AN^{15} pool for L-SRSM slowly increased up to 2.5 h, then decreased for the next 1.5 h, and then rapidly increased until the end of incubation. The AN^{15} pool for U-SRSM remained constant for 1 h, then rapidly decreased up to 4 h, and then decreased at a slower rate up to 10 h. When the model presented in Figure 1 was fitted to these observations, the overall performance of the model was very good with only small deviations between the predicted and observed patterns being evident. The mean rate of SRSM nitrogen degradation estimated using the model was 0.126/h with a standard deviation between runs of 0.0499/h. In conclusion, the current results indicate that the *in vitro* study involving combined use of labelled and unlabelled protein provided sufficient amounts of information to allow accurate parameter estimation for a complex dynamic model.

Effects of particle size and dry matter content of a total mixed ration on intraruminal transport and net portal absorption of VFA in lactating dairy cows

A.C. Storm and N.B. Kristensen
Faculty of Agricultural Sciences, Aarhus University, DK-8830 Tjele, Denmark; nbk@agrsci.dk

Introduction

Volatile fatty acids (VFA) of forestomach origin are absorbed across the rumen epithelium and from compartments distal to the rumen for the VFA being washed out of the rumen (Peters *et al.*, 1990). The relative importance of the possible physiological limitations to VFA absorption from the rumen are sparsely described and understood. Three processes could limit VFA absorption across the rumen epithelium: (a) intraruminal transport of VFA from the site of production to the site of absorption, (b) transport across the rumen epithelium, or (c) limited epithelial blood flow limiting removal of VFA from the serosal side of the epithelium.

In acute rumen acidosis the frequency of primary reticulorumen contractions (FPRC) is reduced or even absent reducing the mixing and passage of rumen contents to the omasum (Crichlow, 1988). Also under common feeding situations, rumen pH and VFA concentrations differ between the medial rumen matt and the ventral rumen sac showing lowest pH and highest VFA concentrations in the matt region (Tafaj *et al.*, 2005). This indicates that the rumen contents are not perfectly mixed and imperfect intraruminal VFA transport increases VFA concentration in the rumen matt. Our hypothesis was that large particles in the rumen matt and high dry matter content of the rumen matt after feeding a dry total mixed ration (TMR) would decrease efficiency of intraruminal VFA transport and rumen VFA absorption efficiency expressed as fractional rumen VFA absorption rate. The objective of the study was to investigate the effect of physical changes in the rumen matt on intraruminal transport- and net portal flux of VFA in lactating dairy cows.

Material and methods

Four Danish Holstein cows (parity 2, 121±17 day in milk, 591±24 kg BW) surgically fitted with rumen cannulas and permanent indwelling catheters in the major splanchnic blood vessels, were used. The experimental design was a 4×4 Latin square with 14 d periods and a 2×2 factorial design of treatments containing corn silage, grass hay, rolled barley and rapeseed cake (35, 30, 20 and 11% of DM, respectively). The diets differed in hay particle size (PS) (30 mm and 3.0 mm) and dry matter (DM) content (42.5 and 50.0%) resulting in four treatments (SM=short/moist, SD=short/dry, LM=long/moist, LD=long/dry). Cows were fed three times daily and feed intake was restricted to 20 kg DM/d. Eight sets of rumen fluid (medial and ventral rumen) and blood samples (arterial, portal- and rumen vein) were simultaneously drawn at hourly intervals from 30 min before the morning feeding on the 13th day in each period. Reticulorumen motility (FPRC) was monitored continuously in the 8 h sampling window and 24 h rumination behaviour was recorded. On day 14 of each period all cows were rumen evacuated before and after feeding. Rumen samples were analysed for pH. Plasma and rumen concentrations of VFA were analysed by gas chromatography (Kristensen, 2000; Kristensen *et al.*, 1996). FPRC was analysed by visual counts of primary contractions from 5 min before to 10 min after the blood sample. Fractional absorption rates of VFA were calculated as (net portal absorption/rumen pool). Data were analysed using a mixed model with cow × treatment as random and sample time as repeated measurements including cow, PS, DM, sample time, and period as class variables.

Results

The treatments were visually very different and differed on PS and DM. DMI was affected by DM×PS ($P<0.003$) because DM of the TMR were not equally adjusted, resulting in 1.16, 0.81, 0.19, 1.64 kg DM/d in excess feeding of SM, SD, LM and LD, respectively. DMI relative to DM fed was not affected by treatments nor were milk yield variables relative to DMI. Total rumen pools of dry matter, fluid and total contents increased ($P=0.01$, $P=0.006$, $P=0.005$) with long compared to short particles (1.7, 6.9, 8.6 kg), respectively. PS affected rumination time (long=490, short=374 min/d) and rumination time/kg DMI (both $P=0.001$). However, ventral rumen pH was only affected by DM×sample time ($P=0.01$) whereas medial rumen pH only was affected by sample time ($P<0.0001$). On average a 0.66 pH unit difference between medial (5.97±0.28) and ventral (6.63±0.38) rumen was observed. Ventral rumen molar proportions of VFA (mol/100 mol) were affected by DM for isobutyrate ($P=0.01$) and DM×PS×sample time for butyrate, valerate, and carporate ($P=0.02$, $P=0.004$, $P=0.03$, respectively). Medial rumen molar proportions of acetate were affected by DM ($P=0.02$), isovalerate by PS ($P=0.05$), and isobutyrate were affected by DM×PS×sample time. No effect was observed on the FPRC or on net portal flux of VFA except for sample time ($P<0.001$). The fractional rumen absorption rates of acetate and propionate were not affected by PS or DM (0.44±0.1 and 0.46±0.12/h, respectively).

Discussion

Even though particle size was reduced extensively in the TMR, which was reflected in the decrease in rumination time and rumen fill, the FPRC, -VFA and net portal flux of VFA were not affected. Even with the reduced rumen fill, fractional rumen absorption rates of acetate and propionate were not affected. This indicates that the physical changes of the rumen contents did not directly affect the absorption kinetics by reducing the rate of intraruminal transport of VFA with long particles or dry rations as was hypothesised. However, since the FPRC was not affected, the intraruminal transport was not evaluated in relation to rumen motility. The pH and VFA concentration differences from medial to ventral rumen indicate a limitation to intraruminal VFA transport that we were unable to manipulate by the applied treatments. In conclusion, the hay particle size and dry matter content of TMR did not affect the intraruminal transport and rumen absorption of VFA in lactating dairy cows.

References

Crichlow, E.C., 1988. Ruminal lactic-acidosis - forestomach epithelial receptor activation by undissociated volatile fatty-acids and rumen fluids collected during loss of reticuloruminal motility. Res. in Vet. Sci. 45: 364-368.

Kristensen, N.B., 2000. Quantification of whole blood short-chain fatty acids by gas chromatographic determination of plasma 2-chloroethyl derivatives and correction for dilution space in erythrocytes. Acta Agric. Scand. Sect. A 50: 231-236.

Kristensen, N.B., A. Danfaer, V. Tetens and N. Agergaard, 1996. Portal recovery of intraruminally infused short-chain fatty acids in sheep. Acta Agriculturae Scandinavica Section A-Animal Science 46: 26-38.

Peters, J.P., R.Y.W. Shen and S.T. Chester, 1990. Propionic acid disappearance from the foregut and small intestine of the beef steer. J. Anim. Sci. 68: 3905-3913.

Tafaj, M., Q. Zebeli, B. Junck, H. Steingass and W. Drochner, 2005. Effects of particle size of a total mixed ration on in vivo ruminal fermentation patterns and inocula characteristics used for *in vitro* gas production. Anim. Feed Sci. and Technol. 123-124: 139-154.

Use of polyethylene glycol (PEG) to assess the effect of condensed tannins on nitrogen balance and digestibility in sheep fed fresh sainfoin (*Onobrychis vicifolia*)

K. Theodoridou, J. Aufrère, D. Andueza and R. Baumont
INRA Unité de Recherches sur les Herbivores, Centre de Clermont-Fd / Theix, 63122 Saint-Genès-Champanelle France; theodori@clermont.inra.fr

Introduction

Proteins in many forage legumes are poorly utilised by ruminants because of their extensive, rapid degradation to ammonia in the rumen and excretion as urea in urine causing nitrogen (N) losses (Min *et al.*, 2003). Improving N utilisation is needed for more efficient and environmental friendly feeding systems in ruminants. Sainfoin is a temperate legume plant containing condensed tannins (CT). Condensed tannins are polyphenols able to bind proteins and to reduce their degradation in the rumen (Waghorn, 2008). The aim of this study was to assess the effect of the CT contained in the sainfoin on its digestibility and nitrogen balance in sheep. Measurements were made without and with polyethylene glycol (PEG), a compound which inactivates CT throughout the entire digestive tract.

Material and methods

Sainfoin (*Onobrychis vicifoilia*) cv. Perly was grown in Clermont-Ferrand (France) and was studied in a first experimental period at the end of flowering in the first growth cycle and in a second experimental period at mid bloom in the second growth cycle. Two groups of 6 Texel, male castrated sheep (12 months old, 60±3 kg live weight) fitted with rumen cannula were used for digestibility, nitrogen (N) balance measurements and sampling of ruminal fluid. One group was dosed daily with 60 g of PEG (2× 30 g in 200 ml water) delivered through the ruminal cannula at 2 times, and the other group was dosed with 200 ml of water (without PEG). The sheep were offered daily 60g dry matter of fresh sainfoin per kg metabolic weight ($W^{0.75}$). Sheep were placed in metabolism crates and urine and faeces samples were collected during six consecutive days. On two occasions, rumen fluid was sampled before the morning meal (T 0h), 1.5, 3 and 6 h after feeding. Total nitrogen content (tN) and ammonia nitrogen (NH3-N) were determined in the rumen fluid. Tannins in sainfoin extracts were determined using the Radial Diffusion Assay as described by Makkar *et al.* (2000). Data were subjected to analysis of variance using the MIXED Procedure by SAS® (2000).

Results and discussion

Organic matter digestibility (OMD) was significantly higher at the mid bloom stage than at flowering, due to a higher cell-wall digestibility ($P<0.001$), but was not significantly affected by the addition of PEG ($P=0.156$) in agreement with Aufrere *et al.* (2008) (Table 1). The addition of PEG significantly increased ($P<0.001$) the total tract N digestibility, revealing a negative effect of tannins on N digestibility. N Digestibility increased with the growth cycle ($P<0.001$) in relation to the higher N intake for the second growth cycle, despite a higher percentage of tannins in the mid bloom (1.58 eq tannic acid) compared to the end of flowering (1.02). The inactivation of tannins by PEG tended to increase N excretion in urine ($P=0.0754$), but this was compensated by significantly lower faecal N excretion. Body N-retained (g/g N intake) was not affected significantly by the addition of PEG. Total N content and NH3-N in the rumen fluid was increased with PEG ($P<0.05$) and the values were higher for the second growth cycle ($P<0.001$).

Table 1. N intake (g/d), total tract organic matter (OM) and nitrogen (N) digestibility, N losses and N retention, total nitrogen (tN) and ammonia nitrogen (NH3-N) in the rumen for sheep fed fresh sainfoin at flowering or mid bloom stage and with or without addition of polyethylene glycol (PEG) in the rumen.

	Flowering		Mid bloom		S.E	S.E	T	GC	T*GC
	PEG	No PEG	PEG	No PEG	GC	T			
N intake (g/d)	29.9	28.3	42.9	41.5	0.005	0.007	ns	***	ns
OM digestibility	0.581	0.607	0.648	0.654	0.006	0.007	ns	***	ns
N digestibility	0.677	0.601	0.775	0.665	0.006	0.007	***	***	*
N(g/gNintake)									
In faeces	0.32	0.40	0.22	0.34	0.003	0.004	***	***	ns
In urine	0.45	0.41	0.48	0.37	0.007	0.001	+	ns	ns
	0.227	0.191	0.298	0.294	0.019	0.025	ns	***	ns
N retained									
tN (mg/g)	0.226	0.193	0.276	0.217	0.008	0.010	*	***	
NH3-N (mg/g)	0.198	0.167	0.261	0.212	0.006	0.008	*	***	

Standard errors for growth cycle (GC) and PEG treatment (T) are given and significant effects (T×GC) are quoted as + = P<0.1, *=*P*< 0.05, **=*P*<0.01, ***=*P*<0.001.

Conclusion

Using PEG to inactivate CT, this study shows that the presence of CT in the sainfoin does not affect the OMD, the N intake and the N-retention. The presence of CT in the sainfoin decreased total tract N digestibility, increased N losses in the faces, but tended to decrease N losses in the urine in accordance with the lower NH3-N and tN contents in the rumen fluid. Work is under progress to investigate further N digestion of sainfoin respectively in the rumen and in the intestine by *in situ* measurements.

Acknowledgement

The authors thank the Marie Curie Research Training, the EU sainfoin project 'Healthy Hay', for supporting this program, and the staff who care the animals and help in the laboratory.

References

Aufrere, J., M. Dudilieu and C. Poncet, 2008. *In vivo* and *in situ* measurements of the digestive characteristics of sainfoin in comparison with lucerne fed to sheep as fresh forages at two growth stages and as hay. Animal 9: 1331-1339.

Makkar, H., I. Mueller-Harvey and A. Hagerman, 2000. Quantification of tannins in tree foliage: A laboratory manual for the FAO/IAEA, p12.

Min, B., T. Barry, T. Attwood and W. McNabb, 2003. The effect of condensed tannins on the nutrition and health of ruminants fed fresh temperate forages. Anim. Feed Sci Technol. 106: 3-19.

Waghorn, G., 2008. Beneficial and detrimental effects of dietary condensed tannins for sustainable sheep and goat production – Progress and challenges. Anim. Feed Sci Technol. 147: 116-139.

Influence of progressive faunation with *Entodinium caudatum*, *Epidinium ecaudatum* and *Eudiplodinium maggii* on ruminal fermentation and total tract digestibility in sheep

J.O. Zeitz[1], S.L. Amelchanka[1], T. Michałowski[2], K. Wereszka[2], M. Kreuzer[1] and C.R. Soliva[1]
[1]ETH Zurich, Department of Agricultural and Food Science, Switzerland; [2]Kielanowski Institute of Animal Physiology and Nutrition, Jabłonna, Poland; johanna.zeitz@inw.agrl.ethz.ch

Introduction

Protozoa can decrease nitrogen utilisation and support methane formation in the rumen, but total tract nutrient digestibility is often improved in the presence of natural ruminal fauna. Differentiation among protozoal species suggesting differences in digestibility as well as in nitrogen utilisation and duodenal N flow was only addressed in a few studies (Koenig *et al.*, 2007; Ivan, 2009). In particular, studies on the role of particular ciliate species in ruminal methanogenesis are scarce. Thus, in the present study, the effect of defaunation and progressive faunation of sheep with 3 species of entodiniomorphid protozoa on ruminal fermentation, methane formation and total tract digestibility in sheep was investigated.

Material and methods

Four rumen-fistulated adult Merino wethers (72.7±5.2 kg) were defaunated (no protozoa; 0P) by a modified rumen washing procedure followed by progressive refaunation first with *Entodinium caudatum* (EC), then *Epidinium ecaudatum* (EC+EE) and lastly *Eudiplodinium maggii* (EC+EE+EM). The diet consisted of hay, ground barley, and soybean meal in a ratio of 0.4:0.4:0.2, and was fed twice daily on a maintenance requirement level. After each refaunation step, protozoa were allowed to propagate for ≥3 wk, followed by an 8-d sampling period. The quantitatively collected samples of feed and faeces were analysed for dry matter, organic matter (OM), neutral and acid detergent fibre (NDF, ADF), and nitrogen (N). Crude protein (CP) was calculated as 6.25 × N. The quantitatively collected urine samples were analysed for N content. Three times a day, ruminal gas was collected through the rumen fistula equipped with a rubber plate using a syringe, and the CH_4/CO_2 ratio was determined by gas chromatography. Ruminal fluid was collected on a daily basis 4 h after morning feeding and ruminal pH and ammonia concentrations were determined using appropriately sensitive electrodes. Short chain fatty acids (SCFA) in ruminal fluid were determined by HPLC. Bacterial counts were determined with a Bürker counting chamber while, to determine the concentration of protozoa in ruminal fluid, all individuals present in a 0.1 ml sample were enumerated. Data were analysed using the MIXED procedure of SAS®.

Results

The highest concentration of protozoal cells was found in the EC treatment (8.8×10^5/ml). The EC species was also predominant in wethers faunated with EC + EE (3.2×10^5/ml; 78.5% EC, 21.5% EE), and with EC + EE + EM (4.9×10^5/ml; 84.3% EC, 14.4% EE, 1.3% EM). In all faunated sheep, the bacterial numbers were reduced (Figure 1, left), indicating engulfment of bacteria by protozoa, and/or competition for nutrients. Digestibility of OM and NDF were unaffected by the ruminal protozoa status, whereas digestibility of ADF was the highest in defaunated (0P) sheep and sheep faunated with EC only (Figure 1, right). Within the faunated sheep, digestibility of CP was numerically the lowest in 'EC' resulting in a non-significant trend to a lower percentage of dietary N excreted with urine (69% in EC as compared to 73.3 and 72.7% with EC + EE and EC

+ EE + EM, respectively). The lowest ammonia concentration was found in 0P sheep as expected, followed by sheep faunated with EC, and, significantly, by the sheep inoculated with EC + EE and with EC + EE + EM (Figure 1, left). Neither ruminal CH_4/CO_2 ratio nor ruminal fluid pH (0.48 and 6.2 on average over all protozoal status, respectively) was affected by the protozoal treatments. Acetate and propionate concentrations were similar in all sheep, but A/P ratio was the highest with 0P (4.21) while it was the lowest with EC (3.42), and EC + EE + EM (3.47) (P<0.1). Butyrate concentrations were the lowest with EC, but increased significantly when EE and EE + EM were added step-wise to EC.

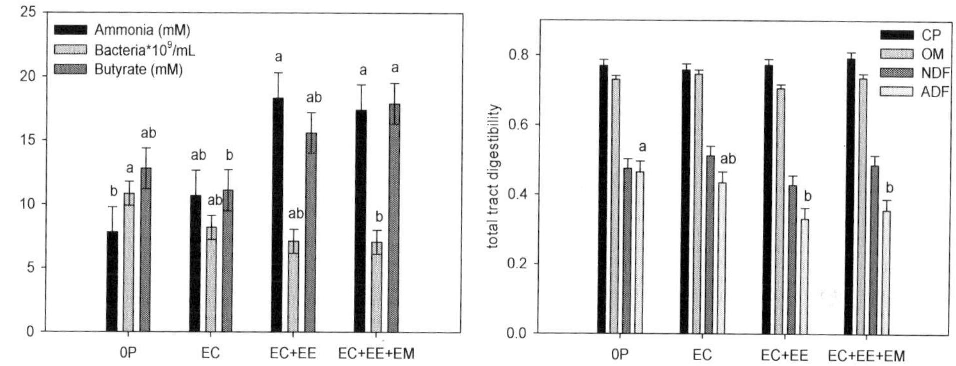

Figure 1. Differences in ruminal fluid variables (left side) and digestibility (right side) of sheep having no (0P), one (EC), two (EC + EE) or three (EC + EE + EM) protozoal species inhabiting the rumen; means without a common letter are different at P<*0.05.*

Conclusion

The results indicate that ruminal N metabolism, bacterial counts and SCFA production depend on the composition of the ruminal fauna population. Especially *Entodinium caudatum,* when present alone, seems to have a less negative influence on N utilisation than in combination with the other species tested. Therefore, when a mixed ciliate fauna population is present, considering only total protozoal counts to predict their influence on ruminal fermentation processes might be a simplification which neglects important aspects.

Acknowledgement

This study was supported by the Swiss National Foundation.

References

Ivan, M., 2009. Comparison of duodenal flow and digestibility in fauna-free sheep inoculated with holotrich protozoa, *Entodinium* monofauna or total mixed protozoa population. Br. J. Nutr. 101: 34-40.

Koenig, K.M., M. Ivan, B.T. Teferedegne, D.P. Morgavi., L.M. Rode, I.M. Ibrahim and C.J. Newbold, 2007. Effect of dietary *Enterolobium cyclocarpum* on microbial protein flow and nutrient digestibility in sheep maintained fauna-free, with total mixed fauna or with *Entodinium caudatum* monofauna. Br. J. Nutr. 98: 504-516.

Characterisation of methanogens in the rumen of cattle with different feed efficiency

M. Zhou, E. Hernandez-Sanabria and L.L. Guan
Department of Agricultural, Food and Nutritional Science, University of Alberta, Edmonton, Alberta T6G2P5, Canada; lguan@ualberta.ca

Introduction

Rumen methanogens play important roles in feed digestion, hydrogen pressure maintenance, and methane production of cattle (Zinder, 1993). However, methane emission causes a significant (6%) loss of dietary energy for the animals (Johnson and Johnson, 1995). It is well known that methane gas produced by the ruminants contributes to 13-19% of global green gas emission (Lassey *et al.*, 1997). Therefore, the energy loss and the consequent methane emission arouse both nutritional and environmental concerns in the livestock industry. Recent studies found that efficient animals produce ~20% less methane than the inefficient ones (Hegarty *et al.*, 2007). However, the linkage between the methanogenic ecology and host feed efficiency has not been established. This study was aimed at elucidating the association between methanogenic ecology and the host's feed efficiency by comparing the methanogenic diversity in the rumen of cattle with different residual feed intake (RFI, an measure of the feed efficiency) under two different diets using culture independent PCR-denaturing gradient gel electrophoresis (PCR-DGGE) and sequencing analysis.

Material and methods

Rumen fluid samples were collected within 3 h after feeding by inducing flexible plastic tubing into the rumen and using the suction created with a 50 ml syringe to remove the fluid from the tubing from 58 ten-month old steers. These animals were fed with a low-energy density feedlot diet (74% oats, 20% hay and 6% commercial supplement) first and then switched to high-energy density feedlot diet (28.3% oats, 56.7% barley, 10% alfalfa pellets, 5% feedlot supplement). RFI values (L-RFI (efficient): RFI <-0.5; M-RFI: $-0.5 \leq$ RFI ≤ 0.5; H-RFI (inefficient): RFI >0.5) were ranked under both diets and the numbers of animals in each RFI group is shown in Table 1. A ~190 bp partial methanogen 16S rRNA gene was amplified from total DNA extracted from each animal and the amplicons were subjected to DGGE (35-45%) analysis. The detectable methanogen profiles were compared among animals, among RFI and between diets. Twenty-eight PCR-DGGE bands were cloned and sequenced. A statistical method was developed to correlate the DGGE methanogenic band patterns to RFI. Each DGGE pattern from an individual steer was assigned by Bionumerics® software (Applied Maths, Austin, TX, USA) to generate a calculated best-fit Gaussian curve for each band using 1% tolerance in the migration distance to rectify the shifts among all the bands from all profiles. A binary matrix where all the bands were allocated was classified into new categories, and then a chi-squared model based on maximum likelihood was fit to analyse the interaction of RFI with the PCR-DGGE bands. Using the CATMOD procedure, the effects of RFI and diet on the prevalence of every band was determined.

Results

The detectable methanogenic PCR-DGGE profiles were clustered according to the ranks of RFI within each diet, suggesting probable correlations between methanogenic diversity and cattle's feed efficiency (data not shown). The PCR-DGGE bands representing the predominant species were found to be different between low-energy and high-energy diets. For example, the sequence resembling to *Methanobrevibacter ruminantium* was the predominant methanogen when fed a low-energy diet. While under high-energy diet, the predominant species fluctuated among animals (data not shown). The diet-associated (11 bands) and RFI-associated (5 bands) were identified. Examples of the diet associated, RFI-associated, diet and RFI associated bands are listed in Table 1. *Methanobrevibacter gottschalkii* was only detected when animals were fed a low-energy diet, and *Methanobrevibacter thaueri* was only detected under high-energy diet. Furthermore, some unidentified species such as PCR-DGGE bands a, d and b, c were found to be associated with L-RFI under low-energy and high-energy diet, respectively.

Table 1. Identified methanogenic species associated with diets and residual feed intake (RFI).

Methanogens	Low-energy diet			High-energy diet		
	L-RFI (n=22)	M-RFI (n=16)	H-RFI (n=20)	L-RFI (n=19)	M-RFI (n=24)	H-RFI (n=15)
PCR-DGGE band a	+	-	-	-	-	-
Methanobrevibacter gottschalkii	+	+	+	-	-	-
Methanosphaera stadtmanii	+	+	+	+	+	+
Methanosphaera stadtmanae	+	+	+	+	+	+
PCR-DGGE band b	-	-	-	+	-	-
Methanobrevibacter ruminantium	+	+	+	+	+	+
Methanobrevibacter sp. AbM4	-	+	+	+	+	+
Methanobrevibacter smithii	+	+	+	+	+	+
PCR-DGGE band c	-	-	-	+	-	-
Methanobrevibacter olleyae	+	+	+	-	+	+
PCR-DGGE band d	+	-	-	-	-	-
Methanobrevibacter thaueri	-	-	-	+	+	+
Methanobrevibacter sp. SM9	-	-	-	+	+	+

Conclusion

This study shows that the methanogenic PCR-DGGE profiles in the rumen were associated with RFI and diet. The change of feed component could affect the methanogenic structure significantly, suggesting that specific combination of methanogens correlating to feed regime and host's feed efficiency may be applied for regulation of methane production in cattle. This is the first study to link the methanogen ecology in the rumen to cattle's feed efficiency trait. A future study linking the methanogenic profiles with the methane gas yield from cattle with different RFI will be performed to verify and elucidate the different mechanisms of methanogesis in the animals with high RFI.

Acknowledgement

This study was supported by ALIDF/AARI (2007041R) and NSERC discovery grants for L.L. Guan. The authors thank Dr. B. Irving, and all the staff who helped the rumen sampling.

References

Hegarty, R.S., J.P. Goopy, R.M. Herd and B. McCorkell, 2007. Cattle selected for lower residual feed intake have reduced daily methane production. J. Anim. Sci. 85: 1479-1486.

Johnson, K.A. and D.E. Johnson, 1995. Methane emissions from cattle. J. Anim. Sci. 73: 2483-2492.

Lassey, K.R., M.J. Ulyatt, R.J. Martin, C.F. Walker and I.D. Shelton, 1997. Methane emissions measured directly from grazing livestock in New Zealand. Atmospheric Environment 31: 2905-2914.

Zinder, S.H., 1993. Physiological ecology of methanogens. In: Ferry J.G. (Ed), Methanogenesis: ecology, physiology, biochemistry and genetics. Chapman & Hall Microbiology/CRC, London, UK, pp. 128-206.

The influence of the grape pomace on ruminal parameters and retained nitrogen of sheep

M.J. Abarghuei[1], Y. Rouzbehan[1] and D. Alipour[2]
[1]Animal Science Department, Faculty of Agriculture, Tarbiat Modares University, Tehran, P.O. Box 14115-336, Iran; [2]Animal Science Department, Faculty of Agriculture, Bu-Ali Sina University, Hamadan, Iran; rozbeh_y@modares.ac.ir

Introduction

Iran faces a scarcity in the quantity and quality of consistent year-round supplies of conventional ruminant feeds. Therefore, better utilisation of non-conventional feed resources, which do not compete as human foods, is imperative. Agro-industrial co-products, such as the grape pomace (GP), may form an important component of ruminant diets. In Iran, the production of GP exceeds 50,000 t/year (Alipour and Rouzbehan, 2006). However, this by-product contains anti-nutrients, condensed tannins, which have a negative effect on nutrient availability (protein, carbohydrates and minerals). Polyethylene glycol (PEG), which possesses a very high affinity for tannins, has been used to deactivate them (Makkar, 2003). Hence, the effect of replacing alfalfa forage with GP with or without PEG on the ruminal parameters of sheep was assessed.

Material and methods

Six fistulated sheep with average body weight of 61.8 kg (SD ± 2.9), were used in a 3×3 change-over design experiment (n=6) made up of three 3-wk periods. The three dietary treatments were the control (39% alfalfa hay, 25% barley grain, 8.3% wheat chaff and 27.7% wheat straw), GPO (72% grape pomace, 13.8% barley grain and 12.8% wheat chaff, 0.4% urea) and GPO + PEG (80 g/day). Animals were held individually in metabolism cages and were adapted to experimental conditions for 10 days. Samples of feed offered, feed refusal, faeces and urine were collected every morning. Conventional chemical composition and phenolic content of the treatments were determined (Makkar, 2000). Retained nitrogen was measured. A bucket containing 100 ml of sulfuric acid solution (containing 10 ml of concentrated sulfuric acid in 100 ml of distilled water), to keep the final pH below 3, was placed below the urine outlet in the metabolic cages for urine collection. Additionally, from urine samples of sheep, the amount of microbial protein produced daily was estimated (Chen and Gomes, 1995), and ruminal parameters, pH and ammonia were measured. Data obtained in experiments were analysed by the GLM procedure of SAS® (2001). Multiple comparisons among means were performed with the Duncan method.

Results

The chemical composition and phenolic compounds of the diets are shown in Table 1. Ruminal pH was in the range of the optimal value and was not significantly different between experimental diets. Ruminal ammonia concentrations were significantly lower for sheep fed the GP diet than the control diet ($P<0.05$). The GPO diet caused a reduction in microbial protein yield and retained nitrogen. The addition of PEG, however, increased those parameters ($P<0.05$).

Table 1. Chemical composition (g/kg DM), phenolic compounds (g/kg DM), ruminal parameters and retained nitrogen (g N/d) of the experimental diets.

	Diets			SE	Diet effect P-value
	Control	GPO	GPO+PEG		
DM, g/kg fresh weight	942	965	933		
OM	965	965	969		
Ash	34.5	34.5	31.5		
CP	115	109	90		
NDF	443	502	438		
TP	7.2	40.7	23.0		
TT	1.9	28.7	11.5		
CT	0	24.3	24.3		
HT	0	40	40		
Ruminal parameters					
pH	6.27	6.03	6.10	0.053	>0.05
Ammonia, mg/dL	18.99^{b}	15.85^{c}	29.91^{a}	0.835	<0.05
Microbial protein, g/d	19.69^{b}	5.5^{e}	11.06^{d}	1.04	<0.05
Nitrogen retained	2.85^{b}	-0.57^{c}	2.06^{b}	0.33	<0.05

OM: organic matter; CP: crude protein; NDF: Neutral-detergent fibre; TP: total phenolic compound; TT: total tannins; CT: condensed tannin; HT: hydrolysable tannins.

^{a,b,c} Means within rows with same superscript letters are not significantly different (P>0.05).

Conclusion

GP induced a severe negative effect on ruminal parameters and nitrogen retained in animals. Addition of PEG improved the ruminal parameters and retained nitrogen. Therefore, such an inclusion of GP in the diet of sheep is not recommended.

References

Alipour, D. and Y. Rouzbehan, 2006. Effects of ensiling grape pomace and addition of polyethylene glycol on *in vitro* gas production and microbial biomass yield. Anim. Feed Sci. Technol. 137: 138-149.

Chen, X.B. and M.J. Gomes, 1995. Estimation of microbial protein supply to sheep and cattle based on urinary excertion of purine derivatives an overview of the technical details. Rowett Research Institute, Bucksburn, Aberdeen AB2 9SB, UK.

Makkar, H.P.S., 2000. Quantification of tannins in tree foliage. A laboratory manual for the FAO/IAEA co-ordinated research project on use of nuclear and related techniques to develop simple tannin assays for predicting and improving the safety and efficiency of feeding ruminants on tanniniferous tree foliage. Joint FAO/IAEA of Nuclear Techniques in Food and Agriculture. Animal Production and Health Sub-programme, FAO/IAEA Working Document. IAEA, Vienna, Austria.

Makkar, H.P.S., 2003. Effects and fate tannins in ruminant animals, adaptation to tannins, and strategies to overcome detrimental effects of feeding tannin-rich feeds. Small Rum. Res. 49: 241-256.

Statistical Analysis System®, 2001. User's Guide: Statistics, Version 8.2. SAS Institute, Carry, NC, USA.

Ammonia inhibits urea transport across the isolated rumen epithelium by modulating cellular extrusion of protons

K. Abdoun[1], F. Stumpff[2], K. Wolf[2] and H. Martens[2]
[1]Faculty of Veterinary Medicine, University of Khartoum, 13314 Shambat, Sudan; [2]Institute of Veterinary Physiology, Free University of Berlin, 14163 Berlin, Germany; abdounn@yahoo.com

Introduction

SCFA, NH_4^+ and CO_2 significantly increase electroneutral Na transport via Na^+/H^+-exchange (NHE), indicating additional intracellular supply of H^+, decrease of intracellular pH (pH_i) and hence, enhanced activity of NHE (Gäbel *et al.*, 1991; Abdoun *et al.*, 2005). Surprisingly, despite the similarly directed effect of SCFA, CO_2 and NH_4^+ on NHE (↑), opposed consequences are observed regarding urea transport. Urea recycling is increased by SCFA, CO_2 (Thorlacius *et al.*, 1971) or intake of concentrate (Theurer *et al.*, 2002) and decreased by ruminal ammonia (Kennedy and Milligan, 1978). Therefore, the aim of this study was to characterise the mechanism by which ruminal ammonia regulates urea recycling.

Material and methods

Isolation and handling of the isolated rumen epithelium have been described in detail by Abdoun *et al.* (2005). The mounted tissues were incubated with standard electrolyte solution (mmol/l): 110 Na^+, 5 K^+, 1 Ca^{+2}, 2 Mg^{+2}, 30 $NMDG^+$, 8 MOPS, 106 Cl^-, 1 $H_2PO_4^-$, 2 HPO_4^{-2}, 25 acetate, 10 propionate, 5 butyrate, 10 glucose, 1 urea, and 1 phenyl phosphorodiamidate (urease inhibitor, to protect urea from microbial hydrolysis) and gassed with O_2. Nmdgcl was replaced by NH_4Cl in ammonia containing buffers. Flux rates of ^{14}C-urea were assayed using a Liquid Scintillation Counter (Wallace-Perkin-Elmer). Statistical evaluations were carried out with the Sigma Plot program version 8.0 for Windows using the Student t-test. N or n refers to number of animal or tissues.

Results

Luminal ammonia did not change urea flux rates at pH 7.4 (Table 1). A decrease of luminal pH (6.4; without ammonia), which enhances the activity of NHE by a decrease in pH_i, induced a ca. 5 fold increase of flux rates (Table 1), which supports the assumption that urea transport is modulated by the decrease of pH_i. However, ammonia reduced urea transport by approximately 67% at pH 6.4 (Table 1), despite a predominant uptake as NH_4^+ at pH 6.4, intracellular release of H^+ (pH_i↓) and stimulation of NHE (Abdoun *et al.*, 2005).
In a second step, we reduced the concentration of ammonia to 5 mmol/l NH_4^+ at a mucosal pH of 6.4. Again, this concentration was found to inhibit serosal to mucosal urea transport (J_{sm}) by some 50% in a manner responsive to amiloride (Table 2).

Discussion and conclusion

It is well known that ruminal ammonia decreases recycling of urea (Kennedy and Milligan, 1978). This finding was confirmed by our data; even low concentrations of ammonia (5 mmol/L) caused a strong reduction of urea transport rates. The underlying mechanism of this inhibition is a little bit obscure, because influx of NH_4^+ and release of H^+ stimulates NHE as SCFA and CO_2 do, both of which are known to increase urea transport. Even more surprising is the effect of amiloride on the transport of urea. In a previous study, we observed that exposure to ammonia stimulates NHE activity to a much larger extent than can be explained by H^+ influx via NH_4^+ uptake alone

(Abdoun *et al.*, 2005). If ammonia shifts the pH set point of the NHE (Aronson *et al.*, 1982) to a more alkalic value, NH_4^+ should alkalinize the cytosol via stimulation of NHE and thus inhibiting urea transport. The presence of 0.8 mM amiloride inhibits the Na^+/H^+ exchanger, hence acidifying the cytosol and counteracting the effect of ammonia. However, direct measurements of pH_i are needed to validate this hypothesis.

From a functional point of view, it appears important to point out that the inhibitory effect of ammonia on urea transport depends on the ruminal pH. At a pH of 7.4 even 30 mmol/l ammonia did not change urea transport (Table 1).

Table 1. Effect of ammonia (30 mmol/l) on mucosal to serosal (Jms) and serosal to mucosal (Jsm) flux rates of urea across the rumen epithelium (Means±SEM).

Luminal pH (NH_4Cl)	J_{sm}^{urea} ($nmol/cm^2/h$)	J_{sm}^{urea} ($nmol/cm^2/h$)	J_{sm}^{urea} ($nmol/cm^2/h$)	N/n
7.4 (0 mmol/l)	22.56[a]±03.00	23.16[a]±02.93	-0.60[a]±02.20	3/9
7.4 (30 mmol/l)	22.69[a]±02.10	21.45[a]±01.53	1.24[a]±02.98	3/9
6.4 (0.0 mmol/l)	122.53[b]±09.47	117.66[b]±09.04	4.87[a]±08.62	4/12
6.4 (30 mmol/l)	39.78[c]±03.77	47.60[c]±03.41	-7.82[a]±05.47	4/12

[a, b, c] Means within column with different superscript letters are significantly different (P<0.05).

Table 2. Effect of 5 mmol/l mucosal ammonia (pH 6.4) on J_{sm} urea (N = 2; n = 7).

	J_{sm}^{urea} ($nmol/cm^2/h$)	Amiloride (mmol/l)
Control	184±14[a]	0
NH_4^+ (5 mmol/l)	93±6[b]	0
NH_4^+ (5 mmol/l)	145±17[c]	0.8

[a,b,c] Means within column with different superscript letters are significantly different (P<0.05)

Acknowledgement

This work has been supported by Alexander von Humboldt Foundation (Georg-Forster scholarship for Dr. K. Abdoun) and by the Margaret-Markus-Charity.

References

Abdoun, K., F. Stumpff, K. Wolf and H. Martens, 2005. Modulation of electroneutral Na transport in sheep rumen epithelium by luminal ammonia. Am. J. Physiol. Gastrointest. Liver Physiol. 289: G508-G520.

Aronson P.S., J. Nee and M.A. Suhm, 1982. Modifier role of internal H+ in activating the Na+-H+ exchanger in renal microvillus membrane vesicles. Nature 299: 161-163.

Gäbel, G., S. Vogler and H. Martens, 1991. Short-chain fatty acids and CO_2 as regulators of Na^+ and Cl^- absorption in isolated sheep rumen mucosa. J. Comp. Physiol. B 161: 419-426.

Kennedy, P.M. and L.P. Milligan, 1978. Transfer of urea from the blood to the rumen of sheep. Br. J. Nutr. 40: 149-154.

Theurer, C.B., G.B. Huntington, J.T. Huber, R.S. Swingle and J.A. Moore, 2002. Net absorption and utilization of nitrogenous compounds across ruminal, intestinal, and hepatic, tissues of growing beef steers fed dry-rolled or steam-flaked sorghum grain. J. Anim. Sci. 80: 525-532.

Thorlacius, S.O., A. Dobson and A.F. Sellers, 1971. Effect of carbon dioxide on urea diffusion through bovine ruminal epithelium. Am. J. Physiol. 220: 162-170.

Comparison of *in sacco* degradability of wheat straw treated in different ways

A. Aghazadeh[1], D. Ghorbannejad[1], N. Maheri-Sis[1] and S. Razzagzadeh[2]
[1]Department of Animal science, Islamic Azad University-Shabstar Branch, Shabstar, Iran; [2]The agricultural and Natural Resources Research Center, Urmia, Iran; a.aghazadeh@mail.urmia.ac.ir

Introduction

In many ruminant production systems, cereal straws of low nutritive value are used as forages during part of the annual seasonal cycle. The recent increase in the human population also has put enormous pressure on land that could be used for fodder production and grazing. As a result, ruminant livestock are fed with more crop residues.
Since cell-wall carbohydrates are the most important components of straw, an efficient microbial digestion in the rumen is important for their utilisation in ruminant feeding. Evaluation of this type of feed with the *in sacco* technique has been considered suitable and more relevant for practical use because it provides information on the degradation characteristics including rumen dynamics (Ørskov *et al.*, 1988). It has been shown that the *in sacco* degradability of roughages gives better prediction of voluntary intake (r=0.82) than digestibility *in vivo* (Hovell *et al.*, 1986). Nevertheless *in vitro* work has shown large variations in the feeding value of wheat straw strongly related to genotypes (Habib *et al.*, 1995). The present study was aimed at evaluating rumen degradability characteristics of wheat straw treated in different ways.

Material and methods

In sacco dry matter degradability was determined according to Ørskov *et al.* (1980). The study was carried out in three rumen fistulated native male buffaloes (3 yr; BW: 389±7 kg). They were fed *ad libitum* with a mixture of wheat straw (40%) and alfalfa hay (60%) to meet nutrient requirements for maintenance. The treatments evaluated were the following: untreated straw (US), 2.5% lime treated straw (LTS2.5), 5% lime treated straw (LTS5), 3% urea treated straw (UTS3), 5% urea treated straw (UTS5) and 2.5% lime + 2.5% urea treated straw (LUTS5). Differently treated straw diets were prepared according to conventional methods (Saadullah *et.al.*, 1981; Munoz *et al.*, 1991). Straw samples, previously ground through a 3 mm mesh screen were sieved to remove fine particles and the 3 g sample was weighed into each Dacron bag (5×10 cm) with a pore size of 50 μm. Incubation times were 6, 12, 24, 36, 48, 72 and 96 h. Chemical additions were made on percentage basis. The degradation characteristics of the treatments were described by the exponential equation $p = a + b(1 - e^{-ct})$ (Ørskov and McDonald, 1979) where p is degradation at time t and a, b and c are constants. The effective degradability (ED) of DM in the rumen was calculated as (Ørskov and McDonald, 1979): $ED = a + (bc / c + k)$. The value of k which represents estimated outflow rate of particulates from the rumen was chosen as 2, 5 and 8%/h. The data were analysed by the analysis of variance procedure using the SPSS package (SPSS Inc., 1986, Chicago, IL, USA). In all cases, means were compared with the Duncan Multiple Range Test. Overall differences between treatment means were declared significant at $P<0.5$.

Results and discussion

The a value of untreated straw was numerically increased as a result of chemical treatment with different levels of lime, urea and their mixture (Table 1). However, the greatest effect was found in LTS5 ($P<0.05$). A parallel increase in b value as a result of chemical treatment except in LTS5 was also found. The mixture of LUTS5 treatment caused the greatest b value followed by UTS3

(P<0.05). From a practical point of view the potentially degradable fraction and rate of degradation are most important values since they influence rumen fill and hence feed intake (Ørskov *et al.*, 1988). The degradation rate, c, was not significantly increased by chemical treatment under this study. However, the estimated effective degradability of wheat straw dry matter at rumen outflow rate of 2%/h significantly increased with a combination of LUTS5.

Table 1. Estimates of rumen degradability parameters of dry matter of wheat straw treated in different ways.

	a (%)	B (%)	C (%/h)	ED (%) at different rumen outflow rates (%/h)		
				2	5	8
Treatments[1]						
US	10.8±1.9[b]	50.7±4.0[bc]	0.028±0.002[ab]	41.0±0.9[b]	30.0±6±0.8[c]	25.3±0.9[c]
LTS2.5	11.86±2.5[b]	51±3.1[bc]	0.03±0.005[ab]	42.5±1.2[b]	30.8±1.9[c]	25.8±2.0[bc]
LTS5	18.0±1.9[a]	47.5±3.6[c]	0.03±0.004[ab]	46.7±0.4[a]	36.0±1.0[a]	31.2±1.63[a]
UTS3	11.23±2.5[b]	54.3±1.8[ab]	0.02±0.001[ab]	42.4±0.2[b]	30.8±0.4[c]	25.9±0.4[bc]
UTS5	13.9±1.9[b]	53.1±3.2[bc]	0.03±0.003[a]	45.3±1.3[a]	33.3±1.0[b]	27.6±1.0[bc]
LUTS5	13.1±1.7[b]	59.33±2.4[a]	0.02±0.005[b]	47±0.4[a]	33.7±0.5[b]	27.9±0.8[b]

[1] US = untreated straw; LTS2.5 = lime treated straw (2.5%); LTS5 = lime treated straw (5%); UTS3 = urea treated straw (3%); UTS5 = Urea treated straw (5%); LUTS5 = lime+urea treated straw (5%).
[a,b,c] Means within columns with same superscript letters are not significantly different (P>0.05).

Conclusion

The main impact of chemical treatments in this study was on b value and effective degradability. The maximum increase in b value and effective degradability at rumen outflow rate of 2%/h were found in a mixture of urea and lime treatment.

References

Habib, G., S.B.A. Shah and K. Inayat, 1995. Genetic variation in morphological characteristics, chemical composition and *in vitro* digestibility of straw from different wheat cultivars. Anim. Feed Sci. Technol. 55: 263-274.

Hovell, F.D., D. De, J.W.W. Ngambi, W.P. Baker and D.J. Kyle, 1986. The voluntary intake of hay by sheep in relation to its digestibility in rumen as measured in nylon bags. Anim. Prod. 42: 111-118.

Munoz, F., M. Joy, R. Facci and X. Alibes, 1991. Treatment of ligno-celloulosic residues with urea. Influence of dosage, moisture, temperature and addition of ureases. Ann. Zootech. 40: 215-225.

Ørskov, E.R. and I. McDonald, 1979. The estimation of protein degradability in the rumen from incubation measurements weighted according to rate of passage. J. Agric. Sci. (Camb.) 92: 499-503.

Ørskov, E.R., G.W. Reid and M. Kay, 1988. Prediction of intake by cattle from digestion characteristics of roughages. Anim. Prod. 43: 271-278.

Ørskov, E.R., F.D. Hovell, B. De and F. Mould, 1980. The use of nylon bag technique for the evaluation of feedstuffs. Trop. Anim. Prod. 5 (3): 195-213.

Saadullah, M., M. Hague and F. Dolberg, 1981. Treatment of rice straw with lime. Trop. Anim. Prod. 6 (2): 116-120.

Ruminal microbial protein synthesis of wethers and heifers fed fresh temperate pastures supplemented or not with sorghum grain

M. Aguerre[1], C. Cajarville[2], G.V. Kozloski[3] and J.L. Repetto[1]
[1]Bovine Department, Facultad de Veterinaria, Universidad de la República (UdelaR), Lasplaces 1550, Montevideo, Uruguay; [2]Animal Nutrition Department, Facultad de Veterinaria, Universidad de la República (UdelaR), Lasplaces 1550, Montevideo, Uruguay; [3]Animal Science Department, Universidade Federal de Santa Maria (UFSM), Campus Camobi, Santa Maria 97105-900, RS, Brazil; aguerremartin@gmail.com

Introduction

Temperate pastures are widely used in semi-intensive beef and sheep production systems in Uruguay. Nitrogen fractions of these kind of pastures are rapidly and extensively degraded in the rumen (Repetto *et al.*, 2005), producing NH_3-N. Supplementation of temperate pastures with grains is suggested to improve nitrogen utilisation efficiency, microbial incorporation of N-NH_3 (Bargo *et al.*, 2003) and microbial protein synthesis efficiency (Horadagoda *et al.*, 2008). The aim of this work was to evaluate the effect of an increase in the level of sorghum grain supplementation on ruminal microbial protein synthesis of wethers and heifers consuming a fresh temperate pasture.

Material and methods

The experiment was conducted on the Experimental farm of the Veterinary Faculty of Uruguay, (34° latitude South and 55° latitude West). Twenty-four Corriedale x Milchschaf wethers (45.6±6.2 kg Body Weight (BW)) and 24 crossbred heifers (210.0±42.5 kg BW) were blocked in four groups according to their BW, and, within each group, were randomly assigned to one of four treatments: non-supplemented (G0) or supplemented with sorghum grain at 5, 10 or 15 g/kg of their BW (G5, G10 and G15, respectively).The animals were housed in metabolic crates and fed fresh temperate pasture (*Lotus corniculatus,* dry Matter (DM): 317 g/kg; organic matter: 930 g/kg of DM; crude protein: 126 g/kg of DM; neutral detergent fiber: 416 g/kg of DM) *ad libitum.* Sorghum grain was provided ground, individually, in two equal meals (08:00 and 20:00 h). After 21 days of adaptation, all urine was collected daily for 5 days, in buckets containing 100 ml and 200 ml of H_2SO_4 (100 ml/l) for wethers and heifers, respectively. The volume was measured and a sample of 100 ml/l of total volume was stored at -18 °C until analysis. Ruminal microbial protein synthesis (g microbial N/day) and ruminal microbial protein synthesis efficiency (g microbial N/kg Organic matter apparently fermented in the rumen) was calculated from total urinary excretion of purine derivatives using the method of Chen and Gomes (1995). Analysis of variance included the effects of species, treatments, block, species vs. treatment interaction and experimental error were performed. Treatment effects were also analysed by linear and quadratic regression.

Results and discussion

Ruminal microbial protein synthesis decreased linearly with increased sorghum supplementation in wethers ($P<0.05$), probably due to a reduction in dry matter intake, while it was not affected by treatments in heifers. The efficiency of ruminal microbial protein synthesis was higher for heifers than for wethers ($P<0.01$) whereas it was not affected by sorghum supplementation in any ruminant species, probably because the control groups (G0) already had a high microbial protein synthesis efficiency. Similarly, recent research of our research group found no differences (Tebot, 2008) or a reduction (Azevedo, 2008) on microbial protein synthesis efficiency when different kinds of energy supplements were included in a fresh pasture based diet.

Table 1. Ruminal microbial protein synthesis and efficiency of ruminal microbial protein synthesis of wethers and heifers fed a fresh temperate pasture (Lotus corniculatus) ad libitum and non-supplemented (G0) or supplemented with ground sorghum grain at 5, 10 or 15 g/kg of their BW (G5, G10 and G15, respectively).

	Ruminal microbial protein synthesis, g microbial N/day		Ruminal microbial protein synthesis efficiency, g microbial N/kg DOMR[1]	
	Wethers	Heifers	Wethers	Heifers
G0	17.6	72.3	20.3	29.8
G5	12.7	73.1	20.4	22.8
G10	11.9	84.8	21.5	25.0
G15	10.5	80.7	15.9	23.4
Mean	13.2[b]	77.7[a]	19.5[b]	25.3[a]
S.D.[2]	4.53	16.01	5.96	5.50
L[3]	<0.01	0.20	0.22	0.12
Q[3]	0.32	0.70	0.20	0.26

[a,b] Means without a common letter differ significantly ($P<0.01$)

[1] Organic matter apparently fermented in the rumen, calculated as 0.65 × digestible organic matter intake.

[2] Standard deviation of the means.

[3] Probability of linear (L) or quadratic (Q) treatment effect.

Conclusion

The impact of sorghum grain supplementation on ruminal microbial growth is different between ruminant species fed fresh temperate forages. Increased supplementation did not affect ruminal microbial protein synthesis of heifers while it had a negative effect on wethers.

References

Azevedo, G., 2008. Valor alimentar de dietas com azevém (*Lolium multiflorum,* LAM.) e suplementação nitrogenada ou energética. MSc Tesis. Universidade Federal de Santa Maria, Santa María, RS, Brasil.

Bargo, F., L.D. Muller, E.S. Kolver and J.E. Delahoy, 2003. Invited review: Production and digestion of supplemented dairy cows on pasture. J. Dairy Sci. 86: 1-42.

Chen, X.B. and M.J. Gomes, 1995. Estimation of microbial protein supply to sheep and cattle based on urinary excretion of purine derivatives - an overview of the technical details. International Feed Resources Unit Rowett Research Institute, Bucksburn Aberdeen, UK.

Horadagoda, A., W.J. Fulkerson, I. Barchia, R.C. Dobos and K.S. Nandra, 2008. The effect of grain species, processing and time of feeding on the efficiency of feed utilization and microbial protein synthesis in sheep. Livest. Sci. 114: 117-126.

Repetto, J.L., C. Cajarville, J. D'Alessandro, A. Curbelo, C. Soto and D. Garin, 2005. Effect of wilting and ensiling on ruminal degradability of temperate grass and legume mixtures. Anim. Res. 54: 73-78.

Tebot, I., 2008. Efecto de los suplementos ricos en energía sobre la función ruminal y el metabolismo del nitrógeno en ovinos alimentados con pasto fresco. MSc Tesis. Universidad de la República, Montevideo, Uruguay.

Effect of using *Megasphaera elsdenii* NCIMB 41125 as a probiotic on feed intake and milk production in early lactation dairy cows

P.C. Aikman[1], P.H. Henning[2], C.H. Horn[2] and A.K. Jones[1]
[1]Animal Science Research Group, Department of Agriculture, University of Reading, Earley Gate, Reading RG6 6AR, United Kingdom; [2]KK Animal Nutrition, Erinvale Building, The Greens Office Park, 26 Charles de Gaulle Street, Highveld, Centurion 0046, South Africa; p.c.aikman@reading.ac.uk

Introduction

High-cereal starch-rich diets typically fed to ruminants to increase dietary energy density and improve production levels can cause rumen acidosis, a metabolic disorder that exists in both acute and sub-acute forms, with serious economic and welfare consequences. Acidosis occurs when rapid fermentation of starch and sugars leads to an accumulation of lactic and volatile fatty acids in the rumen, a drop in pH, and a shift in the delicate balance of rumen microorganisms (Krause and Oetzel, 2006). Highly productive dairy cows are particularly susceptible to acidosis immediately *post partum*, a period characterised by the need for increased feed intake with adaptation to a high-starch diet. The hypothesis of the present study was that boosting rumen populations of *Megasphaera elsdenii*, a naturally occurring lactic acid-utilising rumen bacterium, by dosing at critical time points in the *post partum* period would reduce the effects of acidosis and improve feed intake and milk production.

Material and methods

Eighty multiparous Holstein cows were blocked by previous 305-d milk yield (mean ± SEM 9,710±168 kg) and used in a 2×2 factorial design to assess the effect of diet (standard (S), 584:416 forage:concentrate ratio vs. acidosis challenge (Ac), 392:608 forage:concentrate ratio, DM basis) and probiotic supplementation (Placebo (P) *vs M. elsdenii* NCIMB 41125 (M)) on feed intake and milk production from calving until week 14 of lactation. Diets were offered *ad libitum* as total mixed rations and had concentrations (g/kg DM) of CP, NDF, starch and water soluble carbohydrates of 164, 314, 198 and 41 (S) and 176, 260, 299 and 39 (Ac) with respective ME concentrations of 11.8 and 12.2 MJ/kg DM. Cows were orally drenched (250 ml) with either a minimum of 10^{10} cfu *M. elsdenii* (M) or autoclaved *M. elsdenii* (P) on days 3 and 12 of lactation. Daily milk yield and DM intake (DMI) and weekly milk composition were recorded. Data were analysed as repeated measures within week using the Mixed Procedure of SAS® (2003). The model included diet, probiotic supplement, week, and diet by probiotic interaction, with cow as a random factor. Also, data from the highest-yielding animals in each treatment group (305-d milk yield in previous lactation >10,000 litres, n = 8/treatment) were analysed separately as above to assess whether high-yielding animals responded differently to *M. elsdenii* supplementation. Statistical significance was set at $P<0.05$ and trends were discussed at $P<0.1$.

Results

All cows: Animals receiving the Ac diet were on average 25 kg heavier ($P=0.039$), and yielded 3.4 kg/day more milk ($P=0.007$) with more protein (+2.15 g/kg, $P=0.001$) and less fat (-4.0 g/kg, $P=0.001$) than those receiving diet S, whilst DMI (mean 20.3 kg, $P=0.208$) was similar across diets (Table 1). In addition, cows receiving diet Ac tended ($P=0.061$) to gain weight over the 14 week study whilst those on diet S lost weight. No responses to *M. elsdenii* supplementation or its interaction with diet (S or Ac) were noted.

High-yielding cows only: For high-yielding cows alone, responses to diet were in the same direction as those observed for all cows, but the magnitudes of the responses were generally greater (Table 1). A mean increase in DMI (+1.2 kg/d, P=0.019) was also observed in high-yielding cows fed the Ac diet compared to those on diet S. Cows that received *M. elsdenii* tended (P=0.094) to either lose less weight (S) or gain more (Ac) than those that received the placebo (Table 1). *M. elsdenii* supplementation reduced milk protein (P=0.007) and fat (P=0.080) concentrations in high-yielding cows on both diets, whilst *M. elsdenii* tended to reduce milk yield on diet S but increase it by 2.2 kg/d on the Ac diet (diet×probiotic interaction, P=0.076).

Table 1. Feed intake and productive performance between weeks 1 and 14 of lactation in dairy cows fed a standard (S) or an acidosis challenge (Ac) diet and dosed with either M. elsdenii (M) or a placebo (P).

	Treatment				SEM	P =		
	SP	SM	AcP	AcM		Diet	Prob[1]	Diet*Prob[1]
All cows (n=20 per treatment group)								
Dry matter intake, kg/day	20.0	20.1	21.0	20.2	0.43	0.208	0.429	0.288
Bodyweight, kg	629	620	657	643	12.0	0.039	0.354	0.854
Bodyweight change[2], kg	-4.6	-4.1	6.1	19.6	8.9	0.061	0.439	0.470
Milk yield, kg/day	34.8	35.1	38.9	37.7	1.14	0.007	0.814	0.618
Milk fat, g/kg	43.5	41.6	39.0	38.1	1.08	0.001	0.192	0.658
Milk protein, g/kg	29.7	29.4	31.7	31.7	0.34	0.001	0.690	0.659
High yielding cows (n=8 per treatment group)								
Dry matter intake, kg/day	21.4	20.3	22.1	22.0	0.50	0.019	0.203	0.279
Bodyweight, kg	650	596	649	659	15.4	0.034	0.127	0.032
Bodyweight change[2], kg	-22.9	-8.6	2.9	35.3	13.2	0.015	0.094	0.505
Milk yield, kg/day	38.1	34.7	41.1	43.3	1.58	0.001	0.692	0.076
Milk fat, g/kg	43.6	40.6	38.0	35.4	1.63	0.002	0.080	0.876
Milk protein, g/kg	29.9	28.2	31.8	30.4	0.57	0.001	0.007	0.823

[1] Probiotic supplement (Placebo or *M. elsdenii*).
[2] Total change in bodyweight between week 1 and 14 of lactation.

Conclusion

For all cows, the lack of response to *M. elsdenii* supplementation may have occurred because the diets did not provide a sufficient challenge to rumen pH for the *M. elsdenii* to be effective. In contrast, the highest-yielding animals had higher DM and starch intakes, thereby providing the *M. elsdenii* with substrate. An associated study (unpublished data) showed that *M. elsdenii* shifted rumen VFA production from acetate towards propionate, which would support both the reduction in milk fat concentration and improved energy status (as indicated by reduced weight loss/higher weight gain) seen in the high-yielding animals that received *M. elsdenii* in the present study.

References

Krause, K.M. and G. R. Oetzel, 2006. Understanding and preventing subacute ruminal acidosis in dairy herds: a review. Anim. Feed Sci. Tech. 126: 215-236.

SAS, 2003. Version 9.1, SAS Institute Inc., Cary, NC, USA.

The effects of peppermint addition on the *in vitro* hydrogenation of fatty acids of hay

S. Ando[1], T. Yasutake[1], T. Ichinohe[2] and T. Awano[2]
[1]National Agricultural Research center for Western Region, Ooda Shimane Japan; [2]University of Shimane, Matsue Shimane, Japan; ansada@affrc.go.jp

Introduction

Unsaturated fatty acids ingested by ruminants are hydrogenated in the rumen (Dewhurst *et al.*, 2006; Schmid *et al.*, 2006; Scollan *et al.*, 2006). On the contrary, some plant secondary metabolites are known to alter ruminant performance (Jouany and Morgavi, 2007). Peppermint which contains L-menthol as a secondary plant metabolite is used for many aims. In ruminants, Hosoda *et al.* (2005) reported that peppermint feeding reduced methane production, and Ando *et al.* (2003) reported that peppermint feeding reduced ruminal protozoa counts. With the above reports it can be postulated that peppermint would be able to affect ruminal hydrogenation of unsaturated fatty acids. So, in the present study, the effect of peppermint addition on the ruminal biohydrogenation of fatty acids of hay was investigated *in vitro*.

Material and methods

For the incubations, 4 g of Italian-ryegrass hay was used for the incubation. Three hundred ml of mixed solution (Macdogul artificial saliva:rumen fluid=4:1) were used. Incubation was carried out for 0 h, 3 h, 6 h, 12 h and 24 h, with or without 2 g of dry peppermint put into a 15×15 mm polyethylene bag. Incubated hay was extracted with a methanol and chloroform solution. The analysis of fatty acids was conducted by gas chromatography GC1700 (Shimazu Kyoto Japan). For C18:0, C18:1 and C18:2, the relative values (%) at 3 h, 6 h, 12 h and 24 h to the values at 0 h were calculated. The Student t-test was used for statistical treatment (Snedecor and Cochran, 1967).

Results

Table 1 shows the relative values of C18:0, C18:1 and C18:2. With the progress of incubation the C18:0 contents increased both with and without peppermint. In the total incubation time, C18:0 contents showed significantly ($P<0.01$) higher values without peppermint than with peppermint addition. The C18:1 contents differed much both with and without peppermint. With the progress of incubation, the C18:2 content decreased both with and without peppermint. For the incubation times of 3 h, 6 h and 12 h, C18:2 contents showed significantly ($P<0.01$) higher values with peppermint addition than without peppermint addition.

Table 1. Effect of peppermint addition on the fatty acid profile.

		Incubation hours				
		0 h	3 h	6 h	12 h	24 h
C18:0	No addition	100	654.4[a]	977.3[a]	1,118.2[a]	1,200.0[a]
	Peppermint	100	509.1[b]	595.5[b]	777.3[b]	1,063.6[b]
C18:1	No addition	100	193.8	204.7	195.3	231.3
	Peppermint	100	198.4	225	225	217.2
C18:2	No addition	100	53.6[a]	42.0[a]	32.9[a]	26.1
	Peppermint	100	68.1[b]	55.1[b]	45.4[b]	28

Relative values (%) to 0 h; Significant (P<0. 01) in different letters in the same column.

Conclusion

In the present study peppermint addition had an inhibitory effect upon ruminal hydrogenation of unsaturated fatty acids. It can be concluded that peppermint`s inhibitory effect upon ruminal hydrogenation of unsaturated fatty acids might be due to the alternation of ruminal fermentation.

References

Ando, S., T. Nishida, M. Ishida, K. Hosoda and E. Bayaru, 2003. Effect of peppermint feeding on the digestibility, ruminal fermentation and protozoa. Livest. Prod. Sci. 82: 245-248.

Dewhurst, R.J., K.J. Shingfield, M.R.F. Lee and N.D. Scollan, 2006. Increasing the concentration of beneficial polyunsaturated fatty acids in milk produced by dairy cows in high–foragesystems. Anim. Feed Sci. Technol. 131: 168-20.

Jouany, J.P. and D.P. Morgavi, 2007. Use of 'natural' products as alternative to antibiotic feed additives in ruminant production. Animal 1: 1443-1466.

Hosoda, K., T. Nishida, W.-Y. Park and B. Eruden, 2005. Influence of *Mentha piperita* L. (peppermint) supplementation on nutrient digestibility and energy metabolism in lactating dairy cows. Asian-Aust. J. Anim. Sci. 18: 1721-1729.

Schmid, A., M. Collomb, R. Sieber and G. Bee, 2006. Conjugated linoleic acid in meat: a review. Meat Sci. 73: 29-41.

Scollan, N., J.F. Hocquette, K. Nuernbert, D. Dannnenberger, I. Richardson and A. Moleney, 2006. Innovations in beef production systems that enhance the nutritional and health value of beef lipids and their relationship with meat quality. Meat Sci. 74: 17-33.

Snedecor, G.W. and W.G. Cochan, 1967. Statistical method. The Iowa State University press, Ames, USA, pp. 57-61.

Effect of DCAD on performance of high producing dairy cows can be modulated by protein content of diets

E. Apper-Bossard[1,2], *J.-L. Peyraud*[2,3] *and F. Meschy*[4,5]
[1]*Ecole Supérieure d'Agriculture, 55 rue Rabelais, BP 30748, 49007 Angers Cedex 01, France;* [2]*INRA, UMR1080 Dairy Production, 35000 Rennes, France;* [3]*Agrocampus Ouest, UMR1080 Dairy Production, 35000 Rennes, France;* [4]*INRA, UMR Physiology of Nutrition and Feeding, 75231 Paris Cedex 05, France;* [5]*Agro Paris Tech, UMR Physiology of Nutrition and Feeding, 75231 Paris Cedex 05, France; e.bossard@groupe-esa.com*

Introduction

Based on strong ion theory (Stewart, 1978), Na and K absorption increases blood HCO_3, the main blood buffer. Thus, increasing dietary cation anion difference (DCAD, defined as mEq of (Na + K - Cl - S) per kg of DM) can improve performance in high producing dairy cows fed with high concentrate diets. A meta-analysis (Hu and Murphy, 2004) has confirmed that a large positive DCAD enhances intake and milk yield and raises blood pH and HCO_3 concentration. However, few studies have considered possible the interaction between the response to increased DCAD supply and diet characteristics. In particular, protein contained in diets could interact with DCAD. Indeed, amino-acid metabolism and acid-base homeostasis are linked by glutamine flux between the kidney and liver (Guder *et al.*, 1987). Furthermore, when amino acids are metabolised, whether they are oxidised to yield energy or used as precursors for gluconeogenesis or for the biosynthesis of other uncharged products, one HCO_3^- ion is produced for each COO^- group (Atkinson and Bourke, 1995). By these ways, proteins could have a buffer effect and interact with DCAD. This study was aimed at evaluating the effect of increasing DCAD on performance and blood acid-base status of high producing dairy cows fed with two dietary protein concentrations.

Material and methods

Twenty-four Holstein cows were assigned to two groups of 12 each according to parity, stage of lactation, milk production, protein and fat content, BW and DMI. Each group of cows received one of two levels of protein (low protein, LP, and high protein, HP) during the experiment. Within each group, cows were assigned to two planned levels of DCAD (low DCAD, LD or high DCAD, HD) with a reversed design.

The LP diets were formulated to 13.2% DM of CP. The HP diets were formulated to provide 16.2% DM of CP. The diets consisted of 4 different TMR containing 60% corn silage, 40% concentrate and a mineral mixture. For HP diets, the level of protein was increased by substituting 10% of the concentrate by protected soybean meal. The two planned DCAD were 0 (LD) and 300 (HD) mEq/kg DM. LD was obtained by adding 0.8% NH_4Cl to a mineral mixture. HD was obtained by replacing $CaCO_3$ by $NaCO_3$ and Na_2PO_4. The concentrations of other minerals were kept constant to ensure that the observed effects could be attributed to the manipulation of DCAD.

Each period lasted 4 wk, measurements occurring the last wk of each period. The cows were individually fed to achieve *ad libitum* intake (10% orts). Intake and milk production were recorded daily. Fat and protein contents were recorded from a sample concerning 6 consecutive milkings each wk. For blood gases, mineral and metabolite analysis, blood was collected on d 24 by coccygeal puncture. Urine was collected during the last wk of each period by vulva stimulation to evaluate pH and mineral concentrations. Data were analysed using the general linear models of SAS® (SAS Institute Inc., Cary, NC, USA) according to the model for a split-plot design.

Results

Increasing the proportion of protein increased DMI, but only with a low DCAD, and increasing DCAD increased DMI only with the LP diet ($P<0.05$, Table 1). FCM yield (4%) was significantly increased when cows were fed HP diets but was unaffected by increasing DCAD. No DCAD×protein interaction for the blood acid-base parameters was observed. When HP diets were fed, blood pH and SBE tended to be increased whereas blood Cl decreased. Increasing DCAD significantly increased blood pH and SBE and decreased blood Cl. Blood BHBA increased with the HP diet whereas blood urea increased both with increasing dietary protein and DCAD. Blood protein increased with the HP diet and DCAD increased blood protein but only with the LP diet.

Table 1. Intake, 4% fat corrected milk (FCM), blood parameters and urine pH of dairy cows in response to increasing DCAD with two dietary protein concentrations.

	Treatments[1]				SD[2]	SDprot[3]	Effects		
	LPLD	LPHD	HPLD	HPHD			prot.	DCAD	DCAD×prot
DMI, kg/d	20.7	21.6	22.3	21.6	1.29	2.43	NS	NS	*
4% FCM	26.3	26.0	28.9	29.4	1.86	2.44	**	NS	NS
Blood parameters									
pH	7.41	7.44	7.44	7.47	0.030	0.042	†	**	NS
SBE[4], mmol/L	3.73	5.52	4.46	6.79	1.70	1.30	†	***	NS
Cl, mEq/L	97.2	95.0	94.8	93.3	1.94	1.57	**	**	NS
BHBA, mmol/L	0.33	0.35	0.45	0.48	0.100	0.143	*	NS	NS
Protein, g/L	97.9	94.0	99.6	99.6	2.70	5.52	NS	*	*
Urea, g/L	0.22	0.20	0.37	0.35	0.033	0.032	***	*	NS
Urine									
pH	6.81	8.32	7.11	8.36	0.529	0.549	NS	***	NS

[1] LPLD = low protein and low DCAD, LPHD = low protein and high DCAD, HPLD = high protein and low DCAD, HPHD = high protein and high DCAD, DCAD expressed as mEq/kg DM (Na + K) - (Cl + S).
[2] SD = standard deviation of the statistical model.
[3] SDprot = standard deviation to test the protein effect.
[4] SBE = standard base excess; †$P\leq0.10$, *$P\leq0.05$, **$P\leq0.01$, ***$P\leq0.001$.

Conclusion

The study confirmed that deficient dietary protein supply is a limiting factor of intake and showed the role of DCAD and protein in regulating DMI and blood acid-base status. It confirmed the necessity to take in account all components of diets to propose well-adapted recommendations of DCAD.

References

Atkinson, D.E. and E. Bourke, 1995. pH homeostasis in terrestrial vertebrates; ammonium ion as a proton source. In: N. Heisler (Ed.), Advances in Comparative and Environmental Physiology, vol. 22. Springer-Verlag Berlin, Heidelberg, Germany, pp. 3-23.

Guder, W.G., D. Haussinger and W. Gerok, 1987. Renal and hepatic nitrogen metabolism in systemic acid base regulation. J. Clin. Chem. Clin. Biochem. 25: 457-466.

Hu, W. and M.R. Murphy, 2004. Dietary cation-anion difference effects on performance and acid-base status of lactating dairy cows: a meta-analysis. J. Dairy Sci. 87: 2222-2229.

Stewart, P.A., 1978. Independent and dependent variables of acid-base control. Resp. Physiol. 33: 9-26.

Efficacy of the combined use of acids and heat to protect protein from sunflower meal against rumen degradation: metabolisable protein supply

J.M. Arroyo and J. González
Dpto. de Producción Animal, Universidad Politécnica de Madrid, 28040 Madrid, Spain; josemaria.arroyo.martinez@alumnos.upm.es

Introduction

The utilisation efficiency of proteins from sunflower meal (SFM) is largely reduced by their high degradability in the rumen through the following: (1) the final generation of nitrogenous compounds other than amino acids (ammonia, microbial nucleic acids and amino sugars, etc.) and (2) the reduction of intestinal digestibility, because the undegraded protein concentrates all feed undigestible compounds. A combined treatment with acids and heat has been shown promising to protect protein by decelerating protein breakdown (Ouarti *et al.*, 2006). This study was aimed to evaluate, through *in situ* methods, the efficiency of this treatment on SFM using malic or phosphoric acids on the net supply of metabolisable CP. The effect of using the rate of comminution of particles (k_c) in addition to their rumen outflow rate (k_p) to estimate the digestive availability was also examined.

Material and methods

A SFM 35 was treated with 2 N solutions (400 ml/kg) of malic or ortho-phosphoric acids by pulverisation under continuous mixing. After 1 h resting period, meals were dried for 6 h at 150 °C in a forced-air oven. A 3×3 Latin-square design was carried out using 3 rumen and duodenum cannulated wethers and 3 diets, which include the SFM untreated (UT) or treated with malic (MT) or phosphoric (PT) acids. Diets were isoproteic (about 127 g CP/kg DM) and composed of 60:40 (on DM) Italian ray-grass hay: concentrate (composed by wheat grain and SFM). They were distributed in six equal meals at 75 g/kg $LW^{0.75}$. Samples of SFM were marked by immersion with Yb and pulse dosed to the animals to determine from duodenal samples k_c and k_p rates as described by Ouarti *et al.* (2006). *In situ* studies in the rumen and intestine were carried out as indicated by Arroyo *et al.* (2009). Ruminal studies included 2 incubations with duplicate bags in different days. One bag from each incubation was oven dried for 48 h at 80 °C and used to determine DM degradation kinetics. The remaining bag was freeze-dried and used to generate a representative sample of the outflow from the rumen of undegraded feed DM. These samples were composited by pooling these residues in proportions pre-determined in accordance with the flow of undegraded particles. This flow was determined in turn from the DM degradation and particle transit kinetics using only k_p or both k_c and k_p. The analysis for CP or OM of these composited samples allows determining their effective degradabilities (ED), whereas their incubation in mobile bags (8 sub-samples of about 200 mg) allows determining the intestinal effective digestibility (IED) of CP (Arroyo *et al.*, 2009). These results were corrected for the microbial contamination taking place in the rumen using ^{15}N infusion techniques as indicated by these authors. Variance analyses were performed considering animals as blocks. A split-plot design was used for *in situ* studies with protecting treatment as the main plot.

Results and discussion

Consistent reductions (P<0.05) of k_p values were shown for MT and additionally for PT, whereas similar k_c values were observed in all samples (Table 1). To consider the particle comminution increased the mean retention time in the rumen by 10.5, 10.7 and 8.70% in UT, MT and PT, respectively, in relation to the time determined using k_p. As a consequence, ED was increased for

OM (P<0.001) as well as for CP (P=0.020). The associated diminution of the undegraded CP also led to a reduction (7.7%; P=0.072) in the intestinal digested (ID) CP.

The combined treatment with acids and heat reduced (P<0.001) the ED of CP from 83.0% in UT to 41.2% in MT and 40.5% in PT. In addition, these treatments increased (P<0.05) the IED of the undegraded CP from 72.4% (UT) to 84.1% (MT) and 79.8% (PT). As a consequence of both effects, the ID CP was almost 4 times higher (P<0.001) in the treated meals (Table 1). However, these treatments also reduced the ED of OM in 31.2% (MT) and 40.5% (PT), and, therefore, the microbial synthesis in the rumen associated with the fermentation of these meals. Thus, using these values and based also on an efficiency of 145 g microbial CP/ kg rumen fermented OM and on a content of 64% of intestinal digested true protein in microorganisms (Vérité *et al.*, 1987), the net supply of metabolisable protein (g/kg of feed DM) was 94.5, 209.9 and 200.2 for UT, MT and PT, respectively. Furthermore, the associated digestive efficacy was the following: 25.0, 59.0 and 55.5%, respectively.

Table 1. Effects of treatments of sunflower meal on the rates of comminution (k_c) and outflow (k_p) from the rumen of its particles and on its ruminal and intestinal digestion in accord with the considered transit model.

Item[2]	Transit model				Protection treatment[1]				
	k_p	k_p, k_c	SEM	P	UT	MT	PT	SEM	P
k_p (%/h)					6.23[a]	6.00[b]	5.77[c]	0.024	0.011
k_c (%/h)					56.2	66.3	54.0	5.40	0.493
ED of OM (%)	42.1	44.4	0.06	<0.001	56.8[a]	39.1[b]	33.8[c]	0.61	<0.001
ED of CP (%)	53.2	56.6	0.75	0.020	83.0[a]	41.2[b]	40.5[b]	0.75	<0.001
IED of CP (%)	79.6	77.9	0.78	0.156	72.4[b]	84.1[a]	79.8[a]	1.53	0.021
ID CP (%)	37.9	35.0	0.95	0.072	12.4[b]	49.4[a]	47.5[a]	0.51	<0.001

[1] UT = Untreated; MT = Malic acid treated; PT = Phosphoric acid treated.

[2] ED = Rumen effective degradability; IED = Intestinal effective digestibility; ID = Intestinal digested fraction.

[a,b,c] Means within rows with different superscripts are different at P< 0.05.

Conclusion

The study shows the convenience of considering the rate of particle comminution in a correct *in situ* feed evaluation, as well as a large increase of the protein digestive efficiency of the protective treatments with either of these acids.

Acknowledgement

This work was supported by the CICYT funded Project AGL 2006-08300.

References

Arroyo, J.M., J. González, C.A. Rodríguez, M.R. Alvir and M. Ouarti, 2009. *In situ* evaluation of the protein value of wheat grain corrected for ruminal microbial contamination. J. Sci. Food Agric. 89: 731-734.

Ouarti, M., J. González, L.F.J. Fernandes, M.R. Alvir and C.A. Rodríguez, 2006. Protection of protein from soybean meal by treatments with malic acid solutions and heat. Anim. Res. 55: 165-175.

Vérité, R., P. Chapoutot, B. Michalet-Doreau, J.L. Peyraud and C. Poncet, 1987. Révision du système des protéines digestibles dans l'intestin (PDI), Bull. Techn. CRZV, Theix, INRA. 70: 19-34.

Effects of undernutrition on digestibility and live weight changes in Barbarine ewes

N. Atti[1], *M. Doreau*[2], *M. Mahouachi*[3] *and F. Bocquier*[4]

[1]*Laboratoire des Productions Animale et Fourragère, INRAT 2080 Ariana, Tunisia;* [2]*INRA Theix, UR1213 Herbivores, 63122 Saint-Genès Champanelle, France;* [3]*ESA Kef, Le Kef, Tunisia;* [4]*INRA, Montpellier SupAgro, 2 Place Viala, 34060, Montpellier, France*

Introduction

In ruminants, a negative relationship has been clearly established between the level of intake and digestibility; most experiments in this area were conducted at levels of intake above maintenance (Chilliard *et al.*, 1995). However, when animals are fed at under-maintenance level, the effect of a decrease in intake is variable (Doreau *et al.*, 2003). It has been already shown that the Barbarine genotype is particularly well adapted to restricted feeding mobilising its body reserves (Atti and Bocquier, 1999). The objective of this work was to study the effects of a long term and very marked undernutrition followed by refeeding of Barbarine ewes on the live weight changes, food intake, nitrogen balance and digestibility.

Material and methods

Twenty adult, non pregnant and non lactating ewes of the Barbarine breed were separated into 3 homogeneous groups according to body weight (mean 49 kg). Ewes of the first group, which was considered as the control (C) were fed 0.88 kg DM of vetch-oat hay per head daily throughout the experiment (315 d). In the second (M) and third (L) groups, ewes received only 0.36 and 0.18 kg DM of hay daily, respectively, for 161 d (undernutrition sequence). When refed (refeeding sequence), M and L groups received 1.1 kg DM of hay for 70 days then 0.88 kg DM of hay and 0.45 kg of barley grain for 84 days. Energy supply of the diets, expressed in% of maintenance requirements (MR) calculated from initial body weight, corresponded to 100% (C), 40% (M) and 20% (L) and then 130% (L and M). Diet digestibility and N balance were measured for 7 days in 5 ewes per group by total collection of faeces and urine. Measurements took place at the beginning (P1 at day 24) and the middle (P2 at day 95) of the undernutrition sequence and at the beginning (P3 at day 21) and the middle (P4 at day 98) of the refeeding sequence. Faeces were weighed, dry matter (DM) was determined by drying for 24 h at 105 °C, organic matter (OM), crude protein (CP) and crude fibre (CF) were determined on fresh samples. A two-way analysis of variance, for period and diet level effects on live weight, digestibility and nitrogen balance using the GLM procedure of SAS® (1989), was applied.

Results and discussion

For the whole experiment, total intake for each ewe was 274, 202 and 173 kg DM of hay for C, M and L ewes, respectively; moreover, L and M ewes consumed 38 kg of barley. Undernutrition led to a decrease ($P<0.01$) in body weight (BW). During this period, M and L ewes lost 12.6 and 17.2 kg BW, respectively, which represent 27 and 35% of their initial BW during 160 days. The refeeding sequence resulted in a total recovery of BW.

During the first 2 periods, DM, OM, CP, and CF digestibility was lower at low intake ($P<0.001$, Table 1). Our results confirmed the decrease in digestibility at very low intake (Doreau *et al.*, 2003), on the contrary to the negative relationship between intake and digestibility observed when animals are not underfed (Chilliard *et al.*, 1995). Furthermore, digestibility in undernourished sheep decreased for P2 compared to P1, particularly for L level. With a moderate level of undernutrition, a digestive adaptation occurred; this is confirmed by CP digestibility improvement for M sheep (29% in P2 vs. 16% in P1), it is not the case for a severe undernutrition (-80% for L ewes). During the undernutrition sequence, faecal N decreased with intake ($P<0.001$); the lowest N excretion was observed in P2. Urinary N did not vary with intake during P1 but decreased in P2 suggesting a metabolic adaptation. This confirmed the results on uraemia reduction in undernourished Barbarine ewes (Atti *et al.*, 2002). The N balance was negative in P1 for undernourished groups, it was improved in P2 (Table 1). At the beginning of the refeeding sequence, the N balance was +1.3 g for ewes of all diets; then it increased to reach +5 g for L and M groups as a result of barley incorporation. The N deficit led to a negative N balance elicited amino acid catabolism (Doreau *et al.*, 2003).

Table 1. N balance of ewes fed at control (C), medium (M) or low (L) intake (Int) and digestibility of DM, OM, CP and CF in 4 period (Pi) measurements.

	P1			P2			P3			P4			Stat[1]	
	C	M	L	C	M	L	C	M	L	C	M	L	P	I
Digestibility,%														
DM	62	52	46	49	46	33	54	57	55	52	69	67	*	*
OM	69	61	55	53	54	43	56	59	57	55	71	69	*	*
CP	33	16	10	24	29	10	30	34	32	35	53	51	*	*
N balance, g/d														
N intake	9.0	3.5	1.8	9.8	3.8	2.0	9.4	10.1	10.2	7.9	14.3	14.5	**	*
Faecal N	6.0	2.9	1.7	7.6	2.7	1.8	6.5	6.6	6.9	5.9	6.7	6.4	ns	*
Urinary N	1.9	1.6	1.9	1.3	0.9	1.0	1.6	2.1	1.9	1.0	2.2	2.6	ns	ns
N retention	1.1	-1.1	-1.8	1.0	0.2	-0.8	1.3	1.3	1.4	0.9	5.4	5.4	*	*

[1] P: period; I: Intake level.

Conclusion

This experiment shows that metabolism of Barbarine sheep can face long and severe periods of underfeeding with limited body weight losses and can fully recover when refed. With a moderate level of undernutrition, a digestive and nitrogen balance adaptation occurred; but this is not the case for severe undernutrition since, under a threshold, the digestibility of diet is lowered and the nitrogen balance is negative at a very low level of intake.

References

Atti, N. and F. Bocquier, 1999. Adaptation des brebis Barbarine à l'alternance sous-nutrition – réalimentation, effet sur les tissus adipeux. Ann. Zootech. 48: 189-198.

Atti, N., C. Kayouli, M. Mahouachi, A. Guesmi and M. Doreau, 2002. Effect of a drastic and extended underfeeding on digestion in Barbarine ewe. Anim. Feed Sci. Technol. 100: 1-14

Chilliard, Y., M. Doreau, F. Bocquier and G.E. Lobley, 1995. Digestive and metabolic adaptations of ruminants to variations in food supply. In: M. Journet, E. Grenet, M.H. Farce, M. Thériez and C. Demarquilly (Eds), Recent developments in the Nutrition of Herbivores. INRA, Paris, pp. 329-360.

Doreau, M., B. Michalet-Doreau, P. Grimaud, N. Atti and P. Nozière, 2003. Consequences of underfeeding on digestion and absorption in sheep. Small Rum. Res. 49: 289-301.

Evaluation of DNA extraction methods from rumen contents for gut microbiota studies

G. Balmes[1], A. Serrano[1], A. Bach[1,2], M. Terre[1] and A. Aris[1]
[1]Ruminant Unit, IRTA-Torre Marimon, Caldes de Montbui, España [2]ICREA, Barcelona, Spain; anna.aris@irta.cat

Introduction

The animal gut represents one of the most complex microbial ecosystems with a large degree of microbial biodiversity. Digesta microbiota analyses have been perfomed traditionally by microbiology cultures based on the use of selective media. Molecular biology based methods offer increased sensitivity, specificity and reduced time for bacterial identification with accurate and reliable results. Extraction of highly pure DNA from the rumen community is complex. The high disparity in the structural and morphological features of the microorganisms and the physicochemical nature of these types of samples make DNA extraction procedures particularly difficult. In this context the comparison and the setting up of an efficient DNA extraction methodology is necessary. The nucleic acid purification procedures have to be able to extract DNA from hard-to-lyse bacteria without damaging the DNA from more vulnerable microorganisms. In this study we describe the comparison of four DNA extraction methods for the purification of microbial DNA from ruminal contents. We compared their efficiencies in terms of quantity and quality of extracted DNA. The DNA was used as the template for the quantification of four bacterial species by RT-PCR in order to evaluate whether there was any bias in the DNA extraction yield depending on the microbial structural features.

Material and methods

A rumen digesta sample was collected from a slaughtered one year old calf of 510 kg and fed a diet low in fibre composed of 1kg of straw for every eight kilograms of concentrate. The rumen sample was mixed well, centrifuged and the homogenised pellet distributed in 0.25 g aliquots that were frozen at -80 °C for further DNA extraction. The evaluated DNA extraction methods were (1) the RBB +C (Yu *et al.*, 2004), (2) the phenol/chloroform extraction method combined with bead beating (Whitford *et al.*, 1998), (3) the Power Soil kit (MO BIO Laboratories, Inc) and (4) the Fast DNA® SPIN Kit (Qbiogene, Inc.). DNA extractions were performed by sextuplicate according to the methodology previously cited or the manufacturer's instructions in the commercial kits. DNA concentration was assessed spectrophotometrically in duplicate, DNA purity was analysed by absorbance at 230 nm and 280 nm and DNA integrity was evaluated by 0.8% agarose gel electrophoresis. Rumen microbial DNA was used as a template in RT-PCR assays to amplify specific regions of 16S rDNA of the Gram-negative *Prevotella ruminicola* and *Selenomonas ruminantiu* and the Gram-positive *Ruminococcus flavefaciens* and *Streptococcus bovis.* The absolute quantification of 16S rDNA molecules/ug total DNA was performed using plasmid DNA standards generated by cloning procedures.

Results

Microbial DNA was obtained with all evaluated methods but the classical methodology based on phenol-chloroform extraction resulted in a 4-8 fold increase DNA yield when compared to the other methods used. DNA degradation was detected in all cases but at minor extent with the RBB + C method. The A_{260}/A_{230} ratios were always low, indicating volatile acid contamination in all cases. The RBB + C method offered the lower protein contaminated DNA giving a A_{260}/A_{280}

ratio of 1.8. The results of microbial PCR quantification indicate that the DNA of Gram-negative microorganisms was extracted with greater efficiencies by Fast DNA® SPIN Kit (Figure 1) while a greater proportion of Gram-positive *Ruminococcus flavefaciens* and *Streptococcus bovis* was found in the DNA extracted by the phenol-chloroform method (Figure 1). The Power Soil method yielded the worst results indicating that it was not a suitable kit for ruminal DNA extractions. Although in all tested methods the DNA extraction yield was affected by the microbial structural properties, the phenol-chloroform and the RBB + C method gave acceptable yields for all the microorganisms evaluated in this study.

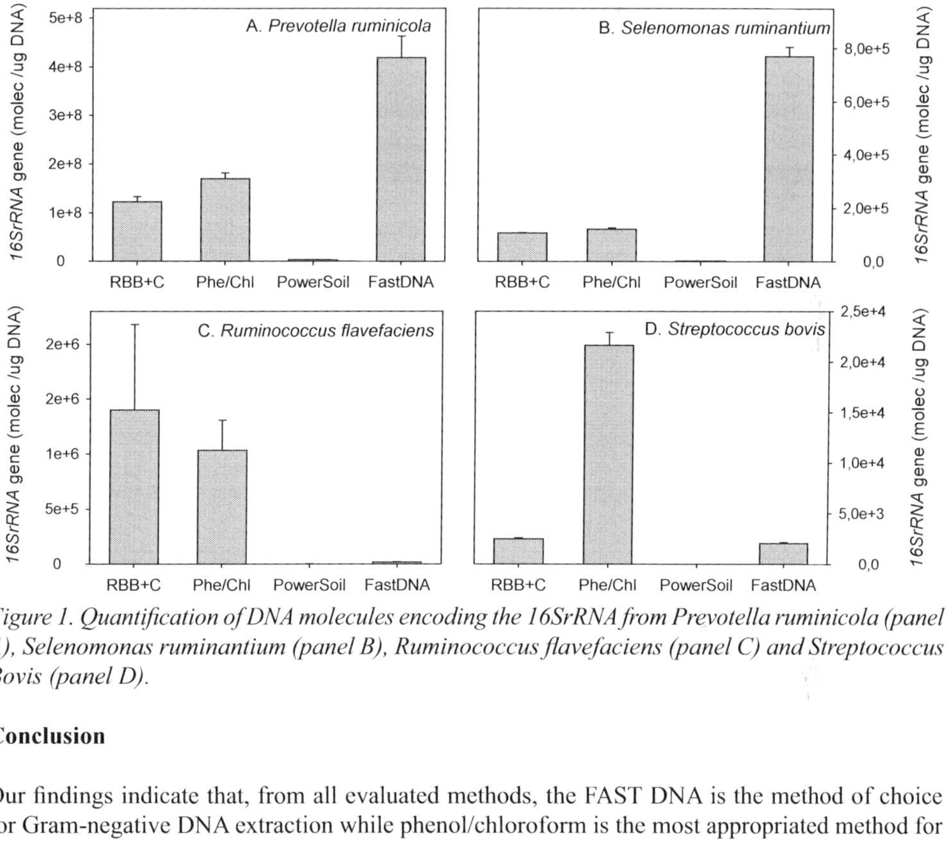

Figure 1. Quantification of DNA molecules encoding the 16SrRNA from Prevotella ruminicola (panel A), Selenomonas ruminantium (panel B), Ruminococcus flavefaciens (panel C) and Streptococcus Bovis (panel D).

Conclusion

Our findings indicate that, from all evaluated methods, the FAST DNA is the method of choice for Gram-negative DNA extraction while phenol/chloroform is the most appropriated method for Gram-positive microorganisms. However, when the combined yield for Gram-positive and Gram-negative bacteria is evaluated, phenol/chloroform and RBB + C method provided the best results.

References

Whitford, M.F., R.J. Forster, C.E. Beard, J. Gong and R.M. Teather, 1998. Phylogenetic analysis of rumen bacteria by comparative sequence analysis of cloned 16S rRNA genes. Anaerobe. 4: 153-163.

Yu, Z. and M. Morrison, 2004. Improved extraction of PCR-quality community DNA from digesta and fecal samples. Biotechniques. 36: 808-812.

Estimating digesta kinetics of large and small particles in dairy cows fed primary growth and regrowth grass silages

A.R. Bayat[1,2], M. Rinne[2], K. Kuoppala[2], S. Ahvenjärvi[2] and P. Huhtanen[3]
[1]Animal Science Department, Shiraz University, 7144165186, Shiraz, Iran; [2]MTT Agrifood Research Finland, Animal Production Research, 31600, Jokioinen, Finland; [3]Dept. of Agricultural Research for Northern Sweden, Swedish University of Agricultural Sciences (SLU), Umeå, Sweden; bayat@shirazu.ac.ir

Introduction

In order to increase digestibility, large particles are selectively retained in the rumen while the small ones have greater possibility to escape (Allen and Mertens, 1988). Neutral detergent fibre (NDF) is not a homogenous entity and it can be divided into indigestible NDF (iNDF) and potentially digestible NDF (pdNDF) (Huhtanen *et al.*, 2006). Rumen evacuation, wet sieving and iNDF determination techniques can be used to study the passage and digestion kinetics of iNDF and pdNDF in both escapable and non-escapable ruminal particles (Huhtanen *et al.*, 2007). The purpose of this study was to compare ruminal pool sizes and digesta kinetics of iNDF and pdNDF in escapable and non-escapable particles in dairy cows fed primary and regrowth grass silages harvested at two stages of growth.

Material and methods

Two primary growth (PG) and two regrowth (RG) silages were prepared from mixed timothy-meadow fescue sward in Jokioinen, Finland. PG silages were harvested on 5 June at early and on 17 June at late growth stage. Early and late RG silages were harvested on 29 July (42 growth days) and on 12 August (68 growth days). The four silages were fed in a 4×4 Latin Square design to four ruminally fistulated cows (75 days in lactation and 622 kg live weight) in four periods (16 d adaptation and 5 d sample collection). Silages and concentrate (8 kg/d) were offered four and two times daily, respectively. Rumen evacuations were conducted on 13 d (prior to morning feeding) and 15 d (6 h after morning feeding) to represent the minimum and maximum ruminal contents. Total faecal collection was performed on 18-21 d. Particle size distribution of ingested silage was adopted from Bayat *et al.* (unpublished data). The particle size distribution of samples was determined by wet sieving using 1.25 and 0.04 mm sieves. The NDF concentration was determined by ANKOM[220] Fiber Analyser. For iNDF determination, samples were incubated for 12 d in the rumen of two cows (Huhtanen *et al.*, 2007). The pdNDF was calculated as NDF-iNDF. Calculations of ruminal pool sizes and kinetics are described by Bayat *et al.* (2007). The following model was used to analyse the data using the SAS® mixed model procedure: $Y_{ijk} = \mu + C_i + P_j + T_k + e_{ijk} + S_l + T_k{*}S_l + \varepsilon_{ijkl}$ where μ is the overall mean, P_i, T_j, S_k and $T_j{*}S_k$ are the fixed effects of period, treatment, particle size and treatment*particle size, respectively. The treatment effect was split into cut, maturity and their interaction using orthogonal contrasts.

Results

The NDF and iNDF concentrations of early and late PG silages, and early and late RG silages were 498, 589, 513 and 538, and 52, 97, 61 and 93 g/kg DM, respectively. The ruminal iNDF content was greater with RG compared to PG grass (P<0.001) and it increased with advancing maturity (P<0.001; Table 1). Faecal iNDF excretion was considerably greater in small (SP, 1.25-0.04 mm) than large (LP, >1.25 mm) particles and it increased in SP with advancing maturity (P<0.001). The passage rate (k_p) of iNDF was greater for SP compared to LP and it was greater in PG compared to RG (P<0.05). The particle breakdown rate (k_r) was faster for PG compared to RG and it increased with advancing maturity only in PG silage (interaction $P<0.01$). The k_p of pdNDF for SP was faster than that for LP ($P<0.001$). The k_p of pdNDF for SP decreased with advancing maturity in PG but it increased with advancing maturity in RG (interaction $P<0.05$). The digestion rate (k_d) of pdNDF decreased with advancing maturity ($P<0.01$) and it was faster in SP compared to LP ($P<0.001$).

Table 1. Ruminal pools and faecal excretion of DM, iNDF and pdNDF, and ruminal rates of intake (k_i), passage (k_p), particle breakdown (k_r) and digestion (k_d) of large (LP) and small (SP) particles of dairy cows fed diets based on different grass silages.

	Primary growth grass				Regrowth grass				SEM[1]	Orthogonal contrasts[2]		
	Early		Late		Early		Late					
	LP	SP	LP	SP	LP	SP	LP	SP		C	M	S
Ruminal digesta, kg												
DM	4.83	3.64	4.80	5.29	4.58	3.80	5.40	4.33	0.405	ns	**	*
iNDF	0.77	0.78	0.99	1.35	0.93	0.96	1.35	1.20	0.102	***	***	*
pdNDF	3.20	1.90	3.09	2.68	2.93	1.70	3.08	1.89	0.242	*	†	***
Faeces, kg/d												
DM	0.42	3.14	0.49	3.86	0.33	2.60	0.43	3.41	0.180	*	***	***
iNDF	0.10	0.96	0.14	1.40	0.11	0.92	0.16	1.19	0.054	ns	***	***
pdNDF	0.29	1.60	0.31	1.80	0.20	1.13	0.24	1.55	0.100	**	**	***
iNDF kinetics												
k_i,1/h	0.036	0.022	0.047	0.014	0.029	0.018	0.028	0.016	0.0022	***	ns	***
k_p,1/h	0.006	0.052	0.006	0.044	0.005	0.041	0.005	0.042	0.0023	*	ns	***
k_r,1/h	0.030		0.040		0.024		0.024		0.0020	***	**	
pdNDF kinetics												
k_i,1/h	0.076	0.038	0.078	0.028	0.070	0.040	0.062	0.035	0.0043	ns	†	***
k_p,1/h	0.004	0.035	0.004	0.028	0.003	0.028	0.003	0.034	0.0020	ns	ns	***
k_d, 1/h	0.042	0.054	0.033	0.045	0.043	0.054	0.035	0.039	0.0038	ns	**	***

[1] SEM, standard error of means, n = 32.
[2] C, cut effect; M, maturity effect; S, particle size effect; ns, non significant; †, $P<0.10$; * $P<0.05$; ** $P<0.01$; *** $P<0.001$.

Conclusion

The passage rate of iNDF in SP and breakdown rate were faster for PG compared to RG while digestion rate was not different. The effect of first and second cuts of forages and maturity stage should be considered in the models predicting feed intake and fibre digestibility.

References

Allen M.S. and D.R. Mertens, 1988. Evaluating constraints on fiber digestion by rumen microbes. J. Nutr. 118: 261-270.

Bayat, A.R., M. Rinne, H. Khalili, R. Valizadeh and P. Huhtanen, 2007. Estimation of digesta kinetics of different particle size fractions using rumen evacuation technique in dairy cows fed red clover-grass silage. J. Anim. Feed Sci. 16: 538-554.

Huhtanen, P., U. Asikainen, M. Arkkila and S. Jaakkola, 2007. Cell wall digestion and passage kinetics estimated by marker and *in situ* methods or by rumen evacuation in cattle fed hay 2 or 18 times daily. Anim. Feed Sci. Technol. 133: 206-227.

Huhtanen, P., J. Nousiainen and M. Rinne, 2006. Recent developments in forage evaluation with special reference to practical applications. Agric. and Food Sci. 15: 293-323.

Effect of supplementation with high levels of soybean oil to rams fed dehydrated lucerne on digestibility and energy valorisation of the diets

R.J.B. Bessa[1,2] *and A.V. Portugal*[2]
[1]*CIISA, Faculdade de Medicina Veterinária, Universidade Técnica de Lisboa, Polo Universitário do Alto da Ajuda, 1300-477, Lisboa, Portugal;* [2]*Estação Zootécnica Nacional, INRB, Fonte Boa, 2005-048, Vale de Santarém, Portugal; rjbbessa@fmv.utl.pt*

Introduction

Supplementation of ruminant diets with polyunsaturated oils has been extensively explored in the last decade in order to modify the fatty acid pattern of their edible fats. It has been known for a long time that feeding high concentrations of polyunsaturated oils to ruminants might reduce intake and digestibility of fibre and hence reduce its productive performance. Our team has been using diets based on dehydrated lucerne with oil inclusions up to 10% in order to increase CLA concentration on lamb meat without observing any major impact on lamb growth (Bessa *et al.*, 2005). In the present experiment, we added up to 120 g/kg dry matter (DM) of soybean oil to dehydrated lucerne basal diet in order to explore the effects on apparent digestibility and energy evaluation of the diets.

Material and methods

Three rams (57±1.2 kg BW) equipped with rumen cannula were used in a 3×3 latin square trial. The animals were fed 4% of wheat straw and 96% of one of the following pelleted diets: dehydrated lucerne (L), dehydrated lucerne with 66 g of soybean oil per kg of DM (LO7) or with 120 g of soybean oil per kg DM (LO12). Animals were fed twice a day aiming a target intake of 40 g of fat-free DM per $kg^{0.75}$ of BW. Experimental periods included 6 weeks of adaptation to allow full adaptation to the diets and avoid carry-over effects. The digestibility trial and N balance were conducted in the first 8 d after the adaptation period followed by 15 days more for rumen sampling procedures (not reported here). Faeces and urine were collected and weighed daily during the digestibility trial period, and 10% of each sampling were pooled and analysed. Feeds and faeces were analysed for ash, N, neutral detergent fibre (NDF), acid detergent fibre (ADF), gross energy (GE) and fatty acids (FA). Urine was analysed for N and GE. Methane production was estimated according to a model that accounts for the inhibitory effect of fat on methane production (Pelchen and Peters, 1998). Data were submitted to analysis of variance using the GLM procedure of SAS® (SAS Institute, Inc., Cary, NC, USA).

Results

Soybean oil inclusion to diets increased intake of FA and decreased intake of ADF (Table 1). The intake of organic matter (OM), N and GE were not different between diets, although a trend to reduce intake of NDF (P=0.098) and OM without fatty acid (P=0.120).
The apparent digestibility of OM, OM without FA and GE were not affected by treatments, but the digestibility of ADF was depressed by oil inclusion. The digestibility of NDF was reduced by intermediate oil supplementation (L7) but not by the highest level of oil inclusion (L12). One single observation was responsible for the lower NDF digestibility in the L7 diet. The apparent digestibility of FA increased with oil supplementation probably by diluting the important endogenous lipid faecal fraction mainly composed of microbial lipids synthesised in the rumen and hindgut. The true digestibility of FA computed as regression of FA intake and FA apparently absorbed was 86% (Absorbed FA = -2.1 (±1.44) + 0.86 (±0.019) FA intake; r^2 = 0.996; n = 9).

Oil supplementation increased DE and ME and metabolisability (q) of diets. Urinary energy losses averaged 0.62 MJ/d and were not affected by treatments.

Table 1. Intake, digestibility and energy evaluation of the diets measured on rams fed dehydrated lucerne and increasing levels of soybean oil.

	Diet[1]			S.E.	*P*-value
	L	L7	L12		
Intake (g/d)					
Organic matter (OM)	806	838	763	29.4	0.384
OM without FA	788	764	662	24.7	0.120
Fatty acids (FA)	17^a	75^b	102^c	4.7	0.012
NDF	468	431	388	13.1	0.098
ADF	365^b	304^a	276^a	8.2	0.031
GE (MJ/d)	16.0	17.1	17.7	0.67	0.365
ME (MJ/d)	7.8	9.3	9.2	0.44	0.172
Digestibility (%)					
OM	62.0	62.8	62.3	0.76	0.815
OM without FA	61.9	60.7	59.1	0.86	0.288
FA	70.5^a	84.2^b	83.1^b	0.73	0.009
NDF	50.7^b	44.4^a	49.7^b	0.65	0.036
ADF	48.6^b	38.6^a	43.1^a	0.77	0.024
GE	58.4	60.3	61.4	0.69	0.173
Energy evaluation (MJ/kg DM)					
DE	10.3^a	11.4^b	12.4^b	0.17	0.026
ME	8.3^a	9.8^b	10.9^b	0.17	0.019
q (ME/GE)	0.47^a	0.52^b	0.54^b	0.007	0.042

[1] L = dehydrated Lucerne pellets; L7 and L12 = dehydrated lucerne pellets including 66 and 120 g of soybean oil per kg of dry matter, respectively.
[a,b,c] Means within the same row with different superscript letters differ significantly ($P<0.05$).

Conclusion

In spite of the very high inclusion of soybean oil to dehydrated lucerne (up to 12% of dry matter), the apparent digestibility of OM was not depressed and DE and ME of diets increased. Nevertheless, fibre digestibility, particularly ADF, was depressed by oil supplementation. Overall, these results support the usefulness of dehydrated lucerne as a basal diet for high oil supplementations in lambs.

References

Bessa, R.J.B., P.V. Portugal, I.A. Mendes and J. Santos-Silva, 2005. Effect of lipid supplementation on growth performance, carcass and meat quality and fatty acid composition of intramuscular lipids of lambs fed dehydrated lucerne or concentrate. Livestock Prod. Sci. 96: 185-194.

Pelchen, A. and K.J. Peters, 1998. Methane emissions from sheep. Small Rum. Res. 27: 137-150.

A meta-analysis of the satiating effect of VFA absorbed in the rumen and glucose absorbed in the intestines of ruminants

A. Boudon[1,2], *J. Juton*[1,2], *L. Delaby*[1,2] *and P. Faverdin*[1,2]
[1]*INRA, UMR 1080 Dairy Production, 35590 St-Gilles, France;* [2]*Agrocampus Ouest, UMR 1080 Dairy Production, 35000 Rennes, France; Anne.Boudon@rennes.inra.fr*

Introduction

In Western European dairy systems, the use of grazed herbage can reduce feed costs, limit the impact of livestock breeding on the environment and improve the quality of animal products. Highly digestible herbage, such as perennial ryegrass generally has a high nutritive value, equivalent to that of a total mixed ration based on corn silage. In spite of this, ray-grass intake remains generally low and variable compared with those of a total mixed ration based on corn silage. A hypothesis to explain this, at least partly, is that the satiating effect of energy nutrient absorbed from herbage, i.e. mainly VFA digested in the rumen, is higher than that of glucose absorbed from the intestines that can represent a substantive part of the energy nutrient absorbed when maize is fed. Our objective was to evaluate the relative satiating effect of VFA and glucose in ruminants.

Material and methods

Ninety published experiments were included in the meta-analysis and fulfilled the following conditions: (1) feed was offered *ad libitum* and intake was precisely given, (2) the nutrient tested was infused in the rumen for VFA, or in the abomasum or duodenum for glucose, (3) a control was included. The final database contained 199 comparisons between a nutrient infusion and a control. The explicated variable was the decrease of NE intake with infusion compared to a control expressed in MJ/MJ of daily NE requirement of the animal. The explicative variables included the nature of the nutrient infused (acetate, propionate, butyrate, mixture of VFA or glucose), the quantity of NE and moles infused per MJ of NE daily requirement and the nature of the infusion protocol. The quantity of NE infused was calculated considering that the NE contents of acetate, propionate, butyrate and glucose were 0.387, 0.980, 1.434 and 2.175 MJ/mole respectively. The infusion protocols were 'continuous infusions' when infusions lasted at least 14 h per day, 'infusions synchronous to ingestion' when infusions were started or stopped when animals started or stopped to eat respectively, and 'infusions during the first meal' when infusions started when diet was fed and lasted less than 6 h. Statistical analyses were performed using the MIXED procedure of SAS® with a statistical model including both quantitative (quantity of nutrients infused) and qualitative (infused nutrient, infusion protocol) variables as described by St-Pierre (2001). The model testing the impact of the infusion protocol was run on the whole database. The model testing the impact of the infused nutrient was run on 2 subdatabases for continuous infusions and infusions synchronous to the meal, respectively.

Results and discussion

When considering the whole database whatever the nutrients infused, intake decreased when the quantity infused increased ($P<0.0001$) but this decrease was quicker when infusions were synchronous to ingestion (interaction slope × infusion protocol, $P<0.0001$; Figure 1).
For infusions synchronous to ingestion, intake decreased with increasing infused quantity ($P<0.05$), but the slopes did not differ between VFA ($P>0.20$; Figure 2). For continuous infusions, intakes decreased with increasing infused quantity when propionate was infused ($P<0.01$), but the slopes did not differ from 0 with the other nutrients ($P>0.20$; Figure 2).

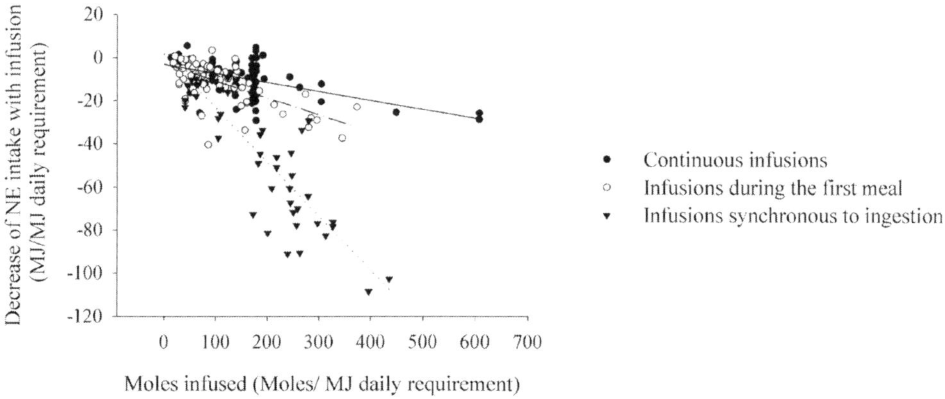

Figure 1. Effect of the infusion protocol on the decrease of NE intake with nutrient infusions compared to a control.

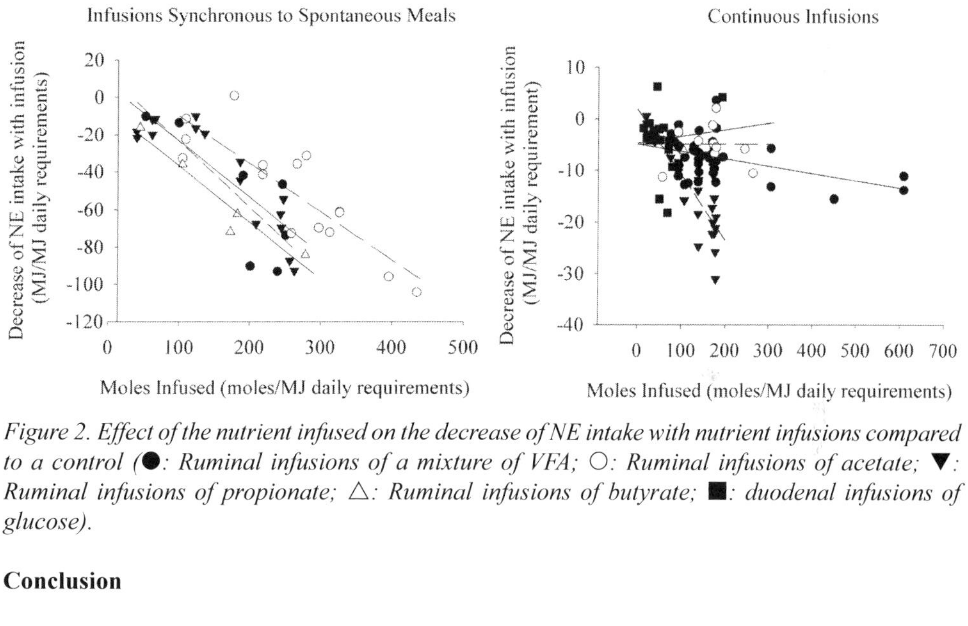

Figure 2. Effect of the nutrient infused on the decrease of NE intake with nutrient infusions compared to a control (●: Ruminal infusions of a mixture of VFA; ○: Ruminal infusions of acetate; ▼: Ruminal infusions of propionate; △: Ruminal infusions of butyrate; ■: duodenal infusions of glucose).

Conclusion

This meta-analysis showed that the satiating effect of VFA strongly depends on their dynamics of absorption by the ruminant. It also showed that propionate has a specific satiating effect when it is continuously infused but not with infusions synchronous to ingestion. With infusions synchronous to ingestion, the effect of VFA infusions on ruminal osmotic pressure is likely to be responsible of satiation.

References

St-Pierre, N.R., 2001. Invited review: Integrated quantitative findings from multiple studies using mixed model methods. J. Dairy Sci. 84: 741-755.

Effects of the methionine analogue isopropyl ester of 2-hydroxy-4-methylthio-butanoic acid (HMBi) on rumen parameters

A. Brisson[1], A. Marquet[2], P. Mosoni[2], D.P. Morgavi[2], E. Forano[2], C. Martin[2] and E. Devillard[1]
[1]Adisseo, 03600 Commentry, France; [2]INRA Theix, 63122 St Genes Champanelle, France; Estelle.Devillard@adisseo.com

Introduction

Balancing rations for metabolisable methionine (Met) improves milk production. To achieve this balance, dairy rations can be supplemented with synthetic sources of Met such as isopropyl ester of 2-hydroxy-4methylthio-butanoic acid (HMBi), which increases milk production and protein concentration and significantly improves N efficiency (St Pierre and Sylvester, 2005). The particularity of HMBi is its double action: around 50% of the supplemented Met is quickly absorbed through the rumen wall and provides Met to the liver and the mammary gland for milk and protein synthesis (Graulet *et al.*, 2005). It has been suggested that the remaining part of HMBi could increase utilisation of the ration feedstuffs, by improving organic matter digestibility (OMD), in particular fibre degradation (Robert *et al.*, 2003, Noftsger *et al.*, 2005). The objectives of the present study were to estimate the effects of HMBi on (a) *in vitro* OMD of different feedstuffs and (b) rumen fermentation products and microbial fibrolytic populations.

Material and methods

Six dry Holstein cows were randomly assigned to 2 dietary treatments in a cross-over design. Cows received a ration of 50% wheat, 10% barley, 35% hay, and 5% straw, supplemented or not with HMBi (20 g/d supplemented via rumen cannula). Dry matter intake was limited and was distributed into 2 equal meals (7.5 kg/day). Each experimental period lasted 8 wk, starting with a 3-wk-adaptation phase. Gas test measurements were carried out to estimate OMD of 7 feedstuffs (maize grain, barley grain, maize silage, grass silage, grass hay, soybean meal, sugar beet pulp), as described by Robert *et al.* (2003). Rumen content samples were also taken for analysis of fermentation products (volatile fatty acids (VFA), ammonia and lactate) and microbial quantification. Protozoa were stained and counted by microscopy. Total and fibrolytic bacteria (*F. succinogenes*, *R. flavefaciens*, *R. albus*) were estimated by qPCR. Bacterial colonisation of feedstuffs was also estimated. Maize grain (MG) and maize silage (MS), introduced into nylon bags (pore size: 48 μm), were incubated for 24 h in the rumen. After washing, bacteria adhering to substrates were quantified by qPCR. Data were analysed using the Proc Mixed procedure of SAS®. The model included run (OMD data) or period (qPCR data), cow, substrate (S), treatment (T), the interaction S×T and the residual error. For the fermentation parameters, the model included cow, period, treatment and residual error. For both models, cow was the random effect.

Results

The OMD of all feedstuffs tested was higher with rumen fluid from supplemented cows, compared to control cows (Table 1). Indeed the OMD of the two cereals (maize and barley grains), the three forages (maize silage, grass silage, grass hay), and sugar beet pulp were improved significantly (+2.4%, 1.3% and 1.3% respectively). In contrast, the OMD of soybean meal was not affected by HMBi treatment.
HMBi increased the rumen concentration of total VFA (from 98 to 110 mM, $P<0.01$), acetate (from 63 to 73 mM, $P<0.01$) and isovalerate (from 1.5 to 2.1 mM, $P<0.001$). Concentration of lactate and $N\text{-}NH_3$ were not affected by HMBi supplementation. Protozoal counts and quantification by

qPCR on rumen contents of total bacteria and of the three fibrolytic bacterial species were similar between cows supplemented or not with HMBi. Total bacteria adhering to MG were not affected by HMBi supplementation, whereas HMBi slightly decreased (by 1.5 fold, $P<0.05$) total bacteria adhering to MS. HMBi had no effect on substrate colonisation by the two main fibrolytic species, *F. succinogenes* and *R. flavefaciens*, but it lowered (~ 2.5 fold, $P<0.05$) *R. albus* populations adhering to MG and MS substrates.

Table 1. OMD (%) of different feedstuffs, estimated by gas test using rumen fluid from cows supplemented or not with the methionine analogue HMBi.

	Control	HMBi	*P*-values
Maize silage	74.8	75.8	<0.001
Grass silage	55.9	56.5	0.02
Grass hay	63.1	64.0	<0.001
Maize grain	89.8	92.1	<0.001
Barley grain	86.7	88.6	<0.001
Soybean meal	92.4	93.1	0.20
Sugar beet pulp	85.6	86.7	0.02

Conclusion

This study showed that HMBi can increase OMD of feedstuffs, corroborating the positive effects of HMBi on rumen fermentation parameters. This effect was not related to an increase in the targeted fibrolytic bacterial populations as hypothesised. However, our *in vitro* OMD and total VFA concentration measurements indicate that the positive effect of HMBi supplementation might be due to an improvement in the degradation of starch-rich (and sugar-rich) substrates, compared to fibrous substrates. This would suggest that HMBi could have a positive effect on amylolytic microbial populations. Further investigations are needed to support this hypothesis.

References

Graulet B., C. Richard and J.C. Robert, 2005. Methionine availability in plasma of dairy cows supplemented with methionine hydroxyl analog isopropyl ester. J. Dairy Sci. 88: 3640-3649.

Noftsger S.M., N.R. Saint-Pierre and J.T. Sylvester, 2005. Determination of rumen degradability and ruminal effects of three sources of methionine in lactating cows. J. Dairy Sci. 88: 223-237.

Robert J.C., S. Paquet, C. Richard and B. Bouza, 2003. Influence of HMBi concentration on *in vitro* estimated organic matter digestibility of diets varying in proportion of corn silage relative to concentrate. J. Dairy Sci. (Suppl. 1) 86: 276.

Saint-Pierre N.R. and J.T. Sylvester, 2005. Effects of 2-hydroxy-4-(methylthio)butanoic acid (HMB) and its isopropyl ester on milk production and composition by Holstein cows. J. Dairy Sci. 88: 2487-2497.

Integrated model of omasal bicarbonate transport in sheep: interactions with SCFA and Na^+/H^+ exchange

D. Caushi[1], M. Beisele[2], K. Wolf[2] and H. Martens[2]
[1]University of Pristina, Faculty of Agriculture and Veterinary Medicine, Kosovo; [2]Freie Universität Berlin, Dept. of Veterinary Physiology, Oertzenweg 19b, 14163 Berlin, Germany; dritoncaushi@gmail.com

Introduction

The omasum has the functions (a) to pass the ingesta from reticulo-rumen into the abomasum (in- and outflow) (Ehrlein and Hill, 1969), (b) to prevent the outflow of long particles out of the reticulo-rumen (Kaske *et al.*, 1992), and (c) to absorb electrolytes and water (v. Engelhardt and Hauffe, 1975; Edrise *et al.*, 1986). The transport mechanisms of Na (Schultheiss and Martens, 1999; Gäbel and Martens, 1988), SCFA (Ali *et al.*, 2006), Cl (Tiling, 1997) and bicarbonate (Wegeler, 2007) have been characterised.
The steady state transcellular transport of HCO_3^- (absorption) depends on a relatively constant intracellular pH, pH_i, and hence, those factors that may influence pH_i should have consequences on bicarbonate transport. The aim of the present experiments was therefore to study the absorption of bicarbonate under conditions which may have consequences for pH_i and hence bicarbonate transport: Transport of SCFA (influx of protons), inhibition of apical Na^+/H^+ exchanger (NHE) by amiloride (reduced efflux of protons) or inhibition of carboanhydrase by ethoxyzolamide (change of intracellular buffer capacity).

Material and methods

Experimental animals: The omasal epithelial tissues were obtained from sheep (*Ovis aries*) fed hay *ad libitum*. The animals had free access to a lick stone and to tap water.
Isolation, preparation and handling of the epithelial tissues: The preparation and incubation of the omasum epithelia have been described in detail by Ali *et al.* (2006). Six to eight large leaves were removed from the wall of the omasum and mucosal sheets on the two surfaces of the leaves were cautiously separated by blunt dissection and cut into pieces ready to be used in Ussing chambers.
Experimental design: All experiments were conducted under short-circuit conditions in an Ussing chamber and were started after an equilibration period for not less than 30 min.
pH-stat method: The pH-stat method carries out continuous measurements of the pH and automatic titration on the serosal side. By titrating acid, the pH will be kept constant at 7.4 and the amount of acid is proportional to HCO_3^- transport. A solution of 0.01 M H_2SO_4 was used for titration.

Results

The isolated omasal epithelium transported HCO_3^- (absorption) under control conditions (50 mmol/l HCO_3^- mucosal side) with a rate of 4-8 μeq/cm^2/h. Experimental conditions that suggest a change of pH_i cause effects on HCO_3^- transport (Table 1). Inhibition of NHE by amiloride (mucosal 1 mmol/l) or high mucosal acetate (100 mmol/l) significantly reduced HCO_3^- transport. A physiological mixture of SCFA (mmol/l: acetate 40; propionate 16; butyrate 8) generally reduced the transport of HCO_3^- and the addition of amiloride further decreased the transport of the HCO_3^-. Inhibition of carboanhydrase by ethoxzolamide significantly decreased HCO_3^- transport rates.

Table 1. Effects of amiloride and/or SCFA and ethoxyzolamide on in vitro HCO_3^- port rates in sheep omasum (x ± SD; N = number of sheep; n = number of tissues).

Treatment	Fluxes J_{ms} HCO_3 (µeq/cm^2/h)	N/n
Control	6.26±0.70	5/7
Amiloride (1 mmol/l)	4.56±0.54[a]	5/8
Control	8.43±2.34	3/4
Acetate (100 mmol/l)	3.42±1.77[a]	3/4
SCFA (64 mmol/l)	5.67±0.94	3/9
SCFA + amiloride (1mmol/l)	2.32±0.87[a]	3/6
SCFA (64 mmol/l)	7.33±1.24	3/7
SCFA + ethoxyzolamide (0.1 mmol/l)	5.53±1.01[a]	4/11

[a] Significantly different from control.

Conclusion

The transcellular transport of HCO_3^- requires as a precondition a tight regulation of pH_i and in particular the avoidance of a decrease of pH_i, which would promote the conversion of $H^+ + HCO_3^-$ to H_2O and CO_2 and hence, cause a decrease of HCO_3^- transport. pH_i is challenged by luminal uptake of HSCFA and intracellular H^+ release, which are recycled via NHE. The results obtained support this model of HCO_3^- transport and the possible interference of this transport by rapid uptake of HSCFA and underline the involvement of NHE in regulation of pH_i.

References

Ali, O., Z. Shen, U. Tietjen and H. Martens, 2006. Transport of acetate and sodium in sheep omasum: mutual, but asymmetric interactions. J. Comp. Physiol. B 176: 477-487.

Edrise, B.M., R.H. Smith and D. Hewitt, 1986. Exchanges of water and certain water-soluble minerals during passage of digesta through the stomach compartments of young ruminating bovines. Br. J. Nutr. 55: 157-167.

Ehrlein, H.J. and H. Hill, 1969. Motility and nutrional transport in the omasum of sheep. Zbl. Vet. Med. A. 16: 573-596.

Engelhardt, W. v. and R. Hauffe, 1975. Funktionen des Blättermagens bei kleinen Haus-wiederkäuern. IV: Resorption und Sekretion von Elektrolyten. Zbl. Vet. Med. A 22: 363-375.

Kaske, M., S. Hatiboglu and W. v. Engelhardt, 1992. The influence of density and size of particles on rumination and passage from the reticulo-rumen of sheep. Br. J. Nutr. 67: 235-244.

Martens, H. and G. Gäbel, 1988. Transport of Na and Cl across the epithelium of ruminant forestomachs: rumen and omasum. A review. Comp. Biochem. Physiol. A 90: 569-575.

Schultheiss, G. and H. Martens, 1999. Ca-sensitive Na transport in sheep omasum. Am. J. Physiol. 276: G1331-G1344

Tiling, C., 1997. *In vitro* Untersuchungen zum Chloridionentransport des Blättermagenepithels von Schafen. Thesis., Berlin - Freie Universität.

Wegeler, C., 2007. Transport von HCO_3^- am isolierten Psalterepithel des Schafes. Thesis, Berlin – Freie Universität.

Effects of urea treated *Leucaena leucocephala* leaves and supplements on *in vitro* fermentation characteristics

Z.P. Chen[1], Y.H. Yang[1], Z.S. Wang[1], A.G. Zhou[1], B. Xue[1] and Y.M. Cai[2]
[1]Animal Nutrition Institute, Sichuan Agricultural University, Ya'an 625014, China; [2]National Institute of Livestock and Grassland Science, Tokyo 329-2793, Japan; wangzs007@yahoo.com.cn

Introduction

Leucaena leucocephala is a tropical leguminous shrub with 22 to 34% CP and its leaf is considered to be a promising protein source for ruminants, however, the utilisation of the tree leaf is limited by mimosine (Jones, 1979). Recently, Liu *et al.* (2009) found that the concentration of mimosine had a reducing tendency after the tree leaf was treated with urea. Urea treatment supplied the nitrogen needed by the rumen microorganism, but, destroyed the nutritive elements such as rumen fermentable carbohydrates and vitamins (Van Soest, 2006), and the degradation parameters of urea treated *Leucaena* leaves may be change. Moreover, the urea treated *Leucaena* leaves could possibly be used efficiently for rumen microbial production if it is mixed with other ingredients. So the objective of this study was to evaluate the effects of urea treatment on *in vitro* fermentation characteristics of the tree leaf, and the tree leaf supplementation in the diet.

Material and methods

Fresh *Leucaena* plant material was harvested in August 20, 2008, at Suining, Sichuan province of China. Urea treated *Leucaena* was prepared according to Liu *et al.* (2009). The treatments were the following : untreated *Leucaena* (LL); urea treated *Leucaena* (ULL); a corn silage and concentrate mixture (LL, 200; maize, 488; rice bran, 13.8; alfalfa meal, 85.3; wheat bran, 46; soybean meal, 28.5; rapeseed meal, 96.3; salt, 4 and premix mixture, 38.1.g/kg DM) in a 50:50 ratio (LLC); and a corn silage and concentrate mixture (ULL, 200; maize, 500; rice bran, 18.5; alfalfa meal, 79.6; wheat bran, 44; soybean meal, 25; rapeseed meal, 90.6; salt, 4 and premix mixture, 38.1g/kg DM) in a 50:50 ratio (ULLC). The compound diets (LLC and ULLC) were considered to have the same ration of energy and nitrogen.

Ruminal fluid was obtained from four ruminally fistulated Nanjiang goats fed a maintained diet before the morning feeding. The fermentation parameters were determined at 24 h incubation of 200 mg substrate in triplicate into 100 ml calibrated glass syringes, fitted with plungers as described by Menke and Steingass (1988). The gas production (GP) was recorded at 2, 4, 6, 8, 12, 24, 48 and 72 h. After terminating the incubation at 24 h, the fermentation was stopped by swirling the syringes in ice water. About 30 ml mixed fermentation medium was used for the analysis of ammonia nitrogen (NH_3-N), volatile fatty acids (VFA) and microbial crude protein (MCP) according to Hu *et al.* (2005). Treatment was the experimental unit for GP parameters, MCP, NH_3-N and VFA. Data were analysed using the general linear model procedure of SPSS 16.0 software (SPSS Inc., Chicago, IL, USA). Overall differences between treatment means were declared significant at $P<0.05$.

Results

The potential GP was higher for ULLC than for the others ($P<0.05$, Table 1), and ULL's rate of GP increased significantly after being combined with others (ULLC) ($P<0.05$). Meanwhile, microbial crude protein (MCP) concentration for ULLC elevated 4.24% ($P<0.05$), 16.04% ($P<0.05$) and 4.24% ($P<0.05$) compared to LL, ULL and LLC respectively. The concentrations of total VFA, acetate, propionate and butyrate had the same significant response to compound diets (LLC and ULLC) ($P<0.05$), moreover, ULLC had a greater tendency than LLC; and the proportion of acetate:propionate was altered (P<0.05). This evolution was less marked for ammonia nitrogen, when ULLC was compared to LL and ULL respectively.

Table 1. Effects of urea treated Leucaena *on* in vitro *GP parameters at 72 h and the concentration of NH_3-N, MCP and VFA at 24 h of incubation.*

	Treatment[1] least square mean				SEM	P-value
	LL	ULL	LLC	ULLC		
GP parameters						
Potential GP, ml/g	46.30^{c}	40.74^{d}	56.34^{b}	59.78^{a}	1.181	<0.001
Rate of GP, ml/h	0.72^{a}	0.57^{b}	0.56^{b}	0.68^{a}	0.004	0.009
Microbial crude protein, mg/ml	1.18^{b}	1.06^{c}	1.18^{b}	1.23^{a}	0.008	<0.001
Ammonia nitrogen, mg/dl	23.21^{b}	25.16^{a}	19.70^{c}	23.77^{ab}	0.253	<0.001
VFA, mmol/l						
Acetate	51.95^{c}	49.53^{c}	62.91^{b}	72.51^{a}	4.152	0.002
Propionate	17.79^{b}	17.99^{b}	26.61^{a}	29.14^{a}	1.834	<0.001
Butyrate	4.41^{b}	4.65^{b}	9.44^{a}	9.83^{a}	0.612	<0.001
Total	74.14^{b}	72.17^{b}	98.96^{a}	111.48^{a}	6.394	<0.001
Acetate:propionate	2.92^{b}	2.77^{b}	2.36^{a}	2.49^{a}	0.120	0.005

[1] LL = untreated *Leucaena*, ULL = urea treated *Leucaena*, LLC and ULLC = diets including 20% LL and ULL, respectively.
[a,b,c,d] Means with different letters in the same row differ significantly (P<0.05).

Conclusion

This study shows that urea treated *Leucaena* (ULL) is not ideal, however, combined with other feeds (ULLC) it enhances the digestibility of the diet and the production of microbial crude protein *in vitro* fermentation.

Acknowledgement

This work is jointly supported by grants from the Chunhui Project by the Ministry of Education of China (No.Z2005-2-61004), and supported by the Educative Department of the Sichuan Province.

References

Hu, W.L., J.X. Liu, J.A. Ye, Y.M. Wu and Y.Q. Guo, 2005. Effect of tea saponin on rumen fermentation *in vitro*. Anim. Feed Sci. Technol. 120: 333-339.

Jones, R.J., 1979. The value of *Leucaena leucocephala* as a feed for ruminants in the tropics. World Anim. Rev. 31: 13-23.

Liu, L.P., Z.S. Wang, Y.M. Cai and W.M. Huang, 2009. The nutritional evaluation of *Leucaena leucocephala* from different areas and season in Sichuan province. New Feed. 1: 24-26.

Menke, K.H. and H. Steingass, 1988. Estimation of the energetic feed value obtained by chemical analysis and *in vitro* gas production using rumen fluid. Anim. Res. Develop. 28: 7-55.

Van Soest, P.J., 2006. Rice straw, the role of silica and treatments to improve quality. Anim. Feed Sci. Technol. 130: 137-171.

Evaluating anaerobic fermentation profiles of corn milling co-products

M.L. Chizzotti[1,2], L.O. Tedeschi[2] and P.J. Kononoff[3]
[1]Department of Animal Science, Universidade Federal do Vale do São Francisco, Brazil; [2]Department of Animal Science, Texas A&M University, USA; [3]Department of Animal Science, University of Nebraska-Lincoln, USA; mario.chizzotti@univasf.edu.br

Introduction

The biofuel industry is currently experiencing tremendous growth. The ethanol obtained from corn produces large quantities of several different corn co-products that can be successfully utilised by ruminants as a relatively inexpensive feed alternative. The *in situ* rumen disappearance and *in vitro* gas production techniques are useful for rapid access of ruminal utilisation. The objective of this study was to provide information on rumen degradability, based on *in vitro* gas production and *in situ* degradability profiles of corn milling co-products.

Material and methods

Composite samples of six corn milling co-products were investigated in this study, including Dakota Gold corn germ dehydrated (Germ), Dakota pelleted bran cake (Bran), wet corn gluten feed (WCGF), wet distillers grains (WDGS), a high protein dried distillers grains (HPDDG), and a dried distillers grains plus solubles (DDGS).

Three ruminally-cannulated Angus steers, fed *ad libitum*, twice daily, a 50% of timothy hay and 50% of alfalfa hay plus 2 kg/d of concentrate diet, were used for determination of *in situ* rumen DM and NDF degradation according to Mertens (1993). Samples were incubated in each animal for 0, 2, 4, 6, 8, 12, 16, 24, and 48 h using 3 replicates. Kinetics of DM and NDF degradation were calculated using the PROC NLIN procedure of SAS® (SAS Institute Inc., Cary, NC, USA) according to the following model (Mertens, 1993): $Y = a + b \times (1-e^{(-c \times (t-L))})$; if t>L, and Y=0; if t≤L, where Y is DM disappearance loss at time t (%); a is intercept of the fitted curve (%); b is asymptote of the curve (%); c is fractional degradation rate constant (%./h); t is time of incubation (h); and L is lag time (h). Effective degradability (ED) was calculated using the above parameters and using rumen outflow rate (kp) of 5%, according to the following model: $ED = a + [b \times c / (c + k)] \times e^{-(c + kp) \times L}$.

In vitro fermentation was determined for each feed using the gas production technique (Schofield and Pell, 1995), using a computerised system to record pressure. Gas production data were fitted to the following equation using the PROC NLIN procedure of SAS®: $Y = b \times (1-e^{(-c \times (t-L))})$; if t>L, and Y=0; if t≤L, in which Y is the accumulated gas (ml/100 g of incubated DM); b is the asymptote of gas production (ml/100 g of incubated DM); c is the fractional rate of gas production (h/1); t is the time (h); and L is the lag time (h).

The difference among corn co-products was analysed using PROC MIXED of SAS®, assuming feeds as fixed effects and steers as random effects, and comparing the least squares means of parameters using 0.05 as significance level.

Results and discussion

The highest (P<0.05) rate of *in situ* degradation of DM coincides with the highest (P<0.01) asymptote of gas production, rate of gas production and IVDMD of Germ and likely are related to its low NDF and high NFC (30.1 and 30.3%, respectively). The fat content of Germ (17.7%) impact the lag time of *in vitro* fermentation, which was the highest (P<0.05), but it did not affect the *in situ* lag time when it was incubated in the rumen. The gas production technique might be more effective than the *in situ* technique for determining inhibitory effects of feed compounds, such as fat, because effects such as toxicity to microbes would be diluted and difficult to detect *in situ*. Bran Cake had the highest ($P<0.01$) zero time intercept from the *in situ* trial, and the largest gas volume. Similar to Germ, the high EE level (12.1%) probably inhibits the adhesion of microbes and increases the lag time in the *in vitro* fiber fermentation. Among corn milling co-products, WCGF has an intermediary rate of DM fermentation and degradability. HPDDG was the least ruminally degradable corn co-product evaluated. DDGS had a good availability of both fiber and non-fiber carbohydrates.

Table 1. In situ *DM, NDF and NDS degradability of six corn milling co-products.*

	Corn milling co-product						SEM	P
	Germ	DDGS	HPDDG	Bran	WDGS	WCGF		
In situ DM degradation								
a	0.496	0.532	0.3	0.6	0.462	0.502	0.017	***
b	0.406	0.364	0.581	0.331	0.446	0.428	0.015	***
c, /h	0.178	0.04	0.059	0.039	0.046	0.065	0.021	***
L, h	0	0	1.698	0	1.279	0	0.756	***
ED	0.813	0.694	0.561	0.745	0.651	0.744	-	-
In situ NDF degradation								
A	0.131	0.067	0.05	0.01	0.05	0.047	0.019	***
B	0.717	0.894	0.893	0.925	0.892	0.844	0.017	***
c, /h	0.077	0.034	0.093	0.033	0.052	0.05	0.003	***
L, h	0	2.137	1.005	10.57	3.179	1.971	0.373	***
ED	0.566	0.369	0.553	0.163	0.379	0.394	-	-
In vitro gas production of DM								
b, ml	48.4	34.6	34	52.4	32.3	42.1	1.73	***
c, h^{-1}	0.196	0.132	0.16	0.161	0.123	0.119	0.02	***
L, h	1.59	0	0.62	1.26	0.89	0.68	0.39	0.03
IVDMD	0.915	0.801	0.841	0.88	0.759	0.888	0.02	0.01
In vitro gas production of NDF								
b, ml	16.7	17.9	9.99	15.3	19.1	25.4	1.18	***
c, h^{-1}	0.081	0.065	0.129	0.055	0.073	0.073	0.01	***
L, h	0.6	1.37	0.98	3.34	2.35	1.9	0.28	***
IVNDFD	0.854	0.804	0.927	0.865	0.787	0.815	0.02	0.01

a: intercept; b: asymptote; c: rate of degradation (/h); L: lag time (h); ED: effective degradability assuming 0.05 of rumen outflow rate; IVDMD: *in vitro* DM digestibility; IVNDFD: *in vitro* NDF digestibility; ***$P<0.001$.

Conclusion

The ruminal degradation profile is variable among different corn milling co-products and our study provides new information.

References

Mertens, D.R., 1993. Rate and extent of digestion. In: J.M. Forbes and J. France (Eds.), Quantitative aspects of ruminant digestion and metabolism. CABI University Press, Cambridge, UK, pp. 13-52.

Schofield, P. and A.N. Pell, 1995. Validity of using accumulated gas pressure readings to measure forage digestion *in vitro*: A comparison involving three forages. J. Dairy Sci. 78: 2230-2238.

Two different drying methods of bovine faeces for estimating *n*-alkane concentration, intake and digestibility: a comparison

F. Sánchez Chopa[1,2], *L.B. Nadin*[2] *and H.L. Leandro*[1]
[1]*Facultad de Ciencias Veterinarias – UNCPBA, Tandil, Argentina;* [2]*CONICET, Argentina; fsanchez@vet.unicen.edu.ar*

Introduction

Many different techniques have been used to estimate herbage intake. The *n*-alkane technique (Mayes *et al.*, 1986) indicates that the material to be analysed should be freeze-dried, although the routine analysis would be simplified if the samples could be oven-dried. Dove and Mayes (1991) indicated that oven-drying was found to reduce the *n*-alkane concentration. However, Oliván and Osoro (1995) and Elwert *et al.* (2006) concluded that oven-dried samples (up to 65 ºC) yield a good estimation of different *n*-alkane concentrations. There is still lack of information about the effect of different drying methods on *n*-alkane concentration of faeces samples. The objective of the present work was to evaluate if the oven-drying method could replace the freeze-drying method in the determination of *n*-alkane concentration in faeces, the estimation of intake, faecal output and forage digestibility.

Material and methods

Four Holstein Friesian steers (182.2±12.1 kg LW) were fed twice a day (8:30 and 16:30 h), for 21 d with perennial ryegrass hay (*Lolium perenne* L.), ground to 12 mm. On the last 10 d, animals were dosed twice a day at feeding time with a cellulose pellet containing *n*-alkanes C_{32} (88.49 mg/pellet) and C_{36} (87.12 mg/pellet). During the last 5 d, total faeces were collected using harnesses 3 times a day (8:00, 16:00 and 22:30 h). Each time, the faeces were weighed and homogenised, and a sample (ca. 400g each) was kept at -20 ºC until analysis. Diet samples (n=4) and faeces samples (n=64) were prepared for the *n*-alkane analysis by two different methods: freeze-drying or oven-drying at 60 ºC. After water removal the samples were treated in pairs. The *n*-alkane analysis was performed as described by Mayes *et al.* (1986). The daily dry matter intake was calculated individually by difference (offered – rejected). The estimations of intake and faecal output by the *n*-alkane technique were calculated according to Dove and Mayes (2006). Estimated intake – using either pairs C_{32}:C_{31} or C_{32}:C_{33} – , estimated faecal output – using C_{36} – , and estimated apparent digestibility by *n*-alkanes, all calculated by both drying methods, were compared against the observed values. The data were statistically analysed by ANOVA. The mean comparisons were made by the Duncan multiple range test.

Results and discussion

The oven-drying method reduced the amount of all the *n*-alkanes present in the faeces, except for C_{36} (Table 1). All the *n*-alkanes, except for C_{23} (P>0.08), were affected (P<0.01) by the drying method (Table 1). The ratios between pairs C_{32}:C_{31} and C_{32}:C_{33} were also affected by the drying method (P<0.01) (Table 1).
Intake, fecal output and digestibility estimated by the *n*-alkane technique are shown in Table 2. When using the C_{32}:C_{33} ratio, there were no differences in estimating intake neither with oven nor freeze-drying methods *versus* observed intake (P>0.14), although there were differences between oven-dried and observed when using the C_{32}:C_{31} ratio (P<0.05). There were no differences in estimating the faecal output (P>0.30) by both of the methods applied in this work *versus* the observed values. There were differences (P<0.05) using the C_{32}:C_{31} ratio between the oven-dried method and the observed method, although no differences were detected using the C_{32}:C_{33} ratio (P>0.15) when compared to the observed value. It would be preferable to use C_{33} instead of C_{31}

to estimate intake and faecal output due to a similar recovery between C_{32} and C_{33} in comparison with recovery of C_{32} and C_{31} (data not shown).

The drying method affects the *n*-alkane concentration in faeces but, when using the C_{32}:C_{33} ratio to estimate intake and digestibility (C_{36}/C_{32}:C_{33}), the results did not statistically differ from the observed values. However, numerical differences between oven-dried and observed intake and digestibility values should not be underestimated.

Table 1. N-alkane concentration in faeces (mg/kg DM). Mean of 64 samples (± standard error) dehydrated by oven-drying at 60 °C or freeze-drying.

N-alkane	Oven-dried	Freeze-dried	P
23	11.35 (±4.28)	12.28 (±5.00)	0.08
25	67.69 (±16.93)	76.51 (±15.40)	<0.001
27	129.15 (±33.24)	154.48 (±27.58)	<0.001
29	302.75 (±94.48)	390.87 (±68.17)	<0.001
31	497.34 (±136.72)	606.51 (±52.36)	<0.001
32	79.13 (±16.68)	85.90 (±18.46)	<0.001
33	106.86 (±25.75)	128.67 (±13.25)	<0.001
35	11.90 (±2.35)	13.25 (±2.66)	<0.001
36	74.41 (±17.16)	70.99 (±15.10)	<0.001
32:31	0.41 (±0.06)	0.38 (±0.04)	<0.001
32:33	0.87 (±0.11)	0.82 (±0.07)	<0.001

Table 2. Intake, fecal output and digestibility estimated by the n-alkane technique by oven of freeze-drying and observed values.

		Oven-dried	Freeze-dried	Observed	P
Intake	C_{32}:C_{33}	4.65	5.34	5.48	0.14
Intake	C_{32}:C_{31}	4.21[a]	4.90[a,b]	5.48[b]	0.04
Faecal output	C_{36}	2.44	2.48	2.37	0.30
Digestibility (%)	C_{36}/C_{32}:C_{33}	47.09	53.48	56.29	0.15
Digestibility (%)	C_{36}/C_{32}:C_{31}	41.49[a]	49.27[a,b]	56.29[b]	0.04

[a,b] Different superscript letters within the same row indicate significant differences (P<0.05).

References

Dove, H. and R.W. Mayes, 1991. The use of plant wax alkanes as marker substances in studies of the nutrition of herbivores: a review. Aust. J. Agric. Sci. 42: 913-952.

Dove, H. and R.W. Mayes, 2006. Protocol for the analysis of *n*-alkanes and other plant-wax compounds and for their use as markers for quantifying the nutrient supply of large mammalian herbivores. Nature protocols Vol. 1, Nº 4: 1680-1696.

Elwert, C., H. Dove and M. Rodehutscord, 2006. Effect of roughage species consumed on fecal alkane recovery in sheep, and effect of sample drying treatment on alkane concentrations. Aust. J. Exp. Agric. 46: 771-776.

Mayes, R.W., C.S. Lamb and P.M. Colgrove, 1986. The use of dosed and herbage *n*-alkanes as markers for the determination of herbage intake. J. Agric. Sci. 107: 161-170.

Oliván, M. and K. Osoro, 1995. The effect of drying treatment on the *n*-alkane analysis. Ann. Zootech. 44 Suppl.: 238.

Increasing alfalfa non structural carbohydrates through genetic selection and cutting management

C. Chouinard-Michaud[1,2], R. Michaud[2], Y. Castonguay[2], A. Bertrand[2], G. Bélanger[2], G.F. Tremblay[2], R. Berthiaume[3] and G. Allard[1]
[1]Université Laval, Québec, QC, G1K 7P4, Canada; [2]Agriculture and Agri-Food Canada, Québec, QC, G1V 2J3, Canada; [3]Agriculture and Agri-Food Canada, Sherbrooke, QC, J1M 1Z3, Canada; caroline.chouinard-michaud@agr.gc.ca

Introduction

Increasing the concentration of non structural carbohydrates (NSC) in forages improves N utilization by ruminants (Brito *et al.*, 2009) and their performance (Brito *et al.*, 2008). Plant NSC concentration increases during the day (Burns *et al.*, 2005) and it can be improved by genetic selection (Humphreys, 1989). Our objective was to determine, under field conditions, the effect of selected populations of alfalfa for high and low concentrations of NSC cut in the morning or the afternoon on NSC concentration and other attributes of nutritive value.

Material and methods

Three populations of alfalfa derived from the cultivar AC Caribou were tested. Two populations were obtained by intercrossing 10 genotypes selected for high (NSC+) or low (NSC-) NSC concentrations from 500 genotypes. A third population (NSC0) was obtained from intercrossing 10 genotypes randomly chosen. The populations were compared in a field experiment established in 2006 near Quebec City, Canada (46°48'N; 71°23'W). A split plot design was used with time of cutting (AM: 9h00, and PM: 15h00) as main plots and populations as subplots. Populations were harvested at the early flowering stage of development three times in both 2007 and 2008. Forage samples were dried at 55 °C immediately after each harvest, ground using a Wiley mill to pass through a 1 mm screen, and scanned using near infrared reflectance spectroscopy. Calibration and validation samples were analysed for water soluble carbohydrates (WSC) by HPLC, starch by colorimetry (Bertrand *et al.*, 2008), crude protein (CP), acid detergent fibre (ADF), neutral detergent fibre (NDF), *in vitro* true digestibility (IVTD), and *in vitro* digestibility of NDF (dNDF) (Goering and Van Soest, 1970). Concentration of WSC was estimated by the sum of sucrose, glucose, fructose, and pinitol. Concentration of NSC was estimated by the sum of WSC and starch. An analysis of variance was performed using the MIXED procedure of SAS® with harvests as random effects, and time of cutting and populations as fixed effects. The LSMEANS test with a Tukey-Kramer adjustment was used to compare treatment means.

Results and discussion

In both years, PM-cutting of alfalfa increased WSC, starch, NSC concentrations, IVTD and dNDF, and decreased ADF and NDF concentrations (Table 1). Averaged across years, the PM-cut alfalfa had a greater concentration than AM-cut alfalfa for WSC (+13%), starch (+52%), NSC (+23%), IVTD (+2%), dNDF (+4%), and a lower concentration for ADF (-6%), and NDF (-4%). The NSC+ population had a greater NSC concentration than the NSC- population in 2007 (+10%) and 2008 (+5%). The increase in NSC was due mainly to an increase in concentrations in starch (+21%) in 2007 and in WSC (+4%) in 2008. In both years, the NSC+ and NSC- populations did not differ for CP, ADF, and NDF concentrations, and for IVTD, dNDF, and DM yield. The difference in NSC concentration between the two populations was probably not large enough to affect the other attributes of nutritive value. There was no interaction between populations and time of cutting.

Table 1. Effect of time of cutting (AM, PM) and selected populations (NSC-, NSC0, NSC+) on attributes of nutritive value of alfalfa (in mg/g DM except for dNDF in mg/g NDF and yield in g DM/m²; means of three harvests per year).

	2007					2008				
	AM	PM	NSC-	NSC0	NSC+	AM	PM	NSC-	NSC0	NSC+
WSC[1]	61.0[b]	70.2[a]	64.4[b2]	65.6[ab]	66.8[a]	86.4[b]	96.3[a]	89.7[b]	91.4[ab]	92.8[a]
Starch	32.2[b]	46.0[a]	34.9[b]	40.1[a]	42.2[a]	19.7[b]	33.0[a]	25.2[a]	26.0[a]	27.9[a]
NSC	93.2[b]	116.2[a]	99.3[b]	105.8[a]	109.0[a]	106.1[b]	129.3[a]	114.9[b]	117.4[ab]	120.7[a]
CP	151[a]	149[a]	150[ab]	153[a]	147[b]	175[a]	170[b]	173[ab]	176[a]	168[b]
ADF	325[a]	308[b]	326[a]	305[b]	319[ab]	287[a]	265[b]	282[a]	271[a]	275[a]
NDF	392[a]	377[b]	394[a]	372[b]	387[a]	349[a]	332[b]	344[a]	336[a]	340[a]
IVTD	759[b]	777[a]	760[b]	780[a]	765[b]	826[b]	840[a]	831[a]	835[a]	834[a]
dNDF	390[b]	412[a]	393[b]	413[a]	398[b]	514[b]	532[a]	522[a]	523[a]	523[a]
Yield	405[a]	383[a]	400[a]	372[a]	411[a]	420[a]	412[a]	416[a]	399[a]	433[a]

[1] WSC = water soluble carbohydrates; NSC = non structural carbohydrates; CP = crude protein; ADF = acid detergent fibre; NDF = neutral detergent fibre; IVTD = *in vitro* true digestibility; dNDF = *in vitro* digestibility of NDF; AM = AM-cutting; PM = PM-cutting; NSC- = population selected for low NSC concentration; NSC0 = population randomly chosen; NSC+ = population selected for high NSC concentration.

[2] The effect of populations on WSC concentration in 2007 is significant at P=0.06; [a-b]For each year and factor, means within a row followed by the same letter are not significantly different (P>0.05).

Conclusion

Alfalfa NSC concentration can be increased by cutting in the afternoon and via genetic selection; this increase was more important with time of cutting (+25%) than with selection (+10%). The increase in NSC concentration with PM-cutting was associated with a decrease in ADF and NDF concentrations, and an increase in IVTD and dNDF. Additional cycles of selection may further increase NSC concentration and affect digestibility; we are currently testing this hypothesis.

References

Bertrand, A., G.F. Tremblay, S. Pelletier, Y. Castonguay and G. Bélanger, 2008. Yield and nutritive value of timothy as affected by temperature, photoperiod and time of harvest. Grass Forage Sci. 63: 421-432.

Brito, A.F., G.F. Tremblay, A. Bertrand, Y. Castonguay, G. Bélanger, R. Michaud, H. Lapierre, C. Benchaar, H.V. Petit, D.R. Ouellet and R. Berthiaume, 2008. Alfalfa cut at sundown and harvested as baleage improves milk yield of late-lactation dairy cows. J. Dairy Sci. 91: 3968-3982.

Brito, A.F., G.F. Tremblay, H. Lapierre, A. Bertrand, Y. Castonguay, G. Bélanger, R. Michaud, C. Benchaar, D.R. Ouellet and R. Berthiaume, 2009. Alfalfa cut at sundown and harvested as baleage increases bacterial protein synthesis in late-lactation dairy cows. J. Dairy Sci. 92: 1092-1107.

Burns, J.C., H.F. Mayland and D.S. Fisher, 2005. Dry matter intake and digestion of alfalfa harvested at sunset and sunrise. J. Anim. Sci. 83: 262-270.

Goering, H.K. and P.J. Van Soest, 1970. Forage fiber analysis (apparatus, reagents, procedures, and some applications). USDA Agric. Handb. 379. U.S. Government Printing Office, Washington, DC, USA.

Humphreys, M.O., 1989. Water-soluble carbohydrates in perennial ryegrass breeding. II. Cultivar and hybrid progeny performance in cut plots. Grass Forage Sci. 44: 237-244.

Effect of induction of sub-acute ruminal acidosis (SARA) on milk fat profile and rumen parameters

E. Colman[1], W. Fokkink[2], M. Craninx[1], J.R. Newbold[2] and V. Fievez[1]
[1]Laboratory for Animal Nutrition and Animal Product Quality, Ghent University, Proefhoevestraat 10, 9090 Melle, Belgium; [2]Provimi Research & Innovation Centre, Lenneke Marelaan 2, Sint-Stevens-Woluwe, Belgium; veerle.fievez@ugent.be

Introduction

Acidosis occurs when cattle consume amounts of fermentable carbohydrates sufficient to cause a nonphysiological accumulation of volatile fatty acids in the rumen, with a concurrent reduction in pH (Nagaraja and Lechtenberg, 2007). High-concentrate diets, which can lead to SARA, are known to result in changes of the fermentation pattern in the rumen, with possible changes in the mammary production of fatty acids (Enjalbert *et al.*, 2008). The focus of this experiment was odd and branched chain fatty acids (OBCFA), which mainly originate from rumen bacteria rather than from animal synthesis and that reflect rumen volatile fatty acid proportions (VFA) (Craninx *et al.*, 2008). The objective of this study was to describe the modifications of milk fat content and rumen parameters during a SARA induction protocol.

Material and methods

Twelve multiparous, rumen fistulated lactating cows followed a SARA induction protocol: every week 25% of the 'normal-pH' concentrate (sugar beet pulp) was replaced by 'low-pH' concentrate with more rapidly fermentable carbohydrates (wheat). The forage/concentrate ratio remained 1:1 during the first 5 wk of the experiment. In wk 6 the total amount of concentrate was increased varying from 2 to 4 kg, depending on the cow's milk yield. Milk samples and rumen fluid were collected every week at d 2 and 7 (day after and before diet change respectively). Analysis of milk fat and quantification of rumen volatile fatty acids and lactic acid were performed. Rumen pH was measured continuously from d 1 until d 43 using an indwelling pH-probe.
Statistical analysis was performed with SPSS 15.0 for Windows (SPSS Inc., 1986-2006, Chicago, IL, USA). Principal component analysis (PCA) was performed based on rumen parameters and milk fatty acids of all data points (% of milk fat) (pH min, pH max, pH fall per hour, area under curve pH <5.6 or 6.0, time pH <5.6 or 6.0, molar acetate, butyrate and propionate proportions, total VFA (mmol/L), *ante* C13:0, *iso* C13:0, *iso* C14:0, C15:0, *iso* C15:0, *iso* C16:0, C17:0+C17:1 *cis9*, C17:1 *cis9*/C17:0, C18:1 *trans10*, C18:1 *trans11*, C18:2 *cis9 trans11*, C18:2 *trans10 cis12*, C18:2 *trans11 cis15*). Data of all measuring days, except for d 7 of wk 6, were included in the PCA.

Results and discussion

Figure 1, showing the score and loading plots based on the first and second principal component (PC1 and PC2), indicates that variance among cows, mostly reflected in PC1, was larger than the variance between induction weeks, mainly according to the PC2. The fatty acids with the highest loading on PC1 that negatively correlate with each other were C18:1 *trans10* and *iso* C14:0. This indicates a major difference between cows in milk fat *iso* C14:0 and C18:1 *trans10* concentration. The score plot of Figure 1B shows standard deviations of PC1 means increased with induction week. This was mainly due to C18:1 *trans10*, of which the variation coefficient increased from 18 (wk 1) to 75% (wk 6). Variation during the first 4 wk of the experiment (i.e. increasing proportions of 'low-pH' concentrate) could be exclusively described by a decrease in the PC2 with highest loadings for fatty acids *anteiso* C13:0 and C18:2 *cis9 trans11*, that negatively correlate with each

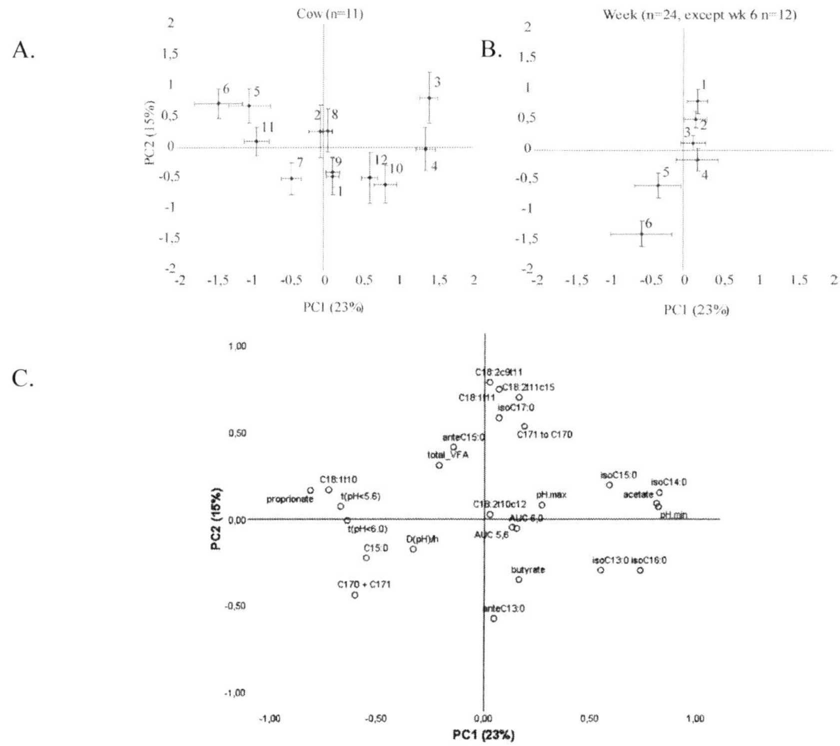

Figure 1. Principal component analysis based on rumen parameters and milk fatty acids as indicated in the loading plot (Figure 1C) of components 1 and 2. Figures 1A and 1B present mean principal component 1 and 2 scores of cows (n=11) and weeks (n=12,) respectively.

other. However, scores of the last 2 wk of the experiment with a final increase of the proportion (wk 5) and total amount of 'low-pH' concentrate (wk 6) deviated from the PC2 axis.

Conclusion

C18:1 *trans10* and *iso* C14:0 explained most of the variation between cows while part of the variation between the first 4 wk of the experiment can be explained by *anteiso* C13:0 and C18:2 *cis9 trans11*. Wk 5 and 6 scores are moving towards PC1 indicating that individual relative changes in milk fatty acid proportions instead of absolute threshold values could be more able to predict ruminal acidosis.

Acknowledgement

The author would like to acknowledge the IWT - Institute for the Promotion of Innovation by Science and Technology in Flanders - for their financial support.

References

Craninx, M., V. Fievez, B. Vlaeminck and B. De Baets, 2008. Artificial neural network models of the rumen fermentation pattern in dairy cattle. Comput. Electron. Agric. 60: 226-238.

Enjalbert, F., Y. Videau, M.C. Nicot and A. Troegeler-Meynadier, 2008. Effects of induced subacute ruminal acidosis on milk fat content and milk fatty acid profile. J. Anim. Physiol. Anim. Nutr. 92: 284-291.

Nagaraja, T.G. and K.F. Lechtenberg, 2007. Acidosis in feedlot cattle. Vet. Clin. North Am. Food Anim. Pract. 23: 333-350.

Protein fermentation characteristics in rumen fluid determined with the gas production technique

J.W. Cone[1], M.A.M. Rodrigues[2], C.M. Guedes[2] and M.C. Blok[3]
[1]Animal Nutrition Group, Wageningen University, P.O. Box 338, 6700 AH Wageningen, the Netherlands; [2]CECAV – Universidade de Trás-os-Montes e Alto Douro, Vila Real, Portugal; [3]CVB, Product Board Animal Feed, The Hague, The Netherlands; john.cone@wur.nl

Introduction

The gas production technique was developed to determine the fermentation kinetics of organic matter in rumen fluid. This technique can also be adapted for the determination of protein fermentation characteristics. To do that, the buffer must be N-free. All the N coming with the rumen fluid must be incorporated into microbial mass. This can be done by supplying the buffered rumen fluid with an excess of fast fermentable carbohydrates. This makes N the limiting factor for fermentation and the obtained gas production profiles reflect the availability of N from the feed samples. The aim of the present study was to investigate if the adapted gas production technique is suitable to determine differences in protein availability in rumen fluid. The fermentation characteristics of N of 19 feed samples were determined using the adapted gas production technique. The amount of dry matter of the samples incubated differed as each sample contained exactly 15 mg N. The results were compared with data of N degradation obtained with the nylon bag technique.

Material and methods

Gas production incubations were performed as described by Cone *et al.* (1996), with the exception that ammonium carbonate was replaced by $NaHCO_3$ (Cone *et al.*, 2005) and that 1 volume of rumen fluid was added to 19 volumes of buffer/mineral solution instead of 2 volumes of buffer/mineral solution. The rumen fluid was obtained from 2 non lactating Holstein Friesian cows receiving 1 kg of concentrate in the morning and *ad libitum* grass silage. To bind all N coming with the rumen fluid, 3.33 g/l glucose, 3.33 g/l xylose and 3.33 g/l soluble starch were added to the buffered rumen fluid and gas production was recorded. After 4 h of incubation gas production was ceased, indicating that no N was available for further fermentation. After 4 h an amount of sample providing 15 mg N for each of the 19 different samples was added to the bottles in duplicate, retaining an excess of quickly fermentable carbohydrates available. Under these conditions the availability of N from the samples determines the rate and extent of the gas production. The N degradation in the rumen for the 19 samples was also determined with the nylon bag technique (Ørskov and McDonald, 1979). Samples were incubated in triplicate for 0, 3, 8, 16, 48 and 336 h in three lactating Holstein Friesian cows receiving a standard ration with 720 g/kg forage (grass and maize silage) and 280 k/kg concentrate. The washout (W), undegradable (U) and degradable (D) fractions of N were determined as well as the rate of degradation (kd/h) and the amount of rumen escape protein (REP) was calculated, assuming a rumen passage rate of 0.06/h. The results of both techniques were compared using the Student t-test.

Results and discussion

Figure 1 shows the gas production profiles determined by the N availability of some of the investigated samples. It is shown that there were large differences in both rate and extent of gas production. The W fraction, determined with the nylon bag technique ranged from 0 for linseed hulls to 0.65 for maize gluten feed, U ranged from 0 for soybean meal to 0.23 for lucerne meal, D ranged from 0.32 for maize gluten feed to 0.96 for linseed hulls and kd ranged from 0.010/h for a protected rapeseed meal to 0.151/h for wheat meal. The calculated amount of REP ranged from 197 g/kg CP for lupin to 840 g/kg CP for the protected rapeseed meal. Comparing the gas production results with the nylon bag data showed that there was a moderate relationship between gas production after 5 h (GP5) and W ($r^2 = 0.63$) and between GP25 and kd ($r^2 = 0.52$). However, there was a rather good relationship between the gas production after 12, 15 and 20 h and the calculated amount of REP ($r^2 = 0.83$). It can be concluded that the adapted gas production technique can be used as a rapid screening methodology to determine differences in N availability between samples and provides a good estimate of the amount of rumen escape protein.

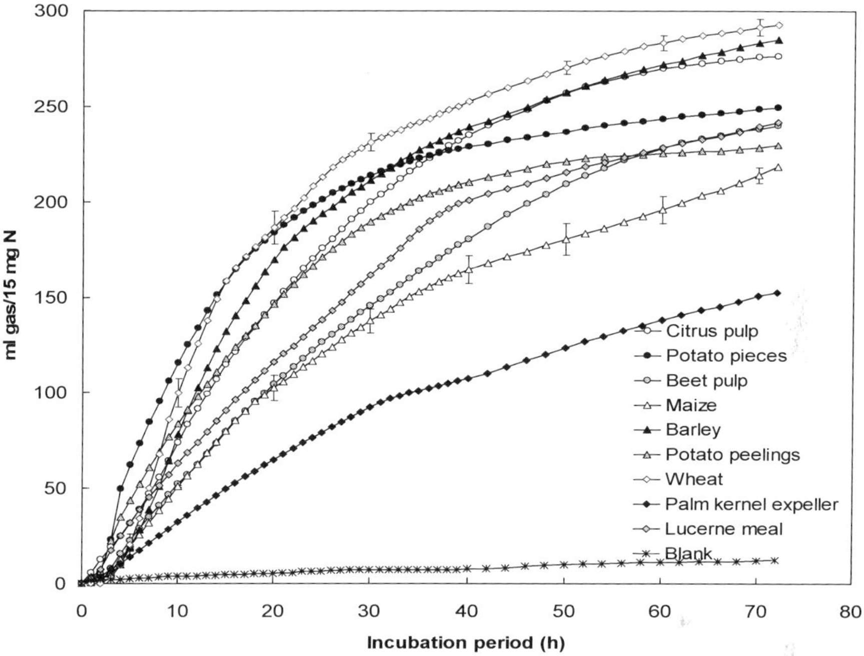

Figure 1. Gas production profiles of fast fermentable carbohydrates in an N-free medium, supplemented with different amounts of samples, all containing exactly 15 mg N. The blank was not supplemented with N.

References

Cone, J.W., A.H. van Gelder, G.J.W. Visscher and L. Oudshoorn, 1996. Influence of rumen fluid and substrate concentration on fermentation kinetics measured with a fully automated time related gas production apparatus. Anim. Feed Sci. Technol. 61: 113-128.

Cone, J.W., A.W. Jongbloed, A.H. van Gelder and L. de Lange, 2005. Estimations of protein fermentation in the large intestine of pigs with the gas production technique. Anim. Feed Sci. Technol. 123-124: 463-472.

Ørskov, E.R. and I. McDonald, 1979. The estimation of protein degradability in the rumen from incubation measurements weighted according to rate of passage. J. Agric. Sci. Camb. 92: 499-503.

Performance and ruminal protozoa in lambs with chromium supplementation

B.S.L. Dallago, C.M.M. Pimentel, D.F. Caldeira, A.C. Lopes, T.P. Paim, E. Franco, B.O. Borges and H. Louvandini
Faculdade de Agronomia e Medicina Veterinária, Universidade de Brasília, Distrito Federal, 70910-900, Brazil; hlouvand@unb.br

Introduction

Chromium picolinate (CrPic) is a mineral chelate used as a nutritional supplement for humans and livestock. The main action of CrPic involves the potentiation of insulin signaling by an interaction between a chromium-containing molecule, insulin and the insulin receptor that facilitates the response of insulin-sensitive tissues (Striffler *et al.*, 1993). This role, in turn, can modify the general body metabolism acting in growth, deposition of fat, or often, in voluntary food intake. On the contrary, some species of heavy metals are toxic, especially to simple life forms such as the ruminal protozoa. This study was aimed at investigating the effects of chromium supplementation on lamb performance and the ruminal protozoa population.

Material and methods

Twenty-four Santa Inês male lambs were used. The initial mean body weight was 22.89±2.23 kg. The lambs were assigned to four treatments with different levels of chromium picolinate: placebo, 0.250, 0.375 and 0.500 mg of chromium/animal/day. The lambs were kept in individual pens during two weeks for adaptation and 84 days for chromium supplementation and were fed with *Panicum maximum* cv Massai hay and concentrate (85% of cassava flour, 11.5% of mineral salt and 3.5% of urea). The control of the feed intake was taken three times a week. Water was drunk *ad libitum*. Lambs were weighed every two weeks. Ruminal content was sampled in the first, 21st, 43rd, 63rd e 84th days of experimental phase to quantify the ruminal protozoa. On these days, three hours after the morning feed, approximately 20 ml of ruminal content were sampled by an esophagus probing coupled in a collecting bomb. The liquid was filtered in sterile gauze and its pH was measured with a pHmeter. After that, the ruminal content was diluted 1:5 (v/v) in a TBSF (Trypan Blue Formol Salin) and then stored at 4 °C. The ruminal protozoa counting was done by optical microscopy in a 400× zoom using a Neubauer chamber. All quantitative analysis of Cr was done by inductively coupled plasma atomic emission spectrometry (model Thermo Jarrell Ash IRIS/AP®). The samples were incinerated at 450 °C and than were submitted to a chloridric digestion according to Silva and Queiroz (2006). Data were analysed by the polynomial regression of the Statistical Analysis System (SAS®, 1999) in a completely randomised design with Cr levels as a source of variation.

Results

The initial mean body weight (BWi), dry matter and Cr intake, final body weight (BWf), dairy gain (DG), total body weight gain (GT) in the different levels of chromium supplementation are presented in Table 1. No difference was recorded in any parameter listed above.
The ruminal pH was not affected by the treatments: the average ruminal pH held stable over normal range (6.49±0.28). The mean concentration of ruminal protozoa population in the treatments is shown in Figure 1. A negative linear relationship was observed between Cr consumption and the protozoa count.

Table 1. Cr and dry matter (DM) intake, initial body weight (BWi), final body weight (BWf), daily gain and total body weight gain in the different levels of chromium supplementation.

Variables	Treatments (mg of Cr/day)				SE	Regression ($P\leq0.05$)
	0.000	0.250	0.375	0.500		
Cr intake (mg/kg of DM)	0.014	0.264	0.389	0.514		
Dry matter intake (kg of DM)	0.923	0.798	0.863	0.797	0.04	ns
BWi (kg)	23.60	22.57	22.20	23.18	0.94	ns
BWf (kg)	30.17	29.06	29.37	29.60	1.00	ns
Daily gain (kg)	0.078	0.077	0.085	0.076	0.007	ns
Total gain (kg)	6.567	6.500	7.167	6.417	0.59	ns

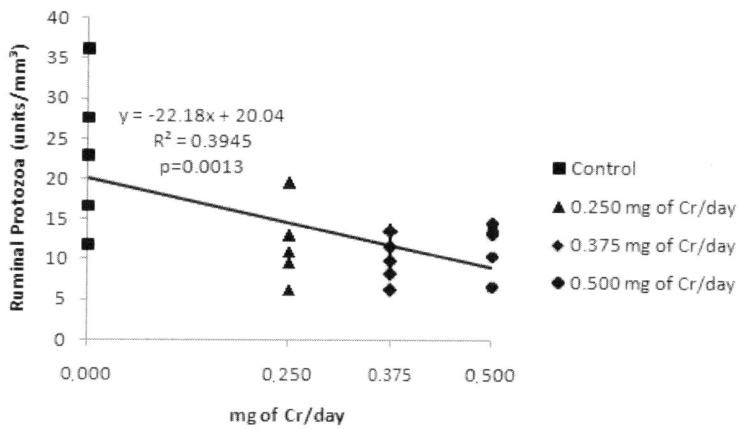

Figure 1. Relationship between Cr supplementation and ruminal protozoa count.

Conclusion

The chromium oral supplementation as CrPic did not show a positive effect in growth performance of lambs, but it impairs the ruminal protozoa population acting as a toxic agent for this population.

Acknowledgement

To CNPq for scholarships and Finatec for financial support.

References

SAS Institute, 1999. SAS statistical package for Windows v. 8.0. SAS Institute, Cary, NC, USA.

Silva, D. J. and A.C. Querioz, 2002. Análise de alimentos: métodos: químicos e biológicos. 3 ed. Viçosa, MG: Editora UFV, 235 p.

Striffler, J.S., M.M. Polansky and R.A. Anderson, 1993. Dietary chromium enhances insulin secretion in perfused rat pancreas. J. Trace Elem. Exp. Med. 6: 75-81.

The effect of grain sources on *in vitro* rumen acid load of close-up dry cow diets

S. Danesh Mesgaran, A. Heravi Moussavi, H. Jahani-Azizabadi, A.R. Vakili, F. Tabataiee and M. Danesh Mesgaran
Dept. Animal Science, Faculty of Agriculture, Ferdowsi University of Mashhad, P.O. Box 91775-1163, Mashhad, Iran; sadjaddm@yahoo.com

Introduction

Over the last two decades, it has become common practice to feed rations of higher grain density during the close-up period (Penner *et al.*, 2007). It has been proposed that the feeding of a high non-fibre carbohydrate diet virtually always results in higher *pre partum* dry matter intake and frequently results in more positive effects on energy metabolism. However, there is a need to manage the inclusion of these feeds to avoid reduced performance due to subacute ruminal acidosis (Garrett *et al.*, 1999) and an increased incidence of clinical acidosis and related disorders. An *in vitro* work has assessed and evaluated a new technique (acidogenicity value; AV) for studying the production of acids during rumen fermentation (Wadhwa *et al.*, 2001). It was proposed that the high concentration of starch in wheat grain is a critical point in AV of a dairy cow diet. Moreover, the impact of different close-up dry cow diets on rumen acid load has not been evaluated. The objective of this experiment was to evaluate the effect of grain sources and combination on the AV of a series of close-up dry cow diets.

Material and methods

Various diets were provided using different concentrations of grain sources including barley, maize and wheat. A basal diet (BD) was provided including barley and maize grains, then wheat grain (WG) was replaced in the basal diet as a part of wheat straw (BDWG1 and BDWG2), or substituted for maize or both barley and maize grains (BDWG3 and BDWG4), or added to the basal diets (BDWG5 and BDGW6). The diets are shown in Table 1. The acidogenicity values of the diets were determined using the procedure as described by Wadhwa *et al.* (2001). Samples were oven-dried (48 h, 68 °C) and ground through a 1 mm screen on a laboratory mill. One-gram (DM) samples were weighed and incubated, in triplicate, with 30 ml of buffered rumen liquor comprising 60% buffer and 40% rumen liquor. The buffer was made up at 20% of the strength of the Tilley-Terry (1963) buffer. Cysteine hydrochloride monohydrate (0.025% wt/vol) was added just prior to incubations. Rumen fluid was collected, 3 h after morning feeding, from four fistulated sheep that were maintained on lucerne hay and concentrates (70 to 30% in the DM). The incubations were carried out in 100-ml bottles held in a water bath at 38.7 °C. Samples (2 ml) were withdrawn from bottles after 24 h and transferred to 2-ml micro tubes containing 50 mg (excess) of $CaCO_3$ powder. The mixture was shaken manually for 5 s and then centrifuged at 4,000 rpm for 10 min before analysis of Ca content in the supernatant using Atomic Absorption. The AV was calculated as the product of Ca concentration (from the analysis) and fluid volume (30 ml) divided by the sample weight. Data were analysed using the completely randomised design of the GLM procedure of SAS® (1999).

Results

The results indicate that the AV of BDWG1, BDWG2 and BDGW6 (11.1, 11.2 and 11.9, respectively) were significantly ($P<0.05$) higher than those of BD, BDWG3, BDWG4 and BDWG5 (10.1, 10.1, 10.1 and 10.7, respectively).

Table 1. The close- up dry cow experimental diets (%DM) and the acidogenicity values.

Items	Diets						
	BD	BDWG1	BDWG2	BDWG3	BDWG4	BDWG5	BDWG6
Maize silage	27.5	27.5	27.5	27.5	27.5	26.5	25.5
Lucerne hay	19.9	19.9	19.9	19.9	19.9	19.1	18.4
Wheat straw	14.6	10.7	6.7	14.6	14.6	14.0	13.5
Wheat bran	4.6	4.6	4.6	4.6	4.6	4.4	4.3
Barley grain	8.6	8.6	8.6	8.6	0.0	8.3	8.0
Maize grain	8.6	8.6	8.6	0.0	0.0	8.3	8.0
Wheat grain	0.0	3.9	7.9	8.6	17.2	3.8	7.4
Soybean meal	4.6	4.6	4.6	4.6	4.6	4.4	4.3
Cottonseed meal	5.3	5.3	5.3	5.3	5.3	5.1	4.7
Rape seed meal	2.1	2.1	2.1	2.1	2.1	2.1	2.0
Anionic salts	2.5	2.5	2.5	2.5	2.5	2.4	2.3
Vitamin & mineral premix	0.4	0.4	0.4	0.4	0.4	0.4	0.4
Protected fat	1.3	1.3	1.3	1.3	1.3	1.3	1.3
Acidogenicity values	10.1	11.1	11.2	10.1	10.1	10.7	11.9

Discussion and conclusion

The present results indicate that the AV of the diets evaluated was influenced by the source and concentration of the grains. In addition, the decrease in wheat straw, as observed in diets BDWG1 and BDWG2, resulted in an increasing AV. This result was in accordance with the finding of a previous study (Wadhwa *et al.,* 2001) in which the forages generally had AV that were lower than that of starchy feeds. In addition, wheat grain had the highest AV of the grains, while wheat straw had the lowest value of the forages. It was concluded that the low values for wheat straw may reflect its resistance to fermentation. Overall, the result of the present study provides additional information for addressing issues of rumen acid load when different sources and concentrations of NFC are included in a dry cow diet formulation.

References

Garrett, E.F., M.N. Pereira, K.V. Nordlund, L.E. Armentano, W.J. Goodger and G.R. Oetzel, 1999. Diagnostic methods for the detection of subacute ruminal acidosis in dairy cows. J. Dairy Sci. 82: 1170-1178.

Penner, G.B., K.A. Beauchemin and T. Mutsvangwa, 2007. Severity of ruminal acidosis in primiparous Holstein cows during the periparturient period. J. Dairy Sci. 90: 365-375.

Tilley, J.M.A. and R.A. Terry, 1963. A two stage technique for the *in vitro* digestion of forage crops. J. Br. Grassl. Soc. 18: 104-111.

Wadhwa, D., N.F.G. Beck, L.P. Borgida, M.S. Dhanoa and R.J. Dewhurst, 2001. Development of a simple *in vitro* assay for estimating net rumen acid load from diet ingredients. J. Dairy Sci. 84: 1109-1117.

In vitro first order dry matter disappearance kinetics of guar meal

M. Danesh Mesgaran, H. Jahani-Azizabadi, M. Vatandoost, M. Mojtahedi, E. Abdi Ghezeljeh, A.R. Vakili and A. Fanaie-Nokar
Departement Animal Science, Excellence Centre for Animal Science, Faculty of Agriculture, Ferdowsi University of Mashhad, P.O. Box 91775-1163, Mashhad, Iran; danesh@um.ac.ir

Introduction

The guar plant is grown for the guar bean's gum that has many food and industrial applications. Isolation of gum (a galactomannan) from guar seed led to yield a high protein by-product containing 55 to 60 percent crude protein (CP), which has been used as a protein source in ruminant and non-ruminant livestock feeding. However, information about the kinetics of nutrient disappearance of guar meal in ruminants is very low. The aim of the present study was to evaluate *in vitro* first order dry matter (DM) disappearance kinetics of guar meal.

Material and methods

Samples of guar meal as non-heat processed (CP: 566 and 580 g/kg DM; GM566 and GM580, respectively) and heat processed (CP: 594 g/kg DM; GM594P) were provided. They were dried using a forced-air oven at 60 °C for 48 h. All feed samples were ground to pass through a 2-mm screen and then analysed for crude protein (CP), ether extract (EE) and ash (AOAC, 1995). Neutral detergent fibre (NDF) and acid detergent fibre (ADF) were determined using the method of Van Soest *et al.* (1991). Samples were incubated in a medium prepared as described by Arroquy *et al.* (2005). Fifty-five ml of medium were supplied into 100 ml bottles that approximately contained 0.45 g of feed (6 replicate for each feed). Then, each bottle was inoculated under carbon dioxide with 5 ml of mixed rumen microbes. Rumen fluid was obtained from three sheep (49.5±2.5 kg) fitted with rumen fistulae, before the morning feeding, and immediately strained through four layers of cheesecloth. The animals fed 1 kg/d of DM of lucerne hay and 0.3 kg/d of DM concentrates (165 g CP/ kg of DM). The bottles were incubated for 4, 8, 16 and 24 h at 39 ºC. Then, bottle content was filtered through a 42 μm filter, and DM of the unfiltered medium was determined. Non-linear first order model was used to estimate the digestion kinetic parameters of DM. The model was the following: $D_{(t)} = D_{(i)} . exp\ (-kd.\ time) + I$; Where, $D_{(t)}$ is the residual DM at any time, $D_{(i)}$ is the potentially digestible fraction, *kd* is the fractional rate constant of digestion (/h) and *I* is the indigestible fraction.

Results

The chemical compositions of the samples are shown in Table 1. Non-linear first order parameters of *in vitro* DM digestion of the samples are presented in Table 2. The results of the present study indicate that DM of various guar meals, except for GM60P, were completely digestible. The *kd* was significantly (P<0.05) higher in GM566 compared with G580 and GM594P.

Table 1. Chemical compassion of the samples (g/kg).

Feed	Nutrients			
	EE	NDF	ADF ADF	Ash
GM566	36.6	218.0	131.0	52.0
GM580	75.2	205.0	144.0	50.9
GM594P	71.9	238.0	140.0	51.7

Table 2. In vitro first order DM disappearance parameters of guar meal.

Feed	Parameters			
	Kd[1]	I[2]	Di[3]	R^2
GM566	0.13±0.007	0.0	1.02±0.02	0.99
GM580	0.108±0.005	0.0	1.04±0.017	0.99
GM594P	0.101±0.011	0.07±0.03	0.91±0.035	0.95

[1] Fractional rate constant of digestion (/h).
[2] Indigestible fraction.
[3] Potentially digestible fraction.

Conclusion

The results of the present study demonstrated that *in vitro* DM indigestible fraction of various guar meals is influenced by heat processing. While the DM of GM566 and GM580 completely disappeared after 24 h of incubation, about 7% of DM of GM594P remained in the medium. It was also concluded that the rate of DM disappearance of GM566 was higher than that of the other samples.

Acknowledgement

The authors wish to acknowledge the financial support received from Aria Shirin Nosh Company and Ferdowsi University of Mashhad.

References

AOAC, 1995. Official Methods of Analysis, 16th ed. Official Methods of Analysis of AOAC International, Gaithersburg, MD, USA.

Arroquy, J.I., R.C. Cochran, T.G. Nagaraja, E.C. Titgemeyer and D.E. Johnson, 2005. Effect of types of non-fiber carbohydrate on *in vitro* forage fiber disappearance of low-quality grass hay. Anim. Feed Sci. Technol. 120: 93-106.

Van Soest P. J., J.B. Robertson and B.A. Lewis, 1991. Methods for dietary fibre, neutral detergent fibre, and nonstarch polysaccharides in relation to animal nutrition. J. Dairy Sci. 74: 3583-3597.

Effects of *Saccharomyces cerevisiae* from rice distiller's by-product on *in sacco* degradability kinetics of dry matter constituents

S. Das, P. Biswas and A.K. Patra
Department of Animal Nutrition, West Bengal University of Animal and Fishery Sciences, 37 K. B. Sarani, Belgachia, Kolkata 700037, India; drpbiswas56@yahoo.com

Introduction

The ethanol industry is expanding rapidly and coproducts such as corn and rice distiller's by-products are becoming widely available. The rice distiller's by-product (RDB) contains a high percentage of crude protein (CP) and soluble nutrients. In addition to a good source of protein, corn distiller solubles have been shown to stimulate cellulose digestion *in vitro* (Chen *et al.*, 1976), which have not been exploited suitably in animal feeding systems, which was not due to the presence of soluble protein, vitamin mix and amino acids (Chen *et al.*, 1976). The RDB inherently contains live cells of *Saccharomyces cerevisiae* that might improve fibre digestibility. The present study was aimed at evaluating *S. cerevisiae* from RDB for *in sacco* nutrient degradation.

Material and methods

An *in sacco* trial for determination of degradability parameters of feeds was conducted largely following the procedures described by Vanzant *et al.* (1998). One rumen fistulated indigenous steer (250 kg of body weight) was fed a diet containing 21.9% CP, 46.5% neutral detergent fiber (NDF) and 63.5% total digestible nutrients to meet its maintenance requirements. Three diets were tested for three periods of 18 days: control diet (RDB0) (period 1), control diet supplemented with 286 g (RDB2) containing 2×10^9 live yeast cells (period 2) and 571 g (RDB3) of RDB containing 4×10^9 live yeast cells (period 3), respectively. For each period, 3 g of ground rice straw (D1), concentrate (D3) and straw and concentrate (1:1, D2) were introduced in standard nylon bags (11×6 cm, and mean pore size 50 μm). They were incubated in the rumen in duplicate for 2, 6, 12, 24, 48 and 72 h after 14 days of feeding for each period. Dry matter (DM), NDF, acid detergent fibre (ADF) and CP disappearance data were fitted to the model of Ørskov and McDonald (1979) to estimate the degradability parameters. The effective degradability (ED) of nutrients was calculated using the equation: $ED = a + [(b \times c)/(c + r)]$ with rumen fractional outflow rates (r) of 0.05 h/1, where a = rapidly degradable fraction (%), b = insoluble but potentially degradable fraction (%), c = rate of degradation (h/1). Data were analysed with the one way ANOVA procedure with RDB as the main effect and linear contrast was used to determine the effect of different level of RDB.

Results and discussion

The 'a' values were similar among the levels of RDB for all nutrients (data not presented). The 'b' and 'c' values for DM in D1 did not show statistical significance due to RDB addition but ED increased linearly with increasing RDB feeding (Table 1). The ED of DM also increased linearly with increasing RDB due to greater 'b' and 'c' values in D2, and 'b' value in D2. Although the 'b' values of NDF in D1 and D2 were similar due to RDB feeding, the 'c' value and ED in D1, and ED in D2 increased with increasing levels of RDB. For D3, both 'b' and 'c' of NDF showed a linear response, which resulted in a linear increase in ED of NDF. With regards to ADF, the 'b', 'c' and ED in D1, ED in D2 and 'c' and ED in D3 increased linearly with greater levels of RDB. The ED of CP in different feeds increased with increasing levels of RDB, but 'b' and 'c' values did not differ significantly among the feeds except that the 'b' value in D2 increased linearly with supplementation of RDB. The feeding of RDB increased the *in sacco* degradability of fibre and CP of feeds probably because of the presence of *S. cerevisiae* in RDB, and hence it might improve nutrient utilisation in ruminants.

Table 1. Effects of rice-distiller's by-products on in sacco degradation kinetics of feeds.

Nutrient	Feed	Degradation parameter	RDB0	RDB1	RDB2	SEM	Linear effect (*P*)
DM	D1	b,%	62.1	63.0	64.1	1.24	0.60
		c, h/1	0.038	0.041	0.039	0.002	0.53
		ED,%	31.5	34.4	35.4	0.42	0.01
	D2	b,%	50.4	50.4	53.0	0.65	0.06
		c, h/1	0.046	0.052	0.051	0.001	0.08
		ED,%	41.1	44.1	44.8	0.48	0.01
	D3	b,%	47.6	52.0	54.2	1.21	0.07
		c, h/1	0.042	0.043	0.042	0.002	0.86
		ED,%	47.1	49.4	50.5	0.45	0.01
NDF	D1	b,%	55.9	54.4	54.7	0.96	0.58
		c, h/1	0.029	0.040	0.049	0.003	0.05
		ED,%	26.1	29.2	31.1	0.45	0.01
	D2	b,%	55.3	56.4	56.5	1.81	0.63
		c, h/1	0.047	0.056	0.058	0.004	0.35
		ED,%	32.7	35.7	36.1	0.35	0.01
	D3	b,%	58.6	63.1	62.3	1.06	0.09
		c, h/1	0.068	0.077	0.076	0.002	0.05
		ED,%	41.6	43.9	44.4	0.44	0.02
ADF	D1	b,%	51.1	55.5	54.8	0.49	0.01
		c, h/1	0.036	0.041	0.046	0.003	0.11
		ED,%	21.2	23.9	25.3	0.55	0.01
	D2	b,%	52.5	52.6	54.7	1.01	0.23
		c, h/1	0.044	0.053	0.054	0.004	0.16
		ED,%	24.5	27.4	27.2	0.41	0.02
	D3	b,%	56.4	59.6	60.4	1.41	0.14
		c, h/1	0.062	0.073	0.079	0.003	0.02
		ED,%	29.8	32.7	33.5	0.42	0.01
CP	D1	b,%	45.7	47.0	47.3	0.61	0.27
		c, h/1	0.039	0.043	0.049	0.005	0.27
		ED,%	37.8	41.5	42.1	0.65	0.04
	D2	b,%	42.6	45.4	49.6	0.88	0.01
		c, h/1	0.053	0.059	0.062	0.003	0.28
		ED,%	44.3	46.5	47.3	0.52	0.03
	D3	b,%	45.7	49.5	48.3	2.26	0.47
		c, h/1	0.057	0.058	0.064	0.003	0.45
		ED,%	48.8	51.8	52.9	0.50	0.01

References

Chen, M.C., W.M. Beeson and T.W. Perry, 1976. *In vitro* studies on the effect of screening processed corn distillers solubles and centrifuge processed corn distillers solubles on cellulose digestion and microbial protein synthesis. J. Anim. Sci. 43: 1280-1285.

Ørskov, E.R. and I. McDonald, 1979. The estimation of protein degradability in the rumen from incubation measurements weighed according to rate of passage. J. Agric. Sci. (Camb.) 92: 499-505.

Vanzant, E.S., R.C. Cochran and E.C. Titgemeyer, 1998. Standardization of *in situ* techniques for ruminant feedstuff evaluation. J. Anim. Sci. 76: 2717-2779.

The effect of pH and osmolality on the level and composition of soluble N in untreated legumes for ruminants

L.H. de Jonge[1], W. Spek[1], H. van Laar[2] and J. Dijkstra[1]
[1]Animal Nutrition Group, Wageningen University, P.O. Box 338, 6700 AH Wageningen, the Netherlands; [2]Nutreco R&D, Ruminant Research Centre, the Netherlands; leon.dejonge@wur.nl

Introduction

In current protein evaluation systems, feed protein fractionation is based on *in situ* techniques and on solubility in buffers and detergent solutions. Some systems, including the DVE/OEB (Tamminga *et al.*, 1994), NRC (2001) and PDI (Vérité *et al.*, 1979) systems, assume that soluble N is very fast fermented in the rumen. The amount of soluble N is determined under fixed conditions using tap water or a borate phosphate solution at pH 6.8 as solvent. Rumen conditions in terms of pH and osmolality are, however, not constant but influenced by feed, feed intake pattern and other feed and animal related characteristics, which may lower rumen pH to values considerably lower than 6.8 (Bach *et al.*, 2007), and can increase the osmotic pressure to 400 mOsm/l (Giger-Reverdin, 2000). The aim of this study was to investigate the influence of pH and osmolality within the physiological rumen range on the amount of soluble N, and on the composition of this fraction in untreated legumes

Material and methods

Feedstuffs in the experiment were soybeans, peas, faba beans, and lupins, which were ground through a 3-mm sieve (Retsch ZM100, Haan, Germany). Nitrogen solubility was determined by extracting 0.5 g feedstuff with 25 ml buffer solution in a beaker under constant mechanical stirring during one hour at 38 °C. After centrifuging, the absolute amount of N in the residue was determined by the Kjeldahl method. The buffer solutions used cover 4 pH values (5.0, 5.6, 6.2 and 6.8) and 2 osmolality values (300 and 400 mOsm/l). The non-protein N (NPN) was estimated after addition of 2 ml of 40% (w/v) TCA to the supernatant, and analysing the residue for N. The molecular weights of the solubilised proteins at the different pH values were measured after denaturing with 3.5% DDT during 5 min at 90 °C. Analysis of variance was conducted using the GLM procedure of SAS® 9.1 (2002). The effect of pH, osmolality and their interaction on N solubility was tested and if significant ($P<0.05$), the Tukey test was used to test for pairwise comparisons.
Nitrogen solubility was determined by extracting 0.5 g feedstuff with 25 ml buffer solution in a beaker under constant mechanical stirring during one hour at 38 °C. After centrifuging, the absolute amount of N in the residue was determined by the Kjeldahl method. The buffer solutions used cover 4 pH values (5.0, 5.6, 6.2 and 6.8) and 2 osmolality values (300 and 400 mOsm/l). The non-protein N (NPN) was estimated after addition of 2 ml of 40% (w/v) TCA to the supernatant, and analysing the residue for N. The molecular weights of the solubilised proteins at the different pH values were measured after denaturing with 3.5% DDT during 5 min at 90 °C. Analysis of variance was conducted using the GLM procedure of SAS 9.1 (2002). The effect of pH, osmolality and their interaction on N solubility was tested and if significant ($P<0.05$), the Tukey test was used to test for pairwise comparisons.

Results and discussion

Nitrogen solubility was significantly affected by pH of the solvent used (Table 1). A decline of the pH from 6.8 to 5.0 led to a decrease of the average N solubility from 48 to 21%. Especially between pH 6.2 and 5.6, a pronounced decrease of the N solubility was noticed. The effect of the osmolality was smaller compared to the effect of pH, and varied between the different feedstuffs. The relative

amount of NPN in the total N fraction of the feedstuff was not significantly affected by decreasing the pH value from 6.2 to 5.0. Neglecting the effect of reduced pH values on the solubility of N in untreated legumes may lead to an overestimation of their rumen degradation.
Electrophoresis results showed that the composition of individual proteins in the soluble fraction was also affected by the pH value of the solvent used. Especially the solubility of storage globulins (7S and 11S), which were the most abundant soluble proteins at pH 6.2 and 6.8, strongly decreased at lower pH levels (i.e. 5.0 and 5.6). The solubility of the 2S albumins was much lesser affected by the decrease in pH. Although albumins are considered to be more resistant against rumen degradation than globulins (Spencer *et al.*, 1998), further investigation is needed to evaluate this effect on the availability of protein for rumen degradation.

Table 1. Relative N solubility (in % total N) of raw materials as affected by pH and osmolality of the solvent at 38 °C (n = 4).

Feed	Osm/l	pH				SE	*P*-value		
		5.0	5.6	6.2	6.8		pH	Osm	pH×osm
Faba beans	300	21.1^{a}	22.8^{ax}	43.7^{bx}	51.6^{c}	0.44	<0.01	<0.01	< 0.01
	400	21.3^{a}	26.2^{by}	46.3^{cy}	50.7^{d}	0.44			
Lupins	300	15.3^{ax}	22.7^{b}	44.0^{c}	49.3^{c}	1.55	<0.01	<0.01	0.03
	400	23.7^{ay}	26.6^{a}	45.6^{b}	47.9^{b}	1.55			
Peas	300	26.1^{a}	30.4^{bx}	43.7^{c}	52.9^{d}	0.51	<0.01	<0.01	0.02
	400	28.3^{a}	33.4^{by}	44.3^{c}	52.7^{d}	0.51			
Soybeans	300	16.6^{a}	20.8^{bx}	33.1^{cx}	39.5^{d}	0.42	<0.01	<0.01	0.38
	400	17.8^{a}	23.4^{by}	35.3^{cy}	41.3^{d}				

[a, b,c,d] Means in the same row with different superscripts differ (*P*<0.05).
[x,y] Means in the same column within a feedstuff with different superscripts differ (*P*<0.05).

Conclusion

A decrease of the pH from 6.8 to 5.0 greatly reduces the amount of soluble N in untreated legumes. Increasing osmolality from 300 to 400 mOsm/l increased N solubility. The decrease in pH also changed the composition of the soluble N and solubility of globulins in particular decreased. Current protein evaluation systems do not include such changes and this may affect accuracy of protein availability.

References

Bach, A., C. Iglesias and M. Devant, 2007. Daily rumen pH pattern of loose-housed dairy cattle as affected by feeding pattern and live yeast supplementation. Anim. Feed Sci. Technol. 136:146-153.

Giger-Reverdin, S., 2000. Characterisation of feedstuffs for ruminants using some physical parameters. Anim. Feed Sci. Techn. 86: 53-69.

National Research Council, 2001. Nutrient requirements of dairy cattle, 7th revised ed. National Academy Press, Washington DC, USA.

Spencer, D., T.J.V. Higgins, M. Freer, H. Dove and J.B. Coombe, 1988. Monitoring the fate of dietary proteins in rumen fluid using gel electrophoresis. Br. J. Nutr. 60: 241-247.

Tamminga, S., W.M. van Straalen, A.P.J. Subnel, R.M.G. Meijer, A. Steg, C.J.G. Wever and M.C. Blok, 1994. The Dutch protein evaluation system: The DVE\OEB system. Livest. Prod. Sci. 40: 139-155.

Vérité, R., M. Journet and R. Jarrige, 1979. A new system for the protein feeding of ruminants: the PDI system. Livest. Prod. Sci. 6: 349-367.

Effect of the rumen environment and type of supplemented nitrogen on the predation of rumen bacteria by protozoa *in vitro*

G. de la Fuente, A. Belanche, J. Balcells and M. Fondevila
Depatamento de Producción Animal y Ciencia de los Alimentos, Universidad de Zaragoza, M. Servet 177, 50013 Zaragoza, Spain

Introduction

The presence of protozoa and their engulfment and digestion of rumen bacteria is the most important activity regulating the efficiency of microbial growth (Leng and Nolan, 1984) and the bacterial N turnover (Coleman, 1975). Dietary nitrogen competes with engulfed bacteria as a nutritional substrate for protozoa. Such an effect may differ among protozoa species and is therefore determined by the rumen environmental conditions (Dehority, 2003). This work studies *in vitro* the effect of the type of nitrogen availability on the predatory activity of different ciliate population in the rumen under two rumen environments.

Material and methods

Rumen contents were extracted from cannulated sheep (n= 4) given two alfalfa hay and ground barley grain proportions (100:0, FOR and 33:67, MIX). Rumen contents were filtered twice through two layers of gauze and washed with Simplex type solution (Williams and Coleman, 1992) and allowed to decant for separation of the protozoal extract. This was divided by filtration into the large protozoal fraction (LP) (remaining after filtering at 45 μm) and small protozoal fraction (SP) (remaining after filtering at 10 μm). Protozoa from both fractions were counted before the incubation. Thirty-five ml of both protozoal fractions (LP and SP) were incubated with 35 ml of ^{15}N-labelled bacteria. Either urea or casein peptone were added as sources of non-protein or protein nitrogen, respectively. Rumen bacteria were ^{15}N enriched using [$(^{15}NH_4)_2SO_4$; 7.68 mg/l isotope] in a free-protozoa filtrate from rumen contents.

Protozoal incubation media were sampled (20 ml) at 1, 2 and 4 h, centrifuged (500 x *g*, 5 min), and 15 ml of the supernatant were replaced with saline solution and the process was repeated twice. The supernatant of the first centrifugation was preserved as a blank of bacteria. After the third centrifugation, 3ml of the residue (protozoal fraction) were frozen in liquid N and lyophilised. Isotope ^{15}N abundance was determined by IRMS and protozoa isotopic N enrichment (taken as an index of bacterial predation) was calculated by difference with the natural abundance from rumen liquid samples from the same sheep.

Results and discussion

Total protozoal numbers and their group types are shown in Table 1. Counts of *Dasytricha* were higher in MIX than FOR (P<0.05) and *Entodinium* spp. was higher in SP than LP fractions (P<0.001). No further differences were detected between protozoal fractions (P>0.10).

The source of supplemented N did not affect microbial ^{15}N enrichment at any level (P>0.10). Protozoa isolated from animals fed with MIX had a higher (P<0.01) enrichment in ^{15}N than samples isolated from FOR animals (Table 2). No significant differences were recorded through incubation times, but the interaction between time and diet (P<0.001) and the time, diet and fraction triple interaction (P<0.10) showed a higher enrichment in SP fraction from MIX than FOR diet at 2 and 4 h, but not at 1 h. This behaviour leads to analyse the enrichment rate within each group, assuming a linear relationship throughout the incubation time (Table 3). There was a higher enrichment rate in the samples from MIX than from FOR fed animals (P<0.001). A trend in the diet by fraction interaction was also observed (P<0.10), where the SP fraction (protozoa smaller than 45 μm diameter) from the diet MIX showed a higher enrichment slope than the same fraction from diet FOR.

Table 1. Total protozoal counts (log no. cells m/l) of LP and SP fractions from both rumen environments at initial time (n=4).

	MIX		FOR		S.E.M.
	LP	SP	LP	SP	
Total	4.61	4.73	4.23	4.67	0.147
Entodinium	4.11^{ab}	4.64^{a}	3.78^{b}	4.57^{a}	0.132
Sub. Diplodiniinae	3.59	2.75	3.42	2.78	0.347
Sub. Ophryoscolecinae	3.77	2.34	3.51	2.94	0.446
Isotricha	3.64	3.14	3.23	2.84	0.252
Dasytricha	3.52	3.59	3.21	3.32	0.123

Within lines, letters indicate significant differences ($P<0.05$).

Table 2. Enrichment values (% of isotope increase in total N present in the sample of LP and SP fractions from both rumen environments at the different incubation times (n = 8).

	MIX		FOR		S.E.M.
	LP	SP	LP	SP	
H1	0.0210	0.0249	0.0156	0.0124	0.00392
H2	0.0318^{ab}	0.0497^{a}	0.0210^{b}	0.0174^{b}	0.00602
H4	0.0571^{ab}	0.0997^{a}	0.0274^{b}	0.0242^{b}	0.01319

Within lines, letters indicate significant differences ($P<0.05$).

Table 3. Slopes of enrichment in ^{15}N (n = 8). R.S.D. = 0.0092.

	MIX	FOR	Overall
LP	0.0121^{b}	0.0038^{b}	0.0080^{A}
SP	0.0249^{a}	0.0037^{b}	0.0150^{A}
Overall	0.0149^{A}	0.0037^{B}	–

Different letters (a, b or A, B) indicate significant differences ($P<0.05$).

Conclusion

Small protozoa, mainly belonging to the genus *Entodinium* seem to show a high predation rate in a rumen environment induced by a mixed diet (MIX) than similar populations under all forage environment (FOR). No effect was detected in the large protozoal population.

References

Coleman, G.S., 1975. The interrelationship between rumen ciliate protozoa and bacteria. In: McDonald, I.W. and A.C.I. Warner (Eds.), Digestion and metabolism in the ruminant. Univ. New England Publishing Unit, Sydney, Australia, pp. 149-164.

Dehority, B.A., 2003. Rumen microbiology. Nottingham University Press, Thrumpton, Nottingham, UK, pp. 372.

Leng, R.A. and J.V. Nolan, 1984. Nitrogen metabolism in the rumen (Review). J. Dairy Sci. 67: 1072-1089.

Williams, A.G. and G.S. Coleman, 1992. The rumen protozoa. Springer-Verlag, New York, USA, pp. 441.

Endogenous phosphorus flow in ruminants

R.S. Dias[1], T. Silva[2], R.M.P. Pardo[3], J.C. Silva Filho[4], D.M.S.S. Vitti[2], E. Kebreab[5], S. Lopez[6] and J. France[1]
[1]Centre for Nutrition Modelling, Department of Animal and Poultry Science, University of Guelph, N1G 2W1, Canada; [2]Animal Nutrition Laboratory, Centro de Energia Nuclear na Agricultura, CEP 13400-970, Brazil; [3]Facultade de Ciências Agropecuarias, University of Sucre, Carrera 28 5-267, Sincelejo, Sucre, Colombia; [4]Federal University of Lavras, Animal Research Laboratory, CEP 37200-000, Brazil; [5]Department of Animal Science, University of Manitoba, R3T 2N2, Canada; [6]Departamento de Producción Animal, Universidad de León, 24071 León, Spain; rsouza@uoguelph.ca

Introduction

The essentiality of P to animal health and production is well known, leading producers to overfeed their herd to guarantee its performance. Surplus P excreted in faeces leads to harmful consequences to the environment, besides wasting money in P supplementation. Endogenous P represents a considerable part of total P excreted in faeces. This fraction of faecal P is deemed an inevitable loss and derives from animal metabolism, mainly from saliva (Bravo *et al.*, 2003), therefore knowledge of how P ingestion affects endogenous flow is necessary for precise determination of maintenance P requirement. It is of interest to know how increasing levels of bicalcium phosphate affect endogenous P flow in ruminants so as to better understand the role of endogenous P and determine P requirements more accurately.

Material and methods

Twenty-four growing male sheep were fed 0.14, 0.32, 0.49 and 0.65%P in DM (dry matter). The treatments consisted of a basal diet supplemented with 0, 1.5, 3 and 4.5 g/kg DM of bicalcium phosphate to provide increasing P levels, representing treatments T0, T1, T2 and T3. After a 20 d adaptation period, the animals were placed in metabolism cages and during the first 7 d, samples of blood, faeces, saliva and ruminal contents were taken for P analysis. The animals then received a single dose of ^{32}P (7.4 MBq) injected into the jugular vein, and during the next 7 d, samples of blood and faeces were taken. Mixed saliva samples, likewise rumen fluid samples, were taken for specific activity measurement at days 4, 5 and 6 after injection in the morning before feeding. On the last day of collection, the animals were sacrificed, and rumen content was weighed and volume was estimated. Secretion rate of endogenous P into the rumen (S_r, g/d) was calculated according to Smith *et al.* (1955):

$$S_r\,(\text{g/d}) = [R \times SA_r(t) + D \int_0^t SA_r dt] \,/\, [\int_0^t SA_p dt - \int_0^t SA_r dt]$$

where R = grams of P in the rumen, SA_r = specific activity in rumen (MBq); t = time (6 d) after ^{32}P injection, D = dietary P intake (g/d); SA_p = specific activity plasma (MBq). Experimental measurements of P intake, rumen P and saliva P concentration, endogenous P in faeces (endPfec) as well as specific activity (SA_p) in plasma, rumen (SA_r) and saliva (SA_s) were analysed as a completely random design. Calculated endogenous P entering the rumen (S_r) was analysed in the same fashion. The data were taken from 24 animals, 6 for each treatment (P level). Comparison of means was performed using the GLM procedure, with source of variation being P concentration. Standard error of means was obtained using the SAS® LSMEANS procedure.

Results

Mean values and differences between treatments for P intake, P in saliva, P in ruminal liquid, endogenous P entering the rumen, endogenous P in faeces and specific activities (SA_r) in rumen, (SA_s) saliva and (SA_p) plasma are summarised in Table 1. The flow of endogenous P entering the rumen (S_r) was positively related to endogenous P in faeces (endPfec), as given by the linear equation: $S_r = 0.61$ endPfec $+ 0.11$ ($r^2 = 0.75$).

Table 1. Values of P intake, P in saliva, P in ruminal liquid, endogenous P entering the rumen (S_r), endogenous P in faeces (endPfec), specific activities in the rumen (SA_r), saliva (SA_s) and plasma (SA_p).

Measurements	Diets[1]				SE
	T0	T1	T2	T3	
P intake, g/d	1.4[a]	3.5[b]	5.6[c]	7.5[d]	0.06
P saliva, mg/dl	50.5	55	69.3	62.6	5.95
P rumen, mg/dl	53.8[a]	71.6[b]	84.6[b]	85.9[b]	5.17
S_r, g/d	0.54[a]	1.00[a]	1.53[b]	1.92[b]	0.161
endPfec, g/d	0.63[a]	1.65[b]	2.56[c]	2.58[c]	0.181
SA_r, MBq	0.019[a]	0.008[b]	0.006[bc]	0.005[c]	0.0006
SA_s, MBq	0.031[a]	0.019[b]	0.014[bc]	0.010[c]	0.0013
SA_p, MBq	0.040[a]	0.023[b]	0.014[c]	0.012[c]	0.0020

[1]T0, T1, T2 and T3 correspond to treatments providing 0.14; 0.31; 0.49 and 0.65%P in DM respectively.

[a,b,c] Within a row means without a common superscript letter differ ($P < 0.05$).

Discussion and conclusion

Increasing ingestion of P led to higher concentrations of P in ruminal fluid but not in saliva. However the amount of endogenous P entering the rumen increased, suggesting an increase in salivary flow. The amount of P entering the rumen was lower than endogenous P excretion suggesting that some of the endogenous P excreted is released post rumen and it depends on P intake. In summary, higher P intake elicited higher endogenous P excretion that was wasted in faeces, likewise for dietary P.

Acknowledgement

The authors thank Fundacao de Amparo a Pesquisa do Estado de Sao Paulo (FAPESP Proc. 2004/14532-5) for financial support. Funding was provided, in part, by the Canada Research Chairs Program.

References

Bravo, D., D. Sauvant, C. Bogaert and F. Meschy, 2003. III. Quantitative aspects of phosphorus excretion in ruminants. Repr. Nutr. Develop. 43: 285-300.

Smith, A.H., M. Kleiber, A.L. Black and C.F. Baxter, 1955. Transfer of phosphate in the digestive tract. II. Sheep. J. Nutr. 45: 507-527.

Application of the Weston model to predict feed intake in calves

R.S. Dias[1], H. Patino[2], E. Prates[2] and J. France[1]
[1]Centre for Nutrition Modelling, Department of Animal and Poultry Science, University of Guelph, N1G 2W, Canada; [2]Laboratório de Nutrição de Ruminantes, Departamento de Zootecnia, Universidade Federal de Rio Grande do Sul, 9154000, Porto Alegre, RS, Brazil; rsouza@uoguelph.ca; harold.patino@ufrgs.br

Introduction

Feed intake has great impact on performance and therefore its accurate prediction is important. With this aim, mathematical models have been used to predict feed intake in ruminants encompassing different concepts. Many factors may influence intake regulation such as animal features, management, environment and feed properties. In the Weston model (Weston, 1996) not only physical and chemical properties of feed are considered but also energy transactions in the ruminant. Essentially this model is based on the observation that there is an inverse relationship between NE (net energy) intake and rumen fill in sheep. Poppi (2008) simplified the process and the equations proposed by Weston to demonstrate how the model may help understand feed intake regulation. In the present study, data on rumen fill, retention time and feed intake was used to evaluate application of the Weston model to the prediction of feed intake in calves.

Material and methods

The present study was carried out at the Federal University of Rio Grande do Sul, Brazil. Four Hereford fistulated male calves weighing 133 kg were weaned and castrated in the same time period. The animals were fed four levels of oats hay DM (dry matter) (*Avena strigosa* L.) representing the following treatments: T1 = 1.5% of liveweight (% LW); T2 = 2.0% LW; T3 = 2.5% LW and T4 = *ad libitum*. The metabolic energy content of the diets was estimated from OM (organic matter) digestibility, which was measured *in vivo* on the four calves of the experiment. The rumen was emptied manually before morning feeding and 2 and 10 h after it. Thereafter rumen content was weighed and further analysed for DM to calculate rumen fill (F). Retention time (T) was calculated using a marker (Cr) after determining its concentration in faeces. Predicted rumen fill (pF) was calculated using the respective formula (Poppi, 2008): $pF = (c \times T \times MW)/(T \times MW + 0.38 D \times W \times m)$, where $c = 50.67$ and $m = 28.6$ are values from Gherardi and Black (1989), T = retention time (h), MW = metabolic weight (kg$^{0.75}$), D = DM digestibility coefficient and W = animal weight (kg). Predicted DM intake ($pDMin$) was calculated as follow: $pDMin = 24 \times pF/T$, where pF is calculated rumen fill (g DM/LW) and T is retention time (h). The data were analysed as a Latin square design (4×4). The Tukey test was used to compare means between treatments. PROC REG was used to evaluate the linear relationship between the parameters (SAS®, SAS Institute Inc. 2003, Cary, NC, USA). The concordance correlation coefficient (CCC) (Lin, 1989) was used to compare the observed and predicted values of F and DM intake.

Results

Rumen fill (F) was positively related to ME intake ($MEin$) and DM intake ($DMin$), as shown by the equations: $F = 163\ (\pm 162.4) MEin + 14\ (\pm 27.2)$ and $F = 1.9\ (\pm 0.30) DMin + 0.66\ (\pm 7.0)$. However, the relationship to $DMin$ was stronger (R^2=0.77) than to $MEin$ (R^2=0.34). Metabolic energy intake was inversely related to T and its relationship was stronger (R^2=0.63) than the relationship between T and $DMin$ (R^2=0.31) as shown in the following equations: T (h) = $-320\ (\pm 68.7) MEin + 109\ (\pm 11.5)$ and T (h) = $0.66\ (\pm 0.271) DMin + 71\ (\pm 6.5)$. According to CCC, the predicted and measured values of *DMin* were concordant (ρ_c=0.86), presenting a very small scale shift (1.05) and location shift (0.47). The *DMin* prediction was less accurate (0.90) than precise (0.95). The CCC value for the comparison between *F* observed and predicted was also concordant (ρ_c=0.83), with a small scale shift (1.02) and location shift (0.55), however with lower accuracy 0.87 and precision 0.94. Table 1 gives the mean values for predicted and measured DM intake and rumen fill and the measured values for retention time.

Table 1. Mean values of DM intake, retention time (T) and rumen fill (F).

	Diets[1]				SEM
	T1	T2	T3	T4	
Measured					
DM intake, g/kg LW	15.4[a]	20.9[b]	26.6[c]	27[c]	1.05
Rumen fill, g DM/LW	34.4	42.5	49.9	51	3.80
Retention time, h	68[a]	57[b]	49[b]	50[b]	1.79
Predicted					
DM intake, g/kg LW	13.4[a]	15.6[b]	17.9[c]	17.8[c]	0.36
Rumen fill, g DM/LW	38.3[a]	37.5[ab]	36.7[b]	37.3[ab]	0.31

[1] T1, T2, T3 and T4 correspond to treatments providing 1.5, 2.0, 2.5% LW and *ad libitum* of oats hay DM.
[a,b,c] Within a row means without a common superscript letter differ (P<0.05).

Discussion and conclusion

Rumen fill was better related to DM intake while retention time was better related to ME intake. Both relationships should be considered when adjusting a model that views to predict feed intake based on feed properties. Although DM intake and rumen fill were underestimated by the Weston model, CCC showed good accuracy and precision between measured and predicted values, which would likely be higher if all values used in the formulae were obtained from studies using calves. The application of the Weston model shows potential in predicting feed intake for calves.

Acknowledgement

The authors thank the Fundação de Amparo a Pesquisa do Rio Grande do Sul (FAPERGS) and National Counsel of Technological and Scientific Development (CNPq) for financial support. Funding was provided, in part, by the Canada Research Chairs Program.

References

Gherardi, S.G. and J.L. Black, 1989. Influence of post-rumen supply of nutrients on rumen digesta load and voluntary intake of roughage by sheep. Brit. J. Nutr. 62: 589-599.

Lin, L.I., 1989. A concordance correlation coefficient to evaluate reproducibility. Biometrics 45: 225-268.

Poppi, D.P., 2008. The dilemma in models of intake regulation: mechanistic or empirical. In: J. France and E. Kebreab (Eds.), Mathematical Modelling in Animal Nutrition. CAB International, Wallingford, UK, pp. 121-141.

Weston, R.H., 1996. Some aspects of constraint of forage consumption by ruminants. Aust. J. Agric. Res. 47: 175-197.

Effect of coconut oil supplementation on methane emission from grazing yak (*Bos grunniens*) in winter pasture on the Tibetan plateau

X. Ding, R.J. Long, J. Mi and B. Yang
International Centre for Tibetan Plateau Ecosystem Management, Lanzhou University, No.768 Jiayuguan West Rd. Lanzhou, 730020, Gansu China; longrj@lzu.edu.cn

Introduction

Methane emissions from enteric fermentation of ruminants contribute much to agricultural greenhouse gas emissions. Various CH_4 mitigation strategies have been reported, and the magnitude of reduction in CH_4 output following dietary supplementation of fats/oils is source dependent, with coconut oil identified as being very effective (Machmüller and Kreuzer, 1999). But this effect has never been confirmed on grazing yaks, consequently this experiment sought to investigate the effect of increasing dietary levels of coconut oil on enteric CH_4 output from yaks grazing on a winter pasture on the Tibetan Plateau.

Material and methods

An investigation was carried out at the Datong Yak Breeding Farm in Qinghai Province in the People's Republic of China. The area is 2900 to 4800 m above sea level. The average annual temperature is 2.4 °C ranging from 24 °C in August to -31 °C in January, respectively. Generally speaking, the growth of grass changes monthly, which has a direct effect on animal nutrition status. February to May is a severe period, in which most animals are absolutely malnourished. In June the nutritional content of the grass is high but of insufficient growth to support large herds. From July to September there is sufficient forage to support all animals and the condition of all animals reaches a peak in this period. In October there is less new growth of grass and in November only dried grass remains, so the nutritional status of the animals starts to decline. By May the remaining dry grass has been consumed (Wang, 2000).
Three mature female yaks (weight 178±5 kg) were randomly assigned to winter pasture for grazing, and supplemented with coconut oil at three levels of 0 g/d, 60 g/d and 120 g/d whilst each animal was fed 1 kg dry oat hay per day in a Latin-square design experiment (three periods). Each period was extended to 18 d to complete (Sutton *et al.*, 1983). Days 1 to 12 were designed to allow the rumen to adjust to the coconut oil exposure environment, from d 13 to 18, the yaks were subjected to measured methane output. Methane emissions were measured using a modification of the SF_6 tracer gas technique (Johnson *et al.*, 1994). The herbage intake was determined using the chromium oxide. Each day the yaks were grazed on separate fenced pastures and dosed with two gelatin capsules containing 10 g powdered chromium oxide. Daily faecal output of each yak in the individual fenced lots in the pasture was collected by hand, and the chromium oxide content in the faeces was determined by atomic absorption spectroscopy according to the method of Williams and David (1962).

Results

A linear reduction in methane output occurred (145, 117 and 88 l/d) as the levels of coconut oil in the diet increased (0, 60 and 120 g/d) ($P<0.01$) with the greatest reduction at the 120 g/d (Table 1). As the level of coconut oil increased, dry matter intake (DMI) decreased, however these differences were not statistically significant at the various levels ($P>0.05$). The proportional reduction in CH_4 output was greater than the proportional reduction in DMI and hence CH4 l/kg DMI decreased from 25.9 l/kg when no coconut oil was given to 17.1 l/kg when 120 g/d coconut oil was given.

Compared to the control (0 g coconut oil) the CH_4 of 60 g and 120 g coconut oil inclusion decreased 15.9% and 37.3%, respectively. Moreover the methane yield (Y_m) declined from 5.6% to 3.7%.

Table 1. Effect of coconut oil supplementation on methane output from the grazing yak.

	Coconut oil (g per head per day)		
	0	60	120
DMI, kg/d	5.6	5.3	5.2
CH_4, l/d	145^a	117^a	88^b
CH_4, l/kg DM	25.9	22.1	17.1
CH_4, g/kg $W^{0.75}$	2.01^a	1.69^{ab}	1.26^b
Ym, %	5.6^a	4.8^{ab}	3.7^b

[a,b,c] Means in the same column with different letters are significantly different ($P<0.05$); Ym (Methane yield) refers to most feed contain 18.4 MJ gross energy/kg DM, methane energy content 55.65 MJ/kg, so that a typical Ym value of 6% corresponds to 19.8 g CH_4/kg DM (27.72 l CH_4/kg DM) intake.

Conclusion

The inclusion of coconut oil would largely reduce CH_4 production with no adverse effect on DMI of grazing yak. However, structure of rumen microorganism need to be further investigated when coconut oil is added under grazing conditions.

Acknowledgement

The authors thank the staff of the Datang Yak Farm, Qinghai Province.

References

Johnson, K., M. Huyler, H. Westberg, B. Lamb and P. Zimmerman, 1994. Measurement of methane emissions from ruminant livestock. Using a SF6 tracer technique. Env. Sci.Technol. 28: 359-362.

Machmüller, A. and M. Kreuzer, 1999. Methane suppression by coconut oil and associated effects on nutrient and energy balance in sheep. Can. J. Anim. Sci. 79: 65-72.

Sutton, J.D., R. Knight, A.B. McAllan and R.H. Smith, 1983. Digestion and synthesis in the rumen of sheep given diets supplemented with free and protected oils. Br. J. Nutr. 49: 419-432.

Wang, M.Q., 2000. Study on some germplasm characteristics of Datong yak. Doctor Dissertation of Northwest Science & Technology, University of Agriculture and Forestry, P.R. China.

Williams, C.H. and D.J. David, 1962. The determination of chromic oxide in feces samples by atomic absorption spectropotometry. J. Agric. Sci. 59: 381-385.

Effect of extrusion and lignosulfonate treatment of canola seed on feed intake and digestibility of dairy cows

W.B.R. dos Santos[1], C.A. Neves[1], G.T. dos Santos[1], D.C. da Silva[1], A.F. Branco[1], F.S. dos Santos[1] and H.V. Petit[2]

[1]Universidade Estadual de Maringá, Departamento de Zootecnia, Avenida Colombo, 5790, CEP 87020-600, Maringá, Paraná, Brazil; [2]Dairy and Swine Research and Development Centre, Agriculture and Agri-Food Canada, Succ Lennoxville, Sherbrooke, QC, J1M 1Z3, Canada; gtsantos@uem.br

Introduction

The inclusion of fat sources in dairy cow diets increases the energy density. However, fat supplementation can modify digestibility, feed intake, and ruminal fermentation. Canola seed is a source of protein (21 to 25%) and fat (30 to 50%) but its hard seed coat is difficult to digest unless it is broken. Once broken, the seed is rapidly degraded in the rumen and processing that ease this process is very important to maximise the nutritional value of canola (Wang *et al.*, 1999). Lignosulfonate, which is a byproduct of the wood industry, decreases ruminal degradability of protein (Petit *et al.*, 1999). Rumen protection of crude protein usually parallels fat (Khorasani *et al.*, 1992), which may overcome negative effects of supplemental fat on ruminal digestion. Therefore, this experiment was aimed at evaluating feed intake and total apparent digestibility of nutrients in lactating dairy cows fed ground canola seed, extruded or not, and treated or not with lignosulfonate.

Material and methods

Eight Holstein multiparous cows averaging 538 kg of BW and 62 days in milk were utilised in a double 4×4 Latin square design with four 21-day periods. The four total mixed diets consisted of supplements based on: ground canola seed (GC), extruded ground canola seed (EGC), ground canola seed with 50 g/kg DM of lignosulfonate (GCL) and extruded ground canola seed with 50 g/kg DM of lignosulfonate (EGCL). The forage to concentrate ratio was 57 to 43 and the forage was corn silage. Feed consumption was recorded daily. Diets were fed twice daily at 0800 and 1600 h and adjusted for 100 g of orts/kg as fed. Samples of each diet were collected daily from day 15 to 20, frozen, and pooled on a period basis. Samples of feces were collected for 6 consecutive days at 8:00 h on d 15, 10:00 h on d 16, 12:00 h on d 17, 14:00 h on d 18, 16:00 h on d 19, and 18:00 h on d 20 of each experimental period. Samples of feces, orts, and diets were analysed for DM, N, acid detergent fiber, neutral detergent fiber (NDF), and indigestible NDF (the NDF remaining after 144 h of *in vitro* fermentation), which was used as an internal marker to estimate apparent nutrient digestibility and fecal output (Cochran *et al.*, 1986). All results were analysed using the MIXED procedure of SAS® (2000) within a 2×2 factorial arrangement of treatments. Data were analysed using a replicated 4×4 Latin square design and treatments were compared to provide factorial contrasts: (1) extruded *versus* non-extruded canola seed, (2) lignosulfonate-treated canola seed versus untreated canola seed, and (3) the interaction between extrusion and lignosulfonate treatment. Significance was declared at $P<0.05$.

Results

There was no interaction between lignosulfonate and extrusion for feed intake and total apparent digestibility although ADF digestibility tended ($P=0.06$) to be the lowest for cows fed EGC. Intake of DM and nutrients (CP, EE, ADF, and NDF) and total apparent digestibility of DM, CP, EE, ADF, and NDF were similar among treatments.

Table 1. Intake and total apparent digestibility from Holstein cows fed ground canola seed (GC), extruded ground canola seed (EGC), ground canola seed treated with 50 g/kg DM of lignosulfonate (GCL) or extruded ground canola seed treated with 50 g/kg DM of lignosulfonate (EGCL)[1].

	Treatments				SE	Probability[2]		
	GC	EGC	GCL	EGCL		E	L	E x L
Intake (kg/d)								
DM	14.50	14.87	14.20	14.34	0.34	0.48	0.27	0.75
Crude protein	2.58	2.72	2.59	2.62	0.48	0.67	0.07	0.95
Ether extract	1.03	1.22	1.09	1.07	0.07	0.25	0.57	0.18
Acid detergent fibre	3.36	3.43	3.34	3.38	0.07	0.46	0.63	0.82
Neutral detergent fibre	5.67	5.87	5.71	5.39	0.23	0.81	0.39	0.30
Total apparent digestibility (%)								
DM	67.57	64.94	63.73	66.18	1.30	0.95	0.36	0.09
Crude protein	75.23	72.91	71.88	73.37	1.48	0.80	0.37	0.25
Ether extract	90.64	87.67	90.96	90.96	1.59	0.40	0.49	0.29
Acid detergent fibre	46.11	42.75	43.13	44.34	3.77	0.79	0.86	0.57
Neutral detergent fibre	47.06	42.28	45.96	45.96	1.98	0.68	0.90	0.06

[1] Least squares means with pooled standard error (SE).
[2] E: extrusion, L: lignosulfonate.

Conclusion

Processing of canola seeds with treatment of extrusion and addition of lignosulfonate did not enhance feed intake and total apparent digestibility in Holstein lactating cows.

References

Cochran, R.C., D.C. Adams, J.D. Wallace and M.L. Galyean, 1986. Predicting digestibility of different diets with internal markers: Evaluation of four potential markers. J. Anim. Sci. 63: 1476-1483.

Khorasani, G.R., G. De Boer, P.H. Robinson and J.J. Kennelly, 1992. Effect of canola fat on ruminal and total tract digestion, plasma hormones, and metabolites in lactating dairy cows. J. Dairy Sci. 75: 492-501.

Petit, H.V., G.F. Tremblay, M. Turcotte and R. Audy, 1999. Degradability and digestibility of full-fat soybeans treated with different sugar and heat combinations. Can. J. Anim. Sci. 79: 213-220.

SAS®, 2000. Statistical analysis system, Release 8.02.2000. SAS® Institute Inc. Cary, NC, USA.

Wang Y., G. Nowak, D. Culley, L.A. Hadwiger and B. Fristensky, 1999. Constitutive expression of pea defense gene DRR206 confers resistance to blackleg (*Leptosphaeria maculans*) disease in transgenic canola (*Brassica napus*). Mol.Plant-Microbe Interact. 12: 410-418.

Effects of beta acid extracts of hops on ruminal metabolism and apparent total tract digestibility by steers fed high concentrate diets

J.S. Drouillard[1], S. Uwituze[1], M.K. Shelor[1], J.J. Higgins[2] and S. Garden[3]
[1]Department of Animal Sciences and Industry; [2]Department of Statistics, Kansas State University, Manhattan, KS 66506, USA; [3]John I. Haas, Inc., Washington, DC 20016-3341, USA; jdrouill@ksu.edu

Introduction

Feed additives are widely used in many food animal production systems to inhibit or eliminate specific ruminal gram positive organisms such as *Streptococcus bovis* and *Methanobrevibacter ruminantium,* the major ruminal lactate and methane producing organisms, respectively (Martin, 1998; Russell and Strobel, 1989). Use of these antimicrobials improves efficiency of feed utilisation, and decreases incidence of digestive disturbances that are a major cause of morbidity and mortality in cattle feeding operations. However, natural alternatives to antibiotics and other growth promoting substances for producers of natural beef products are not widely available commercially. Hops (*Humulus lupulus*) have been used for centuries to control bacterial contamination in beer production. The beta acid fraction of hops can selectively inhibit Gram-positive species such as *S. bovis* and *M. ruminantium*. Consequently, utilising byproducts of hops may make it possible to selectively control ruminal microbial populations without using feed additive drugs.

Material and methods

Angus cross steers (n = 14; 410±8 kg BW) fitted with ruminal cannulas (Bar Diamond Inc., Parma, ID; dorsal sac) were used to evaluate the effects of beta acid extracts derived from hops plants on ruminal fermentation and apparent total tract digestibility of feedlot diets. The study was conducted as a replicated, balanced incomplete block design with seven treatments. Treatments consisted of a negative control (Control; no additives); monensin fed at 300 mg/d (Monensin); or beta acid extract of hops fed at 10, 80, 160, 240, or 300 mg/day (approximately 1, 8, 16, 24, or 30 ppm). Monensin and beta acids were pulse-dosed intra-ruminally once daily immediately prior to feeding. Steers were randomly assigned to treatments and individual, slatted-floor pens equipped with individual feed bunks and water fountains that allowed access to feed and clean water *ad libitum*. Diet was based on steam-flaked corn and contained 10% DM alfalfa hay and 15% DM dried distiller's grains. Diet was mixed, proportioned, and delivered to each pen once daily at 08:00 h. Each morning before feeding, orts were weighed and dried to determine actual dry matter intake. Four experimental periods were used, each consisting of a 21-day acclimation phase followed by a 3-day collection phase. Starting 96 hours before the collection phase of each period, chromic oxide (10 g) in gelatine capsules (Torpac Inc., Fairfield, NJ, USA) was placed into the rumen prior to feeding each day to estimate total faecal output. Ruminal digesta and faecal samples were collected at 2-h intervals post feeding during the collection phase of each period. Ruminal pH was measured immediately after sampling. Concentrations of ruminal ammonia, and volatile fatty acids were determined. Estimation of total eubacteria, *Streptococcus bovis* and total methanogens was performed using quantitative rt-PCR. Faecal samples were composited by animal within period and used to determine total tract digestibilities of DM, OM, NDF, CP, starch, and ether extract. Volatile fatty acid profiles, pH, and ammonia concentration were analysed as repeated measures using the Proc MIXED procedure of SAS® (version 8.1; SAS Inst; Cary, NC, USA, 2002). Treatment means were calculated using the least-square means option and separated using a protected *F*-test. Digestibility characteristics and bacterial DNA concentrations were analysed using the PROC MIXED procedure of SAS® (version 8.1), with fixed effect of treatment and with animal, period, animal x period as random effects.

Contrasts included linear and quadratic effects of beta acids, the comparison between Control and Monensin treatments, and the comparison between Monensin and beta acid treatments. Significance was declared at $P<0.05$.

Results and discussion

There were no significant effects of treatment on counts of ruminal total bacteria, *Streptococcus bovis,* or methanogens. Treatments did not significantly alter profile or concentrations of major VFA ($P\geq0.20$), or acetate:propionate ratio ($P\geq0.50$). Lactate concentration was not affected ($P\geq0.30$) by treatments but steers dosed with monensin tended ($P=0.12$) to have lower ruminal pH compared to the control group. Cattle that received hops acids tended ($P=0.11$) to have higher ruminal ammonia concentrations compared to steers fed monensin. Bohnert *et al.* (2000) observed increases in ruminal ammonia concentrations in cattle fed laidlomycin with a diet high in concentrate content and low in crude protein. A higher concentration of isobutyrate was observed in steers fed hops acids ($P=0.03$) compared with the control group, but was not different from the monensin-fed group ($P=0.26$). Feeding hops acids also resulted in numerically higher concentrations of isovalerate and valerate relative to the control group, but the increases were not significant. It is thus conceivable that the hops acids either enhanced ruminal protein degradation or reduced the bacterial uptake of recycled urea nitrogen. In situations where requirements of ruminal nitrogen are not being met, this effect would be very useful and could potentially reduce the need for ingredients like urea in diet formulation. However, the increased proteolysis of dietary nitrogen could reduce the amount of feed protein reaching the small intestine, thus requiring a higher proportion of ruminally undegraded protein sources. There was no treatment effect on intake or total tract digestibility of DM, OM, NDF, starch, CP, or ether extract ($P>0.20$).

Conclusion

This study did not demonstrate any significant effect of extracts of hops plants on ruminal fermentation or diet digestibility characteristics. However, it is apparent that hops acids have some biological activity in the rumen, and this activity might have commercial application. More work is encouraged.

References

Bohnert, D.W., D.L. Harmon, K.A. Dawson, B.T. Larson, C.J. Richards and M.N. Streeter, 2000. Efficacy of laidlomycin propionate in low-protein diets fed to growing beef steers: effects on steer performance and ruminal nitrogen metabolism. J. Anim. Sci. 78: 173-180.

Martin, S.A., 1998. Manipulation of ruminal fermentation with organic acids: a review. J. Anim. Sci. 76: 3123-3132.

Russell, J.B. and H.J. Strobel, 1989. Mini-Review: The effect of ionophores on ruminal fermentation. Appl. Environ. Microbiol. 55: 1-6.

Effect of a blend of essential oils on the fermentation of starch-rich substrate as estimated by its gas production profile

S.M. Duval and C.J. Newbold
Institute of Rural Science, University of Wales, Aberystwyth, SY23-3AL, United Kingdom; stephane.duval@dsm.com

Introduction

Recently and in response to concerns raised by both the scientific and consumer communities about the use of antibiotics in animal production, research on the use of plant extracts such as essential oils (EO) as rumen-manipulating agents has intensified. The effects of EO on the metabolism of protein-rich substrates have been investigated in several studies; however, little attention has been paid so far to their effects on starch-rich substrates. Yet, a supplementation of commercial blend of EO affected to a greater extent bacterial populations colonising starch-rich substrates than bacterial populations colonising protein-rich substrates (Duval *et al.*, 2004), possibly because of the ability of starch granules to sorb some components of the EO (Misharina, 2002). The aim of this study was to evaluate the effect of EO on the fermentation of barley as estimated by its gas production profiles, and to investigate the effect of co-incubating the starch-rich substrate and the EO.

Material and methods

The rumen simulation technique (Rusitec) was used as described by Czerkawski and Breckenridge (1977). The fermentation vessels were fed daily with 14 g of basal diet (80% chopped hay, 10% soybean meal, 10% molasses) and 6 g of rolled barley. Four vessels were used as a negative control and no EO were added to these vessels (treatment CR). Four vessels were supplemented with 15 mg/d of EO (mixed with basal diet) and 7 mg were added to 200 ml of artificial saliva in which the bags containing barley were left for 24 h (treatment PR). The last four vessels were also supplemented with 15 mg/d of EO, and the bags containing barley were incubated for 24 h in 200 ml of artificial saliva. However, five minutes before the transfer of these bags to the fermentation-vessels, 7 mg of EO were added to the saliva (treatment ER). The Rusitec was allowed 21 d for adaptation. The fluid from the fermentation-vessels was used to carry out a gas-production experiment. Briefly, 300 mg of rolled barley was put into Wheaton bottles which were divided into three groups. The first group (CR) received 5 ml of water 24 h before the experiment, 5 ml of water just before the experiment, and then 40 ml of fluid from CR fermentation-vessel. The second group (PR) received 5 ml of a solution of EO (0.2 /mg.ml) 24 h before the experiment, 5 ml of water just before the experiment, and 40 ml of fluid from PR fermentation-vessels. The third group (ER) received 5 ml of water 24 h before the experiment, 5 ml of a solution of EO (0.2 /mg.ml) just before the experiment, and 40 ml of fluid from ER fermentation-vessels. The gas production was followed over 88 h and the curves were fitted onto the France model (France *et al.*, 1993).

Results and discussion

Figure 1 presents the gas production curves obtained when the experiment was carried out in the same condition as the Rusitec experiment. The fermentation parameters as determined by the France Model are presented in Table 1. In the present study, a strong effect of EO was observed on the extent of fermentation of the barley. Indeed, this parameter was maximal in the absence of EO and minimal when the substrate had been pre-incubated with EO for 24 h. When EO had been added extemporaneously, the difference was not significant. Moreover, the addition of EO led to a significant increase in the lag phase of fermentation and to a significant decrease in the rate of fermentation. Nevertheless, pre-incubation had the opposite effect. Yet, one could point out that, although statistically significant, the differences observed on the lag phases may be too small to have any real biological effect on the overall microbial digestion in the rumen.

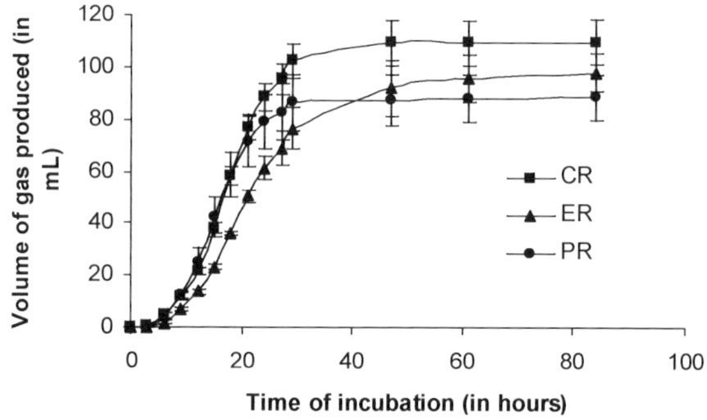

Figure 1. Volume of gas, produced by fluid recovered from Rusitec fermentation-vessel, when 300 mg of rolled barley was fermented, in the absence of EO (CR), or in the presence of EO at a concentration of 0.02 mg.ml^{-1} with (PR) or without (ER) pre-incubation in an EO solution for 24 h.

Table 1. Parameters of the fermentation. CR, no EO; ER, EO at 0.02 mg/ml; PR, EO at 0.02 mg/ml^{1} pre-incubated for 24 h with the substrate.

	CR	ER	PR	s.e.d.	*P*-Value
Extent (ml)	115.5^{a}	102.5^{ab}	92.6^{b}	6.73	0.024
Rate (ml/min) at Vol=0.5 Vol_{max}	0.071^{a}	0.053^{b}	0.084^{c}	0.002	>0.001
Lag (min)	5.825^{a}	6.450^{b}	5.385^{c}	0.217	0.003

$^{a, b, c}$ Means within rows with same superscript letters are not significantly different (P>0.05).

Conclusion

This experiment demonstrated that EO could modify the fermentation parameters of a starch-rich substrate. However, the nature and the extent of this modification were dependent on the time allowed to the EO to bind to the starch. Further work is required to identify the EO compound bound by the starch, and the bacterial species that are affected.

References

Czerkawski, J.W. and G. Breckenridge, 1977. Design and development of a long-term rumen simulation technique (Rusitec). Br. J. Nutr. 38: 371-384.

Duval, S.M., C.J. Newbold, N.R. Mcewan, R.C. Graham and R.J. Wallace, 2004. Effect of a specific blend of essential oils on the colonization of substrates by rumen microorganisms. Reprod. Nutr. Dev. 44: S35.

France, J., M.S. Dhanoa, M.K. Theodorou, S.J. Lister, D.R. Davies and D. Isac, 1993. A Model To Interpret Gas Accumulation Profiles Associated With *In-Vitro* Degradation of Ruminant Feeds. J. Theor. Biol. 163: 99-111.

Misharina, T.A., 2002. Sorption of monoterpenoids and sesquiterpenoids by natural polysaccharides. Appl. Biochem. Microbiol. 38: 480-486.

Prediction of methane production by cattle in some current whole farm models

J.L. Ellis[1], *A. Bannink*[2], *J. Dijkstra*[3], *E. Kebreab*[4] *and J. France*[1]
[1]*Centre for Nutrition Modelling, Department of Animal and Poultry Science, University of Guelph, Guelph, ON, Canada;* [2]*Animal Sciences Group, Division Animal Production, Wageningen University and Research Centre, Lelystad, the Netherlands;* [3]*Animal Nutrition Group, Wageningen Institute of Animal Sciences, Wageningen University, Wageningen, the Netherlands;* [4]*Department of Animal Science, University of Manitoba, Winnipeg, MB, Canada; jellis@uoguelph.ca*

Introduction

The importance of evaluating nutrient emission from dairy cows and their associated mitigation strategies, within the whole farm setting, is being realised as more important than evaluating these emissions in isolation. Methane (CH_4) production from cows is an area that has received much attention as emissions can be significant depending on feeding strategy, and because CH_4 is the major contributor to total greenhouse gas emissions on the farm. Evaluating CH_4 mitigation strategies in isolation provides no information about the financial cost or benefit of making such changes, effect on production level or the effect on other environmental pollutants such as nitrogen. Many current whole farm models that predict CH_4 production have a tendency to use simple regression equations to predict CH_4, and this might compromise their accuracy of prediction. Therefore, the objective of the current paper was to evaluate the performance of several CH_4 prediction equations that are currently being used in whole farm models.

Material and methods

The data used to evaluate CH_4 prediction equations comprised two databases of dairy cow data with diets ranging from 40-70% forage and 10-30 kg DMI/d. The first database is a literature derived treatment average dataset containing 37 data points from 7 studies. The second database is an individual animal dataset of 171 data points from 9 studies. Each study measured CH_4 directly from the animals using whole animal calorimetry, hood calorimetry or the sulphur hexafluoride tracer gas technique.

The predictive ability of the following equations, used in several whole farm models, were evaluated: IPCC (1997) Tier 1 and 2, Schils *et al.* (2005; 2006), Giger-Reverdin *et al.* (2003), Blaxter and Clapperton (1965), Moe and Tyrrell (1979) and two equations by Kirchgeßner *et al.* (1995).

Equation performance was evaluated with mean square prediction error (MSPE) and concordance correlation coefficient (CCC) methods. Square root of the MSPE (RMSPE), expressed as a proportion of the observed mean, gives an estimate of the overall prediction error. The RMSPE can be decomposed into random error, error due to deviation of the regression slope from unity, and error due to overall bias (Bibby and Toutenburg, 1977). The CCC statistic (Lin, 1989) is the product of 2 components: (1) the correlation coefficient, which is a measure of precision (deviation of observations from the best fit line), and (2) a bias correction factor, which indicates how far the regression line deviates from the line of unity (accuracy).

Results and discussion

The equations with the lowest RMSPE value and highest CCC value for the individual cow database were that of Moe and Tyrrell (1979) (RMSPE% = 20.2, CCC = 0.46) and IPCC (1997) Tier 2 (RMSPE% = 20.9, CCC = 0.49). The equation by Moe and Tyrrell (1979) includes the variables non-fibre carbohydrate, cellulose and hemicellulose, while the IPCC (1997) Tier 2 is based on gross energy intake. The equations with the highest RMSPE and lowest CCC values were the IPCC (1997) Tier 1 equation (RMSPE% = 27.6, CCC = 0.00) and the equation by Schils *et al.* (2006) (RMSPE% = 39.9, CCC = 0.254). The IPCC (1997) Tier 1 equation is a country wide estimate per animal, while the Schils *et al.* (2006) equation is based on total intake of concentrate, grass and maize silage with specific methane emissions for each category. The Moe and Tyrrell (1979) equation again performed best on the treatment average literature database (RMSPE% = 24.0, CCC = 0.260), while the IPCC (1997) Tier 1 and Schils *et al.* (2005) equations performed the worst (RMSPE% = 24.3, CCC = 0.00 and RMSPE% = 38.2, CCC = 0.037, respectively). The Schils *et al.* (2005) equation is based simply on milk production.

In general, results show that the simple more generalized equations performed worse than those that attempted to represent important aspects of the diet, but in general significant amounts of bias and deviation of the regression slope from unity existed for all equations. This implies that whole farm models used to estimate environmental impact of dairy farming do not give an accurate prediction of changes in CH_4 production in various mitigation strategies. Improvements could be made by moving towards more mechanistic representation of CH_4 production within these whole farm models.

References

Bibby, J. and H. Toutenburg, 1977. Prediction and Improved Estimation in Linear Models. John Wiley & Sons, Chichester, UK.

Blaxter, K.L. and J.L. Clapperton, 1965. Prediction of the amount of methane produced by ruminants. Br. J. Nutr. 19: 511-521.

Giger-Reverdin, S., P. Morand-Fehr and G. Tran, 2003. Literature survey of the influence of dietary fat composition on methane production in dairy cattle. Livest. Prod. Sci. 82: 73-79.

IPCC, 1997. Revised 1996 IPCC Guidelines for National Greenhouse Gas Inventories: Reference Manual. OECD, Paris.

Kirchgeßner, M., W. Windisch and H.L. Müller, 1995. Nutritional factors for the quantification of methane production. In: Engelhart, W.V., S. Leonhard-Marek, G. Breves and D. Giesecke (Eds.), Ruminant Physiology: Digestion, Metabolism, Growth and Reproduction. Proceedings VIII International Symposium on Ruminant Physiology, pp. 333-348.

Lin, L.I.K., 1989. A concordance correlation coefficient to evaluate reproducibility. Biometrics 45: 255-268.

Moe, P.W. and H.F. Tyrrell, 1979. Methane production in dairy cows. J. Dairy Sci. 62: 1583-1586.

Schils, R.L.M., A. Verhagen, H.F.M. Aarts and L.B.J. Šebek, 2005. A farm level approach to define successful mitigation strategies for GHG emissions from ruminant livestock systems. Nutr. Cyc. Agro. 71: 163-175.

Schils, R.L.M., A. Verhagen, H.F.M. Aarts, P.J. Kuikman and L.B.J. Šebek, 2006. Effect of improved nitrogen management on greenhouse gas emissions from intensive dairy systems in the Netherlands. Global Change Biol. 12: 382-391.

Effect of energy intake on splanchnic net flux and whole body balance of nitrogen in mature sheep fed lucerne hay cubes

M. EL-Sabagh, T. Sugino, T. Obitsu and K. Taniguchi
Graduate School of Biosphere Science, Hiroshima University, Higashihiroshima-shi, 739-8528, Japan; kohzo@hiroshima-u.ac.jp

Introduction

Improvements to feeding systems in ruminants depend on our ability to quantify the exchange and use of nutrients between and by tissues (Loncke *et al.*, 2009). Several studies have considered the effects of metabolisable energy intake (MEI) level on nitrogen (N) metabolism across the portal-drained viscera (PDV). Recent data showed that the conversion of total tract digestible N to PDV net amino acid (AA) supply is varied among steers fed with different dietary ratios of maize silage and lucerne hay (EL-Sabagh *et al.*, 2009). This may be due to differences in N and energy intake from the mixed forage diets. We investigated the effect of increasing levels of MEI on N balance and N metabolites net fluxes across portal and splanchnic tissues in sheep fed lucerne hay cubes.

Material and methods

Four Suffolk mature sheep (61.4±3.6 kg BW) were surgically fitted with catheters in the hepatic portal vein, hepatic vein, mesenteric vein and caudal aorta. They were fed (at 2-h intervals using an automatic feeder) with lucerne hay cubes (19% CP) in a Latin square design with 4 MEI levels ranging from 0.4 to 1.6 fold the maintenance requirements. ME intake was modified by variation in dry matter supply of lucerne hay cubes. The adaptation peroid lasted 7 d. Nitrogen balance was determined by urine and faeces collection over 5 d and Kjeldahl N analyses were performed. Six sets of blood samples were simultaneously collected from arterial and venous catheters at 30-minute intervals, and were analysed for ammonia, urea and 22 individual AA, using *p*-aminohippuric acid as a marker of blood flow. Data were analysed by the Mixed Procedure of SAS® according to a 4×4 Latin square design. The model included the treatment and period as the fixed effect and sheep as the random effect. Treatment means were compared with linear and quadratic contrasts. Significance was declared at $P<0.05$, and tendency at $P<0.1$.

Results

Increasing MEI level induced linear increases in faecal and urinary N losses, digestible N and N retention (Table 1). N balance was negative at the lowest MEI level. Net PDV recovery of total AAN to digestible N varied with MEI increments accounting for 22, 39, 44 and 41% at 0.4, 0.8, 1.2 and 1.6M, respectively. Conversly, the MEI increments linearly reduced ($P<0.05$) the proportion of digestible N recovered as PDV ammonia N release and urea N uptake. As a percentage of digestible N, PDV N output (ammonia N + total AAN) linearely increased ($P<0.05$) with increasing MEI level, but did not vary between treatments relative to gut available N (digestible N + urea N) (Figure 1). Net hepatic removal of ammonia N and total AAN, and net hepatic urea N release linearly increased ($P<0.05$) with inceasing MEI level, but there was no change in the proportion of ammonia N contribution to hepatic urea synthesis. Splanchnic net release of total AAN tended to increase linearly ($P<0.1$) with increasing MEI level.

Table 1. Effect of increments of MEI level on N balance in mature sheep.

	MEI levels (times of maintenance)				SEM[1]	*P*-value[2]	
	0.4M	0.8M	1.2M	1.6M		L	Q[3]
Nitrogen, g/d							
Intake	12.2	23.3	34.5	47.0	1.65	<0.001	0.402
Fecal	2.6	4.8	7.1	10.4	0.45	<0.001	0.167
Urinary	11.4	14.9	19.9	28.1	1.71	<0.001	0.091
Retained	-1.8	3.6	7.4	8.5	1.12	<0.001	0.083

[1] Standard error of means; [2] L=Linear effect; [3]Q=Quadratic effect.

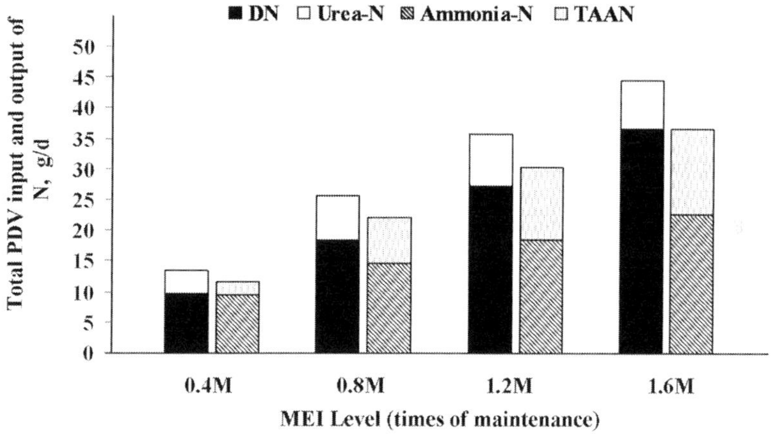

Figure 1. Effect of increments of MEI level on total PDV N input and output in mature sheep (DN=apparently digestible N, TAAN=total amino acid N).

Discussion and conclusion

The remarkably low recovery of digestible N as total AAN by PDV accompanied with high ammonia N recovery at 0.4 M indicate that AAN was used as an energy substrate by PDV. On the contrary, the relatively constant recovery of total AAN between 0.8 and 1.6 M may imply that PDV releases a constant proportion of AAN across levels of energy supply above maintenance. Considering the MEI and PDV recovery of AA can improve the current prediction models of absorbed AA.

References

EL-Sabagh, M., S. Imoto, K. Yukizane, A. Yokotani, T. Sugino, T. Obitsu and K. Taniguchi, 2009. Net portal fluxes of nitrogen metabolites in Holstein steers fed diets containing different dietary ratios of whole-crop corn silage and alfalfa hay. Asian-Aust. J. Anim. Sci. 22: 371-377.

Loncke, C., I. Ortigues-Marty, J. Vernet, H. Lapierre, D. Sauvant and P. Nozière, 2009. Empirical prediction of net portal appearance of volatile fatty acids, glucose, and their secondary metabolites (β-hydroxybutyrate, lactate) from dietary characteristics in ruminants: A meta-analysis approach. J Anim Sci. 87: 253-268.

Methane production by growing bulls fed diets supplemented or not with extruded linseed

M. Eugène[1], *C. Martin*[1], *M.M. Mialon*[1], *D. Krauss*[2], *G. Renand*[3] *and M. Doreau*[1]
[1]*INRA, UR1213 Herbivores, 63122 Saint-Genès-Champanelle, France;* [2]*INRA, UE232 Domaine de la Sapinière, 18390 Osmoy, France;* [3]*INRA, UMR1313 Génétique Animale et Biologie Intégrative, 78352 Jouy en Josas Cedex, France; meugene@clermont.inra.fr*

Introduction

The production of greenhouse gases (GHG) from livestock and their impact on climate changes are a major concern worldwide. It has been reported that enteric methane (CH_4) is the most important GHG emitted (50-60%) at the farm scale in ruminant production systems. Among nutritional strategies proposed to reduce methane emission in ruminants, concentrate and lipid supplementation of diets seem to be the most interesting (Eugène *et al.*, 2008; Martin *et al.*, 2007). The aim of this study was to determine methane production from feedlot Charolais bulls fed a high-concentrate diet supplemented or not with extruded linseed.

Material and methods

Fifty-six Charolais bulls (8.5±0.3 mo of age, bodyweight = 339±8.2 kg at the beginning of the experiment) were allocated to two treatment groups and housed in 8 pens of 7 animals each. The control diet (CTL) contained a concentrate mainly made of cereal by-products (40%), dehydrated lucerne (22%), beet pulp (21%), rapeseed and palm kernel meal (7.5%), and barley straw. The linseed diet (LS) contained a concentrate mainly made of cereals (46%), beet pulp (6%), rapeseed meal (21%) and an extruded mixture containing 50% linseed (12%), and barley straw. The content in crude protein, neutral detergent fibre, acid detergent fibre and ether extract averaged 16, 48, 25 and 2.7% DM for the CTL concentrate and 21, 23, 11 and 5.0% DM for the LS concentrate. In both diets, concentrate and straw were offered *ad libitum* once a day at 08.00 h. The ether extract content of CTL and LS diets was 2.4 and 4.4% DM, respectively. After 4 wk of feeding these diets, methane production was measured individually (9.3±0.3 months of age and 368±29 kg of live weight on average) on 4 consecutive days using the SF6 gas tracer method (Martin *et al.*, 2008). Data were analysed using the PROC GLM procedure of SAS® (SAS Institute Inc., Cary, NC, USA). The statistical model included diet and pen within diet as fixed effects. Bull live weight at the beginning of the experimental period was used as a covariate. Significance was declared at $P<0.05$.

Results and discussion

The respective percentages of concentrate and straw were 87:13% of DM for both diets. Bulls fed the LS diet had lower DM intake ($P<0.01$) and higher net energy intake ($P<0.01$) than bulls fed the CTL diet (Table 1). Bulls fed the LS diet had lower CH_4 production expressed as L/d or as L/net energy intake than bulls fed the CTL diet (-19% and -31% on average, respectively; $P<0.001$). There was no effect of the diet on CH_4 expressed as L/kg DM intake.
Although the control diet was very rich in concentrate (87% DM), daily CH_4 production was high. Indeed, Martin *et al.* (2007) found that bulls fed 6.3 kg DM of a high concentrate diet (86% DM) had lower CH_4 emission (93 L/d). This discrepancy may be due to the nature of the concentrate used in the present study, feedstuffs rich in fibre *vs* starch. Lipid supplementation decreased daily CH_4 emission as reported by Eugène *et al.* (2008) with dairy cows. Our results are in agreement with Martin *et al.* (2008) who observed a decrease in daily CH_4 emission by 38% in dairy cows fed a diet supplemented with 4.4% lipids from extruded linseed. The effect of linseed may be due

to the decrease in organic matter fermented in the rumen and/or to changes of the rumen microbial ecosystem.

Table 1. Effect of linseed supplementation on intake and methane emission by bulls.

	Diet[1]		SEM	Diet effect
	CTL	LS		*P*<
Intake				
DM intake, kg/d	8.1^{a}	7.1^{b}	1.19	0.01
Net energy intake, UFV[2]/d	7.0^{a}	8.1^{b}	1.31	0.01
Methane				
L/d	283.9^{a}	228.9^{b}	8.20	0.001
L/kg DM intake	35.4	32.8	1.40	NS
L/UFV intake	41.6^{a}	28.6^{b}	0.25	0.001

[1] CTL = control diet, LS = diet containing 6% of extruded linseed.
[2] UFV = French net energy intake unit for fattening.
[a,b] Means within rows with same superscript letters are not significantly different (P>0.05).

Conclusion

These results indicate that CH_4 production of bulls, at the beginning of the fattening period, decreased in response to a diet rich in concentrate and supplemented with extruded linseed. These results have to be confirmed on a longer fattening period, and need further highlight by ruminal fermentative and microbial ecosystem parameters.

Acknowledgement

This experiment was supported by a grant from the regional council of the French 'Centre' region, by McKey Food Services (Fleury les Aubrais, France) and by Valorex (Combourtillé, France).

References

Eugène, M., D. Massé, J. Chiquette and C. Benchaar, 2008. Meta-analysis on the effects of lipid supplementation on methane production in lactating dairy cows. Can. J. Anim. Sci. 88: 331-334.

Martin, C., H. Dubroeucq, D. Micol, J. Agabriel and M. Doreau, 2007. Methane output from beef cattle fed different high-concentrate diets. Annual Conference of the British Society of Animal Science, 2-4 April 2007, Dublin, p 46.

Martin, C., J. Rouel, J.P. Jouany, M. Doreau and Y. Chilliard, 2008. Methane output and diet digestibility in response to feeding dairy cows with crude linseed, extruded linseed or linseed oil. J. Anim. Sci. 86: 2642-2650.

Improvement of *in vitro* ruminal fermentation of ensiled peppermint (*Mentha piperitae*) byproduct when combined with alfalfa hay or corn silage

J.-S. Eun[1], *D.R. ZoBell*[1] *and Suhubdy*[2]
[1]*Department of Animal, Dairy and Veterinary Sciences, Utah State University, Logan, UT 84322-4815, USA;* [2]*Department of Animal Feed and Nutrition Sciences, University of Mataram, Mataram, NTB 83125, Indonesia; jseun@usu.edu*

Introduction

The beef and dairy cattle industries use many non-traditional feeds, including by-products from other agricultural industries. These feeds may provide significant economic advantages in ration formulation. However, the nutritional quality of the feedstuffs and their feeding strategies must be considered when fed to ruminants. Mint is a perennial plant, and peppermint *(Mentha piperitae)*, a cultivated mint type, is grown mainly for mint oil. The peppermint is persistent in the Intermountain West area in the U.S., where the right amount of daylight contributes to good yield of peppermint and quality of its oil. The peppermint byproduct after extraction of its oil is high in moisture, fibre and protein (Djouvinov *et al.*, 1997), and it can be conserved as silage to be fed to ruminants. However, the literature is devoid of comparative evaluation of the nutritional value of ensiled peppermint byproducts. Our objective was to determine the effects of the peppermint byproduct silage (PMBS) on *in vitro* fermentation by mixed cultures of ruminal microbes compared to alfalfa hay (AH) and corn silage (CS). We were particularly interested if combining PMBS with AH or CS would improve fibre degradability of PMBS.

Material and methods

The *in vitro* procedures used in this study were the same as those described by Eun and Beauchemin (2007) for forage degradability bio-assay *in vitro*. The incubation consisted of a 24 h *in vitro* batch culture fermentation with treatments applied in a completely randomised design with four replications. Forage samples were dried at 55 °C for 48 h and were milled to pass a 1-mm screen. For combination treatments of the AH and PMBS (AHPMBS) and the CS and PMBS (CSPMBS), equal amounts of the component forages (DM basis) were thoroughly mixed. The PMBS contained (DM basis) the following: 14.8% CP, 54.0% NDF and 42.5% ADF. The AH used contained 17.9% CP, 42.5% NDF, and 31.5% ADF (DM basis), whereas the CS contained 6.8% CP, 41.3% NDF, and 21.7% ADF. Approximately 1.0 g (DM) of ground forages were weighed into acetone-washed and preweighed filter bags (pore size of 25 μm; Ankom Technology, Macedon, NY, USA). The bags were heat-sealed and placed in gas-tight culture vials, and anaerobic buffer medium (40 ml; pH of 6.0) was added to each vial. Strained ruminal fluid (pH of 6.5) through polyester material (pore size of 355 μm) was collected from two ruminally cannulated beef cows fed orchardgrass hay and was dispensed (10 ml per vial) into the culture vials, and then they were stored at 39 °C in an incubator for 24 h. At the end of incubation, the bags were dried at 55 °C for 48 h, and degradabilities of DM and NDF were sequentially determined. Profiles of VFA and ammonia-N (NH_3N) were measured using 5 ml of the fermentation contents added to 1 ml of 25% meta-phosphoric acid and 1% sulphuric acid, respectively. All data were analysed separately as compared when PMBS was combined with AH or CS using the mixed procedures of SAS® with the model that included the fixed effect of forage treatment. Significance was declared at $P<0.05$.

Results

Degradabilities of DM and NDF increased with AHPMBS compared with PMBS ($P<0.02$), whereas DM degradability, but not NDF degradability, increased with CSPMBS compared with PMBS ($P<0.01$; Table 1). The discrepancy on NDF degradability between AHPMBS and CSPMBS may be a result from better N degradation by microbes when PMBS was combined with AH compared with CS. Increased NH_3N concentration when PMBS was combined with AH, but not with CS, supported the improved CP degradation, resulting in increased fibre degradation. Total VFA production increased by combining PMBS with AH or CS ($P<0.01$), but acetate production increased only by combining PMBS with AH ($P<0.01$), which further supports the effect of combining PMBS with AH to improve fibre degradability of PMBS. Propionate production increased by combining PMBS with AH or CS ($P<0.01$).

Table 1. In vitro degradability, volatile fatty acid (VFA) production, and ammonia N (NH_3N) concentration from alfalfa hay, corn silage, peppermint by-product silage and their combinations during 24 h of incubation with ruminal fluid.

	Compared to alfalfa hay[1]			SE	Compared to corn silage[2]			SE
	AH	PMBS	AHPMBS		CS	PMBS	CSPMBS	
Degradability,%								
DM	45.5^a	35.7^c	42.9^b	0.62	41.4^a	35.7^b	41.1^a	0.87
NDF	33.1^a	23.1^c	29.3^b	0.89	39.0^a	23.1^b	25.4^b	1.21
NH_3N, mg/l	612^a	421^c	515^b	6.7	442^a	421^b	418^b	5.7
Total VFA, mM	101^a	66.0^c	83.9^b	1.74	98.6^a	66.0^c	86.4^b	1.65
Acetate, mM	54.5^a	41.5^c	48.5^b	0.81	42.8	41.5	43.1	0.72
Propionate, mM	23.8^a	14.2^c	18.8^b	0.46	29.3^a	14.2^c	24.1^b	0.60

[1] AH = 100% alfalfa hay, PMBS = 100% peppermint by-product silage, AHPMBS = 50% AH and 50% PMBS (DM basis).

[2] CS = 100% corn silage, CSPMBS = 50% CS and 50% PMBS (DM basis).

[a,b,c] Means within a row that do not have a common superscript differ at $P<0.05$.

Discussion and conclusion

Increased NDF degradability and NH_3N concentration when PMBS was combined with AH imply that low degradability of PMBS because of its high acid detergent insoluble protein content (Mustafa *et al.*, 2001) can be improved by incorporating the soluble fraction of N from AH. In order for PMBS to be fed to ruminants, an appropriate source of N must be provided.

References

Djouvinov, D., D. Pavlov, A. Ilchev and E. Enev, 1997. Peppermint (*Mentha piperita* Huds.) and basil (*Ocimum basilicurn* L.) etheric oil by-products as roughages for sheep feeding. Anim. Feed Sci. Technol. 68: 287-294.

Eun, J.-S. and K.A. Beauchemin, 2007. Enhancing *in vitro* degradation of alfalfa hay and corn silage using feed enzymes. J. Dairy Sci. 90: 2839-2851.

Mustafa, A.F., J.J. McKinnon and D.A. Christensen, 2001. Effects of feeding ensiled spearmint (*Mentha spicata*) byproduct on nutrient utilization and ruminal fermentation of steers. Anim. Feed Sci. Technol. 92: 33-43.

Kinetics of *in vitro* ruminal fermentation of glycerol, propylene glycol, molasses and their drenching effect in blood concentrations of glucose and insulin in ewes

S.M. Ferraro[1], G.D. Mendoza[2], L.A. Miranda[3] and C.G. Gutiérrez[1]
[1]Universidad Nacional Autónoma de México, Facultad de Medicina Veterinaria y Zootecnia, Av. Universidad 3000, Mexico DF. 04510; [2]Universidad Autónoma Metropolitana, Xochimilco, Mexico DF; [3]Universidad Autónoma de Chapingo, Chapingo, Mexico; ggcarlos@servidor.unam.mx

Introduction

Glycerol, propylene glycol and molasses are substances often used as energy additives in ruminant nutrition. Glycerol and propylene glycol in particular, are used for their glycogenic and anti-ketogenic properties (Rémond *et al.*, 1993; Kristensen and Raun, 2007). In dairy cows, their addition in early lactation decreased non-esterified free fatty acids and β-hydroxybutyrate whilst increasing blood glucose concentrations. (Nielsen and Ingvartsen, 2004; Chiofalo *et al.*, 2005). Oral administration of propylene glycol increases plasma insulin within 30 min after drenching and increases plasma glucose, although the response is limited, probably because of the large increase in insulin (Nielsen and Ingvartzen, 2004). Molasses is a byproduct of the sugar cane industry used widely in ruminants as a source of soluble carbohydrates rapidly fermentable in the rumen (Wiedmeier *et al.*, 1992). The aim of this study was to determine the gas production and *in vitro* ruminal fermentation of glycerol, propylene glycol, molasses and their drenching effect on blood concentrations of glucose and insulin in ewes.

Material and methods

Experiment 1: Alfalfa (0.5 g), corn silage (0.5 g), glycerol (320 and 640 µl), propylene glycol (320 and 640 µl) and molasses (320 µl) were fermented *in vitro* as described by Menke and Steingass (1988). Gas production was recorded for 92 h. Volatile fatty acid production was measured by gas chromatography. Data were analysed as a completely randomised block design, using the PROC GLM from SAS (1999). Means were compared with the Tukey test. Maximum volume, rate of gas production and lag time *in vitro* were estimated with the PROC NLIN BEST (SAS®, 1999).
Experiment 2: Sixteen ewes (BC 2) were randomly assigned to four treatments. Ewes were maintained in pasture containing alfalfa and ryegrass. The control group received no supplementation; the other three groups received isoenergetic amounts of glycerol (300 ml), propylene glycol (270 ml) or molasses (406 ml) as a single oral drench. Blood insulin and glucose concentrations were measured in samples collected by jugular venipuncture at -30, 0, 30, 90, 120, 150, 180, 210, 360, 540, 720, 1080, 1440 min postdrenching. Glucose was analysed from plasma by the glucose oxidase procedure. Insulin concentration was analysed from serum by radioimmunoassay (RIA). The intra-assay coefficient of variation was 5.31%. Data on plasma glucose and insulin were evaluated using the MIXED procedure for repeated measures by SAS® (1999). The statistical model included: animal, nested within treatment, treatment, time and interaction time*treatment. Overall differences between treatment means was significant at $P<0.05$.

Results

Experiment 1: Glycerol, at both concentrations tested, showed the lowest ($P<0.05$) rate of gas production (0.0242; 0.0192 ml/h) and the largest ($P<0.05$) lag time (10.23; 12.85 h). In contrast, propylene glycol presented the lowest volume of gas production (22.60; 20.54 ml/g) and was rapidly

metabolised with short lag time (1.81; 2.66 h). Molasses showed a fastest rate of gas production (0.0854 ml/h) (P<0.05) and the shortest lag time (1.86 h). Glycerol fermentation resulted in reduced acetate and increased butyrate concentration (P<0.001), while propionate showed a quadratic (P<0.001) effect, increasing at 12 h and decreasing at 26 h of incubation. Inclusion of propylene glycol reduced acetate and butyrate with time (P<0.001), and increased propionate (P<0.001). Molasses fermentation reduced acetate and increased propionate and butyrate (P<0.001).
Experiment 2: Ewes in the control and molasses group showed no increase (P>0.05) in glucose concentrations. In contrast, glycerol and propylene glycol caused an increase in glucose concentrations 30 to 60 min after drenching and remained high until 180 min after treatment. After dosing, insulin blood concentrations increased from 30 to 720 min in the glycerol and from 60 to 720 min in the propylene glycol group. In contrast, in the molasses group insulin concentrations increased only between 30 to 150 min after treatment.

Discussion and conclusion

Molasses are rapidly fermented by rumen microbes, whilst glycerol and propylene glycol are fermented slowly. The long lag time of fermentation for glycerol allows for its passage to the lower tract and its absorption through a long period giving the liver a glycogenic substrate that will increase glucose and thus insulin concentrations. Propylene glycol is basically not metabolised by rumen microbes and acts as a propionate and as a glycogenic precursor. Glycerol may increase liver glucose output by either being converted from propionate or directly from absorbed glycerol.

Acknowledgement

This research was partially supported by UNAM, Papiit IN223907

References

Chiofalo, V., M. Todaro, L. Liotta, S. Margiotta, T. Manzo and G. Leto, 2005. Effect of propylene glycol on pre- and postpartum performance by dairy ewes. Small Ruminant Res. 58: 107-114.

Kristensen, N. and B. Raun, 2007. Ruminal and intermediary metabolism of propylene glycol in lactating Holstein cows. J. Dairy Sci. 90: 4707-4717.

Nielsen, N. and K. Ingvartsen, 2004. Propylene glycol for dairy cows. A rewiew of the metabolism of propylene glycol and its effects on physiological parameters, feed intake, milk production and risk of ketosis. Anim. Feed Sci. Technol. 115: 191-213.

Rémond, B., E. Souday and J.P. Jouany, 1993. *In vitro* and *in vivo* fermentation of glycerol by rumen microbes. Anim. Feed Sci. Technol. 41: 121-132.

Wiedmeier, R., B. Tanner, J. Bair, H. Shenton, M. Arambel and J. Walters, 1992. Effects of a new molasses byproduct. Concentrated separator byproduct, on nutrient digestibility and ruminal fermentation in cattle. J. Dairy Sci. 70: 1936-1940.

Effect of sodium butyrate feed additive in milk replacer and/or starter mixture on mRNA expression of IGF-1, IGF-2 and ghrelin in GIT of neonatal calves

J. Flaga[1], P. Górka[1], Z.M. Kowalski[1], U. Kaczor[2], A. Grzegorzewska[3], M. Jaworski[1], P. Pietrzak[4], A. Kotunia[5] and R. Zabielski[4]

[1]Department of Animal Nutrition, Agriculture University in Krakow, Cracow, Poland; [2]Department of Sheep and Goat Breeding, Agriculture University in Krakow, Cracow, Poland; [3]Department of Animal Physiology, Agriculture University in Krakow, Cracow, Poland; [4]Department of Physiological Sciences, Faculty of Veterinary Medicine, Warsaw University of Life Sciences, Warsaw, Poland; [5]The Kielanowski Institute of Animal Physiology and Nutrition, Polish Academy of Science, Jablonna, Poland; rzkowals@cyf-kr.edu.pl

Introduction

The stimulatory effect of butyric acid on gastrointestinal tract (GIT) development in neonatal calves is quite well documented. Its sodium salt supplementation (SB) in calf milk replacer positively affects villus height, brush border enzyme activity, pancreatic exocrine secretion which in consequence results in better digestion efficiency and enhanced animal growth (Guilloteau *et al.*, 2004 and 2009). Sodium butyrate also shows a stimulatory effect on rumen mucosa development (Mentschel *et al.*, 2001). However, the controlling mechanisms are unclear. Such effects may be mediated by the action of insulin-like growth factor-1 (IGF-1), insulin-like growth factor-2 (IGF-2) and ghrelin which belong to the key regulatory peptides controlling GIT development in newborn mammals. Additionally, the best way of SB supplementation in preruminant calves seems to be controversial. Therefore the aim of present study was to determine the effect of SB supplementation of a milk replacer and/or starter mixture on mRNA expression of IGF-1, IGF-2, ghrelin in GIT of neonatal calves.

Material and methods

Twenty-four bull calves (5-day-old) were randomly allocated to 4 groups: (1) control (milk replacer and starter mixture without SB), (2) milk replacer + SB, (3) starter mixture + SB, and (4) milk replacer + SB and starter mixture + SB. Animals were slaughtered at 26 days of their life, and tissue samples (ruminal, abomasal, duodenal and jejunal mucosa, and liver) were taken for analyses. Expression of mRNA of IGF-1, IGF-2 and ghrelin was evaluated using the semi-quantitative RT-PCR method. The β-actin expression was taken as a housekeeping gene. The PCR products were run in a 2.0% agarose gel and stained with ethidium bromide for visualisation and the density of the gel band was determined using the Scion Image for Windows (Scion Corporation, Maryland, USA). The effect of SB was analysed by two way analysis of variance using PROC GLM of SAS® (8.01). The results were considered significant at $P<0.05$ and as a tendency at $P<0.15$.

Results

Expression of IGF-1 and ghrelin mRNA was different at different sites of GIT and well marked in all GIT segments. The highest level of IGF-1 mRNA expression was observed in the duodenum and jejunum, whereas ghrelin mRNA expression was the highest in the abomasum and duodenum. Expression of IGF-2 mRNA was more or less equal in all tissues. Supplementation of milk replacer with SB induced IGF-2 mRNA expression in the jejunum ($P=0.04$), and tended to stimulate ghrelin mRNA expression in the abomasum ($P=0.07$) and cranial ventral sac of the rumen ($P=0.13$). On the contrary, addition of SB to the milk replacer decreased the level of ghrelin transcript in the cranial dorsal sac of the rumen (*P*=0.01) and tended to reduce IGF-1 mRNA expression in the cranial ventral sac of the rumen (*P*=0.06). The addition of SB in the starter mixture had no significant effect on mRNA expression of the peptides examined in the stomach and small intestine walls. However, this way of its supplementation tended to reduce IGF-1 expression in the liver (*P*=0.11). An interaction between the effects of starter and milk replacer supplementation on the IGF-1 transcript amount in abomasum (*P*=0.09) and ghrelin mRNA level in the cranial dorsal sac (*P*=0.11) and duodenum (*P*=0.06) was observed.

Table 1. Summary of results.

	Milk replacer + SB			Starter mixture + SB		
	IGF-1	IGF-2	Ghrelin	IGF-1	IGF-2	Ghrelin
Rumen (dorsal sac)	NS	ns	↓↓	ns	ns	ns
Rumen (ventral sac)	↓	ns	↑	ns	ns	ns
Abomasum	ns	ns	↑	ns	ns	ns
Duodenum	ns	ns	ns	ns	ns	ns
Jejunum	ns	↑↑	ns	ns	ns	ns
Liver	ns	ns	ns	↓	ns	ns

NS = not significant; ↑↑ = effect in plus, *P*<0.05; ↑ = effect in plus, *P*<0.15; ↓↓ = effect in minus, *P*<0.05; ↓ = effect in minus, *P*<0.15

Discussion and conclusion

In conclusion, this study proved that expression of IGF-1, IGF-2 and ghrelin is site-dependent and differs in different parts of the digestive system. Moreover, milk replacer and/or starter supplementation with SB may affect both positively and negatively, on the mRNA expression of growth factors. It seems that SB supplementation of a milk replacer exerts a more pronounced effect on GIT development, particularly on the abomasum and small intestine, which may be controlled by IGF-2 and ghrelin. More studies are needed to fully understand the effect of SB on GIT development including other regulatory peptides.

Acknowledgement

This investigation was supported by Ministry of Science and Higher Education, Poland (Grant no. 1164/P01/2006/31).

References

Guilloteau, P., V. Romé, L. Le Normand, G. Savary and R. Zabielski, 2004. Is Na-butyrate a growth factor in the preruminant calf? Preliminary results. J. Anim. Feed Sci., 13, Suppl. 1: 393–396.

Guilloteau, P., R. Zabielski, J.C. David, J.W. Blum, J.A. Morisset, M. Biernat, J. Woliński, D. Laubitz and Y. Hamon, 2009. Sodium-butyrate as a growth promoter in milk replacer formula for young calves. J. Dairy Sci. 92: 1038-1049.

Mentschel, J., R. Leiser, C. Mülling, C. Pfarrer and R. Claus, 2001. Butyric acid stimulates rumen mucosa development in the calf mainly by a reduction of apoptosis. Arch. Tierernahr. 55: 85-102.

Activities of microbial fibrolytic enzymes in ten herbivore microbial ecosystems

F.N. Fon and I.V. Nsahlai
School of Agricultural Sciences and Agribusiness, University of KwaZulu-Natal, P/Bag X01, 3209 Scottsville, South Africa; 203514166@ukzn.ac.za

Introduction

Herbivores lack endogenous enzymes for hydrolysing plants cell wall polysaccharides, and therefore depend on a symbiotic relationship with microbes (Lee *et al.*, 2002). Sustainable microbial population is highly dependent on the availability and type of forages. Fibrolytic enzymes are made up of polysaccharidases playing a key role in the hydrolyse of cellulose (exo- and endo-cellulase) and hemicellulose (xylanase) into short chain oligosaccharides. These last ones are then hydrolysed by oligosaccharidases into oses used by microbes for energy generation yielding by-products such as volatile fatty acids which are metabolic substrates in herbivores. Microbial ecosystems were selected based on the fact that the competition for different substrate utilisation especially in the wild might have influenced the evolution of their fibrolytic competence. This study will quantify the microbial cellulase and hemicellulase enzyme activities from non-fractionated ecosystems. Secondly, enzyme efficiencies will be calculated from their kinetic parameters.

Material and methods

The cow, elephant, camel, sheep and horses were fed entirely on hay. The zebra, wildebeest, impala, buffalo and giraffe were grazing on a dry land in an open field where kikuyu hay and other fibres were dominant at the Tala Game Reserve, Umbumbulu. The buffalo often migrated to the valleys where they consumed standing hay while the giraffe browsed on a variety of tree leaves. Samples of each ecosystem were collected from at least two animals of each species (no sex preference) and pooled to constitute a single sample. Rumen digesta (200 ml) was collected (cow, sheep), strained through four layers of cheese cloth and treated with 0.1 mM phyenylmethylsulfonyl fluoride. Faeces were collected in situ from the non-fistulated herbivores (horse, camel, elephant, zebra, wildebeest, giraffe and buffalo). Crude protein was precipitated with ammonium sulfate after sonification and the specific activities of xylanase and cellulases (exo- and endo-cellulase) were determined (Gerrit *et al.*, 1984). Bradford assay and dinitrosalicylic acid was used to determine protein concentration and reducing sugars respectively. All enzyme assays, exo- (1% (m/v) crystalline cellulose), endo-cellulase (0.5% (m/v) carboxymethylcellulose) and xylanase (0.1% (m/v) xylan) were done at 39 °C and pH 6.2 (optimised), however the incubation periods was 1 h for xylanase, 2 h for endo-, and 48 h for exo-cellulase. Each ecosystem was represented by three samples, each of which was analysed in triplicate. Specific activity was defined as μg of xylose or glucose/mg crude protein/min. The kinetic parameters of fibrolytic enzymes were calculated as described by Eisenthal and Cornish-Bowden (1974) using the HYPER software programme. Enzyme catalytic rate (Kcat) is equal to Vmax at constant enzyme concentration. The ratio Kcat/Km is a measure of the catalytic efficiency of an enzyme-substrate pair. SAS® software (2001) was used to perform statistical analysis. The model included the effect of the ecosystem (E), season effect within feed type (S), the interaction ExS and the residual error.

Results

Xylanase and cellulass specific activities differed ($P<0.0001$) among the sampled herbivore microbial ecosystems (Table 1). Xylanase was the highest for the horse and zebra and intermediate for the elephant, cow and sheep. The horse and zebra recorded the highest exo-cellulase activities while the wildebeest and impala had intermediate activities. The horse, zebra and elephant had the highest endo-cellulase activities while those of the cow, wildebeest, impala and buffalo were intermediate (Table 1). It was noted that exo- and endo-cellulase activities were strongly correlated ($r = 0.90$; $P<0.0001$). The sheep and horse had the highest exo-cellulase catalytic efficiencies while the horse and wildebeest showed the highest endo-cellulase catalytic efficiencies. The wildebeest showed the highest efficiency in hydrolysing xylan.

Table 1. Specific activity and efficiency of fibrolytic enzymes of 10 herbivore microbial ecosystems.

Enzyme source	Xylanase specific activity[1]	Exocellulase specific activity[2]	Endocellulase specific activity[2]	Xylanase efficiency[3]	Exocelluase efficiency [3]	Endocellulase efficiency[3]
Cow[R]	38.05	1.83	2.93	67.77	105.40	16.03
Sheep[R]	35.96	1.49	1.77	49.98	848.06	11.81
Horse[f]	46.38	5.39	6.47	155.20	525.83	80.50
Camel[f]	10.13	1.58	1.63	9.07	9.75	11.43
Elephant[f]	34.48	1.96	5.49	244.26	2.98	9.02
Zebra[f]	43.59	4.39	5.97	28.99	ND	1.23
Wildebest[f]	8.00	3.15	3.72	317.50	9.82	40.64
Giraffe[f]	7.73	1.65	1.80	0.93	ND	5.43
Impala[f]	10.62	2.28	2.62	1.66	131.55	3.46
Buffalo[f]	7.64	1.74	2.83	3.68	ND	ND
SED	0.38	0.14	0.15	-	-	-
P<	0.0001	0.0001	0.0001	-	-	-

[R] rumen fluid; [f] feces; SED = standard error deviation; ND = not determined; [1,2] enzyme specific activity = µg xylose (xylanase) and glucose (cellulases) released/mg crude protein/min. [3] enzyme efficiency = ml/mg crude protein/min.

Conclusion

The 10 herbivore microbial ecosystems were classified into two groups based on their fibrolytic (cellulase and hemicellulase) specific activities and catalytic efficiencies. Group A with high enzyme activities comprised of horse, zebra, wildebeest, impala and elephant. Group B with intermediate activities comprised the cow, sheep, giraffe, camel and buffalo.

Acknowledgement

The authors would like to thank NFR for funding the research and Tala Game Reserve.

References

Eisenthal, R. and A. Cornish-Bowden, 1974. The direct linear plot. A new graphical procedure for estimating enzyme kinetics parameters. Biochem. J. 139: 715-720.

Gerrit, B., F. Marjo, M. Frank and G. Fons, 1984. The cellulase of *Trichoderma viride*. Purification characterization and comparison of all detectable endoglucanases, exoglucanases and β-glucosidases. Eur. J. Biochem. 146: 301-308.

Lee, R., J. Paul, H. Willem and S. Isak, 2002. Microbial cellulose utilization: fundamentals and biotechnology. Microbiol. Mol. Biol. Rev. 66: 506-577.

Retinol-Binding-Protein 4 (RBP4) and abomasal displacement (DA)

M. Fürll[1], B. Fürll[1], L. Locher[1] and J. Raila[2]
[1]Faculty of Veterinary Medicine, University of Leipzig, An den Tierkliniken 11, 04103 Leipzig, Germany; [2]Institute for Nutrition Sciences, University of Potsdam, 14558 Nuthetal, Germany; mfuerll@rz.uni-leipzig.de

Introduction

DA is regarded as one of the most common diseases in cattle. The development of a left sided (lDA) or right sided (rDA) DA is caused by various factors (Constable *et al.*, 1992; Fürll, 2001). The lDA results from disturbances in the energy metabolism (free fatty acids [FFA], bilirubin, ß-0H-Butyrate [BHB]) starting during the dry period and increasing *post partum* (p.p.) (Fürll *et al.*, 1999, Le Blanc *et al.*, 2005). lDA's association with fat mobilisation syndrome is likely, however lDA is not always shown through body condition scoring (BCS) or back fat thickness (BFT), indicators of subcutaneous fat. Retinol-Binding-Protein 4 (RBP4) is an indicator for visceral fat as well as insulin resistance in humans (Klöting *et al.*, 2007).
The aim of this study was therefore to check for signs indicating an irregular distribution of subcutaneous (BFT) and visceral fat (RBP4) *ante partum* (a.p.) and p.p. in cows with lDA during early lactation compared to healthy cows.

Material and methods

In a dairy farm, 969 cows and heifers were examined clinically, including BFT, and biochemically (FFA, bilirubin, BHB) in blood serum 4 weeks (W) a.p., 1-2 W a.p., 3 days (d) p.p. as well as 4 W p.p.. Out of those cows, 44 healthy and 11 cows with lDA were selected and comprehensively analysed biochemically (Hitachi 912). RBP4 was determined by means of Western-Blot analyses after isolation in 12% SDS-PAGE. We checked the data using the Wilcoxon-Mann-Whitney-test and observed significant differences between simultaneous samples from lDA and healthy cows, shown in Figure 1 and Table 1 with $P<0.05$.

Results

The main results of the energy metabolism are shown in Table 1 and Figure 1.

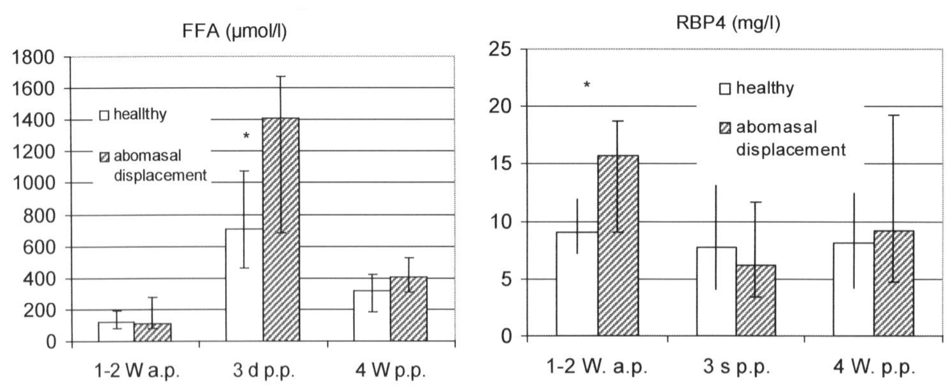

Figure 1. FFA and RBP4 concentrations in healthy cows and cows with abomasal displacement (quartiles).

Table 1. BFT, bilirubin and BHB concentrations in healthy cows and cows with abomasal displacement (2nd, 1st, 3rd quartiles).

		4 W a.p.	1-2 W a.p.	3 d p.p.	4-5 W p.p.
BFT, mm	healthy	**20**-18-25	**21**-19-27	**21**-19-26	**12**-10-17
	IDA	**20**-17-26	**20**-18-26	**19**-17-27	**11**-10-15
Bilirubin,	healthy	**1.8**-1.2-2.4	**1.6**-1.2-3.0	**5.3**[a]-3.9-7.6	**3.7**-2.6-5.0
μmol/l	IDA	**1.3**-1.3-1.7	**1.9**-1.4-3.2	**12.4**[a]-5.6-13.9	**4.7**-3.8-7.4
BHB, mmol/l	healthy	**0.51**-0.43-0.61	**0.46**-0.37-0.57	**0.84**[a]-0.66-0.97	**0.72**-0.61-0.97
	IDA	**0.51**-0.49-0.64	**0.49**-0.42-0.59	**1.02**[a]-0.79-2.05	**0.69**-0.49-0.89

[a] $P \leq 0.05$ in simultaneous samples.

Comparing BFT as an indicator for peripheral fat in both groups, no significant differences were found. The RBP4 concentration representing visceral fat is significantly higher a.p. ($P<0.05$) (Figure 1) in the IDA group. In IDA cows, the concentrations of FFA, bilirubin and BHB increased significantly 3 d p.p. as signs of disturbed energy metabolism.

Discussion and conclusion

Because of studies in cattle in which RBP4 is also reflecting the amount of visceral fat (Locher *et al.*, 2009), the results indicate that cows with IDA p.p. have stored significantly higher visceral fat before calving. Because visceral fat also produces cytokines, the motility of the abomasum can be negatively influenced.

Cows with abomasal displacement have around parturition a disturbed energy metabolism e.g. FFA, bilirubin, BHB. Compared with healthy cows the BFT (subcutaneous fat) does not differ in cows with IDA ($P>0.05$). Increased RBP4 concentrations, being an indicator for visceral fat as well as insulin resistance, before parturition indicate that cows with left side abomasal displacement have stored significantly more visceral fat a.p. than healthy cows and this seems to be involved in the development of insulin resistance with stimulated lipolysis in cattle too.

References

Constable, P.D, G.Y. Miller, G.F. Hoffsis, B.I. Hull and D.M. Rings, 1992. Risk factors for abomasal volvolus and left abomasal displacement in cattle. Am. J. Vet. Res. 53: 1184-1192.

Fürll, M., 2001. Dislocatio abomasi (DA) beim erwachsenen Rind. In: Wiesner E. (Ed.), Handlexikon der Tierärztlichen Praxis, Enke Verlag Stuttgart (GE), 181 xa-181 xw.

Fürll, M., M.N. Dabbagh and L. Jäkel, 1999. Körperkondition und Dislocatio abomasi (DA). Vergleichende Untersuchungen zum Verhalten der Rückenfettdicke sowie weiterer Kriterien bei Rindern. Dtsch. Tierärztl. Wschr. 106: 5-9.

Klöting, N., T.E. Graham, J. Berndt, S. Kralisch, P. Kovacs, C.J. Wason, M. Fasshauer, M.R. Schön, M. Stumvoll, M. Blüher and B.B. Kahn, 2007. Serum retinol-binding protein is more highly expressed in visceral than in subcutaneous adipose tissue and is a marker of intra-abdominal fat mass. Cell. Metab. 6: 79-87.

Le Blanc, S.J., K.E. Leslie and T.F. Duffield, 2005. Metabolic predictors of displaced abomasum in dairy cattle. J. Dairy Sci. 88: 159-170.

Locher, L., L. Zapfe, M. Kern, N. Klöting, M. Blüher, J. Raila and M. Fürll, 2009. Expression of BP4-mRNA in adipose tissue and RBP4 in serum of healthy dairy cows. In: Chilliard, Y., F. Glasser, Y. Faulconnier, F. Bocquier, I. Veissier and M. Doreau (Eds.), Ruminant Physiology. Digestion, metabolism and effects of nutrition on reproduction and welfare. Wageningen Academic Publishers, Wageningen, the Netherlands. p. 580-581.

Inoculants for low-dry matter corn crop ensilage: an ongoing question

A. Ghaempour[1], G.R. Ghorbani[1], M. Khorvash[1] and A. Nikkhah[1,2]
[1]Department of Animal Science, Isfahan University of Technology, Isfahan 84156, Iran; [2]Department of Animal Sciences, College of Agriculture, Zanjan University, Zanjan, Iran; anikkha@yahoo.com

Introduction

In many regions including central Iran, optimum DM content of corn crop before ensilage cannot be ensured if corn is planted as the second crop in the summer. Corn crop may thus be mostly ensiled with <25% DM (Ranjbari *et al.*, 2007). Corn silage makes up approximately half of the dietary forage in most commercial dairy farms in Iran (Kowsar *et al.*, 2008). The common ensilage of low-DM corn crops in bunker silos emphasises the need for effective management strategies to maintain high quality preservation (McDonald *et al.*, 1991). Under such climate and farming conditions, large dairy holders need economical justifications if applying microbial inoculants to low DM corn crop, which can improve silage quality and cattle performance. The objective was to determine the effects of microbial inoculants during low DM corn crop ensilage on silage quality, rumen fermentation, and milk production.

Material and methods

The corn plant was harvested at the milk stage of maturity with 204 g DM per kg of fresh crop, cut for a theoretical particle length of 2 cm, filled in four 60-tone bunker silos, and treated with (1) no inoculants (control), (2) inoculant A (Ecosyl®, Ecosyl Products Ltd. Stokesley, UK) containing *Lactobacillus plantarum*, (3) inoculant B (Biotal®, Lallemand Animal Nutrition, UK) containing *Pediococcus pentosanus*, *Lactobacillus plantarum* and *Propionibacter freudenreichii* or (4) the mixture of inoculants A and B. Eight multiparous lactating Holstein cows at 100±20 days in milk were used in a 4×4 Latin square design with four 20 d periods including 14 d of adaptation and 6 d of sampling. Dietary treatments were TMR containing corn silages with or without inoculants as above. The basal diet contained 329 g corn silage, 143 g alfalfa hay, and 528 g concentrate per kg DM. Rumen fluid was sampled by rumenocentesis at 4 h post-feeding. Acid detergent insoluble ash was measured in composited feed and feces and used as an internal marker to calculate apparent total tract nutrient digestibility.

Results

Inoculants did not affect (P>0.10) silage pH and content of DM, CP, ammonia, lactate, acetate, ash and total VFA. Applying Biotal to the corn crop resulted in greater (P<0.05) silage water soluble carbohydrates (WSC, 47.4 vs. 29.8 g/kg) and lower NDF (494.1 vs. 464.0 g/kg). The mixed inoculants increased (P<0.05) silage butyrate percentage relative to the other treatments. The mixture of Ecosyle and Biotal moderately decreased rumen pH and increased rumen VFA concentrations (P<0.05, Table 1). Dry matter intake increased when corn crop was ensiled with Ecosyle or Ecosyle + Biotal, but inoculants did not significantly affect rumen propionate and branched-chain fatty acids, total tract nutrient digestibility and milk production (P>0.10, Table 1).

Discussion

The absence of treatment effects on silage pH and content of DM, CP, ammonia, and organic acids agrees with earlier studies using high-DM crops (Kung *et al.*, 1993; Merry *et al.*, 1995), suggesting that microbial inoculants did not alter the fermentation products of the low-DM corn crop. The relatively greater rumen VFA concentration in cows on inoculant-treated silage was consistent with the lower pH compared to other groups. These changes could in part be related to altered post-feeding patterns in DMI and rumen kinetics of digestion and passage among treatments. Due to reduced ECM by some inoculants, feed efficiency decreased. The high milk fat and protein were suggestive of healthy rumen conditions.

Table 1. Milk production and composition and rumen fermentation.

Item	Corn silages with or without microbial additives[1]				SE	*P*-value
	Control	Ecosyle (E)	Biotal (B)	E + B		
DMI, kg/d	21.5[b]	22.8[a]	22.4[ab]	21.7[b]	0.5	0.02
4% Fat corrected milk yield, kg/d	33.7	32.8	32.3	33.6	0.7	0.07
Energy corrected milk yield, kg/d	35.6[ab]	34.8[ab]	34.2[b]	35.7[a]	0.7	0.05
Milk fat, g/kg	40.2	39.9	39.0	39.7	1.1	0.77
Fat yield, kg/d	1.35[a]	1.31[ab]	1.27[b]	1.33[ab]	0.03	0.04
Milk protein, g/kg	30.5	30.8	30.6	30.8	0.30	0.52
Protein yield, kg/d	1.02	1.01	1.02	1.04	0.02	0.59
ECM/DMI	1.66[a]	1.52[b]	1.54[b]	1.65[a]	0.04	0.01
Rumen pH	6.04[ab]	6.24[a]	5.94[b]	5.85[b]	0.15	0.08
Total VFA, mmol/l	104.6[b]	108.3[b]	111.7[ab]	123.7[a]	4.7	0.08
Acetate, mmol/l	61.1[b]	61.8[b]	68.8[a]	72.3[a]	2.2	0.007
Propionate, mmol/l	22.8	22.9	22.9	27.5	1.8	0.10
Total tract DM digestibility,%	65.3	69.2	66.7	66.9	2.3	0.63

[1] Control = no inoculant, E = Ecosyl (Ecosyl®, Products Ltd. Stokesley, UK) containing *Lactobacillus plantarum*, B = Biotal (Biotal®, Lallemand Animal Nutrition, UK) containing *Pediococcus pentosanus*, *Lactobacillus plantarum* and *Propionibacter freudenreichii*. [a,b]Means with different superscripts differ statistically within each row.

Conclusion

Applying microbial inoculants to low-DM corn crop (i.e. 20% DM) increased silage WSC and decreased its NDF but did not affect other biochemical properties. Overall, despite some effects on DMI, rumen pH and VFA, inoculants did not improve nutrient dgestibility, and lactation performance of midlactation cows.

References

Kowsar, R., G.R. Ghorbani, M. Alikhani, M. Khorvash and A. Nikkhah, 2008. Corn silage partially replacing short alfalfa hay to optimize forage use in total mixed rations for lactating cows. J. Dairy Sci. 91: 4755-4764.

Kung, L, J.R. Chen, J.H. Kreck and E.M. Knutsen, 1993. Effect of microbial inoculants on the nutritive value of corn silage for lactating dairy cows. J. Dairy Sci. 76: 3763-3770.

McDonald, P., A.R. Henderson and S.J.E. Hesron, 1991. The biochemistry of silage. 2nd ed. Marlow, UK, Chalcombe Publications.

Merry, R.J., A.L. Winters, P.L. Thomas, M. Muller and T. Muller, 1995. Degradation of fructans by epiphytic and inoculated lactic acid bacteria and plant enzymes during in silage of normal and sterile hybrid ryegrass. J. Appl. Bacteriol. 79: 583-591.

Ranjbari, M.,G.R. Ghorbani, A. Nikkhah, and M. Khorvash, 2007. Chemical composition, crude protein fractions, and ruminal degradation of maize silage produced in Isfahan. Int. J. Dairy Sci. 2: 66-72.

Use of chitosans to modulate digestion and ruminal fermentation in sheep

I. Goiri, L.M. Oregui and A. Garcia-Rodriguez
Neiker-Tecnalia, Animal production, Granja Modelo Arkaute, E-01080 Vitoria-Gasteiz, Spain; igoiri@neiker.net

Introduction

As a non-toxic, biodegradable carbohydrate polymer, chitosan has received much attention for its potential applications in medicine and food preservation, due to its antimicrobial properties (Cuero, 1999) against bacteria, moulds, and yeast. Chitosan may provide an alternative to antimicrobial growth promoters in ruminant diets as suggested by Goiri *et al.* (2009a,b). The benefits observed *in vitro* seem to be caused by changes in ruminal fermentation, in particular by increased propionate and decreased methane productions, which in turn lead to energetically more efficient fermentation patterns, and increased diet protein escaping rumen degradation (Goiri *et al.*, 2009b). Nevertheless, these authors, reported negative effects *in vitro* such as decreased dry matter (DM) digestibility, especially in fibre-rich mixtures (Goiri *et al.*, 2009b). However, at present only *in vitro* research has been conducted, and there is no information about the effects of this additive *in vivo*. Therefore the objective of this study was to evaluate the effects of chitosan on rumen degradation and rumen fermentation end products, and on apparent digestibility in sheep.

Material and methods

In a first trial, four ruminally fistulated dry sheep were used. Animals were fed and alfalfa hay-concentrate diet formulated to meet 1.2 their calculated maintenance energy requirements. Sheep were assigned to two treatments (no additive: CTR; 136 mg of chitosan/kg body weight: CHI) in a 2×2 AB/BA 19 day-crossover design. After an adaptation period, an *in sacco* disappearance trial was performed, followed by sampling of ruminal fluid for volatile fatty acids (VFA) and ammonia-N (NH_3) determination. In a second trial, four sheep fed the same diet were used. Sheep were assigned to the same two treatments in a 2×2 AB/BA 14 day-crossover design. Apparent total tract digestibility and sheep nitrogen balance was studied. *In sacco* disappearance data were adjusted to the model:
$p = a + b \times (1 - e^{-ct})$
where p is the disappearance (%) at time t (h), a is an intercept representing the fraction of the constituent that is rapidly degradable (%), b is the fraction of the constituent that is slowly degradable (%), and c is the fractional degradation rate of disappearance of fraction b in the rumen (%/h). Data from the cage-digestibility trial and *in sacco* disappearance experience were analysed using the MIXED procedure of SAS® (2002) according to the following statistical model:
$Y_{ijk} = \mu + S_i + T_j + P_k + \varepsilon_{ijk}$
in which Y_{ijk} represents the value of each individual observation, μ the average, S_i the effect of the ith sequence of treatments (i= AB, BA), T_j the effect of the jth treatment (j = CTR, CHI), P_k the effect of the kth period (k= 1, 2) of the crossover, and ε_{ijk} the residual error. The same model was used for the statistical analysis of ruminal fermentation characteristics (VFA, pH, NH_3), using the MIXED procedure of SAS® (2002) for repeated measures with random (sheep) and repeated (time) statements.

Results

In the analysis of the results obtained with the two cross-over designs, no sequence or carry-over effect was found. Therefore, data from the two periods of the cross-over design were used in the analysis. CHI did not affect diet OM, CP or NDF rapidly (a) and slowly (b) degradable fractions.

CHI decreased NDF degradation rate (c) (9.0 vs. 6.0%/h; P=0.07), but it did not affect OM or CP degradation rate. No differences were observed between CHI and CTR in total VFA concentration or acetate proportion. However, CHI increased propionate proportion (P<0.05) and propionate-to-acetate ratio (P<0.05) and decreased branched-chain VFA (P<0.05) proportion and rumen NH_3 concentration (P<0.05) (Table 1).
Apparent digestibilities of OM, CP and EE averaged, 0.73, 0.76, and 0.78 g/kg, respectively, and were not influenced by treatment, but apparent NDF digestibility was decreased (0.55 vs. 0.47 g/kg; P=0.09) in CHI treated sheep. Outputs of nitrogen in faeces were similar in both treatments, but CHI treatment decreased (2.8 vs. 2.3 g/d; P=0.08) nitrogen output in urine. Retention of nitrogen averaged 20 g/d and was not affected by CHI additive.

Table 1. Effects of chitosan on fermentation characteristics in the ruminal fluid of sheep.

	Treatment[1]		s.e.m.
	CTR	CHI	
Intake, g DM/d	1,134	1,140	18
pH	7.05	6.99	0.08
NH_3, mg/100ml	41.7	35.2	2.72
Total VFA, mmol/L	54.3	62.2	4.52
Individual, mmol/100mmol			
Acetate	70.5	67.6	1.48
Propionate	12.1	14.7[a]	0.67
Butyrate	11.3	12.5	0.93
BCFVA	4.9	3.9[a]	0.33
C3:C2	0.17	0.22[a]	0.02

[1] CTR: control treatment; CHI: chitosan treatment; DM: dry matter; VFA: volatile fatty acids; BCVFA: branched-chain volatile fatty acids (includes isobutyrate and isovalerate); C3:C2: propionate-to-acetate ratio.
s.e.m.: standard error of the mean;
[a] Significant at P<0.05.

Conclusion

Although chitosan reduced NDF apparent digestibility, chitosan shifted ruminal fermentation towards energetically more efficient routes without reducing total VFA production. Moreover, chitosan led to a lower rumen NH_3 concentration and lower urinary nitrogen losses.

References

Cuero, R.G., 1999. Antimicrobial action of exogenous chitosan. In: P. Jolles and R.A.A. Muzarelli (Eds.) Chitin and chitinases. Birkhauser-Verlag Publishing Co, Basel, Switzerland, pp. 315-333.

Goiri, I., A. Garcia-Rodriguez and L.M. Oregui, 2009a. Effect of chitosans on *in vitro* rumen digestion and fermentation of maize silage. Anim. Feed Sci. Technol. 148: 276-287.

Goiri, I., L.M. Oregui and A. Garcia-Rodriguez, 2009b. Dose–response effects of chitosans on *in vitro* rumen digestion and fermentation of mixtures differing in forage-to-concentrate ratios. Anim. Feed Sci. Technol. 151: 215-227.

Efficacy of the combined use of acids and heat to protect protein from sunflower meal against rumen degradation: 2. Feed amino acid supply

J. González[1], J.M. Arroyo[1], M. Ouarti[1] and C. Centeno[2]
[1]Dpto. de Producción Animal, Universidad Politécnica de Madrid, 28040 Madrid, Spain; [2]Instituto de Nutrición y Bromatología, CSIC, 28040 Madrid, Spain

Introduction

To improve protein efficiency in high productive ruminants implies maximising the amino acid (AA) rumen outflow. This objective may be attained by protecting proteins against rumen degradation, in special those of high degradable feeds as sunflower meal (SFM), once the supply of degradable protein to rumen micro-organisms is assured. The digestion may alter the profile of intestinal digested AA from the feed, which may also be affected by this protection. A two fold increase of the SFM metabolisable protein was shown in the associated communication (Arroyo and González, 2009) using a combined treatment with malic or phosphoric acids and heat. In this study, we evaluated the feed content of total and lysine and S-containing AA digested in the intestine as well as the changes on the AA profile with the digestive process.

Material and methods

A SFM 35 was treated with 2 N solutions (400 ml/kg) of malic acid or ortho-phosphoric acid and dried for 6 h at 150 °C in an oven. A 3×3 Latin-square design was carried out using 3 rumen and duodenum cannulated wethers and 3 diets, which include the SFM untreated (UT) or treated with malic (MT) or phosphoric (PT) acids. Rumen studies included 2 incubations on different days and the generation of representative samples of the undegraded feed DM outflow from the rumen. These samples were obtained by pooling the resulting residues in proportions pre-determined in accordance with the flow of undegraded particles defined taking into account the rates of comminution (k_c) and outflow (k_p) of particles from the rumen. The analysis for AA of these composited samples allowed determining their undegraded fraction whereas their intestinal effective digestibility (IED) was determined after incubation in mobile bags. The results were corrected for the microbial contamination taking place in the rumen using ^{15}N infusion techniques. Additional information on diets and procedures was shown in the associated communication (Arroyo and González, in press). Analyses of AA were performed after acid hydrolysis (preceded by oxidation with formic acid and hydrogen peroxide for methionine and cystine), using norleucine as an internal standard. Therefore, the total AA did not include tryptophan. Variance analyses were performed considering animals as blocks. Changes in the AA profiles vs. the original protein were studied by a *t* test of the differences.

Results and discussion

The protective treatments increased the undegraded fraction of total AA by 3.5 times and also its IED, although higher IED values ($P<0.05$) were shown in MT than in PT (Table 1). A similar behaviour was shown for the undegraded fractions of lysine, methionine and cystine. These increases were higher in PT than in MT but this difference was only significant ($P<0.05$) for methionine. The IED of these AA was higher in general in treated meals, although the methionine value in PT did not reach significance ($P>0.05$) in relation with UT. Also, no differences between meals for IED were shown for cystine, although they showed a large variability (Table 1). A lower degradability reduces the concentration of indigestible compounds in the undegraded protein and justifies the increased IED values shown in the treated meals. The ruminal degradation as well as the intestinal digestion altered the essential AA profile (plus cystine) of the SFM (Figure 1). In UT changes with both process ($P<0.05$) were shown for phenylalanine, which decreased, and for histidine, lysine and threonine, which increased. Hull proteins should constitute a large proportion of UT undegraded proteins due to the high degradability of this meal. Hull proteins have respectively a very lower or higher concentration of these AA than SFM proteins (Novus, 1992). In general, the protective treatments mitigated these variations and changed the sign of the lysine variation in relation to the original protein profile ($P<0.05$). Also, a significant change ($P<0.05$) was shown for histidine in PT in both undegraded and intestinal digested proteins.

Table 1. Effects of treatments of sunflower meal on the ruminal undegraded fraction and its intestinal effective digestibility (IED) of total amino acids, lysine, methionine and cystine.

Item	Undegraded fraction (%)					IED (%)				
	UT[1]	MT	PT	SEM	P	UT	MT	PT	SEM	P
TAA[2]	17.3[b]	58.8[a]	62.7[a]	2.92	<0.001	81.3[c]	89.2[a]	85.7[b]	0.56	0.002
Lys	20.9[b]	58.5[a]	63.3[a]	3.37	0.001	81.9[b]	87.7[a]	85.1[ab]	1.08	0.048
Met	17.5[c]	52.0[b]	69.0[a]	3.28	0.002	79.9[b]	86.7[a]	85.9[a]	1.67	0.003
Cys	21.8[b]	61.0[a]	65.3[a]	7.26	0.019	67.2	73.9	68.1	8.48	0.840

[1] UT = untreated; MT = malic acid treated; PT = phosphoric acid treated.
[2] TAA = total amino acids (except tryptophan).
[a, b, c] means within rows with different superscript are different at $P<0.05$.

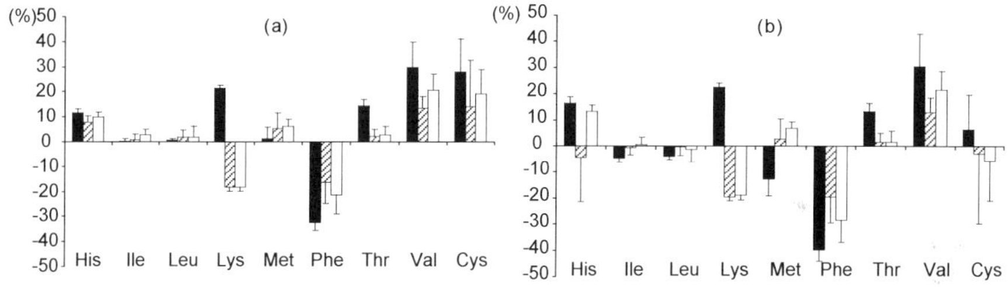

Figure 1. Changes (% in relation to the intact protein) of the essential amino acid (plus cystine) profile of the undegraded (a) and intestinal digested (b) protein of untreated ■, malic acid treated ▨ and phosphoric acid treated □ sunflower meal. Bars are s.e.d.

Conclusion

The combined treatments with acids and heat allowed a large increase of the supply from SFM of total digestible AA as well as of lysine and S-containing AA. The digestion altered the digested AA profile, but these treatments mitigated these variations.

Acknowledgement

This work was supported by the CICYT funded Project AGL 2006-08300.

References

Arroyo, J.M. and J. González, 2009. Efficacy of the combined use of acids and heat to protect protein from sunflower meal against rumen degradation: 1. Metabolizable protein supply. In: Chilliard, Y., F. Glasser, Y. Faulconnier, F. Bocquier, I. Veissier and M. Doreau (Eds.), Ruminant Physiology. Digestion, metabolism and effects of nutrition on reproduction and welfare. Wageningen Academic Publishers, Wageningen, the Netherlands. p. 116-117.

Novus, 1994. Raw material compendium. Novus International, INC.

Two hour chamber measurement provides a useful estimate of daily methane production in sheep

J.P. Goopy, R.S. Hegarty and D.L. Robinson
New South Wales Department of Primary Industries, Beef Industry Centre, Armidale, New South Wales 2351, Australia; john.goopy@dpi.nsw.gov.au

Introduction

In Australia and New Zealand, the significance of enteric methane (CH_4) production is greater than in many industrialised countries, due to large livestock numbers (AGO, 2005). Consequently, there is considerable interest in both countries in breeding livestock which produce relatively less CH_4 per unit of production, as a mitigation strategy. A necessary step in achieving this aim is acquiring the capacity to quickly and reliably estimate the daily methane production (DMP) of ruminants. Mechanistic models used to predict CH_4 production from animals (e.g. Blaxter and Clapperton, 1965) do not reflect differences that have been observed in individual animals on similar intakes (Goopy *et al.*, 2006); 24 h calorimetry remains the only proven method for discerning these observed differences. Unfortunately, this is difficult, resource intensive, and impractical for screening large numbers of animals. Thus, there is considerable impetus to develop alternative methods which will facilitate estimation of DMP in large numbers of animals. This study assessed six (direct and derived) estimates of enteric methane production against 22 h open circuit calorimetry measurements (CHB_{22}) to determine if a reliable, but less demanding estimate of DMP could be found.

Material and methods

Rumen-cannulated Merino-cross hogget ewes (n=12), were assigned on a stratified random basis (by weight) to one of three management groups in a latin-square design. Each animal had DMP_{22h} measured by open-circuit calorimetry at 3 levels of FI (0.7, 1.05, 1.4× maintenance ME requirements, based on LW at entry to the trial), fed for 10 d prior to each measurement. The sheep were fed lucerne chaff (9.0 MJ/kg DM ME), provided in 8 equal portions, four morning (8, 9, 10, 11 am) and four afternoon feeds (4, 5, 6, 7 pm) to simulate morning and evening grazing that typifies the feeding pattern of sheep (Lynch *et al.*, 1992).
A single measure of DMP_{22h} was made for each sheep at each feeding level, with the daily ration being provided as for the training period and refusals collected at the conclusion of the measurement. The construction and operation of the chambers has been described in detail previously (Hegarty *et al.*, 2008). DMP_{22h} was calculated as air flow multiplied by $[CH_4]$ in the effluent air, adjusted for $[CH_4]$ of the incoming air and temperature and atmospheric pressure in each chamber. Additionally, continuous subsamples over the first 3 h and 22 h of the chamber air were collected in Tedlar® bags using peristaltic pumps and analysed for $[CO_2]$ and $[CH_4]$ using a gas chromatograph (Varian CP4900 Micro-GC). At the conclusion of the chamber measurement, sheep were removed and a short, timed, total collection of exhaled air ($BREATH_{3min}$) was taken. Finally, a 3 h subsample of exhaled air drawn from above immediately above the nose ($NOSE_{3h}$), was collected before animals were offered further feed. These both were also analysed for $[CH_4]$ and CO_2:CH_4 ratio.
These five identified measures (1) $BREATH_{3min}$ CO_2:CH_4, (2) CHB_{22} CO_2:CH_4, (3) $NOSE_{3h}$ CO_2:CH_4, (4) $BREATH_{3min}$ and (5) $NOSE_{3h}$ $[CH_4]$ were regressed against DMP_{22h} to determine which best predicted DMP. Additionally, 2 h CH_4 production ($CH4_{2h}$) during sequential periods of the CHB_{22} was regressed against DMP_{22}.

Results

Means and variances of DMP_{22}, the variance due to FI and CV% of DMP_{22} adjusted for FI are shown for each feeding level (Table 1). FI explained 74% of variation in DMP_{22}. $[CH_4]$ and $[CO_2]$ were highly correlated, with CHB_{22} $[CO_2]$, explaining 73% of variation in DMP_{22}. Consequently, CH4:CO2 ratios (measures 1-3) explained 0% of variation in DMP_{22} as did measure (4), and were of no use in predicting DMP. Measure (5) explained 9% of variation, but the correlation was negative – higher $[CH_4]$ being associated with lower DMP_{22}, an effect most likely due to chance. In contrast, $CH4_{2h}$ measured over much of the day explained from 50-82% of the variation in DMP_{22} ignoring FI, but a lesser proportion of variance within feed levels (between 13-55%; Table 2), with the largest amount of variation explained from 4-8 h after start of measurement.

Table 1. Mean CH_4 production (DMP_{22}) and FI of sheep during chamber measurement.

Feed level, x ME_M	Feed intake (FI), g	CH_4, l/d	CH_4 l/d, Variance	ADJ VAR[1] (% explained)	ADV CV%[2]
0.7	624	19.3	2.52	2.13 (15%)	7.6
1.05	960	24.3	9.98	5.96 (40%)	10.0
1.4	1249	29.0	9.29	9.90 (0%)	10.9
All	945	24.2	23.01	5.93 (74%)	10.1

[1] ADJ VAR = variance adjusted for feed intake. [2] ADV CV% = Coefficient of variation adjusted for feed intake.

Table 2. Variance (%) in daily CH_4 production (DMP_{22}) explained by 2 h measures of CH_4.

Period[1]	0	2	4	6	8	10	12	14	16	18	20
VAR[2]	50	57	80	78	82	53	62	77	61	63	64
VAR[2] within feed level	21	41	45	46	55	23	15	35	13	20	23

[1] Period= h after commencement of chamber measurement (08:00 h). [2] VAR =% Variation explained.

Conclusion

Of the range of possible predictors of DMP evaluated, only $CH4_{2h}$ was effective in predicting DMP_{22}. More work is needed to ascertain if shorter periods can be used to predict DMP.

References

AGO, 2005. National Greenhouse Gas Inventory 2003. Australian Greenhouse Gas Office. Commonwealth of Australia.

Blaxter, K.L. and J.L. Clapperton, 1965. Prediction of the amount of methane produced by ruminants. Br. J. Nutr. 19: 511-522.

Goopy, J.P., R.S. Hegarty and R. Dobos, 2006. The persistence over time of divergent methane production lot-fed cattle. In: Proc. 2nd Internat. Congress on Greenhouse Gases. Zurich, Switzerland, 20-24 Sept. 2005. ICS Series 1293 (July), pp. 111-114.

Hegarty, R.S., S.H. Bird, B.A. Vaneslow and R. Woodgate, 2008. Effects of the absence of protozoa from birth or from weaning on the growth and methane production of lambs. Br. J. Nutr. 100: 1220-1227.

Lynch, J.J., G.N. Hinch and D.B. Adams, 1992. The behavior of sheep – Biological Principles and Implications for Production. CAB International, Wallingford, Oxon, UK, pp. 12-13.

The effect of type of liquid feed on small intestine development in newborn calves

P. Górka[1], P. Pietrzak[2], A. Kotunia[3], J. Flaga[1], Z.M. Kowalski[1] and R. Zabielski[2]
[1]Department of Animal Nutrition, Agriculture University in Krakow, Krakow, Poland; [2]Department of Physiological Sciences, Faculty of Veterinary Medicine, Warsaw University of Life Sciences, Warsaw, Poland; [3]The Kielanowski Institute of Animal Physiology and Nutrition, Polish Academy of Science, Jabłonna, Poland; gorka-pawel@wp.pl

Introduction

Gastrointestinal tract development is an important factor determining health and performance of calves in the first weeks of life. Fast rumen development would be desired especially in dairy calves. It is well known that this process is highly dependent on solid feed intake and composition which may substantially accelerate it (Heinrichs, 2005). However, the development of the small intestine (SI) is even more critical for preruminant calf growth and health during the first weeks of life. This may affect animal growth and health and thus also solid feed intake and rumen development. Commonly used milk replacers (MR) slow down SI development (Seegraber and Morrill, 1986). The addition of sodium butyrate (SB) to MR may partially allow overcoming this problem (Kotunia *et al.*, 2004; Guilloteau *et al.*, 2009). However, most studies have not included a group of animals fed with whole milk (WM) (positive control) for comparison. We hypothesise that feeding calves with MR instead of WM will slow down SI development, and SB supplementation will reverse this negative effect. Thus the aim of the present study was to determine the effect of different liquid feeds (WM, MR and MR supplemented with SB) on calf SI development.

Material and methods

Twenty-one bull calves (5-day-old) were randomly allocated to three groups: WM, MR and MR with SB (0.3% as fed) (SBMR). Up to 25% of MR and SBMR (22% CP, 17% fat) composition was based on soy protein concentrate. Liquid feed DMI was equal between treatments and amounted to 1% of initial BW. Animals were slaughtered at 26 days of their life. The duodenum, jejunum and ileum were separated and weighed individually. One cm^2 of whole thickness samples from the middle jejunum were immediately fixed in 4% buffered formaldehyde and then embedded in paraffin. Serial histological sections were stained with hematoxylin and eosin for morphometric analysis (villi length, crypt depth, mucosa thickness and muscular thickness) under light microscopy (OLYMPUS BX 61). Whole thickness middle jejunum samples were fixed in embedding medium, frozen and labelled with a specific set of antibodies. For mitosis, anti-Ki-67-FITC-conjugated (BD Pharmingen), and for apoptosis anti-Cpp32 (DAKO) and secondary Alexa Flour 488 chicken anti-rabbit antibodies (Molecular Probes) were used. Cell nuclei were stained with 7-amino-actinomycin D. Confocal microscopy (OLYMPUS FV500) and the Microimage system (OLYMPUS) were employed for in-tissue-cytometry analysis of mitotic and apoptotic indexes in the small intestinal epithelium. Brush border enzyme activity (lactase, maltase, aminopeptidase A and aminopeptidase N, and dipeptidase IV) in the proximal, distal, and middle jejunum, and ileum were analysed according to Kotunia *et al.* (2004). Data were subjected to one-way analysis of variance using PROC GLM (SAS® 8.01). Pre-planned orthogonal contrast (WM vs. MR and SBMR, MR vs. SBMR) was used. The significance was declared at $P<0.05$ and tendencies at $P<0.15$.

Results

The empty jejunum and ileum mass, crypt depth and mucosa thickness in the middle jejunum were higher in the WM group as compared to the MR and SBMR groups ($P\leq0.13$) (Table 1). The

WM group also tended to have higher maltase activity in the distal jejunum (P=0.08), had higher aminopeptidase A activity in the middle jejunum (P=0.03) and lower apoptic index as compared to MR and SBMR groups (P=0.04). The SB supplementation had no effect on structural SI development. On the contrary, the mitotic index was increased and apoptotic index was reduced in the SBMR group as compared to the MR group (P≤0.01). Additionally, lactase activity tended to be higher in the middle jejunum (P=0.14), and aminopeptidase A and aminopepidase N tended to be higher in the distal jejunum (P≤0.15) in the SBMR group as compared to the MR group.

Table 1. The effect of liquid feed on structural development of the small intestine in calves.

Item	Group[1]			SE[3]	Contrasts[2]	
	WM	MR	SBMR		1	2
Duodenum,% of body mass	0.13[4]	0.12	0.12	0.01	NS[5]	NS
Jejunum,% of body mass	2.64	2.19	2.26	0.08	0.02	NS
Ileum,% of body mass	0.21	0.15	0.13	0.02	0.05	NS
Villi length, µm	437	392	359	19	NS	NS
Crypt depth, µm	223	179	196	12	0.03	NS
Mucosa thickness, µm	671	584	563	28	0.13	NS
Muscularis thickness, µm	355	340	323	13	NS	NS

[1] WM – whole milk, MR – milk replacer, SBMR – milk replacer with sodium butyrate (0.3% as fed).
[2] 1 = WM vs. MR and SBMR, 2 = MR vs. SBMR.
[3] SE = standard error.
[4] Mean.
[5] NS = not significant.

Discussion and conclusion

The present study confirmed a negative effect of feeding calves with MR instead of WM on SI development. The SB did not affect duodenum, jejunum and ileum mass and middle jejunum morphology. It seems that its well documented positive effects on calf performances (Guilloteau *et al.*, 2009) are mediated by enhanced crypt cell proliferation, reduction of apoptosis and stimulation of some brush border enzyme activities. In conclusion, the addition of SB to the MR improves small intestine development.

Acknowledgement

Supported by grant no. 1164/P01/2006/31 and 3196/B/P01/35, Ministry for Research and Higher Education, Poland.

References

Guilloteau, P., R. Zabielski, J.C. David, J.W. Blum, J.A. Morisset, M. Biernat and J. Woliński, 2009. Sodium-butyrate as a growth promoter in milk replacer formula for young calves. J. Dairy Sci. 92: 1038-1049.

Heinrichs, J., 2005. Rumen development in calf. Adv. Dairy Tech. 17: 179-187.

Kotunia, A., J. Woliński, D. Laubitz, M. Jurkowska, V. Rome, P. Guilloteau and R. Zabielski, 2004. Effect of sodium butyrate on the small intestine development in neonatal piglets feed by artificial sow. J. Physiol. Pharmacol. 55(Suppl. 2): 59-68.

Seegraber, F.J. and J.L. Morrill, 1986. Effect of protein source in calf milk replacers on morphology and absorptive ability of small intestine. J. Dairy Sci. 69: 460-469.

Effects of supplementation timing on ruminal digestion and fermentation pattern during continuous culture fermentation of grass herbage

P. Gregorini[1,2] *and K.J. Soder*[1]
[1]*USDA-ARS, Bldg. 3702, Curtin Rd University Park, PA 16802, USA;* [2]*DairyNZ, Ltd. Private Bag 3221, Hamilton, New Zealand; Pablo.Gregorini@dairynz.co.nz*

Introduction

Herbage intake can be concentrated in intensive grazing bouts. However, short and intensive grazing bouts reduce ruminal digestion of herbage (Gregorini *et al.*, 2008). This reduction may be offset by proper supplementation, preparing the rumen for enhanced rates of herbage intake. Several studies have investigated the effects of supplementation levels on herbage digestion (Bargo *et al.*, 2003); but few studies (i.e. Mitani *et al.*, 2005) focused on the effects of timing of supplementation. Continuous culture fermentations have been successfully used to evaluate rumen response to hypothetical feeding scenarios. Using a dual flow continuous culture fermenter system designed to simulate ruminal digestion and nutrient outflow (Soder *et al.*, 2007), this study evaluated the effect of maize silage supplementation, either 9 or 1 h before a hypothetical short and intensive herbage meal, on ruminal digestion and nutrient flows.

Material and methods

Fermenters were operated over four 10 d periods. Treatments included 28 g DM maize silage 9 (9BH) or 1 h (1BH) before providing 42 g DM herbage and control (CTL, no feeding of maize silage, 70 g DM herbage). Herbage was fed as follows; 66% of the total herbage meal (16:00 h), 22% (1720 h) and the remaining 22% at 18:40 h. Effluent sampling and bacterial harvest followed Soder *et al.* (2007) methodology. The effluent was analysed for OM, CP and NDF. Purine concentrations in effluent and bacterial isolates were used to partition effluent N flow into bacterial and non-bacterial fractions, and to calculate true OM digestibility and flows. Fermenters were sampled for pH, volatile fatty acids (VFA), and NH_3-N at 07:30, 11:00, 15:30, 16:00, 17:20, 18:40 and 20:00 h. Data were analysed as a 3×4 incomplete Latin square. True OM, CP, and NDF digestibilities and flows were analysed using a mixed model. Least-squares means were compared by LSD. pH, VFA and NH_3-N pattern was analysed as repeated measures within the linear mixed model framework. The interaction (treatment by sampling time) was analysed by ANOVA. GenStat11.1. was used for the statistical analysis.

Results

True OM (61.1% at mean; P=0.66) and CP (84.6% at mean, P=0.83) digestibilities were not affected (P>0.05) by treatment. Apparent NDF digestibility was the greatest in CTL (84.4%) and least (78.1%) for 9BH (P<0.05). These differences could relate to the results of pH and microbial synthesis. The pH was the greatest (P>0.05) for CTL and the lowest for 9BH (6.6 and 5.6, respectively). Mean bacterial efficiency was 19.6, 17.6 and 15.4 g N/kg OM truly digested for CTL, 9BH and 1BH, respectively. Diurnal patterns of pH, N-NH_3, and acetic/propionic ratio are presented in Figure 1., 9BH maintained the lowest pH values except for the first and the last two samplings. 9BH also maintained the lowest NH_3-N values expect in the first sampling while CTL showed the highest and 1BH the intermediate values. Treatments affected (P<0.05) mean N-NH_3 (18.7, 8.8, and 11.4 mg/100 ml for CTL, 9BH and 1BH, respectively) and mean effluent of N-NH_3 (0.42, 0.19 and 0.26 g N/d for CTL, 9BH and 1BH, respectively). Treatments did not affect (P>0.05) effluent of dietary N (0.32 g/d at mean), or total VFA or acetic acid concentration (91.3 and 44.9 Mmol/ml at mean, respectively). Propionic acid concentration was the greatest ($P<0.05$) for 9BH.

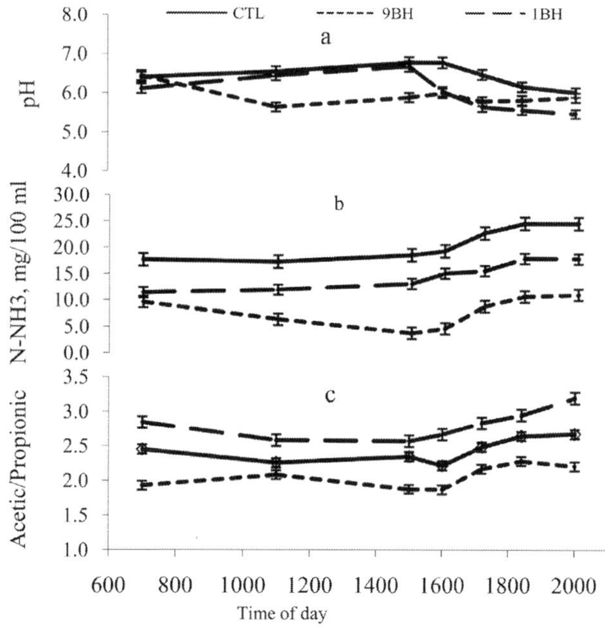

Figure 1. Effect of timing of maize silage supplementation, 9 (9BH) or1 (1BH) h before a single herbage meal on ruminal pH (a), NH_3-N (b) and acetic/propionic VFA patterns.

Conclusion

Strategically timed supplementation with maize silage (9BH vs. 1 BH) improved bacterial efficiency, N utilisation and increased the supply of glucogenic nutrients. Therefore, under the same resource allocation (pasture plus supplement), a simple change in timing of maize silage allocation may improve herbage digestion, whilst probably reducing the negative environmental impact of N losses.

References

Bargo, F.L., D. Muller, E.S. Kolver and J.E. Delahoy, 2003. Invited Review: Production and Digestion of Supplemented Dairy Cows on Pasture. J. Dairy Sci. 86: 1-42.

Gregorini, P., S. Gunter and P. Beck, 2008. Matching plant and animal processes to alter nutrient supply in strip grazed cattle: Timing of herbage and fasting allocation. J. Anim. Sci. 86: 1006-1020.

Mitani, T., M. Takahashi, K. Ueda and S. Kondo, 2005. Effects of pre-feeding of a corn silage-based supplement on the feed intake, milk production and nitrogen utilization of grazing dairy cows. Anim. Sci. J. 76: 453-460.

Soder K., L. Saporito and R. Stout, 2007. Effect of by-product gypsum application rate to grass pasture, grass hay, and corn silage on fermentation by rumen microorganisms in continuous culture. Anim. Feed Sci. Technol. 132: 331-340.

In-series tension receptors and epithelial receptors in the omasum of sheep

W.L. Grovum
Department of Biomedical Sciences, Ontario Veterinary College, University of Guelph, Guelph, N1G 2W1, Ontario, Canada; lgrovum@uoguelph.ca

Introduction

Knowing about the sensory receptors in the ruminant forestomach enables us to understand how its contents can impact salivation, food intake and aspects of motility such as mixing, emptying, eructation and rumination. Identifying and characterising these receptors is difficult enough but determining their functions with certainty is an even greater challenge. In the long term, the possibility exists that their activities and sensitivities may be manipulated with drugs or by changing the diet to alter vago-vagal reflex activity in desirable ways and hence increase the productivity of farm animals. Most neural research on the forestomach has so far been directed at the reticulorumen because its epithelium and muscular walls are relatively easy to expose in anaesthetised animals. In-series tension receptors sense muscle tension and distension whereas epithelial receptors sense tactile stimulation (Leek and Harding, 1975). Recently, a gas pressure sensor was identified in the dorsal regions of the rumen (Grovum and Shaik Mossadeq, 2002). Despite these advances, the sensory innervation of the omasum has not been investigated in large part because of its intractable location ventral to the liver, caudal to the diaphragm and between a small reticulo-omasal orifice on the left and the ribcage on the right. This paper demonstrates how these anatomical roadblocks to investigating omasal sensory physiology were overcome by everting the omasum into the reticulum. Exposing the interior of the omasum in this manner enabled the present finding that in-series tension receptors and epithelial receptors exist in the omasum.

Material and methods

Thirty-four intact Arcott rams ranging in body weight from 31 to 96 kg were fed mixed hay at a research station before being transported to the Veterinary College where they were given only water overnight. Once in the laboratory, each ram was anaesthetised with an injection of sodium pentobarbital via a jugular catheter, and intubated before being laid in dorsal recumbancy on a heated surgical table. A ventilation pump in line with a closed circuit gaseous anaesthetic machine (halothane off) was connected to override the animal's inherent respiratory activity. A rectal probe was inserted to monitor body temperature and the femoral artery was cannulated to record blood pressure. A surgical plane of anaesthesia was maintained using a variable speed peristaltic pump to deliver sodium pentobarbital into the jugular vein. A drip of normal saline was also administered. The reticulorumen was emptied through a rumenotomy before the last 7 ribs and the associated tissues in the thoracic wall on the left side were removed to expose the reticulum. It was then incised upwards from its apex on its left and medial walls to expose the reticulo-omasal orifice. The cut edges of the reticulum were stapled to the skin to close the thoracic and abdominal cavities. After dilating the reticulo-omasal orifice, the small particles between the omasal leaves were eluted (into the reticulum) with warm normal saline to shrink the omasum so that it could be everted it into the reticulum. Strands of nerve fibres were then dissected from the left cervical vagus in a pool of warm silicone oil and draped over a bipolar electrode to intercept mixed afferent nerve activity while the omasal leaves and body wall were stretched and stroked lightly. The strands were severed centrally to eliminate vagal efferent activity. The impulses from the strand were routed through a differential amplifier and then displayed on an oscilloscope and an MSD (Multi Spike Detector) data acquisition system (Alpha Omega, Israel). Single fibre activity was produced from the mixed record using the MSD's template matching system. Once an omasal mechanoreceptor was identified, its receptor field was stretched and stroked lightly to differentiate between in-series and epithelial receptor activity.

Results

The responses from an epithelial receptor (1 of 2) and an in-series tension receptor (1 of 14) are illustrated in Figure 1. Repeated light brushing over the epithelial receptor elicited spike frequencies of up to 150 Hz whereas sustained stretch produced an on-off response with rapid adaptation in between. In contrast, the in-series tension receptor held its elevated activity during sustained stretching (slowly adapting response) and did not respond to light brushing. Such responses are typical for these receptors in the reticulorumen (Leek and Harding, 1975).

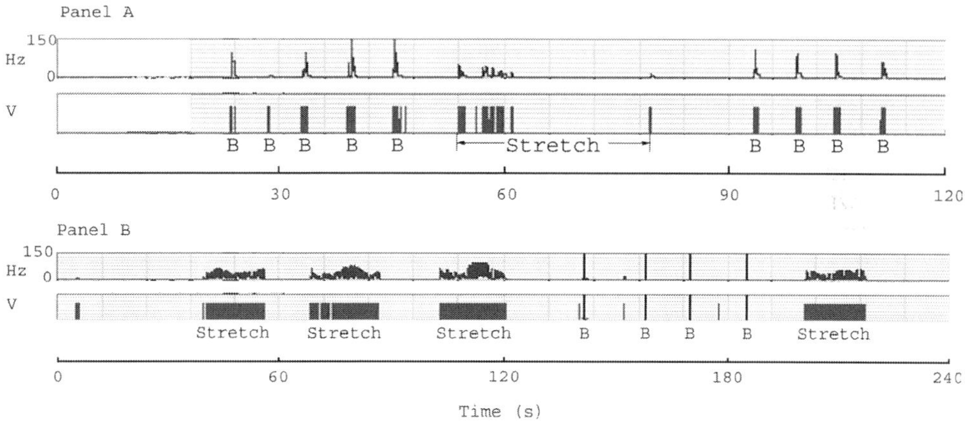

Figure 1. Activity from an epithelial receptor (Panel A) and an in-series tension receptor (B) in the omasum while stretching and brushing (B) over their fields. In each panel, TTL pulses (V) depict impulses. Hz depicts frequency. The bars above the B's in Panel B are markers.

Discussion and conclusion

The inhibition of reticular motility by omasal distension and its stimulation by tactile input (Bost, 1970) may be mediated reflexly respectively by activation of the in-series tension receptors and epithelial receptors reported. Their effects on food intake remain to be studied.

Acknowledgement

Funding was supplied by the Ontario Ministry of Agriculture and Food.

References

Bost, J., 1970. Omasal physiology. In: Phillipson A.T. (Ed.), Physiology of digestion and metabolism in the ruminant. Oriel Press, Newcastle Upon Tyne, England, pp. 52-65.

Grovum, W.L. and W.M. Shaik Mossadeq, 2002. A study of gas pressure receptors in the reticulorumen which may evoke eructation in sheep. Proc. Nutr. Soc. 61: 104A.

Leek, B.F. and R.H. Harding, 1975. Sensory nervous receptors in the ruminant stomach and their reflex control of reticulo-ruminal motility. In: McDonald, I.W. and A.C.I. Warner (Eds.), Digestion and metabolism in the ruminant. UNE Publishing Unit, Australia, pp. 60-76.

Changes in methanogenic populations residing in the rumen of dairy cows in response to a sainfoin (*Onobrychis viciifolia* Scop.) based diet

A. Guglielmelli[1,3], *O. Perez*[2], *F. Tiemessen*[1], *M. Domenis*[1], *R. Albanese*[1,3], *S. Calabrò*[3], *H.S. Smidt*[2] *and W.F. Pellikaan*[1]
[1]*Animal Nutrition Group, Wageningen University, 6700 AH Wageningen, the Netherlands;* [2]*Laboratory of Microbiology, Wageningen University, 6700 AH Wageningen, the Netherlands;* [3]*Dipartimento di Scienze Zootecniche e Ispezione degli Alimenti, Faculty of Veterinary Medicine, Federico II University, 80137 Naples, Italy; Wilbert.Pellikaan@wur.nl*

Introduction

Methanogenesis in the rumen is a natural metabolic process which keeps the fermentation on by continuously utilising the hydrogen (H2) released by fermentation of carbohydrates. Methane emissions by ruminants have implications for (1) animal productivity because they constitute a significant loss of dietary energy, and (2) the environment because methane is considered to be one of the more powerful greenhouses gases involved in global warming. Several plant secondary metabolites can be used for selective inhibition against the methanogens in the rumen. An antimethanogen activity has been attributed to tannins, mainly condensed tannins, present in sainfoin (*Onobrychis viciifolia* Scop.) (Waghorn *et al.*, 2002). These polyphenolic compounds can reduce methane emission by reducing protozoa number and changing the rumen fermentation parameters. A reduction in methane production could be associated with a change in the number and/or in the diversity of the bacterial community in the rumen. Polyethylene glycol (PEG) has a strong affinity with condensed tannins present in sainfoin and it is commonly used to block their action. The current trial involving dairy cows was conducted to study the effect of sainfoin tannins on rumen fermentation and on methanogen functioning, and to assess adaptive behaviour of rumen microbiota when exposed to sainfoin tannins for a prolonged period.

Material and methods

Prior to receiving the experimental diet three rumen fistulated Holstein Friesian dairy cows (DIM = 226±64 d; milk production = 29.4±6.1 kg/ d; mean ± se) were placed on a lucerne based 'uniformity diet' for a 2-wk period to allow animals to adapt to a tannin free legume-based diet. After 2 wk, the lucerne was exchanged for sainfoin. During the first 5 d of the sainfoin feeding polyethylene glycol (PEG4000; 975 g/cow/d) was administered in three portions per day through the fistula. After the PEG treatment the animals remained on their sainfoin based diet for 7 more weeks. Rumen content samples were taken three times per week during the experimental period. Samples were collected before the morning feeding and were prepared for analyses on fermentation end-products: volatile fatty acids (VFA) by gas chromatography and ammonia (NH_3) by a colorimetric method. Ciliate protozoa were microscopically enumerated in rumen fluid samples using a Sedgewick Rafter counting chamber (Dehority, 1984). DNA was extracted from rumen fluid collected and the number of methanogenic Archea were determined in the sample using the SYBR green qPCR assay; *Methanobrevibacter smithii* was used as the reference strain. The PCR conditions and primers used were as reported by Denman *et al.* (2007). In addition, samples were taken to determine the diversity of the rumen methanogen population by denaturing gradient gel electrophoresis (DGGE, Skillman *et al.*, 2006). Data were analysed using the GLM procedure of SAS® 9.1 (SAS Institute Inc., Cary, NC, USA). Differences between dietary treatments (lucerne, sainfoin+PEG, sainfoin) with time were tested using a repeated measures statement.

Results and discussion

The results show a significant ($P<0.05$) decrease in the number of protozoa, and a non- significant decrease in the total number of Archea in the first week after changing to the sainfoin diet (Table 1). The Archea followed a similar tendency as the protozoa but animal variation seemed to be considerably higher. The decrease in Archea numbers already started during the PEG administration period. This suggests that PEG may not have been fully successful in completely blocking the effect of tannins. Total VFA and NH_3 followed a pattern similar to the protozoa numbers. The VFA profile shows some significant differences among diet treatments. The results show no differences in total VFA between the uniformity diet and the sainfoin + PEG diet, but a distinct and significant ($P<0.05$) drop in VFA when PEG treatment was stopped (122.5 vs. 95.67 mmol/l). In contrast, NH_3 shows a significant ($P<0.05$) decrease when animals changed from uniformity diet to the sainfoin + PEG diet, followed by a further drop in the first few days after stopping the PEG treatment; further supporting our observation that PEG was not fully effective. The difference in the DGGE animals profile through time involving all methanogenic groups shows an inhibition in the whole methanogen community.

Table1. Dietary treatment effects on rumen fluid parameters and microbial cell count.

Treatment	Protozoa (log_{10}/ml)	Number of Archea (log_{10}/ml)	Total VFA (mmol/l)	NH_3 (mg/l)
Lucerne	5.79^a	8.14	124.5^a	138.2^a
Sainfoin+PEG	5.78^a	7.69	122.5^a	85.4^{bc}
Sainfoin 1st wk	5.65^c	7.46	95.7^b	54.4^b
Sainfoin 2nd wk	5.72^b	7.57	114.7	109.1^{ac}
Sainfoin 3rd wk	5.73^b	7.41	105.9	94.4^c
MSE	0.0006	0.45	152.1	439.5
P-value	0.0002	0.69	0.08	0.01

Means within columns followed by different superscripts are significantly different at $P<0.05$; MSE: mean square error.

Conclusion

The tannin content in *Onobrychis viciifolia* Scop. resulted in the inhibition of protozoa and number of methanogens. The tendency of the different parameters suggests that the microbial population (protozoa, Archea, other bacteria) responds in different ways to the sainfoin diet over time, suggesting that rumen microbiota adapt to the dietary conditions. Further work is required to evaluate the relation between the microbial inhibition and methane production.

References

Dehority, B.A., 1984. Evaluation of subsampling and fixation procedures used for counting rumen protozoa. Appl. Environ. Microbiol. 48: 182-185.

Denman, S.E., N.W. Tompkins and C.S. McSweeney, 2007. Quantification and diversity analysis of ruminal methanogenic populations in response to the antimethanogenic compound bromochloromethane. FEMS Microbiol. Ecol. 62: 313-322.

Skillman, L.C., P.N. Evans, C. Strompl and K.N. Joblin, 2006. 16S rDNA directed PCR primers and detection of methanogens in the bovine rumen. Lett. Appl. Microbiol. 42: 222-228.

Waghorn, G.C., M.H. Tavendale and D.R. Woodfield, 2002. Methanogenesis from forages fed to sheep. Proc. N.Z. Grassl. Assoc. 64: 167-171.

Virginiamycin supplementation has a selective effect on rumen bacterial population of Chinese Luxi steers

T.J. Guo, J.Q. Wang, D.P. Bu, K.L. Liu, J.P. Wang, D. Li, S.Y. Luan, J. Wang and X.K. Huo
State Key Laboratory of Animal Nutrition, Institute of Animal Science, Chinese Academy of Agricultural Science, 100193, Beijing, China; wang-jia-qi@263.net

Introduction

Virginiamycin (VM) is an antimicrobial feed additive approved by the US Food and Drug Administration for use in cattle to improve performance, digestibility and energy utilisation of grains, and reduce lactic acidosis risk (Clayton *et al.*, 1999; Ives *et al.*, 2002). *In vitro* studies have shown that VM is a potent inhibitor of lactic acid production due to an effect on lactic acid-producing bacteria, including *Streptococcus bovis* and *Lactobacillus* spp. (Nagaraja and Taylor, 1987). VM has an antimicrobial spectrum similar to that of monensin in that gram positive bacteria are susceptible and gram negative bacteria are generally resistant (Nagaraja and Taylor, 1987). However, little is known about the effects of VM on specific ruminal population. The present study was aimed at investigating *in vivo* the effect of VM feeding on the populations of specific ruminal bacteria.

Material and methods

Four Chinese Luxi steers (BW 559.4±30.1 kg) with permanent rumen cannulas were used in a cross-over design experiment. Each period was of 28 d. The basal diet had a forage-to-concentrate ratio of 35:65 (DM basis). The experimental treatments were (1) control; (2) control diet plus VM 30 mg/kg concentrate (DM basis).

Ruminal fluid samples were collected at 7:30 h prefeeding, at 11:30 h and 17:30 h post-feeding on 27 and 28 d of each period from the anterior, dorsal, and mid-ventral regions of the rumen by hand, and were squeezed through four layers of sterile cheesecloth. Part of the pooled rumen fluid was transferred to anaerobic culture (Hungate, 1969). DNA were extracted by the beater-bead method (Yu and Morrison, 2004) for real-time PCR quantification (Koike *et al.*, 2007).

The data were transformed into logarithmic form, and analysed by the MIXED procedure of SAS® 9.0 (SAS Institute Inc., 2004). Significance was declared at $P<0.05$, and tendencies at P value between 0.05 and 0.1.

Results

Steers supplemented with VM had lower ($P<0.01$) amylolytic bacteria (8.41 vs. 8.72 $\log_{10}$ cfu/ml) and proteolytic bacteria (8.60 vs. 8.83 $\log_{10}$ cfu/ml) counts compared with the control (Table 1). VM supplementation led to an increase of *Selenomonas ruminantium* ($P=0.05$), *Anaerovibrio lipolytica* ($P=0.09$) and a decrease of *Ruminococcus albus* ($P=0.07$) and *Streptococcus bovis* ($P=0.10$). VM decreased the relative abundance of *Lactobacillus* spp., but this decrease was not statistically significant.

As far as counts of total viable bacteria ($P=0.36$) and cellulolytic bacteria ($P=0.44$), Genus *Prevotella*, *Prevotella ruminicola*, *Butyrivibrio fibrisolvens*, Genus *Ruminococcus*, *Ruminococcus flavefaciens* and *Megasphaera elsdenii* were concerned, no difference between the control and VM treatment were observed.

Table 1. Quantification of ruminal bacteria in steers fed a control diet or treatment diet.

	Diet[1]		SEM	Diet effect
	control	treatment		*P*-value
Colony counts of ruminal microorganisms ($\log_{10}$ cfu/ml)				
Amylolytic bacteria	8.72	8.41	0.098	<0.01
Proteolytic bacteria	8.83	8.60	0.069	<0.01
Cellulolytic bacteria	9.59	9.56	0.045	0.44
Total viable bacteria	10.91	10.87	0.102	0.36
Quantification of ruminal bacteria by real time PCR ($\log_{10}$ copies/μl)				
Selenomonas ruminantium	4.46	5.55	0.354	0.05
Ruminococcus albus	7.39	6.95	0.165	0.07
Anaerovibrio lipolytica	7.93	8.12	0.120	0.09
Streptococcus bovis	7.74	6.67	0.420	0.10
Genus Prevotella	10.04	9.99	0.029	0.25
Butyrivibrio fibrisolvens	4.74	4.44	0.215	0.35
Lactobacillus spp.	5.22	4.57	0.518	0.41
Ruminococcus flavefaciens	4.00	3.77	0.245	0.52
Prevotella ruminicola	4.32	4.79	0.482	0.54
Genus Ruminococcus	8.45	8.36	0.141	0.67
Megasphaera elsdenii	2.27	2.26	0.231	0.97

[1] Control diet = 35% forage + 65% concentrate; Treatment diet = control diet with 30 mg/kg concentrate of virginiamycin.

Discussion and conclusion

Supplementation of VM reduced the counts of amylolytic bacteria and proteolytic bacteria, and had an increased trend on quantification of lactic acid-utilising bacteria (*Selenomonas ruminantium* and *Anaerovibrio lipolytica*), and a decreased trend on quantification of lactic acid-producing bacteria (*Streptococcus bovis* and *Lactobacillus* spp.). These implied that VM has selective effects on ruminal bacteria, and influences ruminal bacterial populations by changing part of special ruminal bacteria populations.

References

Clayton E.H., I.J. Lean, J.B. Rowe and J.W. Cox, 1999. Effects of feeding virginiamycin and sodium bicarbonate to grazing lactating dairy cows. J. Dairy Sci. 82: 1545-1554.

Hungate R.E., 1969. A roll tube method for cultivation of strict anaerobes. In: J.R. Norris and D.W. Ribbons (Eds.), Methods in Microbiology. Academic Press, London and New York, Vol.3B, p.117-132.

Koike, S., H. Yabuki and Y. Kobayashi, 2007. Validation and application of real-time PCR assays for representative rumen bacteria. J. Anim. Sci. 78: 135-141.

Ives S.E., E.C. Titgemeyer, T.G. Nagaraja, A. del Barrio, D.J. Bindel and L.C. Hollis, 2002. Effects of virginiamycin and monensin plus tylosin on ruminal protein metabolism in steers fed corn-based finishing diets with or without wet corn gluten feed. J. Anim. Sci. 80: 3005-3015.

Nagaraja, T.G. and M.B. Taylor, 1987. Susceptibility and resistance of ruminal bacteria to antimicrobial feed additives. Appl. Environ. Microbiol. 53: 1620-1625.

Yu, Z.T. and M. Morrison, 2004. Improved extraction of PCR-quality community DNA from digesta and fecal samples. BioTech. 36: 808-812.

Dietary protein and carbohydrate alter ruminal fermentation, digesta characteristics and behaviour in lactating dairy cattle

M.B. Hall
U.S. Dairy Forage Research Centre, USDA-ARS, 1925 Linden Dr., Madison, WI 53706 USA; marybeth.hall@ars.usda.gov

Introduction

Diet can affect ruminal fermentation with potential impact on animal productivity, efficiency, and health. Most frequently, direct effects of dietary components on ruminal characteristics are evaluated, but results of some studies suggest that the interaction of protein and carbohydrate supplementation can alter organic acid concentrations and ruminal pH (Aldrich *et al.*, 1993, Carruthers and Neil, 1997) even when no difference in ruminally fermented organic matter is observed (Aldrich *et al.*, 1993). The present study was aimed at evaluating the effects of carbohydrate and ruminally degradable protein supplementation as they affected ruminal fermentation products, ruminal digesta, and animal behaviour in lactating dairy cattle.

Material and methods

Eight ruminally cannulated Holstein cows (average milk yield, BW, and Day in milk of 37.9 kg, 652 kg, and 165 d, respectively) were randomly assigned to series of treatments in a partially balanced incomplete Latin square design with a 2×2 factorial arrangement of treatments. Dietary treatments were supplementation with non fibre carbohydrates (NFC) as sucrose (SUC) or maize starch (STA) at 10% of diet DM, and relative level of ruminal protein degradability (RDP) by supplementation with 48% soybean meal (+RDP) or expeller soybean meal +48% soybean meal (-RDP). All diets contained 33% maize silage and 22% lucerne silage, on a DM basis and were formulated to contain 30% NDF, 17.3% CP, and 41% NFC. Cows were offered the diets *ad libitum* (~5% refusals).
Experimental periods were 21 d with the last 7 d for sample and data collection. Feed offered and refused was measured daily. Rumen digesta evacuation to determine digesta weight and DM% was performed on 2 days with 4 cows sampled at 2 h pre-feeding and 4 at 2 h post-feeding. Animal behaviour was recorded every 5 min for 48 h. Rumen fluid was sampled hourly for 6 h on 1 day for determination of pH, organic acids, and NH_3.
Data were analysed using the MIXED procedure of SAS® (1999). Models included cow, period, treatment, their interactions, and residual error. Time course ruminal data were analysed as repeated measures. Fixed effects included period and treatment; cow or cow within period and treatment were used as random variables. Treatment differences were declared at $P \leq 0.05$, with trends towards significance considered at $P \leq 0.15$.

Results

Although feed intake did not differ among treatments ($P>0.55$), ruminal organic acid concentrations were greater for cows consuming diets SUC than STA (Table 1). Concentration of carbon in organic acids tended to be greater for -RDP with SUC, and for +RDP with STA (interaction $P=0.08$). Average ruminal pH showed the inverse of this pattern with the lower pH for SUC found with -RDP, and for STA with +RDP (interaction $P<0.01$). The lowest pH for the dietary treatments followed a similar numeric pattern, but there was only a tendency for SUC to have a lower pH than STA ($P=0.13$). Although NH_3 was greater for SUC + RDP than -RDP, the + or -RDP values differed little for STA (interaction $P=0.04$). Daily time spent ruminating did not differ by treatment ($P>0.21$). However, -RDP gave greater values than +RDP ($P \leq 0.02$) for time spent eating and standing, and there were greater increases with -RDP for SUC than for -RDP with STA (interaction $P \leq 0.04$). For ruminal digesta characteristics, cows consuming SUC had a greater amount of liquid ($P<0.01$) and tended to have a lower DM% ($P=0.06$) than cows consuming STA.

Table 1. Intake, rumen fluid, rumen digesta, and behaviour measures.

NFC treatment	Diet[1]					Diet effect[2]		
RDP treatment	SUC	SUC	STA	STA		*P*-values		
	+RDP	-RDP	+RDP	-RDP	SED	NFC	RDP	N x R
DM intake, kg/d	22.7	23.4	23.6	23.7	1.1	0.56	0.69	0.74
Rumen pH - mean	5.84	5.69	5.85	5.97	0.06	0.01	0.39	<0.01
- nadir	5.61	5.55	5.69	5.76	0.06	0.13	0.76	0.20
OA[3], m*M*	164	168	157	154	4	<0.01	0.95	0.20
C in OA, m*M*	421	441	396	389	10	<0.01	0.36	0.08
NH_3, m*M*	8.0	5.6	6.0	6.1	0.7	0.18	0.05	0.04
Digesta - DM,%	14.6	14.4	14.8	15.2	0.4	0.06	0.77	0.37
- DM, kg	13.6	13.4	13.1	12.4	0.8	0.20	0.49	0.71
- liquid, kg	80.6	77.9	74.4	70.0	3.4	<0.01	0.19	0.72
Rumination, min/d	404	472	435	468	49	0.71	0.22	0.63
Eating, min/d	197	261	191	216	8	<0.01	<0.01	0.02
Standing, min/d	540	780	617	655	50	0.52	0.02	0.04

[1] SUC = sucrose, STA = maize starch, + RDP = protein supplemented with 48% soybean meal, -RDP = protein supplemented with expeller soybean meal + 48% soybean meal.
[2] NFC = nonfiber carbohydrate, RDP = ruminally degradable protein, N×R = treatment interaction.
[3] OA = organic acids; sum of acetate, propionate, butyrate, lactate, and valerate.

Discussion and conclusion

The effects of carbohydrate source and protein degradability noted in this study are not presently accounted for in our predictions of how diet alters ruminal fermentation and nutrient yield to the animal and animal behaviour. Changes in digesta DM% and liquid amount by NFC source recommend further evaluation of the influence of NFC on passage kinetics as it affects nutrient supply to the animal. The effect of the interaction of carbohydrate and protein on ruminal measures suggests that these dietary components may be important for their impact on the capture of fermentation products in a form useful to the cow as well as on risk of depressing ruminal pH. The effect of this interaction on time standing has implications for its potential influence on hoof health.

References

Aldrich, J.M., L.D. Muller, G.A. Varga and L.C. Griel Jr., 1993. Nonstructural and carbohydrate effects on rumen fermentation, nutrient flow, and performance of dairy cows. J. Dairy Sci. 76: 1091-1105.

Carruthers, B.R. and P.G. Neil, 1997. Milk production and ruminal metabolistes from cows offered two pasture diets supplemented with non-structural carbohydrates. N. Z. J. Agric. Res. 40: 513-521.

Heat stress alters ruminal fermentation and digesta characteristics and behaviour in lactating dairy cattle

M.B. Hall
U.S. Dairy Forage Research Centre, USDA-ARS, 1925 Linden Dr., Madison, WI 53706 USA; marybeth.hall@ars.usda.gov

Introduction

In a feeding study designed to assess the effects of dietary nonfiber carbohydrates (NFC) and ruminally degradable protein (RDP), animals experienced moderate heat stress (HS) in the first period, and no/greatly reduced heat stress (NHS) in the second period which allowed the evaluation of the environment as a period effect. Heat stress has been shown to increase or decrease rumen organic acid concentrations, but the basis for these responses is not clear (Kadzere *et al.*, 2002). Decreased rumen motility (Attebery and Johnson, 1969) and intake, and increased ruminal retention time (Warren *et al.*, 1974) have been implicated. Working from the hypothesis that the period effects were predominantly due to heat stress, the present analysis evaluated the effects of period/heat stress as they affected ruminal fermentation products, ruminal digesta characteristics, and animal behaviour in lactating dairy cattle.

Material and methods

Eight ruminally cannulated Holstein cows (average milk yield, body weight, and days in milk of 37.9 kg, 652 kg, and 165 d, respectively) were randomly assigned to series of treatments in a partially balanced incomplete Latin square design with a 2×2 factorial arrangement of treatments. Dietary treatments were supplementation with NFC as sucrose or maize starch at 10% of diet DM, and relative level of RDP by supplementation with 48% soybean meal (+RDP) or expeller soybean meal +48% soybean meal (-RDP). All diets contained 33% maize silage and 22% lucerne silage, on a DM basis and were formulated to contain 30% NDF, 17.3% CP, and 41% NFC. Cows were offered the diets *ad libitum* (~5% refusals). Mean, maximum, and minimum temperatures for HS were 27.8, 32.6, and 21.8 °C and for NHS were 22.9, 27.7, 18.3 °C, respectively, with a mean relative humidity of 78%.

Experimental periods were 21 d with the last 7 d for sample and data collection. Feed offered and refused was measured daily. Rumen emptying via the rumen cannula to determine digesta weight and DM% was performed on 2 days with 4 cows sampled at 2 h pre-feeding and 4 at 2 h post-feeding. Visual observations of animal behaviour were recorded every 5 min for 48 h. Rumen fluid was sampled hourly for 6 h on 1 day for determination of pH, organic acids, and protein degradation products. Respiration rate and rectal temperature were measured on 2 days at 5:00 and 14:30.

Data were analysed using the MIXED procedure of SAS® (1999). Models included cow, period, treatment, their interactions, and residual error. Time course ruminal data were analysed as repeated measures. Fixed effects included period and treatment; cow or cow within period and treatment were used as random variables.

Results

Heat stressed cows had greater respiration rates ($P<0.01$) and rectal temperatures ($P<0.01$), and tended to have lower feed intake ($P=0.07$) than NHS cows (Table 1). Rumen pH was greater ($P<0.01$), and organic acid concentration smaller ($P<0.01$) with HS. Except for butyrate, molar% of organic acids produced predominantly from carbohydrates differed or tended to differ between HS and NHS (Table 1). In contrast to the organic acids, rumen NH_3 and amino acids were at 1.3-fold

(P=0.02) and four-fold (P<0.01) greater concentrations, respectively, in HS cows. Rumen digesta of HS cows had a lower DM% (P<0.01) and tended to have a greater amount of liquid (P=0.08) even though time spent drinking did not differ for HS and NHS (P=0.58). Adjustment for differences in liquid volume did not equalise the concentrations of organic acids between HS and NHS (data not shown). Digesta DM amount was lower for HS (P=0.02). Compared to NHS cows, HS cows spent fewer minutes per day eating (P<0.01) and tended to spend less time ruminating (P=0.07).

Table 1. Respiration rate, rectal temperature, feed intake, rumen fluid, rumen digesta, and behaviour measures by period[1].

	HS	NHS	SED	P		HS	NHS	SED	P
Respiration rate, breaths/min	68.7	40.9	3.7	<0.01	Millimolar ruminal concentrations[2]				
Rectal temperature, °C	39.0	38.4	0.1	<0.01	OA	147	174	3	<0.01
DM intake, kg/d	21.9	24.8	0.4	0.15	Max. lactate	14	16	5	0.77
Rumination, min/d	403	486	34	0.07	NH_3	7.2	5.7	0.5	0.02
Eating, min/d	190	242	5	<0.01	BCVFA	4.6	4.7	0.3	0.78
Standing, min/d	625	672	30	0.20	Amino acids	3.6	0.8	0.6	<0.01
Drinking, min/d	90	99	11	0.58	*Molar % of OA*				
Digesta - DM,%	13.8	15.6	0.2	<0.01	Acetate	64.8	59.8	1.0	<0.01
- DM, kg	12.5	13.7	0.5	0.02	Propionate	19.8	22.5	0.9	0.01
- liquid, kg	77.6	73.8	2.0	0.08	Butyrate	12.5	13.1	0.8	0.44
Rumen pH, mean	6.03	5.82	0.05	<0.01	Valerate	2.0	2.6	0.2	0.01
Rumen pH, nadir	5.78	5.52	0.04	0.07	Lactate	0.9	2.0	0.7	0.12

[1] HS = heat stress period, NHS = non-heat stress period.
[2] OA = organic acids, sum of acetate, propionate, butyrate, valerate, and lactate; Max. lactate = maximum detected lactate concentration; BCVFA = branched chain fatty acids.

Discussion and conclusion

As compared to non-heat stressed animals, heat stress altered ruminal digesta characteristics, concentrations of fermentation products, and behaviour in lactating dairy cows. Under heat stress, the decrease in organic acid concentration and increased ruminal pH could be related to the tendencies for greater amounts of ruminal liquid and lower intakes of potentially fermentable material. Increased ruminal amino acid concentrations found with HS suggest a greater degree of ruminal protein degradation than production. Increased ruminal NH_3 may reflect reduced ruminal capture as protein and/or increased endogenous contributions of NH_3 if animals were mobilising tissue due to weight loss at decreased DM intake.

References

Attebery, J.R. and H.D. Johnson, 1969, Effects of environmental temperature, controlled feeding and fasting on rumen motility. J. Anim. Sci. 29: 734-737.

Kadzere, C.T., M.R. Murphy, N. Silanikove and E. Maltz, 2002. Heat stress in lactating dairy cows: a review. Livest. Prod. Sci. 77: 59-91.

Warren, W.P., F.A. Martz, K.H. Asay, E.S. Hilderbrand, C.G. Payne and J.R. Vogt, 1974. Digestibility and rate of passage by steers fed tall fescue, alfalfa and orchardgrass hay in 18 and 32 C ambient temperatures. J. Anim. Sci. 39: 93-96.

Effects of replacing soya with *Vicia faba* beans on fermentation in the rumen of the ovine Sicilo-Sarde breed

M. Hammami, H. Rouissi, H. Selmi, B. Rekik and A. Ben Gara
Ecole Supérieure d'Agriculture, 7030 Mateur, Tunisia; hamadi.rouissi@iresa.agrinet.tn

Introduction

Under current international economic conditions, the replacement of the imported soya by locally produced *Vicia faba* beans may be an appealing alternative for feeding the Tunisian Sicilo-Sarde dairy sheep. Rouissi *et al.* (2008) reported that *Vicia faba* may be used in dairy sheep rations without compromising milk performances. The objective of this study was to evaluate fermentation parameters as a function of the protein source in the rumen of Sicilo-Sarde rams.

Material and methods

Four rams of the Sicilo-Sarde breed with permanent canulas were used in this experiment. Rams (mean age = 4.75±0.5; mean live weight = 53.275±6.6 kg) were in individual boxes. The basal ration was 1.5 kg DM oat hay per day distributed in two separate meals complemented with 500 g/day of concentrates. Two concentrates were tested. The first concentrate (RS) was 82.5% barley, 13.5% Soya, and 4% mineral mixture, while the second concentrate (RV) included 71.5% barley, 17.5% *Vicia faba*, 7% Soya and 4% mineral mixture. The energy values were 0.54, 0.96 and 0.96 UFV/kg DM and the crude protein contents were 5.20, 16.8 and 16.2% for hay, RS and RV, respectively. Water was *ad libitum*. The experimental design was constituted of two consecutive experimental 45 day periods separated by a two week adaptation period. Samples of 50 ml of rumen juice were taken before the morning meal and after 2, 5 and 8 hours to determine pH and ammonia nitrogen concentration. Types and counts of protozoa were determined on unfiltered rumen juice taken 2 hours after the morning meal and kept in a 100 ml fixing mixture (50 ml of glycerol + 2 ml of formaldehyde + 48 ml of distilled water). Protozoa types were counted by means of a HAWSKLEY counter, following multiple dilutions of the rumen juice, using a x100 microscope (Ogimoto and Imai, 1981). To determine gas production (CO_2 and CH_4), rumen content was collected in 100 ml plastic syringes before the morning meal and was filtered through four surgical gas layers. Rams were deprived from drinking water during the night before samples for gas determination were taken. Each syringe contained 0.5 g of crushed hay, 10 ml of rumen juice and 40 ml of artificial saliva and the mixture was incubated during 48 h at 39 °C (Menke and Steingass, 1988). DM disappearance was determined using the *in sacco* method using nylon bags with 50 micron diameter. Each bag contained 3 g of crushed hay. The bags were incubated in the rumen for 48 hours. Diet effects were tested using a one way analysis of variance (SAS®, 1989).

Results and discussion

The rumen juice pH was comparable (P>0.05) for the RS and RV concentrates (Table 1). In fact, pH values were 6.18±0.17 and 6.15±0.16 for RV and RS regimens, respectively. These pH levels are favourable for rumen flora proliferation. The proportion of ammonia nitrogen tended to be higher for the RV compared to the RS regimen (P>0.05). This result was in agreement with previous reports by Rouissi and Guesmi (2004). This may be explained by the protein quality of *Vicia faba* grains (rich in methionine and lysine) and also by the size of the ciliate protozoa population in the rumen. Total counts of protozoa types showed that animals fed the RV concentrate had higher protozoa population in the rumen (6.07 vs. 5.24×10^5/ml). However, proportions of protozoa types were (*Entodinium, Isotricha, Ophryoscolex, Polyplastron*) similar for both regimens (P<0.05).

Table 1. Fermentation parameters in the rumen of Sicilo-Sarde rams.

		Time, h			
		0	2	5	8
pH	RV regimen	6.38(0.15)	6.19(0.12)	6.02(0.18)	6.08(0.14)
	RS regimen	6.39(0.12)	6.15(0.13)	6.00 (0.12)	6.06(0.12)
		NS	NS	NS	NS
$N-NH_{3,}$ g/100ml	RV regimen	11.25(2.19)	15.28(4.76)	13.33(4.2)	11.05(2.4)
	RS regimen	12.01(1.49)	14.6(3.28)	12.32(2.27)	10.16(1.6)
		NS	NS	NS	NS

(): Standard deviation; NS: Not significant ($\alpha = 5\%$, $P>0.05$); RS: Soya concentrate, RV: *Vicia faba* concentrate.

This result confirmed the positive correlation between protozoa counts and ammonia nitrogen concentration (Jouany and Senaud, 1982). Total gas production was 53.5 and 51.8 ml for the RV and RS regimens, respectively. Values of *in sacco* disappearance of DM were higher ($P<0.05$) for the RS (34.05%) than for the RV (32.87%) regimen because of differences in chemical composition and quality of the protein source. Elevated protozoa counts when using proteins in *Vicia faba* resulted in increased CO_2 and CH_4.

Conclusion

The use of *Vicia faba* in replacement of soya beans may be a good alternative in the formulation of concentrates for the Tunisian Sicilo-Sarde dairy sheep. Fermentation parameters (pH, $N-NH_3$, protozoa and gas) from the concentrate including *Vicia faba* beans in addition to barley were comparable to those obtained with a concentrate including soya beans as a protein source.

References

Jouany, J.P. and J. Senaud, 1982. Influence des ciliés du rumen sur la digestion des différents glucides chez le mouton. I. Utilisation des glucides pariétaux (cellulose et hémicellulose) et de l'amidon. Reprod. Nutr. Dev. 22: 735-752.

Menke, K.H. and H. Steingass, 1988. Estimation of the energetic feed value obtained from chemical analysis and *in vitro* gas production using rumen fluid. Anim. Res. Develop. 28: 7-55.

Ogimoto, K. and S. Imai, 1981. Atlas of rumen microbiology. Jap. Sci. Soc. Press, Tokyo, Japan, 231 pp.

Rouissi, H. and A. Guesmi, 2004. Etude comparée de la population des protozoaires ciliés dans le rumen des ovins et caprins. Opt. Medit. A, 59: 57-59.

Rouissi H., B. Rekik, H. Selmi, M. Hammami and A. Ben Gara, 2008. Performances laitières de la brebis Sicilo-Sarde Tunisienne complémentée par un concentré local. Livestock Research for Rural Development 20: #110. *Retrieved* from http://www.lrrd.org/lrrd20/7/roui20110.htm

SAS®, 1989. User's Guide, Version 6.10 for Windows. SAS Inst. Inc., Cary, NC, USA.

Diet selection and rumen fermentation parameters of sheep grazing four subtropical pastures during the summer

A. Hassen and W.A. van Niekerk
Department of Animal and Wildlife Sciences, University of Pretoria, Pretoria 0002, South Africa; Abubeker.hassen@up.ac.za

Introduction

Nutritive value of grazed pasture is governed by complex relationships between the pasture and the animal. Due to this, sampling of herbage to predict the nutritive value provides little information regarding grazed pastures. This is due to the influence of selective grazing on pasture quality and the spatial distribution of the preferred herbage within the sward. The present study was aimed at comparing the nutritive value of four sub-tropical pastures through assessment of diet quality and rumen fermentation parameters in sheep grazing the pastures.

Material and methods

The study was conducted at the Hatfield Experimental Farm of the University of Pretoria, South Africa. Four tropical grass pastures (*Panicum maximum* cv. Gatton, *Anthephora pubescens* cv. Wollie, *Digitaria eriantha* ssp. *eriantha* and *Chloris gayana* cv. Katambora) were established each in a 0.4 ha paddock. The pasture was not replicated.
Sixteen mature South African Mutton Merino wethers equipped with ruminal and abomasal cannulae were randomly allocated to four pasture treatments. An additional 4 oesophageal fistulated wethers were allocated randomly to each pasture treatment to estimate the diet quality. Dry matter, ash and N concentrations were determined according to AOAC (2000), neutral detergent fibre (NDF) and acid detergent fibre (ADF) concentration according to the method of Robertson and Van Soest (1981) and acid detergent lignin (ADL) according to Goering and Van Soest (1970). *In vitro* digestible organic matter (IVDOM) was determined according to the method of Tilley and Terry (1963) as modified by Engels and Van der Merwe (1967). Ruminal fluid was collected at each sampling and the supernatants were used for determination of rumen ammonia nitrogen (NH_3-N) and volatile fatty acid (VAF) concentrations as described by Morgan *et al.* (1976).
All data from the experiments were analysed using Proc GLM of SAS® (2001). The statistical model included pasture species and error, and differences ($P<0.05$) between means were tested using the Bonferroni test according to Samuels (1989).

Results and discussion

Sheep grazing *A. pubescens* selected a diet low in NDF and high in N compared to those sheep grazing the other pastures (Table 1). The IVDOM of diets for sheep grazing *A. pubescens* was higher compared to those sheep grazing *C. gayana*. The ADL concentration of the selected diet was lower in *P. maximum* compared to *D. eriantha*.
Rumen ammonia nitrogen (NH_3-N) concentration was high for sheep grazing *A. pubescens* while the lowest was recorded for sheep grazing a *D. eriantha* pasture (Table 2). The level of NH_3-N recorded for *A. pubescens* pasture (26.7 mg/100 ml) was higher while that of *D. eriantha* pasture (16.8 mg/ 100 ml) was slightly lower than the optimum level (20-24 mg/100 ml rumen fluid) suggested by Orskov (1982) for maximum fermentation rate. The highest and lowest total volatile fatty acids (VFA) were recorded for sheep grazing *P. maximum* and *C. gayana* pasture, respectively. The pasture species did not differ in terms of the molar proportion of acetic acid, propionic acid, butyric acid or acetic acid to propionic acid ratio.

Table 1. Diet composition of lambs grazing Panicum maximum, Anthephora pubescens, Digitaria eriantha *and* Chloris gayana *pasture during the summer.*

Species	Chemical composition (% DM)					IVDOM
	N	Ash	NDF	ADF	ADL	
Panicum maximum	3.2[b]	11.7[ab]	56.2[c]	28.1	4.1[b]	62.7[ab]
Anthephora pubescens	3.4[a]	12.1[a]	54.1[d]	28.7	4.3[ab]	66.2[a]
Digitaria eriantha	2.7[c]	11.6[ab]	61.4[b]	27.8	4.4[a]	62.6[ab]
Chloris gayana	2.6[c]	10.6[b]	63.6[a]	29.7	4.3[ab]	60.7[b]
SE	0.03	0.4	0.6	1.2	0.1	1.6

[ab] Means in the same column followed by different superscripts differ (P<0.05).

Table 2. Rumen parameters for sheep grazing four sub-tropical grass pasture species.

Species	NH_3-N (mg/ 100 ml)	VFA (mmol/ 100 ml)	Molar proportion			Acetic/ Propionic acid
			Acetic acid	Propionic acid	Butyric acid	
Panicum maximum	19.8[b]	19.2[a]	0.69	0.21	0.09	3.3
Anthephora pubescens	26.7[a]	18.3[ab]	0.69	0.22	0.09	3.2
Digitaria eriantha	15.9[c]	17.7[ab]	0.69	0.20	0.09	3.4
Chloris gayana	18.9[bc]	16.8[b]	0.70	0.19	0.09	3.6
SE	1.2	0.7	0.007	0.007	0.006	0.2

[ab] Means in the same column followed by different superscripts differ (P<0.05).

Conclusion

This study shows that diets selected from *A. pubescens* pasture are less fibrous, more nutritious and could likely result in better animal performance during the summer season followed by *P. maximum* and *D. eriantha*.

References

AOAC, 2000. Official methods of analysis of the Association of Official Analytical Chemists. 15th Edition, AOAC, Washington D.C., USA.

Engels, E.A.N. and F.J. Van der Merwe, 1967. Application of an *in vitro* technique to South African forages with special reference to the effect of certain factors on the result. S. Afr. J. Agr. Sci. 10: 983-995.

Goering, H.K. and P.J. Van Soest, 1970. Forage fiber analysis (apparatus, reagents, procedures and some applications). U.S.D.A.-A.R.S. Agriculture handbook no. 379, Government Printing Office, Washington, D.C., USA.

Morgan P.J.K., J.P. Pienaar and R.A. Clark, 1976. Animal based methods of determining herbage intake and quality under grazing conditions. Proc.the Grassl. Soc. Southern Afr. 11: 73-76.

Ørskov E.R., 1982. Protein Nutrition in Ruminants. Academic Press, New York, USA.

Robertson J.B. and P.J. Van Soest, 1981. The detergent system of analysis and its application to human foods. In: James, W.P.T. and Theander, O. (Eds), The Analysis of Dietary Fiber in Food. Marcel Dekker, New York. pp. 123-158.

Samuels M.L., 1989. Statistics for the Life Sciences. Collier MacMillan Publishers, London, UK.

SAS®/STAT, 2001. Statistical Analysis Systems User's Guide Version 8.2. SAS® Institute Inc., Cary, NC, USA.

Effect of different inclusion levels of oil palm fronds on *in vitro* rumen fermentation with adapted and non-adapted rumen fluid

H.A. Hassim[1], M. Lourenço[2], G. Goel[2], Y.M. Goh[1] and V. Fievez[2]
[1]Department of Preclinical Sciences, Faculty of Veterinary Medicine, Universiti Putra Malaysia, 43400, UPM Serdang, Selangor, Malaysia; [2]Laboratory for Animal Nutrition and Animal Product Quality, Faculty of Bioscience Engineering, Ghent University, Proefhoevestraat 10, 9090, Melle, Belgium; veerle.fievez@ugent.be

Introduction

Oil palm fronds (OPF) are one of the oil palm by-products and their abundance has resulted in major interest into their potential use as livestock feed in Malaysia. These OPF cannot be used as a sole source of animal feed mainly because of its poor caloric value, 5.65 MJ/kg DM. There has been limited research concerning the inclusion of OPF in ruminant diets. One study suggested an optimal inclusion level of dietary OPF in the diet of 50% and 30% for beef cattle and dairy cows, respectively (Ishida *et al.*, 1997), yet little information is available on the effects of OPF inclusion, in the diet, on animal productivity. Thus, it is important to develop viable strategies to formulate diets containing OPF to achieve optimum growth and productivity for ruminants. The present study was aimed at assessing *in vitro* the effects of different inclusion levels of OPF on rumen fermentation using adapted and non-adapted rumen fluid to OPF diet.

Material and methods

In vitro incubations were performed to assess the effect of OPF inclusion levels by adding OPF (0, 0.025, 0.050, and 0.075 g/incubation) to a standard dairy concentrate (0.250, 0.225, 0.200, and 0.175 g/incubation) for OPF 0, OPF 10, OPF 20 and OPF 30, respectively. Non-adapted rumen fluid was collected from two rumen fistulated Holstein dairy cows receiving a diet based on 8.83 kg (DM) corn silage, 0.2 kg soybean meal, 0.05 kg mineral mix (AVEVE, Belgium) and grass silage *ad libitum*, whereas the adapted rumen fluid was collected from eight four-month old Barbados Black Belly x Malin crossbred lambs that were being fed the same inclusion levels of OPF as for the *in vitro* experiments. Incubations were completed as described by Van Nevel and Demeyer (1996), and done in quadruplicate for non-adapted rumen fluid and in triplicate for adapted rumen fluid.
Non-adapted rumen fluid was collected before the morning feeding (08:00 h) whereas the adapted rumen fluid was collected during slaughter. At 0 h of incubation, samples of both adapted and non-adapted rumen fluid were collected for short chain fatty acid (SCFA) analysis. After 24 h, incubation fluid with non-adapted (0.5 ml) and adapted (2 ml) rumen fluid were recovered for SCFA analysis. Net production of SCFA was calculated by subtracting the amounts at the 0h time point from amounts obtained after 24 h of incubation. Fermentable organic matter (FOM) was calculated as FOM = Acetate/2 + Propionate/2 + Butyrate, with acetate, propionate and butyrate expressed as net molar productions.
Data were analysed using general linear model (univariate) according to $Y_{ij} = \mu + A_i + B_j + \xi_{ij}$, where Y_{ij} is the response; μ the overall mean; A_i the effect of different inclusion levels of OPF (fixed factor); B_j the effect of incubation series (random factor); and ξ_{ij} the residual error. Linear and quadratic effects of the OPF inclusion levels were evaluated through polynomial contrasts.

Results and discussion

In both incubation series, increasing amounts of OPF resulted in a linear decrease in the production of total SCFA and FOM (Table 1 and 2). In incubation with non-adapted rumen fluid, proportions of acetate and propionate linearly increased and decreased, respectively, which is in line with the increasing amounts of fibre at higher OPF inclusion levels. However, in incubation with adapted rumen fluid, the effects on the fermentation pattern are less consistent, which might be explained by the confounding effect between OPF inclusion and inoculum of the donor animal.

Table 1. Effect of increasing inclusion levels of OPF on total SCFA production (µmol/incubation), relative proportions of individual SCFA (mmol/mol total SCFA) and FOM (mg) after 24 h incubation using non-adapted rumen fluid (n=4).

	OPF 0	OPF 10	OPF 20	OPF 30	SEM	OPF inclusion	
						Linear	Quadratic
Total SCFA	1,569	1,506	1,415	1,338	24.3	<0.001	0.775
Acetate	656	658	671	681	3.61	<0.001	0.255
Propionate	244	248	234	221	5.76	0.011	0.176
Butyrate	98.5	92.8	93.8	96.3	4.25	0.771	0.364
FOM	140	134	125	119	1.88	<0.001	0.918

Table 2. Effect of increasing inclusion levels of OPF on total SCFA production (µmol/incubation), relative proportions of individual SCFA (mmol/mol total SCFA) and FOM (mg) after 24 h incubation using adapted rumen fluid (n=3).

	OPF 0	OPF 10	OPF 20	OPF 30	SEM	OPF inclusion	
						Linear	Quadratic
Total SCFA	1,309	1,192	1,093	931	22.9	<0.001	0.369
Acetate	518	549	566	545	6.36	0.013	0.07
Propionate	243	318	231	303	4.18	0.003	0.676
Butyrate	114	75.9	93.0	79.8	1.51	<0.001	<0.001
FOM	105	98.4	87.1	76.1	1.80	<0.001	0.263

Conclusion

Increasing the inclusion levels of OPF reduced rumen fermentation *in vitro*, both using non-adapted and adapted rumen fluid, with concomitant changes in the fermentation pattern.

Acknowledgement

This research was funded by the EU-community (Project TH/Asia-Link/014-141-176). Postdoctoral grant of M. Lourenço funded by the Special Research Fund, Ghent University, Belgium.

References

Ishida, M. and O. Abu Hassan, 1997. Utilization of oil palm fronds as cattle feed. Asian-Australas. J. Anim. Sci. 31: 41-47.

Van Nevel, C.J. and D.I. Demeyer, 1996. Influence of pH on lipolysis and biohydrogenation of soybean oil by rumen contents *in vitro*. Reprod. Nutr. Dev. 36: 53-63.

Identification of novel biohydrogenation intermediates formed during incubations of linoleic acid with rumen microbiota *in vitro*

A.M. Honkanen[1], J.M. Griinari[1], V. Toivonen[2], A. Vanhatalo[1] and K.J. Shingfield[2]
[1]Department of Animal Science, P.O. Box 28, FIN-00014 University of Helsinki, Finland; [2]MTT Agrifood Research Finland, FIN-30100 Jokioinen, Finland; anne.z.honkanen@helsinki.fi

Introduction

On entering the rumen, dietary lipids are exposed to microbial lipases and the fatty acids liberated are subjected to biohydrogenation (BH). Even though ruminal metabolism of dietary polyunsaturated fatty acids to 18:0 is extensive, BH intermediates can accumulate and escape the rumen. Characterising the formation of specific BH metabolites is important with respect to the nutritional quality of ruminant-derived foods and also in understanding physiological responses to lipid supplements in growing and lactating ruminants. Numerous *in vitro* and *in vivo* studies have allowed the major pathways of linoleic acid (LA) metabolism to be elucidated (Harfoot and Hazlewood, 1988), while more recent experiments have provided evidence that several minor BH intermediates are also formed (Wallace *et al.*, 2007). In this study, an *in vitro* batch culture method was used to characterise the formation of BH intermediates during incubations of incremental amounts of LA with rumen microbiota.

Material and methods

An *in vitro* batch culture method was developed to examine LA metabolism. Rumen fluid was collected before morning feeding from a non-lactating rumen-fistulated Ayrshire cow fed a grass silage-based diet (forage:concentrate ratio 70:30). Incremental amounts of LA (1, 2.5, 5, and 10 mg/flask), were added as a suspension in water, Tween80 and 2 M NaOH to incubation flasks containing hay, McDougall buffer, strained rumen fluid and nonadecanoic acid as an internal standard. Flasks were gassed with CO_2, sealed and incubated at 39 °C for 0, 1.5, 3, 4.5, 6, and 9 h on a rotating platform. Metabolism was arrested by placing flasks in ice-cold water and freezing immediately. Incubation samples were freeze-dried and lipid was extracted using a mixture of hexane and isopropanol, and converted to fatty acid methyl esters and 4,4-dimethyloxaline derivatives. BH intermediates were identified using a combination of Ag+ thin-layer chromatography, gas chromatography mass spectrometry and Ag+-high-performance liquid chromatography. Experimental data were analysed by ANOVA for repeated measures (mixed model procedure of SAS®, version 9.1) using a model that included the effects of LA addition, incubation time and incubation day. Polynomial contrasts were used to evaluate the linear, quadratic and cubic effects of incremental amounts of LA on the abundance of BH intermediates. Least squares means are presented and treatment effects were declared significant at $P<0.05$.

Results and discussion

LA was extensively metabolised to 18:0 during incubations with rumen microbiota (Table 1). However, at the highest level, LA inhibited 18:0 production. Incremental addition of LA increased linearly ($P<0.001$) the formation of *cis*-9, *trans*-11 CLA (18.9, 20.5, 24.4 and 64.9 µg/flask) and *trans*-11 18:1 (40.8, 96.6, 219 and 530 µg/flask). In addition to enhancing the formation of known metabolites, inclusion of LA also resulted in the formation and accumulation of several novel BH intermediates, including *trans*-5, *cis*-12 18:2, *trans*-8, *cis*-12 18:2, *cis*-7, *cis*-12 18:2, *trans*-6, *cis*-12 18:2 and *cis*-8, *cis*-12 18:2 (Table 1). It has often been considered that LA metabolism proceeds via isomerisation of the *cis*-12 double bond. Recent studies have demonstrated that LA metabolism occurs via two distinct mechanisms, one involving direct isomerisation yielding 10,12 conjugated products the other leading to the formation of 9,11 18:2 intermediates (Wallace *et al.*, 2007), possibly mediated by a hydrogen-abstraction catalysed reaction (McIntosh *et al.*, 2009). The occurrence of several 18:2 BH intermediates containing a *cis*-12 double bond in this experiment highlights that metabolism of LA can also proceed via action on the *cis*-9 double bond. The mechanisms involved are not known but may involve the formation of a radical intermediate as proposed for LA metabolism proceeding via the *cis*-12 double bond (McIntosh *et al.*, 2009).

Table 1. Effect of incremental levels of linoleic acid on the formation of biohydrogenation intermediates during incubations with rumen microbiota in vitro.

Production (µg/flask)	Linoleic acid (mg/flask)				SEM[1]	P^2	
	1	2.5	5	10		L	Q
cis-7, *cis*-12 18:2	7.20	17.3	60.1	227	11.95	<0.001	0.007
cis-8, *cis*-12 18:2	18.5	28.6	95.4	338	19.50	<0.001	0.009
trans-5, *cis*-12 18:2	20.2	12.4	20.1	73.1	9.43	<0.001	0.054
trans-6, *cis*-12 18:2	11.7	11.8	19.0	22.1	2.39	0.004	0.481
trans-8, *cis*-12 18:2	8.68	24.8	60.3	75.0	14.12	0.004	0.243
trans-9, *cis*-12 18:2	0	7.29	35.5	108	6.14	<0.001	0.078
Σ18:2[3]	789	128	322	963	51.53	<0.001	0.020
Σ CLA	138	164	184	305	24.26	<0.001	0.441
Σ 18:1	291	795	1855	3517	149.8	<0.001	0.505
LA	252	624	1444	4964	204.7	<0.001	<0.001
18:0	402	804	1193	344	187.6	0.289	<0.001

[1] SEM = standard error of the mean, error degrees of freedom 12.
[2] Significance of linear (L) and quadratic (Q) components of the response to linoleic acid addition. Cubic responses to linoleic acid were not significant (P>0.05).
[3] Total 18:2 excluding isomers of conjugated linoleic acid (CLA) and linoleic acid (LA).

Conclusion

A range of novel BH products were found to be formed during LA metabolism *in vitro*, including *cis*-7, *cis*-12 18:2 and *cis*-8, *cis*-12 18:2. The underlying mechanisms are not known, but current data suggests that metabolism of linoleic acid by rumen microbiota may also proceed via action on the *cis*-9 double bond possibly mediated via the formation of a radical intermediate.

References

Harfoot, C.G. and G.P. Hazlewood, 1988. Lipid metabolism in the rumen. In: Hobson P.N. (Ed), The Rumen Microbial Ecosystem. Elsevier Science, New York, NY, USA, pp. 285-322.

McIntosh F.M, K.J. Shingfield, E. Devillard, W.R. Russell and R.J. Wallace, 2009. Mechanism of conjugated linoleic acid and vaccenic acid formation in human faecal suspensions and pure cultures of intestinal bacteria. Microbiol. 155: 285-294.

Wallace, R.J., N. McKain, K.J. Shingfield and E. Devillard., 2007. Isomers of conjugated linoleic acids are synthesized via different mechanisms in ruminal digesta and bacteria. J. Lipid Res. 48: 2247-2254.

The effect of subacute ruminal acidosis induction and recovery on rumen methanogen density in dairy cattle

S.E. Hook[1], M.A. Steele[1], K.S. Northwood[2], A.-D.G. Wright[2] and B.W. McBride[1]
[1]Department of Animal and Poultry Science, University of Guelph, Guelph, Ontario, Canada N1G 2W1; [2]CSIRO Livestock Industries, Queensland Bioscience Precinct, 306 Carmody Rd., St. Lucia, Queensland 4067, Australia; shook@uoguelph.ca

Introduction

Ruminant production systems commonly feed high-concentrate diets during lactation to meet energy requirements, but the feeding of highly fermentable carbohydrates can cause a decline in ruminal pH, resulting in subacute ruminal acidosis (SARA). Alteration of microflora during a diet transition has been previously investigated (Tajima *et al.*, 2001), but little is known about the effect of pH on methanogens *in vivo*. Increased acidity is suspected to cause a loss of methanogens from the rumen, and has been found *in vitro* to inhibit methanogenesis at a pH less than 6.0 (Van Kessel and Russell,1996). The effect of pH on methanogens may differ *in vivo* from *in vitro* though, as methanogens are able to associate with feed particles and protozoa, possibly making them more acid-tolerant, and potentially allowing them to produce methane at a lower pH than expected. Additionally, research suggests that low pH in culture does not kill methanogens, but instead inhibits methane production until the pH recovers to a more tolerable level (Van Kessel and Russell, 1996). These experiments have not been validated *in vivo* so the objective of this research is to measure the effect of rumen pH during SARA and recovery on the density of rumen methanogens.

Material and methods

Four mature dairy cattle (760±30 kg) fitted with a rumen cannula were followed over a five week experimental period. Cattle were fed a baseline diet of 11.0 kg chopped hay and a rumen sample was taken at the end of week 0. SARA was induced over four days and maintained from week 1-3 by feeding 4.5 kg of chopped hay and 8.3 kg mixed grain daily and a rumen sample was collected on the last day of week 1. Cattle were then transitioned back to 11.0 kg chopped hay and a rumen sample was collected at the end of week 4. Indwelling pH measurements were also recorded over 48 hours immediately prior to sampling.

Total DNA was extracted from 1 ml of filtered rumen fluid using cetyltrimethylammonium bromide (CTAB) (Wright *et al.*, 1997). Quantification of methanogens by real-time PCR was performed using the primers and following the methods of Denman *et al.* (2007). ABI SDS software was used to determine the quantity of methanogens in the unknown samples.

Repeated measures analysis of methanogens per gram wet weight (mpgww) was performed with log transformed measurements at weeks 0, 1, and 4. The random effect of animal was used and week was included as a fixed effect. Methanogen density at week 0 and average pH over the experimental period were used as covariates. The analysis was performed as a mixed procedure using the SAS® statistical package (version 9.1.3: Cary, NC) (Wang and Goonewardene, 2004).

Results

The PCR efficiency of the mcrA real-time PCR was found to be 2.01 with a standard R-squared of 0.99. Baseline, week 1, and week 4 methanogen density measurements are shown in Table 1. Repeated measures analysis of mpgww with baseline (week 0) values as the covariate did not find any significant effect of baseline mpgww (P=0.30) or week (P=0.67). When repeated measures analysis was performed for mpgww with average pH (weeks 0, 1, and 4 mean ± SE of 6.40±0.04,

5.86±0.05, and 6.55±0.06, respectively) as the covariate, there was no significant effect of average pH (P=0.19) or week (P=0.30).

Table 1. Methanogen density from the rumen fluid of each animal at week 0, 1, and 4.

	Methanogens per gram wet weight		
	Week 0	Week 1	Week 4
Animal 1	1.14×10^7	1.81×10^6	2.32×10^6
Animal 2	1.27×10^6	1.32×10^6	4.97×10^6
Animal 3	5.35×10^6	1.56×10^6	2.55×10^6
Animal 4	1.79×10^7	3.43×10^6	3.50×10^4
Mean±SE	$8.98\times10^6\pm3.63\times10^6$	$2.03\times10^6\pm4.76\times10^5$	$2.47\times10^6\pm1.01\times10^6$

Discussion and conclusion

Our research found that dietary SARA induction does not significantly decrease the number of methanogens in the rumen of dairy cattle. This suggests that low pH associated with SARA may not kill methanogens, but could suppress their activity, as previously suggested by *in vitro* experiments. Future research will be conducted to determine the effect of SARA on methanogen diversity to determine whether some methanogen species are more acid tolerant than others.

Acknowledgement

The authors wish to acknowledge Dr. Ousama AlZahal (University of Guelph) and Dr. Stuart Denman (CSIRO Livestock Industries) for technical assistance, Dr. Margaret Quinton (University of Guelph) for statistical analysis support, and the barn staff.

References

Denman, S.E., N.W. Tomkins and C.S. McSweeney, 2007. Quantitation and diversity analysis of ruminal methanogenic populations in response to the antimethanogenic compound bromochloromethane. FEMS Microbiol. Ecol. 62: 313-322.

Tajima, K., R.I. Aminov, T. Nagamine, H. Matsui, M. Nakamura and Y. Benno, 2001. Diet-dependent shifts in the bacterial population of the rumen revealed with real-time PCR. Appl. Environ. Microbiol. 67: 2766-2774.

Van Kessel, J.A.S. and J.B. Russell, 1996. The effect of pH on ruminal methanogenesis. FEMS Microbiol. Ecol. 20: 205-210.

Wang, Z. and L.A. Goonewardene, 2004. The use of MIXED models in the analysis of animal experiments with repeated measures data. Can. J. Anim. Sci. 84: 1-11.

Wright, A.-D.G., B.A. Dehority and D.H. Lynn, 1997. Phylogeny of the rumen ciliates *Entodinium, Epidinium* and *Polyplastron* (Litostomatea: Entodiniomorphida) inferred from small subunit ribosomal RNA sequences. J. Eukaryot. Microbiol. 44: 61-67.

Selective enrichment, isolation and characterisation of fast-growing acid-tolerant lactate utilisers from rumen contents of animals on high-energy diets

C.H. Horn[1], *A. Kistner*[2] *and G. Fouche*[2]
[1]*KK Animal Nutrition, PO Box 10520, Centurion 0046, South Africa;* [2]*Agricultural Research Council, Private Bag X2, Irene 0062, South Africa; chorn@kkan.com*

Introduction

Lactate utilising bacteria in the rumen can be a practical tool to prevent lactic acidosis in ruminants fed high energy diets. The present study was targeted at the isolation, screening and selection of such strains of lactate-utilising bacteria, starting from rumen contents of cattle that had successfully adapted to high-concentrate diets. The strategy was aimed at enrichment, from the naturally pre-selected microbial population present in rumen contents of animals on high-concentrate diets, those strains which could thrive under sustained, simultaneous pressure of a combination of discriminatory factors. The application of continuous culture systems is highly suitable for this purpose (Veldkamp, 1970; Krieg, 1981). However, by opting for the pH-auxostat version of continuous culture (Martin and Hempfling, 1976) and keeping the pH differential between control point and the incoming poorly-buffered medium small, a further selection pressure was applied. This was singling out, from the heterogeneous population, the fastest-growing individuals under the given, constant environmental conditions. These individuals will determine the rate at which lactic acid is converted to weaker acids, and thus the rate of deviation from pH set point. This, in turn, will determine the rate at which fresh medium will be added, i.e. the dilution rate of the system. In time, slower growing species and strains in the population will be washed out of the bioreactor.

Material and methods

Samples of rumen contents were collected and were transferred directly into the bioreactor. A New Brunswick Scientific Bioflo 1 continuous culture system was modified into a pH-auxostat by converting the pH-dosing pump to a medium addition pump. Poorly buffered medium was added whenever the pH increased below the set value until the desired value was reached. The maximum dilution rate obtained for a given organism during auxostat cultivation is a measure of the maximum growth rate of that organism during that condition.
Filtered rumen fluid was used to fill the bioreactor (270 ml) initially and the titrator activated to add sterile medium (Medium 1) to the culture proportionally to the increase in pH of the culture. Continuous culturing followed until a pure culture was observed microscopically. The growth rates of the isolates were also verified using the batch cultivation technique and monitoring the increase in optical density over time.

Results

pH auxostat isolations: Seven pH auxostat enrichment experiments were conducted, with the control point set initially at pH 5.3 and later at 5.0. In most of the runs large cocci, occurring in pairs and short chains, became dominant within 48 h to the extent that no other morphological types could be detected microscopically.
Effect of pH on specific growth rates of the pH-auxostat isolates and the type strain: Specific growth rates of four of the pH-auxostat isolates, as well as those of the type strain of *M. elsdenii*, ATCC 25940, at pH values between 4.5 and 6.5 were determined. They represent averages between

values determined with the pH-auxostat and values calculated from turbidity measurements on batch cultures at the same pH values. The growth rate of *M. elsdenii* ATCC 25940 rose slowly from about 0.07/h at pH 4.5 to a peak of 0.66/h at pH 6.0 and then declined rapidly towards pH 6.5. In comparison, the isolates from the pH-auxostat enrichments, CH3, CH4, CH6 and CH7 attained maximum growth rates of 0.86, 0.94, 0.93 and 0.66/h respectively, all at pH 5.5. Of these isolates, CH4 was the most acid tolerant, with a growth rate of 0.39/h at pH 4.5, while the organism with the second best acid tolerance, CH6, attained a growth rate of only 0.19/h at this pH value. Isolates CH3, CH4 and CH6 showed a sharp decline in growth rate between pH 5.5 and 6.0, whereas, in the case of CH7, the effect of pH on growth rate in the range pH 5.0 to 6.0 was relatively small. The same applied to ATCC 25940 in the range pH 5.5 to 6.5.

Conversion of lactate by isolate CH4: Isolate CH4 was grown in the chemostat at three dilution rates, namely 0.94, 0.83 and 0.75/h on lactate medium. Samples were taken during steady state and analysed for utilisation of lactic acid and production of volatile fatty acids (VFA's). The same was done for a sample from a batch culture collected during stationary phase. The analyses were corrected for the concentrations of these acids present in sterile medium.

Although a relatively narrow range of dilution rates was covered in this experiment, a marked shift in the relative production of fatty acids occurred. The proportions of n-butyric and n-valeric acids declined sharply with increasing dilution rate; right down to zero at dilution rate (D) 0.94/h in the case of the latter. Acetate and propionate concentrations peaked at D=0.83/h. As was to be expected, lactate utilisation decreased with increase in dilution rate. During the cultivation D=0.75/h more than 40% of the lactate was converted to VFA's and although lactate utilisation was high a large proportion of the available energy was wasted. When CH4 was cultivated in batch it produced mainly acetate and propionate. The concentrations of VFA's produced during batch cultivation were much lower than expected and a possible explanation is that CH4 utilises the VFA's when lactate is depleted.

Presumptive identification of isolates: The isolates obtained were presumptively identified as strains of *Megasphaera elsdenii*.

Discussion and conclusion

The isolates could be grown on a relatively simple semi-defined medium. However, despite of the high maximum specific growth rates of the isolates, the maximum biomass output rates in continuous culture, expressed in grams per litre culture per hour, were low. This can be ascribed to the low yields per gram of substrate metabolised, a situation common to most anaerobic organisms. A favourable observation with regards to practical applications was that the bacteria survived for up to 20 d at 25 °C on a semi defined medium.

References

Martin, G.A. and W.P. Hempfling, 1976. Method for the regulation of population density during continuous culture at high growth rates. Arch. Microbiol. 107: 41-47.

Krieg, R., 1981. Enrichment and isolation. In: P. Gerhardt (Ed.), Manual of Methods for General Bacteriology. American Society for Microbiology, Washington, pp. 112-114.

Veldkamp, H., 1970. Enrichment cultures of prokaryotic organisms. In: J.R. Norris and D.W. Ribbons (Eds.), Methods in Microbiology, vol. 3A. Academic Press, London, pp. 305-361.

Vitamins D and E are not metabolised in the rumen of high yielding dairy cows

L. Hymøller and S.K. Jensen
Department of Animal Health, Welfare and Nutrition, Aarhus University, Faculty of Agricultural Sciences, Blichers Allé 20, Box 50, DK-8830 Tjele, Denmark; lone.hymoller@agrsci.dk

Introduction

The utilisation of fat-soluble vitamins in ruminants is low compared to monogastric animals and the rumen with its fermentative environment has been traditionally regarded as a major site of loss of vitamins. Today it is accepted that α-tocopherol (vitamin E; α-TOC) is stable in the rumen and a low utilisation of α-TOC may therefore be caused by low absorption. A prerequisite for absorption of α-TOC from *all-rac*-α-tocopheryl acetate (α-TOC-Ac) is hydrolysis of the α-TOC-Ac to α-TOC by pancreatic carboxylester hydroxylase and bile acids (Hidiroglou *et al.*, 1989); incomplete hydrolysis of α-TOC-Ac will cause insufficient absorption of α-TOC. Absorption of ergocalciferol (vitamin D_2; ERG) and cholecalciferol (vitamin D_3; CHO) is not dependent on hydrolysis but their utilisation might be affected by ruminal degradation (Sommerfeldt *et al.*, 1979).

Material and methods

The 5 high yielding Holstein dairy cows used in the study were fed a mixed ration of barley (30%), maize (40%) and grass (30%) silage twice a day. Ten to fifteen kg of rumen content were taken from each cow through a rumen fistula. A sample was taken out (0 hour sample) and the remaining content was mixed with 4 000 mg (α-TOC-Ac) and 250 mg of both ERG and CHO. After mixing the rumen content was returned to the respective cows. Samples of rumen content were collected at 0, 1, 2, 4, 6, 24, and 30 hours (*in vivo*). From the 1 hour sample, 6 sub-samples from each cow were incubated at 37 °C and taken out at 2, 4, 6, 12, 24, and 30 hours (*in vitro*). Samples were analysed by HPLC for content of ERG and CHO as described by Hymøller (2009) and for α-TOC, α-TOC-Ac, β-carotene (β-CAR) and lutein (LUT) as described by Jensen *et al.* (1998). The results from the *in vitro* study were analysed using the GLM procedure of SAS® system (SAS Institute Inc., Cary, NC, USA) testing if curve slopes were different from 0.

Results and discussion

Results from the rumen content regarding α-TOC, α-TOC-Ac, CHO, ERG, β-CAR, and LUT from the *in vivo* and *in vitro* studies are shown in Table 1. *In vivo,* the concentration of the added α-TOC-Ac, CHO, and ERG declined due to the dilution effect of subsequent eating and disappearance out of the rumen. In contrast, LUT and β-CAR were found at constant levels since they originate from the natural content in the feed. Surprisingly also α-TOC was found at constant levels *in vivo*, indicating that α-TOC also originated from the natural content in the feed and not from the added α-TOC-Ac. *In vitro* both the added vitamins and the feed derived β-CAR and LUT were found at constant levels. In the present research there were no indications of any degradation of ERG and CHO in intact rumen content. Previous results (Sommerfeldt *et al.*, 1979) with radio labelled ERG or CHO incubated in rumen fluid from cattle showed that as little as 10 to 25% of the added label was present in ERG or CHO after 24 hours of incubation. Earlier results regarding the ruminal fate of α-TOC showed substantial degradation of the vitamin (Alderson *et al.*, 1971).

In the present research, α-TOC originating from the feed as well as the added α-TOC-Ac was stable in rumen content. Consistent with the present results, other research on the stability of α-TOC in the rumen has shown no degradation of α-TOC in the rumen or in rumen content (Weiss *et al.*,

1995). The stability of α-TOC-Ac in the rumen is, however, shown here for the first time, and it can be speculated if the low utilisation of α-TOC-Ac in cattle is associated with limited hydrolysis of α-TOC-Ac to α-TOC in the duodenum (Hidiroglou *et al.*, 1989). Natural carotenes from feed have, in agreement with the present results, been shown to be stable in incubation studies with rumen fluids from cattle and goats (Mora *et al.*, 1999).

Table 1. Vitamin concentration (μg/g dry matter) during in vivo *and* in vitro *studies of their stability in rumen content from high yielding dairy cows.*

in vitro							*in vivo*						
Time, h	α-TOC	α-TOC-Ac	CHO	ERG	β-CAR	LUT	Time, h	α-TOC	α-TOC-Ac	CHO	ERG	β-CAR	LUT
							0	19.1	31.3	0.4	0.4	31.0	102.0
2	21.8	93.7	12.3	10.4	30.8	96.4	1	18.8	83.9	8.7	7.6	30.6	100.1
4	21.0	95.0	13.4	11.1	32.9	103.6	2	20.0	95.3	11.4	9.7	32.2	104.2
6	23.2	106.8	12.7	10.7	33.7	104.4	4	22.8	100.4	9.0	8.0	32.8	105.4
12	24.2	111.2	15.1	13.1	35.0	109.4	6	21.9	84.6	8.3	6.8	33.4	106.0
24	22.1	108.9	13.1	11.6	34.4	107.2	24	19.0	46.7	1.1	1.3	29.7	96.9
30	22.2	122.7	13.1	11.2	32.7	101.7	30	20.3	42.3	0.4	0.7	31.4	101.6

Conclusion

The present experiment showed that ERG, CHO, α-TOC, α-TOC-Ac, β-CAR and LUT are stable in rumen content both *in vivo* and *in vitro*. The shown resistance of α-TOC-Ac against hydrolysis might cause reduced absorption in the duodenum.

Acknowledgement

The authors wish to thank N.B. Kristensen for providing the cows, 'Landbrugets Hudefond' for financial support, and laboratory staff E. Lyng and M. Würtz.

References

Alderson, N.E., G.E. Mitchell, C.O. Little, R.E. Warner and R E. Tucker, 1971. Preintestinal disappearance of vitamin E in ruminants. J. Nutr. 101: 655-660.

Hidiroglou, N., L.R. McDowell and O. Balbuena, 1989. Plasma tocopherol in sheep and cattle after ingesting free or acetylated tocopherol. J. Dairy Sci. 72: 1793-1799.

Hymøller, L., 2009. Vitamin D status and supply in organic dairy cattle. Ph.D.-thesis. University of Copenhagen, Faculty of Life Sciences. 143 pp.

Jensen, S.K., C. Jensen, K. Jakobsen, R.M. Engberg, J.O. Andersen, C. Lauridsen, P. Sørensen, P. Henckel, L.H. Skibsted and G. Bertelsen, 1998. Supplementation of broiler diets with retinol acetate, ß-carotene or canthaxanthin: Effect on vitamin and oxidative status of broilers in vivo and meat stability. Acta Agric. Scand. A Anim. Sci. 48: 28-37.

Mora, O., J.L. Romano, E. Gonzáles, F.J. Ruiz, and A. Shimada, 1999. *In vitro* and *in situ* disappearence of β-carotene and lutrin from lucerne (*Mecicago sativa*) hay in bovine and caprine ruminals fluids. J. Sci. Food Agric. 79: 273-276.

Sommerfeldt, J.L., R.L. Horst, E.T. Littledike and D.C. Beitz, 1979. *In vitro* degradation of cholecalciferol in rumen fluid. J. Dairy Sci. 62 (suppl. 1): 192-193.

Weiss, W.P., K.L. Smith, J.S. Hogan and T.E. Steiner, 1995. Effect of forage to concentrate ratio on disappearance of vitamin A and E during *in vitro* ruminal fermentation. J. Dairy Sci. 78: 1837-1842.

Effects of dietary linoleic and linolenic acids on the rumen population of cellulolytic bacteria and ciliate protozoa in dairy cows

M. Ivan[1,2], *J. Chiquette*[1], *H.V. Petit*[1] *and A.R. Alimon*[2]
[1]*Dairy and Swine Research and Development Centre, Agriculture and Agri-Food Canada, 2000 College Street, P.O Box 90 STN Lennoxville, Sherbrooke, Quebec, J1M 1Z3, Canada;* [2]*Institute of Tropical Agriculture, University Putra Malaysia, 43400 UPM Serdang, Selangor, Malaysia; michael.ivan@agr.gc.ca*

Introduction

The negative effects of the presence of ciliate protozoa species in the rumen on the efficiency of ruminal microbial synthesis of protein and on the duodenal flow of non-NH_3-N components are well established (Ivan *et al.*, 2000). Dietary fats and oils are toxic to protozoa, but contain mid-chain fatty acids that are also toxic to cellulolytic bacteria (Henderson, 1973). However, dietary supplementation with high linoleic acid – variety of sunflower seeds reduced the protozoa concentration in rumen fluid and the protein requirement in the diet, while producing higher ruminal pH and higher digestion of fibre in lambs (Ivan *et al.*, 2004). It was postulated that the increased dietary concentration of linoleic acid (C-18:2) had positive effects on the cellulolytic ruminal bacteria. This was tested in the present experiment with dairy cows fed diets with oilseed supplements containing high concentration of linoleic (Linola) or linolenic (Nulin) acids.

Material and methods

Sixteen Holstein rumen – cannulated dairy cows were assigned to four dietary treatments following a 4-wk feeding of a control diet based on corn silage and soybean meal. The four diets were control or diets formed by replacement of soybean meal with high C-18 containing oilseeds Linola (high in C-18:2), Nulin (high in C-18:3), or Linola plus Nulin (equal concentration of C-18:2 and C-18:3). All diets contained (dry matter basis) approximately 16% of crude protein and similar concentration of dietary fibre. The 3 oilseed diets each contained 5% of C-18 fatty acids (6% of oil). The cows were fed *ad libitum* twice daily and had a free access to drinking water. Rumen contents were taken from different parts of the rumen and accumulated as a daily sample 2 h after the morning feeding, on days 0, 5, 10, 15, and 20 relative to the start of feeding of the experimental diets. A portion of the rumen contents was squeezed through one layer of cheesecloth and rumen fluid was preserved for enumeration of protozoa. The rest of the contents were used for quantification of cellulolytic bacteria by real-time PCR. The results were analysed statistically with the MIXED procedure of SAS® as repeated measurements in time. Tukey's adjustment was used to further compare treatments on a 2×2 basis.

Results

The present experiment clearly established a positive effect of the dietary linoleic acid (C-18:2) on the total number of the cellulolytic species of bacteria, and a decreasing effect on the total numbers of ciliate protozoa in rumen fluid of dairy cows. In fact, on the average, the linoleic acid supplementation increased the population of cellulolytic bacteria by 43% (Table 1), while decreasing the protozoa population in the rumen by 88%. The effects were already apparent from day 5. The present results explain the increased digestibility of the dietary acid detergent fibre and neutral detergent fibre by 26% and 19%, respectively, and the decreased requirement for dietary protein by 30%, due to increased dietary linoleic acid that was obtained in the study of Ivan *et al.* (2004). It is evident, however, that there was no significant positive effect of the dietary linolenic acid (C-18:3) alone or in combination with linoleic acid on the rumen population of cellulolytic bacteria, while the decrease in the rumen protozoa population due to linolenic acid was less than half (37%) of that of linoleic acid (88%).

Table 1. Effects of dietary linoleic, linolenic, or combination of the two fatty acids on the mean concentration of total cellulolytic bacteria ($n \times 10^8$/ml) and protozoa ($n \times 10^3$/ml) in rumen fluid during the 20 d of dietary supplementation.

Day	Control		Linoleic acid		Linolenic acid		Combination	
	Mean	SE	Mean	SE	Mean	SE	Mean	SE
Bacteria								
0	1.95	0.40	1.87	0.67	2.39	0.82	2.63	0.94
5-20[1]	2.78[a]	0.56	3.94[b] (42)[2]	0.72	3.14[ab] (13)	0.35	3.43[ab] (23)	0.57
Protozoa								
0	504	119	379	116	341	136	307	55
5-20[1]	633[a]	84	78[bc] (88)	25	400[b] (37)	48	45[c] (93)	12

[a,b,c]Means within the same line followed by the same superscript are statistically not different (P>0.05).
[1]Average for days 5 to 20.
[2]Values in brackets are % increase for bacteria and % decrease for protozoa as compared to the Control.

Discussion and conclusion

The present experiment confirmed the postulation by Ivan *et al.* (2004) that the dietary supplements (5% of DM) of linoleic acid have positive effects on the rumen population of cellulolytic bacteria and on the digestion of dietary fibre. Linoleic acid is a precursor for ruminal synthesis of human health promoting conjugated linoleic acids (CLA). Moreover, the decreased protozoa populations due to the dietary linoleic acid decrease the availability of hydrogen for ruminal methanogenesis and the requirements for dietary protein. Therefore, dietary supplements of linoleic acids are also expected to increase the CLA concentrations in milk and meat and decrease the excretion of methane and nitrogen to alleviate the environmental impact of ruminant production.

References

Henderson, C., 1973. The effects of fatty acids on pure cultures of rumen bacteria. J. Agric. Sci. Cambr. 81: 107-112.

Ivan, M., L. Neill and T. Entz, 2000. Ruminal fermentation and duodenal flow following progressive inoculations of fauna-free wethers with major individual species of ciliate protozoa or total fauna. J. Anim. Sci. 78: 750-759.

Ivan, M., P.S. Mir, Z. Mir, T. Entz, M.L. He and T.A. McAllister, 2004. Effects of sunflower seeds on rumen protozoa and growth of lambs. Br. J. Nutr. 92: 303-310.

Effect of maturity stage at harvest on the number of large particles in faeces from pregnant ewes fed grass silage

A.R. Jalali[1], P. Nørgaard[1] and E. Nadeau[2]
[1]Deptartment of Animal and Veterinary Basic Sciences, Faculty of Life Sciences, University of Copenhagen, Grønnegårdsvej 3, 1870 Frederiksberg C, Denmark; [2]Deptartment of Animal Environment and Health, Swedish Univiversity of Agricultural Sciences, Skara, Sweden; alja@life.ku.dk

Introduction

The rumen, as a system, has the capacity to selectively retain large particles (LP). Poppi *et al.* (1980) defined critical particle size for retention to be 1.18 mm, based on an observation that only 5% of the particle dry matter (DM) of faeces from ruminants was retained on a sieve with a pore size of 1.18 mm. Jalali *et al.* (2008) observed, in the faeces of ewes fed grass silage, that the proportion of LP retained on a 1 and 2.36 mm pore size sieve decreased with advancing stage of maturity of grass at harvest. Whilst in cattle, fed a forage diet, Nørgaard (2006) using image analysis (IA), defined critical particle length (PL) to be 5 mm. Subsequently, Nørgaard *et al.* (2007) proposed a simple and fast 'barn sieving' technique for cattle capable of recording faeces LP longer than 2×critical PL. This technique, which acts as an indicator of the capacity of the rumen to retain LP, has not, however, been applied to faeces from small ruminants. The present study therefore sought to validate a new simple 'barn sieving method' for the measurement of the number of LP in faeces, using pregnant ewes fed grass silage harvested at three stages of maturity.

Material and methods

Eighteen late pregnancy ewes bearing twins were randomly assigned to three dietary treatments; namely *ad libitum* feeding with either early cut (EC), medium cut (MC) or late cut (LC) round bale grass silage. The content of crude protein, neutral detergent fibre and acid detergent lignin in EC, MC and LC grass silage (DM basis) were 17, 11 and 8%; 45, 58, and 63% and 2, 4 and 5%, respectively. Samples of faeces (50 g) were collected daily over 4 d in the morning and afternoon. Faeces (15 g) from the pooled daily samples were washed in nylon bags, freeze dried, termed particle DM% (PDM) and sieved using square holes of 2.36 mm, before they were scanned. The length (PL) and width (PW) of the scanned particles was measured using IA and the accumulated distribution for PL values was estimated using a composite function (C(PL)) (Nørgaard, 2006). The mass proportions of particles longer than 7 or 10 mm was estimated from the C(PL). Long (PL >7 mm) and wide particles (PW >1 mm) were defined as the double critical PL and PW respectively, estimated on the basis of the C(PL) and C(PW) functions. A sample of 50 g of faeces was used for the 'barn sieving' technique. The sample was placed on a sieve with a pore size of 2.36 mm and washed with running tap water for about 10 min as described by Nørgaard *et al.* (2007). The long particles were manually selected and sorted into two groups based on their length (7-10 mm; >10 mm) and two width groups (<1 mm; >1 mm). The number of particles assigned to each length and width group combination was counted and the amount of particle DM was measured after drying (105 °C). The proportion of particle DM to PDM (PPDM) was estimated for each size group. After sieving, the number of particles in each size group was analyzed using the Kruskal-Wallis Chi-square procedure. The PPDM values were validated against the corresponding C(PL) values obtained using IA according to an analysis where DM intake was a covariate, using the MIXED procedure of SAS® (SAS Institute Inc. 2002, Cary, NC, USA).

Results

Feeding EC silage to ewes resulted in a significant increase in the number of LP >7, between 7 and 10, and >10 mm, as well as higher PPDM values for particles longer than 7 or 10 mm in faeces, compared to MC or LC silage (Table 1). A linear correlation was observed between PPDM values obtained by barn sieving and the (1-C(PL))% values obtained by IA, for particles longer than 10 mm compared to other particle length criteria (r^2=0.62; Table 2).

Table 1. Number of large faecal particles retained on a 2.36 mm sieve (mean±s.d.) and proportion of particle DM (%) for the different particle length and width groups.

Particle length (PL) group, mm	Particle width	EC	MC	LC	SEM	*P*-value
Number of particles per 50 g						
PL >10	Wide	138±37[a]	78±27[b]	46±14[b]	11	**
	Thin	318±134[a]	127±51[b]	57±22[b]	34	***
7<PL<10	Thin and wide	817±381[a]	517±196[b]	323±59[b]	102	**
PL >7	Thin and wide	1273±485[a]	721±251[b]	426±45[b]	129	**
Proportion of particle DM,%						
PL >10	Wide	0.23[a]	0.11[b]	0.08[b]	0.03	**
	Thin	0.05[a]	0.02[b]	0.01[b]	0.01	**
7<PL<10	Thin and wide	0.41	0.26	0.21	0.07	NS
PL >7	Thin and wide	0.69[a]	0.38[b]	0.30[b]	0.10	*

[a] Means within rows without common superscripts differ (*P*<0.05), NS = not significant (*P*>0.05), ****P*<0.001, ** *P*<0.01, * *P*<0.05.

Table 2. Validation of the PPDM (X) obtained by 'barn sieving' against the C(PL) (Y) values measured by image analysis for the different particle length groups[1].

Particle length (PL) group, mm	Intercept	Slope	*P*-value[2]
PL >10	0.07±0.04	1.15±0.22	***
7<PL<10	0.39±0.12**	0.54±0.20	*
PL >7	0.38±0.10**	0.81±0.19	***

[1] DM intake was included as covariate.
[2] *P*-value = probability for linearity, ****P*<0.001, ***P*<0.01, **P*<0.05.

Conclusion

The LP number in faeces from ewes fed grass silage decreased with advancing stage of maturity of grass at harvest. Moreover, the C(PL) measured using IA can be accurately predicted by 'barn sieving' due to a linear relation between the values obtained by the two techniques used to measure LP in faeces particles.

References

Jalali, A., P. Nørgaard, E. Nadeau, A. Arnesson and M.R. Weisbjerg, 2008. Faecal particle size distribution from ewes fed grass silages harvested at different stages of maturity. Grassl and Science in Europe 13: 822-824.

Nørgaard, P., 2006. Use of image analysis for measuring particle size in feed, digesta and faeces. In: K. Sejrsen, T. Hvelplund and M.O. Nielsen (Eds.), Ruminant physiology. Digestion, metabolism and impact of nutrition on gene expression, immunology and stress, Wageningen Academic Publishers, Wageningen, the Netherlands, pp. 579-585.
Nørgaard, P., E. Nadeau and M. Nordqvist, 2007. Distribution of particle size in manure from cattle – a barn sieving technique. Nordic Association of Agricultural Scientists 3: 293-294.
Poppi, D.P., B.W. Norton, D.J. Minson and R.E. Hendricksen, 1980. The validity of the critical size theory for particles leaving the rumen. J. Agric. Sci. 94: 275-280.

Degradation of lignocellulose and methane production by anaerobic fungal monoculture and their natural co-cultures with methanogens obtained from different herbivores

W. Jin, Y.F. Cheng and W.Y. Zhu
Laboratory of Gastrointestinal Microbiology, College of Animal Science and Technology, Nanjing Agricultural University, 210095, Nanjing, China; zhuweiyun@njau.edu.cn

Introduction

In the rumen, the conversion of lignocellulose to methane by anaerobic fungi and methanogens is efficient. If this conversion could be simulated *in vitro*, it could provide a promising way to produce methane. Interactions between anaerobic fungi and methanogens have been investigated by co-cultures (Bauchop *et al.*, 1981) or triple-cultures (Joblin *et al.*, 2002). However, most of these studies have been focussed on the mixture of pure methanogen and fungal strains. In the present study, natural co-cultures of anaerobic fungi with methanogens were evaluated for their ability to degrade agricultural residues and produce CMCase and xylanase.

Material and methods

A total of eight natural co-cultures of anaerobic fungi and methanogens were estimated in the present study. Two (G1, G2) of them were from the goat rumen, three (C1, C2, C3) from camel faeces, one (B2) from buffalo faeces and two (M1, M2) from mule faeces. The fungi in the co-cultures were *Piromyces* except M1 (*Anaeromyces*) and C2 (*Neocallimastix*). The methanogens in all the co-cultures were *Methanobrevibacter* spp.. Amounts of 10 ml co-cultures were inoculated into 90 ml Orpin medium C containing penicillin (1,600 IU/ml) and streptomycin (2,000 IU/ml) and grown with 8 different substrates (0.8 g) (Table 1) at 39 °C without shaking. Total gas and methane production were determined by the transducer method and GC respectively. Dry matter (DM) loss and fibrolytic enzyme activities (xylanase and CMCase) were determined at the end of fermentation by the methods described by Lowe *et al.* (1987). Cellulase and xylanase activities were expressed as micromoles of glucose (for cellulases) or xylose (for xylanase) released per min per ml of supernatant. Means of the data in triplicate were compared using a single factor ANOVA test at 95% significant level (SPSS 16.0). Regression analysis was carried out to correlate means of total gas production with the means of corresponding DM loss and methane production.

Table 1. Cell wall carbohydrate composition of lignocellulosic substrates.

Substrate	% of dry matter		
	Cellulose	Hemicellulose	Lignin
Wheat straw	35.28	31.90	22.31
Corn stem	36.84	27.77	20.66
Corn core	36.46	40.64	14.06
Bagasse	26.85	46.71	15.63
Sawdust	38.17	22.24	32.86
DDGS[1]	4.63	42.20	7.02
Wheat bran	2.82	51.02	14.33
Rice straw	21.30	30.87	24.86

[1] Distiller Dried Grains with Solubles.

Results

With the same substrate rice straw, the eight co-cultures showed different fermentation profiles. Co-culture B2 produced the largest amount of total gas (150.2 ml) and methane (21.4 ml), and had the highest DM loss (48.2%) and xylanase (17.22 U/(ml·min)) activities ($P<0.05$). Co-cultures M1 and M2 from mule faeces had the lowest performances ($P<0.05$).
Further studies with co-cultures B2, G1 and C3 growing on different substrates showed that the highest DM loss, total gas production and methane production were obtained when growing on corn core ($P<0.05$). It was followed in order by bagasse, wheat straw, and corn stem and wheat bran. But the fibrolytic activities with corn core as substrate were not significantly higher than that with the other substrates. Positive correlations were found between total gas production and the corresponding methane production ($R^2>0.98$) as well as the DM loss ($R^2>0.90$), while no correlation was found between gas production and the fibrolytic activities. To further study the relationship of anaerobic fungi and methanogens in co-cultures, antibiotic chloromycetin (50 mg/l) was added to remove methanogens. Five monocultures of anaerobic fungi were obtained (G1, G2, C1, C2, and C3), the others could not survive in their monocultures of anaerobic fungi.

Discussion and conclusion

Eight lignocellulose-degrading and methane-producing natural co-cultures were obtained from the goat rumen, and faeces of the camel, buffalo and mule. Growing on the same substrate rice straw, co-culture B2 had the best performance on DM loss, gas and methane production and xylanase activity among the 8 co-cultures isolated from 4 different herbivores. Among different substrates, corn core had the highest DM loss, methane and gas production. The natural co-cultures may have a potential for industrial use in effective conversion of lignocellulose to useful compounds. The present approach using natural co-culture of anaerobic fungi and methanogens may also provide useful means to gain insight into the *in vivo* relationship between anaerobic fungi and methanogens.

Acknowledgement

Supported by NSFC of China (30530560).

References

Bauchop, T. and D.O. Mountfort, 1981. Cellulose fermentation by a rumen anaerobic fungus in both the absence and the presence of rumen methanogens. Appl. Environ. Microbiol. 42: 1103-1110.

Joblin, K.N., H. Matsui, G.E. Naylor and K. Ushida, 2002. Degradation of fresh ryegrass by methanogenic co-cultures of ruminal fungi grown in the presence or absence of *Fibrobacter succinogenes*. Curr. Microbiol. 45: 46-53.

Lowe, S.E., M.K.Theodorou and A.P.J.Trinci, 1987. Growth and fermentation of an anaerobic rumen fungus on various carbon sources and effect of temperature on development. Appl. Environ. Microbiol. 53: 1210-1215.

The effects of a trehalose-producing *Saccharomyces cerevisiae* strain on rumen fermentation in sheep

V. Jurkovich[1], H. Fébel[2], J. Kutasi[3], A. Harnos[4], P. Kovács[1], L. Könyves[1] and E. Brydl[1]
[1]Szent István University, Faculty of Veterinary Science, Department of Animal Hygiene, Herd Health and Veterinary Ethology, Budapest, Hungary; [2]Research Institution for Animal Breeding and Nutrition, Herceghalom, Hungary; [3]Dr. Bata Ltd, Ócsa, Hungary; [4]Department of Biomathematics and Informatics, Budapest, Hungary; jurkovich.viktor@aotk.szie.hu

Introduction

Viable *Saccharomyces cerevisiae* yeast strains are widely used as zootechnical additives in ruminant feedstuffs for their favourable impact on milk production and the prevalence of rumen disorders. However, the *in vivo* responses of yeast culture supplementation are highly variable and apparently influenced by the composition of the diet. We suppose that the reason behind this is that the yeast strains adapt differently to rumen conditions. In a previous *in vitro* study, we concluded that a higher trehalose content in the cell wall of yeast cells improves their rate of survival in the rumen fluid (Kutasi *et al.*, 2005). The aim of our study was to investigate the effects of trehalose-producing *Saccharomyces cerevisiae* strain supplementation on rumen fermentation of sheep.

Material and methods

Three rumen cannulated Merino wethers were used in a self controlled experimental design, meaning that all three animals were used as both control and experimental animals. An experimental phase of feeding a high-energy diet (HE) was followed by a phase of a high-fibre diet (HF). The total mixed ration consisted of corn silage, meadow hay, barley grain, corn grain, extruded sunflower hull and extruded soybean meal, meaning 60.2% concentrate and 39.8% forages in the HE diet and 51.3% concentrate and 48.7% forages in the HF diet (on a DM basis). Daily amount of total mixed ration (TMR) was 1.2 kg of DM/d/animal. Both phases (HE and HF) consisted of three successive periods of 3 weeks each during which animals received the following: (1) no yeast supplementation during the control period (CO); (2) 2.5 g/d trehalose non-producing *Saccharomyces cerevisiae* (LS); and (3) 2.5 g/d trehalose-producing *Saccharomyces cerevisiae* (LSDairy). The yeast preparation was mixed in the daily amount of TMR. Rumen fluid samples were taken on the first, third and fifth day of the last week of each period, 3 hours after the morning feeding and were measured for pH and volatile fatty acid concentration (VFA). Degradability of NDF (alfalfa) and starch (corn and barley) was also estimated via the *in sacco* method (Ørskov and McDonald, 1979) in the last week of each period (NDF degradability 72 h of incubation in two repetitions, starch degradability: 8 h of incubation in four repetitions). For hypothesis testing, a general linear mixed model was fitted to the studied parameters. P-values were adjusted by Tukey-Kramer correction at multiple testing. Statistical significance was set at $P<0.05$.

Results

The main results of the experiment are summarised in Tables 1 and 2. Rumen fluid pH in the control period was lower in the HE diet compared to the HF diet ($P<0.05$). Rumen fluid pH of LS and LSDairy periods of HE and HF diets did not show a significant difference. In the HE diet, the LSDairy supplementation resulted in a higher rumen fluid pH compared to CO and LS ($P<0.05$). Total VFA concentration measured for the HE diet was significantly higher in LSDairy than in CO and LS. For the HF diet, total VFA concentration measured in LS and LSDairy were both significantly higher compared to CO ($P<0.05$). Molar proportions of acetate and butyrate were

Table 1. Rumen fermentation parameters during the experiment (Mean ± SD).

	HE diet			HF diet		
	CO	LS	LSDairy	CO	LS	LSDairy
pH	5.5±0.3[a1]	5.5±0.3[a]	5.7±0.2[b]	5.8±0.2[a2]	5.5±0.3[b]	5.6±0.2[a]
Total VFA, mmol/L	91.7±11.3[a]	94.3±14.1[a1]	101.2±19.9[b]	87.4±17.8[a]	112.6±17.8[b2]	109.7±22.1[b]
Acetate, molar%	59.3±8.7	59.4±3.9	60.2±2.3	62.7±3.9	61.0±4.8	61.9±4.7
Propionate, molar%	18.9±5.9[a]	20.5±2.3[a]	22.2±2.1[b]	19.1±4.1	18.7±2.3	20.2±2.8
Butyrate, molar%	14.9±4.1	15.8±1.8	16.1±1.7	15.2±2.1	15.8±1.8	15.4±2.6

[a,b] Means with different superscripts within a diet are significantly different ($P<0.05$).
[1,2] Means with different superscripts mark significant difference between the items HE and HF diets ($P<0.05$).
HE: high-energy; HF: high-fibre; CO:control, LS: trehalose non-producing *Saccharomyces cerevisiae* supplementation, LSDairy: trehalose producing *Saccharomyces cerevisiae* supplementation.

Table 2. In sacco degradability of hay and cereals (Mean ± SD).

	HE diet			HF diet		
	CO	LS	LSDairy	CO	LS	LSDairy
Pelleted alfalfa NDF,%	19.7±1.3[a1]	17.8±2.8[a]	21.7±0.6[b]	23.7±1.1[a2]	20.5±2.9[b]	21.4±1.4[b]
Corn grain starch,%	53.5±15.0	49.5±8.6	50.7±9.0	54.1±7.9	50.6±8.5	50.7±9
Barley grain starch,%	84.9±8.7[a]	84.2±3.9[a]	93.7±2.8[b1]	89.1±2.7[a]	85.9±5.0[b]	89.4±3.4[a2]

[a,b] Means with different superscripts within a diet are significantly different ($P<0.05$).
[1,2] Means with different superscripts mark significant difference between the items HE and HF diets ($P<0.05$).

similar between treatments ($P>0.05$) whereas the molar proportion of propionate was higher using LSDairy compared to CO and LS in the HE diet ($P<0.05$).
The NDF degradability of forage was significantly higher in LSDairy compared to CO and LS in the HE diet ($P<0.05$), but significantly lower in LSDairy and LS compared to CO in the HF diet. There was no difference in the starch degradability of corn, regarding both phases and all periods. Degradability of barley starch was significantly better with LSDairy compared to CO and LS in the HE diet. In HF, the relevant LS value is significantly lower than that of CO and LSDairy, yet CO and LSDairy did not show statistical difference.

Conclusion

As an effect of trehalose-producing *Saccharomyces cerevisae* supplementation, rumen fermentation in sheep was improved as shown by a significant increase in total VFA concentration and in starch degradability entailing an increase in the molar proportion of propionate. When feeding a diet with high fermentable carbohydrate content, the NDF degradability of hay was also improved significantly.

Acknowledgement

Our research was supported by the National Office for Research and Technology (grant no.: OMFB-1213/2004).

References

Kutasi, J., G. Both, Á. Bata and V. Jurkovich, 2005. Induction of trehalose synthesis in *Saccharomyces cerevisiae* yeast strains. In: Proceedings of 16th Hungarian Buiatrics Congress, 5-8 October 2005, Balatonfüred, Hungary (in Hungarian).

Ørskov, E.R. and Y. McDonald, 1979. The estimation of protein degradability in the rumen from determining the digestibility of feeds in the rumen. J. Agric. Sci., 92: 499-503.

Effects of rapeseed lipids in the diet on ruminal lipid metabolism and milk fatty acid composition in cows fed grass silage based diets

P. Kairenius[1], V. Toivonen[1], S. Ahvenjärvi[1], A. Vanhatalo[2], D.I. Givens[3] and K.J. Shingfield[1]
[1]MTT Agrifood Research Finland, 31600 Jokioinen, Finland; [2]University of Helsinki, 00014 University of Helsinki, Finland; [3]University of Reading, Reading, RG6 6AR, United Kingdom; piia.kairenius@mtt.fi

Introduction

Ruminant derived foods are an important source of lipid in the human diet (Givens and Shingfield, 2004) and therefore there has been intense interest in altering the composition of milk and meat to improve long-term human health. Nutritional strategies for enhancing the nutritional value of milk fat have been directed towards reducing 12:0, 14:0 and 16:0 saturated fatty acids (SFA), and increasing *cis* monounsaturated fatty acid (MUFA) and polyunsaturated fatty acid (PUFA) concentrations in milk fat. This can be achieved by using plant oils in the diet, but this often results in an accompanying increase in milk fat *trans* fatty acids (Chilliard *et al.*, 2000). There is evidence that supplementing the diet with intact oilseeds may allow more strategic changes in milk fatty acid (FA) composition compared with plant oils, due to the seedcoat protecting lipids from ruminal metabolism (Chilliard and Ferlay, 2004). To test this hypothesis, the effects of different forms of rapeseeds in the diet on ruminal lipid metabolism and milk fat composition were examined.

Material and methods

Four multiparous Finnish Ayrshire cows each fitted with a rumen cannula were randomly assigned to four dietary treatments in a 4×4 Latin square with 21 d periods. Treatments consisted of a control diet (C) or the same basal ration containing 50 g/kg dry matter (DM) of additional lipid in the form of rapeseed oil (RO), whole rapeseeds (WR) or rapeseeds milled with wheat (MR). Diets were offered *ad libitum* as total mixed rations with lipid sources replacing concentrate ingredients to minimise selection of dietary components and maintain the forage: concentrate ratio 60:40 (DM basis). All the data were analysed by ANOVA using the GLM procedure of SAS 9.1® using a statistical model that included the random effects of cow and fixed effects of period and treatment.

Results

Supplementing the diet with rapeseed feeds had no effect (P>0.05) on DM intake (19.3, 18.8, 21.8 and 20.5 kg/d, SEM 0.69) for C, RO, WR and MR, respectively), milk yield (28.5, 28.0, 28.3 and 26.9 kg/d, SEM 0.72) or milk fat (41.2, 42.2, 43.0 and 43.0 g/kg, SEM 2.09), protein (31.3, 31.0, 31.1 and 30.5 g/kg, SEM 0.28) or lactose content (48.9, 49.0, 49.6 and 48.6 g/kg, SEM 0.59). Relative to the control, rapeseed supplements increased (P<0.05) the amount of 18:0 leaving the rumen (Table 1). Inclusion of whole or milled rapeseeds in the diet enhanced (P<0.05) *cis*-9 18:1 flow at the omasum, whereas, rapeseed oil had no effect (P>0.05) compared with the control (Table 1). Rapeseed ingredients altered the flow of *trans* FA at the omasum (Table 1) due, in the most part, to increased ruminal outflow of *trans* 18:1 intermediates (41, 162, 86 and 103 g/d, SEM 15.0). Supplementing the diet with whole rapeseeds resulted in the largest increase (P<0.05) in *cis* 18:1 (17, 45, 102 and 57 g/d, SEM 9.60) and 18:2n-6 (7.97, 8.14, 26.2 and 10.6 g/d, SEM 1.92) flow at the omasum, supporting the view that the seedcoat affords some protection of unsaturated FA from ruminal metabolism. Inclusion of rapeseed lipids in the diet altered milk FA composition, resulting in a reduction (P<0.05) in milk fat SFA concentration (70.3, 56.3, 63.4 and 59.2 g/100g FA, SEM 0.86) and an increase (P<0.05) in milk fat MUFA concentrations (26.6, 40.6, 33.8 and

38.1 g/100g FA, SEM 0.83), responses that were higher for RO compared with WR and MR. Reductions in milk saturates were largely confined to decreases in the secretion of FA synthesised de novo, while the increase in milk MUFA was due, in the most part, to increases in milk fat *cis*-9 18:1 content (Table 1). Inclusion of rapeseed oil in the diet increased ($P<0.05$) milk total *trans* FA content compared with other treatments (Table 1). Rapeseed lipids in the diet had relatively minor effects on 18:2 concentrations, other than causing a reduction ($P<0.05$) in 18:2*n*-6 concentrations (1.11, 0.97, 0.98 and 0.91 g/100g FA, SEM 0.02), while rapeseed oil enhanced ($P<0.05$) milk fat *cis*-9, *trans*-11 conjugated linoleic acid content (343, 562, 274 and 325 mg/100g FA, SEM 18.07).

Table 1. Effects of rapeseed lipids in the diet on the flow of fatty acids at the omasum (g/d) and milk fatty acid composition (g/100 g fatty acids).

	Treatment[1]				SEM[2]
	C	RO	WR	MR	
Omasal canal flow, g/d					
18:0	293[b]	963[a]	787[a]	1020[a]	69.7
cis-9 18:1	10.8[c]	28.3[bc]	87.8[a]	44.6[b]	8.10
Σ *Trans* fatty acids	46.6[c]	172[a]	93.3[bc]	111[b]	16.2
Milk fatty acid composition, g/100 fatty acids					
Σ ≤14	26.3[b]	17.5[c]	21.7[b]	19.3[c]	0.58
16:0	32.0[a]	18.9[c]	24.4[b]	20.8[c]	0.82
18:0	9.91[b]	17.8[a]	15.5[a]	17.3[a]	0.48
cis-9 18:1	19.0[c]	31.0[a]	26.2[b]	30.2[a]	0.94
Σ *Trans* fatty acids	3.49[c]	6.49[a]	4.18[bc]	4.70[b]	0.33

[1] Refers to grass silage diets containing none (C) or 50 g/kg diet dry matter of supplemental lipid in the form of rapeseed oil (RO), whole rapeseeds (WR) or milled rapeseeds (MR); [2]Standard error of the mean; degrees of freedom of error 6.
[a,b,c] Means within row not sharing common roman superscripts differ significantly ($P<0.05$).

Conclusion

In conclusion, all rapeseed feeds increased flow of 18:0 at the omasum, reduced milk fat saturates and enhanced milk *cis*-9 18:1 concentrations compared with the control. Supplementing the diet with rapeseed oil increased the flow of *trans* FA at the omasum compared with the other treatments, while whole rapeseeds resulted in the largest increase in ruminal outflow of *cis*-9 18:1. However, the increases in unsaturated FA at the omasum did not result in similar increases in milk fat *cis*-9 18:1 and PUFA concentration, suggesting that the digestion and absorption of lipid contained in intact rapeseeds is lower compared with rapeseed oil or when the seedcoat is ruptured during milling.

References

Chilliard, Y. and A. Ferlay, 2004. Dietary lipids and forages interactions on cow and goat milk fatty acid composition and sensory properties. Reprod. Nutr. Dev. 44: 467-492.

Chilliard, Y., A. Ferlay, R.M. Mansbridge and M. Doreau, 2000. Ruminant milk fat plasticity: nutritional control of saturated, polyunsaturated, *trans* and conjugated fatty acids. Ann. Zootech. 49: 181-205.

Givens, D.I. and K.J. Shingfield, 2004. Food derived from animals: the impact of animal nutrition on their nutritive value and ability to sustain long-term health. Nutr. Bull. 29: 325-332.

The effects of garlic oil on *in vitro* rumen fermentation and methane production are influenced by the basal diet

C. Kamel[1], *H.M.R. Greathead*[1], *M.J. Ranilla*[2], *M.L. Tejido*[2], *S. Ramos*[2] *and M.D. Carro*[2]
[1]*Faculty of Biological Sciences, University of Leeds, Leeds LS2 9JT, United Kingdom;* [2]*Departamento de Producción Animal, Universidad de León, 24071 León, Spain; mdcart@unileon.es*

Introduction

Previous studies (Busquet *et al.*, 2005; Calsamiglia *et al.*, 2007) have shown that garlic oil (GO) supplementation to *in vitro* cultures of ruminal micro-organisms leads to reduced methane production and acetate: propionate ratios, but the effects are related to the dose. Our hypothesis was that GO effects might depend on the rumen microbial populations, and thus on both the incubated substrate and the type of diet fed to donor animals. The aim of this study was therefore to analyse the effects of different doses of GO on *in vitro* fermentation of two diets in batch cultures of mixed rumen micro-organisms from the rumen of sheep fed the same diets.

Material and methods

Eight Merino sheep (56.1±2.80 kg) fitted with permanent ruminal cannulas were used as rumen fluid donors. Four sheep received a medium-concentrate (MC; 500:500 alfalfa hay: concentrate) and the other 4 were fed a high-concentrate (HC; 150:850 barley straw: concentrate) diet for 10 d before starting the *in vitro* incubations. Samples (300 mg) of MC and HC diets were incubated with 30 ml buffered ruminal fluid in batch cultures. GO was added to the cultures at 0, 20, 60, 180 or 540 mg/l of incubation medium. After 16 h of incubation at 39 ºC, the main fermentation variables were determined. Incubations were repeated on four non-consecutive days. Data were analysed as a mixed model using the MIXED procedure (SAS® Inst. Inc., Cary, NC, USA). Five concentrations of GO, diet, and the interaction of GO x diet were included in the model as fixed effects, and incubation day was considered as a random effect. Nonorthogonal polynomial contrasts were used to test for linear effects of GO. Differences among treatments were declared at $P<0.05$, and P-values of 0.05 to 0.10 were considered as trends.

Results and discussion

There were no effects ($P=0.16$ to 0.86) of GO on pH or concentration of ammonia-N and total lactate (results not shown). As shown in Table 1, GO x diet interactions ($P=0.06$ to <0.001) were observed for most of the measured parameters. Total VFA production was not affected by any dose of GO with the HC diet, but it was reduced ($P=0.02$) by GO540 for MC diet indicating inhibition of rumen fermentation. Increasing doses of GO reduced linearly ($P<0.001$) the proportion of acetate and acetate:propionate ratio, and increased ($P<0.001$) the proportion of propionate; whereas GO showed significant effects at 60 mg/l with MC diet, a minimum concentration of 180 mg/l was required with the HC diet. The addition of GO to MC did not modify the proportion of butyrate, but in agreement to previous results (Busquet *et al.*, 2005), GO at 60, 180 and 540 mg/l increased ($P<0.05$) butyrate proportion with the HC diet. For the HC diet, GO20 reduced CH_4 and CH_4/VFA ratio by 9.6 and 12.1%, respectively, but no effects ($P>0.05$) were observed for the MC diet. GO at 60, 180 and 540 mg/l decreased CH_4 production to 87, 58 and 36% of the CON values for MC diet, and to 91, 75 and 38% of CON values for the HC diet. Hydrogen recovery, calculated from the stoichiometric relationships between the end products formed (Demeyer, 1991), was linearly lowered ($P<0.001$) by increasing GO supplementation. This could indicate accumulation of hydrogen or a reduced end product other than CH_4, hydrogen, propionate, butyrate and valerate, since these are involved in the calculation of the hydrogen balance.

Table 1. Effects of five doses of garlic oil (GO; 0, 20, 60, 180 and 540 mg/l for CON, GO20, GO60, GO180 and GO540, respectively) on total volatile fatty acid production (VFA; µmol), molar proportions (mol/100 mol) of acetate (C2), propionate (C3) and butyrate (C4), acetate:propionate ratio (C2:C3; mol/mol), CH_4 production (µmol), CH_4/VFA ratio (mol/mol) and hydrogen recovery (HR; %) after in vitro fermentation of diets (300 mg) with medium (MC) and high (HC) concentrate content by mixed rumen micro-organisms for 16 h (n = 4).

	Diet	VFA	C2	C3	C4	C2:C3	CH_4	CH_4/VFA	HR
CON	MC	2069	62.9	21.5	10.6	2.93	541	0.262	90.2
GO20	MC	2068	62.5	21.8	10.8	2.87	532	0.257	89.7
GO60	MC	2083	60.8*	23.3*	11.2*	2.62*	472*	0.227*	85.5
GO180	MC	1993	56.9*	25.6*	12.4*	2.23*	316*	0.159*	75.6*
GO540	MC	1929*	54.7*	26.8*	13.8*	2.05*	197*	0.102*	65.7*
CON	HC	2196	52.5	30.9	14.0	1.70	467	0.214	92.2
GO20	HC	2258	53.3	30.6	13.5	1.75	422*	0.188*	84.4*
GO60	HC	2233	52.9	30.9	13.5	1.71	424*	0.192*	87.8
GO180	HC	2178	50.8*	32.6*	14.0	1.57*	349*	0.160*	83.8*
GO540	HC	2203	47.9*	35.0*	14.5	1.37*	179*	0.081*	71.2*
SEM		57.4	0.53	0.60	0.29	0.066	0.29	0.0117	2.61
Significance (*P*=)									
GO[1]		0.26	<0.001	<0.001	<0.001	<0.001	<0.001	<0.001	<0.001
Diet		<0.001	<0.001	<0.001	<0.001	<0.001	<0.001	<0.001	0.02
GO x Diet		0.45	<0.001	0.06	<0.001	<0.001	0.001	0.003	0.06

* Within columns and for each diet, values differ ($P<0.05$) from CON.
[1] Linear effects of GO dose.

Conclusion

The results of this study indicate that GO may show different effects on *in vitro* rumen fermentation depending on the administered dose and the substrate composition and microbial populations in the inoculum, which could help to explain the variation in the response observed in different studies.

Acknowledgement

This research was supported by the CICYT of Spain (AGL2008-04707-C02-02) and the Consejería de Educación de la Junta de Castilla y León (Ref. GR158).

References

Busquet M., S. Calsamiglia, A. Ferret, M.D. Carro and C. Kamel, 2005. Effect of garlic oil and four of its compounds on rumen microbial fermentation. J. Dairy Sci. 88: 4393-4404.

Calsamiglia, S., M. Busquet, P.W. Cardozo., L. Castillejos and A. Ferret, 2007. Invited Review: Essential oils as modifiers of rumen microbial fermentation. J. Dairy Sci. 90: 2580-2595.

Demeyer, D.I., 1991. Quantitative aspects of microbial metabolism in the rumen and hindgut. In: Jouany, J.P. (Ed.), Rumen Microbial Metabolism and Ruminant Digestion. INRA Ed., Paris, France, pp. 217-237.

Effects of yellow grease supplementation with two levels of forage to concentrate ratios on digestion and milk production of lactating dairy cows

S. Kargar, G.R. Ghorbani and M. Alikhani
Department of Animal Sciences, Isfahan University of Technology, Isfahan 84156, Iran; kargar@ag.iut.ac.ir

Introduction

Technological processing increases the cost of the ruminally inert fat products such as hydrogenated palm oil. Because yellow grease is an inexpensive source of fat, there is great interest in maximising its utilisation by dairy cows. Differential intake, digestion and production responses to such fats may be attributed to the level of dietary fat, fatty acid profile of the fat source, or interactions between fat source and feed ingredients of the basal diet. To be beneficial, supplemental fat should not compromise rumen fermentation and, concomitantly, DMI and milk fat percentage (Onetti *et al.*, 2001). Hydrogenated palm oil and yellow grease differ in the ratio of unsaturated to saturated fatty acids (0.2 vs. 5.1, respectively). Unsaturated fatty acids are toxic for rumen microbes and decrease fibre digestion (Palmquist and Jenkins, 1980) and evidence suggests that negative responses to fat occur likely when corn silage is the major source of forage (Smith and Harris, 1993). The objective of this study was to investigate the effects of supplementing yellow grease at two forage:concentrate (F:C) ratios and to compare the effect of yellow grease to hydrogenated palm oil at a lower forage ratio, on performance and total tract digestion of nutrients, when alfalfa was the sole source of forage to avoid probable detrimental effects of yellow grease on rumen fermentations.

Material and methods

Eight multiparous Holstein dairy cows averaging 55±9 d in milk (DIM; mean ± SD) and 40.4±2.4 kg milk yield were used in a replicated 4×4 Latin square design with 21-d periods. The four diets were the following: (1) no supplemental fat and 34:66 F:C ratio (Control; DM, 74.80%; CP, 18.03%; NDF, 31.56%; NE_l, 1.55 Mcal/kg), (2) 2% hydrogenated palm oil (C16:0, 49.34%; C18:0, 30.31%; C18:1, 15.39%; C18:2, 1.03%) and 34:66 F:C ratio (HPO; DM, 74.57%; CP, 18.35%; NDF, 31.80%; NE_l, 1.63 Mcal/kg), (3) 2% yellow grease (C16:0, 11.85%; C18:0, 4.05%; C18:1, 25.68%; C18:2, 51.59%) and 34:66 F:C ratio (YG; DM, 75.81%; CP, 18.36%; NDF, 31.53%; NE_l, 1.62 Mcal/kg), and (4) 2% yellow grease and 45:55 F:C ratio (YGHF; DM, 75.00%; CP, 18.13%; NDF, 34.57%; NE_l, 1.60 Mcal/kg). Alfalfa hay (DM, 95.21%; CP, 14.23%; NDF, 43.43%) was the sole source of forage in all diets. The diets were fed as TMR, and *ad libitum* consumption was allowed.

Feed was supplied twice daily at 07:30 and 15:30 h in amounts that allowed 5 to 10% orts. Orts and the TMR amounts offered and refused were quantified on 6 consecutive days to determine intake. Cows were milked three times daily at 03:30, 13:00 and 21:30 h and milk yield was recorded and sampled at each milking during the last 7 d of each period.

The mixed procedure of SAS® (SAS Institute Inc. 2003, Cary, NC, USA) was used to analyse data. The model included the fixed effects of square, period, and treatment and cow within square was the random effect. Preplanned statistical contrasts were used to test the effect of fat supplementation (Control vs. HPO + YG); the effect of source of fat supplement (HPO vs. YG); and the effect of forage to concentrate ratio within diets supplemented with yellow grease (YG vs. YGHF). Least squares means are reported throughout. Statistical significance was set at a P-value≤0.05.

Results and discussion

Feeding YGHF significantly decreased DM and OM intake (Table 1). Digestibilities of DM and OM were not affected by fat supplementation, but were decreased in HPO fed cows compared to cows fed YG ($P<0.04$) and tended to decrease with a high forage diet. Milk yield was significantly decreased with HPO and YGHF diets compared to other diets. In addition, the yield of 4% FCM tended to increase ($P=0.07$) in fat supplemented diets. Fat supplementation and source of fat significantly increased milk fat yield and high ratio of F:C tended ($P=0.08$) to decrease it. Avila *et al.* (2000) found that linear increasing of yellow grease had no effect on DMI, OM digestion, milk production and milk fat yield but feeding fat significantly increased milk production and milk fat yield.

Table 1. Effects of different dietary fat supplements and forage:concentrate ratio on intake, digestibility, and milk yield and composition.

	Diet				SE	*P*-value		
	Control	HPO	YG	YGHF		Fat supplementation	Source of fat	F:C
Intake								
DM, kg/d	27.3	26.5	27.2	25.3	1.11	0.81	0.76	<0.01
OM, kg/d	24.7	24.0	24.7	22.7	1.00	0.53	0.19	<0.01
Digestibility								
DM,%	75.7	70.5	75.8	71.0	2.55	0.24	0.04	0.06
OM,%	76.9	71.7	77.1	72.4	2.53	0.25	0.04	0.07
Milk yield								
Actual, kg/d	38.2	37.4	40.1	38.5	1.75	0.54	<0.01	0.03
4% FCM, kg/d	36.0	36.0	39.1	37.1	2.22	0.07	<0.01	0.03
Fat,%	3.62	3.77	3.86	3.75	0.19	0.07	0.48	0.37
Fat, kg/d	1.38	1.41	1.54	1.45	0.10	0.05	0.02	0.08

Conclusion

Milk yield and fat improved due to the increased energy intake when cows consumed fat supplemented diets and when cows were fed the YG diet compared with the YGHFdiet. Based on the source of supplemental fat inclusion in the diet used in the current experiment, yellow grease had a greater feeding value and was more cost effective relative to hydrogenated palm oil for lactating dairy cows.

Acknowledgement

The authors thank the personnel of the farm animal research and teaching unit of Lavark.

References

Avila, C.D., E.J. DePeters, H. Perez-Monti, S.J. Taylor and R.A. Zinn, 2000. Influences of saturation ratio of supplemental dietary fat on digestion and milk yield in dairy cows. J. Dairy Sci. 83: 1505-1519.

Onetti, S.G., R.D. Shaver, M.A. McGuire and R.R. Grummer, 2001. Effect of type and level of dietary fat on rumen fermentation and performance of dairy cows fed corn silage-based diets. J. Dairy Sci. 84: 2751-2759.

Palmquist, D.L. and T.C. Jenkins, 1980. Fat in lactation rations. J. Dairy Sci. 63: 1-14.

Smith, W.A. and B. Harris Jr, 1993. The influence of type of forage on the production response of lactation dairy cows supplemented with different types of fat. Prod. Anim. Sci. 8: 7-21.

Improved methodology for estimating rumen protein degradation using the *in vitro* gas production technique

L. Karlsson[1], M. Hetta[1], P. Udén[2] and K. Martinsson[1]
[1]Departement of Agricultural Research for Northern Sweden, Swedish University of Agricultural Sciences, 901 83 Umeå, Sweden; [2]Department of Animal Nutrition and Management, Swedish University of Agricultural Sciences, Kungsängen Research Centre, 753 23 Uppsala, Sweden; linda.karlsson@njv.slu.se

Introduction

Rumen protein degradation is an important feed characteristic, often estimated with the *in sacco* technique (Ørskov and McDonald, 1979). However, the method is laborious, expensive and has problems with loss of undegraded N from the bags used in the technique (López, 2005). Hence, simple, inexpensive, and reliable alternative methods are required. One potential method is the gas production (GP) technique (Raab *et al.*, 1983) where *in vitro* degradable N (IVDN) is determined via ammonia-N and gas production measurements. The aims of the presented study were to improve the convenience and reliability of the GP technique for estimating rumen protein degradability of feeds and to compare the results obtained with the improved method and the *in sacco* method.

Material and methods

Cold-pressed hempseed and rapeseed cakes, rapeseed expeller, heat-treated rapeseed meal, and soybean meal were used as test feeds. The feeds were incubated with buffered rumen fluid and carbohydrates (at four concentrations), and amounts of ammonia-N and gas produced were measured after 4, 8, 12, 16, 24, and 30 h. The IVDN for each feed were estimated after each incubation period via linear regression of ammonia-N vs. GP, as described by Raab *et al.* (1983). The improved GP technique included the following changes: (1) samples for ammonia-N analysis were collected by a novel sampling device from the same incubation flask over time, and (2) the rumen fluid was pre-incubated with carbohydrates to enhance microbial activity and to reduce background ammonia-N levels. The *in sacco* method was used as a reference method. The effective protein degradation (EPD) was estimated after fitting the IVDN and *in sacco* N disappearance data to a non-linear equation (Ørskov and McDonald, 1979). The results yielded by the *in vitro* and *in sacco* techniques were compared by subjecting the EPD values to analysis of variance using the MIXED procedure in SAS®. The linear relationships between IVDN and *in sacco* N disappearance were determined by linear regression at 4, 8, 12 and 24 h of incubation.

Results

Comparisons of EPD showed differences in the protein degradability estimates due to feed and method but there was also a feed x method interaction ($P<0.001$). The GP method gave lower EPD values for the cold pressed feed cakes ($P<0.001$), while no differences were found for the other feeds (Table 1). There were linear relationships ($P<0.01$) between IVDN and *in sacco* N disappearance when data for cold-pressed cakes and expeller and meals were analysed separately with R^2 values of 0.87 and 0.93, respectively (Figure 1).

Discussion and conclusion

The modified GP technique can provide convenient estimates of IVDN that enable EPD values of protein feeds to be calculated. We believe that the technique has the potential to be developed into a useful method for estimating protein degradation in the rumen. However, additional studies are required to further improve its reliability and to evaluate the differences between the results obtained with the improved GP method and the *in sacco* method.

Table 1. Least squares means of effective protein degradation (EPD) calculated at a passage rate of 0.08/h from results obtained with the gas production (GP) and in sacco methods.

	GP method			*In sacco* method			Effects (*p*)			
	n	EPD	SD	n	EPD	SD	M	F	M×F	M within F
Hempseed cake	3	0.33	0.03	3	0.84	0.00	<0.001	<0.001	<0.001	<0.001
Rapeseed cake	4	0.59	0.05	3	0.89	0.01				<0.001
Rapeseed expeller	4	0.46	0.06	3	0.40	0.04				0.696
Rapeseed meal	4	0.36	0.06	3	0.37	0.03				1.000
Soybean meal	4	0.67	0.04	3	0.65	0.04				1.000

n = number of *in vitro* runs resulting in a calculated EPD value; SD = standard deviation; F = feed; M = method.

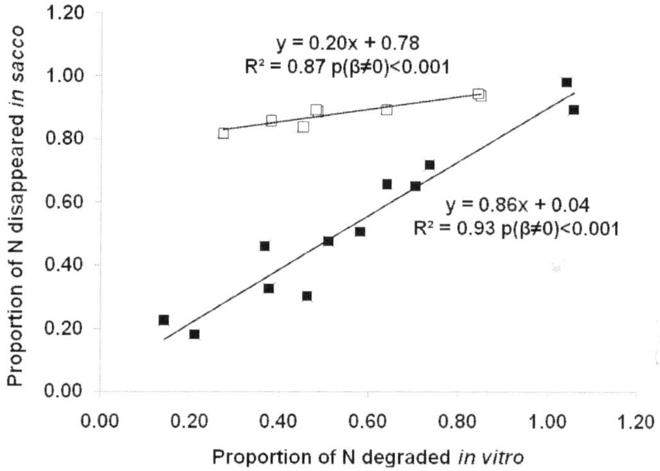

Figure 1. Relationships between in vitro degradable N and in sacco N disappearance measurements obtained in experiments with hempseed cake and rapeseed cake (□), and rapeseed expeller, rapeseed meal, and soybean meal (■) after 4, 8, 12, and 24 h incubation.

References

Lopéz, S., 2005. *In vitro* and *in situ* Techniques for estimating digestibility. In: Dijkstra J., Forbes J.M. and J. France (Eds), Quantitative aspects of ruminant digestion and metabolism. CAB Publishing, UK, Chapter 4, pp. 87-121.

Ørskov, E.R. and I. McDonald, 1979. The estimation of protein degradability in the rumen from incubation measurements weighted according to rate of passage. J. Agric. Sci. Camb. 92: 499-503.

Raab, L., B. Cafantaris, T. Jilg and K.H. Menke, 1983. Rumen protein degradation and biosynthesis 1. A new method for determination of protein degradation in rumen fluid *in vitro*. Br. J. Nutr. 50: 569-582.

Relationship between ruminal mat characteristics and chewing activity in Holstein dry cows fed beet pulp and alfalfa or grass hay

K. Izumi
Research farm, Rakuno Gakuen University, Ebetsu, Japan; izmken@rakuno.ac.jp

Introduction

In general, although by-product feeds have relatively high fiber content, they do not strongly stimulate chewing activity because the nonforage fiber they contain is less effective physically and therefore an insufficient ruminal mat may be formed. The lack of a sufficient ruminal mat in cows offered by-product feed in large quantities may carry a risk of disorder of ruminal fermentation and increasing rapid escape of small particles from the rumen. In our previous study (Izumi *et al.*, 2008), rumen fill was evaluated by measuring the penetration resistance value (PRV) and the depth of the rumen digesta. In addition, the ruminal mat could be defined as the layer over the intersection point on two line charts of the relationship between PRV and depth of the rumen digesta. In the present study, we focus on beet pulp since it is one of the typical fiber-rich by-product feeds and regarded as a nonforage fiber source. The aim of this study was to evaluate ruminal mat characteristics by using PRV, as well as to investigate the influence of the ratio of beet pulp in the diet and the quality of the combined forage hays on rumen fill and chewing activity in Holstein dry cows.

Material and methods

Four rumen-cannulated Holstein dry cows were randomly assigned to a 4×4 Latin square design. Cows were fed four different diets: the ratios of alfalfa hay (AH) to beet pulp (BP) were 8:2 (dry matter basis, A8B2) and 2:8 (A2B8), and those of grass hay (GH) to BP were 8:2 (G8B2) and 2:8 (G2B8). All diets were fed once daily at a maintenance level of TDN. In the experimental period, intake, chewing activity, and rumen fill were measured. The rumen fill was evaluated by measuring PRV and depth of the rumen digesta and ruminal mat characteristics. The penetration resistance tests to evaluate PRV and the depth of the rumen digesta were conducted at 0, 2, 4, 6, 9, 12, and 18 h after feeding. The trial was conducted with the device we previously developed (Izumi *et al.*, 2008). The device consisted of a rod inserted into the rumen, a control box with a load cell, a displacement transducer and a computer. The ruminal mat was detected according to the method of Izumi *et al.* (2008), where PRV was averaged every 10 cm and plotted on a scatter chart. Where the two regression lines intersect on the chart, the ruminal mat is defined as the part to the upper side of the intersection point and nonmat material is defined as the part to the lower side.

Results

Total eating time decreased with increasing BP content ($P<0.01$; Table 1). Rumination time per day for AH was shorter than that for GH ($P<0.01$), and it decreased with increasing BP content ($P<0.01$). The number of rumination periods was greater with the lower-BP diets ($P<0.01$). The duration of the rumination period for GH was longer than that for AH ($P<0.05$). The PRV of the ruminal mat was the highest with the G8B2 diet and decreased with increasing BP content ($P<0.05$) and feeding AH ($P<0.05$). The thickness of the ruminal mat did not differ for AH and GH, but was greater for decreasing BP content ($P<0.05$).

Table 1. Chewing activity and characteristics of ruminal mat and nonmat material in cows.

	Diets				SEM	Effects, *P*		
	A8B2	A2B8	G8B2	G2B8		Hay	Hay:BP[1]	Hay× Hay:BP
Chewing activity								
Eating time, min/d	208.3	88.7	286.6	120.0	20.8		**	
No. of meals, /d	5.6	3.4	5.3	3.6	0.7			
Duration of meal, min	55.6	37.2	76.9	45.0	0.7			
Rumination time, min/d	309.7	197.0	470.5	338.2	23.5	**	**	
No. of rumination periods, /d	13.5	11.5	14.8	12.5	0.4		**	
Duration of rumination periods, min	23.1	17.7	32.1	27.4	1.9	*		
Ruminal mat								
PRV, N	7.3	6.3	10.5	7.4	0.4	*	*	
Thickness, cm	32.3	25.6	33.2	26.5	1.7		*	

[1] Feeding ratio of hay to BP; **P*<0.05; ***P*<0.01.

Conclusion

Increasing the PRV of the ruminal mat stimulated rumination activity. The ruminal mat could be formed, although it was soft even when cows were offered a large quantity of BP; however, rumination activity for diet A2B8 was extremely low. From the standpoint of maintaining long-term rumen health, GH would be more desirable forage combined with a high ratio of BP compared with AH.

References

Izumi, K, T. Sakamoto, S. Shibayama and H. Tsuji, 2008. Evaluation of penetration resistance value for the physical factor of rumen digesta in lactating dairy cows. Nihon Chikusan Gakkaiho. 79: 361-368.

Stability of fatty acids in grass and maize silages after exposure to air during the feed out period

N.A. Khan, J.W. Cone and W.H. Hendriks
Animal Nutrition Group, Department of Animal Sciences, Wageningen University, P.O. Box 338, 6700 AH Wageningen, the Netherlands; John.Cone@wur.nl

Introduction

The growing interest in forage lipids stems from their potential to modulate the milk fatty acid (FA) profile of dairy cows favourably to benefit human health. Lipids in the photosynthetic tissues of plants are rich in polyunsaturated FA (PUFA; 0.74±0.09 g/g total FA), particularly in C18:3*n*-3 (0.61 ± 0.07 g/g total FA) and offer a natural resistance against biohydrogenation in the rumen through their complex chemical structure and protection by cell walls. However, the feeding value of herbage lipids is significantly reduced by ensiling. Compared to grazing, fresh grass and hay, the transfer of PUFA from silage into milk is lower with silages. The latter is related to the extensive hydrolysis of forage lipids in the silo. Wilted silages result in a higher milk PUFA content compared to unwilted silages despite lower C18:3 concentrations (Noci *et al.*, 2007). Wilting increases the dry matter (DM) content at ensiling which restricts the fermentation, reducing the in silo lipolysis (Van Ranst *et al.*, 2009). The degree of in silo lipolysis and the fermentation quality therefore effect the post ensiling stability of the herbage lipids. Although in a well sealed and compacted silo the concentration of FA remains stable. During opening and feeding of silages the free fatty acids (FFA) are exposed to air. In the presence of oxygen oxidation can be induced by light, microbes and plant lipoxygenases. The objective of the present study was to evaluate the stability of FA in grass and maize silages with varying ensiling qualities, over the length of the feeding period.

Material and methods

Eight commercially prepared maize silages were chosen on the basis of dry matter (DM) contents, ranging from very wet (DM<280 g/kg), wet (DM=280-320 g/kg), normal (DM=320-360 g/kg) to dry (DM>360 g/kg). Eight commercially prepared grass silages (mainly *L. perenne*) were chosen on the basis of pH and NH_3 level (in g NH_3/100 g total N) and categorised as low NH_3 low pH (NH_3<8, pH<4.5), high NH_3 low pH (NH_3>11, pH<4.5), low NH_3 high pH (NH_3<8, pH>4.8) and high NH_3 high pH (NH_3>11, pH>4.8). From each silage, approximately 700 g of sample was taken from the middle (in height) at the back of the silage clamp with a hollow drill (2 cm diameter). The samples were transported anaerobically to the laboratory in a cooled box. Samples forming individual silages were mixed thoroughly and divided into three subsamples. The three subsamples were exposed to air for 0,12 and 24 h respectively. Lipids from the freeze dried samples were extracted with chloroform-methanol (2:1 v/v; Folch *et al.*, 1957). FA in the residual lipids were esterified or transesterified using both acid and base catalysed methods. The fatty acid methyl esters (FAME) were quantified using GC (Carlo Erba 8560 HRGC, Rodano, Italy) with a fused silica capillary column (100 m x 0.250 mm and 0.2 μm film thickness; Superlco; SP2560, St. Louis MO, USA) using helium as a carrier gas. Repeated measures of analysis of variance (PROC MIXED of SAS® Version 9.1; SAS Institute, Inc., Cary, NC) were performed for the concentrations and proportions of FA. Silage was considered as a repeated effect.

Results

Exposure to air significantly lowered (P<0.01) the mean concentrations (g/kg DM) of C18:3, C18:2, C18:1 and total FA in maize silages (Table 1). Similarly, in grass silages exposure to air decreased

($P<0.05$) the mean concentrations of C18:3, C18:2 and total FA (Table 1). In both grass and maize silages the decrease in the concentrations of the major unsaturated FA was associated with an increase ($P<0.01$) in the proportion (g/g total FA) of C16:0. The relative decrease in total FA after 24 h exposure to air was higher in the very wet maize silages compared to more dry silages. In contrast, pH and NH_3 levels of grass silages had no effect ($P>0.05$) on the stability of FA during the feed out period. There was a positive relationship between the DM contents of maize silages and the concentrations of C18:2 (C18:2 = 1.48 + 0.288xDM, $R^2=0.67$, $P<0.01$). Conversely, the concentration of C18:3 decreased (C18:3 = 2.85 - 0.005xDM, $R^2=0.44$, $P<0.05$) with an increasing DM content in the maize silages. Among grass silages there were large differences in concentrations of total and individual FA. These differences in FA concentrations were associated with variations in the proportion of C18:3 from 0.44 to 0.66 g/g total FA.

Table 1. Mean concentrations (g/kg DM) of total and major individual FA in grass and maize silages after exposure to air for 0,12 and 24 h during the feed out period.

ET	Maize silages					Grass silages			
	Total FA	C16:0	C18:1	C18:2	C18:3	Total FA	C16:0	C18:2	C18:3
0	20.36[a]	3.20	4.19[a]	10.96[a]	1.28[a]	17.61[a]	3.29	3.10[a]	10.2[a]
12	19.73[b]	3.22	4.06[b]	10.51[b]	1.21[b]	16.93[b]	3.26	2.89[b]	9.77[b]
24	19.47[b]	3.22	4.03[b]	10.35[b]	1.14[c]	16.72[b]	3.29	2.83[b]	9.61[b]
SEM	0.762	0.069	0.318	0.397	0.11	1.376	0.193	0.253	0.985
Significance									
Q	NS	NS	NS	NS	NS	NS	NS	NS	NS
ET	**	NS	**	**	**	*	NS	*	*
Q x ET	NS	NS	NS	NS	NS	NS	NS	NS	NS

[a,b,c] Means with different superscripts within column are significantly different ($P<0.05$); * = $P<0.05$; ** = $P<0.01$; NS = non significant $P>0.05$; ET= exposure time; Q = quality (DM in maize silages or NH_3 and pH level in case of grass silages).

Conclusion

Although quantitatively small (<0.06 g/g total FA), there was a consistent decline in the concentrations of major unsaturated and total FA in both maize and grass silages after 24 h exposure to air. This decline in the concentration of UFA was associated with a concomitant increase in C16:0 per unit total FA. The present study demonstrates that silages should not be exposed to air for longer periods than needed to avoid losses of unsaturated FA.

References

Folch, J., M. Lees and G.H. Sloane-Stanley, 1957. A simple method for the isolation and purification of total lipids from animal tissues. J. Bio. Chem. 226: 497-509.

Noci, F., F.J. Monahan, N.D. Scollan and A.P. Moloney, 2007. The fatty acid composition of muscle and adipose tissue of steers offered unwilted or wilted grass silage supplemented with sunflower oil and fishoil. Br. J. Nutr. 97: 502-513.

Van Ranst, G., V. Fievez, J. De Riek and E. Van Bockstaele, 2009. Influence of ensiling forages at different dry matters and silage additives on lipid metabolism and fatty acid composition. Anim. Feed Sci. Technol. in press.

Effects of a diverse high altitude forage in comparison with a total mixed ration on ruminal nutrient fermentation and methanogenesis *in vitro*

R. Khiaosa-ard, F. Leiber, M. Kreuzer and C.R. Soliva
ETH Zurich, Department of Agricultural and Food Science, 8092 Zurich, Switzerland; ratchaneewan.khiaosa-ard@inw.agrl.ethz.ch

Introduction

Alpine pastures are known for their high biodiversity of plant species. Alpine herbs and leguminous forages may contain substantial amounts of secondary plant metabolites (SPM). These SPM may affect the ruminal microbial population and activity resulting in changes in ruminal nutrient degradation and decreasing ruminal methanogenesis as shown before for specific SPM additions (e.g. Khiaosa-ard *et al.*, 2009, Wang *et al.*, 2009). The aim of the present study was to evaluate the effects on ruminal nutrient fermentation of different forage sources and methanogenesis *in vitro* when the ruminal fluid donor cow's diet changed from a total mixed ration (TMR), typical for lowland feeding, to alpine pasture grass. Further, linseed oil was supplemented to all the feeds to evaluate their methane decreasing effect.

Material and methods

The *in vitro* Hohenheim gas test apparatus was used to incubate in triplicate 200 mg dry matter (DM) of four different forages for 24 h either without or with 5% of linseed oil (DM basis). The forages were pure ryegrass hay (R), assumed to be poor in SPM, alpine pasture hay (A), assumed to contain substantial amounts of SPM, grass silage (G), another typical lowland feed, and pure hemicellulose (H), a purified fibre easily digestible for ruminal microbes. Ruminal fluid used for the incubations was collected from a fistulated Brown Swiss cow. In period 1 (P1), the cow was located in the lowlands (400 m.a.s.l.) receiving a conventional diet consisting of silage and concentrate (TMR). In period 2 (P2), the cow stayed on an alpine site (2000 m.a.s.l) and ruminal fluid was collected after 10 wk of alpine grazing. The sward on the alpine site consisted of 71 plant species with 36% grass, 23% legumes, and 40% herbs (Leiber *et al.*, 2005). In period 3 (P3), the cow had returned to the lowlands again, and ruminal fluid was collected after 6 wk of adaptation to TMR. After 24 h of incubation at 39 °C in 30 ml incubation liquid (ruminal fluid:buffer solution; 1:3), fermentation gas volume was recorded and gas was analysed for its concentration of CH_4, CO_2, and H_2, pH, ammonia concentration, and microbial counts were determined after each run in the incubation liquid. Data were analysed by the Mixed procedure of SAS® (version 9.1 for windows, SAS® Institute Inc., Cary, NC, USA) with periods (P1-P3), experimental forages, and linseed oil supplementation being considered as effects, as well as the respective interactions. The comparisons among means were performed with the Tukey method.

Results

Changing the donor cows' diet from TMR to alpine grass and back (P1-P3) resulted in significant effects on the *in vitro* fermentation traits (Table 1). Compared to P1, bacterial and holotrich protozoal counts increased in P2 while entodinium protozoal counts decreased. Ammonia concentration in the incubation liquid and CH_4 in fermentation gas were also lower in P2 than in P1, whereas ammonia concentration, similar to entodinium protozoal counts, became even lower when returning to TMR (P3). Hemicellulose resulted in the lowest values of all fermentation traits (except bacterial number) compared with the other three forages. The use of the alpine pasture hay resulted in lower ammonia concentration and lower daily CH_4 emissions than ryegrass hay and grass silage. Linseed oil level did not reveal an effect regarding methane mitigation in this study (data not shown).

Table 1. Ruminal nutrient fermentation and methanogenesis in vitro as affected by differing ruminal fluid origin (lowland P1, alpine P2, and lowland P3) and different forages (ryegrass hay R, alpine pasture hay A, grass silage G, and hemicellulose H).

Traits	Periods (n=12)			SE	Forages (n=9)				SE
	P1	P2	P3		R	A	G	H	
pH	6.81^{c}	6.88^{b}	7.04^{a}	0.010	6.95^{a}	6.98^{a}	6.94^{a}	6.78^{b}	0.011
NH_3, mmol/L	11.05^{a}	9.78^{b}	9.44^{c}	0.079	11.77^{b}	10.55^{c}	12.56^{a}	5.48^{d}	0.091
Bacteria, x10^8/ml	18.5^{c}	34.3^{a}	22.2^{b}	0.99	29.0^{a}	21.5^{b}	22.4^{b}	27.0^{a}	1.14
Protozoa, x10^4/ml									
Holotrichs	0.11^{b}	0.50^{a}	0.40^{a}	0.052	0.44^{ab}	0.49^{a}	0.30^{b}	0.12^{c}	0.060
Entodinium	5.77^{a}	4.43^{b}	2.02^{c}	0.224	3.20^{bc}	3.74^{b}	6.47^{a}	2.88^{c}	0.259
CH_4 ml/24 h	6.92^{a}	5.90^{b}	6.11^{b}	0.109	6.81^{a}	5.98^{b}	6.84^{a}	5.63^{b}	0.125
CH_4:total gas,%	12.2^{a}	11.3^{b}	12.0^{b}	0.16	12.71^{a}	12.20^{a}	12.71^{a}	9.78^{b}	0.187
CH_4:CO_2,%	14.7^{a}	13.1^{c}	13.9^{b}	0.11	15.1^{a}	14.4^{b}	15.0^{a}	11.2^{c}	0.13

^{a,b,c,d} Mean values within the same row sharing no common superscript are significantly different at $P<0.05$.

Discussion and conclusion

Ruminal fluid collected from a cow while grazing on alpine pasture yielding a high biodiversity and a high proportion in dicotyledons led to a clear increase of the bacterial population and of the proportion of bacteria to protozoa, compared to a silage-concentrate diet with a very low share of herbs. Also, *in vitro* fermentation properties of ruminal fluid were altered by the alpine sojourn of the donor cow, leading to a certain decrease in methane formation. These properties, however, resided also after re-changing the cow's diet to TMR. The contrasting effect on bacteria of alpine pasture hay used in the *in vitro* test show, that complex interdependency of feed and ruminal fluid properties exist, and that the different timeframes and environments *in vivo* and *in vitro* certainly play a role.

References

Khiaosa-ard, R., S.F. Bryner, M.R.L. Scheeder, H.-R. Wettstein, F. Leiber, M. Kreuzer and C.R. Soliva, 2009. Evidence for the inhibition of the terminal step of ruminal α-linolenic acid biohydrogenation by condensed tannins. J. Dairy Sci. 92: 177-188.

Leiber, F., M. Kreuzer, D. Nigg, H.-R. Wettstein and M.R.L. Scheeder, 2005. A study on the causes for the elevated n-3 fatty acids in cows' milk of alpine origin. Lipids 40: 191-202.

Wang, C.J., S.P. Wang and H. Zhou, 2009. Influences of flavomycin, ropadiar, and saponin on nutrient digestibility, rumen fermentation, and methane emission from sheep. Anim. Feed Sci. Technol. 148: 157-166.

Microbial outflow determined from reticular or omasal sampling of dairy cows fed grass silage with different neutral detergent fibre content at two levels of concentrate supplementation

S.J. Krizsan[1], S. Ahvenjärvi[2], S.K. Nes[1] and H. Volden[1]
[1]Department of Animal and Aquacultural Sciences, Norwegian University of Life Sciences, Ås, Norway; [2]MTT-Agrifood Research Finland, Animal Production Research, Jokioinen, Finland; sophie.krizsan@umb.no

Introduction

Different methods, using ruminally and/or duodenally cannulated animals and marker techniques, have been used to determine microbial protein synthesis experimentally. Increased accuracy of digesta flow measurements using a triple marker system (Ahvenjärvi *et al.*, 2003), and less interference from endogenous N flow and abomasal digestive processes are the advantages with the omasal sampling technique when determining microbial N outflow (Ahvenjärvi *et al.*, 2000). However, the composition of omasal samples compared to samples collected from the duodenum have deviated more from calculated true digesta and omasal large particle phase consists of heterogeneous particulate matter (Ahvenjärvi *et al.*, 2000). This could increase errors in cases of unrepresentative digesta sampling. Flow of particulate matter from the reticulum through the omasal canal depends on reductions in particle size and proportion of digestible material remaining in the particles. The objective of this study was to compare microbial N outflow of the rumen determined using the omasal sampling technique or from selected reticular material of lactating dairy cows fed grass silage differing in NDF content and at two levels of concentrate feeding.

Material and methods

Six Norwegian Red cows with mean BW 594±62 kg, parity 2.7±0.8, 41±20 DIM and producing 34±5 kg milk/d at the start of the trial, were used in an incomplete Latin square design with four 24-d periods. Dietary treatments consisted of 3 grass silages and 2 levels of concentrate feeding (5 and 9 kg) in a 3×2 factorial arrangement. The grass silages; G1, G2 and G3 from the same field (mainly timothy) contained (DM basis): 922±2, 923±2 and 934±2 g OM, 412±10, 530±12 and 639±21 g NDFom, 215±6, 127±5 and 125±5 g CP, and 40.2±3.5, 79.4±5.7 and 126±8 g INDF, respectively. Dry matter content of G1, G2 and G3 were 310±14, 349±10, and 242±9 g/kg, respectively. The composition of the concentrate used was (DM basis): 947±4 g OM, 3.92±1.29 g NDFom, 158±11 g CP, and 733±30 g starch.

Sampling from the omasum or reticulum combined with 3 markers (Co + Yb + INDF) using the reconstitution system were used to quantify nutrient flow. Continous infusion of +10 atom% $(^{15}NH_4)_2SO_4$ from >60 h in advance of the first sampling occasion until the last were used to label microbial protein. Samples from the reticulum and omasum were collected during 3 d at 4 different occasions each day with 6 h interval between sampling sites. On the last 2 d, sampling occasions were moved 2 h later than on the previous day. Bacterial samples from reticulum and omasum were separated in liquid and particle associated bacteria primarily by differential centrifugation. Large particles were separated by filtering a 3 d composite digesta sample through a 100 μm fabric. The filtrate was centrifuged at 1000× g at 5 °C for 10 min to separate fluid and the small particle phase. Reticular samples were sieved through a 1 mm sieve before isolation of bacteria and digesta phases. This value was based on determination of particle distribution in reticular and omasal samples collected the last 2 days in the first adaptation period. Functional specific gravity was also determined, but did not differ (1.07 g/ml) between reticular and omasal digesta.

All data was analysed using the MIXED procedure of SAS® (2002). The model included fixed effects of period, sampling site, level of concentrate fed, silage type and the interaction between concentrate level and silage type. Cow within site was treated as a random effect.

Results

Total DMI and intake of N were different ($P \leq .04$) between diets, ranging from 15.2 to 17.7 kg/d and 326 to 583 g/d, respectively. Nonammonia N flow, microbial N flow parameters and OM truly digested in the rumen within diet are presented in Table 1. There were clear dietary effects of all parameters ($P \leq 0.01$; Table 1) except OM truly digested in the rumen ($P=0.99$; Table 1). Further, there was no significant effect of sampling site ($P>0.05$) for any of the given parameters in Table 1. However, microbial efficiency tended ($P=0.07$; Table 1) to be affected by sampling site indicating a 1.74 g of NAN/kg of OM truly digested in the rumen higher efficiency estimated from reticular compared with omasal digesta.

Table 1. Effect of sampling site, concentrate level and silage source on microbial flow parameters and OM truly digested in the rumen of dairy cows.

Silage[1]	G1		G2		G3		SE	*P*-value[2]			
Concentrate level	5	9	5	9	5	9		S	C	G	C×G
Item[3]											
NAN flow, g/d	395	274	318	311	263	314	25	0.19	0.24	0.11	0.01
Microbial NAN flows											
LAB, g/d	110	81.9	78.6	65.0	68.8	85.6	8.1	0.32	0.27	<0.01	<0.01
PAB, g/d	134	122	141	96.8	120	116	8.0	0.22	<0.01	0.04	<0.01
Total, g/d	243	204	219	164	189	201	13	0.77	0.01	<0.01	<0.01
Microbial efficiency, g of NAN/kg of OMTDR	18.9	17.4	21.3	15.7	19.5	19.5	1.1	0.07	0.02	0.27	0.01
OMTDR, kg/d	13.1	12.8	10.8	10.6	10.7	10.5	0.6	0.96	0.63	<0.01	0.99

[1] Chemical characterisation of the silages given in text.
[2] Probability of a significant effect of S = sampling site, C = concentrate level, G = silage type, C×G = interaction between concentrate level and silage type.
[3] NAN = nonammonia N, LAB = liquid-associated bacteria, PAB = particle-associated bacteria, OMTDR = OM truly digested in the rumen.

Conclusion

This study shows that digesta collected from the reticulum and selected with regards to a cut-off particle size of 1 mm is representative of material entering the omasum over a range of diets with regards to microbial flow parameters, NAN flow and OM truly digested in the rumen.

References

Ahvenjärvi, S., A. Vanhatalo, P. Huhtanen and T. Varvikko, 2000. Determination of reticulo-rumen and whole-stomach digestion in lactating cows by omasal canal or duodenal sampling. Br. J. Nutr. 83: 67-77.

Ahvenjärvi, S., A. Vanhatalo, K.J. Shingfield and P. Huhtanen, 2003. Determination of digesta flow entering the omasal canal of dairy cows using different marker systems. Br. J. Nutr. 90: 41-52.

Potassium transport across gastrointestinal epithelia of ruminants

N. Kronshage and S. Leonhard-Marek
Department of Physiology, School of Veterinary Medicine Hannover, Bischofsholer Damm 15, 30173 Hannover, Germany; nina.katrin.kronshage@tiho-hannover.de

Introduction

The small intestine is assumed to be the main site of gastrointestinal potassium absorption. However, *in vivo* studies have shown that preintestinal regions are likewise able to contribute to K absorption, if the luminal availability of K is increased (Greene *et al.* 1983, Wylie *et al.* 1985, Khorasani *et al.* 1997). Cows that suffer from abomasal displacement, a disease mainly found in high yielding dairy cows around parturition, often show a (sometimes severe) hypokalemia. This drop in plasma K might be the cause as well as the consequence of abomasal displacement. Since our knowledge about the mechanisms and regulation of gastrointestinal K absorption is very limited, the aim of the present study was to investigate K absorption across different gastrointestinal epithelia from ruminants.

Material and methods

Parts of the intestinal tract were isolated from goats, sheep and cattle immediately after slaughter. Isolated epithelia were prepared from the ventral rumen wall, the abomasal corpus, mid jejunum and proximal colon and were incubated in Ussing chambers under short circuit conditions. Epithelia were bathed in a standard bicarbonate buffer. On the mucosal (luminal) side the potassium concentration was varied between 4 and 100 mmol/l. Potassium was osmotically replaced by the impermeable N-Methyl-D-glucamine. The short circuit current (Isc) was taken as a measure for electrogenic transport. To discriminate between K related currents and an indirect effect of K on Na-dependent currents, the effects of increasing K concentrations were compared in the presence and absence of sodium. Potassium absorption, which can include electrogenic and electroneutral transport mechanisms, was calculated from the change in potassium concentration on the serosal side. For this purpose samples were taken from the original buffer and after 30 min of incubation the potassium concentration was measured by atomic absorption spectroscopy. To investigate a potential regulatory effect of hypokalemia on gastrointestinal K transport, the K concentration on the serosal (blood) side of the epithelia was changed from 4 mmol/l to 2 mmol/l. Data are presented as means ± SEM. Statistical significance was evaluated using analysis of variance and the Student t-test or Wilcoxon signed rank test.

Results

Electrogenic transport: Increasing the mucosal potassium concentration from 4 to 100 mmol/l in the absence of sodium resulted in an increase of current (ΔIsc in μeq/cm^{2}h) across the rumen of goats (1.0±0.1, n=16), sheep (2.5±0.2, n=14) and cattle (1.8±0.6, n=6), as well as across the colon of goats (2.5±0.4, n=8) and sheep (3.3±0.4 n=11). In the presence of Na, the basal currents were higher across rumen and colonic epithelia (measured in goats and cattle). The potassium dependent current remained in the same order of magnitude across rumen epithelia, while it increased across the goat colon (Figure 1). No K dependent increase of current could be shown for the abomasum. In the jejunum, an increasing K concentration (from 4 to 100 mmol/l) increased Isc by 1.5±0.3 μeq/cm^{2}h (n=11) in goats, but had no effect in sheep. Decreasing the concentration of potassium on the serosal side (hypokalemia) had no effect on any current.
Potassium absorption: Gastrointestinal potassium absorption was measured in sheep. A luminal increase in potassium concentration from 4 to 100 mmol/l augmented K absorption (in μmol/cm^{2}h)

by 2.5±0.2 (rumen, n=14), 2.9±0.4 (colon, n=6), 3.5±0.5 (abomasum, n=6) and 4.5±0.4 (jejunum, n=14). This increase in K absorption was significantly higher in the jejunum than in the rumen and colon ($P<0.05$), while the other absorption rates did not differ from each other. In the rumen and colon, the increase in K absorption was equivalent to the parallel increase in current (ΔIsc). In the abomasum and jejunum, the increase in K absorption was much higher than the increase in current.

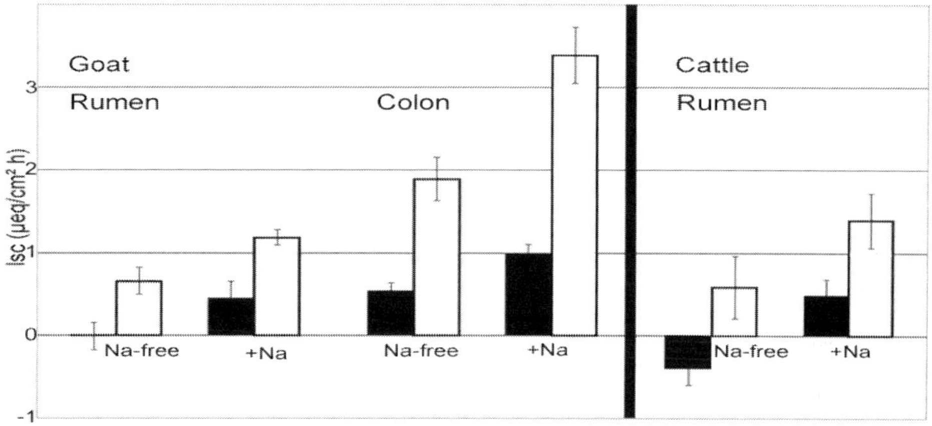

Figure 1. Effect of potassium and sodium on short circuit current (Isc) across different epithelia. ■ mucosal K^+ concentration = 4mmol/l □ mucosal K^+ concentration = 50mmol/L. The sodium concentration in the +Na buffers was 141.4 mmol/l in the presence of 4 mmol/l K^+, and 95.4 mmol/l in the presence of 50 mmol/l K^+. n=8 (rumen), n=4 (colon). All K-dependent increases of Isc were significant; the K-dependent increase of Isc across the goat colon was higher in the presence of Na.

Discussion and conclusion

A luminal increase in K concentration increased K absorption across the different gastrointestinal segments with the same order of magnitude. However, the mechanisms of potassium absorption are apparently different. While electrogenic K transport seems to dominate in the rumen and colon, K absorption across the abomasum and jejunum appears mainly electroneutral. These results form the basis for further studies on mechanisms and regulation of K absorption. Hypokalemia per se, tested as a first approach to regulation, was not able to increase (electrogenic) K absorption.

References

Greene, L.W., P.J. Fontenot and K.E. Webb, 1983. Site of magnesium and other mineral absorption in steers fed high levels of potassium. J. Anim. Sci. 57: 503-510.

Khorasani, G.R., R.A. Janzen, W.B. McGill and J.J. Kennelly, 1997. Site and extend of mineral absorption in lactating cows fed whole-crop cereal grain silage of alfalfa silage. J. Anim. Sci. 75: 239-248.

Wylie, M.J., J.P. Fontenot and L.W. Greene, 1985. Absorption of magnesium and other macrominerals in sheep infused with potassium in different parts of the digestive tract. J. Anim. Sci. 61:1291-1229.

Rumen pH and function in dairy cows of the South Island of New Zealand

J. Laporte-Uribe and J. Gibbs
Lincoln University, P.O. Box 84, Canterbury, New Zealand; Jim.Gibbs@lincoln.ac.nz

Introduction

Environmental conditions in the rumen allow the growth of a myriad of symbiotic microorganisms that in exchange generate large amounts of nutrient for the ruminant host. The pastoral system in the South Island of New Zealand is unique, because the sole feed is high quality pasture maintained throughout lactation by highly managed grazing, irrigation and fertilisation practices. In this system, cows have been selected for high intakes of fresh pasture. For the last four years Lincoln University Rumen Studies Group has conducted research seeking to describe the rumen environment of such high intake cows on a large, commercial farm. Over this period, our understanding of this unusual rumen environment has been developed by the use of *in situ* rumen recordings of more than 300 d of pH and temperature. The work described here was designed to assess other rumen metabolite parameters associated with the described diurnal patterns of pH and temperature.

Material and methods

Six ruminally fistulated dairy cows from a highly managed commercial herd grazing high quality pasture were equipped with dataloggers to continuously record pH and temperature. Staggered sampling of rumen content was conducted every 4 h for 48 h to provide samples for every 2 h interval across a 24 h diurnal period. Samples were analysed for short chain fatty acids (SCFA) using gas chromatography (HP6890 series, Hewlett Packard, USA). A linear mixed model was build to asses relationship between variables, time of sampling was used as repeated measurement, and animal as a subject, the fixed model was constructed using time of the day as the main effect and the animal as a covariant, proportions of SCFA are expressed as mean ± standard error and pH as mean values every 10 min (SPSS Statistics 17.0©, 2009, SPSS Inc., Chicago, IL, USA).

Results

Grazing management was the largest influence on the rumen parameters assessed; there is substantial diurnal variation in every parameter with the largest and most rapid changes occurring immediately after access to the daily pasture allowance at 18:00 h. Rumen pH nadir was between 20:00-22:00 h with a second minor drop at 12:00-13:00 h, while the zeniths were at 08:00-10:00 h and 16:00-18:00 h after morning and afternoon milkings (Figure 1). Acetic acid dominates the SCFA profile; however pH appeared to be driven by the increase in the proportions of propionic and butyric acid (Figure 1).

Discussion

The dairy grazing management system of the South Island of New Zealand uses once daily allocation of fresh pasture and a highly disciplined approach of enforcing very low residual pasture covers to ensure high utilisation and high quality regrowth. It demands a diurnal intake pattern of marked 'feast and famine', which combined with pastures high in rapidly fermentable carbohydrate, produces an unusual rumen environment characterised by strong variation in pH and SCFA proportions. The pH of these rumens is clearly dependent on grazing management, and the pattern of zeniths and nadirs appears to be driven by the accumulation of propionate (R^2=-0.48) and butyric (R^2=-0.51) acids, as acetic acid proportions followed positively pH curve (R^2=0.64) throughout the day (Figure 1).

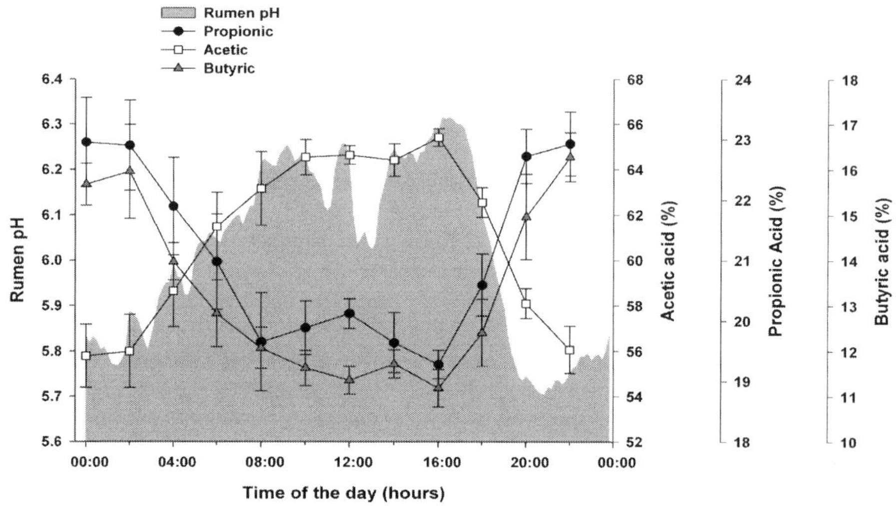

Figure 1. Diurnal pattern of pH and proportions of main SCFA in pasture fed dairy cows.

A ratio of acetic/propionic acid below 3 has been suggested in pasture fed animals to supply an adequate amount of energy in the form of water soluble carbohydrates (Corbett *et al.*, 1969). In this study a ratio below 3 was observed for approximately three quarters of the day (data not shown), suggesting a high degree of synchrony between longer and shorter term carbohydrate metabolism, which could benefit microbial growth. The diurnal patterns of the parameters observed in this study show that the rumen environment and function in the unusual grazing management systems of the South Island is atypical when compared with existing studies (Carruthers *et al.*, 1997), and more dynamic than typical TMR fed systems (Schwartzkopf-Genswein *et al.*, 2003). Rumen pH assessment alone in these systems could not be a good indicator of rumen function, and further consideration of the dynamism of the rumen is required in the future to develop better indicators of rumen dysfunction for these systems.

References

Carruthers, V.R., P.G. Neil and D.E. Dalley, 1997. Effect of altering the non-structural:structural carbohydrate ratio in a pasture diet on milk production and ruminal metabolites in cows in early and late lactation. Anim. Sci. 64: 393-402.

Corbett, J.L., R.A. Leng and B.A. Young, 1969. Measurement of energy expenditure by grazing sheep and the amount of energy supplied by volatile fatty acids produced in the rumen. In: K.L. Blaxter, J. Kielanowski and G. Thorbek (Eds.), Energy metabolism of farm animals. Oriel Press Ltd, Newcastle-upon-Tyne, UK, pp. 177-186.

Schwartzkopf-Genswein, K.S., K.A. Beauchemin, D.J. Gibb, D.H. Crews Jr, D.D. Hickman, M. Streeter and T.A. McAllister, 2003. Effect of bunk management on feeding behavior, ruminal acidosis and performance of feedlot cattle: A review. J. Anim. Sci. 81: E149-E158.

Influence of various concentrations of triticale meal in cattle feed on rumen protozoa

O. Latal[1], J. Pozdisek[1] and A. Pechova[2]
[1]Agrorisearch Rapotin Ltd., Vyzkumniku 267, 788 13, Vikyrovice, Czech Republic; [2]University of Veterinary and Pharmaceutical Sciences Brno, Palackeho 1/3, 612 42, Brno, Czech Republic; latal.oldrich@post.cz

Introduction

Rumen ciliate protozoa play diverse and important roles in ruminant metabolism (Hristov *et al.*, 2001). Few studies have investigated the effects of grain diets (high/medium) on rumen protozoa (Hristov *et al.*, 2001, Franzolin and Dehority, 1996 in Dehority, 2003). Triticale is an important feeding grain. The aim of this study was to assess the possible influence of various concentrations of grain (triticale meal (TM)) as one concentrate in cattle feed on the total number of rumen protozoa of the *Isotrichidae* family and selected rumen characteristics (pH, total acidity and volatile fatty acids (VFA)). Our question was whether it was possible to range TM in cattle as one concentrate in the feeding ration.

Material and methods

This study was monitored at the same time as an *in vivo* balance trial at the Research Institute for Cattle Breeding, Ltd., Rapotin (RICB) at an accredited balance stable in February 2006. The balance trial using the regression method described in Pozdíšek and Vaculová (2008). In this paper there was an evaluated trial with various percentages of TM in ration dry matter (4.8-41.2%). The study was carried out on 12 heifers (Czech Fleckvieh, 241-338 kg live weight (LW). The preparatory period was 2 wk, followed by the experimental period of 10 days. The sample of TM (particle 4 mm size, taken in 2005), was added to the feed of clover/grass (C/G) silage (was produced by RICB using the silage space). The rumen fluids were removed from the heifers 3.5 h after the end of the main period using a stomach tube for rumen fluid removal. Protozoa from the family *Isotrichidae* were preserved in 10% methanal, dyed with methylene blue and counted in a Fuchs-Rosenthal counting chamber according to the method of VFU Brno and Dehority (2003, in Dehority, 2003). VFA was analysed by gas chromatography, pH using a standard pH meter and total acidity by the titration method. The samples (TM and G/C silage) were determined according to the Czech State Standard (CSS) 46 7092 'Method for Feed Testing' (Weenden method): crude protein (CP), crude fibre (CF), fat (F), ash (A), nitrogen free ext. (NFE) and organic matter (OM). All results were adjusted to g/kg DM. The data were analysed using descriptive statistics and general linear regression models (GRM) with one fixed effect (rate of TM in diet) using the SAS® (2007). The primary statistical parameters for the model equation, coefficients of correlation and determination and the level of significance were set at $P<0.05$.

Results and discussion

Weende analysis of TM (g/kg DM) gave the following: -100.2 (CP), 34.6 (CF), 21.2 (A), 10.8 (F), 833.2 (NFE) and 978.8 (OM); C/G silage (g/kg DM) -143.0 (CP), 273.2 (CF), 99.8 (A), 28.5 (F), 455.5 (NFE) and 900.2 (OM). The daily intake of total DM was balanced at the level of 20 gDM/kgLW. The main results for rumen characteristic are shown in Table 1. There was a significant correlation between percentage of TM (x) and (1) number of rumen protozoa ($y=2.9087 + 0.1499x$, $r=0.7266$: $P<0.05$), (2) pH ($y=6.9519 - 0.0117x$, $r=-0.6887$: $P<0.05$) and (3) VFA ($y = 85.4076 + 0.7754x$, $r = 0.6929$: $P<0.05$) but not in total acidity ($y = 11.9805 + 0.1291x$, $r = 0.5584$: $P>0.05$).

Total number of protozoa ranged from 332,000-1,360,000/ml, pH 6.95-6.2, family *Isotrichidae* ranged from 3,730-16,260/ml and VFA 82.0-132.5 mmol/l. Dehority (2003) and other authors have described normal values of rumen protozoa as ranging from 200,000-400,000/ml, pH 6.2-6.8 and VFA 80-120 mmol/l. Our hypothesis was statistically confirmed. It is possible to feed TM in cattle as one concentrate.

Dehority (2003), and Hristov *et al.* (2001) describe maximum protozoal number as between 40 – 60% (grain – concentrate) of DM. When high concentrate rates are fed and ruminal pH decreases to below 6.0, ruminal numbers are decreased (Dehority, 2003) or can be totally lost (Eadie *et al.*, 1970 in Hristov *et al.* 2001).

Table 1. Main results of rumen characteristic in 12 heifers.

Heifer	Rate of TM in diet, % of total DM	pH	Average number of rumen prot., 10^5	Family *Isotrichidae*, 10^3	Total acidity, tit. un.	Volat. fatty acids (VFA), mmol/l
1	5.5	6.91	5.44	5.69	13.6	91.9
2	10.5	6.74	6.64	6.85	17.8	106.7
3	19.4	6.78	3.84	4.11	11.6	93.8
4	27.3	6.56	6.36	7.25	16.4	104.2
5	38.8	6.40	8.30	9.61	18.0	125.3
6	40.8	6.21	13.60	16.26	20.2	132.5
7	19.4	6.62	4.56	5.56	16.8	97.6
8	29.2	6.43	4.64	5.43	17.0	123.0
9	41.2	6.81	8.12	9.60	15.2	100.1
10	37.5	6.73	8.48	9.95	13.4	99.7
11	10.9	6.95	4.36	5.46	10.0	89.3
12	4.8	6.95	3.32	3.73	10.6	82.0

Conclusion

The percentage of TM 4.8-41.2% DM in cattle feed in our study had a positive effect on No. of rumen protozoa of the *Isotrichidae* family. pH and VFA had near normal values. Our results confirm the value of including TM in cattle feed.

Acknowledgement

The work was partly supported by Ministry of Education CR, Prague No. LA 332 and MSM 2678846201.

References

Dehority, B.A., 2003. Rumen microbiology. Nottingham University Press, Nottingham, UK, 372 pp.

Hristov, A.N., M. Ivan, L.M. Rode and T.A. McAllister, 2001. Fermentation characteristic and ruminal ciliate protozoal populations in cattle fed medium or high concentrate barley based diets. J. Anim. Sci. 79: 515-524.

Pozdíšek, J. and K. Vaculová, 2008. Study of wheat (*Triticum aestivum* L.) quality for feeding ruminants using *in vitro* and *in vivo* methods. Czech J. Anim. Sci. 53: 253-264.

Strategies of forage supplementation to increase dry matter intake and rumen outflow rate in heifers fed low-quality hay of tropical grass

J.D. Latorre, A.J. Ayala and J.C. Ku
Campus de Ciencias Biológicas y Agropecuarias, UADY, Mérida, Yucatán, México; kvera@uady.mx

Introduction

Decreased DM intake during the dry season is one of the main constraints for ruminant production in the tropics. This is due to the low concentration of CP and high content of NDF in the forage, resulting in low digestibility of DM and reduced weight gain. It has been shown that the use of forage supplements may increase the amount of fermentable N or easily degradable fiber promoting an increase in cellulolytic microorganisms which improve the utilisation of fiber (Manyuchi *et al.*, 1997). Another alternative is the use of chemical treatments that promote delignification of the cell wall allowing a better utilisation of the nutritional components of the forage (Klopfenstein *et al.*, 1972). Changes in rumen osmotic pressure may lead to a faster outflow of particles from the rumen. The aim of this experiment was to investigate the effect of different forage supplementation strategies on DM intake, digestibility and rumen parameters in heifers fed mature tropical grass.

Material and methods

Four crossbreed heifers (*Bos indicus* × *Bos taurus*) fitted with permanent rumen cannulae and weighing 273±58 kg were housed in individual stalls. A 4×4 Latin square design was used with 10 days of adaptation and 9 days of measurements. Treatments were the following: (CT) control consisting of *Brachiaria brizantha* hay (BbH) only, (HT) 80% BbH plus 20% of hay soaked in NaOH (10%), (LT) 80% BbH plus 20% of chopped forage of *Leucaena leucocephala* and (PT) 80% BbH plus 20% of chopped young forage of *Pennisetum purpureum*. The basal diet (BbH) for all treatments was provided *ad libitum* and included a mixture of cane molasses-urea. Urea was provided in different amounts to make the rations isonitrogenous. DM intake was measured by weighing the feed offered and that refused next day. *In situ* rumen degradation was evaluated with the nylon bag technique (Orskov *et al.*, 1980). Six incubation times were used for BbH, treated hay (NaOH), *Pennisetum purpureum* and *Leucaena leucocephala*. Osmotic pressure was measured in rumen liquor at 0, 3, 6 and 9 h postpandrium the 4th day. Rumen outflow rate was estimated the 5th day of measurements through the infusion via rumen cannulae of 150 g of PEG and sampling at 0, 3, 6, 12 and 18 h post-infusion. Total collection of faeces was carried out on 5 consecutive days to determine total tract apparent digestibility of DM and ADF. The last day of every period, heifers were emptied of rumen contents to determine DM and ADF loads. All data were analysed with the GLM procedure of SAS® (2000).

Results and discussion

Non significant differences were found among treatments for apparent digestibility of DM (43.5% at mean; P=0.19) and ADF (49.1% at mean; P=0.09) nor for rumen parameters and total DM content in the rumen (Table 1). However, a trend was observed for a higher osmotic pressure (P=0.19) in the rumen for the NaOH treated hay. Intake of DM and ADF was increased by 20% (P<0.01) with PT and 13% (P<0.01) with LT without substitution of the basal diet. Although a significant response was found for *in situ* DM degradation of supplemented forage, with ED (k=0.02) 29% (P<0.01) higher for HT compared to CT; DM intake was decreased for this treatment. The highest rate of degradation of DM was found for LT (P<0.01). DM intake was improved with PT, but there were neither reductions in DM digestibility nor changes in rumen load (DM and ADF), which could be the result of a faster rate of passage from the rumen.

Table 1. Intake, digestibility, in situ rumen degradation, rumen content and rumen parameters in heifers fed low quality hay of tropical grass and supplemented with different forages.

Item	Treatments[1]				SEM	P
	CT	HT	LT	PT		
Intake						
gDM/kg^{0,75}	68.9^{bc}	61.3^{c}	78.1^{ab}	82.5^{a}	2.61	<0.01
gADF/kg^{0,75}	35.3^{ab}	31.0^{b}	37.4^{ab}	41.0^{a}	1.36	<0.01
Basal diet intake						
gDM/kg^{0,75}	68.9^{a}	47.4^{b}	58.8^{ab}	62.1^{a}	2.61	<0.01
Digestibility (%)						
DM	43.3	39.9	44.6	46.2	1.83	0.19
ADF	51.9	46.8	46.0	51.8	1.73	0.09
Ruminal degradation (DM)						
Supplements						
A+B	63.2^{b}	92.8^{a}	71.3^{b}	75.6^{b}	2.75	<0.01
C	0.027^{b}	0.032^{b}	0.073^{a}	0.052^{ab}	0.007	<0.01
ED (k=0,02)	45.5^{c}	74.5^{a}	64.7^{ab}	60.4^{b}	2.23	<0.01
Rumen contents						
DM, kg/animal	8.0	7.8	8.1	7.8	0.61	0.98
ADF, kg/animal	4.2	4.2	4.3	4.0	0.33	0.93
Rumen parameters						
OP[2] (mOsm/l)	261.5	282.0	266.8	261.5	6.53	0.19
Outflow rate (%/h)	5.7	7.3	6.6	7.0	0.57	0.31

[1] CT = Control treatment, *Brachiaria brizantha* hay (Bbh); HT = 80% Bbh plus 20% of treated hay (NaOH) at 10%; LT = 80% Bbh plus 20% of *Leucaena leucocephala*; PT = 80% Bbh plus 20% of *Pennisetum purpureum*.
[2] Osmotic pressure; $^{a, b}$ Means within rows with same superscript letters are not significantly different (P>0.05).

Conclusion

DM intake was increased in response to forage supplements particularly with *Pennisetum purpureum* without reduction of the intake of BbH. Although effective rumen degradation of low quality hay was improved by treatment with NaOH, DM intake was reduced and digestibility was not affected by alkali treatment.

References

Klopfenstein, T.J., V.E. Krause, M.J. Jones and W. Woods, 1972. Chemical treatment of low quality roughages. J. Anim. Sci. 35: 418-422.

Manyuchi, B., F.D. Deb Hovell, L.R. Ndlovu, J.H. Topps and A. Tigere, 1997. Napier or ground hay as supplements in diets or sheeps consuming poor quality natural pasture hay. Livest. Prod. Sci. 49: 43-52.

Orskov, E.R., F.D. DeB Hovell and F. Mould, 1980. The use of the nylon bag technique for the evaluation of feedstuffs. Trop. Anim. Prod. 5: 195-213.

Effect of feeding grain on ruminal acidosis in cattle: a pilot study

I.J. Lean and A.R. Rabiee
SBScibus, 2 Broughton Street, Camden, NSW 2570, Australia; ianl@dairydocs.com.au

Introduction

Ruminal acidosis induced by feeding large amounts of rapidly fermentable substrate is a well recognised clinical condition of cattle. However, sub-clinical acidosis and secondary effects such as depressed milk production, milk fat content and laminitis have not been well documented until recently in predominantly pasture-fed animals. Bramley *et al.* (2008) developed a method to identify sub-clinical acidosis based on rumen concentrations of volatile fatty acids, ammonia, pH and lactic acid concentrations. The index developed was strongly influenced by rumen volatile fatty acids and ammonia concentrations, whereas lactic acid and pH, were less important predictors. It may not be possible to eliminate all acidosis and still maintain economic levels of milk production, but the feeding parameters must be managed and controlled. The capacity to screen grains for the potential to cause acidosis provides a possible method to avoid acidosis. The objective of this study was to determine the amount of the grains required to create large, but not life-threatening, changes in ruminal conditions following the oral administration of grains.

Material and methods

Eight Holstein heifers, <18 mo of age, and 8 Holstein non-lactating cows were randomly selected from a group of cattle and allocated to 2 challenge diets, fed at four different levels. The body weights (BW) of heifers and cows at the beginning of the experiment ranged from 410 to 650 kg and 650 to 800 kg, respectively. Cattle were accustomed to the feeding system during a 7-d pre-adaptation period during which cattle were fed 1 kg of mixed grain and grass silage *ad libitum*. During a 5-d challenge period, animals received 1 kg per day of rolled mixed grains (n=8) or 1 kg rolled triticale (n=8) and grass silage *ad libitum*. During this period, cattle were withheld from all feed for a period of 14 h and then fed challenge diets. Cattle were fed first with 1 kg of wet ryegrass silage to reduce the saliva contamination during the rumen sampling at time 0. Cattle were challenged with each of the two diets at four levels; 0%, 0.4%, 0.8% or 1.2% of BW, respectively. The mixed grains were an equal mixture of oats, wheat, barley, triticale and sorghum cultivars. Following the initial rumen sampling at time 0, rumen samples were collected 1 h after feeding grain, and every 45 min for the 5 subsequent samples. Rumen pH was measured immediately in samples of the raw rumen fluid. Rumen fluid was analysed for volatile fatty acids using gas chromatography. The coefficients of variation for assay of propionic, acetic, isobutyric, butyric, isovaleric, valeric and caproic acids were 3.8%, 5.1%, 3.2%, 4.0%, 3.4%, 3.7% and 3.2% respectively. Concentrations of rumen ammonia, D- and L-lactate were analysed using enzymatic methods.
Estimated marginal mean responses were calculated using generalised linear models with repeated measures. Further analyses to determine concordance with groups reflecting acidosis status were performed using K-means cluster analysis and discriminant analysis (SPSS, v.12, SPSS Inc., Chicago, IL, USA). These groups had been previously defined by using cluster analysis and discriminant analysis (Bramley *et al.*, 2008) in a large study involving 100 herds and 800 lactating cows.

Results

Cattle fed at 1.2% of BW had a significant reduction in pH (P=0.01), a significant increase in concentrations of rumen valerate (P=0.004), propionate (P=0.01), and ammonia (P=0.04) over controls (Table 1). Key indicators of the change in rumen function were rumen pH, propionate and valerate concentrations that varied with the amount of grain fed. Of these, only valerate also significantly differed between grain types, for time × group and grain type × time effects ($P<0.05$). Concentrations of valerate were significantly higher in cattle fed triticale (0.69 mM) than in those fed the mixed grains (0.61 mM). Concordance with the acidotic group described by Bramley *et al.* (2008) using K-mean cluster analysis showed cattle fed at the rate of 1.2% of BW had the highest average acidosis rank, and cattle in the control (0 kg grain) and 0.4% groups had the lowest average rank.

Table 1. Estimated marginal means of rumen fermentation products for cattle fed at 0%, 0.4%, 0.8% and 1.2% of BW. Mean values for the two diets, i.e. mixed gains and triticale.

	Estimated marginal means, mean ± SD			
Percentage of grain fed	Control (0 kg)	0.4% of BW (2-3 kg)	0.8% of BW (4-6 kg)	1.2% of BW (6-8 kg)
pH	7.30±0.04	7.12±0.03	7.03±0.03	6.99±0.03
Total VFA, mM	59.06±8.24	62.0±5.82	74.70±5.82	76.1±5.82
Acetate, mM	45.14±6.27	45.64±4.43	53.96±4.43	54.45±4.43
Propionate, mM	7.99±1.11	8.85±0.78	11.63±0.78	12.77±0.78
A/P ratio	5.67±0.37	5.22±0.27	4.64±0.27	4.28±0.27
Butyrate, mM	4.24±1.12	5.67±0.79	6.52±0.79	6.20±0.79
Isobutyrate, mM	0.53±0.06	0.49±0.04	0.62±0.04	0.63±0.04
Isovalerate, mM	0.76±0.15	0.78±0.11	0.98±0.11	1.05±0.11
Valerate, mM	0.31±0.11	0.46±0.07	0.81±0.07	0.88±0.07
Caproate, mM	0.08±0.06	0.12±0.05	0.17±0.05	0.12±0.05
L-lactate, mM	0.16±0.06	0.08±0.04	0.07±0.04	0.01±0.04
D-lactate, mM	0.13±0.06	0.12±0.04	0.16±0.04	0.08±0.04
Ammonia, mM	3.96±1.29	4.47±0.91	6.75±0.91	6.09±0.91

Discussion and conclusion

The 1.2% BW feeding rate created a substantial change in some rumen indicators of acidosis. The acidosis index previously developed (Bramley *et al.*, 2008) was useful for ranking grain and challenge level groups. Valerate concentrations were the single most useful indicator of an acidotic challenge.

Acknowledgement

We thank J.L. Black, R. King and S. Nielson for their support with this study.

References

Bramley, E., I.J. Lean, W.J. Fulkerson, M.A. Stevenson, A.R Rabiee and N.D. Costs, 2008. The definition of acidosis in dairy herds predominantly fed on pasture and concentrates. J. Dairy Sci. 91: 308-321.

Effect of feeding grain on ruminal acidosis in cattle: acidosis indices

I.J. Lean and A.R. Rabiee
SBScibus, 2 Broughton Street, Camden, NSW Australia 2570; ianl@dairydocs.com.au

Introduction

Variation in fibre digestion and in rates of fermentation has been noted between grain species (Opatpatanakit *et al.,* 1994). It is also likely that these differences extend beyond grain species to differences among cultivars within a grain. The potential to screen grains fed for acidosis risk is one method that may reduce risk of acidosis in cattle. A near infrared reflectance spectroscopy method (NIR) developed to rank grains for acidosis risk based on feed chemistry results (Grains Research and Development Corporation, Australia, unpublished) required validation *in vivo.* Our objective was to rank 20 different grain cultivars for acidosis risk, based on an index of acidosis previously developed using rumen volatile fatty acid (VFA) analysis, pH and ammonia concentrations from 800 cattle in 100 dairy herds exposed to a wide range of diets (Bramley *et al.,* 2008). A challenge model (Lean and Rabiee, 2009) in which cattle were fed grains at 1.2% of bodyweight was used to test the differences in acidosis risk among grains.

Material and methods

Forty-two Holstein heifers were assigned in a randomised, controlled and partially blinded clinical trial. Determinations were balanced for day of testing and for grains to provide 3 determinations per grain. Additional determinations were made for the control which was an equal mix of the 20 grains. Cattle were used twice in the trial, after 14 d of rest. The body weights of heifers at entry ranged from 230 to 500 kg. Cattle were accustomed to the feeding system using a pre-adaptation period of 7 d, when cattle were fed 0.5 kg of triticale and 0.5 kg of the test grain and grass silage; a 4 d period when 1 kg per day of test grain and grass silage was fed. During the challenge period, cattle had all feed withheld for 14 h. Cattle were then fed 0.3 kg of ryegrass silage to reduce the saliva contamination during rumen sampling at time 0, then 1.2% of BW of the test grain. The test grains were oats (n=3), wheat (n=6), barley (n=4), triticale (n=4) and sorghum (n=3). Grains were processed by dry rolling. Following the initial rumen sampling at time 0, rumen samples were collected 1 h after feeding grain, and every 45 min for 5 subsequent samples. Rumen pH was measured immediately in samples of the raw rumen fluid. Rumen fluid was analysed for VFA using gas chromatography and concentrations of rumen ammonia, D- and L-lactate were analysed using enzymatic methods. Estimated marginal mean responses were calculated using generalised linear models with repeated measures. The rumen data were used to calculate concordance with the 'acidotic' group cattle identified by Bramley *et al.* (2008) using cluster, and discriminant analysis. Correlations between ranking of acidosis risks based on findings from the mixed model using all variables, increased valerate concentrations, or the Bramley *et al.* (2008) index were calculated using the Spearman rank test.

Results

Rumen concentrations of propionate, isovalerate, valerate and ammonia significantly differed among grains (Table 1). Concentrations of valerate were significantly increased following the challenge (P<0.001), and varied among grains (P<0.001). These data strongly indicated that changes in concentrations of valerate following challenge were more discriminatory than other single measure. Rankings of acidosis indices for some of the different grains and control, using the 3 methods outlined above are given in Table 1. Correlations between acidosis scores of different grains from cluster analysis and valerate (r 0.852; P<0.0001), and NIR rankings were significant (r 0.681; P=0.0007). Those for the mixed model and NIR were significant (r 0.466; P=0.033). The acidosis ranks in Table 1 are 1 to 21, with rank 1 highest, 21 least acidotic.

Table 1. Estimated marginal means (± SE) of rumen metabolites, and acidosis ranks, for 10 examples from the 20 grains. The control grain was an equal mixture of all 20 test grains.

Grain	pH	Mean ranks cluster mixed, NIR	Acetate	Propionate	Butyrate	Valerate	Lactate		Ammonia
							D-	L-	
Control (mixed)	6.94 ±0.02	10,12,14	52.3 ±1.81	12.1 ±0.42	7.9 ±0.55	1.07 ±0.07	0.07 ±0.01	0.06 ±0.01	7.57 ±1.16
Barley1	6.98 ±0.05	16,11,16	55.4 ±3.64	12.1 ±0.04	8.4 ±1.11	1.14 ±0.14	0.10 ±0.02	0.09 ±0.02	9.52 ±2.34
Barley3	6.96 ±0.06	12,15,6	58.3 ±4.30	13.6 ±0.99	6.3 ±1.31	0.88 ±0.16	0.06 ±0.02	0.03 ±0.03	3.69 ±2.76
Oats1	6.89 ±0.05	18,14,19	52.5 ±3.64	11.9 ±0.84	9.0 ±1.11	0.83 ±0.14	0.07 ±0.02	0.05 ±0.02	4.45 ±2.34
Sorghum2	7.0 ±0.06	19,21,13	47.4 ±4.20	10.3 ±0.97	7.2 ±1.29	0.76 ±0.16	0.07 ±0.02	0.06 ±0.03	3.34 ±2.70
Triticale2	6.88 ±0.05	1,4,4	56.7 ±4.29	14.9 ±0.99	9.5 ±1.31	1.43 ±0.16	0.10 ±0.02	0.09 ±0.03	11.20 ±2.76
Triticale4	6.94 ±0.06	9,9,7	55.8 ±4.26	12.9 ±0.98	10.5 ±1.31	1.33 ±0.16	0.08 ±0.02	0.07 ±0.03	7.91 ±2.74
Wheat1	6.93 ±0.65	8,2,11	55.7 ±3.61	13.6 ±0.83	9.4 ±1.11	1.61 ±0.14	0.12 ±002	0.11 ±0.02	13.96 ±2.32
Wheat6	6.88 ±0.06	4,7,2	56.6 ±4.17	13.2 ±0.96	9.5 ±1.28	1.29 ±0.16	0.10 ±0.02	0.05 ±0.02	5.87 ±2.68

Discussion and conclusion

Grains did differ in acidosis risk based on the cluster analysis index. Prediction of group allocation using the statistical methods developed by Bramley *et al.* (2008) was strongly based on propionate, valerate and ammonia concentrations. These were, similarly, the most significantly different variables influenced by grain cultivar in this study (Table 1). This observation is consistent with the conversion of lactic acid to valerate and propionate in the rumen, an essential function to sequester hydrogen ions in safe compounds.

Acknowledgement

We thank J.L. Black, R. King and S. Nielson for their support with this study.

References

Bramley, E., I.J. Lean, W.J. Fulkerson, M.A. Stevenson, A.R. Rabiee and N.D. Costs, 2008. The definition of acidosis in dairy herds predominantly fed on pasture and concentrates. J. Dairy Sci. 91: 308-321.

Opatpatanakit, Y., R.C. Kellaway, I.J. Lean, G. Annison and A. Kirby, 1994. Microbial fermentation of cereal grains *in vitro*. Aust. J. Agr. Res. 45: 1247-1263.

Lean, I.J. and A.R. Rabiee, 2009. Effect of feeding grain on ruminal acidosis in cattle: A pilot study. In: Chilliard, Y., F. Glasser, Y. Faulconnier, F. Bocquier, I. Veissier, M. Doreau (Eds.), Ruminant Physiology. Digestion, metabolism and effects of nutrition on reproduction and welfare. Wageningen Academic Publishers, Wageningen, the Netherlands. p. 256-257.

Using Grazplan software to estimate annual methane outputs of grazing Merino ewes having different lifetime reproductive performances

G.J. Lee
NSW Department of Primary Industries, Orange Agricultural Institute, NSW 2800, Orange, Australia; greg.lee@dpi.nsw.gov.au

Introduction

There is a large degree of within-flock variation in the lifetime reproductive performance of the Australian Merino (Lee *et al.*, 2009). For example, the difference in net reproduction rate (NRR) between ewes in the top and bottom quartiles is 3.6-6.4 times, which represents at least one additional lamb annually for each ewe mated. The nutrient requirements of the different NRR quartiles of the ewe flock will reflect the large variation in lifetime productivity.
Methane (CH_4) emissions from grazing ruminant industries are seen as making an important contribution to anthropogenic greenhouse gas (GHG) emissions, with consequences for climate change. As CH_4 production is positively related to feed intake (Molano and Clark, 2008), the animals within a flock having the highest productivity will also have higher nutrient requirements, and hence intakes (feed costs). This study sought to estimate the relative annual CH_4 production of breeding Merino ewes with different lifetime NRR, within a wool producing enterprise grazing in the wheat-sheep belt of southern New South Wales.

Methodology

The availability of pasture and its quality were simulated on a daily basis throughout the year for a mixed (perennial and annual grass and legume species) pasture grazed by a flock of breeding Merino ewes in southern NSW, using *GrasGro®3* software (Moore *et al.*, 1997). The median values from the simulation of dead and green matter availability, and their digestibility, and the proportion of legume throughout the year were used as the feed base in subsequent calculations of the daily pasture intakes, liveweight changes and CH_4 outputs for ewes of each birth type/rearing type (BT/RT). The mean birth type was 0.99 lambs per ewe joined, of which 0.78 survived to weaning.
It was assumed that initial ewe liveweight was 44 kg at 1 January, that the ewes were mated on 1 March and lambed on 30 July, that litter size was limited to 0, 1 or 2 lambs, that lamb losses occurred in the first week after birth, and that lambs were weaned at 12 weeks (21 October).The daily intake of pasture, rates of liveweight change and CH_4 production of each BT/RT were calculated at intervals throughout the year using *GrazFeed* (Freer *et al.*, 1997). Liveweights were revised for each run using the liveweight change estimates from the previous period for the respective BT/RT. Daily intakes and methane production were calculated for the ewe-lamb unit for the period between birth and weaning, and per ewe at other times. From the daily estimates, the total annual metabolisable energy intake (MEI) as pasture and CH_4 production were calculated for each of the NRR quartiles, based on the proportions of each BT/RT within each quartile. Table 1 shows the annual intakes and CH_4 production by ewes of each BT/RT.

Results and discussion

The 25% of ewes with the highest lifetime NRR (median = 1.2 lambs weaned / ewe mated) consumed 37% more energy annually than the lowest performers, produced 12% more CH_4 and weaned more than 5 times as many lambs over their lifetime (Table 2). However, the top 25% of ewes used 75% less energy than ewes in the lowest NRR quartile and produced 80% less CH_4 to wean each lamb.

Table 1. GrazFeed estimates of the annual pasture intakes (as metabolisable energy, MEI) and CH_4 production of the ewe:lamb unit for each birth type/rearing type.

Lambs born:	0	1	1	2	2	2
Lambs weaned:	0	0	1	0	1	2
MEI pasture of ewe and lamb (GJ/yr)	3.60	3.67	4.75	3.68	4.73	6.98
CH_4 output of ewe and lamb (kg/yr)	9.15	9.29	10.71	9.31	10.60	11.39

Table 2. Net reproduction rate (lambs weaned/ewe mated) and GrazFeed estimates of the annual pasture intakes (as metabolisable energy) and CH_4 production, expressed as either annual total or per lamb weaned, within the lifetime net reproductive rate quartiles.

	1st quartile	2nd quartile	3rd quartile	4th quartile
Net reproduction rate	0.24	0.66	0.92	1.31
MEI pasture of ewe and lamb (GJ/yr)	3.95	4.41	4.77	5.41
MEI/lamb weaned (GJ)	16.27	6.71	5.21	4.14
CH_4 output of ewe and lamb (kg/yr)	9.59	10.10	10.38	10.74
CH_4 output/lamb weaned (kg)	39.55	15.35	11.35	8.22

The *GrazFeed* model (Freer *et al.*, 1997) estimates CH_4 production from non-linear relationships with feed intake and diet quality (Blaxter and Clapperton, 1965). More recently Molano and Clark (2008) suggested that CH_4 production per unit of dry matter intake is constant on forage diets, in which case the CH_4 production estimates for higher performing ewes will have been underestimated to some extent, and the relative CH_4 production differences will be more accurately reflected by differences in MEI.

Retaining poor performing ewes in the breeding flock increases the cost of production through the higher pasture intake (MEI) required to wean each lamb, and those increased costs are unlikely to be offset by the value of a slightly higher fleece weight (Lee *et al.,* 2009). In addition, there are also increased GHG emissions per unit output.

References

Blaxter, K.L. and J.L. Clapperton, 1965. Prediction of the amount of methane produced by ruminants. Br.J. Nutr. 19: 511-522.

Freer, M., A.D. Moore and J.R. Donnelly, 1997. GRAZPLAN: Decision support systems for Australian grazing enterprises. II. The animal biology model for feed intake, production and reproduction and the GrazFeed DSS. Agric. Syst. 55: 77-129.

Lee G.J., K.D. Atkins and M.A. Sladek, 2009. Heterogeneity of lifetime reproductive performance, its components and associations with wool production and liveweight of Merino ewes. Anim. Prod. Sci. 49: in press.

Molano, G. and H. Clark, 2008. The effect of level of intake and forage quality on methane production by sheep. Aust. J. Exp. Agric. 48: 219-222.

Moore, A.D., J.R. Donnelly and M. Freer, 1997. GRAZPLAN - decision support systems for Australian grazing enterprises. 3. Pasture growth and soil moisture submodels, and the Grassgro DSS. Agric. Syst. 55: 535-582.

Substrate oriented rumen fermentations in sheep during provoked acidosis

A. Lettat[1,2], P. Nozière[1], M. Silberberg[1], D.P. Morgavi[1] and C. Martin[1]
[1]INRA, UR1213 Herbivores, Site de Theix, 63122 Saint Genès Champanelle, France; [2]Danisco France SAS, Zone d'Activités de Buxières, BP 10, 86220 Dangé Saint Romain, France; cecile.martin@clermont.inra.fr

Introduction

Latent rumen acidosis is a major problem in high producing ruminants. It decreases feed efficiency and production and predisposes to disease (Stock et al., 1990). This rumen disorder is due to excessive consumption of readily fermentable carbohydrates (RFC), which induces a reduction in pH and affects the microbial ecosystem. During latent acidosis, microbial fermentations are unstable and variably oriented to lactate, propionate or butyrate at the expense of acetate according to the severity of acidosis (Martin *et al.,* 2006), and existing mechanistic rumen models fail to predict rumen VFA profiles (Offner and Sauvant, 2004). The aim of this study was to develop an experimental model for rumen acidosis in sheep for monitoring the fermentation pathways (lactic, propionic and butyric) and studying the changes in the microbial ecosystem.

Material and methods

Three ruminally cannulated Texel wethers, averaging 63.4±5.4 kg BW at the start of the experiment, were used. Wethers were housed in individual stalls (1.00×1.50 m) with individual feed-bunks, and were randomly assigned to three dietary treatments in a 3 x 3 Latin square design. Each period lasted 14 d with 11 d where sheep were fed (1.2% of BW) in two equal portions at 8:00 and 16:00 h with 80% hay + 20% wheat based concentrate, and 3 d of acidosis induction by a drastic change of diet (challenge). During the challenge periods, morning feeding was suppressed and sheep were intraruminally dosed with wheat (readily fermentable starch), maize (slowly fermentable starch) or beet pulp (easily digestible fibres). The acidosis challenges were repeated 3 times on 3 consecutive days (1 challenge per day). Animals had free access to water and mineralised salts blocks. Intake of feed and water were recorded daily. During the acidosis induced periods, rumen liquid samples were taken from the ventral sac one hour before (-1) and 1, 3, 5 and 6 h after challenge induction. Rumen pH, lactic acid and volatile fatty acids were measured on each sample. All data were analysed in repeated time using the MIXED procedure of SAS®, with period, substrate (S, wheat vs. maize vs. beet pulp), challenge (C, 1 to 3), time (T, -1 vs. +1 vs. +3 vs. +5 vs. +6 h) and S×C, S×T, C×T, S×C×T interactions as fixed effects, and animal as random.

Results

Rumen pH (Figure 1A) was affected by substrate ($P<0.01$) and S×T interaction ($P<0.001$), indicating that difference among substrates was enhanced by time. Mean rumen pH was the lowest for a wheat challenge (5.71), intermediate for maize (5.97) and highest for beet pulp (6.13). At the end of the acidosis induced period, total VFA concentration (data not shown) was affected by time ($P<0.001$) but not by substrate. Molar proportion of propionate (Figure 1B) was influenced by substrate ($P<0.01$), time, and S × T interaction ($P<0.001$). Beet pulp had the highest propionate proportion while it was intermediate for maize and lowest for wheat (21.6%, 16.7% and 12.1%, respectively). Molar proportion of butyrate (Figure 1C) varied with time ($P<0.001$) and tended ($P=0.09$) to be higher for maize (16.5%) than for the other substrates (9.5% on average). Lactate concentration (Figure 1D) in rumen fluid was affected by time ($P= 0.002$) and S×T interaction ($P<0.001$). It reached 45.5 mM for wheat and was lower for maize and beet pulp (8.3 mM on average).

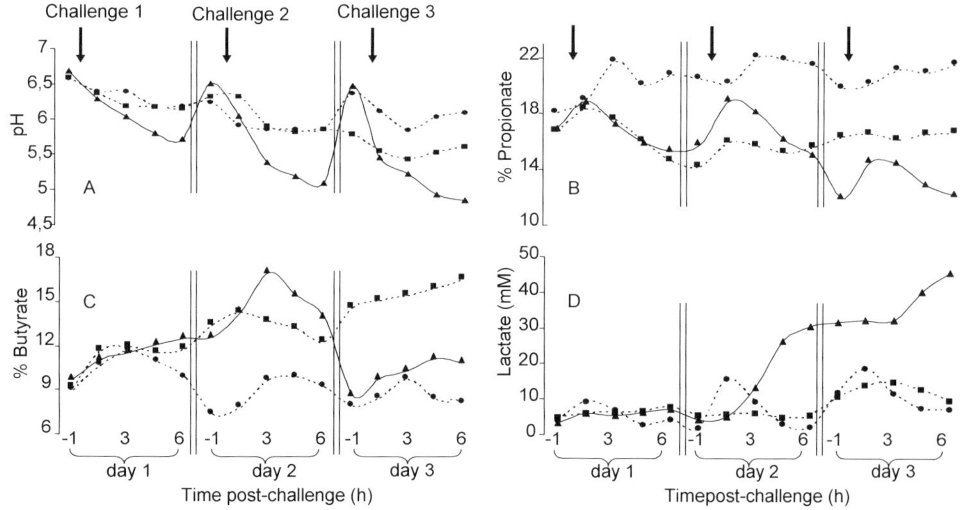

Figure 1. Effect of three different feed challenges based on wheat (▲), maize (■) and beet pulp (●) on pH (A), molar proportion of propionate (B) and butyrate (C), and lactate concentration (D) during rumen induced acidosis in sheep.

Conclusion

We developed a model which allowed us to obtain differentiated rumen fermentations in sheep. The rumen pH and fermentation parameters obtained with the wheat challenge were indicative of a lactic acidosis, whereas butyric and propionic latent acidosis were observed for maize and beet pulp challenges, respectively. This model should be confirmed before beginning the microbial ecosystem characterisation.

Acknowledgement

The authors thank Danisco SAS for their financial support and S. Alcouffe, M. Fabre and D. Roux for the care of the animals.

References

Martin, C., L. Brossard and M. Doreau, 2006. Mécanismes d'apparition de l'acidose latente et conséquences physiopathologiques et zootechniques. INRA Prod. Anim. 19: 93-108.

Offner, A. and D. Sauvant, 2004. Comparative evaluation of the Molly, CNCPS, and LES rumen models. Anim. Feed Sci. Technol. 112: 107-130.

Stock, R.A., M.H. Sindt, J.C. Parrot, and F.K. Goedeken. 1990. Effect of grain type, roughage level and monensin level on finishing cattle performance. J. Anim. Sci. 68: 3441-3455.

Effects of levels and combinations of fish oil and sunflower oil inclusion in the diet on rumen fermentation and total tract digestibility in China Nooxi steers

S. Liang, D.P. Bu, J.Q. Wang, Khas-Erdene, S.J. Liu, H.Y. Wei and L.Y. Zhou
State Key Laboratory of Animal Nutrition, Institute of Animal Science, Chinese Academy of Agricultural Sciences, Beijing 100193, P.R. China; wang-jia-qi@263.net

Introduction

It is well established that adding fish oil (FO) and sunflower oil (SFO) to animal diets is one of the best strategies to enhance CLA content in ruminant food products (Cruz-Hernandez *et al.*, 2007). However, it is generally accepted that adding unsaturated fatty acid to ruminant diets may exert negative effects on the rumen ecosystem and decrease digestion (Jenkins, 1993; Havartine and Allen, 2006), and these negative effects may be ameliorated by a high forage diet (Bateman and Jenkins, 1998; Ueda *et al.*, 2003). To the best of our knowledge, the effects of supplementing both FO and SFO simultaneously on nutrient digestion have not been evaluated. The present study evaluated the effects of including various combinations of FO and SFO on rumen fermentation and digestion in China Nooxi beef steers fed a high forage diet.

Material and methods

Four China Nooxi steers fitted with ruminal cannulas were used in a 4×4 Latin Square experimental design. Treatments were the control diet (14% alfalfa hay, 51% Chinese wildhey, and 35% concentrate on a DM basis), or a basal diet with 3% SFO plus 1% FO (SF1), or a basal diet with 2.5% SFO plus 1.5% FO (SF2), or a basal diet with 2% SFO plus 2% FO (SF3). Experimental diets were isoenergetic. Steers were allowed access to a quantity of feed that just met maintenance requirements: DMI was restricted to 8.01 kg/head/d. Each experimental period lasted for 21 d and measurements occurred during the last 7 d of each period. Total tract digestibility was estimated using a dual-phase marker system (Faichney, 1975). Ruminal fluid was collected every 2 h from 07:30 to 19:30 h on d 15 of each experimental period, to assay volatile fatty acid (VFA) and ammonia concentrations.

All data were analysed by the MIXED procedure of SAS® 8.2 (SAS Institute Inc.). The statistical model included cow and period as random effect, and treatment as the fixed effect. Least Squares Means and pooled SEM are reported for all data. Orthogonal contrasts were used to test the effect of control versus oil-supplementation and linear and quadratic effect of FO. Treatment differences were determined using the PDIFF option of SAS®. Significance was declared at $P<0.05$, and tendencies at P value between 0.05 and 0.1.

Results

Compared with controls, including FO and SFO decreased ($P<0.01$) rumen fluid acetate and butyrate concentrations, while propionate content increased ($P<0.01$). Accordingly, the ratio of acetic to propionic acid decreased ($P<0.01$) when oil was added. Butyrate acid content in the rumen fluid decreased linearly ($P=0.05$) as dietary FO increased, and acetic acid decreased quadratically (Table 1), which indicated that fish oil had a strong influence on rumen fermentation. Oil supplementation tended to increase ammonia N in rumen fluid ($P=0.06$). Total tract digestibility of DM, OM, NDF and ADF was not affected ($P>0.05$) by supplemental fat (Table 1).

Table 1. Ruminal VFA and ammonia N concentrations, and total tract digestibility in beef cattle fed different various combinations of SFO and FO.

Item	Diets				SEM	*P*-Value		
	Control	SF1	SF2	SF3		Control *vs* oil	Linear	Quadratic
Ruminal VFAs and ammonia N								
Acetate, mmol/L	63.4^{a}	57.8^{b}	54.7^{bc}	52.2^{c}	1.54	<0.001	0.13	0.02
Propionate, mmol/L	15.7^{c}	18.5^{b}	19.8^{a}	21.1^{a}	1.05	<0.001	0.39	0.16
Acetate/propionate	4.2^{a}	3.2^{b}	2.9^{b}	2.7^{c}	0.15	<0.001	0.21	0.09
Butyrate, mmol/L	10.2^{a}	8.5^{b}	7.7^{b}	7.5^{b}	0.39	<0.001	0.05	0.12
Ammonia N, mg/dL	7.2	9.7	8.8	9.0	0.82	0.06	0.49	0.86
Total tract digestibility, %								
DM	65.6	68.0	65.6	72.0	4.21	0.52	0.67	0.30
OM	63.1	66.4	63.2	70.4	4.35	0.44	0.56	0.26
NDF	51.5	57.5	58.7	63.7	5.60	0.66	0.28	0.17
ADF	50.9	56.8	58.8	62.6	7.38	0.70	0.14	0.22

^{a,b,c} $P<0.05$.

Discussion and conclusion

Combining dietary FO and SFO reduced rumen fluid acetic and butyric acid contents as well as the acetic to propionic acid ratio, while propionic acid content was increased. Combining dietary FO and SFO did not decrease nutrient digestion when China beef steers were fed a high forage ration. These results should be interpreted with caution, because the steers used in this study had a relative lower DMI and thus may not be a perfect model for dairy cows. Nonetheless, a high forage diet might mask some of the negative effects supplemental oils have on digestion kinetics.

References

Bateman, H.G. and T.C. Jenkins, 1998. Influence of soybean oil in high fiber diets fed to nonlactating cows on ruminal unsaturated fatty acids and nutrient digestibility. J. Dairy Sci. 81: 2451-2458.

Cruz-Hernandez, C., J.K.G. Kramer, J.J. Kennelly, D.R. Glimm, B.M. Sorensen, E.K. Okine, L.A. Goonewardene and R.J. Weselake, 2007. Evaluating the conjugated linoleic acid and *trans* 18:1 isomers in milk fat of dairy cows fed increasing amounts of sunflower oil and a constant level of fish oil. J. Dairy Sci. 90: 3786-3801.

Faichney, G.J., 1975. The use of markers to partition digestion within the gastro-intestinal tract of ruminants. In: McDonald I.W. and A.C.I. Warner (Eds), Proc. IV Intl Symp. on Ruminant Physiology. Univ of New England Publ., Armidale, Australia, pp. 277-291.

Harvatine, K.J. and M.S. Allen, 2006. Effect of fatty acid supplements on ruminal and total tract nutrient digestion in lactating dairy cows. J. Dairy Sci. 89: 1092-1103.

Jenkins, T.C., 1993. Lipid metabolism in the rumen. J. Dairy Sci. 76: 3851-3863.

Ueda, K., A. Ferlay, J. Chabrot, J.J. Loor, Y. Chilliard and M. Doreau, 2003. Effect of linseed oil supplementation on ruminal digestion in dairy cows fed diets with different forage:concentrate ratios. J. Dairy Sci. 86: 3999-4007.

Effect of combined ensiling of sorghum and soybean with or without molasses and lactobacilli on *in vitro* rumen fermentation

R. Lima[1], *M. Lourenço*[2], *R.F. Díaz*[1,3], *A. Castro*[3] *and V. Fievez*[2]
[1]*Central University of Las Villas (UCLV), Carretera a Camajuaní km 5½, Santa Clara, Cuba;* [2]*Ghent University, LANUPRO, Proefhoevestraat 10, 9090 Gontrode, Belgium;* [3]*UCLV, CIAP, Carretera a Camajuaní km 5½, Santa Clara, Cuba; veerle.fievez@ugent.be*

Introduction

Compared to fresh material, ensiling might result in a lower content of rumen fermentable organic material due to the loss of quickly fermentable carbohydrates and soluble protein (González *et al.*, 2007). The addition during the ensiling process of molasses, as a source of water soluble carbohydrates, and lactate producing bacteria, *Lactobacillus* spp. (LAB) as an inoculant, could limit these losses as it accelerates a pH drop in the silo. The objective of this study was to assess changes in rumen fermentable substrate when ensiling a sorghum-soybean mixture with or without molasses and lactobacilli.

Material and methods

Two sorghum (*Sorghum bicolor* (L.) Moench) varieties (CIAP 2E-95 & CIAP 49V-95), with the former showing higher grain and total DM yield, but also increased tannin contents (Castro *et al.*, 2000) and one soybean (*Glycine max* (L.) Merr.) variety (INCASOY-35) were sown, harvested (at pasty grain state), chopped and ensiled (CIAP, Cuba) in the following combinations: two sorghum varieties with soybean in two proportions (0.4 and 0.6) and ensiled with or without molasses (3.5% of fresh material) and *Lactobacillus* sp. as inoculant (3×10^5 colony forming units/g). The experiment consisted of two types of *in vitro* incubations, each performed in triplicate: in gastight culture flasks (24 h) using 20 ml of a phosphate buffer and in syringes (96 h) using 20 ml of a buffer mixture (phosphate/bicarbonate), according to Fievez *et al.* (2005). Further, 0.5 g of dried material of each treatment and 5 ml of rumen contents were added. After 24-h incubation in culture flask, the contents were acidified, centrifuged, filtered and used for SCFA and NH_3 determination. The apparent ruminal degradable organic matter (ARDOM) was estimated according to Demeyer (1991). The gas production kinetics were estimated as:

$V_t = V_f\{1+\exp[2-4 \times k_f \times (t-L_n)]\}^{-1}$,

with V_t, the gas production (ml) at time (t), V_f, the asymptotic gas volume corresponding to maximum digestion of the incubated material (ml), k_f, the fractional fermentation rate (h^{-1}) and L_n, the lag time before gas production starts (h). A general linear model was used for statistical analysis (using SPSS 15.0, SPSS Inc., Chicago, IL, USA):

$Y = \mu + SP_{i=1\text{-}2} + SV_{j=1\text{-}2} + EE_{k=1\text{-}3} + R_{t=1\text{-}3} + SP_i \times EE_k + SV_j \times EE_k + \varepsilon$,

with $SP_{i=1\text{-}2,}$ the sorghum proportion, $SV_{j=1\text{-}2,}$ the sorghum varieties, $EE_{k=1\text{-}3,}$ the ensiling effect (fresh forage *vs* ensiled without molasses and inoculum *vs* ensiled with molasses and inoculum), $R_{l=1\text{-}3,}$ the 3 incubation days, $SP_i \times EE_{k,}$ and $SV_j \times EE_{k,}$ the interaction between the indicated factors. Further, a post-hoc Duncan test was performed.

Results and discussion

Ensiling resulted in lower acetate and higher propionate proportions (Table 1). However, ensiling without molasses and inoculant also reduced SCFA production and hence ARDOM and V_f as well as k_f. As expected, a higher proportion of sorghum increased the molar propionate proportion (mmol/mol total SCFA) (387 vs. 370), k_f(h/1) (0.09 vs. 0.07) and L_n (h) (3.94 vs. 3.07), whereas ammonia (mmol/l) concentrations (1.91 vs. 2.09) were reduced. CIAP 2E-95 silages produced higher molar propionate proportions (398 vs. 360), induced higher V_f(ml/g OM) (131 vs. 117), lower acetate proportions (472 vs. 520 mmol/mol total SCFA) and ammonia (1.82 vs. 2.18 mmol/l) concentrations.

Table 1. Total net production of SCFA (mmol/g OM) and proportions of individual SCFA (mmol/mol total SCFA), NH_3 content (mmol/l), ARDOM (%) and parameters of curve fitting (Vf (ml/g OM), kf (h/1), L_n (h)) of combinations of sorghum and soybean incubated in flasks (24 h) or syringes (96 h) either as fresh forage (F) or as silages (E), ensiled with (+) or without (-) molasses and Lactobacillus sp. inoculum (n=3).

	SCFA	C2	C3	C4	NH_3	ARDOM	V_f	k_f	L_n
Treatments									
Sorghum-soybean F	4.50^a	546^a	336^c	118	1.95^b	40.8^{ab}	135^a	0.09^a	3.90
Sorghum-soybean E-	4.30^b	477^b	386^b	124	2.11^a	38.8^b	116^b	0.07^b	3.19
Sorghum-soybean E+	4.57^a	465^b	414^a	118	1.93^b	41.4^a	121^b	0.09^a	3.42
SEM	0.08	12.2	11.0	7.72	0.04	0.73	3.02	0.01	0.27
Statistics (*P*-values)									
EE	0.078	<0.001	<0.001	0.611	0.001	0.040	0.002	0.012	0.139
SP	0.148	0.108	0.021	0.106	<0.001	0.325	0.749	<0.001	0.006
SV	0.131	<0.001	<0.001	0.008	<0.001	0.390	0.002	0.025	0.039
SP×EE	0.529	0.690	0.027	0.764	0.964	0.630	0.727	0.332	0.357
SV×EE	0.005	0.706	0.086	0.922	0.006	0.002	0.002	0.291	0.782

^{a,b,c} Different superscripts within a column indicate significant differences ($P<0.05$).

Conclusion

Molasses and *Lactobacillus* spp. added to sorghum-soybean prior to ensiling promote rumen propionate production and did not affect the rumen degradable organic matter as compared to fresh forages. Despite the higher tannin content of the fresh material, CIAP 2E-95 was preferred in this respect.

Acknowledgement

Research financed by the VLIR (Belgium) IUC-Programme. M. Lourenço is post-doctoral grant-holder of the UGent Special Research Fund.

References

Castro, A., L. Marrero and A. Arias, 2000. Potencial productivo y composición nutritiva del grano de cuatro variedades cubanas de sorgo cultivadas en condiciones de bajos insumos, In: Libro de resúmenes, IV Simposio de Agricultura Sostenible, INCA (RF-P-34), p. 146.

Demeyer, D., 1991. Quantitative aspects of microbial metabolism in the rumen and hindgut, In: J.P. Jouany (Ed.), Rumen microbial metabolism and ruminant digestion. INRA Editions, Paris, France, pp. 217-237.

Fievez, V., O.J. Babayemi and D. Demeyer, 2005. Estimation of direct and indirect gas production in syringes: A tool to estimate short chain fatty acid production that requires minimal laboratory facilities. Anim. Feed Sci. Technol. 123: 197-210.

González, J., J. Faría-Mármol, C.A. Rodríguez and A. Martínez, 2007. Effects of ensiling on ruminal degradability and intestinal digestibility of Italian rye-grass. Anim. Feed Sci. Technol. 136: 38-50.

In vitro gas production measurements to evaluate interactions among corn, soybean meal and distillers grain

Y. Lin[1]*, Z.S. Wang*[1]*, S.J. Lai*[2] *and G.Y. Yang*[2]
[1]*Animal Nutrition Institute, Sichuan Agricultural University, Ya'an 625014, China;* [2]*Animal Science Institute, Sichuan Agricultural University, Ya'an 625014, China; wangzs@sicau.edu.cn*

Introduction

Distiller grains (DG) of white spirit are a by-product of ethanol production. With recent increases in ethanol production, DG is increasingly available for livestock feed. This feedstuff provides a substantial amount of energy, phosphorus and may serve as well as a supplement for beef cattle. But DG provides a little amount of protein and starch. The associative effects (AE) describe the nonlinear response in nutrient utilisation when two feedstuffs are combined (Moe, 1979). AE are important because diets are typically balanced using linear equations. The deviations in the digestibility may be positive or negative as compared with calculated digestibility. The nutritional balance of the feedstuffs may be the main reason for interactions in ruminants. There are few data focussing on the use of DG of white spirit as a supplement for ruminants, by using the method of AE. The objectives of this study were to quantify the magnitude of AE among DG, corn and soybean meal (SM) by using the gas production (GP) technique.

Material and methods

In Trial 1, corn was incubated together with SM in proportions of 0, 20, 40, 60, 80, and 100% respectively. In Trial 2, corn and SM were used as a mixture (corn:SM = 6:4 from Trial 1) with DG in proportions of 0, 20, 40, 60, 80 and 100% respectively. *In vitro* GP was determined according to Menke and Steingass (1988). Rumen fluids were collected from steers (450±25 kg). All laboratory handling of rumen fluid was carried out under a continuous flow of CO_2. Samples (200±10 mg) of the air-dry feedstuffs were accurately weighed into 100-ml glass syringes. The GP was recorded after 2, 4, 6, 8, 12, 24, 36, 48, and 72 h of incubation. About 30 ml of mixed fermentation medium was collected for analysis of ammonia nitrogen (NH3-N), volatile fatty acids (VFA) and microbial nitrogen (MN) according to Hu *et al.* (2005). Data were analysed using the general linear model procedure of SPSS16.0 software (SPSS Inc., Chicago, IL, USA). Overall differences between treatment means were declared significant at $P<0.05$.

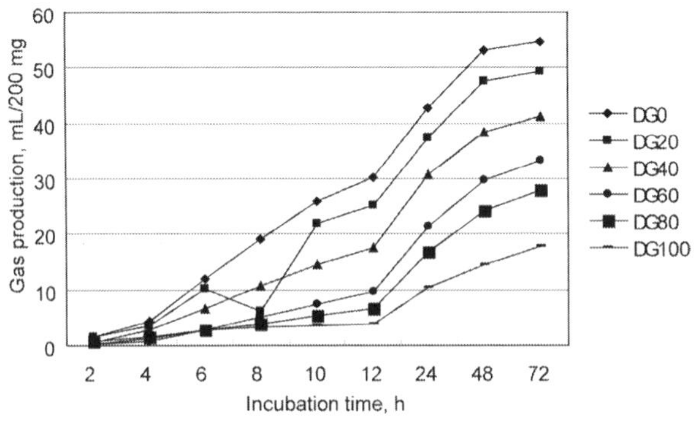

Figure 1. Cumulative gas production of individual feedstuffs at different times of incubation. DG0, 20, 40, 60, 80, 100 mean that the proportion of distiller's grains are 0, 20, 40, 60, 80, 100 in feedstuff of mixture and distillers grain.

Results

In Trial 1, the positive AE between corn and SM existed at 24 h of SM40, but the AE of other treatments were negative. So the corn and SM were mixed at the ratio of 6:4 as a feedstuff in Trial 2. The GP curves of corn with different levels of SM and DG at different times of incubation are given in Figure 1. The cumulative GP at all incubation times were the highest for treatment DG0 (P<0.05), the amount of cumulative GP reduced with the decreasing level of DG. The AE of feedstuffs of the GP observed for the mixtures of SM, corn and DG are given in Table 1. The AE of DG40 was greater than other treatments (P<0.001). The concentrations of N-NH3 were improved from 25.2 to 30.84 mg/100 ml when the DG level reduced (P<0.001). The MN production increased with the decreasing level of DG *(P*=0.02). The level of acetate, propionate, and butyrate were reduced with the decreasing level of DG (P<0.001, P=0.002 and P<0.001 respectively). The ratio acetate:propionate differ between DG and other treatments (P<0.001).

Conclusion

In vitro GP showed that the positive AE existed and was greater than that with the other treatments among corn, SM and DG at the ratio of 3.6:2.4:4, and at this proportion, the ratio of acetate:propionate decreased which may be one of the reasons why there were positive AE.

Table 1. Difference (%)[1] among the gas production observed for the different mixtures of soybean meal, corn and DG and the effect of mixture of different DG proportion on MN, NH_3-N, acetate, propionate, and butyrate.

	Difference (%)[1] of different incubation time, h			MN mg/100 ml	NH_3-N mg/100 ml	Acetate mmol/l	Propionate mmol/l	Acetate/ Propionate mol/mol	Butyrate mmol/l
	24	48	72						
DG0				6.82^b	30.84^b	0.25^b	0.149^b	1.678^c	0.042^b
DG20	3^b	4^c	4^d	5.88^c	29.81^c	0.202^c	0.111^c	1.82^d	0.037^c
DG40	4^b	5^c	22^b	5.13^d	28.35^d	0.187^{cd}	0.124^{cd}	1.508^b	0.034^c
DG60	-9^d	-0.4^d	-3^e	4.15^e	26.94^e	0.163^d	0.098^d	1.663^c	0.032^c
DG80	-6^c	9^b	12^c	3.34^f	26.26^e	0.157^e	0.084^d	1.869^e	0.026^d
DG100				2.61^g	25.2^f	0.143^f	0.084^d	1.702^{cd}	0.028^d
SEM	0.52	0.45	0.52	0.4	0.05	0.01	0.07	0.03	0.02
P-value	<0.001	<0.001	<0.001	<0.001	0.02	<0.001	0.002	<0.001	<0.001

[1] Difference (%) = [(Observed gas production - predicted gas production)/ predicted gas production] × 100.
[b-g] Significant differences (P<0.05).

Acknowledgement

This work was supported by the program for public welfare behaviour (Agriculture) research fund of the Agricultural Department of China (No. nyhyzx07-036-03).

References

Hu, W.L., J.X. Liu, J.A. Ye, Y.M. Wu and Y.Q. Guo, 2005. Effect of tea saponin on rumen fermentation *in vitro*. Anim. Feed Sci. Technol. 120: 333–339.

Moe, P.W., 1979. Associated effects of feedstuffs. Pacific Northwest Anim. Nutr. Conf., p 177.

Menke, K.H. and H. Steingass, 1988. Estimation of the energetic feed value obtained from chemical analysis and *in vitro* gas production using rumen fluid. Anim. Res. Dev. 28: 7-55.

Effect of acetate kinase gene deletion engineering bacteria of *Selenomonas ruminantium* on propionate metabolism *in vitro*

M. Long, X. Xing, L. Liu, X.Y. Pang, Z. Wang and G.W. Liu
Laboratory of Animal Nutritional and Metabolic Diseases, College of Animal Science and Veterinary Medicine, Jilin University, Changchun, Jilin, China; liuguowen2008@163.com

Introduction

The metabolic diseases of periparturient dairy cows such as ketosis and fatty liver caused by negative energy balance are due to propionate deficiency. The traditional preventive and therapeutic measures involve supplying dairy cows with precursors of glucose such as propionate, but they are not economical and practical. Recently, ionophore antibiotics, such as monensin and lasalocid, have been shown to have a consistent and well-documented positive effect on propionogenesis, but their use as feed additives in animal production is forbidden in Europe since 2006. So it is important to find a new way to regulate the quantity of propionate instead of using ionophore antibiotics at peripartum. *Selenomonas ruminantium* is one of the dominant bacteria in the rumen which can ferment lactate into acetate, propionate and butyrate. We constructed acetate kinase gene deletion engineering bacteria, but did not evaluate whether it could affect propionate metabolism. So the purpose of this study was to investigate the effects of adding it into rumen fluid on propionate metabolism *in vitro*.

Material and methods

Selenomonas ruminantium K6 (K6) was isolated from the rumen of a healthy dairy cow, identified by 16S rDNA PCR, and then its acetate kinase gene deletion engineering bacteria (TnK6) was constructed in our lab. Rumen fluid was collected from a rumen cannulated dairy cow 2 h after feeding, and 60 ml were equally divided and put into three flasks. The rumen fluid of each flask was added with substrates (2 g/l lactic acid and 2 g/l pyruvate) and then incubated without added bacterial strains (control group), with K6 bacteria (8.2×10^6 organisms/ml, group K6) and TnK6 bacteria (8.2×10^6 organisms/ml, group TnK6) at 39 °C for 24 h under anaerobic conditions. Triplicate flasks were prepared for each treatment. The concentrations of lactate, acetate and propionate in the rumen fluid were measured by Waters-Baseline 810 liquid chromatography at 0, 4, 8, 10, 12, and 24 h after *in vitro* incubation. The results are given as means ± SD (n=18) and differences between groups were analysed by ANOVA for the mean of all sampling times.

Results

The lactate concentration of groups K6 and TnK6 decreased with incubation time whereas it increased in the control group (Table 1). Lactate concentration was significantly lower in group TnK6 than in group K6 and in the control group ($P<0.01$). The acetate concentration of the three groups increased with incubation time to reach a peak at 12 h for the control group and at 10 h for groups K6 and TnK6. Mean acetate concentration was higher in group K6 than in the other two groups ($P<0.01$) and similar between the control and TnK6 group ($P>0.05$).The propionate concentration of the three groups increased with incubation time to reach a peak at 10 h for the control group and at 12 h for groups K6 and TnK6. Mean propionate concentration was higher in group K6 than in the control group and in group TnK6 ($P<0.01$). The concentration ratio of acetate to propionate of the three groups decreased with incubation time to reach the lowest point at 10 h for the control group and at 12 h for groups K6 and TnK6. Mean concentration ratio of acetate to propionate was lower in group TnK6 than in group K6 and in the control group ($P<0.01$).

Table 1. Effect of adding Selenomonas ruminantium *(K6) or its acetate kinase gene deletion engineering bacteria (TnK6) on the lactate, acetate and propionate concentrations of the rumen fluid incubated* in vitro.

Group	Time of sampling (h)						Significance test (n=18)
	0	4	8	10	12	24	
Lactate (mmol/L)							
control	11.13±0.31	11.90±0.40	14.47±0.49	15.77±0.15	16.07±0.35	12.10±0.44	13.01±2.22[a]
K6	11.13±0.31	11.53±0.25	9.20±0.26	8.13±0.31	6.63±0.21	9.17±0.25	9.78±1.76[b]
TnK6	11.13±0.31	10.10±0.50	8.17±0.25	7.27±0.15	8.63±0.15	10.17±0.32	9.59±1.59[c]
Acetate (mmol/L)							
control	50.07±0.25	50.87±0.25	59.10±0.36	57.33±0.38	59.23±0.38	53.83±0.35	54.00±4.28[a]
K6	50.07±0.25	53.80±0.30	64.07±1.27	69.60±0.92	69.13±0.47	61.17±0.76	59.45±7.55[b]
TnK6	50.07±0.25	52.17±0.25	59.23±0.35	61.00±0.53	58.47±0.59	55.70±0.26	55.12±4.18[a]
Propionate (mmol/L)							
control	18.10±0.44	19.67±0.85	22.17±0.31	22.77±0.38	21.27±0.59	20.40±0.78	20.20±2.13[a]
K6	18.10±0.44	21.80±0.36	27.17±0.25	29.80±0.56	33.07±0.38	28.23±0.55	25.07±5.17[b]
TnK6	18.10±0.44	19.80±0.53	23.07±0.15	25.13±0.45	26.13±0.21	24.83±0.25	22.04±3.12[a]
Acetate / Propionate							
control	2.77±0.07	2.59±0.13	2.67±0.03	2.52±0.06	2.79±0.09	2.64±0.09	2.68±0.14[a]
K6	2.77±0.07	2.64±0.06	2.57±0.03	2.43±0.03	2.24±0.03	2.24±0.03	2.53±0.20[b]
TnK6	2.77±0.07	2.47±0.04	2.36±0.03	2.34±0.02	2.09±0.02	2.17±0.06	2.41±0.23[c]

[a-c] Means with different superscript letters in the same column and for the same parameter are significantly different at $P<0.05$.

Discussion and conclusion

This *in vitro* study showed that the acetate kinase gene deletion engineering bacteria of *Selenomonas ruminantium* could ferment lactate from ruminal juice into volatile fatty acids. Compared to the ruminal juice with the strain K6, the acetate concentration and the ratio of acetate to propionate were obviously reduced in the ruminal fluid incubated with the strains TnK6, even though the concentration of propionate was slightly reduced. This demonstrated that the engineering bacteria could turn fermentation towards propionate, and it also gave us a new idea that using engineering bacteria can regulate the fermentation of ruminal microorganisms. But the *in vivo* use of such genetic modified microorganisms needs further research and will require regular constraints.

Acknowledgement

This study was funded by the China National Science Foundation Committee (30600441).

Comparisons in bacteria community changes in the rumen and in *in vitro* cultures as revealed by denaturing gradient gel electrophoresis

X.J. Lv, S.Y. Mao and W.Y. Zhu
Laboratory of Gastrointestinal Microbiology, College of Animal Science and Technology, Nanjing Agricultural University, Nanjing, 210095, China; zhuweiyun@njau.edu.cn

Introduction

As a convenient method, an *in vitro* fermentation system has been used to investigate the ruminal microbial metabolism for long history, but this system is limited to a certain set of micro-organisms which could be cultured. Thus, the microbial community in the rumen must be different from that obtained *in vitro*. The aim of the present study was to compare the changes of the bacterial community in the rumen (rumen bacterial community, RBC) to that of the bacterial community obtained from the same rumen but cultured *in vitro* (cultured rumen bacterial community, CRBC) using the PCR-DGGE method.

Material and methods

The animals were fed 500 g of forage and 500 g of concentrate (70% corn and 30% soybean) once a day at 8.00 h in the morning. After an adaptation period to the diet of 13 d, rumen contents including rumen fluid and solid were obtained from two fistulated male goats before (0 h) and 2, 6, 12 and 24 h after feeding. For the *in vitro* fermentation system, mixed rumen contents (equal volume) from both goats were collected at 0 h, transported to the laboratory in a thermos within 10 min of sampling and mixed with dilution buffer (Longland *et al.*, 1995) in a proportion of 1:9 (v/v) at 39 °C under continuous flushing with CO_2. And then a 100 ml mixture was transferred to 160 ml serum bottles containing 1 g of substrate (the same diet as above). A total of 12 bottles were prepared and incubated at 39 °C without shaking. Samples including fluid and solid were collected at 2, 6, 12 and 24 h (three replicates for each).

DNA was isolated according to a bead-beating method. Primers U968-GC and L1401 were used to amplify the V6–V8 regions of the bacterial 16S rRNA gene (Nübel *et al.*, 1996). Amplicants were checked by agarose gels and then used for sequence-specific separation by DGGE (Zoetendal and Akkermans, 1998). The gel was stained with $AgNO_3$ according to Sanguinetti *et al.* (1994). Similarity analyses were determined by calculating the band similarity (Dice) coefficient SD (Konstantinov *et al.*, 2003).

Results and discussion

The microbial community changes among samples are shown in Figure 1A. Generally, the DGGE similarity between RBC and CRBC remained high (more than 88%, Figure 1B). The Shannon index (SI) of diversity for RBC was in the range of 0.82 to 1.21, with 0 and 24 h sample at 1.03 and 1.21, respectively, while SI for CRBC was in a range of 0.99 to 1.10, with a 24 h sample at 1.02, suggesting RBC had relatively more fluctuation than CRBC. Many bands were common for both RBC and CRBC. However, bands A and B were only detected in RBC which suggested that they may not be culturable or outgrown *in vitro*. Band C could only be detected at 6 and 12 h in CRBC, which indicated that the corresponding bacterial population was increasing during this period. Band D was only present in CRBC suggesting its enrichment *in vitro*. Bands E, F, G were detected at 24 h in RBC, and band H was present at 12 and 24 h in RBC, suggesting that these bacteria might become more dominant in the community when feed was used towards the end.

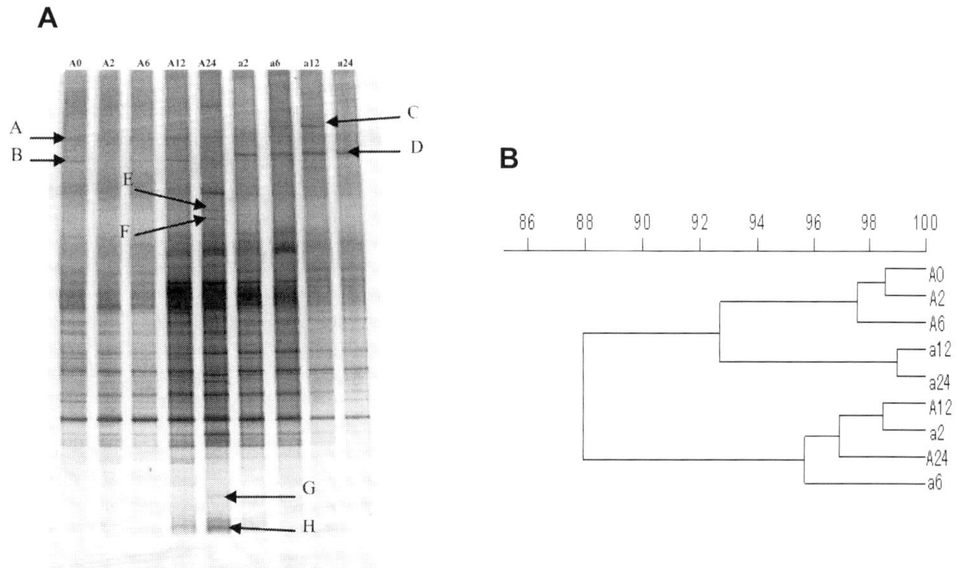

Figure 1. (A) DGGE profiles of bacteria: A0, A2, A6, A12, A24 = RBC at 0, 2, 6, 12 and 24 h; a2, a6, a12, a24 = CRBC at 2, 6, 12 and 24 h; (B) Similarity analysis of DGGE profiles: A = similarity of RBC; a = similarity of CRBC.

Conclusion

The overall microbial community was highly similar between RBC and CRBC, but several rumen microbial populations changed while incubated *in vitro*, with some species becoming not detectable and some enriched. But, the species corresponding to these changes need to be identified and their function to be valuated.

Acknowledgement

Supported by Natural Science Foundation of China (30530560).

References

Konstantinov, S.R, W.Y. Zhu, B.A. Williams, S.E. Tamminga, W.M. de Vos and A.D.L. Akkermans, 2003. Effect of fermentable carbohydrates on piglet faecal bacterial communities as revealed by denaturing gradient gel electrophoresis analysis of 16S ribosomal DNA. FEMS Microbiol. Ecol. 43: 225-235.

Longland, A.C., M.K. Theodorou, R. Sanderson, S.J. Lister, C.J. Power and P. Morris, 1995. Non-starch polysaccharide composition and *in vitro* fermentability of tropical forage legumes varying in phenolic content. Anim. Feed Sci. Technol. 55: 161-177.

Nübel, U., B. Engelen, A. Felske, J. Snaidr, A. Wieshuber, R.I. Amann, W. Ludwig and H. Backhaus, 1996. Sequence heterogeneities of genes encoding 16S rRNAs in *Paenibacillus polymixa* detected by temperature gradient gel electrophoresis. J. Bacteriol. 178: 5636-5643.

Sanguinetti, C.J., N.E. Dias and A.J. Simpson, 1994. Rapid silver staining and recovery of PCR products separated on polyacrylamide gels. Biotechniques 17: 914-921.

Zoetendal, E.G. and A.D.L. Akkermans, 1998. Temperature gradient gel electrophoresis analysis from human fecal samples reveals stable and host-specific communities of active bacteria. Appl. Environ. Microbiol. 64: 3854-3859.

Effect of feed intake on the intestinal supply of N fractions in dairy cows

G.L. Lynch[1], T.H. Klusmeyer[2], I.R. Ipharraguerre[3] and J.H. Clark[4]
[1]Land O'Lakes Purina Feed, 100 Danforth Dr. Gray Summit, MO, 63039, USA; [2] Monsanto Company, 800 N. Lindbergh Blvd, St. Louis, MO 63167, USA; [3]Lucta S.A., Ctra. de Masnou a Granollers, km 12.400, Montornés del Vallés 08170, Spain; [4]ASL 315, University of Illinois, 1207 W. Gregory Dr., Urbana, IL 61801, USA; ignacio.ipharraguerre@lucta.es

Introduction

Dry matter intake (DMI) is a major determinant of the amount and nature of N fractions that reach the intestines of dairy cows (Clark *et al.*, 1992). This notion is embedded in most empirical systems currently used to evaluate and (or) formulate dairy diets. For instance, estimation of microbial protein synthesis (MPS) by the NRC (2001) model is indirectly discounted for the expected decline in digestibility of organic matter (OM) caused by increases in DMI. Likewise, the supply of rumen undegradable protein (RUP) is adjusted for predicted changes in the passage rate of ruminal contents in response to alterations in DMI. Although empirically appropriate, these models may not accurately represent the contribution of microbial and feed protein to the supply of metabolisable protein (MP) (Huhtanen, 2005). Mounting evidence suggests that more research on the impact of DMI on N transactions in the gastrointestinal tract (GIT) of cows is required to improve the accuracy of current MP supply estimates (Schadt *et al.*, 2007; Seo *et al.*, 2006). Our objectives were to study nutrient digestion and duodenal flow of N fractions in dairy cows fed varying amounts of the same diet during late lactation.

Material and methods

Four multiparous Holstein cows fitted with ruminal and duodenal cannulas were used in a 4×4 Latin square design (mean milk yield, BW, and days in milk were 20.2 kg/d, 581 kg, 201 d, respectively). Treatments were 100, 88, 76, and 64% of *ad libitum* DMI. The *ad libitum* DMI was established for each cow during the week preceding the onset of the trial. On DM basis, the experimental diet consisted of 40% alfalfa silage, 40% corn silage, 15% ground corn, 3.5% soybean meal, and 1.5% minerals plus vitamins. Experimental periods lasted 14 d, the first 9 d were used to adapt cows to treatments and the last 5 d were used to collect data. Cows were housed individually and fed and milked twice daily. Samples of rumen fluid (50 ml) and duodenal digesta (500 ml) were collected during the last 3 d of each period. The sampling time was adjusted ahead 1 h daily so that a sample was obtained for each 1-h interval of the day (24 total samples). Ruminal bacteria were isolated from 6 samples of whole reticular content (500 ml) taken during the last 3 d of each period. Nutrient digestibility and passage of N fractions to duodenum were estimated using chromic oxide as digesta marker and purines as microbial marker. Data were analysed as a 4×4 Latin square using the MIXED procedure of SAS® (2000), with treatment and period treated as fixed effects and cow treated as a random variable. The degrees of freedom for treatment were partitioned into four single-degree-of-freedom orthogonal contrasts: linear, quadratic, and cubic.

Results

Feed intake ranged from 18.1 to 12.1 kg of DM/d (Table 1). Because cows were fed the same diet, decreasing DMI linearly decreased OM and N intake and the amount of OM truly digested in the rumen (OMTDR). As a result, the concentration of ruminal ammonia (9.9 to 8.1 mg/dl) decreased linearly ($P<0.05$) whereas ruminal pH (6.2 to 6.6) increased linearly ($P<0.01$). Passage to duodenum of OM and N of microbial and feed origin paralleled the decreases in DMI. However, flows of microbial N and RUP showed a quadratic response and the proportion of OMTDR as well as the efficiency of MPS were not affected by treatments. The NRC model overpredicted the supply of RDP and underpredicted the flow of microbial N and RUP at all levels of DMI.

Table 1. Digestibility and duodenal flow of N fractions in dairy cows in response to varying levels of feed intake.

	DM intake, kg/d				SEM	*P*>F
	18.1	16.0	14.2	12.1		
OM intake^{L}, kg/d	16.4	14.4	12.8	11.0	0.65	0.0001
OMTDR^{1L}, kg/d	8.4	6.9	6.1	5.4	0.51	0.01
OMTDR,% of OM intake	51.7	48.6	48.0	49.2	5.15	0.76
N intake^{L}, g/d	453	397	356	303	18.2	0.0001
RDP^{2} supply^{L}, g/d	1871	1697	1530	1112	78.9	0.0005
Flow to duodenum						
NAN^{3L}, g/d	475	425	372	322	32.4	0.0002
NANMN^{4lQ}, g/d	188	156	138	148	17.7	0.02
Microbial N^{Lq}, g/d	287	269	234	184	16.9	0.0004
Microbial N, g/kg OMTDR	34.0	39.1	38.9	34.6	1.10	0.40
RUP^{5LQ}, g/d	962	785	693	783	104	0.05
NRC predictions^{6}						
RDP supply, g/d	2052	1828	1634	1404		
Microbial N, g/d	235	209	186	160		
RUP, g/d	777	673	586	487		

[1] OM truly digested in the rumen.
[2] Rumen degradable protein.
[3] Non-ammonia N.
[4] Non-ammonia non-microbial N.
[5] Rumen undegradable protein.
[6] Calculated using LS means for production and intake data. ^{L}Linear effect ($P<0.05$);^{Q}Quadratic effect ($P<0.05$); ^{l} Linear trend ($P<0.0.1$); ^{q}Quadratic trend ($P<0.1$).

Conclusions

This study supports the views (1) that DMI plays a central role in dictating the passage of N fractions to the intestines and (2) that unbiased predictions of MP supply to dairy cows will require more complex approaches to account for variability in N transactions within the GIT caused by changes in DMI.

References

Clark, J.H., T.H. Klusmeyer and M.R. Cameron, 1992. Microbial protein synthesis and flows of nitrogen fractions to the duodenum of dairy cows. J. Dairy Sci. 75: 2304-2323.

Huhtanen, P., 2005. Critical aspects of feed protein evaluation systems for ruminants. J. Anim. Feed Sci. 14: 145-170.

National Research Council, 2001. Nutrient Requirements of Dairy Cattle. 7th rev. ed. Natl. Acad. Sci., Washington, DC, USA.

Schadt, I., G. Azzaro, R. Petriglieri, P.J. Van Soest, K.-H. Sudekiim and G. Licitra, 2007. Influence of level of intake upon rumen degradability of protein sources. J. Dairy Sci. 90 (Suppl.1): 682.

Seo, S., L.O. Tedeschi, C.G. Schwab, B.D. Garthwaite and D.G. Fox, 2006. Evaluation of the passage rate equations in the 2001 dairy NRC model. J. Dairy Sci. 89: 2327-2342.

Is the *trans*-10 shift that sometimes occurs in the ruminal biohydrogenation of linoleic acid caused by low pH or starch? A Rusitec study

M.R.G. Maia[1,2], R.J.B. Bessa[1,3] and R.J. Wallace[4]
[1]Unidade de Produção Animal, Instituto Nacional de Recursos Biológicos, Fonte Boa, 2005-048 Vale de Santarém, Portugal; [2]REQUIMTE- GABAI, Campus Agrário de Vairão, Rua Padre Armando Quintas, 4485-661 Vairão VC, Portugal; [3]CIISA- Faculdade de Medicina Veterinária, UTL, Pólo Universitário do Alto da Ajuda, 1300-477 Lisboa, Portugal; [4]University of Aberdeen Rowett Institute of Nutrition and Health, Bucksburn, Aberdeen AB21 9SB, United Kingdom; mrgmaia@mail.icav.up.pt

Introduction

Conjugated linoleic acids (CLA) are formed during the biohydrogenation of linoleic acid (*cis*-9,*cis*-12-18:2; LA) in the rumen. The main product is usually *cis*-9,*trans*-11-CLA (rumenic acid), but many other CLA are formed too and found subsequently in milk, including *trans*-9,*trans*-11-18:2 and *trans*-10,*cis*-12-18:2 (Shingfield *et al.*, 2003). Occasionally, when a high-concentrate diet containing fish or vegetable oil is fed, the *trans*-10,*cis*-12 isomer becomes a major intermediate, giving rise to high concentrations of *trans*-10-18:1 in digesta (the '*trans*-10 shift'; Griinari and Bauman, 1999). This results in milk fat depression, which has serious consequences to the animal (Chouinard *et al.*, 1999). The aim of this study was to identify whether the starch in the diet was the main cause of the *trans*-10 shift, or if the lower pH resulting from starch fermentation caused the change in the biohydrogenation pathway.

Material and methods

An *in vitro* study, using the RUSITEC system with 4 fermenters, was conducted to study the effects of pH and starch on rumen biohydrogenation. Three rams were used as rumen content donors. Two diets including 8% soybean oil were used: low starch (LS; lucerne hay) or high starch (HS; 50% lucerne hay and 50% starch). The artificial saliva continuously infused was buffered either at low (6.0) or high (7.0) pH. The 4 treatments were: LS diet and low pH (LSLpH); HS diet and low pH (HSLpH); LS diet and high pH (LSHpH); and HS diet and high pH (HSHpH). Each run lasted for 15 days and was repeated twice. Fatty acids (FA) in effluents were analysed by gas chromatography, using a 100 m CP-Sil 88 (Varian Inc., Walnut Creek, CA, USA) capillary column. The MIXED procedure of SAS® (SAS Institute, Inc., Cary, NC) was used for statistical analysis considering the starch and pH as fixed effects and the run as random. Day was considered a repeated measurement nested within each run.

Results

The low pH treatments produced lower concentrations of stearic acid (18:0) than high pH treatments and higher residual concentrations of LA and *cis*-9,*cis*-12,*cis*-15-18:3 (Table 1). Vaccenic acid (*trans*-11-18:1, VA) accumulated to a greater extent at high pH, while the concentrations of *trans*-10-18:1 and CLA were unchanged. Thus, low pH appeared to slow biohydrogenation but did not alter the route, which was predominantly *via cis*-9,*trans*-11-CLA and VA (Table 1). Starch, however, had no effect on the extent of biohydrogenation but increased *trans*-10-18:1 regardless of pH. VA formation was the highest on LSHpH and decreased both in the presence of starch and low pH, which had an additive effect.

Table 1. Biohydrogenation profile (g/100 g C18 fatty acids) of fatty acids in effluents from Rusitec vessels in response to pH and starch supplementation.

	Treatments[1]				SEM	*P* value[2]		
	LSLpH	HSLpH	LSHpH	HSHpH		pH	S	pH x S
18:0	23.8	19.0	31.7	30.6	3.17	0.028	0.395	0.581
trans-10-18:1	3.5	10.2	2.9	8.1	1.22	0.343	0.013	0.589
trans-11-18:1	9.2^{b}	3.0^{a}	21.7^{c}	11.5^{b}	0.82	<0.001	<0.001	0.039
cis-9,*trans*-11-18:2	2.0	1.4	1.7	1.3	0.26	0.467	0.082	0.668
trans-10,*cis*-12-18:2	0.6	0.6	0.5	0.9	0.13	0.591	0.177	0.197
cis-9,*cis*-12-18:2	23.8	29.4	8.8	16.1	3.96	0.019	0.167	0.837
cis-9,*cis*-12,*cis*-15-18:3	3.7	3.2	2.9	2.2	0.34	0.049	0.143	0.854

[1] LSLpH = low starch diet at low pH; HSLpH = high starch diet at low pH; LSHpH = low starch diet at high pH; HSHpH = high starch diet at high pH.
[2] pH = pH level effect; S = starch supplementation effect; pH×S = pH level by starch supplementation interaction effect.
[a,b] Means within rows with different superscripts are statistically different ($P<0.05$).

Discussion and conclusion

These results suggest that the *trans*-10 shift that occurred during a 15-day adaptation of the ruminal microbiota was caused by the presence of starch rather than the culture pH. This contrasts with the results of Choi *et al.* (2005), who found that lowering pH in a 6-h incubation caused the isomers of CLA accumulating to switch from *cis*-9,*trans*-11-CLA to *trans*-10,*cis*-12-CLA. Starch presumably enriches, over time, for the bacteria that form *trans*-10,*cis*-12-CLA.

Acknowledgement

This work was supported by the Fundação para a Ciência e a Tecnologia (FCT), Portugal (POCTI/CVT/2000/95/2002) and the Scottish Government Rural and Environmental Research Directorate (RERAD). M.R.G.M. was also supported by FCT with a PhD grant (SFRH/BD/6976/2001).

References

Choi, N.J., J.Y. Imm, S. Oh, B.C. Kim, H.J. Hwang and Y.J. Kim, 2005. Effect of pH and oxygen on conjugated linoleic acid (CLA) production by mixed rumen bacteria from cows fed high concentrate and high forage diets. Anim. Feed Sci. Technol. 123-124: 643-653.

Chouinard, P.Y., L. Corneau, D.M. Barbano, L.E. Metzger and D.E. Bauman, 1999. Conjugated linoleic acids alter milk fatty acid composition and inhibit milk fat secretion in dairy cows. J. Nutr. 129: 1579-1584.

Griinari, J.M. and D.E. Bauman, 1999. Biosynthesis of conjugated linoleic acid and its incorporation into meat and milk in ruminants. In: M.P. Yurawecz, M.M. Mossoba, J.K. Kramer, M.W. Pariza and G.J. Nelson (Eds.), Advances in conjugated linoleic acid research, volume 1. AOCS Press, Champaign, Illinois, pp. 180-200.

Shingfield, K.J., S. Ahvenjarvi, V. Toivonen, A. Arola, K.V.V. Nurmela, P. Huhtanen and J.M. Griinari, 2003. Effect of dietary fish oil on biohydrogenation of fatty acids and milk fatty acid content in cows. Anim. Sci. 77: 165-179.

Variation in sire genetics is an irrelevant determinant of digestibility in supplemented crossbred sheep fed canola and lupins

A.E.O. Malau-Aduli, R.E. Walker, J.M. Sykes, C.F. Ranson and C.W. Bignell
Animal Production & Genetics, School of Agricultural Science, University of Tasmania, Sandy Bay, Hobart, TAS 7001, Australia; Aduli.MalauAduli@utas.edu.au

Introduction

The efficiency with which sheep produce meat and/or wool relies on a combination of available high quality nutrition and good genetics, hence the constant quest for sheep breed combinations that best utilise feeds to the maximum. High digestibility and nutrient retention of feed on offer are important indices of protein and energy available for wool fibre synthesis or muscle accretion in sheep. Pitchford (1992) stated that both wool fibre number and diameter were strongly genetically determined. To our knowledge, it has not yet been established if sire genetics alone influences the digestibility and nutrient retention of dietary energy and protein in supplemented crossbred sheep. Therefore, the objective of this study was to evaluate the influence of sire genetics, dietary protein supplement type, level of feeding, sex and their second order interactions on metabolisable energy and nitrogen digestibility in first cross progeny of Merino dams sired by 5 ram breeds.

Material and methods

Weaner sheep (40) sired by 5 ram breeds (Poll Dorset, Coopworth, Texel, East Friesian and White Suffolk), were balanced for liveweight (30 kg) and body condition score (3.0) on the average, before being randomly assigned to two supplementary feeds (canola or lupins) and fed at two levels (1% or 2.0% BW). The feeding trial lasted for six weeks including an initial adjustment period of 3 weeks and the final 1 wk of faecal and urinary collection. All treatment groups received a daily allocation of an isocaloric and isonitrogenous diet comprising 0.5 kg of barley, 0.1 kg molasses-treated straw and 0.001 kg vitamin-mineral mix at 10:00 h. Each sheep had *ad libitum* access to clean, drinking water. DM intake and output, body weight, and change in wool fibre diameter (difference in wool microns at the beginning and end of the feeding trial) were measured. DM digestibilities were measured and data subjected to a general linear models procedure (PROC GLM) of the Statistical Analysis System® (SAS Institute, 2007) and significance established using orthogonal contrasts and Tukey pairwise comparisons. The model included sire breed, supplement, feeding level, sex as main effects and their second order interactions.

Results and discussion

Regardless of sire genetics, feeding level or gender, sheep supplemented with canola consumed 4.5% more feed (DMI 163.5 vs. 149.2 g/day), voided 17% more faeces (51.08 vs. 35.97 g/day), digested 8.5% more ME (52.23 vs. 44.23%) and had 4% heavier liveweights (40 vs. 36.9kg) than those supplemented with lupins. While lupins are the major legume grain fed to ruminants in Australia, their value as a source of undegraded protein is limited by their high rumen degradability and low methionine content (White *et al.*, 2000). Canola meal on the other hand, is partially protected from degradation in the rumen with estimates of degradability ranging from 0.6 to 0.8 (Hill, 1991). These trends conform to our study's findings. Feeding supplements at 1% of body weight triggered higher ME (49.9% vs. 46.5%) and N (64.9% vs. 63.2%) digestibility responses than feeding at 2%. There was a tendency for females to eat more than males (161.8 vs. 149.6 g/day DMI) and N digestibility was 2% higher in males (65%) than females (63%). Sire genetics × level of feeding interactions significantly influenced ME and N digestibility (P<0.05) whereby Coopworth-sired sheep supplemented at 1% of their body weight recorded the highest ME and N digestibility of 54% and 67% compared to 42% and 62% respectively, than their counterparts fed at 2% of body weight. There was a highly significant (P<0.01) effect of type of supplement × level of feeding interaction on wool fibre diameter because sheep fed canola supplements at 1% of body weight had finer wool (22.1 microns) than their 2%-fed counterparts (25.4 microns). Canola meal has been reported to contain more methionine (approximately 2 g/100 g CP) than lupins (0.4-1.0 g/100 g CP; Hill, 1977) or rumen digesta (1.2-1.6 g/100 g amino acids; Wallace, 1994) and is therefore expected to provide a more appropriate source of amino acids for wool growth as demonstrated in our study and Masters and Mata (1996). Regression of wool fibre diameter on digestibility revealed that there was no correlation between the two (R^2 =0.0087-0.169).

Conclusion

The postruminal delivery of nutrients in sheep supplemented with canola was more efficient than in lupin-supplemented sheep, hence their higher energy and protein digestibility and retention. Furthermore, variation due to sire genetics alone was insufficient in accounting for differences in digestibility and wool fibre diameter, but significantly interacted with type of supplement, level of feeding and sex. Finally, sire breed variation in digestibility is unlikely to be a useful predictor of genetic merit for wool fibre diameter in first cross sheep.

Acknowledgement

The authors are grateful to the Australian Wool Education Trust for funding this research and for the provision of postgraduate and honours research scholarships. We are thankful to Dr Barrie Wells - UTAS Animal Welfare Officer for his veterinary care of the experimental sheep and blood sampling.

References

Hill, G.D., 1977. The composition and nutritive value of lupin seed. Nutr. Abstr. Rev. B: Livestock Feeds and Feeding 47: 511-529.

Hill, R., 1991. Rapeseed meal in the diets of ruminants. Nutr. Abstr. Rev. B: Livestock Feeds and Feeding 61: 139-154.

Masters, D.G. and G. Mata, 1996. Responses to feeding canola meal or lupin seed to pregnant, lactating, and dry ewes. Aust. J. Agric. Res. 47: 1291-1303.

Pitchford, W.S., 1992. Effect of crossbreeding on components of hogget wool production. Aust. J. Agric. Res. 43: 1417-1427.

Wallace, R.J., 1994. Amino acid and protein synthesis, turnover and breakdown by ruminal microorganisms. In: J.M. Asplund (Ed), Principles of Protein Nutrition of Ruminants. CRC Press Inc., Boca Raton, FL, USA, pp. 71-111.

White, C.L., P. Young, N. Phillips and M. Rodehutscord, 2000. The effect of dietary protein source and protected methionine (Lactet) on wool growth and microbial protein synthesis in Merino wethers. Aust. J. Agric. Res. 51: 173-183.

Carbonic anhydrase II is secreted into whole saliva of three ruminating species

M. Mau[1], *T.M. Kaise*[2] *and K.-H. Südekum*[1]
[1]*Institute of Animal Science, University of Bonn, Endenicher Allee 15, 53115 Bonn, Germany;* [2]*Zoological Institute and Museum, University of Hamburg, Martin-Luther-King-Platz 3, 20146 Hamburg, Germany; mmau@itw.uni-bonn.de*

Introduction

Other than humans, ruminants secrete large volumes of alkaline saliva mostly for lubricating food particles and providing a constant pH for rumen microbes (Kay, 1960). In this context, salivary carbonic anhydrase (CA) is of major interest, because it participates in basic processes such as local pH regulation of the oral cavern and the alimentary tract, in bicarbonate transport and electrolyte balance (Parkkila and Parkkila, 1996). Since the initial discovery of the 29-kDa CA-II, different CA isoenzymes have been found in numerous mammalian tissues (Sly and Hu, 1995). The only enzyme yet known to be secreted into saliva, is the CA-VI, which is exclusively characterised by an apparent molecular weight of 42 kDa (Sly and Hu, 1995). Recently, the enzyme was shown to be a part of the bovine *in vitro* salivary pellicle (Mau *et al.*, 2006). Since salivary glands are highly variable in the composition of their secretions, Salivary proteins could be a primary way by which species adapt or react to their diets. Thus, feeding adaptation may correlate with saliva composition. Hence, animals using identical feeding niches should possess similar salivary proteins. In the present study we compared salivary protein concentrations and patterns of three grass-eating species (cattle: *Bos primigenius* f. *taurus*; goat: *Capra aegagrus* f. *hircus*; camel: *Camelus* f. *bactrianus*). The knowledge of specific functional adaptations to grass diets in the oral milieu of ruminating animals could help to elucidate the maintenance of function in their specialised digestive system.

Material and methods

Saliva samples of nine female goats (*Capra aegagrus* f. *hircus*) and four cows (*Bos primigenius* f. *taurus*) were obtained from the Research Institute for the Biology of Farm Animals (FBN) in Dummerstorf, Germany. Saliva samples of camels (*Camelus ferus* f. *bactrianus*; two male, one female) were provided by the Zoological Garden Rheine, Germany. All salivary samples were taken according to Mau *et al.* (2009) and compared by SDS-gel electrophoresis and immunoblotting. For immunohistochemistry, serial sections of 6 µm were cut from bovine parotid glands and incubated with primary antibodies raised against bovine or human CA-II.

Results

Salivary protein patterns were similar among individuals of the same species but varied largely among species (not shown). However, all samples showed proteins of apparently 29 and 42 kDa, identified as carbonic anhydrases (CA) by immunoblotting (Figure 1). CA-VI (42 kDa) was highly expressed in cattle and camel saliva but showed lower expression in goat saliva. CA-II (29 kDa) was also found in saliva of all species tested (Figure 1) and was proven not to originate from cellular debris of the oral mucosa or ingested food (not shown). Additionally, cross reactivity of the antibody with plant CA was excluded using 50 µg crude protein isolated from meadow grass (not shown). Immunohistochemistry on bovine parotid glands revealed intense positive staining for CA-II in the ductal cells (not shown). Furthermore, positive staining was observed inside the duct lumen in the form of clear brinks (not shown). Additionally, CA-II protein was present in protein extracts from bovine parotid tissue using immunoblot analysis (not shown).

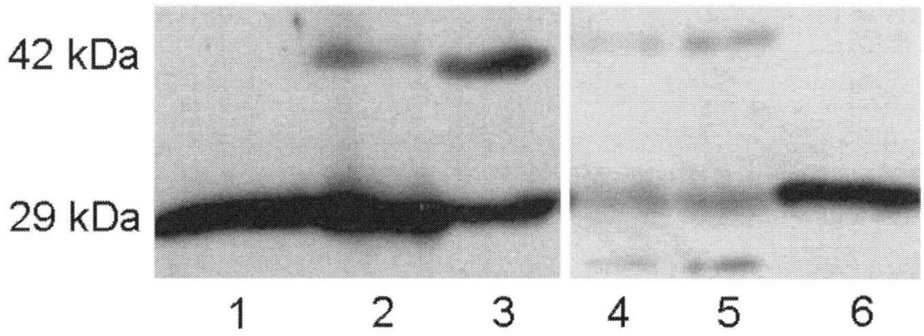

Figure 1. Immunoblot of bovine, camel and goat whole saliva to detect salivary CA using a rabbit anti-bovine CA antibody (1:400; AbD Serotec). All samples were positive for CA-II (29 kDa) and showed an additional band at 42 kDa, representing CA-VI. 1,6: bovine CA-II (positive control); 2: bovine saliva; 3: camel saliva; 4,5: goat saliva.

Discussion and conclusion

The present study shows that besides CA-VI, CA-II is another CA isoform secreted in the saliva of ruminating species. The results further demonstrate that CA-II is expressed in bovine parotid glands and is secreted from parotid ductal cells. We conclude that the two CA isoenzymes detected may form a complementary system, protecting mucosa from acidity and helping to maintain a constant bicarbonate concentration in the animal's mouth and digestive tract. Because both isoenzymes are highly active and supply the alimentary tract with bicarbonate, a decrease in the expression of either CA-II or CA-VI might lead to severe disturbances of digestion and to increased susceptibility to acidosis in ruminating species.

Acknowledgement

The study was supported by grant DFG SU 124/15-1.

References

Kay, R.N.B., 1960. The rate of flow and composition of various salivary secretions in sheep and calves. J. Physiol. 150: 515-537.

Mau, M., T.M. Kaiser and K.-H. Südekum, 2009. Evidence for the presence of carbonic anhydrase 29-kDa-isoenzyme in salivary secretions of three ruminating species and the gelada baboon. Arch. Oral. Biol. 54: 354-360.

Mau, M., C. Müller, J. Langbein, C. Rehfeldt, J.P. Hildebrandt and T.M. Kaiser, 2006. Adhesion of bovine and goat salivary proteins to dental enamel and silicate. Arch. Anim. Breed. 49: 439-446.

Parkkila, S. and A.K. Parkkila, 1996. Carbonic anhydrase in the alimentary tract. Roles of the different isozymes and salivary factors in the maintenance of optimal conditions in the gastrointestinal canal. Scand. J. Gastroenterol. 31: 305-317.

Sly, W.S. and P.Y. Hu, 1995. Human carbonic anhydrases and carbonic anhydrase deficiencies. Annu. Rev. Biochem. 64: 375-401.

Effects of forage to concentrate ratio on rumen fermentation pattern in buffaloes

A. Aghazadeh[1], N. Parvishi[1] and H. Mansoury[2]
[1]Department of Animal Science, Urmia University, Urmia, Iran; [2]Institute of Animal Science Research, Karaj, Iran; a.aghazadeh@mail.umia

Introduction

Buffaloes in Iran are mostly reared under a mixed farming system, in which crop residues and by-products are used as livestock feeds, and in turn manures and draft power from livestock are used in crop production. This system is also most prevalent in Asia, in which crop residues are used as livestock feeds (Borghese, 2005). Nevertheless, supplementary feed is usually required in order to sustain adequate milk production. Milk production may vary according to the husbandry and production system. Milk production is also related to the quality of forage and the forage to concentrate (F: C) ratio. Many studies have been conducted on the effect of forage to concentrate ratio on the rumen fermentation pattern of dairy cows (Lee *et al.*, 2006; Loor *et al.*, 2004). However, little information is available on the effect of F:C ratio on the rumen fermentation of native buffalo. The objective of this study was to investigate the effects of different F: C ratio on rumen fermentation of the Azarbaijani buffalo.

Material and methods

Three adult male ruminally fistulated buffaloes (average body weight 450 ± 20 kg, average age 3 yrs) where housed in an enclosed barn in 1×2×1.7 m individual tie stalls. The experimental design was a 3×3 Latin square. The diets were based on 35:65, 45:55 and 55: 45 for low, medium and high F: C ratio respectively. The low F:C diet consisted of approximately 25% alfalfa hay, 10% wheat straw and 65% concentrate, medium F:C diet consisted of 23.17% alfalfa hay, 31.83% wheat straw and 45% concentrate and high F:C diet consisted of 17.83% alfalfa hay, 47.7% wheat straw and 35% concentrate (DM basis). Animals were kept for 15 d on each diet for microbial adaptation. The pH was measured soon after collection of rumen liquor at each time with a Toshniwal pH meter. The total VFA were determined by the method of Kroman *et al.*, 1967. The effect of dietry treatment on PH and total VFA were analysed according to the following: $y_{ijk}=\mu + T_i + P_j + C_k + e_{ijk}$ where y_{ijk} is the individual observations, μ is the overall means, T_i is the effect of dietry treatment, P_j is the effect of experimental period, C_k is the effect of cow, and e_{ijk} is the residual error. Effect of cow was tested as a random effect. The data were subjected to statistical analyses using SPSS software. Significant difference between means was tested with the Duncan multiple range tests (Duncan, 1955).

Results

Ruminal PH increased significantly ($P<0.05$) as the ratio of F:C decreased. Total VFA concentration in the rumen increased in buffaloes fed the LF:C diet. Overall ruminal PH of the diets varied between 6.31-7.02. Ruminal PH values lower than 6.00 are known to depress cellulolytic activity, thus reducing fibre digestion (Kennelly *et al.*, 1999). The reason for a decrease of the ruminal PH with a decrease in F:C ratio is that high concentrate diets containing high amounts of non structural carbohydrate are quickly fermented by ruminal microbes, resulting in a declining ruminal PH and increasing ruminal VFA (Lee *et al.*, 2006). The effect of concentrate in decreasing PH and increasing total VFA have been reported by other workers. However the size of the decline in rumen PH when increasing the composition of concentrate in the diet is highly variable. The results of this study were in agreement with those of Kennelly *et al.* (1994) (LF:C=25:75 and HF:C=50:50) and Murphy *et al.* (2000) who fee a high proportion (>0.70) concentrate in the diet and in contrast to those of Khorasani *et al.* (2001) (LF:C=35:65 and HF:C= 65:35), Ueda *et al.* (2003) (LF:C=65:35 and HF:C=35:65), Loor *et al.* (2004) (LF:C=35:65 and HF:C=65:35) and Lee *et al.* (2006) (LF:C=20:80 and HF:C=80:20) who did not observe a significant reduction in ruminal PH as a result of increasing concentrate level in the diet.

Table 1. Changes in pH and total volatile fatty acids in rumen liquor of buffaloes fed different F: C ratio diets (Mean±S.E.).

Hours after feeding	TVFA (mmol/l)			pH		
	LF:C	MF:C	HF:C	LF:C	MF:C	HF:C
0	90.65±1.3[c]	93.34±0.7[b]	93.6±1.0[a]	7.25±0.04[a]	6.93±0.05[b]	6.72±0.05[c]
3	107.09±2.4[c]	120.91±0.6[b]	132.37±2.4[a]	6.92±0.04[a]	6.68±0.10[b]	6.28±0.15[c]
6	106.32±1.8[c]	117.21±1.5[b]	126.79±0.8[a]	6.89±0.05[a]	6.44±0.12[b]	5.93±0.10[c]
9	102.92±2.0[c]	108.37±2.9[b]	119.04±2.0[a]	6.97±0.04[a]	6.58±0.03[b]	6.24±0.67[c]
12	94.40±1.2[c]	99.22±1.7[b]	106.78±0.9[a]	7.08±0.01[a]	6.71±0.03[b]	6.36±0.08[c]
Overall mean	100.8[c]	107.81[b]	115.71[a]	7.02[a]	6.67[b]	6.31[c]

LF:C = concentrate: forage35:65, MF:C = concentrate: forage45:55, HF:C = concentrate: forage55:45. Means with different superscripts in the same row are significantly different *(P<0.01)*.

Conclusion

Increasing the concentrate component in the diet from 0.45 to 0.65 significantly reduced rumen PH and total VFA. However, the possible effect of other factors such as buffering capacity of the feed, saliva production, nature of the forage used should not be over looked. Additional research is needed to study the impact of different F: C diets on the rumen kinetics of buffaloes and possibly compare them with cattle under the same feeding conditions.

References

Borghese, A., 2005. WAAP Book of the year 2005. Wageningen Academic Publishers, Wageningen, the Netherlands, pp. 189-201.

Duncan, D.B., 1955. Multiple range and multiple F test. Biometrics 11: 1-42.

Kennelly, J.J., B. Robinson and G.R. Khorasani, 1999. Influence of carbohydrate source and buffer on rumen fermentation characteristics, milk yield, and milk composition in early-lactation Holstein cows. J. Dairy Sci. 82: 2486-2496.

Khorasani, G.R., E.K. Okine and J.J., Kennelly, 2001. Effects of forage source and amount of concentrate on rumen and intestinal digestion of nutrients in late-lactation cows. J. Dairy Sci.84: 1156-1165.

Kroman, R.P., J.P. Mayer and W.G. Stielan, 1967. Steam distrillation of volatile fatty acids in the rumen digesta. J. Dairy Sci. 50: 75-76.

Lee, M.R.F., J.K.S. Tweed, R.J., Dewhurst and N.D. Scollan, 2006. Effect of forage: concentrate ratio on ruminal metabolism and duodenal flow of fatty acids in beef steers. Br. Anim. Sci. 82: 31-40.

Loor, J.J., K. Ueda, A. Ferlay, Y. Chilliard and M. Doreau, 2004. Biohydrogenation, duodenal flow, and intestinal digestibility of *trans* fatty acids and conjugated linoleic acids in response to dietary forage:concentrate ratio and linseed oil in dairy cows. J. Dairy Sci.87: 2472-2485.

Murphy, M., M.A. Kerlind and K. Holtenius, 2000. Rumen fermentation in lactating cows selected for milk fat content fed two forage to concentrate ratios with hay or silage. J. Dairy Sci. 83:756-764.

Ueda, K., A. Ferlay, J., Chabrot, J.J. Loor, Y. Chilliard and M. Doreau, 2003. Effect of linseed oil supplementation on ruminal digestion in dairy cows fed diets with different forage:concentrate ratios. J. Dairy Sci. 86: 3999-4007.

Mid to long term stability of ruminal physicochemistry in dairy cows fed a fibre- or a starch-based diet

V. Monteils, M. Rey and T. Gidenne
INRA, Université de Toulouse, INPT-ENSAT, ENVT, UMR 1289 TANDEM, Tissus Animaux, Nutrition, Digestion, Ecosystème et Métabolisme, F-31326 Castanet-Tolosan, France; valerie.monteils@ensat.fr

Introduction

Ruminal physicochemistry is the result of the biocenosis activity in interaction with the host (absorption of fermentation terminal products through the ruminal wall, motricity and transit, etc). It characterises the biotope and also influences the microbiota. In young ruminants, the establishment of microbiota is gradual and relatively quick (Fonty *et al.*, 1987). Ruminal fermentation and degradation vary with age (Rotger *et al.*, 2005). Are the ruminal parameters time-stable in the adult cow? If variability is observed, is this variability linked to the diet composition? This study addresses these questions.

Material and methods

Six ruminal cannulated Holstein heifers without previous acidosis status were randomly assigned to one of the two groups, and fed either a fibre- or a starch-based diet, offered twice daily in equal portions. The fibre-diet was composed of 50.2% of dry matter (DM) of dehydrated alfalfa, 20.8% DM of wheat straw, 25.5% DM of ground corn and 3.5% DM of minerals. The starch-diet was composed of 40.2% DM of corn silage, 43.4% DM of ground corn, 13.2% DM of soybean meal and 3.1% DM of minerals. After a 6-wk adaptation period to the diet, the measurements and sampling were done individually once a week during 15 consecutive weeks. Both pH and redox potential were done according to the ex-vivo design described by Marden *et al.* (2005) and recorded hourly from the morning to the evening feeding (i.e. during 8 hours). A correction of the redox potential was done (+199 mV) according to the reference electrode used (Ag-AgCl). Ruminal fluid samples were collected just before and 2, 3, 4, 6 and 8 hours after feeding and conserved with the addition of mercuric chloride (2% wt/vol). Volatile fatty acid (VFA) concentration was determined using gas chromatography with an adaptation of the method of Playne (1985). The determination of ammonia was based on the modified Berthelot reaction with the Skalar Method. Data were analysed with analysis of variance using linear mixed-effects and repeated-measures models of R. The statistical model included diet, time, week, cow effects, the diet x week interaction, and the residual error. Differences between diets or weeks effects were assessed by pairwise comparisons (Tukey test). Differences were declared significant at $P<0.05$.

Results

The daily mean ruminal pH was 0.5 points higher ($P<0.01$) for the fibre diet than the starch-diet all along the study (Table 1). However the starch diet involved a higher sporadic variability of pH values between weeks (SE= 0.008 and 0.015 for fibre and starch diet respectively). At the beginning of the trial, the redox potential differed sharply among the two diets (78 mV, $P<0.05$), then the values became progressively similar from the 19th week on (Figure 1A). The ammonia concentration was 50% lower for starch compared to fibre diet ($P<0.001$) along the whole study, and the values linked to the fibre diet were less time-stable (SE= 2.38 and 1.91 for fibre and starch diet respectively). The total VFA concentrations were 13% higher in the starch diet, with a propionate proportion that was sporadically variable along time for the starch diet, and particularly stable for the fibre diet (Figure 1B).

Table 1. Ruminal physicochemistry[1] according to diet composition along a 15 wk period.

	Diet		SE	Effects		
	Fibre	Starch		Diet	Week	Diet x Week
pH	6.63[a]	6.16[b]	0.011	<0.01	<0.001	<0.001
Redox potential (mV)	-210[a]	-171[b]	1.15	<0.001	<0.001	<0.001
Ammonia (mg/l)	64.9[a]	33.9[b]	1.84	<0.001	<0.001	<0.001
Total VFA (mmol/l)	75.1[b]	86.5[a]	0.69	<0.05	<0.001	<0.05

[1] Mean values for a 15 weeks period with a daily measurement after the morning feeding.

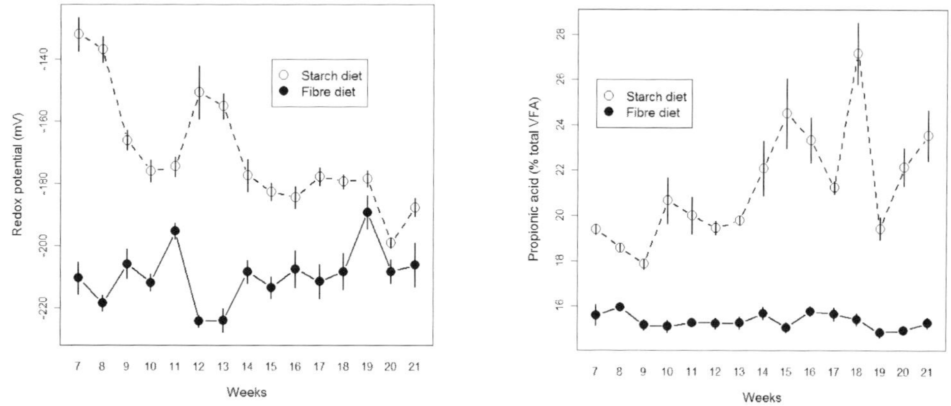

Figure 1. Variation of the Redox potential (A) and propionic acid (B) in the rumen[1] according to the diet composition and along a 15 wk period.
[1] Mean values of a daily measurement after morning feeding.

Conclusion

After a dietary treatment of 21 weeks, the ruminal biotope is unstable particularly for the starch diet that involved strong variations of ruminal parameters among weeks. These sporadic variations suggest disturbances in the biocenosis balance and probably consequences for the host digestive health and efficiency. This study does not allow concluding about the existence of a stable state.

References

Fonty, G., P. Gouet, J.P. Jouany and J. Senaud, 1987. Establishment of the microflora and anaerobic fungi in the rumen of lambs. J. Gen. Microbiol. 133: 1835-1843.

Marden, J.-P., C. Bayourthe, F. Enjalbert and R. Moncoulon, 2005. A New Device for Measuring Kinetics of Ruminal pH and Redox Potential in Dairy Cattle. J. Dairy Sci. 88: 277-281.

Playne, M.J., 1985. Determination of ethanol, volatile fatty acids, lactic acid and succinic acid in fermentation liquids by gas chromatography. J. Sci. Food Agric. 36: 638-644.

Rotger, A., A. Ferret, S. Calsamiglia and X. Manteca, 2005. Changes in ruminal fermentation and protein degradation in growing Holstein heifers from 80 to 250 kg fed high-concentrate diets with different forage-to-concentrate ratios. J. Anim. Sci. 83: 1616-1624.

Fatty acid profile and fermentation characteristics of ruminal fluid of dairy cows fed TMR complemented with different grazing times

E. Morales-Almaráz, F. Vicente, A. González, A. Soldado, A. Martínez-Fernández and B. de la Roza-Delgado
Department of Nutrition, Grasslands and Forages, Agro-Food Research and Development Institute (SERIDA), P.O. Box 13; E-33300 Villaviciosa (Asturias), Spain; fvicente@serida.org

Introduction

Polyunsaturated fatty acids (PUFA) of the diet suffer a ruminal biohydrogenation (BH) process, considered as a way to protect against the toxic effects of rumen bacteria. Microorganisms transform linoleic acid to stearic acid in three steps, with conjugated linoleic acid (*c*9*t*11 CLA) and vaccenic acid (VA) being the intermediate products. These FA have received special attention due to their potential beneficial effects on human health, inducing the interest in increasing their concentration in dairy products. Feeds have an influence on ruminal BH, because the substrate available is associated with changes in the rumen environment and it will lead to changes in microbial activity that correspond to altered end-product fermentation (Martin and Jenkins, 2002). Diets based on forage, besides supplying PUFA, can promote the proliferation of rumen bacteria during the BH process being able to biohydrogenate PUFA and synthesise these beneficial intermediary products. The aim of this work was to evaluate the influence of feeding dairy cows with TMR complemented with different grazing times on rumen fermentation parameters and FA content in ruminal fluid.

Material and methods

Three Holstein cows (833 kg of live weight) fitted with a simple cannula in the rumen were grouped randomly in three groups with four Holstein cows (648 kg) in production (38.1 l/d). They were randomly assigned to three feeding systems based on TMR plus different grazing times during three experimental periods of three weeks, including two adaptation weeks and one for milk and feed sampling. Ruminal fluid was sampled twice the last day of each period: 7:00 (AM) and 19:00 (PM). Feeding systems were the following: TMR00, feeding only TMR; TMR06, six hours grazing (12:30-19:30) and TMR12, twelve hours grazing (7:30-19:30). During the hours without grazing access, the animals had free access to TMR offered *ad libitum* including maize silage, cereal straw, grass hay and concentrate with a 70:30 F:C ratio. Five adjacent paddocks were used for grazing. Milk yield and TMR intake were registered and sampled for near infrared spectroscopy (NIRS) analysis. In addition, pH ruminal fluid was recorded immediately and sampled for analysis of NH_3, volatile fatty acids (VFA) and FA (Sukhija and Palmquist, 1988) by gas chromatography. Data were analysed as a 3×3 latin square design.

Results

Both lactating and cannulated cows showed DMI (kg/d) of TMR higher in TMR00 (23.5 vs. 12.1) and TMR06 (22.4 vs. 11.6) than in TMR12 (17.5 vs. 8.32, respectively). Pasture intake estimation was 16.5 kg DM/d for TMR12 and 5.4 for TMR06 (P<0.05). NH_3 ruminal fluid concentration increased in PM samples (20.85 vs. 31.18 mg/dl to AM and PM samples respectively, P<0.05; Table 1). There was no difference between treatments on VFA, however, acetic acid concentration was higher in AM (69.94 mol/100 mol) than PM (64.86 mol, P<0.05). Short and medium chain FA were not affected by treatments. TMR00 showed higher concentration of C18:0 in AM than in PM samples, but there was no difference between treatments. PM samples contained higher concentrations of C18:1*c*9 and C18:2 in each treatment. C16:0 presented a linear decrease (33.21,

32.43 and 30.33 g/100 g FA to TMR00 TMR06 and TMR12 respectively, $P<0.05$) while VA and C18:3 increased with an increasing grazing time (4.10, 4.94 and 7.88 g VA/100 g FA and 0.07, 0.26 and 0.53 g C18:3/100 g FA to TMR00 TMR06 and TMR12 respectively ($P<0.05$) especially in PM.

Table 1. Fermentative products and fatty acid contents of ruminal fluid of dairy cows fed TMR complemented with different grazing times.

Time (T)	AM			PM			SEM	Significance		
Feeding (F)	TMR00	TMR06	TMR12	TMR00	TMR06	TMR12		T	F	T*F
pH	6.87	6.56	6.34	6.46	6.43	6.26	0.085	NS	NS	NS
NH_3, mg/dL	19.63	26.20	16.73	25.76	31.76	36.03	1.372	*	NS	NS
Total VFA, m*M*	82.2	96.4	117.0	97.9	117.8	114.4	5.694	NS	NS	NS
VFA, mol/100mol										
Acetate	72.8	71.5	65.5	65.4	63.5	65.7	1.045	*	NS	NS
Propionate	17.4	17.9	18.1	19.8	18.9	19.7	0.413	NS	NS	NS
Butyrate	9.8	10.6	16.5	14.7	17.6	14.6	1.300	NS	NS	NS
Fatty Acids, g/100g FA										
C6:0 – C15:0	4.87	5.42	6.31	7.67	7.34	4.64	0,489	NS	NS	NS
C16:0	32.93	32.25	29.35	33.48	32.61	31.31	0.374	NS	*	NS
C16:1	0.00	0.00	0.03	0.05	0.04	0.00	0.009	NS	NS	NS
C17:0	0.23	0.21	0.15	0.15	0.15	0.13	0.020	NS	NS	NS
C17:1	0.03	0.08	0.05	0.04	0.03	0.11	0.022	NS	NS	NS
C18:0	48.97[a]	46.64[ab]	46.49[ab]	38.58[b]	39.76[b]	40.05[b]	1.237	*	NS	*
C18:1*c*9	5.47[b]	6.18[ab]	5.70[b]	9.75[a]	8.86[a]	9.46[a]	0.508	*	NS	*
C18:1*t*11,VA	3.96	4.72	8.06	4.25	5.16	7.71	0.289	NS	*	NS
C18:2	3.29[b]	4.11[ab]	3.50[b]	5.78[a]	5.48[a]	5.46[a]	0.259	*	NS	*
C18:3	0.06	0.17	0.20	0.09	0.35	0.86	0.062	NS	*	NS
C20:0	0.18	0.20	0.16	0.13	0.15	0.20	0.019	NS	NS	NS

[a,b] Means rows with different letters differ significantly by T and F interaction; * $P<0.05$; NS, Non Significant.

Conclusion

The general characteristics of rumen fermentation were not affected by the feeding system, but long chain FA content in ruminal fluid was modified, especially VA, from cows grazing 12 hours due to higher availability of C18:3 and C18:2 from pasture. Both FA are main precursors of *c*9*t*11 CLA and VA, however, this CLA isomer was not quantified probably, due to the methodology of FA extraction (Sukhija and Palmquist, 1988) and which can transform it partially to other isomers (Harvatine and Allen, 2006). Moreover, ruminal pH values are slightly lower than the optimum pH (7.0) for *c*9*t*11 CLA production (Nam and Garnsworthy, 2007).

Acknowledgement

Work supported by INIA project RTA2007-00058-C02-01. Stage of E. Morales-Almaráz is supported by a grant of CONACYT-México.

References

Harvatine, J.K.J. and M.S. Allen, 2006. Fat supplements affect fractional rates of ruminal fatty acid biohydrogenation and passage in dairy cows. J. Nutr. 136: 677-685.

Martin, S.A. and T.C. Jenkins, 2002. Factors affecting conjugated linoleic acid and *trans*-C18:1 fatty acid production by mixed ruminal bacteria. J. Anim. Sci. 80: 3347-3352.

Nam, S. and C. Garnsworthy, 2007. Factors influencing biohydrogenation and conjugated linoleic acid production by mixed rumen fungi. J. Microbiol. 45: 199-204.

Sukhija, P.S. and D.L. Palmquist, 1988. Rapid method for determination of total fatty acid content and composition of feedstuffs and feces. J. Agric. Food Chem. 36: 1202-1206.

Relationship between degradation characteristics of canola and pasture hays and milk production characteristics of late lactation dairy cows

S.K. Muir, J. Hill, Phanchung, J. Tharmaraj and D.F. Chapman
Department of Agriculture and Food Systems, Melbourne School of Land and Environment, The University of Melbourne, Parkville, Victoria 3010, Australia; s.muir2@pgrad.unimelb.edu.au

Introduction

The Australian dairy industry has traditionally had a high reliance on home grown supplements (hay, pasture silage). Supplements are used in dry periods, such as summer feed gap and under drought conditions. Conserved forage (hay and silage) are the traditional supplements used in Victoria (Stockdale *et al.,* 1997), however feeding concentrates or grains during periods of low pasture availability is becoming more common (ABARE, 2005). The use of alternative feed sources becomes more common as costs limit the use of traditional supplements, particular under periods of low availability. Demand for canola hay as a supplement for dairy cows has increased in recent years, however there is little knowledge of the effects that feeding canola hay has on milk production of dairy cows. The objective of the study was to evaluate the milk production response to feeding canola hay vs. pasture hay and investigate the degradation characteristics of canola hay and relationship to milk production and milk urea nitrogen.

Material and methods

The study was conducted at the DemoDAIRY research farm, in SW Victoria, Australia. The experiment was divided into a 20 d acclimatisation period and a 10 d measurement period. Twenty-four Holstein dairy cows were allocated to four dietary treatments. Diets contained basal forage of pasture or canola hay *ad libitum*, supplemented with 4 kg dry matter (DM) whole crop cereal silage and concentrate. The concentrate was (6 kg DM) comprised of wheat grain/canola meal blend and palm kernel expeller (PKE) with the latter replacing 0 or 50% of the concentrate. Pasture and canola hay was offered to each treatment group in stand off areas on pasture with herbage mass of <1,100 kg DM/ha. Whole crop cereal silage and PKE were offered via individual feed stalls. Milk samples were collected during the measurement period and analysed for fat and protein using NIR. Subsamples of milk were de-fatted and urea content was measured via an enzymatic assay of Nousiainen *et al.* (2004).

The degradation characteristics of canola and pasture hay were determined following the experiment. Dry samples of feed were ground to 1mm sieve. Triplicate samples of 50 mg were incubated with fungal cellulase (*Trichoderma viride*) at 37 °C for up to 36 h. Dry matter disappearance was measured at 2, 4, 6, 8, 12, 24 and 30 h. Additional triplicate samples were incubated for 10 min in sodium acetate buffer to determine the soluble fraction. Incubation was stopped by the addition of sodium hydroxide. Residue was collected by filtration with glass filter paper and dried at 100 °C for 24 hours (Lopez *et al.*,1998).

Results

No differences (P>0.05) in voluntary feed intake (VFI) were observed between the two forage types. The impact of PKE on total VFI reflected a depression in intake when PKE level in the diet was increased.

Milk yield, fat and protein concentrations and fat and protein outputs were not affected (P<0.05) by feeding canola hay or increasing the proportion of PKE in the ration. Canola hay was associated with an increase (P>0.05) in milk yield and a slight increase in milk protein output. Milk urea output was increased ($P<0.001$) by feeding canola hay and significantly decreased ($P<0.001$) by increasing the level of PKE in the ration.

Milk urea was increased ($P<0.001$) by feeding canola hay, suggesting increased soluble fraction ($P<0.05$) and fractional degradation rate ($P>0.05$), confirming the degradation study. Lopez *et al.* (1998) reported fractional degradation rates (c values) in the range of 0.039 to 0.059 for grass based hays. The average fractional degradation rate for canola hay of 0.056 indicates that it is similar to grass hays in terms of rumen degradation.

Table 1. Milk production characteristics of late lactation dairy cows fed pasture or canola hay and two levels of grain replacement with PKE.

	Forage type				Main effect[a]	Interaction
PKE level	Pasture hay		Canola hay		s.e.d	s.e.d
	0	50	0	50		
Milk yield (kg/d)	14.80	13.84	16.02	16.45	1.310^{NS}	1.853 ^{NS}
Fat (g/kg)	50.5	47.1	49.6	47.3	2.44 ^{NS}	3.45 ^{NS}
Protein (g/kg)	38.89	36.15	38.49	36.64	1.373 ^{NS}	1.942 ^{NS}
Total milk solids (g/d)	1,310	1,154	1,382	1,379	98.4 ^{NS}	139.2 ^{NS}
Milk urea (mmol/l)	7.71	6.14	8.07	7.59	0.240***	0.340*

NS = not significant, Significance levels denoted as *= $P<0.05$, **= $P<0.01$, ***=$P<0.001$.
[a] Main effect s.e.d = forage s.e.d and PKE s.e.d.

Table 2. Average degradation parameters for canola and pasture hays, determined by cellulase incubation.

	Pasture hay	Canola hay	s.e.d
A (soluble)	3.87	6.2	1.34*
B (insoluble but degradable)	54.1	64.8	3.95**
C (fractional degradation rate)	0.051	0.056	0.023^{NS}

NS = not significant, Significance levels denoted as *=$P<0.05$; **=$P<0.01$; ***=$P<0.001$.

Conclusion

The results of this study suggest that canola hay can be used to replace pasture hay in the diets of late lactation dairy cows without negatively impacting milk production. Canola hay has a greater soluble fraction than pasture hay, correlated with an increased milk urea.

References

ABARE, 2005. Australian dairy: Production systems 05.1, Productivity and Profit. Australian Bureau of Agricultural and Resource Economics.

Lopez, S., M.D. Carro, J.S. Gonzalez and F.J. Ovejero, 1998. Comparisons of different *in vitro* and *in situ* methods to estimate the extent and rate of degradation of hays in the rumen. Anim. Feed Sci. Technol. 73: 99-113.

Nousiainen, J.I., K.J. Shingfield and P. Huhtanen, 2004. Evaluation of milk urea nitrogen as a diganostic of protein feeding. J. Dairy Sci. 87: 386-398.

Stockdale, C.R., D.W. Dellow, C. Grainger, D. Dalley and P.J. Moate, 1997. Supplements for dairy production in victoria. Department of Natural Resources and Environment, Tatura.

Effect of feeding different levels of banana peelings on the rumen environment, degradability and digesta kinetics of cattle fed a basal diet of elephant grass

J. Nambi-Kasozi[1], F.B. Bareeba[1], E.N. Sabiiti[1] and E. Spörndly[2]
[1]Makerere University, P.O. Box 7062, Kampala, Uganda; [2]Swedish University of agricultural Sciences, P.O. Box 7070, Uppsala, Sweden; fbareeba@agric.mak.ac.ug

Introduction

Peri-urban dairy farmers in Uganda use elephant grass (EG) and banana peels (BP) as the main basal feeds supplemented with locally available protein supplements. However there is limited information on CP degradation characteristic of these feeds. Kabi *et al.* (2005) studied the CP and NDF degradability of elephant grass, cotton seed cake and *Gliricidia sepium* and found that different diets affected the degradability of the feeds studied. The present study was aimed at evaluating the effects of feeding different levels of banana peelings and elephant grass supplemented with a mixture of cotton seed cake (CSC), maize bran (MB) and *G. sepium* (GS) on the rumen environment, degradability and digesta kinetics of feeds.

Material and methods

Three rumen fistulated Friesian steers (average BW of 445±15 kg) were used in a completely randomised block design with feeding periods of four weeks on each diet after which they were switched to another diet. Four different levels of the banana peelings (0, 20, 40 and 60% DM basis) were given with elephant grass and each was supplemented with a mixture of cotton CSC, MB and GS. The quantities of GS, CSC and MB were offered in a ratio of 1:2:2 and were increased as the level of banana peelings in the diets increased to make the diets iso-nitrogenous at 12-13% CP. The first fourteen days on each diet were for adaptation, the fifteenth day for rumen fluid collection and the next six days were for incubation of the nylon bags while the last seven days were for the rate of passage studies. Rumen fluid was sampled from each animal at about 6 hours after the morning feeding during each treatment for pH and VFA determination. The nylon bag technique was used to determine the CP degradability of the feed ingredients. Three nylon bags (9×16 cm and 40 μm pore size) for each feed sample(5 g) were incubated in each of the animals for 0, 6, 12, 24, 48, 72, 96 and 120 hours. The exponential model according to McDonald (1981) was fitted to disappearance data to derive degradation parameters of CP. The animals were given 80 g of chromium-mordanted hay via the rumen cannula. Faecal grab samples were collected 0, 12, 24, 30, 33, 36, 48, 54, 60, 72, 96, 120 and 144 hours after consumption of the marked feed. The passage rate along the digestive tract and other dynamic variables were determined according to Huhtanen and Kukkonen (1995). Rumen pH and VFA data were subjected to one way ANOVA in randomised blocks using the Genstat statistical package. Crude protein degradability characteristics were analysed according to the general linear models (GLM) procedure of the statistical analysis system (SAS®, 1990) as described by Littell *et al.* (1996) with periods being included in the model as independent variables.

Results

The chemical composition of ingredients was 5.8%, 10.6%, 26.5% CP and 45.6%, 73.0% and 33.8% NDF for BP, EG and concentrate mixture respectively.
The diet without banana peelings had higher ($P<0.05$) pH than those with banana peelings and ranged from 6.7-5.9. Addition of banana peelings reduced ($P<0.05$) the acetate (54.63-46.81

molar%) and propionate (26.73-19.97 molar%) levels. Effective CP degradability of BP was higher ($P<0.05$) at the 0 and 20% levels of BP and ranged between 62.9% and 80.2%. Similarly effective CP degradability of EG was higher ($P<0.05$) at the 0 and 20% levels of BP (Table 1). Total tract retention time was the lowest at the higher BP levels. Effective CP degradability was 64.6%, 82.8% and 72% for MB, CSC and GS respectively with the 40% BP in the diet.

Table 1. Crude protein degradability (%) and total tract retention time of elephant grass in steers fed varying levels of banana peelings.

Parameter[1]	Percentage of banana peelings in the diets				SEM
	0	20	40	60	
a	32.34	32.57	32.68	32.68	0.20
b	65.29[a]	64.06[a]	45.66[b]	46.66[b]	0.88
a+b	97.63[a]	96.64[a]	78.22[b]	79.34[b]	0.77
c	0.018[b]	0.018[b]	0.030[a]	0.027[a]	0.001
Tl	1.46	1.55	1.80	1.61	0.391
ED	56.46[a]	56.91[a]	55.26[b]	54.75[b]	0.349
TRMT, h	56.22[b]	63.72[a]	64.89[a]	47.97[c]	1.39

[a,b,c] Means within a row with different superscripts are significantly different ($P \leq 0.05$).
[1] a = zero time intercept; b = insoluble but slowly degradable fraction; a+b = potential degradability; c = the rate of degradation of the b component; tl = time lag (h); ED = effective degradability; TRMT= total tract retention time; SEM = Standard error of the least square means.

Conclusion

This study shows that feeding high levels of banana peelings negatively affects rumen environment and CP degradability. The 20% BP level is recommended for inclusion as a basal diet.

References

Huhtanen, P. and U. Kukkonen, 1995. Comparison of methods, markers, sampling site and models for estimating digesta passage kinetics in cattle fed at two levels of intake. Anim. Feed Sci. Technol. 52: 141-158.

Kabi, F., F.B. Bareeba, Ø. Havrell and I.D.T. Mpofu, 2005. Evaluation of protein degradation characteristics and metabolisable protein of elephant grass (*Pennisetum purpureum*) and locally available protein supplements. Livest. Prod. Sci. 95: 143-153.

Littell, R.C., G.A. Milliken, W.W. Stroup and R.D. Wolfinger, 1996. SAS® systems for mixed models, SAS Institute, Inc., Cary, NC, 633 pp.

McDonald, I., 1981. A revised model for the estimation of protein degradation in the rumen. J. Agric. Sci. Camb. 96: 251-252.

Effect of associating ryegrass to lucerne or sainfoin on rumen digestion *in vitro*

V. Niderkorn, R. Baumont, A. Le Morvan, R. Bergeault, Y. Papon and D. Macheboeuf
INRA, UR1213 Herbivores, Site de Theix, F-63122 Saint-Genès Champanelle, France; vincent.niderkorn@clermont.inra.fr

Introduction

Ruminant feeding with multi-species grasslands rather than monocultures could contribute to sustainable breeding systems integrating productivity and environmental requirements. However, digestive interactions between forage species need to be better understood. Indeed, they could modify the efficiency of the digestion in terms of use of nutrients by animals and reduction of methane (CH_4) and nitrogen emissions (Niderkorn and Baumont, 2009). The present study was aimed at evaluating the associative effects between plants from mixtures containing one grass species and one legume species on rumen digestion *in vitro*.

Material and methods

We studied the effects of associating to ryegrass (RG, *Lolium perenne L.*), two different legumes, either lucerne (LU, *Medicago sativa* L.) or sainfoin (SF, *Onobrychis Vicifolia*) that contain condensed tannins (CT), on ruminal fermentation parameters *in vitro* according to Theodorou *et al.* (1994). Freeze-dried forages (0.6 g/40 ml) sampled at a vegetative stage were fermented alone or mixed in equal proportions (1:1). Incubations were done anaerobically at 39 °C during 24 h in culture bottles containing buffered rumen juice from sheep. Gas production, composition of gases (H_2, CO_2 and CH_4), true organic matter digestibility (TOMD), and ammonia and volatile fatty acids (VFA) in the fermentation medium were measured at 3.5 h and at 24 h. For each measuring period, the CO_2/CH_4 ratio, the partitioning factor (PF, mg OM truly degraded/ml gas) and the acetate/propionate ratio were calculated. Values were adjusted by subtracting at each collection point the values from blanks without substrate at T=0. Data were submitted to analysis of variance using a mixed procedure with repetition considered as the random effect. For each association, the three levels (0, 0.5 and 1) of legume-ratio in the mixture were used to test linear and quadratic contrasts.

Results and discussion

At the early stage of the fermentation, the TOMD was significantly different among the three pure substrates tested: LU>SF>RG ($P<0.05$), whereas the gas production was similar ($P>0.05$) (Table 1). Hence, the PF values were inversely correlated to TOMD. The SF showed a particular fermentation profile at 3.5 h of incubation: the production of CH_4 was lower than the other substrates, and the ammonia content in the medium was null. At the end of the fermentation, few differences were observed on the parameters measured. However, the ammonia content in the presence of SF remained the lowest among the three species.

Among the quadratic effects observed at 3.5 h of incubation, gas production from mixtures was higher than the balanced median value calculated from the pure substrates. Interestingly, the associative effects between RG/LU and RG/SF on the CO_2/CH_4 ratio ($P<0.01$) were opposite: positive and negative, respectively. This observation suggests that some components of SF depress the CH_4 production from RG, whereas that LU amplifies it. The ammonia content in the medium in presence of RG/SF was near zero. The property of CT for binding proteins and hence reducing their degradation in the rumen and the ammonia content in the rumen fluid is well known (Barry and McNabb, 1999). In this study, the negative associative effect observed on ammonia ($P<0.001$)

indicates that CT from SF also reduce the ammonia production from RG. At 24 h, the associative effects seemed to be attenuated, indicating a possible delay in the degradation of fibre and proteins.

Table 1. Total gas productions, gas compositions, truly degraded organic matter (TOMD), partition factors (PF), ammonia (NH_3-N) and volatile fatty acids (VFA) obtained from the early and late fermentation stages of ryegrass, lucerne, sainfoin and the two grass/legume associations.

Parameter	RG[1]	LU[1]	SF[1]	RG-LU[2]			RG-SF[2]		
				Mean	L	Q	Mean	L	Q
Incubation time: 3.5 h									
Gas (mmol/batch)	2.695	2.778	2.601	2.839	NS	*	2.688	**	*
CH_4 (mmol/batch)	0.331^{ab}	0.348^{a}	0.292^{b}	0.380	NS	**	0.303	NS	**
CO_2/CH_4	7.07^{b}	7.32^{ab}	8.23^{a}	6.56	NS	**	8.16	NS	**
TOMD (%)	55.79^{c}	70.65^{a}	60.67^{b}	61.36	***	NS	57.52	NS	NS
PF (mg TOMD/ml gas)	4.647^{b}	5.868^{a}	5.455^{a}	4.905	***	NS	5.090	NS	*
NH_3-N (mmol/batch)	0.100^{b}	0.449^{a}	-0.004^{c}	0.324	***	*	0.016	NS	***
VFA (mmol/batch)	1.797^{b}	2.017^{a}	1.697^{b}	1.988	**	NS	1.841	*	NS
Acetate/propionate	2.04	2.77	2.36	2.58	***	*	2.18	**	***
Incubation time: 24 h									
Gas (mmol/batch)	5.637	5.291	5.289	5.487	**	NS	5.504	NS	NS
CH_4 (mmol/batch)	1.036	1.060	1.041	1.088	NS	NS	1.025	NS	NS
CO_2/CH_4	4.36	4.05	4.34	4.15	NS	NS	4.38	NS	NS
TOMD (%)	80.62	80.22	78.08	79.98	NS	NS	79.31	NS	NS
PF (mg TOMD/ml gas)	3.355	3.553	3.438	3.433	*	NS	3.450	NS	NS
NH_3-N (mmol/batch)	0.375^{b}	1.140^{a}	0.230^{b}	0.812	***	NS	0.307	NS	*
VFA (mmol/batch)	3.691	3.716	3.723	3.743	NS	NS	3.705	NS	NS
Acetate/propionate	2.62^{b}	3.05^{a}	2.86^{a}	2.92	***	NS	2.72	NS	*

[1] For pure forages, means within rows with different superscript letters are significantly different ($P<0.05$).
[2] For mixtures, data shown are means, linear (L) and quadratic (Q) contrasts: NS, *$P<0.05$, **$P<0.01$, ***$P<0.001$.

Conclusion

Digestive interactions mainly occurred during the early stage of the fermentation. In particular, SF allows reducing the production of CH_4 and ammonia from the fermentation of RG, probably due to the CT content of SF. The clarification of the exact role of CT from SF in reducing detrimental fermentation end-products and *in vivo* investigations are needed to evaluate the possible benefit of forage mixtures containing SF on the environmental footprint.

References

Barry, T.N. and W.C. McNabb, 1999. The implications of condensed tannins on the nutritive value of temperate forages fed to ruminants. Br. J. Nutr. 81: 263-272.

Niderkorn, V. and R. Baumont, 2009. Associative effects between forages on feed intake and digestion in ruminants. Animal. In press.

Theodorou, M.K., B.A. Williams, M.S. Dhanoa, A.B. McAllan and J. France, 1994. A simple gas production method using a pressure transducer to determine the fermentation kinetics of ruminant feeds. Anim. Feed Sci. Technol. 48: 185-197.

Methane production and microbial profile in the rumen from three high water-soluble carbohydrate perennial ryegrass monocultures differing in their heading dates using RUSITEC

V. Niderkorn[1], E.J. Kim[2], F.J. Hou[3], C.J. Newbold[2] and N.D. Scollan[2]
[1]INRA, UR1213 Herbivores, Site de Theix, 63122 Saint-Genès Champanelle, France; [2]IBERS, Aberystwyth University, Gogerddan, Aberystwyth, Ceredigion, SY23 3EB, United Kingdom; [3]Key Laboratory of Grassland Agro-Ecosystem, College of Pastoral Agriculture Science and Technology, Lanzhou University, Gansu, China; nigel.scollan@aber.ac.uk

Introduction

Readily available energy in the form of water-soluble carbohydrate (WSC) has beneficial effects on animal productivity and environmental footprint. Elevating the level of readily available energy in the rumen can increase the capture of rumen ammonia and hence increase microbial protein synthesis (Lee *et al.*, 2003). In addition, increasing WSC in perennial ryegrass has been noted to reduce CH_4 production (Lovett *et al.*, 2006). This work sought to provide further evidence that WSC content in grass influences rumen efficiency impacting on rumen microbial protein synthesis and methane production. Three perennial ryegrass monocultures bred for higher WSC content and differing in their heading dates were assessed using an *in vitro* semi-continuous culture system (RUSITEC).

Material and methods

Two identical eight-vessel RUSITEC were randomly allocated to four dietary treatments: a control WSC perennial ryegrass variety 'Premium', and three high WSC perennial ryegrass varieties AberStar, AberMagic and AberAvon differing in their heading dates (Table 1). All varieties were harvested at 3 wk of regrowth. On the first day, each vessel was charged with buffered rumen fluid from three ruminally fistulated Holstein-Friesian cows, one dacron bag containing 40 g of rumen digesta solids and another containing fresh forage harvested at 09:00 h (9.5 g DM/bag). The population of rumen micro-organisms was maintained and mixed in an anaerobic environment. The vessels were immersed in a water bath maintained at 39 °C and continuously infused with artificial saliva (dilution rate = 0.65/d). After 24 h, the rumen digesta solids were removed and a new feed bag was inserted. On subsequent days, the feed bag that had been in the vessel for 48 h was replaced in a similar way as to the rumen digesta solids. The effluent liquors and fermentation gases were collected in effluent bottles and gas collection bags respectively. The experiment was run for 16 d. After an adaptation period of 10 d, gas production and composition (CO_2 and CH_4) were determined daily. Sampling for rumen parameters (volatile fatty acids, VFA; ammonia, NH_3-N) occurred on d 11 and 12. The DM disappearance was estimated on d 14 and 15. On d 16, vessel content was sampled for microbial profiling by qPCR (total bacteria). Data were submitted to analysis of variance using a mixed procedure from GenStat (VSN International Ltd, Hemel Hempstead, UK) with the RUSITEC unit considered as the random effect.

Results and discussion

The WSC level of AberStar, AberMagic and AberAvon were significantly higher than the control ($P<0.05$; Table 1). The DM disappearance at 48 h of AberMagic and AberAvon tended to be higher in comparison with the control and AberStar. Although the differences between treatments were not significant, AberAvon, the more digestible variety, also contained the highest content of total bacteria. The latter suggests that this variety provided an improved balance of nutrients for assimilation by the rumen microbial population. However, no difference was observed on the production of VFA (or individual VFA, data not shown) and NH_3-N in the effluents (P>0.05). AberAvon also produced significantly less CH_4 per gram of degraded DM than the other varieties (P<0.05).

Table 1. Heading date, chemical composition and dry matter disappearance, volatile fatty acids (VFA) and ammonia concentrations in effluents, methane production and microbial profiling from the RUSITEC fed with three ryegrass high WSC varieties and control.

	Control	AberStar	AberMagic	AberAvon
Heading date	29 May	20 May	29 May	5 June
WSC, g/kg	269^a	300^b	302^b	307^b
Crude protein, g/kg	101	99	97	101
DM disappearance at 48 h,%	52.8	51.5	57.7	59.3
VFA, mmol/L	32.3	31.2	31.3	30.7
NH_3-N, mmol/L	3.38	3.38	3.16	3.43
CH_4, mmol/g DM degraded	3.80^b	3.28^b	3.60^b	1.90^a
Total bacteria, μg DNA/g DM residue	557	499	521	669

Data shown are means from 4 vessels. ^{a,b} Values with a different letter indicate a significant difference between treatments (P<0.05).

Conclusion

The results of this study provide further evidence that breeding grasses for WSC content may impact on rumen efficiency. However, factors other than simply WSC content must underpin the reduction in methane (mmol/g DM degraded) observed on AberAvon, since the WSC content of the three high WSC varieties were similar, but methane production was not different between the control, AberStar and AberMagic.

Acknowledgement

This study was associated with a larger project funded by the Department for Environment, Food and Rural Affairs, UK. We are grateful to all staff who assisted with laboratory analysis at IBERS, Aberystwyth University.

References

Lee, M.R.F., R.J. Merry, D.R. Davies, J.M. Moorby, M.O. Humphreys, M.K. Theodorou, J.C. MacRae and N.D. Scollan, 2003. Effect of increasing availability of water-soluble carbohydrates on *in vitro* rumen fermentation. Anim. Feed Sci. Technol. 104: 59-70.

Lovett, D.K., D. McGilloway, A. Bortolozzo, M. Hawkins, J. Callan, B. Flynn and F.P. O'Mara, 2006. *In vitro* fermentation patterns and methane production as influenced by cultivar and season of harvest of *Lolium perenne* L. Grass Forage Sci. 61: 9-21.

The relationship between soluble and total faecal phosphorus excretion in lactating dairy cows of the Swedish Red and White breed

M. Nordqvist, R. Spörndly and K. Holtenius
Swedish University of Agricultural Sciences, Department of Animal Nutrition and Management, Kungsängen Research Centre, SE-753 23, Uppsala, Sweden; Maria.Nordqvist@huv.slu.se

Introduction

Total phosphorus (P) losses from agricultural operations have increased markedly during recent years. More than 2/3 of the P consumed by cows is generally excreted in manure (Ekelund *et al.*, 2005). Soluble P in faeces is the most vulnerable P fraction with regards to potential runoff losses from farm land which causes eutrophication in water, a major threat to aquatic ecosystems worldwide (Kleinman *et al.*, 2002). When P is provided in excess, dairy cows have a profound ability to excrete the surplus with faeces. The total P determination of faeces consists in both a largely insoluble fraction (plant cell wall residues, microbial residues, sloughed gastric intestinal cells and digestive secretions) and a soluble P fraction, mainly representing surplus P. Determination of the acid soluble P fraction in faeces is a simple and useful tool to identify excessive P feeding (Dou *et al.*, 2007). This study focusses on investigating the relation between different P fractions in faeces from dairy cows fed a range of rations with varying P content.

Material and methods

In this trial, 30 lactating dairy cows of the Swedish Red and White Breed were studied. The cows were fed different combinations of concentrate and silage (~40-90% silage of total DM) giving rise to a variation in P intake. When the cows had adapted to the diet (after at least 14 days on the specific diet) sample collection was performed during a five day period. Samples of feed offered and grab samples of faeces were collected daily during the collection period. The samples were pooled and frozen, with a subsample being freeze dried for mineral analysis. The total P content in dried feed and faeces was analysed by plasma emission spectrophotometry by an accredited commercial laboratory.

Acid soluble P in faeces was determined using a modified method by Dou *et al.* (2007). Faeces samples were thawed and 95 ml of an HCl solution (0.1%) was added to 5 g of the sample. The mixture was then transferred to a shaker for 60 min, after which the solution was centrifuged and filtered. The P content in the solution was measured using a commercially available colorimetric kit. The residual P fraction was calculated by subtracting the acid soluble fraction from the total P. This fraction largely consisted of insoluble plant cell wall residues and microbial residues, sloughed gastric intestinal cells and digestive secretions respectively.

Results

The phosphorus concentration in the diet ranged from 2.6 to 4.2 g/kg DM. The total P and extractable P varied in a similar fashion with P intake. The residual P fraction was virtually not affected by P intake (Figure 1).

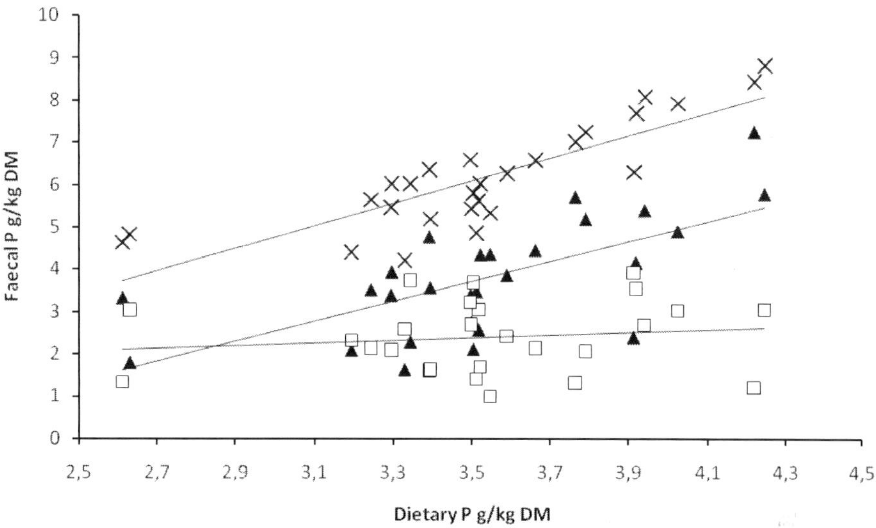

Figure 1. Relationship between dietary P and 3 different fractions of faecal P; total P(X), acid soluble P(▲), and the difference between faecal total P and acid soluble P(□).

Conclusion

The present results achieved in dairy cows of the Swedish Red and White Breed agrees with those from previous studies of the Holstein breed (Dou *et al.*, 2007). The variation in total P excretion in faeces was almost entirely explained by variations in soluble P, which is the most vulnerable P fraction with regards to potential runoff losses. Our results indicate, in line with the results of Dou *et al.* (2007), that determination of soluble P in faeces can be used as a tool to assess P nutrition in dairy cows.

References

Dou, Z., C.F. Ramberg, Jr., L. Chapuis-Lardy, J.D. Toth, Y. Wang, R.J. Munson, Z. Wu, L.E. Chase, R.A. Kohn, K.F. Knowlton and J.D. Ferguson, 2007. A novel test for measuring and managing potential phosphorus loss from dairy cattle feces. Environ. Sci. Technol. 41: 4361-4366.

Ekelund, A., R. Spörndly, H. Valk and M. Murphy, 2005. Effects of varying monosodium phosphate intake on phosphorus excretion in dairy cows. Livest. Prod. Sci. 96: 301-306.

Kleinman, P.J.A., A.N. Sharpley, B.G. Moyer and G.F. Elwinger, 2002. Effect of mineral and manure phosphorus sources on runoff phosphorus. J. Environ. Qual. 31: 2026-2033.

Prediction of digestibility and intake of mixed diets in dairy cows from faecal samples with near infrared reflectance spectroscopy (NIRS)

L. Nyholm[1], J. Nousiainen[1], M. Rinne[2], S. Ahvenjärvi[2] and P. Huhtanen[2]
[1]Valio Ltd., P.O. Box 30, FIN-00039 Valio, Finland; [2]MTT Agrifood Research Finland, FIN-31600 Jokioinen, Finland; laura.nyholm@valio.fi

Introduction

The capacity of an individual cow to consume and utilise forages is the main factor limiting milk production, and it is altered by many factors and their interactions. Dry matter intake (DMI) and organic matter digestibility (OMD) are important parameters to estimate when evaluating differences between individual animals for example for feeding and breeding purposes. DMI and OMD are laborious to measure with traditional methods. A reliable, accurate and rapid method is needed to predict DMI and OMD of individual cows. Near infrared reflectance spectroscopy (NIRS) has been used to evaluate forage chemical composition and digestibility with an adequate level of accuracy. The faecal NIR scans have also been used to predict forage and diet DMI and OMD and chemical composition of the faeces of ruminants (Dixon and Coates, 2009). The objective of this study was to create a direct faecal NIRS calibration to estimate differences between the individual animals' capacity to ingest and digest the mixed diets based on forage (grass silage) and concentrates.

Material and methods

The faecal data set (n=198) was obtained from nine trials where DMI and diet digestibility of dairy cows had been measured individually by a total faecal collection method. The cows were fed grass silage ad libitum supplemented with cereal based concentrates and protein supplements of varying quantity and quality. Dried and ground faecal samples were scanned with NIR Systems 6500 instrument at 2 nm intervals over the wavelength range of 400-2500 nm. The whole faecal data set was split into calibration (n=168) and validation (n=30) sets and faecal calibrations were tested with an internal cross validation method and external validation samples. NIR calibrations to faecal nitrogen (N), neutral detergent fibre (NDF) and indigestible neutral detergent fibre (iNDF) concentration, diet OMD and DMI were calculated with WINISI II v1.50 (Foss NIRSystems, Infrasoft International) software using modified partial least squares regression method (MPLS). Different calibrations were calculated using a spectral range of 400-2500 and 1100-2500 nm, and different mathematical treatments for the spectra: first and second derivative and standard normal variate and detrend (SNV-D) as a scatter correction technique. The best models were selected based on standard error of cross validation (SECV), R^2 values of the calibrations and standard error of prediction (SEP).

Results and discussion

The best NIRS calibrations to faecal N, NDF and iNDF were achieved using a spectral range of 1,100-2,500 nm and mathematical treatments 1,4,4,1 (Table 1). Using a spectral range of 400-2,500 nm improved N calibration slightly, but improvement was not notable since the number of samples in the final 400-2500 nm calibration was smaller. Validation samples predicted almost the same accuracy. The best calibrations were achieved using the whole spectral range of 400-2,500 nm with first and second derivative to OMD and DMI respectively (Table 2). OMD of the validation samples were predicted with the same accuracy. DMI predictions for the validation samples were unsatisfactory (SEP 2.61 kg/d and R^2 0.29). The difference between SEC and SEP indicates that the calibration dataset does not cover all the variability in faecal composition due to variance in the measured intake. The results agreed with earlier publications with the bovine faecal NIRS calibration (Boval *et al.*, 2004, Dixon and Coates, 2009).

Table 1. Calibration statistics to faecal composition with different faecal NIRS calibrations.

Constituent	Spectral region (nm)	Math treatment	n	Mean	SD	SEC	R^2	SECV	1-VR	SD/SECV
Nitrogen	400-2,500	1,4,4,1	157	28.70	2.55	0.62	0.94	0.67	0.93	3.8
(g/kg DM)	400-2,500	2,4,4,1	161	28.67	2.56	0.68	0.93	0.75	0.91	3.4
	1,100-2,500	1,4,4,1	160	28.67	2.56	0.64	0.94	0.69	0.93	3.7
	1,100-2,500	2,4,4,1	161	28.66	2.56	0.67	0.93	0.72	0.92	3.6
NDF	400-2,500	1,4,4,1	166	519.7	47.2	14.6	0.90	18.2	0.85	2.6
(g/kg DM)	400-2,500	2,4,4,1	166	519.4	46.9	16.5	0.88	19.7	0.82	2.4
	1,100-2,500	1,4,4,1	165	520.1	47.0	15.3	0.89	16.9	0.87	2.8
	1,100-2,500	2,4,4,1	162	519.6	46.9	15.8	0.89	17.5	0.86	2.7
iNDF	400-2,500	1,4,4,1	163	213.0	40.0	12.7	0.90	16.3	0.83	2.5
(g/kg DM)	400-2,500	2,4,4,1	162	211.6	38.8	8.6	0.95	15.0	0.85	2.6
	1,100-2,500	1,4,4,1	164	212.2	40.1	11.7	0.91	14.2	0.87	2.8
	1,100-2,500	2,4,4,1	161	210.9	38.6	10.4	0.93	14.5	0.86	2.7

Table 2. Calibration statistics for diet OMD and DMI with different faecal NIRS calibrations.

Constituent	Spectral region (nm)	Math treatment	n	Mean	SD	SEC	R^2	SECV	1-VR	SD/SECV
OMD	400-2,500	1,4,4,1	163	732.2	28.3	12.9	0.79	15.8	0.69	1.8
(g/kg)	400-2,500	2,4,4,1	164	733.1	28.8	17.6	0.62	18.8	0.57	1.5
	1,100-2,500	1,4,4,1	162	731.7	27.7	13.4	0.77	16.7	0.64	1.7
	1,100-2,500	2,4,4,1	163	731.7	27.4	16.5	0.64	19.3	0.51	1.4
DMI	400-2,500	1,4,4,1	164	19.99	2.22	1.63	0.46	1.84	0.31	1.2
(kg/d)	400-2,500	2,4,4,1	165	20.03	2.29	1.43	0.61	1.78	0.40	1.3
	1,100-2,500	1,4,4,1	162	19.96	2.21	1.78	0.35	1.93	0.23	1.1
	1,100-2,500	2,4,4,1	162	20.00	2.20	1.56	0.50	1.84	0.30	1.2

Conclusion

NIRS analysis enables an accurate prediction of faecal chemical composition and diet digestibility. DMI was not predicted accurately enough with a direct NIRS calibration of faecal samples. Therefore wider calibration data would be needed to create a robust NIR calibration of DMI. Alternatively an external indigestible marker would be needed to evaluate DMI with satisfactory accuracy from NIR spectra of faecal samples.

References

Boval, M., D.B. Coates, P. Lecomte, V. Decruyenaere and H. Archimède, 2004. Faecal near infrared spectroscopy (NIRS) to assess chemical composition, *in vivo* digestibility and intake of tropical grass by Creole cattle. Anim. Feed Sci. Technol. 114: 19-29.

Dixon, R. and J. Coates, 2009. Review: Near infrared spectroscopy of faeces to evaluate the nutrition and physiology of herbivores. J. Near Infrared Spectrosc. 17: 1-31.

Bioavailability of soil-bound persistent organic pollutants in dairy ruminants: a review

F. Ounnas, G. Rychen, C. Feidt and S. Jurjanz
UR AFPA, Nancy University, INRA, 2 avenue de la forêt de Haye, F-54505 Vandœuvre-lès-Nancy, France; faycal.ounnas@ensaia.inpl-nancy.fr

Anthropogenic activities are at the origin of emission in the environment of Persistent Organic Pollutants (PoPs), compounds which may contaminate environmental matrices like soil or plants. PoPs are characterised by strong persistence in the environment, and they have raised concern about the risk of transfer through the food chain via the animal product. PoPs are listed in several international conventions dealing with their potential toxicity for humans and the environment (Stockholm convention in 2001).

Soil is identified as an important reservoir for PoPs (Armitage *et al.*, 2006). Indeed, PoPs can be accumulated in the soil organic matter over a long period ('memory effect of soil') and therefore much higher concentrations of pollutants are found in soil than in grass (Rychen *et al.*, 2008). The concentrations varied from µg to mg/kg of soil depending on the PoPs family.

By grazing, cows may ingest up to 1 kg of soil per day, depending on numerous factors such as season and climate (Laurent *et al.*, 2005). Indeed, the pasture on sparse vegetation even leads to higher daily intake of soil (Mayland *et al.*, 1977). Thus the PoPs contained in soil could enter into the food chain to several extents via oral intake on pasture. However, few studies have been focussed on the soil-bound PoPs bioavailability in ruminants. In the available studies, PoPs were directly administered to the ruminant animals via contaminated oil or via feed (Lyche *et al.*, 2004; Grova *et al.*, 2006). The highest transfer rates 'feed-milk' were found for polychlorinated dibenzo-dioxins/furans (PCDD/Fs; 1 to 53%) and for polychlorinated biphenyls (PCB; 5-90%). Polycyclic aromatic hydrocarbons (PAH) were found to be significantly absorbed but very weakly excreted in milk (less than 1%), due to the high metabolisation of these compounds (Lapole *et al.*, 2007). To our knowledge, no study investigated the bioavailability of soil-bound PoPs in ruminants. Nevertheless, numerous studies have been carried out in rodents and mini-pigs in order to estimate the oral bioavailability of PoPs from soil by determination of DNA adducts or ethoxyresorufin-*O*-deethylase (EROD) activity or by detection of urinary metabolite excretion (Fouchécourt *et al.*, 1999; Roos *et al.*, 2004). In these studies, the bioavailability of soil-bound PoPs was significantly reduced when compared to the bioavailability of PoPs administered via feed or oil. Indeed, in soil the PoPs interact with the organic matter leading to the formation of covalent bonds (Gevao *et al.*, 2000). After soil ingestion, the contaminants may be partially or totally released from the soil particles during the digestion process. Hence, bioavailability of soil-bound PoPs is the result of various processes including soil ingestion, bioaccessibility, absorption, and first-pass effect.

Bioaccessible soil-bound PoPs are considered as the fraction that is mobilised in the digestive juice. *In vitro* digestion models based on human physiology have been developed in order to estimate the fraction of soil-bound PoPs mobilised under gastro-intestinal conditions (Khan *et al.*, 2008). Liquid/solid ratio, incubation time and bile salts appear to be the most determinants on PoP bioaccessibility (Lu *et al.*, 2008). On the contrary, the sequestration of PoPs by soil seems to be specific depending on the nature of the compounds (number of rings and chloration degree) and on physical and chemical soil parameters such as the content of clay and organic carbon (Tang *et al.*, 2006).

In recent years, studies related to bioavailability of soil-bound PoPs in lactating ruminants have been initiated in our research unit. Regarding the transfer of PoPs from soil to milk, halogenenated compounds and PAHs need to be distinguished: PCDD/F and PCB are generally considered as bioaccumulable in livestock products whereas PAH are largely metabolised. Different strategies must be led to investigate these compounds (PCB; PAH).

A typical grassland soil was spiked with 3 PAH model compounds (Pyrene, Phenanthrene and Benzo[a]pyrene). The estimation of the relative bioavailability (RB) was performed by comparing urinary and milk excretion of the hydroxylated-metabolite compounds after ingestion of soil-bound PAH incorporated in feed (test material) and PAH introduced in feed via a contaminated oil matrix (control material). The urinary excretion of PAH prevailed compared to the milk excretion, with a 33-fold higher value. Bioavailability of Pyr was estimated to 34%, this ratio corresponds to a minimal absorption rate by lactating goats. The bioavailability of soil-bound Pyr was reduced to 20% compared to Pyr directly incorporated in feed via oil. No differences were found for Phe.

For halogenated compounds, an agricultural PCB contaminated soil was collected and chronically fed to lactating ruminants after incorporation of the contaminated soil in feed. Carry-Over Rates (COR) of PCB toward milk will be calculated and compared with PCB contaminated hay COR (Costera *et al.*, 2006) obtained in very similar experimental conditions.

These different studies will make it possible to determine the effect of soil on PoP bioavailability.

References

Armitage, J., M. Hanson, J. Axelman and I. Cousins, 2006. Levels and vertical distribution of PCBs in agricultural and natural soils from Sweden. Sci. Total Environ. 371: 344-352.

Costera, A., C. Feidt, P. Marchand, L. Le Bizec and G. Rychen, 2006. PCDD/F and PCB transfer to milk in goats exposed to a long-term intake of contaminated hay. Chemosphere 64: 650-657.

Fouchécourt, O., M. Arnold, P. Berny and L. Rivière, 1999. Assessment of the bioavailability of PAHs in rats exposed to a polluted soil by natural routes. Environ. Res. 80: 30-39.

Gevao, B., K.T. Semple and K.C. Jones, 2000. Bound pesticide residues in soils: a review. Env. Poll. 108: 3-14.

Grova, N., G. Rychen, F. Monteau, B. Le Bizec and C. Feidt, 2006. Effect of oral exposure to polycyclic aromatic hydrocarbons on goat's milk contamination. Agr. Sust. Develop. 26: 195-199.

Khan, S., Q. Cao, J. Lin and G. Zhu, 2008. Concentrations and bioaccessibility of polycyclic aromatic hydrocarbons in wastewater-irrigated soil using *in vitro* test. Env. Sci. Pollut. Res. 15: 344-353.

Lapole, D., G. Rychen, N. Grova, F. Monteau, B. Le Bizec and C. Feidt, 2007. Milk and urine excretion of polycyclic aromatic hydrocarbons and their hydroxylated metabolites after a single oral administration in ruminants. J. Dairy Sci. 90: 2624-2629.

Laurent, C., C. Feidt and F. Laurent, 2005. Contamination des sols. EDP/ADEME, France, 215 pp.

Lu, M., D. Yuan, Q. Li and T. Ouyang, 2008. Application of response surface methodology to analyze the effects of soil/liquid ratio, pH, and incubation time on the bioaccessibility of PAHs from soil in *in vitro* method. Water Air Soil Poll. 200: 387-397.

Lyche, J.L., J.U. Skaare, H.J. Larsen and E. Ropstad, 2004. Levels of PCB 126 and PCB 153 in plasma and tissues in goats exposed during gestation and lactation. Chemosphere 55: 621-629.

Mayland, H.F., G.E. Shewmaker and R.C. Bull, 1977. Soil ingestion by cattle grazing crested wheatgrass. J Range Manage. 30: 264-265.

Roos, P.H., S. Tschirbs, A. Hack, P. Welge and M. Wilhelm, 2004. Different mechanisms of ingested polycyclic aromatic hydrocarbons in mammalian species. Xenobiotica 34: 781-795.

Rychen, G., S. Jurjanz, H. Toussaint and C. Feidt, 2008. Dairy ruminant exposure to persistent organic pollutants and excretion to milk. Animal 2: 312-323.

Tang, X., L. Tang, Y. Zhu, B. Xing, J. Duan and M. Zheng, 2006. Assessment of the bioaccessibility of polycyclic aromatic hydrocarbons in soils using an *in vitro* test. Environ. Pollut. 140: 279-285.

Effect of medium-chain fatty acids from coconut oil or krabok oil on *in vitro* rumen biohydrogenation

P. Panyakaew[1,2], *G. Goel*[1], *M. Lourenço*[1], *C. Yuangklang*[2] *and V. Fievez*[1]

[1]*Laboratory for Animal Nutrition and Animal Product Quality, Ghent University, Proefhoevestraat 10, 9090 Melle, Belgium;* [2]*Department of Animal Science, Faculty of Natural Resources, Rajamangala University of Technology Isan, Sakon-Nakhon Campus, Sakon Nakhon 47160 Thailand; veerle.fievez@ugent.be*

Introduction

The ruminal biohydrogenation (BH) process is responsible for alteration of dietary fatty acids (FA) in the rumen. Medium chain fatty acids (MCFA) from different oils are reported to change the rumen fermentation pattern due to their antimicrobial activity. Coconut oil, rich in saturated fatty acids such as lauric (C12:0) and myristic (C14:0) acid has been extensively investigated for its anti-methanogenic activities (Soliva *et al.*, 2003). Oil from krabok (*Irvingia malayana*) seeds is also rich in C12:0 and C14:0 (44.4 and 43.7 g/100 g total FA, respectively). The MCFA from these oils were investigated for their antimicrobial activity against biohydrogenating bacteria. The present study reports the effect of coconut oil and krabok oil on biohydrogenation of linolenic and linoleic acids *in vitro*.

Material and methods

Standard Thai concentrate (0.2 g) was used as a basal substrate with four treatments of 20 and 30 mg doses of C12:0+C14:0 from either coconut oil (C20 and C30) or krabok oil (K20 and K30). A flask without any MCFA was taken as the control. A mixture of sunflower oil (10 mg) and linseed oil (10 mg) were added as polyunsaturated fatty acid (PUFA) sources. An *in vitro* incubation was carried out according to Boeckaert *et al.* (2007). Mixed rumen contents were collected before the morning feeding from two rumen fistulated cows. The strained mixed rumen fluid (5 ml) and phosphate buffer (20 ml; pH 6.8) were added (1:4, v/v) in each incubation flask. The flasks were incubated in a shaking water bath at 39 °C for 24 h. After incubation, extraction of fatty acids was done according to Boeckaert *et al.* (2007). All the incubations were completed in three runs on separate days.

Apparent biohydrogenation values for C18:2 n-6 and C18:3 n-3 were calculated as the proportional loss of these fatty acids after 24 h of *in vitro* incubation, i.e. $(FA_{0h} - FA_{24h})/FA_{0h}$ with FA_{0h} and FA_{24h} representing the amount of C18:2 n-6 and C18:3 n-3 in the incubation flask (mg/flask) at the start of the incubation and after 24 h incubation, respectively.

All data were statistically evaluated using the general linear model procedure (univariate) according to $Y_{ij} = \mu + A_i + B_j + \xi_{ij}$, where Y_{ij} is the response; μ the overall mean; A_i the effect of treatment (fixed factor); B_j the effect of run (random factor) and ξ_{ij} the residual error. Treatments were compared to the control using a 2-sided Dunnett as the post-hoc test. All statistical analyses were performed using SPSS (SPSS software, 16.0., SPSS Inc., Chicago, IL, USA).

Results and discussion

The effect of coconut oil and krabok oil on the individual C18 fatty acid profile (mg/flask) is shown in Table 1. Both oils resulted in higher accumulation of C18:2 t11c15 compared to the control. Moreover, coconut or krabok supplementation dose dependently decreased all C18:1-isomers, except C18:1 t11 and C18:1 t12-t14. Krabok oil at 30 mg level resulted in lower biohydrogenation of C18:2 n-6 and C18:3 n-3 compared to coconut oil. Higher biohydrogenation was observed for C18:2 n-6 and C18:3 n-3 with both MCFA rich oils.

Table 1. Effect of coconut oil and krabok oil on C18 fatty acids contents (mg/flask) and biohydrogenation of linoleic and α-linolenic acid (n=3).

	Control	C 20	K 20	C 30	K30	SEM	Dose effect		Oil effect	
							C	K	Dose 20	Dose 30
C18:1 t6-t8	0.41	0.26[a]	0.27[a]	0.19[a]	0.24[a]	0.020	*	*	+	+
C18:1 t9	0.35	0.29[a]	0.23[a]	0.23[a]	0.21[a]	0.011	*	*	*	*
C18:1 t10	0.54	0.37[a]	0.35[a]	0.28[a]	0.29[a]	0.039	*	*	NS	NS
C18:1 t11	3.14	2.88	3.38	2.71	3.94	0.268	NS	NS	NS	NS
C18:1 t12-t14	0.27	0.26	0.27	0.21	0.28	0.022	*	*	NS	NS
C18:1 c9	10.4	9.93	8.39[a]	7.94[a]	6.73[a]	0.245	*	*	*	*
C18:1 c11	0.32	0.32	0.29	0.20[a]	0.31	0.020	*	*	NS	NS
C18:1 c15	0.13	0.05[a]	0.06[a]	0.02[a]	0.04[a]	0.003	*	*	*	*
C18:2 t11c15	0.51	0.99[a]	1.33[a]	1.20[a]	1.66[a]	0.104	NS	NS	*	*
CLA c9t11	0.09	0.12	0.11	0.15[a]	0.11	0.010	NS	NS	NS	NS
Biohydrogenation,%										
C18:2 n-6	63.3	65.1	73.5[(a)]	80.4[a]	75.4[a]	2.60	*	*	NS	NS
C18:3 n-3	56.8	61.8	76.8[a]	84.6[a]	78.8[a]	3.82	*	*	*	*

[a,(a)] Mean values are significantly different or have tendency to be different from the control value presented in the same row; SEM = standard error of mean; $*P<0.05$; $+0.05<P<0.10$, NS = not significant.

Conclusion

MCFA from coconut and krabok oil affected the rumen biohydrogenation to a similar extent, generally increasing linoleic and linolenic acid biohydrogenation and reducing the amount of accumulating biohydrogenating intermediates, except for C18:2 t11c15.

Acknowledgement

This research was funded by the EU-projects (TH/Asia-Link/014-141-176 and Food-CT-23006-36241, Prosafebeef).

References

Boeckaert, C., B. Vlaeminck, J. Mestdagh and V. Fievez, 2007. *In vitro* examination of DHA-edible micro algae: 1. Effect on rumen lipolysis and biohydrogenation of linoleic and linolenic acid. Anim. Feed Sci. Technol. 136: 63-79.

Soliva, C.R., I.K. Hindrichsen, L. Meile, M. Kreuzer and A. Machmüller, 2003. Effects of mixtures of lauric and myristic acid on rumen methanogens and methanogenesis *in vitro*. Lett. Appl. Microbiol. 37: 35-39.

Evaluation of the nutritive value of processed barley grain with different methods using an *in vitro* gas production technique with two sources of inocula

E. Parand and A. Taghizadeh
Department of Animal Science, Faculty of Agriculture, Tabriz University, Iran; ehsanparand@gmail.com

Introduction

Whole barley grain with an intact pericarp is largely resistant to digestion because a fibrous hull which has a low digestibility surrounds barley grain besides, the intact pericarp of kernels is resistant to bacterial attachment in the rumen. So feeding whole kernels to dairy cattle leads to a considerable excretion of them in the faeces (Yang *et al.*, 2000). Since barley starch, once exposed to micro-organisms in the rumen, is readily degradable, the effects of processing on the ruminal degradability of it will be important due to disorders associated with rapid carbohydrate fermentation. In addition processing may change the digestibility of the protein matrix in barley and may limit the availability of starch granules for rumen bacteria. In this study, the effects of three different barley processing methods were evaluated in a completely random design using an *in vitro* gas production technique. Also this study was carried out to evaluate the accuracy of sheep faeces as the source of microbial enzyme for gas production technique compared to rumen fluid.

Material and methods

The grains received three different processings (roasting, steam flaking and microwave processing) and were compared with a control group (not processed barley). Barley grain was steam flaked by exposing the grains to low pressure steam for 40 min, attaining temperatures of 95-99 °C. Then steamed grains were immediately moved to a flaker. Roasting was accomplished in a spinning pot, made of cast iron on butane flame for 10 min. The dry matter (DM) content of the barley grain was determined (from duplicate 1 g samples) and water was added to increase its moisture content up to 300 g/kg. Three samples (100 g each) were subjected to microwave irradiation (Butane microwave BC380W) at a power of 900 W for 3 min. Rumen fluid was provided by two fistulated sheep, fed an alfalfa hay (400 g/kg) plus commercial sheep concentrate (600 g/kg) twice daily. The rumen fluid was collected 2 h after the morning feed and was strained through four layers of cheesecloth. Fifty g of wet sheep faeces were collected within one hour of voiding from the same donor animals. The faeces were macerated and mixed with 50 ml of McDougall (1948) buffer solution, which had previously been saturated with CO_2. The mixture was filtered and then made up to 300 ml with buffer solution, and the pH was adjusted to 6.8. Gas production was measured using the method of Fedorak and Hrudey (1983). Triplicate grounded and dried samples of each treatment (approximately 300 mg) were weighed and placed in 50 ml capacity serum bottles; then bottles were incubated in 20 ml of buffered rumen fluid/faecal liquor (buffer:rumen fluid, 2:1, v/v) for 48 h. Gas production was recorded at 2, 4, 6, 8, 12, 16, 24, 36 and 48 h of incubation. The data at the different times was analysed using completely randomised design.

Results

The results showed that the microwave processing and control group had more total gas production (volume) than the others (Table 1). Also we found a relationship between *in vitro* gas production volume (ml/g DM), using rumen liquor (x), and *in vitro* gas production volume (ml/g DM) using faecal liquor (y) at different times (Table 2) which suggests that faeces can be used successfully in gas production technique to evaluate concentrate feed stuff.

Table 1. The effect of four different processing methods on in vitro gas production with rumen liquor (ml/g DM).

Time	Processed barley grain				s.e
	Roasted	Microwaved	Steam flaked	Control	
2 h	13.0[a]	13.0[a]	12.3[a]	15[a]	3.5
4 h	35.6[a]	24.0[b]	25.6[b]	26.3[b]	3.4
6 h	53.8[a]	35.0[b]	25.5[b]	36.6[b]	4.7
8 h	68.1[a]	44.0[b]	32.3[b]	47.0[b]	5.4
12 h	96.3[a]	77.3[ab]	53.6[b]	80.5[a]	7.9
16 h	129.44[a]	119.3[a]	82.8[b]	121.4[a]	3.8
24 h	184.7[a]	185.7[a]	153.2[b]	185.4[a]	3.6
36 h	220.0[ab]	246.3[a]	212.5[b]	245.1[a]	3.9
48 h	239.7[b]	275.9[a]	239.4[b]	275.3[a]	3.3

[a,b] Means in the same row with different letters differ ($P<0.05$).

Table 2. The relationship between in vitro gas production volume (ml/g DM), using rumen liquor (x), and in vitro *gas production volume (ml/g DM) using faecal liquor (y) at different times.*

Processing method	Variance accounted for (r^2)	
	Incubation time: 2 h-48 h	Incubation time:8 h, 12 h, 24 h and 36 h
Control	0.94	0.80
Microwave processing	0.97	0.98
Roasting	0.91	0.89
Steam flaking	0.98	0.96

Conclusion

Because of considerations about disorders like acidosis or bloat, use of processing like steam flaking and roasting seems to be more reasonable in order to bypass some starch to the small intestine while leaving a sufficient amount of degradable starch in the rumen. It seems that the heat during these processing methods turns starch to a more resistant form against digestion. Denaturation of the protein matrix reduces the starch degradability in the rumen. Using microwave processing is questionable due to costs of this method and difficulties in using it on a large scale. Our results suggest that faecal liquor can successfully be used to evaluate concentrate feed stuff with the *in vitro* gas production technique instead of rumen liquor and may reduce the costs of this method. The total gas production volume by faeces liquor is lower than those with rumen liquor and mathematical correction is needed.

References

Fedorak, P.M. and D.E. Hrudey, 1983. A simple apparatus for measuring gas production by methanogenic cultures in serum bottles. Environ. Technol. Lett. 4: 425-432.

McDougall, E.I., 1948. Studies on ruminant saliva. 1. The composition and output of sheep saliva. Biochem. J. 43: 99-109.

Yang, W.Z., K.A. Beauchemin and L.M. Rode, 2000. Effects of barley grain processing on extent of digestion and milk production of lactating cows. J. Dairy Sci. 83: 554-568.

Comparison of transcriptome and proteome expressions in the anaerobic rumen fungus *Neocallimastix frontalis* PMA02 under different substrate conditions

M.-A. Park[1,3], J. Song[1,3], M. Kwon[2], J.K. Ha[3] and J. Chang[1]
[1]Department of Agricultural Science, Korea Open National University, Seoul, Korea; [2]Department of Forest Products, Kookmin University, Seoul, Korea; [3]Department of Food and Animal Biotechnology, Seoul National University, Seoul, Korea; jschang@knou.ac.kr

Introduction

Currently, six genera of anaerobic fungi were isolated from the digestive tract of herbivorous animals and belong to family *Neocallimatigaceae* (NCBI, 2009). In the rumen, the biomass of anaerobic fungi is about 5% of total microbial biomass, however, anaerobic fungi play important roles in feed digestion with their diverse enzyme system. Unlike other microbes in the rumen, anaerobic fungi degrade feed particles both with chemical reaction and mechanical action. During fermentation, they produce volatile fatty acids, ethanol, CO_2 and hydrogen. The amount and ratio of each fungal metabolite can be shifted by a fermentation condition such as substrate types. The change in substrate during fermentation might cause a change in fungal enzyme production and this might cause a change in fungal physiology and metabolic profile. The present study was conducted to compare the transcriptome and proteome changes in anaerobic rumen fungus *Neocallimastix frontalis* PMA02 fermented under different substrate conditions.

Material and methods

The anerobic fungus *Neocallimastix frontalis* PMA02 was isolated from the rumen of Holstein steer and identified by analysing both morphological characteristics and sequences of rRNA ITS1 region (Brookman *et al.*, 2000). Fungal strain was cultivated with modified Lowe medium (Lowe *et al.*, 1985) containing 2% of glucose, starch, filter paper or Sigmacell cellulose powder as carbohydrate source at 39 °C for 72 h under anaerobic conditions. After incubation, fungal cells were harvested and homogenised under liquid nitrogen. Total RNA and cellular proteins were sequentially extracted using Trizol™ reagent (Invitrogen, USA) according to the manufacture's instruction. The protein samples from different substrate conditions were compared using the 2D-DIGE method using Cy3™, Cy5™ and Cy2™ as the pre-staining dye. After iso-electric focussing (IEF) and subsequent sodium dodecyl sulfate-polyacryl amide gel electrophoresis (SDS-PAGE), the gels were scanned with Typhoon 9400 system (GE healthcare, USA) for image capturing and Decyder software was used for image analysis. After quantification of total extracted RNA, cDNA were synthesised with 1.5 μg of RNA using a cDNA synthesis kit (Roche, Germany). The primers for real time PCR were designed based upon the results of *N. frontalis* PMA02 EST data base (Kwon *et al.*, 2009) using Vector NTI software (Infomax, USA). The $2^{-\Delta\Delta CT}$ method (Livak *et al.*, 2001) was used for comparing the RNA expression using rRNA ITS1 as an internal control.

Results

The protein expression levels of enolase, triose phosphate isomerase and phosphoglycerate kinase in culture with soluble carbohydrate substrate were significantly ($P<0.05$) higher than those obtained with insoluble carbohydrate substrate (Table 1). The protein expression levels of malate dehydrogenase, phosphoenol pyruvate carboxykinase, malic enzyme and glyceraldehyde 3-phosphate dehydrogenase were significantly ($P<0.05$) higher in culture with insoluble carbohydrate substrate than in those with soluble carbohydrate substrate. The RNA expression levels of enolase and triose phosphate isomerase were higher in culture with soluble carbohydrate substrate than in those with insoluble carbohydrate substrate, which matched with protein expression pattern.

Table 1. Results of proteome and transcriptome expression in N. frontalis *under different carbohydrate substrate conditions.*

Protein Name	Protein Express.	RNA Express.
Enolase	S > G > C > F	S > G > F> C
Triose phosphate isomerase (TIM)	S > G > F > C	S > G > F > C
Phosphoglycerate kinase (PGK)	S > G > C > F	S > F > G > C
Succinyl CoA synthetase α subunit (SCSA)	C > S > F > G	S > F > C > G
Succinyl CoA synthetase β subunit (SCSB)	C > S > F > G	S > F > G > C
Pyruvate formate lyase (PFL)	C > S > F > G	S > F > C > G
Malate dehydrogenase (MDH)	C > F > G > S	S > F > G > C
Glyceraldehyde 3-phosphate dehydrogenase (GAPDH)	C > F > S > G	S > F > G > C
Phosphoenolpyruvate carboxykinase (PEPCK)	C > F > S > G	S > F > G > C
Malic enzyme (ME)	C > F > S > G	F > S > G > C

Carbohydrate substrate: C: Sigmacell cellulose; F: filter paper; G: glucose; S: starch.

Conclusion

The protein expression levels of MDH, GAPDH, PDPCK and ME were higher in culture with insoluble carbohydrate substrate than in those with soluble carbohydrate substrate. The protein expression levels of enolase, TIM, and PGK were higher with soluble carbohydrate substrate than with insoluble ones. On the contrary, the protein expression levels of SCSA, SCSB and PFL were not consistent among the carbohydrate types. In addition, the protein expression levels were not matched with RNA expression levels. The existence of isozymes and post-transcriptional modification might cause the inconsistency between protein and RNA expression levels. To solve this question, further studies are required.

References

Brookman, J.L., G. Mennim, A.P.J. Trinci, M.K. Theodorou and D.S. Tuckwell, 2000. Identification and characterization of anaerobic gut fungi using molecular methodologies based on ribosomal ITS1 and 18S rRNA. Microbiol. 146: 393-403.

Kwon, M., J. Song, J.K. Ha, H.-S. Park and J. Chang, 2009. Analysis of functional genes in anaerobic rumen fungus *Neocallimastix frontalis* PMA02. Asian-Australasian J. Anim. Sci. in press.

Livak, K.J. and T.D. Schmitten, 2001. Analysis of relative gene expression data using real-time quantitative PCR and the $2^{-\Delta\Delta CT}$ method. Methods 25: 402-408.

Lowe, S.E., M.K. Theodorou, A.P.J. Trinci and R.B. Hespell, 1985. Growth of anaerobic rumen fungi on defined and semi-defined media lacking rumen fluid. J. Gen. Microbiol. 131: 2225-2229.

Effects of crude glycerin on ruminal metabolism and digestibility when fed in combination with steam-flaked corn

G.L. Parsons and J.S. Drouillard
Kansas State University, Manhattan, KS, USA; jdrouill@ksu.edu

Introduction

Expansion of the biodiesel industry has created increased supplies of crude glycerin that may have direct applications in livestock feeding. Catalysed reactions between methanol and triglycerides from vegetable oils, such as soybean oil, yield biodiesel and the co-product, crude glycerin (Van Gerpen, 2005). Approximately 10% of the weight of soybean oil used to produce biodiesel becomes glycerin. Recent increases in feed costs have induced livestock producers to seek cost-effective alternatives to traditional feed ingredients. Several studies have evaluated the use of glycerin in poultry (Cerrate *et al.*, 2006), swine (Kijora *et al.*, 1995), and dairy cattle (DeFrain *et al.*, 2004). Limited work has been conducted to understand digestion and metabolism of glycerin in ruminant livestock. Upon ingestion, glycerol is converted to glucose via phosphorylation to glycerol-3-phosphate (G3P), which is catalysed by glycerol kinase and enters gluconeogenesis in the liver (Mourot *et al.*, 1994). Trabue *et al.* (2007) reported that approximately 80% of glycerol is metabolised in the rumen within 24 hours, and Schröder and Südekum (2007) observed lower acetate:propionate ($P<0.05$) ratio with glycerol administration. Glycerin has been demonstrated to be useful as an energy source in several species, and may yield significant improvements in beef cattle performance (Parsons *et al.*, 2009) when administered at low levels. Availability of information pertaining to the effects of crude glycerin on ruminal fermentation and diet digestibility in high-concentrate diets is limited.

Material and methods

Crossbred steers (n = 9; 624 ± 80 kg) fitted with ruminal cannulae (Bar Diamond Inc., Parma, ID; dorsal sac) were arranged in a 3×3 Latin square design. Treatments consisted of steam-flaked corn diets containing 0, 2, and 4% crude glycerin (DM basis). Steers were allowed *ad libitum* access to finishing diets fed once daily. Diets contained 6% alfalfa hay and 82.6, 80.2, and 77.8 steam-flaked corn respectively; provided 14% crude protein, 3.5% protein equivalent as non-protein nitrogen, 300 mg/d monensin, 90 mg/d tylosin, 1000 IU/lb vitamin A, 0.3% salt, 0.7% calcium, and 0.7% potassium. Periods consisted of a 10-day acclimation phase followed by a 3-day collection phase. Chromic oxide (10 g) was used as an indigestible marker to estimate total faecal output, and was dosed intraruminally prior to feeding each day beginning 7 days prior to the sampling phase. Starting on day 11 of each period, ruminal digesta were collected throughout a 3-day collection phase. Collection times were scheduled at the following times relative to feeding each day: day 1 at 0, 6, 12, 18 and 24 h post-feeding; day 2 at 2, 8, 14, and 20 h post feeding; day 3 at 4, 10, 16 and 22 h post-feeding. Digesta was removed from the rumen via the ruminal cannula and strained through eight layers of cheesecloth. Analyses of strained ruminal fluid consisted of the determination of volatile fatty acid profiles, pH, ammonia concentration, and apparent total tract digestibilities of dry matter, organic matter, NDF, starch, protein, and ether extract. Data were analysed using the Proc MIXED procedure of SAS® (version 8.1; SAS Inst; Cary, NC 2002).

Results

Dry matter intake was similar among treatments ($P>0.98$). Faecal output was ($P=0.74$) 1.21, 1.27, and 1.28 kg/day when glycerin was fed at 0, 2, and 4%, respectively. Dry matter digestibilities were similar ($P=0.84$) for cattle for each of the dietary treatments. Feeding glycerin linearly increased the mean ruminal pH from 5.61 in control steers to 5.67 and 5.73 when glycerin was added at 2 and 4%, respectively ($P<0.053$). No treatments by time interactions were observed. Acetate concentrations were linearly decreased with increasing glycerin concentrations. Concentrations of butyrate and valerate significantly decreased as crude glycerin increased in the diet (Table 1).

Table 1. Ruminal concentrations of volatile fatty acids and ammonia and acetate:propionate ratio in steers fed 0, 2, or 4% crude glycerin.

VFA, mM	Glycerol,% of diet dry matter			SEM	Probability values		
	0	2	4		0 vs Gly	Linear	Quadratic
Acetate	52.8	50.7	50.1	1.60	0.06	0.06	0.56
Propionate	50.8	48.7	50.7	3.50	0.81	0.76	0.32
Acet:Prop ratio	1.16	1.20	1.14	0.14	0.82	0.68	0.25
Butyrate	14.4	13.5	12.5	1.05	0.05	0.03	0.99
Isobutyrate	0.88	0.85	0.84	0.05	0.38	0.34	0.92
Valerate	5.45	4.85	3.74	0.73	0.01	< 0.01	0.52
Isovalerate	2.25	1.85	2.35	0.37	0.67	0.80	0.20
Total VFA	120	116	116	4.7	0.13	0.23	0.31
Ammonia	7.99	7.80	7.67	1.33	0.74	0.72	0.97

Conclusion

Adding 2 to 4% crude glycerin to cattle finishing diets containing steam-flaked corn does not affect dry matter intake, faecal output, or nutrient digestibility. Modest changes in pH occurred with addition of glycerin, as well as changes in ruminal concentrations of some VFA.

References

Cerrate, S,F. Yan, Z. Wang, C. Coto, P. Sacakli and P. Waldroup, 2006. Evaluation of glycerin from biodiesel production as a feed ingredient for broilers. Int. J. Poult. Sci. 5: 1001-1007.

DeFrain, J.M., A.R. Hippen, K.F. Kalscheur and P.W. Jardon, 2004. Feeding glycerol to transition dairy cows: effects on blood metabolites and lactation performance. J. Dairy Sci. 87: 4195-4206.

Kijora, C., R.D. Kupscy and L. Hagemann, 1995. Glycerol as a feed component in diets of fattening pigs. Arch. An. Nutr. 47: 345-360.

Mourot, J., A. Aumaitre, A. Mounier, P. Peiniau and A.C. Fracois, 1994. Nutritional and physiological effects of dietary glycerol in growing pigs. Consequences on fatty tissues and post mortem muscular parameters. Livest. Prod. Sci. 38: 237-244.

Parsons, G.L., M.K. Shelor and J.S. Drouillard, 2009. Performance and carcass traits of finishing heifers fed crude glycerin. J. Anim. Sci. 87: 653-657

Schröder, A. and K.H. Südekum, 2007. Glycerol as a by-product of biodiesel production in diets for ruminants. Institute of Animal Nutrition, Physiology and Metabolism, University of Kiel. http://regional.org.au/au/gcirc/1/241.htm. Accessed July 23 2007.

Van Gerpen, J.V., 2005. Biodiesel processing and production. Fuel Process. Technol. 86: 1097-1107.

Evaluation of models for prediction of voluntary feed intake in beef steers

H. Patino[1], K. Swanson[2], J. France[2] and E. Prates[1]
[1]LANUR, Dept. de Zootecnia, Universidade Federal de Rio Grande do Sul, Brazil; [2]Centre for Nutrition Modelling, DAPS, University of Guelph, Canada; harold.patino@ufrgs.br; hpatino@uoguelph.ca

Introduction

Prediction of voluntary feed intake in ruminants has been a major challenge for nutritionists for more than 50 years. The National Research Council uses an empirical model for feed intake prediction based on the relationship between feed intake and dietary energy concentration (NRC, 2000). However, in situations where diets are predominantly based on forages, most models have generated estimates that are not adequate, apparently because there are interactions between energy balance and ruminal fill that govern voluntary feed intake and are not included for those models. The Standing Committee on Agriculture uses an empirical model that predicts feed intake in grazing conditions as the product of the potential intake of food by the specific animal and the proportion of that potential that the animal can obtain from the available feed supply (Freer *et al.*, 1997). Some studies have shown that forage ruminal degradation characteristics can be used to predict intake, digestibility and weight gain. Madsen *et al.* (1994) proposed a model to estimate feed consumption based on the prediction of ruminal fill, knowing the animal's rumen capacity and assuming that the physical limitation is the determining factor in consumption. The present study was aimed at evaluating four models to estimate the intake of dry matter (DM) and neutral detergent fiber (NDF) of forage based diets.

Material and methods

The data set for model evaluation was obtained from an experiment with four-rumen fistulated Hereford steers with an average BW of 150 kg in a 4×4 Latin square design. Treatments were four offered levels of oat (*Avena strigosa, L.*) hay DM (9.1%CP; 76.8%NDF; 46.4%ADF; 5.2%ADL): 1.5% BW, 2.0% BW, 2.5% BW and *ad libitum* (defined to ensure a leftover of food equal to 15% of offered). Feed intake and digestibility were measured in a conventional digestibility trial and metabolisable energy content in diets was estimated. The intake and net energy value for maintenance (NEm) were predicted using the equation for growing calves of NRC (2000). The intake was also predicted using the model described by Freer *et al.* (1997). The relative availability was calculated for each animal considering a value 0.60, 0.80, 1.0 and 1.20 for animals in treatments with hay DM offered at 1.5, 2.0, 2.5% BW and *ad libitum*, respectively. To evaluate the model of Madsen *et al.* (1994), ruminal degradability data were obtained by placing 5 g hay samples inside nylon bags (10×20 cm and pore size of 40 μ) and incubating sequentially for 3, 6, 9, 12, 24, 48, 72 and 96 hours. The residue contained in the bags was ground to a particle size of 1 mm and analysed for NDF. Passage rate was calculated using 50 g Cr-mordant per animal as an indicator and adjusting the values of fecal concentration indicator for a two compartment model. The ruminal capacity was determined by manual emptying of rumen contents before the meal in the morning, 2 and 10 hours afterwards. Samples of rumen contents were collected during evacuation and were analysed for NDF, incubated in the rumen for 132 hours and the resulting residues were analysed for NDF. These results were used to calculate the amounts indigestible NDF (INDF) present in the rumen, estimate ruminal fill (RF) and NDF intake (NDFI). Evaluations of the accuracy and precision of models were conducted by regressing observed DM intake on predicted DM. Data were analysed using the Model Evaluation Systems v. 3.1.1. (Tedeschi, 2006).

Results

The models evaluated had intermediate values of concordance correlation coefficient, indicating an intermediate account in assessing both accuracy and precision. The NRC and Freer models were not accurate and overestimated ($P < 0.001$) and underestimated ($P = 0.03$) intake, respectively. The Madsen model was accurate without mean bias ($P = 0.32$) and with a high value of the bias factor correction factor (Cb=0.86). The Freer and Madsen models were 38% more precise than the NRC model for intake prediction (MSEP: 0.66 vs. 0.41). The NRC model showed inaccuracy and imprecision to predict the consumption of forage-based diets, probably because only 1% of the data were used to calculating NEm with diets with ME concentration smaller than 1.9 Mcal/kg. The Freer model was more accurate than precise with 60% of MSEP due to variation not explained by the model. The Madsen model was accurate and precise but 65% of MSEP was due to errors not explained by the regression, because of the small number of observations used (n=16).

Table 1. Comparison of different models for prediction of DM intake for steers.

Parameters[1]	Models		
	NRC	Freer	Madsen
CCC	0.58	0.60	0.69
MB	-0.72	0.34	-0.17
Cb	0.65	0.75	0.86
MSEP	0.66	0.40	0.42
Bias,%	78.39	29.19	6.66
Regression errors,%	2.35	11.07	64.46
Random errors,%	19.26	59.74	28.88

[1] CCC: concordance correlation coefficient; MB: mean bias; Cb: model accuracy; MSEP: mean square error of prediction.

Conclusion

The Madsen model had a smaller MSPE and it was the only one to be accurate without bias according to the concordance correlation coefficient. This shows the need to include forage and animal characteristics related to digestive kinetics in models of intake prediction.

References

Freer, M., A.D. Moore and J.R. Donelly, 1997. GRAZPLAN: decision support systems for Australian grazing enterprises- II. The animal biology model for feed intake, production and reproduction an the GrazFeed DSS. Agric. Syst. 54: 77-126.

Madsen, J., T. Stensing, M.R. Weisbjerg and T. Hvelpund, 1994. Estimation of the physical fill of feedstuffs in the rumen by in sacco degradation characteristics. Livest. Prod. Sci. 39: 43-47.

NRC, 2000. Nutrient Requeriments of beef Cattle: Update 2000. 7th rev ed. Natl. Acad. Press, Washington DC, USA.

Tedeschi, L.O., 2006. Assessment of the adequacy of the mathematical models. Agric. Syst. 89: 225.

Effect of different combining ratios of high-quality and poor-quality roughage on rumen fermentation parameters *in vitro*

D.Y. Peng, Z.S. Wang, B. Xue, L.Z. Wang and A.Q. Lai
Animal Nutrition Institute, Sichuan Agricultural University, Ya'an 625014, China; wangzs007@yahoo.com.cn

Introduction

Alfalfa shortage and its high cost are the major limitation for dairy production in the south of China. Roughage (corn silage or rice straw) is an abundant feedstuff resource in southern China. However, because of low feed intake, digestibility and utilisation by ruminants, roughage is not utilised effectively and is often burnt, contributing to waste of resources and air pollution. To improve the nutritional value of roughage, it is common to combine straw with an alfalfa pellet to provide a more balanced diet for ruminants. Positive associative effects for alfalfa addition to low quality forages were observed on *in vitro* fermentation (Haddad, 2000). In recent years, it has been shown that *in vitro* gas production (GP) is an effective technique to evaluate associated effects of ruminant feedstuff. Thus, the objective of the experiment was to study the associative effects of rice straw, corn silage and alfalfa, aiming to provide an optimal ratio for the practical diet formulation of dairy cows in southern China.

Material and methods

Fresh rice straw was harvested at Ya'an City, Sichuan province of China, on August 29, 2008. Corn silage and alfalfa pellets were obtained from the Newhope natural dairy farm. A compound forage mixture (CR) was formulated by corn silage and rice straws with the ratio of 50:50 (dry matter as basis). CR was mixed with alfalfa according to the ratio of 0:100 (CR0), 40:60 (CR40), 75:25 (CR75) and 100:0 (CR100) to constitute roughage. These four mixed roughages were then combined with concentrate at the proportion of 70:30 to constitute four complete diets. Ruminal fluid was obtained from Holstein cows fed twice daily on a mixed diet of alfalfa hay and mixed concentrates (7:3, w/w). For the *in vitro* gas production test, the medium consisted of buffer and rumen liquor in a 2:1 ratio (Menke and Steingass, 1988), and each treatment was replicated. After 48 h incubation, the gas production mixed fermentation medium was used for the analysis of ammonia nitrogen, volatile fatty acids (VFA), pH and microbial crude protein (MCP). VFA were estimated by gas chromatography (CP-3800GC, Varian). The MCP was determined by the method of Zinn and Owens (1986). To describe the dynamics of GP over time, equations of GP parameters were based on those of McDonald (1981).
Statistical analysis was performed using SPSS16.0 (SPSS Inc., Chicago, IL, USA). Data were analysed by one-way ANOVA. LSD multiple comparisons were used to test the differences between treatments, which are denoted by different letter superscripts. Statistical significance was accepted at $P<0.05$.

Results

The cumulative GP and rate of GP at all incubation times were the highest for treatment CR0 ($P<0.05$). The amounts of cumulative GP decreased with the decreasing level of alfalfa. The decrease may be quadratic, which could be tested. Concentration of ammonia nitrogen also decreased with the decreasing level of alfalfa ($P<0.01$). MCP concentration decreased with the decreasing proportion of alfalfa in the forage, even though the decrease of MCP was irregular between CR40 and CR75. CR40, CR75 and CR100 had higher potential GP than CR0 ($P<0.01$). Lag time decreased from 2.95

h for CR0 to 1.83 for CR75 and then increased to 2.58 for CR100 ($P<0.05$). The acetate:propionate ratio did not differ between CR0, CR40 and CR75 and increased with CR100 ($P<0.05$).

Table 1. The effects of different ratio and quality of roughage on in vitro gas production (GP) parameters at 72 h, the concentrations of NH3-N, MCP and VFA at 48 h (n=6 per treatment).

Items	Diet[1]				SE	*P*-value
	CR0	CR40	CR75	CR100		
GP parameters						
Cumulative GP, ml	40.77[a]	39.80[ab]	36.83[b]	32.45[c]	1.338	0.025
Potential GP, ml/g	33.39[b]	40.75[a]	42.80[a]	43.82[a]	0.888	<0.001
Rate of GP, ml/h	0.72[a]	0.70[a]	0.60[ab]	0.56[b]	0.052	0.04
Lag time, h	2.95[a]	2.04[bc]	1.83[c]	2.58[ab]	0.292	0.019
Microbial crude protein, mg/dl	80.40[a]	69.33[bc]	76.71[ab]	64.91[c]	1.475	<0.001
Ammonia nitrogen, mg/dl	18.86[a]	16.96[b]	13.36[c]	10.72[d]	0.276	<0.001
pH	6.6	6.68	6.73	6.57	0.057	0.079
VFA, mmol/l						
Acetate	6.10[a]	3.95[b]	4.33[b]	4.52[ab]	0.698	0.063
Propionate	3.86[a]	2.70[bc]	2.87[b]	2.01[c]	0.313	0.003
Butyrate	1.19[b]	1.07[b]	1.27[ab]	1.36[a]	0.118	0.167
Acetate: propionate	1.58[b]	1.47[b]	1.52[b]	2.27[a]	0.073	0.049

[1] The ratios of CR to alfalfa in CR0, CR40, CR75 and CR100 are 0:100, 40:60, 75:25 and 100:0, respectively.

[a,b,c,d] Means with different letters in the same row differ significantly ($P<0.05$).

Conclusion

In the present study, no negative effects on rumen fermentation parameters were observed with the CR to alfalfa ratio of 75:25, thus this ratio could be recommended to save alfalfa and prevent air pollution induced by burning of rice straw.

Acknowledgement

This work was supported by the program for public welfare behaviour (Agriculture) research fund of the Agricultural Department of China (No. nyhyzx07-036-03).

References

Haddad, S.G., 2000. Associative effect of supplementing barley straw diets with alfalfa hay on rumen environment and nutrient intake and digestibility for ewes. Anim. Feed Sci. Technol. 87: 163-171.

McDonald, I., 1981. A revised model for the estimation of protein degradability in the rumen. J. Agric. Sci. (Camb.) 96: 251-252.

Menke, K.H. and H. Steingass, 1988. Estimation of the energetic feed value obtained by chemical analysis and *in vitro* gas production using rumen fluid. Anim. Res. Develop. 28: 7-55.

Zinn, R.A. and F.N. Owens, 1986. A rapid procedure for purine measurement and its use for estimating net ruminal protein synthesis. Can. Anim. Sci. 66: 157-166.

Mucosal acidification and hyperosmolarity differentially affect the barrier function of the isolated ovine ruminal epithelia

G.B. Penner[1], J.R. Aschenbach[2], G. Gäbel[2] and M. Oba[1]
[1]Department of Agricultural, Food and Nutritional Science, University of Alberta, Edmonton, Alberta, T6G 2P5, Canada, [2]Institute for Veterinary Physiology, University of Leipzig, An den Tierkliniken 7, 04103 Leipzig, Germany; gpenner@ualberta.ca

Introduction

Sub-acute ruminal acidosis (SARA) is a persisting digestive disorder in ruminant production that is characterised by temporal decreases in ruminal pH below pH 5.8 (Penner *et al.*, 2007). Accompanying the decrease in ruminal pH is increased osmolarity which may cause increased water inflow into the rumen (Gaebel *et al.*, 1987) and decreased barrier function (Schweigel *et al.*, 2005). Current data demonstrating the effect of SARA on the barrier function of ruminal epithelia are limiting. The present study aimed to determine the effect of a single episode of SARA in vivo on the barrier function of the isolated ovine ruminal epithelia.

Material and methods

Twenty-four German Merino sheep (72.3±10.1 kg of BW) were fed an all-hay diet and were assigned to either the glucose (GLU; n=17) or sham (SHAM, n=7) treatment. The GLU sheep were orally dosed with a 2.2 M glucose solution to supply 5 g glucose/kg BW; whereas, SHAM sheep received an equal volume of water. Ruminal pH was measured for 3 h after the oral dose. Sheep were then euthanised and ruminal epithelia from the ventral sac were collected.
Epithelia were mounted in Ussing chambers under short-circuit conditions. The mucosal (pH 6.1) and serosal (pH 7.4) solutions contained (mmol/l) 15.6 NaCl, 5.5 KCl, 1.3 MgCl, 0.6 NaH_2PO_4, 2.4 Na_2HPO_4, 1.0 L-glutamine, 10.0 HEPES free acid, 24.0 $NaHCO_3$, 5.0 Na-D/L-lactate, 10.0 Na-acetate, 10.0 Na-propionate, 10.0 butyric acid, 120.0 mannitol, and 10 NaOH. Radioactivity (^{3}-H mannitol, 100 kBq/15 ml) was added 20 min after mounting and 40 min were allowed for isotope equilibration. Epithelia were exposed to three 1-h periods to measure the serosal-to-mucosal mannitol flux (SM), as an indicator of barrier function (Lodemann and Martens, 2005). The flux periods consisted of pre-challenge, challenge, and recovery periods and were interspersed by 30-min periods for treatment equilibration. During the challenge period, epithelia were exposed to either an acidotic challenge by adding 150 ml of 1.5 mol/l gluconic acid (final pH 5.2, and osmolarity 293 mOsm/l), or an osmotic challenge through the addition of 0.405 g of mannitol (final pH 6.1 and osmolarity 450 mOsm/l), while a third group served as control (final pH 6.1 and osmolarity 293 mOsm/l). The mucosal buffer solution was replaced for the recovery period.
Data were analysed using the Proc Mixed procedure of SAS® (version 9.1.3, SAS Institute Inc.) as a split-plot design. The model included the fixed effects of in vivo treatment, *in vitro* treatment, block, and *in vitro* measurement period. The sheep nested within block × *in vivo* treatment was included as a random effect. Further, the model included *in vitro* measurement period as a repeated measure. Significance was declared when $P<0.05$.

Results and discussion

Sheep receiving GLUC had lower mean pH (6.67±0.09 and 5.77±0.06; $P<0.001$), and increased duration (111.3±11.2 vs. 0.4±17.4 min/180 min; $P<0.001$) and area under the curve (26.7±4.9 vs. 0.04±7.6 pH × min/180 min; $P=0.007$) for pH below 5.8 compared to SHAM sheep, indicating the successful induction of SARA with the oral glucose challenge.

The SM flux rate of mannitol and the tissue conductance were not affected by in vivo treatment (P>0.05). However, interactions between the *in vitro* treatment and measurement period were detected for the SM flux, and tissue conductance (Table 1). Collectively these data demonstrate that low mucosal pH does not affect epithelial barrier function during the challenge but increases permeability post-challenge. In contrast, hyper-osmolarity increased the permeability of the ruminal epithelia during the challenge period; however, the SM mannitol flux and tissue conductance were not different from the control during the recovery period.

Table 1. Interactions between in vitro treatment and in vitro period for the serosal-to-mucosal mannitol flux rate and tissue conductance (G_T).

		Pre-challenge	Challenge	Recovery	SEM
Mannitol flux, μmol/cm²/h	Control	0.64^{d}	0.94^{c}	1.04^{bc}	0.07
	Acidotic	0.63^{d}	0.99^{bc}	1.40^{a}	
	Osmotic	0.66^{d}	1.14^{b}	1.08^{b}	
G_T, mS/cm²	Control	2.07^{cd}	2.29^{cd}	2.42^{bc}	0.19
	Acidotic	2.00^{d}	2.03^{d}	3.51^{a}	
	Osmotic	2.04^{d}	2.72^{b}	2.33^{cd}	

[a,b,c,d] Means within a dependent variable with different superscripts differ significantly ($P < 0.05$).

Conclusion

The results of this study indicate that a single exposure to SARA in vivo does not have persistent effects on the barrier function of the isolated ruminal epithelia. However, a more severe acidotic challenge as imposed *in vitro* (pH 5.2) may have persistent effects. In contrast to the acidotic challenge, effects of increased osmolarity appear to be fully reversible up to a luminal osmolarity of 450 mOsm/L.

References

Gaebel, G., H. Martens, M. Suendermann and P. Galfi, 1987. The effect of diet, intraruminal pH and osmolarity on sodium, chloride and magnesium absorption from the temporarily isolated and washed reticulo-rumen of sheep. Quat. J. Exp. Physiol. 72: 501-511.

Lodemann, U. and H. Martens. 2006. Effects of diet and osmotic pressure on Na+ transport and tissue conductance of sheep isolated rumen epithelium. Exp. Physiol. 91: 539-550.

Penner, G.B., K.A. Beauchemin and T. Mutsvangwa, 2007. The severity of ruminal acidosis in primiparous Holstein cows during the periparturient period. J. Dairy Sci. 90: 365-375.

Schweigel, M., M. Freyer, S. Leclercq, B. Etschmann, U. Lodemann, A. Böttcher and H. Martens, 2005. Luminal hyperosmolarity decreases Na transport and impairs barrier function of sheep rumen epithelium. J. Comp. Physiol. B. 175: 575-591.

Effect of supplemental yeast (*Saccharomyces cerevisiae*) and fat level on feed intake and nutrient digestion in beef cattle

W. Polviset[1], *C. Yuangklang*[2], *C. Wachirapakorn*[3] *and S. Chumpawadee*[4]
[1]*Faculty of Agricultural and Technology, Rajabhat Mahasarakham University Thailand;* [2]*Department of Animal Science, Faculty of Natural Resources, Rajamangala University of Technology Isan, Sakon Nakhon Campus, Phangkhon, Sakon Nakhon, 47160, Thailand;* [3]*Department of Animal Science, Faculty of Agriculture, Khon Kaen University, Khon Kaen 40002, Khon Kaen, Thailand;* [4]*Division of Animal Science Faculty of Veterinary and Animal Sciences, Mahasarakham University, Mahasarakham, 44000, Thailand; polviset@hotmail.com*

Introduction

Currently, the use of antibiotics in animal production is attracting much interest. Alternative use of feed additives for ruminants especially *Saccharomyces cerevisiae* have been utilised as an antimicrobial feed additive for over ten years (Lynch and Martin, 2002). *Saccharomyces cerevisiae* increases feed efficiency, rumen pH and total volatile fatty acids (Quigley, 1992). Supplemental fat is typically used as an energy source. However, it has also been reported that high fat diet reduces fibre digestion. The explanation is that a high fat diet decreases the number of cellulolytic bacteria in the rumen. Currently, yeast supplementation has been promisingly used in beef production to improve fibre digestion by stimulating cellulolytic bacteria in the rumen, so that supplemental yeast would counteract the inhibitory effect of high fat on fibre digestion in beef cattle. Our hypothesis was that supplemental yeast (*Saccharomyces cerevisiae*) would reduce the inhibitory effect of a high fat diet on fibre digestion in beef cattle.

Material and methods

Six crosbred Brahman steers were used in a replicate 3×3 Latin square design. The dietary treatments were low fat diet and high fat with or without yeast supplementation. Steers were housed in individual pens (8 m^2). Water and mineral blocks were available at all times. During each period, the animals received 1.5%BW concentrate and 2.0% ruzi hay (*Brachiaria ruzizeinsis*) as a roughage source. Steers were fed twice a day at 7:00 and 15:00 h. Each experimental period lasted for 28 days: 21 days for feed intake measurements following by the last 7 days for sample collection. During the last five days of each period, faeces sample were quantitatively collected. Rumen fluid was obtained via a stomach tube from each cattle approximately 2 h after the morning feeding. Rumen fluid samples were immediately measured for pH and filtered through four layers of cheesecloth. Feed and faeces samples were dried at 60 °C for 72 h and ground and analysed for dry matter (DM), crude protein (CP) and ash by the method of AOAC (1984). Neutral detergent fibre (NDF), Acid detergent fibre (ADF) and Acid detergent lignin (ADL) were measured by the method of Goering and Van Soest (1970). Digestibility of nutrients was calculated as nutrient intake – nutrient in faeces/nutrient intake × 100 (Schneider and Flatt, 1975).

Results and discussion

Ruzi hay intake was not significantly different (P>0.05) among treatments. There were no significant differences in nutrient digestion. Yeast supplementation has been reported to improve nutrient digestion, particularly fibre digestion. In this trial, however, supplemental yeast in a high fat diet did not improve fibre digestion when compared with no yeast supplementation. It might be possible that ruzi hay is a good quality roughage source. The proportions of acetate, propionate and butyrate were not significantly different (P>0.05) among treatments. Propionate tended (P=0.94) to increase when yeast was added to a high fat diet. Thus, the experimental data did not agree with our hypothesis.

Table 1. Effect on feed intake, nutrient digestion, rumen pH and volatile fatty acids.

Item	Fat 3%	Fat 6%	Fat 6%+yeast	SEM
Ruzi hay intake,% BW	1.14	1.13	1.09	0.07
Concentrate intake,% BW	1.50	1.50	1.50	-
Total intake,% BW	2.64	2.63	2.59	0.13
Digestion coefficient,%				
DM	77.13	77.60	77.84	1.60
CP	84.76	86.66	85.10	1.26
EE	92.97	89.62	88.21	3.62
NDF	72.24	68.97	68.22	2.23
ADF	62.21	64.26	58.44	4.08
Rumen pH	6.73	7.00	6.75	0.10
Volatile fatty acids,				
Acetate,%	62.28	55.42	61.18	4.66
Propionate,%	25.95	31.92	33.54	4.02
Butyrate,%	11.35 [a,b]	12.67 [a]	9.66 [b]	0.71

[a,b] P<0.05.

Conclusion

Very few significant effects on rumen fermentation and fibre digestion were observed in this experiment.

Acknowledgement

The authors would like to sincerely thank Rajabhat Mahasarakham University for financial and facilities supports.

References

AOAC, 1984. Offical Methods of Analysis. 14th ed. The Association of Official Analytical Chemists, Washington, D.C., USA.

Goering, H.K. and P.J. Van Soeast, 1970. Forage Fiber Analyses (apparatus,reagents,Procedures and some applications). Agric.Handbook No.379. Washington, D.C.ARS, USDA, USA.

Lynch, H.A. and S.A. Martin, 2002. Effects of *Saccharomyces cerevisiae* culture and *Saccharomyces cerevisiae* live cells on *in vitro* mixed ruminal microorganism fermantation. J. Dairy Sci. 85: 2603-2608.

Quigley, J.D., L.B. Wallis, H.H. Dowlen and R.N. Heitmann,1992. Sodium bicarbonate and yeast culture effects on ruminalfermentation, growth, and intake in dairy calves. J. Dairy Sci. 75: 3531-3538.

Schneider, B.H. and W.P. Flatt, 1975. The Evaluation of Feed through digestibility Experiment. Athens. The Univ. of Gorgia Press., Gorgia, U.S.A.

Methanogenesis kinetics and fermentation patterns in the rumen of sheep with or without protozoa

M. Popova, C. Martin, Y. Rochette, D. Graviou and D.P. Morgavi
INRA, UR1213 Herbivores, Site de Theix, 63122 Saint-Genès-Champanelle, France; diego.morgavi@clermont.inra.fr

Introduction

Methane (CH_4) is an important greenhouse gas that has a global warming potential 21 times higher than carbon dioxide. Livestock, and ruminants in particular, have been identified as the largest source of anthropogenic CH_4 worldwide. In addition, CH_4 production by ruminants represents an energy loss to the animal. Therefore mitigating CH_4 emissions from livestock has economic and environmental benefits. Some of the current approaches aiming to reduce enteric CH_4 emissions from ruminants consist in improving nutrition and/or manipulating the rumen microbiota. Thus, diet composition, feeding frequency, level of intake or elimination of protozoa from the rumen (termed defaunation) were shown to influence rumen fermentation patterns and CH_4 production (Iqbal *et al.*, 2008). The effect of these factors on CH_4 production has been described, but the dynamics of methanogenesis throughout the day is still poorly understood. The aim of the present work was to monitor the rate of CH_4 production during the day in sheep harbouring or not protozoa. Changes in microbial population dynamics will also be studied to understand the relationship with the extent and kinetics of CH_4 production.

Material and methods

Four rumen-fistulated Texel wethers (61.5±1.3 kg) were used in this study. They were fitted with rumen gas cannulae allowing frequent gas sampling. Animals were fed once daily (0800 h) a diet composed of 700 g alfalfa pellet, 200 g prairie hay and 300 g cracked corn grain. The study consisted of two experimental periods: 'faunated' and 'defaunated', each lasting 4 weeks: 2 weeks of adaptation to the diet, 1 week of gas sampling and 1 week of rumen content sampling. After the first 'faunated' period, animals were defaunated by rumen emptying and washing.

Kinetics and total CH_4 production were determined using the sulfure hexafluoride (SF_6) tracer technique. For kinetics measures, gas samples were taken on 5 consecutive days over a 14-h period, every hour from 08:00 to 12:00 h and then every two hours up to 22:00 h. In order to quantify overall CH_4 production over the 14-h period, representative breath samples from each animal were sampled into pre-evacuated PVC collection devices as described by Pinares-Patiño *et al.* (2003). All gas samples were analysed by gas chromatography.

Rumen content samples were taken at 08:00, 10:00, 13:00 and 18:00 h, on three non-consecutive days. Volatile fatty acid (VFA) concentration of rumen fluid was analysed by gas chromatography. Data were analysed using the MIXED Procedure of SAS® v9. Kinetics data were analysed in repeated time with sheep, treatment (Tr), sampling day (Sd), sampling time (St), Sd(Tr) and Tr*St interaction as fixed effects. For total CH_4 production data, the model included sheep (S), treatment (Tr), day (D), D(Tr) and S*Tr interaction as fixed effects.

Results

The absence of protozoa reduced CH_4 production daily from 31.6 to 28.4 l ($P<0.05$) as compared to the same wethers before defaunation. Total VFA and propionate concentrations increased in wethers without protozoa (from 91.0 to 99.5 mM and from 15.8 to 19.8 mM, respectively $P<0.05$), whereas those of acetate and butyrate remained the same (65.7 and 9.0 mM on average, respectively). The acetate/propionate ratio decreased from 4.04 to 3.57 in wethers with and without protozoa respectively ($P<0.001$). The presence/absence of protozoa did not influence rumen pH (6.2 on average).

The kinetics of CH_4 production and VFA concentration are shown in Figure 1. The curves remained the same independently of the presence or not of protozoa ($P>0.05$). Methane production peaked 1 h after feeding with a three-fold increase as compared to production before feeding. Hourly emissions slowly decreased afterwards reaching the initial level 14 h later. For total VFA concentration, a two-fold increase was observed 2 h after feeding.

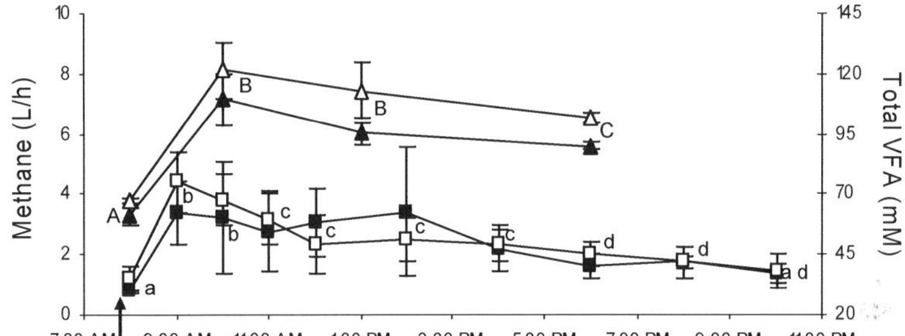

Figure 1. Kinetics of CH_4 production (□,■) and of volatile fatty acids concentration (△,▲) in sheep with (filled symbols) and without (open symbols) protozoa. Capital and lower case letters are common for VFA and CH_4 curves respectively. Time points with a different letter, within the same curve, are significantly different (P<0.05). Arrow indicates time of feeding.

Conclusion

This study confirmed that methane production in sheep decreases in the absence of protozoa, whereas the kinetics of CH_4 production and total VFA concentration remain unchanged. In the absence of protozoa, rumen fermentation was oriented towards the production of propionate. Further work is in progress to monitor microbial community structure and dynamics in order to make the link to CH_4 production kinetics. The comprehension of this relationship seems necessary to devise methane-abatement strategies based on feeding.

Acknowledgement

The authors thank M. Fabre and S. Alcouffe for taking care of animals and their help in sampling.

References

Iqbal, M., Y.-F. Cheng, W.-Y. Zhu and B. Zeshan, 2008. Mitigation of ruminant methane production: current strategies, constraints and future options. World Journal of Microbiology and Biotechnology 24: 2747-2755.

Pinares-Patiño, C.S., R. Baumont and C. Martin, 2003. Methane emissions by Charolais cows grazing a monospecific pasture of timothy a four stages of maturity. Can. J. Anim. Sci. 83: 767-777.

Methodological aspects of quantitative analysis of ruminant faeces for adenosine triphosphate by the firefly luciferin-luciferase system

M. Predotova, A. Sundrum, R.G. Joergensen and E. Schlecht
Faculty of Organic Agricultural Sciences, University of Kassel, Steinstrasse 19, 37213 Witzenhausen, Germany; tropanimals@uni-kassel.de

Introduction

Microbial activity in ruminant faeces receives little attention so far, although it is an indicator of hindgut fibre degradation and may serve as an indicator for the speed of nutrient release from excreta. The adenylates ATP, ADP and AMP and the adenylate energy charge (AEC) are determined as quantitative measures of microbial activity by established soil microbiological approaches. These often rely on ion-paired-reverse-phase HPLC (Dyckmans and Raubuch, 1997) which is sensitive to disturbance from sample preparation. Therefore a more robust method was sought for analysing ATP in ruminant faeces.

Material and methods

Faecal samples (30 g) from 3 *Bubalus bubalis* heifers (503±60.3 kg) fed 62% Rhodes grass hay, 35% corn and 3% soy bean meal (intake 4 kg DM/d) were used for a duplicate experiment repeated within 6 d. Each time, 5 frozen sample replicates of 0.5±0.05 g were thawed at room temperature (3 h). In analogy to Tate and Jenkinson (1982) a 2 l aq. solution containing 11.7 g EDTA (20 mM), 11.2 g KOH (0.1 M) and 7.6 g Na_3PO_4 (10 mM) was prepared. Thirteen ml H_2O demin. were added to 1 l of this solution (extractant-A), and 1 l was mixed with 1.63 ml 0.1 mM ATP (spike-solution) and 11.37 ml H_20 (extractant-B). Tris-buffers (-I, -II, -III) were prepared following the Rothamsted Protocol (2004) but adjusting pH to 7.75 instead of 7.5. Luciferin (2 mg) were dissolved in 5 ml iced Tris-I and stored on ice (light-protected) for 4 h max. Twenty-five µl of luciferase solution (1 mg luciferase in 1 ml Tris-I) were added to 5 ml iced Tris-III and stored on ice. A matrix was prepared of 20 ml dimethylsulfoxide (DMSO), 80 ml extractant-A and 100 ml BC-buffer; the latter consisted of 250 µl benzalkoniumchloride solution mixed with 249.75 ml Tris-II. Fifty ml of this matrix were mixed with 1 ml spike solution to yield a concentration of 2 pmol ATP/µl. From this, standards containing 0, 0.02, 0.04 and up to 0.4 pmol ATP/µl were prepared. Instantly before sample reading, 4 ml DMSO were added to the thawed replicates. The mixture was stirred (120 s, 150 rmp) and 16 ml of extractant-A or -B were added to 5 replicates each and stirred again (120 s). Suspensions were ultratrasonified for 120 s to disrupt microbial cells, and 0.5 ml of the suspension were then added to 0.5 ml BC-buffer. After ultratrasonification (5 s), 5 µl of the suspension were pipetted into a vial containing 0.5 ml Tris-II and 50 µl of the luciferase-Tris-III solution. The vial was inserted into a Promega luminometer (setting: kinetics off, frequency 1/1) where 100 µl luciferin were routinely added. The luminescence was taken after a lag phase of 2 s and an integration time of 10 s. Before and after a measuring cycle all standards were read; after every 5th sample a randomly taken standard was also measured. Background readings originated from three blanks (extractant-B, DMSO, BC-buffer solution).

The difference in luminescence between 'sample×extractant-A' and 'sample×extractant-B' was divided by the luminescence of the blanks. Correcting for sample weight and water content, sample ATP concentration (nmol/g DM) was subjected to analysis of variance (SAS®, 2000) to determine the influence of animal and day.

Results and discussion

The four start- and end-calibration curves (Figure 1) obtained on the two days yielded correlation coefficients of $R^2 \geq 0.97$. However, R^2 values were higher for the range of ATP concentration tested here (0 to 0.4 pmol ATP/μl) than when standards of up to 2 pmol ATP/μl were included in the calibration. For faeces of a higher ATP concentration than those used here, the number of standards needed and the luminometer settings should therefore be tested first so as to obtain linearity of the calibration curve beyond 120,000 relative units of luminescence.

Differences in replicate mean ATP concentrations between days 1 and 2 were ≤12% of the mean across all 10 replicates (P=0.36), while the coefficient of variation (CV) within 5 replicates measured on one day was ≤25% (Figure 2). ATP concentrations differed between animals (P<0.001), while sample x day interactions were insignificant (P=0.92). Since the method detects differences in faecal ATP between identically fed individuals, it should allow detecting ration-specific ATP concentrations. This could be used in quality assessments of ruminant dung. Unfavourably, the enzymatic test in comparison to the HPLC method only assesses ATP but not ADP and AMP. On the contrary, our own HPLC analysis of frozen and dewed replicates (n=6) of buffalo faeces yielded a CV of 108% for ATP while CV for ADP (54%) and AMP (78%) were clearly reduced (data not shown).

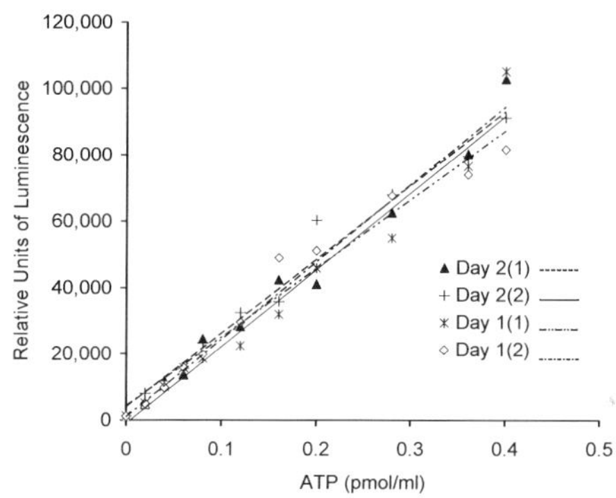

Figure 1. Four calibration curves showing the correlation between ATP concentration in standard solutions (x) and relative units of luminescence (y) from luciferin decay.

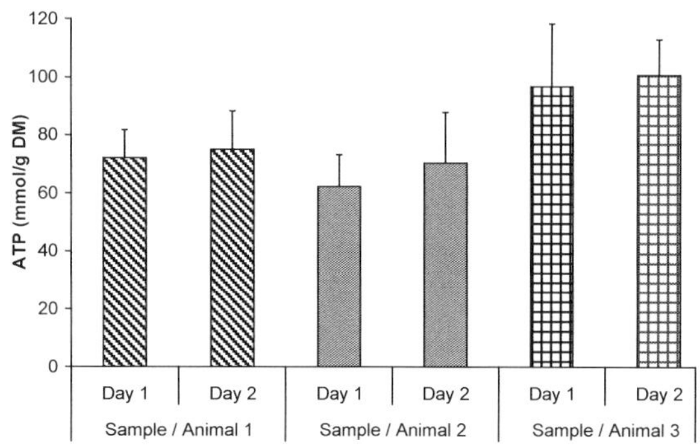

Figure 2. ATP concentration in three faecal samples measured by the luciferin-luciferase method in 5 replicates on two different days (Means and SD).

Conclusion and outlook

The luciferin-luciferase method used for soils can be adjusted for a repeatable determination of ATP in ruminant faeces. Further tests should address (1) sample storage prior to analysis (fresh / refrigerated / frozen / shock-frozen), (2) storage duration (days / weeks / months) and (3) sample processing (way and duration of dewing, homogenisation).

Acknowledgement

We are grateful to Mrs. G. Dormann, Department of Soil Biology and Plant Nutrition, Kassel University, and Mr. J.J. Melapie for substantial support with the analytics and thank Deutsche Forschungsgemeinschaft (Research Training Group 1397) for funding.

References

Dyckmans, J. and M. Raubuch, 1997. A modification of a method to determine adenosine nucleotides in forest organic layers and mineral soils by ion-paired reversed-phase high-performance liquid chromatography. J. Microbiol. Methods 30: 13-20.

Rothamsted Protocol, 2004. Soil Microbial Biomass Research Group Protocol; http://www.rothamsted.bbsrc.ac.uk/aen/smbweb1/methods.php?id=1042.

Tate, K.R. and D.S. Jenkinson, 1982. Adenosine triphosphate measurements in soil: an improved method. Soil Biol. Biochem. 14: 331-335.

Effects of increasing dietary protein on intake and total tract apparent digestibility in dairy crossbred heifers

M.F.S. Queiroz, T.T. Berchielli and R.D. Signoretti
UNESP, Faculdade de Ciências Agrárias e Veterinárias, Jaboticabal, SP, Brazil; mfernanda_queiroz@yahoo.com.br

Introduction

Dairy replacements are the foundation of any dairy enterprise. In Brazil, dairy replacements have an important participation in the cost of this business. Improvement of a herd is possible when older cows are replaced by well fed, healthy and genetically superior cows. Dairy replacement heifers require adequate amounts of dietary protein to support growth (Hoffman *et al.*, 2001) and must be healthy so they can be bred at 15 and 18 mo of age and maintain their pregnancy.
Most forage can be offered to heifers and a large percentage of farmers use pasture for this animal category throughout the year. Nevertheless, in dry parts of the year, the winter season in Brazil, there is a shortage of pasture for rearing heifers and sugarcane is a great option due to its high productivity in this season. Sugarcane alone cannot adequately provide nutritional requirements of heifers and they therefore need a concentrate supplementation in their diets. Gabler and Heinrichs (2003) found no difference on dry matter intake and apparent total tract digestibility evaluating four dietary crude protein (CP) levels for Holstein heifers and 16.7% CP was recommended for prepubertal heifers. Limited data are available for growing crossbred dairy heifers and this current study was aimed at evaluating the effect of increasing levels of dietary protein on intake and total tract apparent digestibility in dairy crossbred Holstein × Gir heifers.

Material and methods

Four dairy crossbred heifers Holstein × Gir, averaging 240±20 kg of body weight, and fitted with rumen cannulae were used to evaluate increasing levels of dietary protein intake (13, 15, 19 and 22% CP) on total tract apparent digestibility during July to November (winter/spring). The diet was made up of sugarcane (IAC 862480) and a concentrate with ground corn grain, soybean meal, urea, ammonia sulphate and mineral mix, in a 70:30 forage:concentrate ratio. Heifers were individually housed in pens with access to water at all times and *ad libitum* access to the diets (10% refusals). Offered and refused feed were quantified daily during the marker adaptation period to determine intake. Collection of faeces were in 2 d at hours 2, 5, 8 and 11 am and pm to determine total tract digestibility. They were randomly assigned to one of four treatment rations in a 4×4 Latin square design with 14 d periods, consisting of a 7 d adaptation period to diet, 5 d adaptation to marker (CrEDTA) and a 2 d faeces sampling period. Marker (CrEDTA) was infused directly in the rumen two times per day, 8 am and 5 pm, at the same time that heifers were individually fed. All data from the experiment were analysed as a 4×4 Latin square using the GLM procedure of SAS® (SAS Institute Inc., Cary, NC, USA). Overall differences between treatment means were declared significant at $P<0.05$.

Results

The dry matter (5.0 kg/d), organic matter (4.8 kg/d) and nonstructural carbohydrate (2.0 kg/d) intake presented a linear trend to increase whenever dietary protein level increased while carbohydrate intake (1.6 kg/d) was similar ($P>0.05$) among treatments. However, the crude protein intake as well as neutral digestible fibre (1.8 kg/d) were linearly different ($P<0.001$) among treatments. Protein digestibility was 26.8% lower at level 13% CP than the 22% dietary crude protein level and dry matter and organic matter total tract digestibility increased with an increasing protein content of the diet (Table 1).

Table 1. Nutrient intake and total tract digestibility by heifers fed increasing protein dietary levels.

	Dietary protein level (% CP)				CV[3]	Diet effect *P*-value		
	13	15	19	22		Linear	Quadratic	Cubic
Intake[1]								
DM, kg/d	4.8 [a]	4.8 [a]	5.2 [a]	5.3 [a]	10.2	0.1538	0.7450	0.6060
% BW[2]	2.0 [a]	2.0 [a]	2.1 [a]	2.2 [a]	12.8	0.4757	0.8563	0.8713
OM	4.6 [a]	4.6 [a]	5.0 [a]	5.1 [a]	10.6	0.1914	0.8141	0.5212
% BW	2.0 [a]	1.9 [a]	2.0 [a]	2.1 [a]	13.7	0.7256	0.6637	0.7844
CP	0.7 [b]	0.8 [b]	1.1 [a]	1.3 [a]	10.3	<0.0001	0.1843	0.1238
% BW	0.3 [c]	0.3 [c]	0.5 [b]	0.6 [a]	9.4	<0.0001	0.0972	0.1936
NDF	1.7 [a]	1.8 [a]	1.9 [a]	1.9 [a]	7.8	0.0293	0.7358	0.5504
% BW	0.7 [a]	0.7 [a]	0.8 [a]	0.8 [a]	13.8	0.4591	0.8090	0.9137
NSC	2.2 [a]	2.0 [a]	1.9 [a]	1.8 [a]	13.6	0.0742	0.7231	0.8735
% BW	0.9 [a]	0.9 [a]	0.8 [a]	0.7 [a]	18.2	0.0897	0.8725	0.9427
C	3.9 [a]	3.8 [a]	3.8 [a]	3.7 [a]	10.5	0.5562	1.0000	0.7054
% BW	1.6 [a]	1.6 [a]	1.6 [a]	1.5 [a]	14.1	0.4469	1.0000	0.9223
Digestibility (%)								
DM	47.3 [a]	50.1 [a]	57.8 [a]	59.0 [a]	13.3	0.0359	0.8425	0.4975
OM	49.3 [a]	51.6 [a]	59.3 [a]	60.3 [a]	12.6	0.0388	0.8521	0.4666
CP	55.7[b]	57.8[b]	70.5 [ab]	76.1 [a]	10.2	0.0033	0.2866	0.1268
C	53.0 [a]	52.4 [a]	58.1 [a]	56.8 [a]	15.8	0.1876	0.7764	0.6001

[a.,b,c] Superscripts that differ are significant at $P<0.05$.
[1] kg/DM; DM = dry matter; OM = organic matter; CP = crude protein; NDF = neutral digestible fibre; NSC = nonstructural carbohydrates; C = carbohydrates.
[2] BW = Body weight.
[3] CV = Coefficient of variation (%).

Conclusion

Increasing dietary protein levels on diets compounded by sugarcane and concentrate in a 70:30 forage:concentrate ratio increased dry matter, organic matter and crude protein total tract digestibility and showed a trend towards increased nutrient intake by dairy heifers.

Acknowledgement

The authors thank FAPESP and APTA-Colina milking workers and experimental staff.

References

Gabler, M.T. and A.J. Heinrichs, 2003. Effects of increasing dietary protein on nutrient utilization in heifers. J. Dairy Sci. 86: 2170-2177.

Hoffman, P.C., N.M. Esser, L.M. Bauman, S.L. Denzine, M. Engstrom and H. Chester-Jones, 2001. Short Communication: Effect of dietary protein on growth and nitrogen balance of Holstein heifers. J. Dairy Sci. 84: 843-847.

Modulation of Na transport by heat shock proteins in sheep rumen epithelium

I. Rabbani, U. Tietjen and H. Martens
Institute of Veterinary Physiology, Oertzenweg 19b, 14163, Freie Universität Berlin, Germany; martens.holger@vetmed.fu-berlin.de

Introduction

Subacute ruminal acidosis (SARA) results as a consequence of feeding large amounts of rapidly fermentable carbohydrates in conjunction with inadequate fibre and is characterised by high ruminal concentrations of short chain fatty acids (SCFA), periods of low ruminal pH, depressed feed intake and subsequent health problems (Kleen *et al.*, 2003). Low pH and high SCFA concentrations cause stress to the ruminal epithelium which has to cope with these challenges. It is known that heat shock proteins (Hsp), also called *stress proteins*, are induced when a cell undergoes various types of environmental stresses (Anckar and Sistonen, 2007). However, no study has been performed regarding the possible protective role of Hsp against low pH and high SCFA. It was therefore the aim of the present study to investigate the physiological significance of Hsp in the ruminal epithelium.

Material and methods

Ruminal tissues from hay fed sheep were obtained and cut into 3×3 cm pieces to be used in the Ussing chamber for the induction of Hsp by various stress factors like temperature, pH or osmotic stress, etc. A reduction in pH from 7.4 to 6.4 and an increase in SCFA concentration up to 70 mM were applied at the mucosal side to induce Hsp synthesis. Tissues were further processed for western blotting. After establishing a protocol for the induction of Hsp, further experiments were carried out to determine Na flux rates. Radioactive isotope (^{22}Na) was used for flux studies. Electrophysiological data were recorded at all times from a computer controlled voltage clamp device (AC Micro-Clamp, Aachen, Germany) and all the experiments were carried out under short circuit conditions. The data were analysed using Microsoft Excel and Sigma Plot 8.0

Results

The western blot revealed that Hsp70 were generally expressed in the rumen epithelium and were over expressed when the tissues were incubated at 70 mM SCFA, pH 6.4 or 380 mOsmol/l buffer solution on mucosal side (Table 1).
Induction of Hsp 70 is accompanied by a significant increase of electroneutral Na transport (J_{ms}Na and J_{net}Na; Table 2). Inhibition of Hsp synthesis by cycloheximide (CHX) significantly reduced Na transport. CHX treatment in control tissues did not influence basal Na transport rates.

Conclusion

Adaptation of the rumen epithelium to changes of the diet is well known (for details see Etschmann *et al.*, 2009) and includes morphological, biochemical and physiological alterations, which require days (transport properties) or even weeks (number and size of papillae) and are probably mediated by luminal factors as SCFA or hormones (Baldwin *et al.*, 2004). However, diurnal variations of concentrations of SCFA, pH, temperature and osmotic pressure are remarkable and may interfere with the integrity and function of the rumen epithelium (Gaebel *et al.*, 1989; Schweigel *et al.*, 2005). The present study tested the hypothesis that heat shock proteins may be involved in short term adaptation of the rumen epithelium. The data obtained clearly show that Hsps are acutely induced by changes of usual rumen parameters (pH, SCFA, temperature and osmotic pressure). Induction of Hsp 70 is accompanied by an increase of electroneutral Na transport via Na^+/H^+ exchange (NHE) which recycles H^+ taken up by non-ionic diffusion of undissociated SCFA (HSCFA) into the epithelium. The increase of NHE activity can be considered as a first line of defence against rapid HSCFA and hence H^+ uptake.

Table 1. Expression of Hsp as integrated density values (IDV) $\times 10^3$ with SCFA, pH stress and high osmotic pressure.

Integrated Density Values (IDV)					
Control	15.36±1.49	pH 7.4	28.17±5.10	Control	6.70±0.54
70 mM SCFA	51.52±1.31*	pH 6.4	65.71±2.22*	High osmolarity	13.25±0.85*

* Describes the significance at $P<0.05$.

Table 2. Interaction of Hsp with Na transport in isolated rumen epithelium of sheep.

Treatment	J^{Na}_{ms} (μeq/cm²/h)	J^{Na}_{sm} (μeq/cm²/h)	J^{Na}_{net} (μeq/cm²/h)	I_{sc} (μeq/cm²/h)	Hsp: IDV×10^3	N/n
Control	4.61±0.50	1.16±0.32	3.11±0.62	1.25±0.17	5.07±0.50	3/9
Challenged	9.32±0.74*	1.63±0.38	7.70±0.82*	1.10±0.33	27.12±1.85*	3/9
Control CHX	4.04±1.17	1.69±0.38	2.35±1.27	1.69±0.14	7.76±1.07	3/9
Challenged CHX	6.32±0.88*	1.49±0.39	4.83±1.02*	1.31±0.23	9.50±0.78*	3/9

* Describes the significance (P<0.05); N = Number of animals; n = Number of epithelia.

Acknowledgement

This study was supported by a scholarship of the Government of Pakistan and DAAD. The running costs were provided by a grant of the MMC.

References

Anckar, J. and L. Sistonen, 2007. Heat shock factor 1 as a coordinator of stress and developmental pathways. Adv. Exp. Med. Biol. 594: 78-88.

Baldwin, R.L., K.R. McLeod and A.V. Capuco, 2004. Visceral tissue growth and proliferation during the bovine lactation cycle. J. Dairy Sci. 87: 2977-2986.

Etschmann, B., A. Suplie and H. Martens, 2009. Change of ruminal Na transport in sheep during dietary adaptation. Arch. Anim. Nutr. 63: 26-38.

Gaebel, G., M. Bell and H. Martens, 1989. The effect of low mucosal pH on sodium and chloride movement across the isolated rumen mucosa of sheep. Q. J. Exp. Physiol. 74: 35-44.

Kleen, J.L., G. A. Hooijer, J. Rehage and J.P. Noordhuizen, 2003. Subacute ruminal acidosis (SARA): A review. J. Vet. Med. A. Physiol. Pathol. Clin. Med. 50: 406-414.

Schweigel, M., M. Freyer, S. Leclercq, B. Etschmann, U. Lodemann, A. Bottcher and H. Martens, 2005. Luminal hyperosmolarity decreases Na transport and impairs barrier function of sheep rumen epithelium. J. Comp. Physiol. [B] 175: 575-591.

Electrogenic transport of SCFA anions in sheep rumen epithelium

R. Rackwitz, J.R. Aschenbach, P. Philipp and G. Gäbel
Institute of Veterinary Physiology, An den Tierkliniken 7, 04103 Leipzig, Germany; rackwitz@vetmed.uni-leipzig.de

Introduction

In the rumen, huge amounts of short-chain fatty acids (SCFA) are produced by microbial fermentation. For the uptake of these high-energy substrates, sufficient transport mechanisms are necessary. In their protonated form, SCFA are able to pass biological membranes by lipophilic diffusion. However, at the physiological pH in the rumen, SCFA are mainly available in their dissociated form. The transport of these SCFA anions ($SCFA^-$) is particularly mediated by electrically silent mechanisms. But electrophysiological data suggest that electrogenic mechanisms are also involved. The aim of this study was to enlighten this latter aspect of $SCFA^-$ transport.

Material and methods

Stripped sheep rumen epithelia were mounted in Ussing chambers with an exposed area of 3.14 cm^2. On both sides, a chloride- and bicarbonate-free buffer solution adjusted to pH 7.4 and gassed with oxygen was used. After equilibration under open-circuit conditions, the transepithelial potential difference (PD_t) was clamped to either -40 or +40 mV with the mucosal side as reference. Before measuring acetate uptake at one side of the epithelium, the buffer at the contralateral side was exchanged with a solution containing 100 mM potassium and 0.5 μM nigericin.
Uptake measurements were conducted by addition of 40 mM acetate to either the mucosal or the serosal side. Five minutes after acetate application, ^3H-labelled acetate was added and samples of the labelled buffer solution were taken. One minute later the uptake was stopped by washing the epithelia three times with ice-cold buffer solution. The epithelia were dismounted and lysed in 0.1 N NaOH. Samples from the buffer solution and lysate were measured in a scintillation counter. Uptake (U) was calculated by a simple ratio equation and corrected for protein content of the epithelia. Following Stumpff *et al.* (2009), we used 1 mM diisothiocyanatostilbene 2,2'-disulfonic acid (DIDS) or 1 mM p-hydroxymercuribenzoic acid (pHMB) as anion channel blockers. The blockers were added simultaneously with acetate on the uptake side.

Results

Without potassium treatment of the contralateral side, mucosal and serosal uptakes of 40 mM acetate showed a slight but not statistically significant PD_t dependence (data not shown). After depolarisation of the contralateral side by 100 mM potassium and 0.5 μM nigericin, mucosal acetate uptake was again not affected by PD_t. In contrast, the serosal acetate uptake showed significantly higher values at -40 mV PD_t than at +40 mV PD_t (Figure 1). DIDS (1 mM) affected neither the mucosal nor the serosal uptake. Addition of 1 mM pHMB had no effect on mucosal uptake. On the serosal side, however, 1 mM pHMB reduced the uptake at -40 mV PD_t to the level of the uptake at +40 mV PD_t (data not shown). Under all conditions tested, serosal uptake was approximately twice as high as mucosal uptake (Figure 1).

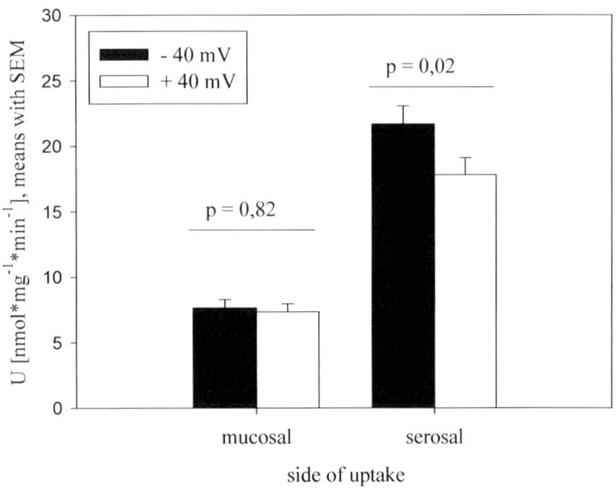

Figure 1. Mucosal and serosal uptake (U) of 40 mM acetate at transepithelial potential differences of -40 mV (black bars) and +40 mV (white bars) after depolarisation of the contralateral side by 100 mM potassium and 0.5 μM nigericin. Means with SEM, n=9, paired t-test.

Conclusion

The PD_t dependence of acetate uptake supports the idea that an electrogenic mechanism is involved in $SCFA^-$ transport. Our data suggest the existence of a basolateral located, acetate-permeable anion channel in sheep rumen epithelium. Driven by serosal membrane potential this conductance may present a supporting pathway for $SCFA^-$ efflux.

Acknowledgement

The authors thank the 'Deutsche Forschungsgemeinschaft' (DFG) for financial support.

References

Stumpff, F., H. Martens, S. Bilk, J.R. Aschenbach and G. Gäbel, 2009. Cultured ruminal epithelial cells express a large-conductance channel permeable to chloride, bicarbonate, and acetate. Pflugers Arch – Eur. J. Physiol. 457: 1003-1022.

Effects of the ratio of nonfibre carbohydrates to rumen degradable protein on feed intake and digestibility in mid-lactation Holstein cows

H. Rafiee
Departments of Animal Science, University of Tehran, Aboureihan Campus, Tehran, Iran; harafiee@yahoo.com

Introduction

Recent increases in feed costs have led livestock producers to seek diet formulations which optimise microbial synthesis in the rumen and nutrient supplies to the ruminants. Energy and rumen degradable protein (RDP) supply in an appropriate time and amounts are means to modify the efficiency and synthesis of microbial mass (Hoover and Stokes, 1991). Currently, nonfibre carbohydrate (NFC) is used as a source of energy in dairy diets. However, dietary protein is changed by replacing high-protein supplements with cereal grains, sources of NFC. Consequently, in these experiments the amount and form of N and carbohydrate available in the rumen and intestines might have differed between treatments. Therefore, it remains unclear whether the reported effects on cow performance were achieved by alterations in rumen metabolism or postruminal supply of N and carbohydrates (Ipharraguerre and Clark, 2005). Although numerous experiments have been conducted to optimise the energy and nitrogen utilisation in dairy cows, little information is available when low quality nitrogen sources, e.g. urea, and rapidly degradable carbohydrate source, e.g. barley grain, are used in low producing cows. Our hypothesis was that finding an optimum NFC: RDP ratio in diets of mid-lactating dairy cows would be valuable particularly when diets are balanced with an emphasis on minimising costs and maximising rumen fermentation.

Material and methods

Nine multiparous Holstein cows (171±17 DIM and 24.1±3.3 kg/d of milk) were used in a 3×3 Latin square design with 21-d periods with 14 d of adaptation and 7 d for sampling and data collection. Based on the NRC (2001) recommendations, 3 isocaloric total mixed rations (TMR) were prepared for cows. Diets were composed of fixed level of NFC (40% of DM) and ground barley was the sole source of cereal. Expeller Cottonseed meal was used as the main protein source in the concentrate portion and percentage of RDP was increased through urea supplementation. Three diets were tested differing by RDP% and NFC: RDP ratio. They were 9.8 and 4.1 (diet 1), 10.8 and 3.7 (diet 2), and 11.8 and 3.3 (diet 3). Feed intake and orts were measured and sampled daily for each cow during the collection period. Food and orts samples were analysed for ash and CP (Kjeldahl N × 6.25). NFC calculated as 100 − CP − NDF − ash − ether extract (NRC, 2001). Fecal grab samples were taken 4 h after feeding and analysed for DM, OM and CP and AIA. Digestibility of diets was measured by acid-insoluble ash (AIA) method for a collection period of 4 d (Van keulen and Young, 1977). Milk yield was determined on 7 consecutive days. Data were analysed using Proc Mixed of SAS® (SAS®, 2002). The model included square, period, and diet. Cow within square was the term of the RANDOM statement. Values reported as least squares means. Diet effect (i.e. different NFC: RDP ratios) was partitioned into linear and quadratic contrasts. Significance was declared at $P \leq 0.05$, and a trends was noted if $0.05 < P \leq 0.10$.

Results

Data are presented in Table 1. The DMI and OMI decreased as the ratio of NFC: RDP decreased and this decrease is linear for OMI. As expected, intake of CP increased linearly ($P<0.01$) from 2.64 to 3.05 kg/d for diets 1 to 3. Milk yields ($P=0.36$) were similar among treatments. The proportion of DM, OM and CP apparently digested in the total tract were unaffected by dietary NFC: RDP ratio. There was a trend (P=0.06) towards lower amounts (Kg/d) of DM apparently digested in the total tract in diet 3. Treatments had significant effect on OM digestibility (Kg/d) (P<0.05), and diet 3 had lower digestibility versus diets 1 and 2. The same pattern was observed for CP digestibility where diet 3 was of lower CP digestibility when compared to other diets, though treatments were only numerically different (P>0.05). The reasons of differences between digestibilities (kg/d) were disparity in DMI, OMI and percentage of digestibility. These variations could be reported to variations in intake, lower for DM and OM and higher for CP. Totally, the apparent digestibility of diet 2 was higher from other diets and the NFC: RDP ratio equal to 3.7 was better than that of other ratios.

Table 1. Effect of dietary NFC: RDP ratio on intake and digestibility.

Items	Dietary treatments[1]			SE	P-value		
	1	2	3		Diet	Linear	Quadratic
DM Intake, kg/d	21.83[a]	21.29[b]	21.08[b]	0.59	<0.01	<0.10	0.74
ADTT[2],%	69.8	70.4	67.7	1.16	0.24	0.20	0.26
ADTT, kg/d	15.06	15.23	14.25	0.55	<0.10	0.30	0.39
OM Intake, kg/d	19.79[a]	19.30[b]	19.00[b]	0.53	<0.01	<0.05	0.87
ADTT,%	67.7	68.2	65.5	1.15	0.24	0.19	0.27
ADTT, kg/d	13.24[a]	13.37[a]	12.44[b]	0.48	<0.05	0.24	0.36
CP Intake, kg/d	2.64[b]	2.91[a]	3.05[a]	0.10	<0.01	<0.05	0.62
ADTT,%	51.9	53.2	49.6	3.09	0.67	0.58	0.51
ADTT, kg/d	1.36	1.35	1.49	0.08	0.35	0.28	0.34
Milk yield, kg/d	22.9	23.0	23.4	0.92	0.36	0.48	0.83

[a,b] Least squares means within the same row without a common superscript differ (P<0.05).
[1] Diet 1 = 9.8 (% DM); diet 2 = 10.8 (% DM); diet 3 = 11.8% (% DM). [2] ADTT = apparently digested in the total gastrointestinal tract.

Conclusion

Results confirming that low producing cows are less likely to respond to altering NFC: RDP ratio. Diets containing 15.3% CP and 10.8% RDP with 40% NFC supported maximal production and digestibility in mid-lactation dairy cows compared with diets with higher CP content.

References

Hoover, W.H. and S.R. Stokes, 1991. Balancing carbohydrates and proteins for optimum rumen microbial yield. J. Dairy Sci. 74: 3630-3644.

Ipharraguerre, I.R. and J.H. Clark, 2005. Varying protein and starch in the diet of dairy cows. II. Effects on performance and nitrogen utilization for milk production. J. Dairy Sci. 88: 2556-2570.

NRC, 2001.Nutrients Requirements of Dairy Cattle. 7th Revised edition. National Academy of Sciences. Washinton D.C., USA. 381p.

Van Keulen, J. and B.A. Young, 1977. Evaluation of acid-insoluble ash as a natural marker in ruminant digestibility studies. J. Anim. Sci. 44: 282-287.

Influence of diet and detachment procedure on recovery of solid-associated microbes from sheep ruminal digesta

S. Ramos, M.L. Tejido, M.E. Martínez, M.J. Ranilla, C. Saro and M.D. Carro
Departamento de Producción Animal, Universidad de León, 24071 León, Spain; mjrang@unileon.es

Introduction

Detachment procedures (DP) commonly used to remove solid-associated microbes (SAM) from ruminal digesta have a low efficiency, and total recovery of detached microbes have often been reported to be lower than 35% (Craig *et al.*, 1987; Martín-Orúe *et al.*, 1998). We hypothesised that efficiency of DP might be influenced by the diet of the animal host, but only one diet has been investigated in most of the studies. The aim of this work was to evaluate the effectiveness of three different DP to recover SAM from ruminal digesta from sheep fed four diets differing in forage:concentrate (F:C) ratio and type of forage.

Material and methods

Six rumen fistulated Merino sheep (58.6±4.46 kg body weight) were fed four diets over four 21-d experimental periods. Dietary treatments were a 2×2 factorial arrangement of forage type (grass hay (G) and alfalfa hay (A)) and F:C ratio (high (HF; 70:30) and low (LF; 30:70). ^{15}N was used to label microbial biomass. Samples of solid ruminal digesta were taken 4 h after feeding and subjected to three DP: (1) MET: digesta was incubated (38 °C, 15 min) with saline solution (0.9% NaCl) containing 0.1% methylcellulose; (2) STO: digesta was mixed with cold saline solution and detached with a stomacher (230 rpm, 5 min); (3) FRE: digesta was immediately frozen at -20 °C for 72 h, thawed at 4 °C, mixed with saline solution and subjected to the STO procedure. Following all treatments, samples were stored at 4 °C for 24 h after the treatment, homogenisation, filtration, and resuspension of digesta two times in the treatment solutions before isolating SAM (Ranilla and Carro, 2003). The proportion of detached microbes (% detachment) for each DP was calculated as [1 – (μg of ^{15}N in digesta after treatment/μg of ^{15}N in digesta before treatment)×100]. The proportion of recovery from detached microbes (% recovery) was calculated as [μg of ^{15}N in microbial pellet/(μg of ^{15}N in digesta before treatment – μg of ^{15}N in digesta after treatment) × 100]. The total recovery was calculated as [(% detachment × % recovery) / 100]. Data were analysed according to a mixed model using the MIXED procedure of SAS®. The effects of F:C, type of forage, DP, period, and the interactions F:C × DP and forage × DP were considered fixed, and sheep effect was considered random. Mean effects were declared significant at $P<0.05$.

Results

Detachment efficiency was greater (P=0.002) for LF compared to HF diets (56.4 and 54.3%, respectively; Table 1), in agreement with Ranilla and Carro (2003). In contrast, the recovery of detached was greater (P<0.001) for HF (74.1%) compared to LF diets (63.0%) which would indicate that microbes attached to forage particles were more resistant to cell lysis than those attached to particles of concentrate (Ranilla and Carro, 2003). Total recovery was greater (P=0.04) for HF diets compared to LF, which supports previous studies (Ranilla and Carro, 2003; Martínez *et al.*, 2009). In contrast, the type of forage did not affect any of the variables analysed (P=0.50 to 0.80). Detachment values for STO were on average 1.5 and 1.2 times greater (P<0.001) than those obtained by MET and FRE, respectively. There were no differences between MET and STO in the values of recovery (P=0.54), but FRE produced lower (P<0.001) values which would indicate adverse effects of this treatment on microbial cell integrity., Values of total recovery averaged across diets were 48.2, 31.4 and 26.1% for STO, FRE and MET, respectively, with the values being greater

(P<0.001) for STO compared to the other two methods. There were no F:C × DP or forage x DP interactions for any variable (P=0.09 to 0.88).

Table 1. Percentage of detachment, recovery of detached and total recovery of solid-associated microbes from ruminal digesta in sheep fed different diets after applying different detachment procedures (DP) as determined using ^{15}N as microbial marker (n = 6).

	Diet	DP[1]			SEM	Significance (P =)[2]		
		MET	STO	FRE		F:C	Forage	DP
Detachment,%	HFA	44.5[a]	61.9[c]	52.3[b]	1.86	0.002	0.50	<0.001
	LFA	42.9[a]	68.4[c]	58.3[b]				
	HFG	44.0[a]	66.1[c]	57.2[b]				
	LFG	42.6[a]	67.4[c]	58.5[b]				
Recovery,%	HFA	79.8[b]	81.8[b]	62.2[a]	4.18	<0.001	0.80	<0.001
	LFA	67.7[b]	65.6[b]	52.9[a]				
	HFG	86.9[b]	77.6[b]	56.7[a]				
	LFG	68.7[b]	70.8[b]	52.2[a]				
Total recovery,%	HFA	35.1[a]	50.5[b]	32.4[a]	2.31	0.04	0.71	<0.001
	LFA	29.3[a]	44.9[b]	30.7[a]				
	HFG	38.2[a]	50.8[b]	31.9[a]				
	LFG	29.2[a]	46.7[b]	30.8[a]				

[a, b] Means within a row with unlike superscripts differ (P<0.05); [1]See text for description of DP and diets; [2]F:C: forage:concentrate; No significant F:C × DP and Forage × DP interactions were detected (P>0.05).

Conclusion

The results indicate that the effectiveness of DP is influenced by the F:C ratio in the diet of the animal but not by the type of forage. Treatment with STO was the most effective method to remove SAM for all diets, but further research is needed to decrease microbial losses during the isolation process in order to increase total recovery of SAM.

Acknowledgement

This research was supported by the C.I.C.Y.T. of Spain Project AGL2004-04755-C02-01). S. Ramos, M.E. Martínez and C. Saro gratefully acknowledge the receipt of scholarships from the M.E.C. of Spain (BES-2005-6842, AP2005-1797 and AP 2006-03049, respectively).

References

Craig, W.M., G.A. Broderick and D.B. Ricker, 1987. Quantization of microorganisms associated with the particulate phase of ruminal ingesta. J. Nutr. 117: 56-62.

Martín-Orúe, S.M., J. Balcells, F. Zakraoui and C. Castrillo, 1998. Quantification and chemical composition of mixed bacteria harvested from solid fractions of rumen digesta: effect of detachment procedure. Anim. Feed Sci. Technol. 71: 269-282.

Martínez, M.E., M.J. Ranilla, S. Ramos, M.L. Tejido, C. Saro and M.D. Carro, 2009. Evaluation of procedures for detaching particle-associated microbes from forage and concentrate incubated in Rusitec fermenters: efficiency of recovery and representativeness of microbial isolates. J. Anim. Sci. (In press).

Ranilla, M.J. and M.D. Carro, 2003. Diet and procedures to detach particle-associated microbes from ruminal digesta influence chemical composition of microbes and estimation of microbial growth in Rusitec fermenters. J. Anim. Sci. 81: 537-544.

Mode of action of *Chrysanthemum coronarium* as a modulator of biohydrogenation of fatty acids in the rumen

E. Ramos Morales[1], N. McKain[1], C. Atasoglu[2], T.A. Wood[1] and R.J. Wallace[1]
[1]University of Aberdeen Rowett Institute of Nutrition and Health, Bucksburn, Aberdeen AB21 9SB, United Kingdom; [2]Canakkale Onsekiz Mart Universitesi, 17100 Canakkale, Turkey; e.morales@abdn.ac.uk

Introduction

Ruminant milk and meat contain conjugated linoleic acids (CLA), mainly the *cis*-9,*trans*-11-18:2 isomer, and vaccenic acid (VA; *trans*-11-18:1), which are associated with putative health benefits such as cancer prevention and improved immune response. CLA and VA are intermediates in the biohydrogenation sequence that converts unsaturated fatty acids, particularly linoleic acid (LA; *cis*-9,*cis*-12-18:2), in the diet to saturated fatty acids in the rumen and beyond. The inclusion of the daisy plant, *Chrysanthemum coronarium*, in a dairy sheep diet results in higher concentrations of CLA and VA in milk (Cabiddu *et al.*, 2006). Previous studies in our laboratory have shown that *C. coronarium* contains a very high content of linolenic acid (LNA; *cis*-9,*cis*-12,*cis*-15-18:3, 878.8 μg/100 mg). In addition, it is known that *C. coronarium* contains an unusual epoxy fatty acid, coronaric acid (*cis*-9,10-epoxy,*cis*-12-18:1; Earle, 1970). The aims of the present study were to determine if the reported change in milk fatty acid composition was the result of the effects of *C. coronarium* on ruminal biohydrogenating bacteria, and to investigate which constituent of *C. coronarium* may be responsible for the observed effects. Coronaric acid and LNA were investigated as the main compounds that could alter LA biohydrogenation.

Material and methods

Four sheep receiving a mixed hay-concentrate (30:70) diet were used to provide an inoculum for *in vitro* incubations of ruminal digesta with LA in the presence and absence of *C. coronarium* var. Primrose Gem. Additionally, rumen fluid was incubated either with LA or with a combination of LA and coronaric acid ((±)-*cis*-9, 10-epoxy, *cis*-12-18:1). One ml of strained rumen fluid was added under CO_2 to Pyrex tubes containing one of the following: 0.2 ml distilled water; 5 mg ground plant and 0.2 ml distilled water; 0.1 ml of 2 mg LA/ml and 0.1 ml distilled water; 5 mg ground plant, 0.1 ml of 2 mg LA/ml and 0.1 ml distilled water, or 0.1 ml of 2 mg LA/ml and 0.1 ml of 2 mg coronaric acid/ml. Suspensions were incubated under CO_2 at 39 °C for up to 24 h. Samples were analysed for fatty acids by GC of methyl esters. Statistical significance was determined at each time point by a randomised block analysis of variance with sheep as blocks. Genstat 10th edition (VSN International, UK) was used.

Results

The rate of metabolism of LA and rate of production of stearic acid (SA) decreased as a result of *C. coronarium* addition to mixed ruminal microorganisms (results not shown). VA accumulated during the incubation, and *C. coronarium* increased the accumulation. An inhibition ($P<0.05$) of the metabolism of LA was observed when coronaric acid was added (Figure 1a), as well as an inhibition ($P<0.05$) of the accumulation of CLA and a slowdown in its metabolism (Figure 1b). Decreases ($P<0.05$) in VA (Figure 1c) and SA (Figure 1d) metabolism were also observed.

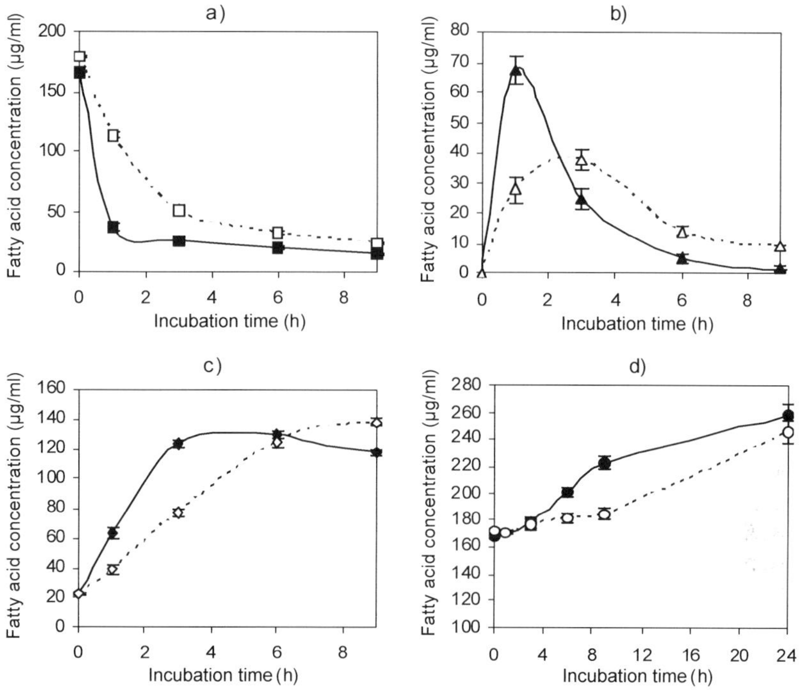

Figure 1. Influence of coronaric acid on metabolism of LA in ruminal fluid from four sheep receiving a mixed hay-concentrate diet (30:70). LA and coronaric acid were added to an initial concentration of 200 µg/ml. (a) LA (■,□). (b) CLA cis-9, trans-11 (▲,Δ). (c) VA (♦,◊). (d) SA (●,○). Closed symbols are from incubations with LA alone; open symbols are from incubations with LA + coronaric acid. Results are mean ± SE from four sheep. Statistical significance was determined at each time point by a randomised block analysis of variance with sheep as blocks.

Conclusion

As LNA inhibits biohydrogenation at high concentrations (Wąsowska *et al.*, 2006) and coronaric acid showed an inhibitory effect on the metabolism of LA, it was concluded that the combined effect of LNA and coronaric acid in *C. coronarium* putatively caused an increase in VA and decrease in SA *in vitro*. This effect would lead to an increased flow of VA from the rumen, which after absorption from the small intestine would be converted back to CLA by Δ^9-desaturase in mammalian tissue, which in turn would lead to an increased CLA concentration in milk from animals receiving *C. coronarium*.

References

Cabiddu, A., M. Addis, G. Pinna, S. Spada, M. Fiori, M. Sitzia, A. Pirisi, G. Piredda and G. Molle, 2006. The inclusion of a daisy plant (*Chrysanthemum coronarium*) in dairy sheep diet. 1: Effect on milk and cheese fatty acid composition with particular reference to C18:2 *cis*-9, *trans*-11. Livest. Sci. 101: 57-67.

Earle, F.R, 1970. Epoxy oils from plant seeds. J. Amer. Oil Chem. Soc. 47: 510-513.

Wąsowska, I., M.R.G. Maia, K.M. Niedzwiedzka, M. Czauderna, J.M.C. Ramalho Ribeiro, E. Devillard, K.J. Shingfield and R.J. Wallace, 2006. Influence of fish oil on ruminal biohydrogenation of C18 unsaturated fatty acids. Br. J. Nutr. 95: 1199-1211.

Portal absorption of ethanol and propanol in early lactating dairy cows

B.M.L. Raun and N.B. Kristensen
Faculty of Agricultural Sciences, Aarhus University, DK-8830 Tjele, Denmark; nbk@agrsci.dk

Introduction

Field studies on Danish corn silage have shown that 35% of the dairy farms included were feeding corn silage that in part of the year contained 20-24 g/kg DM of ethanol. 20% of dairy farms fed corn silages that contained 5-9 g/kg DM of propanol. When selecting farms experiencing problems with the quality of corn silage, concentrations as high as 34 g/kg DM of ethanol and 25 g/kg DM of propanol have been observed. In some respects, silages with high levels of propanol seem favourable because this correlates with acetic acid content and increases the aerobic stability in the silage (Danner *et al.*, 2003). However, based on reports from practice, we hypothesised that cows in early lactation have problems tolerating high levels of propanol, but very little is known about the absorption of dietary alcohols in early lactation. The aim of the present experiment was to investigate portal absorption and ruminal metabolism of ethanol and propanol in early lactating dairy cows.

Material and methods

Eight lactating Holstein cows implanted with a ruminal cannula and permanent indwelling catheters in major splanchnic blood vessels were used to investigate metabolism of ethanol and propanol at days 4, 15 and 29 in lactation (DIM). Cows were randomly allocated to a 2 by 2 factorial design with one factor being level of branched chain alcohol (isopropanol from HMBi; 0.26% of dry matter; Adisseo, France) compared with no addition of isopropanol (calcium carbonate) and the second factor being the source of straight chain ethanol alcohols (1.9% of dry matter) compared with propanol (1.6% of dry matter). Cows were fed the experimental rations from time of calving. The rations were fed in three equally sized portions daily at 8 h intervals. Eight hourly sets of ruminal, arterial, and portal vein samples were collected starting 30 min before the morning feeding. Net portal metabolite fluxes were calculated as described by Kristensen *et al.* (2007). Data were analysed as a 2 by 2 factorial design with samples within day as repeated measurements using the mixed procedure of SAS®. Only data for straight chain alcohol levels are reported. Data are presented as means ± SEM.

Results and discussion

Dry matter intake and milk yield increased (P >0.01) with DIM, but were not affected by treatment (Table 1). Total ruminal VFA concentration was not affected by DIM or treatment, but the proportion of acetate of total VFA decreased from 60 to 57% with propanol and the ruminal propionate proportion increased (P<0.05) with propanol. Ruminal concentrations of ethanol and propanol increased according to the dietary intake of the respective alcohol, but the ruminal concentrations of ethanol and propanol were not affected by DIM even though alcohol intake increased with 25 and 33% from 4 to 29 DIM for ethanol and propanol, respectively. The net portal flux of ethanol and propanol increased in association with the increased ruminal concentrations. However, the net portal flux accounted for a decreasing fraction of ethanol and propanol intake with increasing DIM that might be due to the adaptation of rumen microbial or rumen/gut epithelium alcohol metabolism.

Table 1. Effects of ethanol and propanol intake on dry matter intake (DMI), energy corrected milk (ECM) production and metabolic variables in lactating dairy cow samples at 4, 15, and 29 d in milk (DIM).

	Treatment						SEM	*P*-value	
	4 DIM		15 DIM		29 DIM				
	Ethanol	Propanol	Ethanol	Propanol	Ethanol	Propanol		Trt	DIM
DMI, kg/d	14.7	13.8	16.5	17.5	19.7	20.6	0.6	0.52	<0.01
ECM, kg/d	30	32	36	33	38	37	3	0.93	<0.01
Ruminal concentration, mmol/l									
Ethanol	3.2	1.5	3.3	1.7	2.8	1.6	0.3	<0.01	0.64
Propanol	0.5	1.9	0.5	1.9	0.5	1.8	0.3	<0.01	0.87
Total VFA	110	105	111	106	108	105	5	0.49	0.85
mol/100 mol total VFA									
Acetate	59.6	57.9	60.4	57.1	60.3	57.5	0.4	0.01	0.94
Propionate	22.2	25.4	21.3	24.8	22.0	23.2	0.8	0.04	0.15
Arterial concentration, mmol/l									
Ethanol	0.046	0.027	0.073	0.036	0.074	0.048	0.023	0.30	0.46
Propanol	0.009	0.011	0.012	0.013	0.011	0.019	0.005	0.53	0.33
Net portal flux, mmol/h									
Ethanol	71	27	97	41	114	40	15	0.02	0.04
Propanol	15	62	17	62	17	77	6	<0.01	0.31
Portal recovery, %									
Ethanol	39	47	28	37	24	31	4	0.06	<0.01
Propanol	73	44	51	33	43	29	7	0.02	0.03

Arterial concentrations of ethanol and propanol were not affected by treatment or by DIM and thereby indicate an efficient hepatic extraction of the alcohols.

Conclusion

Dietary supplementation with ethanol and propanol led to increased ruminal concentrations and increased net portal fluxes of both alcohols, but the arterial concentrations were not affected. Portal recoveries of ethanol and propanol were greater on day 4 of lactation compared with later sampling times, but portal absorption accounted for at least 24% of dietary alcohol intake throughout the period studied. Alcohols from silage will be partly absorbed by the blood of lactating dairy cows with the potential of affecting intermediary metabolism.

References

Danner, H., M. Holzer, E. Mayrhuber and R. Braun, 2003. Acetic acid increases stability of silage under aerobic conditions. Appl. Environ. Microbiol. 69: 562-567.

Kristensen, N.B., A. Storm, B.M.L. Raun, B.A. Rojen and D.L. Harmon, 2007. Metabolism of silage alcohols in lactating dairy cows. J. Dairy Sci. 90: 1364-1377.

Effects of high non-structural carbohydrate concentration in lucerne on feeding behaviour and ruminal pH of early lactating cows

G. Régimbald[1], V. Girard[2], A.F. Brito[3], G. Allard[1], D. Pellerin[1], G.F. Tremblay[4] and R. Berthiaume[3]
[1]Université Laval, Québec, Canada; [2]Université de Montréal, St-Hyacinthe, Canada; [3]Agriculture and Agri-Food Canada, Sherbrooke, Canada; [4]Agriculture and Agri-Food Canada, Québec, Canada; regimbaldg@agr.gc.ca

Introduction

Previous studies have shown that forages containing high concentration of non-structural carbohydrates (NSC) increase milk yield (Miller *et al.*, 2001; Brito *et al.*, 2008) and N utilisation (Brito *et al.*, 2008) of late-lactating cows fed silage only. Spot sampling of ruminal fluid (Brito *et al.*, 2008) revealed that the mean pH was higher for cows fed high- vs. low-NSC forage. Although high levels of NSC in the diet generally elicit a decrease of ruminal pH, we observed the opposite (Brito *et al.*, 2008). We hypothesized that cows fed the high-NSC forage were ruminating longer and were producing more saliva as shown by Gregorini *et al.* (2005). The objective of this study was to determine if an increase in ruminal pH potentially associated with high-NSC forage may be caused by an augmentation in the rumination activity.

Material and methods

Four Holstein cows in early lactation (average initial BW 640 ± 44 kg and daily milk production 39±5 kg) were randomly assigned to one of two dietary treatments in a crossover design: a high-NSC (67.2 g/kg of DM) vs. a low-NSC diet (36.6 g/kg of DM). Diets were comprised of high or low NSC lucerne baleage plus a common concentrate (60:40 DM basis, forage:concentrate ratio). Lucerne (*cv* AC Caribou) was grown on three fields in Normandin, Québec. Each field was separated in two; the first half was cut at sundown (18:00 to 20:00) after a sunny day and the second half was cut on the following morning (06:00 to 08:00) at late bud stage of development. Forages were harvested as large rectangular bales, wrapped in stretch plastic, and then transported at the Research Centre of Agriculture and Agri-Food Canada in Sherbrooke, Québec. Each bale was analysed for NSC and DM concentrations. Bales were paired to have similar DM but the greatest difference in NSC concentrations. Cows were fed *ad libitum* (allowing 10% feed refusals). Forage was offered in a single meal at 8:30 h whereas concentrate was offered in three meals at 10:00 h, 15:00 h, and 20:00 h. Animals were housed in tie stalls and had free access to fresh water. Each experimental period lasted 24 d. Feeding behaviour and ruminal pH were continuously monitored during the last three days of each period. Each cow was wearing a halter with a magnetic switch to follow the activity of chewing during eating and rumination. When a cow is chewing, an elastic strap under the jaw of the animal opens the circuit and counts one bite. Data were compiled to calculate the average number of bites per minute during rumination and the time spent ruminating. A continuous recording system using a heavy-duty pH electrode inserted in the ventral sac of the rumen and a wireless data logger was used (AlZahal *et al.*, 2007). This system recorded ruminal pH data every five minutes. Data were summarised by calculating the time ruminal pH was below 6.0 and below 5.5, respectively, for each 24 h period. Data were analysed using the MIXED procedure of SAS® for a crossover design. Considering the small number of cows used in this trial, significance was declared at $P<0.10$.

Results

There was no difference in DMI and milk yield and composition between treatments (Table 1). However, there was a noticeable, although not statistically significant, difference between the high-

and low-NSC treatments for the period of time when the pH is under 6.0 (Table 2). In the present trial, mean pH was not higher with the high-NSC treatment. This might be due to the presence of concentrate (40% of DMI) in the diet. The time spent ruminating when cows were fed the high-NSC lucerne was longer than on the low-NSC treatment (P=0.09).

Table 1. Intake, milk yield and milk composition of dairy cows fed a high- vs. *a low-NSC diet.*

Item	Treatment		SED[1]	$P>F$[2]
	High-NSC	Low-NSC		
DMI, kg/d	23.3	23.2	0.48	0.75
Milk yield[3], kg/d	35.5	35.1	0.70	0.62
Milk yield:DMI	1.53	1.51	0.03	0.73
4% FCM, kg/d	31.3	31.7	0.86	0.69
4% FCM:DMI	1.35	1.37	0.05	0.73
ECM[4], kg/d	33.0	33.4	0.54	0.53
ECM:DMI	1.42	1.44	0.04	0.67
Milk fat,%	3.22	3.36	0.10	0.31
Milk fat, kg/d	1.14	1.18	0.04	0.44
Milk protein,%	2.60	2.64	0.08	0.66
Milk protein, kg/d	0.92	0.93	0.03	0.86
Milk lactose,%	4.45	4.50	0.04	0.36
Milk lactose, kg/d	1.58	1.57	0.04	0.95
MUN, mg/dL	9.29	9.62	0.78	0.72

[1] SED = standard error of the least squares means difference.
[2] Probability of treatment effect (High- vs. Low-NSC).
[3] Intake measured during the 3d of feeding behaviour measurements.
[4] ECM = [0.327 × milk yield (kg/d)] + [12.95 × fat yield (kg/d)] + [7.2 × protein yield (kg/d)] (Orth, 1992).

Table 2. Ruminal pH and ruminating activity of dairy cows fed a high- vs. *a low-NSC diet.*

Item	Treatment		SED	P-value	
	High-NSC	Low-NSC		NSC	NSC x Day
Ruminal pH below 5.5, min/d	0.00	1.33	1.54	0.51	0.36
Ruminal pH below 6.0, min/d	314	210	237	0.70	0.75
Number of rumination bouts/day	13.5	13.5	1.27	0.99	0.90
Bite rate, bites/min of each rumination bout	61.9	64.8	3.12	0.46	0.42
Rumination time, min spent/rumination bout	27.4	25.1	0.74	0.09	0.28
Rumination bites, bites/rumination bout	1,696	1,625	85.3	0.49	0.62
Total number of bites, bites/d	23,010	21,547	1,958	0.53	0.64

Conclusion

This study suggests that increasing the NSC concentration in lucerne increases the time spent ruminating in early lactating cows. The lack of an effect on ruminal pH might be due to the large proportion of concentrates (40%) in the diet.

References

AlZahal, O., B. Rustomo, N.E. Odongo, T.F. Duffield and B.W. McBride, 2007. Technical note: A system for continuous recording of ruminal pH in cattle. J. Anim. Sci. 85: 213-217.

Brito, A.F., G.F. Tremblay, A. Bertrand, Y. Castonguay, G. Bélanger, R. Michaud, H. Lapierre, C. Benchaar, H.V. Petit, D.R. Ouellet and R. Berthiaume, 2008. Alfalfa cut at sundown and harvested as baleage improves milk yield of late-lactation dairy cows. J. Dairy Sci. 91: 3968-3982.

Gregorini, P., M. Eirin, R. Refi, M. Ursino, O.E. Ansin and S.A. Gunter, 2006. Timing of herbage allocation in strip grazing: Effects on grazing pattern and performance of beef heifers. J. Anim. Sci. 84: 1943-1950.

Miller, L.A., J.M. Moorby, D.R. Davies, M.O. Humphreys, N.D. Scollan, J.C. MacRae and M.K. Theodorou, 2001. Increased concentration of water-soluble carbohydrate in perennial ryegrass (*Lolium perenne* L.): milk production from late-lactation dairy cows. Grass For. Sci. 56: 383-394.

Orth, R., 1992, Sample Day and Lactation Report. DHIA 200 Fact Sheet A-2. Mid-States DRPC, Ames, IA, USA.

Comparison of marker infusion techniques to determine the clearance of ruminal volatile fatty acids

J.C. Resende Júnior, J.L.P. Daniel, F.C. Meireles, M.B. Moreira, R.F. Lima and M.G. Cardoso
Departamento de Medicina Veterinária da Universidade Federal de Lavras, MG, Brazil; joaocrj@ufla.br

Introduction

Ruminal acidosis is a metabolic disorder triggered when the production of volatile fatty acids (VFA) is higher than its absorption through the rumen wall and passage with the rumen fluid through the reticulum-omasum orifice. The VFA accumulate in the reticulorumen causing deleterious effects on performance and health of ruminants. Some studies have used different techniques for measuring ruminal clearance, including pulse-dose ruminal infusion of valeric acid linked to the Co-EDTA in evacuated digesta (Resende Júnior *et al.*, 2006) or infusion without evacuation (Allen *et al.*, 2000). The rumen evacuation technique destabilises the rumen environment, through the destruction of the mat, loss of stratification, an increase in oxygen tension and reduction of the digesta temperature, and it also changes the motility and blood flow of the rumen. However, when rumen evacuation is not used, the rumen environment can also suffer disturbance, because a good mix mechanism is necessary. Moreover, without rumen evacuation, the necessary volume of marker solution may be very high (Allen *et al.* 2000) and could interfere in the VFA clearance. Daniel *et al.* (2007) have proposed a technique that uses a low volume of homogenised marker solution to intact digesta without artificial mixture. The purpose of the current experiment was to validate the technique of Daniel *et al.* (2007) by comparing it to the technique of pulse-dose ruminal infusion of valeric acid linked to the Cr-EDTA to evacuated digesta (Resende Júnior *et al.*, 2006).

Material and methods

Four not lactating and not pregnant, rumen-cannulated Jersey cows, housed in a Tie Stall on a sand bed, were placed in a sequence of four treatments in a split plot design, conducted simultaneously with periods of 18 days. The four treatments were arranged in a 2×2 factorial arrangement considering the method of marker solution infusion and diet : T1: Corn silage diet and marker solution infusion to the intact diet (ID); T2: Corn silage diet and marker solution infusion to the evacuated digest (ED); T3: Corn silage plus concentrate (50% on dry matter basis) and marker solution infusion to ID; T4: Corn silage plus concentrate (50% on dry matter basis) and marker solution infusion to ED. The infusions to ID were performed on day 14 and the infusions to ED were performed on day 18. The marker solution had a volume of 4.0 l and pH adjusted to 6.5±0.1, containing 700 mg/l of Cr, as Cr-EDTA (Binnerts *et al.*, 1968) and 75g/l of valeric acid. In ID treatments, the marker solution was introduced to the rumen through a perforated tube being ¼ of the volume in the cranial sac and ¾ of the volume in different places of the ventral sac. There was no artificial mechanism of agitation in the rumen and the mixture was done only by rumen motility. So the first time of fluid collection was 90 minutes from the marker infusion and was considered as time zero in an exponential curve of the marker decay. In the ED treatment, the marker solution was mixed to evacuate digesta by hand. So the time zero of the fluid collect was immediately after returning the evacuated ruminal content. At times zero, 15, 30, 60, 120 and 240 minutes about 20 ml of ruminal fluid were collected from ventral sac samples and were divided into two aliquots, one for analysis of Cr, by means of atomic absorption spectrophotometry, and the other for analysis of VFA by means of gas-liquid chromatography. The fractional rate of VFA absorption by the rumen wall was calculated by the difference between the fractional rate of total clearance and fractional rate of passage of fluid, as described by Resende Junior *et al.* (2006).

Results

The estimate of total fractional rate of clearance using the ID infusion technique (38%/h) did not differ (P=0.39) from that obtained by the ED infusion technique (30%/h). Absorption fractional rates were not different also, showing that the techniques are equivalent. The fractional passage rate of fluid (12%/h) estimated by the ID technique was higher (P=0.06) than that estimated by the ED technique, probably reflecting the destabilisation of the ruminal digesta, which could have affected the rumen motility. The fractional rate of VFA absorption tended (P=0.14) to be higher in the diet with forage plus concentrate (30%/h) than in the diet with only forage (17%/h), probably reflecting a greater absorptive surface in the diet with concentrate.

Conclusion

The results indicate that the marker infusion techniques in intact digesta as purposed by Daniel *et al.* (2007) are secure to estimate the ruminal VFA clearance.
The intact digesta infusion technique was better than the evacuated infusion technique because it minimises interference in the stratification of ruminal digesta.
Diets with quickly fermentable carbohydrates tended to provide higher fractional rates of ruminal VFA absorption, probably reflecting the larger absorptive surface of the ruminal epithelium.

References

Allen, M.S., L.E. Armentano, M.N. Pereira, Y. Ying and J. Xu, 2000. Method to measure fractional rate of volatile fatty acid absorption from the rumen. Proc. 25th Conf. rumen Function, Chicago, IL., p. 26.

Binnerts, W.T., A.T Van' t Klooster and A.M. Frens. 1968, Soluble chromium indicator measured by atomic absorption in digestion experiments. Vet. Rec. 82: 470.

Daniel, J.L.P., J.C. Resende Junior, L.Q. Melo, R.L.M. Souza and M.C. Guerreiro, 2007. Funcionalidade do sulco do reticulo em bovinos adultos e avaliação de técnicas alternativas para preparação, infusão e coleta de marcadores da fase fluida ruminal. Ciênc. Agrotec. 31: 1850-1856.

Resende Júnior, J.C., M.N. Pereira, H. Bôer and S. Tamminga, 2006. Comparison of Techniques to determine the Clearance of ruminal Volatile Fatty Acids. J. Dairy Sci. 89: 3096-3106.

Differences of bacterial communities in the rumen liquor and faeces of steers fed on alfalfa or sainfoin silage and under arctic production

G.A. Romero-Perez[1], K.H. Ominski[2], T.A. McAllister[3] and D.O. Krause[2]
[1]Creative Research Institution 'Sosei', Hokkaido University, 001-0021, Sapporo, Japan; [2]Department of Animal Science, University of Manitoba, MB, R3T 2N2, Winnipeg, Canada; [3]Lethbridge Research Centre, AB, T1J 4B1, Lethbridge, Canada; gromprz@cris.hokudai.ac.jp

Introduction

Faeces shedding and manure from cattle production are considered important sources of bacterial contamination of food and the environment. The activity of bacterial populations in the rumen can be negatively affected if diets contain tannins (Jones *et al.*, 1994) or the ambient temperature is very low (Von Keyserlingk and Mathison, 1993). However, bacteria can develop tolerance to tannins (Nelson *et al.*, 1998) and remain active even when low temperatures are induced (Jones *et al.*, 1987). Little is known about the combined effects of arctic temperatures and fodder tannins on the rumen and hindgut bacterial populations. The objective of the present work was to characterise the differences of bacterial communities in the rumen liquor and faeces of beef steers fed a tanniferous diet and under arctic conditions.

Material and methods

Individual oral probe-aspirated rumen liquor and rectal faecal samples from donor steers (10× four pens) given water and either alfalfa silage or sainfoin silage at 100% of *ad libitum*, were collected weekly in the first three sampling days and fortnightly afterwards for 10 wk. Samples were divided by sample type, pooled per pen and sub-sampled. Daily ambient temperature (range: -4.6 ~ -30.0 °C) was registered and averaged to weekly mean temperature. The DNA was extracted with ZR Faecal DNA kits, as per manufacturer protocol. The 16S ribosomal DNA was amplified using primers blue-fluorescent 27f, 5′-GAAGAGTTTGATCATGGCTCAG-3′, and 1100r, 5′-CTGCTGCCTCCCGTAG-3′. The PCR amplicons were produced in 35 cycles (denaturation: 94 °C, 0.5 min; annealing: 61 °C, 1.0 min; and extension: 72 °C, 0.5 min). Amplicons were digested (Bhandari *et al.* 2008), cleaned with EtOH and re-suspended in a sample loading solution. The terminal restriction fragment length was estimated by the Beckman-Coulter CEQ 8800 Genetic Analysis System. Combined profiles of T-RFLP data of at least two sample replicates were built and compared with an online sequence database. Bacterial community proportions of the total population in the sequences were analysed by Linear Mixed Models (GenStat, 2007). Fixed effects included sample type, diet and ambient temperature and the random effects were week × pen. Significance levels ($P<0.05$; trends towards significance: $P<0.10$) and standard errors of the means were calculated with natural-logged data and tested using the Wald test.

Results

There were differences in the bacterial community proportions in phylum *Bacteriodetes* ($P=0.053$) and order *Lactobacillales* ($P=0.057$), as they were higher in the faecal samples than in the rumen liquor samples (Table 1). In contrast, the bacterial community proportions in phylum *Tenericutes* ($P<0.05$) and order *Bacillales* ($P<0.01$) were significantly higher in the rumen liquor than in the faeces (Table 1). Ambient temperature significantly affected ($P<0.05$) the bacterial community proportions in phylum *Firmicutes* and tended to cause differences ($P=0.088$) in those in order *Lactobacillales*. The sample type x diet interaction caused significant changes ($P<0.05$) in the bacterial proportions in phyla *Proteobacteria* and *Verrucomicrobia* and a tendency in phyla

Bacteriodetes, *Fusobacteria* and *Tenericutes* (P=0.061, P=0.065 and P=0.063, respectively), as sainfoin silage affected primarily bacterial communities in the faeces but alfalfa silage mainly those in the rumen liquor (Table 1).

Table 1. Effect of sample type, diet and ambient temperature on the bacterial communities in the faecal and rumen liquor samples of steers fed on sainfoin silage (SainSil) or alfalfa silage (AlfSil) (proportion of total bacterial communities in sequences ×100).

Phyla[1]	Order[2]	Rumen liquor		Faeces		s.e.	Level of significance			
		AlfSil	SainSil	AlfSil	SainSil		Sample	Diet	Temp.[3]	Sample x Diet
BD		0.37	0.14	0.40	0.48	0.286	0.053	n.s.	n.s.	0.061
FM		86.93	87.80	93.23	90.47	0.038	n.s.	n.s.	*	n.s.
	BAC	45.84	44.04	26.98	30.42	0.133	**	n.s.	n.s.	n.s.
	LB	46.76	43.60	67.70	61.26	0.158	0.057	n.s	0.088	n.s.
FB		0.06	0.05	0.04	0.07	0.190	n.s.	n.s.	n.s.	0.065
PB		4.64	3.30	3.11	4.37	0.172	n.s.	n.s.	n.s.	*
TN		0.15	0.18	0.15	0.12	0.092	*	n.s.	n.s.	0.063
VM		0.83	0.10	0.08	0.46	0.756	n.s.	n.s.	n.s.	*

[1] Bacterial phyla: BD = *Bacteroidetes*; FM = *Firmicutes*; FB = *Fusobacteria*; PB = *Proteobacteria*; TN = *Tenericutes* and VM = *Verrucomicrobia*.
[2] Firmicutes orders: BAC = *Bacillales*; LB = *Lactobacillales*.
[3] Temp. = Ambient temperature.
* P<0.05; ** P<0.01 = Statistical significance; n.s. = not significant.

Conclusion

The results showed that sample type and its interaction with diet had a larger effect on bacterial communities than diet and temperature alone. All this may be due to an influence of the bio-geographical location of the digestive tract site, the formidable survivability that certain bacterial communities show in adverse surroundings, including extreme temperature changes (Jones *et al.*, 1987) and presence of anti-nutrients such as tannins (Nelson *et al.*, 1998), or a combination of all. Further study of environmental variables affecting bacteria in rumen and ruminant hindgut and characterisation of their species are highly recommended.

References

Bhandari, S.K., B. Xu, C.M. Nyachoti, D.W. Giesting and D.O. Krause, 2008. Evaluation of alternatives to antibiotics using an Escherichia coli K88+ model of piglet diarrhea: Effects on gut microbial ecology. J. Anim. Sci. 86: 836-847.

Jones, G.A., T.A. McAllister, A.D. Muir and K.J. Cheng, 1994. Effects of sainfoin (*Onobrychis viciifolia* Scop.) condensed tannins on growth and proteolysis by four strains of ruminal bacteria. Appl. Environ. Microbiol. 60: 1374-1378.

Jones, P.G., R.A. VanBogelen and F.C. Neidhardt, 1987. Induction of proteins in response to low temperature in Escherichia coli. J. Bacteriol. 169: 2092-2095.

Nelson, K.E., M.L. Thonney, T.K. Woolston, S.H. Zinder and A.N. Pell, 1998. Phenotypic and phylogenetic characterization of ruminal tannin-tolerant bacteria. Appl. Environ. Microbiol. 64: 3824-3830.

Von Keyserlingk, G.E. and G.W. Mathison, 1993. The effect of ruminal escape protein and ambient temperature on the efficiency of utilization of metabolizable energy by lambs. J. Anim. Sci. 71: 2206-2217.

Molecular diversity of the bacterial community in the rumen of the feral dromedary camel

A.A. Samsudin[1,2], *A.-D.G. Wright*[2] *and R.A.M. Al Jassim*[1]
[1]*School of Animal Studies, Faculty of Natural Resources, Agriculture and Veterinary Science, The University of Queensland, Gatton Campus, QLD 4343, Australia;* [2]*CSIRO Livestock Industries, Queensland Bioscience Precinct, 306 Carmody Road, St Lucia, Queensland 4067, Australia; anjas.samsudin@csiro.au*

Introduction

The dromedary camel is known for its ability to efficiently utilise the vegetation of arid and semi-arid land. Under desert conditions, camels browse on a range of forage plants that are of little nutritional value or are not palatable to cattle, sheep, goats and other herbivores (Kayouli *et al.*, 1993). Australia has the world's second largest area of arid and semi-arid land and the largest population of feral dromedary camels. In Australia, feral camels feed mainly on tannin-rich shrubs and trees, which contain anti-nutritional compounds that can cause illness in other ruminants (i.e. cattle, sheep). In contrast, camels thrive well on tannin-rich plants and increase in numbers suggesting the ability to break down these anti-nutritional compounds and to alleviate their effect. The present study was aimed at investigating the bacterial community diversity of the rumen of the dromedary camel using culture-independent techniques.

Material and methods

Rumen contents from 12 feral dromedary camels were collected after being slaughtered and genomic DNA was extracted. The samples were taken from the digestive compartment 1, also known as the rumen, and DNA was PCR amplified using bacterial-specific 16S rRNA primers (27f and 1492r). PCR products were cloned with suitable vector and were sequenced using an ABI Prism 3730 48 capillary sequencer using Big Dye Terminator and Taq FS. The sequence data were initially analysed using Genbanks' BLAST program to find percent identity to validly describe bacteria and unculturable environmental samples.

Results

A total of 181 sequences ranging in length from 1,250 to 1,450 base pairs were analysed. Eleven of the 181 sequences (6%) represent strains closely related to *Ruminococcus flavefaciens*, *Brevundimonas* sp., *Butyrivibrio fibrisolvens*, *Prevotella* sp., *Succiniclasticum ruminis* and *Anaerovibrio lipolytica*. However, 94% of the 181 sequences (170 sequences) may represent new species. Of the 181 sequences, 144 sequences may represent new genera, whereas 18% and less than 1% of the sequences may represent new families and a new order respectively (Table 1). At the phylum level, there is much similarity of the rumen bacterial population between dromedary camel and cattle (Figure 1). The *Firmicutes* (58.6%) and the *Bacteroidetes* were the most abundant phyla represented in the rumen of the dromedary camel.

Table 1. Breakdown of 181 sequences based on nearest valid species.

Sequence total	98-99.5%	<98%	<93%	<85%	<80%
	'Strain'	'New species'	'New genus'	'New family'	'New order'
181	11 (6%)	170 (94%)	144 (80%)	33 (18%)	1 (0.5%)

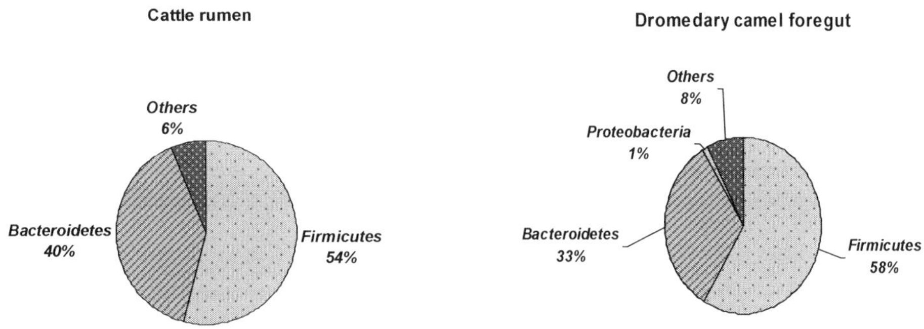

Figure 1. Comparison of bacterial phyla between the rumen of the cattle (Edward et al., 2004) and the rumen of the dromedary camel.

Conclusion

This study presents important preliminary information on the bacterial diversity in the camel rumen, the first such study ever reported. Further, this study provides data that will enhance our understanding of the taxonomy and function of a consortium of a poorly understood microbial group. The use of clone library analysis in this research uncovered many novel non-culturable species that represent a large proportion of the bacteria in the rumen of feral dromedary camels.

References

Edward, J.E., N.R. McEwan, A.J. Travis and R.J. Wallace, 2004. 16S rDNA library-based analysis of ruminal bacterial diversity. Antonie van Leeuwenhoek 86: 263-281.

Kayouli, C., J.P. Jouany, D.I. Demeyer, Ali-Ali, H.Taoueb and C. Dardillat, 1993. Comparative studies on the degradation and mean retention time of solid and liquid phases in the forestomachs of dromedaries and sheep fed on low-quality roughages from Tunisia. Anim. Feed Sci. Technol. 40: 343-355.

Variations in the production of CH_4 per unit of digestible organic matter intake

D. Sauvant and S. Giger-Reverdin
AgroParisTech - INRA, 16 rue Claude Bernard, 75231 Paris Cedex 05, France; sauvant@agroparistech.fr

Introduction

Production of CH4 is frequently considered as proportional to the fermented organic matter in the rumen, or to the digestible OM in the whole tract (DOM). However, Sauvant and Giger-Reverdin (2007) observed that variations in the feeding level (FL) of dry matter intake in% Body Weight (DMI% BW) and in the dietary concentrate (0<PCO<1) could induce contradictory relationships between digestible energy and CH4 energy. To go further on, the present study was performed to elucidate and quantify the variations of the CH4 production per unit of digestible organic matter intake (g/kg DOMI).

Material and methods

Two data bases were built from results of calorimetric studies. The 1st base pooled experiments focussed on the influences of FL (103 experiments, 295 treatments, tr, FL=1.61 ± 0.64, min=0.56, Max=4.01). The 2nd base pooled experiments focussed on the impact of PCO (91 experiments, 275 treatments, PCO=0.39±0.26, min=0.0, Max=0.87). For the bases 1 and 2 respectively, the measures were performed either on growing (143 and 94 tr) or on lactating (0 and 67 tr) cattle, growing sheep (150 and 85 tr) and goats (2 and 31 tr). After a first separate analysis performed to study the marginal effects, the two bases were pooled since the meta-design exhibited a large independence between FL and PCO variations (Figure 1). The two variables studied were CH_4 production (26.3±5.5 g/kg DOMI, min=8.3, Max=36.2) and digestible organic matter (DOM: 66.3±7.5% dry matter intake, min=43.5, Max=88.9). CH_4 was also explained by DOM. Meta-analyses were performed to split among and within experiment influences, only within-experiment relationships were considered. Outliers treatments were excluded for normalised residues >3.

Results

Marginal influences of FL and PCO: FL significantly depressed CH_4 production (33.4 – 4.10 FL; nexp=102, n=292, RMSE=3.0) and dietary DOM (70.0 – 2.27 FL; nexp=94, n=292, RMSE=1.6). The influence of PCO on CH_4 was curvilinear (27.8 + 11.85 PCO – 26.6 PCO^2, nexp=82, n=240, RMSE=2.4). PCO linearly influenced DOM but with a significant interaction exp×covariable (58.4 + 20.0 PCO + Exp × PCO, nexp=91, n=275, RMSE=1.9). This interaction was caused by a logic negative relationship between values of intercept and slope of the experiments (R=-0.66).
Combined influences of FL and PCO: When the two bases were pooled, significant interactions appeared between FL and PCO. For CH_4, all the quadradatic terms were significant (37.6 - 7.6 FL + 1.1 FL^2 + 18.3 PCO – 29.4 PCO^2 - 3.4 FL × PCO; nexp=144, n=430, RMSE=2.5). Figure 2 presents the mapping trace of this equation, the interacting influences of both FL and PCO to reduce CH_4 production appears particularly when FL>2. For DOM, PCO presented a quadratic influence and there was no interaction (64.1 – 2.76 FL + 24.39 PCO - 4.4 PCO^2; nexp=152, n=455, RMSE=2.3).
Relationships between CH4 and DOM: With experiments focussed on FL, there was a linear relationship between CH_4 and DOM (-12.7 + 0.59 DOM, nexp=102, n=292, RMSE=3.8). In contrast, with experiments focussed on PCO, the expression of CH_4 in function of DOM was curvilinear (-2.4 + 1.05 DOM – 0.009 DOM^2, nexp=158, n=467, RMSE=3.7). Despite the significance of

this last equation, its residual variation was significantly higher than that obtained from combined FL and PCO (Figure 2) and was significantly explained by these two items. An analysis of CH_4 production, pooling DOM, FL and PCO was performed, in this case the coefficients of DOM and DOM^2 were not significant while those of FL and PCO were significant, coming back to the equation illustrated by Figure 2.

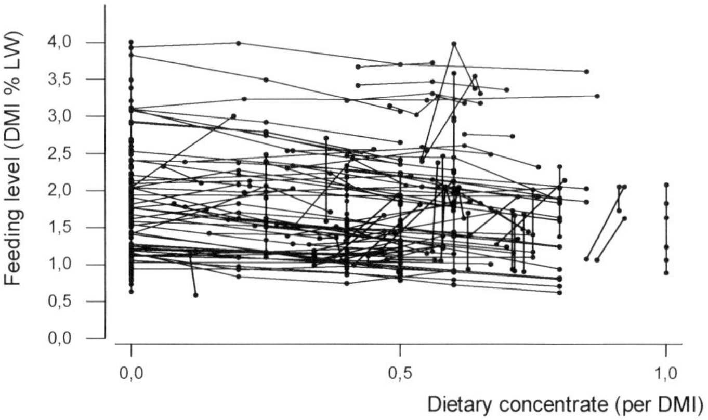

Figure 1. Structure of the meta-design.

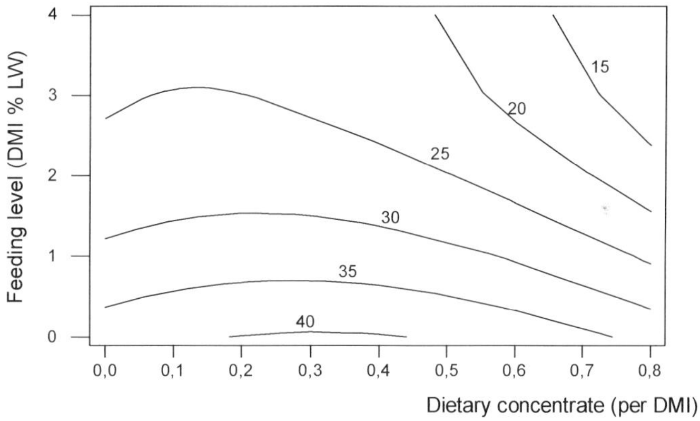

Figure 2. Production of CH_4 (g/kg DOM) as a function of feeding level and dietary concentrate.

Conclusion

Stoechiometry of CH_4 production from DOM was not constant. It can be partly and curvilinearly explained by the dietary DOM. However FL and PCO were more accurate predictors when they were combined. The results stressed that CH_4 production per unit of DOM is particularly inhibited with diets having a high level of DOM or with diets rich in concentrate given at a high level of intake.

References

Sauvant, D. and S. Giger-Reverdin, 2007. Empirical modelling by meta-analysis of digestive interactions and CH_4 production in ruminants. In: I. Ortigues-Marty (Ed.), 2nd EAAP international symposium on energy and protein metabolism and nutrition. Wageningen Academic Publishers, Wageningen, the Netherlands, pp. 561-563.

Trace elements of gastrointestinal tract contents of the European moose (*Alces alces*)

A. Scopin and T. Rukavishnikova
Laboratory of Animal Ecology, Russian Research Institute of Game Management and Fur Farming, Kirov, 610000, 79 Engels St., Russia; scopin@bk.ru

Introduction

Studies on domestication and free-range breeding of many wild ungulates are carried out on the territory of the Russian Federation. The moose is the largest ungulate of the taiga region in Northern Eurasia. Its domestication is quite a difficult task since the biology of this animal has not been well studied (Glushkov, 2001; Heptner and Nasimowisch, 1974). One of the most complicated questions here is the peculiarities of moose nutrition and digestion in nature and in captivity (Weber *et al.*, 1992). In particular, there are extremely few data about the concentration of chemical elements in moose fodder and their metabolic modifications in the organism of this species. Thus, the aim of our research was to evaluate the concentration of trace element concentrations in gastrointestinal tract contents.

Material and methods

All the material was collected on the territory of the Kirov Region located in the southern taiga of the European part of the Russian Federation. During selective shooting in November 2007, eight adult animals at the ages of 1.5 to 4.0 years were taken. Samples were collected during 30 min after shooting. For the analyses, samples from the rumen and rectum of moose were drawn, packed in plastic bags and frozen at -2 °C. Later, the samples were dried at 60 °C. Eight trace elements (Cd, Cr, Cu, Fe, Mn, Ni, Pb, Zn) were determined. To estimate trace elements, an atomic-absorption method based on the standard procedure of forage plant analyses was used (Samokhvalov *et al.*, 2002; Welz and Sperling, 1999).

Results

Among the evaluated elements taken from the intestinal tract, maximum concentrations were found for Mn, Fe, Zn, and minimum concentrations for Cr, Pb, Cd. When the food transits through the gastrointestinal tract, most trace elements change concentration., There were no reliable differences in the amount of Cr and Mn at $P<0.05$ in digested food from the rumen and rectum. The concentration of other elements in food decreases during digestion. The amount of trace elements is considerably higher in rumen contents as compared with faeces (Table 1). The observed differences were reliable at $P<0.05$. The most significant differences were found for concentrations of Pb and Zn ($P<0.01$).

Conclusion

The differences in concentrations of trace elements directly depend on the predominance of certain species of plants in a moose diet (Ohlson and Staaland, 2001). The moose is a browser consuming in the winter mostly shoots and the bark of aspen, birch, mountain ash and willow as well as some of coniferous trees (pine, fir). A characteristic feature of this ungulate is a gradual weakening of digestion and absorption processes during the winter (Weber *et al.*, 1992). However, early in the winter, the decreased amount of microelements in moose faeces is an effect of a strong absorption of chemical elements from food by an organism. Probably, a strong absorption of trace elements in the autumn and at the beginning of winter makes it possible to maintain the metabolism of elements in a moose organism during the whole winter.

Table 1. Concentration of trace elements in gastrointestinal tract contents of the moose.

	Concentration of trace elements, mg/kg		*P*-value
	Rumen contents	Faeces	
Cd	0.30±0.08	0.13±0.01	0.043
Cr	1.56±0.19	2.11±0.28	>0.05
Cu	6.08±0.87	3.24±0.43	0.011
Fe	91.18±8.94	66.47±3.40	0.021
Mn	99.78±21.05	83.18±13.62	>0.05
Ni	2.59±0.36	1.53±0.26	0.033
Pb	1.91±0.18	1.18±0.08	0.002
Zn	39.35±2.83	29.78±1.13	0.007

References

Glushkov, V.M., 2001. The Moose. Ecology and Management of Populations. VNIIOZ RAAS Publ., Kirov, the Russian Federation, 318 pp. (In Russian).

Heptner, W.G. and A.A. Nasimowisch, 1974. Der Elch. A. Ziemsen Verlag, Wittenberg, Lutherstadt, the Germany, 239 pp.

Ohlson, M. and H. Staaland, 2001. Mineral diversity in wild plants: benefits and bane of moose. Oikos. 94: 442-454.

Samokhvalov, S.G., N.A. Chebotareva, G.I. Gorshkova, V.A. Chuikov and Kh.K. Khudyakova, 2002. Vegetable feeds. Methods for determination of microelements. In: Vegetable feeds. Methods of analyses. Russian State Standards, IPK Standards Publ., Moscow, the Russian Federation, p. 13-44. (In Russian).

Weber, A.E., A.F. Simakov, N.I. Chuviyurova, A.V. Chalyshev, L.N. Badlo, T.I. Kochan and N.I. Mochalov, 1992. The Physiology of Nutrition and Metabolism of Moose. Komi Scientific Centre Publ., Syktyvkar, the Russian Federation, 123 pp. (In Russian).

Welz, B. and M. Sperling, 1999. Atomic Absorption Spectrometry. Wiley-VCH Verlag GmbH, Weinheim, the Germany, 965 pp.

Effect of low dietary P on rumen microbial P metabolism and synthesis

J. Sehested, P. Lund, M.R. Weisbjerg and T. Hvelplund
Faculty of Agricultural Sciences, Aarhus University, P.O. Box 50, 8830, Tjele, Denmark; jakob.sehested@djf.au.dk

Introduction

Phosphorus (P) is an essential nutrient for the dairy cow as well as for the ruminal microbes. The microbial metabolism and degradation of feed implies that a significant part of dietary P will be incorporated into e.g. microbial phospholipids and nucleotides before it reaches the intestine. Ruminants are dependent on ruminal fermentation and microbial deficiency of P is likely to reduce ruminal feed degradation, and thereby feed intake and animal performance. According to Durand and Komisarczuk (1988) the requirement of the rumen microbes is about 5 g available P per kg feed digestible organic matter, including the contribution of P in saliva. The recommendations of P for dairy cows have been reduced significantly to about 3.4 g per kg feed dry matter (DM) in recent years due to environmental concerns regarding surplus P in animal slurry (Sehested, 2004). There is limited knowledge on the microbial P requirement in cows at a high feeding and production level, and on the effects of a reduced P allocation on ruminal metabolism and feed degradation. The aim of the present experiment was to determine the effect of reducing dietary P supplementation below current estimates of dairy cow P requirement on rumen microbial P metabolism and synthesis and degradation of NDF.

Material and methods

The original design was a 4×4 Latin square experiment, including two levels of forage digestibility and P-supplementation. Due to a missing observation an additional cow was included in the last three periods; in total, 19 observations were made with rumen and duodenal fistulated lactating Holstein Friesian dairy cows. The experiment complied with the guidelines of the Danish Ministry of Justice. Each period consisted of 21 d. Cows were fed a TMR consisting of (%DM) soy bean meal (12), dried beet pulp (31), molasses (14), maize silage (21), grass silage (21), and NaH_2PO_4 (0.5) for cows on High P only. The content of P in the non-supplemented (Low P) and supplemented (High P) feeds were 2.2 and 3.2 g/kg DM, respectively. Feed offer and refusals were recorded daily. Cows were fed twice daily (08.30 h; 17.30 h), and ten grams of chromic oxide was administrated via the rumen cannula as a flow marker before each of the two daily feedings. During the collection procedure on day 10 to 14, 12 samples of duodenal chyme were taken from each cow and pooled to give a representative sample of the diurnal flow. A rumen evacuation was performed at 12.00 h on day 21, and bacteria were isolated from the liquid and particle fractions. Rumen bacteria were isolated by sequential centrifugation and microbial synthesis was determined based on RNA in isolated bacteria and duodenal chyme. Statistical analysis was done using the GLM procedure in SAS® 9.1 using cow, period and treatment as class variables. Only effects of dietary P level are reported here, since there were no statististically significant interactions between forage digestibility levels and dietary P levels.

Results and discussion

The results are shown in Table 1. Feed intake was unaffected by treatments, whereas dietary P intake differed significantly as planned from 3.2 g P/kg DM (High P) which is close to the physiological requirement, to 2.2 which is significantly below requirement (Sehested, 2004).

Table 1. Intake of feed dry matter (DM) and phosphorus (P). Content of P in ruminal pools of DM and microbes. Flux of P at duodenum and estimated recycling of P to the rumen. Ruminal NDF digestibility and microbial synthesis of organic matter (OM), protein and P-containing compounds. Data are given as least squares means with standard error (SE).

	Diet[1]		SE	Diet effect
	High P	Low P		*P*-value
Feed intake, kg DM/d	19.9	19.9	0.5	0.94
P intake, g/d	63	44	2	<0.0001
Duodenal P, g/d	100	74	3	<0.0001
Minimum P recycling to the rumen, g/d[2]	37	30	4	0.22
P in ruminal content,% of DM	0.53	0.44	0.01	0.0004
P in ruminal pool, g	52	42	2	0.001
P in rumen microbes from fluid,% of DM	1.52	1.44	0.02	0.02
P in rumen microbes from particles,% of DM	1.62	1.73	0.09	0.42
Microbial OM synthesis, kg/d	2.9	3.1	0.1	0.24
P in microbial net synthesis, g/d	54	52	2	0.48
Microbial protein synthesis, g N/d	260	270	10	0.48
Ruminal NDF digestibility,%	65	68	1	0.11

[1] Diets with 2.2 (Low P) and 3.2 (High P) g P/kg feed DM.
[2] Calculated as P flow at duodenum - dietary P intake.

Low P decreased duodenal P flow significantly as compared to High P, whereas recycling of P to the rumen (estimated as P flow at duodenum - dietary P intake) was unaffected. The ruminal allowance of P with feed and saliva can thus be estimated to 5 g P/kg feed DM at High P, which is close to the microbial requirements (Durand and Komisarczuk, 1988), whereas the ruminal allowance of P was significantly lower (3.7 g P/kg feed DM) at Low P. Low P significantly decreased the total ruminal pool and concentration (g/kg DM) of P. The content of P in microbial DM from the ruminal fluid fraction was significantly decreased with Low P, whereas the content of P in microbes from the particle fraction was unaffected. Treatments did not affect ruminal pools of DM, organic matter (OM), or NDF (data not shown). The microbial net syntheses of OM, protein and P in microbial net synthesis were unaffected by treatments. Accordingly, duodenal digestibility of NDF was not affected by treatments.

Conclusion

It is concluded that reduction of dietary P-supplementation from 3.2 to 2.2 g P/kg feed DM reduced ruminal P concentration, the total ruminal P pool, and the content of P in ruminal microbes (from fluid fraction) without affecting salivary recycling of P to the rumen, microbial synthesis and ruminal degradation of NDF.

References

Durand, M. and S. Komisarczuk, 1988. Influence of major minerals on rumen microbiota. J. Nutr. 118: 249-260.
Sehested, J., 2004. Feeding phosphorus (P) to cattle: A new basis for nordic recommendations. Review article. Acta Agricultura Scandinavica, Section A 54: 169-180.

Effects of synchronisation of energy and nitrogen supply on ruminal fermentation and microbial protein synthesis

J.K. Seo, H.J. Kim, J.K. Baek and J.K. Ha
Department of Agricultural Biotechnology, College of Agriculture & Life Sciences, Seoul National University, Seoul, Korea; jongha@snu.ac.kr

Introduction

Proteins reaching small intestine for absorption are composed of three sources: microbial, undegraded dietary and endogenous proteins. Amongst these sources, microbial proteins are by far the most important for ruminant animals in terms of both quantity and quality. Microbial protein that is synthesised in the rumen may provide from 50% to nearly all amino acids required for beef cattle (NRC, 1996). Many factors have been known to influence microbial protein synthesis (MPS) in the rumen and synchronisation of energy and nitrogen supply has been suggested as one of the key factors (Cabrita *et al.*, 2006). The majority of past studies reached energy and nitrogen synchronisation by exchanging feed ingredients (Kaswari *et al.*, 2007) or supplementation of energy or nitrogen sources to animals, so that the effects of synchrony were compounded with that of dietary ingredient. The objectives of the present studies were to assess the effects of synchronisation of energy and nitrogen supply on MPS in steers and to determine the different synchrony indexes (SI) for diets formulated with the same ingredients on rumen fermentation and MPS under *in vitro* condition.

Material and methods

Three Holstein steers (385±20.95 kg) fitted with a rumen cannula were fed three diets, which contained the same level of protein and energy, but different SI in a 3×3 Latin square design. SI of each feed was calculated as suggested by Sinclair *et al.* (1993) from data on fractionation of carbohydrate and crude protein (the Cornell Net Carbohydrate and Protein System (CNCPS) developed in 2004) in residues after nylon bag studies. Total faeces and urine were collected for 3 d and ruminal fluids were sampled 8 times a day after a 14 d adaptation period to determine total tract digestibility, and concentrations of ruminal ammonia N, VFA, urine N and urinary purine derivatives. The modified Tilly and Terry (1963) *in vitro* fermentation system was used to evaluate the effects of synchronisation on ruminal fermentation and MPS. Soybean meal (S) was treated with 0.2% alcalase (ES) or 0.5% formaldehyde (FS) and 3 different combinations of corn(C)-SBM(S) were prepared (C+S; C+ES; C+FS). DM digestibility, pH and ammonia N concentration were measured at 0, 3, 6, 12, 24 h. Purine base contents were measured at 0, 12 h to estimate MPS. An *in vitro* study was conducted as a randomised complete design. All data were statistically analysed by the MIXED procedure of SAS® (SAS Institute Inc. 2000, Cary, NC, USA).

Results

Feeding three different diets having various SI resulted in similar DM digestibility (Table 1). However, steers on the diet having higher SI (0.83) excreted more purine derivatives in urine ($P<0.05$) than those on a lower SI diet, which may indicate that the synchronised diet improved MPS in the rumen. Steers receiving a low SI diet had lower total VFA ($P<0.05$) and a tendency of higher ruminal ammonia N concentrations. Different combinations of corn and soybean meal in an *in vitro* study resulted in different SI values (Table 2). As observed in the *in vivo* study, higher concentrations of purine base suggested more MPS with high SI. Corn-soybean meal combination with higher SI also improved *in vitro* DM digestibility.

Table 1. In vivo *total tract digestibility, ruminal VFA, ammonia N and urinary purine derivative contents as influenced by synchrony index.*

	Synchrony index			SEM
	0.77	0.81	0.83	
Total tract DM digestibility,%	73.53[a]	75.01[a]	73.66[a]	0.56
Total VFA, mM	51.10[b]	97.27[a]	81.07[a]	5.98
Total purine derivative, mg/d	9.76[a]	11.07[ab]	12.46[b]	0.42
Ruminal ammonia N, mg/100 ml	16.18[a]	14.34[a]	13.26[a]	0.62

[a,b] Within the same row, means without common superscript are significantly different (P<0.05).

Table 2. In vitro *ruminal DM digestibility, ammonia N concentration and purine base contents as influenced by synchrony index.*

	Treatments[a]			SEM
	C+FS	C+S	C+ES	
Synchrony index	0.71	0.85	0.95	
DM degradability,%	21.84[c]	31.00[b]	32.27[b]	1.66
Ammonia N, mg/100 ml	2.66[c]	11.27[b]	11.96[b]	1.50
Total purine base, ug/ml	19.52[d]	30.09[c]	33.98[b]	2.16

[a] C = corn, S = soybean meal, ES = soybean meal treated with alcalase (0.2%), FS = soybean meal treated with formaldehyde (0.5%).
[b,c,d] Within the same row, means without common superscript are significantly different (P<0.05).

Conclusion

The present results of an *in vivo* and an *in vitro* experiment reveal that synchronisation of energy and nitrogen supply may improve microbial protein synthesis and dry matter digestion. Since microbial protein synthesis was measured by an indirect method in the present study, the concept may have to be confirmed using direct measurement of microbial protein synthesis.

Acknowledgement

This study was carried out with the support of Cooperative Research Program for Agricultural Science & Technology Development (Project No. 20090101-030-166-001-03-00), RDA, Republic of Korea.

References

Cabrita, A.R.J., R.J. Dewhurst, J.M.F. Abreu and A.J.M. Fonseca, 2006. Critical evaluation of the effects of synchronising the availability of N and energy in the rumen. Anim. Res. 55: 1-24.

Kaswari, T., P. Lebzien and G. Flachowsky, 2007. Studies on the relationship between the synchronization index and the microbial protein synthesis in the rumen of dairy cows. Anim. Feed Sci. Technol. 139: 1-22.

NRC, 1996. Nutrient Requirements of Beef Cattle, 7th edition. National Academy Press, Washington, DC, 17 pp.

Sinclair, L.A., P.C. Garnsworthy, J.R. Newbold and P.J. Buttery, 1993. Effect of synchronizing the rate of dietary energy and nitrogen release on rumen fermentation and microbial protein synthesis in the sheep. J. Agri. Sci. 120: 251-263.

Tilly, J.M. and R.A. Terry, 1963. A two-stage technique for the *in vitro* digestion of forage crops. J. Br. Grassl. Soc. 18: 104-111.

Influence of tree leaf supplementation on nutrient utilisation, rumen fermentation and digesta kinetics in sheep fed *Cenchus ciliaris* grass based diets

S. Singh
Plant Animal Relationship Division, Indian Grassland and Fodder Research Institute, Jhansi, India; singh.sultan@rediffmail.com

Introduction

Sheep are primarily fed on crop residues and grasses by grazing on fallow, community and roadside lands in India and they are usually reared by landless and resource poor sections of the community, sometimes being supplemented with kitchen waste. In the dry season, both the quantity and quality of available crop residues and grasses is poor particularly in available N and minerals. Tree foliage usually rich in protein and minerals is the only supplement to natural pasture, grazing lands and crop residues to increase dietary protein mineral supply to rumen microorganisms (Osuji *et al.*, 1995) to improve rumen function. The effect of tree leaf supplementation with *Pennisetum* hay on sheep rumen fermentation and digesta kinetics has been demonstrated (Navas-Camacho *et al.*, 1993). The present work assesses the response of foliage supplementation to a *Cenchrus ciliaris* grass diet in sheep on nutrient digestion, rumen fermentation and digesta kinetics.

Material and methods

Four adult local male sheep (21-30 kg) were fed *Cenchrus ciliaris* (CC)-*Leucaena leucocephala* (LL) and *Cenchrus ciliaris* (CC)- *Grewia optiva* (GO) diets (75:25 ratios on% DM basis) for 3 months on each diet in a switch-over experiment. Half grass and *LL* and *GO* leaves (collected from whole plant canopy and wilted for 24 h before feeding) were offered once at 9:00 a.m. and the remaining half at 12:00 noon. Feed intake and nutrient digestibility were determined at more than 2 months of feeding by conducting a 7 day metabolism trial using metabolic cages, where 24 h faecal collection was made for each individual animal. More than 1 month feeding rumen liquor samples were collected at 1 month intervals for 2 consecutive days from animals at 0 and 4 h feeding to estimate pH, TVFA and rumen metabolites on each diet using standard methods. Rumen liquid digesta kinetics (volume, out flow rate and dilution rate) was determined by the method of Hyden (1961) using PEG-6000. Data were analysed in cross over design.

Results

CP contents were 5.02, 20.10 and 16.87% in *CC, LL* and *GO* respectively (Table 1).
NDF, ADF and cellulose contents were 10.48, 6.19 and 6.43 units more in GO than LL. Sheep tended to have higher DMI (g/kg $W^{0.75}$) on the GO supplemented diet (Table 2). DM, CP, NDF, ADF, cellulose

Table 1. Chemical composition (% DM basis) of C. ciliaris, L. leucocephala *and* G. optiva.

Nutrients	*Cenchrus ciliaris*	*Leucaena leucocephala*	*Grewia optiva*
Crude protein	5.02	20.10	16.87
Organic matter	92.53	89.10	88.21
NDF	88.76	30.70	41.18
ADF	49.05	19.30	25.49
Cellulose	39.31	11.80	18.13
Hemicellulose	31.71	11.40	15.69
Lignin	4.05	8.10	2.76

and hemicellulose digestibility were ($P<0.05$) higher in sheep on the *CC-GO* than *CC-LL* diet. Tree leaf supplementation did not influence sheep rumen volume and its dilution rate. However, outflow rate was relatively higher in sheep on the *CC-LL* diet. Supplementation of *GO* resulted in a higher TVFA (meq/l) concentration in sheep rumen liquor than with the *LL* supplemented diet (Table 3). On the contrary, *LL* supplementation resulted in higher concentrations of N metabolites than the *GO* supplement.

Table2. DMI and nutrient digestibility in sheep fed grass-tree leaf diets.

		CC-LL	CC-GO
Intake	% body weight	2.57±0.30	2.76±0.17
	g/kg $W^{0.75}$	58.23±4.28	60.21±3.41
Digestibility	DM*	47.98±3.02	56.54±1.45
	OM*	49.67±2.72	59.30±1.34
	CP*	44.02±1.23	57.60±2.03
	NDF*	47.42±3.80	56.80±1.36
	ADF*	40.47±5.00	50.54±1.28
	Cellulose*	50.11±3.89	64.74±1.46
	Hemi-cellulose*	53.53±3.30	63.05±1.45
Liquid digesta	Rumen volume, L	6.52±0.07	7.00±0.05
	Dilution rate,%/h	5.12±0.02	5.14±0.03
	Out flow rate, L/d	8.10±0.04	6.12±0.02

* Differ significantly at $P<0.05$.

Table 3. TVFA and N metabolites in rumen liquor of sheep fed grass-tree leaf diets.

	CC-LL		CC-GO	
	0 h	4 h	0 h	4 h
pH	7.06±0.04	6.98±0.14	6.41±0.04	6.57±0.01
TVFA, meq/l	67.3±10.02	87.5±2.5	91.5±8.29	107.2±7.95
TN, mg/100 ml	92.7±9.42	105.8±6.50	68.5±7.77	99.05±14.43
NH3-N, mg/100 ml	13.06±0.61	18.5±1.05	14.5±0.72	16.4±0.83
TCA-N, mg/100 ml	72.8±5.6	74.2±2.68	51.20±4.71	71.71±5.81

Conclusion

Sheep had higher DMI, nutrient digestibility and TVFA on the *GO* supplemented diet, while N metabolite concentrations and rumen outflow rate were higher with the *LL* supplemented diet. *GO* seems to be a better supplement for improved nutrient utilisation.

References

Hyden, S., 1961. The use of reference substances and measurement of out flow rate in alimentary tract. In: Digestive Physiology and Nutrition of Ruminants. Butterworth, London, UK, p 37.

Navas-Camacho., A., M.A. Laredo., A. Cuesta., H. Anzola and J.C. Leon, 1993. Effect of supplementation with tree legume forage on rumen function. Livestock Research for Rural Development 5: #. Retrieved from http://www.lrrd.org/lrrd5/2/navas.htm

Osuji, P.O., S. Fernandez-Rivara and A. Odenyo, 1995. Improving fiber utilization and protein supply in animals fed poor quality roughages. In: Wallace R.J. and K.A. Lahlou (Eds.), Rumen Ecology Research Planning. Proc. Workshop held at ILRI, Ethiopia, 13-18 March 1995, pp. 1-22.

A new method for simultaneous recording of methane eructation, reticulo-rumen motility and jaw movements in rumen fistulated cattle

A.-K. Skovsted Koch[1], P. Nørgaard[1] and K. Hilden[2]
[1]Department of Basic Animal and Veterinary Sciences, Faculty of Life Sciences, University of Copenhagen, 1870 Frederiksberg C, Denmark; [2]Department. of Basic Sciences and Environment, Faculty of Life Sciences, University of Copenhagen, 1870 Frederiksberg C, Denmark; anne-k@dsr.life.ku.dk

Introduction

Increasing focus on global warming has increased the interest in methane (CH_4) emission from cattle. Total daily CH_4 production from ruminants has been recorded using different respiration chamber techniques and tracer techniques within time intervals up to 24 h. It is well known that CH_4 production and release vary throughout the day depending on level of feed intake, fermentation process, and ruminal environment, and that eructation of CH_4 occurs during secondary contraction cycles of the rumen. To our knowledge, only few studies have recorded CH_4 eructations during eating, rest and rumination, and simultaneous recording of CH_4 eructations and measurement of reticulo-rumen motility has not yet been done. That combination will enable investigation of the eructation mechanisms in ruminants, variation in CH_4-eructation during periods of eating, ruminating or resting and how it can be affected by different feeding. The sulphur hexafluoride (SF_6) tracer technique (Johnson *et al.*, 1994) estimates total CH_4 release by eructation and respiration under different production conditions by sampling air above the nostrils of the animal. Inspired by that technique, new CH_4 censoring equipment was developed. The aim of the present study was to perform a trial using a new non-invasive technique for investigating ruminant CH_4 eructation and its correlation to reticulo-rumen motility.

Material and methods

Four rumen fistulated tie-stalled non-lactating Jersey cows were used in the experiment. They were fed at maintenance level with a hay based diet supplemented with 2 kg concentrate. Jaw movement oscillations (JMO) were recorded using a Hall sensor (Nørgaard and Hilden, 2004). Motility of the reticulum (Re) and ventral rumen sac (VS) was recorded by two pressure transducers (Honeywell, 24PCAFA6D) connected to a logger by electric wires through the rumen fistula. Eructated CH_4 from the rumen was continuously recorded by a CH_4-specific electrochemical TGS sensor (Figaro TGS2611C00) placed underneath a nose-screen just above the nostrils (Figure 1), ensuring capture of the respired and eructated air without preventing the cow's regular behavioural patterns such as eating, drinking, etc. The signal from the four sensors were digitised and sampled at 20 Hz. The experimental setup was tested through 8 to 24 h periods on four cows while eating, drinking, ruminating and resting behaviour was observed. The sensor signals were plotted using the Gplot procedure (SAS® 9.1) and time of primary (PC) and secondary contractions (SC) of the Re and VS was identified according to reticulo-rumen motility patterns found by Backus *et al.* (1993).

Figure 1. Sketch of the position of the CH_4-specific TGS sensor under a nose-screen.

Results

Concentration of CH_4 under the nose-screen, JMO, and pressure variation in the Re and VS during ruminating are shown in Figure 2, where time of PC, SC, and regurgitation of rumen content boluses (B) are marked. Recorded concentration of CH_4 increased about five fold (400-2200 ppm) within 2-3 s after a SC, and then decreased until the next eructation. This indicates a short response time of the TGS sensor and suitability of the CH_4 censoring equipment for identifying occurring CH_4 eructations and SC. Variation along the decreasing CH_4 curve was interpreted as inspiration and expiration of eructated CH_4 through the lungs. All sensor outputs could be retrieved and interpreted nicely during eating, resting or ruminating when the cows were standing. Lying position gave disturbance in the Re and VS outputs, while CH_4 detection was unchanged if not clearer. JMO were unaffected by position.

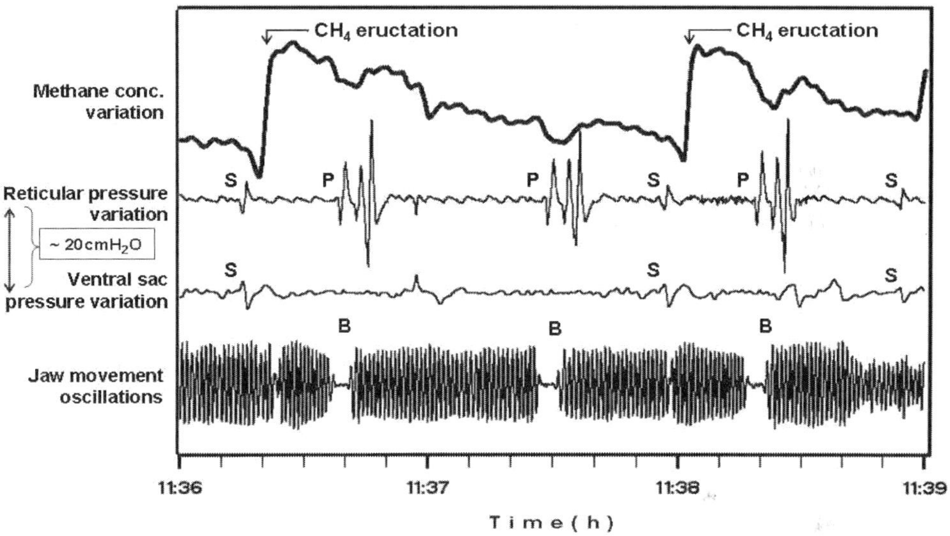

Figure 2. Simultaneous recording of methane concentration in expired air under the nose-screen, reticular pressure, ventral sac pressure, and jaw movement oscillations during 3 min of rumination. P=primary contraction, S=secondary contraction, B=bolus regurgitation.

Conclusion

The present technique appears to have potential for simultaneously studying CH_4 eructations, diurnal variations in rumen fermentation, rumen pH and reticulo-rumen motility during eating, ruminating and resting in ruminants exposed to different dietary treatments. The technique can be applied on different ruminant species kept tied, loose or grazing.

References

Backus, R.C., H.W. Colvin and E.J. DePeters, 1993. The effect of elevation of intrarumen pressure by nitrogen insufflation on eructation in cattle (*Bos taurus*). Comp. Biochem. Physiol. 106A: 731-738.

Johnson, K., M. Huyler, H. Westberg, B. Lamb and P. Zimmerman, 1994. Measurement of methane emissions from ruminant livestock using a SF_6 tracer technique. Environ. Sci. Technol. 28: 359-362.

Nørgaard, P. and K. Hilden, 2004. A new method for recording mastication during eating and ruminating in sheep. J. Anim. Feed Sci. 13 (supplement 1): 171-174.

Dramatic shifts in rapidly fermentable carbohydrates influence mRNA expression of IGFBP3, IGFBP5 and IGFBP6 in rumen papillae

M.A. Steele, S.E. Hook, S.L. Greenwood, O. Al Zahal and B.W. McBride
Department of Animal and Poultry Science, University of Guelph, Guelph, Ontario, Canada; msteele@uoguelph.ca

Introduction

It has been suggested that the rumen adapts to rapidly fermentable (high-grain) diets by increasing the size of rumen papillae, thus maximizing the surface area for nutrient absorption (Gäbel *et al.*, 2002). Increasing dietary grain has been shown to rapidly increase cellular proliferation in the rumen epithelium (Goodlad, 1981). However the molecular mechanisms governing these changes are unknown. It has recently been shown that rumen papillae proliferation is associated with more IGF-1 receptors and increased plasma IGF-1 concentrations (Shen *et al.*, 2004). Since IGF binding proteins (IGFBP) modulate IGF stimulated cellular events (Firth *et al.*, 2002), they may play a role in the regulation of rumen papillae adaptation. Therefore, the objective of this study was to characterise the mRNA expression of three IGFBP (IGFBP3, IGFBP5 and IGFBP6) in rumen papillae during the transition from a high fiber to a rapidly fermentable carbohydrate diet.

Material and methods

Four mature, non-lactating Holstein dairy cattle (760±30 kg) fitted with ruminal cannulae were utilised to meet these objectives. The experimental protocol was conducted over seven weeks. Two diets; high fiber (HF) and a high-grain diet (HG) and the transition from HF to HG were tested. The HF consisted of 100% chopped hay and was fed at 1.4% of BW. The HG consisted of 35% chopped hay and 65% mixed grain and was fed at 1.7% of BW. Cattle were maintained on the HF for several months before a baseline measurement at week 0. The transition to the HG took place over four days and the HG diet was fed for three experimental weeks (HG period; experimental weeks 1 to 3) before switching back to the original HF (HF period; experimental weeks 4 to 6). Rumen papillae were biopsied from the ventral sac on the last day of experimental weeks 0, 1, 3, 4 and 6. In brief, the reticulorumen contents were partially evacuated to facilitate the retraction of the ventral sac. Rumen papillae were excised, washed in ice-cold PBS, frozen in liquid nitrogen and stored at -80 °C until mRNA isolation. Total RNA was isolated and assessed for quality before performing quantitative real-time PCR (qRT-PCR) in duplicate. *GAPDH* was used as the house-keeping gene to calculate Delta threshold cycle values as described by Loor *et al.* (2005). Previously reported IGFBP3 primers (Loor *et al.*, 2005) were utilised and primers were designed for IGFBP5 (5'-CTACAAGAGAAAGCAGTGCAAACC-3', 5'-TCCACGCACCAGCAGATG-3') and IGFBP6 (5'-CGCAGAGACCAACAGAGGAACT-3',5'-GGGACCCATCTCAGTGTCTTG-3') using Primer Express 3.0 (Applied Biosystems). qRT-PCR results are presented as fold-change relative to week 0. A mixed model procedure with repeated measures in SAS® was used to determine differential mRNA expression between weeks (P<0.05).

Results

Differential mRNA expression was observed for all three IGFBP during the HG diet and the expression patterns are displayed in Figure 1. IGFBP5 mRNA was up-regulated during weeks 1 and 3 to 1.60 and 1.79±0.21 fold, respectively. In contrast, IGFBP3 mRNA was down-regulated by 0.53±0.08 fold during week 1 of the HG period and 0.70±0.08 fold by week 3. IGFBP6 mRNA expression followed a similar expression pattern as IGFBP3 as mRNA levels were depressed 0.70

and 0.56±0.13 fold during weeks 1 and 3 of the HG period. The mRNA expression of all three IGFBP during weeks 4 and 6 of the HF diet were not significantly different from the baseline measurement (week 0, HF).

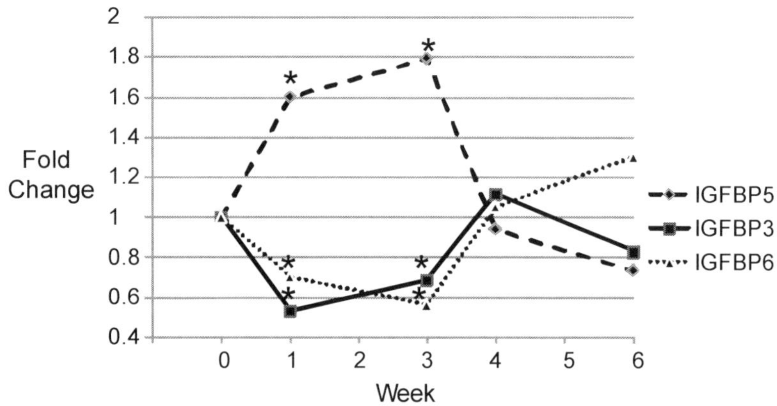

*Figure 1. Patterns of IGFBP3, IGFBP5 and IGFBP6 mRNA expression in rumen papillae throughout the experiment (Baseline, week 0; HG feeding, weeks 1 to 3; HF feeding, weeks 4 to 6). Values are means (n=4) and are expressed as fold-change from the baseline measurement (week 0; CON). Pooled SEM are IGFBP3, 0.08; IGFBP5, 0.21; IGFBP6, 0.13 (** P<0.05).*

Conclusion

These results suggest that mRNA expression levels of IGFBP3, IGFBP5 and IGFBP6 are responsive to shifts in dietary fermentable carbohydrates. In many cell models an up-regulation of IGFBP5 and a down-regulation of IGFBP3 and IGFBP6 correspond with increased cellular proliferation (Firth and Baxter, 2002). Since IGFBP can modulate cellular functions independently of IGF-1, our results suggest that IGFBP are involved in the adaptive response of rumen papillae. These results warrant further investigation of how IGFBP elicit cellular activities in combination with the IGF-axis during the adaptation of the rumen epithelium to rapidly fermentable carbohydrate diets.

References

Firth, S.M. and R.C. Baxter, 2002. Cellular actions of the insulin-like growth factor binding proteins. Endocrine Reviews 23: 824-854.

Gäbel, G., J.R. Aschenbach and F. Müller, 2002. Transfer of energy substrates across the ruminal epithelium: Implications and limitations. Anim. Health Res. Rev. 31: 15-30.

Goodlad, R.A., 1981. Some effects of diet on the mitotic index and the cell cycle of the ruminal epithelium of sheep. Q. J. Exp. Physiol. 66: 487-499.

Loor, J.J., H.M. Dann, R.E. Everts, R. Oliveira, C.A. Green, N.A. Janovick-Guretzky, S.L. Rodriguez-Zas, H.A. Lewin and J.K. Drackley, 2005. Temporal gene expression profiling of liver from periparturient dairy cows reveals complex adaptive mechanisms in hepatic function. Physiol. Genomics 23: 217-226.

Shen Z.M., H.M. Seyfert, B. Lohrke, F. Schneider, R. Zitnan, A. Chudy, S. Kuhla, H.M. Hammon, J.W. Blum, H. Martens, H. Hagemeister and J. Voigt, 2004. An energy-rich diet causes rumen papillae proliferation associated with more IGF type 1 receptors and increased plasma IGF-1 concentrations in young goats. J. Nutr. 134: 11-17.

Effects of various linseed treatments on biohydrogenation of C18:3n3 *in vitro*

A. Sterk[1,2], R. Hovenier[1], B. Vlaeminck[3], A.M. van Vuuren[1] and J. Dijkstra[1]
[1]*Animal Nutrition Group, Wageningen University, Marijkeweg 40, 6709 PG Wageningen, the Netherlands;* [2]*CCL Research, N.C.B.-laan 52, 5462 GE, Veghel, the Netherlands;* [3]*Laboratory for Animal Nutrition and Animal Product Quality, Ghent University, Proefhoevestraat 10, 9090 Melle, Belgium; attje-rieke.sterk@wur.nl*

Introduction

The fatty acid (FA) profile of bovine milk fat can be altered by manipulating dietary FA profile and metabolism in the rumen. Since lipids are extensively altered in the rumen through the processes of lipolysis and biohydrogenation, the FA profile of lipids leaving the rumen (mostly saturated FA (SFA)) differs greatly from the FA profile in the diet (mostly unsaturated FA (UFA)) (Jenkins *et al.*, 2008). Protecting dietary FA to bypass these processes will result in a higher proportion of UFA in the lipids leaving the rumen. Three main protection processes can be distinguished: (1) chemical protection, e.g. through encapsulation in a protein matrix followed by formaldehyde treatment of oilseeds; (2) alterations of FA structure through formation of calcium salts and amides of FA; and (3) technological treatments of oilseeds, such as extrusion, cracking, etc. (Fievez *et al.*, 2007). The objective of this study was to investigate whether various chemically or technologically treated linseed products can effectively change the rumen biohydrogenation kinetics of C18:3n3 *in vitro* compared to pure linseed oil.

Material and methods

Mixed ruminal fluid was collected before the morning feeding from 4 lactating Holstein Friesian cows fitted with a rumen cannula and fed a total mixed ration (TMR) diet. The TMR contained (fresh weight basis) 35.9% ryegrass silage, 54.5% maize silage, 1.1% straw, 0.4% minerals, and 8.2% concentrate (containing 32.8% soybean meal, 32.8% wheat, 32.8% rapeseed meal and 2.0% cane molasses). The mixed ruminal fluid was strained through a double layer of cheese cloth and continuously flushed with CO_2. The ruminal fluid was mixed (1:4, v/v) with a phosphate buffer (per l distilled water: 28.8 g $Na_2HPO_4.12H_2O$; 6.1 g $NaH_2PO_4.H_2O$; 1.4 g NH_4Cl and adjusted to pH 6.8 by adding NaOH solution). Linseed products consisted of pure linseed oil (LO), linseed crushed in a roller mill (0.25 mm; CL), linseed extruded as a mixture of whole linseed and wheat bran (70:30; 127 °C for 20-30 seconds; EL), linseed extruded as a mixture of crushed linseed and wheat bran (70:30; 115 °C for 20-30 seconds; ECL), micronised crushed linseed (115-120 °C for 90 seconds; MCL), and formaldehyde treated crushed linseed (4.5 g formaldehyde per kg crushed linseed; FCL). A freeze-dried and ground (1 mm) sample of the TMR was used as the incubation substrate. The treatments comprised 1 g TMR with 0.06 g LO, 0.89 g TMR with 0.17 g CL, MCL or FCL, and 0.82 g TMR with 0.24 g EL or ECL, thereby providing equal amounts of supplemented FA and fermentable substrate. Substrate and 50 ml of the rumen fluid-phosphate buffer mixture were added to glass gastight incubation flasks (150 ml) under anaerobic conditions. Flasks were flushed with CO_2 before incubation started in a shaking water bath at 39 °C for 0.5, 1, 2, 4, 6, 12, and 24 hours. All incubations were repeated in a second incubation run on separate days. At the end of the incubation periods, flasks were removed and placed on ice. Zero hour samples were not incubated, but immediately placed on ice. After incubation, pH was measured and the incubation residues were then freeze dried. The FA were extracted, methylated and analysed by gas chromatography (Carlo Erba 8560 HRGC). Methylated FA were separated using a fused silica capillary column (100 m, 0.25 mm, i.d. 0.2μm thickness, Supelco; SP2560). Disappearance of C18:3n3 from the incubation flasks at each sampling time was used to estimate the fractional biohydrogenation rate and lag time according to the exponential model of Ørskov and McDonald (1979). Effective C18:3n3 biohydrogenation was calculated according to Dhanoa *et al.* (1999) assuming a fractional passage rate of 6%/h.

Results

After 24-h incubations *in vitro* the amount of C18:3n3 was higher in the EL and FCL treatments compared to the LO, ECL and MCL treatments (Table 1). The amount of C18:3n3 in the flasks after 24-h incubations in the EL treatment was also higher compared to the CL treatment. The estimated fractional rates of biohydrogenation and lag time did not differ between the various treatments of linseed. The calculated effective biohydrogenation differed between the linseed treatments: EL had a lower effective biohydrogenation compared to ECL; EL tended ($P<0.07$) to have a lower effective biohydrogenation compared to LO, and FCL tended ($P<0.07$) to have a lower effective biohydrogenation compared to ECL and LO.

Table 1. Estimated biohydrogenation for C18:3n3 for linseed treatments incubated in vitro.

Treatment	LO	CL	EL	ECL	MCL	FCL	SEM	*P*-value
C18:3n3 24-h,%[1]	36.7^{c}	39.6^{bc}	65.6^{a}	30.0^{c}	35.3^{c}	59.4^{ab}	4.5	<0.001
Kh,%/h[2]	6.65	4.13	1.79	5.93	4.53	2.47	1.31	0.223
Lag time, h	1.49	2.21	0.25	2.00	2.01	2.99	0.50	0.104
Eff. bh,%[3]	43.5^{ab}	35.6^{ab}	22.7^{b}	43.7^{a}	38.1^{ab}	24.3^{ab}	3.5	0.026

[1] C18:3n3 24-h =% of C18:3n3 of total C18 FA after 24-h incubations relative to 0-h incubation.
[2] Kh = fractional rate of biohydrogenation; [3]eff. bh = calculated effective biohydrogenation.

Conclusion

The results of this study confirm that extrusion of whole linseed and the chemical treatment of crushed linseed with formaldehyde result in protection of C18:3n3 against rumen microbial biohydrogenation. This was shown by the higher residue of C18:3n3 after 24-h incubations and the lower calculated effective biohydrogenation.

References

Dhanoa, M.S., J. France, S. López, J. Dijkstra, S.J. Lister, D.R. Davies and A. Bannink, 1999. Correcting the calculation of extent of degradation to account for particulate matter loss at zero time when applying the polyester bag method. J. Anim. Sci. 77: 3385-3391.

Fievez, V., B. Vlaeminck, T.C. Jenkins, F. Enjalbert and M. Doreau, 2007. Assessing rumen biohydrogenation and its manipulations *in vivo*, *in vitro* and *in situ*. Eur. J. Lipid Sci. Technol. 109: 740-756.

Jenkins, T.C., R.J. Wallace, P.J. Moate and E.E. Mosley, 2008. Board-invited review: Recent advances in biohydrogenation of unsaturated fatty acids within the rumen microbial ecosystem. J. Anim. Sci. 86: 397-412.

Ørskov, E.R. and J. McDonald, 1979. The estimation of protein degradability in the rumen from incubation measurements weighed according to rate of passage. J. Agric. Sci. (Camb.) 92: 499-503.

The ruminal anion channel: a pathway for the efflux of SCFA

F. Stumpff, M. Georgi and H. Martens
Institute of Veterinary Physiology, Free University of Berlin, 14163 Berlin, Germany; stumpff@zedat.fu-berlin.de

Introduction

Rapid absorption of SCFA from the rumen is of crucial importance for the prevention of ruminal acidosis, but our understanding of the absorptive pathway is incomplete. Uptake of SCFA leads to a stimulation of sodium absorption via NHE (Gäbel *et al.*, 1991), suggesting that SCFA acidify the intraepithelial space either by uptake of the undissociated SCFA, or via exchange of the anion for HCO_3^- (Aschenbach *et al.*, 2009). Conversely, basolateral efflux of acetate probably occurs through a recently characterised large conductance anion channel (Stumpff *et al.*, 2009) that has previously been postulated to mediate efflux of chloride (Sehested *et al.*, 1999; Leonhard-Marek *et al.*, 2006). However, acidification of the ruminal epithelium by SCFA has never been directly demonstrated, and the basolateral efflux pathway for SCFA other than acetate remains unclear. In the current study, the response of intraepithelial pH to the application of SCFA was studied in intact tissue using pH sensitive microelectrodes. The permeability of the anion channel for propionate was studied on isolated cells from the ruminal epithelium using the patch clamp technique.

Material and methods

After isolation (Abdoun *et al.*, 2005), rumen epithelium was incubated with standard electrolyte solution (mmol/l): 140 Na^+, 5 K^+, 1 Ca^{+2}, 1 Mg^{+2}, 104 Cl^-, 1 $H_2PO_4^-$, 2 HPO_4^{-2}, 10 Hepes and 10 glucose with either 40 gluconate$^-$ or 40 SCFA$^-$ (25 acetate, 10 propionate, 5 butyrate) added as indicated. Piggy back double barrelled microelectrodes were prepared from filamented GC120F10 and GC150F15 borosilicate glass tubing (Harvard Apparatus, Kent, UK). The ion selective channel was silanised with dichloromethylsilane and filled with Hydrogen Ionophore I – Cocktail A (both Sigma Aldrich) and 100 mmol/l KCl, 20 mmol/l Hepes, pH 7.4. The reference channel was filled with 0.5 mmol/l KCl.
Patch clamp experiments were performed as described previously (Stumpff *et al.*, 2009). The pipette solution contained (mmol/L) 131 Na^+, 121.2 gluconate$^-$, 1.1 Mg^{2+}, 2 Ca^{2+}, 5 K^+, 1 $H_2PO_4^-$, 20 Cl^-, 5 EGTA and 10 Hepes (7.2/Trizma). Extracellular solutions contained 131 Na^+, 130 of the anion as indicated and 0.9 Mg^{2+}, 1.7 Ca^{2+}, 5 K^+, 1 $H_2PO_4^-$, 10.2 Cl^- and 10 Hepes (7.4/Trizma). Osmolarity was adjusted to 290 mOsmol/l with the dominant salt. Results are given as means ± SEM, significance levels were tested with the student t-test if normally distributed or the Wilcoxon test otherwise.

Results

After calibration, microelectrodes were lowered onto the epithelium. A drop in pH from 7.4 to a pH of 7.2±0.5 (P=0.05, n=7) signalled proximity of the transporting layer of cells. Impalement of the cytosolic space of the tissue was judged to have taken place when the potential dropped to negative values, combined with a change in fractional apical resistance from 0% to 50 ± 10%. Cytosolic pH was 7.08±0.09 (n=7, P=0.002 versus pH of bulk solution (7.4)). When perfusion was switched to a solution with pH 6.4, cytosolic pH declined slowly (-0.05±0.02 pH/min) to 6.8±0.2 (n=7, P=0.004 vs. original level). Following addition of SCFA, the rate of acidification increased markedly (to -0.10±0.01 pH/min, P=0.02 vs. acidification before addition of SCFA) with cytosolic pH dropping to 6.60±0.15 (n=5). Return to a pH of 7.4 resulted in a return of pH to 6.9 ± 0.1 (+0.17±0.08 pH/min, P= 0.005 vs. previous).

In the patch clamp experiments, cells in NaCl solution had a reversal potential of -29 ± 2 mV in NaCl solution (n=14), indicating a conductance for chloride. Exposure to Na-gluconate solution depolarised the cells to 2.5±1.5 mV (n=10, P<0.001). Replacement with Na-proprionate resulted in a decrease in reversal potential to -9.6±1.3 mV (P<0.001), suggesting electrogenic influx propionate$^-$ in the form of the anion. DIDS (200 μmol/L), which is known for its voltage dependent block of anion channels, blocked outward current to 36±6% of the level without DIDS (n=14, P<0.001), but had no significant effect on reversal potential (-5.1±1.5 mV). After return to NaCl solution, reversal potential returned to -28±2 mV (n=12, P<0.001). The relative conductance of propionate versus chloride was 0.45±0.03, significantly higher than that for gluconate (0.33±0.05, P=0.007), but slightly though not significantly lower than the corresponding value for acetate (0.53±0.03 (n=28)) from a previous study (Stumpff *et al.*, 2009).

Conclusion

The microelectrode experiments confirm the presence of an apical pH microclimate within the stratum corneum and the acidifying effects of SCFA as suggested previously (Gäbel *et al.*, 1991). The patch clamp data show that the ruminal anion channel conducts not only acetate as recently reported (Stumpff *et al.*, 2009), but also propionate. We suggest that in the intact tissue, this channel is basolaterally located and mediates efflux of anions such as chloride, acetate and propionate, driven by the potential generated by the Na^+-K^+-ATPase. This mechanism should be helpful in maintaining cellular volume and thus, contribute to the ability of the ruminal epithelium to absorb SCFA.

Acknowledgement

The authors wish to thank the Deutsche Forschungsgemeinschaft (DFG STU 258/4-1) and the Margarete-Marcus-Charity for their financial support.

References

Abdoun, K., F. Stumpff, K. Wolf and H. Martens, 2005. Modulation of electroneutral Na transport in sheep rumen epithelium by luminal ammonia. Am. J. Physiol. 289: G508-20.

Aschenbach, J.R., S. Bilk, G. Tadesse, F. Stumpff and G. Gabel, 2009. Bicarbonate-dependent and bicarbonate-independent mechanisms contribute to non-diffusive uptake of acetate in the ruminal epithelium of sheep. Am. J. Physiol. (Epub).

Gäbel, G., S. Vogler and H. Martens, 1991. Short-chain fatty acids and CO_2 as regulators of Na^+ and Cl^- absorption in isolated sheep rumen mucosa. J. Comp. Physiol. B 161: 419-426.

Leonhard-Marek, S., G. Breves and R. Busche, 2006. Effect of chloride on pH microclimate and electrogenic Na+ absorption across the rumen epithelium of goat and sheep. Am. J. Physiol. Gastrointest. Liver Physiol. 291: G246-252.

Sehested, J., L. Diernaes, P.D. Møller and E. Skadhauge, 1999. Transport of butyrate across the isolated bovine rumen epithelium--interaction with sodium, chloride and bicarbonate. Comp. Biochem. Physiol. A Mol. Integr. Physiol. 123: 399-408.

Stumpff, F., H. Martens, S. Bilk, J.R. Aschenbach and G. Gäbel, 2009. Cultured ruminal epithelial cells express a large-conductance channel permeable to chloride, bicarbonate, and acetate. Pflugers Arch. 457: 1003-1022.

Novel technique for tracing ingestive and ruminative behaviours

Suhubdy, B.A. Young[1,2], D.R. ZoBell[3] and F.D. Provenza[4]
[1]Research Centre for Tropical Rangeland and Grazing Animal Production Systems, Faculty of Animal Science, University of Mataram, Mataram NTB, Indonesia; [2]School of Animal Studies University of Queensland, Gatton Campus, Gatton Qld 4343 Australia; [3]Depatment of Animal, Dairy, and Veterinary Sciences, Utah State University, Logan UT, USA; [4]Department of Wildland Resources, Utah State University, Logan UT, USA; suhubdy1960@gmail.com

Introduction

Eating and/or rumination are sequential processes. Eating is initiated by prehending feed, followed by mastication to form a bolus, and swallowing to cease the activity. Rumination begins with aspiration or regurgitation of the ingested bolus, re-swallowing the liquid and fine materials, re-mastication of the rough materials, and finally re-swallowing of the finely ground regurgitated bolus. Clearly identifying ingestive and ruminative behaviors would help to determine and predict how feed quality, palatability, diet selection, and feed intake of ruminant livestock are interrelated (Beauchemin and Iwaasa, 1993; Suhubdy, 2002). Past research to measure ingestive and/or ruminative behaviours relay on tracing only the pattern of movements of the jaw during eating and/or ruminating (Beauchemin *et al.*, 1989; Gordon, 1995). However, the limitation of tracing only the pattern of jaw movement cannot capture and distinguish when swallowing occurs. A limited effort has been done to distinguish the behaviour of swallowing (Suhubdy *et al.*, 1999; Suhubdy, 2002). This paper depicts the traces of jaw and laryngeal movements during eating and rumination and describes the importance of traces for identifying eating and ruminating behaviours of ruminants.

Material and methods

Four 2-3 year old Merino whether sheep were used in this trial. The sheep were kept indoor and fed a basal diet consisting of 700 g alkali-treated bagasse pellets (91% DM, 1.68% CP, and 70% NDF) and 300 g of cottonseed meal (90% DM, 38.59% CP, and 32% NDF). Drinking water and mineral lick were available at all times. A sensitive and robust *Suhubdy-collar* sensor was used (Suhubdy *et al.*, 1999; Suhubdy, 2002). For capturing the jaw and laryngeal movement patterns during eating and rumination, two similar sensors were attached to each sheep. Each sensor was connected to the Neotrace chart recorder (Neomedix System Pty Ltd., Sydney, Australia) and analogical signals were recorded on chart paper.

Results

Figure 1 shows the traces of jaw and laryngeal movement patterns during eating (A) and ruminating (B). It was observed that the laryngeal movement patterns were consistent when eating and ruminating. Patterns were consistent except when swallowing, which shows amplitudes. Jaw movement patterns were different during eating and ruminating activities. During eating, there were two types of patterns expressing the event of nibbling and mastication ingested feed. Sheep use their lip when prehending and nibbling patterns were short in amplitude and faster in rates. When masticating, sheep used their molars to grind the feed and patterns were higher in amplitude and consistent in rates. In addition to rumination, the jaw movement patterns were consistent in amplitude. Since swallowing is the end process of eating and ruminating activities, the inclusion of the measurement of laryngeal movement patterns could give more meaning to the jaw movement patterns since they were not shown in the measurement of eating and ruminating behaviour using only a jaw movement indicator.

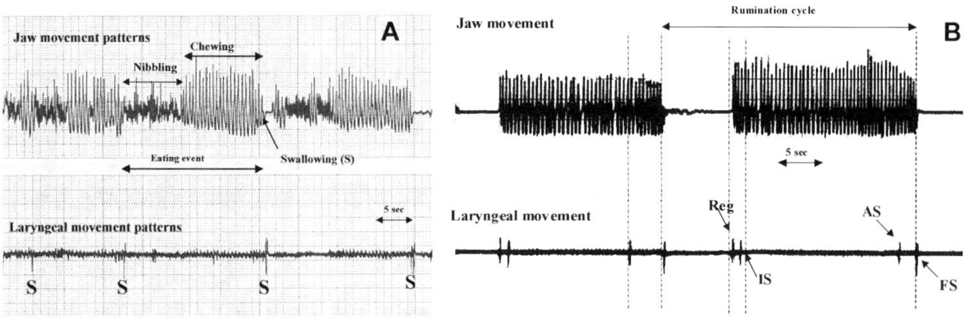

Figure 1. A combined jaw and laryngeal movement patterns during eating (A) and ruminating (B). The letters S, AS, FS, IS, and Reg denote swallowing, additional swallowing, final swallowing, immediate swallowing, and regurgitation.

Conclusion

Simultaneously capturing the pattern of jaw and laryngeal movements using two sensors allowed us to clearly identify the eating and ruminating behaviours of sheep and/or other ruminants. For long research observation, we need to develop a data logger that can record data for 24 h measurements.

Acknowledgement

The first author is Fulbright Senior Visiting Scholar at the Department of Wildland Resources, Utah State University, Logan UT 84322 USA. Thank to A. Goodwin, R. Englebright and K. Vockenson for assistance during the experiments in The School of Animal Studies, The University of Queensland, Gatton Campus, Gatton - Queensland 4343 Australia.

References

Beauchemin, K.A., S. Zelin, D. Genner and J.G. Buchanan-Smith, 1989. An automatic system for quantification of eating and ruminating activities of dairy cattle. J. Dairy Sci. 72: 2746-2756.

Beauchemin, K.A. and A.D. Iwaasa, 1993. Eating and ruminating activities of cattle fed alfalfa or orchardgrass harvested at two stages of maturity. Can. J.Anim.Sci. 73: 79-88.

Gordon, I.J., 1995. Animal-based techniques for grazing ecology research. Small Rum. Res. 16: 203-214.

Suhubdy, B.A. Young, R.S Copland and L. Blagbrough, 1999. Jaw and laryngeal movement patterns during eating and drinking. South African J. Anim. Sci. (Special Issue ISRP) 29: 48-49.

Suhubdy, B.A., 2002. Ingestive behaviour in sheep. Ph.D Thesis, The University of Queensland, Brisbane Qld, Australia, 160 pp.

Postnatal changes in the expression of the ruminal monocarboxylate transporter 1

F. Taifour[1], J. Steinhoff[2], H. Pfannkuche[1], H.M. Hammon[2] and G. Gäbel[1]
[1]Institute of Veterinary Physiology, An den Tierkliniken 7, 04103, Leipzig, Germany; [2]FBN Dummerstorf, Wilhelm-Stahl-Allee 2, 18196, Dummerstorf, Germany; gaebel@vetmed.uni-leipzig.de

Introduction

As previously shown (Müller *et al.*, 2002), a monocarboxylate transporter 1 (MCT1) is located in the *Str. basale* of the rumen epithelium and facilitates transport of monocarboxylic acids from cytosol of the ruminal epithelial cells into the blood. Therefore MCT1 can be regarded as a bottleneck for the transepithelial transfer of short chain fatty acids (SCFA: mainly acetate, propionate and n-butyrate) and their intraepithelial catabolites. Recent studies in the colon support the view that MCT1 is subjected to regulation on the transcriptional level (Cuff *et al.*, 2002). SCFA produced in the lumen are regarded as the main cause for the changes of MCT1 expression in the colon (Cuff *et al.*, 2002). However, systemic influences on MCT1 expression cannot be ruled out. The objective of the present study was to investigate the effects of systemic influences on MCT1 expression in the forestomach in the absence of SCFA. To achieve the absence of SCFA, we looked at the early postnatal period. The forestomach is not colonised during this time. SCFA are not produced and can, therefore, be ruled out as triggers for changes in the expression of MCT1.

Material and methods

Twenty-eight Holstein Frisian calves were divided into four groups according to their feeding and age (n=7 per group). Treatment of the calves is summarised in Table 1. After slaughtering, tissue samples were collected from the atrium ruminis, the ventral and dorsal sac of the rumen, the reticulum and the omasum. MCT1 expression and distribution was evaluated by indirect immunofluorescence on cryostat sections of the samples. Expression rate was electronically quantified on photomicrographs (three per section, data were pooled per animal) by measuring mean pixel intensity in the stained regions. To exclude variations between the sections, the difference between labelled regions of MCT1 within the *Str. basale* and the background fluorescence of the epithelium was calculated.

Table 1. Postnatal treatment of calves.

Group	Parturition	Feeding	Slaughter time
1	Natural	Colostrum	4 d postnatal
2	Natural	Colostrum based formula diet[1]	4 d postnatal
3	Caesarean section, ~9 d before calculated parturition	24 h fasted, 1× colostrum	1 d postnatal
4	Natural	24 h fasted, 1× colostrum	1 d postnatal

[1] Composition of formula diet was daily adjusted to the mineral, nutrient and energy composition analysed in the colostrum fed in group 1.

Results

Immunohistochemical staining confirmed that MCT1 was expressed in all of the sections examined from every group. Immunoreactivity was solely expressed in the *Str. basale* in the forestomach epithelia. In the atrium ruminis and in the ventral and dorsal rumen sac, MCT1 expression was higher in the older groups (1 and 2) than in the younger groups (3 and 4) (Figure 1). Developmental changes could not be observed in the reticulum and the omasum (Figure 1).

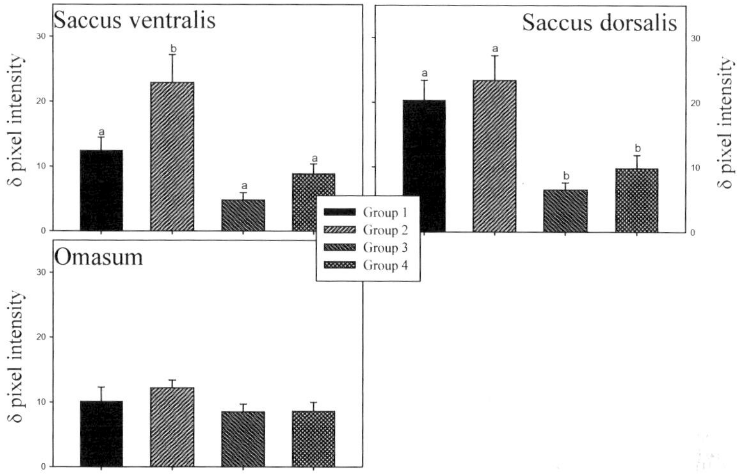

Figure 1. Difference of fluorescence intensity (measured as difference [δ] of pixel intensity) between MCT1-labelled regions and unlabelled regions in various parts of the forestomach. Changes in the atrium ruminis and in the reticulum are described in the text. Means ± SEM, n=7. a, b: different letters indicate significant differences within one graph, ANOVA with subsequent Student-Newman-Keuls-Test.

Conclusion

MCT1 is expressed in the *Str. basale* of the rumen, reticulum and omasum. Expression rate in the rumen strongly increases after birth. Since the forestomach is not colonised at that time, induction by luminal substrates like SCFA does not seem to be responsible for the upregulation. In contrast, systemic stimuli occurring during or shortly after birth probably have a regulatory influence on MCT1 expression.

References

Cuff, M.A., D.W. Lambert and S.P. Shirazi-Beechey, 2002. Substrate-induced regulation of the human colonic monocarboxylate transporter, MCT1. J. Physiol. 539: 361-371.

Müller, F., K. Huber, H. Pfannkuche, J.R. Aschenbach, G. Breves and G. Gäbel, 2002. Transport of ketone bodies and lactate in the sheep ruminal epithelium by monocarboxylate transporter 1. Am. J. Physiol. Gastrointest. Liver Physiol. 283: G1139-1146.

Digestion site and extent of nitrogen fractions in growing steers fed maize silage and lucerne hay with different ratios

K. Taniguchi, K. Yukizane, T. Obitsu and T. Sugino
Graduate School of Biosphere Science, Hiroshima University, Higashihiroshima, 739-8528, Japan; kohzo@hiroshima-u.ac.jp

Introduction

Maize silage (MS) and lucerne hay (LH) among various forages are characterised by high-energy and high-CP forage, respectively, and their combination is important to formulate high forage diets. The intestinal digestibility of rumen undegradable protein from forages is less than that of concentrate feeds, estimated by the mobile bag technique (Frydrynch, 1992). However, the intestinal digestibility using mobile bags varies with the ruminal incubation time and the assumed passage rate (Haugen *et al.*, 2006). Furthermore, although nitrogen (N) bound to cell wall may relate to intestinal N availability, the small residual samples from mobile bags make the chemical analysis difficult. Therefore, we investigated N utilisation, especially on N bound to the cell wall in the rumen and intestine of growing steers fed MS and LH in different ratios.

Material and methods

Four Holstein steers (initial BW: 236±7.2 kg) fitted with duodenal cannulae were used in a 4×4 Latin square design. Diets consisted of MS and LH in the ratios of 80:20 (M8L2), 60:40 (M6L4), 40:60 (M4L6) and 20:80 (M2L8), which contained 10.5, 12.0, 13.5 and 15.0% CP in DM, respectively. Maize silage containing 9.0% CP and 56.1% NDF in DM was harvested at dough stage and prepared with a roll-baler. Lucerne hay containing 16.6% CP and 52.6% NDF was a commercial product. Each experimental period consisted of a minimum of 14 days for adaptation and 4 days for collection of total faeces and twelve duodenal samples. Feeding level was restricted at 95% of *ad libitum* intake to measure N digestion. Steers were offered the diets and orally dosed 2 g of chromic oxide twice daily (08:00 and 20:00 h). Neutral detergent insoluble nitrogen (NDIN) was assayed by Kjeldahl N after ND solution treatment without amylase and sodium sulfite. Acid detergent insoluble nitrogen (ADIN) was assayed after AD solution.

Duodenal DM flow was estimated by dividing faecal excretion of chromic oxide and acid detergent lignin by their duodenal concentration, and averaged. All data were analysed as a 4×4 Latin square using the GLM procedure of SAS® (2000). The model included period, steer, treatment and residual error. Significance of treatment effects was declared at $P<0.05$, and tendency at $P<0.10$.

Results and discussion

Intake of DM, total N, NDIN and ADIN increased linearly with increased LH levels (Table 1). Proportion of NDIN to total N intake also increased from 0.35 to 0.46, but that of ADIN decreased from 0.15 to 0.10 with LH levels. Duodenal flow of NDIN was not affected by MS and LH ratios. Consequently, the ruminal digestibility of NDIN linearly increased from 40 to 74% with LH levels. Duodenal flow of ADIN slightly increased with LH levels, and the ruminal digestibility was similar among diets (35% at mean). Both faecal excretions of NDIN and ADIN were not affected by the diets, and their duodenal flows were almost the same. Because the proportion of NDIN in non-ammonia N entered the duodenum decreased from 0.17 to 0.14 with LH levels ($P<0.05$), the intestinal digestibility of non-ammonia N tended to slightly increase ($P=0.07$), ranging from 60 to 64%. To exclude the influence of NDIN, when intestinal available N fraction was calculated by subtracting faecal NDIN from duodenal non-ammonia N, the digestibility slightly decreased with

LH levels but averaged 74%. Assuming that 20% of duodenal microbial N was excreted in faeces, however, the intestinal digestibility of non-ammonia non-microbial N including NDIN varied 29 to 42% among diets.

Table 1. Intake, duodenal flow and faecal excretion of nitrogen (N) fractions in steers fed maize silage and lucerne hay diets varying the ratio.

	Diet[1]				SE	Linear effect (*P*)
	M8L2	M6L4	M4L6	M2L8		
Intake						
DM, kg/d	4.49	4.94	4.90	5.44	0.282	<0.05
Total N, g/d	75.6	95.6	104.0	130.0	5.79	<0.01
NDIN[2], g/d	26.0	36.5	43.7	58.4	2.50	<0.01
ADIN[3], g/d	11.2	12.4	12.6	14.1	0.81	<0.05
Duodenal flow						
Microbial N, g/d	54.5	55.4	59.3	60.7	4.80	0.34
NANMN[4], g/d	33.4	43.6	37.5	46.8	3.31	0.06
NDIN, g/d	15.3	17.3	14.4	15.0	1.30	0.56
ADIN, g/d	6.9	8.0	7.9	8.6	0.34	<0.05
Faecal excretion						
Total N, g/d	34.6	36.3	36.7	39.9	2.47	0.24
NDIN, g/d	16.5	16.3	15.2	15.6	1.23	0.53
ADIN, g/d	7.5	8.1	8.8	8.4	0.62	0.25

[1] Diet consisting maize silage and lucerne hay in the ratio of 80:20, 60:40, 40:60 and 20:80, respectively.
[2] Neutral detergent insoluble nitrogen.
[3] Acid detergent insoluble nitrogen.
[4] Non-ammonia non-microbial nitrogen.

Conclusion

Ruminal digestibility of NDIN in lucerne hay was greater than that of maize silage because of the lower proportion of ADIN to NDIN. Ruminal digestibility of NDIN in forage could be estimated from intake and faecal excretion of NDIN. Additionally, ruminal indigestible NDIN of forage largely affects the intestinal digestibility of non-ammonia non-microbial N.

References

Frydrynch, Z., 1992. Intestinal digestibility of rumen undegraded protein of various feeds as estimated by the mobile bag technique. Anim. Feed Sci. Technol. 37: 161-172.

Haugen, H.L., S.K. Ivan, J.C. MacDonald and T.J. Klopfenstein, 2006. Determination of undegradable protein digestibility of forages using the mobile nylon bag technique. J. Anim. Sci. 84: 886-893.

Ruminal fatty acid profile and fermentation characteristics in ewes fed sunflower and fish oils

P.G. Toral[1], G. Hervás[1], K.J. Shingfield[2], V. Toivonen[2], A. Belenguer[1] and P. Frutos[1]
[1]Instituto de Ganadería de Montaña (CSIC-ULE), 24346 Grulleros, León, Spain; [2]Animal Production Research, MTT Agrifood Research Finland, 31600 Jokioinen, Finland; pg.toral@eae.csic.es

Introduction

Fish oil (FO) in the diet alters milk fatty acid (FA) composition, enhances milk fat conjugated linoleic acid content (CLA), and also induces milk fat depression (MFD) in lactating cows (Loor *et al.*, 2005). Even though ewes appear less prone to MFD, there is some evidence that supplementing high-concentrate diets with 10 g FO and 20 g sunflower oil (SO)/kg resulted in transient increases of milk fat content (+12%) after 3 d of supplementation, followed by a sustained decrease (up to -27%; Toral *et al.*, unpublished). Under these circumstances, MFD in ewes was not explained by an increase in milk fat *t*10 *c*12-CLA concentration. Changes in milk FA composition were characterised by a severe reduction in 18:0 together with an increase in *trans*-18:1 content. Furthermore, the addition of SO and FO decreased DM intake (-19%) and reduced milk yield (-14%) which may indicate a negative impact of lipid supplements on ruminal fermentation in sheep fed high-concentrate diets. The current experiment was therefore designed to examine the effect of a mixture of SO and FO in the diet on ruminal FA profile and fermentation characteristics in ewes.

Material and methods

Five cannulated Merino ewes (BW: 63±6.0 kg) were fed a total mixed ration (control diet; 35% dehydrated alfalfa hay and 65% concentrate; 41 g DM/kg $BW^{0.75}$) for 11 d (adaptation period). Thereafter, the diet was supplemented with 20 g SO plus 10 g FO/kg DM (SFO treatment) and offered to the sheep for 11 more days. This experimental design was used to avoid carry over effects of oils on ruminal microbiota and biohydrogenation (BH). At the end of the adaptation period (Control) and after 3 (SFO_3) and 10 (SFO_{10}) d on the SFO diet, *in vivo* pH and lactate, ammonia and total VFA concentrations, and *in situ* alfalfa hay DM, NDF and CP disappearance after 24 h incubation (DMD, NDFD, and CPD, respectively) were measured (Toral *et al.*, 2009). Ruminal digesta collected at 0, 3, 6, 12, 18, and 24 h post-feeding was composited to provide single daily samples per ewe and submitted for FA analysis (Shingfield *et al.*, 2003). *In vivo* data taken over time (hours post-feeding) were analysed by repeated measures, and *in situ* measurements and FA composition data by one-way analysis of variance using the MIXED procedure of SAS® (Version 9.1).

Results

Ewes consumed all the feed offered throughout the experiment. The SFO supplement had no effect (P>0.10) on ruminal fermentation characteristics, DMD or NDFD, but increased (P<0.01) marginally CPD (Table 1). However, it resulted in a decrease (P<0.01) in the proportion of polyunsaturated FA and a gradual increase (P<0.01) in relative concentrations of monounsaturated FA (Table 2), due to progressive accumulation of *trans*-18:1 BH intermediates (*t*11-18:1 accounting for 71% of the increase). Consistent with this, the relative proportions of 18:0 were substantially reduced (P<0.01) after 10 d on the SFO treatment. SFO tended (P=0.07) to increase the relative amount of CLA, but no *t*10 *c*12-CLA was detected.

Table 1. Changes in ruminal fermentation characteristics and degradation of alfalfa hay in sheep fed a mixture of 20 g sunflower oil and 10 g fish oil (SFO)/kg DM, on days 0 (Control), 3 (SFO_3) and 10 (SFO_{10}) on diet.

	pH	Lactate, mmol/l	Total VFA, mmol/l	Ammonia, mg/l	DMD, %	NDFD, %	CPD, %
Control	6.3	0.96	113.0	325.9	57.9	30.2	72.7^b
SFO_3	6.2	0.81	118.6	403.8	61.6	36.0	76.7^a
SFO_{10}	6.3	0.75	120.7	425.0	61.0	35.7	75.3^a
s.e.d.	0.07	0.256	8.10	50.88	1.68	2.78	1.08

[a,b] Different superscripts within a column indicate significant differences (P<0.05).

Table 2. Partial FA profile (g/100 g total FA) of ruminal digesta lipids in sheep fed 20 g sunflower oil and 10 g fish oil (SFO)/kg DM, on days 0 (Control), 3 (SFO_3) and 10 (SFO_{10}) on diet.

	Control	SFO_3	SFO_{10}	s.e.d.		Control	SFO_3	SFO_{10}	s.e.d.
18:0	43.6^a	28.9^b	10.4^c	4.68	22:6 *n*-3	$<0.1^b$	0.6^a	0.9^a	0.16
*t*10-18:1	0.5^c	1.5^b	2.1^a	0.25	*cis*-18:1	6.1	7.7	9.5	1.36
*t*11-18:1	4.1^c	14.3^b	28.3^a	2.38	*trans*-18:1	7.1^c	26.6^b	41.3^a	2.73
*c*9-18:1+*t*14-18:1	5.0	6.3	7.8	1.31	Total CLA	0.4	0.7	0.8	0.15
*c*9 *t*11-CLA	0.1	0.3	0.4	0.15	Saturated FA	69.0^a	50.3^b	31.3^c	4.58
*c*9 *c*12-18:2	11.3^a	5.5^b	5.1^b	1.45	MUFA	14.7^c	37.8^b	54.0^a	3.28
24:0+20:5 *n*-3	0.5^b	0.6^a	0.6^a	0.04	PUFA	15.3^a	9.6^b	10.2^b	2.02

[a,b,c] Different superscripts within a row indicate significant differences (P<0.05).

Conclusion

Relatively high proportions of *trans*-18:1 together with the previously reported decrease in 18:0 and the lack of effects on *in vivo* and *in situ* ruminal parameters, suggests that supplementing ewe diet with 20 g SO plus 10 g FO/kg DM inhibits complete BH of unsaturated FA in the absence of apparent detrimental effects on ruminal fermentation.

Acknowledgement

P.G. Toral received an I3P predoctoral grant from the CSIC. This work was supported by Projects AGL2008-04805-C02-02 (MICINN) and CSI01B08 (Council of Castille and Leon).

References

Loor, J.J., M. Doreau, J.M. Chardigny, A. Ollier, J.L. Sebedio and Y. Chilliard, 2005. Effects of ruminal or duoenal supply of fish oil on milk fat secretion and profiles of *trans*-fatty acids and conjugated linoleic acid isomers in dairy cows fed maize silage. Anim. Feed Sci. Technol. 119: 227-246.

Shingfield, K.J., S. Ahvenjärvi, V. Toivonen, A. Ärölä, K.V.V. Nurmela, P. Huhtanen and J.M. Griinari, 2003. Effect of fish oil on biohydrogenation of fatty acids and milk fatty acid content in cows. Anim. Sci. 77: 165-179.

Toral, P.G., A. Belenguer, P. Frutos and G. Hervás, 2009. Effect of supplementation of a high-concentrate diet with sunflower and fish oils on ruminal fermentation in sheep. Small Ruminant Res. 81: 119-125.

Nutritive value attributes in timothy and alfalfa as affected by sample preparation treatments

G.F. Tremblay, S. Pelletier, A. Bertrand, G. Bélanger, Y. Castonguay and R. Michaud
Agriculture and Agri-Food Canada, Québec, QC, G1V 2J3, Canada; Gaetan.Tremblay@agr.gc.ca

Introduction

Non structural carbohydrate (NSC) concentration in forages provides an estimate of the energy readily available to rumen microbes. Proper sample preparation is crucial for precise estimation of NSC. Enzymatic activity that occurs after cutting may affect the final NSC concentration in forages. Freezing followed by lyophilisation often represents the reference method to conserve nutrients in forage samples but it is difficult to use with large samples under field conditions. Other methods used for large samples exist but their effectiveness in preserving NSC and other attributes of nutritive value remains to be determined. Our objective was to assess the effect of sample preparation treatments on concentrations of NSC, N, and fibres, and on *in vitro* digestibilities of DM and NDF in timothy and alfalfa. We hypothesised that high temperature or freezing, immediately after harvest, limits enzymatic activity and preserves NSC in grass and legume forage samples.

Material and methods

The experiment was conducted on spring growth and summer regrowth of timothy and alfalfa in 2007 and 2008. Around 7 kg of each pure field-grown alfalfa and timothy were cut between heading and anthesis for timothy and between early budding and early seed pod for alfalfa, separated in 5 replication sets of 5 sub-samples of 250 g each, put on ice in 5 coolers, and kept in a refrigerator at 4 °C. As soon as possible after harvest (<1 h), the first set of 5 sub-samples of 250 g each was treated as described in Table 1. The 4 other replication sets of 5 sub-samples were subsequently treated in the same way. There were 100 samples of each species (2 years × 2 growth periods/year × 5 sample preparation treatments × 5 replications). Dried samples were 1-mm Wiley milled and analysed for water soluble carbohydrates (WSC) and starch by colorimetric methods, and for pinitol and high degree of polymerisation fructans (HDPF) by HPLC (Bertrand *et al.*, 2008). There were no fructans of low degree of polymerisation in the timothy samples. Concentration of WSC was estimated by the sum of sucrose, glucose, fructose, and pinitol. Concentration of NSC was estimated by the sum of WSC and starch. Concentrations of N, fibres, and NDIN, and digestibilities of DM and NDF (Table 1) were also determined (Licitra *et al.*, 1996; Bertrand *et al.*, 2008). For each species, the analysis of variance was performed using the MIXED procedure of SAS® with years and growth periods as random effects. The Tukey-Kramer test was used to compare treatment means.

Results and discussion

As compared to the reference method of lyophilisation, drying forage samples at 55 °C for 48 h resulted in lower concentrations of NSC, WSC, starch, glucose, and fructose, lower values of IVTD and dNDF, and higher concentrations of total N, NDF, and NDIN in both species (Table 1). Concentrations of HDPF in timothy summer regrowth and of pinitol in alfalfa were also lower in samples dried at 55 °C than lyophilised. Heating timothy and alfalfa samples in a microwave oven for one minute followed by drying at 55 °C for 48 h resulted in similar NSC, WSC, starch, HDPF, pinitol, total N, ADF, IVTD, and dNDF values than with lyophilisation. Drying at 100 °C for 1 h and then at 55 °C for 48 h, or freezing at -20 °C for 1 month followed by drying at 55 °C for 48 h often gave intermediate results between drying at 55 °C and lyophilisation.

Table 1. Effect of five sample preparation treatments on attributes of nutritive value (in mg/g DM except for dNDF which is in mg/g NDF) of timothy (n=100) and alfalfa forage (n=100).

Nutritive value attributes[2]	Sample preparation treatments[1] applied to Timothy					Alfalfa				
	55	100	Free	M-O	Lyo	55	100	Free	M-O	Lyo
NSC	44.6[a]	55.0[b]	62.1[c]	62.2[c]	63.1[c]	54.1[a]	76.5[b]	78.4[b]	88.5[c]	89.3[c]
WSC	38.5[a]	47.5[b]	55.0[c]	54.4[c]	55.5[c]	42.4[a]	50.6[b]	53.7[b]	62.2[c]	61.7[c]
Starch	6.1[a]	7.5[b]	7.1[b]	7.9[b]	7.6[b]	11.7[a]	25.9[b]	24.7[b]	26.3[b]	27.7[b]
Sucrose	28.6[c]	20.9[b]	3.5[a]	28.5[c]	17.6[b]	18.8[cd]	13.9[b]	4.2[a]	22.0[d]	16.1[bc]
Glucose	5.7[a]	16.2[b]	27.6[c]	17.3[b]	24.3[c]	8.5[a]	12.9[b]	17.7[c]	15.4[bc]	18.0[c]
Fructose	4.2[a]	10.4[b]	23.9[d]	8.5[b]	13.5[c]	2.8[a]	9.8[b]	17.6[d]	10.3[b]	12.8[c]
HDPF[3]	32.9[a]	36.2[a]	34.6[a]	44.8[ab]	52.3[b]	-	-	-	-	-
Pinitol	-	-	-	-	-	12.3[a]	13.9[ab]	14.2[ab]	14.6[b]	14.8[b]
Total N	17.1[b]	16.7[ab]	16.5[ab]	16.2[a]	16.2[a]	32.6[b]	32.9[b]	32.4[ab]	32.2[ab]	31.0[a]
ADF	355[c]	353[bc]	355[c]	346[ab]	341[a]	309[b]	293[ab]	293[ab]	281[a]	293[ab]
NDF	613[bc]	616[cd]	625[d]	605[b]	576[a]	369[b]	346[a]	351[ab]	343[a]	348[a]
NDIN	4.8[b]	6.9[c]	7.4[c]	7.0[c]	2.1[a]	3.3[b]	6.4[d]	4.4[c]	8.1[e]	2.2[a]
IVTD	788[a]	801[ab]	802[ab]	801[ab]	818[b]	810[a]	828[b]	823[ab]	831[b]	831[b]
dNDF	658[a]	681[ab]	687[ab]	672[ab]	687[b]	488[a]	508[ab]	506[ab]	517[b]	520[b]

[1] 55 = dried at 55 °C for 48 hrs; 100 = dried at 100 °C for 1 hr and at 55 °C for 48 hrs; Free = frozen in a plastic bag at -20 °C for about 1 month and dried at 55 °C for 48 hrs; M-O = heated in a microwave oven (maximal electrical power = 1100 W) during 1 min at maximum intensity to reach around 70 °C and dried at 55 °C for 48 hrs; Lyo = frozen in a plastic bag at -20 °C for about 1 month and lyophilised.
[2] NSC = non structural carbohydrates; WSC = water soluble carbohydrates; HDPF = high degree of polymerisation fructans; ADF = acid detergent fibres; NDF = neutral detergent fibres; NDIN = neutral detergent insoluble nitrogen; IVTD = *in vitro* true digestibility; dNDF = *in vitro* digestibility of NDF.
[3] Means are for summer regrowth samples only (n = 50); there were no HDPF in spring growth samples.
[a-e] For each species, means within rows followed by the same superscript letters are not significantly different (P>0.05).

Conclusion

The standard procedure of drying forage samples at 55 °C for 48 h, compared to lyophilisation, reduced NSC, WSC, starch, glucose, and fructose concentrations in both forage species as well as their digestibilities while increasing their total N, NDF, and NDIN concentrations. A high temperature treatment with a microwave oven (250 g at maximum intensity during 1 min to reach 70 °C) applied immediately after harvest limits enzymatic activity and prevents the loss of NSC in forage samples while maintaining most other attributes of nutritive value. This high temperature treatment, however, would denature proteins and render them insoluble since it increased the NDIN concentration.

References

Bertrand, A., G.F. Tremblay, S. Pelletier, Y. Castonguay and G. Bélanger, 2008. Yield and nutritive value of timothy as affected by temperature, photoperiod and time of harvest. Grass Forage Sci. 63: 421-432.

Licitra, G., T.M. Hernandez and P.J. Van Soest, 1996. Standardization of procedures for nitrogen fractionation of ruminant feeds. Anim. Feed Sci. Technol. 57: 347-358.

Intake, growth, ciliate protozoa and extra cellular microbial enzyme status of lambs on different yeast culture feeding

M.K. Tripathi and S.A. Karim
Division of Animal Nutrition, Central Sheep and Wool Research Institute, Avikanagar 304 501 (Via- Jaipur), Rajasthan, India; mktripathi@gmail.com

Introduction

Recent concern of researchers has been the exploration of non-antibiotic alternatives to manipulating the rumen microbial ecosystem in order to promote fermentation and growth by reducing energy and nutrient losses from the ruminant production system. Among the alternatives, microbial feed additives have shown such capabilities, however results on improved rumen function and production efficiencies are inconsistent. Yeast probiotics are more potent to maintaining a higher rumen pH and exert inhibitory activity on different pathogenic bacterial strains (Chaucheyras-Durand *et al.*, 2008). Performance responses of ruminants including lambs on yeast and yeast cultures have been variable (Kawas *et al.*, 2007; Stella *et al.*, 2007), are dose and diet and culture strain dependent. Since different yeast strains differ in their ability to produce favourable responses, the present study evaluated the suitability of three dairy yeast strains and a mixed culture on performance and rumen fermentation of lambs, to select a promising yeast culture for lamb feeding for fattening.

Material and methods

Sixty weaned lambs (90±3.5 d old and 15.9±0.50 kg BW) were randomly assigned to five equal groups. Animals were individually fed *ad libitum* a composite feed (TMR) with R:C 25:75, for 91 d. The control group was supplemented sterilised culture medium while other groups were supplemented with one of the three or a mixed yeast culture orally each day just after offering the fresh feed of the day. The mixed live yeast culture had *Kluyveromyces marximanus* NRRL3234 (KM), *Saccharomyces cerevisiae* NCDC42 (SC) and *Saccharomyces uvarum* ATCC9080 (SU) cultures in a ration of 1:1:1 ratio. The cultures were supplemented with 1 ml/kg live weight, having $1.5\text{-}2\times10^9$ live cells/ml. Three yeast stains were cultivated separately and mixed before feeding. Lamb BW were recorded for two consecutive days every 7 d to determine BW gain (ADG). On six randomly selected lambs, a metabolism trial was carried out. To estimate microbial protein synthesis, the procedure of urinary purine derivate excretion was used. Rumen fluid samples were collected at 3 h post feeding using a stomach tube to monitor the ciliate protozoa population and hydrolytic rumen microbial enzymes. The results were analysed for statistical significance using general linear mathematical model and significant means were separated using the Duncan Multiple Range Test.

Results

Yeast culture supplementation increased ADG (P=0.002) and feed efficiency (P=0.002) but feed intake did not improve. Efficiency of ME, DCP and absorbed N for live weight gain increased ($P\leq0.001$) by yeast culture supplementation. The ciliate population was similar while microbial protein synthesis increased (P=0.033) by yeast culture supplementation (Table 1). Extra cellular microbial enzyme activities though were not statistically different but were higher in SC and mixed culture supplemented lambs compared to the control. The SC culture supplementation improved activity of proteases 434.4, α-amylase 10.4, α-glucosidase 28.6, Xylanase 1.5 and Coarboxymethyl cellulase activity 19.7 units/100 ml rumen fluid in comparison to control lambs. Among the three cultures, *Saccharomyces cerevisiae* improved intake 5.9%, growth 21%, feed efficiency 16.5% and microbial protein synthesis by 45% to that in control lambs.

Table 1. Intake, growth, feed and nutrient efficiency, extra cellular enzymes and microbial protein synthesis in lambs supplemented with yeast culture.

	Dietary treatments (type of yeast culture[1])					SEM	*P* values
	Control	KM	SC	SU	Mixed yeast[2]		
Feed intake and average daily gain (g/d)							
Feed intake	978.7^a	771.1^b	1040.2^a	924.8^{ab}	954.8^a	28.368	0.036
Average daily gain	145.4^b	136.2^b	184.0^a	162.4^{ab}	172.8^a	4.554	0.002
Efficiency (per kg live weight gain)							
Feed (kg)	6.76^b	5.67^a	5.64^a	5.71^a	5.59^a	0.117	0.002
ME (MJ)	61.69^b	50.09^a	47.72^a	51.78^a	48.64^a	1.138	<0.001
N absorbed (g)	107.6^b	89.5^a	86.2^a	92.3^a	86.7^a	1.914	<0.001
Microbial protein synthesis (g/d)	27.8a	39.2ab	50.5b	30.7a	39.2ab	2.547	0.033
Extra cellular microbial enzyme activities (units/ 100 ml rumen fluid)							
Protease	367.67	377.46	802.10	403.04	613.08	90.905	0.515
α-amylase	45.25	47.97	55.64	30.80	46.67	3.374	0.221
α-glucosidase	40.98	60.91	69.59	37.53	11.24	11.52	0.581
β-glucosidase	113.19	70.16	108.12	68.09	97.43	18.175	0.810
Xylanase	16.97	16.18	18.44	13.34	14.34	1.042	0.589
Coarboxymethyl cellulase activity	37.04	47.68	56.75	36.56	90.52	16.747	0.874

[1] KM, *Kluyveromyces marximanus;* SC, *Saccharomyces cerevisiae*; SU, *Saccharomyces uvarum.*
[2] Mixed culture of three yeasts in equal proportion.
[a,b] Means within rows with same superscript letters are not significantly different.

Conclusion

Among the three, *Saccharomyces cerevisiae* have a potential as a growth promoting feed additive for feedlot lamb production, being an alternate to antibiotics or ionophores since culture supplementation improved intake 5.9%, growth 21%, feed efficiency 16.5% and microbial protein synthesis by 45% in fattening lambs.

References

Chaucheyras-Durand, F., N.D. Walker and A. Bach, 2008. Effects of active dry yeast on the rumen microbial ecosystem: Past, present and future. Anim. Feed Sci. Technol. 145, 5-26.

Kawas, J.R., R. García-Castillo, F. Garza-Cazares, H. Fimbres-Durazo, E. Olivares-Sáenz, G. Hernández-Vidal and C.D. Lu, 2007. Effects of sodium bicarbonate and yeast on productive performance and carcass characteristics of light-weight lambs fed finishing diets. Small Rumin. Res. 67: 157-163.

Stella, A.V., R. Paratte, L. Valnegri, G. Cigalino, G. Soncini, E. Chevaux, V. Dell'Otro and G. Savoini, 2007. Effect of administration of live Saccharomyces cerevisiae on milk production, milk composition, blood metabolites and faecal flora in early lactating dairy goats. Small Rumin. Res. 67: 7-13.

Enzymatic approach of linoleic acid ruminal biohydrogenation

A. Troegeler-Meynadier[1,2], *M.C. Nicot*[1,2] *and F. Enjalbert*[1,2]
[1]*Université de Toulouse; INPT, ENVT; UMR 1289 Tandem, F-31076 Toulouse, France;* [2]*INRA; UMR 1289 Tandem, F-31326 Castanet-Tolosan, France; a.troegeler@envt.fr*

Introduction

Ruminal biohydrogenation (BH) corresponds to a microbial reduction of dietary unsaturated fatty acids. The linoleic acid (C18:2) BH is divided into three steps: first an isomerisation into conjugated linoleic acids (CLA), then a reduction producing mainly *trans*-octadecenoic acids (*trans*-C18:1), and a final reduction producing stearic acid (C18:0). Isomerisations of CLA and *trans*-C18:1 can lead to a number of positional and geometrical isomers. The control of BH reactions is of interest for researchers because BH directly affects the composition of fatty acids of milk and meat. In order to better understand C18:2 BH and its variations, the development of an enzymatic approach is necessary to ascertain if the action of modulators affects the bacterial enzyme activity or ruminal bacteria. The aim of this study was to investigate the C18:2 BH capacity of ruminal content after inactivation of bacteria by chloramphenicol (Cm), an inhibitor of protein synthesis in prokaryotes.

Material and methods

Rumen fluid was collected from a dry dairy cow equipped with a ruminal canula, and strained on a metal sieve (1.6 mm). This strained ruminal fluid was mixed with 1 mg Cm/ml (adapted from Allison *et al.* (1962) and Rocha *et al.* (1996)). Then the mixture was gassed with CO_2 and placed at 39 °C during 5 h for a complete inhibition of bacterial growth (adapted from Allison *et al.*, 1962). Afterward, 1 ml of the mixture was incubated with 1 ml of a bicarbonate buffer solution in cell culture vials containing 1 mg of C18:2 (purity ≥99%, Sigma). Filled vials were gassed with CO_2, placed in a waterbath at 39 °C and agitated during 0, 1, 2 or 3 h, in 3 replicates. After incubation, reactions were stopped by placing the vials in ice. Then, vial contents were frozen and lyophilised. Fatty acids were quantified by gas chromatography. Fatty acid amounts in vials (mg) were analysed by the General Linear Model of SYSTAT with time as a categorical variable, followed by a pairwise comparison (Tukey test).

Results

The amount of C18:2 significantly ($P<0.01$) decreased during incubation: 30.3% disappeared after 1 h, 44.9% after 2 h and 52.7% after 3 h of incubations. The *trans*10,*cis*12-CLA and *cis*9,*trans*11-CLA were numerically the most abundant CLA isomers (Table 1). The amount of *trans*10,*cis*12-CLA and *trans*9,*trans*11-CLA increased throughout incubation, the amount of *cis*9,*trans*11-CLA was maximum after 1 h of incubation. The *trans*10-C18:1 and *trans*11-C18:1 were numerically the most abundant *trans*-C18:1 isomers. They increased throughout incubation, as *cis*12-C18:1 and *trans*6+7+8-C18:1. The amount of C18:0 was only significantly ($P<0.05$) increased after 3 h of incubation.

Discussion and conclusion

The BH of C18:2 produced mainly *cis*9,*trans*11-CLA and *trans*10,*cis*12-CLA, and *trans*11-C18:1 and *trans*10-C18:1 isomers, as previously described (Jouany *et al.*, 2007). The increase in *cis*12-C18:1 came from reduction of *trans*10,*cis*12-CLA, that of *trans*6+7+8-C18:1 from the reduction of minor CLA isomers not quantified in this study, like *trans*8,*trans*10-CLA (Shingfield *et al.*, 2008). The *trans*11 pathway was rapid: the *cis*9,*trans*11-CLA production was maximal at about 1 h of incubation while *trans*11-C18:1 accumulated throughout incubation. On the contrary, the *trans*10 pathway was slow: *trans*10,*cis*12-CLA regularly increased during incubation, so that it was more abundant than *cis*9,*trans*11-CLA after 3 h incubation, and *trans*10-C18:1 only began to increase after 2 h of incubation. The amount of C18:0 began to increase in the media when *trans*11-C18:1 concentration was over 0.05 mg/ml. Such evolution of fatty acids involved in C18:2 BH was similar to that reported *in vitro* with living ruminal microorganisms by Harfoot *et al.* (1973) and Jouany *et al.* (2007). So, this enzymatic approach using Cm could be an interesting and valid method to study C18:2 BH, however 3 h of incubation were not sufficient to study the final reduction.

Table 1. Evolution of amounts (mg) of CLA and C18:1 isomers involved in BH during incubation of C18:2 with inactivated mixed ruminal bacteria.

Duration of incubation	0h	1h	2h	3h	SEM	P
*cis*11-C18:1	0.005	0.005	0.005	0.006	0.001	0.661
*cis*12-C18:1	0.002[a]	0.004	0.006	0.007[b]	0.001	0.031
*cis*15-C18:1	0.001	0.001	0.000	0.000	0.000	0.131
*trans*6+7+8-C18:1	0.004[a]	0.004	0.005[b]	0.005	0.000	0.013
*trans*9-C18:1	0.002	0.002	0.002	0.002	0.000	0.936
*trans*10-C18:1	0.003[a]	0.009	0.025	0.032[b]	0.006	0.047
*trans*11-C18:1	0.056[a]	0.087	0.105[b]	0.115[b]	0.007	0.002
*trans*12-C18:1	0.009	0.007	0.005	0.007	0.001	0.560
*trans*13+14-C18:1	0.000	0.000	0.000	0.000	0.000	-
*trans*15-C18:1	0.004	0.004	0.004	0.004	0.000	0.712
*trans*16-C18:1	0.007	0.006	0.006[a]	0.007[b]	0.000	0.036
*trans*10,*cis*12-CLA	0.001[a]	0.028[b]	0.047[c]	0.074[d]	0.003	<0.001
*cis*9,*cis*11-CLA	0.000	0.000	0.000	0.000	0.000	-
*cis*9,*trans*11-CLA	0.003[a]	0.034[b]	0.024	0.019	0.006	0.029
*trans*9,*trans*11-CLA	0.000[a]	0.004[b]	0.005[c]	0.008[d]	0.000	<0.001

[a,b] Means in the same row with different superscripts are significantly different ($P<0.05$).

References

Allison, J.L., R.E. Hartman, R.S. Hartman, A.D. Wolfe, J. Ciak and F.E. Hahn, 1962. Mode of action of chloramphenicol. J. Bacteriol. 83: 609-615.

Harfoot, G.C., R.C. Noble and J.H. Moore, 1973. Factors influencing the extent of biohydrogenation of linoleic acid by rumen microorganisms *in vitro*. J. Sci. Food Agric. 24: 961-970.

Jouany, J.P., B. Lassalas, M. Doreau and F. Glasser, 2007. Dynamic features of the rumen metabolism of linoleic acid, linolenic acid and linseed oil measured *in vitro*. Lipids 42: 351-360.

Rocha, E.R., T. Selby, J.P. Coleman and C.J. Smith, 1996. Oxidative stress response in an anaerobe, *Bacteroides fragilis*: a role for catalase in protection against hydrogen peroxide. J. Bacteriol. 178: 6895-6903.

Shingfield, K.J., S. Ahvenjärvi, V. Toivonen, A. Vanhatalo, P. Huhtanen and J.M. Griinari, 2008. Effect of incremental levels of sunflower-seed oil in the diet on ruminal lipid metabolism in lactating cows. Br. J. Nutr. 99: 971-983.

Intake and partial digestion of nitrogen by sheep grazing four subtropical pastures during the summer

W.A. van Niekerk and A. Hassen
Department of Animal and Wildlife Sciences, University of Pretoria, Pretoria 0002, South Africa; willem.vanniekerk@up.ac.za

Introduction

The selection of a cultivated pasture species depends mainly on the adaptability of the species to a specific area, its dry matter yield, persistency and nutritive value of the forage. This study was set out to compare the nutritive potential of four sub-tropical grass pastures during the summer period and endeavoured to explain the difference between the species through the measurement of feed intake and partial digestibility of the nitrogen component by sheep.

Material and methods

The study was conducted at the Hatfield Experimental Farm of the University of Pretoria, South Africa.
Four tropical grass pastures (*Panicum maximum*, *Anthephora pubescens*, *Digitaria eriantha* and *Chloris gayana*) were established each in a 0.4 ha paddock.
Sixteen mature South African Mutton Merino wethers equipped with ruminal and abomasal cannulae were randomly allocated to four pasture treatments. An additional 4 oesophageal fistulated wethers were allocated randomly to each pasture treatment to estimate the diet quality. The wethers were fitted with faecal collection bags to determine voluntary intake. Pasture intake was estimated from the ratio faeces OM voided in the collection bags (Langlands, 1975) and the indigestibility of oesophageal samples by converting the in vitro digestibility to in vivo digestibility, according to Engels *et al.* (1981). The double marker technique, with continuous infusion and sampling at pre-determined times, as described by Faichney (1975), was used to determine the partial digestibility of the grass.
All data from the experiments were analysed using Proc GLM of SAS® (2001). The statistical model included pasture species and error and differences ($P<0.05$) between means were tested using the Bonferroni test according to Samuels (1989).

Results

Sheep grazing *A. pubescens* pasture had a higher organic matter intake (OMI) than those sheep grazing the other three pastures (Table 1). The lowest intake was recorded for sheep grazing *C. gayana*. The lower level of intake in this species corresponded well with the higher NDF, lower N and lower IVDOM value of the selected diet (Hassen and Van Niekerk 2009, this proceeding). The N intake (g/d) and the total N, NH_3-N or non-ammonia nitrogen (NAN) (g/d) flow in the abomasum and ileum were high for sheep grazing *A. pubescens* compared to for those sheep on *C. gayana* pasture (Table 1). The NAN disappearance (g/d) in the small intestine followed the same trend, but when the NAN disappearance is compared as a percentage of N intake, the difference between the pasture species was not significant ($P>0.05$). The pasture species also did not differ ($P>0.05$) in terms of NAN digestibility in the gastrointestinal tract, as the highest and lowest N intake and faecal NDF-N was recorded respectively for sheep grazing *A. pubescens* and *C. gayana* pasture.

Table 1. Intake and N utilisation of sheep grazing different sub-tropical grass pasture.

Parameters[1]	Pasture type				SE
	P. maximum	*A. pubescens*	*D. eriantha*	*C. gayana*	
Intake					
OMI (g/d)	961^{b}	1174^{a}	1018^{b}	792^{c}	40.1
OMI (g/kg $W^{0.75}$/d)	31.4^{b}	40.3^{a}	33.8^{b}	25.0^{c}	0.9
N intake (g/d)	30.8^{b}	39.9^{a}	27.5^{b}	20.6^{c}	1.3
Abomasum					
Digesta flow (l/d)	23.1^{b}	25.5^{a}	20.1^{c}	18.3^{d}	0.4
Total N flow (g/d)	25.3^{b}	31.6^{a}	22.5^{b}	17.2^{c}	1.1
NH_3-N flow (g/d)	5.6^{b}	6.9^{a}	5.6^{b}	4.2^{c}	0.2
NAN flow (g/d)	19.7^{b}	24.7^{a}	16.9^{b}	13.0^{c}	1.1
NAN flow/N intake	0.64	0.62	0.61	0.63	0.02
Ileum					
Digesta flow (l/d)	5.1^{ab}	4.9^{ab}	4.6^{b}	5.8^{a}	0.4
Total N flow (g/d)	6.2^{b}	8.2^{a}	5.6^{b}	4.2^{c}	0.4
NH_3-N flow (g/d)	1.5^{a}	1.9^{a}	1.4^{ab}	1.1^{b}	0.1
NAN flow (g/d)	4.7^{b}	6.3^{a}	4.2^{b}	3.1^{c}	0.3
NANdisappearance (g/d)	15.0^{b}	18.4^{a}	12.7^{bc}	9.9^{c}	0.5
NAN disappearance (% N intake)	49	46	46	48	1.4
NAN digestibility	76	75	75	76	0.9
Faecal NDF-N (g/d)	4.2^{b}	5.8^{a}	4.2^{b}	2.8^{c}	0.3

[1] OMI=organic matter intake; N= nitrogen; NAN= non-ammonia nitrogen; NDF-N= nitrogen bind to the NDF.

[a,b,c] Means in the same row followed by different superscripts differ significantly (P<0.05).

Conclusion

This study shows that sheep grazing *A. pubescens* pasture had higher intake and higher NAN disappearance in the small intestine that could likely result in better animal performance during the summer season followed by *P. maximum* and *D. eriantha*.

References

Engels, E.A.N., H.O. De Waal, L.C. Biel and A. Malan, 1981. Practical implications of the effect of drying and treatment on nitrogen content and *in vitro* digestibility of samples collected by oesophageally fistulated animals. S. Afr. J. Anim. Sci. 11: 247-254.

Faichney, G.J., 1975. The use of markers to partition digestion within the gastro-intestinal tract of ruminants. In: McDonald I.W. and Warner A.C.T. (Eds), Proceedings of the 4th International Symposium on Ruminant Physiology. University of New England Publ. Unit, Armidale, N.S.W., Australia. pp. 277-291.

Langlands, J.P., 1975. Techniques for estimating nutrient intake and its utilization by the grazing ruminant. In: McDonald I.W. and Warner A.C.T. (Eds), Proceedings of the 4th International Symposium on Ruminant Physiology. University of New England Publ. Unit, Armidale, N.S.W., Australia. pp. 320-332.

Samuels, M.L., 1989. Statistics for the Life Sciences. Collier MacMillan Publishers, London.

SAS®/STAT 2001. Statistical Analysis Systems User's Guide Version 8.2. SAS® Institute Inc., Cary, NC, USA.

Effects of extruded linseed, a mixture of C8:0 and C10:0 fatty acids, and diallyldisulfide on methane emission in dairy cows

S.M. van Zijderveld[1,2], *W.J.J. Gerrits*[1], *J. Dijkstra*[1], *J.R. Newbold*[2], *D. Deswysen*[2] *and H.B. Perdok*[2]
[1]*Animal Nutrition Group, Wageningen University, Marijkeweg 40, 6709 PG, Wageningen, The Netherlands;* [2]*Provimi Research and Innovation Centre, Lenneke Marelaan 2, Sint-Stevens-Woluwe, B1923, Brussels, Belgium; svanzijderveld@nl.provimi.com*

Introduction

In recent decades, methane has been implicated as a greenhouse gas and the global population of ruminants contributes significantly to the overall production of methane from anthropogenic sources (Steinfeld *et al.*, 2006). In addition, methane represents a potential loss of dietary energy to the animal. Both topics have led to a widespread interest among nutritionists to reduce methane emissions from ruminant animals.

Extruded linseed has been shown to decrease methane emission by 38% from dairy cows (Martin *et al.*, 2008) when included in the diet at a high level. The reduction in methane emission was, however, confounded with a decreased DMI, digestibility and animal performance. Also, the inclusion of dietary lipids will dilute the content of fermentable organic matter in the diet, which in itself will be partly responsible for the reduction in methane production. Diallyldisulfide and the fatty acids C8:0 and C10:0 have been shown to decrease methane emissions *in vitro* by 68.5% and 20-60% respectively (Busquet *et al.*, 2005; Ajisaka *et al.*, 2002), but to the authors' knowledge this has never been demonstrated *in vivo* in dairy cows. This experiment was designed to examine methane mitigating properties of extruded linseed, a mixture of C8:0 and C10:0 fatty acids and diallyldisulfide *in vivo* at practical inclusion levels without the confounding effect of a reduced DMI on methane emissions in dairy cattle.

Material and methods

The experiment was designed as a randomised block experiment with four treatments. Each cow was offered the same total mixed ration (41% grass silage, 35% maize silage and 14% concentrates on a DM-basis). The dietary additives were added to this diet as a premix at 10% of the ration DM. The treatments involved a control treatment (CON), 100 g/kg DM of an extruded linseed product (50% extruded linseed, 30% wheat bran and 20% sunflower meal; LIN), 25 g/kg DM of a mixture of C8:0 and C10:0 fatty acids (50:50% (w/w); MCFA) and 0.2 g/kg DM of diallydisulfide (DADS). Diets were formulated to be isolipidic and animals were fed restrictedly to avoid confounding effects of DMI on methane emission.

The experiment was conducted in 10 successive trial periods of 17 d each. Forty Holstein-Friesian dairy cows (27.9±7.0 kg milk, 167±99 days in milk) were blocked in 10 blocks of 4 cows each based on milk yield, days in milk and parity. Within each period, a pair of cows from one block was assigned to one of two of the dietary treatments, housed in a tie-stall facility for 12 d and subsequently housed in pairs in one of two identical climate-controlled respiration chambers for 5 d for the measurement of methane emission. In addition, digestibility measurements and a complete energy balance were performed. During the initial 8 d, cows were fed *ad libitum* and after 8 days cows received 95% of the total feed intake of the cow within that block consuming the lowest amount of feed.

Results and discussion

Dry matter intake was similar among treatments as a result of the experimental design (Table 1). Milk production was unaffected by treatment. Milk fat content on MCFA was significantly higher than all other treatments. Methane production was unaffected by treatment (Table 2).

Table 1. Dry matter intake, milk production and milk composition (n=10/treatment).

	CON	LIN	MCFA	DADS	s.e.d.	P-value
DMI (kg/day)	16.6	16.9	16.8	16.8	0.31	0.784
Milk production (kg/day)	24.6	25.3	22.5	24.7	1.46	0.260
Milk protein content (%)	3.40	3.34	3.58	3.41	0.109	0.170
Milk fat content (%)	4.82[a]	4.47[a]	5.38[b]	4.52[a]	0.214	<0.001
Milk lactose content (%)	4.58	4.62	4.58	4.54	0.062	0.603
Milk urea content (mg/dl)	25.2	27.5	23.3	25.5	1.51	0.075

[a,b] Means in the same row followed by different superscripts differ significantly.

Table 2. Methane production (n=5/treatment).

	CON	LIN	MCFA	DADS	s.e.d.	P-value
Methane production (g/d)	340	359	355	352	32.6	0.94
Methane production (g/kg DMI)	20.6	21.2	21.2	21.0	1.52	0.97
Methane production (g/kg milk)	14.4	14.7	16.6	14.3	2.3	0.74

In contrast with previously observed reduction in methane emission by extruded linseed, diallyldisulfide and MCFA, the results of the present experiment show that, in the dosage included in the ration, they exert no effect on methane emission in dairy cows *in vivo*. The experimental power was sufficient to statistically show treatment differences in methane emission of 9%. It is concluded that previously reported methane reductions *in vivo* are quite often related to reductions in feed intake or a dilution of fermentable organic matter with lipids, which hampers conclusions with regards to the efficacy of the additives tested.

Conclusion

The addition of extruded linseed, a mixture of C8:0 and C10:0 fatty acids, and DADS to restrictedly fed dairy cattle did not affect methane emission.

References

Ajisaka, N., N. Mohammed, K. Hara, K. Mikuni, K. Hara, H. Hashimoto, T. Kumata, S. Kanda and H. Itabashi, 2002. Effects of medium-chain fatty acid-cyclodextrin complexes on ruminal methane production *in vitro*. Anim. Sci. J. 73: 479-484.

Busquet, M., S. Calsamiglia, A. Ferret, M.D. Carro and C. Kamel, 2005. Effect of Garlic Oil and Four of its Compounds on Rumen Microbial Fermentation. J. Dairy Sci. 88: 4393-4404.

Martin, C., J. Rouel, J.P. Jouany, M. Doreau and Y. Chilliard, 2008. Methane output and diet digestibility in response to feeding dairy cows crude linseed, extruded linseed, or linseed oil. J Anim Sci 86: 2642-2650.

Steinfeld, H., P. Gerber, T. Wassenaar, V. Castel, M. Rosales and C. De Haan, 2006. Livestock's Long Shadow. Food and Agriculture Organization of the United Nations.

Ruminal bacteria, protozoa and fatty acid profile in sheep and goats supplemented with tannins

V. Vasta[1], D.R. Yáñez-Ruiz[2], M. Mele[3], A. Serra[3], G. Luciano[1], M. Lanza[1] and A. Priolo[1]
[1]DACPA, Sezione di Scienze delle Produzioni Animali, via Valdisavoia 5, 95123, Catania, Italy; [2]Estación Experimental del Zaidín, CSIC, profesor Albareda, 18008, Granada, Spain; [3]Dipartimento di Agronomia e Gestione dell'Agro Ecosistema, via del Borghetto 80, 56124, Pisa, Italy

Introduction

Meats and milk fatty acid composition are affected by the biohydrogenation (BH) of polyunsaturated fatty acids (PUFA). From the isomerisation of C18:2 *cis*-9, *cis*-12 is produced forming 18:2 *cis*-9, *trans*-11 (CLA), a fatty acid active in the prevention of chronic diseases (Ip *et al.*, 1991). The saturation of CLA brings C18:1 *trans*-11 (vaccenic acid, VA), which is then saturated to C18:0 (stearic acid, SA). One of the bacterial strains mostly active in the conversion of PUFA to CLA is *Butyrivibrio fibrisolvens*, while other distinct groups of bacteria carry out the conversion of VA to SA (SA producers) (Paillard *et al.*, 2007). Also rumen protozoa seem to play a role in the BH of PUFA (Yáñez-Ruiz *et al.*, 2007). It has been shown *in vitro* that the BH, in particular the conversion of VA to SA, is strongly reduced by tannins (Vasta *et al.*, 2009). The present study was aimed at investigating, *in vivo*, if dietary tannins could affect some rumen microbial groups and the BH in sheep and goats.

Material and methods

Thirteen lambs and nine kids weaned at the age of 45 days were assigned to two dietary treatments. Six lambs and 5 kids (control) received a concentrate containing (as fed basis): barley (55.1%), alfalfa hay (30.0%), soybean meal (13.0%) and vitamin and mineral premix (1.9%). The remaining animals received the same concentrate with a supplementation of quebracho tannins (8.44% as fed). After 70 d of the experiment, the animals were slaughtered and the rumen content was immediately sampled. An aliquot of the rumen content was immediately frozen at -80 °C and then freeze-dried prior to DNA extraction. Another aliquot was filtered through cheesecloth and stored at -80 °C for fatty acid analyses. Protozoal and bacterial rDNA concentration was determined (Yáñez-Ruiz *et al.*, 2007) and the relative populations of *B. fibrisolvens* and of those bacteria which produce stearic acid (SA producing bacteria) were quantified according to Stevenson and Weimer (2007) and Paillard *et al.* (2007), respectively. Ruminal fluid fatty acids were analysed as described in Vasta *et al.* (2009). Data were analysed by two-way ANOVA.

Results

Tannin supplementation increased the proportion of *B. fibrisolvens* in rumen bacteria ($P<0.05$) and total protozoa ($P<0.0005$) in both sheep and goats. Total bacteria concentration was not affected by tannin supplementation, while the relative proportion of SA producing bacteria was reduced by tannins both in sheep and goats. Tannin supplementation increased by twofold the concentration of VA ($P<0.0005$) and reduced the concentration of SA ($P<0.005$) in the ruminal fluid.

Conclusion

Tannin supplementation leads to the accumulation of VA and reduces the production of SA in the ruminal fluid through changes in microbial groups involved in different BH steps. This is probably due to the reduced concentration of SA producing bacteria in the presence of tannins, which causes a lower rate/extent of the terminal step of ruminal BH. Considering that CLA is also produced in the muscle and in the mammary gland from the desaturation of VA (Corl *et al.*, 2001), our results suggest that supplementing tannins is a good strategy to enhance CLA concentration in ruminant products.

Table 1. Effect of tannin on ruminal microorganisms and fatty acid profile in rumen fluid of sheep and goats.

	Goats		Sheep		SEM	*P*-value		
Tannin Supplementation	None	Yes	None	Yes		S[1]	T[1]	S × T
Protozoa (log 10 copies/g FM)	6.44	9.55	5.08	7.82	0.433	0.019	<0.0005	0.993
Bacteria (log 10 copies/g FM)	10.03	9.49	9.50	9.57	0.110	0.320	0.300	0.175
B. fibrisolvens (% of total bacteria)	10.55	15.53	4.22	8.76	1.041	<0.0005	0.005	0.709
SA producers (% of total bacteria)	6.03	4.06	3.99	2.77	0.399	0.046	0.032	0.743
Fatty acids (% of total fatty acids) in rumen fluid								
C18:2 *cis*-9, *cis*-12	0.49	0.67	1.18	1.83	0.215	0.04	0.33	0.56
C18:2 *cis*-9, *trans*-11	0.003	0.13	0.03	0.41	0.097	0.45	0.23	0.53
C18:1 *trans*-11	1.03	2.44	1.48	3.09	0.234	0.10	<0.0005	0.75
C18:0	56.65	40.88	50.85	44.39	1.920	0.74	0.005	0.18

[1] S = species (sheep or goat); T = tannin supplementation (yes vs. none).

Acknowledgement

This research was funded by the Italian Ministry of Science and Research (PRIN 2007, main investigator A. Priolo).

References

Corl, B.A., L.H. Baumgard, D.A. Dwyer, J.M. Griinari, B.S. Phillips and D.E. Baumann, 2001. The role of Δ9-desaturase in the production of *cis*-9, *trans*-11 CLA. J. Nutr. Biochem. 12: 622-630.

Ip, C., S.F. Chin, J.A. Scimeca and M.W. Pariza, 1991. Mammary cancer prevention by conjugated dienoic derivative of linoleic acid. Cancer Res. 51: 6118-6124.

Paillard, D., N. McKain, M.T. Rincon, K.J. Shingfield, D.I. Givens and R.J. Wallace, 2007. Quantification of ruminal Clostridium proteocalsticum by real-time PCR using a molecular beacon approach. J. Appl. Microbiol. 103: 1251-1261.

Stevenson, D.M. and P.J. Weimer, 2007. Dominance of *Prevotella* and low abundance of classical ruminal bacterial species in the bovine rumen revealed by relative quantification real-time PCR. Appl. Microbiol. Biotenchnol. 75: 185-174.

Vasta, V., H.P.S. Makkar, M. Mele and A. Priolo, 2009. Ruminal biohydrogenation as affected by tannins *in vitro*. Br. J. Nutr. doi:10.1017/S0007114508137898.

Yáñez-Ruiz, D.R., S. Williams and C.J. Newbold, 2007. The effect of the absence of protozoa on rumen biohydorgenation and the fatty acid composition of lamb muscle. Br. J. Nutr. 97: 938-948.

Effects of dietary concentrate to forage rate on microbial protein recycling in the rumen of goats

M.Z. Wang, H.R. Wang and G.X. Li
College of Animal Science and Technology, Yangzhou University, 225009, Jiangsu, China; hongrongwang@sina.com

Introduction

Bacterial ingestion by protozoa in the rumen will cause bacterial protein recycling, and leads to a low utilisation efficiency of N (Wallace and McPherson, 1987). Dietary structure might influence the protozoal population, and the community structure (Brown *et al.*, 2006). These changes might not only directly affect MCP yield; but also indirectly affect MCP yield and N utilisation efficiency, through affecting the micro-recycling. However, reports on rumen micro-recycling have seldom been seen. Thus, currently, the Fluorescence-labelled Bacteria Technique (FLB) was used to investigate the effects of dietary structure on rumen MCP recycling. We also hope to offer some references for research on nitrogen regulating techniques and how to save feed protein.

Material and methods

Four Xuhuai goats (26.2±1.6 kg), fitted with rumen cannulas, were randomly assigned to four dietary treatments (Table 1) in a 4×4 Latin square design. Each experimental period consisted of a 14-day preliminary period, and then followed by 7 days for rumen fluid sampling. Bacteria and protozoa enumeration were conducted using a direct microscopic count technique; a grazing experiment was carried out according to the method of Wang *et al.* (2008). Statistical analysis was done using SPSS, v11.5.

Table 1. Dietary composition and nutrient levels of experimental sheep in vivo.

Ingredient, %	A	B	C	D
Corn	0.97	17.39	38.65	60.87
Soybean meal	8.70	11.59	9.66	6.76
Straw	86.96	67.63	48.31	28.99
Urea	0.97	0.97	0.97	0.97
Dicalcium phosphate	1.06	1.06	1.06	1.06
Salt	0.77	0.77	0.77	0.77
Premix	0.58	0.58	0.58	0.58
Total	100.00	100.00	100.00	100.00
Nutrient level				
Metabolism energy (ME), MJ/kg	5.78	7.17	8.62	10.14
Dry matter (DM),%	91.84	91.26	90.63	89.99
Crude protein (CP),% DM basis	9.20	10.91	11.03	10.82
Non-structure carbohydrate (NSC)[1],%	23.71	33.64	45.89	58.56
Structure carbohydrate (SC)[1],%	54.87	44.23	33.72	23.23
NSC/SC	0.43	0.76	1.36	2.52

[1] SC=NDF-NDFN; NSC=100-((NDF–NDFN)+CP+ EE+ASH).

Results

To ensure the validity of the testing results, it was important to determine the time of the ingestion experiment. Referring to previous studies, the regression analysis time was set to 25 min. The fluorescent microscopy photo of protozoa shows that the 25-min photo was clear enough to count FLB cells ingested by protozoa (Figure 1A); however, the longer the experimental time was, the blurrier the FLB cells in the protozoal body was, and in Figure 1B (50-min photo) it was difficult to count FLB cells inside the protozoal body.
Table 2 shows that density of protozoa and bacteria was the highest in C and B respectively; while both protozoa and bacteria density of A was the lowest. Grazing rates varied with diets (P=0.000), and the highest peak was found in A. Further extrapolating the assimilation quantity of bacteria N by protozoa per goat per day, there were 136.49, 369.02, 485.99, and 440.56 mg N/(d capita) respectively. And the estimation of bacterial protein recycling showed that C recorded the highest recycling quantity of bacterial protein (3.37%).

Table 2. Effect of dietary concentrate to forage rate on rumen micro-ecosystem.

Items	A	B	C	D	SEM	P
Protozoa density, $\times 10^5$ cells/ml	6.13^a	19.41^b	24.07^c	21.13^{bc}	1.69625	0.000
Bacteria density, $\times 10^9$ cells/ml	8.92^a	35.18^b	30.58^c	31.27^{bc}	2.06950	0.000
Grazing rates, cells/cell h	429.5^d	366.74^b	389.48^b	402.2^c	2.73453	0.000
Grazing quantity, $\times 10^5$ cells/ml h	2632.84	7118.42	9374.78	8498.49	–	–
Bacterial recycling rates, $\times$100%	2.95	2.02	3.07	2.72	–	–
Bacterial recycling time, h	33.9	49.4	32.6	36.8	–	–
N recycling rates, pg/cell h	2.319	1.98	2.103	2.172	–	–
N recycling quantity, mg/d captia	136.49	369.02	485.99	440.56	–	–
Pr recycling quantity, g/d captia	0.853	2.306	3.37	2.754	–	–

[a,b,c] The same superscript within the same row were not significantly different (P>0.05); Neighbouring letters were significantly different (P<0.05); Parted letters indicate that the means are extremely different (P<0.01).

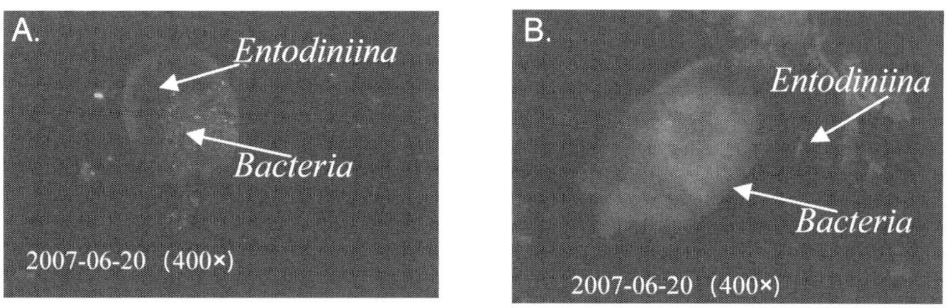

Figure 1. The fluorescent photo of protozoa with FLB celsl inside its body. A: The 25-min fluorescent photo; B: The 50-min fluorescent photo.

Conclusion

In conclusion, microbial density, protozoa grazing rates, and bacterial protein recycling in the rumen were modified by dietary structure. Diet C showed the highest recycling quantity of bacterial protein.

References

Brown, M.S., C.H. Ponce and R. Pulikanti, 2006. Adaptation of beef cattle to high-concentrate diets: Performance and ruminal metabolism. J. Anim. Sci. 84(Suppl.): E25-E33.

Wallace, R.J. and C.A. McPherson, 1987. Factors affecting the rate of breakdown of bacterial protein in rumen fluid. Br. J. Nutr. 59: 313-323.

Wang, M.Z., H.R. Wang, G.X. Li, H.C. Cao and Z.J. Lu, 2008. The preliminary report on rumen protozoa grazing rate on bacteria with fluorescence-labeled technique. Scientia Agricultura Sinica 7: 101-105.

In vitro effects of phlorotannins from *Ascophyllum nodosum* (brown seaweed) on rumen bacterial populations and fermentation

Y. Wang, T.W. Alexander and T.A. McAllister
Agriculture and Agri-Food Canada Research Centre, Lethbridge, T1J 4B1, Alberta, Canada; yuxi.wang@agr.gc.ca

Introduction

Tannins from terrestrial plants exert varying effects on rumen microbial activity, depending on their concentration and source and on the microbial species involved. Phlorotannins (PT) from marine brown algae also exhibit strong inhibitory activity against ruminal microbes (Wang *et al.,* 2008) but information on specific interaction(s) of PT with principal rumen bacteria in mixed culture is lacking. We used real-time polymerase chain reaction (PCR) to investigate the effects of PT extracted from brown seaweed (*Ascophyllum nodosum*) on seven major rumen bacterial species in batch culture ruminal incubation. Polyethylene glycol, which inactivates PT (Jones and Mangan, 1977), was included as a control to isolate PT effects.

Material and methods

Phlorotannins isolated from *A. nodosum* were included in a batch culture incubation at 0 (control or CON) or 500 µg/ml (PT), alone or with polyethylene glycol (PEG) at 0 or 1.0 mg/ml, in a 2×2 factorial arrangement of treatments. Substrate comprised a barley silage/alfalfa hay/grass hay mixture, and inoculum was prepared (Wang *et al.,* 2008) using ruminal fluid from two steers fed a 60:40 alfalfa hay:barley silage diet. Gas production was measured after 3, 6, 9, 12 and 24 h in 3 vials/treatment. Triplicate vials were removed from the incubation after 0, 6 and 24 h and centrifuged (20,000× *g*; 30 min; 4 °C). Supernatants were analysed for volatile fatty acids (VFA) by gas chromatography (Wang *et al.,* 2000). The pellets, which comprised feed residues and mixed microbes, were washed 2 times with phosphate buffer, lyophilised and ground to a powder, and DNA was extracted using a Qiagen QIAamp DNA Stool Mini Kit.
Plasmid standards were created for three cellulolytic: (*Fibrobacter succinogenes* subsp. *succinogenes* S85, *Ruminococcus albus* 7 and *Ruminococcus flavefaciens* C94) and four non-cellulolytic (*Prevotella bryantii* $B_1$4, *Streptococcus bovis* 45S1, *Ruminobacter amylophilus* 70, and ATCC 19205 *Selenomonas ruminantium* subsp. *lactilytica*) bacterial reference strains. DNA extracted from the incubation pellets was analysed by quantitative real-time PCR (SYBR green) to determine relative quantities of each of the seven species, expressed as copy number of species-specific16S rDNA. Analysis of variance was conducted using the MIXED procedure of SAS®, with individual vial as the random factor. Significance of differences among treatments was tested using LSMEANS with the PDIFF option.

Results

Phlorotannins (without PEG) markedly reduced gas production ($P<0.05$) and VFA accumulation ($P<0.001$) at 6, 12, and 24 h, as compared with other treatments (data not shown). For 16S rDNA copy numbers, a PT×PEG interactive effect ($P<0.01$) was observed for *F. succinogenes* at 6, 12, and 24 h, for *R. albus* at 24 h, and for the non-cellulolytic bacteria at 12 and 24 h. Including PT without PEG reduced ($P<0.001$) *F. succinogenes* 16S rDNA copy numbers by 78, 83, and 65% at 6, 12, and 24 h, respectively, and those of *R. albus* by 42% at 24 h, compared with CON. Among the non-cellulolytic species, 16S rDNA copy numbers were reduced ($P<0.01$) by PT for *S. ruminantium* at 6 h, but were increased ($P<0.01$) for *S. ruminantium*, *P. bryantii* and *R. amylophilus* at 12 and

24 h, and for *S. bovis* at 24 h only compared with CON. Analysis of data grouped by cellulolytic capacity revealed a PT-associated reduction ($P<0.01$) in total 16S rDNA copy numbers of the three cellulolytic species, by 63, 65, and 58% at 6, 12, and 24 h, respectively (Figure 1). In contrast, 16S rDNA copy numbers of the non-cellulolytic species were reduced by PT (by 39%; $P<0.01$) only at 6 h. After 12 and 24 h, they were increased by 40 and 190%, respectively ($P<0.001$). Overall (i.e. for all seven species studied), total 16S rDNA copy numbers tended ($P=0.068$) to be reduced, were unaffected, and were markedly increased ($P<0.01$) by PT, at 6, 12 and 24 h of incubation, respectively.

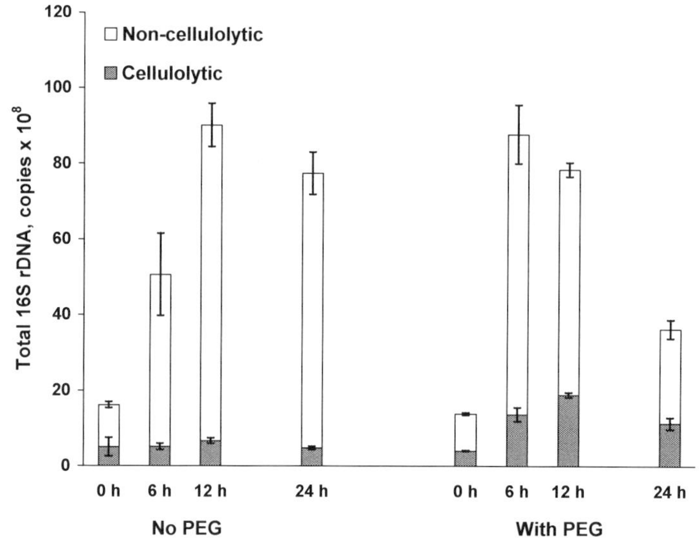

Figure 1. Total 16S rDNA of three cellulolytic and four non-cellulolytic bacterial species during 24 h of batch culture ruminal incubation of forage-based substrate with phlorotannins (PT; 500 µg/ml) and with or without polyethylene glycol (PEG; 0 or 1.0 mg/ml) to inactivate the PT. Bars indicate SE (n=3). See text for species names.

Conclusion

At 500 µg/ml, phlorotannins inhibited *in vitro* ruminal fermentation overall, and affected rumen bacteria in a species-specific manner. PT inhibited growth of *F. succinogenes*, had a minimal effect on *R. flavefaciens* and *R. albus*, and promoted growth of *S. ruminantium, S. bovis*, *R. amylophilus,* and *P. bryantii.* Overall, PT supplementation reduced the population of cellulolytic bacteria but increased total and non-cellulolytic bacterial populations.

References

Jones, W.T. and J.L. Mangan, 1977. Complexes of condensed tannins of sainfoin (*Onobrychis viciifolia* Scop.) with fraction 1 leaf protein and with submaxillary mucoprotein, and their reversal by polyethylene glycol and pH. J. Sci. Food Agric. 28: 126-136.

Wang, Y., T.A. McAllister, L.J. Yanke, Z.J. Xu, P.R. Cheeke and K.J. Cheng, 2000. *In vitro* effects of steroidal saponins from *Yucca schidigera* extract on rumen microbes. J. Appl. Microbiol. 88: 887-896.

Wang, Y., Z. Xu, S.J. Bach and T.A. McAllister, 2008. Effects of phlorotannins from *Ascophyllum nodosum* (brown seaweed) on *in vitro* ruminal digestion of mixed forage or barley grain. Anim. Feed Sci. Technol. 145: 375-395.

Study on the measurement of fluorescence-labelled technique for protozoa predation rate on bacteria in the rumen

M.Z. Wang, H.R. Wang and L.H. Yu
College of Animal Science and Technology, Yangzhou University, 225009, Jiangsu, China; hongrongwang@sina.com

Introduction

The bacterial ingestion by protozoa in the rumen is a negative factor affecting MCP yields, due to bacterial protein recycling, and resultantly leads to a low utilisation efficiency of nitrogen in ruminants (Wallace and McPherson, 1987). However, because there is no powerful method to investigate the grazing rate of rumen protozoa on bacteria, there has been little research on rumen micro-recycling. Therefore, the recycling rule of the rumen microbe is not clear at all even now. Currently, the fluorescence-labelled bacteria (FLB) technique was introduced, in the light of a measurement method of protozoal grazing rate in the marine ecosystem, and was aimed at developing new approaches in predation rate measurement of rumen protozoa, and promoting research in the field of rumen micro-recycling.

Material and methods

Four Xuhuai goats (23.7±2.6 kg), fitted with rumen cannulas, were used to provide rumen liquor. Two groups were designed as follows: one group was the whole bacteria which were labelled using fluorescence by removing free bacteria from rumen fluid (regarded as WFLB group); the other group was bacteria which were labelled using fluorescence without removing free bacteria from rumen fluid (regarded as FLB group). An ingestion experiment was carried out according to the method of Sherr *et al.* (1987), with a slight modification. One ml of subsamples were taken at 5-min intervals for 40 min after ingestion in the experiment, and the time series samples were fixed immediately and inspected by light microscopy at first (100× to 400×), to find and identify the target protozoa; and then converted visible light to fluorescent light, to count FLB numbers within the protozoa body (1000×). Six protozoa per slide were measured, and 3 repeats were set for each sampling-time point.

Number of FLB were converted to N biomass using an average bacterial biovolume of 0.1 μm^3, and a N conversion factor of 0.054 pg N of μm^3 (Sherr *et al.*, 1987; Li *et al.*, 2005). Statistical analysis was done using SPSS, ver. 11.5.

Results

The results presented in Table 1 showed that, the FLB numbers engulfed by protozoa increased with time within 25 min, and did not reach a plateau period. The linear regression analysis between the FLB numbers engulfed by protozoa and time within 25 minutes was carried out (Figure 1), and the regression equations were the following: for WFLB, $Y = 21.27 + 6.64X$ ($r = 0.971$); for FLB, $Y = 4.87 + 3.84X$ ($r = 0.977$); where Y – FLB numbers grazed by protozoa (cells/cell), X – time (min). The r value of the linear equtions were all higher than 0.97, indicating a good linear relationship between FLB number and time. According to the regression equations, the grazing rates were 398.4 and 230.4 cells/(cell h) for WFLB and FLB, respectively.

Through conversion, the bacterial N predation rates were 2.15 pg N/(cell h), and 1.24 pg N/(cell h) for WFLB and FLB respectively. And further extrapolation of the assimilation quantity of bacterial N by protozoa was through an estimation of 4 l of goat's rumen content and protozoa density level of about 5×10^5/ml; there were 103.2 mg N/(d capita), and 59.5 mg N/(d capita) respectively.

Furthermore, the estimations of recycling quantity of bacterial protein by protozoa per day per goat, respectively, were 0.645 g Pr/(d capita), and 0.372 g Pr/(d capita) for WFLB and FLB.

Table 1. The quantity and rate of protozoa grazing on rumen bacteria.

Time point, min	WFLB Number of FLB, cells/cell	WFLB Grazing rate of protozoa, cells/(cell h)	FLB Number of FLB, cells/cell	FLB Grazing rate of protozoa, cells/(cell h)
2	18±1.00	480	6±0.00	180
5	53±2.52	636	22±1.53	264
10	107±2.31	642	49±2.08	294
15	137±2.65	548	71±1.73	284
20	146±1.15	438	86±1.53	258
25	178±1.53	427.2	91±3.06	218.4
30	182±2.89	364	98±2.52	186
35	182±1.53	312	97±2.00	166.8
40	191±2.08	286.5	102±2.52	153

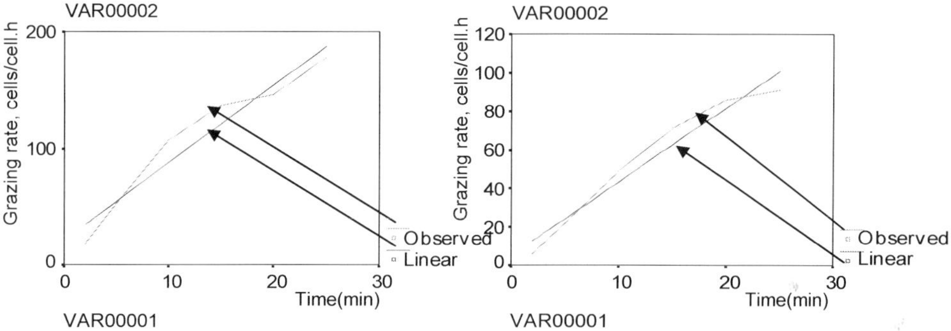

A (WFLB group) *B (FLB group)*

Figure 1. Regression analysis between bacterial numbers engulfed by protozoa with time within 25 minutes. (A) WFLB group; (B) FLB group.

Conclusion

It could be concluded that, bacterial predation rates of rumen protozoa were 398.4 cells/(cell h) and 230.4 cells/(cell h) for WFLB, and FLB, respectively; MCP recycling quantities were 0.645 g Pr/(d capita) and 0.372 g Pr/(d capita), respectively. And finally, the fluorescence-labelled technique would be a powerful method for testing protozoan grazing rate in the rumen.

References

Li, H.B., T. Xiao, S.J. Zhao and H.D. Yue, 2005. Measuring and estimating parameters of the heterotrophic bacterioplankton in marine ecosystem. Marine Sci. 29: 58-63 (in Chinese).

Sherr, B.F., E.B. Sherr and R.D. Fallon, 1987. Use of monodispersed, fluorescently labeled bacteria to estimate in situ protozoan bacterivory. Appl. Environ. Microbiol. 53: 958-965.

Wallace, R.J. and C.A. McPherson, 1987. Factors affecting the rate of breakdown of bacterial protein in rumen fluid. Br. J. Nutr. 59: 313-323.

Effect of freeze-thaw treatment of herbage on the biohydrogenation of α-linolenic acid

D. Warner[1,2], A. Elgersma[2] and R.J. Dewhurst[1]
[1]Lincoln University, P.O. Box 84, Lincoln, Canterbury, New Zealand; [2]Wageningen University, Department of Plant Sciences, P.O. Box 16, 6700 AA Wageningen, the Netherlands

Introduction

Grazed grass is an important source of the n-3 α-linolenic acid (C18:3) in ruminant diets and products. However, exploitation of C18:3 has been limited because rumen biohydrogenation (BH) is often most extensive for high-forage diets. Consequently, we investigated plant-based mechanisms that might decrease BH. Disruption to plant cells causes rapid release of volatile plant defence compounds such as methanol and C6 oxygenates including hexenal ('green odour'). Lee *et al.* (2007) identified individual effects of some components or analogues of 'green odour' on BH. This study adopted a complementary approach, using the rapid release of compounds when herbage is frozen and then thawed. One limitation of this model is that the composition of green odour released after freeze-thaw differs from that due to physical damage (Fall *et al.*, 2001). Since herbage fatty acid levels increase in the autumn, particularly when temperatures are low (Witkowska *et al.*, 2008), a further objective of this work was to investigate the effects of autumn management of pasture on BH.

Material and methods

A batch incubation study was conducted to investigate the effects of freeze-thawing of herbage on BH. Measurements of gas production (GP) provided a parallel assessment of the effects on overall fermentation activity. Ryegrass/white clover herbage that had regrown for 4 wk was harvested on May 23 (late autumn). Adjacent plots had either a grazing rotation omitted (RO) or not (C) so that the preceding rotation had either 6- or 3-wk regrowth respectively. Herbage was carefully harvested and immediately frozen. Half of the herbage was then freeze-dried and ground (<1 mm) and then weighed into fermentation vessels (1 g DM), whilst the remaining herbage was held in the freezer. On the day of the experiment, the frozen herbage was quickly weighed (8 g fresh weight) into fermentation vessels, sealed with parafilm and placed in an incubator at 39 °C. There were 3 replicates of a 2×2 arrangement of previous field management (RO *vs* C) and sample processing (freeze/thaw (FT) *vs* freeze-dry and grind (FD)). Rumen incubations were conducted in 260 ml fermentation vessels of an automated GP system (Ankom, Macedon, NY, USA). When the FT herbage reached 39 °C (90 min), 80 ml of a 50/50 (v/v) mixture of buffer solution (pH 8.1; Lee *et al.*, 2007) and rumen fluid was added and the incubations started. The buffer was pre-warmed and gassed with CO_2, whilst the rumen fluid was collected from two grazing cows, blended and strained through 4 layers of muslin. Incubations of the FD herbage were conducted in parallel. GP was recorded at 5 min intervals via a radio frequency transmitter. At the end of the 6-h incubation, the fermentations were stopped and the bottle contents were frozen until analysis. The chemical composition of the herbage was analysed using near-infrared (NIR) analysis and fatty acid analysis used base methylation (Lee *et al.*, 2007). BH of C18:3 was calculated as the proportional loss of the fatty acid over 6 h (Lee *et al.*, 2007). Gas pressures were converted to volumes and expressed in relation to the amount of OM incubated. Two-factor analysis of variance was performed with the SPSS 17.0 statistical package, SPSS Inc., Chicago, IL, USA.

Results

The concentrations (g/kg DM) of crude protein, water-soluble carbohydrates, NDF, total fatty acids and C18:3 were 236 and 262, 171 and 176, 303 and 277, 65 and 84, and 47 and 63 for RO and C pasture respectively, demonstrating the exceptionally high quality of this late autumn herbage. The effects of treatments on GP and BH of C18:3 are shown in Table 1.

Table 1. Effects of previous pasture management and processing method on gas production (GP) and biohydrogenation (BH) of C18:3.

	Rotation omitted (RO)		Control pasture (C)		S.E.M.	*P*-value	
	FD[1]	FT[2]	FD	FT		Management	Processing
GP, ml/g OM	36.5	45.6	35.4	43.9	0.64	n.s.	<0.001
BH of C18:3, g/g	0.75	0.27	0.77	0.26	0.06	n.s.	<0.001

[1] FD = Freeze-dry and ground (<1 mm).
[2] FT = Freeze-thaw.

Conclusion

It is possible that the reduced BH of FT herbage resulted from its different physical form. However, Kim *et al.* (2005) found extensive BH when crushed herbage, which is physically similar to FT herbage, was incubated *in sacco*. The freeze-thaw treatment increased GP relative to freeze-drying and grinding, confirming the extensive release of cell contents. The results are consistent with the hypothesis that green odour selectively inhibits bacteria involved in BH. Herbage fatty acid levels were exceptionally high at this time, probably as a result of the low mean daily temperature (6.5 °C; Witkowska *et al.*, 2008). Previous pasture management affected fatty acid level, with exceptionally high levels for autumn pasture that had been managed intensively, but there was no effect on BH.

Acknowledgement

The financial support of the Grassland Science Foundation is gratefully acknowledged.

References

Fall, R., T. Karl, A. Jordon and W. Lindinger, 2001. Biogenic C5 VOCs: release from leaves after freeze-thaw wounding and occurrence in air at a high mountain observatory. Atmos. Environ. 35: 3905-3916.

Kim, E.J., R. Sanderson, M.S. Dhanoa and R.J. Dewhurst, 2005. Fatty acid profiles associated with microbial colonization of freshly ingested grass and rumen biohydrogenation. J. Dairy Sci. 88: 3220-3230.

Lee, M.R.F., S.A. Huws, N.D. Scollan and R.J. Dewhurst, 2007. Effect of fatty acid oxidation products (green odor) on rumen bacterial populations and lipid metabolism *in vitro*. J. Dairy Sci. 90: 3874-3882.

Witkowska, I.M., C. Wever, G. Gort and A. Elgersma, 2008. Effects of nitrogen rate and regrowth interval on perennial ryegrass fatty acid content during the growing season. Agron. J. 100: 1371-1379.

Estimation of the fractional rate of forage NDF digestion by *in vitro* gas production or *in situ* methods

M.R. Weisbjerg[1], M. Rinne[2] and P. Huhtanen[3]
[1]Faculty of Agricultural Sciences, Aarhus University, Denmark; [2]MTT Agrifood Research Finland, Animal Production Research, Jokioinen, Finland; [3]Dept. of Agricultural Research for Northern Sweden, Swedish University of Agricultural Sciences (SLU), Umeå, Sweden; martin.weisbjerg@djf.au.dk

Introduction

Forages form the basis of ruminant nutrition and are highly variable in digestion properties depending on species, growth stage etc. The current and future models and rationing systems describing nutrient availability and production of ruminants require information on the rate of digestion of feeds. The *in vitro* gas production and the *in situ* methods are both used to determine this rate. The objective of this study was to compare the fractional rate of digestion (k_d) of forage neutral detergent fibre (NDF) determined using *in vitro* gas production and *in situ* methods, and to compare their ability to predict the *in vivo* NDF digestibility.

Material and methods

Samples comprised 2 barley straw, 3 festulolium, 2 field bean, 7 grass-clover, 1 grass-clover silage, 4 green barley whole crop, 7 lucerne, 2 lupin whole crop, 4 maize silage, 4 pea whole crop, 1 pea straw, 5 red clover, 1 red fescue straw, 4 perennial ryegrass, 1 ryegrass straw, and 2 white clover resulting in a total of 50 samples. Samples were divided into grasses and legumes, and in case of mixtures this was done according to the dominating species based on botanical analysis. *In vivo* digestibility was measured in 4 mature castrated male sheep fed at maintenance level using the total faecal collection method. *In vitro* gas production measurements were made by an automated system (Huhtanen *et al.*, 2008). Samples of NDF residues were incubated in three replicates for 4 d in the presence of rumen fluid and buffer. A two-pool Gompertz model was fitted to the cumulative gas curves. *In situ* NDF digestion was measured using 3 dry Holstein cows fed 0.67 hay and 0.33 concentrate at maintenance level. Bag pore size was 37 μm and incubation times 0, 2, 4, 8, 24, 48 and 96 h. NDF digestion was calculated relative to the residue after 0 h (washing in washing machine) to account for particle losses. Rate of digestion was estimated using exponential models without lag time. Indigestible NDF (iNDF) was measured as NDF residue after 288 h rumen incubation in 12 μm bags.
Predicted *in vivo* NDF digestibility values were estimated based on iNDF, and fractional rate of NDF digestion (k_d) from gas production and *in situ* using a two pool model (Huhtanen *et al.*, 2006) with an assumed 50 h mean retention time in the rumen, divided with 40% in the first pool and 60% in the second pool.

Results and discussion

The samples covered a broad range of chemical composition and digestibility (Table 1). Estimated k_d from *in vitro* gas production varied between 0.023-0.101/h, and from *in situ* between 0.010-0.137/h. Mean k_d was slightly higher for gas production (0.070/h) than for *in situ* (0.066/h) (Figure 1a). Compared to gas production, the *in situ* method underestimated k_d for low digestibility feeds, and overestimated that of high digestibility feeds.
In Figures 1b and 1c predictions of *in vivo* digestibility based on k_d estimated by the two methods were examined. Generally, the predictions seemed to be less precise for legumes compared to

grasses. However, neither slope nor intercept differed significantly between legumes and grasses. Simple regression equations of predicted NDF digestibility on observed NDF digestibility for gas production data were the following: 0.005 + 1.005 X (R^2 0.90) and for *in situ* data: 0.092 + 0.901 X (R^2 0.93). The mean prediction errors (MPE) were 0.048 and 0.053 for the gas production and *in situ* data. The mean bias and slope errors were greater for the *in situ* data, whereas random error was greater for the gas production data. Correlation between the NDF digestibility values estimated with the two methods was high (R^2 0.95).

Table 1. Chemical composition and digestibility of the forage samples (n=50).

	Concentration in DM (%)				Digestibility	
	CP	NDF	ADL	iNDF	OM	NDF
Mean	15.5	43.3	4.5	12.8	0.704	0.640
S.d.	6.1	14.6	2.8	9.4	0.131	0.152
Min	3.5	21.4	1.1	1.9	0.269	0.312
Max	26.0	83.1	14.3	43.7	0.858	0.900

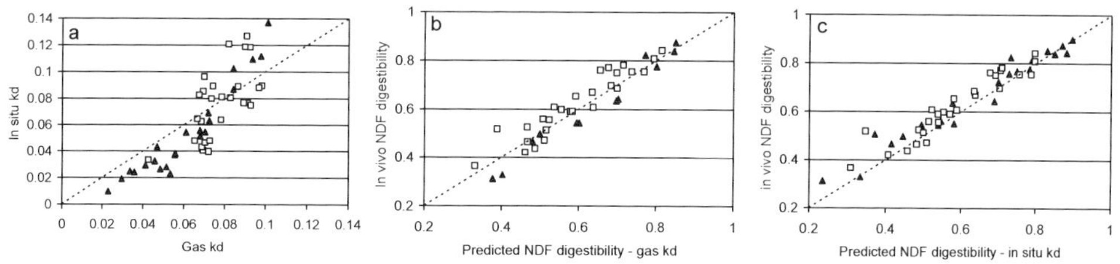

Figure 1. Plot of k_d estimated using in situ vs. in vitro gas (a), and plot of measured digestibility of NDF vs. k_d based predicted NDF digestibility for in vitro gas production (b) and in situ (c). Broken line y=x, □ = legumes, ▲ = grasses.

Conclusion

The *in vitro* gas production method was more accurate and biologically the relationship was more correct, but the *in situ* method was more precise in predicting *in vivo* digestibility for NDF from the k_d estimated by the respective methods. Predictions seem to be less accurate for legumes compared to grasses.

References

Huhtanen, P., S. Ahvenjärvi, M.R. Weisbjerg and P. Nørgaard, 2006. Digestion and passage of fibre in ruminants. In: Sejrsen K., T. Hvelplund and M.O. Nielsen (Eds.), Ruminant physiology. Digestion, metabolism and impact of nutrition on gene expression, immunology and stress. Wageningen Acedemic Publishers, Wageningen, the Netherlands, Proceedings Xth ISRP, pp. 87-135.

Huhtanen, P., A. Seppälä, M. Ots, S. Ahvenjärvi and M. Rinne, 2008. *In vitro* gas production profiles to estimate extent and effective first-order rate of neutral detergent fibre digestion in rumen. J. Anim. Sci. 86: 651-659.

Gastrointestinal calcium (Ca) transport in sheep as affected by dietary Ca and treatment with 1.25-dihydroxyvitamin D_3

M. Wilkens, N. Mrochen, G. Breves and B. Schröder
Department of Physiology, School of Veterinary Medicine, Bischofsholer Damm 15/102, 30173, Hannover, Germany; mirja.wilkens@tiho-hannover.de

Introduction

It is generally accepted that gastrointestinal Ca absorption is increased by 1.25-dihydroxyvitamin D_3, the biologically active metabolite of vitamin D, in monogastric animals (Hoenderop *et al.*, 2005). For ruminants, marked differences to monogastric species have been described concerning the localisation and vitamin D sensitivity of Ca absorption along the gastrointestinal axis. Functional studies could not clearly demonstrate regulation of active Ca transport by 1.25-dihydroxyvitamin D_3 in sheep and goats (Schröder *et al.*, 1997), although the expression of proteins required for the expected mechanism has been shown in ovine small intestine (Schröder *et al.*, 2001; Wilkens *et al.*, 2008). However, active Ca transport mechanisms have been demonstrated in ovine and caprine rumen. Whereas long-term Ca depletion of growing goats led to elevated concentrations of plasma calcitriol and increased Ca net flux rates across ruminal epithelia as measured in Ussing chambers, vitamin D_3 treatment or Ca depletion followed by an increase in plasma calcitriol levels did not alter ruminal net flux rates of Ca in sheep (Schröder *et al.*, 2001; Schröder *et al.*, 1997; Schröder *et al.*, 1999).

It was the aim of the present study to clarify whether an alimentary Ca depletion, a treatment with 1.25-dihydroxyvitamin D_3 in pharmacological dosage or the combination of depletion and treatment can stimulate gastrointestinal Ca transport in sheep.

Material and methods

Twenty female Suffolk sheep aged six months were randomly divided into two groups, which were fed different diets either deficient or adequate in Ca content. Refusals were weighed to calculate the average daily intake of nutrients. After feeding the diets for at least four weeks, five sheep of each group received treatment with 1.25-dihydroxyvitamin D_3 (0.5 µg per kg body weight i.v.). Twelve hours later, the animals were sacrificed. Samples were taken from the ventral rumen sac, duodenum, jejunum and colon. Epithelia were stripped immediately from the underlying muscle layers and mounted into Ussing chambers.

All buffer solutions used in the study were modified Krebs-Henseleit-buffers (300 mosm/kg, pH adjusted to 7.4 when tempered at 38 °C and aerated with carbogen). The serosal solutions were enriched with 5 mmol/l glucose. The mucosal buffer for incubation of ruminal samples contained short chain fatty acids at physiological molar proportions. The chambers were kept under short-circuit conditions and electrophysiological parameters (tissue conductance and short-circuit current) were measured continuously. A radioisotope tracer technique was used to determine unidirectional flux rates of Ca (J_{ms} mucosal-to-serosal flux rate, J_{sm} opposite direction).

To test whether Ca net flux rates ($J_{net}=J_{ms}-J_{sm}$) were significantly different from zero, the Student *t*-test was used. Differences between the four groups were verified by means of two-way-ANOVA (diet and treatment).

Results

In the rumen, significant Ca net flux rates of approximately 10 nmol/cm^2/h were determined. However, neither 1.25-dihydroxyvitamin D_3 nor alimentary Ca depletion had any effect on the Ca transport.

In all treatment groups, no significant Ca net flux rates could be found in duodenal tissues. This is in contrast to monogastric animals investigated so far, e.g. the rat (Jungbluth and Binswanger, 1989). In the jejunum, a significant Ca absorption was only found in epithelia from animals treated with 1.25-dihydroxyvitamin D_3 in supraphysiological dosage. The net flux rates ranged around 10 nmol/cm^2/h, which is very low compared to duodenal flux rates of monogastric animals (Jungbluth and Binswanger, 1989).

Ca net flux rates determined in colon preparations were difficult to interpret due to high variances. A rather low, but statistically significant Ca net flux rate of approximately 3 nmol/cm^2/h could only be confirmed in the alimentary Ca depleted and 1.25-dihydroxyvitamin D_3 treated group.

Conclusion

Ca transport in the ovine rumen does not seem to be regulated by 1.25-dihydroxyvitamin D_3. Against the background of several *in vivo* studies which demonstrated that ruminal Ca absorption is dependent on sufficient amounts of Ca intake as well as the presence of short chain fatty acids, these characteristics suggest that the transport mechanism differs from that described for intestinal tissues which is upregulated by low dietary Ca and via increased plasma 1.25-dihydroxyvitamin D_3. Furthermore, our results indicate that intestinal Ca absorption via active mechanisms does not have the same relevance in sheep as in monogastric animals, where the duodenum was identified as the major site of active and regulated Ca absorption.

References

Hoenderop, J.G., B. Nilius and R.J. Bindels, 2005. Calcium absorption across epithelia. Physiol. Rev. 85: 373-422.

Jungbluth, H. and U. Binswanger, 1989. Unidirectional duodenal and jejunal calcium and phosphorus transport in the rat: effects of dietary phosphorus depletion, ethane-1-hydroxy-1,1-diphosphonate and 1,25-dihydroxycholecalciferol. Res. Exp. Med. 189: 439-449.

Schröder, B., W. Goebel, K. Huber and G. Breves, 2001. No effect of vitamin D 3 treatment on active calcium absorption across ruminal epithelium of sheep. J. Vet. Med. A Physiol. Pathol. Clin. Med. 48: 353-363.

Schröder, B., I. Rittmann, E. Pfeffer and G. Breves, 1997. *In vitro* studies on calcium absorption from the gastrointestinal tract in small ruminants. J. Comp. Physiol. B 167: 43-51.

Schröder, B., S. Vössing and G. Breves, 1999. *In vitro* studies on active calcium absorption from ovine rumen. J. Comp. Physiol. B 169: 487-494.

Wilkens, M.R., C. Kunert-Keil, H. Brinkmeier and B. Schroder, 2008. Expression of calcium channel TRPV6 in ovine epithelial tissue. Vet. J., Aug. 2008 (Epub ahead of print).

Effects of propionate-producing bacteria on propionate metabolism *in vitro*

X. Xing, M. Long, X.Y. Pang, Z. Wang and G.W. Liu
College of Animal Science and Veterinary Medicine, Jilin University, 5333 Xi'an Road, Changchun, Jilin, 130062, China; liuguowen2008@163.com

Introduction

Under normal circumstances, carbohydrates in the rumen of ruminants are fermented to volatile fatty acids (VFA), mainly in acetic, propionic and butyric acids, by a variety of bacteria that contain the corresponding enzymes. Absorbed VFA are precursors of the synthesis of fat and glucose. Approximately 90% of glucose required by ruminant animals is derived from gluconeogenesis. The major glycotropic precursor is propionic acid. Ketosis is one of the most harmful cow diseases, being caused by decreased dry matter intake and insufficient propionic acid intake (Oba and Allen, 2003). The traditional preventative and therapeutic measures involve supplying cows with precursors of glucose, such as propionate. Successful production of propionic acid via ruminal propionate-producing bacteria has not yet been reported. The aim of this study was to investigate the effect of propionate-producing bacteria on ruminal propionic acid metabolism *in vitro* and, thus, to demonstrate the possibility of increasing the level of propionate *in vivo* via genetic engineering or by adding propionic bacteria to the diet.

Material and methods

Bacterial strains, *Prevotella ruminicola* (C11), *Megasphaera elsdenii* (H6), *Veillonella parvula* (H2), *Acidaminococcus fermentans* (H18), were identified by 16S rRNA PCR. For experimentation, small Tail Han sheep from different nurseries (23-28 kg BW and 0.8-1 years old) with good body condition were used. The sheep were given prepared food in accordance with China Merino husbandry standards. Rumen fluids were collected within 2 h after feeding from the rumen of three sheep and were filtered through four gauzes. The fluids were then gassed with CO_2 for 5 min, and quickly transferred to serum bottles (20 ml/bottle) which had been heated and filled with CO_2. For *in vitro* experiments, ruminal fluid was incubated with the primary substrates (lactic acid and pyruvate) and a bacterial strain, C11, H6, H2 or H18 called group I, II, III and IV, respectively. The samples were cultured at 37 °C in anaerobiosis. Culture fluids were sampled at 0, 4, 8, 12, 24 and 48 h of incubation for measurement of lactate, acetate and propionate concentrations. The results are given as means ± SD (n=30), and differences between groups were analysed by ANOVA.

Results

We found that the lactate concentrations of different strains decreased with culture time (Table 1). The concentration of group II decreased significantly, and reached the lowest point at 24 h. Between 0 and 48 h, there were no significant differences between groups I and IV (P>0.05); however, there were highly significant differences between groups I, II and III, and between groups IV, II and III (all: P<0.01). The acetate concentrations of the different cultures increased with incubation time. Between 0 and 48 h, there were significant differences between group II and the others (P<0.01). The propionate concentrations of the different cultures increased with incubation time. In particular, the concentration of propionate was markedly increased in groups II and III. Between 0 and 48 h, there were significant differences between groups (P<0.01).

Table 1. Effects of propionic acid-producing bacteria on the evolution of VFA concentrations in vitro.

Group	Time of sampling (h)						Significance[1]
	0	4	8	12	24	48	test (n=30)
Lactate (mmol/l)							
I	20.30±0.10	19.23±0.38	19.50±0.27	19.20±0.27	18.50±0.10	18.73±0.50	19.16±0.64^{a}
II	20.83±0.55	19.17±0.55	16.53±0.25	12.97±0.47	10.53±0.25	11.40±0.10	15.22±4.05^{c}
III	20.60±0.20	19.30±0.36	17.83±0.31	16.80±0.30	16.13±0.31	14.03±0.21	16.96±3.05^{b}
IV	20.53±0.25	19.57±0.15	19.03±0.40	18.73±0.35	18.50±0.10	18.20±0.10	18.93±0.93^{a}
Acetate (mmol/l)							
I	23.37±0.67	31.83±0.25	48.93±0.25	51.83±0.38	58.63±0.75	53.17±0.31	42.70±11.94^{B}
II	23.20±0.78	27.83±0.25	35.53±0.21	37.50±0.10	40.57±0.75	39.37±0.15	32.87±5.72^{C}
III	23.10±0.20	35.07±0.25	50.43±0.35	55.10±0.44	52.60±2.00	50.43±0.15	43.17±10.48^{B}
IV	23.17±0.35	32.07±0.25	48.93±0.25	52.23±0.32	58.40±0.44	56.30±0.40	44.71±12.64^{A}
Propionate (mmol/l)							
I	7.40±0.10	8.43±0.15	10.63±0.15	11.50±0.10	12.60±0.36	15.80±0.27	10.76±2.57^{3}
II	7.53±0.06	12.43±0.15	19.53±0.15	23.77±0.35	28.97±0.32	26.73±0.21	19.18±6.98^{1}
III	7.43±0.06	11.30±0.46	20.53±0.15	26.20±0.44	26.43±0.55	21.73±0.32	17.23±6.58^{2}
IV	7.33±0.06	7.83±0.06	9.37±0.15	10.13±0.21	12.10±0.30	10.00±0.44	8.86±1.48^{4}

[1] Means with different superscript letters or numbers in the same column and the same VFA are significantly different at $P<0.01$.

Conclusion

This study showed that the ability of strains C11 and H18 to consume lactate were weaker than strains H6 and H2, and likely produced acetate. The propionate production by strains H6 and H2 increased with time, and they used the propionic acid pathway of fermentation. Strain H6 had the most powerful ability to produce propionate. We successfully obtained H6, the highest-performing propionate-producing strain, and provided a foundation for increasing the level of propionate *in vivo.*

Acknowledgement

This study was funded by the China National Science Foundation Committee (30600441).

References

Oba, M. and M.S. Allen, 2003. Extent of hypophagia caused by propionate infusion is related to plasma glucose concentration in lactating dairy cows. J. Nutr. 133: 1105-1112.

Comparison of passage rate, structure and motility of the reticulo-rumen in two sheep breeds

A. Yamazaki, S. Choki, T. Kakizaki, A. Matsuura, M. Irimajiri and K. Hodate
Department of Animal Science, School of Veterinary Medicine, KITASATO University, Towada, Aomori, Japan; yamazaki@vmas.kitasato-u.ac.jp

Introduction

Breeds of domestic animals that have been genetically improved have many differences in many aspects. In ruminants like sheep, each breed has a different ability for production, for example feed digestibility. Digesta passage rate is an important factor for feed usability. Goats known as 'browsers' have higher passage rates than sheep known as 'grazers' (Katoh *et al.*, 1988). There is a very strong relationship between feed utilisation and digesta passage rate, and passage rate is mainly regulated at the reticulo-rumen (Poppi *et al.*, 1985).
Therefore it is considered that the structure and function of the rumen are at the origin of many differences seen in breeds. In these experiments, digesta passage rates were compared in two sheep breeds and also the relationship between the passage rate and the structure and motility of the reticulo-rumen were examined.

Material and methods

Mature Corriedale and Suffolk breed sheep were used.
Experiment 1: Plastic particles (2 mm diameter, 4 mm length) of seven specific gravities (SG; 0.90 to 1.84) were put into the rumen from the mouth through the gullet. Excreted plastic particles were collected from faeces each day over a 10 d period, and the numbers of particles that were excreted and ruminated were counted.
Experiment 2: Computed tomographies of the sheep digestive organ were taken every 2.5 mm or 5.0 mm interval. Nine Corriedale (68.0±11.2 kg) and nine Suffolk (75.3±14.0 kg) sheep were used. Feed calculated from body weight was given just 3 h before photography. The images were processed by a computer and software. Gullet cardia, reticulo-ruminal capacity, reticulo-ruminal fold, position and size of reticulo-omasal orifice were compared by the image of digestive organs provided by photography.
Experiment 3: Ruminal contraction amplitude and the number of primary contractions were measured in two breeds. Each animal was fitted with a cannula on the dorsal sac of the rumen and an electrode on the reticulum. Ruminal motility was measured over 24 h. Student t-tests were used for statistical evaluation.
This study is conformed to the Kitasato University Ethical Code.

Results

Experiment 1: The total passage rates of the seven SG of plastic particles showed a peak on d 2 in both breeds, but the Suffolk breed showed a significantly higher value ($P<0.05$). The Suffolk breed took 1.99 d to excrete 50% of administered particles whereas the Corriedale breed took 4.83 d.
Experiment 2: The ruminal fold was significantly longer and the gullet cardia diameter tended to be smaller in the Corriedale breed. Significant difference was not seen in the reticulo-ruminal capacity (l/kg BW) between the two breeds.
Experiment 3: The rumen contraction amplitudes were significantly higher in the Suffolk breed compared with the Corriedale breed, but the number of contractions was significantly smaller in the Suffolk breed.

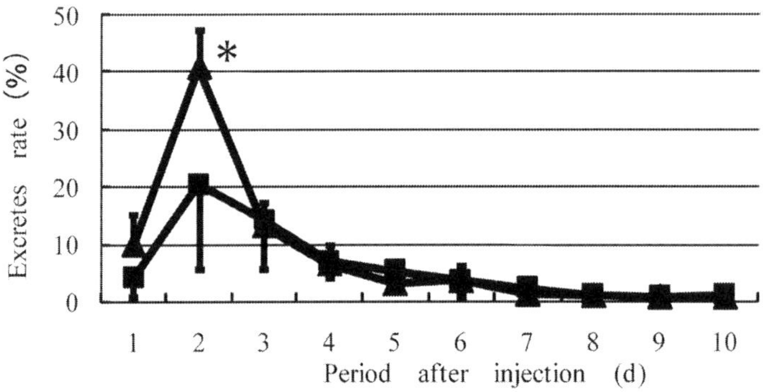

*Figure 1. Excretion rates of plastic particles in two breeds (▲: Suffolk breed, ■: Corriedale breed). * Means are significantly different between the two breeds (P<0.05).*

Table 1. Alimentary canal morphological characteristic and ruminal motility in the two sheep breeds.

	Corriedale breed	Suffolk breed
Reticulo-ruminal capacity, l	12.1±1.8	13.4±2.6
Reticulo-ruminal capacity, l/kg BW	0.18±0.03	0.18±0.03
Gullet cardia diameter, mm	26.5±3.9	32.4±4.4
Reticulo-ruminal fold, mm	29.7±9.6	26.1±12.4
Reticulo-ruminal contraction, times/d	1,609.7±161.1	1,382.6±88.6[a]
Reticulo-ruminal contraction amplitude, mmHg	14.0±1.7	21.2±1.7[a]

[a] Means are significantly different between the two sheep breeds (P<0.05).

Conclusion

The Suffolk breed has a higher alimentary canal passage rate of digesta compared with the Corriedale breed. High amplitude of the ruminal contractions might promote the passage rate in the Suffolk breed. The Suffolk breed had a small number of primary contractions, however, it was suggested that the digesta are removed faster from the reticulo-rumen to lower the alimentary canal by stronger ruminal contractions.

References

Katoh, K., F. Sato, A. Yamazaki, Y. Sasaki and T. Tsuda, 1988. Passage of indigestible particles of various specific gravities in sheep and goats. Br. J. Nutr. 60: 683-687.

Poppi, D.P., R.E. Hendricksen and D.J. Minson, 1985. The relative resistance to escape of leaf and stem particles from the rumen of cattle and sheep. J. Agric. Sci. 105: 9-14.

Effect of dietary vitamin E supplementation on dietary nutrient digestibility in the Boer goat

L. Yan, H. Meng, H. Luo and H. Zhu
State Key Lab of Animal Nutrition, College of Animal Science and Technology, China Agricultural University, Beijing 100193, PR China; luohailing@cau.edu.cn

Introduction

Since the discovery of vitamin E 80 years ago, the effects of vitamin E on animals have been described in several studies (Cheah *et al.*, 1995; Politis *et al.*, 1995; Sahin and Kucuk, 2001a; 2001b). However, limited research has evaluated the effects of vitamin E supplementation on nutrient digestibility in the goat. Nutrient utilisation can influence the deposition of nutrients in the organism, resulting in changes of performance, slaughter performance and meat quality of animals. Macit *et al.* (2003) reported that vitamin E supplementation of lambs improves the feed conversion efficiency and daily weight gain. Sahin and Kucuk (2001a) found digestibility of nutrients is greater with higher dietary vitamin E supplementation. We suppose that vitamin E supplementation can influence the nutrient digestibility of goats, thus affecting performance and meat quality. The objective of the present study was to determine the effects of dietary vitamin E level on nutrient digestibility in the Boer goat.

Material and methods

A total of 24 healthy Boer goats of 3-months age with similar body weights (BW) were selected and randomly divided into four groups. Each group was supplemented with vitamin E at 0, 80, 320 and 880 IU/kid/d for 5 months. Treatments will be referred to as Groups 1, 2, 3 and 4, respectively. The vitamin E used in this study was DL-alpha-tocopheryl acetate (1 mg contains 1 IU vitamin E). The kids were fed a diet consisting mainly of forage and concentrate. The ratio of forage to concentrate was 7/3. All kids were penned individually and housed indoors during the feeding period.

During the last week of the experiment, three goats from each treatment with similar feed intake were selected for collection of faeces. After a 2-day adapted period, total faeces were collected for 3 days for each goat and weighed daily. Ten percent of the faeces were dried. Crude protein (CP) and ether extract (EE) content of feed and faeces were analysed according to the methods of the Association of Official Analytical Chemist (AOAC, 1995). Neutral detergent fibre (NDF) and Acid detergent fibre (ADF) were determined using the method of Van Soest *et al.* (1991).

The univariate of the general linear model of SPSS 13.0 was applied in this study to analyse the data. The results are expressed as mean and standard error. Values of $P<0.05$ were considered significant.

Results

The results in Table 1 show that the nutrient digestibility was affected by dietary vitamin E supplementation on different levels.

In comparison with Group 1, the supplementation with vitamin E results in a higher CP, EE and ADF digestibility in the experimental groups, but there was no significant differences among them. Vitamin E supplementation also had beneficial effects on NDF digestibility, this latter one being increased significantly in Group 3 and Group 4 compared to Group 1, however, no differences between vitamin E-supplemented groups were shown.

Table 1. Effect of dietary vitamin E supplementation on nutrient digestibility in the Boer goat.

Digestibility,%	Control	Group 2	Group 3	Group 4
CP	60.69±1.17	65.10±3.26	66.16±1.86	66.79±1.78
EE	72.35±0.03	74.34±2.58	73.97±1.31	75.97±0.89
NDF	45.89±1.64[a]	55.03±3.35[ab]	58.39±4.34[b]	57.65±1.67[b]
ADF	39.38±2.11	58.41±11.71	52.51±5.63	55.34±2.27

[a,b] Means with common or no superscript under the same classification do not differ (P >0.05).

Conclusion

The study showed that there was a growing tendency for digestibility of CP, EE, NDF and ADF with the increasing level of vitamin E; the supplementation of vitamin E at 320 and 880 IU/kid/d level increased the digestibility of NDF significantly. So, dietary vitamin E supplementation results in improvement of nutrient utilisation in the Boer goat. Further work is in progress to evaluate the effect of vitamin E supplementation on performance and meat quality in goats.

Acknowledgement

The authors thank experimental sheep farm workers and staff of the laboratory. The present study is supported by the project of State Key Laboratory of Animal Nutrition of China (2004DA125184-0803) and the grants from the Trans-Century Training Programme Foundation for the Talents by the Ministry of Education of China (NCET-05-0136).

References

AOAC, 1995. Official Methods of Analysis (16th Edition.) Association of Official Analytical Chemists, Arlington, VA, USA.

Cheah, K.S., A.M. Cheah and D.I. Krausgrill, 1995. Effect of dietary supplementation of vitamin E on pig meat quality. Meat Sci. 39: 255-264.

Macit, M., V. Aksakal, E. Emsen, M.I. Aksu, M. Karaoglu and N. Esenbuga, 2003. Effects of vitamin E supplementation on performance and meat quality traits of Morkaraman male lambs. Meat Sci. 63: 51-55.

Politis, I., M. Hidiroglou, T.R. Batra, J.A.Gilmore, R.C. Gorewit and H. Scherf, 1995. Effects of vitamin E on immune function of dairy cows. Am. J. Vet. Res. 56: 179-184.

Sahin, K. and O. Kucuk, 2001a. Effects of vitamin C and vitamin E on performance, digestibility of nutrients, and carcass characteristics of Japanese quails reared under chronic heat stress. J. Anim. Physiol. A. Anim. Nutr. 85: 335-341.

Sahin, K. and O. Kucuk, 2001b. Effects of vitamin E and selenium on performance, digestibility of nutrients, and carcass characteristics of Japanese quails reared under heat stress. J. Anim. Physiol. A. Anim. Nutr. 85: 342-348.

Van Soest, P.J., J.B. Robertson and B.A Lewis, 1991. Methods for dietary fiber, neutral detergent fiber, and non-starch polysaccharides in relation to animal nutrition. J. Dairy Sci. 74: 3583-3597.

Site and extent of feed digestion in the digestive tract of beef cattle fed high-grain diet supplemented with cinnamaldehyde or eugenol

W.Z. Yang[1], *C. Benchaar*[2], *M.L. He*[1] *and K.A. Beauchemin*[1]
[1]*AAFC, Research Centre, 5403 – 1 Ave. S. PO Box 3000, Lethbridge, AB, T1J4B1, Canada;* [2]*AAFC, Dairy and Swine R&D Centre, 2000 College street - PO Box 90, STN-Lennoxville, Sherbrooke, QC, J1M 1Z3 Canada; wenzhu.yang@agr.gc.ca*

Introduction

Essential oils (EO) from plant extracts have been reported to have an anti-bacterial activity against Gram-negative and Gram-positive bacteria. Studies have revealed that the EO such as cinnamaldehyde oil (CIN), carvacrol, eugenol (EUG) have the potential to favourably alter rumen metabolism such as a reduction in the rate of deamination of amino acids, and inhibition of the production of methane, which is beneficial for improving feed utilisation and for reducing environmental impact (McIntosh *et al.*, 2003). The CIN, a phenylpropanoid with antimicrobial activity, is the main active component of cinnamon (*C. cassia*) oil, accounting for up to 75% of its composition (Calsamiglia *et al.*, 2007). The EUG is a phenolic compound with wide-spectrum antimicrobial activity, and it is one of the main active components in clove bud and cinnamon oils. The effects of supplementation of finishing beef diet with CIN or EUG on feed intake, site and extent of digestion were evaluated in two digestion studies.

Material and methods

Two studies were conducted using eight spayed beef heifers with ruminal and duodenal cannulas. In each study, four animals were randomly assigned to a single 4×4 Latin square design with treatments: control (no addition of EO), 400, 800 and 1600 mg/head/d CIN or EUG, respectively, for study 1 or 2. Each experimental period lasted 21 d, and the animals were housed in individual tie stalls. A total mixed ration containing 15% barley silage, 80% dry-rolled barley grain and 5% supplement (DM basis) was fed twice daily at 8:30 a.m. and 5:30 p.m. Measurements of intake and digestibility occurred the last week of each period. External marker of $YbCl_3$ and ^{15}N were used, respectively, to label digesta and rumen microbes. Data were analysed using the MIXED procedure of SAS® (2007) with model including period and treatment as fixed effects and animal as random effect. Linear and quadratic contrasts were tested using the CONTRAST statement of SAS®.

Results

Intake of DM was quadratically changed with increasing CIN supplementation to be higher (P<0.10) with the low CIN but lower (P<0.05) with high CIN dosage compared with the control (Table 1). Digestibility (truly) of DM in the rumen was quadratically affected by CIN with no effect of the low and medium CIN supplementation, but with a reduction of 11% by the high CIN supplementation compared with the control. Intestinal digestibility of DM was not affected whereas that of DM in the total tract linearly (P<0.05) decreased with increasing CIN supplementation. Microbial N production was not affected by treatments.

For study 2, DM intake (averaged 9.6 kg/d) was not affected by EUG supplementation. Digestibility of DM in the rumen, intestine and total tract was not affected by treatments. However, digestibility of ADF tended to linearly (P<0.10) reduce from 40 to 31% in the rumen and from 51 to 43% in the total tract without affecting intestinal digestion with increasing dosage of EUG. Microbial N production was quadratically affected (P<0.08) to be the highest with the medium EUG and the lowest with the control.

Table 1. Effect of CIN or EUG supplementation on intake and digestibility.

	EO, mg/head/day				SE[1]	Effect	
	0	400	800	1600		Linear	Quadratic
Cinnamaldehyde							
Intake of DM, kg/d	9.7	10.7	10.1	8.7	0.6	0.04	0.04
Digestibility of DM,% of intake							
Rumen (apparent)	50.6	54.2	51.4	44.3	3.9	0.07	0.15
Rumen (true)[2]	62.7	64.3	63.3	55.6	3.8	0.03	0.09
Intestine	27.2	25.6	24.7	28.4	2.3	0.63	0.27
Total tract	77.8	79.8	76.1	72.7	3.2	0.05	0.41
Microbial N, g/d	110.1	115.6	116.7	103.3	9.5	0.37	0.21
Eugenol							
Intake of DM, kg/d	9.4	9.6	9.9	9.7	0.7	0.57	0.44
Digestibility of DM,% of intake							
Rumen (apparent)	49.0	39.2	43.1	43.6	5.3	0.85	0.24
Rumen (true)[2]	66.0	57.0	62.2	62.6	5.5	0.84	0.24
Intestine	28.5	37.2	33.8	31.0	4.5	0.97	0.09
Total tract	77.5	76.4	76.9	74.6	3.4	0.20	0.79
Microbial N, g/d	110.5	114.0	128.4	120.8	11.3	0.09	0.08

[1] SE = standard error.

[2] Corrected for rumen microbial portion.

Conclusions

Supplementation of high-grain diets with CIN affected feed intake and ruminal digestion of feeds in a dose-dependent manner. A low dose of CIN increased nutrient availability in the rumen due to increased feed intake and higher ruminal digested DM. In contrast, feed intake and ruminal digestion of feeds was adversely affected when a high dose of CIN was supplemented. Lowering ruminal CP degradation can provide more rumen undegraded protein to the small intestine. Supplementation of a feedlot finishing diet with EUG had minimal effect on site of DM digestion. However, it would be detrimental to ruminal microbial activity on reducing fibre digestion in the rumen. The results suggest that the CIN and EUG have different modes of action in the digestive tract of ruminants.

Acknowledgement

The CIN product was provided by the Pancosma S.A. (Bellegarde-sur-Valserine, France).

References

Calsamiglia, S., M. Busquet, W. Cardozo, L. Castillejos and A. Ferret, 2007. Invited Review: Essential oils as modifiers of rumen microbial fermentation. J. Dairy Sci. 90: 2580-2595.

McIntosh, F.M., P. Williams, R. Losa, R.J. Wallace, D.A. Beever and C.J. Newbold, 2003. Effects of essential oils on ruminal microorganisms and their protein metabolism. Appl. Environ. Microbiol. 69: 5011-5014.

Effects of aniso-prescription of chinese herbal medicine on the main digestive enzymes in the jejunum of growing cattle

W.R. Yang, Y.H. Cui, Z.B. Yang, S.Z. Jiang and P. Wang
Department of Animal Sciences and Technology, Shandong Agricultural University, Taian, Shandong 271018, P. R. China; wryang211@163.com

Introduction

There has been an increasing interest in exploiting natural products as feed additives to solve problems in animal nutrition and livestock production. Herbs have been evaluated for their ability to alter ruminal fermentation and improve nutrient utilisation in ruminants (Hristov *et al.*,1999). In addition, supplementation with herbs (lemongrass powder) improved digestibilities of nutrients, rumen microbial population, and microbial protein synthesis efficiency in beef cattle (Wanapat *et al.*, 2008). But to our knowledge, the effect of mixed-herbs on the activity of digestive enzymes in the jejunum of growing cattle has never been confirmed. The present study was aimed at evaluating the effects of herbs on the main digestive enzymes in the jejunum of growing cattle.

Material and methods

Twenty crossbred beef cattle (Limousin × Lu-xi, BW: 199.37±8.67 kg) were randomly divided into five treatments with a control group (no herb added) and four tested groups (A, B, C, D; Table 1). Cattle had *ad libitum* access to diets.

Table 1. Composition of all mixed-herbs.

	Herbs included
A	Tang shen, Largehead Atractylodes Rhizome, Indian Bread, Common Yan Rhizome, Lablab, Dried Tangerine peel, Platycodon Root, Villous Amomum Fruit, Coix Seed
B	Baical Skullcap Root, Cape Jasmine Fruit, Oriental Waterplantain Rhizome, Akebiae, Chinese Angelica, Bupleurum chinense, Radix Glycyrrhizae, Rehmannia Dride Rhizome, Radix Gentianae, Plantain Seed
C	Bupleurum chinense, Chinese Angelica, White peoney Alba, Largehead Atractylodes Rhizome, Indian Bread, Cape Jasmine Fruit, Cortex Moutan, Prepared Dried Ginger, Peppermint, Prepared Liauorice Root
D	Massa Medicata Fermentata, Milkvetch Root, Hawthorn Fruit, Immature Orange Fruit, Atractylodes Rhizome, Perillaseed, Twotoothed Achyranthes Root, White Mulberry Root-bark, Radish Seed, Stir-baked

The experimental period lasted 90 days from 7 to 10 months of age. Cattle were slaughtered at the end of the experiment. The abdominal cavity was opened and the entire gastrointestinal tract was removed. The small intestine was carefully dissected free. The intestinal tissue was collected and immediately frozen in liquid nitrogen for analysis. Intestinal tissue was thawed at 4 °C and homogenised in 1 wt/vol of ice-cold NaCl for 15 s. The homogenate was centrifuged (15,000× g for 10 min at 4 °C), and aliquots of the supernatant were stored at -20 °C until analysis. The substrate used for trypsin determination was benzoyl DL-arginine p-nitroanilide (B 4875, Sigma, St. Louis, MO), and succinyl ala-ala-pro-phe p-nitroanilide (S 7388, Sigma) was used as a substrate to measure chymotrypsin activity. The activity of lipase was measured at pH 6.5 with tributyrin

(T8626, Sigma Chemical, St. Louis, MO, USA) as a substrate using a Titralab™ (Radiometer, Copenhagen, Denmark). Amylase activity was measured according to procedures described by Ceska *et al.* (1969) using Phadebase amylase reagent as a substrate (Pharmacia Diagnostics, Uppsala, Sweden) (Jensen *et al.*, 1997). All data from the experiment were analysed by the procedure of SPSS. The statistical model included cattle, period, treatment and residual error. Fixed effects included period and treatment. Cattle were the random effect. Trends towards significance were considered at $P<0.05$.

Results

The activities of trypsin, chymotrypsin and lipase in jejunum were significantly stimulated by the aniso-prescription with Chinese herbal medicine ($P<0.05$). The content of amylase was similar between groups ($P>0.05$) (Table 2).

Table 2. The activities of trypsin, chymotrypsin, lipase and amylase in the jejunum of growing-cattle in response to aniso-prescription with Chinese herbal medicine.

	Trypsin	Chymotrypsin	Lypase	Amylase
Control	0.82[c]	0.47[d]	66.74[c]	3.15[a]
A	1.06[b]	0.6[abc]	74.82[bc]	2.69[ab]
B	1.06[b]	0.62[ab]	159.76[a]	3.02[ab]
C	1.18[a]	0.7[a]	115.27[b]	2.87[ab]
D	1.16[a]	0.59[abc]	127.4[b]	2.72[ab]
SEM	0.017	0.015	3.32	0.16
P-value	<0.05	<0.05	<0.05	>0.05

[a,b] Means within arrays with same superscript letters are not significantly different ($P>0.05$).

Conclusion

This study shows that the aniso-prescription with Chinese herbal medicine increased the activities of trypsin, chymotrypsin and lipase in the jejunum of growing cattle. But the amylase activity was not significantly affected by different mixed-herbs.

References

Ceska, M., K. Birath and B. Brown, 1969. A new and rapid method for the clinical determination of α-amylase activities in human serum and urine. Clin. Chim. Acta 26: 437-444.

Hristov, A.N., T.A. McAllister, F.H. Van Herk, K.J. Cheng, C.J. Newbold and P.R. Cheeke, 1999. Effect of *Yucca schidigera* on ruminal fermentation and nutrient digestion in heifers. J. Anim. Sci. 77: 2554-2563.

Jensen, M.S., S.K. Jensen and K. Jakobsen, 1997. Development of digestive enzymes in pigs with emphasis on lipolytic activity in the stomach and pancreas. J. Anim. Sci. 75: 437-445.

Wanapat, M., A. Cherdthong, S. Wanapat and P. Pardee, 2008. Manipulation of rumen ecology by dietary lemongrass (*Cymbopogon citratus Stapf.*) powder supplementation. J. Anim. Sci. 86: 3497-3503.

Effects of dietary energy intake and ruminal SCFA on mRNA expression of Na^+/H^+ exchangers in rumen epithelium of goats

W. Yang[1], H. Martens[2] and Z. Shen[1]
[1]Key Laboratory of Animal Physiology and Biochemistry, College of Veterinary Medicine, Nanjing Agricultural University, China; [2]Institute of Veterinary Physiology, Free University, Berlin, Germany; zmshen@njau.edu.cn

Introduction

Na^+/H^+ exchangers (NHEs) are ubiquitous plasma membrane transporters in mammalian cells. The mRNA of NHE1 and NHE3 were detected in the rumen epithelium of sheep, goats and cattle (Shen *et al.*, unpublished). Electroneutral Na transport in rumen epithelium is mediated by NHE (Martens *et al.*, 1991). Recent studies have demonstrated the stimulating effects of rumen microbial fermentation products like short-chain fatty acids (SCFA) (Uppal *et al.*, 2003), ammonia (Abdoun *et al.*, 2003), as well as dietary energy intake (Shen *et al.*, 2004) on rumen epithelial NHE activity. However, the underlying mechanism of this stimulation is not clear. Musch *et al.* (2001) reported an increased colonic NHE3, but not NHE2, protein, mRNA, and brush-border activity of rats fed soluble fibre pectin and suggested that luminal SCFA are important physiological cues for regulating Na absorption. In a previous study (Shen *et al.*, 2004), we observed that an energy-rich diet caused an increase of butyric acid concentration in rumen fluid and enhanced NHE activity in the rumen epithelium. But the causal relationship between diet, ruminal SCFA and NHE in the rumen epithelium is not fully understood. We therefore hypothesised that diets and feeding strategy induce the alteration of rumen fermentation products and subsequently, the fermentation products since SCFA exerts its effect on NHE. In this paper we studied the following: (1) The effects of high or low energy intake on ruminal SCFA and on mRNA expression of NHE1, 2 and 3, and the effects of feeding or fasting on rumen SCFA and mRNA expression of NHE1, 2 and 3. (2) Correlation of altered SCFA concentration with NHE1, 2 and 3 gene expression in the rumen epithelium.

Material and methods

Two experiments, with weanling goats (70-d old, BW 14.27±0.69 kg), were performed with different feed intake and feeding methods before slaughter (fasting or feeding). In EXP A ten goats were assigned into two groups (n=5). Animals were fed peanut straw *ad libitum* (LLD-A; energy consumed 0.6 $MJ/kg^{0.75}/d$; nitrogen intake 1.0 $g/kg^{0.75}/d$) or peanut straw supplemented with 400 g/d concentrate (HLD-A, 1.0 $MJ/kg^{0.75}/d$ and 2.4 $g/kg^{0.75}/d$) at 8:00, 11:00, 14:00 and 17:00. Water was always available. The experimental period lasted 42 d. Before slaughter the goats were fasted for 16 h. In EXP B six goats were assigned into LLD-B and HLD-B groups (n=3) and fed according to the feeding program of EXP A. The animals were slaughtered 7 h after regular feeding. For both groups the rumen fluid was sampled to determine the ruminal pH and SCFA and, rumen epithelia were taken from the ventral rumen sac for NHE mRNA measurement. The qualitative analysis of mRNA expression was measured by semi-quantitative reverse transcription-polymerase chain reaction (RT-PCR). All statistical analyses were performed using One-Way ANOVA, LSD and Bivariate Correlations of SPSS 12.0 for Windows. Treatment effects were considered significant at $P<0.05$ and a tendency to be significant at $P<0.10$.

Results

The mRNA of NHE1, 2 and 3 were detected in all collected rumen epithelium. The sequence homology between the reported bovine sequences of NHE3 (*Bos taurus* U49432), the NHE1 (*Bos taurus* AJ131764.1) and our present PCR products were 95% and 91%, respectively.
In EXP A (16 h fasting before slaughtering) the mRNA abundance of NHE1 (P=0.64), NHE2 (P=0.44) and NHE3 (P=0.43) did not differ between HLD-A and LLD-A, nor the total SCFA (TSCFA) concentration (P=0.18). There was no correlation between NHE mRNA and rumen TSCFA nor pH. In EXP B the mRNA abundance of the NHE1 (P<0.05), NHE3 (P=0.05) and TSCFA concentrations were higher (P<0.05) in HLD-B than in LLD-B, indicating the effects of dietary energy intake on NHE1 and NHE3 gene expression. Furthermore, correlations between the ruminal TSCFA (r=0.93) and the mRNA abundance of NHE3 (P<0.05), between the ruminal pH (r=-0.85) and the mRNA abundance of NHE1 (P=0.07) were observed. These data agreed with those of a previous study on colonocytes of pectin-fed rats (Musch *et al.*, 2001), in which the authors reported that luminal microbial fermentation products of SCFA regulate NHE3 expression in colonic epithelium. The data obtained from EXP A demonstrated that fasting abolished the differences of ruminal TSCFA and NHE gene expression in the rumen epithelium despite the different intake of energy.

Conclusion

mRNA expression of NHE3 and NHE1 in the rumen epithelium is positively correlated to the higher concentration of ruminal TSCFA and lower pH, caused by a energy-rich diet. The adaptation of NHE3 and NHE1 gene expression to ruminal TSCFA, induced by diet intake, could occur acutely within several hours.

Acknowledgement

This study is financially supported by Chinese National Nature Science Foundation (No. 30771568).

References

Abdoun, K., K. Wolf, G. Arndt and H. Martens, 2003. Effect of ammonia on Na+ transport across isolated rumen epithelium of sheep is diet dependent. Brit. J. Nutr. 90: 751-758.

Martens, H., G. Gabel and B. Strozyk, 1991. Mechanism of electrically silent Na and Cl transport across the rumen epithelium of sheep. Exp. Physiol. 76: 103-114.

Musch, M.W., C. Bookstein, Y. Xie, J.H. Sellin and E.B. Chang, 2001. SCFA increase intestinal Na absorption by induction of NHE3 in rat colon and human intestinal C2/bbe cells. Am J Physiol Gastrointest. Liver. Physiol. 280: 687-693.

Shen, Z., H. Syfert, B. Loehrke, F. Schneider, R. Zitnan, A. Chudy, S. Kuhla, H.M. Hammon, J.W. Blum, H. Martens, H. Hagemeister and J. Voigt, 2004. An energy-rich diet caused rumen papillae proliferation associated with more IGF type 1 receptors and increased plasma IGF-1 concentrations in young goats. J. Nutr. 134: 11-17.

Uppal, S.K., K. Wolf, S.S. Khahra and H. Martens, 2003. Modulation of Na+ transport across isolated rumen epithelium by short-chain fatty acids in hay- and concentrate-fed sheep. J. Anim. Physiol. Anim. Nutr. (Berl). 87: 380-388.

Effect of calcium level on feed intake, nutrient digestion, fecal microbial population and growth performance of dairy calves

C. Yuangklang[1], C. Wachirapakorn[2] and A.C. Beynen[1]
[1]Department of Animal Science, Faculty of Natural Resources, Rajamangala University of Technology Isan, Sakon Nakon Campus, Phangkhon, Sakon Nakon 47160 Thailand; [2]Department of Animal Science, Faculty of Agriculture, Khon Kaen University, Khon Kean, 40002 Thailand; chayua@hotmail.com

Introduction

Feeding the dairy calf during the transition period from birth to weaning is a critical time for the dairy farmer, because the calf's diet is changed from liquid to solid feeds, which may cause some diseases. During the transition period, calves are more susceptible to diseases such as scouring. High calcium intake has been shown to decrease fat digestion and to enhance fecal bile acid excretion in veal calves, which can also be explained by the formation of extra calcium phosphate sediment (Xu *et al.*, 1998). Moreover, Yuangklang *et al.* (2004) demonstrated that high calcium intake decreased fat digestion and increased calcium and phosphorus excretion in faeces. Moreover, Yuangklang *et al.* (2006) found that calves fed a high calcium diet were heavier than calves fed a low calcium diet when fed ruzi hay as a roughage source. Furthermore, dietary high calcium intake depresses the number of *Escherichia coli* in faeces when compared with low calcium intake in rats (Bovee-Oudenhoven *et al.,* 1999). The objective of the present experiment was aimed at studying the effect of calcium level on feed intake, nutrient digestion, growth performance and faecal microbial population in dairy calves.

Material and methods

Forty male Thai Friesian calves about one week old with an average BW of 33 kg were used in a randomised completed block design. Treatments were 0.5, 1.0, 1.5 and 2.0% calcium in concentrate diet. Calves were fed concentrate at 1.0% of body weight and calves received pangola hay ad libitum. Calves consumed reconstituted milk at 10% of body weight. Animals were weighed weekly. The experiment lasted for 90 days. On d 85 to d 90, faeces samples were quantitatively collected. On d 89 to d 90, feces samples were directly collected from the rectum for microbial population. Experimental diets and faeces samples were analysed for dry matter, crude protein and ash (AOAC, 1990). Calcium and phosphorus were analysed according to Yuangklang et al. (2004).

Results and discussion

The initial body weight of calves was similar among treatments. High calcium intakes linearly enhanced ($P<0.05$) final body weight of calves. Average daily gain was linearly increased ($P<0.05$) with an increasing calcium intake. Fecal score measurement was 1.25, 1.30, 1.30 and 1.30 in 0.5, 1.0, 1.5 and 2.0% Ca in concentrate diet, respectively ($P>0.05$). The fecal characteristics were strict and dark-yellow color. Fecal *E. coli* number population was not significantly different ($P>0.05$) among treatments, but fecal *lactobacilli* population increased ($P<0.05$) with an increasing calcium level. Dietary calcium phosphate stimulates intestinal lactobacilli in rats (Bovee-Oudenhoven *et al.* 1999). Digestion of nutrients were not significantly different ($P>0.05$) among treatments. Yuangklang *et al.* (2004) found that high calcium intake reduces phosphorus absorption. In conclusion, this study indicates that feeding high calcium increases the *lactobacilli* population in feces and improves average daily gain. Increasing the level of calcium did not influence nutrient digestion.

Table 1. Effect of calcium level on body weight, average daily gain, fecal score, feed intake, nutrient digestion and fecal microbial population.

	Calcium level (%)				SEM	Contrast		
	0.5	1.0	1.5	2.0		L	Q	C
Initial BW, kg	32.6	34.8	34.6	30.6	1.27	ns	ns	ns
Final BW, kg	97.8	109.2	113.2	114.0	1.44	*	ns	ns
ADG, g/d	724.4	826.7	873.3	926.7	12.64	*	ns	ns
Total intake, gDM/d	2857	3008	3045	2996	51.53	ns	ns	ns
Fecal score (1-5)	1.25	1.30	1.30	1.30	0.01	ns	ns	ns
Fecal *E. coli*, $\log_{10}$ CFU/g	1.98	2.08	2.12	1.91	0.04	ns	ns	ns
Fecal *Lactobacillus* sp. $\log_{10}$ CFU/g	1.36	1.71	1.75	2.00	0.02	*	ns	ns
Digestion,% of intake								
DM	89.0	87.6	90.0	90.9	0.93	ns	ns	ns
OM	90.8	89.9	93.4	91.1	0.62	ns	ns	ns
CP	96.8	96.4	97.4	96.6	0.27	ns	ns	ns
EE	86.5	83.7	91.7	84.5	1.66	ns	ns	ns
Ca	89.3	90.2	92.4	93.8	1.78	ns	ns	ns
P	90.4	90.2	88.2	87.9	1.65	ns	ns	ns

[a,b,c] Means within rows with same superscript letters are not significantly different $P<0.05$.

Acknowledgement

The authors would like to sincerely thank the Higher Commission of Education and Thailand Research Fund through the Young Scientist Award for financial support and Faculty of Natural Resources, Rajamangala University of Technology Isan for facility supports.

References

AOAC, 1990. Official Methods of Analysis. AOAC, 1990. Official Methods of Analysis. 15th ed. Association of Official Analytical Chemists, Arlington, VA, USA.

Bovee-Oudenhoven, I.M., M.L. Wissink, J.T. Wouters and R. Van der Meer, 1999. Dietary calcium phosphate stimulates intestinal *Lactobacilli* and decreases the severity of a *Salmonella* infection in rats. J. Nutr. 129: 607-612.

Xu, C., T. Wensing and A.C. Beynen, 1998. Effects of High Calcium Intake on Fat Digestion and Bile Acid Excretion in Feces of Veal Calves. J. Dairy Sci. 81: 2173-2177.

Yuangklang, C., K. Vasupen, S. Wonsuthavas and P. Srenanual, 2006. Effect of calcium level on feed intake, rumen fermentation and nutrient digestibility in dairy calves. In: Proceeding of Animal Science Meeting, Chiang Mai University, Chiang Mai, Thailand.

Yuangklang, C., Th. Wensing, L. Van den Broek, S. Jittakhot and A.C. Beynen, 2004. Fat digestion in veal calves fed milk replacers low or high in calcium and containing either casein or soy protein isolate. J. Dairy Sci. 87: 1051-1056.

Effect of protein level on feed intake, nutrient digestibility and blood urea nitrogen in crossbred Brahman heifers

C. Yuangklang, K. Vasupen, S. Wongsuthavas, S. Bureenok and J. Khotsakdee
Department of Animal Science, Faculty of Natural Resources, Rajamangala University of Technology Isan, Sakon Nakon Campus, Phangkhon, Sakon Nakon 47160 Thailand; chayua@hotmail.com

Introduction

It is well known that protein is an essential nutrient for all animals. Its activities are involved in many biological functions such as muscle development. Protein is also an expensive nutrient in animal rations so thus it is important to understand protein requirement to meet the animal's need. In Thailand, Brahman cattle have been imported from India and Pakistan for many years. The objectives are to improve the performance of Thai native cattle. Presently, there are numerous crossbred Brahman cattle, which have been distributed throughout the whole country. The knowledge of protein requirement of Thai crossbred Brahman cattle is scarce. Especially, Thai crossbred Brahman cattle are fed low quality roughage such as rice straw, particularly in the northeast of Thailand. Kearl (1982) reported that the protein requirement for the growing crossbred Brahman was 12.0% crude protein. Thus, the present experiment was aimed at investigating the effect of protein level on feed intake, average daily gain, nutrient digestion and feed conversion ration of Thai crossbred Brahman heifers.

Material and methods

Sixteen crossbred Brahman heifers were used in a randomised completed block design. The initial body weight was 258±29 kg (mean ± SD). The experiment lasted for 120 days. The animals were fed dietary crude protein (CP) levels of 8.0, 10.0, 12.0 and 14.0% with similar amounts of metabolisable energy (2.6 Mcal per kg of dry matter) at 3.0% body weight (BW) as dry matter (DM) intake. The animals were weighed every two weeks. On day 115-120 of the experimental period, the faeces samples were collected for five consecutive days. Feed samples were collected weekly and pooled for analysis. Blood was taken from jugular rena-puncture for blood urea nitrogen analysis. Feed and faeces samples were dried at 60 °C for 72 h; ground and analysed for dry matter (DM), crude protein (CP), as hand acid insoluble ash (AIA) by the method of AOAC (1990). Digestibility of nutrients was calculated as nutrient intake – nutrient in feces × nutrient intake/1 × 100 (Schneider and Flatt, 1977).

Data were subjected to analysis of variance using the GLM procedure for orthogonal polynomial contrast analysis (SAS®, 1996) according to a randomised complete block design (RCBD) using the initial body weight as the block. Significance was shown at $P<0.05$ unless otherwise noted.

Results and discussion

The crude protein contents of the experimental diets were 8.1, 10.4, 12.3 and 14.2%, respectively. Feed intake did not differ among treatments (Table 1). Average daily gain was 450, 465, 478 and 498 g/d in heifers fed on 8.0, 10.0, 12.0 and 14.0% CP, respectively ($P<0.05$). It has been known that nitrogen is needed for animal growth (Yuangklang, 2007). Digestibility of DM, OM and CP did not significantly differ ($P>0.05$) among treatments. Although protein digestion was not significantly different among treatments, but it was increased as protein level increased. In a previous trial, Yuangklang *et al.* (2008) reported that the protein digestion increases with increasing protein levels. Blood urea nitrogen was linearly increased as protein level increased ($P<0.05$). This might explain that energy intake was restricted so that protein is used as an energy source leading to increased breakdown of protein into ammonia nitrogen and consequently increased urea nitrogen concentration in the blood. This finding was in accordance with Chantiratikul *et al.* (2009) who demonstrated that increased dietary protein level is linearly increased with blood urea nitrogen in Thai-Indigenous heifers. From the experimental data, it can be concluded that increased protein level did improve growth performance of crossbred Brahman heifers fed a total mixed ration with rice straw as a roughage source.

Table 1. Feed intake, average daily gain, digestion coefficient of nutrient and blood urea nitrogen.

	Dietary protein level (%CP)				SEM
	8.0	10.0	12.0	14.0	
Feed intake,					
kgDM/d	9.00	9.00	10.00	10.00	0.12
%BW	3.0	3.0	3.0	3.0	
g/kgBW$^{0.75}$	120.9	120.4	117.4	121.2	0.46
Average daily gain, g/d	450^{b}	465^{b}	478^{ab}	498^{a}	27
Digestion coefficient,%					
Dry matter	71.3	75.4	77.9	79.8	2.45
Organic matter	74.5	77.8	80.2	82.6	3.34
Crude protein	65.1	69.2	71.9	74.8	3.83
Blood urea nitrogen	12.9^{b}	13.4^{b}	14.5^{ab}	16.1^{a}	0.25

^{a,b,c} Means within rows with same superscript letters are not significantly different $P<0.05$.

Acknowledgement

The authors would like to sincerely thank Japan International Research Center for Agricultural Sciences (JIRCAS) for financial support and Faculty of Natural Resources, Rajamangala University of Technology Isan for facilities supports.

References

AOAC, 1990. Official Methods of Analysis. AOAC, 1990. Official Methods of Analysis. 15th ed. Association of Official Analytical Chemists, Arlington, VA, USA.

Chantiratikul, A., S. Chumpawadee, W. Kanchanamayoon and P. Chantiratikul, 2009. Effect of dietary protein on nutrient digestibility and nitrogen metabolism in Thai-Indigenous heifers. J. Anim. Vet. Adv. 8: 297-300.

Kearl, L.C., 1982. Nutrient Requirements of Ruminants in Developing Countries. International Feedstuff Institute Publication, Utah State University, Logan, USA.

SAS, 1996. Statistical Analysis Systems Institute Inc., Cary, NC, USA.

Schneider, B.H. and W.P. Flatt, 1977. The Evaluation of Feeds through Digestibility Experiments. The University of Georgia Press, Athens, Georgia, USA, 423 pp.

Yuangklang, C., 2007. Ruminant Nutrition and Feeding. Rajamangala University of Technology Isan, Department of Animal Science, Faculty of Natural Resources, Sakon Nakhon Campus, Sakon Nakhon, 179 pp.

Yuangklang, C., K. Vasupen, S. Wongsuthavas and P. Panyakaew, 2008. Effect of increasing protein level on feed intake and nitrogen utilization in purebred Brahman cattle. In: Proc. 13th AAAP, Hanoi, Vietnam.

Effect of fat level and supplemental yeast (*Saccharomyces cerevisiae*) on voluntary feed intake, digestion coefficient of nutrients and growth performance in meat goats

C. Yuangklang and J. Khotsakdee
Department of Animal Science, Faculty of Natural Resources, Rajamangala University of Technology Isan, Sakon Nakhon Campus, Phangkhon, Sakon Nakhon 47160 Thailand; chayua@hotmail.com

Introduction

Typically, goats are raised in natural fields to reduce feed cost. But the performances of goats are not well improved so that supplementation is needed. Fat is normally used in ruminant diets to enhance energy density. Yuangklang *et al.* (2005) demonstrated that supplemental fat improved growth of meat goats, in agreement with other researchers who showed that fat supplementation increased body weight gain and nutrient utilisation in goats. However, added fat in ruminant diets has been shown to reduce fibre digestion (Devendra and Lewis, 1974). The native effect of added fat must be considered before feeding. Yeast (*Saccharomyces cervesiae*) is frequently incorporated into ruminant diets to improve animal performance. In sheep, yeast supplementation decreased rumen protozoa count when compared with the control. Newbold *et al.* (1996) demonstrated that yeast culture enhanced cellulotytic bacteria. The present experiment studied the effect of fat level and yeast supplementation on feed intake, digestion coefficient of nutrient digestibility and growth performance in meat goats.

Material and methods

Eighteen meat goats were used in a randomised complete block design. Goats received 3.0% fat and 6.0% fat with or without yeast supplementation. Goats were fed concentrate *ad libitum* and grass silage was fully supplied. Goats were weighed weekly. The experiment lasted for 10 days. On d 115 to d 120, faeces samples were quantitatively collected. Feed and faeces samples were dried at 60 °C for 72 h; ground and analysed for dry matter (DM), crude protein (CP) and ash by the AOAC method (1990). Neutral detergent fibre (NDF), Acid detergent fibre (ADF) and Acid detergent lignin (ADL) were measured by the method of Goering and Van Soest (1970). Digestibility of nutrient was calculated as nutrient intake – nutrient in faeces × nutrient intake/1 × 100 (Schneider and Flatt, 1975). Various data were subjected to analyses of variance (ANOVA) according to a randomised complete block design using the General Linear Models (GLM) of the SAS® System for Windows (SAS, 1996). Treatment means were compared using the Duncan new multiple range test (Steel and Torrie, 1980).

Results and discussion

Silage and total intakes were not significantly different (P>0.05) among treatments. But concentrate intake was significantly different (P<0.05). High fat intake lowered concentrate intake and yeast supplementation reduced concentrate intake. Belewu *et al.* (2004) found that yeast supplementation had improved feed intake and nutrient digestion in goats. Goats fed high fat diets had higher average daily gain than goats fed a low fat diet. Supplemental yeast increased average daily gain in goats fed a high fat diet. This was in accordance with Yuangklang *et al.* (2005) who found supplemental fat improved the growth of meat goats. Dry matter, ether extract and protein digestibilities were not significantly different (P>0.05) among treatments. Goats fed diet containing yeast had enhanced organic matter, neutral detergent fibre and acid detergent fibre digestibilities when compared with goats fed diets containing low and high fat. Ahmed *et al.* (2007) found that supplemental yeast enhanced cellulose activity in goats fed Berseem hay.

Table 1. Effect on feed intake, average daily gain (ADG) and digestion coefficient nutrients.

	Fat 3%	Fat 6%	Fat 6%+yeast	SEM
Silage intake,% BW	1.37	1.42	1.39	0.05
Concentrate intake,% BW	1.27^{a}	1.22^{ab}	1.19^{b}	0.01
Total intake,% BW	2.64	2.65	2.58	0.13
Average daily gain, g/d	41.67^{b}	66.67^{ab}	80.56^{a}	6.40
Digestion coefficient,%				
DM	79.24	82.87	86.99	1.89
OM	80.25^{b}	85.74^{ab}	90.81^{a}	1.50
EE	60.02	79.81	69.49	4.51
CP	76.50	83.37	84.77	2.01
NDF	59.36^{b}	61.57^{b}	72.84^{a}	2.16
ADF	50.70^{b}	56.35^{b}	70.42^{a}	3.90

^{a,b,c} Means within the same row different superscripts differ significantly (P<0.05)

Conclusion

In conclusion, yeast supplementation improved average daily gain and nutrient digestibility in goats fed a high fat diet.

Acknowledgement

The authors would like to sincerely thank the Faculty of Natural Resources, Rajamangala University of Technology Isan for financial and facility support.

References

Ahmed, M.El.W. and H.R. Ibrahim, 2007. Effect of *Saccharomyces cerevisiae* of yeast on fiber digestion in sheep fed Berseem (*Trifolium alexandrinum*) hay and cellulase activity. Austr. J. Basic Appl. Sci. 1: 379-385.

AOAC (Association of Official Analytical Chemists), 1990. Official Methods of Analysis, vol. I, 15th ed. AOAC, Arlington, VA, USA.

Belewu, M.A., D.J. Ologunla and K.Y. Belewu, 2004. Effect of yeast culture (*Saccharomyces cerevisiae*) on feed intake and digestibility coefficient of west african dwarf goat fed sawdust-cassava waste based diet. Bulg. J. Agric. Sci. 10: 239-242.

Devendra, C. and D. Lewis, 1974. Fat in the ruminant diet: a review. Ind. J. Anim. Sci. 44: 917-938.

Goering H.K. and P.J. Van Soest, 1970. Forage fiber analyses. In: Agricultural Handbook n°379, US Department of Agriculture, Washington DC, USA, pp. 1-20.

Newbold, C.J., R.J. Wallace and F.M. McIntosh, 1996. Mode of action of the yeast *Saccharomyces cerevisiae* as a feed additive for ruminants. Br. J. Nutr. 76: 249-261.

Schneider R.H. and W.P. Flatt, 1975. The evaluation of feeds through digestibility experiments. The University of Georgia Press, Athens, USA, 168 pp.

Steel, R.G.D. and J.H. Torrie, 1980. Principles and Procedures of Statistics with Special Reference to the Biological Sciences. McGraw Hill, New York, USA, 451pp.

Yuangklang, C., S. Wittayakun, K. Vasupen, C. Sarnklong, S. Wonsutthavas, P. Srenanul, R. Kamdaw and S. Chaiphokhum, 2005. Effect of substitution of tallow by soybean oil on feed intake, rumen fermentation and nutrient digestibility in dairy steers fed with urea-treated rice straw as roughage source. In: Proceeding of Integrating Livestock-Crop System to meet the challenges of globalization. AHAT/BSAS International Conference, 14-18 November 2005, Khon Kaen, Thailand. T29.

The ruminal ratio of *trans*-10/*trans*-11 fatty acids obtained *in vitro* reflects *in vivo* values and strongly depends on the diet of the donor cow

A. Zened[1,2], A. Troegeler-Meynadier[1,2], M.C. Nicot[1,2] and F. Enjalbert[1,2]
[1]Université de Toulouse; INPT, ENVT; UMR 1289 Tandem, 31076 Toulouse Cedex 3, France;
[2]INRA, UMR 1289 Tandem, 31326 Castanet-Tolosan, France; f.enjalbert@envt.fr

Introduction

The composition of fatty acids (FA) of ruminant products has a potential impact on human health. Among them, *trans* FA, which are intermediates of the ruminal biohydrogenation of dietary unsaturated FA, deserve interest, in particular *trans*-10 and *trans*-11 isomers, which would have negative and positive effects on human health, respectively (Tricon *et al.*, 2004). A large variability in the ratio of *trans*-10 to *trans*-11 isomers (t10/t11) has been observed (Shingfield *et al.*, 2006). Some dietary factors shifting from t11 to t10 have been identified, like the proportion of concentrate (Griinari *et al.*, 1998), or the addition of oil (Roy *et al.*, 2006).
The use of *in vitro* systems to simulate rumen fermentation presents technical, economical and ethical advantages compared to *in vivo* experiments, and allows screening studies. The aim of the present study was to investigate if the ruminal fluid of a cow having the t11 to t10 deviation results in the same deviation during *in vitro* batch incubation and if the pathway of biohydrogenation *in vitro* depends mainly on the donor cow or on the fermentative substrate.

Material and methods

Four Holstein dry cows receiving 12 kg of dry matter per day were fed four different diets based on corn silage during two successive periods. The control diet (C, 20% of starch, <3% of crude fat, 15% of crude protein) was based on corn silage, the starch diet (S, 40% of starch) was supplemented with barley and wheat, the oil diet (O) was supplemented with 5% of sunflower oil and the starch plus oil diet (SO) was rich in both starch and sunflower oil. Each period consisted of 3 wk with the C diet, followed by 2 wk with C, S, O or SO diet. Cows were assigned to different diets during periods 1 and 2. On the last day of each period, ruminal fluid of each donor cow was incubated for 5 h with the four diets used as substrates, replacing sunflower oil by pure linoleic acid. Five hours after the morning meal, rumen fluid was taken from each cow (*in vivo* data). FA of *in vivo* and *in vitro* samples were analysed by gas chromatography. The t10/t11 ratios observed *in vivo* were compared to the ratios obtained *in vitro* with the same donor cow and the same diet by a paired Student t test. Effects of cow's diets and culture substrates were analysed by the GLM procedure of Systat®.

Results

As expected, the t10/t11 ratio was much higher when the diet of the cow was supplemented with both starch and sunflower oil. There was no significant (P=0.344) difference in the t10/t11 ratio of the ruminal fluid between *in vivo* experiments and incubations conducted *in vitro* (Fig.1). There was no effect (P=0.289) of the *in vitro* substrates on the t10/t11 ratio (Fig. 2), but the *in vitro* ratio was strongly affected by the diet of the donor cow (P<0.001).

Conclusion

This study showed that the t10/t11 ratio of the ruminal fluid after 5 h *in vitro* incubation reflects the *in vivo* values. This ratio *in vitro* did not depend on culture substrates, but was related to the diet of the donor cows, suggesting a major importance of the ruminal inoculum on the biohydrogenation pathway. This might be due to the short incubation time preventing the bacterial communities to evolve according to the *in vitro* substrate. As a consequence, short duration batch *in vitro* cultures cannot be used to study the dietary conditions of the t11 to t10 shift. Nevertheless, they could possibly be used for the study of the effects of feed additives on the t10/t11 ratio.

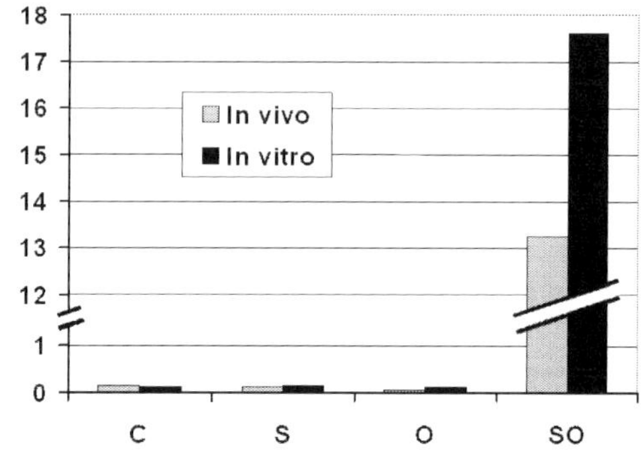

Figure 1. Comparison of in vivo and in vitro t10/t11 ratios obtained with control (C), added starch (S), added oil (O) and added starch + oil (SO) diet or substrate.

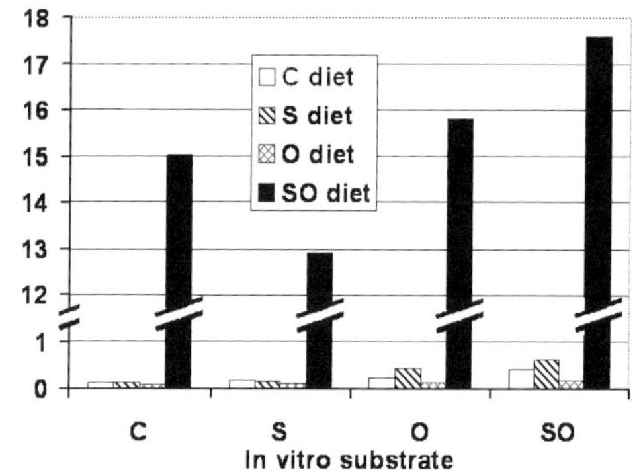

Figure 2. Effect of the diet of donor cow on the t10/t11 ratio obtained in vitro with control (C), added starch (S), added oil (O) and added starch + oil (SO) substrate.

References

Griinari, J.M., D.A. Dwyer, M.A. McGuire, D.E. Bauman, D.L. Palmquist and K.V.V. Nurmela, 1998. *Trans*-octadecenoic acids and milk fat depression in lactating dairy cows. J. Dairy Sci. 81: 1251-1261.

Roy A., A. Ferlay, K.J. Shingfield and Y. Chilliard, 2006. Examination of the persistency of milk fatty acid composition responses to plant oils in cows given different basal diets, with particular emphasis on *trans*-C18:1 fatty acids and isomers of conjugated linoleic acid. Anim. Sci. 82: 479-492.

Shingfield K.J., C.K. Reynolds, G. Hervas, J.M. Griinari, A.S. Grandison and D.E. Beever, 2006. Examination of the persistency of milk fatty acid responses to fish oil and sunflower oil in the diet of dairy cows. J. Dairy Sci. 89: 714-732.

Tricon S., G.C. Burdge, S. Kew, T. Banerjee, J.J. Russell, E.L. Jones, R.F. Grimble, C.M. Williams, P. Yaqoob and P. Calder, 2004. Opposing effects of *cis*-9,*trans*-11 and *trans*-10,*cis*-12 conjugated linoleic acid on blood lipids in healthy humans. Am. J. Clin. Nutr. 80: 614-620.

Short communications
Metabolism and endocrinology

Effects of extracellular essential amino acid deprivation on protein synthesis signaling in bovine mammary epithelial cells *in vitro*

J.A.D.R.N. Appuhamy[1], A.L. Bell[1], J. Escobar[2] and M.D. Hanigan[1]
[1]Department of Dairy Science, Virginia Polytechnic Institute and State University, Blacksburg, VA 24060, USA; [2]Department of Animal and Poultry Science, Virginia Polytechnic Institute and State University, Blacksburg, VA 24060, USA; appuhamy@vt.edu

Introduction

It has been shown that transgenic cows having additional copies of casein genes produced 10-20% more milk protein, whereas the transgenic mice with the same genes expressed casein at much higher levels (Brophy *et al.*, 2003) suggesting a limitation on protein synthesis in the bovine mammary gland (Moshel *et al.*, 2006). Protein synthesis efficiency depends primarily on translation initiation and elongation rates and responds to various signals such as hormones, amino acids, and energy supply. There has been a large amount of work demonstrating the positive effects of amino acids, particularly Leu, on protein synthesis stimulation in skeletal muscle and liver. These signals have been shown to converge at mammalian target of rapamycin (mTOR). Phosphorylation of mTOR, in turn, stimulates phosphorylation of ribosomal protein S6 (rpS6), 4E-binding protein 1 (4E-BP1) and dephosphorylation of eukaryotic elongation factor 2 (eEF2). Phosphorylation of rpS6 and 4E-BP1 enhances translation initiation whereas phosphorylated eEF2 impairs elongation. Moshel *et al.* (2006) reported that protein synthesis in bovine mammary epithelial (BME) cells is more sensitive to the elimination of all amino acids than to Leu alone. This suggests that amino acids other than Leu could have stronger effects on protein synthesis signaling in BME cells. Moreover, many studies have shown that Lys and Met are the most limiting amino acids for milk production in dairy cows fed corn based diets. The objective of this study was to investigate the total and individual effects of essential amino acid (EAA) on protein synthesis signaling in BME cells.

Material and methods

An established BME cell line, Mac-T, was used as a model. Cells were grown to 85% confluence, serum-starved overnight, and treated for one hour with media containing all amino acids (+EAA), media lacking all EAA (-EAA) or media deprived independently of Arg, His, Ile, Leu, Lys, Met, Phe, Thr, Trp, and Val. All treatment media contained insulin (1 μg/ml) and glucose (3.5 g/l). The cells were then lysed in the presence of protease and phosphatase inhibitors. Cell lysates were subjected to western immunoblotting using antibodies against phosphorylated rpS6 ($Ser^{235/236}$), mTOR (Ser^{2448}), and eEF2 (Thr^{56}). Membranes were stripped and reprobed with antibodies against the total forms of each signaling protein. The ratio of the two forms (phosphorylated:total) represented the phosphorylation state (PS) for each signaling protein. Significance of the effects of individual EAA deprivations and -EAA on PS, compared to that of +EAA, were obtained from a statistical model accounting for fixed effects of experiments and treatments.

Results

Deprivation of all EAA (-EAA) reduced (P<0.05) PS of mTOR and rpS6 by 86% and 63% respectively. In the absence of Arg, Ile, Phe, Leu, and Trp the PS of rpS6 was reduced (P<0.05) by 60%, 60%, 58%, 46% and 43% respectively. In response to Arg deprivation, PS of mTOR tended (P=0.15) to decrease by 70%. The -EAA did not affect PS of eEF2 (P=0.57). The PS of mTOR was positively correlated with PS of rpS6 (r = 0.68, P<0.05) and negatively correlated with PS of eEF2 (r=0.69, P<0.05). The PS of rpS6 had a negative correlation (r = -0.55, P=0.06) with PS of eEF2.

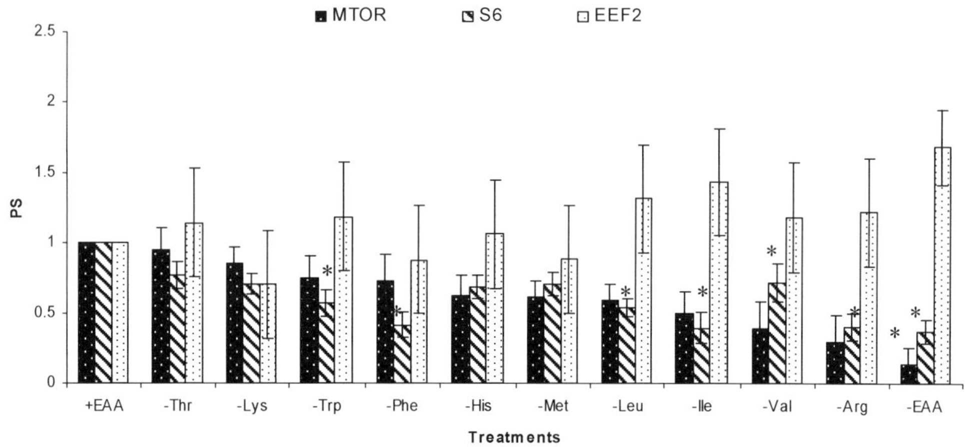

*Figure 1. Mean phosphorylation state (PS) of mTOR, rpS6, and eEF2 for individual and total EAA (-EAA) deprivation treatments. All PS of deprivation treatments were standardised to the PS of +EAA. * PS is significantly (P<0.05) different compared to +EAA.*

Conclusion

Essential amino acid deprivation, particularly Arg, Phe, Trp, and branched-chain amino acids significantly reduced signaling for translation initiation via mTOR and rpS6 in BME cells. Elongation was not significantly affected by EAA deprivation despite its high correlation with mTOR PS. Thus, phosphorylation of eEF2 appears to be regulated through mTOR in BME cells.

References

Brophy, B. and G. Smolenski, 2003. Cloned transgenic cattle produce milk with higher levels of beta-casein and kappa-casein. Nat. Biotechnol. 21: 138-139.

Fingar, D.C. and J. Blenis, 2004. Target of rapamycin (TOR): an integrator of nutrient and growth factor signals and coordinator of cell growth and cell cycle progression. Oncogene 23: 3151-3171.

Moshel, Y., R.E. Rhoads and I. Barash, 2006. Role of amino acids in translational mechanisms governing milk protein synthesis in murine and ruminant mammary epithelial cells. J. Cell Biochem. 98: 685-700.

Plasma angiopoietin-like protein 4 concentration is decreased by energy restriction in lactating dairy cattle

B.J. Bradford[1], L.K. Mamedova[1], K.J. Harvatine[2] and Y.R. Boisclair[2]
[1]Department of Animal Sciences & Industry, Kansas State University, 127 Call Hall, Manhattan, Kansas 66506, USA; [2]Department of Animal Science, Cornell University, 259 Morrison Hall, Ithaca, New York 14853, USA; bbradfor@ksu.edu

Introduction

Angiopoietin-like protein 4 (ANGPTL4; also known as fasting-induced adipose factor) is a 55-kDa glycosylated protein secreted by multiple tissues in the bovine, with greatest expression in adipose tissue and liver (Mamedova *et al.*, 2008). Among its many emerging roles, ANGPTL4 inhibits lipoprotein lipase activity, modulates angiogenesis, and regulates systemic carbohydrate and lipid metabolism (Hato *et al.*, 2008), although its receptor has not yet been identified. Expression of ANGPTL4 is often increased during periods of energy deficit in rodents, and this response is thought to contribute to nutrient repartitioning through inhibition of lipoprotein lipase. The objective of this study was to determine the effects of energy restriction and recombinant bovine somatotropin (bST) administration on liver and adipose tissue expression and plasma concentration of ANGPTL4 in lactating dairy cattle.

Material and methods

The experimental design has been described previously (Rhoads *et al.*, 2007). Briefly, 6 non-pregnant, late-lactation cows were fed at 120% of energy requirements (well-fed) for 14 d, followed by a 3-day intervening period and a second treatment period during which cows were fed at 30% of maintenance energy requirements (underfed) for 14 d. Subplot treatments were daily injections of either saline or bST (40 mg/d) administered during d 5-8 and 11-14 of each treatment period. Plasma samples and biopsies of liver and adipose tissue were collected at the end of each subplot period. Total RNA was extracted using a commercial kit (Qiagen RNAeasy Kit, Qiagen, Inc., Valencia, CA, USA) and reverse-transcribed to cDNA with random primers. Relative ANGPTL4 mRNA abundance was determined by real-time qPCR using the $2^{-\Delta Ct}$ method. Amplification was performed in triplicate using the SYBR Premix Ex Taq Kit (Takara, Madison, WI, USA). Primer sequences (5'-3') were as follows: ANGPTL4, forward: GATGGCTCCGTGGACTTTAACC, reverse: GGATGTGATGCACCTTCTCCAG; ribosomal protein subunit 9 (control gene), forward: GAACAAACGTGAGGTCTGGAGG, reverse: ATTACCTTCGAACAGACGCCG. Cellular extracts were prepared for Western blotting as described (Rhoads *et al.*, 2007). Forty μg total protein or 1 μl plasma were separated by SDS-PAGE on a 4-12% Tris-HCl gel and dry-transferred onto nitrocellulose membranes (iBlot; Invitrogen, Carlsbad, CA, USA). Membranes were blocked and then incubated with polyclonal goat anti-ANGPTL4 antibody (1:1000 dilution; Santa Cruz Biotechnology, Santa Cruz, CA, USA) using a One-Step Western Kit (GenScript Corp., Piscataway, NJ, USA). Blots were quantified by scanning densitometry. Results were analyzed using mixed models including fixed effects of nutrition, injection, and their interaction, and the random effect of cow. Data were log-transformed when necessary and reported means were back-transformed; data points with studentised residuals >2.5 or <-2.5 were omitted. Significance was declared at $P<0.05$, and tendencies at $P<0.10$.

Results

As reported previously, under-fed cows had significantly lower energy balance than well-fed cows (Rhoads *et al.*, 2007). Under-feeding significantly decreased mRNA abundance of ANGPTL4 in adipose tissue (P=0.04, Table 1); however, at the protein level, ANGPTL4 was decreased by bST (P=0.01), but not by the feeding protocol. Nutrition and bST interacted to influence mRNA abundance in the liver (P<0.01), with the greatest abundance in under-fed cows administered saline. As in adipose tissue, protein concentration did not correlate with mRNA abundance, and no treatment effects were detected. Relative plasma ANGPTL4 concentration was decreased by under-feeding (P=0.05).

Table 1. Effects of nutrition and bST administration on mRNA and protein abundance of ANGPTL4 in lactating dairy cattle.[1]

Relative ANGPTL4	Well-fed		Under-fed		SEM	Treatment
Abundance	Saline	bST	Saline	bST		effects[2]
Adipose mRNA	1.648	1.372	0.775	0.585	0.445	Nutr.
Adipose protein	3.05	2.45	3.35	2.61	0.23	bST
Liver mRNA	0.060	0.150	0.574	0.212	0.079	Nutr. × bST
Liver protein	1.76	1.64	1.75	1.68	0.27	-
Plasma protein	3.03	2.65	2.21	2.54	0.28	Nutr.

[1] Values are LS means, n = 6 per group; outliers removed from adipose mRNA (1) and protein (2) analyses.

[2] Significant treatment effects (P<0.05) are denoted as follows: Nutr., nutrition effect; bST, injection effect; Nutr. × bST, interaction of main effects.

Conclusion

Under-feeding lactating dairy cattle significantly decreased plasma ANGPTL4 concentration, and the primary tissue responses to under-feeding were decreased adipose tissue mRNA abundance and increased liver mRNA abundance in saline-treated cows. Liver is not likely the primary source of circulating ANGPTL4 (Kersten *et al.*, 2000), but protein abundance in adipose tissue was also unrelated to plasma concentration. These data suggest that (1) translation and/or secretion of ANGPTL4 is likely regulated, (2) other tissues, such as gastrointestinal tissue, may contribute substantially to circulating ANGPTL4 in cattle (Mamedova *et al.*, 2008), and (3) bovine ANGPTL4 is not 'fasting-induced' as originally described in rodents (Kersten *et al.*, 2000).

References

Hato, T., M. Tabata and Y. Oike, 2008. The role of angiopoietin-like proteins in angiogenesis and metabolism. Trends Cardiovasc. Med. 18: 6-14.

Kersten, S., S. Mandard, N.S. Tan, P. Escher, D. Metzger, P. Chambon, F.J. Gonzalez, B. Desvergne and W. Wahli, 2000. Characterization of the fasting-induced adipose factor FIAF, a novel peroxisome proliferator-activated receptor target gene. J. Biol. Chem. 275: 28488-28493.

Mamedova, L., B. Bradford, K. Robbins and B. Johnson, 2008. Tissue expression of fasting-induced adipocyte factor in cattle. FASEB J. 22: 1116.1112 (Abstr.).

Rhoads, R.P., J.W. Kim, M.E. Van Amburgh, R.A. Ehrhardt, S.J. Frank and Y.R. Boisclair, 2007. Effect of nutrition on the GH responsiveness of liver and adipose tissue in dairy cows. J. Endocrinol. 195: 49-58.

Thiazolidinediones increase lipogenic enzyme activity in internal and external adipose tissue depots in sheep

F.T. Fahri[1,2,3], I.J. Clarke[4], D.W. Pethick[1,2], B.G. Tatham[3], R.D. Warner[2,3] and F.R. Dunshea[1,3,5]
[1]School of Veterinary and Biomedical Sciences, Murdoch University, 6150, Murdoch, WA, Australia; [2]Australian Sheep Industry CRC, University of New England, 2350, Armidale, NSW, Australia; [3]Department of Primary Industries, 3030, Werribee, VIC, Australia; [4]Department of Physiology, Monash University, Clayton, VIC, 3800, Australia; [5]Melbourne School of Land and Environment, The University of Melbourne, 3010, Melbourne, VIC, Australia; fdunshea@unimelb.edu.au

Introduction

Thiazolidinediones (TZD) are synthetic compounds that are currently used clinically as oral antidiabetic drugs. These synthetic PPARγ ligands offer a relatively new line of therapy for the treatment of many human conditions including impaired glucose tolerance type 2 diabetes mellitus and hyperinsulinemia (Willson *et al.*, 2000). Studies using an *in vitro* cell culture model of 3T3-L1 adipocyte cells, have suggested the potential of utilising TZD in animal production for the promotion of adipocyte differentiation (Gerhold *et al.*, 2002) and for the possible manipulation of fat mass development (Aleo *et al.*, 2003). Due to their agonistic behaviour towards PPARγ, TZD are also known potent inducers of adipocyte differentiation (Morrison, 2000). The action of TZD on adipose tissue is associated with increased glucose and FA uptake, lipogenesis, and glucose oxidation (Hauner, 2002). Studies both *in vivo* and *in vitro*, suggest that PPARγ activation by TZD ligands possess the ability to increase and improve insulin sensitivity in tissues which are insulin responsive. In the ruminant, the process of lipogenesis occurs almost entirely in the adipose tissue and is under the control of insulin. The present study was conducted to determine the effect of TZD on fat metabolism in sheep.

Methods

Thirty Poll Dorset × Merino × Border Leicester 5 mo old lambs (15 ewes and 15 wethers) were blocked on liveweight and body fat (as evaluated by Dual X-Ray absorptiometry) and allocated to a 2×3 factorial design with the respective factors being sex (ewe or wether) and dose of TZD (0, 8 or 24 mg/d of rosiglitazone maleate (Avandia™)). Lambs were fed commercial pellets (Rumevite™) *ad libitum* for 8 wk. TZD tablets or placebo were administered via gavage daily for the duration of the study. At slaughter, adipose tissue samples were obtained from the neck (N), the rump (SC), peri-renal (PR), and omental (OM) areas of the animal, and snap frozen before enzymatic activity assays. Homogenised and extracted samples were analysed for glycerol-3-phosphate dehydrogenase (G3PDH), glucose-6-phosphate dehydrogenase (G6PDH), and fatty acid synthase (FAS). As there were no effects of sex or any difference between 8 and 24 mg/d of TZD the data were pooled across sex and TZD groups.

Results

Activity of all lipogenic enzymes was higher ($P<0.001$) in the two external adipose tissue depots (Neck and SC) than in the internal fat depots (PR and OM) (Table 1). Also, the activity of all enzymes was increased ($P<0.002$) in all adipose tissue depots by dietary TZD treatment. However, there were significant interactions which suggested that the absolute (but not proportionate) increase in enzyme activity was greater in the external adipose depots.

Table 1. Effect of TZD treatment and site on glycerol-3-phosphate dehydrogenase (G3PDH), glucose-6-phosphate dehydrogenase (G6PDH), and fatty acid synthase (FAS) activity. All enzyme activities are expressed as nmoles NADH (or NADP or NADPH) oxidised (or reduced) per min per mg of cytosolic protein. Data are pooled across sex and TZD doses.

	TZD	Site				sed	Significance		
		NECK	SC	PR	OM		TZD	Site	T x S
G3PDH	- (n=10)	1,666	2,181	809	703	244.7	0.002	<0.001	0.10
	+ (n=20)	2,376	3,109	1,010	1,225				
G6PDH	- (n=10)	434	593	248	165	73.5	<0.001	<0.001	0.044
	+ (n=20)	754	868	335	350				
FAS	- (n=10)	52.4	67.8	33.8	19.7	9.28	0.002	<0.001	0.021
	+ (n=20)	84.4	104.4	38.2	39.6				

Discussion

This experiment evaluated the enzymatic activity of three lipogenic enzymes within two external and two internal adipose tissue depots in sheep which had been treated with TZD. These data clearly indicate lipogenic enzyme activity is greater in external than in internal adipose tissue depots. Previously we showed that dietary TZD were absorbed in ruminants (F.T. Fahri, unpublished observations) and these data demonstrate that dietary TZD have metabolic effects in sheep. In this context, dietary TZD increased lipogenic enzyme activity in all adipose tissue depots with the absolute increase being the greatest in external adipose tissue sites which are already most lipogenically active. The observed effects were consistent with human studies and clearly indicate that manipulation of adipose tissue through PPARγ can alter fat metabolism. Further research is needed to target adipose tissue sites of economic importance such as intramuscular adipose tissue.

References

Aleo, M.D., G.R. Lundeen, D.K. Blackwell, W.M. Smith, G.L. Coleman, S.W. Stadnicki and W.M. Kluwe, 2003. Mechanism and implications of brown adipose tissue proliferation in rats and monkeys treated with the thiazolidinedione darglitazone, a potent peroxisome proliferator-activated receptor-{gamma} agonist. J. of Pharmacol. and Exp. Ther. 305: 1173-1182.

Gerhold, D.L., F. Liu, G. Jiang, Z. Li, J. Xu, M. Lu, J.R. Sachs, A. Bagchi, A. Fridman, D.J. Holder, T.W. Doebber, J. Berger, A. Elbrecht, D.E. Moller and B.B. Zhang, 2002. Gene expression profile of adipocyte differentiation and its regulation by peroxisome proliferator-activated receptor-g agonists. Endocrinology 143: 2106-2118.

Hauner, H., 2002. The mode of action of thiazolidinediones. Diabetes-Metab. Res. 18: S10-S15.

Morrison, R.F. and S.R. Farmer, 2000. Hormone signalling and transcriptional control of adipocyte differentiation. Proceedings Experimental Biology 00 Symposium, San Diego, CA: 3116S-3121S.

Willson, T.M., P.J. Brown, D.D. Sternbach and B.R. Henke, 2000. The PPARs: from orphan receptors to drug discovery. J. Med. Chem. 43: 527-550.

Selection for muscling reduces muscle response to adrenaline

G.E. Gardner[1,2,3], *P. McGilchrist*[1,3], *J.M. Thompson*[1,2] *and K.M. Martin*[2]
[1]*Australian Cooperative Research Centre for Beef Genetic Technologies, Australia;* [2]*Cooperative Research Centre for Sheep Industry Innovation, Australia;* [3]*School of Veterinary & Biomedical Science, Murdoch University, 6150, WA, Australia; G.Gardner@murdoch.edu.au*

Introduction

Carcass lean meat yield is a key profit driver within the sheep and beef industries, and is improved by selection for muscling. This selection causes muscle hypertrophy which is associated with greater proportions of fast-glycolytic type IIX myofibres in both cattle and sheep (Wegner *et al.*, 2000; Greenwood *et al.*, 2007). Muscle tissue that is high in type IIX myofibres will also have increased glycolytic and glycogenolytic capacity (Wegner *et al.*, 2000), which is likely to result in a greater potential for stress/adrenaline induced muscle glycogen depletion. This has great relevance to industry, as low muscle glycogen at slaughter will result in high ultimate pH (>5.7) carcases leading to dark firm dry (DFD) meat. However, in contradiction to this theory Martin *et al.* (2004) found more muscle glycogen in sheep selected for muscling. This was most evident when energy intake was high, implicating greater insulin sensitivity which may counteract the hypothesised greater adrenaline response. Therefore a link between selection for muscling and response to stress/adrenaline remains to be confirmed. In this study, we tested the hypothesis that selection for muscling will increase the muscle response to adrenaline.

Material and methods

This paper details two experiments where adrenaline challenges were administered to a group of 20 sheep at 4 and 16 months of age, as well as 12 cattle at 15 and 36 months of age. The sheep were the progeny of Merino and Poll Dorset sires selected for a diverse range in Australian Sheep Breeding Values for yearling eye muscle depth (YEMD). The steers were the progeny of Piedmontese (heavy muscling genotype), or Angus sires. Prior to administering adrenaline challenges, animals were habituated in individual pens for 2 weeks on an *ad-libitum* grain-based diet. Adrenaline challenges were then administered via indwelling jugular catheters at 7 levels (2/day) ranging between 0.1-3.0 μg/kg liveweight. Sixteen blood samples were taken between -30 and 130 minutes relative to adrenaline administration. Plasma was analysed for lactate concentration which reflects the muscle response to adrenaline. In both experiments plasma lactate area under curve (AUC) for the first 10 minutes following adrenalin challenge was analysed using a linear mixed effects model. Fixed effects included age and sire-breed (cattle experiment only), covariates were adrenaline challenge and YEMD (sheep experiment only), and animal within sire was used as the random term.

Results

Following adrenaline challenge the plasma lactate AUC was about 30% greater ($P<0.01$) for the 16 month lambs compared to their response at 4 months of age (Figure 1a). Selection for muscling reduced ($P<0.01$) lactate AUC by at least 50% across the range of YEMD (Figure 1a), with this effect apparent at both ages and across all levels of adrenaline challenge. In cattle, age also impacted ($P<0.01$), with the 36 month cattle having plasma lactate AUC about double that of their 15 month responses (Figure 1b). While there was no difference between breeds at 15 months, the increase in responsiveness in the Angus sired cattle at 36 months was almost twice that of the heavily muscled Piedmontese genotype, although only at adrenaline doses above 0.9 μg/kg liveweight.

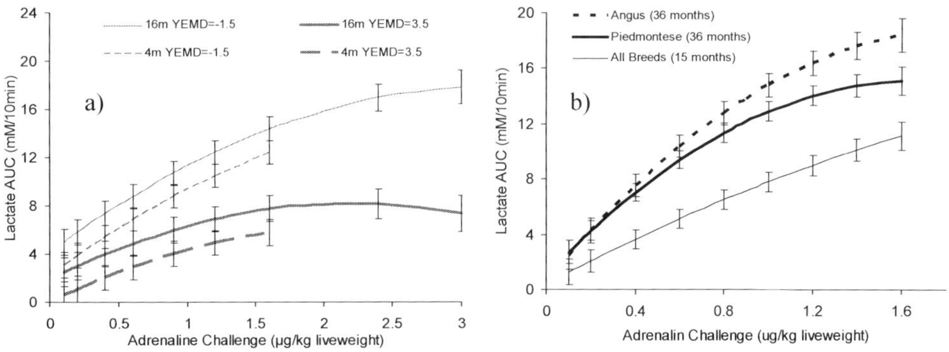

Figure 1. The effect of adrenaline challenge on plasma lactate area under the curve (mM/10 min) for (a) sheep with high and low YEMD and (b) Angus, and Piedmontese sired cattle at 15 and 36 months of age. Values are least squares means ± sem.

Conclusion

In both sheep and cattle the progeny of heavily muscled sires were less responsive to adrenaline, a finding contrary to our initial hypothesis, however fitting well with the observed greater muscle glycogen levels in these genotypes (Martin *et al.,* 2004). An explanation for this reduced muscle responsiveness may be found in work by Martin *et al.* (1989) who showed greater density of β_2-adrenoreceptors in rats with more oxidative myofibres (type I), resulting in increased glycogenolysis. Given that selection for muscling in sheep and cattle reduces the proportion of type I myofibres, this is likely to lower the density of β_2-adrenoreceptors thus reducing the response to adrenaline. The reverse may apply to older animals as muscle tissue becomes more oxidative as animals age (Greenwood *et al.,* 2007) which explains why the muscle of older animals was more responsive to adrenaline than younger animals. Ultimately selection for muscling is likely to lower the response to adrenaline in muscle and reduce the incidence of DFD meat.

References

Greenwood, P.L., S. Harden and D.L. Hopkins, 2007. Myofibre characteristics of ovine *longissimus* and *semitendinosus* muscles are influenced by sire breed, gender, rearing type, age and carcass weight. Aust. J. Exp. Agric. 47: 1137-1146.

Martin, K.M., G.E. Gardner, J.M. Thompson and D.L. Hopkins, 2004. Nutritional impact on muscle glycogen metabolism in lambs selected for muscling. J. Anim. Feed Sci. 13 (Suppl. 1): 639-642.

Martin, W., S. Murphree and J. Saffitz, 1989. Beta-adrenergic receptor distribution among muscle fibre types and resistance arterioles of white, red, and intermediate skeletal muscle. Circulation Res. 64: 1096-1105.

Wegner, J., E. Albrecht, I. Fiedler, F. Teuscher, H.J. Papstein and K. Ender, 2000. Growth- and breed-related changes of muscle fibre characteristics in cattle. J. Anim. Sci. 78: 1485-1496.

Glucose metabolism in neonatal calves: dependence on postnatal maturation

H.M. Hammon[1], J. Steinhoff[1], S. Goers[1], E. Kanitz[1], R.M. Bruckmaier[2] and C.C. Metges[1]
[1]Research Institute for the Biology of Farm Animals, Dummerstorf (FBN), Wilhelm-Stahl-Allee 2, 18196 Dummerstorf, Germany; [2]Veterinary Physiology, Vetsuisse Faculty, University of Bern, Bremgartenstr. 109a, 3012 Bern, Switzerland; hammon@fbn-dummerstorf.de

Introduction

In the foetus, only negligible endogenous glucose production occurs, but endocrine changes close to term lead to maturation of hepatic gluconeogenic enzymes to ensure postnatal endogenous glucose production (Fowden *et al.*, 1993), when the continuous glucose supply via placenta ceased. Neonatal calves, especially prematurely born calves, show hypoglycaemia after birth, and besides oral glucose intake, endogenous glucose production is important to maintain plasma glucose concentrations (Girard *et al.*, 1992; Bittrich *et al.*, 2002). Preliminary data also indicate that colostrum feeding might stimulate endogenous glucose production (Hammon *et al.*, 2003). Therefore, we tested the hypothesis that ontogenic maturation affects gluconeogenesis (GNG) and postprandial hepatic gluconeogenic activities in neonatal calves.

Material and methods

Calves (n=7 per group) were either born preterm (PT; delivered by section 9 d before term), or at term (T; spontaneous vaginal delivery). A third group of calves was fed colostrum (C; spontaneous vaginal delivery at term) for 4 d. Group C received colostrum of milkings 1, 3 and 5 after calving in amounts (per meal) of 4% body weight (BW) on d 1 and 5% BW on d 2 and 3. Calves were fed twice daily on d 1 and 2 and only once on d 3. Immediately after birth (groups PT and T; without feeding) or on d 3 of life (group C; 12 h without feed intake) calves were gavaged with deuterium-labelled water (2×10 g D_2O (70 atom% D)/kg BW within 4 h) and received [U^{13}C]glucose (99 atom% ^{13}C) i.v. (prime: 4.3 μmol/kg BW; infusion: 6.4 μmol/[kg BW × h] for 4 h) to measure plasma glucose appearance rate (Ra_{Gluc}) and gluconeogenesis (GNG) by isotope dilution. Blood samples were taken immediately after birth and during tracer tests. On the following d colostrum of milking 5 was fed at 5% of BW to all calves and blood samples were taken before and 2 h after feeding. Plasma concentrations of glucose, insulin, glucagon and cortisol were measured in plasma samples. Calves were slaughtered 2 h after feeding and liver samples were snap frozen in liquid nitrogen to measure glycogen content and mRNA levels, using real-time RT-PCR, and enzyme activities of pyruvate carboxylase (PC; EC 6.4.1.1), cytosolic phosphoenolpyruvate carboxykinase (PEPCK; EC 4.1.1.32) and glucose-6-phosphatase (G6-Pase; EC 3.1.3.9). Body weight, plasma glucose and hormones and isotopic data were analysed by the Mixed Model of SAS® with gestation length as fixed effects and individual calves as random effects. Liver data were analysed using the General Linear Model of SAS® with different gestation length as the fixed effect.

Results

Body weight at birth was lower ($P<0.05$) in PT than in T and C. Plasma glucose concentrations were similar among groups immediately after birth, decreased ($P=0.1$) from birth to d of slaughter only in PT and increased ($P<0.05$) after feed intake in PT and C. On d of slaughter glucose concentrations were the highest ($P<0.05$) in C. Plasma insulin concentrations were higher ($P<0.05$) in C than PT and T and the postprandial insulin increase was the highest ($P<0.01$) in C. Plasma glucagon concentration was higher ($P<0.05$) in PT than in T and C and decreased ($P<0.05$) after feeding in

PT. Plasma cortisol concentrations after birth was lower ($P<0.05$) in PT than in T and decreased ($P<0.05$) during the experimental period in T and C and after feeding in PT and T. On d of slaughter C-calves showed the lowest ($P<0.05$) cortisol concentrations. The Ra_{Gluc} was higher ($P<0.05$) in C and T than in PT (Table 1). The percentage of GNG at Ra_{Gluc} was higher ($P < 0.05$) in C than in PT and T, whereas absolute GNG was the highest ($P<0.05$) in C and higher ($P<0.05$) in T than in PT (Table 1). Hepatic glycogen content was the highest ($P<0.05$) in C and higher ($P<0.05$) in T than PT. In C-calves, mRNA levels of PC were lower ($P<0.05$) than in T and mRNA levels of G6Pase were lower than in PT. On the contrary, PEPCK enzyme activities were the highest ($P<0.05$) in C and higher in ($P<0.05$) T than in PT.

Table 1. Glucose appearance rate (Ra_{Gluc}) and gluconeogenesis (GNG) in preterm (PT), term (T) and 4-d old calves (C).

	Groups			SE	Ontogenic effect
	PT	T	C		*P*-value
Ra_{Gluc} μmol/(kg × min)	10.3[b]	20.8[a]	23.3[a]	1.1	<0.01
GNG,% of Ra_{Gluc}	30.7[b]	36.3[b]	49.5[a]	2.7	<0.01
GNG, μmol/(kg × min)	3.2[c]	7.6[b]	11.4[a]	0.6	<0.01

Data are presented as LSMeans and pooled standard error (SE). [a,b,c] LSMeans within a row with different uppercase letter are different ($P<0.05$).

Conclusion

This study indicates an impaired glucose metabolism and reduced endogenous glucose production in calves immature at birth. Obviously, GNG was not stimulated by endocrine changes in PT, although, hepatic gluconeogenic enzymes are expressed in PT. Apart from oral glucose intake, the improved glucose status in C might be a result of enhanced endogenous glucose production with about 50% resulting from GNG.

References

Bittrich, S., C. Morel, C. Philipona, Y. Zbinden, H.M. Hammon and J.W. Blum, 2002. Physiological traits in preterm calves during their first week of life. J. Anim. Physiol. Anim. Nutr. 86: 185-198.

Fowden, A.L., J. Mijovic and M. Silver, 1993. The effects of cortisol on hepatic and renal gluconeogenic enzyme activities in the sheep fetus during late gestation. J. Endocrinol. 137: 213-222.

Girard, J., P. Ferré, J.P. Pégorier, and P.H. Duée, 1992. Adaptions of glucose and fatty acid metabolism during perinatal period and suckling-weaning transition. Physiol. Rev. 72: 507-562.

Hammon, H.M., S.N. Sauter, M. Reist, Y. Zbinden, C. Philipona, C. Morel, and J.W. Blum, 2003. Dexamethasone and colostrum feeding affect hepatic gluconeogenic enzymes differently in neonatal calves. J. Anim. Sci. 81: 3095-3106.

The effects of beta-adrenergic agonist (BA) and growth hormone (GH) on metabolic characteristics and factors involved in determining skeletal muscle fibre type in growing lambs

K. Hemmings, T. Parr, Z. Daniel, P. Buttery and J. Brameld
Division of Nutritional Sciences, University of Nottingham, School of Biosciences, Sutton Bonington, Leicestershire, LE12 5RD, United Kingdom; tim.parr@nottingham.ac.uk

Introduction

Anabolic agents, such as growth hormone (GH) and beta-adrenergic agonists (BA), have hypertrophic effects on skeletal muscle. In lambs, we have shown that BA had a greater effect than GH on muscle growth and myosin heavy chain (MyHC) isoform mRNA expression, following short term exposure, causing a switch in expression toward a faster MyHC fibre type (Hemmings *et al.*, 2009). The development and differentiation of muscle is dependent on the myogenic regulatory factors (MRFs), myoD, myf-5, myogenin and MRF4, but there are also suggestions that the relative ratio of these factors may also influence fibre type (Te Kronnie and Reggiani, 2002). In addition, signalling enzymes such as calmodulin kinase II (CaMKII) and calcineurin (CaN) have been implicated in muscle fibre type transitions (Bassel-Duby and Olson, 2006). The aim of this study was to determine whether the transition toward a faster MyHC fibre type found when lambs were treated with anabolic agents was associated with changes in these MRFs, and by measuring lactate dehydrogenase (LDH) and isocitrate dehydrogenase (ICDH) activity whether treatment was associated with changes in muscle oxidative or glycolytic capacity respectively.

Material and methods

The experimental details of the treatments are described in another abstract submitted to the conference (Hemmings *et al.*, 2009). In summary, Mule×Charolais male twin lambs were split into d 60 (D60) and d 120 (D120) age groups, which were further split into three treatment groups: group CO (n=11) was the control; group BA (n=10) were fed the beta-agonist cimaterol at 10 ppm and group GH (n=10) were administered prolonged release bovine GH at 3.75 mg/kg BW. After treatment for 6 d, lambs were slaughtered and a whole transverse section sample of the *longissimus dorsi* (LD) muscle from the region of the 10th rib (*lumbarum et thoracis)* which was snap frozen in liquid nitrogen and stored at -80 °C. The frozen LD were crushed and mixed, and a sample taken for determination of LDH and ICDH activity using the methods described by Brandstetter *et al.* (1998). CaMKII and CaN activities were determined as described by the manufacturer of the assay kits (Promega and Biomol International). In addition, LD total RNA was extracted (Trizol) then first strand cDNA was generated using reverse transcriptase and random primers. The relative level of mRNA expression was determined using quantitative RT-PCR analysis (Roche). Real-time PCR primers and probes specific for ovine MRFs (myoD, myf-5, myogenin and MRF4) were designed using Primer Express (Applied Biosystems). Data were analysed by ANOVA (GenStat, VSN International Ltd, Hemel Hempstead, UK) and the *post-hoc* Dunnett test.

Results

At both D60 and D120, ICDH activity was significantly decreased in BA treated lambs relative to CO (Table 1), but there were no effects of either anabolic agent on LDH, CaMKII or CaN activities. The mRNA expression of some of the MRFs was significantly affected by BA treatment, but not GH. BA treatment significantly decreased myoD expression in D120 lambs and increased

Myogenin expression at both ages. There were no significant effects of either treatment at either age on MRF4 and Myf-5 expression.

Table 1. Effects of treatment of lambs with BA and GH for a 6 d period.

Age	D60					D120				
Treatment	CO	BA	GH	SED[1]	P	CO	BA	GH	SED	P
Enzyme activity										
ICDH, mOD/min/mg	49.3	33.9*	49.4	4.8	0.004	37.3	28.2[a]	33.1	3.3	0.041
LDH, mOD/min/mg	510	489	519	86	0.928	637	618	651	114	0.957
CaMK II, pmol/min/ug protein	11.5	14.3	11.0	2.1	0.246	13.8	14.9	11.5	1.8	0.160
Calcineurin, Arbitrary units	49.5	43.0	44.4	15.5	0.901	34.1	27.1	22.5	11.1	0.562
mRNA expression, Arbitary units										
MyoD	2.95	2.66	2.73	0.40	0.735	3.00	2.01[a]	3.28	0.39	0.008
Myogenin	3.34	4.65*	3.65	0.47	0.023	3.15	4.54[a]	3.23	0.40	0.002
MRF4	3.02	4.47	3.59	1.36	0.552	4.08	4.14	5.06	0.63	0.230
Myf-5	2.71	3.69	3.22	0.40	0.059	4.33	3.71	3.71	0.57	0.436

[1] SED = standard error of the differences of the means.
[a] Values significantly different from control (P<0.05).

Conclusion

Despite only a short-term (6 d) treatment, BA significantly reduced ICDH activity, which we suggest is associated with reduced oxidative capacity, indicating metabolic transition to a more glycolytic fibre type, even though LDH activity was unaffected. This agrees with the increase in MyHC IIX and induction of MyHC IIB mRNA expression seen in this same trial (Hemmings *et al.*, 2009). Increased CaMKII and CaN activities have been associated with glycolytic to oxidative (or fast to slow) changes in fibre type (Bassel-Duby and Olson, 2006). However, no changes in CaMKII and CaN activities were observed, suggesting that they are not involved in transitions toward a glycolytic fibre type. It has previously been reported that myoD mRNA expression is higher in glycolytic muscle fibres, whilst myogenin expression is higher in slow oxidative muscles (Hughes *et al.*, 1993). Our observations appear to contradict this, with an increase in myogenin and decrease in myoD expression associated with a decrease in oxidative metabolism. In contrast to BA, GH treatment had no effect on the enzyme activities or mRNA expression measured.

References

Bassel-Duby, R. and E. Olson, 2006. Signaling pathways in skeletal muscle remodelling. Annu. Rev. Biochem. 75: 19-37.

Brandstetter, A.M., B. Picard and Y. Geay, 1998. Muscle fibre characteristics in four muscles of growing bulls I. Postnatal differentiation. Livest. Prod. Sci. 53: 15-23.

Hemmings, K., T. Parr, Z. Daniel, P. Buttery and J. Brameld, 2009. The effects of beta-adrenergic agonist (BA) and growth hormone (GH) on lamb growth characteristics and myosin heavy chain expression. International Symposium on Ruminant Physiology, these proceedings.

Hughes, S.M., J.M. Taylor, S.J. Tapscott, C.M. Gurley, W.J. Carter and C.A. Peterson, 1993. Selective accumulation of MyoD and myogenin mRNAs in fast and slow adult skeletal muscle is controlled by innervation and hormones. Development 118: 1137-1147.

Te Kronnie, G. and C. Reggiani, 2002. Skeletal muscle fibre type specification during embryonic development. J. Muscle Res. Cell Motil. 23: 65-69.

Recovery of α-linolenic acid in milk fat of dairy cows fed flowering forage plants

T. Kälber, M. Kreuzer, H.R. Wettstein and F. Leiber
ETH Zurich, Department of Agricultural and Food Science, 8092, Zurich, Switzerland; fleiber@ethz.ch

Introduction

One of the most important functional fatty acids in milk is α-linolenic acid (ALA). It is a plant fatty acid, and its carry-over from feed to milk largely depends on the proportion which escapes ruminal biohydrogenation (Chilliard *et al.,* 2007). It was shown that high alpine pastures, where herbs are frequent and plants are often in their flowering stage, facilitate a much higher ALA carry-over compared to lowland ryegrass pastures (Leiber *et al.*, 2005). These authors therefore hypothesised that plant secondary compounds, predominant especially in the flowering stage of the herbs may inhibit ruminal biohydrogenation thus increasing ALA carry-over. In order to develop strategies for lowland farms to increase the fat quality of milk, some flowering plants rarely used as forage were evaluated for their use as feed for dairy cows and particularly for their influence on ALA concentrations in milk.

Material and methods

Three flowering catch crop plants (*Fagopyrum esculentum* [buckwheat; FE], *Phacelia tanacetifolia* [phacelia; PT] and *Trifolium alexandrinum* [berseem clover; TA] were tested for their influence on ALA carry-over from feed to food during 20 days. Additionally, a non-flowering forage herb (*Cichorium intybus* [chicory; CI]) and a forage grass (*Lolium multiflorum* var. [Italian ryegrass; LM]) were used as controls. Test plants were sown in mixture with ryegrass (20%) and fed in the flowering stage. Thirty dairy cows (Holstein Friesian and Brown Swiss, mid lactation, fed a highly digestible silage-concentrate diet before) were allocated to five feeding treatments (n=6). Test forages were fed *ad libitum*. Only 1 kg/d respectively of energy and protein concentrate per cow was fed to prevent masking of forage effects. All cows received 2.5 kg/d of pure ryegrass hay to provide enough structured fibre. Data on intake and milk yield, and samples of 15-20 experimental days were used for calculations and analyses. Lipids from milk and forage samples were extracted, transesterified and gas chromatographically analysed by standard methods as described in Leiber *et al.* (2005). Data were analysed with a linear ANOVA model and a Levene-T-Test with the SPSS 14.0 software.

Results

Total fatty acid contents of CI, FE, PT, TA, and LM were 22.3, 15.4, 20.1, 21.2, and 22.8 g/kg dry matter, respectively. The corresponding concentrations of ALA were 44.9, 35.3, 40.5, 49.9, and 47.0 g/100g fatty acid methyl esters (FAME). This resulted in a significantly lower intake of total lipids and ALA with FE compared to PT, TA, and LM (Table 1). Milk fat yield was in the same range for all groups. Milk ALA yield was the highest with PT and TA and the lowest with CI and LM, as was also the case for milk ALA concentration. The resulting ratio of ALA excreted in milk to ALA ingested from feed was significantly (P<0.01) lower in the LM group compared to all groups fed flowering herbs (FE, PT and TA; Figure 1).

Conclusion

The results show that the three tested flowering catch crop plants increase the carry-over of ALA from feed to milk, compared to a ryegrass diet. With phacelia and berseem clover, the resulting

ALA concentrations in milk fat are as high as in milk from high alpine pastures (Leiber *et al.*, 2005). The fact that buckwheat resulted in somewhat lower ALA concentrations and yields was due to the particularly low concentrations of fat and ALA in this plant. This supports the hypothesis that plant secondary compounds related to flowering processes inhibit ALA biohydrogenation in the rumen. Chicory was not flowering when used in the present experiment. Although it is relatively high in phenolics, the ALA carry-over was not as high as with the flowering plants. This again underlines the possible role of the flowering stage. Identification of the involved plant secondary compounds requires further research.

Table 1. Intake rates and excretion of lipids and ALA in milk (n=6 per group).

Dietary treatment	C. intybus	F. esculentum	P. tanacetifolia	T. alexandrinum	L. multiflorum
Feed intake (kg DM/d)	14.6[b]	17.3[a]	17.7[a]	18.5[a]	16.9[a]
Lipid intake (g/d)	359[ab]	330[b]	399[a]	430[a]	415[a]
ALA intake (g/d)	130[b]	97[c]	135[b]	176[a]	164[a]
Milk fat yield (g/d)	914	951	1069	1017	971
Milk ALA yield (g/d)	7.8[bc]	8.9[abc]	11.5[ab]	12.4[a]	7.3[c]
ALA in milk fat (g/100g FAME)	0.86[c]	0.94[bc]	1.07[ab]	1.22[a]	0.75[c]

[a,b,c] Means within row without common superscript are significantly different (P<0.05).

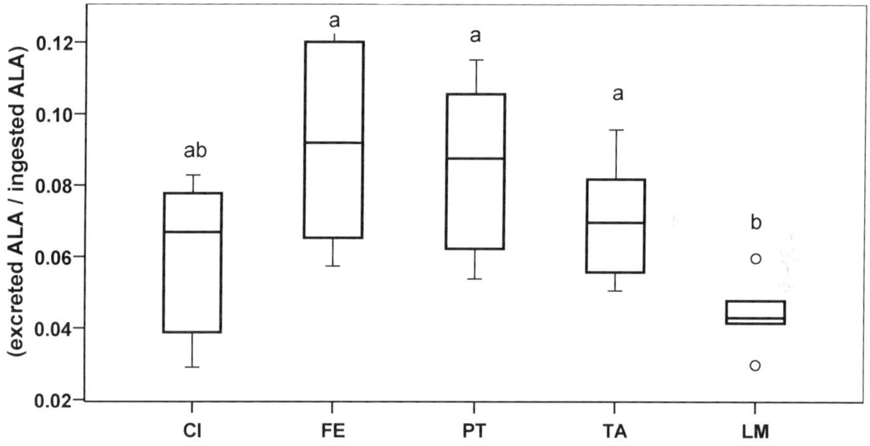

Figure 1. Recovery of ALA from feed in milk (n=6 per group).

Acknowledgement

This study was supported by the Swiss Grassland Society (AGFF), Zurich.

References

Chilliard, Y., F. Glasser, A. Ferlay, L. Bernard, J. Rouel and M. Doreau, 2007. Diet, rumen biohydrogenation and nutritional quality of cow and goat milk fat. Eur. J. Lipid Sci. Technol. 109: 828-855.

Leiber, F., M. Kreuzer, D. Nigg, H.R. Wettstein and M.R.L. Scheeder, 2005. A study on the causes for the elevated n-3 fatty acids in cows' milk of alpine origin. Lipids 40: 191-202.

Effect of diet and breed on fatty acid composition of beef steers

E.J. Kim[1], R.I. Richardson[2], K. Gibson[2], D. Coulmier[3] and N.D. Scollan[1]
[1]Institute of Biological, Environmental and Rural Sciences, Aberystwyth University, Aberystwyth, SY23 3EB, United Kingdom; [2]Department of Clinical Veterinary Science, University of Bristol, Langford, BS40 5DU, United Kingdom; [3]Desialis - France Luzerne, Complexe Agricole Mont Bernard, Route de Suippes, BP 124, 51007 Chalons en Champagne, France; nigel.scollan@aber.ac.uk

Introduction

Differences in the fatty acid (FA) composition of beef have been noted between breeds. However, these effects are often confounded with differences in growth rate and body composition (Scollan *et al.*, 2006). Feeding diets differing in *n*-3 (i.e. grass, linseed) and *n*-6 (i.e. grain, soya) polyunsaturated fatty acids (PUFA) produce meat with higher concentrations of 18:3*n*-3 and 18:2*n*-6, respectively and their long chain derivatives. Such diet-induced differences in FA composition occur despite high levels of biohydrogenation of dietary PUFA in the rumen. Feeding ruminally protected lipids deliver more PUFA through to muscle (Scollan *et al.*, 2006). This study examined Charolais × Friesian and British Shorthorn steers growing at similar rates and fed either all-forage or a cereal-based diet. The effects of including a PUFA-rich lucerne-based plant extract (PX) were also assessed within the Charolais cross only.

Material and methods

Thirty Charolais × Friesian (CF) and twenty British Shorthorn (BS) steers (~22 weeks of age) were allocated to one of three dietary treatments: (1) forage (F: grass/white clover swards during grazing followed by grass silage in the winter period), (2) restricted barley straw and concentrate (C: 40:60 on a DM basis), and (3) restricted barley straw and concentrate with 'PX' (PXC: 40:60 on a DM basis; Charolais × Friesian breed only). The 'PX' is a PUFA-rich lucerne-based plant extract (Désialis – France Luzerne, France) developed from the liquid fraction extracted from fresh lucerne (*Medicago sativa* L.), and then heat-treated and dried. Liveweight was monitored every 28 days and gains were calculated for individual animals. These data were used to modify feed intake of the C and PXC animals to achieve a similar growth rate to those fed on *ad libitum* F. Animals were slaughtered when they achieved fat class 4L and samples of *longissimus thoracis et lumborum* were taken at 48 h post-mortem for FA analysis. An analysis of variance was conducted with various contrasts including, overall interaction, diet × breed interaction, PX supplement effect within CF breed only using GenStat (117 edition) statistical software.

Results

Total fat was not different across breeds and diets (Table 1). In neutral lipid (NL), 18:1*n*-9 was lower in BS vs. CF ($P<0.05$). A similar trend was noted for 18:1*n*-9 in phospholipid (PL). Also in PL, the concentration and proportion of 18:3*n*-3, 20:5*n*-3 and 22:6*n*-3 were higher in BS vs. CF ($P<0.01$). Diet F resulted in higher levels of 18:3*n*-3 and longer chain derivatives in muscle lipids whilst C increased 18:2*n*-6 and associated derivatives. PXC resulted in higher 18:2*n*-6 and 18:3*n*-3 relative to F or C, resulting in a higher P:S ratio. Longer-chain PUFA (20:5*n*-3 and 22:6*n*-3) were up to 3× higher ($P<0.001$) in F-fed animals relative to C or PXC.

Table 1. Fatty acid (FA) composition (mg/100 g muscle) of neutral and phospholipid fraction, and total fatty acid of longissimus thoracis *and* lumborum.

Breed	BS		CF			sem[1]	D	B	Sig.[2]	PX	I
Diet	F	C	F	C	PXC				D×B		
Neutral lipid											
18:0	406	366	375	404	313	0.1	NS	NS	NS	NS	NS
18:1*n*-9	986	783	1147	1,155	891	0.1	NS	*	NS	NS	0.09
18:2*n*-6	23.6	33.1	21.9	41.2	46.3	0.11	***	NS	NS	NS	***
18:3*n*-3	15.8	7.2	13.4	5.3	25.0	0.13	***	NS	NS	***	***
18:2*c*-9, *t*-11	23.4	11.5	11.1	10.0	9.4	0.13	**	**	*	NS	***
Total	2,905	2,329	2,921	3,177	2,406	0.1	NS	NS	NS	0.09	NS
Phospholipid											
18:0	48.0	48.5	51.5	50.4	51.0	0.04	NS	NS	NS	NS	NS
18:1*n*-9	109	91	124	98	64	0.1	***	0.08	NS	***	***
18:2*n*-6	42.4	75.6	38.4	77.1	99.2	0.05	***	NS	NS	***	***
18:3*n*-3	20.9	6.0	18.5	4.8	23.1	0.07	***	*	NS	***	***
18:2*c*-9, *t*-11	1.87	0.93	0.92	0.74	0.75	0.081	***	***	*	NS	***
20:4*n*-6	28.3	42.0	34.3	49.6	42.2	0.05	***	***	NS	*	***
20:5*n*-3	17.3	6.9	17.2	4.7	13.1	2.85	***	**	**	***	***
22:5*n*-3	24.5	12.6	25.4	10.7	19.6	0.04	***	NS	*	***	***
22:6*n*-3	3.5	1.5	4.15	1.0	1.9	0.06	***	0.06	***	***	***
Total	499	488	531	521	523	0.04	NS	NS	NS	NS	NS
Total muscle FA	3,422	2,833	3,461	3,721	2,960	0.1	NS	NS	NS	NS	NS
n-6:*n*-3 ratio	1.2	4.7	1.3	6.9	2.4	0.05	***	***	**	***	***
P:S ratio	0.07	0.10	0.07	0.08	0.16	0.074	***	NS	NS	***	***

[1] sem (standard error of mean) is on log scale.
[2] *, ** and *** for $P<0.05$, $P<0.01$ and $P<0.001$, respectively; D=diet, B=breed, D×B=diet × breed interaction, PX=PXC effect within CF breed, I=overall interaction.

Conclusion

The increased content of 18:3*n*-3 and longer chain PUFA in the Shorthorns may reflect differences in underlying gene expression or enzymes involved in FA synthesis. Forage feeding increased the content of *n*-3 PUFA of the meat. PXC resulted in the highest level of total PUFA in meat relative to F or C and this probably relates to the rumen protected nature of the lipid in this concentrated plant extract.

Acknowledgement

This work was supported by the Department for Environment, Food and Rural Affairs, UK.

References

Scollan, N., J.-F. Hocquette, K. Nuernberg, D. Dannenberger, I. Richardson and A. Moloney, 2006. Innovations in beef production systems that enhance the nutritional and health value of beef lipids and their relationship with meat quality. Meat Sci. 74: 17-33.

Effects of abomasal infusion of tallow and camelina oil on responses to glucose and insulin in dairy cows during late pregnancy

T. Kokkonen, S. Salin, J. Taponen, K. Elo and A. Vanhatalo
University of Helsinki, 00014 University of Helsinki, Finland; tuomo.kokkonen@helsinki.fi

Introduction

The abrupt increase of nutrient demand at the onset of lactation represents a major challenge to the dairy cow. The decreased circulating insulin concentrations and reduced responsiveness of skeletal muscle and adipose tissue to insulin promote increased nutrient partitioning to the mammary gland. Reflecting increased whole-body insulin resistance, the clearance rates of glucose in intravenous glucose tolerance tests have been shown to decrease between late pregnancy and early lactation (Holtenius *et al.*, 2003, Bossaert *et al.*, 2008). The elevation of plasma non-esterified fatty acids (NEFA) during the periparturient period may enhance insulin resistance. In nonlactating, nongestating cows induction of hyperlipidemia by infusion of tallow impaired glucose clearance (Pires *et al.*, 2007), whereas infusion of linseed oil, rich in C18:3 n-3 (50-60% of total FA), had an insulin sensitising effect (Pires *et al.*, 2008). The current study was carried out to test the hypothesis that an elevated NEFA concentration of blood induces whole-body insulin resistance in dairy cows during late pregnancy, and that camelina oil, rich in C18:3 n-3 (35-40% of total FA), attenuates insulin resistance.

Material and methods

Six nonlactating, primiparous Finnish Ayrshire cows, approximately 6 wk prior to the predicted parturition date, were allocated at random to experimental treatments according to a replicated 3×3 Latin square with 5-day periods. Treatments were abomasal infusion of water (CONT), tallow (TAL) or camelina oil (CAM). The amount of water/fat administered was 500 ml/d, which was infused via abomasal lines in ten equal portions (50 ml) using 100 ml syringes every second hour between 6:00 and 24:00 h. To reduce carry-over effects, 5-d washout period was allowed between experimental periods. Intravenous glucose tolerance test (IVGTT) was performed at 09:00 h on day 5 of each period by administering 0.25 g/kg body weight of glucose, and insulin challenge (IC) was performed at 19:00 h on day 5 of each period by administering 0.1 IU of insulin/kg body weight. Blood samples were collected at -15, -5, 5, 10, 15, 20, 30, 40, 50, 60, 90 and 120 min relative to glucose infusion and insulin challenge, and in addition at 150 and 180 min relative to glucose infusion. During experimental periods energy intake was restricted to 95% of calculated energy requirements for pregnant cows. Energy content of fat infusion was taken into account in the calculation of individual energy allowances. Diet was composed of 80% grass silage and 20% grass hay. Data were analysed by the mixed procedure of SAS® (version 9.1) with a model that included the random effects of cow and fixed effects of square, period within the square and treatment.

Results

Metabolisable energy (ME) intakes were similar between the treatments, when ME content of lipid infusion was taken into account. As intended, infusion of lipids increased basal NEFA concentrations (P<0.01) (Table 1). Lipid infusion reduced the glucose clearance rate during IVGTT and IC (P<0.05), and glucose nadir was lower (P<0.05) in CONT after IC. Treatments did not influence glucose response measured as the area under the curve (AUC) during IVGTT, whereas the absolute value of glucose AUC was greater in CONT after IC (P<0.05). Nadir of NEFA concentration was lower in CONT during IVGTT and IC (P<0.001), but no differences between the treatments were observed in NEFA AUC.

Table 1. The effect of abomasal infusions of water, tallow or camelina oil on plasma glucose and NEFA responses to glucose and insulin challenges.

	Treatment [1]			SEM	P-value	
	CONT	TAL	CAM		Fat vs. CONT	TAL vs. CAM
Glucose tolerance test						
Glucose CR_{60} [2]	1.74	1.48	1.34	0.128	0.03	0.35
Glucose AUC_{120} [3]	561.6	557.0	598.0	31.07	0.69	0.39
NEFA basal, mmol/l	0.11	0.21	0.23	0.026	0.005	0.53
NEFA nadir, mmol/l	0.04	0.10	0.11	0.010	0.0007	0.37
NEFA, AUC_{120} [3]	-5.30	-6.16	-4.65	2.24	0.97	0.65
Insulin challenge						
Glucose nadir, mmol/l	2.05	2.30	2.22	0.099	0.04	0.39
Glucose CR_{60} [2]	1.02	0.66	0.83	0.074	0.01	0.12
Glucose, AUC_{120} [3]	-158.9	-127.0	-137.1	11.45	0.02	0.37
NEFA basal, mmol/l	0.07	0.19	0.21	0.023	<0.0001	0.25
NEFA nadir, mmol/l	0.04	0.11	0.12	0.016	0.0002	0.46
NEFA, AUC_{60} [3]	-0.40	-2.39	-1.03	0.796	0.23	0.27

[1] CONT = Control; TAL = Tallow; CAM = Camelina oil.
[2] CR = Clearance rate during the first 60 min,%/min.
[3] AUC = Area under the curve during the first 60 or 120 min, mmol/l × 60 min or mmol/l × 120 min.

Conclusion

Induction of hyperlipidemia by infusion of tallow or camelina oil impaired glucose clearance during glucose and insulin challenges. This indicates that hyperlipidemia reduces insulin sensitivity and responsiveness in dairy cows during late pregnancy. This study gives no support to the hypothesis that increased supply of C18:3 n-3 in camelina oil would enhance the response to insulin. Further work is in progress to study the effect of lipid infusions on the development of insulin resistance in dairy cows.

References

Bossaert, P., J.L.M.R. Leroy, S. De Vliegher and G. Opsomer, 2008. Interrelations between glucose-induced insulin response, metabolic indicators and time of first ovulation in high-yielding dairy cows. J. Dairy Sci. 91: 3363-3371.

Holtenius K., S. Agenäs, C. Delavaud and Y. Chilliard, 2003. Effects of feeding intensity during the dry period. 2. Metabolic and hormonal responses. J. Dairy Sci. 86: 883-891.

Pires, J.A.A., A.H. Souza and R.R. Grummer, 2007. Induction of hyperlipidemia by intravenous infusion of tallow emulsion causes insulin resistance in Holstein cows. J. Dairy Sci. 90: 2735-2744.

Pires, J.A.A., J. Pescara, A. Brickner, N. Rio, A.P. Cunha and R.R. Grummer, 2008. Effects of abomasal infusion of linseed oil on responses to glucose and insulin in Holstein cows. J. Dairy Sci. 91: 1378-1390.

Mammary amino acid metabolism in response to increased energy and protein supply in lactating dairy cows

S. Lemosquet[1,2], *H. Lapierre*[3], *H. Rulquin*[1,2] *and J. Guinard-Flament*[2,1]
[1]*INRA, UMR1080 Dairy Production, 35590, Saint-Gilles, France;* [2]*Agrocampus ouest, UMR1080 Dairy Production, 35000, Rennes, France;* [3]*Agriculture and Agri-Food Canada, J1M 1Z3, Sherbrooke, Canada; sophie.lemosquet@rennes.inra.fr*

Introduction

Milk protein yield usually increases in response to increased supply of energy (E) and protein (P), alone or in combination, in cows fed under requirements. Does this increment occur through a similar mechanism in terms of mammary metabolism of amino acids (AA)?

Material and methods

Four multiparous dairy cows received 4 diets providing 70% (LP: 1324 g PDI/d) or 125% (HP: 2247 g PDI/d) of protein requirements combined with 70% (LE: 22.8 Mcal NEL/d) or 100% (HE: 30.9 Mcal NEL/d) of energy requirements, in a Latin Square design with 14-d periods. Requirements (INRA, 1989) were determined based on milk yields predicted during experimental periods using a milk persistency of 98% per week. The 4 diets were composed of maize silage (taken at 12.2, 8.7, 12.2, 12.9 kg MS/d in LELP, LEHP, HELP, HEHP, respectively), of energy concentrate (0.45, 0.33, 5.7, 2.4 kg MS/d), soybean meal (1.4, 4.9, 0.15, 3.7 kg MS/d) and urea (160, 40, 180, 140 g/d). Cows were adapted to feed components 3 weeks before the experiment. Cows were fitted with an ultrasonic flow probe (Transonic type A Systems Inc., Ithaca, NY) to measure blood flow on left udder and with two catheters in a carotid artery and in the left mammary vein. On day 12 of each period, blood was collected every 2 h between the 12 h milking interval to analyse plasma AA on Biotronik LC 3000 (GmbH, Maintal, Germany). Data were analysed using the MIXED procedure of SAS® (2000) with treatment, period and cow as fixed effects. Significance level was set at $P \leq 0.05$, tendency at $P \leq 0.10$. During the 4th period, a flow probe connector did not work as a consequence n = 3 for HEHP. In Table 1, AA were grouped as in Mepham (1982): Phe+Tyr and Met were included in Group 1; branched-chain AA (BCAA) and Lys in Group 2.

Results

Supply of E and P increased milk yield by 10.7% and 16.6%, and protein yield by 16.6% and 17.0%, respectively, with no interaction (E×P). Only E tended to increase milk protein content. Increasing E decreased arterial concentration of Arg. Increasing P increased arterial concentrations of almost all EAA (Table 1) and of Arg but decreased the arterial concentrations of Gln + Glu and Ala. Half mammary plasma flow (MPF) increased with E but remained unchanged with P. Increasing E tended to increase the uptake of Met + Phe + Tyr and the uptake of Ala + Glu + Gln. Increasing P increased BCAA + Lys and Arg uptakes but decreased Ala + Glu + Gln uptakes. The uptake to output (U:O) ratio of Met + Phe + Tyr did not change with all treatments and was not significantly different from 1. The U:O ratio of BCAA + Lys and Arg decreased with E. At the opposite, increasing P supply increased the U:O ratio of BCAA + Lys and Arg, but with a larger effect on the BCAA + Lys U:O ratio when E supply was low (LEHP; *ExP: P* = 0.07), whereas it decreased the ratio of Ala + Glu + Gln.

Table 1. Half udder milk production and AA metabolism in dairy cows receiving increased energy (E) or protein (P) supply.

	Treatments[1]				SEM[2]	*P*-values[3]		
	LELP	LEHP	HELP	HEHP		E	P	E×P
Milk Yield (kg/12 h)	5.6	6.5	6.2	7.2	0.20	0.01	0.01	0.68
Milk Protein (g/12 h)	149	169	172	199	6.7	0.02	0.01	0.58
Protein content (g/kg)	26.3	26.4	27.9	27.4	0.70	0.08	0.74	0.58
MPF (L/h)4	260	280	353	305	22.4	0.03	0.49	0.13
A (μM)[5]								
Met+Phe+Tyr	111	122	110	138	9.0	0.38	0.04	0.31
BCAA+Lys	529	1029	544	939	117	0.71	<0.01	0.62
Arg	84	106	67	98	5.6	0.04	<0.01	0.38
Ala+Glu+Gln	461	399	468	411	13.9	0.45	<0.01	0.80
Uptake (mmol/h)								
Met+Phe+Tyr	11	11	12	15	1.2	0.06	0.13	0.23
BCAA +Lys	38	55	44	53	2.6	0.40	0.03	0.16
Arg	6	8	6	8	0.7	0.46	<0.01	0.93
Ala+Glu+Gln	22	17	30	18	2.6	0.10	<0.01	0.22
Uptake: Ouput[6]								
Met+Phe+Tyr	1.07	0.98	1.00	1.09	0.09	0.78	0.96	0.30
BCAA +Lys	1.27	1.61	1.24	1.33	0.06	0.04	0.01	0.07
Arg	2.43	2.95	1.93	2.41	0.24	0.05	0.05	0.92
Ala+Glu+Gln	0.95	0.61	1.07	0.57	0.09	0.62	<0.01	0.34

[1] LELP: Low Energy and Low Protein; LEHP: Low Energy and High Protein; HELP: High Energy and Low Protein; HEHP: High Energy and High Protein.
[2] The higher standard error of means is given (HEHP, n=3).
[3] Effects: E: increasing energy supply; P: increasing protein supply; ExP: interaction.
[4] Mammary Plasma Flow.
[5] Arterial concentration. [6] Ratio of uptake to output in milk protein.

Conclusion

Energy and protein supplies increased milk protein yield as expected even if treatments were applied during relative short periods (14 d). It occurred through different effects on mammary AA metabolism as observed in Raggio *et al.* (2006). Increased energy supply increased milk protein secretion through an increased uptake of Group1-AA (mainly used, on a net basis, for milk protein synthesis) and an increased uptake of non essential AA (e.g. Ala + Glu + Gln). There was, however, no paralleled increment in the uptake of the BCAA+Lys and Arg. On the contrary, increasing protein supply mainly increased the uptake of the AA already taken in excess relative to milk protein output, i.e. BCAA + Lys and Arg, providing C skeletons and N for non essential AA synthesis and possibly, energy through oxidation.

References

Mepham, T.B., P. Gaye and J.C. Mercier, 1982. Biosynthesis of milk proteins. In: Developments in Dairy chemistry: 1 Proteins. In: Fox, P.F. (Ed.), Applied Science Publishers, London, UK. pp 115-116.

Raggio, G., S. Lemosquet, G.E. Lobley, H. Rulquin and H. Lapierre, 2006. Effect of Casein and propionate supply on mammary protein metabolism in lactating dairy cows. J. Dairy Sci. 89: 4340-4351.

Empirical prediction of net splanchnic release of ß-hydroxybutyrate in ruminants

C. Loncke[1], P. Nozière[1], J. Vernet[1], H. Lapierre[2], D. Sauvant[3] and I. Ortigues-Marty[1]
[1]INRA, UR 1213, Theix, France; [2]Agriculture and Agri-Food Canada, STN Lennoxville, Sherbrooke, QC, Canada; [3]INRA-Agro Paris Tech, UMR 791, Paris, France; isabelle.ortigues@clermont.inra.fr

Introduction

To adapt feeding strategies for ruminants to the new demands of animal production, feed evaluation systems have to evolve toward nutrient-based systems where energy would no longer be considered as an aggregated unit. In a first step, prediction equations of net portal appearance (NPA) of energetic nutrients were developed based on dietary characteristics (Loncke *et al.*, 2009). The second part of this work was to establish predictions of net splanchnic release (NSR) of energetic nutrients towards peripheral tissues. This study presents preliminary results focussed on NSR of ß-hydroxybutyrate (BOH).

Material and methods

The FLORA (Vernet and Ortigues-Marty, 2006) database was used. All publications presenting results on NPA and NSR of BOH were selected except studies including infusions. All feed ingredients and diets were characterised according to the INRA feed evaluation system with calculation of the ruminal fermentable organic matter intake (RfOMI = digestible organic matter – fat – undegradable crude protein and starch – fermentation products). Animal energy status was defined as the quantity of energy retained (ES>0) or mobilised (ES<=0). It was calculated for non productive or growing animals as the net energy for production (NEP) expected from their ME intake, or for lactating animals as the difference between the expected and the observed NEP, the latter being calculated from the reported animal performances. The NSR of BOH was defined as the sum of its NPA and net hepatic release (NHR). Potential predictors of NHR of BOH were the NPA of butyrate (C4), acetate (C2) and BOH, the arterial concentration of non-esterified fatty acids (NEFA) and the calculated ES. Data were submitted to meta-analyses (Sauvant *et al.*, 2008), considering the study effect as fixed (Loncke *et al.*, 2009). Relationships between NHR (Y) and predictors (X) were studied using a covariate GLM model nested within physiological status (PHY; non-productive adults, growth, lactation): $Y = \alpha + \beta_i X_i + PHY + study\ (PHY) + PHY \times X_i + error$. Post analyses were carried out to check the absence of interfering factors (species included) on slopes, LSMeans and residuals.

Results and discussion

The studies (n=24) selected out of FLORA included 24 treatments on non-productive adults, 22 on growing/finishing animals and 12 on lactating dairy cows. Animals ate 21±7.6 g DM/d/kg BW and diets contained 37±35 g concentrate/100 g DM, which indicates a large range of intake level and diet composition (fasting included). The RfOMI averaged 11±4.2, 10±3.7 and 14±1.8 g/d/kg BW whereas the ES averaged 4.9±1.5 (range = -37.8 to 26.2), 10.5±2.1 (-41.4 to 21.8) and -0.5±1.9 (-17.0 to 14.1) kcal/d/kg BW, for non-productive, growing/finishing animals and lactating cows respectively. The NHR of BOH averaged 6.48±4.7 mmol/d/kg BW independently of PHY (P=0.48) or species (P=0.99). Successive steps of the meta-analyses showed that the amount of OM fermented in the rumen and the total inflow of NEFA to the liver were the major drivers of the NHR of BOH. Indeed, the NHR of BOH was positively correlated to RfOMI (P<0.001), NPA of C2 (P=0.013), C3 (P<0.001), C4 (P=0.021), but was not correlated to NPA of BOH (P=0.18).

Moreover, the NPA of C2, C3 and C4 were positively correlated to RfOMI ($P<0.001$). This was consistent with an increased supply of carbon chains to the liver, particularly of BOH precursors, when RfOMI increases. Principal component analyses confirmed the influence of RfOMI on NHR of BOH and showed that NHR of BOH was positively correlated to arterial NEFA concentrations, reflecting supply of BOH precursors from mobilisation of body reserves. Because in the selected publications, the dataset was not complete for relevant nutrients (C2, C4, NEFA, BOH) it was not possible to elaborate a model which took into account all these parameters. Because arterial NEFA concentrations were strongly and negatively linked to ES (R^2=0.90, RMSE=0.058 mM in a within study GLM analysis), ES was considered as a predictor of NEFA hepatic inflow. The RfOMI was considered as an index of NPA of C2 and C4. The prediction equation of NHR of BOH was thus established (Table 1) in response to variations in RfOMI (P=0.005) and ES^2 ($P<0.001$). The ES^2 variable was influenced by the physiological status reflecting a higher NHR of BOH in lactating dairy cows during mobilisation. No other interfering factors were observed.

Table 1. Response of net hepatic release of ß-hydroxybutyrate (mmol/h/kg BW) to variations in rumen fermentable OM intake (RfOMI) and energy status (ES).

N	Intercept	RfOMI (g/d/kg BW)	ES (kcal/d/kg BW)	ES^2	RMSE	R^2
58	-0.0078^{NS}	0.0188	NS	0.00081 + Δ^{δ}	0.041	0.91

$^{\delta}\Delta$ lactation = 0.00105±0.00022 ($P<0.001$); Δ non productive adults = -0.00043±0.00012 ($P<0.001$); Δ growing = -0.00062±0.00016 ($P<0.001$); NS = non significant.

Prediction of NSR of BOH was obtained by combining the prediction models of NPA (Loncke *et al.*, 2009) and NHR of BOH: NSR = 0.138 + (0.174 × RfOMI) × (0.08 – 0.073 × rumen digested NDF/RfOM) + 0.0158 × RfOMI – $0.000813 \pm \Delta_{PHY} \times ES^2$. A GLM analysis against the observed NSR of BOH (R^2=0.87; RMSE=0.081) indicated an intercept (0.016±0.032) and a slope (0.90±0.2) non different from 0 (P=0.58) and 1 (P=0.35) respectively.

Conclusion

This study proposes a prediction model of NSR of BOH and suggests that NSR of BOH is not solely driven by nutrient supply but also by the metabolic demand of the animal. The subsequent challenge will be to establish predictive models of the NSR of the other nutrients.

References

Loncke, C., I. Ortigues-Marty, J. Vernet, H. Lapierre, D. Sauvant and P. Nozière, 2009. Empirical prediction of net portal appearance of volatile fatty acids, glucose, and their secondary metabolites (ß-hydroxybutyrate, lactate) from dietary characteristics in ruminants: A meta-analysis approach. J. Anim. Sci. 87: 253-268.

Vernet, J. and I. Ortigues-Marty, 2006. Conception and development of a bibliographic database of blood nutrient fluxes across organs and tissues in ruminants: data gathering and management prior to meta-analysis. Reprod. Nutr. Develop. 5: 257-546.

Sauvant, D., P. Schmidely, J.J. Daudin and N.R. St-Pierre, 2008. Meta-analyses of experimental data in animal nutrition. Animal 2: 1203-1214.

Postweaning adaptation of liver activity to solid diet in goat kids

D. Magistrelli, A.A. Aufy and F. Rosi
Department of Animal Science, University of Milan, via G. Celoria 2, 20133 Milan, Italy; damiano.magistrelli@unimi.it

Introduction

Weaning is the process of transitioning young mammals from milk to solid diet. In ruminants, milk is progressively substituted by forage and concentrate or grain-based diets, which are processed in the developing rumen. The pattern of nutrients absorbed changes from glucose, long-chain fatty acids, and milk-derived amino acids to volatile fatty acids, ketones and amino acids from feed and microbial sources. The change requires the adaptation of liver metabolism and activity that shift from glycolytic to gluconeogenic pathways (Baldwin *et al.*, 2004). Moreover, the requirement of amino acids by the digestive apparatus is high and may affect amino acid availability for growth (Leat, 1970), so the first days after the removal of milk represent a critical phase during which kids can experience growth stasis or weight loss.
The aim of the present study was to investigate changes in liver composition and activity in Saanen goat kids, during the first days after the completion of weaning.

Material and methods

Two trials were conducted on male Saanen goat kids, reared by two different weaning programs. In the first trial, 11 3-day old kids were divided into two homogeneous groups: MILK (6 animals), that received goat milk *ad libitum* for the entire study period, and WEAN2 (5 animals) that was subjected to weaning. In the second trial, 12 3-day old kids were assigned to one of two homogeneous groups (6 animals per group): MILK, that was fed goat milk *ad libitum*, and WEAN8, that was weaned. The weaning programme was based on the progressive replacement of milk by a weaning mixture constituted of grass hay (30%), dehydrated alfalfa hay (10%), steam-flaked corn (19%), corn gluten meal (3%), dried sugar beet pulp (8%), soybean meal (15%), sunflower seeds (4%), sugar cane molasses (4%) and mineral/vitamin supplement (7%). The experimental feeds were administered twice a day (9:00 am and 7:00 pm).
All the kids studied in the trials were slaughtered (5 hours after the meal) 2 days (in the first trial) or 8 days (in the second trial) after the complete removal of milk from the diet of WEAN2 and WEAN8 groups, respectively. At slaughtering, body, carcass and liver weights were recorded and samples of liver were taken and analysed for DNA and RNA (Munro and Fleck, 1966), soluble protein (Pierce Biotecnology Inc.), ether extract (AOAC method 960.39.), phospholipids (Folch *et al.*, 1957) and glycogen (Kemp and Kits van Heijningen, 1954) content, and alanine aminotransferase (ALT) and aspartate aminotransferase (AST) activity (Boehringer Mannheim GmbH).
Differences between milk-fed and weaned kids, recorded 2 and 8 days after complete weaning, were evaluated by ANOVA using the GLM procedure of SAS® (SAS Institute Inc.).

Results

Body weight did not differ between weaned and milk-fed kids, nor 2 days, nor 8 days after weaning, but WEAN8 kids exhibited a lower carcass weight than milk-fed ones. Two days after weaning, liver weight was similar in the experimental groups, but 8 days after weaning, liver weight significantly decreased in WEAN8 kids, in comparison to the MILK ones.
DNA and phospholipid contents did not differ between groups of the first trial, but DNA and phospholipids were higher in the weaned kids (WEAN8) of the second trial.

A reduction of 25% in the glycogen content of the liver of WEAN2 kids was noted in the first trial. This reduction was greater in WEAN8 kids, compared to MILK ones.

Two days after weaning, no difference was observed in aminotransferase activity between the experimental groups, but 8 days after weaning, both ALT and AST activity were higher in WEAN8 kids, compared to milk-fed ones.

Table 1. Body and carcass weights and liver analysis.

	First trial				Second trial			
	Milk	Wean2	SEM	P	Milk	Wean8	SEM	P
Body weight (kg)	16.2	15.3	1.56	0.72	15.0	13.0	0.69	0.20
Carcass weight (kg)	10.7	9.54	0.74	0.45	9.68	6.70	0.67	0.02
Liver weight (g)	377	330	18.2	0.25	372	220	31.7	0.03
DNA (mg/g)	2.54	2.96	0.18	0.19	2.88	5.57	0.51	0.01
Total DNA (g)	0.96	0.98	0.08	0.86	1.07	1.23	0.24	0.33
RNA (mg/g)	10.6	10.9	0.34	0.58	9.66	10.2	0.19	0.26
SP (mg/g)[1]	238	223	8.90	0.84	223	250	10.2	0.28
Glycogen (mg/g)	55.3	40.9	4.77	0.08	40.6	13.8	4.46	0.00
Ether extract (mg/g)	381	363	42.2	0.82	453	587	36.6	0.17
P-lipids (mg/g)[2]	3.87	3.80	0.22	0.86	3.79	4.84	0.20	0.02
ALT (U/g)	8.22	7.71	0.97	0.79	6.36	8.84	0.60	0.08
AST (U/g)	229	200	17.9	0.39	169	224	8.66	0.00

[1] SP = soluble protein.

[2] P-lipids = phospholipids.

Conclusion

Weaning alters liver composition and activity, without affecting the number of cells within the tissue. Liver activity shifts from glycolytic to gluconeogenic pathways and the effects of this transition are significant 8 days after the complete removal of milk from the diet. The reduction in hepatic glycogen content is accompanied with the increase in amino acid transamination, which is the first step in the catabolism of amino acids (Hagopian *et al.*, 2003).

References

Baldwin, R.L.VI, K.R. McLeod, J.L. Klotz and R.N. Heitmann, 2004. Rumen development, intestinal growth and hepatic metabolism in the pre- and postweaning ruminant. J. Dairy Sci. 87: E55-E65.

Folch, J., M. Lees and G.H.S. Stanley, 1957. A simple method for the isolation and purification of total lipides from animal tissues. J. Biol. Chem. 226: 497-509.

Hagopian, K., J.J. Ramsey and R. Weindruch, 2003. Caloric restriction increases gluco-neogenic and transaminase enzyme activities in mouse liver. Exp. Gerontol. 38: 267-278.

Kemp, A. and A.J.M. Kits van Heijningen, 1954. A Colorimetric micro-method for the determination of glycogen in tissues. Biochem. J. 56: 646-648.

Leat, W.M.F., 1970. Carbohydrate and lipid metabolism in the ruminant during post-natal development. In: A.T. Phillipson (Eds), Digestive physiology and metabolism in the ruminant. Oriel Press, Newcastle Upon Tyne, UK, pp. 211-222.

Munro, H.N. and A. Fleck, 1966. The determination of nucleic acids. Methods Biochem. Anal. 14: 113-176.

Quantitative estimation of the endogenous synthesis of rumenic acid in goats fed lipid supplements

J. Mouriot[1,2], L. Bernard[1], P. Capitan[1], C. Joly[2], O. Loreau[3], J.M. Chardigny[2] and Y. Chilliard[1]
[1]INRA, UR1213 Herbivores, Site de Theix, F-63122 Saint-Genès-Champanelle, France; [2]INRA, UMR 1019 Nutrition Humaine, F-63122 Saint-Genès-Champanelle, France; [3]CEA, IBITECS, Chimie Bioorganique et de Marquage, F-91191 Gif sur Yvette, France; laurence.bernard@clermont.inra.fr

Introduction

Rumenic acid (RA, *cis*-9,*trans*-11-18:2) in milk, the main conjugated linoleic acid isomer, is derived from ruminal metabolism of 18:2*n*-6 and from endogenous synthesis via the action of Δ-9 desaturase on vaccenic acid (VA, *trans*-11-18:1). VA is an intermediate of linoleic (18:2*n*-6) and linolenic (18:3*n*-3) acid metabolism in the rumen (Bauman and Griinari, 2003) and its conversion in RA in the udder represents the major source of milk RA in the bovine (Mosley *et al.*, 2006; Shingfield *et al.*, 2007). However, no data are available so far in goats on the endogenous synthesis of RA from VA in the mammary gland, even though a close relationship between milk RA and VA has been established in this species (Chilliard *et al.*, 2003). The objectives of this study were to use ^{13}C-labelled VA to determine the uptake and conversion of VA into RA in the udder and to quantify the fraction of milk RA originating from VA, in lactating goats fed lipid supplements.

Material and methods

Eight multiparous Alpine goats in late-lactation (266 (SD 7) d in milk) were fed 2 diets during a 21 d period in 2 groups of 4 animals. Diets consisted of natural grassland hay and a concentrate mixture (in a 68/32 ratio) and included a lipid supplement of 90 g/d of sunflower-seed oil (SO) or a combination of 60 g/d sunflower-seed oil and 30g/d of fish oil (SFO), offered as two equal meals. On d 18, a single bolus of 1.5 g of 1-^{13}C-VA as free fatty acid (CEA, France), was delivered by jugular injection after morning milking. After the dose, the goats were completely milked every 4 h for 24 h and for each milking, milk yield was recorded, and samples of milk were collected for analysis of milk fat content, fatty acid composition and ^{13}C-enrichment. ^{13}C-VA and ^{13}C-RA enrichments were determined respectively on dimethyl disulfide (DMDS) adducts as described by Mosley *et al.* (2006) and directly from FAME, by analysis on GC–MS (Agilent model 7890A GC system attached to an Agilent model 5975C inert XL EI/CI mass detector). The tracer:tracee ratio (TTR) was calculated for VA and RA from the mass abundance of the ^{13}C and ^{12}C fragments using the equation TTR = ^{13}C/^{12}C. Total grams of VA and RA at each time point for 24 h post-bolus were calculated in milk fat, and respectively separated into grams of ^{13}C- and ^{12}C-VA, and ^{13}C- and ^{12}C-RA and summed for each goat. These values were used to calculate VA% converted to RA in the udder and RA% in milk originating from VA (Table 1). Data were statistically evaluated using the nonparametric Wilcoxon U-test. Treatment effects were considered significant at $P<0.05$.

Results

The conversion of VA to milk RA was estimated to be 31.7 and 31.6% ($P=0.89$) for SO and SFO treatments, respectively. From these conversion factors it was calculated that 73.1 and 62.9% ($P=0.47$) of milk RA came from VA, for SO and SFO treatments, respectively.

Table 1. Calculations of percentages of vaccenic acid (VA) converted to rumenic acid (RA) in the udder and of milk RA originating from VA, in lactating goats fed a hay-based diet with sunflower-seed oil alone or in combination with fish oil.

Variables calculated over 24 h post-bolus	Diet[1] SO	SFO	Calculation
% VA converted to RA	31.7±3.35	31.6±2.82	(g ^{13}C-RA) / (g ^{13}C-VA + g ^{13}C-RA) x 100
Secreted VA, g	3.1±0.52	2.9±0.95	(g ^{13}C-VA secreted + g ^{12}C-VA secreted)
Uptake of VA, g	4.4±0.57	4.1±1.23	(g secreted VA x 100)/(100 - % VA converted to RA)
VA converted to RA, g	1.4±0.11	1.2±0.28	(total uptake g VA - total secreted g VA)
Secreted RA, g	1.9±0.28	1.9±0.35	(g ^{13}C-RA secreted + g ^{12}C-RA secreted)
% RA from VA	73.1±7.82	62.9±3.45	(g VA converted to RA)/ g secreted RA

[1] Hay-based diet supplemented with 90 g/d sunflower-seed oil (SO) or with 60 g/d of sunflower-seed oil and 30 g/d of fish oil (SFO); values are means (n=4) ± SEM.

Conclusion

This study provides quantitative estimations, using ^{13}C-labelled VA, on VA metabolism and its desaturation into RA in the mammary gland of the goat. It demonstrates that the endogenous synthesis of RA from VA in the caprine mammary gland is the major source of milk RA with 63-73% of milk fat RA originating from VA. Further work is in progress to evaluate if these estimations were related to the other parameters of mammary Δ-9 desaturation (mammary level of transcripts and/or *ex vivo* activity of Δ-9 desaturase).

Acknowledgement

The authors thank goat unit workers and staff of the laboratories involved. This research was supported by the French National Agency for Research within the 'Transqual' (projet ANR-PNRA 2006-08) research program.

References

Bauman, D.E. and J.M. Griinari, 2003. Nutritional regulation of milk fat synthesis. Annu. Rev. Nutr. 23: 203-27.

Chilliard, Y., A. Ferlay, J. Rouel and G. Lamberet, 2003. A review of nutritional and physiological factors affecting goat milk lipid synthesis and lipolysis. J. Dairy Sci. 86: 1751-70.

Mosley, E.E., B. Shafii Dagger, P.J. Moate and M.A. McGuire, 2006. *Cis*-9, *trans*-11 conjugated linoleic acid is synthesized directly from vaccenic acid in lactating dairy cattle. J. Nutr. 136: 570-575.

Shingfield, K.J., S. Ahvenjarvi, V. Toivonen, A. Vanhatalo and P. Huhtanen, 2007. Transfer of absorbed *cis*-9, *trans*-11 conjugated linoleic acid into milk is biologically more efficient than endogenous synthesis from absorbed vaccenic acid in lactating cows. J. Nutr. 137: 1154-1160.

Recycling of phosphate is not affected by P intake in lactating dairy cows

L. Puggaard, N.B. Kristensen and J. Sehested
Department of Animal Health, Welfare and Nutrition, Faculty of Agricultural Sciences,Aarhus University, 8830, Tjele, Denmark; nbk@agrsci.dk

Introduction

Phosphorus (P) plays an important role in the production and health of cattle and is essential for rumen microbial fermentation. Salivary P secretion is an important P input to the rumen, but salivary secretion and fecal output are also the quantitatively most important excretion route for P in cattle (Scott *et al.*, 1985). A significant part of the P secreted in saliva is reabsorbed, thereby creating a system of P recycling, but there is a shortage on quantitative data of salivary secretion of P. In order to increase P utilisation of dairy cattle, it is desirable to have a high salivary P secretion. This provides rumen microbes with the highest possible amount of substrate which, in combination with an efficient intestinal reabsorption of P, ensures that the cow loses as little P in faeces as possible. The present study was aimed at testing the hypothesis that a reduction in P intake increases the efficacy of intestinal P absorption and increases P recycling. The objective of the present study was to quantify salivary secretion and portal absorption of P in dairy cows fed two levels of dietary P (at and below recommendations).

Material and methods

Five Holstein Dairy cows, 597±31 kg of body weight and 145±55 d in milk were used in an incomplete cross-over experiment. The cows were rumen cannulated and fitted with permanent indwelling catheters into an artery, mesenteric vein, and the hepatic portal vein (Kristensen *et al.*, 2007). P concentrations in the total mixed rations were 2.4 g/kg DM (LP) and 3.4 g/kg DM (HP). HP corresponded to Danish recommendations (Sehested, 2004). The rations were based on maize silage, clover grass silage, sugar beet molasses and sugar beet pulp. Samplings were conducted for 8 h at d 14 in each period. To allow for an 8-h sampling window, the cows were fed three times per d (8:00, 16:00, and 24:00 h). Diets were offered at 20 kg DM/d. Portal blood flow was measured by down stream dilution of *p*-aminohippuric acid infused into the mesenteric vein. Sampling included simultaneous sampling of arterial and portal blood as well as ruminal fluid, urine, faeces, and milk. Data were analysed as repeated measurements in a cross-over design by the MIXED procedure in SAS® version 9.2. Treatment, period, sequence and sample were used as fixed effects. The effect of cow within sequence was included as a random effect. Data are presented as means ± SE. Significance was declared at $P \leq 0.05$.

Results and discussion

Dry matter (DM) intake and milk production in kg/d were not affected by treatment ($P>0.67$; Table 1). Intake of P and excretion of P in faeces were significantly higher ($P<0.024$) in treatment HP, whereas excretion of P in urine was not affected by treatment. The feed – fecal P difference was significantly higher in treatment HP compared with LP ($P=0.01$). P balance, calculated as P input – P output in milk, urine and faeces, was significantly lower ($P=0.05$) and negative in LP compared with HP. The P concentration in ruminal fluid and arterial plasma concentration of P was significantly lower in LP ($P<0.0014$). Net portal flux of P and recycling of P were not affected by treatment ($P=0.20$).
The results indicate that recycling of P was not affected by a reduced dietary supply of P thus explaining that the cows were unable to maintain the ruminal fluid concentration of P when P intake

decreased. Moreover, the apparent digestibility (feed-fecal difference/P intake) was similar in both treatments (not shown), which indicates that efficiency of P digestion does not increase when P level in the diet is decreasing. A constant blood to gut flux of P and a constant P digestibility means that cows on LP had a similar P output from the intercellular compartment, but a lower input from the gut. The results indicate that the cows on LP were mobilising P from bones in order to maintain the same salivary P flux as HP cows. The negative P balance with LP was in agreement with the reduced arterial blood plasma concentration of P. We could not detect (P=0.20) any difference in net portal flux of P between treatments, but the flux was numerically higher with HP. However, when relating net portal flux of P to P intake the net portal flux was relatively higher with LP compared with HP (1.19 vs. 0.98), indicating that P was more efficiently recycled in LP.

Table 1. Present dry matter intake and milk production from the dairy cows together with the intake, excretion, absorption, balance, recycling and net fluxes of P.

	Treatment		SE	Treatment effect
	LP[1]	HP[2]		*P*-value
DM intake, kg/d	19.8	19.6	0.37	0.67
Milk production, kg/d	30.6	30.8	2.69	0.71
P intake, mmol/h	64.4	90.4	1.58	<0.0001
P in feces, mmol/l	29.0	41.5	2.32	0.024
P in urine, mmol/l	0.5	0.5	0.02	0.32
Feed – fecal P difference, mmol/h	35.4	48.9	2.80	0.01
[3]P balance, mmol/h	-6.2	7.1	3.27	0.05
Ruminal fluid P, mmol/l	3.26	5.33	0.68	0.0001
Arterial plasma P, mmol/l	1.15	1.85	0.10	0.0014
[4]Net portal flux of P, mmol/h	76.8	88.2	6.39	0.20
[5]Recycling of P, mmol/h	40.0	38.5	5.66	0.86

[1] Treatment LP contains 2.4 g P/kg DM.
[2] Treatment HP contains 3.4 g P/kg DM.
[3] P balance is calculated from the assumption that P content in milk is 1 g P/kg milk (Sehested, 2004).
[4] Net portal flux of P is calculated as follows: portal blood flow x (portal conc. – arterial conc.).
[5] Recycling of P is calculated as net portal flux – (feed – fecal P difference).

Conclusion

The present experiment shows that recycling of P was not affected by a reduced dietary P intake in lactating dairy cows. Cows were unable to fully compensate for a reduced dietary P intake by increasing intestinal reabsorption of P and reduced P intake forced cows into negative P balance.

References

Kristensen, N.B., A. Storm, B.M.L. Raun, B.A. Røjen and D.L. Harmon, 2007. Metabolism of silage alcohols in lactating dairy cows. J. Dairy Sci. 90: 1364-1377.

Scott, D., F.G. Whitelaw, W. Buchan and L.A. Bruce, 1985. The effect of variation in phosphorus intake on salivary phosphorus excretion in sheep. J. Agric. Sci. 105: 271-277.

Sehested, J., 2004. Feeding phosphorus (P) to cattle: A new basis for Nordic recommendations. Acta Agric. Scan. Sec. A-Animal Sci. 54: 169-180.

Regulation of dairy cattle adipose tissue metabolism by adrenergic control systems and gene transcription mechanisms dictating increased overall efficiency

J.M. Sumner, C. Schachtschneider, A. Hutjens, A. Youngquist, G. Duncan, S. Rocco, J. Miller, J.L. Vierck and J.P. McNamara
Washington State University, Pullman WA, USA; mcnamara@wsu.edu

Introduction

Metabolic adaptations in adipose tissue are a critical part of establishment and maintenance of lactation and lactational efficiency. Adipose tissue stores and releases energy and secretes metabolic regulators and cytokines. Previous work (McNamara, 2004; Sumner and McNamara, 2007) suggested that lipogenesis and lipolysis are controlled differently by gene transcription for enzyme synthesis or post-translation responses. Our objective was to obtain a more quantitative understanding of the regulation of gene transcription and energy flux underlying the adipose response to lactation.

Methods

Two related studies were conducted, for the first, 20 nulliparous Holstein heifers had biopsies from subcutaneous adipose tissue taken at 30 days *pre partum* and 14 days of lactation. All animals were on the same diet, fed to requirements. Diet and milk composition, were determined by AOAC methods. Changes in body fat and protein were determined from body weight, fat cell size and condition score by validated published equations (Waltner *et al.*, 1994). Adipose tissue was extracted for RNA by the RNA-easy mini-kit (Qiagen 75842), and a total of 12 pairs of samples (30 days *pre partum* and 14 days of lactation, 2 samples each from 12 animals; 10 ug of RNA per sample) were hybridised to the Affymetrix Genechip® Bovine Genome Array. The array data were normalised to a signal of 125 using the Affymetrix GCOS analyser, and the relative expression signal data were analysed by ANOVA using both GeneSpring and SAS in a one-way analysis to measure the change from *pre partum* to *post partum*.
In the second study, 48 Holsteins were blocked by parity and were either from High Genetic (HG, Predicted transmitting ability for milk (PTAM)=870 kg), or Low Genetic (LG, PTAM = 378) sires; and half of each genetic group was fed either to requirements (HE) or to 90% of energy requirements (LE), other components fed to requirements. Feed intake (daily), milk (daily) and component (monthly) production, body weight and body condition score (weekly) were measured. Biopsies of subcutaneous adipose tissue were taken from 32 animals (16 each diet group; 8 each Genetic/Diet group) at –21, -7, 7, and 28 days around parturition. Lipogenesis was measured by incorporation of ^{14}C-Acetate into fatty acids in tissue slices for 2 hours *in vitro*. Lipolysis was measured by release of glycerol from adipose tissue slices for 2 hours *in vitro*.

Results

In Study 1, milk yield was 29.8 (SEM=1.3 kg/d for the first 60 DIM (range 18.6 to 44.8 kg/d)), animals lost 42.6 kg of BW (SEM 8.4, range +9.1 to -113.6) and 0.38 BCS units (SEM 0.10, range 0 to -1.0) from 0 to 14 DIM. Using the relative expression signal strength from the Affymetrix Bovine Gene Array (normalised to 125 on the GCOS system), Gene expression for enzymes controlling anabolic pathway decreased from 30 days *pre partum* to 14 days of lactation (P<0.05), including (% change, SEM): SREBP, -25.1%, (6.2); GLUT1, - 57.3% (14.1); THRSP14, -30.8%

(7.4); LPL, -48.4% (7.7) and AcCoA Carboxylase, -60.6% (13.0). The regression (R^2) of transcript change on milk production was 0.18 for AcCoA carb and 0.26 for ATP-CL (P<0.05). Genes that code for lipolytic control elements increased, with much variation among animals, including Ca channel subunit 338% (203); beta-2 adrenergic receptor 52.0% (8.8); PKC receptor 10.1% (2.6) and HSL mRNA 23.0% (17.9). The regression of transcript change on milk yield was 0.30 and 0.25 for B2AR and HSL mRNA.

In Study 2, feed intake in the dry period (-21 to -1 d *pre partum*) was 13.6 and 12.7 kg DMI/d for HE and LE (SE=1.5), during lactation it was 21.2 and 17.4 kg/d (SE=1.4) for the first 56 DIM. Feed intake was 36.1 and 33.3 kg/d for HG and LG for 27-56 DIM (P<0.05). Milk production was 28.6, 26.0, 38.1 and 38.0 (SE 1.2) kg/d for HG and LG in parity 1 and parity 2. For the LGHE, HGHE, LGLE and HGLE groups, milk production was 33.7, 32.8, 31.7, 31.5 kg/d, HE>LE, P<0.05). Although the HG animals produced slightly less milk than LG when well fed, the reduction due to LE was slightly less in the HG than LG group. Loss of BW, BCS and body fat were greater (P<0.05) on LE diet and in parity 2. Adipose tissue lipogenesis decreased (P<0.05) from 7 d *pre partum* to 28 d *post partum* (day -7: Lipogenesis (nm/g/2h=846 ln [Ac], mM -62; Day 28 lipogenesis=199 ln [Ac] -167. At d 28, lipogenesis was lower (P< 0.05) in animals fed LE (LE lipog=686 ln [Ac] -85; E lipog=314 ln[Ac] -34.

Conclusion

From these studies we have extended our quantitative description that reductions in adipose tissue lipogenesis during lactation are primarily due to a reduction in transcription of genes for the enzymes leading to a reduction in pathway flux. The control of lipolysis includes both increases in gene expression of beta-receptors and hormone-sensitive lipase and increases in activation (post-translational) of the enzyme by adrenergic stimulation. These data have identified the mechanisms regulating adipose tissue metabolic flux. They can be incorporated into models of metabolism to help identify patterns and mechanisms of metabolic regulation in the most efficient dairy animals.

References

McNamara, J.P., 2004. Research, improvement and application of mechanistic, biochemical, dynamic models of metabolism in lactating dairy cattle. Animal Feed Sci. Techn. 112: 155-176.

Sumner, J.S. and J.P. McNamara, 2007. Expression of key genes controlling lipolysis in adipose tissue of Holstein dairy cattle during the dry period and lactation. J. Dairy Sci. 90: 5237-5246.

Waltner, S.S., J.P. McNamara and J.K. Hillers. 1994. Validation of indirect measures of body fat in lactating cows. J. Dairy Sci. 77: 2570-2579.

Evaluation of response to insulin infusion in Holstein cattle undergoing an extended lactation

L.C. Marett[1], K.L. Macmillan[2], C. Grainger[3], C.V.C. Phyn[4], F.R. Dunshea[1] and B.J. Leury[1]
[1]Department of Agriculture and Food Systems, University of Melbourne, Australia; [2]Department of Veterinary Science, University of Melbourne, Australia; [3]Department of Primary Industries, Ellinbank, Australia; [4]Dairy NZ Ltd, Hamilton 3240, New Zealand; brianjl@unimelb.edu.au

Introduction

Recent research has shown variation in the ability of dairy cows to extend lactation (EL) beyond the traditional 300 days (Auldist *et al.*, 2007; Kolver *et al.*, 2006). Some cows may be more efficient at partitioning nutrients to milk production rather than body gain in a pasture-based system. Kolver *et al.* (2006) found that different combinations of diets affected a cow's ability to maintain milk production for extended periods. Their results suggest that lactation persistency is not only influenced by the rate of secretory cell death within the mammary gland (Capuco *et al.*, 2003), but also by nutritional or animal factors.
Whilst the biological basis for the variation in persistency of milk production is not fully known, it may be partly associated with variation in whole animal and/or tissue specific insulin sensitivity. Insulin is essential for the regulation of glucose and lipid metabolism and affects the uptake of nutrients into insulin dependant tissues. In this study we used the insulin tolerance test (ITT) to investigate changes in whole animal insulin sensitivity, estimated by changes in plasma concentrations of glucose and non-esterified fatty acids (NEFA) (Kaneko, 1997) at different stages of an EL.

Material and methods

Holstein-Friesian cows grazed pasture throughout a 670 day lactation following calving in Winter 2006. Cows were milked twice daily at 07:00 h and 16:00 h. At 100, 240, 420, and 580 days in milk (DIM), twelve cows were housed undercover and individually fed for a period of five weeks. Six cows were offered a ration of 160 MJ ME/cow/day of fresh cut pasture or pasture silage and the other six were offered 215 MJ ME/cow/day comprising pasture/silage plus 5 kg grain. During the fifth week, cows were fitted with jugular catheters and housed in metabolism stalls. Feed was removed at 20:00 h, prior to the ITT at 08:00 h the following morning. Basal blood samples were collected at -20,-10 and -1 minutes prior to intravenous injection of 0.12 U/kg liveweight insulin (Actrapid, Novo Nordisk). Blood samples were then taken every 5 minutes for 2 hours and then at 150, 180, 210, 240, 270, and 300 min post-insulin infusion. Plasma was collected and stored at -20 °C until analysis for glucose and NEFA. The trapezoidal method was used to calculate the area under the curve (AUC). Fractional rate constants (K) were calculated for the decrease in plasma glucose and NEFA for the interval 5-30 minutes post insulin. Data were analysed for differences in response measures across the lactation and between diets using Genstat REML. Basal and recovery values are the average of the first and final three samples taken, respectively.

Results

Milk yields followed the typical bi-peak extended lactation curve. There was no significant diet effect on yield. Average daily yield (pooled) was 23.80, 13.59, 18.30, and 11.06 l at day 100, 240, 420 and 580, respectively. Basal plasma glucose was not affected by stage of lactation (SOL). Rate of glucose clearance ($K_{Glucose}$) 5-30 min post infusion was significantly reduced at 240 DIM compared with 100, 420 and 580 DIM, and was numerically fastest at 420 DIM, but there was no effect of SOL on Glucose AUC_{0-30}. Recovery plasma glucose was significantly higher at 100 DIM compared with 240, 420 and 580 DIM.

Table 1. Plasma glucose and NEFA responses to infusion of insulin across an extended lactation. Data pooled across diets.

	Days in milk (DIM)				sed[1]	*P*-value
	100	240	420	580		
Basal glucose (mM)	3.63	3.66	3.54	3.41	0.15	0.29
$K_{Glucose}$ 5-30 (%/min)	-2.12	-1.59	-2.52	-2.33	0.23	0.002
Glucose AUC_{0-30} (mM/min)	-29.2	-30.40	-24.82	-27.77	4.24	0.611
Recovery glucose (mM)	4.40	3.90	3.88	3.83	0.16	0.009
Basal NEFA (μM)	520.0	152.0	252.4	133.3	39.7	<0.001
K_{NEFA} 5-30 (%/min)	-4.2	-2.2	-3.2	-1.8	0.4	<0.001
NEFA AUC_{0-30} (mM/min)	-5943	-693	-712	-395	696	<0.001
Recovery NEFA (μM)	643.5	208.9	411.2	219.6	51.4	<0.001

[1] Standard error of the difference for DIM.

Basal plasma NEFA concentration was the highest at 100, intermediate for 420 and lowest at 240 and 580 DIM. The variation in basal plasma NEFA corresponded to a similar pattern in the variation in milk yield throughout the EL, suggesting an increase in adipose tissue mobilisation at these times. Rate of NEFA disappearance (K_{NEFA} 5-30) was higher, and AUC_{0-30} greater, at 100 and 420 DIM compared with 240 and 580 DIM suggesting that cows may be more sensitive to the anti-lipolytic actions of insulin at these stages. Recovery plasma NEFA was significantly affected by SOL and reflected the variation in basal plasma NEFA.

Conclusion

Insulin sensitivity, estimated by an ITT, in Holstein-Friesian cows managed in a pasture-based system changes throughout an extended lactation. The highest milk yields (100 and 420 DIM) were associated with enhanced insulin sensitivity as indicated by increased glucose and NEFA clearance following exogenous insulin administration. Further studies are required to investigate the role of seasonality and/or nutrition in this response.

Acknowledgement

The authors thank P. Gore, P. Cakebread and K. Watkins for expert technical assistance.

References

Auldist, M.J., G.N. O'Brien, D.J. Cole, K.L. Macmillan and C. Grainger, 2007. Effects of varying lactation length on milk production capacity of cows in pasture-based dairying systems. J. Dairy Sci. 90: 3234-3241.

Capuco, A.V., S.E. Ellis, S.A. Hale, E. Long, R.A. Erdman, X. Zhao and M.J. Paape, 2003. Lactation persistency: Insights from mammary cell proliferation studies. J. Anim. Sci. 81: 18-31.

Kaneko, J.J., 1997. Carbohydrate metabolism and its diseases. In: J.J. Kaneko (Ed.) Clinical Biochemistry of Domestic Animals. Academic Press: San Diego, CA, USA, pp. 45-81.

Kolver, E.S., J.R. Roche, C.R. Burke and P.W. Aspin, 2006. Effect of genotype and diet on milksolids production, body condition, and reproduction of cows milked continuously for 600 days. Proc. N. Z. Soc. Anim. Prod. 66: 245-251.

Nutritive value and silage characteristics of partly stoned olive cakes treated with molasses

M.J. Abarghuei[1], Y. Rouzbehan[1] and D. Alipour[2]
[1]Animal Science Department, Faculty of Agriculture, Tarbiat Modares University, Tehran, P.O. Box14115-336, Iran; [2]Animal Science Department, Faculty of Agriculture, Bu-Ali Sina University, Hamadan, Iran; rozbeh_y@modares.ac.ir

Introduction

Each ton of olives produces 800 kg of olive cake (OC) (Alburquerque *et al.*, 2004). Efficient utilisation of the OC in ruminant feeding requires appropriate preservation and upgrading of its nutritional value (Hadjipanayiotou, 1994). Storage of OC is difficult due to its high moisture and fat content. Ensiling is a suggested technique for long-term preservation of OC alone or together with other by-products such as wheat bran or molasses (Hadjipanayiotou, 1994). Little information is available on the nutritive value of these olive by-products as ruminant feeds in Iran. This experiment was conducted to evaluate nutritive value and silage quality of Shirazi partly stoned OC with and without molasses.

Material and methods

The samples of partly stoned OC were prepared from the Golden olive factory in Shiraz (Fars Province). Molasses was added or not up to 5% of wet weight to the fresh samples. Samples were ensiled in 3-kg bags, and were opened after 80 days of fermentation. Dry Matter (DM), ash, crude protein (CP), neutral detergent fibre (NDF), acid detergent fibre (ADF) and acid detergent lignin (ADL) were determined according to AOAC (1990). Water soluble carbohydrates (WSC) were measured using the anthrone method (MAFF, 1982). Ammonia in silages was determined using phenol-hypochlorite (Broderick and Kang, 1980). Lactic acid and VFA were measured by high performance liquid chromatography (HPLC). Gas production parameters from fermentation samples were measured as described by Menke and Steingass (1988). Differences between values of ammonia-N, VFA and lactic acid were examined with the *t*-test. Other data were subjected to analysis using the GLM procedure of SAS®, based on the statistical model: $Y_{ijk}=\mu + A_i + B_j + AB_{ij} + e_{ijk}$, where Y_{ijk} is the measured parameter, μ the general mean, A_i the effect of molasses level, B_j the effect of ensilage treatment, AB_{ij} is the interaction between factors and e_{ijk} the error term.

Results

NDF, ADF and ADL content of ensiled OC was higher than that of fresh OC (Table 1) and so caused a decrease in its nutritive value. Ensiling increased cell wall constituents (NDF and ADF), but decreased DM and CP ($P<0.05$). The WSC content significantly decreased after ensiling ($P<0.05$). Ensiling caused a significant decrease in pH of samples ($P<0.05$; Table 2). Addition of molasses had no significant effect on the pH of fresh OC ($P>0.05$), but decreased pH of ensiled samples ($P<0.05$). OMD and ME were low and were decreased during ensiling ($P<0.05$). The c value was increased by ensiling ($P<0.05$). Addition of molasses into the fresh and ensiled samples resulted in an increase in DM and WSC contents, and a decrease in OM, NDF, ADF and lignin contents ($P<0.05$). Adding molasses increased lactic acid and propionic acid, but decreased ammonia-N, acetic and butyric acids ($P<0.05$). OMD, ME and c increased with the addition of molasses ($P<0.05$).

Table 1. Chemical composition and gas production parameters, before and after ensiling (g/kg DM).

	M[1]	DM[2]	OM[3]	CP[4]	NDF[5]	ADF[6]	ADL[7]	IFF[8]	Rate[9]	OMD[10]	ME[11]
Fresh	0	428[b]	964[a]	86.7[a]	648[b]	468[b]	236[a]	21.1[b]	0.054[b]	375[b]	5.7[b]
	50	445[a]	957[b]	82.7[a]	582[c]	425[c]	202[c]	23.2[a]	0.053[b]	398[a]	6.1[a]
Ensiled	0	389[d]	966[a]	76.1[b]	733[a]	532[a]	241[a]	21.1[b]	0.063[a]	356[d]	5.4[d]
	50	403[c]	942[c]	85.3[a]	650[b]	472[b]	225[b]	23.2[a]	0.046[c]	364[c]	5.6[c]
SEM[12]		0.413	2.34	0.98	2.12	1.85	1.32	0.02	0.002	0.06	0.009
P-value		<0.01	<0.05	<0.05	<0.01	<0.01	<0.01	<0.05	<0.05	<0.01	<0.01

[1] M, molasses level (g/kg fresh weight); [2] DM, dry matter; [3] OM, organic matter; [4] CP, crude protein; [5] NDF, neutral detergent fibre; [6] ADF, acid detergent fibre; [7]ADL, acid detergent lignin; [8] IFF, insoluble but fermentable fraction (ml); [9] Rate, rate constant of gas production during incubation (ml/h); [10] OMD, organic matter digestibility (g/kg DM); [11] ME, metabolisable energy (MJ/kg DM); [12] SEM: standard error of the means; [a,b,c,d] Means in the same column with different superscripts differ ($P<0.05$).

Table 2. Fermentation characteristics of OC, before and after ensiling.

	M[1]	pH	WSC	NH_3	Acetate	Butyrate	Propionate	Lactate
fresh	0	5.3[a]	36.3[b]	-	-	-	-	-
	50	5.2[a]	69.9[a]	-	-	-	-	-
ensiled	0	4.9[a]	12.6[d]	46.0[a]	26.9[a]	2.7[a]	0.30[b]	16.2[b]
	50	4.4[b]	23.5[c]	35.8[c]	6.9[b]	2.3[b]	0.34[a]	27.5[a]
SEM[2]		0.051	0.016	0.005	0.031	0.003	0.003	0.121
P-value		<0.05	<0.01	<0.01	<0.01	<0.5	<0.05	<0.01

[1] M, molasses level (g/kg fresh weight); WSC, acetate, butyrate, propionate and lactate (g/kg DM); NH_3-N (g/kg N); SEM: standard error of the means.
[a,b,c,d] Means in the same column with different superscripts differ ($P<0.05$).

Conclusion

It can be concluded that this by-product has a potential value as animal feedstuff. Also, ensiling can be used to preserve these by-products for long time. The addition of molasses improved the quality of ensiled OC.

References

Alburquerque, J.A., J. Gonzalvez, D. Garciaand and J. Cegarra, 2004. Agrochemical characterization of 'alpurejo', a solid by-product of two-phase centrifugation method for olive oil extraction. Bioresource Technol. 91: 195-200.

AOAC, 1990. Offical methods of analysis.15th edition. Association of Official Analytical Chemists.Washington, D.C., USA.

Broderick, G.A. and J.H. Kang, 1980. Automated simultaneous determination of ammonia and total amino acids in ruminal fluid and *in vitro* media. J. Dairy Sci. 63: 64-75.

Hadjipanayiotou, M., 1994. Laboratory evaluation of ensiled olive cake, tomato pulp and poultry litter. Livest. Res. Rural Develop. 6: #12. Retrieved from http://www.lrrd.org/lrrd6/2/cyprus1.htm.

MAFF (Ministry of Agriculture, Fisheries and Food), 1982. The Analysis of Agricultural Materials, 2nd ed. MAFF, London, UK.

Menke, K.H and Y.H. Steingass, 1988. Estimation of energetic feed value obtained from chemical analysis and *in vitro* gas production using rumen fluid. Anim. Res. Dev. 28: 7-55.

Expression of adipogenic genes in *longissimus* muscle and different adipose tissues of cattle representing either the accretion or the secretion type

E. Albrecht[1], J.X. Xu[2], T. Viergutz[1], G. Nürnberg[1], R.Q. Zhao[2] and J. Wegner[1]
[1]Research Institute for the Biology of Farm Animals, Dummerstorf, Germany; [2]Nanjing Agricultural University, Nanjing, PR China; elke.albrecht@fbn-dummerstorf.de

Introduction

Cattle of the secretion type metabolise their energy intake predominantly into milk and store excess energy in times of positive energy balance in the form of fat in adipose tissues, which is mobilised at times of negative energy balance. Cattle of the accretion type metabolise their energy intake into muscle, resulting in a lower body fat content. Holstein and Charolais cattle were used as representatives of the respective breed types, having a similar body frame size and economical importance (Kühn *et al.*, 2002). The study focusses on fat distribution among different tissue types and mRNA abundance of adipogenic genes, namely perilipin, CCAAT/enhancer binding protein (C/EBP) α and C/EBPβ. Perilipin is a protein that coats lipid droplets in adipocytes and is an important regulator of lipid storage (Souza *et al.*, 2002). The transcription factors C/EBPα and C/EBPβ are known to directly influence adipocyte development and regulate PPARγ expression, which is responsible for expression of adipose genes like adiponectin and perilipin (Park *et al.*, 2004). The objective of the study was to find tissue specific and/or breed specific differences in the expression of these genes.

Material and methods

Holstein and Charolais calves, 18 of each breed, were raised on a milk replacer diet until 120 d of age. From 5 mo to slaughter at 18 mo of age, animals were tethered on individual feeding places. The diet was offered *ad libitum* and was composed of a concentrate and chaffed hay in the proportion of 3:1 (energy content: 11.7 MJ ME/kg dry matter of feed).

Longissimus muscle (LM) tissue and omental, perirenal and subcutaneous adipose tissues were collected within 30 min after slaughter and immediately frozen in liquid nitrogen. The weights of different fat depots were determined. The intramuscular fat (IMF) content of LM samples was obtained in triplicate via the Soxhlet-extraction method with petroleum ether. Total cellular RNA from muscle was extracted using TRIzol reagent (Invitrogen); total RNA from adipose tissues was extracted using the RNeasy Lipid Tissue Kit (Qiagen), according to the manufacturer's instructions. The iScriptTM cDNA Synthesis Kit (BioRad) was used to synthesise cDNA from 100 ng of total RNA. The abundance of RPS18, perilipin, C/EBPα and C/EBPβ mRNA was quantified by real-time RT-PCR (iCycler, BioRad) and calculated as pg/μg total RNA, using the known concentration of a standard and amplification efficiency.

Statistical analysis was performed using the SAS® statistical software (Version 9.1). Data were analysed by ANOVA using the MIXED procedure, with fixed factors breed and tissue, and their interaction. Data were normalised introducing RPS18 mRNA abundance as the covariable in the model. The t-test was used as the *post hoc* test.

Results

Holstein and Charolais bulls, reared under the same conditions, reached different body weights and carcass compositions at 18 mo of age (Table 1). As expected, the body weight was greater (P<0.01) in Charolais bulls and Holstein bulls stored more fat (P<0.01) in different depots, also in the LM (P=0.02). However, the proportion of subcutaneous fat was not different (P=0.51) between the breeds. Variance analysis of perilipin, C/EBPα and C/EBPβ mRNA abundance revealed a significant influence of tissue (P<0.01), but no influence of breed (P≥0.71) and breed x tissue interaction

($P \geq 0.30$). Comparing the breeds, only the C/EBPβ mRNA level in perirenal fat tended to be greater ($P=0.09$) for Charolais bulls (Figure 1). Comparing the tissues, the omental fat had greater ($P<0.05$) mRNA levels of perilipin, C/EBPα and C/EBPβ than the LM and subcutaneous fat in both breeds. The mRNA levels of omental fat were greater or in tendency greater compared to perirenal fat, except for C/EBPα in Holstein bulls.

Table 1. Body weight and composition of Holstein and Charolais bulls at 18 months of age.

Trait	Holstein, LSMean	Charolais, LSMean	SE	*P*-value
Body weight, kg	664	747	11	<0.01
Body fat content,%	26.4	19.2	0.9	<0.01
Omental fat,%BW	3.06[a]	1.80[a]	0.12	<0.01
Perirenal fat,%BW	2.87[a]	1.87[a]	0.18	<0.01
Subcutaneous fat,%BW	2.92[a]	2.80[b]	0.18	0.51
IMF content of LM,%	4.06	2.62	0.40	0.02

[a,b] Means within a breed with same superscript letters are not significantly different between fat depots ($P>0.05$).

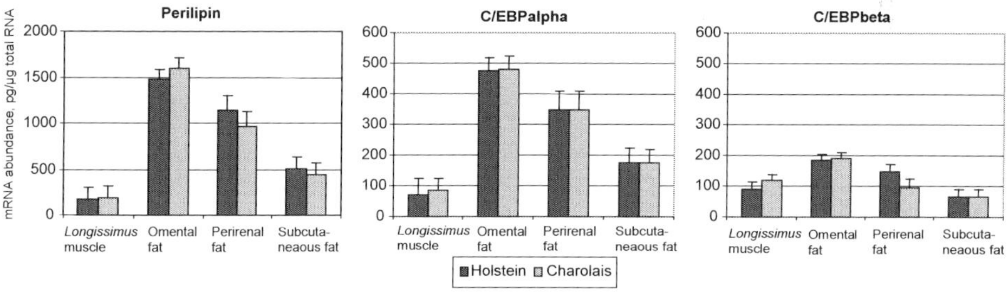

Figure 1. Abundance of perilipin, C/EBPα and C/EBPβ mRNA in muscle and adipose tissues of Holstein and Charolais bulls at 18 months of age.

Conclusion

The regulation of the investigated genes was obviously more depot- than breed-specific. Differences between these cattle breeds in fat deposition were not accompanied by differences in mRNA abundance of perilipin, C/EBPα and C/EBPβ. The greater transcriptional activity in omental tissue was not associated with greater fat deposition. The results suggest that adipocytes are still differentiating and maturing in all adipose tissues of cattle at 18 mo of age.

References

Kühn, C., O. Bellmann, J. Voigt, J. Wegner, V. Guiard and K. Ender, 2002. Experimental approaches for studying genetic and physiological background of nutrient transformation in cattle with respect to secretion and accretion type. Arch. Tierz. 45: 317-330.

Park, B.H., L. Qiang and S.R. Farmer, 2004. Phosphorylation of C/EBPβ at a consensus extracellular signal-regulated kinase/glycogen synthase kinase 3 site is required for the induction of adiponectin gene expression. Mol. Cell. Biol. 24: 8671-8680.

Souza, S.C., K.V. Muliro, L. Liscum, P. Lien, M.T. Yamamoto, J.E. Schaffer, G.E. Dallal, X. Wang, F.B. Kraemer, M. Obin and A.S. Greenberg, 2002. Modulation of hormone-sensitive lipase and protein kinase A-mediated lipolysis by perilipin A in an adenoviral reconstituted system. J. Biol. Chem. 277: 8267-8272.

Influence of lipid sources on the fatty acid composition of *longissimus* muscle of heifers finished in a feedlot[1]

T.T. Berchielli[2,3], G. Fiorentini[2,3] and R.A. Reis[2]
[1]Research funded by FAPESP, São Paulo, Brazil; [2]Faculdade de Ciências Agrárias e Veterinárias/ UNESP, Jaboticabal, São Paulo, Brazil; [3]Fundação de Amparo à Pesquisa do Estado de São Paulo–FAPESP, Brazil; ttberchi@fcav.unesp.br

Introduction

Nowadays, there is concern with the alimentary health of the human being, not only on the sanitary quality of foods, but mainly, in relation to the possible effect (deleterious or beneficial) of certain foods or nutrients on the health of consumers.
Fatty acids directly affect the quality of meat (Wood *et al.,* 2002). The variation on fatty acid composition has an important effect on the firmness and softness of meat fat, especially the subcutaneous and intramuscular fat. According to these authors, the presence of unsaturated fatty acids, especially those with more than two double bonds, has a great importance on the rapid oxidation of meat, decreasing shelf time (rancification and deterioration of colour). In addition, this propensity to oxidation is important in the development of flavor during cooking.
The objective of this study was to evaluate the effect of different lipid sources on the composition of fatty acids in the *longissimus* muscle of heifers finished in a feedlot system.

Material and methods

Twenty-one crossbred heifers (Nellore × Santa Gertrudis × Braunvieh) with 300 kg of body weight were fed 60% of roughage basis of corn silage and 40% concentrate, with 5.8% of ether extract level. The fat sources were the following: soybean grain, protected fat (Megalac-E®) and soybean oil. After an adaptation period to the experimental facilities, management and diets, the animals were confined for 68 d.
The samples of meat were obtained at slaughter of the animals. Approximately 2.5 cm of thickness was obtained from a perpendicular cut in the *longissimus* muscle, at the 12th rib, used for subsequent extraction of lipids, transesterification and methylation of fatty acids in the *longissimus* muscle. The extraction and evaluation of the total lipids of samples were performed according to the modified method of Hara and Radin (1978), who used about 3.0 g sample of *longissimus* muscle for the extraction of fat, and hexane/isopropanol 3:2 (v/v).
The methyl esters of fatty acids were separated on a capillary column of 100 m of fused silica (SP-2560) with hydrogen as a carrier gas (1.8 ml/min) and flame-ionisation detector (FID).
The fatty acid composition of each fat source evaluated is presented in Table 1.

Table 1. Fatty acid composition (%) of each fat supplement.

Fatty acid[1]	Supplement fat		
	SG	ML	OS
C18:0	10.72	5.29	7.50
C18:1 c9	27.78	21.29	28.48
18:2 c9 c12	31.04	42.82	23.76
18:3 n3	1.20	2.94	0.76

[1] % of total fatty acids; [2] SG = Soybean grain, ML = Protected fat (Megalac-E), OS= Soybean oil.

The experiment was conducted in a completely randomised design with three treatments and seven replicates. Data were analysed by the Proc GLM of SAS® program (2004) and the averages were compared by the Tukey test 5%.

Results

The results of the meat composition fatty acids on sample steaks of crossbred fed heifers with different lipid sources are presented in Table 2.

Concentration of c9 c12 of 18:2 (linoleic) and 18:3 n3 (linolenic) were different in the *longissimus* muscle ($P<0.05$). The other fatty acids were not affected ($P>0.05$) by lipid sources. Animals fed with soybean grain had higher concentrations of fatty acids 18:2 c9 c12 and 18:3 n3 in the *longissimus* muscle than those fed with protected fat, and the diet with soybean oil did not differ from the other treatments. There was no difference in the proportion of other fatty acids in samples of muscle.

Table 2. Fatty acid composition (%) of longissimus *muscle samples of heifers finished in a feedlot fed with different lipid sources.*

	Diets[2]			P[3]	CV[4]
Fatty acid[1]	SG	ML	OS		
C16: 0	25.58	26.39	25.42	0.38	5.20
C17: 0	0.87	0.88	0.88	0.98	12.04
C18: 0	14.63	15.40	13.68	0.24	12.53
C18: 1 c9	37.78	38.70	38.82	6.68	0.74
18:2 c9 c12	3.84[a]	2.33[b]	2.90[ab]	0.04	33.18
18:3 n3	0.21[a]	0.12[b]	0.17[ab]	0.01	26.89
18:2 c9 t 11	0.49	0.50	0.56	0.73	34.59

[1] % of total fatty acids;

[2] SG = Soybean grain; ML = Protected fat (Megalac-E); OS= Soybean oil. Means, within a row, followed by different superscript letters are different ($P<0.05$) by Tukey test.

[3] $P< 0.05$.

[4] Coefficients of variation.

Conclusion

The use of soybean grain in the diet increased the concentration of 18:2 c9 c12 and 18:3 n3 in meat, and can improve beef cattle quality, making it a food containing an appropriate fat profile which can contribute to prevention and development of some diseases.

References

Hara, A. and N.S. Radin, 1978. Lipid extraction of tissues with a low-toxicity solvent. Anal. Biochem. 90: 420-426.

Wood, J.D., R.J. Richardson, G.R. Nute, A.V. Fisher, M.M. Campo, E. Kasapidou, P. R. Sheard and M. Enser, 2002. Effects of fatty acids on meat quality: a review. Meat Science 66: 21-32.

The effects of condensed tannins in *Lotus corniculatus* on valine kinetics in the mammary gland of the ewe

E.N. Bermingham, W.C. McNabb, B.R. Sinclair, M. Tavendale and N.C. Roy
Food, Metabolism & Microbiology, AgResearch Limited, Grasslands Research Centre, Tennent Drive, Private Bag 11008, 5301, Palmerston North, New Zealand; nicole.roy@agresearch.co.nz

Introduction

Condensed tannins (CT) in the diet can increase the apparent absorption of amino acids (AA), in particular the branched-chained amino acids (BCAA) from the small intestine (Bermingham *et al.*, 2001). This may be responsible for the increased lactational performance observed in ruminants fed fresh CT containing forages (Woodward *et al.*, 1999). These results suggest that these legumes may have a potential as improved forages for dairy ruminants. However, it is unclear as to what are the consequences of changes in absorption of AA from the small intestine on the flux and use of BCAA to the mammary gland.
Our hypothesis was that CT in *Lotus corniculatus* increase the partitioning of BCAA to the mammary gland and is used for protein synthesis in late lactating ewes. The effects of CT on BCAA flux across the mammary gland were elucidated by comparing animals fed Sulla and orally supplemented with polyethylene glycol (PEG; CT inactive) with animals that had not received PEG (Control; CT active). PEG preferentially binds and inactivates CT.

Material and methods

Ten days post-lambing 12 ewes were prepared with an indwelling catheter in the mesenteric artery and a permanent cannula in the abomasum. A transonic flow probe was fitted around the pudic artery to measure mammary blood flow. Three days prior to the start of the sampling periods, a temporary catheter was inserted into the caudal superficial epigastric (mammary) vein for blood sampling. All ewes were offered fresh *L. corniculatus* (2,000 g DM/d; 68 g CT/d) from 25 days after the start of lactation. Half the ewes were orally drenched (4 times per d) with polyethylene glycol (PEG; 160 g/d in water) to remove the effects of the CT and the remaining ewes (Control) received a drench of water (day 0 of the experimental period). The treatments were applied according to a completely randomised block design with the block representing the week the ewes underwent surgery (4 ewes per week – 2 ewes per treatment). On day 46 of lactation, a 12 h continuous infusion of [1-^{13}C]-valine into the abomasum was used to quantify the kinetics of valine across the mammary gland. Blood was obtained by 3 consecutive 2 h continuous blood samples (from 6 to 12 h) from the mesenteric artery and mammary vein. Mammary blood flow was measured continuously during this period. Blood samples were centrifuged and the plasma was removed and stored at 85 °C until assayed for valine concentration and isotopic enrichment and activity. Isotopic enrichment (IE) of valine in the plasma and mammary intracellular pool and mammary valine irreversible loss rate (ILR; Biolo *et al.*, 1994) were calculated. Data were analysed using the GLM procedure of Minitab using a block design. Covariates (DM intake and milk yield) were used as co-variates but were not significant, therefore being removed from the final analysis.

Results

The CT in the diet did not affect (P=0.33) feed intake (Control: 1779 vs. PEG: 1919 (SED 90) g DM/d) but tended (P=0.06) to decrease plasma flow to the mammary gland (Control: 126 vs. PEG: 203 (SED 24) ml/min). Daily milk yield was similar (P=0.98) between treatments (ca. 890 g/d). Valine kinetics are shown in Figure 1, and only estimates of mammary valine ILR (Fom; P=0.04) and valine released from protein degradation (Fmo; P=0.08) were decreased by the presence of CT in the diet.

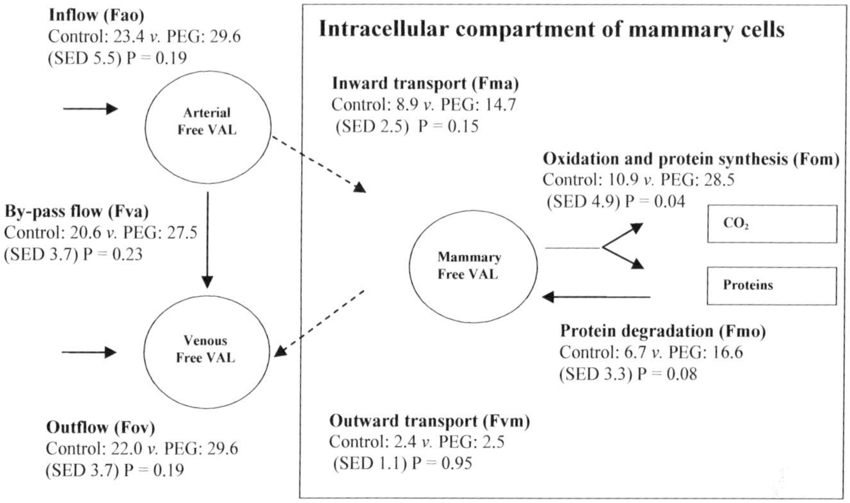

Figure 1. Least square means and standard error of the difference (SED) of valine kinetics (µmol/min) across the mammary gland of lactating ewes fed fresh Lotus corniculatus *either orally supplemented with polyethylene glycol (PEG; CT inactive), or not supplemented with PEG (Control; CT active).*

Conclusion

Despite a decreased plasma flow to the mammary gland, the presence of CT in *L. corniculatus* did not affect daily milk yield. All the valine kinetics presented are within the range reported by Bequette *et al.* (2002). There was no effect of CT on the inflow of valine (Fao) to the mammary gland. The ILR of valine across the mammary gland was lower in the control group suggesting that protein synthesis (and/or oxidation) in this tissue was decreased by the presence of CT. However, protein degradation was also decreased resulting in similar transport of valine out of the mammary gland, indicating no overall change in valine uptake by the gland. These results contradict those reported for ewes fed fresh Sulla (80 g CT/d; Roy *et al.*, 2004) suggesting that not all CT act in a similar manner.

References

Bermingham, E.N., K.J. Hutchinson, D.K, Revell, I.M. Brookes and W.C. McNabb, 2001. The effect of condensed tannins in sainfoin (*Onobrychis viciifolia*) and sulla (*Hedysarum coronarium*) on the digestion of amino acids in sheep. Proc. NZ Soc. Anim. Prod. 61: 116-119.

Bequette, B.J., C.E. Kyle, L.A. Crompton, S.E. Anderson and M.D. Hanigan, 2002. Protein metabolism in lactating goats subjected to the insulin clamp. J. Dairy Sci. 85: 1546-55.

Biolo, G., A. Gastaldelli, X.J. Zhang and R.R. Wolfe, 1994. Protein synthesis and breakdown in skin and muscle: a leg model of amino acid kinetics. Am. J. Phys. 267: E467-E474.

Roy N.C., B.R. Sinclair, B. Treloar, A. Kirk, J. Peters, M. Tavendale and W.C. McNabb, 2004. The effects of condensed tannins in sulla (*Hedysarum coronarium*) on valine kinetics in the ovine mammary gland. Anim. Prod. Aust. 25: 148-151.

Woodward, S.L., M.J. Auldist, P.J. Laboyrie and E.B.L. Jansen, 1999. Effect of *Lotus corniculatus* and condensed tannins on milk yield and milk composition of dairy cows. Proc. NZ Soc. Anim. Prod. 59: 152-5.

Lipid supplements rich in n-3 polyunsaturated fatty acids deeply modify *trans* 18:1 isomers in the *longissimus thoracis* muscle of finishing cattle

E. Bispo Villar[1,2], A. Thomas[1], B. Lyan[3], D. Gruffat[1], D. Durand[1] and D. Bauchart[1]
[1]INRA, Herbivore Research Unit, 63122 St-Genès-Champanelle, France; [2]Centro de Investigacións Agrarias de Mabegondo, 15080 A Coruña, Spain; [3]INRA, Human Nutrition Unit, 63122 St-Genès-Champanelle, France; dominique.bauchart@clermont.inra.fr

Introduction

Trans fatty acids are notably formed during partial hydrogenation of polyunsaturated fatty acids (PUFA) provided by vegetable oils in the rumen and then deposited in tissues of ruminants especially in muscle fats. Among *trans* 18:1 isomers found in ruminant products, 18:1Δ11*trans* (vaccenic acid) is generally the most abundant (Wolff, 1995) and considered to be innocuous or even protective against cardiovascular disease (CHD) for consumers whereas 18:1Δ9 *trans* (elaidic acid) and 18:1Δ10 *trans* are detrimental to health by favouring atherosclerosis and CHD (Jacobsen *et al.*, 2006), but also inflammation, diabetes and infant development (Dalainas and Ioannou, 2008). The objective of the study was to determine the impact of lipid supplements rich in n-3 PUFA and/or rich in 18:1n-9*cis* on *trans* 18:1 isomers in the *longissimus thoracis* muscle of finishing bovine.

Material and methods

Animals and diets: The experiment was performed with 19 Normand culled cows (48-60 months old, mean live weight 642 kg) for a 100 d finishing period. Animals were assigned at random to three iso-energetic and iso-nitrogenous rations composed of straw (30% diet DM) and concentrate (70%). Animals were given the basal diet (diet C) or the same diet supplemented with extruded linseeds (diet L) or with a mixture of extruded rapeseeds (2/3) and linseeds (1/3) (diet RL). Lipid supplements amounted to 40 g lipid/kg diet DM for a mean DM intake of 10.5 kg/d. Animals were slaughtered at a mean live weight of 787 (SD 66) kg for the three diets with a mean body fat score of 3.53 and BWG of 1.56 kg/d for the 100 d finishing period. Samples (150 g) of *longissimus thoracis (*LT) muscle were collected 1d *post mortem,* mixed in N_2 liquid as a fine powder and stored at -20 °C until analysis.
Beef trans 18:1 analysis: Total beef lipids were extracted by chloroform /methanol 2/1 and their fatty acids (FA) extracted and transmethylated as FA methyl esters (FAME) with sodium methanolate and BF3-methanol. Total FA composition was determined by GLC in a CP Sil 88 glass capillary column (100 m × 0.25 mm, Varian). Total *trans* 18:1 were isolated from FAME by preparative reversed-phase HPLC using a series of two Kromasil KR100-5C18 inverse phase columns (5 μm, 250×10 mm) with acetonitrile as the eluting solvent (4 ml/min) (Juaneda, 2002) and detected at 206 nm. Specific distribution of *trans* 18:1 isomers, converted into dimethyl disulfide (DMDS) adducts, was achieved by GLC-MS (Figure 1) in the Agilent 7890A GC (HP5 MS, 30 m × 25 mm, carrier gas: He) linked to a mass spectrometer Agilent 5975E (ionising energy 70 eV), allowing the structural characterisation and quantification of individual 18:1 *trans* isomer (Figure 2).

Results and discussion

GLC analysis of beef FA showed that *trans* monounsaturated (MUFA) FA represented 2.5% of total FA in the control group (diet C). Lipid supplements increased *trans* MUFA by 33.7% in cows given the linseed diet (L diet) and by 105.5% in cows given the mixture linseed/rapeseed diet (diet LR) ($P<0.05$). Total *trans* 18:1 isomers were well separated from other FA (especially from *cis* 18:1 isomers) by preparative HPLC allowing the specific analysis of their isomers by GLC-MS (Figure

1). Mass spectrum analysis of each *trans* 18:1 isomer (Figure 2) determined their chemical structure and their relative importance in total *trans* 18:1 isomers. It showed that each isomer was eluted as a single peak, excepted for Δ6tr to Δ8tr, Δ13tr-Δ14 tr, and Δ15 tr (contaminated by 18:1Δ9*cis*). Analysis of specific distribution of *trans* 18:1 isomers in LT samples by GLC-MS showed that in the control diet, *trans* 18:1 isomers of LT muscle were dominated by the Δ10t (33.7%) and Δ11t (36.1%) forms that had respectively negative and positive health values for humans. The other isomers from Δ12t to Δ16t represented, each, less than 4.5% of total isomers and from Δ6t to Δ8t less than 2.0% in agreement with data previously reported in beef (Wolff, 1995). L diet rich in n-3 PUFA (31.6%) had improved the health values of *trans* 18:1 in meat by decreasing both Δ9t (-41.2%) and Δ10t (-53.7%) isomers to the benefit of Δ12t to Δ16t isomers. On the contrary, RL diet lower in n-3 PUFA (19.7%) and richer in 18:1 n-9*cis* (28.5% vs. 18.0% for C and L groups) altered the health value of LT muscle by both decreasing the level of Δ11t (-30.7%) and increasing that of Δ10t (+22.0%).

These results clearly showed the beneficial effect for consumers of lipid supplements rich in 18:3n-3 (linseed) on beef *trans* 18:1 isomers when given as a unique source of unsaturated FA supplement in a concentrate-based diet. On the contrary, they showed the negative effect on the nutritional and health value of beef *trans* isomers (favouring deposition of Δ9*trans* and Δ10 *trans*) of linseed when associated to rapeseed rich in 18:1n-9*cis*. Indeed, we hypothesised that the beneficial effect of L diet on beef *trans* 18:1 isomers should be associated to its high value of the 18:3 n-3/18:1 n-9*cis* ratio (1.77) compared to the negative effect of the mixture linseed + rapeseed in the RL diet having a low value of the 18:3 n-3/18:1 n-9*cis* ratio (0.69).

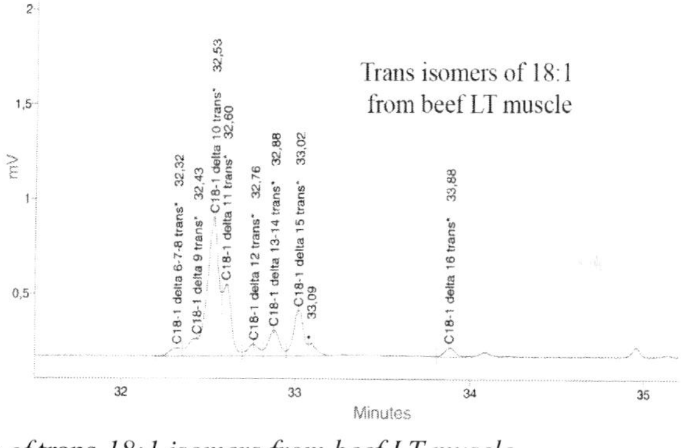

Figure 1. GLC profile of trans *18:1 isomers from beef LT muscle.*

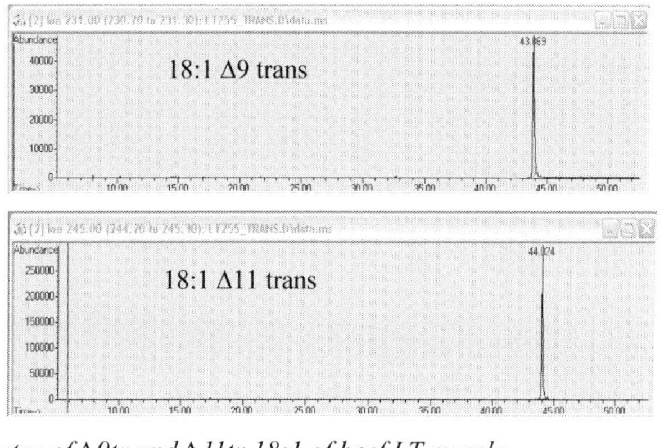

Figure 2. Mass spectra of Δ9tr and Δ11tr 18:1 of beef LT muscle.

References

Dalainas I. and H.P. Ioannou, 2008. The role of *trans* fatty acids in atherosclerosis, cardiovascular disease and infant development. Int. Angio. 27: 146-156.

Jacobsen, M.U., A. Bysted, N.L. Andersen, B.L. Heitmann, H.B. Hartkopp, T. Leth, K. Overvad and J. Dyerberg, 2006. Intake of ruminant *trans* fatty acids and risk of coronary heart disease. An overview. Atherosclerosis Supplements, 7: 9-11.

Juanéda, P., 2002. Utilisation of reversed-phase high-performance liquid chromatography as an alternative to silver-ion chromatography for the separation of *cis*- and *trans*-C18:1 fatty acid isomers. J. Chromatography A 954: 285-289.

Wolff, R.L., 1995. Content and distribution of *trans*-18:1 acids in ruminant milk and meat fats. Their importance in European diets and their effect on human milk. J. Am. Oil Chemist Soc. 72: 259-272.

Serum IGF-I concentration from birth to slaughter in calves under different management strategies analysed with a spline model

M. Blanco[1], I. Casasús[1] and D. Villaba[2]
[1]Unidad de Tecnología en Producción Animal, CITA, Avda, Montañana 930, 50059 Zaragoza;
[2]Departament de Producció Animal, Universitat de Lleida, Avda, Rovira Roure 191, 25198 Lleida, Spain; mblanco@aragon.es

Introduction

Early weaning could be useful to improve the match between herd requirements and forage resources in dry mountain areas, but it implies changes in calf nutrition since milk can be substituted by forage or concentrates. These changes have a strong influence on the expression/secretion of a variety of growth-related hormones, including Insulin-like growth factor-I (IGF-I), which plays an important physiological role in growth and development of mammals by acting in specific organs or systemically (Hossner *et al.*, 1997).
Longitudinal data obtained to describe IGF-I evolution from birth to slaughter involve a collection of data at different time points for several subjects. The statistical analysis is complicated because factors such as age and weight at the collection date interfere. Spline curves constructed from pieces of low degree random effect polynomials could be used for a more accurate analysis of data. In this paper an spline curve with random coefficients and heterogeneous residual variances was used to analyse IGF-I concentration patterns using repeated measurements on calves under different management strategies.

Material and methods

Thirty-nine Parda de Montaña calves were born in the autumn in La Garcipollera Research Station, in the central Pyrenees. At birth, calves were randomly assigned to early weaning (EW) at 90 d or traditional weaning (TW) at 150 d. During lactation, half of the calves of each age at weaning were offered a starter concentrate mix (11.55 MJ ME/kg DM, 17.5% CP) (S) while the remainders received no concentrate (NS).
During the nursing period, calves were allowed to suckle their dams 30 min twice a day. After weaning, calves were individually fed concentrates on an *ad libitum* basis. From weaning until they reached 350 kg BW, calves received a fattening concentrate (11.47 MJ ME/kg DM, 14.9% CP); thereafter and until slaughter at 450 kg (age 313±26 d), they received a finishing concentrate (11.65 MJ ME/kg DM, 13.7% CP).
Calves were weighed weekly and they were bled monthly to determine IGF-I concentration using a commercial EIA kit (OCTEIA® IGF-I, IDS, Boldon, UK).
Serum IGF-I concentration, live weight and concentrate intake were described as an spline curve that consists of two third-degree polynomials partially overlapped around weaning age. The model also included random regression coefficients and was implemented using the MIXED procedure of the SAS® System (Littell *et al.*, 1996). Age differences at samplings were considered. Average daily gain (ADG) at age day was calculated as the first derivative of body weight model. Comparisons between estimated least squares means were performed using a t-test with the Kenward-Roger adjusted degrees of freedom. Co-variance parameters were tested using a Wald test. Partial correlations were calculated using the animal plus residual deviations obtained from the models.

Results and discussion

The spline polynomial mixed model with random coefficients used in the current study allowed the modelling of IGF-I evolution over a long period with different age of animals at controls including age and days from a specific event (weaning). Actually, the model could be described as the simplest spline polynomial model with only one knot placed for each animal at the moment of maximum change in the growth, this is the weaning.

Pre-weaning concentrate feeding did not affect IGF-I concentration until 90 d (Figure 1) because concentrate intake was negligible during the first 2 mo. From 90 to 150 d, IGF-I concentration was the greatest in EWS, intermediate in EWNS and TWS and lowest in TWNS calves. These differences were related with the energy intake of the different management strategies (Elsasser *et al.*, 1989) because EWS and EWNS calves were fed concentrates, TWS calves were fed milk and concentrates and TWNS calves only milk. After weaning, EWS calves had greater IGF-I concentration than EWNS calves until 180 d (after 4 mo of concentrates) and TWS calves had greater IGF-I concentration than TWNS calves until 180 d (after only 2 mo). It took shorter for TWNS to reach the IGF-I concentrations of their TWS counterparts because they had greater concentrate intake, while both groups of EW calves had similar concentrate intake. IGF-I concentration increased until a maximum, which was attained at 150, 180, 210 and 240 d for EWS, TWS, EWNS and TWNS, respectively. Early weaning advanced the peak 30 d whereas pre-weaning concentrate feeding advanced the peak 60 d. Concentrate intake and ADG were positively correlated with the corresponding IGF-I concentration. However, no consistent correlation was found between IGF-I concentration at younger ages with post-weaning performance thus it could not be used as a predictor of future performance.

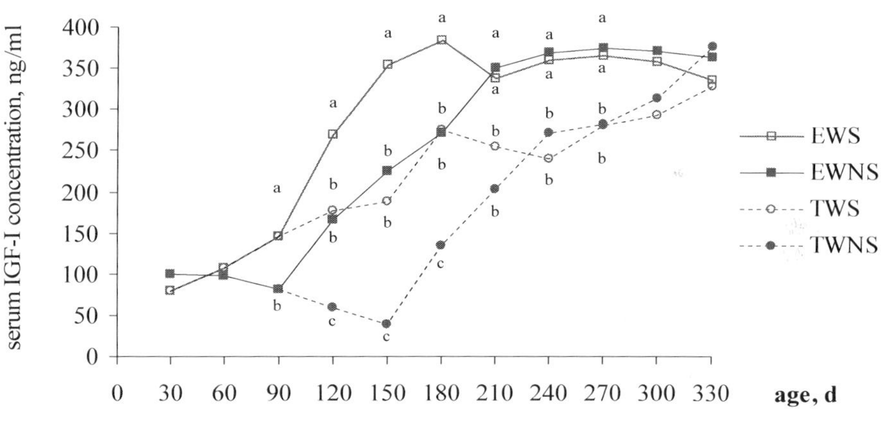

Figure 1. Evolution of IGF-I concentration according to management strategy (EW = early weaned; TW = traditionally weaned; S = pre-weaning concentrate feeding; NS = without pre-weaning concentrate feeding). Within an age, means with different letter differ at P<0.05.

References

Elsasser, T.H., T.S. Rumsey and H.M. Hammon, 1989. Influence of diet on basal and growth hormone-stimulated plasma concentrations of IGF-1 in beef cattle. J. Anim. Sci. 67: 128-141.

Littell, R.C., G.A. Milliken, W.W. Stroup and R.D. Wolfinger, 1996. SAS system for mixed models, volume 1. SAS Institute Inc., Cary, NC, USA.

Hossner, K.L., R.H. McCusker and M.V. Dodson, 1997. Insulin-like factors and their binding proteins in domestic animals. Anim. Sci. 64: 1-15.

Effect of crown daisy (*Chrysanthemum coronarium*) and ricinoleic acid on sheep milk production and quality

R. Bodas[1], S. Andrés[1], A.B. Rodríguez[1], J. Romero[1], R.J. Wallace[2], F.J. Giráldez[1] and S. López[1]
[1]Instituto de Ganadería de Montaña (CSIC-ULE), 24346 Grulleros, Spain; [2]University of Aberdeen, Rowett Institute of Nutrition and Health, Aberdeen AB21 9SB, United Kingdom; s.lopez@unileon.es

Introduction

It is known that diet plays an important role in modulating the fatty acid composition of milk, and one of the objectives of the 'Replace' Project (FP6) was to study the possible impact of plants and plant extracts as feed additives on food quality (especially fatty acids, FA) and human health. After identifying promising candidates from a collection of 500 plants and plant extracts, a small number of samples were tested in animal trials. *Chrysanthemum coronarium* flowers (crown daisy, CF) inhibited C18:0 production and promoted accumulation of *trans*-11 C18:1 in cultures of mixed ruminal microorganisms *in vitro* (unpublished data). Previous studies had shown that sheep grazing pasture containing CF during the springtime consumed the flowers and produced milk with increased concentrations of rumenic and vaccenic acids (Cabiddu *et al.*, 2006). One aim of the present *in vivo* trial was therefore to investigate further the impact of *C. coronarium* on milk production and composition.

A second observation, also investigated further here, was that ricinoleic acid (12-OH *cis*-9 C18:1, RC) inhibited the biohydrogenation process *in vitro*. It had been suggested that RC might be an intermediate in the biohydrogenation of linoleic acid. However, RC was metabolised more slowly than linoleic acid, forming 10-OH C18:0 rather than non-hydroxylated C18:1 or C18:2 intermediates (Wallace *et al.*, 2007), and inhibited C18:0 formation. Thus, RC was incorporated into the ration of lactating sheep in a combined trial comprising commercial crown daisy flowers (*C. coronarium*) at different doses (50 and 100 g/kg concentrate) and commercial ricinoleic acid (30 g/kg concentrate) to determine their effects on milk yield and composition.

Material and methods

Thirty-two Merino lactating ewes in early lactation were allocated to four groups balanced by body weight (79.1±2.30 kg) and milk yield (1,648±129.5 ml/day). All the groups received 1.2 kg of concentrate (with 30 g sunflower oil added per kg concentrate) and 2 kg of lucerne hay per animal and per day in separate feeding troughs. Each group received one of the following concentrates: no additive (Control), 50 or 100 g CF/kg concentrate (50CF, 100CF), or 30 g ricinoleic acid/kg concentrate (RC). Lambs remained with their dams throughout the experiment. Once a week, ewes were separated from their lambs at 09:30 h, injected with 5 IU of oxytocin and milked until the udder was empty. After 8 h a second injection and milking were carried out. Milk collected from each animal was weighed and a sample was taken for analysis of milk composition. Samples collected on week 0 (just before the administration of an additive) and week 4 after receiving the additives were analysed to determine FA composition. After the second milking, lambs were allowed to suckle until the next milking day. Fat, protein, and total solid concentrations were determined by infrared spectrophotometry. Methyl esters of FA were obtained as described by Nudda *et al.* (2005) and analysed by GC. Data were subject to one way ANOVA, using the GLM procedure (SAS® Inst. Inc., 1999), with the additives tested as the only source of variation.

Results

The inclusion of either CF or RC acid did not affect milk yield and its macro-composition (Table 1). Treatment 100CF tended to decrease the proportion of C18:0 ($P<0.10$). However, regardless of the dose, CF tended ($P=0.10$) to increase *trans*-11 C18:1 (VA), and caused a significant 44% increase in *cis*-9, *trans*-11 C18:2 (conjugated linoleic acid, CLA, $P<0.05$) in milk. This increase in CLA resulted in a tendency to a higher CLA to VA ratio ($P<0.10$). With RC, a numerical but not significant decrease of C18:0 of around 7% was observed. The proportion of saturated FA in milk tended to decrease in response to 50CF and RC ($P<0.10$), while polyunsaturated FA tended to increase over 7% in milk from ewes receiving the 50CF concentrate ($P<0.10$).

Table 1. Milk yield and composition in sheep receiving 50 and 100 g of Chrysanthemum coronarium *flowers (50CF, 100CF) and 30 g of ricinoleic acid (RC) per kg of concentrate.*

	Control	50CF	100CF	RC	RSD	*P*-value
Milk yield, g/d	1116	1069	813	1150	510.5	0.60
Fat yield, g/d	89	86	82	86	42.7	0.99
Protein yield, g/d	56	56	53	57	23.9	0.99
Total solids yield, g/d	211	205	189	214	94.8	0.97
Fatty acids, g/kg total fatty acids						
C18:0	131	134	107	123	20.6	0.09
trans-11 C18:1 (VA)	36.4	46.6	47.8	42.0	9.50	0.10
cis-9, trans-11 C18:2 (CLA)	14.6^{b}	21.0^{a}	21.0^{a}	17.7^{ab}	4.70	0.04
Saturated fatty acids	669	633	657	629	33.1	0.08
Polyunsaturated fatty acids	58.2	68.4	64.7	61.2	7.73	0.08
CLA / VA	5.10	6.77	6.89	5.97	1.366	0.06

^{a,b} Means in the same row with different superscript differ significantly (P<0.05).

Conclusion

The inclusion of 50 or 100 g CF/kg concentrate slightly affected milk fatty acid composition, enhancing RA and VA concentrations without any effects on sheep performance. 30 g RC/kg concentrate, however, tended to decrease the proportion of saturated FA.

Acknowledgement

European Union FP6 Project 'Replace' (Contract no.: FOOD-CT-2004-506487).

References

Cabiddu, A., M. Addis, G. Pinna, S. Spada, M. Fiori, M. Sitzia, A. Pirisi, G. Piredda and G. Molle, 2006. The inclusion of a daisy plant (*Chrysanthemum coronarium*) in dairy sheep diet. 1: Effect on milk and cheese fatty acid composition with particular reference to C18:2 *cis-9, trans-11*. Livest. Sci. 101: 57-67.

Nudda, A., M.A. McGuire, G. Battacone and G. Pulina, 2005. Seasonal variation in conjugated linoleic acid and vaccenic acid in milk fat of sheep and its transfer to cheese and ricotta. J. Dairy Sci. 88: 1311-1319.

Wallace, R.J., N. McKain, K.J. Shingfield and E. Devillard, 2007. Isomers of conjugated linoleic acids are synthesized via different mechanisms in ruminal digesta and bacteria. J. Lipid Res. 48: 2247-2254.

Heat production of dairy cows under acute and chronic heat load

A. Brosh[1], A. Asher[1], J. Miron[1], A. Shabtay[1], G. Adin[1], U. Moalem[1], E. Tahar[2], S. Abboud[3] and Y. Aharoni[1]

[1]Institute of Animal Science, Department of Ruminant Science & Genetics Agricultural Research Organization, P.O. Box 1021, 30095, Ramat Yishay, Israel; [2]Veteix Ltd, Veterinary Expert systems, P.O.Box 36, 30600, Or Aqiva, Israel; [3]Department of Medical Engineering, Tel Aviv University, P.O.Box 39040, 69978, Tel Aviv, Israel; brosha@volcani.agri.gov.il

Introduction

According to the classic thermoregulation literature, when animals are exposed to heat load above the upper critical temperature, they activate evaporative cooling mechanisms. Consequently the animals' heat production (HP) has to increase. Most of the classic thermal regulation studies in the past were carried out under constricted conditions, in respiration chambers. The heart rate O_2 pulse method (Brosh *et al.*, 1998, Brosh, 2007) allows to measure the animals' HP throughout days and even weeks, while animals are maintained in their natural habitat (feedlot, cowshed, grazing). This method enables to follow animals' long term behaviour and physiological responses to the changes in the environmental conditions. The following presentation describes these responses in lactating dairy Holstein cows exposed to heat load.

Material and methods

The first presentation (Figure 1) describes changes in heart rate (HR) of high yielding cows in the hot Jordan Valley. The cows were subjected to two feeding regimes, night and day feeding, in order to test the effect of time of feeding on heat load throughout the day. During the trial the cooling system of the night fed cows was shut down by accident. The second presentation describes the HP of high yielding dairy cows in the summer in two feeding treatments, during which the cows were cooled several times during the day by showers and fans, and when the cooling was stopped. The third presentation describes recorded data of the following physiological parameters: HR, respiration rate (RR), reticulum temperature (Rt) and interval between rumen constrictions. The data were recorded by newly invented technology reticulum bolus, Veterix®, Israel. These data were transmitted by Radio Frequency communication to a central computer unit that stores and interprets the data to produce a set of alerts for controlling the cows' welfare and production management.

Results and discussion

In the first presentation (Figure 1), it is shown that the first response of night fed cows to the inactivity of the cooling system and consequently the increasing heat load was an increased HR, which was recorded at 12:00 18/5. The second response, from 18:00 18/5 to the end of the measurement process, was an adaptation to the heat load by a significant reduction of HR, i.e. reducing HP. The daily HR pattern of the day fed cows, who received cooling, did not change. In the second presentation dairy cows were fed on two diets being either cooled or not cooled. Milk production of the cooled cows was 39.0 kg/d but decreased to 31.5 kg/d when cows were not cooled. The interval between the two measurements was 64 days; the expected decrease in their HP in respect to their day in milking was 5%, (Aharoni *et al.*, 2005). However, the HP of the cooled cows on diets 1 and 2 were 1,122 and 1,092 $kJ \cdot kgBW^{-0.75} \cdot day^{-1}$, and HP decreased by 25 and 22%, respectively, when cows were not cooled. The third presentation describes a sequence of changes in several parameters throughout exposure to the external heat load. The cows' first response was a considerable increase in HR; later, HR decreased to a lower level than the initial one, with a simultaneous sharp increase of RR and a moderate increase of Rt, which was interrupted by Rt drops due to more frequent drinking. The overall conclusion of the above described three trials is that the first response of a cow to heat load is to increase blood flow in order to increase heat dissipation, including activation of evaporative cooling. When this cooling was not sufficient, HR decreased as a result of the animals' effort to reduce the intrinsic heat load by decreasing the production rate. The respiration rate was still increased to enhance evaporative heat dissipation from the lungs, but the overall production rate and consequently the HP is adjusted to balance the heat dissipation and heat load from the intrinsic and the extrinsic sources.

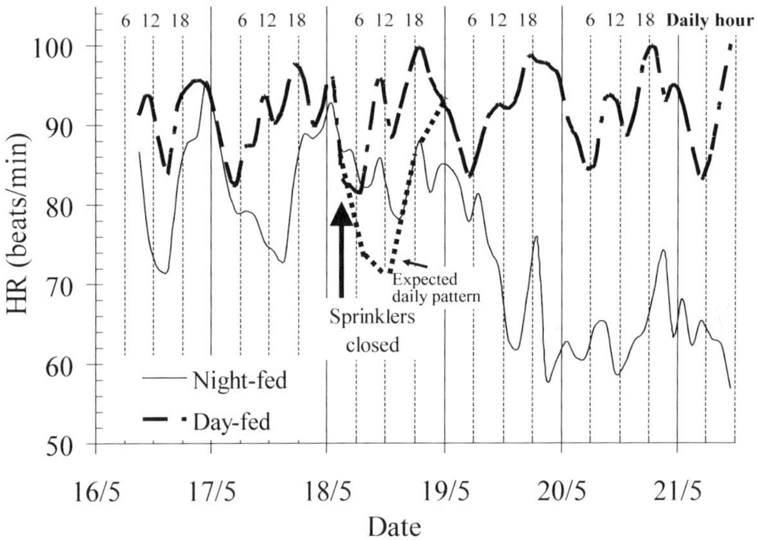

Figure 1. Heart rate of dairy cows fed at day hours (cut line) and at night hours (solid line), before and after the sprinkler cooling system of night fed cows was closed.

References

Aharoni, Y., A. Brosh and Y. Harari, 2005. Night feeding for high-yielding dairy cows in hot weather: effects on intake, milk yield and energy expenditure Livest. Prod. Sci. 92: 207-219.

Brosh, A., Y. Aharoni, A.A. Degen, D. Wright and B.A. Young, 1998. Estimation of energy expenditure from heart rate measurements in cattle maintained under different conditions. J. Anim. Sci. 76: 3054-3064.

Brosh, A., 2007. Heart rate measurements as an index of energy expenditure and energy balance in ruminants: A review. J. Anim. Sci. 85: 1213-1227.

Effect of different supply and source of polyunsaturated fatty acid on milk fat synthesis of grazing dairy sheep

A. Cabiddu, M. Addis, S. Spada, M. Acciaro, M. Sitzia, M. Decandia and G. Molle
AGRIS Sardegna, Dipartimento per la Ricerca nelle Produzioni Animali, 07040 Olmedo, Italy; acabiddu@agrisricerca.it

Introduction

Ruminant dietary lipid supplementation has been considered as a strategy to improve animal product healthiness. Unfortunately when supplementation levels are high, this strategy may influence milk fat synthesis. A previous study reported that the ewe mammary gland may be less sensitive to the effect of CLA t-10 c-12 on milk fat synthesis than that of the cow (Reynolds *et al.* 2006) in agreement with the theory that regulation of milk fat synthesis differs between species. No studies are available on the combined effect of C18:1 and CLA isomers on milk fat content (MFC) in dairy sheep. The aim of the present study was to evaluate the effect of different supplementations of PUFA on milk fatty acids involved in MFC regulation.

Material and method

Forty-eight dairy sheep (DIM 88 + 10, BCS 2.42 + 0.24, BW 42.9 + 4.73 kg, and milk yield 1624 + 235 ml) were randomly assigned to the following dietary treatments: only pasture (PAS, sheep grazing for 24 h with the exception of milking times), non-fat enriched supplementation (NFS, sheep grazing for 3 h, supplemented with 900 g of cereal based concentrate), high linolenic acid supplementation (C183H, sheep grazing for 3 h, supplemented with 900 g of linseed based concentrate) and moderate linolenic acid supplementation (C183M, sheep grazing for 3 h, supplemented with 900 g of linseed and sunflower-seed based concentrate). In C183H linseed and sunflower seed was 13% and 4% whereas in C182M concentrate was 7% and 0.35% respectively. Herbage intake was measured by the n–alkane method, concentrates were individually measured at the milking parlour (data not shown). Milk yield and composition were measured weekly. Fatty acid methyl esters from milk fat were obtained using base-catalysed methanolysis. Separation and quantification of FAME was performed using a gas chromatograph. The ANCOVA model with orthogonal contrast was used to test the effect of dietary treatments on milk fat, milk fat yield and the main isomers involved in MFC regulation.

Results

Concentrate supplementation (TR contrast, Table 1) decreased milk fat content compared to pasture treatment, whereas no difference was found between supplemented groups (Fat contrast). Fat yield on the contrary was increased with C183 supply compared to NFS (Fat contrast). The best predictors of MFC were C18:1 t-10, C18:2 t-9 t-12, and CLA t-9 c-11 isomers. The increase of C18:1 t-10 was about 3 fold in 183M with 25% of a detrimental effect on milk fat content. The source of lipid supplementation strongly influenced the level of C18:1 t-10, CLA t-9 c-11 and C18:2 t-9 t-12. On the contrary, a CLA t-10 c-12 isomer was not related to MFC probably because its level was very low, eightfold lower than in cow's milk when it is associated with milk fat depression. These results were in agreement with Cabiddu *et al.* (2006) who found similar levels of CLA t-10 c-12 without any effect on MFC. Furthermore, there were negative relationships between CLA t-10 c-12 and C18:1 t-10 ($r^2 = 0.38$; $P<0.01$) or CLA t-9 c-11 ($r^2 = 0.44$; $P<0.01$). The negative relationships between different isomers and MFC also showed a high individual variability (Figure 1).

Table 1. Effect of treatments on milk fat content, milk fat yield and the main isomers putatively involved in milk fat content decrease.

	Treatment				SEM	Contrast *P*-value				
	PAS	NFS	C183H	C183M		TR	FAT	PF	PN	TF
Milk fat content,%	7.00	5.33	5.18	5.39	0.19	**	ns	**	**	ns
Milk fat yield, g/d	85	82	89	95	3.54	ns	*	ns	ns	ns
C18:1 t-4,%FAME	0.017	0.016	0.025	0.023	0.002	ns	0.07	ns	ns	ns
C18:1 t-5, "	0.014	0.016	0.027	0.029	0.002	0.09	**	*	ns	ns
C18:1 t-6 t-8, "	0.223	0.364	0.512	0.465	0.02	**	**	**	**	ns
C18:1 t-10, "	0.600	1.225	1.45	2.610	0.25	*	ns	**	ns	*
C18:2 t-9 t-12, "	0.030	0.060	0.060	0.134	0.01	0.06	ns	*	ns	*
CLA t-10 c-12, "	0.063	0.044	0.055	0.052	0.01	**	0.09	0.06	**	ns
CLA t-9 c-11, "	0.030	0.033	0.039	0.077	0.002	ns	ns	ns	ns	**

TR = PAS vs. NFS+C183H+C183M; FAT = C183H+C183M vs. NFS; PF = PAS vs. C183H+C183M; PN = PAS vs. NFS; TF = C183H vs. C183M. *P*-value <0.05 was considered significant.

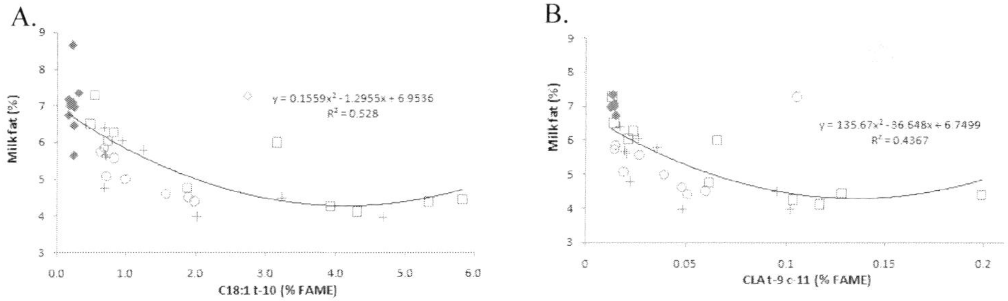

Figure 1. Relationship between C18:1 t-10 (A), CLA t-9 c-11(B) and milk fat content (each point is an individual value, ♦ PAS, ○ NFS, + C183H, □ C183M).

Conclusion

This study shows that the levels of isomers linked to MFC increase when sheep are fed with fat supplements. In particular the highest levels of these isomers are found when feeding sunflower linseed based concentrate (C183M).

Acknowledgement

This work was partially funded by the Sardinia Regional Government (OVISAN-Project).

References

Cabiddu, A., G. Molle, M. Decandia, M. Sitzia, A. Cannas, N. Fois, G. Piredda, A. Pirisi and M. Addis, 2006. Effect of linseed supplementation on milk fatty acid composition of dairy sheep grazing an organic grass–based pasture. Proc. 4th Euro Fed Lipid Congress, 1-4 October 2006, Madrid, Spain, p. 580 (Abstr.).

Reynolds, C.K., V.L. Cannon and S.C. Loerch, 2006. Effects of forage source and supplementation with soybean and marine algal oil on milk fatty acid composition of ewes. Anim. Feed Sci. Technol. 131: 333-357.

Effect of roughage diet type and NaCl addition on the milk urea content in dairy cows

S. De Campeneere, J.M. Vanacker and D.L. De Brabander
ILVO, Animal Sciences Unit, Scheldeweg 68, 9090 Melle, Belgium;
sam.decampeneere@ilvo.vlaanderen.be

Introduction

Several models (De Brabander *et al.*, 1999; Nousiainen *et al.*, 2003) are used to predict N excretion of dairy cows from milk urea content (MUC). These models only include nutritional factors, like protein intake and dietary protein/energy ratio. However, De Campeneere *et al.* (2006) concluded from a feeding trial that a MUC value of a 100% maize silage based diet (100MS) (217 mg/l) was significantly higher than that of a 100% prewilted grass silage based diet (100PGS) (146 mg/l), after correction for small differences in energy and protein supply (De Brabander *et al.*, 1999). Additional N-balance trials confirmed that although total N-excretion was similar (392 and 389 g/d), the MUC content for 100MS was 84 mg/l higher than for 100PGS. De Campeneere *et al.* (2007) found that adding KCl to 100MS could not explain the difference found by the former while adding NaCl could explain part of it. The present trial studied the effect of NaCl addition (as a compensation for the lower Na content in MS *versus* PGS) further and tests if that effect is the same for a 100% maize silage based diet (MS) and a 50/50 maize silage/prewilted grass silage based diet (MS/PGS).

Material and methods

A Latin square trial with 4 periods was set up with 20 Holstein cows involving 2 treatments with MS and 2 with MS/PGS. As MUC reflects dietary changes very rapidly, periods of 2 weeks were used. Within each roughage type (R), there was one control treatment, the other was supplemented with 0.5 kg NaCl. Each period consisted of 1 week of adaptation and 1 week of sampling. After each period, groups switched treatments. All cows were housed individually in a tie-stall with rubber beddings and individual mangers. At the start of the trial, concentrate and soybean meal supply was calculated individually to cover 105% of the net energy and digestible protein requirements based on expected milk yield and composition. Administered feed and milk production (MP) was recorded daily and milk of the last 4 milkings of each collection week was sampled for analysis on fat, protein and urea. N-excretion was calculated as N intake minus N retained in milk and LW change (14 g N/kg LW change). Statistical analysis was done as a GLM procedure with as model $Y_{ijk} = \mu + P_i + C_j + R_k + NaCl_l + R_k \times NaCl_l + e_{ijkl}$; using factors period (P), cow (C), R (MS or MS/PGS) and NaCl (0 or 500g) in Statistica (2007).

Results and discussion

DM-intake significantly differed between MS and MS/PGS, but was not influenced by the NaCl factor (Table 1). Adding the salt significantly increased MP, with a stronger effect for the MS diet (1.3 kg) than for the MS/PGS diet (+0.4 kg). This was also found by De Campeneere *et al.* (2006) and is probably linked with an increased water intake. Milk fat and protein content were not affected by NaCl. After correction of the MUC values for differences in energy and protein supply (De Brabander *et al.*, 1999), MUCcorr was 28% and 11% lower when NaCl was added to the MS and the MS/PGS diets, respectively. NaCl significantly interacted with R, having a larger effect for the MS diet than for the MS/PGS diet. Possibly, the higher Na content in the PGS reduced the effect of the NaCl addition. N-excretion only decreased with 5% and 3% for these diets, respectively. MUCcorr was significantly lower for MS/PGS than for MS, although N-excretion was systematically higher, indicating MUC as an ineffective predictor of N-excretion, when different roughage types are involved. Due to an increased MP, NaCl improved N efficiency, again to a larger extent in MS than in MS/PGS diets. The decreasing effect of NaCl on MUC of the MS diet was comparable with the difference in MUC found by De Campeneere *et al.* (2006) between 100MS and 100PGS. Sehested and Lund (2006) found a decrease in MUC from 4.4 to 2.9 mM/l when increasing the dietary K content (with KCl) from 12 to 35 g/kg DM, but within that high K content, adding Na (from 1 to 10 g/kg DM) did not influence the MUC any further. These authors found no effect of adding NaCl (at high KCl content) on MP or protein content.

The results confirm that MUC is not a good predictor for N-excretion, when different roughage types are involved or NaCl is added. Research should focus on the physiological pathways of milk urea formation, before MUC can be used as a reliable N-excretion management tool.

Table 1. DM-intake, milk yield and composition and N-balance data.

	MS		MS / PGS		SEM	*P*-value		
	-	+NaCl	-	+NaCl		R	NaCl	R×NaCl
DM-intake, kg/d[1]	19.0	19.7	20.4	20.2	0.2	***	ns	†
Milk, kg	26.0	27.3	27.9	28.3	0.3	**	*	ns
Fat,%	4.36	4.23	4.40	4.44	0.05	†	ns	ns
Protein,%	3.47	3.49	3.42	3.39	0.03	†	ns	ns
MUC, mg/l	330	245	234	211	8	***	***	*
MUCcorr, mg/l	257	184	150	134	6	***	***	**
N-intake, g/d	482	486	509	495	4	***	ns	ns
N-excretion, g/d	346	329	356	346	4	**	**	ns
N-efficiency, g/d	28.8	30.9	28.9	29.9	0.4	ns	**	ns

[1] Incl. NaCl added; DM = dry matter; MUCcorr=MUC corrected for differences in energy and protein supply; N-eff = N retained in milk/N-intake; † =$P<0.10$; * = $P<0.05$; ** = $P<0.01$; *** = $P<0.001$; SEM=standard error of the mean; R=roughage effect (MS vs. MS/PGS).

References

De Brabander, D.L., S.M. Botterman, J. Vanacker and Ch.V. Boucqué, 1999. The milk urea concentration as indicator for nutrition and N-excretion. (in Dutch). Report nr 1109. Dep. Animal Nutrition and Husbandry, Centre Agricultural Research Ghent, Belgium, 8 pp.

De Campeneere, S., D.L. De Brabander and J.M. Vanacker, 2006. Milk urea concentration as affected by the roughage type offered to dairy cattle. Livest. Sci. 103: 30-39.

De Campeneere, S., D.L. De Brabander and J.M. Vanacker, 2007. Constraints in estimating N-excretion from the milk urea content in dairy cows. In: Ortigues-Marty I. (Ed), Energy and protein metabolism and nutrition. Wageningen Academic Publishers, Wageningen, the Netherlands, EAAP Series 124, pp. 149-150.

Nousiainen, J., K.J. Shingfield and P. Huhtanen, 2003. Evaluation of milk urea nitrogen as a diagnostic of protein feeding. J. Dairy Sci. 87: 386-398.

Sehested, J. and P. Lund, 2006. NaCl and KCl intake affects milk composition and freezing point in dairy cows. In: Van der Honing Y. (Ed.), Book of Abstracts of the 57th Annual Meeting of the EAAP, Antalya, Turkey, 17-20 Sept. 2006. Wageningen Academic Publishers, Wageningen, the Netherlands, EAAP Book of Abstracts 12, p. 295.

Statistica 8.0, 2007. Statsoft Incl, Tulsa, USA.

Heavy metals in poultry manure, bovine tissues and human kidneys in Yucatán México

A. Castellanos-Ruelas and G. Rosado-Rubio
Fac. de Ing. Química, Univ. Aut. de Yucatán, Campus de Ingenierías y Ciencias Exactas, Periférico Nte. km 33.5. Tablaje Catastral 13615. Col. Chuburná de Hidalgo Inn. C.P. 97203, Mérida, Yuc. México; cruelas@uady.mx

Introduction

The human population in the state of Yucatán México, is the largest in the country affected by lithiasis. This syndrome has different origins: genetic, nutritional, and environmental among others. The exposure to heavy metals is also considered a risk factor for presentation of lithiasis. Poultry manure is frequently used as a feedstuff for ruminants (Castellanos *et al.*, 2000) and it possibly has a large amount of Cu since $CuSO_4$ is commonly used as a growth promoter. The objective of this research project was to analyse Cu content in poultry manure as well as Cu and Pb in bovine tissues and finally in human kidneys of patients presenting lithiasis.

Material and methods

The first experiment was carried out during one year. Samples of poultry manure were collected (n=165) from the three must important avian producing companies (A, B and C) over periods covering the rainy, northwind and dry seasons. Cu content was measured. In the second experiment, bovine tissues (n = 160) were sampled in two abattoirs located in the city of Mérida. Samples of muscle, kidney and liver from each animal were taken during a period of six months for Cu and Pb analysis. Finally, samples from human kidneys (n= 48) were collected in a local hospital. They were chirurgically obtained from patients with lithiasis, since it was prescribed by the physician. Cu and Pb were also analysed. Atomic absorption spectrophotometry method of analysis was used using specific hollow cathodes to detect Cu and Pb. Means and SD of the results were calculated and the effect of the avian producing companies (A, B and C), the season of the year (rainy, northwind and dry season), their interactions and the error were estimated using the least squares method (SAS®, 1988). Data of experiment two had to be previously transformed to its log N since heteroskedasticity was detected.

Results

Cu content detected in the avian companies was 162±64, 29±20 and 23±13 ppm for A, B and C respectively. A large variation was found between companies. Variation should be attributed to the intensity of usage of $CuSO_4$ as a bacteriostatic in the avian feedstuff.
Depending on the season of the year, Cu content was 74±77, 100±89 and 71±64 ppm for the rainy, northwind and dry season respectively. The highest level coincides with the presentation of fungi growth in the digestive tract of birds, due to the accumulation of humidity. Copper sulfate is also used to control this problem. Considering that bovines have a maximum Cu tolerance of 100 ppm (NRC, 1996), the amount found in company A and during the northwind season, is dangerous since it may bioaccumulate in the animals and therefore in humans since they are the last link of the food chain. Sheep are even more sensible to Cu toxicity than bovines, hence the problem can be exacerbated. The results of experiment 2 are shown in Table 1.
Most of the tissue samples had more Cu and Pb than the amount permitted by law in México (NOM 004 ZOO, 1994).
The results of Cu and Pb in human kidneys are shown in Table 2. The range considered normal is (ppm): 0.7-10.3 ppm for Cu and 0.02-0.9 ppm for Pb. No statistical differences were found between

gender (P>0.05). Eleven percent of male samples and 14% of female samples had a higher Cu level than the one considered as normal. The results obtained with Pb were more important: one hundred and 98% respectively were above normality. Obviously these patients have been exposed to large quantities of these metals, with poultry manure possibly being the origin for Cu. The origin of Pb is uncertain, but possibly it could be attributed to industrial contamination of local industries.

Table 1. Cu and Pb content (ppm) in bovine tissues in the Yucatán state.

Parameter	Muscle (n= 159)	Kidney (n= 158)	Liver (n= 160)
Cu			
Mean and SD	4.8±4.1^{b}	14.6±6.2^{b}	187.0±97.4^{a}
Min Max	0.4-27.4	2.4-49.6	5.8-450
Maximun tolerable level according to Mexican legislation	2	10	60
% Samples above max. level	79	79	88
Pb			
Mean and SD	1.6±1.8^{a}	1.6±1.7^{a}	1.3±1.8^{b}
Min-max	0-6.8	0-8.9	0-10.6
Maximun tolerable level according to Mexican legislation	0.5	2	2
% samples above max. level	88	68	40

^{a,b} Means within rows with same superscript letters are not significantly different (P>0.05).

Table 2. Concentration of Cu and Pb (ppm) in kidney tissues obtained by nefrectomy.

	Mean ± Standard Deviation		Min - Max values		% of samples above normality	
	Male (n= 14)	Female (n=34)	M	F	M	F
Cu	10.4±4.5 ^{a}	8.3±4.5 ^{a}	4.5-17.5	2.7-20	11	14
Pb	1.3±1.8 ^{a}	3.5±1.8 ^{a}	0.4-5.6	0.7-9.8	100	98

^{a,a} Means within rows with same superscript letters are not significantly different (P>0.05).

Conclusion

Cu and Pb are consistently present in the food chain in Yucatán, México. The pathway starts with the contamination of poultry manure by the use of $CuSO_4$, which in turn is fed to bovines. The consumption of bovine meat may facilitate its bioaccumulation in the human kidney. Both heavy metals in human tissues may collaborate to induce lithiasis.

References

Castellanos, R.A.F., O.M.L. Murguía and O.Y. Moguel, 2000. Effect of the dehydration process of poultry manure on its nutritional value and microorganisms count. Tec. Pecu. Mex. 38: 219-230.

NOM (Norma Oficial Mexicana) 004-ZOO-1994. Grasa, hígado, músculo y riñón en aves, bovinos, caprinos, cérvidos, equinos, ovinos, y porcinos. Residuos tóxicos. Límites máximos permisibles y procedimientos de muestreo. Modificación de la Norma Oficial Mexicana.

NRC, 1996. The nutrient requirements of beef cattle. 7th revised ed. Washington, DC, USA, National Academy Press.

SAS,1988. User's Guide: Statistics (ver. 6.03) Cary, NC, USA. SAS® Inst. Inc.

Maternal nutritional plane alters ovine jejunal mRNA expression of glucagon like peptide-2 in offspring at 20 and 180 days of age

J.S. Caton[1], L.P. Reynolds[1], J.M. Wallace[2], K.A. Vonnahme[1], A.M. Meyer[1], M.L. Johnson[1] and D.A. Redmer[1]
[1]Center for Nutrition and Pregnancy, Department of Animal Science, North Dakota State University, Fargo 58108-6050, USA; [2]Rowett Research Institute, Bucksburn, Aberdeen AB21 9SB, United Kingdom; joel.caton@ndsu.edu

Introduction

Maternal nutritional status is a major factor implicated in the development and function of fetal organ systems (Wallace *et al.*, 1999; Wu *et al.*, 2006). Prenatal growth is sensitive to maternal dietary intake throughout gestation. Growth restricted offspring are at risk of immediate postnatal complications, may exhibit poor growth and productivity, and in humans, develop significant diseases later in life (Barker, 2004). Glucagon Like Peptide-2 (GLP-2) belongs to the family of glucagon like peptide hormones and is produced in intestinal, pancreatic, and brain tissues of several nonruminant species (Burrin *et al.*, 2003). Research in several laboratories has shown numerous intestinal effects of GLP-2 including stimulation of intestinal growth, regulation of intestinal adaptation, improved intestinal postsurgical recovery, enhanced nutrient absorption, and increased intestinal blood flow. Research investigating GLP-2 in ruminants is very sparse in the literature. Therefore, we hypothesise that the GLP-2 system is present in ovine intestinal tissues and that mRNA expression of GLP-2 and receptor (GLP-2R) in offspring will be responsive to maternal nutrition.

Material and methods

Rambouillet ewe lambs (n=168) were used in two experiments (84 in Exp. 1 and Exp. 2, respectively) to evaluate the effects of maternal nutritional plane and selenium (Se) supply on mRNA expression of GLP-2 and GLP-2R. In institutionally approved experiments, ewes (approximately 240 d of age; 52 kg BW at breeding) were allocated to a 2×3 factorial arrangement of treatments. Factors included Se (adequate Se [ASe, 9.5 (Experience1) or 11.5 (Experience 2) μg/kg BW] or high Se [HSe, 81.8 (Experience 1) or 77.0 (Experience 2) μg/kg BW]) initiated at breeding and nutritional plane (60% [RES], 100% [CON], and 140% [HIGH] of requirements) initiated at d 50 (Experience 1) or d 40 (Experience 2) of gestation. Ewes were fed pelleted diets and housed individually indoors. At parturition, lambs were removed from their dams before suckling, fed artificial colostrum to body weight, and then maintained on common diets. Lambs were euthanised at 180 (Experience 1) or 20 (Experience 2) days of age, and detailed necropsies were performed. Jejunal mucosal scrapes were immediately collected and snap frozen in super-cooled isopentane (submerged in liquid nitrogen) and stored at -80 °C until analysis. Jejunal mucosal scrapes were analysed for mRNA expression of ovine GLP-2 and GLP-2R via quantitative real-time RT-PCR using TaqMan reagents and previously published procedures (Redmer *et al.*, 2005). Data were analysed using the GLM procedure of SAS. Means were separated using the method of least significant difference.

Results

To our knowledge, these data represent the first published report of ovine GLP-2 and GLP-2R expression in sheep intestinal tissues. Maternal nutrient restriction during gestation resulted in greater ($P \leq 0.05$) GLP-2 expression in offspring jejunal mucosal scrapes at 20 d of age compared with offspring from dams fed either control or high planes of nutrition (Table 1). Conversely, at 180 d, offspring from dams fed restricted diets had lower ($P \leq 0.07$) GLP-2 expression compared with those from dams fed control or high planes of nutrition. Maternal Se supply did not alter offspring jejunal mRNA expression of GLP-2. Expression of GLP-2R in offspring jejunal tissues was not impacted by treatments.

Table 1. Effects of maternal nutrition and Se supply on mRNA expression of glucagon like peptide-2 (GLP-2) and receptor (GLP-2R) in offspring jejunal mucosal scrapes.

Item	Nutrition[1]				Selenium[2]			*P*-value[3]		
	RES	CON	HIGH	SEM	ASe	HSe	SEM	Nut	Se	Nut*Se
Offspring day 20										
GLP-2	0.537[a]	0.357[b]	0.401[b]	0.051	0.392	0.472	0.042	0.05	0.19	0.62
GLP-2R	0.799	0.784	0.880	0.060	0.771	0.878	0.049	0.53	0.13	0.92
Offspring day 180										
GLP-2	0.133[a]	0.173[b]	0.151[b]	0.013	0.148	0.156	0.011	0.07	0.61	0.66
GLP2R	0.272	0.312	0.259	0.013	0.279	0.282	0.033	0.61	0.94	0.90

[1] RES = ewes fed to 60% of controls, CON = control ewes fed at requirements, HIGH = ewes fed to 140% of CON.
[2] Ewes fed 11.0 and 9.5 µg/kg BW (ASe; no added Se) vs. 77.0 and 81.8 µg/kg BW (HSe) for day 20 and 180 offspring, respectively.
[3] Probability values for effects of nutrition (Nut), selenium (Se), and the interaction.
[a,b] Means within a row having differing superscripts differ ($P \leq 0.07$).

Conclusion

The results indicate that the GLP-2 system is present in ovine intestinal tissues. This gastrointestinal hormone, which has strong early postnatal intestinal tropic affects in some species, appears to be sensitive to maternal plane of nutrition during gestation in sheep. Observed GLP-2 responses in lambs from dams fed differing planes of nutrition are novel and merit further investigation.

Acknowledgement

Partially supported by USDA-NRI grants No. 2003-35206-13621, 2005-35206-15281, and 2009-35206-05253 from the USDA-CSREES, and by NIH Grant HL 64141.

References

Barker, D.J., 2004. Developmental origins of well being. Phil. Trans. Royal Soc. Lond. 359: 1359-1366.

Burrin, D.G., X. Guan, B. Stoll, Y. Petersen and P. Sangild, 2003. Glucagon-like peptide 2: A link between nutrition and intestinal adaptation in neonates. J. Nutr. 133: 3712-3716.

Redmer, D.A., R.P. Aitken, J.S. Milne, L.P. Reynolds and J.M. Wallace, 2005. Influence of maternal nutrition on messenger RNA expression of placental angiogenic factors and their receptors at midgestation in adolescent sheep. Biol. Reprod. 72: 1004-1009.

Wallace, J.M., D.A. Bourke and R.P. Aitken, 1999. Nutrition and fetal growth: paradoxical effects in the overnourished adolescent sheep. J. Reprod. Fertil. Suppl. 54: 385-399.

Wu, G., F.W. Bazer, J.M. Wallace and T.E. Spencer, 2006. Intrauterine growth retardation: Implications for the Animal Sciences. J. Anim. Sci. 84: 2316-2337.

Long term chronic and oral exposure of dairy goats to mixtures of polycyclic aromatic hydrocarbons: research of potential bioindicators of exposure in milk, urine and blood lymphocytes

A. Chahin, Y. Guiavarc'h, M.A. Dziurla, H. Toussaint, C. Feidt and G. Rychen
URAFPA, Nancy University, INRA, 2 avenue de la forêt de Haye, 54505, Vandœuvre-lès-Nancy, France; abir.chahin@ensaia.inpl-nancy.fr

Introduction

Among environmental and health issues, polycyclic aromatic hydrocarbons (PAH) and their metabolites are the subject of numerous studies in terms of production (Rhead and Hardy, 2003), spatial deposition as well as transfers in native or metabolised forms (Lapole *et al.*, 2007) in the food chain. Considering their toxicity and presence in our environment, the US Environmental Protection Agency identified 16 unsubstituted PAH as priority pollutants. Although no transfer equations from food to milk are available in the literature, urinary 1-hydroxypyrene is considered as a valuable biomarker of exposure of humans to pyrene and other PAH because pyrene is systematically present in PAH mixtures, is almost only metabolised in the 1OH-pyrene form, and because its increase in the atmosphere can be linearly related to other congener concentrations (Ravindra *et al.*, 2006). Faced to the lack of data on exposure of dairy ruminants, Chahin *et al.* (2008) demonstrated that, under oral and chronic exposure of dairy goats to a mixture of pyrene, phenanthrene and benzo(a)pyrene with increasing ingested doses every 7 d, a linear relation was observed between 1OH-pyrene concentrations in milk and urine and the level of exposure to the mixture (1, 7 and 49 mg/d during 1 week each). This linear relationship was recently confirmed including similar experiments on much lower doses of 0.28 and 0.04 mg/d (data not shown) This allowed concluding that monitoring of 1-OH pyrene in milk, which is easy to sample (collecting trucks…), should be considered with interest in order to evaluate a global level of exposure to PAH in a herd of dairy ruminants. Transfer equations could be established. The purpose of the present study was to assess whether this biomarking capability still holds when dairy goats are submitted to long term oral and chronic exposure. We also took advantage of this experiment to evaluate the evolution of EROD activity inside blood lymphocytes, which, to our knowledge, has never been performed on dairy ruminants.

Material and methods

Two groups of three goats were daily and respectively submitted to ingestion of 1 mg or 50 mg of a mixture of pyrene, phenanthrene and benzo(a)pyrene during 40 d. Urine, milk and blood were sampled every 10 d during 50 d. 1OH-pyrene and other PAH metabolites in milk and urine samples were extracted and analysed via a HPLC-fluorescence detector. Lymphocytes were isolated using ficoll gradients for subsequent measurement of ethoxyresorufin-O-deethylase (EROD) in a fluorescence microplate reader. Multilinear regression was used to model the achieved kinetic data with goats being considered as an uncontrolled and possible interacting factor (SAS® 9.1.3. Cary, NC, USA). Residual normality was assessed based on the Shapiro-Wilk test.

Results

After an increase of 1OH-pyrene release into milk and urine, a significant decay was observed between d 20 and d 30 in both matrixes thus revealing clear changes in the metabolism of pyrene and/or transfer of 1OH-pyrene (data not shown). After stopping the exposure to PAH at day 40, no residual 1OH-pyrene was found in milk nor urine. Such results suggest that 1OH-pyrene should actually be used with caution as a biomarker. But its use under accidental and punctual exposures to PAH due to chemical incidents or forest fires seems to be valuable to quickly evaluate the daily ingestion of PAH without using heavy analytical procedures. As depicted in Figure 1, EROD activity measured in isolated blood lymphocytes increased with time exposure thus reflecting strong CYP450 1A1 induction. After 40 d of exposure, EROD activity reached a plateau with a 1 mg/d dose and was still increasing at a 50 mg/d dose, before a sudden decay after the end of exposure (day 40).

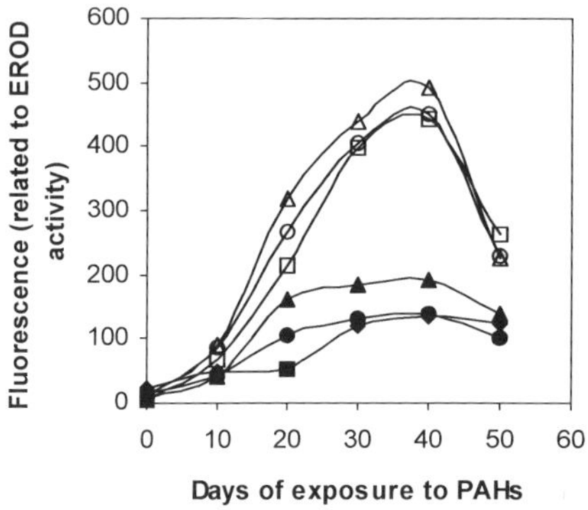

Figure 1. Evolution of EROD activity in blood lympcytes of dairy goats (black symbols correspond to the dose of 1 mg PAHs/d; white symbols correspond to the dose of 50 mg PAHs/d. Each curve corresponds to one goat.

Conclusion

These data indicated that 1OH-pyrene could be a good indicator in the case of exposure of a herd of dairy ruminants to PAH and global pollution. No data on EROD activity in ruminant lympocytes were available up to date. The non specificity of this bioindicator of exposure to xenobiotics must, however, be kept in mind.

References

Chahin, A., Y. Guiavarc'h, M.A. Dziurla, H. Toussaint, C. Feidt and G. Rychen, 2008. 1-hydroxypyrene in milk as a bioindicator of polycyclic aromatic hydrocarbons exposure of ruminants. J. Agric. Food Chem. 56: 1780-1786.

Lapole, D., G. Rychen, N. Grova, F. Monteau, B. Le Bizec and C. Feidt, 2007. Milk and urine excretion of polycyclic aromatic hydrocarbons and their hydroxylated metabolites after a single oral administration in ruminants. J. Dairy Sci. 90: 2624-2629.

Ravindra, K., L. Bencs, E. Wauters, J. de Hoog, F. Deutsch, E. Roekens, N. Bleux, P. Berghmans and R. Van Grieke, 2006. Low pressure gaz chromatography ion-trap mass spectrometry for the fast determination of polycyclic aromatic hydrocarbons in air samples. Atmosph. Env. 40: 771-785.

Rhead, M. and S. Hardy, 2003. The source of polycyclic aromatic compounds in diesel engine emissions. Fuel. 82: 385-393.

Fat body partition in dry Pelibuey ewes fed roughage diets with three levels of energy

A. Chay-Canul[1], A. Ayala-Burgos[1], J. Magaña-Monforte[1], J. Ku-Vera[1] and L.O. Tedeschi[2]
[1]Facultad de Medicina Veterinaria y Zootecnia, Universidad Autónoma de Yucatán, México; [2]Department of Animal Science, Texas A&M University, College Station, TX 77843-2471, USA; aayala@uady.mx

Introduction

In Yucatan, Mexico, sheep production is a growing activity due to the increasing demand for mutton in Central Mexico. The Pelibuey is the main breed hair sheep well adapted to tropical conditions (heat, parasites, low-quality feedstuffs), however, carcass characterisation and nutritional requirements for this breed have not been well established, particularly in adult females. This aspect is critical, considering that adult ewes define the gross efficiency of the flock performance. One strategy of this breed to cope with seasonality is to store fat during the favorable season of the year for its use in seasons of shortage (Ermias *et al.*, 2002). The objective of this work was to evaluate the effect of metabolisable energy intake (MEI) on carcass composition and changes in fat depots of dry Pelibuey ewes, fed roughage diets under tropical conditions.

Material and methods

The work was carried out at the, University of Yucatan, (FMVZ-UADY) in Merida, Yucatan, Mexico. Twenty-four, 3 year old Pelibuey ewes, were divided into 4 groups of six ewes, of similar initial body weight (BW= 37±4.1 kg), condition score (2.5) and assigned to a completely randomised experimental design. The group Base line (BS) was slaughtered initially; and eighteen remaining ewes were individually penned in metabolic crates, and fed at three levels of ME intake: Low (L), Medium (M) and High (H) during 65 days. Metabolisable energy intake was adjusted every two weeks based on theoretical requirements for maintenance (426 kJ ME/kg $BW^{0.75}$; AFRC, 1993). Forage of the basal diet was offered in equal amounts at 08:00 and 15:00 h. Forage composition was the following: DM: 283 g/kg, CP: 31 g/kg DM, NDF: 693 g/kg DM; the forage was offered at 44 g DM/kg $BW^{0.75}$, whereas a supplement (CP: 140 g/kg DM and ME: 11.5 MJ/kg DM) based on ground maize grain, velvet bean fruit (husk and grain), plus minerals and sugar cane molasses was offered at 0, 16 and 32 g DM/kg$BW^{0.75}$ for L, M and H respectively. Dry matter intake (DMI) and apparent DM digestibility were recorded every two weeks during 5 consecutive days. Feed and feces samples per ewe were taken daily, pooled and frozen (-4 °C.) for further analysis (DM, ash). MEI was estimated according to DM intake and the McDonald *et al.* (2002) equation (ME diet MJ/kg DM: DOM in DM $\times$ 0.16). At the end of the experiment the ewes were slaughtered according to regulations of the Ethical Committee of FMVZ-UADY. Data recorded at slaughter were weights of viscera and carcass. Adipose tissue was dissected, weighed and grouped as from organs (pelvic region), mesenteric and inter-intestinal regions. The gastrointestinal tract (GIT) was weighed with digestive content (full) and after emptying. The empty BW (EBW) was computed as the difference between shrunk BW (SBW) and GIT contents. Carcass was split at the dorsal midline in two equal halves, weighed and chilled at 6 °C. After 24 h refrigeration, the left side was dissected into muscle, bone and fat. One ewe of treatment H was removed at the end of the experiment for reasons other than experimental (pneumonia). Statistical analysis was performed with the PROC GLM (SAS), testing the linear (*L*) or quadratic (*Q*) effects of the MEI.

Results

Ewes in the L and M groups lost 7.3 and 1.4 kg respectively, while group H gained about 2.1 kg. Daily mean weight changes were the following: -107, -21 and 30 g/d for L, M and H respectively. Daily DMI and MEI were different among L, M and H groups (Table 1). The EBW and dressing percentages (%SBW) were different (*L*, $P<0.01$) among groups (Table 1). Tissues in carcass were different among treatments (*L*, $P<0.05$). Visceral fat was increased with MEI.

Table 1. DM, ME intakes and carcass characteristics of dry Pelibuey ewes fed three levels of ME.

Item	Initial (n=6)	Low (n=6)	Medium (n=6)	High (n=5)	SEM	*P*-value* L	Q
EBW (%SBW)	79[ab]	74[ac]	78[ab]	80[ab]	0.67	0.016	0.53
Dressing percentage (%SBW)	42[ab]	39[a]	43[b]	43[b]	0.44	0.008	0.20
Intake							
Total DM, g /kg$^{0.75}$/d	-	33[a]	56[b]	65[b]	2.98	<0.0001	0.08
ME, MJ/kg$^{0.75}$/d	-	0.279[a]	0.478[b]	0.573[c]	0.021	<0.0001	0.06
Tissues carcass (kg/kg carcass)							
Muscle	0.590[ab]	0.600[b]	0.640[c]	0.630[bc]	0.004	0.03	0.09
Fat	0.150[c]	0.070[a]	0.100[b]	0.120[bc]	0.44	<0.0001	0.63
Bone	0.260[a]	0.330[b]	0.260[a]	0.250[a]	0.038	0.002	0.08
Fat partition (g/kg EBW)							
Carcass fat	80.4[d]	34.3[a]	53.3[b]	66.9[c]	2.07	<0.0001	0.49
Internal Fat	76.4[bc]	28.8[a]	63.2[b]	99.2[c]	3.86	<0.0001	0.93
MesFat[1]	25.9[b]	12.7[a]	24.1[ab]	45.8[c]	2.06	0.0002	0.37
OrgFat[2]	20.6[bc]	6.0[a]	15.4[b]	26.8[c]	1.07	<0.0001	0.68
InterFat[3]	30.0[b]	10.2[a]	23.7[b]	26.6[b]	1.61	0.0015	0.15

[a-c] Means with different superscript letters in a row differ ($P<0.05$);*Probability of either linear (*L*) or quadratic (*Q*) effect of level of MEI.

[1] MesFat: Mesenteric fat.

[2] OrgFat: Organs fat.

[3] InterFat: Interintestinal fat.

Conclusion

Internal fat deposition in mature dry Pelibuey ewes was increased linearly with increasing MEI. It seems that mature Pelibuey ewes store a larger proportion of absorbed energy in the internal fat depots rather than the carcass. This may represent an adaptative strategy of this breed for energy storage during times of plenty in prevision for the periods of scarcity during the dry season.

References

AFRC, 1993. Technical Committee on Responses to Nutrients. Energy and Protein Requirements of Ruminants. CAB International, Wallingford, UK.

Ermias, E., A. Yami and J.E.O. Rege, 2002. Fat deposition in tropical sheep as adaptive attribute to periodic feed fluctuation. J. Anim. Breed. Genet. 119: 235-246.

McDonald, P., R.A. Edwards, J.F.D. Greenhalgh and C.A. Morgan. 2002. Animal Nutrition. 6th Ed. Prentice Hall. p 693.

Investigation of the potential to use isotopic fractionation between milk and urine as a test for nitrogen use efficiency of dairy cows

L. Cheng and R.J. Dewhurst
P.O. Box 84, Lincoln University, Canterbury, New Zealand; richard.dewhurst@teagasc.ie

Introduction

Conversion of feed protein into milk and meat is the basis of sustainable ruminant production systems from both economic and environmental perspectives. Nitrogen use efficiency (NUE) describes the efficiency of converting feed N into product N (in milk or meat). This project addresses the need for rapid screening tools to support nutritional and animal breeding strategies to improve NUE. The nitrogen balance technique for measuring NUE is difficult to run, particularly with large numbers of animals, and subject to experimental errors and bias (Spanghero and Kowalski, 1997). Studies in monogastrics have shown relationships between N utilisation and the fractionation of N isotopes in the pathways leading to protein and urea (Sick *et al.*, 1997). This study investigated this phenomenon in relation to milk and urine from dairy cows offered a wide range of diets, alongside nitrogen balance measurements.

Material and methods

Nine multiparous Holstein-Friesian cows, in early lactation, were used in an incomplete changeover design with three four-week periods (3 cows/treatment). All cows received 4 kg/d concentrates (4.0% N on a DM basis) and dietary treatments were based on a range of silages offered *ad libitum*: PRG: Perennial ryegrass; TIM: Timothy; TF: Tall Fescue; RC: Red clover; RM: Red clover/maize mixture (40/60 on a DM basis); RO: Red clover/whole-crop oat mixture (40/60 on a DM basis); OR: Red clover/whole-crop oat mixture (25/75 on a DM basis). The remaining treatments were based on RO with feed intake restricted to the level of PRG (ROr) or with a low protein concentrate (50/50 mixture of barley and molassed sugar beet pulp) (ROlp). Feed intake was recorded and total collections of faeces, milk and urine were made over the final 6 d of each period. Urine was analysed for purine derivatives (PD; a marker for rumen microbial protein synthesis) according to Dewhurst *et al.* (1996) and nitrogen-15 ($\delta^{15}N$ units; ‰) by Isotope Ratio Mass Spectrometry. The results were analysed by REML with a Student-Newman-Keuls multiple range test to separate means (Genstat 10, VSN International Ltd, Hemel Hempstead, UK).

Results

There were highly significant effects of the dietary treatments on milk yield, N intake and the output of N in faeces, milk and urine (Table 1).

Table 1. Effects of dietary treatments on feed intake, milk production and N partitioning.

	PRG	TIM	TF	RC	RM	RO	OR	ROr	ROlp	SED	Sig
Silage DMI, kg/d	12.8	13.8	11.6	14.7	15.7	15.6	13.6	14.4	13.3	1.27	NS
Milk yield, kg/d	24.4^{bc}	22.3^{abc}	19.5^{a}	26.1^{c}	27.2^{c}	26.1^{c}	24.8^{bc}	25.2^{bc}	20.5^{ab}	1.58	***
N intake, g/d	605^{c}	547^{bc}	494^{b}	626^{c}	512^{b}	473^{b}	359^{a}	464^{b}	379^{a}	36.4	***
Faecal N, g/d	175	163	151	193	166	148	139	149	141	15.6	NS
Milk N, g/d	115^{bcd}	103^{abc}	90^{a}	121^{cd}	128^{d}	125^{d}	116^{bcd}	115^{bcd}	100^{ab}	6.10	**
Urine N, g/d	302^{c}	259^{c}	256^{c}	275^{c}	181^{b}	148^{ab}	122^{a}	162^{ab}	117^{a}	18.8	***

Retained N averaged 22 g/d and was not significantly different between treatments. There were highly significant effects of diet on ^{15}N in faeces, urine and milk, mirroring feed values. Fractionation of ^{15}N between faeces, milk and urine – expressed as deviations from respective feed values, was also significantly different between diets (Table 2). Milk was enriched in ^{15}N relative to feed, whilst urine was depleted.

Overall, there were no significant relationships between N isotope fractionation and the partitioning of dietary N between urine and milk. However, a significant relationship emerged when values for TIM and TF were excluded:

Urine $\delta^{15}N$ – feed $\delta^{15}N$ = -4.7 (se=0.49)*** + 1.11 (se=0.290) urine N/milk N***
R^2=0.41; n=21; residual sd=0.771; P<0.001

Table 2. Effects of dietary treatments on urinary excretion of purine derivatives and $\delta^{15}N$ contents of feed, faeces, urine and milk.

	PRG	TIM	TF	RC	RM	RO	OR	ROr	ROlp	SED	Sig
Urinary PD, mmol/d	260^{ab}	201^{a}	200^{a}	259^{ab}	312^{b}	275^{b}	276^{b}	254^{ab}	261^{ab}	23.4	***
Feed $\delta^{15}N$	6.59^{c}	8.38^{d}	7.15^{c}	2.27^{a}	4.03^{b}	2.37^{a}	1.82^{a}	2.29^{a}	2.59^{a}	0.348	***
Faecal $\delta^{15}N$	7.95^{cd}	8.76^{d}	7.04^{c}	4.27^{a}	5.75^{b}	4.20^{a}	4.27^{a}	4.62^{a}	4.53^{a}	0.369	***
Milk $\delta^{15}N$	9.09^{c}	9.76^{c}	8.77^{c}	5.32^{b}	6.80^{b}	5.54^{a}	5.59^{a}	6.05^{a}	6.57^{a}	0.378	**
Urine $\delta^{15}N$	4.09^{d}	3.91^{d}	2.62^{c}	1.01^{b}	0.71^{b}	-1.13^{a}	-1.42^{a}	-0.92^{a}	-1.06^{a}	0.584	***
Faecal $\delta^{15}N$ – feed $\delta^{15}N$	1.51^{b}	0.15^{a}	-0.03^{a}	1.69^{b}	1.65^{b}	2.06^{b}	2.61^{c}	2.40^{bc}	1.85^{b}	0.211	***
Milk $\delta^{15}N$ – feed $\delta^{15}N$	2.40^{b}	1.67^{a}	1.43^{a}	3.49^{d}	2.88^{c}	3.38^{d}	3.64^{d}	3.85^{d}	3.77^{d}	0.287	**
Urine $\delta^{15}N$ – feed $\delta^{15}N$	-2.48^{b}	-4.43^{a}	-4.58^{a}	-1.19^{c}	-3.30^{ab}	-3.43^{ab}	-3.12^{ab}	-3.35^{ab}	-3.73^{ab}	0.496	***

[a,b,c,d] Values in the same row with different superscripts are significantly different (P<0.05).

Conclusion

This study generated a wide range of N partitioning with the low protein diet (OR) supporting milk N output almost equal to urinary N. In contrast, the high-protein grass silages led to urinary N exceeding milk N by at least 2.5 times. For most of the diets, the relationship between N isotopic fractionation and N utilisation was consistent with the monogastric study of Sick *et al.* (1997), with less depletion of ^{15}N in urine when protein intake (urinary N output) was the highest. However, the results with TF and TIM diets were not consistent and this may be related to the much lower microbial protein synthesis (urinary PD) with these diets.

References

Dewhurst, R.J., A.M. Mitton, N.W. Offer and C. Thomas, 1996. Effects of the composition of grass silages on milk production and nitrogen utilisation by dairy cows. Anim. Sci. 62: 25-34.

Sick, H., N. Roos, E. Saggau, K. Haas, V. Meyn, B. Walch and N. Trugo, 1997. Amino acid utilization and isotope discrimination of amino nitrogen in nitrogen metabolism of rat liver *in vivo*. Z. Ernährwiss. 36: 340-346.

Spanghero, M. and Z.M. Kowalski, 1997. Critical analysis of N balance experiments with lactating cows. Livest. Prod. Sci. 52: 113-122.

Blocking vasodilatory prostaglandin synthesis by ketoprofen fails to prevent the renal blood flow increase induced by insulin in conscious sheep

A. Cirio[1], I. Tebot[1], J.Y. Ayoub[2], C. Paquet[2], S. Junot[2] and J.M. Bonnet[2]
[1]Area de Fisiología, Facultad de Veterinaria, 11600, Montevideo, Uruguay; [2]Université de Lyon, Ecole Nationale Vétérinaire de Lyon, EA 4173 INSERM ERI 22, 69280 Marcy L'Etoile, France; albertocirio@yahoo.com

Introduction

A circadian rise of renal blood flow (RBF) after meals and independent of blood pressure has been recently reported in sheep (Tebot *et al.*, 2009). In a preliminary work (not published) we found an increase in RBF during insulin perfusion in fasted sheep, which lasted longer than the rise in plasma insulin, suggesting the existence of an insulin-induced mediator of the RBF increase. We also found that this effect was only partially mediated by NO. In non-ruminant animals, vasodilatory prostaglandins (PG) have been invoked to account for the RBF changes caused by pancreatic hormones or aminoacids (reviewed by Tuttle *et al.*, 2002). The aim of this work was to describe in sheep the effect of insulin perfusions on RBF after blockade of PG synthesis by ketoprofen, a cyclooxygenase inhibitor. Systemic arterial pressure (SAP) was also monitored since sympathetically mediated insulin-dependent vasoconstriction in skeletal muscle in humans and rats has been reported (reviewed by Muniyappa *et al.*, 2007), suggesting a role on blood pressure, and variations in renal hemodynamics are partly attributed to changes in arterial pressure (Miller-Craig *et al.*, 1978).

Material and methods

Under general anesthesia, 6 adult Ile de France ewes (54-70 kg BW) were bilaterally implanted with transit-time ultrasonic flow-metering probes (Transonic Systems, Ithaca, NY) around renal arteries, as previously described (Tebot *et al.*, 2009). Four of the sheep were provided with a telemetry measurement system (Physiotel Transmitter, Data Sci. Int., St Paul, MN) inserted into one carotid artery for SAP monitoring. After recovery, sheep were housed in individual pens and data were transmitted to a processing system (Acqknowledge III for MP150WSW, Biopac Systems, Sta. Barbara, CA). Mean values of RBF (right + left flows) and SAP were calculated every 10 min. On the experimental days, 16 h-fasted sheep received constant rate i.v. infusions (0.5 ml/min, 10:00-12:00 h) of ketoprofen (Ketofen, Merial, Lyon, France, 0.2 mg/kg/min, primed with 6 mg/kg) in 0.9% saline and, during the second hour, insulin (Caninsulin, Intervet SA, Beaucouzé, France, 6 mU/kg/min, primed with 70 mU/kg) or saline perfusions. Each sheep received one perfusion per day, on alternate days and random order. During insulin perfusions euglycaemia was maintained and hypokalemia was prevented by perfusing a 40% glucose and a 2.5% KCl solution. On the perfusion days, RBF was recorded between 9:00 h and 14:30 h and blood samples were taken every 30 min (9:30 to 14:00 h) for insulin (RIA kit Insik 5, DiaSorin, Antony, France) and glucose (GOD POD kit, Thermo Fisher Scientific Oy, Vantaa, Finland) analyses. Analysis of variations in RBF, SAP and insulin values was performed by a one-way ANOVA with subsequent Fisher PLSD test.

Results and discussion

The 2 h perfusion of ketoprofen (Figure 1) caused an 11% decrease in RBF ($P<0.05$). When insulin, but not saline, was added, RBF rapidly rose to 14% above pre-perfusion values ($P<0.05$), in spite of the continuous perfusion of ketoprofen and without changes in SAP. The significant rise in RBF lasted 100 min after the end of insulin infusion and was similar to that found after meals (results not published). The progressive RBF reduction during saline perfusion was similar to that described for fasted sheep (Tebot *et al.*, 2009). Owing to the euglycaemic clamp, no changes in plasma glucose during perfusions were observed.

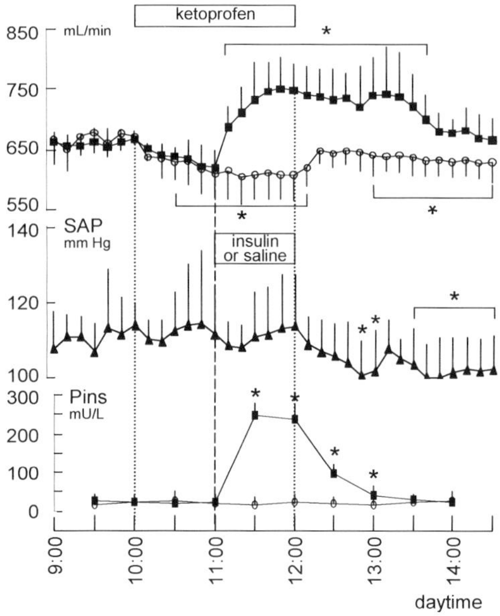

Figure 1. Renal blood flow (RBF), systemic arterial pressure (SAP) and plasma insulin (Pins) during and after perfusions of ketoprofen (between dotted lines) with insulin (■) or saline (○) during the second hour. SAP values correspond to insulin perfusions. RBF and SAP data = 10 min means ± SD. * = P<0.05 *vs. 1 h pre-infusion mean values.*

Conclusion

Blocking PG synthesis with ketoprofen reduced renal perfusion but failed to prevent the RBF increase induced by insulin in conscious sheep, suggesting that the effect of insulin on renal haemodynamics involves vasodilatory agents other than PG.

Acknowledgement

The authors thank the staff of the Endocrinology Laboratory of the Veterinary School of Lyon for RIA analyses. Financial support was provided by the Comité ECOS-Sud, action U05S02.

References

Tebot, I., J.M. Bonnet, S. Junot, J.Y. Ayoub, C. Paquet and A. Cirio, 2009. Roles of eating, rumination, and arterial pressure in determination of the circadian rhythm of renal blood flow in sheep. J. Anim. Sci. 87: 554-561.

Miller-Craig, M.W., C.N. Bishop and E.B. Raftery, 1978. Circadian variation of blood pressure. Lancet 1: 795-797.

Muniyappa, R., M. Montagnani, K. Kon Koh and M.J. Quon, 2007. Cardiovascular actions of insulin. Endocrine Reviews 28: 463-491.

Tuttle, K.R., M.E. Puhlman, S.K. Cooney and R.A. Short, 2002. Effects of amino acids and glucagon on renal hemodynamics in type 1 diabetes. Am. J. Physiol. 282: F103-112.

Blocking NO synthesis by L-NAME perfusion partially prevents the renal blood flow increase induced by insulin perfusion in conscious sheep

A. Cirio[1], I. Tebot[1], C. Paquet[2], J-Y. Ayoub[2] and J.M. Bonnet[2]
[1]Area de Fisiología, Facultad de Veterinaria, 11600, Montevideo, Uruguay; [2]Université de Lyon, Ecole Nationale Vétérinaire de Lyon, EA 4173 INSERM ERI 22, 69280 Marcy L'Etoile, France; albertocirio@yahoo.com

Introduction

A circadian rise of renal blood flow (RBF) after meals and independent of blood pressure has been recently reported in sheep (Tebot *et al.*, 2009). In addition, plasma insulin increases after meals in sheep and cattle (reviewed by Basset, 1972). In a preliminary study (not published) we found an increase in RBF during insulin perfusion in fasted sheep, lasting longer than the rise in plasma insulin and suggesting the existence of an insulin-induced vasodilating mediator of the RBF increase. Nitric oxide (NO) has been proposed as being responsible for the renal haemodynamic changes caused by different vasodilator agents in non-ruminant animals (Rajapakse *et al.*, 2002). The aim of this work was to describe the effect of insulin perfusions on RBF after blockade of NO synthesis by L-NAME in sheep chronically implanted with ultrasonic flow probes around both renal arteries.

Material and methods

Under general anesthesia, 8 adult Ile de France ewes (54-70 kg BW) were bilaterally implanted with transit-time ultrasonic flow-metering probes (4 mm, R-series, Transonic Systems, Ithaca, NY) around renal arteries for RBF measurement, as previously described (Tebot *et al.*, 2009). After recovery, sheep were housed in individual pens for chronic studies and fed on alfalfa pellets, hay and *ad libitum* water. Data were transmitted to a processing system (Acqknowledge III for MP150WSW, Biopac Systems Inc., Sta. Barbara, CA), and mean values of RBF (right + left flows) were calculated every 10 min. On experimental days, 16 h-fasted sheep received constant rate i.v. infusions (0.5 ml/min, 10:00-12:00h) of (a) insulin (Caninsulin, Intervet SA, Beaucouzé, France, 6 mU/kg/min, primed with 70 mU/kg) in 0.9% saline or saline only, (b) N^G-nitro-L-arginine methyl ester (L-NAME, Sigma-Aldrich, Lyon, France, 0.22 mg/kg/min) in saline, and (c) L-NAME + insulin during the second hour (11:00-12:00 h) of L-NAME perfusion. Each sheep received one perfusion per diem, on alternate days and random order. During insulin perfusions euglycaemia was maintained and hypokalemia was prevented by perfusing 40% glucose and 2.5% KCl solutions. The perfusion days and RBF were recorded (8:00 h to 16:00 h) and blood samples were taken every 30 min (9.00 h to 15.00 h) for insulin (RIA kit Insik 5, DiaSorin, Antony, France) and glucose (GOD POD kit, Thermo Fisher Scientific Oy, Vantaa, Finland) analyses. Variations in RBF and insulin values were analysed by a one-way ANOVA and subsequent Fisher PLSD test (StatView, SAS®, Cary, NC, USA).

Results and discussion

Insulin perfusion (Figure 1a) induced an increase in RBF whose maximum value (895±59 ml/min) was reached during the decline in plasma insulin. The progressive RBF reduction during saline perfusion was similar to that described for fasted sheep (Tebot *et al.*, 2009). The 2 h perfusion of L-NAME alone (Figure 1b) caused a 17% decrease in RBF which recovered only after the end of the perfusion. However, when insulin was added during the second hour, RBF rises to pre-perfusion values in spite of the continuous perfusion of L-NAME. Nevertheless, this increase was of less magnitude than the increase observed with the perfusion of insulin alone. Owing to the euglycaemic clamp, no changes in plasma glucose during perfusions were observed.

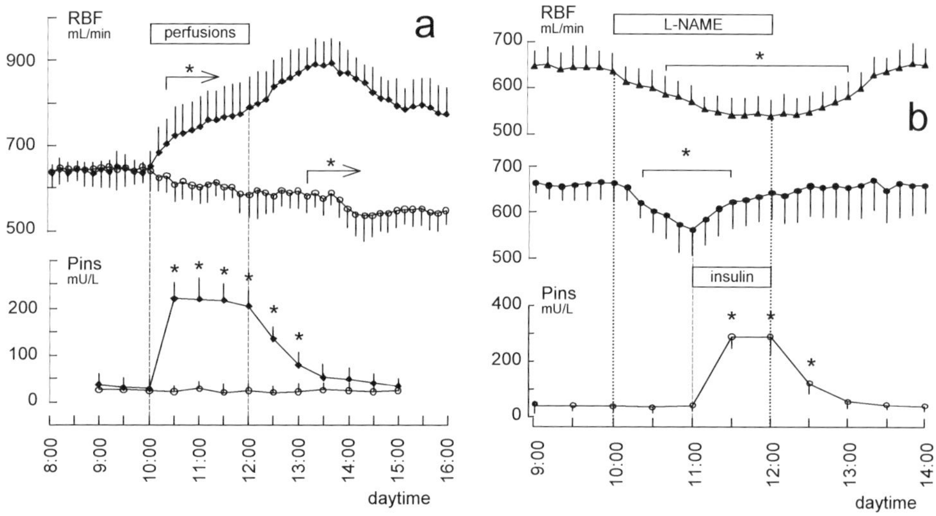

*Figure 1. Renal blood flow (RBF) and plasma insulin concentration (Pins) in 8 sheep during and after (a) perfusions (between dashed lines) of insulin (◆) or saline (○) and (b) perfusion of L-NAME (between dotted lines) alone (▲) or with insulin (●) during the second hour. RBF data are 10 min means ± SD. * = P<0.05 vs. 1 h pre-infusion mean values.*

Conclusion

Blocking NO synthesis with L-NAME reduced renal perfusion in sheep but only partially prevented RBF increase induced by insulin, suggesting the existence of complementary vasodilatory agents accounting for the effect of insulin on renal haemodynamics.

Acknowledgement

The authors thank the staff of the Endocrinology Laboratory of the National Veterinary School of Lyon for biochemical analyses. Financial support was provided by the Comité ECOS-Sud, action U05S02.

References

Basset, J.M., 1972. Dietary and gastro-intestinal control of hormones regulating carbohydrate metabolism in ruminants. In: McDonald, I.W. and A.I. Warner (Eds.), Digestion and Metabolism in the ruminant. University of New England, Armidale, NSW, Australia, pp. 384-398.

Rajapakse, N.W., J.J. Oliver and R.G. Evans, 2002. Nitric oxide in responses of regional kidney blood flow to vasoactive agents in anesthetized rabbits. J. Cardiovasc. Pharmacol. 40: 210-219.

Tebot, I., J.M. Bonnet, S. Junot, J.Y. Ayoub, C. Paquet and A. Cirio, 2009. Roles of eating, rumination, and arterial pressure in determination of the circadian rhythm of renal blood flow in sheep. J. Anim. Sci. 87: 554-561.

Concentrate feeding increases plasma leptin level in mid lactation goats

C. Delavaud[1], J. Rouel[1], E. Bruneteau[2], M. Tourret[1], P. Guillouet[2], A. Ferlay[1] and Y. Chilliard[1]
[1]INRA, UR1213 Herbivores, Theix, F-63122 Saint-Genès Champanelle, France; [2]INRA, UE88 Insémination Caprine et Porcine, F-86480 Rouillé, France; yves.chilliard@clermont.inra.fr

Introduction

Plasma leptin is closely related to energy homeostasis in ruminants (Chilliard *et al.*, 2005) including goats in which plasma leptin is down-regulated by lactation (Bonnet *et al.*, 2005). It was recently shown in lactating goats (Bonnet *et al.*, 2009), that this hypoleptinaemia is regulated by dietary starch degradability, dietary crude fibre or sunflower-seed oil. The aim of this experiment was to study the effect of diet composition (high or low forage-to-concentrate ratio in diets supplemented or not with plant oils) on plasma leptin levels in mid lactation goats.

Material and methods

Six groups of 11-12 Alpine goats (83±12 days in milk) were fed experimental diets during 5 weeks. Diets with a high forage-to-concentrate ratio (F) were based on *ad libitum* alfalfa hay with a limited amount of maize silage (0.34 kg DM/d/goat) and 34% concentrate (barley 13%, soybean meal 13%, sugarbeet pulp 8%) whereas diets with a low forage-to-concentrate ratio (C) contained 72% concentrate (barley 53%, soybean meal 16%, sugarbeet pulp 3%). Each type of diet was supplemented, or not, with 180 g/d (ca. 7%) of either sunflower-seed (SO) or linseed (LO) oil. Lipids were substituted to sugarbeet pulp and soybean meal in order to have isoenergetic diets. At the end of the experimental period, body weight (BW), milk yield and composition were determined. Plasma leptin was measured by disequilibrium double-antibody ovine-specific RIA validated for determination in goat plasma (Delavaud *et al.*, 2000). Milk fatty acid (FA) composition was determined by gas chromatography using a 100 m capillary column (CPSil-88) (Chilliard *et al.*, 2006). Data were analysed using the GLM procedure of SAS (2000). Fixed effects included goat, basal diet, lipid supplement and basal diet × lipid supplement interaction. For BW, milk yield and composition, pre-experimental data were used as the covariate when significant. The least squares means procedure of SAS® (2000) was used for mean comparisons.

Results

Plasma leptin concentrations were significantly higher (+66%, $P<0.001$) in goats fed the C diets (compared to the F diets), for which both energy intake and energy balance were higher (Table 1). Lipid supplementation did not change plasma leptin, whereas it increased milk fat concentration in *trans* 18:1 (SO and LO), and C18:3n-3 (LO). Milk yield was significantly increased (+22%, $P<0.05$) and its concentration in 18:3n-3 was decreased ($P<0.05$) in goats fed the C (compared to the F) diets.

Discussion and conclusion

A positive effect of dietary 18:2n-6 on plasma leptin was shown in lactating goats (Bonnet *et al.*, 2009) but not in the present trial. The higher plasma leptin in goats fed the C diets was probably due in part to the higher energy balance of animals, which is known to increase leptin in ruminants (Chilliard *et al.*, 2005). Furthermore, the lower crude fibre content (107 *vs* 261 g/kg DM) of these diets could also be linked to higher leptin, as shown previously in lactating goats (Bonnet *et al.*, 2009). This remains, however, to be confirmed for identical levels of energy intake.

Table 1. Effect of a basal diet with either a 66% (F) or a 28% (C) forage-to-concentrate ratio, supplemented or not with plant oils, on body weight, plasma leptin and milk production and composition in 6 groups of 11-12 mid lactation goats.

Forage-to-concentrate	F			C		
Lipid supplement[1]	-	SO	LO	-	SO	LO
Goats *n*	11	12	12	12	12	12
Plasma leptin, ng/ml	2.69^{ac}	2.36^{a}	2.53^{a}	4.20^{bc}	3.77^{ab}	4.60^{b}
Body weight, kg	60.7^{a}	60.4^{a}	60.5^{a}	63.2^{b}	60.3^{a}	61.5^{ab}
Milk yield, kg/day	3.40^{a}	3.52^{a}	3.29^{a}	4.33^{b}	4.14^{b}	3.96^{c}
Milk fat, g/day	98^{a}	129^{bc}	114^{ac}	108^{a}	138^{b}	128^{bc}
Milk FA,% total FA						
*trans*10 18:1	0.17^{a}	0.95^{b}	0.41^{a}	0.18^{a}	1.07^{b}	1.42^{b}
*trans*11 18:1	0.51^{a}	12.7^{d}	10.3^{c}	0.89^{a}	6.82^{b}	6.19^{b}
18:2 n-6	1.83^{a}	3.81^{c}	1.48^{a}	2.45^{b}	2.17^{b}	1.76^{a}
18:3 n-3	0.49^{b}	0.39^{b}	1.29^{c}	0.24^{a}	0.13^{a}	0.50^{b}
Energy intake, MJ/day[2]	17.5	18.4	17.6	22.7	22.0	21.7
Energy balance, MJ/day[2]	3.70	2.91	2.70	7.03	4.83	5.54

[1] SO = sunflower-seed oil, LO = linseed oil.

[2] Feed intake was controlled for each group of animals and no statistical analysis was done on these data.

[a,b,c,d] Means within rows with the same superscript letters are not significantly different (P>0.05).

Acknowledgement

The authors thank F. Dugué and N. Riquet for animal care and technical assistance at INRA experimental farm Les Verrines.

References

Bonnet, M., C. Delavaud, J. Rouel and Y. Chilliard, 2005. Pregnancy increases plasma leptin in nulliparous but not primiparous goats while lactation depresses it. Domest. Anim. Endocrinol. 28: 216-223.

Bonnet, M., C. Delavaud, L. Bernard, J. Rouel and Y. Chilliard, 2009. Sunflower-seed oil, rapidly degradable starch and adiposity up-regulate leptin gene expression in lactating goats. Domest. Anim. Endocrinol. 37: 93-103

Chilliard, Y., C. Delavaud and M. Bonnet, 2005. Review. Leptin expression in ruminants: nutritional and physiological regulations in relation with energy metabolism. Domest. Anim. Endocrinol. 29: 3-22.

Chilliard, Y., J. Rouel and C. Leroux, 2006. Goat's alpha-s1 casein genotype influences its milk fatty acid composition and delta-9 desaturation ratios. Anim. Feed. Sci. Tech. 131: 474-487.

Delavaud, C., F. Bocquier, Y. Chilliard, D.H. Keisler, A. Gertler and G. Kann, 2000. Plasma leptin determination in ruminants: effect of nutritional status and body fatness on plasma leptin concentration assessed by a specific RIA in sheep. J. Endocrinol. 165: 519-526.

SAS, 2000. Statistical analysis system, Release 8.02.2000. SAS Institute Inc. Cary, NC.

Milk production and composition of dairy cows fed extruded canola and lignosulfonate

G.T. dos Santos[1], C.A. Neves[1], D.C. da Silva[1], W.B.R. dos Santos[1], J.C. Damasceno[1] and H.V. Petit[2]
[1]Universidade Estadual de Maringá, Departamento de Zootecnia, Avenida Colombo, 5790, CEP 87020-600, Maringá, Paraná, Brasil; [2]Dairy and Swine Research and Development Centre– Agriculture and Agri-Food Canada, Succ Lennoxville, Sherbrooke,QC J1M 1Z3, Canada; gtsantos@uem.br

Introduction

Canola seed is a potentially interesting source of fat for dairy cows with 40 to 50% of fat and 21 to 25% of crude protein. Extrusion, which is a method to heat-process oilseeds, can be used to increase nutrient utilisation by cows, thus enhancing milk production and composition (Whitlock *et al.*, 2002). Lignosulfonate, which is a byproduct of the wood industry containing xylose, has different surface-active properties such as binders, dispersants, emulsifiers, humectants, sequestrates; it also combines with protein (Melbar, 2000). So, the combination of heat and sugar addition can modify milk production and composition. The present study was aimed to evaluate milk production and milk composition of cows fed extruded canola seed treated or not with lignosulfonate.

Material and methods

Eight multiparous cows averaging 538 kg of BW and 62 days in milk were utilised in a double 4×4 Latin square design with four treatments and four 21-d periods. Diets were fed twice daily at 0800 and 1600 h and adjusted for 100 g of orts/kg as fed. The four total mixed diets consisted of supplements based on the following: ground canola seed (GC), extruded ground canola seed (EGC), ground canola seed with 5% of lignosulfonate (GCL) and extruded ground canola seed with 5% of lignosulfonate (EGCL). The forage to concentrate ratio was 57 to 43 and the forage was corn silage. Milk production was recorded twice daily. Milk samples were obtained from 4 consecutive milkings on days 15 and 16 of each experimental period and pooled within cow and period relative to production to obtain one composite sample per cow per period for further chemical analysis. Milk samples were kept at room temperature with a preservative, 2-brome- 2-nitropropane-1,3 diol (Bronopol, D&F Control Systems Inc., San Ramon, CA, USA), for determination of protein, fat, lactose and total solid concentrations by infrared spectroscopy (Bentley model 2000; Bentley Instrument Inc., Chaska, MN, USA).

All results were analysed using the MIXED procedure of SAS® (2000) within a 2×2 factorial arrangement of treatments. Data were analysed using a replicated 4×4 Latin square design with the following general model:

$$Y_{ijklm} = \mu + T_i + P_j + Q_k + TQ_{ik} + A/Q_{lk} + e_{ijklm}$$

where: Y_{ijklm} = the dependent variable, μ = the overall mean, T_i = the effect of treatment, P_j = the effect of period (j = 1 to 4), Q_k = the effect of square (k = 1, 2), TQ_{ik} = the interaction between treatment and square, A/Q_{lk} = cow within square, and e_{ijklm} = the random residual error. Treatments were compared to provide factorial contrasts: (1) extruded versus non-extruded canola seed, (2) lignosulfonate-treated canola seed versus untreated canola seed, and (3) the interaction between extrusion and lignosulfonate treatment. Significance was declared at $P<0.05$.

Results and discussion

There was no interaction between lignosulfonate and extrusion for milk production and composition. Milk production (kg/d) and 3.5% fat-corrected milk yield were similar among treatments ($P>0.05$).

Similarly, Neves *et al.* (2007) reported no difference in milk production between cows fed extruded and non-extruded soybeans treated or not with lignosulfonate. Milk concentrations of protein, lactose and total solids were similar among diets. However, diets containing extruded ground canola seed decreased concentration of milk fat (3.05%) compared with those containing non-extruded seeds (3.42%).

Table 1. Milk production and milk composition of Holstein cows fed ground canola seed (GC), extruded ground canola seed (EGC), ground canola seed treated with 50 g/kg DM of lignosulfonate (GCL) or extruded ground canola seed treated with 50 g/kg DM of lignosulfonate (EGCL)[1].

	Treatments				SE	Probability[2]		
	GC	EGC	GCL	EGCL		E	L	E×L
Production (kg/d)								
Milk	18.80	19.51	18.64	19.01	0.54	0.33	0.55	0.75
3.5FCM[3]	18.45	18.05	18.23	17.55	0.57	0.35	0.54	0.81
Fat	0.64	0.59	0.63	0.58	0.03	0.10	0.64	0.89
Composition,%								
Protein	3.09	3.10	3.10	3.02	0.05	0.46	0.47	0.40
Fat	3.44	3.05	3.39	3.04	0.15	0.02	0.86	0.89
Lactose	4.44	4.40	4.33	4.39	0.06	0.88	0.43	0.45
Total solids	11.88	11.52	11.77	11.32	0.24	0.12	0.54	0.85

[1] Least squares means with pooled standard error (SE).
[2] E = extrusion; L = Lignosulfonate.
[3] 3.5% fat-corrected milk.

Conclusion

Extrusion and lignosulfonate treatment of canola seeds had no effect on milk production and milk concentrations of protein, lactose, and total solids. On the contrary, feeding extruded canola seeds decreased milk fat concentration.

References

Melbar, 2000. Lignosulfonato. 22p. São Paulo. [catalogue].

Neves, C.A., G.T. Santos, M. Matsushita, E.M. Alves, R.L. Oliveira, A.F. Branco, D.C. Silva, A.C. Furlan and H.V. Petit, 2007. Intake, whole tract digestibility, milk production, and milk composition of Holstein cows fed extruded soybeans treated with or without lignosulfonate. Anim. Feed Sci. Technol. 134: 32-44.

SAS, 2000. Statistical analysis system, Release 8.02.2000. SAS Institute Inc. Cary, NC.

Whitlock, L.A., D.J. Schingoethe, A.R. Hippen, K.F. Kalscheur, R.J. Baer, N. Ramaswamy and K.M. Kasperson, 2002. Fish oil and extruded soybeans fed in combination increase conjugated linoleic acids in milk of dairy cows more than when fed separately. J. Dairy Sci. 85: 234-243.

Effects of antioxidant supplementation in the diet on blood parameters and muscle characteristics in fighting bulls during extreme exercise

D. Durand[1], V. Santé Lhoutellier[1], D. Micol[1], N. Mirabeau[2], J. Garcia-Schneider[2], H. Compan[2] and B. Picard[1]
[1]INRA, UR 1213 and UR370, Theix, F-63122 Saint-Genès-Champanelle, France; [2]Ecole Nationale Vétérinaire, F-31076 Toulouse, France; denis.durand@clermont.inra.fr

Introduction

The relationship between physical exercise and oxidative tissue damage has been investigated in numerous studies on men and animals. Exercise induces oxidative stress characterised by a rise in reactive oxygen species (ROS) which contribute to accelerated muscle fatigue leading to poor performances. At the cellular level, oxidation of fatty acids, also referred to as lipoperoxydation, has major deleterious consequences. So, numerous studies have demonstrated that an increase in efficiency of the antioxidant defence system, such as physical training, reduces the exercise-induced oxidative stress (Cooper *et al.*, 2002). The aim of this study was to investigate the effect of antioxidant supplementation on fighting bull performances through muscle characteristics, metabolic orientations and the ability to prevent oxidative stress during extreme exercise such as bullfighting.

Material and methods

Twelve fighting bulls and 12 Brave cows from French farms, were fed with a basal diet (hay and concentrate) supplemented (treated) or not (control) during 8 weeks with minerals (Ca, P, Na, Mg, and Cl), vitamins (A, D3, E, H, B1, C), trace elements (Fe, Cu, Zn, I, Co, Se) and plant extract rich in polyphenols (PERP) (according to bovine recommended allowances (Gladine *et al.*, 2007)). After an extreme exercise of 20 minutes, blood from the jugular vein and muscles (*Semitendinosus*) were collected. Plasma antioxidant markers (vitamin E, total antioxidant status (TAS)), markers of plasma susceptibility to lipoperoxydation (lag phase LP; maximum rate of peroxydation (Tmax) and maximum amount of conjugated diene produced (Qmax)) was measured according to Gladine *et al.* (2007). Plasma markers of metabolic responses (free fatty acids (FFA), urea, glucose, beta-hydroybutyrate (BOH) and lactate) were evaluated according to Ortigues *et al.* (1995). Markers of muscular metabolism and fibre types (I slow oxidative, IIA fast oxido-glycolytic and IIX fast glycolytic) were measured according to Picard *et al.* (1999). All results were expressed as means and standard deviations; the effect of dietary supplementation was tested by 't' test. Significance was declared at $P<0.05$ and $P\leq0.10$ was considered as a trend.

Results

The level of Vit E in plasma of bulls supplemented increased by 70% ($P<0.01$), with the TAS not being affected. Antioxidant supplementation led to an increase of the ability to delay lipid oxidation (LP +22%; $P\leq0.10$) (Table 1). Our results showed, under anaerobic conditions, an efficient utilisation

Table 1. Plasmatic markers of antioxidative status (total antioxydative status, TAS; vitamin E, VitE) and susceptibility to lipoperoxydation (LP, Tmax and Qmax) in fighting bulls receiving a diet without C (Control) or with (Treated) antioxidant supplementation in the diet.

	Vit E	TAS	LP	Tmax	Qmax
Control (n=12)	2.67±0.80[a]	1.63±0.08	27.5±6.4[e]	4.2±1.3	168±34[c]
Treated (n=12)	4.54±1.10[b]	1.60±0.21	33.7±13.0[f]	5.3±2.2	223±60[d]

of energetic molecules, with higher glycogen depletion without any deleterious effect. Indeed, with antioxidant supplementation muscle and blood lactate content remained lower. On the contrary, FFA as the second substrate utilised by muscle tended to be 20% less oxidised in the supplemented group than in the control group (Table 2). The proportion of fast IIX fibres was significantly lower in the treated group comparatively to the control. On the contrary, the proportion of slow type I fibres was increased in the supplemented group (Table 3).

Table 2. Plasmatic markers of energetic metabolism (free fatty acid, FFA; Urea; Glucose; Hydroxybutyrate (BOH) and lactate) in fighting bulls receiving the diet without C (Control) or with (Treated) antioxidant supplementation in the diet.

	FFA	Urea	Glucose	BOH	Lactate
Control	0.9±0.8	32.1±3.7	318±29	202±53^{e}	68.3±5.5^{c}
Treated	0.9±0.6	31.7±3.7	307±43	239±59^{f}	64.0±5.5^{d}

Table 3. Muscular characteristics in fighting bulls receiving the diet without C (Control) or with (Treated) antioxidant supplementation in the diet.

	TBIIX	TBIIA	TBI	Glyc	Lactate
Control	14.5±8.4^{e}	69.0±6.1	16.5±5.6^{c}	33.6±5.3^{a}	42.3±2.5
Treated	8.3±5.3^{f}	69.5±3.4	22.2±3.2^{d}	23.3±5.4^{b}	44.2±6.1

Statistical significant: $^{a,b}P<0.01$; $^{c,d}P<0.05$; $^{e,f}P<0.10$.

Conclusion

This study showed that oral antioxidant supplementation in fighting bulls is able to modify both muscle fibre composition and its ability to 'cope' during exercise, especially aerobic utilisation of glucose without lactate production and delayed plasma peroxydation processes then increasing the physical performances of bulls during extreme exercise.

Acknowledgement

The authors thank the Association of Veterinary and Phytosynthèse Company for their financial support and French fighting Brave breeders (Mrs Fano and Gallon).

References

Cooper, C.E., N.B.J. Vollaard, T. Choueiri and M.T. Wilson, 2002. Exercise, free radical and oxidative stress. Bioch. Soc. Trans. 30: 280-285.

Gladine, C., E. Rock, C. Morand, D. Bauchart and D. Durand, 2007. Biovailability and antioxydant capacity of plant extract rich in polyphenols, given as a single acute dose, in sheep made highly susceptible to lipoperoxydation. Br. J. Nutrition 98: 691-701.

Ortigues, I., C. Martin and D. Durand, 1996. Circadian changes in net nutrient fluxes across the portal-drained viscera, the liver, and the hindquarters in preruminant calves. J. Anim. Sci. 74: 895-907.

Picard, B., C. Barboiron, M.P. Duris, H. Gagniere, C. Jurie and Y. Geay, 1999. Electrophoretic separation of bovine muscle myosin heavy chain isoforms. Meat. Sci. 53: 1-7.

Effects of stage of grass silage maturity and level of concentrate in ewes in late gestation and early lactation on feed intake, blood energy metabolites and the performance of their lambs

M. Eknæs[1], Å.T. Randby[1] and P. Nørgaard[2]
[1]Norwegian University of Life Sciences, Dept. of Animal and Aquacultural Sciences, Ås, Norway;
[2]Department of Basic Animal and Veterinary Sciences, University of Copenhagen, Frederiksberg, Denmark; margrete.eknas@umb.no

Introduction

The pregnancy-lactation cycle represents a challenging physiological period for the ewe, during which energy and nutrient requirements change dramatically. The energy requirements in ewes during late gestation increase 0.5-1.2 folds compared with maintenance level (Robinson, 1999). The objective was to study the effect of *ad libitum* feeding with grass silage with different qualities supplemented with different levels of concentrate in late gestation and early lactation on feed intake, the metabolic blood profile and performance of lambs.

Material and methods

The experiment was conducted with 24 ewes of the Norwegian White Sheep breed allocated into six dietary treatments from six wk before to four wk after parturition based on live weigt and number of foetuses identified by scanning. The dietary treatments included *ad libitum* feeding grass silages with 3 different qualities supplemented at one to three levels of concentrate. The grass silages were made from the primary growth of a mixed sward of timothy, meadow fescue and red clover in the proportions of 0.47, 0.37 and 0.15, respectively and were harvested at Ås, Norway at three stages of maturity: (1) Very early (S1), May 22; (2) Early (S2), June 5; and (3) Normal (S3), June 13. The three silages contained 235, 239, 238 g DM/kg; 156, 125, 105 g CP/kg DM; 440, 550, 594 g NDF/kg DM, respectively. Metabolisable energy (ME) content, determined *in vivo* from digestibility trials with sheep, was 11.8, 10.6 and 9.2 MJ/kg DM for S1, S2 and S3, respectively and 11.6 MJ/kg DM for the concentrate. The three concentrate levels (0, M and H) during the nursing period were 0.0 kg/d, 0.2 kg/d and 0.4 kg/d for ewes with twins, and 0.0 kg/d, 0.4 kg/d and 0.8 kg/d for ewes with triplets. The ewes were fed half of their daily nursing concentrate level during the last three to four wk before lambing. The six dietary treatments included: S1_0, S1_M, S1_H, S2_M, S2_H and S3_H. The ewes were weighed two consecutive d every wk and in addition the second d after lambing. Blood samples were taken from the jugular vein before the morning meal <7 d before lambing, and serum was analysed for glucose, non esterified fatty acids (NEFA) and β-hydroxybutyrate (BHBA). The lambs were weighed at birth and two consecutive d every wk. The statistical model (GLM procedure of SAS®) included block, treatment and residual error. Due to illness and some stillborn lambs, the data set was not completely balanced.

Results

The pregnant ewes on the S1_0 treatment had a higher daily intake of silage DM, a lower content of BHBA in serum (as S2_M) and a higher daily live weight gain (LWG) (as S1_M and S1_H groups) than the other dietary treatments (Table 1). A negative correlation between ME intake and NEFA ($r=-0.48$, $P<0.0001$) and a positive correlation between ME intake and LWG ($r=0.64$, $P<0.0001$) in late gestation was observed. The LWG by the lambs was not significantly affected by the dietary treatments, but later harvesting time resulted in lower total ME intake and also in higher energy mobilisation of ewes during early lactation.

Table 1. Feed intake, blood metabolites, live weight and live weight gain in response to dietary treatments during late gestation and early lactation.

	Dietary treatments[1]						SE	Diet effect
	S1 0	S1 M	S1 H	S2 M	S2 H	S3 H		*P*-value
No. of ewes	3	4	4	4	4	4		
Feed intake, late gestation								
Silage (kg DM/d)	2.28^{a}	2.04^{b}	2.02^{b}	2.02^{b}	2.15^{ab}	1.75^{c}	0.071	<0.001
Concentrate (kg DM/d)	0.00^{d}	0.12^{c}	0.21^{b}	0.13^{c}	0.23^{a}	0.21^{b}		
Total ME[2] (MJ/d)	26.8^{a}	25.5^{a}	26.2^{a}	22.9^{b}	25.4^{a}	18.5^{c}	0.77	<0.001
Feed intake, early lactation								
Silage (kg DM/d)	2.66^{ab}	2.72^{a}	2.61^{ab}	2.39^{b}	2.63^{ab}	2.40^{b}	0.093	0.06
Concentrate (kg DM/d)	0.00^{d}	0.27^{c}	0.59^{b}	0.31^{c}	0.65^{a}	0.59^{b}		
Total ME[2] (MJ/d)	31.4^{b}	35.3^{a}	37.7^{a}	28.9^{b}	35.5^{a}	28.9^{b}	0.98	<0.001
Blood metabolites (mmol/l)[3]								
Glucose	3.08^{bc}	3.29^{b}	3.79^{a}	2.80^{c}	3.35^{ab}	1.97^{d}	0.165	<0.001
NEFA	0.46^{a}	0.77^{a}	0.57^{a}	0.64^{a}	0.70^{a}	1.66^{b}	0.212	<0.01
BHBA	0.32^{c}	0.87^{a}	0.67^{ab}	0.52^{bc}	0.75^{a}	0.83^{a}	0.077	<0.001
Live weight, ewes (kg)								
Start of experiment	93.0	92.0	101.3	97.7	96.9	98.8	3.41	0.40
2 d after lambing	84.2^{b}	87.4^{b}	105.9^{a}	92.8^{b}	90.6^{b}	89.0^{b}	3.11	<0.001
Live weight gain, ewes (g/d)								
Late gestation	516^{a}	507^{a}	458^{a}	251^{c}	379^{b}	286^{c}	25.2	<0.001
Early lactation	-120^{b}	55^{a}	59^{a}	-257^{c}	-172^{bc}	-225^{bc}	36.4	<0.001
No. of lambs (twins_triplets)	8(2_6)	10(4_ 6)	11(2_9)	11(2_9)	12(0_12)	11(2_ 9)		
Live weight, lambs (kg)								
Birth	5.1	5.2	4.2	4.2	4.5	4.9	0.29	0.07
End of experiment	12.9	13.6	12.0	11.8	11.9	12.1	0.60	0.23
Live weight gain, lambs (g/d)	273	295	273	260	257	243	13.0	0.12

[1] S1, S2 and S3 = very early, early and normal harvesting time, respectively; 0, M and H = zero, medium and high concentrate level, respectively.
[2] Metabolisable energy, determined *in vivo* from digestibility trials with sheep.
[3] Early morning samples, <7 d before lambing.
[a,b] Means within rows with different superscript are significantly different (P<0.05).

Conclusion

Very early harvested (S1) or early harvested (S2) grass silage fed with none or moderate concentrate supplementation gave similar or higher energy intake than normal harvested silage with high concentrate supplementation. The high energy intake was in line with the levels of blood metabolites. No significant difference in lamb performance was found.

References

Robinson, J.J., K.D. Sinclair, R.D. Randel and A.R. Sykes, 1999. Nutritional management of the female ruminant: mechanistic approaches and predictive models. In: H.J.G. Jung and G.C.Jr. Fahey (Eds.), Nutritional Ecology of Herbivores. Proceedings of the Vth International Symposium on Nutrition of Herbivores. American Society of Animal Science, Savoy, Illinois. pp. 550- 603.

Influence of intensive nursing and feeding during early growth stage on growth and muscle physiology in grass-fattening Japanese Black cattle (Wagyu)

K. Etoh[1], K. Metoki[2], S. Kaneda[2], T. Abe[2], T. Etoh[1], K. Hayashi[1], Y. Nakamura[1], F. Ebara[1], J. Wegner[3] and T. Gotoh[1]

[1] Kuju Agricultural Research Center, Faculty of Agriculture, Kyushu University 8780201, Japan; [2]Kuju, Japan, National Livestock Breeding Center (NLBC), 961-8581, Fukushima, Japan; [3]Research Institute for Biology of Farm Animals, 18196 Dummerstorf, Germany; gotoh@farm.kyushu-u.ac.jp

Introduction

Grain is food that human beings can eat. Recently, there has been a global shift to produce agro-energy, i.e. ethanol from grain because of oil shortage. The price of grain has increased. In some developed countries, a considerable amount of grain feeds are fed to cattle. In Japan, Wagyu (Japanese Black cattle) are known for their excellent marbled beef which is achieved by feeding them with a considerable amount of concentrate (4,000 to 5,000 kg altogether, until slaughter at 28-30 mo of age). Cattle are ruminants, which means they have an important ecological niche that capitalises on the symbiotic relationship between fibre fermenting ruminal microbes and mammalian demand for usable nutrients. We aim at producing a high-quality safe beef product while maximising the use of domestic grass resources. This would contribute to the beef production in lowlands and uplands of South East Asian countries. It has been shown, however, that alterations in foetal and early postnatal nutrition and endocrine status may result in developmental adaptations that permanently change the structure, physiology, and metabolism in the adult life of rats, mice, domestic species and humans (Levin *et al*, 2000). This phenomenon is referred to as 'metabolic imprinting' based on medical research regarding 'the developmental origins of health and disease (DOHaD)'. However, there are very few reports of metabolic imprinting in cattle.
Japanese beef production has many problems such as food safety, animal welfare, excreta from livestock and the sudden price of grain jumping because we depend much on grain for livestock. Accordingly, we aim at producing a high-quality safe beef product while maximising the use of domestic grass resources. In this study, we investigated, using molecular biology and histochemistry methods, whether the metabolic imprinting effect of differences in feeding during early growth influences the muscle fibre of Japanese Black steers.

Material and methods

Japanese Black steers were randomly allocated to 2 groups. The high energy group (HE: n=12) underwent intense nursing (maximum intake of 1.8 kg per day) at 3 months of age and was fed a high-concentrate diet for 3 to 10 months of age. On the contrary, the Roughage group (R: n=11) underwent normal nursing and was fed only roughage (orchardgrass hay) *ad libitum* from 3 to 10 months of age. Furthermore, nursing was used for each quality milk replacer in every group. After feeding at 10 months of age, both groups were fed only roughage (orchardgrass hay) *ad libitum* from 10 to 14 months of age. Subsequently all animals were put out to the same pasture and grazed until 20 months of age. Samples of tissues from the *longissimus* muscles (LM) in all animals were collected and biopsied at four times: the finished point just after nursing (T1, 3 mo of age), the end point of different nutritional treatment (T2, 10 mo of age), the time point just after feeding only hay (T3, 14 mo of age) and the end point of grazing (T4, 20 mo of age). Gene expressions related to adipogenesis, fatty acid synthesis (PPARγ2, C/EBPα, Leptin, G6P, SCD) and IGF- I receptor in muscle were measured by real-time PCR. The microarray study aimed at investigating which genes were differentially expressed between the HE and the R groups among growth treatment at T2. The composition of myofiber type and the myofiber diameter was calculated by the histochemical data based on acid-preincubation ATPase acitivity and NADH dehydrogenase activity in LM. The adipocyte size was measured histochemically using Oil-red O and HE staining. Moreover, blood samples were taken from the jugular vein and serum IGF-1 concentration was measured by RIA at four times in every animal.

Results and discussion

The average live weight was always significantly higher in the HE group than in the R group after 4 weeks of age. Serum IGF-1 concentration indicated significantly higher levels in the HE group than in the R group at T1, T2 and T3. Gene expressions related to muscle growth, adipogenesis and fatty acid synthesis were significantly higher in the HE group than in the R group at T2. In addition, diameters of every myofiber type and diameter of adipocytes were higher in the HE group than in the R group at T2. Percentages of type I and IIA myofibers were also significantly different between the HE group and R group. Microarray analysis showed differentially expressed genes related to fatty acid synthesis, PPAR signalling and insulin signalling pathways were significantly higher in the HE group than in the R group at T2.

These results indicate that the high energy treatment during the early growth phase influenced live weight, gene expression and myofiber classification and morphometry. However, no parameters were significantly different between the HE and R groups at T3 and T4, except for live weight, diameter of adipocyte and IGF-I receptor expression.

Conclusion

This study did not provide clear evidence of differences between the two treatment groups at T4. We, however, describe in detail the sequential change of meat-quality related factors influenced by high energy treatment during the early growth phase in Japanese Black steers.

References

Levin, B.E., 2000. The obesity epidemic: metabolic imprinting on genetically susceptible neural circuits. Obes. Res. 8: 342-347.

Transcriptomic profile in adipose tissues is modified by nutrition in lactating goats

Y. Faulconnier, J. Domagalski, M.B. Montazer Torbati, Y. Gaudron, D. Bany, Y. Chilliard and C. Leroux
INRA, UR1213 Herbivores, Site de Theix, F-63122 Saint Genès-Champanelle, France; christine.leroux@clermont.inra.fr

Introduction

The major role of ruminant adipose tissue (AT) is to synthesise fatty acids and store triglycerides for use in the support of productive functions. In particular, body fat mobilisation is required at the beginning of lactation corresponding to a period of negative energy balance (Vernon, 2005). Similarly, changes in the amount and metabolism of AT occur in underfed ruminants. Some reports have shown that undernutrition inhibits the enzymatic activity and expression of some genes encoding the key lipogenic enzymes in ruminant AT (Chilliard *et al.*, 2000; Faulconnier *et al.*, 2001). However, the gene expression profiling of ruminant AT in response to nutritional restriction have not been studied. Recently, bovine microarray accessibility allowed the simultaneous analysis of the expression of several thousand genes (Ollier *et al.*, 2007; Taniguchi *et al.*, 2008). Thus, the objective of this study was to investigate the molecular mechanisms of the response to food deprivation (FD) in AT of lactating goats by identifying differentially expressed genes using bovine microarray. We also compared the FD impact on gene expression in two different AT (perirenal: PR and omental: OM) to investigate the mechanism underlying their different deposits in response to dietary manipulations (Faulconnier *et al.*, 2007).

Material and methods

Ten peak-lactating Alpine goats were assigned to 2 groups differing in their feeding level (control diet *ad libitum* vs. 48-h FD). At the end of the experiment, all goats were slaughtered and immediately after death PR and OM AT were collected in sterile conditions. Total RNA was extracted using TRIZOL Reagent and DNA contamination was removed by DNase I treatment (Invitrogen Life Technologies). RNA concentration, purity and integrity were determined by a Nanodrop. Transcriptomic analyses were performed using a bovine oligochip containing 8379 probes (Operon Biotechnologies). Five comparisons were performed and for each comparison, data from the 4 replicated spots per oligonucleotide (2 intra- and 2 inter-slides) were normalised, averaged and after filtering statistically analysed with a standard Student *t* test to detect differentially expressed genes between the two groups. Probability values were adjusted using the Benjamini & Hochberg correction (at 5%). Differentially expressed genes were classified according to their biological process ontology determined by bioinformatical tools. RT-PCR validations were performed for 9 genes for OMAT and 10 genes for PRAT, using the Step one plus^{TM} Real-Time PCR System with Power SYBR Green kit (Applied Biosystems) according to the manufacturer's instructions, with 10 pmol of a primer pair designed from a caprine or bovine sequence close to that of the microarray probe.

Results

Hybridisation of AT RNA on the bovine oligochip followed by statistical analyses allowed the identification of 456 differentially expressed genes in OM and 199 in PR AT, between the 2 feeding levels studied. Among these differentially expressed genes, 211 and 91 genes were down regulated whereas 245 and 108 genes were upregulated in FD compared with control goats, in OM and PR AT, respectively, with 97 genes identical in both AT. We confirmed by real-time RT-PCR a significant effect of the FD for 6 genes (including SCD and FABP4) in both AT, 3 genes (LPL, ACSL1 and GPAM) in PR AT and a tendency was shown for 2 genes (LPL and GPAM) in OM AT. Only one gene (PKP4) was not validated by real-time RT-PCR in both AT. The differentially expressed genes were classified into 10 functional categories according to Gene Ontology annotation in both AT sites. Our results highlight genes involved in the cell cycle, proliferation, differentiation and death as the major altered class as well as an average of 6.5% of genes involved in lipid metabolism and transport. In particular, among the 97 genes identified in both AT, 13% are involved in lipid metabolism with 10 genes downregulated (including LPL, SCD and GPAM) and 3 upregulated (including FABP4).

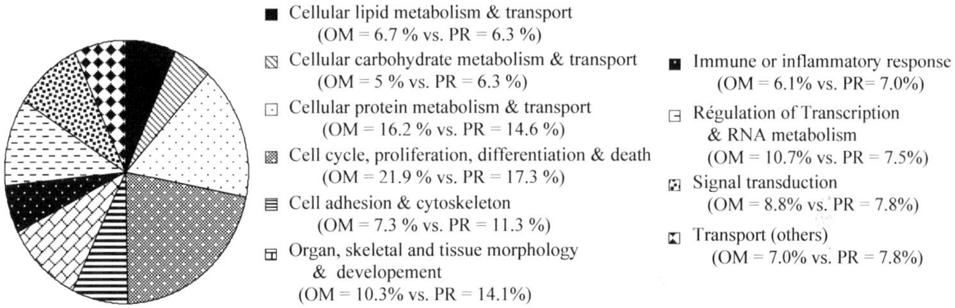

Figure 1. Biological process of the 456 genes altered by food-deprivation in omental AT.
In legend% of biological process in omental adipose tissue (OM) compared with its counterpart in perirenal adipose tissue (PR).

Conclusion

To our knowledge, this is the first transcriptomic analysis studying the impact of nutrition on ruminant adipose gene expression, which demonstrates that short term FD greatly modifies adipose transcriptomes. Globally similar biological processes were altered in PR and OM AT.

References

Chilliard, Y., A. Ferlay, Y. Faulconnier, M. Bonnet, J. Rouel and F. Bocquier, 2000. Adipose tissue metabolism and its role in adaptations to undernutrition in ruminants. Proc Nutr Soc 59: 127-134.

Faulconnier, Y., M. Bonnet, F. Bocquier, C. Leroux and Y. Chilliard, 2001. Effects of of photoperiod and feeding level on adipose tissue and muscle lipoprotein lipase activity and mRNA level in dry non-pregnant sheep. Br. J. Nutr. 85: 299-306.

Faulconnier, Y., I. Ortigues-Marty, C. Delavaud, D. Dozias, R. Jailler, D. Micol and Y. Chilliard, 2007. Influence of the diet and grazing on adipose tissue lipogenic activities and plasma leptin in steers. Animal 1: 1263-1271.

Ollier, S., C. Robert-Granié, L. Bernard, Y. Chilliard and C. Leroux, 2007. Mammary transcriptome analysis of food deprived lactating goats highlights gene involved in milk secretion and programmed cell death. J. Nutr. 137: 560-567.

Taniguchi, M., L.L. Guan, J.A. Basarab, M.V. Dodson and S.S. Moore, 2008. Comparative analysis on gene expression profiles in cattle subcutaneous fat tissues. Comp. Bioch. Physiol. Part D, 3: 251-256.

Vernon, R.G, 2005. Lipid metabolism during lactation: a review of adipose tissue-liver interactions and the development of fatty liver. J. Dairy Res. 72: 460-469.

Glucose release in response to adrenaline is lower in Merino ewes bred for lower fatness

M.B. Ferguson[1,2], J.R. Briegel[1], D.W. Pethick[1], N.R. Adams[1], H.E. Pugh[1] and G.E. Gardner[1]
[1]School of Veterinary & Biomedical Science, Murdoch University, 6150, WA, Australia; [2]Department of Agriculture and Food Western Australia, South Perth, 6151, WA, Australia;
mark.ferguson@agric.wa.gov.au

Introduction

Genetic selection for lower fatness is commonly practiced in sheep breeding programs to improve consumer appeal for sheep meat products. This selection strategy may result in changes to glucose production, in particular liver glucose output, which is of particular importance for breeding Merino ewes since their ability to build energy stores in nutritionally marginal environments is paramount for reproductive success. Hepatic glucose output can be measured by administering controlled doses of adrenaline which is a strongly catabolic hormone that among other things stimulates the liver to rapidly increase the output of glucose. This release of glucose is the net result of adrenaline-stimulated mobilisation of stored glycogen combined with glucose synthesis from gluconeogenic precursors. The rate of gluconeogenesis in response to adrenaline may be higher in genetically fat sheep as both the basal and adrenaline-stimulated rates of glucose synthesis are higher in genetically fat rats (Rohner-Jeanrenaud *et al.*, 1986; Sánchez-Gutiérrez *et al.*, 2000). Furthermore, obese humans have higher rates of gluconeogenesis and higher concentrations of hepatic glycogen than their lean counterparts, both of which would favour greater glucose release in response to adrenaline (Müller *et al.*, 1997). Thus we hypothesise that glucose output following an adrenaline challenge will be lower in ewes that are genetically leaner.

Material and methods

The blood glucose response to adrenaline was measured in 24 Merino ewes that were 1.5 years old at the commencement of the experiment and had a diverse range of Australian Sheep Breeding Values (ASBV) for subcutaneous fat depth (HFAT) (merged into 3 groups). Five levels of adrenaline (0.2, 0.6, 1.2, 2.0 and 3.0 μg/kg liveweight) were administered to the ewes via indwelling jugular catheters over three days. The experiments were repeated during late-pregnancy, peak lactation and the non-breeding state. The ewes were individually fed at maintenance and were in the fed state when the experiments were conducted. In each experiment, 15 blood samples were collected into fluoride-oxalate blood tubes from a jugular catheter between -30 and 130 min relative to the administration of adrenaline. Blood samples were placed on ice, centrifuged, and the plasma harvested and frozen at -20 °C for later determination of glucose concentrations. Plasma glucose area under the curve for the first 10 minutes following adrenaline challenge (AUC) was analysed using linear mixed effects models (SAS v8.0®) with physiological state (pregnant, lactating, non-breeding) as a fixed effect, and adrenaline dose and HFAT as covariates. Animal was used as a random term.

Results

When glucose AUC was averaged across all adrenaline challenges it was around 20% lower ($P<0.05$) when ewes were pregnant (6.52±0.53 mM/10 min.) than when lactating (8.20±0.53 mM/10 min.) or non-breeding (8.64±0.53 mM/10 min.). Glucose AUC increased ($P<0.001$) by around 500% across the range of adrenaline doses administered and was greater ($P<0.05$) in ewes with higher HFAT ASBV but only when they were pregnant or non-breeding (Figure 1). During pregnancy this HFAT effect was evident at all levels of adrenaline challenge, with this difference as large as 4 mM/10 min between the HFAT extremes. When the ewes were non-breeding the HFAT effect was only evident at adrenaline challenges greater than 2 ug/kg liveweight, and was similar in magnitude to that during pregnancy.

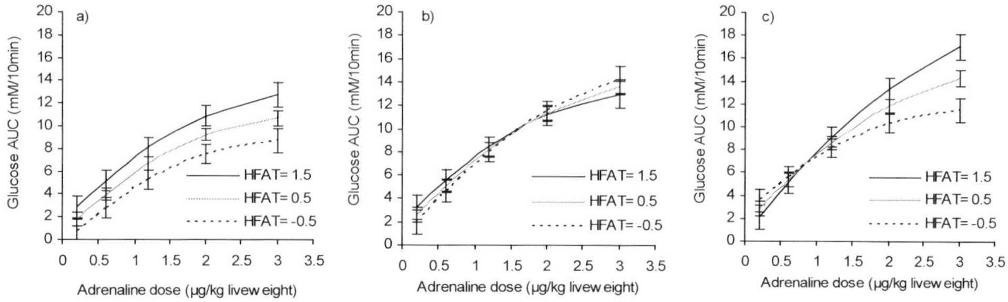

Figure 1. Glucose concentration area under curve between 0 and 10 minutes (AUC) relative to adrenaline dose in: (a) pregnancy;(b) lactation; and c) non-breeding ewes with breeding values for subcutaneous fat at hogget age (HFAT) of -0.5, 0.5 and 1.5 mm.

Discussion and conclusion

These findings align with our original hypothesis, suggesting that selection of ewes for lower fatness reduces the glucose output from the liver in response to stress. This was evident when ewes were pregnant or non-breeding, but not during lactation when hepatic and renal gluconeogenesis are maximised to cope with the high demand for glucose. We propose that the lower glucose output in response to adrenalin is in part due to a reduced gluconeogenic capacity in lean ewes. The link between fatness and gluconeogenesis is potentially provided by likely differences in leptin concentrations since lean individuals have lower circulating leptin concentrations and low leptin concentrations are associated with lower hepatic gluconeogenesis (Frühbeck and Salvador, 2000). If the relationship between genetic fatness and glucose output is confirmed, selection for reduced fatness is likely to have adverse effects on lamb birth weight as glucose is the key limiting factor for fetal growth during pregnancy.

References

Frühbeck, G. and J. Salvador, 2000. Relation between leptin and the regulation of glucose metabolism. Diabetologia 43: 3-12.

Müller, C., E. Assimacopoulos-Jeannet, F. Mosimann, Ph. Schneiter, J.P. Riou, C. Pachiaudi, J.P. Felber, E. Jequier, B. Jeanrenaud and L. Tappy, 1997. Endogenous glucose production, gluconeogenesis and liver glycogen concentration in obese non-diabetic patients. Diabetol. 40: 463-468.

Rohner-Jeanrenaud, F., J. Proietto, E. Ionescu and B. Jeanrenaud, 1986. Mechanism of abnormal oral glucose tolerance of genetically obese (fa/fa) rats. Diabetes 35: 1350-1355.

Sánchez-Gutiérrez, J.C., J.A. Sánchez-Arias, B. Samper and J.E. Feliu, 2000. Modulation of gluconeogenesis by adrenaline in hepatocytes isolated from genetically obese (fa/fa) Zucker rats. Arch. Biochem. Biophys. 373: 249-254.

Glucose uptake in response to insulin is lower in Merino ewes bred for lower fatness.

M.B. Ferguson[1,2], J.R. Briegel[1], N.R. Adams[1], D.W. Pethick[1] and G.E. Gardner[1]
[1]School of Veterinary & Biomedical Science, Murdoch University, 6150, WA, Australia; [2]Department of Agriculture and Food Western Australia, South Perth, 6151, WA, Australia; mark.ferguson@agric.wa.gov.au

Introduction

Muscle and fat are the major insulin-responsive tissues in all mammals including ruminants. Given that muscle uses more glucose than fat, a greater proportion of muscle to fat will increase the uptake of glucose per kilogram liveweight in response to stimulation by insulin (Prior and Smith, 1982). This concept has been demonstrated in sheep where the response to insulin is higher in animals that have less fat as a result of lower nutrition (Bergman *et al.*, 1989). In addition, glucose uptake in response to insulin is affected by differences in the sensitivity of muscle and fat tissue to insulin which is lower in obese animals (Bergman *et al.*, 1989). Since the proportion of fat in the whole body in sheep can be effectively reduced by selective breeding (Hegarty *et al.*, 2006), there is interest in the impact of this selection strategy on glucose metabolism in breeding ewes due to the importance of glucose for foetal development and milk synthesis. Given the impact of decreased fatness on glucose use and whole body responsiveness to insulin, selection strategies aimed at lowering whole body fatness are likely to increase the glucose response to insulin. Therefore, we hypothesise that the uptake of glucose in response to insulin will be higher in ewes bred for lower fatness.

Material and methods

The blood glucose response to insulin was measured in 24 Merino ewes that were 1.5 years old at the commencement of the experiment and had a diverse range of Australian Sheep Breeding Values (ASBV) for subcutaneous fat depth (HFAT). The hyperinsulinaemic-euglycaemic clamp technique (Bergman *et al.* 1989) was used to determine whole body response to insulin infusion. A continuous infusion of insulin was administered via an indwelling jugular catheter and commenced at a rate of 0.6 mU/kg liveweight per minute. Glucose (50% w/v) was infused concurrently and the glucose infusion rate was constantly adjusted until the pre-infusion level of blood glucose was reached and held constant for 1 hr, this glucose rate was defined as the steady state glucose infusion rate (SSGIR). The process was repeated with a rate of insulin infusion of 6 mU/kg liveweight per minute. This protocol was carried out during pregnancy (128-134 days post-conception), peak lactation (33-40 days post-lambing) and the non-breeding state. The ewes were individually fed at maintenance and were in the fed state when experiments were conducted. The SSGIR was analysed using a linear mixed effects model (SAS® v8.0), with physiological state (pregnant, lactating, non-breeding) and insulin infusion rate (0.6 and 6.0 mU/kg liveweight per minute) as fixed effects, HFAT was fitted as a covariate and animal was used as a random term. Models were fitted with and without the inclusion of liveweight or condition score. Animal was used as a random term. In the data set there was no correlation between HFAT and condition score.

Results

The effect of HFAT on SSGIR was only evident at the high insulin infusion rate, showing a positive association (P<0.05) between the two but only for the pregnant and lactating states (Figure 1). The effect during lactation was twice that of the pregnant state, with SSGIR increasing by 65% across the range of HFAT ASBV during lactation compared to only 29% during pregnancy. There was no effect of liveweight on SSGIR, however there was a positive association ($P<0.05$) between condition score and SSGIR across all states, with SSGIR increasing by 3.3±1.3 ml/h for each unit increase in condition score. The inclusion of liveweight or condition score in the final model and no impact on the relationship between HFAT and SSGIR.

SSGIR did not differ between states ($P>0.05$) at the low insulin infusion rate (7.4±1.3, 9.3±1.4 and 9.2±1.5 ml/h for pregnant, lactating and non-breeding states). However, at the high insulin infusion rate the SSGIR was 17% lower ($P<0.05$) in the pregnant state (18.5±1.4 ml/h) than either lactating (23.0±1.5 ml/h) or non-breeding (21.5±1.5 ml/h) states.

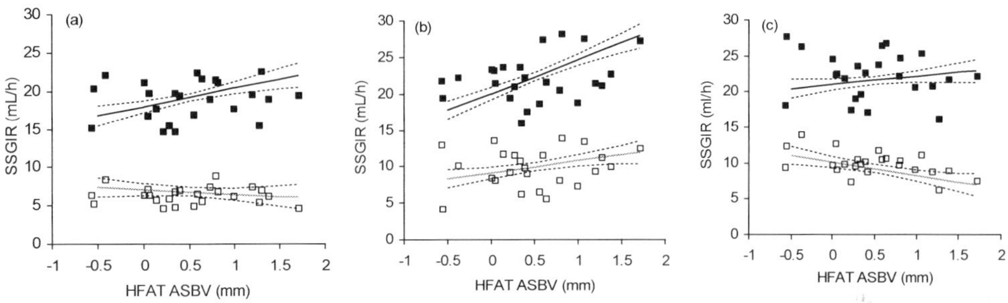

Figure 1. The effect of ewe fat breeding value (HFAT ASBV) on steady state glucose infusion rate (SSGIR; 50% glucose solution). Predicted means ± se at insulin infusion rates of 0.6 (grey line, open squares) and 6.0 mU/kg.min (black line, closed squares) during (a) pregnant, (b) lactating, and (c) non-breeding states. Symbols represent an experiment on a single sheep, values have been adjusted for effects significant in the final model.

Discussion and conclusion

In both pregnant and lactating ewes, those with higher HFAT breeding values had greater uptakes of glucose in response to high insulin. This finding is contrary to our hypothesis, however it is possible that since the ewes were managed in moderate condition, those that had a genetic propensity to be fatter maintained a higher whole body responsiveness to insulin to enhance fat storage, similar to that which occurs in early pregnancy (Ramos *et al.*, 2003). We hypothesise that ewes with higher fat breeding values maintain a state of maternal tissue storage by having a higher uptake of glucose into fat cells when circulating levels of insulin are elevated. Further investigation is required to determine the impact of differences in genetic fatness on response to insulin when ewes are fatter than they were in this experiment.

References

Bergman, E.N., S.S. Reulein and R.E. Corlett, 1989. Effects of obesity on insulin sensitivity and responsiveness in sheep. Am. J. Phys. 257: E772-E781.

Hegarty, R.S., D.L. Hopkins, T.C. Farrell, R. Banks and S. Harden, 2006. Effects of available nutrition and sire breeding values for growth and muscling on the development of crossbred lambs. 2: Composition and commercial yield. Aust. J. Agric. Res. 57: 617-626.

Prior, R.L. and S.B. Smith, 1982. Hormonal effects on partitioning of nutrients for tissue growth: role of insulin. Fed. Proc. Fed. Am. Soc. Exp. Biol. 41: 2545-2549.

Ramos, M.P., M.D. Crespo-Solans, S. del Campo, J. Cacho and E. Herrera, 2003. Fat accumulation in the rat during early pregnancy is modulated by enhanced insulin responsiveness. Am. J. Phys. 285: E318-E328.

Effect of supplementation with different urea levels on young grazing bulls recently weaned in the dry season in tropical conditions

H.J. Fernandes[1,2], M.O. Porto[2], A.A. Rocha[2], J. Cavali[2] and M.F. Paulino[2]
[1]State University of Mato Grosso do Sul – UEMS / FUNDECT, MS. Rod. Aquidauana-Cera km 12. Aquidauana, MS. 79200-000, Brazil; [2]Federal University of Viçosa. Dep. de Zootecnia, UFV. Viçosa, MG. 36570-000, Brazil; ike.fernandes@hotmail.com

Introduction

The first dry season of the animal's life is the most critical phase in grazing production systems of beef cattle in the tropics. The beginning of the dry season results in rapid decline in the nutritional quality of the forage available to recently weaned calves. The reduction in nitrogen (N) levels in plants is the main factor associated with this nutritional restriction (Poppi and McLennan, 1995). The use of concentrated supplements, particularly those rich in non-protein nitrogen (NPN) compounds, has been adopted in order to reduce this problem. However, efficient utilisation of N from supplements and from the pasture depends upon the type of protein supplemented and the availability of energy sources and other nutrients. Inappropriate use of extra N can actually make the system less efficient by the waste of nutrients. There is also an environmental risk due to potential contamination of water reserves by the excreted surplus N. The objective of this experiment was to evaluate the effects of concentrated dietary supplements with different urea levels on protein utilisation in recently weaned grazing bulls during the tropical dry season.

Material and methods

This study was conducted at the Federal University of Viçosa, in Viçosa, Brazil, during the dry season. Forty young bulls, with initial weights of 256±30 kg were randomly distributed into four groups and housed on *Brachiaria decumbens* Stapf. pastures. To avoid possible differences between the pastures, the groups were rotated among the pastures every seven days. Animals within a group were supplemented daily with either mineral salt *ad libtum* (control), or 1.5 kg of one of the three concentrate rations. The concentrates were formulated with corn, soybean meal and urea, according to the requirements of the animals. The same level of crude protein (CP) (32.5%) was maintained in the isonitrogenous supplements, but the non-protein nitrogen (NPN) content varied depending on the amount of urea, which comprised 0, 4, or 8% of the dry matter. Spot urine and blood samples were collected on day 55 of the experimental period, approximately 4 hours after the supplement supply. Urine samples were analysed to determine creatinine, urea and purine derivatives, and blood samples for plasma urea-N (PUN), as described by Pina *et al.* (2009). Urine volume was estimated using creatinine excretion. Microbial crude protein (mCP) synthesis was estimated via total excretion of purine derivatives (Chen and Gomes, 1992). The effect of concentrate supplementation and the linear and quadratic effects of urea levels were evaluated by partitioning the treatment sum of squares in orthogonal contrasts. The significance level used was 5%.

Results and discussion

Protein Concentrate supplementation affected mCP synthesis, PUN levels, and urinary urea excretion (UUE) in young grazing bulls (P<0.01), indicating increases in the protein metabolism of the animals (Table 1). The increase in urea excretion can be associated with the increase of available N, and not to a reduction in the efficiency of N use. In fact, the urea level in different supplements did not affect any of the parameters measured. The similar levels of PUN and UUE among the supplemented animals showed that they had comparable capacities for utilising metabolic N. These results can be partly explained by the ingestive behavior of the animals. Increase in the urea level reduces the velocity of the intake, meaning that the animals fed with concentrate supplement lacking urea consumed this supplement more quickly. Despite the rapid intake of protein, the degradability rate of this one was smaller than urea. This facilitated a better use of protein by the rumen microorganism and by the animal during the day. On the contrary, due to slower consumption by animals that received the concentrate supplement with 8% urea, the urea intake was distributed during the day, which could also allow a more efficient metabolism of the protein content. Animals receiving the supplement with 4% urea also obtained this balance between the velocity of supplement intake and protein degradation.

Table 1. Effects of concentrate supplementation and levels of urea on the ruminal microbial crude protein synthesis (mCP), plasma urea-N (PUN), and urinary urea excretion (UUE).

Item	Supplement				CV (%)	*P*-value (effects)		
	Salt (control)	Concentrate with				Supplementation	Urea level	
		0% urea	4% urea	8% urea			Linear	Quadratic
mCP, g/d	161.8	522.3	398.0	441.4	52.2	0.001	0.383	0.298
PUN, mg/dl	6.2	19.6	21.9	22.0	17.2	<0.0001	0.084	0.340
UUE								
mg/KgBW	74.5	512	490	443	30.5	<0.0001	0.199	0.791
g/d	15.5	126.7	124.4	108.6	29.2	<0.0001	0.156	0.534

Conclusion

Concentrate supplementation increased protein metabolism in young grazing bulls in their first dry season. However, increasing urea (NPN) levels in the isonitrogenous supplement did not affect protein use.

References

Chen, X.B. and M.J. Gomes, 1992. Estimation of microbial protein supply to sheep and cattle based on urinary excretion of purine derivatives – an overview of technical details. Aberdeen: Rowett Research Institute/ International Feed Research Unit. 21p.

Pina, D.S., S.C. Valadares Filho, L.O. Tedeschi, A.M. Barbosa and R.F.D. Valadares, 2009. Influence of different levels of concentrate and ruminally undegraded protein on digestive variables in beef heifers. J. Anim. Sci. 87: 1058-1067.

Poppi, D.P. and S.R. McLennan, 1995. Protein and energy utilization by ruminants at pasture. J. Anim. Sci. 73: 278-290.

Effects of amino acid infusion on ghrelin action in lactating cows

R. Fukumori[1], A. Yokotani[1], T. Sugino[1], F. Itoh[2], H. Shingu[2], N. Moriya[2], Y. Hasegawa[3], M. Kojima[4], K. Kangawa[5], T. Obitsu[1], S. Kushibiki[2] and K. Taniguchi[1]
[1]Graduate School of Biosphere Science, Hiroshima University, 739-8528, Higashi-Hiroshima, Japan; [2]National Institute of Livestock and Grassland Science, 305-0901, Tsukuba, Japan; [3]School of Veterinary Medicine and Animal Sciences, Kitasato University, 034-8628, Towada, Japan; [4]Institute of Life Science, Kurume University, 839-0864, Kurume, Japan; [5]National Cardiovascular Center Researchi Institute, 565-8565, Osaka, Japan; sugino@hiroshima-u.ac.jp

Introduction

Ghrelin, mainly secreted by the abomasums in ruminants (Hayashida *et al.*, 2001), is related to feeding and nutritional status (Sugino *et al.*, 2004), and can stimulate food intake and GH secretion. Although plasma ghrelin concentrations are affected by dietary compositions in wethers (Takahashi *et al.*, 2007), it is unclear whether ghrelin affects nutrient metabolism. We investigated the effects of amino acid infusion on ghrelin action in lactating cows.

Material and methods

Six Holstein dairy cows (calving number: 3.3±0.8, the days in milk: 57.2±2.1, daily milk yield: 32.5±0.4 kg, BW: 687±18 kg) were fed total mixed ration twice daily (9:00 h and 18:00 h) to meet the nutrient requirements of the Japanese feeding standard and were trained to an assigned meal feeding regimen for 10 days. Then, cows were randomly assigned to two infusion treatments in a cross-over design (an interval of 1 week). At 4 hours after feeding (13:00 h), a mixture solution of amino acids [Aminic. (Ajinomoto Pharma Co, Ltd.) plus methionine dissolved into saline, (total amino acid: 91.0 mg/ml, infusion rate: 4.0 mg/kgMBW/min)] (AMI) or saline (CON) was continuously infused into the jugular vein catheter for 4 hours. At 2 h after the initiation of infusion, synthetic bovine ghrelin (1 μg/kgBW) was single injected into the jugular vein through a catheter. Blood samples were taken at -130, -120, -10, 0, 5, 10, 15, 20, 25, 30, 40, 50 and 60 min relative to the ghrelin injection from the catheter of another side jugular vein and collected in heparinised tubes with aprotinine. Plasma ghrelin, GH and insulin concentrations were determined by time-resolved fluoro-immunoassay (TR-FIA, Sugino *et al.*, 2004). Plasma glucagon concentrations were determined by RIA. Plasma metabolites (glucose, NEFA, alpha-amino N and urea-N) were measured by enzymatic methods. All dates were evaluated by a Student *t*-test using a mixed model analysis. $P<0.05$ was considered as significant.

Results

Amino acid infusion increased the concentrations of plasma alfa-amino N and urea-N ($P<0.05$), but not plasma hormones, glucose and NEFA. Ghrelin injection rapidly increased plasma GH with no difference between CON and AMI. Similarly, plasma levels of insulin and glucagon were increased by ghrelin injection in both treatments ($P<0.05$, Figure 1A, B). Threreafter, plasma glucagon and insulin concentrations gradually declined, but were higher in AMI compared with CON ($P<0.05$, Figure 1A, B). On the contrary, the increased plasma glucose and NEFA concentrations by ghrelin injection decreased more rapidly in AMI (Figure 1C, D).

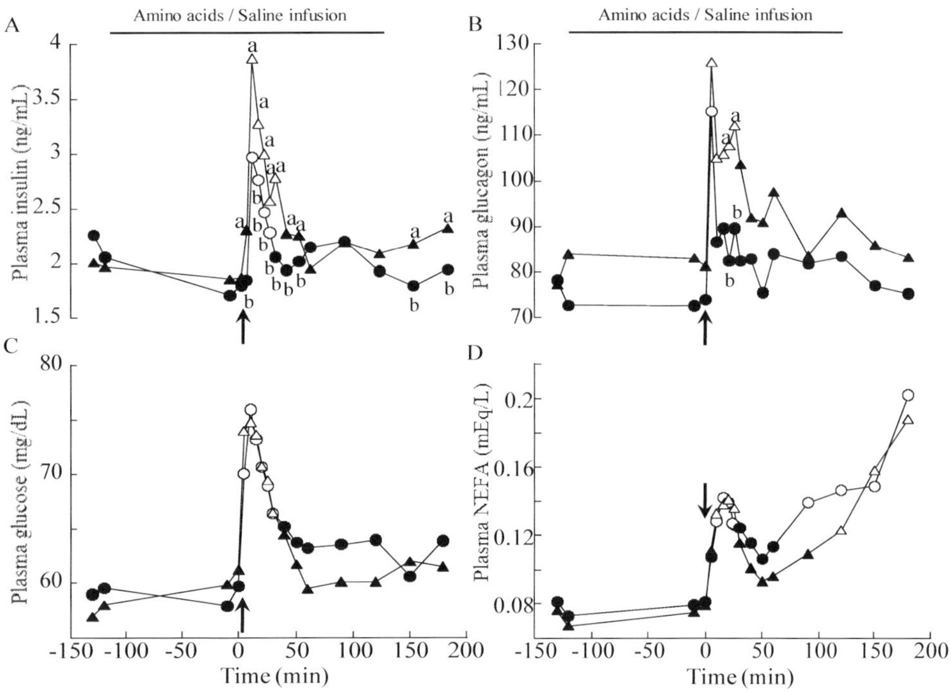

Figure 1. Effects of ghrelin injection on plasma insulin, glucagon, glucose and NEFA concentrations in amino acids/saline infusion of lactating cows. Plots of circle and triangle show the CON and AMI treatments, respectively. a, b: P<*0.05, between CON and AMI at each time. Open plot:* P<*0.05, between pre- (-10 min and 0 min) and post-injection. The arrow and bar show the initiation of ghrelin injection and the period of amino acids/ saline infusion, respectively.*

Conclusion

These results suggest that ghrelin increases plasma pancreatic hormone and metabolite levels. Therefore, ghrelin may directly act as a regulator to nutrient metabolism in lactating cows. Furthermore, amino acid infusion enhanced ghrelin action on pancreatic hormone secretion. Amino acids may modify ghrelin action in lactating cows.

References

Hayashida, T., K. Murakami, M. Nishihara, M. Nakazato, M.S. Mondal, Y. Horii, M. Kojima, K. Kangawa and N. Murakami, 2001. Ghrelin in domestic animals: distribution in stomach and its possible role. Domestic Anim. Endocrinol. 21: 17-24.

Sugino, T., Y. Hasegawa, Y. Kurose, M. Kojima, K. Kangawa and Y. Terashima, 2004. Effects of ghrelin on food intake and neuroendocrine function in sheep. Anim. Reprod. Sci. 82-83: 183-194.

Takahashi, T., Y. Kobayashi, S. Hasegawa, E. Touno, Y. Otani, S. Haga, F. Itoh, K. Katoh and Y. Obara, 2007. Different responses in postprandial plasma ghrelin and GH levels induced by concentrate or timothy hay feeding in wethers. Domestic Anim. Endocrinol. 34: 432-439.

Effect of *Lactobacilli* probiotic supplementation on blood glucose, insulin and NEFA performance of dairy cattle during late pregnancy and early lactation

M.A. Galina[1], V.J. Chavez[1], J. Pineda[2], J.D. Hummel[2], R.M. Ortiz and M. Delgado-Pertiñez[3]
[1]FES Cuautitlan-Universidad Nacional Autónoma de México, km 2.5 carretera Cuautitlan-Teoloyucan. Xhala, Cuautitlan Izcalli, Edo. de México; [2]Facultad de Medicina Veterinaria y Zootecnia Universidad de Colima, México; [3]Departamento de Ciencias Agroforestales, Universidad de Sevilla, Spain; miguelgalina@correo.unam.mx

Introduction

Bacteria of the genus *Lactobacillus* (LAB) have been proven to be beneficial microorganisms of particular interest because of their long successful history in human health (Holzapfel, 2002). Lactobacilli were among the first organisms used by man for processing foodstuffs (Konigs *et al.*, 2000) and for preserving food by inhibiting invasion by some pathogenic microorganisms (Adams, 1999). Supplementation with LAB probiotics as beneficial microorganisms could be an important alternative for feeding dairy ruminants (Galina *et al.*, 2008). The objective of this study was to evaluate the effect of LAB/SIUS supplementation (probiotic) during the *pre* and *post partum* period on transition performance of dairy cattle.

Material and methods

Eighty-four multiparous Jersey cows (540±22 kg) selected by milk performance in previous lactation (±24 kg/d) were divided into two treatments (42 cows each) from 3 wk *pre partum* throughout 10 wk *post partum*. Treatment groups consisted of two *pre partum* diets (with or without SIUS/LAB). After calving, each group was individually fed with a lactation diet (with or without SIUS/LAB). Probiotic supplementation (LAB) contained approximately 4×10^7 cfu of lactic bacteria composed of *Lactobacilli plantarum*; *L. delbrueckii*; *L. helvaticus*; *Lactoccocus lactis*; *Leuconostoc mesenteroides,* and *Bifidus* spp. blend on a mixture of 35% molasses and 65% cheese whey. Two-hundred fifty g supplement/cow per day in both *pre partum* and lactation period. The SIUS formula was published before by Galina *et al.* (2007). *Pre-* and lactation diets were offered twice a day, assuring at least a 10% refusal. Daily DMI per cow were recorded. Daily milk production and weekly composition were measured. Blood samples were taken weekly since 21 d before parturition, continuing daily after parturition until the end of the study. Blood samples were taken before the morning feeding and throughout 24 h to determinate NEFA (Plasma nonesterified fatty acids) and BHBA (Blood β hydroxybutyrate) according to Nocek *et al.* (2003). The results were evaluated by split-plot-in-time for repeated measures procedure of SAS®: dependent variable = μ + treatment + period + cow (treatment) + treatment * period + E, where μ is the overall mean of the population and E is the random error. When treatment was significant ($P<0.05$), the Tukey-Kramer test was used to identify treatment effects within the period as described by Nocek *et al.*, (2003).

Results

Dry matter intake from 21 to 8 d *pre partum* were higher in SIUS/LAB animals, 11.2 kg/d compared to control 9.4 kg/d ($P<0.05$). During the 2nd and 3rd week *post- partum* cows fed LAB had higher intakes ($P<0.05$) than those fed no LAB (15.8 kg/d control 17.4 kg/d SIUS/LAB). On day 8 through 21, cows consuming LAB *pre* and *post- partum* produced more milk than cows receiving no LAB, 21.2 kg control 25.3 kg SIUS/LAB ($P<0.05$). This same general tendency ($P<0.05$) was demonstrated from weeks 3 through 10 *post partum*. Although there were no significant effects of treatment on milk fat percentage, cows that received LAB regardless of stage tended ($P<0.1$) to have numerically higher milk fat percentages than those that did not receive LAB (4.20g% to 4.12% respectively). During the first week *post partum*, milk protein percentage was higher ($P<0.05$) with LAB/SIUS (3.30%) in comparison with 3.05% in the no LAB group. During the first 3 wk *post-partum*, there was a treatment effect on blood BHBA, control 0.81 μM/L; 0.88 μM/l SIUS/LAB and NEFA 635 μM/l control; 589 μM/l LAB/SIUS ($P<0.05$). After 22 d *post- partum* NEFA and BHBA were lower in cow feeding LAB/SIUS *pre partum* and NEFA also remained lower in cows feeding Lab/SIUS *post partum* ($P<0.05$). Blood glucose levels 51.8 μM/l control; 60.4 μM/l LAB/SIUS and insulin 22.5 μM/l control; 24.3 μM/l LAB/SIUS were higher ($P<0.05$) compared to plasma NEFA levels were lower ($P<0.05$) for cows receiving LAB/SIUS during the *post partum* period (635 μM/l control; 589 μM/l LAB/SIUS). Blood metabolite information suggests this response was associated with more glucose being made available and less fatty acids being mobilised from lipid stores. Increased DMI in LAB/SIUS probably resulted from higher cell wall utilisation and development of bacterial protein from non protein nitrogen as previously demonstrated (Galina *et al.*, 2007).

Conclusion

Including LAB/SIUS in the *pre partum* and *post partum* diets did improve quantity and quality of milk production. In addition, there were improved metabolic profiles in cows as reflected by glucose, insulin, NEFA and BHBA status. The performance of cows supplemented with LAB/SIUS was improved.

Acknowledgement

The authors thank PAPIIT IN 200809 UNAM

References

Adams, M.R., 1999. Safety of industrial lactic acid bacteria. J. Biotechnol. 68: 171-178.

Galina, M.A, M.A. Ortiz-Rubio, M. Delgado-Pertiñez and L.J. Pineda, 2008. Effect of a lactic probiotic supplementation on goat kids growth. J. Options Méditerrannées.

Galina, M.A., M. Guerrerro and D.C. Puga, 2007. Fattening Pelibuey lambs with sugar cane tops and corn complemented with or without Slow Intake Urea Supplement. Small Rum. Res.70:101-109.

Holzapfel, W.H., 2002. Appropiate starter culture technologies for small-scale fermentation in developing countries. Int. Food Microbiol. 75: 197-212

Konings, W.N., J. Kok, O.P. Kuipers and B. Poolman, 2000. Lactic acid bacteria: the bug of the new millennium. Curr. Opin. Microbiol. 3: 276-282.

Nocek, J.E., W.O. Kautz, A.Z. Leedle and E. Block, 2003. Direct-fed microbial supplementation on the performance of dairy cattle during the transition period. J. Dairy Sci. 86: 331-335.

An unprotected conjugated linoleic acid (CLA) supplement reduces milk fat synthesis and forage intake in lactating goats

M.A.S. Gama[1], D.E. Oliveira[2], D. Fernandes[2], J. de Souza[2] and J.H. Bruschi[1]
[1]National Dairy Cattle Research Centre, Embrapa, Juiz de Fora, Minas Gerais, Brazil; [2]Santa Catarina State University (UDESC), Chapecó, Santa Catarina, Brazil; gama@cnpgl.embrapa.br

Introduction

Supplements containing *trans*-10,*cis*-12 CLA reduce milk fat synthesis in lactating cows, with the magnitude of response being similar to that observed in lactating ewes when the same CLA dose (on a BW basis) is fed (Lock *et al.*, 2006). In contrast, milk fat content of lactating goats was unchanged when similar doses of *trans*-10,*cis*-12 CLA were fed as lipid-encapsulated (Erasmus *et al.*, 2004) or infused duodenally (Andrade and Schmidely, 2006). This suggests that goats are less sensitive to *trans*-10,*cis*-12 CLA than cows and ewes, corroborating data from a recent dose-response study (Shingfield *et al.*, 2009). Besides, the lack or little response to CLA in some of the above-mentioned studies could be related to a short supplementation period (8 d or less), since Lock *et al.* (2008) found similar milk fat reduction when lactating goats received for two weeks about 50% of lipid-encapsulated *trans*-10,*cis*-12 CLA dose used by Erasmus *et al.* (2004). This study was designed to evaluate the effects of an unprotected CLA supplement on feed intake, milk production and milk composition of lactating goats.

Material and methods

Twenty Toggenburg lactating goats (primiparous, 60 to 110 DIM, average initial milk yield and BW of 2.8±0.4 kg/d and 40±3.7 kg, respectively) were blocked according to milk production and BW and randomly assigned to the following treatments: (a) Control: 30 g/d of calcium salts of soybean oil and (b) CLA: 30 g/d of an unprotected CLA supplement. The experimental design was a cross-over with 14-d treatment periods separated by 6-d washout intervals. The fat supplements were mixed into 1.2 kg of concentrate and fed individually twice a day (0.6 kg per meal) after morning and afternoon milking. The CLA supplement contained about 30% of *trans*-10,*cis*-12 CLA as the methyl-ester. Diet was composed of corn silage and a concentrate mixture which were fed separately. The corn silage was fed *ad libitum* and orts were weighed daily in order to calculate the forage intake. Milk yield was recorded daily and milk samples were collected every 2 days for analysis of its components and SCC. Goats were weighed at the beginning and end of each experimental period (1st and 14th days, respectively). Data were analysed as repeated measures design using the MIXED procedure of SAS® (2000) assuming period and goat within treatment sequence as random effects. The statistical model included treatment, day and interaction treatment vs. day as sources of variation. Differences between treatments were declared significant at $P<0.05$.

Results and discussion

Least squares means for milk yield, milk composition and forage intake from Control and CLA treatments are presented in Table 1. Milk fat content and yield were reduced by 16.3 and 15.5% in response to CLA, respectively ($P<0.01$). The temporal variation showed that milk fat yield was gradually decreased by CLA from the 4th to 10th day of treatment, when it reached a plateau (interaction day vs. treatment, $P<0.01$). Milk protein content and yield were similar between treatments, but lactose content was lower in CLA than in the Control ($P<0.01$). Noteworthy, it was observed that CLA reduced forage intake by 6.4% ($P<0.01$) and tended to increase the milk yield (20 g/d) very slightly ($P=0.07$).

Table 1. Milk yield, forage intake and milk composition from lactating goats receiving Control or CLA supplement for 14 days.

Variable	Treatments[1]		SE	Effects (*P*-value)		
	Control	CLA		treatment	day	interaction
Milk yield, kg/d	2.45[a]	2.47[a]	0.25	0.07	<0.01	<0.01
Forage intake, kg/d (as fed)	4.50[a]	4.21[b]	1.14	<0.01	<0.01	0.88
Milk fat,%	3.00[a]	2.51[b]	0.17	<0.01	<0.01	<0.01
Milk protein,%	2.66[a]	2.68[a]	0.18	0.17	<0.01	0.07
Milk lactose,%	4.17[a]	4.14[b]	0.09	<0.01	<0.01	<0.01
Fat yield, g/d	73.0[a]	61.7[b]	5.78	<0.01	<0.01	<0.01
Protein yield, g/d	63.6[a]	63.9[a]	4.13	0.43	<0.01	<0.01
Lactose yield, g/d	102.0[a]	101.7[a]	9.85	0.67	<0.01	0.65
Linear score for SCC	5.53[a]	5.41[a]	0.79	0.13	0.04	0.89

[1] Control = 30 g of calcium salts of long chain fatty acid; CLA = 30 g of an unprotected CLA supplement.
[a,b] Means within rows with different superscript letters differ significantly ($P<0.05$).

Conclusion

It was concluded that dietary supplementation with a source of unprotected CLA reduces milk fat synthesis and forage intake in lactating goats.

Acknowledgement

The authors thank Fapemig and Agrofuturo for financial support.

References

Lock, A.L., B.M. Teles, J.W. Perfield II, D.E. Bauman and L.A. Sinclair, 2006. A conjugated linoleic supplement containing *trans*-10, *cis*-12 reduces milk fat synthesis in lactating sheep. J. Dairy Sci. 89: 1525-1532.

Erasmus, L.J., Z. Bester, T. Fourie, R.J. Coertze and L Hall, 2004. Effect of level of rumen protected CLA supplementation on milk yield and composition in Saanen goats. S. Afr. J. of Anim Sci., 34 (Suppl.1): 42-45.

Andrade, P.V.D. and P. Schmidely, 2006. Effect of duodenal infusion of *trans*-10, *cis*-12 CLA on milk performance and milk fatty acid profile in dairy goats fed high or low concentrate diet in combination with rolled canola seed. Reprod. Nutr. Dev. 46: 31-48.

Lock, A.L., M. Rovai, T.A. Gipson, M.J. De Veth and D.E. Bauman, 2008. A conjugated linoleic acid supplement containing *trans*-10, *cis*-12 conjugated linoleic acid reduces milk fat synthesis in lactating goats. J. Dairy Sci. 91: 3291-3299.

Shingfield, K.J., J. Rouel and Y. Chilliard, 2009. Effect of calcium salts of a mixture of conjugated linoleic acids containing *trans*-10 *cis*-12 CLA in the diet on milk fat synthesis in goats. Br. J. Nutr. 101: 1006-1019.

The effects of chromium supplementation on blood parameters related to protein and lipid metabolism in early lactating cows

G.R. Ghorbani, M. Khorvash, M. Mirzaee and H.R. Rahmani
Department of Animal Science, Isfahan University of Technology, Isfahan, Iran; mirzaee.1984@gmail.com

Introduction

Insulin is the primary anabolic hormone in the body that manages amino acid direction to muscle cells. Chromium (Cr) is a mediate for insulin anabolic activities (Mowat, 1997). Chromium increases glucose and some amino acid uptake by facilitating conjunction of insulin to its receptor (Pechova *et al.*, 2002). High plasma non-esterified fatty acid (NEFA) concentrations represent high catabolic activities and negative energy balance of dairy cattle in early lactation. If Cr increases insulin sensitivity, lipogenesis should be stimulated and lipolysis inhibited, theoretically resulting in a decrease in NEFA concentrations (Kegley *et al.*, 2000). Decreases in NEFA concentration can increase dry matter intake through the lipoacetate mechanism. This study was conducted to evaluate the effects of Cr supplementation on blood metabolites (protein and lipid metabolism) of Holstein cows during early lactation.

Material and methods

Fifteen Holstein cows (12 multiparous and three primiparous) averaging 38±12 d in milk and 620±45 kg weight were arranged in three dietary chromium treatments of (1) control (no supplemental Cr), (2) 0.05 and (3) 0.10 mg Cr/kg $BW^{0.75}$ as chromium-methionine. The supplement was top dressed with 250 g of ground corn in the morning feed (Sumner *et al.*, 2007). This trial was conducted as a complete randomised design lasting for 65 d during the summer 2007. The amount of Cr-Met applied in this experiment was based on their body weight measured at the beginning of the experiment and biweekly. Cows were housed in individual pens (4×4 m). After the initial two weeks, cows were bled weekly at 11:30 h (2.5 h after the morning feed) by puncture of the coccygeal vein for seven weeks. Blood samples were centrifuged at 2000×*g* for 15 min and obtained serums were preserved at -20 °C. Serum albumin was measured by biconjugate gradient (BCG) using a commercial kit (Pars Azmoon, Tehran, Iran). Concentration of total protein, triglycerides (TG), and high density lipoproteins (HDL) were measured using Biuret, enzymatic and sedimentary methods with commercial kits (Pars Azmoon, Tehran, Iran), respectively. NEFA were determined using a chemical method with a commercial kit (DRG, Co. Germany) as well. Data were analysed using the mixed procedure with time as repeated measures (SAS®, 2005, SAS Institute Inc., Cary, NC, USA). Individual cow was used as the experimental unit for blood data. The model for serum data included treatment, time, treatment×time interaction. Significance was declared at $P \leq 0.05$; trends were declared at $(0.05 < P \leq 0.10)$.

Results

Blood data are presented in Table 1. Chromium supplemented groups had higher serum albumen concentration than the control group but only the difference between 0.05 and control groups was significant. Total protein, globulin concentrations and albumin to globulin ratio were not affected by chromium. Chromium supplementation level of 0.05 tended to reduce significantly serum NEFA concentration compared with the control level. High density lipoprotein, cholesterol and triglycerides were not affected by chromium supplementation.

Table 1. The effects of chromium levels on blood metabolites.

	Treatment (Trt or Cr dose)			SEM	*P*-value
	0	0.05	0.10		Trt
Total proteins, g/dl	7.35	7.40	7.38	0.06	0.45
Albumin, g/dl	3.77[a]	3.91[b]	3.88[ab]	0.05	0.26
Globulin, g/dl	3.58	3.48	3.49	0.06	0.89
Albumin:globulin	1.06	1.14	1.13	0.03	0.71
NEFA, μEq/l	158.9[a]	144.6[b]	157.3[ab]	4.7	0.07
Triglycerides, mg/dl	19.3	18.3	19.9	0.74	0.31
Cholesterol, mg/dl	130.4	133.1	131.2	1.3	0.30
High-density lipoprotein, mg/dl	25.5	27.9	26.6	0.9	0.21

Conclusion

The changes in the albumin concentration with 0.05 Cr supplementation indicate a higher hepatic production of albumin since it is only produced in the liver, suggesting that Cr may improve amino acid synthesis, possibly via insulin. Supplementation with the medium level of chromium mediated hypolipidemic effects on serum metabolites because of reduction in lipolysis of adipose tissue and hepatic ketogenesis. Hypolipidemic effects of chromium supplementation are related to insulin effects on decreased lipolysis and increased fatty acid connection to adipocytes. Decreased NEFA can in turn increase DMI by reversing the lipostatic mechanism, since high circulating NEFA concentrations depress feed intake.

Acknowledgement

The authors are grateful to Isfahan University of Technology for research and teaching services and the Sanadam-Pars Corporation (Tehran, Iran) for partial financial support.

References

Kegley, E.B., D.L. Galloway and T.M. Fakler, 2000. Effect of dietary chromium-L-methionine on glucose metabolism of beef steers. J. Anim. Sci. 78: 3177-3183.

Mowat, D.N., 1997. Organic chromium in animal nutrition. Chromium Books, Guelph, ON, Canada.

Pechova, A., J. Illek, M. Sindelar and L. Pavlata, 2002. Effects of chromium supplementation on growth rate and metabolism in fattening bulls. Acta Vet. Brno. 71: 535-541.

Sumner, J.M., F. Valdes and J.P. McNamara, 2007. Effects of chromium propionate on response to an intravenous glucose tolerance test in growing Holstein heifers. J. Dairy Sci. 90: 3467-3474.

Production and processing studies on calpain-system gene markers in cattle

P.L. Greenwood[1], L.M. Cafe[1], D.W. Pethick[2], D.L. Robinson[1] and J.M. Thompson[3]
Australian Cooperative Research Centre for Beef Genetic Technologies
[1]NSW DPI Beef Industry Centre of Excellence, UNE, NSW 2351, Armidale, Australia; [2]School of Veterinary & Biomedical Science, Murdoch University, 6150, WA, Australia; [3]University of New England, NSW 2351, Armidale, Australia; paul.greenwood@dpi.nsw.gov.au

Introduction

Gene markers for tenderness have been shown to be related to shear force of meat from cattle (Barendse *et al.*, 2008; Page *et al.*, 2002; White *et al.*, 2005). Four tenderness markers currently available commercially are based on polymorphisms in genes controlling components of the *post-mortem* tenderisation process, specifically the calpain proteolytic system. The objectives of this study were to quantify the magnitude of effects of tenderness gene markers on growth, feed efficiency, carcass characteristics and beef quality in Brahman cattle, and to quantify interactions within and between tenderness gene markers, gender, hormonal growth promotant, method of carcass hang, and major muscles.

Material and methods

Bos indicus cattle (n=1,700) in commercial and research herds (n=15 herds) were blood sampled and genotyped for beef tenderness markers (calpastatin, *CAST*; calpain 1, *CAPN1*; and calpain 3, *CAPN3*) at weaning. Two concurrent experiments were conducted, one in New South Wales (NSW, n=167) and one in Western Australia (WA, n=73). Experimental herds were established based on gene marker status. The experimental design in NSW comprised 2 *CAST* status (0 vs. 2 favourable alleles) × 2 *CAPN3* status (0 vs. 2) × HGP treatment (+ or - Revalor-H® during feedlotting) × gender (heifer *vs* castrate male) × method of hang (achilles vs. tenderstretch) × aging period (1 vs. 7 d). The experimental design in WA comprised 3 *CAST* status (0 vs. 1 vs. 2) × 3 *CAPN3* status (0 vs. 1 vs. 2) × HGP treatment × method of hang × aging period. In both herds, groups were balanced for *CAPN1* status. The weaner cattle were backgrounded for 6 months then grain-finished in a feedlot for 82 d (WA) or 116 d (NSW). In the NSW herd, cattle were allocated during feedlotting to pens (10-11 cattle/pen) with automated feed intake recorders to measure individual animal intake. Growth, feedlot intake and efficiency, carcass and Meat Standards Australia (MSA) chiller characteristics, and *longissimus* (LD, striploin) and *semitendinosus* (ST, eye round) objective beef quality characteristics were measured.

Results

A reduction ($P<0.05$) in LD shear force in achilles hung carcasses after 7 d aging of 12.0 N for the NSW herd (Table 1) and 0.93 N for the WA herd was evident for the cattle with the favourable *CAST* and *CAPN3* alleles compared to those with the unfavourable alleles (2_2 vs. 0_0 alleles, respectively). Significant differences (WA, $P<0.05$) or tendencies towards significant differences (NSW, $P<0.10$: Table 1) between the 0_0 and 2_2 cattle were also evident for the LD of the tenderstretched sides after 1 d (results not shown) and 7 d aging. There were no significant associations between the gene markers and LD compression, cooking loss and meat colour, and ST shear force or other meat quality measurements (results not shown). There were no significant effects of the gene markers on carcass and MSA chiller characteristics (Table 1) or on growth, feed intake or feed efficiency (Table 2) for either herd.

Table 1. Effect of calpastatin (CAST) and calpain 3 (CAPN3) markers on longissimus *shear force at 7 d ageing for normally (AT) and tenderstretch (TS) hung carcass sides, and on carcass characteristics of Brahman cattle (steers and heifers) from the NSW herd. Least squares means with different superscripts differ (*$P<0.05$*, italic* $P<0.10$*).*

CAST_CAPN3	n	AT 7d shear (N)	TS 7d shear (N)	HSCW (kg)	EMA (cm^2)	HotP8 (mm)	Uslean (colour)	Temp at pH6	pHu
0_0	38	78.5^{a}	47.3^{a}	246	59.1	12.7	171	22.1	5.49
0_2	26	71.3^{ab}	46.7^{ab}	244	59.0	12.4	171	22.3	5.49
2_0	45	73.8^{ab}	45.1^{ab}	246	60.1	12.5	153	19.6	5.49
2_2	41	66.5^{b}	44.6^{b}	243	59.9	12.3	152	19.8	5.48
sed		4.13	1.33	5.6	1.55	0.6	14.3	1.55	0.011

Table 2. Effect of calpastatin (CAST) and calpain 3 (CAPN3) markers on growth and feed intake and efficiency of Brahman cattle (steers and heifers) from the NSW herd (all $P>0.10$*).*

CAST_CAPN3	n	Background ADG (g)	Feedlot entry wt (kg)	Feedlot ADG (kg)	DMI (kg/day)	FCR (kg DM/ kg gain)	Feedlot exit wt (kg)
0_0	38	737	322	1.22	8.4	7.0	441
0_2	26	752	318	1.20	8.3	7.2	437
2_0	45	719	321	1.14	8.0	7.6	439
2_2	41	734	317	1.13	8.0	7.4	435
sed		21.2	7.2	0.054	0.27	0.48	10.2

Discussion and conclusion

The favourable alleles improved tenderness (reduced shear force) of 7 d aged LD of Brahman cattle. Given the influences of the calpain proteolytic system that drives *post-mortem* tenderisation, observed increases in differences as meat aged would be expected. The positive effect of the favourable alleles was also evident in the tenderstretched LD, but less so compared to normally hung sides due to reduced variation in shear force following tenderstretch. The findings confirm the value of calpain system markers as an option in the production of more tender beef from Brahman cattle, without adverse effects on other production characteristics, at least within the present study. They also suggest potential for processing efficiencies due to different aging rates associated with the markers. Samples and results from the present tenderness gene marker study are also enabling interactions with gene markers to be further assessed, sensory (taste panel) results for beef with the different marker status to be generated for subsequent incorporation into the Meat Standards Australia model for beef eating quality, and for studies on biology underpinning these gene markers.

References

Barendse, W., B.E. Harrison, R.J. Bunch and M.B. Thomas, 2008. Variation at the Calpain 3 gene is associated with meat tenderness in zebu and composite breeds of cattle. BMC Genetics 9: 41.

Page, B.T., E. Casas, M.P. Heaton, N.G. Cullen, D.L. Hyndman, C.A. Morris, A.M. Crawford, T.L. Wheeler, M. Koohmaraie, J.W. Keele and T.P.L. Smith, 2002. Evaluation of single-nucleotide polymorphisms in CAPN1 for association with meat tenderness in cattle. J. Anim. Sci. 80: 3077-3085.

White, S.N., E. Casas, T.L. Wheeler, S.D. Shackelford, M. Koohmaraie, R.D. Riley, C.C. Chase Jr, D.D. Johnson, J.W. Keele and T.P.L Smith, 2005. A new single nucleotide polymorphism in CAPN1 extends the current tenderness marker test to include cattle of *Bos indicus*, *Bos taurus*, and crossbred descent. J. Anim. Sci. 83: 2001-2008.

Effects of grazing time allocation on intake, foraging behaviour and hunger-related hormone and metabolites of dairy cows during the first grazing session

P. Gregorini, C.E.F. Clark, J.G. Jago, C.B. Glassey, K.L.M. McLeod and A.J. Romera
DairyNZ, Ltd. Private Bag 3221, Hamilton, New Zealand; pablo.gregorini@dairynz.co.nz

Introduction

The time available to eat, and the time elapsed since the last meal are key stimuli in determining the degree of hunger (Forbes, 1995), which in turn motivates particular foraging strategies (Pittroff and Soca, 2006). These changes are mediated by the energetic status of the animal, and conveyed by the orexigenic agent ghrelin (Gregorini *et al.*, 2009), which increases with hunger. Cattle consume most of the daily available herbage during the first hours when strip-grazed. Foraging strategies during such a period, thus, have a large impact on daily herbage intake. There are few data on how, or if, the timing of pasture allocation affects blood metabolites before and foraging strategies during the first grazing session of the day (GS). The objective of the present study was to determine the impact of the timing of pasture allocation on dairy cow intake, foraging behaviour and key hunger-related hormone (ghrelin) and metabolites (non-esterified fatty acids, NEFA) during the GS after milking.

Material and methods

Forty-eight Holstein-Friesian cows (470±47 kg BW; 35±9 DIM) were strip-grazed on a perennial ryegrass pasture for 4 h after each milking (2×4), 8 h between am and pm milkings (1×8) or for the 24 h period excluding milking times (CTL). Herbage allowance was the same for all treatments. Cows did not receive supplements. Each treatment had two groups of eight cows. Cows were bled before the GS (08:00-11:30 h); plasma was analysed for ghrelin and serum was analysed for non-esterified fatty acids (NEFA). Herbage mass was measured pre, during, at the end of the GS, and post-grazing. Herbage mass data were fitted to a model (Ørskov and McDonald, 1979) to estimate pasture disappearance rates. Intake and bite mass were calculated based on herbage mass disappearance and behavioural measurement. Eating, searching, rumination, and idling behaviour were recorded during the GS every 2 min. Bite rate was determined during one continuous minute at 0, 30, 60, and 120 min after the GS started. Behavioural times, bite rate and herbage intake were summarised into means of the GS and three periods within the GS: 0-60, 60-120 and 120-210 min. Treatment effects on ghrelin and NEFA, foraging behaviour, intake during the GS, and daily intake were analysed using ANOVA. Repeated measurements through the GS were analysed as a mixed model with treatment, time and treatment × time as fixed effects and cow group as a random effect. A compound symmetry covariance structure was used for the within-group repeated measurements and heterogeneity of variance at the time points was allowed for. Pasture disappearance was analysed by the model parameter estimates and herbage disappearance rates at different times using ANOVA. GenStat 11.1 was used for statistical analysis.

Results

Although daily intake was not affected by treatment ($P>0.05$); intake during the GS (Table 1) was. Bite mass differed ($P<0.05$) between treatments (Table 1) and changed ($P<0.05$) during the GS according to treatments. Bite mass was the smallest for CTL during the first 60 min ($P<0.05$); but it was the greatest ($P<0.05$) during the last 90 min, when cows in 2x4 had the smallest ($P<0.05$) bite mass. During the GS, cows in 1×8 spent the longest ($P<0.05$) time eating and the least ($P<0.05$)

searching and ruminating (Table 1). Eating time was the highest ($P<0.05$) for 1×8 cows during the first 60 and last 90 min. Searching time only differed in the second 60 min, where it was the lowest for cows in 1×8. Cows did not ruminate during the first 120 min. Cows in CTL showed the most ($P<0.05$) rumination time during the last 90 min. The model fitted to represent dynamics of pasture disappearance showed differences ($P<0.05$) in the parameter representing the fractional herbage disappearance rate (0.7, 1.3 and 0.3%/h for 1×8, 2×4 and CTL, respectively). Simulated pasture disappearance rates were the greatest ($P<0.05$) for 1×8 during all GS. Ghrelin and NEFA concentration were affected ($P<0.05$) by treatment (Table 1).

Table 1. Effects of grazing time allocation on ghrelin and NEFA concentrations before, and foraging behaviour of dairy cows during, the first grazing session of the day (08:00-11:30 h).

	Treatment[1]			SED	Treatment effect *P*-value
	1×8	2×4	CTL		
Daily intake, kg DM/cow	12.5	13.9	13.7	1.23	0.528
Intake during GS, kg DM/cow	11.5^{a}	9.2^{b}	9.4^{b}	0.340	0.012
Eating time,% of GS	0.81^{a}	0.68^{b}	0.58^{c}	0.021	0.004
Searching time,% of GS	0.02^{b}	0.05^{a}	0.07^{a}	0.008	0.018
Rumination time,% of GS	0.009^{b}	0.012^{b}	0.112^{a}	0.032	0.079
Idling time,% of GS	0.13	0.21	0.21	0.027	0.119
Bite rate, bites/ min	52.1^{b}	54.9^{a}	46.47^{c}	0.541	0.001
Bite mass, g DM	1.28^{ab}	1.10^{bc}	1.46^{a}	0.064	0.026
Ghrelin, mmol/l	230.40^{a}	152.62^{b}	178.82^{b}	10.055	0.009
NEFA, mmol/l	0.423^{a}	0.197^{b}	0.273^{ab}	0.071	0.10

[a,b,c] Means in the same row with different superscript are different ($P<0.05$).
[1] 1×8, 8 h between milkings; 2×4, 4 h after each milking and CTL, 24 h excluding milking times.

Conclusion

During the first hours of grazing, cows in 1×8 were hungrier than cows in 2×4 and CTL cows, matching the increments in ghrelin and NEFA levels, intake and eating time during the GS. This study presents a link between foraging behaviour and physiological markers of hunger. Alteration of available grazing time sets the cows' reactions to the perception of the same feed.

References

Forbes, J.M., 1995. Voluntary food intake and diet selection in farm animals, 1st ed. CAB International, Wallingford, Oxfordshire, UK, 540 pp.

Gregorini, P., K.J. Soder and R.S. Kensinger, 2009. Effects of rumen fill on short-term ingestive behavior and circulating concentrations of ghrelin, insulin and glucose of dairy cows foraging vegetative micro-swards. J. Dairy Sci. in press.

Ørskov, E.R. and I. McDonald, 1979. The estimation of protein degradability in the rumen from incubation measurement weighed according to the rate of passage. J. Agric. Sci. Camb. 92: 499-508.

Pittroff, W. and P. Soca, 2006. Physiology and models of feeding behavior and intake regulation in ruminants. In: Bels V. (Ed.), Feeding in Domestic Vertebrates: From Structure to Behaviour. CAB International, Wallingford, Oxfordshire, UK, pp. 278-301.

Compared hepatic metabolism of linoleic and linolenic acids of finishing bovines given a n-3 PUFA-rich diet

D. Gruffat, M. Gobert, D. Durand and D. Bauchart
INRA, UR 1213 Herbivores, Site de Theix, F-63122 Saint Genès Champanelle, France; Dominique.Gruffat@clermont.inra.fr

Introduction

Feeding strategies based on the addition of vegetal lipids rich in n-3 polyunsaturated fatty acids (n-3 PUFA) in diets of meat producing bovines especially given during the finishing period are used in routine for (1) a better efficiency of energy supply with a lower cost and a shorter fattening period, (2) a higher nutritional lipid value of meat beneficial for human health (higher level of anti-atherogenic n-3 PUFA), and (3) a more favourable flavour of meat (n-3 PUFA, precursors of aroma) (Scollan *et al.*, 2005). However, following rumen biohydrogenation, these fatty acids could still be largely metabolised (oxidation, elongation, desaturation, esterification, etc.) in different tissues/ organs of animals involved in fatty acid metabolism such as the liver with the intensity of these pathways affecting the fatty acid deposition in muscles. Thus, the objective of the present study was to determine the metabolism of linoleic (18:2 n-6) and linolenic (18:3 n-3) acids in the liver of animals fed a control diet or the same diet supplemented with a mixture of extruded linseed and rapeseed, by using the incubated tissue slice method.

Material and methods

Two groups of Normandy culled-cows were fed for a 100 d finishing period a basal ration composed of concentrate (70%) and straw (30%) without any lipid supplement (control, C, n=5) or supplemented with a mixture of extruded linseed (1/3) and rapeseed (2/3) (40 g lipids / kg diet DM) (LRC, n=6). The diets for both groups were calculated to be iso energetic and iso nitrogenous and to allow a mean live gain weight of 900 g/d. Animals were slaughtered at the experimental abattoir of the INRA Centre of Theix. The metabolism of fatty acids representative of these diets, the linoleic (18:2 n-6) and linolenic (18:3 n-3) acids, was studied in the liver of these animals using the incubated tissue slice method (Graulet *et al.*, 1998). Briefly, livers were taken up just after slaughtering in conditions that preserve tissue functional integrity. Tissue samples were prepared for metabolic labelling in the presence of ^{14}C-linoleic or ^{14}C-linolenic acids) and of a mixture of FA representative of plasma NEFA in the culture medium to be close to physiological conditions. After 17 h of incubation, the extent of uptake, β oxidation (in CO_2 trapped in hyamine hydroxyde and in perchloric acid soluble products), esterification as parts of neutral lipids and phospholipids (separated by liquid chromatography), desaturation (detected by gas-liquid chromatography combined with a flow counter) and secretion as part of VLDL particles (isolated by ultracentrifugal flotation) were measured for the two fatty acids at the end of incubation. Values were means ± SE. All data were treated by ANOVA using the general linear model procedure of SAS®.

Results

Hepatic uptake of the two fatty acids was higher with the LR than with the C diet (+56 and +49% for 18:2 n-6 and 18:3n-3, respectively) (Table 1). Moreover, for the two diets, the uptake of 18:3 n-3 was higher (+46%, P=0.04) than that of 18:2 n-6. The 18:3 n-3 was much more oriented towards mitochondrial β-oxidation (mainly in the form of ketone bodies) than the 18:2 n-6 with the C diet (+79%, P=0.015), this difference being more marked with the LR diet (+246%, P=0.015) (Table 1). This extent of oxidation of 18:3 n-3 corresponded to more than 50% of 18:3 n-3 taken up by hepatocytes. Whatever the diet studied, 18:3 n-3 was not bio-converted into longer (by elongation) and/or more unsaturated fatty acids (by desaturation) while about 14% of 18:2 n-6 were converted into 20:4 n-6 (arachidonic acid). Intensity of esterification was higher (+70%, P=0.004) with the LR than with the C diet, for both fatty acids tested. Finally, hepatic secretion of 18:3 n-3 as part of VLDL, which represents the fraction of 18:3 n-3 that could reach the target tissues such as muscles, was lower than that of 18:2 n-6 (-58 and -23% for C and LR diets, respectively, P=0.02; Table 1).

Table 1. Uptake, β oxidation, conversion, esterification and secretion of 18:2 n-6 and 18:3 n-3 by the liver of Normandy culled-cows fed a control diet (C, n=5) or the same diet supplemented with a mixture of extruded linseed (1/3) and rapeseed (2/3) (LR, n=6).

	C diet		LR diet		Effect of	
	18:2 n-6	18:3 n-3	18:2 n-6	18:3 n-3	Diet	Fatty acids
Uptake (nmol/g liver)	10.5±1.0	15.4±2.8	16.4±2.3	23.0±3.7	0.02	0.04
β-oxidation (nmol/g liver)	3.4±0.38	9.5±1.91	3.7±0.39	12.8±3.39	NS	0.015
into CO_2 (% oxidised FA)	1.6±0.09	0.6±0.08	2.7±0.29	1.2±0.31	0.002	0.0001
into ketone bodies (% oxidised FA)	98.4±0.09	99.4±0.08	97.3±0.29	98.9±0.31	0.002	0.0001
Conversion (% converted FA)	13.1±0.33	ND	13.9±0.45	ND	NS	0.0001
Esterification (nmol/g liver)	6.8±0.81	5.8±1.25	11.4±2.24	10.0±0.97	0.004	NS
into NL (% esterified FA)	51.4±5.7	55.2±11.8	56.5±7.1	43.3±6.5	NS	NS
into PL (% esterified FA)	48.6±5.7	44.8±11.8	43.5±7.1	56.7±6.5	NS	NS
Secretion (nmol/g liver)	0.31±0.09	0.13±0.02	0.26±0.03	0.20±0.05	NS	0.02

ND: non detectable

Conclusion

Our results suggest that the liver is highly active in the 18:3 n-3 catabolism. On the contrary, in our experimental *in vitro* conditions, the bovine liver was not efficient in the conversion of 18:3 n-3 by elongation and desaturation for the synthesis of long chain n-3 PUFA whereas the *de novo* synthesis of arachidonic acid from 18:2 n-6 actively takes place in the liver.

References

Graulet, B., D. Gruffat, D. Durand and D. Bauchart, 1998. Fatty acid metabolism and very low density lipoprotein secretion in liver slices from rats and preruminant calves. J. Biochem. 124: 1212-1219.

Scollan, N., I. Richardson, S. De Smet, A.P. Moloney, M. Doreau, D. Bauchart and K. Nürnberg, 2005. Enhancing the content of beneficial fatty acids in beef and consequences for meat quality. In: Indicators of milk and beef quality, EAAP Publ. n°112, (Eds. J.F. Hocquette and S. Gigli), Wageningen Academic Publishers, Wageningen, the Netherlands, pp. 151-162.

Effect of muscle and animal types on the expression of HSP in cattle muscle

N. Guillemin[1], H. Levéziel[2], C. Jurie[1], J.F. Hocquette[1] and B. Picard[1]
[1]INRA, UR1213 Herbivores, Site de Theix, 63122 Saint-Genès-Champanelle, France; [2]INRA-Université de Limoges, UMR 1061, 87060 Limoges, France; nicolas.guillemin@clermont.inra.fr

Introduction

Beef tenderness is a very important criterion for consumers. Several recent genomics programmes have revealed some potential meat tenderness biomarkers (DNA, RNA, proteins) (Hocquette *et al.*, 2007). Among these, members of the Heat Schok Protein family (HSP): Hsp20, Hsp27, Hsp40, Hsp70-1A/B, Hsp70-8 and αB-crystallin were differentially expressed between extreme groups of tenderness and appeared to be good markers of beef tenderness. The aim of this study was to determine the expression of protein HPS in different muscles or animal types.

Material and methods

The study was conducted on 104 different samples corresponding to 29 Charolais young bulls and 23 Charolais steers from the INRA experimental programme MUGENE (funded by ANR and APIS-GENE through the GENANIMAL call). Animals were slaughtered at the experimental slaughterhouse of our Unit in compliance with the current ethical guidelines for animal welfare. Muscle samples from *longissimus thoracis* (LT, mixt fast oxido-glycolytic) and *semitendinosus* (ST, fast glycolytic) were excised from each animal (29 young bulls and 23 steers) within 15 min of slaughter. Muscle samples were immediately frozen in liquid nitrogen and stored at -80 °C until protein extraction, performed according to Bouley *et al.* (2004) in the denaturation buffer (8.3M urea, 2M thiourea, 1% DTT, 2% CHAPS). The protein concentration was determined by spectrophotometry with the Bradford assay. Protein extracts were stored at -20 °C.
Protein quantification was done by Dot-Blot, as described by N. Guillemin (unpublished data 2009). All samples were replicated four times.
The primary antibodies were used according to the following concentrations, at 37 °C during 90 min: Hsp20 (Santa Cruz) 1/200, Hsp27 (Santa Cruz) 1/3000, Hsp40 (Santa Cruz) 1/250, Hsp70-1A/B (Abnova) 1/2000, Hsp70-8 (Santa Cruz) 1/250 and αB-crystallin (Assay Designs) 1/500. The second fluorescent antibody anti-mouse (LI-COR) was used at 1/20,000 at 37 °C for 30 min. The statistical analysis was performed using the GLM procedures of SAS®. The effects of animal type, muscle and animal type X muscle interaction were tested. When significant effects were detected, differences were evaluated by the PDIFF option of SAS®. No animal type X muscle interactions were observed, so only main effects are reported.

Results

Animal type effect (Table 1): whereas the expression of Hsp40 and Hsp70-1A was not significantly different between bulls and steers, αB-crystallin, Hsp20 and Hsp27 were found to be expressed more in young bulls compared with steers, for the two muscles. These proteins constitute a functional complex, according to bibliographic data available in human databases.
Muscle type effect (Table 1): all the HSP, except Hsp27, were found more expressed in LT than in ST muscle. It is well documented in the literature that the muscles of young bulls are more oxidative than steers, due to the difference in androgenic hormones (higher in young bulls) (Cassar-Malek *et al.*, 1998). Moreover, LT is more oxidative than the ST. Consequently, we can conclude that HSP are associated with oxidative metabolism.

Table 1. Expression of different HSP from Charolais bulls and steers in longissimus thoracis *(LT) and* Semitendinosus *(ST)* muscles.

	Animal type		SE	Effect	Muscle		SE	Effect
	Bulls	Steers		*P*-value	LT	ST		*P*-value
αB-crystallin	16.4	10.6	0.66 (0.74)	0.0001	17.8	9.2	0.66	0.001
Hsp20	16.2	11.3	0.43 (0.48)	0.0001	16.9	10.7	0.38	0.0001
Hsp27	11.7	5.9	0.63 (0.7)	0.0001	8.54	9.15	0.54	0.43
Hsp40	15.4	14.5	0.43 (0.48)	0.15	16.4	13.5	0.43	0.0001
Hsp70-1A	14.2	14.6	0.56 (0.62)	0.63	17.8	11	0.61	0.0001
Hsp70-8	15.1	16.7	0.33 (0.36)	0.0017	16.5	15.4	0.29	0.0122

Protein quantities are given in arbitrary units.
*SE for bulls out of brackets and for steers between brackets.

Discussion and conclusion

All these HSP were found to be expressed more in an androgenic / oxidative metabolism, which was in accordance with human protein databases. These proteins are implicated in chaperone activities and are also apoptose inhibitors. They are induced by a thermal or oxidative stress, produced for example during muscular work.
So, these results show that different beef production systems influence the expression of some potential beef tenderness markers, by muscle effects and animal types. The same experiments will be done on around twenty other proteins, such as metabolic enzymes and structural proteins.

Acknowledgement

The authors thank Bruno Meunier, Isabelle Cassar-Malek, Christiane Barboiron, Nicole Dunoyer, Daniel Taillandier, the INRA experimental farms of Bourges and the experimental slaughter house or our Unit. The authors also thank Gilles Renand (INRA, SGQA UNIT, Jouy-en-Josas) for the conception of the experimental design of the MUGENE program.

References

Bouley, J., C. Chambon and B. Picard, 2004. Mapping of bovine skeletal muscle proteins using two-dimensional gel electrophoresis and mass spectrometry. Proteomics 4: 1811-1824.

Cassar-Malek, I., A. Listrat and B. Picard, 1998. Contrôle hormonal des caractéristiques des fibres musculaires après la naissance. INRA Prod. Anim. 11: 365-377.

Hocquette, J.F., S.A. Lehnert, W. Barendse, I. Cassar-Malek and B. Picard, 2007. Recent advances in cattle functional genomics and their application to beef quality. Animal 1: 159-173.

Effects of terpene oral administration on their transfer in goat milk

I. Hadjigeorgiou[1], *I. Poulopoulou*[1], *E. Zoidis*[1] *and T. Masouras*[2]
[1]*Laboratory of Nutritional Physiology and Feeding, Faculty of Animal Science and Aquaculture;* [2]*Laboratory of Dairy Research, Faculty of Food Science, Agricultural University of Athens, 75 Iera Odos, Athens, Greece; ihadjig@aua.gr*

Introduction

Food traceability methods have appeared recently as a consequence of a rising consumer's interest about their food and in particular that of animal origin. Efforts have been made to identify substances in animal products that act as tracers of the animal's diet. In this direction, terpenes appear as promising candidates to certify animal products from grazing animals (Prache *et al.*, 2005).

Material and methods

In a 20-day experiment, 8 dairy goats were divided into two balanced groups, representing control (C) and treatment (T). Goats were fed a 24 kg/day complete ration, at a 40:60 forage to concentrate ratio. The treatment group orally received 1 g/goat of each of three terpenes: α-pinene, limonene and b-caryophyllene, once a day during the first 18 days. Milk samples were collected regularly and the dosed terpenes were analysed by Solid Phase Micro-extraction Method (SPME) and the use of a GC-MS. Milk samples were also analysed for fat, protein, lactose, total solid and total non fat solid contents through MILKOSCAN. Furthermore, hard cheeses were produced on 3 different dates throughout the experimental period, twice during the administration of terpenes and one week after the end of the experiment. Terpene concentration and chemical composition (dry matter, fat and protein content) of the cheeses produced were also determined through the SPME, GC-MS method and classical chemical analysis respectively. Means were compared by paired t-tests.

Results

Milk production did not differ between the two groups of goats averaging 1.2 kg/day. Moreover, milk chemical composition was not different between the two groups, with the exception of fat content (Table 1).
Milk analysis for terpene contents showed that during the experiment, animals from the Control group had no terpenes detected in their milk. However, in the Treatment group α-pinene and limonene were found in milk samples in increasing concentrations, after the third experimental day (Figure 1) and decreasing after stopping the dosing (18th day), while b-caryophyllene was detected in less than 10% of the samples.
Cheese produced from the Treatment group had consistently lower dry matter, fat and higher total and soluble protein contents than the Control group cheese (Table 2) in all three batches. The profile of the studied terpenes in the goat cheese, after four months of ripening, showed a closer association with that of the orally distributed substances. All three terpenes were detected in cheeses produced at day 15 from milk of the reatment group, but at different proportions as compared to the oral doses. However, cheese from the Control group contained 0.5 to 1.5 ppm of limonene.

Discussion and conclusion

The present results indicate terpenes could act as diet tracers of both milk and cheese, in agreement with Priolo *et al.* (2004) who found volatile compounds to be useful as tracers in animal feeding systems. Since α-pinene and limonene were found in all milk samples, it can be concluded that mono-terpenes can be transferred more easily to the raw animal products and they could have a high capacity as biomarkers. However, the terpene profile of cheeses indicate that terpenes are biochemically active compounds, which easily shift between slightly different chemical forms (i.e. oxidised – reduced forms). The transfer process of such compounds from feed to raw animal products and finally the processed food (cheese) appears to be a complex procedure. Several biochemical paths are involved in their metabolism and interactions occur (Villalba *et al.*, 2006). Certification of the dairy product origin on the basis of their terpene content presents strong challenges to the researcher.

Table 1. Chemical composition of goat milk produced by control and treatment groups.

	Fat,%	Protein,%	Lactose,%	Non fat solids,%	Solids,%
Control	3.92	4.01	4.34	8.64	12.56
Treatment	3.23	4.07	4.01	8.32	11.56

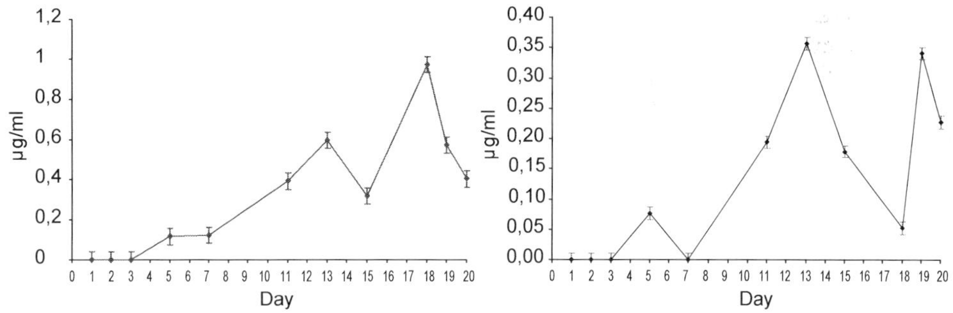

Figure 1. Terpene concentration in milk of goats receiving 1g/d of the substance. Left: α-pinene, right: limonene.

Table 2. Chemical composition of goat cheese produced by control and treatment groups.

	Dry matter,%	Fat,%	Total protein,%	Soluble protein,%
Control	56.4	29.03	24.13	1.60
Treatment	54.7	26.50	26.42	1.82

References

Prache, S., A. Cornu, J.L. Berdague and A. Priolo, 2005. Traceability of animal feeding diet in the meat and milk of small ruminants. Small Rum. Res. 59: 157-168.

Priolo, A., A. Cornu, S. Prache, M. Krogmann, N. Kondjoyan, D. Micol and J.L. Berdague, 2004. Fat volatile tracers of grass feeding in sheep. Meat Sci. 66: 475-481.

Villalba, J.J., F.D. Provenza and K.C. Olson, 2006. Terpenes and carbohydrate source influence rumen fermentation, digestibility, intake, and preference in sheep. J. Anim. Sci. 84: 2463-2473.

Effects of lactogenic hormones on the expression of IGF-binding protein mRNA in cultured bovine mammary epithelial cells

A. Hagino[1], Y. Ohtani[1], S. Oda[2] and K. Katoh[1]
[1]Laboratory of Animal Physiology, Graduate School of Agricultural Science, Tohoku University, Sendai 981-8555, Japan; [2]Laboratory of Animal Nutrition, Faculty of Agriculture, Iwate University, Morioka 020-8550, Japan; hagino@bios.tohoku.ac.jp

Introduction

The insulin-like growth factor (IGF) system regulates the growth and function of the mammary gland through complex interactions with other growth factors and hormones. For example, Cohick (1998) reported that IGF-I possesses the ability to stimulate proliferation and inhibit apoptosis in normal mammary epithelial cells. Furthermore, six IGF-binding proteins (IGFBP-1 to -6), which bind to IGF-I and modulate its action, have been detected in the mammary tissues of several species and their expression levels vary considerably between stages of development. IGFBP are also expressed in cultured mammary epithelial cells and their expression is affected by GH (Sakamoto *et al.*, 2007). These observations suggest that IGFBP may play an important role in the growth, development and involution of the mammary gland. However, very little is known about the control of IGFBP expression by hormones and growth factors. We therefore focussed on the effect of prolactin, insulin and dexamethasone on the expression of IGFBP mRNA in cultured bovine mammary epithelial cells issued from a pregnant heifer.

Material and methods

Bovine mammary epithelial (BME) cells were established from the mammary gland of one 102-day-pregnant Holstein heifer (Sakamoto *et al.*, 2007). The cells were grown in 10% FCS DME to confluence, then washed twice with serum-free DMEM and cultured in serum-free DMEM supplemented with 0.2% BSA. Following a 24 h wash-out period, dexamethasone, prolactin and/or insulin (10 μg/ml each) were added to the culture and incubated for 24 h. The cells were lysed in RNAiso (Takara) and the total RNA was isolated according to the manufacturer's instructions. RNA quality was verified by inspection of the 18S and 28S rRNA bands after agarose gel electrophoresis. DNAase treated RNA were reverse transcribed using a Prime Script RT reagent kit (Takara) with random primers. For quantitative PCR, primer sets for IGFBP-2, IGFBP-3, IGFBP-4 and IGFBP-5 were obtained from Voge *et al.* (2004). The primer set for the type I IGF receptor (IGF-I-R) was from Plath-Gabler *et al.* (2001) and GAPDH was from Yonezawa *et al.* (2008). To determine the gene expression profiles, individual samples were diluted 1:4, and 1 μl were amplified in a 20 μl reaction mixture containing SYBR Premix Ex Taq (Takara) and 0.4 μM each of the forward and reverse gene-specific primer. Quantitative PCR was performed on 96-well plates and fluorescent PCR products were detected using an Opticon-2 real-time PCR detection system (MJ Research). Data were normalised with GAPDH and the relative expression of the treated sample to untreated sample was calculated by the $2^{-\Delta\Delta Ct}$ method (Livak *et al.*, 2001). Data are presented as means ± standard error (S.E.) from three independent observations. Mean comparisons were done using the Student *t*-test with $P<0.05$ as the level of significance.

Results

IGFBP-2 through -5 mRNA expressions were detected in cultured bovine mammary epithelial cells. Treatment with the 3 lactogenic hormones added together for 24 h, dexamethasone, insulin and prolactin (DIP), increased the IGFBP-3 ($P<0.01$) expression but decreased the IGFBP-5 one

(*P*<0.01), whereas the IGFBP-2, -4 and IGF-I-R expressions were not affected by the treatment. Dexamethasone alone caused an increase in IGFBP-3 and a decrease in IGFBP-5 as well as DIP treatment. Insulin decreased the IGFBP-5 expression (*P*<0.05). Prolactin had no effect on the IGFBP and IGF-I-R mRNA expressions in the BME cells.

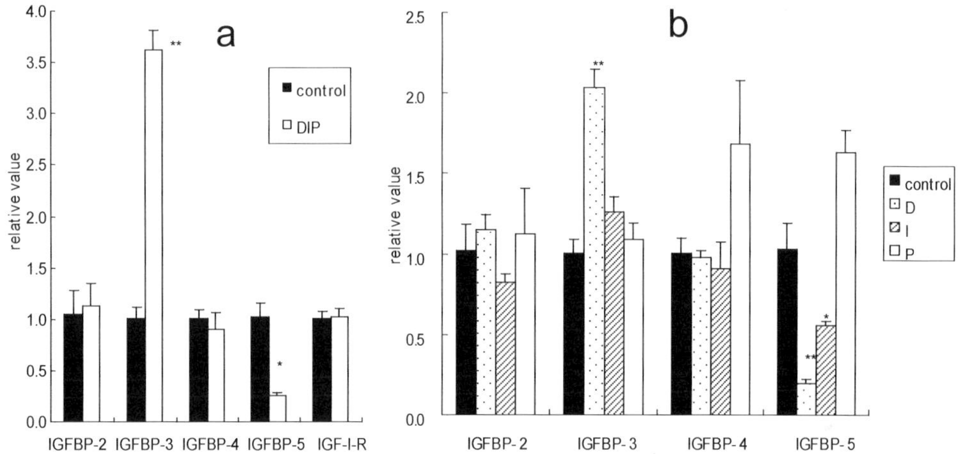

Figure 1. Effects of dexamethasone (D), insulin (I) and prolactin (P) added together (a), and D, I, or P added alone (b) for 24 h on the expression of IGFBP and IGF-I-R. (P<0.05, ** P<0.01 vs. control).*

Conclusion

It is likely that dexamethasone affects the function of BME cells, at least partly, by regulating IGFBP expression. Further study is needed to elucidate the mechanism.

References

Cohick, W.S., 1998. Role of the insulin-like growth factors and their binding proteins in lactation. J. Dairy Sci. 81: 1769-1777.

Livak, K.J. and T.D. Schmittgen, 2001. Analysis of relative gene expression data using real-time quantitative PCR and the $2^{-\Delta\Delta Ct}$ method. Method 25: 402-408.

Plath-Gabler, A., C. Gabler, F. Sinowatz, B. Berisha and D. Shams, 2001. The expression of the IGF family and GH receptor in the bovine mammary gland. J. Endocrinol. 168: 39-48.

Sakamoto, K., T. Yano, T. Kobayashi, A. Hagino, H. Aso and Y. Obara, 2007. Growth hormone suppresses the expression of IGFBP-5, and promotes the IGF-I-induced phosphorylation of Akt in bovine mammary epithelial cells. Domestic Anim. Endocrinol. 32: 260-272.

Voge, J.L., C.A.T. Santiago, P.Y. Aad, D.W. Goad, J.R. Malayer and L.J. Spicer, 2004. Quantification of insulin-like growth factor binding protein mRNA using real-time PCR in bovine granulosa and theca cells: effect of estradiol, insulin, and gonadotropins. Domestic Anim. Endocrinol. 26: 241-258.

Yonezawa, T., S. Haga, Y. Kobayashi, K. Katoh and Y. Obara, 2008. Unsaturated fatty acids promote proliferation via ERK1/2 and Akt pathway in bovine mammary epithelial cells. BBRC. 367: 729-735.

Effect of vitamin E levels in diet on the slaughter performance of the Boer goat

H. Luo, H. Meng, H. Zhu, G. Zhang, L. Yan and D. Yue
State Key Laboratory of Animal Nutrition, College of Animal Science and Technology, China Agricultural University, Beijing 100193, PR China; luohailing@cau.edu.cn

Introduction

The absorption, transport and distribution of Vitamin E within the body are linked to that of dietary fat, and intestinal absorption requires the presence of bile salts, pancreatic enzymes and adequate fat (Kayden and Traber, 1993). Jenkins and Mitchell (1986) found that body weight (BW) increases with increasing Vitamin E supplemented in the rat diet. The ewes BW and birth lamb weights are increased by supplementary vitamin E in the diet (Yaprak *et al.,* 2004). However, no significant effect on the growth performance in the pig was found (Lepine *et al.*, 1990). Limited references on the impact of vitamin E on slaughter performance were found in the goat. The objective of the present study was to determine the effect of different vitamin E levels in the diet on the slaughter performance of the Boer goat.

Material and methods

In Beijing Gaote Animal Husbandry Ltd, a total of 24 healthy Boer male goats, 3-months old, and with similar body weights (BW), were selected and fed a basic ration in this study to investigate the effect of vitamin E levels in the diet on slaughter performance. The forage/concentrate ratio of the diet was 7/3, and the forage was a mixture of silage and clover. The formulation was developed according to the NRC (1985) Lamb Feeding Standard. All kids, during the feeding period, were penned individually and housed indoors and received the same quantities of diet. All procedures involving animals were conducted under the approval of the Chinese Agricultural University Animal Care and Use Committee. Vitamin E powder consisting of Vitamin E acetate (1 mg contains 1 IU Vitamin E) was obtained from the Beijing Chemical Reagents Company. The kids were randomly divided into four groups, and each was supplemented with Vitamin E added to the concentrate directly at 0, 80, 320 and 880 IU/kid/d for 5 months. Treatments will be referred to as Groups 1, 2, 3 and 4, respectively. At the end of the experiment, fasting for 24 hours, three Boer goats in each group were weighed (living body weight) and slaughtered to measure the dressed weight (the weight of the carcass), dressing percentage (the percentage of dressed weight to living body weight) and eye muscle area. The general linear model of SPSS 13.0 was applied to analyse the data. The results are expressed as means and standard deviation. Values of $P<0.05$ were considered significant.

Results

The results in Table 1 indicate a tendency of increasing living body weight, dressed weight, with vitamin E supplemented in the diet. The dressed weight and dressing percentage were increased significantly ($P<0.05$) especially in Group 3 (320 IU/kid/d) compared with the control. Our results were consistent with those of Jenkins and Mitchell (1986) who found that in the rat BW increased with increasing Vitamin E in the diet.
We noticed that all parameters decreased in Group 4 (880 IU/kid/d), compared with Group 3 (320 IU/kid/d), although these parameters were higher in Group 4 than in the control. We presume that the dose of 880 IU/kid/d was too high for the 8-month old goat. Perhaps the negative effect including a toxic affect will be found if the dose continuously increases. However, further work will be done to evaluate the effect of vitamin E supplementation in the goat's diet.

Table 1. Effect of different vitamin E levels in the diet on slaughter performance of the Boer goat.

	Control	Group 2	Group 3	Group 4
Living body weight / kg	16.87±1.12[a]	23.00±3.77[a]	24.25±1.30[a]	21.92±1.66[a]
Dressed weight / kg	6.84±0.86[a]	10.50±1.81[ab]	12.51±1.40[b]	10.28±1.16[ab]
Dressing percentage /%	40.38±2.24[a]	45.56±0.43[ab]	51.27±3.01[b]	46.77±2.58[ab]
Eye muscle area /cm^2	9.65±0.45[a]	10.92±0.58[a]	12.00±0.66[a]	11.26±0.73[a]

[a,b]Means with common or no superscript under same classification do not differ (P >0.05).

Conclusion

At the supplemented dose of 80, 320 and 880 IU/kid/d in the diet, the increasing tendency of the slaughter performance of the Boer Goat was found with increasing levels of vitamin E supplemented in the diet, especially the dressed weight and dressing percentage increased significantly in Group 3 (320 IU/kid/d). However, the parameters decreased in Group 4 (880 IU/kid/d), compared with Group 3. We suggest the dose of 80-320 IU/kid/d is suited for the 8-month old goat.

Acknowledgement

The authors thank the experimental sheep farm workers and staff of the laboratory. The present study is supported by the project of State Key Laboratory of Animal Nutrition of China (2004DA125184-0803) and the grants from the Trans-Century Training Programme Foundation for the Talents by the Ministry of Education of China (NCET-05-0136).

References

Jenkins, Y.M. and G.V. Mitchell, 1986. Effects of dietary protein on growth and on plasma and liver Vitamin A and E levels in young rats. Nutr. Res. 6: 1083-1094.

Kayden, H.J. and M.G. Traber, 1993. Absorption, lipoprotein transport and regulation of plasma concentrations of Vitamin E in humans. J. Lipid Res. 34: 343-358.

Lepine, A.J., B.E. Moore and H.A. Agboola, 1990. Effect of vitamin E, phosphorus and sorbitol on growth performance and serum and tissue cholesterol concentrations in the pig. J. Anim. Sci. 68: 3252-3260.

Yaprak, M., E. Emsen, B. Emsen and M. Macit, 2004. The influence of vitamin E supplementation during late pregnancy on lamb mortality and ewe productivity in Awassi ewes. J. of Anim. Vet. Adv. 3: 190-193.

Effect of plant oils on milk fatty acid composition in cows fed red clover silage based diets

A. Halmemies[1,2], *T. Kokkonen*[1], *S. Jaakkola*[1], *A.-M. Lampi*[1], *V. Toivonen*[2], *K.J. Shingfield*[2] *and A. Vanhatalo*[1]
[1]*University of Helsinki, FIN-00014 Helsinki, Finland;* [2]*MTT Agrifood Research Finland, FIN-30100 Jokioinen, Finland; anni.halmemies@helsinki.fi*

Introduction

Based on the potential benefits to human health there is interest in developing sustainable nutritional strategies to alter the fatty acid composition of bovine milk. Replacing grass silage with red clover silage is known to enhance milk fat polyunsaturated fatty acid (PUFA) content, C18:3n-3 in particular, and decrease saturated fatty acid (SFA) concentrations (Dewhurst *et al.*, 2003, Vanhatalo *et al.*, 2007). Plant oil inclusion in maize or grass silage based diets typically reduces the proportion of SFA and increases C18:0, C18:1 c9, *trans* fatty acid, PUFA and conjugated linoleic acid (CLA) concentrations in milk fat (Dewhurst *et al.*, 2006). Camelina is an ancient oil plant rich in C18:3n-3 (Peiretti and Meineri, 2007), but information on the potential as a feed ingredient to alter milk fat composition is limited. The current study was designed to establish if the milk fatty acid composition of cows fed red clover silage based diets can be improved by moderate amounts of plant oil supplements.

Material and methods

Five multiparous Finnish Ayrshire cows averaging 115±5 d in milk were allocated at random to experimental diets according to a 5×5 Latin square with 21 d periods. Treatments consisted of 5 concentrates containing no additional lipid (C), or 29 g/kg of lipid from rapeseed oil (RO), sunflower-seed oil (SFO), camelina-seed oil (CO) or camelina expeller (CE). Concentrates were comprised of barley, wheat, rapeseed meal, molasses sugar beet pulp, cereal bran and plant oil. For treatment CE, camelina expeller replaced rapeseed meal. Concentrates were fed at a flat rate of 12 kg/d to cows offered pure red clover silage *ad libitum.* Intake and milk data were analysed by the MIXED procedure of SAS® (version 9.1) with a model that included the random effect of cow and fixed effects of period and treatment.

Results and discussion

Treatment had no effect (P>0.05) on silage dry matter intake, milk yield or milk fat content (Table 1). Plant oils in the diet decreased the concentration of SFA in milk synthesised *de novo* and enhanced milk fat unsaturated fatty acid (UFA) content. Concentrates containing RO increased milk fat C18:1 c9 content, whereas concentrations of C18:2n-6 were higher and that of C18:3n-3 were lower for SFO than camelina supplemented diets (P<0.01), reflecting the inherent fatty acid composition of lipid supplements. Relative to CO, milk from CE contained higher (P<0.01) amounts of *trans* fatty acids, C18:1 t10 and t11, CLA c9t11 and C18:2 t11c15 and lower (P<0.01) C18:0 and C18:1 c9 concentrations suggesting that complete biohydrogenation of the lipids was lower for the expeller than for camelina oil. It is possible that the presence of a physically disrupted seed coat offers some protection of UFA against metabolism in the rumen or that other components in camelina oilseeds exert inhibitory effects on ruminal biohydrogenation.
In conclusion, moderate amounts of plant oils in diets based on red clover silage had no effect on silage dry matter intake or milk production and altered milk fatty acid composition with the potential

to improve human health. Camelina expeller was found to be more effective for enhancing milk fat UFA content than RO, SFO or CO.

Table 1. Effect of plant oils on dry matter intake, milk yield, milk fat content and milk fatty acid composition in cows fed red clover silage based diets.

	Treatment [1]					SEM	Significance [2]			
	C	RO	SFO	CO	CE		Lipid	MUFA vs. PUFA	C18:2 vs. C18:3	CO vs. CE
Silage DMI [3], kg/d	12.9	12.8	12.4	12.7	12.2	0.44	0.15	0.12	0.93	0.09
Diet DMI, kg/d	23.3	23.4	23.0	23.3	22.7	0.46	0.31	0.15	0.91	0.07
Milk yield, kg/d	31.1	32.3	32.3	31.2	32.2	2.65	0.26	0.64	0.52	0.33
Milk fat content, g/kg	39.6	38.6	36.4	39.3	36.7	1.70	0.32	0.55	0.44	0.26
Milk fatty acid composition, g/100 g total fatty acids										
C4-C14	28.6	26.2	26.0	25.9	25.8	0.65	<0.01	0.30	0.42	0.78
C16:0	32.4	27.3	26.5	27.1	26.8	1.38	<0.01	0.21	0.38	0.55
C18:0	7.63	10.4	10.9	9.86	7.33	0.468	<0.01	<0.01	<0.01	<0.01
C18:1 c9	13.2	17.3	16.6	16.5	13.5	0.59	<0.01	<0.01	<0.01	<0.01
C18:1 t10	0.36	0.51	0.56	0.42	0.96	0.046	<0.01	<0.01	0.02	<0.01
C18:1 t11	0.96	1.28	1.42	1.21	2.18	0.082	<0.01	<0.01	<0.01	<0.01
C18:2 c9c12	2.08	1.99	2.55	2.10	1.98	0.091	0.30	<0.01	<0.01	0.16
C18:2 t11c15	0.15	0.16	0.11	0.23	0.62	0.024	<0.01	<0.01	<0.01	<0.01
CLA c9t11	0.44	0.56	0.64	0.57	1.02	0.044	<0.01	<0.01	<0.01	<0.01
C18:3 c9c12c15	1.10	1.02	0.99	1.17	1.06	0.049	0.21	0.09	<0.01	<0.01
Trans fatty acids	5.98	7.43	7.77	7.45	12.44	0.375	<0.01	<0.01	<0.01	<0.01
Saturates	71.0	66.1	65.9	65.4	62.6	1.00	<0.01	<0.01	<0.01	<0.01
Monounsaturates	23.2	28.1	27.7	28.2	29.7	0.78	<0.01	0.25	<0.01	<0.01
Polyunsaturates	5.33	5.40	5.99	5.93	7.27	0.240	<0.01	<0.01	<0.01	<0.01

[1] Refers to red clover silage based diets containing no additional lipid (C), rapeseed oil (RO), sunflower-seed oil (SFO), camelina-seed oil (CO) or camelina expeller (CE).
[2] Preplanned orthogonal single degree of freedom comparisons were the following: Lipid = (C vs. RO, SFO, CO and CE), MUFA vs. PUFA (RO vs. SFO, CO and CE), C18:2 vs. C18:3 (SFO vs. CO and CE) and CO vs. CE.
[3] DMI = dry matter intake.

References

Dewhurst, R.J., W.J. Fisher, J.K.S. Tweed and R.J. Wilkins, 2003. Comparison of grass and legume silages for milk production. 1. Production responses with different levels of concentrate. J. Dairy Sci. 86: 2598-2611.

Dewhurst, R.J., K.J. Shingfield, M.R.F. Lee and N.D. Scollan, 2006. Increasing the concentrations of beneficial fatty acids in milk produced by dairy cows in high-forage systems. Anim. Feed Sci. Technol. 131: 168-206.

Peiretti, P.G. and G. Meineri, 2007. Fatty acids, chemical compostion and organic matter digestibility of seeds and vegetative parts of false flax (*Camelina sativa* L.) after different lengths of growth. Anim. Feed Sci. Technol. 133: 341-350.

Vanhatalo, A., K. Kuoppala, V. Toivonen and K.J. Shingfield, 2007. Effects of forage species and stage of maturity on bovine milk fatty acid composition. Eur. J. Lipid Sci. Technol. 109: 856-867.

Myosin heavy chain expression in ovine skeletal muscles

K. Hemmings[1], T. Parr[1], Z. Daniel[1], B. Picard[2], P. Buttery[1] and J. Brameld[1]
[1]Division of Nutritional Sciences, University of Nottingham, School of Biosciences, Sutton Bonington, LE12 5RD, Leicestershire, United Kingdom; [2]INRA UR1213 Herbivore Research Unit, Muscle Growth and Metabolism group, Theix, 63122, Saint-Genès-Champanelle, France; sbxkmh@nottingham.ac.uk

Introduction

Myosin heavy chain (MyHC) isoforms form the basis of the contractile fibre type in skeletal muscle with four adult isoforms (MyHCI, IIA, IIX and IIB) being expressed in the skeletal muscle of many species. In sheep it is unclear which isoforms are expressed, and the expression of the fastest isoform, MyHCIIB, has only recently been identified in bovine skeletal muscle (Picard and Cassar-Malek, 2009). Therefore the current study was aimed at characterising adult MyHC expression in sheep skeletal muscles.

Material and methods

Longissimus dorsi (LD), *supraspinatus* (SS) and *semitendinosus* (ST) skeletal muscle samples were obtained at slaughter from six Mule × Charolais wether lambs at 66±2 days of age. *cutaneus trunci* (CT) and *diaphragma* (Di) muscle samples were obtained from a 19 month-old charolais bull (Picard *et al.*, 1999), and *longissimus thoracis* (LT) from a 15 month old Blonde d'Aquitaine bull (Picard and Cassar-Malek, 2009), were used as controls, since CT contains MyHCIIA and IIX, Di contains MyHCI and IIA and LT contains MyHCI, IIA, IIX and IIB (Picard and Cassar-Malek, 2009). Samples were snap frozen in liquid nitrogen and stored at -80 °C. Protein expression of the MyHC isoforms was assessed by SDS-PAGE. Semi-quantitative PCR (35 cycles) was carried out to examine MyHC IIB mRNA expression using published primers designed to the bovine sequence (Vuocolo *et al.*, 2007). Expression was compared to that of MyHCIIX + IIB combined, using primers designed to the available sequence for the ovine MyHCIIX isoform, but expected to also amplify MyHCIIB, since this region is identical in the porcine sequences for MyHCIIX and IIB.

Results

MyHC isoforms migrate differently in different species and therefore the conditions of electrophoresis have to be optimised. Since cattle and sheep are phylogenetically close it is likely that their MyHC isoforms will migrate similarly (Sayd *et al.*, 1998), therefore we characterised ovine MyHC isoform migration by comparing it to cattle. Only three MyHC bands were observed in the sheep muscles studied (Figure 1), corresponding to MyHCI, IIA and IIX isoforms in cattle, indicating a lack of MyHCIIB protein expression in ovine muscle. Semi-quantitative RT-PCR indicated very low or undetectable levels of expression of MyHCIIB at the mRNA level, with expression predominantly detected in the ST muscle (Figure 2).

Conclusion

In the ovine muscles studied the major MyHC isoforms expressed were MyHCI, IIA and IIX. Expression of the fastest isoform, MyHCIIB, was only observed at the mRNA level, where expression was very low. It therefore appears that MyHC IIX is the predominant isoform expressed in fast glycolytic fibres in sheep muscles. Expression of the MyHCIIB isoform at the protein level has been demonstrated recently in bovine skeletal muscle, but expression is rare, only being observed in some muscles of the Blonde d'Aquitaine breed of bulls (Picard and Cassar-Malek, 2009).

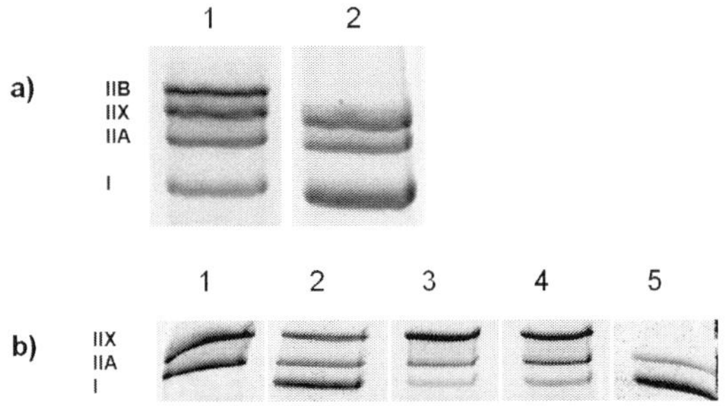

Figure 1. Electrophoretic separation of myosin heavy chain isoforms in (a) bovine LT muscle (1) and ovine SS muscle (2); (b) ovine SS (2), ST (3) and LD (4) muscles and bovine CT (1) and Di (5) muscles. The band for MyHCIIB was absent in all ovine samples.

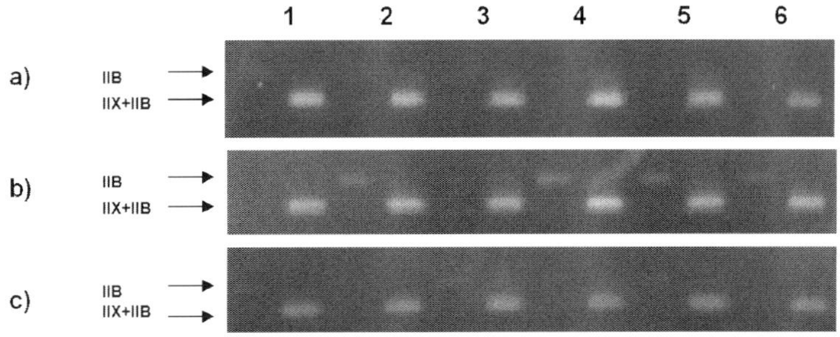

Figure 2. Semi-quantitative PCR for MyHCIIB and MyHCIIB + IIX combined in (a) LD, (b) ST and (c) SS muscles of the 6 lambs (1-6).

References

Picard, B., C. Barboiron, M.P. Duris, H. Gagniere, C. Jurie and Y. Geay, 1999. Electrophoretic separation of bovine muscle myosin heavy chain isoforms. Meat. Sci. 53: 1-7.

Picard B. and I. Cassar-Malek, 2009. Evidence for expression of IIb myosin heavy chain isoform in some skeletal muscles of Blonde d'Aquitaine bulls. Meat. Sci. 82: 30-36.

Sayd, T., T. Mera, V. Martin and E. Laville, 1998. Spatial distribution of myosin heavy chain isoforms and lactate dehydrogenase M-4 in the limb musculature of two crossbred lambs. Comp. Biochem. Physiol. B, Biochem. Mol. Biol. 120: 153-163.

Vuocolo, T., K. Byrne, J. White, S. McWilliam, A. Reverter, N.E. Cockett and R.L. Tellam, 2007. Identification of a gene network contributing to hypertrophy in callipyge skeletal muscle. Physiol. Genomics 28: 253-272.

The effects of beta-adrenergic agonist (BA) and growth hormone (GH) on lamb growth characteristics and myosin heavy chain expression

K. Hemmings, T. Parr, Z. Daniel, P. Buttery and J. Brameld
Division of Nutritional Sciences, University of Nottingham, School of Biosciences, Sutton Bonington, Leicestershire, LE12 5RD, United Kingdom; john.brameld@nottingham.ac.uk

Introduction

In skeletal muscles of sheep and cattle the predominant myosin heavy chain (MyHC) isoforms expressed are types I, IIA and IIX, with the expression of the 'fastest' isoform, type IIB, being rarely detected at the protein level (Picard and Cassar-Malek, 2009). In this study the effects of growth hormone (GH) and the beta-adrenergic agonist (BA), cimaterol, on lamb growth characteristics and mRNA expression of MyHC isoforms were determined.

Material and methods

Mule×Charolais male twin lambs were weaned at 53±5 d of age after being given free access to a standard creep feed for approximately 4 wk. The pairs of twins were then split into d 60 (D60) and d 120 (D120) age groups. These groups were then split into three treatment groups: group CO (n=11) was the control having creep feed *ad libitum*; group BA (n=10) were *ad-libitum* fed the creep diet containing the beta-agonist cimaterol at 10 ppm; and group GH (n=10) were administered prolonged release bovine GH (3.75 mg/kg BW, Monsanto) by a single subcutaneous injection and *ad-libitum* fed the creep diet. Treatments were for 6 d starting at 60±5 or 120±4 d of age. At slaughter plasma was collected as well as a whole transverse section of the *longissimus lorsi* (LD) muscle from the region of the 10th rib (*lumbarum et thoracis)*, which was snap frozen in liquid nitrogen and stored at -80 °C. Whole LD, *semitendinosus* (ST), *vastus lateralis* (VL), *vastus intermedius* (VI) and *supraspinatus* (SS) muscles were dissected from the right side of the carcass and weighed, along with the liver. Plasma IGF-1 and LD glycogen content were measured as described by Sensky *et al.* (2006). Total RNA was extracted (Trizol) from crushed LD samples and first strand cDNA generated using random primers. Relative levels of MyHC isoform mRNA expression were determined using quantitative RT-PCR analysis. Real-time PCR primers and probe specific for MyHC I, IIA and IIX isoforms were designed using published partial length ovine cDNA sequences. Porcine MyHC IIX and IIB cDNA are identical in the region where the IIX primers and probe were designed; therefore these primers are predicted to also detect sheep MyHC IIB mRNA, if present. Specific expression of MyHC IIB mRNA was determined by semi-quantitative RT-PCR using primers previously shown to detect MyHC IIB transcripts in sheep (Vuocolo *et al.*, 2007); whilst combined MyHC IIX + IIB mRNA was determined by semi-quantitative RT-PCR using the Quantitative RT-PCR primers described above. Data were analysed by ANOVA (GenStat, VSN International Ltd, Hemel Hempstead, UK) and the *post hoc* Dunnett test.

Results

As expected, GH treatment significantly increased plasma IGF-1 concentrations and liver weights, while BA treatment significantly decreased LD muscle glycogen content, both at D60 and D120 (Table 1). No change in bodyweight was observed between treatment groups (data not shown). At D60 there were no effects of either treatment on muscle weights whereas at D120 BA treatment significantly increased the SS and ST muscle weights, with a trend for an increase in VL and LD muscle weights. There was a significant effect of BA treatment on the relative expression level (%) of MyHC mRNA in both age groups with a decrease in MyHC IIA and increase in MyHC IIX/B levels, relative to the control, whilst MyHC I levels decreased in the D60 group only (Table 1). MyHC IIB mRNA was barely detectable in CO or GH treated animals, but BA treatment induced MyHC IIB expression (Figure 1). In contrast, the Quantitative RT-PCR primers designed to detect both MyHC IIX and IIB mRNA generated bands in all of the treatment groups at both D60 and D120, suggesting that this signal was predominantly due to MyHC IIX.

Table 1. Effects of treatment of lambs with BA and GH for a 6 d period.

Age		D60					D120				
Treatment		CO	BA	GH	SED	P	CO	BA	GH	SED	P
Plasma LD	IGF-1, ng/ml	375	373	928*	129	<0.001	972	964	2,092*	127	<0.001
	Glycogen, mg/g	25.5	19.0*	23.8	1.6	<0.001	21.6	18.3*	24.2	1.3	<0.001
	Liver, g	475	460	567*	32	0.005	928	926	1,101*	55	0.004
Muscle weights	SS, g	53.5	55.2	55.9	3.9	0.812	98.5	116.4*	98.0	6.3	0.009
	ST, g	63.3	67.1	65.0	4.3	0.667	112.4	131.2*	120.4	6.6	0.023
	VL, g	85.4	88.0	89.3	6.3	0.815	164.5	180.3	163.8	7.5	0.058
	LD, g	305	306	327	27	0.655	592	667	632	35	0.108
	VI, g	28.5	27.5	30.9	2.3	0.331	51.7	55.7	52.4	3.7	0.487
LD mRNA levels	MyHC I,%	7.2	3.9*	6.6	0.8	<0.001	6.8	5.0	8.1	1.9	0.277
	MyHC IIA,%	26.9	7.0*	34.1	3.4	<0.001	21.1	5.2*	25.4	3.6	<0.001
	MyHC IIX/B,%	65.1	89.1*	59.3	3.6	<0.001	72.1	89.8*	66.5	4.2	<0.001

SED = standard error of the differences of the means; * Values significantly different from control ($P<0.05$).

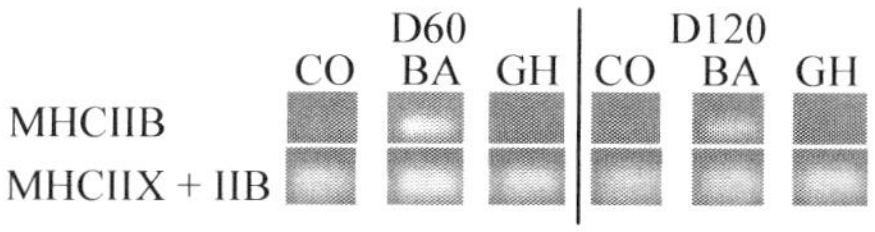

Figure 1. Representative semi-quantitative PCR for MyHC IIB or MyHC IIB+IIX (35 cycles).

Conclusion

Despite only a short-term (6 d) treatment with the anabolic agents there was a significant effect of BA administration on the expression of myosin heavy chain mRNA, driving MyHC expression toward the isoforms associated with fast glycolytic muscles (MyHC IIX), including MyHC IIB, an isoform that has previously been shown to have limited expression in ruminant species. There was also a tendency for muscle weights to increase in response to BA in the older lambs. In contrast, GH treatment had little or no effect on muscle weights or MyHC isoform expression. Overall BA was the more potent anabolic agent (over this short period and dose), increasing muscle weights and inducing expression of MyHC IIB mRNA.

References

Picard, B. and I. Cassar-Malek, 2009. Evidence for expression of IIb myosin heavy chain isofom in some skeletal muscles of Blonde d'Aquitaine bulls. Meat Sci. 82: 30-36.

Sensky, P.L., K.K. Jewell, K.J.P. Ryan, T. Parr, R.G. Bardsley and P.J. Buttery, 2006. Effect of anabolic agents on calpastatin promoters in porcine skeletal muscle and their responsiveness to cyclic adenosine monophosphate-and calcium-related stimuli. J. Anim. Sci. 84: 2973-2982.

Vuocolo T., K. Byrne, J. White, S. McWilliam, A. Reverter, N.E. Cockett and R.L. Tellam, 2007. Identification of a gene network contributing to hypertrophy in callipyge skeletal muscle. Physiol. Genomics 28: 253-272.

The association of interleukin-6 and insulin sensitivity in bovine subcutaneous and perirenal adipose tissue explants treated with propionate

A. Hosseini, M. Mielenz and H. Sauerwein
Institute of Animal Science, Physiology & Hygiene Unit, University of Bonn, D-53115, Bonn, Germany; sauerwein@uni-bonn.de

Introduction

Adipose tissue (AT) was known as an energy store, but it is now considered to be an endocrine organ, too. Interleukin-6 (IL-6), a pro-inflammatory cytokine contributes to the development of insulin resistance in the mouse and human (Pittas *et al.*, 2004). The release of IL-6 in human AT explants is the highest after 24 h incubation, whereas in isolated human adipocytes, the highest expression was observed already after 4 h (Fain *et al.*, 2004). In man, IL-6 secretion of AT follows insulin secretion in the blood (Orban *et al.*, 1999), and is 3-fold higher in omental than in subcutaneous (SC) AT (Fried *et al.*, 1998). Short-chain fatty acids (SCFA) increase the mRNA abundance of IL-6 in intestinal samples of piglets after long-term infusion (Milo *et al.*, 2002). Since little is known about the expression and regulation of IL-6 mRNA in AT and its relation with AT insulin sensitivity in ruminants, we analysed the effect of propionate (C3) on IL-6 mRNA in SC and perirenal (PR) bovine AT *in vitro*.

Material and methods

SC and PR AT were obtained from 7 slaughtered Holstein Friesian cows. The tissues were incubated for 4 h in basal medium (DMEM/Ham F-12 with L-Glutamine) or in medium supplemented with either 100 nM insulin or with 0.5, 1, 2, or 3 mM C3. Total RNA was extracted and transcribed to cDNA. The mRNA abundance of housekeeping genes (HKGs) and IL-6 (Acc. number: BC123577) was quantified by real-time-PCR. Subsequent pairwise analysis of variation using geNorm™ program identified LRP10, HPCAL1 and GAPDH as the most stable HKG for accurate normalisation for SC and PR AT explants out of 7 HKs. Real-time-PCR data were normalised and every C3 dosage was compared against a control using the Paired-samples t-test with the SPSS program (trends towards significance were considered at $P<0.15$).

Results

We observed an increase (1.8-fold vs. control) in PR AT by short-term insulin (100 nM) stimulation ($P=0.13$), which is comparable to a numerical increase by 0.5 mM propionate ($P=0.178$) as a trend. With higher C3 (2 mM) dosage, IL-6 mRNA abundance decreased ($P=0.093$; 2.1-fold vs. control) dose dependently as a trend exclusively in PR AT explants (Figure 1B). No influence of insulin or C3 on the mRNA of IL-6 was observed in SC AT explants ($P\geq0.378$).

Discussion and conclusion

Our results might suggest that potential *in vitro* effects of C3 treatment on IL-6 mRNA abundance seem to be limited to PR AT. As in humans, PR AT might be one of the main sources for IL-6 mRNA expression in bovine. In view of the data known from monogastric species, increased IL-6 mRNA in the presence of high (pharmacological) concentrations of C3 (0.5 mM) might thus increase insulin resistance in PR AT in cattle. Concentration dependent silencing by demethylation might also lead to repression of mRNA abundance of IL-6 (Benjamin and Jost, 2001) at 2 mM C3 in PR AT.

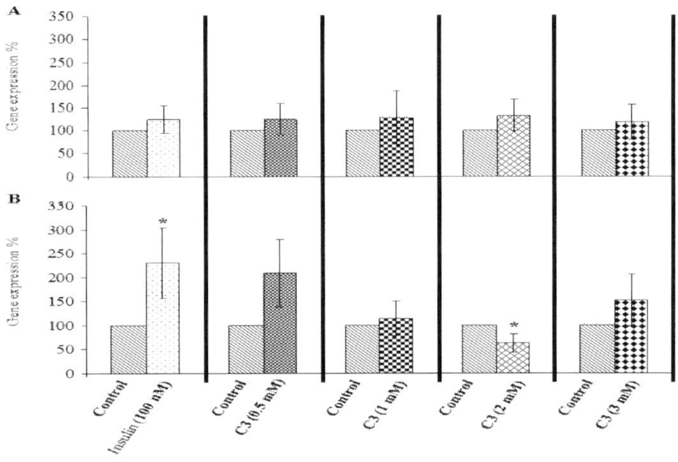

*Figure 1. IL-6 mRNA abundance (means ± SEM) in SC (A) and PR (B) AT explants treated with insulin or different concentrations of C3; * P≤0.13 versus respective fat depot control.*

Acknowledgement

This study was supported by the H. Wilhelm Schaumann foundation.

References

Benjamin D. and J.P. Jost, 2001. Reversal of methylation-mediated repression with short-chain fatty acids: evidence for an additional mechanism to histone deacetylation. Nucleic. Acids Res. 29: 3603-3610.

Fain J.N., A.K. Madan, M.L. Hiler, P. Cheema and S.W. Bahouth, 2004. Comparison of the release of adipokines by adipose tissue, adipose tissue matrix, and adipocytes from visceral and subcutaneous abdominal adipose tissues of obese humans. Endocrinology. 145: 2273-2282.

Fried S.K., D.A. Bunkin and A.S. Greenberg, 1998. Omental and subcutaneous adipose tissues of obese subjects release interleukin-6: depot difference and regulation by glucocorticoid. J. Clin. Endocrinol Metab. 83: 847-850.

Milo L.A., K.A. Reardon and K.A. Tappenden, 2002. Effects of short-chain fatty acid-supplemented total parenteral nutrition on intestinal pro-inflammatory cytokine abundance. Dig. Dis. Sci. 47: 2049-55.

Orban Z., A. Remaley, M. Sampson, Z. Trajanoski and G.P. Chrousos, 1999. The differential effect of food intake and ß-adrenergic stimulation on adipose-derived hormones and cytokines in man. J. Clin. Endocrinol. Metab. 84: 2126-2133.

Pittas, A.G., N.A. Joseph and A. Greenberg, 2004. Adipocytokines and insulin resistance. J. Clin. Endocrinol. Metab. 89: 447-452.

New insights on mammary tissue responses to dietary lipids using transcriptomics

G. Invernizzi[1,2,3], B.J. Thering[2,3], M. Bionaz[2,3], D. Graugnard[2,3], P. Piantoni[2,3], R.E. Everts[3], H.A. Lewin[3], G. Savoini[1] and J.J. Loor[2,3]
[1]Department of Sciences and Technologies for Food Safety, University of Milan, 20133 Milan, Italy; [2]Mammalian NutriPhysioGenomics; [3]Department of Animal Sciences and Division of Nutritional Sciences, University of Illinois at Urbana-Champaign, Urbana, 61801 Illinois, USA; jloor@illinois.edu

Introduction

The underlying genomic and physiological adaptations that occur in mammary tissue in response to dietary lipids remain relatively unknown. Several lines of evidence (e.g. Bionaz and Loor, 2008) have led us to propose that regulation of milk component synthesis in the bovine mammary gland is a complex dynamic process likely controlled by interactions between networks of genes and environmental factors such as nutrition. The primary objective of this study was to determine alterations of mammary tissue gene networks in cows consuming either a twenty-carbon polyunsaturated fatty acid milk fat-depressing diet or a saturated fatty acid milk fat-enhancing diet.

Material and methods

Thirteen Holstein cows [100 d (DIM), 41.7 kg milk/d] were allocated at random to three experimental treatments during a 4-week study. Diets were based on corn silage and alfalfa silage and included a control diet with no added lipid (CTR, n=4), a milk fat-depressing diet (FSO, n=4) containing 10 g/kg DM of fish oil (19% 16:0, 14% 20:5n-3, 15% 22:6n-3) and 25 g/kg DM of soybean oil (20% *cis*9-18:1, 53% 18:2n-6), and a diet containing 35 g/kg DM of saturated lipid supplement (Energy Booster 100; 40% 16:0, 40% 18:0, 6% *cis*9-18:1; n=5, EB100) to enhance milk fat yield. Diets had a forage:concentrate ratio of 70:30, with lipid supplements in FSO and EB100 replacing soybean hulls and were fed *ad libitum* as a total mixed ration (TMR).

Transcriptomics of mammary biopsies obtained on d 0 and 21 from start of treatments was performed using a 13,257 bovine oligonucleotide microarray in a reference sample (i.e. pool of several bovine tissues) dye-swap design (Loor *et al.*, 2007). Data from a total of 52 microarrays were normalised for dye and array effects (i.e. Loess normalisation and array centering) and used for statistical analysis. A repeated measures model with covariate was fitted to the adjusted ratios using Proc MIXED (SAS®). The model consisted of treatment, dye as fixed effects, as well as cow and microarray as the random effects. *P*-values for fixed effects were adjusted with false discovery rate (FDR) and considered significant at $P \leq 0.05$. Post-hoc analyses were performed using Scheffé comparisons. Gene expression network and pathway analysis was performed using Ingenuity Pathway Analysis® (IPA).

Results

FSO led to lower milk fat percentage and yield (CTR=1.22, EB100=1.23, FSO=0.95 kg/d; time × treatment $P \leq 0.01$), whereas milk protein yield did not differ (time × treatment $P=0.11$). EB100 resulted in 1,432 differentially expressed genes (DEG) vs. CTR and 1,137 vs. FSO. FSO resulted in 847 DEG vs. CTR. DEG between EB100 and FSO were involved in lipid metabolism, molecular transport, small molecule biochemistry, carbohydrate metabolism, amino acid metabolism and protein synthesis (Figure 1). Molecular and cellular function analysis of DEG using IPA indicated that EB100 reduced amino acid catabolism, affected transport of cations (i.e. Na^+), and downregulated genes involved in lipid catabolism except those required for synthesising membrane constituents (i.e. LASS5, SPTLC1), transport of lipids, and glyceroneogenesis. Pathway analysis clearly showed that EB100 downregulated energy metabolism (i.e. TCA cycle, oxidative phosphorylation). FSO mostly affected protein synthesis and free radical scavenging. Gene networks focusing on transcription factors (TF) revealed different mechanisms of action on mammary tissue between the two treatments. EB100 upregulated expression of E2F4, NFE2L2, RARA, FOXO1, and NFYA whereas it downregulated expression of BRCA1 and NR5A2. A network of 16 PPARG-regulated genes mostly involved in transport and glyceroneogenesis (i.e. PC, PCK1) were downregulated. Among TF affected by FSO, YY1 (↓), HDAC5 (↑), and NR3C1 (↑), formed larger networks with other DEG. Genes modulated by FSO-network TF were involved in response to stimulus, development, transcription, and programmed cell death.

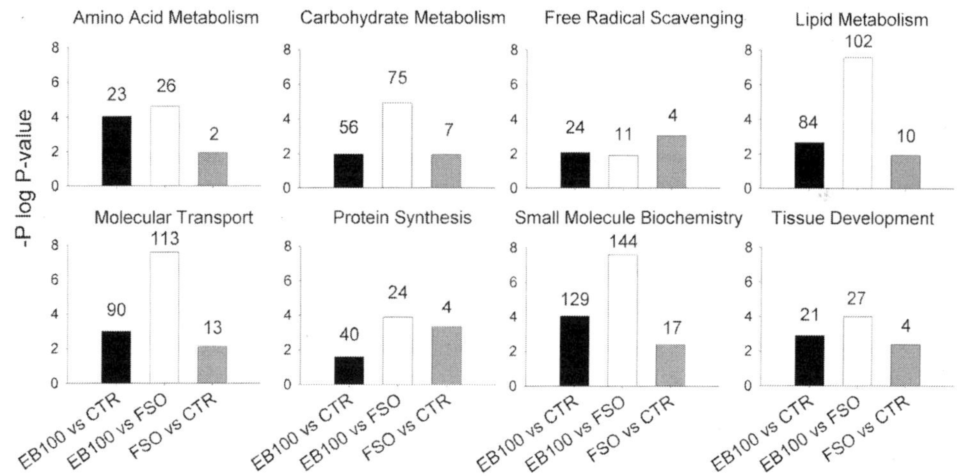

Figure 1. Most significantly enriched IPA functions among DEG consistently UP or DOWN regulated in mammary tissue. The y-axis denotes the –log P-value calculated for each function. The numbers above each column represent the number of DEG.

Conclusion

Transcript profiling uncovered a stronger effect of EB100 on mammary tissue compared with FSO. The results suggest a strong downregulation of energy metabolism in mammary tissue of cows supplemented with palmitate and stearate (EB100), and increased synthesis of the signaling lipid ceramide. The two sources of dietary fat administered affected mammary gene networks through different transcription factors/nuclear receptors.

References

Bionaz, M. and J.J. Loor, 2008.Gene networks driving bovine milk fat synthesis during the lactation cycle. BMC Genomics. 9: 366.

Loor, J.J., R.E. Everts, M. Bionaz, H.M. Dann, D.E. Morin, R. Oliveira, S.L. Rodriguez-Zas, J.K. Drackley and H.A. Lewin, 2007. Nutrition-induced ketosis alters metabolic and signaling gene networks in liver of periparturient dairy cows. Physiol. Genomics 32: 105-116.

Effect of body condition score at parturition on blood glucose and insulin responses during a glucose tolerance test in Estonian Holstein and Estonian Red cows

H. Jaakson, K. Ling, J. Samarütel, A. Ilves, T. Kaart and O. Kärt
Institute of Veterinary Medicine and Animal Sciences, Estonian University of Life Sciences, 1 Kreutzwaldi St., 51006, Tartu, Estonia; hanno.jaakson@emu.ee

Introduction

In dairy cows insulin and tissues sensitivity to insulin play a key role in *post partum* lipid mobilisation. At the beginning of lactation in ruminants, blood insulin concentration decreases compared to *pre partum* (Blum *et al.*, 1973), insulin response to glucose infusion reduces and adipose tissue becomes resistant to the lipogenic effect of insulin (Debras *et al.*, 1989). Being, to a moderate extent, common in dairy cows at the beginning of lactation, insulin resistance enhances lipolysis to cover energy requirements and to support the onset of milk production under conditions of negative energy balance. *Post partum* glucose-induced insulin response is inversely correlated with milk producing ability (Hammon *et al.*, 2007) and body condition score (BCS) at calving (Holtenius *et al.*, 2003). Therefore the aim of the study was to compare glucose and insulin responses in relation to BCS at parturition in Estonian Holstein (EH) and Estonian Red (ER) cows, using the glucose tolerance test (GTT) carried out during the period of negative energy balance.

Material and methods

The study was carried out on 30 dairy cows (EH, n=15; ER, n=15) kept in tie stall barns, fed *ad libitum* TMR, providing on average 11.7 MJ metabolisable energy and 101.1 g metabolisable protein per 1 kg dry matter, and milked thrice a day. 305-day milk yields from the previous lactation in the experimental groups were 8,999±319 and 8,253±287 kg in EH and ER respectively. BCS at calving was evaluated according to Edmondson *et al.* (1989). The GTT was carried out 31±1.9 days *post partum*. Cows were deprived of feed 60 min before and during the GTT. Glucose (0.15 g/kg BW) was infused and blood was sampled using a jugular vein catheter. Samples were collected at the following times: -15, -5, 5, 10, 20, 30, 40, 50 and 60 min relative to the start of infusion. Plasma was separated and kept at -24 °C until analysis for glucose and insulin. The following GTT parameters were calculated: basal concentration (mean of pre-infusion samples) and maximum increase, calculated as the difference between the basal concentration and the highest concentration, and according to Holtenius *et al.* (2003) the area under the curve (AUC) for glucose and insulin and clearance rate (CR) for glucose. The effect of BCS at calving on GTT parameters was analysed using Correlation Analysis with the SAS® System. Significance was declared at $P<0.05$.

Results

BCS at calving in ER (3.55±0.13) compared to EH (3.10±0.12) was higher ($P<0.05$). If grouped according to individual BCS at calving as thin (BCS≤3.0), with optimum BCS (3.25-3.5) and over-conditioned (BCS≥3.75), the distribution of cows into BCS groups had an opposite pattern between breeds: amongst EH nine cows were classified as thin, four with optimum BCS and two over-conditioned; amongst ER there were three thin, seven with optimum BCS and five over-conditioned cows. Least squares means of GTT parameters and correlations between BCS at calving and GTT parameters are presented in Table 1. In EH cows BCS at calving correlated positively with a maximum increase (r=0.51, $P<0.05$) and AUC of insulin (r=0.53, $P<0.05$); correlations with GTT parameters for glucose were weak. Conversely, in ER cows BCS at calving correlated positively with a maximum increase (r=0.62, $P<0.05$) and AUC of glucose (r=0.53, $P<0.05$); correlations with parameters characterising insulin response were weak; at the same time, on the contrary to EH, these were negative.

Table 1. Least squares means (LSM) and pooled standard errors of means (SEM) of glucose tolerance test (GTT) parameters and their correlation (r) with body condition score (BCS) at calving in Estonian Holstein (EH) and Estonian Red (ER) cows.

Parameter	LSM[1]		Pooled SEM	r	
	EH	ER		EH	ER
Glucose					
Basal concentration; mg/dl	91.1	78.2	7.2	0.15	0.08
Maximum increase; mg/dl	103.0	99.8	7.1	0.03	0.62*
Area under the curve; mg/dl×min	2,213.4	1,876.3	187.6	0.01	0.53*
Clearance rate; %/min	1.42	1.52	0.10	0.09	0.34
Insulin					
Basal concentration; μU/ml	2.6	3.0	0.8	0.13	-0.03
Maximum increase; μU/ml	39.3	57.8	9.9	0.51*	-0.22
Area under the curve; μU/ml×min	875.3	1,119.3	220.3	0.53*	-0.17

[1] LSM did not differ significantly in breeds.
* Significant ($P<0.05$) correlation between GTT parameters and BCS at calving.

Conclusion

Glucose and insulin responses were associated with BCS at calving, while the effect of BCS was different between breeds. Higher BCS at calving in EH cows was associated with a stronger insulin response; in ER cows the association was weak and contrary to EH. Higher BCS in ER was associated with stronger glucose response, indirectly suggesting a less pronounced effect of insulin on stabilising blood glucose in cows with high BCS at calving in this breed.

References

Blum, J.W., R.B. Wilson and D.S. Kronfeld, 1973. Plasma insulin concentrations in parturient cows. J.Dairy Sci. 56: 459-464.

Debras, E., J. Grizard, E. Aina, S. Tesseraud, C. Champredon and M. Arnal, 1989. Insulin sensitivity and responsiveness during lactation and dry period in goats. Am. J. Physiol. 256: E295-302.

Edmondson, A.J., L.J. Lean, L.D. Weaver, T. Farver and G.A. Webster, 1989. Body condition scoring chart for Holstein dairy cows. J. Dairy Sci. 73: 68-78.

Hammon, H.M., O. Bellmann, J. Voigt, F. Schneider and C. Kühn, 2007, Glucose-dependent insulin response and milk production in heifers within a segregating resource family population. J. Dairy Sci. 90: 3247-3254.

Holtenius, K., S. Agenäs, C. Delavaud and Y. Chilliard, 2003. Effects of feeding intensity during the dry period. 2. Metabolic and Hormonal Responses. J. Dairy Sci. 86: 883-891.

Effects of feeding rapeseed oil, soybean oil or linseed oil on stearoyl-CoA desaturase expression in the mammary gland of dairy cows

A.A.A. Jacobs[1], A.M. van Vuuren[1], J. van Baal[1], D. van den Hengel[2] and J. Dijkstra[1]
[1]Animal Nutrition Group, Wageningen University P.O. Box 338, 6700 AH, Wageningen, the Netherlands; [2]Cehave Landbouwbelang Voeders bv., Poort van Veghel 4949, 5466 SB, Veghel, the Netherlands; antoon.jacobs@wur.nl

Introduction

Extensive biohydrogenation of dietary fatty acids (FA) occurs in the rumen of dairy cattle, giving rise to a high proportion of saturated FA in milk fat. Saturated FA may contribute to increased risks of cardiovascular disease and the metabolic syndrome (Williams, 2000). Saturated FA, as well as several mono-unsaturated FA, can be desaturated by Δ9-desaturase, also known as stearoyl-CoA desaturase (SCD), present in the mammary gland of dairy cows. It is known that nutrition, especially polyunsaturated FA (PUFA), can affect the expression of SCD in rodents (Ntambi, 1999). Although various FA have been identified which can affect mammary SCD expression in dairy cattle, such knowledge is limited compared with rodents. Therefore, the objective of this study was to investigate the effect of dietary FA supplementation of C18:1 *cis*-9, C18:2 *cis*-9,12 or C18:3 *cis*-9,12,15, by feeding rapeseed oil, soybean oil or linseed oil respectively, or its mixture, on SCD expression in the mammary gland of dairy cows.

Material and methods

Twenty-eight Holstein-Friesian cows in mid-lactation, averaging 153±82 days in milk (DIM), were blocked according to parity, DIM, milk yield, fat content and protein content, and randomly assigned to one of the four dietary treatments. The total mixed ration (TMR) of the four treatments included 2% of either: rapeseed oil (RO), soybean oil (SO), linseed oil (LO) or 2% of a 1:1:1 mixture of the three oils (MO). Cows were milked daily at 06:00 h and 18:00 h and were grazing on pasture from 08:00 h until 16:00 h. At other times, cows were inside the barn and fed the TMR. Treatment period lasted 3 weeks and after the treatment ended, cows were fed a control diet again (without oil supplementation) for 4 weeks. Individual milk production and feed intake for each treatment group were recorded daily. On the last day of the treatment period and control period, a biopsy from the mammary gland was taken, according to the method of Farr *et al.* (1996) with minor modifications. These biopsies were used for analysis of SCD expression by using quantitative RT-PCR, and 18S was used as endogenous control. In addition, on the last day of the treatment period and control period, one milk sample (combined morning and evening milking) was obtained for FA analysis. The FA were extracted (Folch *et al.*, 1957), methylated and analysed by gas chromatography (TRACE GC Ultra™). Methylated FA were separated using a fused silica capillary column (100m, 0.25mm, i.d. 0.2μm thickness; Restek RT-2560).

Results

Milk yield and milk fat, protein and lactose content did not differ (P>0.05) between the four dietary treatments (data not shown). Feeding SO resulted in a significant down-regulation (P=0.003) of SCD compared to RO and LO (Figure 1).
In addition, the desaturase indices calculated from FA pairs in milk, which are frequently used to estimate *in vivo* SCD activity, were lower for SO and this was significant (P=0.038) for C16:1 c9/C16:0 compared to MO (Table 1).

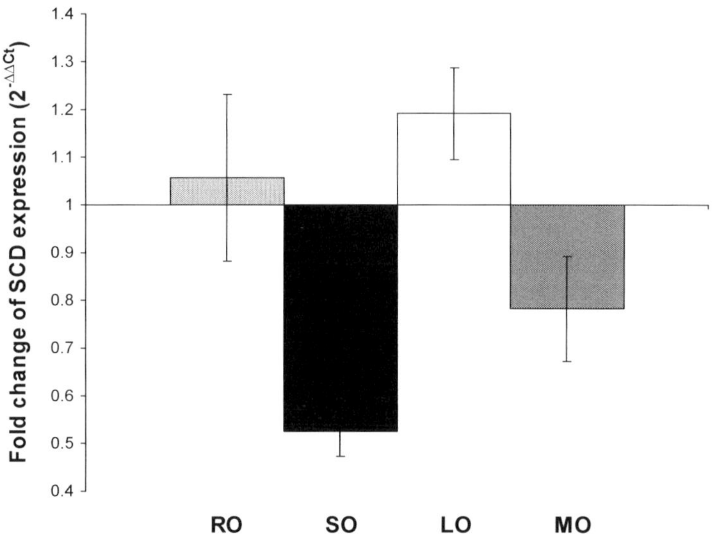

Figure 1. SCD expression of cows fed various oil supplements. Expressed as fold change compared to control ($2^{-\Delta\Delta Ct}$ method). Vertical lines represent the SEM with n=7 per group.

Table 1. Desaturase indices from milk fatty acids of cows fed various oil supplements; n=28 for the control group, and n=7 per treatment group.

Treatment	Control[1]	RO	SO	LO	MO	sem	P
Desaturase indices (calculated from milk FA)							
C14:1 c9/C14:0	0.13	0.13	0.10	0.11	0.15	0.02	0.112
C16:1 c9/C16:0	0.08	0.08^{ab}	0.06^{b}	0.08^{ab}	0.09^{a}	0.01	0.038
C18:1 c9/C18:0	2.35	2.44	1.94	2.19	2.54	0.17	0.068

[1] Control values are shown for comparison only, and are not statistically analysed against the treatments.

[a,b,c,d] Means within a row without common superscript differ ($P<0.05$).

Conclusion

This study shows that mammary SCD expression is significantly down-regulated in dairy cows by feeding soybean oil, and this is partly reflected by the lower desaturase indices in the milk.

References

Farr, V.C., K. Stelwagen, L.R. Cate, A.J. Molenaar, T.B. McFadden and S.R. Davis, 1996. An improved method for the routine biopsy of bovine mammary tissue. J. Dairy Sci. 79: 543-549.

Folch, J., M. Lees and H. Sloane Stanley, 1957. A simple method for the isolation and purification of total lipids from animal tissues. J. Biol. Chem. 226: 497-509.

Ntambi, J.M., 1999. Regulation of stearoyl-CoA desaturase by polyunsaturated fatty acids and cholesterol. J. Lipid Res. 40: 1549-1558.

Williams, C.M., 2000. Dietary fatty acids and human health. Ann. Zootech. 49: 165-180.

Effects of chromium supplementation on production responses and some blood indicators of glucose metabolism in heat stressed dairy cows

M. Khorvash, G.R. Ghorbani, M. Mirzaee and H.R. Rahmani
Department of Animal Science, Isfahan University of Technology, Isfahan, Iran; mirzaee.1984@gmail.com

Introduction

Chromium (Cr) improves action of insulin in animals. All physiological activities of insulin can be affected by chromium supplementation. The ability of insulin to control glucose utilisation determines milk production, fertility and health status of lactating cows (Pechova and Pavlata, 2007). Studies indicated that various stressors (e.g. heat stress, lactation, infection) increased urinary excretion of chromium. During stress, glucose and insulin metabolism are increased, this is likely influenced by increasing cortisol secretion and other glucoregulatory hormones into the blood (Mowat, 1997). The objective of this project was to investigate the effects of Cr as chromium-methionine on the performance, dry matter intake (DMI) and some blood factors related to glucose metabolism of heat stressed cows in early lactation.

Material and methods

This study was conducted as a randomised complete block design with two parities as a block of 12 multiparous and three primiparous Holstein cows (averaging 38±12 days in milk and 620±45 kg weight) and three levels of chromium in five replications. Chromium levels of 0, 0.05 and 0.1 mg of Cr/kg of $BW^{0.75}$ were applied as control (C), moderate (M) and high (H) levels, respectively. This trial lasted for 65 days from the beginning of June to September in 2007. Cows were housed in individual pens (4×4 m). Maximum temperature and minimum relative humidity were recorded daily and maximum temperature-humidity index (THI) was calculated according to Garcia *et al.* (2006). It has been suggested that physiological responses to heat stress such as depressed DMI and milk production and increased maintenance energy expenditure begin to occur at a THI of about 72 (Johnson, 1987). The mentioned Cr levels were daily mixed to 0.25 kg of ground corn and top dressed in the morning feeding. After two weeks of experimental period, dry matter intake and milk yield were recorded daily for each cow. Milk samples were collected weekly from three consecutive milkings, and preserved using potassium dichromate. Milk samples were analysed for fat, protein, lactose and total solid by Milk-O-Scan (134 BN Foss Electric, Hillerod, Denmark). Cows were bled weekly at 11:30 h (2.5 h after the morning feeding) by puncture in the coccygeal vein. An enzymatic method was conducted to determine the concentrations of serum glucose with a commercial kit (Pars Azmoon, Tehran, Iran). Insulin was measured by radio immunoassay (RIA) using a commercial kit (Biosource, Co. Italy). Data were analysed by repeated measures using the Mixed procedure of SAS® (2005). The model included the effects of treatment, block, time, treatment×time interaction. The results are presented as least squares means. Significance was declared at $P \leq 0.05$; trends were declared at ($0.05 < P \leq 0.15$).

Results

The effects of levels of chromium supplementation on production responses, DMI and blood parameters are presented in Table 1. The average of maximum THI in this experiment was 77.7 units that indicate medium degrees of heat stress during the study. Dry matter intake was significantly higher for the M group in comparison with the control group ($P \leq 0.05$). The increasing in DMI of the H group was numerically higher than in the control group ($P \leq 0.15$). Milk yield, fat corrected milk (FCM), milk component concentrations and yields were not affected by treatments, but milk protein and total solid percentages tended to be higher for the M group than cows that were not supplemented by Cr ($P \leq 0.09$ and $P \leq 0.11$, respectively). Supplementation with chromium (M and H groups) significantly decreased insulin concentration in comparison with the control group but glucose serum concentration was not affected by chromium supplementation.

Table 1. The effects of chromium supplementation on production responses and blood parameters.

Response Variable	Treatment (Trt) C	M	H	SEM	*P*-value Trt
DMI, kg/d	20.18[a]	22.51[b]	21.86[ab]	1.06	0.12
Milk, kg/d	36.65	37.89	36.49	1.77	0.80
FCM, kg/d	33.63	36.47	32.76	2.37	0.30
Fat,%	3.50	3.73	3.39	0.19	0.26
Fat, kg/d	1.27	1.41	1.22	0.08	0.17
CP,%	2.72	2.84	2.81	0.05	0.21
CP, kg/d	0.99	1.08	1.01	0.06	0.32
Lactose,%	5.73	5.84	5.64	0.10	0.19
Lactose, kg/d	2.08	2.21	2.03	0.14	0.46
TS,%	12.14	12.61	12.06	0.42	0.14
TS, kg/d	4.41	4.78	4.34	0.22	0.29
Glucose, mg/dl	67.81	66.18	66.78	1.11	0.36
Insulin, µIU/ml	8.90[a]	8.16[b]	8.34[b]	0.26	0.04

Conclusion

The results indicate that DMI was increased, but milk production was not influenced in heat stressed cows supplemented with chromium in early lactation. The main reason for this may be related to a decrease in insulin concentration in cows supplemented with chromium. Concentration of insulin decreased as a result of insulin binding to its receptor leading to an increased insulin signalling and glucose utilisation in peripheral tissues (e.g. muscle and fat). Therefore, a lower amount of glucose may be available for lactogenesis in the mammary glands.

References

Garcıa-Ispierto, I., F. Lopez-Gatius, P. Santolaria, C. Nogareda and F. De Rensis, 2006. Relationship between heat stress during the peri-implantation period and fetal loss in dairy cattle. Theriogenology 65: 799-807.

Mowat, D.N, 1997. Organic chromium in animal autrition. Chromium Books, Guelph, ON, Canada, pages: 1-298.

Pechova, A. and L. Pavlata, 2007. Chromium as an essential nutrient. a review. Vet. Med. 52: 1-18.

Johnson, H.D., 1987, Bioclimates and livestock: bioclimatology and the adaptation of livestock. World Anim. Sci. 21: 140-151.

Effect of late gestation maternal nutrition on leptin, IGF-1, insulin and glucose concentration in suckling lambs

A. Kiani[1], A.H. Tauson[2], A. Chwalibog[2] and M.O. Nielsen[2]
[1]Animal Sciences Group, Faculty of Agricultural Sciences, Lorestan University, Khoramabad, Iran; [2]Department of Basic Animal Veterinary Science, Faculty of Life Sciences, University of Copenhagen, Denmark; arkashkia@gmail.com

Introduction

Maternal undernutrition can cause alterations in endocrine function in the foetus (Fowden *et al.*, 2006), and undernutrition in late gestation decreases concentrations of anabolic hormones (e.g. insulin, IGF-I) in foetal circulation (Gicquel and Le Bouc, 2006). It has, therefore, been proposed that alteration in anabolic hormones during foetal life induced by maternal undernutrition may transit to postnatal life and continue to permanently alter concentrations of circulating hormones. We investigated the effect of late gestational nutrition in twin bearing ewes on plasma concentrations of leptin, IGF-1, insulin and glucose in their offspring.

Material and methods

Fourteen lambs whose mothers were fed either 60% (Restricted; 5 females and 3 males) or 100% (Control; 3 females and 3 males) of their energy requirements during the last six weeks of gestation were used. Lactating ewes were fed *ad libitum* and all lambs were reared by their own dam throughout the experiment. Blood samples from lambs were taken from the jugular vein at the day of birth (day 1) and also at 7, 17 and 32 days of age. Plasma samples were analysed for glucose, insulin, leptin and IGF-1. Data were analysed statistically with the MIXED procedure of SAS® version V8.2. The model included the fixed effects of late gestation maternal nutrition, age and sex of the lamb and samples within each lamb were declared as repeated. Differences between means were declared significant at $P<0.05$.

Results

There were no significant effects of sex in the traits measured in this experiment except for IGF-1, and therefore data from both sexes were analysed together except for IGF-1. Restricted lambs were lighter at 7 and 17 days of age, but at birth and at 32 days of age there was no significant difference in weight between the two groups (Figure 1A). At birth, plasma glucose in Restricted lambs was higher than that in Control lambs. All lambs were hypoglycaemic at birth, but plasma glucose concentration increased steeply by 7 days of age (Figure 1B). Restricted lambs had lower plasma concentration of insulin at 7 days of age (Figure 1C), and this resulted in a higher ratio of glucose-to-insulin in the Restricted lambs (Figure 1D). Plasma leptin concentrations did not differ between Control and Restricted lambs. However, similar to plasma glucose, plasma concentration of leptin increased steeply by 7 days of age (Figure 1E). Plasma concentration of IGF-1 was not different at birth, but at 7 and 32 days of age it was significantly lower in Restricted lambs than in Controls (Figure 1F). Regardless of maternal nutrition, males had higher plasma concentration of IGF-1 than females.

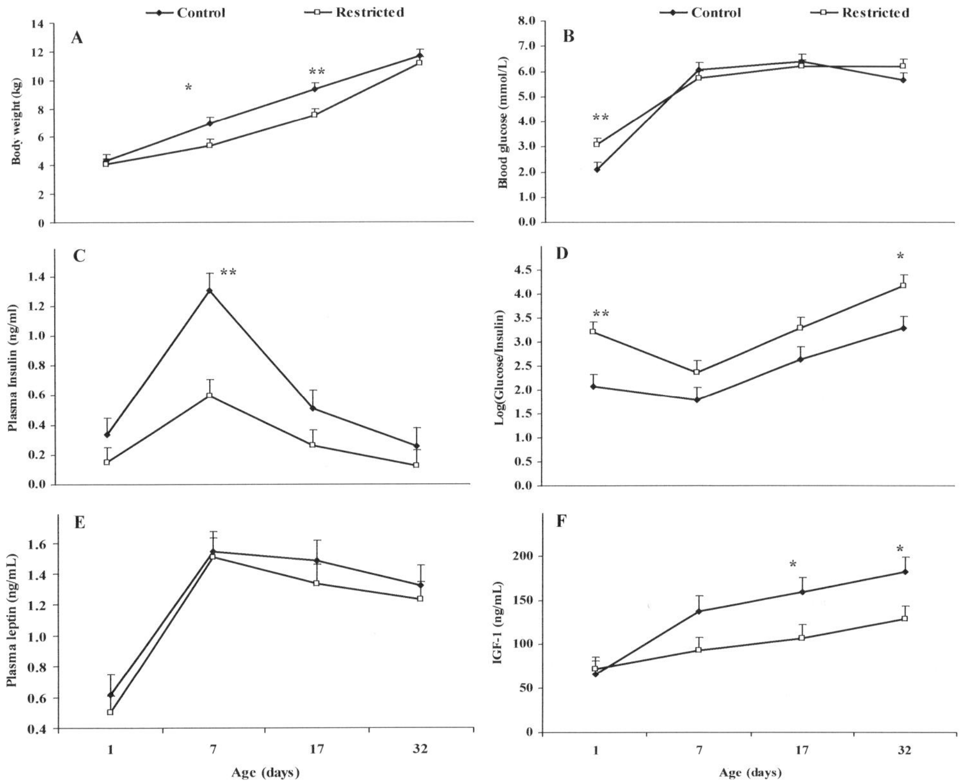

Figure 1. Lamb's body weight (A) and plasma concentration of glucose (B), insulin (C), ratio of glucose to insulin (D), leptin (E), and insulin like growth factor-1 (F) in response to late gestation maternal nutrition. Lambs born from ewes fed either 60% (Restricted; □) or 100% (Control; ♦) of their energy requirements during the last six weeks of gestation. * P< *0.05,* ** P< *0.01.*

Conclusion

Late gestation maternal undernutrition may decrease concentration of insulin in suckling offspring as it does during foetal life. This was also associated with a change in the ratio of glucose to insulin and a decrease in the concentration of IGF-1 in suckling lambs. The results also confirm that circulating glucose rather than insulin would be a determinant of leptin secretion (Tokuda, 2003).

References

Fowden, A.L., D.A. Giussani and A.J. Forhead, 2006. Intrauterine programming of physiological systems: causes and consequences. Physiol. 21: 29-37.

Gicquel, C. and B.Y. Le Bouc, 2006. Hormonal regulation of fetal growth. Horm. Res. 65 Suppl 3: 28-33.

Tokuda, T., 2003. Plasma leptin concentration in pre- and post-weaning lambs. Anim. Sci. 76: 221-227.

Effects of increasing supplementation levels of rice bran on milk production of lactating dairy goats

C.-H. Kim[1], J.K. Park[2], H.J. Choi[2], D.Y. Park[2] and J.D. Kim[3]
[1]School of Animal Life and Environment Science, Hankyong National University, 456-749, Anseong, Korea; [2]Graduate School of Bio-Environment and Information Technology, Hankyong National University, 456-749, Anseong, Korea; [3]Cheonan Yonam College, 330-709, Cheonan, Korea; kimch@hknu.ac.kr

Introduction

In the last year (2008), Korea has produced approximately 4.8 million tonnes of rice. Rice bran constitutes about 10% of the weight of rough rice; therefore, 480,000 tonnes annually were available to be used in animal feeds. Rice bran (RB) is rich in vitamins B and E, and trace minerals, and also protein and fat are of relatively high biological value (Warren and Farrell, 1990; Webb *et al.*, 2001). However, there is limited research on the performance of lactating dairy goats supplemented with rice bran. Therefore, the objective of the following study was to evaluate the effects of rice bran fed with different supplementation levels on milk production of dairy goats.

Material and methods

Four Saanen dairy goats (initial BW 55.0±7.1 kg) in their first or second lactations were used. The goats were 60 to 80 days into their lactation at the start of the experiment. The experimental design was a 4 x 4 Latin square with four animals and four 21-day periods. All experimental diets consisted of dry ground corn, rice bran, soybean meal, cotton seed meal, sugar beet pulp, alfalfa hay and timothy hay. Animals were allocated to the four groups fed diets *ad libitum* and four diets contained 0, 5, 10 and 20% DM of rice bran (RB), respectively, reducing the supplementation levels of corn and cotton seed meal (4.5 and 0.5%, 9 and 1% and 18 and 2%, respectively) while, they were equally formulated to be 10.7 ME/kg, 18.8% CP/kg DM (based on NRC, 1981) (Table 1). Feed intake and milk yield was recorded daily. Feed samples were obtained once weekly and composited across 3-week periods for composition of DM, CP, EE, CF and ash (AOAC, 1990) and NDF and ADF (filter bag technique; ANKOM Technology Corp., Fairport, NY, USA). Milk samples from the last four consecutive milkings in each experimental period were obtained and analysed for fat, protein, lactose and MUN with an infrared spectrophotometer (Milko-scan 4000, Foss Electric Co., Denmark). Data were analysed using the MIXED procedure of SAS® (1996) and the statistical model considered the main effects of period, animal and diet of the latin square design. Differences among diets were determined using the LSMEAN option.

Table 1. Chemical composition of experimental diets.

Item	Rice bran supplementation levels (%)			
	0	5	10	20
DM,%	88.82	88.93	89.03	89.24
CP,%	18.69	18.80	18.79	18.89
EE,%	3.89	4.52	5.17	6.45
ME, MJ/kg	10.69	10.70	10.70	10.72
NDF,%	35.91	33.01	33.17	33.67
ADF,%	20.47	20.61	20.63	20.80

All analyses considered $P<0.05$ as the minimum threshold for statistical significance.

Results

Feed intake and milk production are given in Table 2. Concentrate DM intake was the highest in the 5% treatment compared to the 0 and 20% treatments ($P<0.05$) whereas, forage DM intake was not significantly different among all the treatments. Total DM intake of the 5% treatment was significantly higher than the control ($P<0.05$). Milk yield was the lowest and milk fat content was higher than the control in the 20% treatment ($P<0.05$). Milk protein contents in 0 and 5% treatments were significantly lower than the 10 and 20% treatments ($P<0.05$). However, milk protein production was not significantly different among all the treatments ($P>0.05$). Milk lactose production was lower in the 20% treatment compared to the 0 and 5% treatments ($P<0.05$).

Table 2. Milk production and dry matter intake by supplementation of increased levels of rice bran in dairy goats.

	Rice bran supplementation levels (%)				Significance ($p<$)	SE
	0	5	10	20		
DM intake, g/d						
Concentrate	1,006^{b}	1,159^{a}	1,083^{ab}	1,022^{b}	0.01	36.58
Forage	484	468	463	510	ns	80.43
Total	1,490^{b}	1,627^{a}	1,541^{ab}	1,533^{ab}	0.05	62.63
Milk yield, g/d	2,080^{ab}	2,170^{a}	2,015^{b}	1,826^{c}	0.01	76.90
Milk fat,%	3.31^{b}	3.58^{ab}	3.70^{ab}	4.00^{a}	0.05	0.23
Milk fat, g/d	78.10	76.27	74.13	78.61	ns	6.40
Milk protein,%	2.85^{b}	2.83^{b}	2.95^{a}	2.96^{a}	0.01	0.11
Milk protein, g/d	62.83	61.60	59.06	58.34	ns	4.59
Lactose,%	4.49	4.55	4.52	4.50	ns	0.10
Lactose, g/d	99.03^{a}	99.24^{a}	90.59^{ab}	88.55^{b}	0.05	5.39

Means with different letters (a, b, c) within the same row differ for the reported threshold of significance.

Conclusion

The dietary fat contents were increased by increasing supplementation levels of rice bran though, only the milk yield of the 20% rice bran supplementation group was reduced. The cause of the result is probably due to restriction of the digestibility by supplementation of high dietary fat contents in the rumen. Therefore, the present experiment indicates that rice bran supplementation level should not be more than 10% in diets for milk production in diary goats.

References

AOAC, 1995. Official Methods of Analysis. 16th ed. Association of Official Analytical Chemist, Washington, DC, USA.

NRC, 1981. Nutrient Requirements of goats: Angora, Dairy and Meat Goats in Temperate and Tropical Countries. National Academy Press, Washington, D.C. USA

SAS Institute, 1996. SAS User's Guide: Statistics. SAS Institute, Inc., Cary, NC, USA.

Warren, B.E. and D.J. Farrell, 1990. The nutritive value of full fat and defatted Australian rice bran. I. Chemical composition. Anim. Feed Sci. Technol. 27: 219-228.

Webb, S.M., A.W. Lewis, D.A. Neuendorff and R.D. Randel, 2001. Effects of dietary rice bran, lasalocid, and sex of calf on *post partum* reproduction in Brahman cows. J. Anim. Sci. 79: 2968-2974

Monocarboxylate transporters (MCT1-MCT14) in the ruminant pancreas

D. Kirat and S. Kato
Department of Veterinary Physiology, School of Veterinary Medicine, Rakuno Gakuen University, 582 Bunkyodai-Midorimachi, Ebetsu, Hokkaido 069-8501, Japan; kato@rakuno.ac.jp

Introduction

Because of the uniqueness of ruminant digestion, the regulation of pancreatic secretion in ruminants differs from that in non-ruminants. In ruminants, short chain fatty acids (SCFA) are not only an important energetic and nutritional source but also essential regulators for the pancreatic endocrine and exocrine secretions, and consequently SCFA seem to act directly on the pancreas (Manns and Boda, 1967; Harada and Kato, 1983). Currently, 14 members of the monocarboxylate transporters (MCTs) family are known, seven of which have been functionally characterised. MCT1-MCT4 catalyse proton-coupled transport of metabolically important monocarboxylates such as SCFA, lactate, pyruvate and ketone bodies (Halestrap and Meredith, 2004). MCT6 mediates the transport of certain drugs. MCT8 transports thyroid hormones (Friesema *et al.*, 2003), while MCT10 catalyses the transport of aromatic amino acids (Kim *et al.*, 2001). Among MCT isoforms (MCT1-MCT8) studied in human tissues, MCT7 and MCT8 were detected in the pancreas (Price *et al.*, 1998). In rats, however, the study by Zhao *et al.* (2001) revealed that MCT1 was the only detected isoform among the four studied MCT (MCT1-MCT4). Our work was the first study undertaken to investigate the expression of all MCT family members (MCT1-MCT14) in the bovine pancreas.

Material and methods

Pancreases were collected from six Holstein–Friesian cows (520-600 kg) immediately after slaughter by bleeding from the carotid artery following intravenous injection with sodium pentobarbital (35 mg/kg). mRNA was isolated using MagNA Pure LC mRNA isolation Kit II (Roche Diagnostics GmbH, Germany) and the cDNA was generated by reverse transcription using Transcriptor first strand cDNA synthesis kit (Roche) following the manufacturer's instructions. PCR was performed using an Expand High Fidelity plus PCR system (Roche). PCR condition and primer sequences used for the amplification of MCT genes (MCT1–MCT14) are fully described in Kirat *et al.* (2009). DNA sequencing was performed with the BigDye Terminator v3.1 Cycle Sequencing kit, according to the manufacturer's instructions, on an ABI Prism 3100 automated sequencer (Applied Biosystems Inc.).

Results

Our RT-PCR analysis using the specific primers for MCT (MCT1-MCT14) revealed that eight MCT isoforms, namely MCT1, MCT2, MCT3, MCT4, MCT5, MCT8, MCT13 and MCT14 of the expected size of 1206, 981, 244, 1052, 904, 692, 996, and 834 bp, respectively are expressed in the bovine pancreas (Figure1). These amplified cDNA segments were confirmed by sequence analysis and deposited in GenBank under various accession numbers. The homology search of the identified segments of each gene showed 100% identity with the equivalent MCT of *Bos taurus* (GenBank accession numbers NP_001032396, NP_001069804, XP_612876, NP_001103450, NP_001094593, XP_615986, NP_001069600, and XP_585259 for MCT1, MCT2, MCT3, MCT4, MCT5, MCT8, MCT13 and MCT14, respectively). We could not detect MCT6, MCT7, MCT9, MCT10, MCT11, or MCT12.

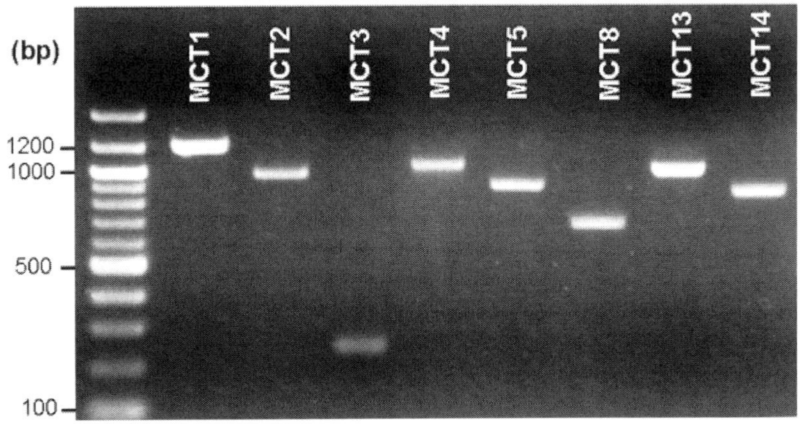

Figure 1. Agarose gel electrophoresis for RT-PCR analysis of the detected MCT isoforms in the bovine pancreas.

Conclusion

Our results present novel data for the expression of eight MCT isoforms (MCT1-MCT5, MCT8, MCT13 and MCT14) in the ruminant pancreas. MCT1-MCT4 may play a role in SCFA transport across the pancreatic cells, while MCT8 can mediate the transport of thyroid hormones to the pancreas. MCT5, MCT13 and MCT14 are not functionally characterised yet.

References

Friesema, E.C., S. Ganguly, A. Abdalla, J.E. Manning Fox, A.P. Halestrap and T.J. Visser, 2003. Identification of monocarboxylate transporter 8 as a specific thyroid hormone transporter. J. Biol. Chem. 278: 40128-40135.

Halestrap, A.P. and D. Meredith, 2004. The SLC16 gene family-from monocarboxylate transporters (MCTs) to aromatic amino acid transporters and beyond. Pflugers Arch-Eur. J. Physiol. 44: 619-628.

Harada, E. and S. Kato, 1983. Effect of short-chain fatty acids on the secretory response of the ovine exocrine pancreas. Am. J. Physiol. 244: G284-290.

Kim, D.K., Y. Kanai, A. Chairoungdua, H. Matsuo, S.H. Cha and H. Endou, 2001. Expression cloning of a Na+-independent aromatic amino acid transporter with structural similarity to H+/monocarboxylate transporters. J. Biol. Chem. 276: 17221-17228.

Kirat, D., K. Sallam, H. Hayashi, T. Miyasho and S. Kato, 2009. Presence of ten isoforms of monocarboxylate transporter (MCT) family in the bovine adrenal gland. Mol. Cell. Endocrinol. 298: 89-100.

Manns, J.G. and J.M. Boda, 1967. Insulin release by acetate, propionate, butyrate, and glucose in lambs and adult sheep. Am. J. Physiol. 212: 747-755.

Price, N.T., V.N. Jackson and A.P. Halestrap, 1998. Cloning and sequencing of four new mammalian monocarboxylate transporter (MCT) homologues confirms the existence of a transporter family with an ancient past. Biochem. J. 329: 321-328.

Zhao, C., M.C. Wilson, F. Schuit, A.P. Halestrap and G.A Rutter, 2001. Expression and distribution of lactate/monocarboxylate transporter isoforms in pancreatic islets and the exocrine pancreas. Diabetes 50: 361-366.

Circadian variation in plasma total antioxidative capacity and levels of ascorbic acid in sheep

S. Kobayashi[1], M. Kumagai[1], Y. Kikuchi[1], A. Hagino[2] and S. Oda[1]
[1]Department of Animal Science, Faculty of Agriculture, Iwate University, Ueda 3-18-8, 020-8550, Morioka, Japan; [2]Laboratory of Animal Physiology, Graduate School of Agricultural Science, Tohoku University, Tsutsumidori-amamiyamachi 1-1, Aoba-ku, 981-8555, Sendai, Japan; soda@iwate-u.ac.jp

Introduction

Reactive oxygen species (ROS) play a dual role as both beneficial and deleterious species. ROS are well recognised for processes of pathogen clearance. Antioxidants as scavenger sof ROS have been shown to play a role in the maintenance of cellular redox homeostasis (Valko *et al.*, 2007). However, overproduction of ROS results in oxidative stress and damage to cells and tissues, leading to progression of diseases (Hogg, 1998). Especially, livestock are susceptible to oxidative stress because of the requirement for high productivity and economic efficiency. In some diseases such as endotoxemia in sheep or pneumonia in calves, overproduction of ROS by endotoxin is one of the causes of tissue damage and inflammatory response (Polla *et al.*, 1991; Ledwozyw and Stolarczyk, 1992). In addition, administration of vitamins C and E, which are antioxidants, showed to improve heat stress-induced oxidant status, leading to better production performance of layers (Panda *et al.*, 2008). Therefore, the assessment of the overall antioxidative capacity is significant from the perspective of production performance and prevention of diseases. However, total antioxidative capacity including circadian rhythm still remains unclear in ruminants. The present study was aimed at evaluating circadian variation in plasma total antioxidative capacity and levels of ascorbic acid as an antioxidant in sheep.

Material and methods

Four mixed-breed wethers were each kept in metabolic cages. Diets were composed of a maintenance ration (orchard-timothy mixed hay and concentrate). Plasma samples that were obtained from wethers were collected 7 times in a 24-h period in November and December 2008. The average weekly temperatures for November and December were about 10 °C and 2 °C, respectively. Plasma total antioxidative capacity was measured with a high-sensitive chemiluminescence assay kit (TK-0005TAO-K, Tokken Co., Ltd., Japan) and auto luminescence analyzer. A type of antioxidant that has superoxide-scavenging ability was evaluated. The results of the samples were calculated with a standard curve of vitamin C levels and expressed in vitamin C equivalent. Plasma ascorbic acid analysis was carried out using high-performance liquid chromatography. Opsonized zymosan (OZ)-stimulated superoxide production in blood was also measured using chemiluminescence with luminol. All the data from experiments are expressed as the mean ± S.E. Statistical analyses of the data for circadian variation were evaluated by ANOVA and the Tukey method. Trends towards significance were considered at $P<0.1$.

Results

Total antioxidative capacity in ovine plasma increased transiently 1 to 3 h after feeding and paralleled variations in OZ-stimulated superoxide production by phagocytes (Tables 1 and 2). Both total antioxidative capacity and superoxide production subsequently remained at basal levels. There was no difference in 24 h variations between November and December. However, mean values of both parameters in December were significantly lower than those in November ($P<0.05$). Ascorbic acid levels in plasma were also increased by feeding, however the basal level tended to be lower in November (2.5±0.2 μg/ml) than in December (4.1±0.6 μg/ml).

Table 1. Circadian variation in total antioxidative capacity in ovine plasma.

Plasma total antioxidative capacity [Vitamin C equivalent (mg/ml)]								mean±S.E.
time (hrs)	pre	feeding	1	2	4	14	22	
Nov.	17.7±3.2	21.4±3.9	24.8±3.9	21.8±5.0	16.8±3.3	18.8±3.0	14.9±3.3	19.5±3.6[a]
Dec.	9.5±4.4	13.2±5.1	15.4±5.9	9.7±5.0	11.7±3.7	13.6±4.5	13.5±5.1	12.4±1.5[b]

[a,b] Different superscript letters indicate a significant difference (P=0.0224).

Table 2. Circadian variation in OZ-stimulated superoxide production by phagocytes.

superoxide production per cell (counts/10 sec)								mean±S.E.
time (hrs)	pre	feeding	1	2	4	14	22	
Nov.	3.7±3.7	3.0±0.1	4.2±1.1	2.6±0.5	3.4±0.9	2.5±0.2	2.4±0.6	2.8±0.6[a]
Dec.	1.5±0.1	1.3±0.2	1.9±0.3	1.8±0.4	1.7±0.2	1.8±0.5	1.0±0.1	1.4±0.2[b]

[a,b] Different superscript letters indicate a significant difference (P=0.0096).

Conclusion

This study revealed that total antioxidative capacity in ovine plasma increased transiently after feeding and paralleled variations in superoxide production by phagocytes. It seemed that an increase in superoxide production by feeding initiated the following upregulation of total antioxidative capacity. There was no correlation between total antioxidative capacity and level of ascorbic acid in plasma, indicating that plasma ascorbic acid was not a main superoxide scavenger. Moreover, our data suggest that both plasma total antioxidative capacity and superoxide production by phagocytes were affected negatively by cold exposure.

References

Hogg, N., 1998. Free radicals in disease. Semin. Reprod. Endocrinol. 16: 241-248.

Ledwozyw, A. and H. Stolarczyk, 1992. The involvement of polymorphonuclear leukocytes in the pathogenesis of bronchopneumonia in calves. VI. Superoxide dismutase and lipoprotein lipase activities. Acta. Vet. Hung. 40: 267-277.

Panda A.K., S.V. Ramarao, M.V. Raju and R.N. Chatterjee, 2008. Effect of dietary supplementation with vitamins E and C on production performance, immune responses and antioxidant status of White Leghorn layers under tropical summer conditions. Br. Poult. Sci. 49: 592-599.

Polla, B.S., J. Clerc, J.F. Pittet and D.R.Morel, 1991. Superoxide production by peripheral blood monocytes during sustained endotoxaemia in sheep. Clin. Sci. 81: 815-822.

Valko, M, D. Leibfritz, J. Moncol, M.T. Cronin, M. Mazur, and J. Telser, 2007. Free radicals and antioxidants in normal physiological functions and human disease. Int. J. Biochem. Cell Biol. 39: 44-84.

Expression of fatty acid and amino acid transporters around differentiation in bovine mammary epithelial cells (BMEC)

Y. Kobayashi[1], K. Higuchi[1], I. Nonaka[1], H. Ohtani[1], N. Kanematsu[1], K. Katoh[2], K. Sato[3], O. Enishi[1] and M. Sutoh[1]

[1]National Institute of Livestock and Grassland Science, Tsukuba, Ibaraki, 305-0901, Japan; [2]Graduate School of Agricultural Science, Tohoku University, Sendai, Miyagi, 981-8555, Japan; [3]Institute of Symbiotic science and technology, Tokyo University of Agriculture and Technology, Fuchu, Tokyo, 183-8509, Japan; kobayou@affrc.go.jp

Introduction

Bovine mammary epithelial cells (BMEC) extensively transport many nutrients to produce a large amount of milk components. BMEC are able to transport some nutrients into the cells via nutrient-specific transporters (Bionaz and Loor, 2008), synthesise milk components, and secrete milk into the luminal side. However, BMEC, when cultured in a medium, have no ability to produce milk components. They have to reconstitute the functional lumen structure such as mammospheres, and need to be exposed to some endocrine signals. There are few reports demonstrating the relationship between cell differentiation with lumen structure and milk synthesis although BMEC cultured on the extracellular matrix preparation (Matrigel) is known to secrete the alfa-casein into the luminal side (Rose *et al.*, 2002). It is also not shown how the regulatory mechanism of nutrient absorption changes after differentiation in BMEC. In the present study, therefore, we analysed mRNA expression of fatty acid and amino acid transporters in BMEC cultured on Matrigel and treated with lactogenic hormones.

Material and methods

BMEC were sampled from the mammary gland of one 102-day-pregnant Holstein cow and were cloned by the limiting dilution method. FACS analysis showed that all of the cells had the existence of cytokeratin. Cells were seeded at the density of 1×10^5 cells/cm^2 in Matrigel (BD biosciences, MA)-coated 6-well plate and cultured in DMEM (Sigma Aldrich) contained 10% fetal bovine serum, glutamine (2mM), penicillin (100 U/ml), streptomycin (100 μg/ml) and lactogenic hormones. Lactogenic hormones consisted of dexamathasone (10 μg/ml), insulin (10 μg/ml) and prolactin (10 μg/ml). Cultured cells were maintained under 5% CO_2 and humidified air at 37 °C. BMEC were cultured for 0, 1, 3, 5, 7 days after seeding, and were subjected to quantitive RT-PCR.

After being cultured for the planned periods, cells were washed twice with ice-cold PBS (-) and were perfused into TRIzol reagent to isolate total RNA according to the manifactur's instructions (Invitrogen, CA). Total RNA was reverse-transcribed using Super Script First-Strand Synthesis System for RT-PCR (Invitrogen, CA, USA). Quanatitive RT-PCR was performed using LightCycler FastStart DNA MsterPLUS SYBR Green I (Roche, Switzerland).

The expression of CD36 was calculated as a fatty acid transporter. Furthermore, those of CAT-1 (cationic amino acid transporter-1) and LAT-1,2 (L-type amino acid transporter-1,2) were calculated as amino acid transporters. The sequence of RT-PCR primers for transporters was previously described (Hayashi *et al.*, 2005, Liao *et al.*, 2008). The expression level of transporters were normalised by that of GAPDH, and are represented as relative mRNA levels.

All experiments were done in triplicate and values are expressed as means ±SEM. Statistical significance was estimated by means of one-way ANOVA followed by the Bonferroni multiple range test.

Results

BMEC functionally differentiated on Matrigel-coated plate under co-stimulation with lactogenic hormones because alfa-s1-casein mRNA expressed at day1 and continued for 7 days (data not shown).

The expression of CD36 gradually and significantly increased with culture, with the value of the expression for day 7 being 10-fold greater than that for day 0 (P<0.01) (Table 1). For amino acid transporters, there were no consistent changes and no significant change in LAT1. However, there was a similar pattern between the time courses for the expression of CAT1 and LAT2 (Table 1).

Table 1. Gene expression of fatty acid and amino acid transporters.

	mRNA levels relative to GAPDH mRNA				
	day0	day1	day3	day5	day7
CD36	0.18 ^{e}	0.47 ^{d}	0.85 ^{c}	1.02 ^{b}	2.00 ^{a}
CAT1	1.03 ^{ab}	0.95 ^{bc}	0.88 ^{c}	1.05 ^{a}	1.08 ^{a}
LAT1	1.12	1.09	1.13	1.28	1.19
LAT2	1.17 ^{ab}	1.16 ^{ab}	1.02 ^{b}	1.47 ^{a}	1.38 ^{a}

BMEC were cultured for 0, 1, 3, 5, 7 days on Matrigel-coated plate under co-stimulation with lactogenic hormones. ^{a,b,c,d,e}; Values within rows with different superscript letters are significantly different (P<0.05).

Conclusion

Our present findings clearly demonstrated that the gene expression of fatty acid and amino acid transporters changes around the formation of lumen structures in BMEC. Because of drastic changes for CD36, it is suggested that fatty acid transport on the membrane surface of BMEC may be activated for production of milk fats. For amino acid transporters, other types of transporters may have a key role for amino acid transport in BMEC. These data would be useful for research on the regulatory mechanisms of milk production in BMEC.

References

Bionaz, M. and J.J. Loor, 2008. ACSL1, AGPAT6, FABP3, LPIN1, and SLC27A6 are the most abundant isoforms in bovine mammary tissue and their expression is affected by stage of lactation. J. Nutr. 138: 1019-1024.

Hayashi, H., T. Yonezawa, T. Kanetani, F. Terada, K. Katoh and Y. Obara, 2005. Expression of mRNA for sodium-glucose transporter 1 and fatty acid translocase in the ruminant gastrointestinal tract before and after weaning. Anim. Sci. J. 76: 339-344.

Liao, S. F., E.S. Vanzant, J. A. Boling and J.C. Mattews, 2008. Identification and expression pattern of cationic amino acid transporter-1 mRNA in small intestinal epithelia of Angus steers at four production stages. J. Anim. Sci. 86: 620-631.

Rose, M.T., H. Aso, S. Yonekura, T. Komatsu, A. Hagino, K. Ozutsumi and Y. Obara, 2002. *In vitro* differentiation of a cloned bovine mammary epithelial cell. J. Dairy. Res. 69: 345-355.

Hepatic acetylation of the blood flow marker *p*-aminohippuric acid affect measurement of hepatic blood flow in cattle

N.B. Kristensen, B.A. Røjen, B.M.L. Raun, A.C. Storm, L. Puggaard and M. Larsen
Faculty of Agricultural Sciences, Aarhus University, DK-8830, Tjele, Denmark; nbk@agrsci.dk

Introduction

For decades the standard procedure for determining portal and hepatic blood flows in cattle implanted with permanent indwelling catheters in an artery and in portal, hepatic, and mesenteric veins has been based on down stream dilution of *p*-aminohippuric acid (pAH) infused into the mesenteric vein (Huntington *et al.*, 1989). This technique is based on the Fick principle (Zierler, 1961) and requires: (a) a quasi steady state, (b) ability to obtain representative samples of the blood, and (c) that the blood flow marker is not metabolised in the isolated tissue beds. Acetylation of pAH is known to occur in the liver of sheep (Katz and Bergman, 1969), but to our knowledge no investigations on hepatic metabolism of pAH exists for cattle. Several observations on hepatic metabolism in cattle point toward a potential bias in estimates of hepatic arterial blood flow: (a) hepatic uptake of glucogenic precursors cannot account for the apparent hepatic glucose output (Larsen and Kristensen, 2009), (b) the carbon source for the apparent hepatic acetate production is unaccounted for (Reynolds, 1995), and (c) hepatic mass balances for [^{13}C]acetate infused into the jugular vein indicates hepatic uptake of acetate (Kristensen and Harmon, 2004) although the activity of acetyl-CoA synthetases in the ruminant liver is very low. For all three entities the inability to balance the liver fluxes of precursors and products could be caused by overestimation of the hepatic arterial flow rate driven by overestimation of hepatic vein flow rate. Based on these findings we hypothesised that the standard procedure for measuring splanchnic blood flows in cattle without a deacetylation step overestimates hepatic arterial blood flow because of acetylation of pAH in the liver. The objective of the present study was to test the effect of deacetylation of plasma pAH on estimates of splanchnic blood plasma flows in lactating dairy cows.

Material and methods

Arterial, portal, and hepatic blood plasma was obtained from eight lactating Danish Holstein cows (parity 2) fitted with permanent indwelling catheters in the vessels sampled as well as in the mesenteric vein. Continuous infusion of pAH (27±0.4 mmol/h) into the mesenteric vein was initiated 1 h before first sampling time. Eight sets of blood samples were obtained at hourly intervals starting 0.5 h before morning feeding. Plasma pAH was analysed with and without deacetylation by the method described by Harvey and Brothers (1962) using a continuous flow analyzer (Autoanalyzer 3, method US-216-72 Rev.1; Seal Analytical Ltd, Burgess Hill, England). For deacetylation of pAH, plasma was deproteinised by combining with an equal volume of 20% trichloroacetic acid (w/v) and the supernatant incubated at 100 °C for 1 h. Data were analysed by a paired t-test using the means procedure of SAS®.

Results and discussion

Deacetylation of plasma pAH markedly increased ($P<0.01$) the measured concentration of pAH in arterial and hepatic blood plasma (Table 1); however, the hepatic concentration was increased relatively more than the arterial concentration leading to a decreased ($P<0.01$) estimate of hepatic vein and hepatic arterial blood flow. The hepatic arterial blood flow was reduced by 50% with deacetylation indicating that hepatic acetylation is affecting measurements of hepatic blood flow in cattle. Based on the findings of the present study it is apparent that the portal vein contributes relatively more to total hepatic blood flow than typically have been observed for cattle and this has large implications on measurements of net hepatic fluxes. The calculated net hepatic flux for all compounds with a detectable hepatic vein – arterial concentration difference will be biased by an overestimation of the hepatic arterial blood flow. Therefore, the current findings imply that we need to reassess numerous aspects of hepatic metabolism in cattle, for example, what we assume to know about hepatic metabolism of glucose, acetate, and oxygen.

Table 1. Effect of pAH deacetylation on splanchnic blood flow variables in lactating dairy cows (mean ± SE; n=8).

Item	Without deacetylation	With deacetylation	*P*-value
Arterial pAH, mmol/l	0.073±0.004	0.085±0.004	<0.01
Hepatic pAH, mmol/l	0.090±0.004	0.106±0.005	<0.01
Hepatic vein plasma flow, l/h	1,550±28	1,303±33	<0.01
Hepatic arterial blood plasma flow, l/h	329±48	162±24	<0.01
Portal vein blood plasma flow in proportion of hepatic vein blood plasma flow, proportion	0.80±0.03	0.88±0.02	0.01

References

Harvey, R.B. and A.J. Brothers, 1962. Renal extraction of para-aminohippurate and creatinine measured by continuous in vivo sampling of arterial and renal-vein blood. Ann. N. Y. Acad. Sci. 102, 46-54.

Huntington, G.B., C.K. Reynolds and B.H. Stroud, 1989. Techniques for measuring blood flow in splanchnic tissues of cattle. J. Dairy Sci. 72: 1583-1595.

Katz, M.L. and E.N. Bergman, 1969. Simultaneous measurements of hepatic and portal venous blood flow in the sheep and dog. Am. J. Physiol. 216: 946-952.

Kristensen, N.B. and D.L. Harmon, 2004. Splanchnic metabolism of VFA absorbed from the washed reticulorumen of steers. J. Anim. Sci. 82: 2033-2042.

Larsen, M. and N.B. Kristensen, 2009. Effect of abomasal glucose infusion on splanchnic amino acid metabolism in periparturient dairy cows. J. Dairy Sci. in press.

Reynolds, C.K., 1995. Quantitative aspects of liver metabolism in ruminants. In: Ruminant physiology: digestion, metabolism, growth and reproduction. W.V. Engelhardt, S. Leonhard-Marek, G. Breves and D. Giesecke (Eds.) Ferdinand Enke Verlag, Stuttgart, Germany, pp. 351-371.

Zierler, K.L., 1961. Theory of the use of arteriovenous concentration differences for measuring metabolism in steady and non-steady states. J. Clin. Invest. 40: 2111-2125.

Magnesium and calcium metabolism in periparturient dairy cows fed different levels of calcium

C. Kronqvist[1], U. Emanuelson[2], R. Spörndly[1], M. Tråvén[2] and K. Holtenius[1]
[1]Swedish University of Agricultural Sciences, Dept. of Animal Nutrition and Management, 753 23, Uppsala, Sweden; [2]Swedish University of Agricultural Sciences, Dept. of Clinical Sciences, Box 7054, 750 07, Uppsala, Sweden; cecilia.kronqvist@huv.slu.se

Introduction

Calving and the onset of lactation is a great challenge for the dairy cow. Inability to maintain a sufficient level of calcium (Ca) in the blood due to losses in milk may result in milk fever or subclinical hypocalcaemia. Parathyroid hormone (PTH) is a key factor regulating calcium homeostasis. Hypomagnesaemia can impair both the release of PTH and the tissue response to the hormone (Goff, 2008). Earlier studies in sheep have shown that an increasing Ca concentration in the rumen may inhibit the uptake of magnesium (Mg) (Care *et al.*, 1984). The Mg homeostasis is not hormonally regulated, and the urinary excretion reflects the uptake (Ram *et al.*, 1998). Urinary calcium output is generally low and not affected by dietary Ca (Goff, 2008). The aim of the present study was to investigate the effects of dietary Ca on the homeostasis of Ca and Mg in periparturient dairy cows.

Material and methods

Dry, pregnant cows of the Swedish Red breed (n=29) in their 2nd to 6th gestation were assigned to blocks according to calving date and adapted to a diet consisting of 6 kg DM of grass haylage and 2.8 kg DM of concentrates. Three weeks before expected calving, the cows were randomly assigned to 3 diets with Ca levels of 4.9 (L), 9.3 (M) and 13.6 g/kg DM (H), achieved by limestone supplement. The Mg concentration in all 3 diets was 1.8 g/kg DM. After calving, all cows were fed similar diets. Blood samples were taken two times weekly until calving, and at 6, 12 and 24 h, 2, 4 and 7 d after calving. Spot samples of urine were taken twice weekly until calving. Samples of silage and concentrates were taken regularly. Ca and Mg concentrations were determined in all samples. PTH in plasma as well as creatinine in urine was analysed. Mineral excretion in the urine was calculated using a daily creatinine output rate of 29 mg/kg BW (Valadares *et al.*, 1999). For the statistical analysis, the time factors (day before calving and day in treatment) were classified in 4-day-periods. The MIXED Procedure of SAS® was used, with treatment, cow, time relative to calving (TRC) and time in treatment as fixed effects and block as a random effect. Cow was treated as subject with repeated measurements over TRC. Differences were considered significant at P<0.05.

Results

There were no significant differences in Ca level in plasma between the experimental diets, but cows fed diet H had significantly lower plasma levels of Mg after calving compared to the two other groups (Figure 1). No significant differences in PTH between treatments were found.
The daily output of Ca in the urine was the highest for cows fed diet L (Table 1). The excretion of Mg in the urine was the lowest for cows fed diet H.

Conclusion

Plasma Mg declined after calving among the cows fed diet H, even though all cows then were given the same diet. The lower urinary Mg output in cows fed diet H indicates that Ca interfered with Mg uptake during the dry period. The decrease in Mg did not seem to affect the Ca metabolism after calving, since there were no significant differences between treatments in Ca or PTH levels. Since an impaired Mg status of cows around calving may increase the risk of developing milk fever, the effect of Ca on the metabolism of Mg needs further studies.

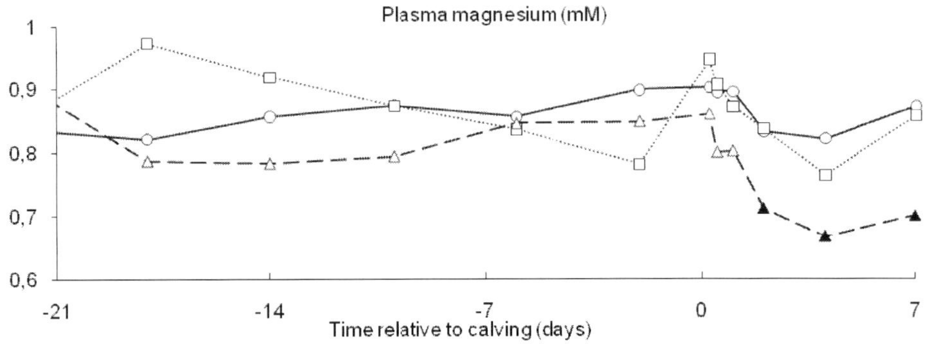

Figure 1. Plasma Mg (least square means) in 29 dairy cows fed different Ca levels, L (○), M (□) and H (△) during the last 21 days of gestation. Filled symbols indicats difference between diets (P<0.05).

Table 1. Urinary output (least squares means) of Ca and Mg in the last 21 days of gestation for cows fed different Ca levels; L (4.9 g Ca/kg DM), M (9.3 g Ca/kg DM) and H (13.6 g Ca/kg DM).

	L	M	H	SEM	*P*-value
Ca, g/day	1.1[a]	0.4[b]	0.5[b]	0.20	0.009
Mg, g/day	2.7[a]	2.7[a]	1.7[b]	0.29	0.004

[a, b] Means in the same row with the same letter are not significantly different (P>0.050).

References

Care, A.D., R.C. Brown, A.R. Farrar and D.W. Pickard, 1984. Magnesium absorption from the digestive tract of sheep. Quarterly J. Exp. Phys. 69: 577-587.

Goff, J.P., 2008. The monitoring, preventing, and treatment of milk fever and subclinical hypocalcemia in dairy cows. Vet. J. 176: 50-57.

Ram, L., J.T. Schonewille, H. Martens, A.T. van't Klooster and A.C. Beynen, 1998. Magnesium absorption by wethers fed potassium bicarbonate in combination with different dietary magnesium concentrations. J. Dairy Sci. 81: 2485-2492.

Valadares, R.F.D., G.A. Broderick, S.D. Valadares Filho and M.K. Clayton, 1999. Effect of replacing alfalfa silage with high moisture corn on ruminal protein synthesis estimated from excretion of total purine derivatives. J. Dairy Sci. 82: 2686-2696.

Insulin resistance after single dose dexamethasone treatment in dairy cows

M. Kusenda, A. Starke, M. Kaske, M. Piechota, M. Hoeltershinken and J. Rehage
Clinic for Cattle, University of Veterinary Medicine, Foundation, Bischofsholer Damm 15, 30173 Hannover, Germany; juergen.rehage@tiho-hannover.de

Introduction

Glucocorticoids such as dexamethasone are widely used in medical therapy (20 to 40 µg/kg bodyweight [BW]) for ketosis and the clinical efficacy is clearly demonstrated in dairy cows. Dexamethasone leads to considerable hyperglycemia which is predominately explained by enhanced hepatic gluconeogenesis (Shpigel *et al.*, 1996, Jorritsma *et al.*, 2004). However, recent studies showed after dexamethasone treatment in calves no increase in key enzyme activities of hepatic gluconeogenesis (Hammon *et al.*, 2003, 2005). In humans it has been shown that dexamethasone leads to insulin resistance and thereby to hyperglycemia (Nicod *et al.*, 2003, Dake *et al.*, 2004). Thus, the aim of the study was to investigate whether dexamethasone affects whole body insulin resistance in early lactating cows.

Material and methods

Twelve clinically healthy Holstein cows (2-4 weeks *post partum*, second and third lactation, five days after right flank omentopexy for correction of left sided abomasal displacement, no treatment with corticosteroids for at least four weeks prior to the start of the study) were assigned randomly and blinded to one of two treatment groups: dexamethasone group (40 µg dexamethasone/kg BW intravenously [IV]; N=6), or control group (corresponding volume of saline IV, N=6). Cows were housed in single pens on straw and were fed a diet based on grass land hay, corn silage and concentrate (roughage to concentrate ratio: 60%:40%) according to maintenance. To characterise the peripheral insulin resistance, euglycemic hyperinsulinemic glucose clamps (EHGC) were performed the day after administration of the drugs. Each cow received five consecutive continuous infusion periods (two hours each) with increasing doses of bovine insulin (0.1, 0.5, 2, 5, 10 mU/kg BW min; IV). Blood glucose concentration was clamped at the basal level by adjusting the intravenous infusion rate of glucose according to blood glucose level measured every 10 min. In order to allow separate insulin and glucose infusions, cows were fitted with two indwelling venous catheters (left and right jugular vein) the day before the EHGC. Before the EHGC basal insulin (analysed by RIA), glucose (commercial enzymatic test kit), and non-esterified fatty acid (Nefa; enzymatic commercial test kits) serum concentrations were measured. During the last 30 minutes of each insulin infusion period the steady state (SS) insulin serum concentration (SSIC; mU/ml), the steady state glucose infusion rate (SSGIR; µmol/kg BW per min) and the steady state Nefa serum concentration (SSNefa; µmol/l) were assessed. The insulin sensitivity ratio (ISR) was calculated from SSGIR and SSIC (ISR=SSGIR/SSIC). The results were statistically evaluated using the Wilcoxon Signed Ranks Test (SAS® statistical package; version 9.1). *P*-values of less than 0.05 were considered significant.

Results

The day after dexamethasone treatment in average serum glucose and insulin concentrations were significantly higher compared to controls ($P<0.001$). Compared to controls dexamethasone led to significantly lower SSGIR during the second ($P<0.01$; 0.5 mU/kg min), and the fourth insulin infusion period ($P<0.05$, 5 mU/kg min). During the first (0.1 mU/kg min) and second insulin infusion periods but not in the following periods of the EHGC the ISR (Figure 1) was significantly lower

($P<0.01$) after Dexamethasone application compared to control cows. The SSNefa dropped during the EHGC in control cows and dexamethasone treated cows to 6%±2 (mean ± sem) and 25% ± 4 of baseline values, resp. ($P<0.05$).

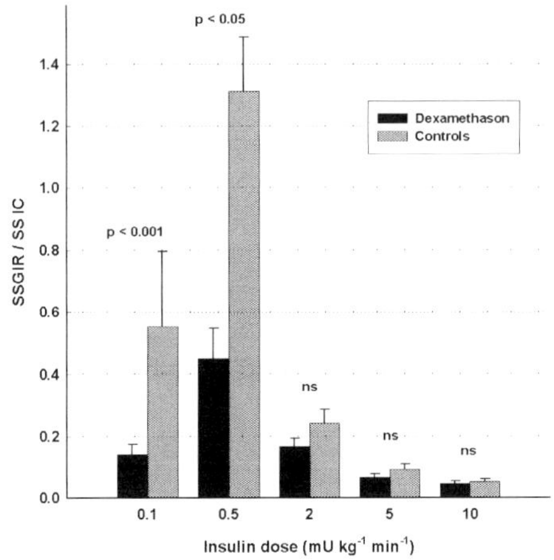

Figure 1. Steady state insulin sensitivity ratio (SSGIR/SSIC; µmol ml/mU kg min) during euglycemic hyperinsulinemic glucose clamps in dairy cows one day after dexamethasone or placebo (control) treatment (N=6 each; mean ± sem; corresponding means differ significantly where indicated, ns not significant).

Conclusion

Dexamethasone treatment leads to increased insulin resistance due to reduced whole body insulin sensitivity. Peripheral insulin response appears not to be affected. Insulin resistance affects whole body glucose and fatty acid metabolism.

References

Dake, Q., T. Pulinikunnil, A. Ding, G. Sanjoy, A. Ashraf, J.A. Pospisilik, R. Brownsey, R. Wambolt, M. Allard and B. Rodrigues, 2004. Single-dose dexamethasone induces whole-body insulin resistance and alters both cardiac fatty acid and carbohydrate metabolism. Diabetes 53: 1790-1797.

Hammon, H.M., S.N. Sauter, M. Reist, Y. Zbinden, C. Philipona, C. Morel and J.W. Blum, 2003. Dexamethasone and colostrum feeding affect hepatic gluconeogenic enzymes differently in neonatal calves. J. Anim. Sci. 81: 3095-3106.

Hammon, H.M., C. Philipona, Y. Zbinden, J.W. Blum and S. Donkin, 2005. Effects of dexamethasone and growth hormone treatment on hepatic gluconeogenic enzymes in calves. J. Dairy Sci. 88: 2107–2116.

Jorritsma, R., J. Thanasak, M. Houweling, J.P. Noordhuisen and K.E. Muller, 2004. Effects of a single dose of dexamethasone-21-isonicotinate on the metabolism of heifers in early lactation. Vet. Rec. 155: 521-523.

Nicod, N., V. Giusti, C. Besse and L. Tappy, 2003. Metabolic adaptations to dexamethasone-induced insulin resistance in healthy volunteers. Obesity research 11: 625-631.

Shpigel, N.Y., R. Chen, Y. Avidar and E. Bogin, 1996. Use of corticosteroids alone or combined with glucose to treat ketosis in dairy cows. J. Am. Vet. Med. Assoc. 208: 1702-1704.

Plasma concentrations of incretins (GIP and GLP-1) did not increase in periparturient cows abomasally infused with glucose

M. Larsen[1], A.E. Relling[2], C.K. Reynolds[3] and N.B. Kristensen[1]
[1]Faculty of Agricultural Sciences, Aarhus University, DK-8830 Tjele, Denmark; [2]Dept. of Animal Science, The Ohio State University, Wooster, 44691,USA; [3]School of Agriculture, Policy, and Development, University of Reading, Reading, United Kingdom; Mogens.Larsen@agrsci.dk

Introduction

Increasing the glucogenic status of the periparturient cow by feeding diets containing bypass starch is, in theory, an attractive strategy to overcome the problems of hypoglycaemia and hyperlipidemia in early lactation. In a study with periparturient cows, Larsen and Kristensen (2009) found the dry matter intake (DMI) to be 37% lower when 1,500 g/d of glucose were abomasally infused compared to non-infused cows. In humans, the incretin peptides, glucagon-like peptide 1(7-36) amide (GLP-1) and glucose-dependent insulinotropic polypeptide (GIP) participate, directly and indirectly, in the hormonal regulation of appetite and glucose metabolism (Holst, 2004). Therefore, we hypothesised an increase in plasma concentrations of GLP-1 and GIP in periparturient cows abomasally infused with glucose.

Material and methods

The study is described in detail by Larsen and Kristensen (2009). In brief, six periparturient Holstein cows fitted with ruminal cannulas and permanent indwelling catheters in major splanchnic blood vessels were used to study the effects of abomasal glucose infusion on splanchnic glucose metabolism. Cows were assigned to one of two treatments: no infusion (CONT) or infusion of 1,500 g/d of glucose into the abomasum (GLUC) initiated at the day of calving. Abomasal glucose infusion was conducted using a device inserted into the abomasum via the rumen cannula. Blood samples were collected at 12±6 days *pre partum* as well as 4, 15, and 29 days in milk (DIM). Concentrations of GIP and GLP-1 were measured in pooled arterial plasma samples using radio-immuno-assays (Relling and Reynolds, 2007). Intraassay coefficients of variation were 10.3% and 6.7% for GIP and GLP-1, respectively. Analysis of variance was conducted using a mixed model including the random effect of cow and the fixed effects of treatment, DIM and the interaction.

Results and discussion

The data on DMI and plasma concentrations of glucose and insulin has been reported previously (Larsen and Kristensen, 2009). From calving to 29 DIM, voluntary DMI increased at a lower rate (P=0.05) with GLUC compared with CONT and the overall treatment means were 10.7 and 16.9 kg/d, respectively. The plasma concentrations of GIP and GLP-1 were unaffected by GLUC (P=0.50 and P=0.76, respectively; Figure 1), but increased as lactation progressed (P=0.01 and P<0.01, respectively). Relling and Reynolds (2007) found a similar pattern of concentrations of GIP and GLP-1 in early lactating dairy cows, and multiple regression analysis showed that the *post partum* increases were related to actual DIM and not to actual DMI. The lower DMI with GLUC might have been caused by impairment of digesta flow, as no blank infusion was conducted with CONT, but a test experiment showed that the infusion device did not induce major negative impact on DMI (Larsen and Kristensen, 2009). The cows with GLUC did not experience the normally observed abrupt fall in plasma concentrations of glucose and insulin at calving (Figure 1), which might be of importance for the endocrine cascade triggering metabolic adaptations to lactation.

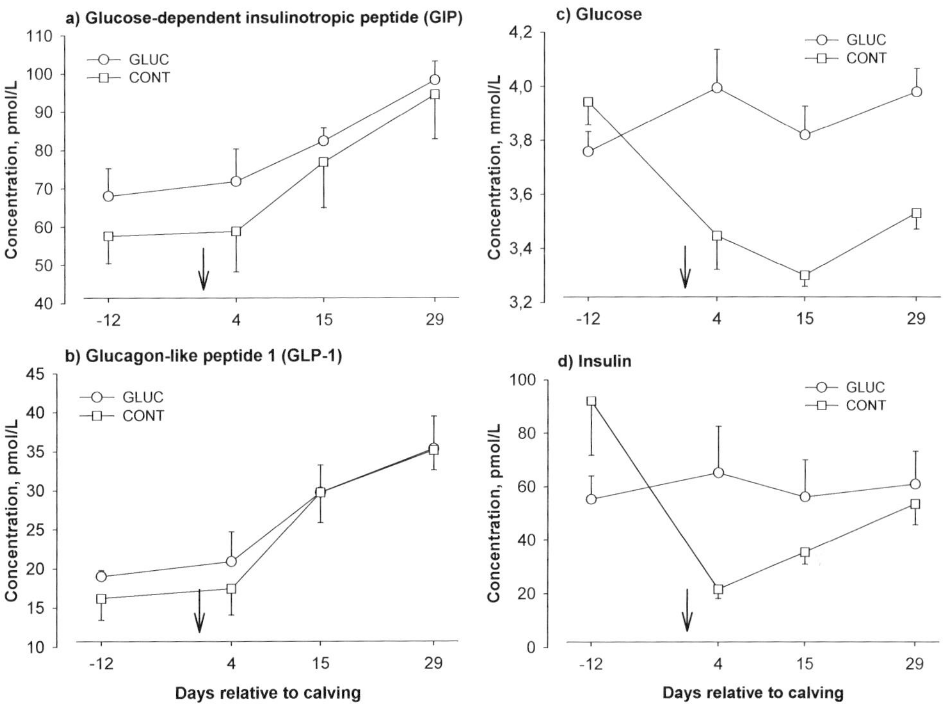

Figure 1. Arterial concentrations of glucose-dependent insulinotropic polypeptide (GIP; a), glucagon-like peptide 1(7-36) amide (GLP-1; b), glucose (c), and insulin (d) in periparturient dairy cows with either no infuson (□, CONT) or infused with 1,500 g/d of glucose into the abomasum (○, GLUC) from the day of calving (arrow). Each point is mean ± SEM (n=3). The concentrations of GIP and GLP-1 were unaffected by GLUC (P=0.50 and P=0.76, respectively), but increased as lactation progressed (P=0.01 and P<0.01, respectively). The concentrations of glucose and insulin decreased more from pre partum to 4 DIM with CONT compared to GLUC (P=0.05 and P=0.03, respectively; Larsen and Kristensen, 2009).

Conclusion

Abomasal glucose infusion in periparturient dairy cows did not affect the plasma concentrations of GIP and GLP-1, whereas the concentration increased as lactation progressed. From this data we cannot confirm that GIP or GLP-1 participated in regulation of appetite or glucose metabolism in periparturient cows abomasally infused with glucose.

References

Holst, J.J., 2004. On the physiology of GIP and GLP-1. Horm. Metab. Res. 36: 747-754.

Larsen, M. and N.B. Kristensen, 2009. Effect of abomasal glucose infusion on splanchnic and whole body glucose metabolism in periparturient dairy cows. J. Dairy Sci. 92: 1071-1083.

Relling, A.E. and C.K. Reynolds. 2007, Plasma concentrations of gut peptides in dairy cattle increase after calving. J. Dairy Sci. 90: 325-330.

Response of plasma ghrelin to growth hormone releasing hormone (GHRH) administration during compensatory growth in steers

H.G. Lee[1], C.H. Lee[2], Z.S. Hong[3], C.X. Xu[2], Y.C. Jin[2], H. Kuwayama[4] and Y.J. Choi[2]
[1]Department of Animal Science and PNU-Special Animal Biotechnology Center, Pusan National University, 627-706, Gyeongnam, Korea; [2]Department of Agricultural Biotechnology, Seoul National University, 441-744, Seoul, Korea; [3]Department of Animal science & Technology, Tianjin Agricultural University, P.R. China; [4]Department of Animal Science, Obihiro University of Agriculture and Veterinary Medicine, 080-8555, Obihiro, Japan; hglee66@pusan.ac.kr

Introduction

During compensatory growth, animals show great body weight (BW) gain, increased efficiency of energy utilisation, reduced maintenance requirements because of depression of the basic metabolic rate, enhanced appetite and feed intake capacity, changes in endocrine status, altered body tissue composition compared with animals fed conventionally (Ryan *et al.*, 1993; Park *et al.*, 1994; Choi *et al.*, 1997; Ahn *et al.*, 1996). Growth hormone (GH) is related to the compensatory growth. Ghrelin is an important regulator of the somatotropic axis in ruminants and its effect might be altered by the physiological status of animals. Therefore this study was conducted to examine the effect of growth hormone releasing hormone (GHRH) administration on plasma active and total ghrelin levels during compensatory growth induced by a stair-step growth pattern in steers.

Material and methods

Ten 6 month-old Korean native steers, averaging 180 kg of BW (SE = 3.22), were used for the experiment. The test group was subjected to a stair-step growth pattern and was fed according to a schedule of 3, 2, 4 and 2 mo. The first stair-step began with feeding for maintenance for 3 mo (20% below the control) followed by feeding for compensatory growth (25% above the control) for 2 mo. The second step was 4 mo of maintenance feeding followed by 2 mo of compensatory feeding. Steers were assigned randomly to injection of vehicle (5 ml 0.1% BSA-saline) and GHRH (0.45 μg/kg BW) on the previous and compensatory growth period. Blood samples were collected each time at -20, -10, 0, 5, 10, 20, 30, 40, 50, 60, 90, 120 and 180 min after injection of GHRH and saline solution for plasma ghrelin assay. Plasma active ghrelin, total ghrelin and GH were measured by double antibody radioimmunoassay (RIA) as described by ThidarMyint *et al.* (2006). The area under the curve (AUC) of ghrelin and GH was calculated for 180 min after GHRH administration and corrected for the basal AUC (-20, -10 and 0 min), which was evaluated by analysis of the student *t*-test.

Results and discussion

In the present study, the response of GH in the stair-step growth group was stimulated by GHRP-2 administration compared to the normal feeding group ($P<0.05$) (Table 1). In addition the steers in the stair-step growth group showed greater body weight (BW) gain and feed efficiency ($P<0.05$). For plasma ghrelin, although total plasma ghrelin response to GHRH administration was increased in the stair-step growth group ($P<0.05$), there was no apparent effect of GHRH treatment on active plasma ghrelin concentrations during the compensatory growth in steers (Table 1). In conclusion, we suggest that the development of growth performance by compensatory growth may increase total plasma ghrelin in steers.

Table 1. Feed efficiency and responses of bovine plasma active and total ghrelin to GHRH administration during compensatory growth in steers.

	Normal	Stair-step Growth
Feed efficiency[1]	7.9 ± 0.21	8.6 ± 0.25*
Active ghrelin		
Tot AUC (ng/min.ml)[2]	154,283.6 ± 60,319.76	132,583.8 ± 86,461.46
AVG (ng/ml)[3]	625 ± 275.4	574 ± 420.4
Total ghrelin		
Tot AUC (ng/min.ml)[2]	696.9 ± 40.90	1156.7 ± 34.21*
AVG (ng/ml)[3]	3.8 ± 2.41	5.7 ± 2.20

Values are expressed as mean ± SEM for 4 steers.

[1] (Daily weight gain/daily feed intake) × 100%; DMI: dry matter intake.

[2] Area calculated for the 180 min (between 0 and 180) periods.

[3] Calculated as the average of concentration for 180 min.

* Significance $P<0.05$.

References

Ahn, K.H., I.K. Han, K. Kwon, J.H. Woo and Y.J. Choi, 1996. Effects of stair-step feeding pattern by manipulating energy contents on the growth rate, mammary gland development and reproductive performance. Kor. J. Anim. Nutr. Feed. 20: 231.

Choi, Y.J., I.K. Han, J.H. Woo, H.J. Lee, K. Jang, K.H. Myung and Y.S. Kim, 1997. Compensatory growth in dairy heifers: The effect of a compensatory growth pattern on growth rate and lactation performance. J. Dairy Sci. 80: 519-524.

Park, C.S., M.G. Baik, W.L. Keller and W.D. Slanger, 1994. Dietary energy restriction-mediated growth and mammary development in rats. J. Anim. Sci. 72: 2319-2324.

Ryan, W.J., I.H. Williams and R.J. Moir, 1993. Compensatory growth in sheep and cattle. I. Growth pattern and feed intake. Aust. J. Agric. Res. 44: 1609-1621.

ThidarMyint H., H. Yoshida, T. Ito and H. Kuwayama, 2006. Dose-dependent response of plasma ghrelin and growth hormone concentrations to bovine ghrelin in Holstein heifers. J. Endocrinol 189: 655-664.

Milk fatty acid profile of cows fed diets supplemented with soybean or fish oil and with two concentrate levels

L.C. Leite[1] and D.P.D. Lanna[2]
[1]Universidade Estadual de Maringá, Paraná, Brazil; [2]Universidade de São Paulo, Escola Superior de Agricultura Luiz de Queiroz, São Paulo, Brazil; dplanna@esalq.usp.br

Introduction

Conjugated linoleic acid (CLA) is a known anticarcinogen and fat supplementation in the ruminant diet has been used to increase its content in milk. The major source of CLA in milk is endogenous synthesis from vaccenic acid (*trans*-11 18:1) (Griinari *et al.*, 2000). Bauman and Lock (2006) have shown that inhibition of the last step of the biohydrogenation pathway can result in the accumulation of *trans*-11 18:1. This can be accomplished by diet manipulations that modify the rumen environment such as changes in forage:concentrate ratios and the use of fish oil. In high-concentrate diets, the normal biohydrogenation metabolism is altered, with the formation of a different isomer, *trans*-10 18:1, instead of *trans*-11 18:1. This reduces substrate availability for endogenous synthesis of *cis*-9, *trans*-11 CLA and can contribute for a reduced secretion of CLA in milk fat (Shingfield *et al.*, 2005). With this background, the objective of this work was to evaluate the effects of concentrate level and source of oil on the fatty acid profile of milk fat.

Material and methods

Four Holstein dairy cows (average initial milk production of 31.24±8.57 kg/d and with 109±10 days in milk) were used in a 4×4 Latin square design to evaluate the effect of oil source and concentrate level on milk fatty acid profile. Treatments in a 2×2 factorial arrangement were the following: two concentrate levels in the total DM (40 and 60%) diet and two sources of oil (soybean oil or fish oil at 2% of the total diet DM). A total mixed ration was fed twice daily with corn silage as roughage source. Fatty acid profiles were analysed according to Chouinard *et al.* (1998). The fatty acid profile was determined by GC with a 200 m capillary column of fused silica (Varian CP-2571). Statistical analysis used the mixed model, considering the cow as random effect.

Results

Interestingly, while oil source had a pronounced effect on fatty acid profiles, concentrate level did not alter them (Table 1). It is also noteworthy the absence of interaction between treatments. Fish oil reduced saturated and increased mono and poly-unsaturated fatty acids. Considering the relationships with the main milk fatty acids, fish oil decreased the levels of 18:0 and increased the levels of total *trans* 18:1, *trans*-10 18:1, *trans*-11 18:1, *cis*-9, *trans*-11 CLA, 20:5n-3 and 22:6n-3, when compared to soybean oil.

Conclusion

This work confirms that different sources of oil have a consistent effect on fatty acid profiles, particularly *trans*-11 18:1. However, this work corroborates previous results in which concentrate levels have inconsistent effects on fatty acid profiles, which need to be predicted more precisely before nutritional strategies can be used to alter milk fatty acid profiles.

Table 1. Milk fatty acid profile (g/100 g) of Holstein cows fed diets supplemented with soybean or fish oil and with two concentrate levels.

	Treatment				SE			
Oil source[1] (O)	Soybean oil		Fish oil			*P*-value		
Concentrate level (C)	60%	40%	60%	40%		O	C	O x C
Summary								
SCFA (Σ C4-10)	9.05	8.41	8.57	8.23	0.56	0.56	0.41	0.80
MCFA (Σ C11-16)	42.80	40.88	42.94	46.15	1.40	0.09	0.66	0.10
LCFA (> C17)	48.19	50.73	48.52	45.67	1.87	0.24	0.93	0.18
SFA	70.73	68.59	58.58	61.52	2.07	<0.01	0.85	0.25
MUFA	24.65	26.73	31.19	28.23	1.59	0.03	0.79	0.15
PUFA	4.37	4.48	6.04	5.66	0.57	0.04	0.82	0.68
Main fatty acids								
14:0	10.78	10.23	10.46	11.35	0.42	0.36	0.69	0.12
16:0	26.66	25.59	26.78	29.02	0.99	0.11	0.57	0.13
18:0	18.71	19.09	6.86	6.77	1.44	<0.01	0.92	0.88
*cis*9-18:1	23.62	25.64	29.61	26.69	1.56	0.05	0.78	0.15
Total *trans*	6.36	7.39	18.09	17.38	1.26	<0.01	0.90	0.51
trans-10	1.62	1.04	4.88	3.78	1.44	0.07	0.58	0.86
trans-11	2.21	3.47	9.52	9.98	1.60	<0.01	0.60	0.81
18:2	3.78	3.88	4.59	4.31	0.50	0.24	0.87	0.71
cis-9, *trans*-11	0.50	0.71	1.44	1.38	0.29	0.02	0.81	0.65
trans-10, *cis*-12	0.02	0.01	0.01	0.02	0.01	0.61	0.99	0.29
18:3n-3	0.27	0.30	0.30	0.27	0.03	0.96	0.96	0.47
20:5n-3	0.04	0.05	0.28	0.28	0.04	<0.01	0.88	0.96
22:6n-3	0.05	0.03	0.26	0.22	0.05	<0.01	0.64	0.80

[1] 2% fish or soybean oil in DM of diet.

References

Bauman, D.E. and A.L. Lock, 2006. Conjugated linoleic acid: biosynthesis and nutritional significance. In: Fox, P.F. and P.L.H. McSweeney (Eds), Advanced dairy chemistry. Springer, New York, p. 93-136.

Chouinard, P.Y., V. Girard and G.J. Brisson, 1998. Fatty acids profile and physical properties of milk fat from cows fed calcium salts of fatty acids with varying unsaturation. J. Dairy Sci. 81: 471-481.

Griinari, J.M., B.A. Corl, S.H. Lacy, P.Y. Chouinard, K.V. Nurmela and D.E. Bauman, 2000. Conjugated linoleic acid is synthesized endogenously in lactating dairy cows by delta-9 desaturase. J. Nutr. 130: 2285-2291.

Shingfield, K.J., C.K. Reynolds, B. Lupoli, V. Toivonen, M.P. Yurawecs, P. Delmonte, J.M. Griinari, A.S. Grandison and D.E. Beever, 2005. Effect of forage type and proportion of concentrate in the diet on milk fatty acid composition in cows given sunflower oil and fish oil. Anim. Sci. 80: 225-238.

Performance, metabolic parameters and fatty acid composition of milk fat due to dietary CLA and rumen-protected fat of dairy cows

T. Liermann[1], J. Groß[1], P. Möckel[2], A.-M. Pfeiffer[3], G. Jahreis[2] and F.J. Schwarz[1]
[1]Technische Universität München, Department of Animal Science, Section of Animal Nutrition, Hochfeldweg 4, 85350 Freising-Weihenstephan, Germany; [2]University Jena, Institute of Nutrition, Dornburgerstr. 24, 07743 Jena, Germany; [3]BASF SE, Charlottenstr. 59, 10117 Berlin, Germany; schwarzf@wzw.tum.de

Introduction

After parturition energy intake by dairy cows is lower than energetic output. Therefore cows mobilise body tissue to compensate their negative energy balance. To lower this lack two possibilities can be discussed: either increasing the energy density of the diet (e.g. supplementing rumen protected fat (RF)) or lowering the energy output (e.g. reducing milk fat yield by conjugated linoleic acid (CLA) in comparison to practical feeding) (Griinari and Bauman, 2006). The purpose of this paper was to highlight the potential effects of feeding CLA alone or in combination with RF with regards to energy partioning and metabolic responses as well as the modulation of the fatty acid (FA) composition of milk fat. Further attention was paid to the period after finishing feeding the supplements.

Material and methods

The trial lasted from the first week p.p. until the 14th wk (period 1, with supplements) and from the 15th to the 26th wk (post-period, without feeding the supplements) of lactation. Dairy cows (Red Holstein (RH)×Simmental) were allocated as follows: I-control (n=17), II-CLA (n=18) and III-CLA+RF (n=18). CLA supplement (Lutrell®, BASF, 40 g/cow/d, lipid-encapsulated, 10.7% CLA c9t11, 10.7% CLA t10c12, 10.0% C16:0, 55.7% C18:0 and 9.7% C18:1 c9) was added with an extra concentrate of 0.2 kg. RF (Dunafat 100, EURODUNA) was given separately (700 g/cow/d). All cows were fed *ad libitum* once a day a partial mixed ration (PMR) (50.0% corn silage, 28.5% grass silage, 15.5% concentrate, 6.0% hay, dry matter basis). Additional concentrate without supplements was offered when daily milk yield was above 21 kg. Milk yield, intake of PMR, concentrate, CLA and RF were recorded daily for each cow. Aliquots of blood samples (taken once per week) were stored at -20 °C until analysis of plasma glucose, NEFA, BHBA, liver enzymes (AST, GLDH, γ-GT) and bilirubin. Milk samples were taken once per week and freeze-dried. The FA composition was analysed in wk 1, 5, 9 and 14 by means of GC-FID (Kraft *et al.*, 2003). Analysis of variance (performance, metabolic parameters) was conducted using a mixed model (SAS JMP® v7.01) with treatment as the fixed effect and week as the random effect. Data of fatty acid composition were subjected to analysis of variance (ANOVA) within week using SAS® (SAS Institute Inc., 1998, Cary, NC, USA).

Results

The effects of treatments on performance and metabolic parameters are presented in Table 1. Daily feed intake was unaffected by treatment. Milk yield tended to be higher in group II and III (35.9 kg and 35.8 kg) in contrast to the control (34.8 kg). Milk fat was significantly depressed by CLA (group II and III: 3.00% and 3.06% vs. group I: 3.67%). Cows of group II and group III in average reached about 2-3 wk earlier a positive energy balance after parturition. After finishing the feeding of the supplements milk fat again raised (3.57% (II, III) *vs* 3.83% (I)) whereas the milk yield stayed on a higher level (27.6 kg (II, III) vs. 25.5 kg (I)). Blood parameters, liver enzymes and bilirubin showed no significant treatment differences.

Table 1. Effect of treatment on performance and metabolic parameters (SD in brackets).

	I-control	II-CLA	III-CLA+RF
DM intake, kg/d			
period 1	19.1 (2.2)	19.0 (2.3)	19.0 (2.3)
post-period	20.0 (0.45)	19.6 (0,57)	19.6 (0.47)
Milk yield, kg/d			
period 1	34.8 (2.6)	35.9 (2.4)	35.8 (2.3)
post-period	25.5 (3.6)	27.6 (3.7)	27.5 (3.8)
Milk fat,%			
period 1	3.67[a] (0.28)	3.00[b] (0.44)	3.06[b] (0.46)
post-period	3.83 (0.16)	3.61 (0.17)	3.53 (0.24)
Energy balance (period 1), MJ NEL/d	-9.3[a] (4.89)	-1.8[b] (4.81)	2.0[b] (4.81)

[a,b] Means within a row with different superscripts differ ($P<0.05$).

Table 2 shows the modulation of milk fat composition. CLA alone did not significantly change the FA pattern. Surprisingly, the further addition of RF in combination with CLA changed the FA composition of the milk fat much more clearly than CLA supplementation alone. In wk 14 of lactation the sum of trans C-18:1 and sum of CLA fatty acids of milk fat significantly increased only in group III (∑C-18:1 *trans* (% of FAME): 1.88 (I), 2.04 (II), 3.64 (III), ∑CLA FAME (% of): 0.46 (I), 0.54 (II), 0.61 (III)). The proportion of CLA c9t11 was on average about 80% whereas CLA t10c12 only amounted to 3-4% of total CLA after CLA supplementation.

Table 2. Milk fat content (%) and FA composition (% FAME) at lactation wk 14.

	Fat	SFA	MUFA	PUFA	∑CLA	c911t	t10c12	C18:1 t10	C18:1 t11
I	3.65[a]	75.0[a]	22.4[b]	2.57[b]	0.46[b]	0.37[b]	0.00	0.24[b]	0.51[b]
II	2.87[b]	74.4[a]	22.7[b]	2.98[a]	0.54[ab]	0.43[b]	0.02	0.23[b]	0.60[b]
III	2.88[b]	72.1[b]	25.4[a]	2.50[b]	0.61[a]	0.51[a]	0.02	0.51[a]	0.73[a]

[a,b] Means within a row with different superscripts differ ($P<0.05$).

Conclusion

Dietary supplementation with CLA clearly reduces milk fat content. As also described in earlier studies (Griinari and Bauman, 2006) there was a potential effect of energy partioning, since supplemented cows in this trial reached a positive energy balance about 2-3 wk sooner than cows fed the control diet. After feeding CLA+RF the sum of trans C18:1 fatty acids and CLA was significantly higher compared to the control diet or feeding CLA alone.

References

Griinari, J.M. and D.E. Bauman, 2006. Milk fat depression: concepts, mechanisms and management. In: K. Sejrsen, T. Hvelplund and M.O. Nielsen (Eds.), Ruminant physiology: digestion, metabolism, and impact of nutrition on gene expression, immunology and stress. Wageningen Academic Publishers, Wageningen, the Netherlands, pp. 389-417.

Kraft, J., M. Collomb, P. Möckel, R. Sieber and G. Jahreis, 2003. Differences in CLA isomer distribution of cow's milk lipids. Lipids 38: 657-664.

A technique to assess internal body fat of dairy goats using real-time ultrasound

L.D. Lima, I.A.M.A. Teixeira, H.G. Silva, K.T. Resende, J.C. Canola and O.B. Neto
UNESP, São Paulo State University, Jaboticabal, SP 14870-000, Brazil; lisidelima@gmail.com

Introduction

The understanding of fat deposition is essential to determine body composition, which is an important indicator of animal nutritional status. One of the most commonly indirect methods to assess body and carcass fat *in vivo* is real-time ultrasound (RTU). It is a noninvasive technique and also has been considered as a practical, accurate and precise method (Ribeiro, 2008). Normally RTU has been used to measure subcutaneous fat thickness and intramuscular fat in cattle. The application of RTU in small ruminants was considered limited at the beginning, because of the presence of hair or wool and also the reduced subcutaneous fat thickness. The last characteristic is especially important in goats, since this species mainly accumulates internal fat (Teixeira *et al.*, 2008). Thus, the prediction of internal fat would be helpful in estimating body fat composition in goats. However, the assessment of internal fat deposition using ultrasound is difficult, due to the complexity of establishing an anatomical reference point to make the evaluations. Therefore, it is imperative to develop methods that can evaluate internal fat depot in goats. The aim of this study was to develop a technique to evaluate internal fat in goats, based on the measurement of kidney fat using real-time ultrasound.

Material and methods

Data for this study were obtained from 48 dairy pregnant female goats (average BW = 67.1±12.5; age = from 3 to 5 years; body condition score = 3.01±0.74), in which 3 does were 80 days pregnant, 14 does were 110 d pregnant and 25 does were 140 d pregnant. They were fed a diet based on hay and concentrate (60:40 roughage: concentrate ratio) formulated according to their requirements. The RTU measurements were collected in a calm relaxed animal stand just before slaughter. Hair was clipped to increase image quality, and gel was used as a coupling agent. The RTU kidney fat image was collected after the 13th rib, as a posterior view (Figure 1) in order to evaluate the kidney fat thickness (mm) in the lower border of the right kidney. RTU measurements were made using an Aquila Vet ultrasound unit (Esaote Pie Medical, the Netherlands) with a 5-MHz convex array transducer. Images were collected and interpreted on site at the ultrasound console. After slaughter, the weights of internal fat depots (kidney, heart, omental, mesenteric and abdominal fats) were properly recorded. Pearson correlations between RTU kidney fat and the weights of internal fat depots were calculated using the CORR procedure and regression equations between those parameters were calculated using the REG procedure of SAS® (SAS® Inst., Inc., Cary, NC).

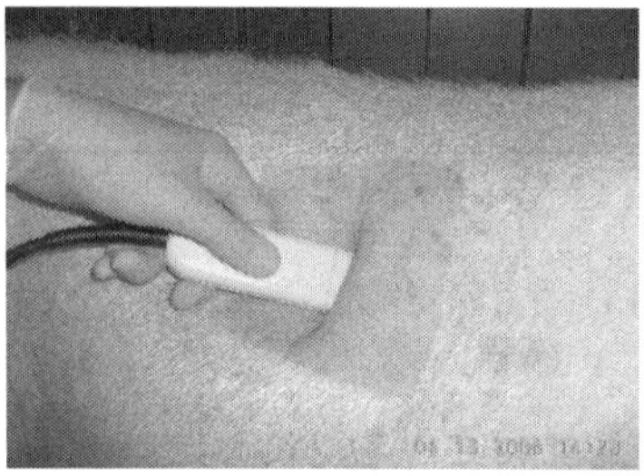

Figure 1. Evaluation of kidney fat using real-time ultrasound.

Results

There was a high correlation between non-carcass fat and omental fat (r = 0.93, $P<0.0001$) and kidney fat (r = 0.84, $P<0.0001$). RTU kidney fat was not correlated to internal fat depots, except to kidney fat (r = 0.33, $P<0.01$), however this relationship presented a low precision. The regression equation between kidney fat and RTU kidney fat presented a low R^2 (0.11; Equation 1; Table 1). These results were probably due to the large variation in the body condition score of the females during pregnancy. It has been reported in previous studies that fat deposition is intensified at the beginning of pregnancy and the fat reserves are used to meet the nutritional requirements at the end of the pregnancy (Scheaffer *et al.*, 2004). This phenomenon has been observed in a large number of animals, chiefly in species for which abdominal fat is the main fat depot (Véras *et al.*, 2001). Since the body condition score is a simple and widely used tool to assess body fat stores, it was added as an independent variable in the equations in order to increase the precision of them (Equations 2 and 3; Table 1). The same procedure was used, adding pregnancy age in Equations 2 and 3. The amount of kidney fat and non-carcass fat decreased with the increasing age of pregnancy. A negative linear relationship was observed between days of pregnancy and amount of internal fat depots, mainly due to the fact that the majority of the evaluated goats were in the final third of pregnancy, a period during which higher nutrient mobilisation occurs. The multiple regression equations were more precise than the simple ones ($R^2 = 0.91$ and 0.55 for kidney fat and non-carcass fat, respectively).

Table 1. Regression equations between real-time ultrasound kidney fat and kidney fat and non-carcass fat depots.

Variable	Regression equation	R^2
Kidney fat (Equation 1)	Y=1203.54497* + 2063.68402 RTU**	0,11
Kidney fat (Equation 2)	Y= 2684.32895 RTU* - 7.10126 D** + 623.25166 BCS*	0,91
Non-carcass fat (Equation 3)	Y= 6330.96827** + 6610.91028 RTU** - 64.33429 D* + 2545.95881 BCS*	0,55

*= ($P<0,01$); ** = ($P<0,05$); RTU= real-time ultrasound kidney fat; D=days of pregnancy; BCS= body condition score, according to the scale of 1 to 5 proposed by Morand-Fehr and Hervieu (1999).

Conclusion

The real time ultrasound can be used as a practical, precise and useful tool to monitor fat body stores in goats during pregnancy, especially associated to body condition score and pregnancy age.

References

Morand-Fehr, P. and J. Hervieu, 1999. Apprécier l'état corporel des chèvres: Intérêt et méthode. Rev. La Chèvre 231: 22-34.

Ribeiro, F.R.B., L.O. Tedeschi and J.R. Stouffer and G.E. Carstens, 2008. Technical note: A novel technique to assess internal body fat of cattle by using real-time ultrasound. J. Anim. Sci. 86: 763-767.

Scheaffer, A.N., J.S. Caton, D.A. Redmer and L.P. Reynolds, 2004. The effect of dietary restriction, pregnancy, and fetal type in different ewe types on fetal weight, maternal body weight, and visceral organ mass in ewes. J. Anim. Sci. 82: 1826-1838.

Teixeira, A., M. Joy and R. Delfa., 2008. *In vivo* estimation of goat carcass composition and body fat partition by real-time ultrasonography. J. Anim. Sci. 86: 1-8.

Véras, A.S.C., S. de Campos Valadares Filho, J.F.C. da Silva, M.F. Paulino, P.R. Cecon, R.F.D. Valadares, M. de Andrade Ferreira and C.M. Sá Fontes, 2001. Efeito do nível de concentrado sobre o peso dos órgãos internos e do conteúdo gastrintestinal de bovinos nelore não-castrados. R. Bras. Zootec. 30, suppl 3: 1120-1126.

Energy expenditure of Angus heifers divergently selected for residual feed intake

D.S. Lines[1], M.L.Wolcott[2], W.S. Pitchford[1], C.D.K. Bottema[1], R.M. Herd[3] and V.H. Oddy[2]
Cooperative Research Centre for Beef Genetic Technologies
[1]Animal Science, The University of Adelaide, 5371, Roseworthy, South Australia; [2]Animal Science, University of New England, Armidale, 2351, New South Wales, Australia; [3]NSW DPI Beef Industry Centre of Excellence, Armidale, 2351, New South Wales, Australia; david.lines@adelaide.edu.au

Introduction

In beef cattle divergently selected for residual feed intake (RFI; a measure of feed efficiency), up to 95% of the variation in RFI could be attributed to differences in energy expenditure (EE) rather than to energy retained in body tissues (Richardson and Herd, 2004). These authors suggested EE associated with whole body protein turnover, tissue metabolism and stress may contribute 37% of the variation in RFI. The literature suggests that the energy costs of protein turnover contribute toward 20-25% of maintenance EE; 15-20% of basal metabolic rate across a range of species; protein synthesis alone contributes 30% maximally toward heat production in cattle as reviewed by Richardson and Herd (2004). Given this evidence, the aim of this study was to assess the contribution of protein synthesis to EE and RFI of beef cattle.

Methods

Sixteen Angus heifers from a research population divergently selected for approximately 3.5 generations for high RFI (low efficiency; n = 8, RFI estimate breeding value (EBV) = 0.64 kg/d) and low RFI (high efficiency; n = 8, RFI EBV = -0.78 kg/d) were used. They were fed 105% maintenance requirements (MEm) and 95% *ad libitum*, equal to approximately 180% maintenance, in a crossover design. First; one half (4 high RFI and 4 low RFI) were fed at 180% MEm and the other half (4 high RFI and 4 low RFI) fed 105% MEm. After adaptation to this feeding level, the following measurements were taken. Ultrasound for subcutaneous rump and rib fat depth, eye muscle area (EMA) and intramuscular content of fat (IMF) were measured. EE was estimated by infusion with ^{13}C-sodium bicarbonate and whole body muscle protein synthesis with ^{13}C-leucine. Then the dietary treatments were switched and the measurements were taken again. The time between first and last ultrasound measurements within each dietary treatment was 35 and 42 d for periods 1 and 2. The animals were fed twice daily of a diet of 50% grain, 40% chopped sorghum hay, 9% Molofos® and 1% minerals. The diet was estimated to contain 11.5MJ ME and 12.5 g of crude protein/kg DM. Feed refusals were weighed twice daily.

Results

No significant (*P*>0.05) interactions were observed between RFI line and feeding level for body composition traits. Feeding level had the largest effect on the change in body composition traits during the measurement periods. Heifers fed at 180% MEm grew faster and laid down more fat over the rump and ribs, and as IMF, than heifers fed 105% MEm (Table 1). Additionally, the low RFI heifers had lower rump and rib fat deposition (*P*<0.05), but not IMF deposition, than the high RFI heifers, regardless of feeding treatment.
Diet affected EE (Figure 1), but there was no significant difference between RFI genotype. There was no effect of diet and feeding level on whole body muscle protein synthesis.

Table 1. Main effects means and SEM for changes in weight and tissue depots for high and low RFI heifers fed at either 105% or 180% maintenance feeding levels.

Main effects	Group	ADG (kg/d)	EMA (cm^2)	Rump fat (mm)	Rib fat (mm)	IMF (%)
RFI line	-0.78±0.09	0.66^a±0.12	4.2^a±0.61	1.0^a±0.30	1.0^a±0.27	0.9^a±0.16
	0.64±0.03	0.70^a±0.11	4.8^a±1.07	1.9^b±0.52	1.4^b±0.38	0.9^a±0.21
Feeding level	180%	0.88^a±0.09	4.8^a±0.85	2.4^a±0.33	2.0^a±0.24	1.3^a±0.16
	105%	0.47^b±0.09	4.3^a±0.89	0.5^b±0.41	0.4^b±0.27	0.5^b±0.15

[a,b] Means with different superscripts differ significantly (P<0.05). Main effect interactions were not significant.

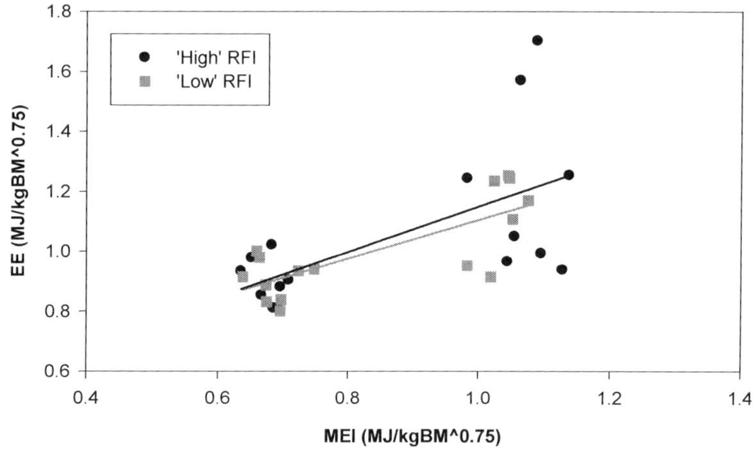

Figure 1. Energy expenditure of high and low RFI heifers against ME-intake. EE overestimated due to method used, outliers gained proportionately more muscle than fat and one was significantly more agitated on day of EE measurement.

Conclusion

Most of the variation in EE could be accounted for by the amount of energy consumed. Mean EE per unit of ME-intake did not differ between the selection lines. There was evidence for differences in fat deposition; the high-RFI animals retained more energy in fat. Muscle protein synthesis was not different between the lines and therefore, appears not to contribute to between RFI line variation in EE by these animals. Modelling energy transactions suggested that there was no difference in efficiency of energy utilisation between the RFI lines.

Acknowledgement

The authors thank the NSW Department of Primary Industries for access to the heifers.

References

Richardson, E.C. and R.M. Herd, 2004. Biological basis for variation in residual feed intake in cattle. 2. Synthesis of results following divergent selection. Aust. J. Exp. Agr. 44: 431-440.

Pattern of change and correlation of blood NEFA and urea with energy balance and related variables in dairy cows the first 21 days post-calving

N.E. Lobos[1], M.A. Wattiaux[1], G.A. Broderick[2] and P.M. Crump[1]
[1]University of Wisconsin-Madison, WI., USA; [2]U.S. Dairy Forage Research Center, Madison, WI, USA; wattiaux@wisc.edu

Introduction

Cows in early lactation rely on body stores of energy that are released as non-esterified fatty acids (NEFA) from adipose tissue triglycerides. However cows can also derive energy from amino acids mobilised from skeletal muscle and other tissue proteins. Thus, body protein mobilisation may impact hepatic deamination and blood urea nitrogen (BUN). Rastani *et al.* (2006) observed a significant positive relationship between NEFA and BUN in wk 2 post-calving. Formigoni *et al.* (1996) found that propylene glycol, a glucose precursor, lowered NEFA and milk urea nitrogen (MUN, a generally accepted proxy for BUN); however, these results were not confirmed by Chibisa *et al.* (2008). Thus, our objectives were (1) to describe the pattern of change in BUN, NEFA, body weight (BW), dry matter intake (DMI), milk yield (MY), and N utilisation efficiency (NUE), and (2) to assess the association of BUN with MUN, energy related measurements and NUE during the first 21 days of lactation.

Material and methods

Sixty Holstein cows were selected for a trial that extended from two weeks prior to calving date until three weeks post-calving. A pre-calving and a post-calving ration were formulated according to NRC (2001) and offered *ad libitum* once daily as a total mixed ration. Post-calving, BW was recorded twice weekly resulting in 6 data points per cow (DPC), DMI and MY were recorded daily (21 DPC), milk samples (for MUN and milk true protein analysis), and blood samples (for NEFA and BUN analysis) were collected three times a week (9 DPC) at 0530 h before a.m. feeding. Milk N divided by intake N was computed as NUE. Data were analysed as repeated measures (with cow as 'subject') using the Mixed Procedure of SAS® and the 'ar(1)' variance-covariance structure to account for auto-correlated errors. To determine the number of inflection points in the response pattern of the measured variables, four models were used for each cow to determine significance of the t (linear), t^2 (quadratic), t^3 (cubic) and t^4 (quartic) effects, where t was days relative to calving. If no model was significant, a cow's pattern was classified as 'complex'. For each cow, Pearson's correlation was used to assess relationships between BUN and MUN, NEFA, EB, BW, DMI, MY and NUE.

Results

Responses were as follows (mean ± SD): DMI 19.4±5.9 kg/d, MY 30.9±9.1 kg/d, NEFA 460±286 μeq/l, BUN 9.5±3.3 mg/dl, NUE 0.28±0.09. Although all cows were fed a 18.3% crude protein diet, BUN ranged from 0.8 to 30.7 mg/dl, and NUE from 0.09 to 0.88.

The complex pattern was the most frequent for BUN, NUE, NEFA and EB, revealing a high degree of dynamic changes (Table 1). Changes in BW and DMI exhibited mostly linear patterns whereas MY clustered around one to three inflection points. Among cows with linear NEFA pattern (n=19), a majority (n=15) had a negative slope, but some (n=4) exhibited an increase in NEFA over time. Among cows with linear BUN pattern (n=11), a majority (n=8) had a positive slope and some (n=3) exhibited a decrease in BUN over time. Among the four cows that showed a linear pattern for both NEFA and BUN, three exhibited a positive correlation whereas one exhibited a negative correlation (P<0.10). For cows with linear NUE pattern (n=14) the slope was consistently but barely negative (-0.011±0.01). There were no clear relationships between type of pattern in NEFA and DMI or MY (data not shown). For the majority of cows, BUN was not related to NEFA, EB, BW, DMI, NUE or MY (Table 2). However, when correlations existed, the relationship was either strongly positive or strongly negative. Concentrations of BUN and MUN were correlated in only 42% of the cows (n=25).

Table 1. Frequency distribution of cows (%) with distinct patterns of change in BUN, NUE, NEFA, BW, DMI, MY, and EB during the first 21days after calving (n=60).

Pattern	Infl. points	BUN	NUE	NEFA	BW	DMI	MY	EB
Linear	0	18.3	23.3	31.7	38.3	40.0	6.7	13.3
Quadratic	1	8.3	10.0	6.7	15.0	16.7	26.7	13.3
Cubic	2	6.7	10.0	5.0	18.3	5.0	25.0	3.3
Quartic	3	6.7	1.7	11.7	11.7	13.3	30.0	10.0
Complex	≥4	60.0	55.0	45.0	16.7	25.0	11.7	60.0

Table 2. Number of cows and range in correlation coefficient – between BUN and NEFA, EB, BW, DMI, MY, NUE, and MUN – categorised as absence of correlation (No corr), positive (+ corr) or negative (- corr) correlation (P<0.10) during the first 21 days after calving (n=60).

	NEFA	EB	BW	DMI	MY	NUE	MUN
No corr, n	45	51	44	46	40	42	35
+ corr, n	7	4	4	5	9	6	25
+ corr, range	0.60 to 0.79	0.69 to 0.93	0.76 to 0.84	0.59 to 0.94	0.62 to 0.98	0.64 to 0.97	0.61 to 0.87
- corr, n	8	5	12	9	11	2	0
- corr, range	-0.89 to -0.58	-0.93 to -0.64	-0.99 to -0.64	-0.84 to -0.59	-0.81 to -0.61	-0.92 to -0.82	n.a.

Conclusion

In this trial, 60% of the cows showed a complex pattern of change in BUN during the first 21 days after calving. Only a minority of cows followed the expected decline in NEFA after calving; 45% showed complex pattern and 7% showed an increase in NEFA over time. The decline in BW and increase in DMI followed mostly linear or quadratic patterns; however, the increase in milk yield showed additional inflection points. The poor relationship between BUN and MUN or NUE observed in this trial contrasted with mid to late lactation data reported in the literature. Any metabolic relationship between BUN and NEFA was difficult to discern. Analyses of patterns of change may be a useful tool in future transition cow studies. The dynamic fluxes that characterise the homeorhesis of cows post-calving followed a simple and easily predicted pattern in only a limited number of cows.

References

Chibisa, G.E., G.N. Gozho, A.G. Van Kessel, A.A. Olkowski and T. Mutsvangwa, 2008. Effects of peripartum propylene glycol supplementation on nitrogen metabolism, body composition, and gene expression for the major protein degradation pathways in skeletal muscle in dairy cows. J. Dairy Sci. 91: 3512-3527.

Formigoni, A., M.C. Cornil, A. Prandi, A. Mordenti, A. Rossi, D. Portetelle and R. Renaville, 1996. Effect of propylene glycol supplementation around parturition on milk yield, reproduction performance and some hormonal and metabolic characteristics in dairy cows. J. Dairy Res. 63: 11-24.

NRC, 2001. Nutrient requirements of dairy cattle 7th rev. ed, Washington, DC.

Rastani, R.R., N.E. Lobos, M.J. Aguerre, R.R. Grummer and M.A. Wattiaux, 2006. Relationships between blood urea nitrogen and energy balance or measures of tissue mobilization in Holstein cows during the periparturient period. The Professional Animal Scientist 22: 382-385.

Expression of RBP4-mRNA in adipose tissue and RBP4 in serum of healthy dairy cows

L. Locher[1], L. Zapfe[1], M. Kern[2], N. Klöting[2], M. Blüher[2], J. Raila[3] and M. Fürll[1]
[1]Department of Internal Medicine, Faculty of Veterinary Medicine, University of Leipzig, An den Tierkliniken 11, 04103 Leipzig, Germany; [2]Internal Medical Clinic III, Faculty of Medicine, University of Leipzig, 04103 Leipzig, Germany; [3]Institute for Nutrition Sciences, University of Potsdam, 14558 Nuthetal, Germany; locher@vetmed.uni-leipzig.de

Introduction

Lipolysis and insulin resistance play a crucial role in predisposition, incidence and outcome of disease in the periparturient period in dairy cows. Retinol-Binding-Protein 4 (RBP4) a transport protein for retinol in serum, is also an adipokine which regulates glucose uptake in adipocytes and impairs insulin action in the muscle and liver. It is synthesised in most tissues, especially in the liver, kidney and fat. Studies on tissue and body fluid distribution and excretion have been done in carnivores (Raila *et al.*, 2000), but to our knowledge there are few investigations in cattle. High serum levels in man reflect an impaired insulin action and can be used as a predictive marker for type II diabetes (Wolf, 2007). Furthermore mRNA expression of RBP4 is higher in visceral fat. In human medicine, it is considered as an indicator for abdominal fat mass (Klöting *et al.*, 2007). This study was aimed at investigating whether mRNA expression of RBP4 of healthy cows is different in subcutaneous in comparison to visceral fat and whether serum RBP4 concentrations in dairy cows change during the periparturient period or correlate with NEFA-concentrations shortly after calving.

Material and methods

Samples were taken at slaughter immediately after death from 12 dairy cows, slaughtered for non metabolic reasons in mid or late lactation. Fat was taken from the omentum, renal capsule, inguinal region (retroperitoneal), hip (subcutaneous) and heart base, immediately quick-frozen and stored at -70 °C. mRNA expression for RBP4 was measured using quantitative RT-PCR. The results were calculated relative to expression of 18S mRNA. Blood samples from 44 healthy, not overconditioned [body fat thickness (BFT) 20-30 mm], Holstein-Friesian/German Black Pied (HF/SB) cows (≥2 lactations) were collected 10 d *ante partum* (a.p.), 3 d *post partum* (p.p.) and 4 wk p.p. and analysed for RBP4 using western blot. Serum was analysed for clinical chemical routine parameters, including BHB and NEFA. BFT was measured by ultrasound. Since data were normally distributed, the Friedman-test for comparison of dependent samples and Spearman Rho for correlation were used.

Results

Expression of RBP4 mRNA in all sample sites could only be detected in 6 of 12 cows analysed. It was higher in retroperitoneal, renal and pericardial than in subcutaneous or omental fat. It was significantly (P=0.019) upregulated in pericardial fat in comparison to omental fat (Table 1). Serum RBP4 levels 3 d p.p were 8 (4.05;13.44) mg/l (median and quartiles) and did not change significantly between the three sampling points (Figure 1). There was a significant correlation (r=0.5) between RBP4 concentrations 10 d a.p. and NEFA-concentration 3 d p.p (results not shown). No relationship between BFT and RBP4 could be shown in this study.

Table 1. Relative mRNA expression in bovine adipose tissue from different fat depots.

	n	median	1st quartile	3rd quartile
RBP4				
Subcutaneous	9	0.01910	0.00804	0.90350
Omental*	9	0.17400	0.06495	0.51100
Renal	12	0.48250	0.32350	1.81325
Pericardial	9	1.93000	0.51400	6.73000
Retroperitoneal*	11	1.12000	0.31600	7.03000

* Indicating significant difference at $P<0.05$

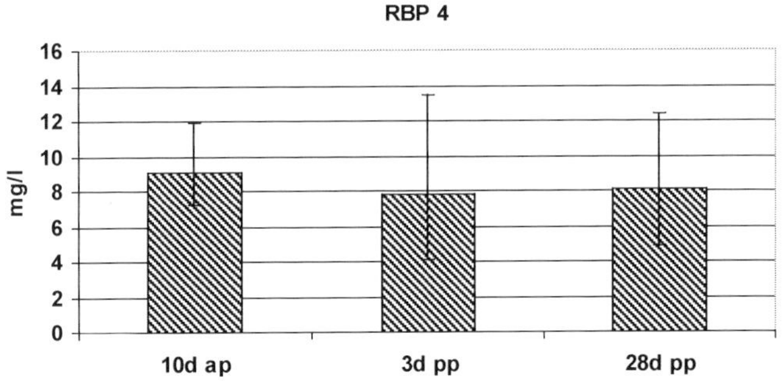

Figure 1. RBP concentrations in serum of normal HF/SB cows (<2 lactations) in the periparturient period (median, 1st and 3rd quartile)

Conclusion

High RBP mRNA expression in pericardial fat is probably due to the high metabolic activity and glucose turnover in the heart muscle. The general mRNA-expression level in visceral fat (except omental fat) was higher than in subcutaneous depots. The constant serum levels of RBP4 show, that in normal conditioned cows with balanced energy metabolism, RBP4 levels were not altered during the periparturient period and did not correlate with BFT. The relationship between RBP4 *ante partum* and NEFA *post partum* suggests RBP4 as a possible predictive marker for fat mobilisation in cattle as well. Therefore further investigations in cows in different body conditions and suffering from fat mobilisation syndrome are needed.

References

Klöting, N., T.E. Graham, J. Berndt, S. Kralisch, P. Kovacs, C.J. Wason, M. Fasshauer, M.R. Schön, M. Stumvoll, M. Blüher and B.B. Kahn, 2007. Serum retinol binding protein is more highly expressed in visceral than in subcutaneous fat and is a marker for intraabdominal fat mass. Cell Metab. 6: 79-87.

Raila, J., I. Buchholz, H. Aupperle, G. Raila, H.A. Schoon and F.J. Schweigert, 2000. The distribution of vitamin A and retinol-binding protein in the blood plasma, urine, liver and kidneys of carnivores. Vet. Res. 31: 541-551.

Wolf, G., 2007. Serum retinol binding protein: A link between obesity, insulin resistance and type 2 diabetes. Nutr. Rev. 65: 251-256.

Comparative aspects of hormone sensitive lipase (HSL), lipoprotein lipase (LPL), adiponectin and leptin mRNA-expression in bovine fat tissue

L. Locher[1], L. Zapfe[1], N. Klöting[2], M. Kern[2], M. Blüher[2] and M. Fürll[1]
[1]Department of Internal Medicine, Faculty of Veterinary Medicine, University of Leipzig, An den Tierkliniken11, 04103 Leipzig, Germany; [2]Internal Medical Clinic III, Faculty of Medicine, University of Leipzig, 04103 Leipzig, Germany; locher@vetmed.uni-leipzig.de

Introduction

Negative energy balance and subsequent fat mobilisation in the periparturient period are almost inevitable in dairy cows, due to high milk yield and lowered dry matter intake around calving. Adipokines seem to play a crucial role in this stage. Leptin and non esterified fatty acids (NEFA) for example are known to influence feed intake directly (Ingvartsen and Andersen, 2000). Adiponectin supports insulin action and decreases lipolysis. Hormone sensitive lipase (HSL) and lipoprotein lipase (LPL) are rate limiting enzymes for lipolysis and triglyceride uptake, respectively. Adipose tissue can be divided into subcutaneous (SC) and visceral fat. Visceral fat contains intraabdominal and retroperitoneal fat (Wajchenberg, 2000). Studies in mice and man proved that adipose tissue acts differently according to its body site and there is evidence that, at least for leptin, similar facts can be assumed for ruminants (Chilliard *et al.*, 2001). The aim of the study was to investigate mRNA expression of leptin, adiponectin, HSL and LPL in bovine fat from different locations.

Material and methods

Samples were taken at slaughter immediately after death from 12 cows, slaughtered for non metabolic reasons in mid or end-lactation. Samples were collected from the omentum, renal capsule, inguinal region (retroperitoneal), hip (subcutaneous) and heart base. They were immediately deep frozen using liquid nitrogen and then stored at -70 °C until analysis. mRNA expression for HSL, LPL, adiponectin and leptin was measured using quantitative RT-PCR method. The results are calculated relative to the expression of the house keeping gene 18S, indicating metabolic activity of the tissue. Differences between fat tissues were statistically verified using Friedmann-test. Correlation was assessed using the Spearman Rho. Significant differences were assumed for $P \leq 0.05$.

Results

HSL mRNA was the highest in omental fat whilst in perirenal fat LPL mRNA was the highest. The highest adiponectin mRNA levels were detected in pericardial fat (Table 1). All these differences where not significant. In contrast, leptin mRNA showed the highest values in retroperitoneal fat (Figure 1) which was significantly higher than SC mRNA values (P=0.05). There was significant positive correlation between HSL and LPL mRNA in subcutaneous (P=0.015; r=0.8), omental (P=0.005; r=0.83) and perirenal (P=0.001; r=0.84) fat. HSL and adiponectin were also significantly (P<0.01) positively correlated (r=0.9) except in omental tissue, where they correlated negatively (r=-0.8).

Discussion and conclusion

Perirenal, pericardial and retroperitoneal fat in healthy cows showed higher levels of candidate mRNA expression (relative to 18s mRNA) whilst SC and omental fat were relatively inert. With HSL and LPL, mRNA expression seems to act in the same direction when there is no lipolysis. The negative correlation of adiponectin and HSL in omental fat in contrast to other locations might indicate higher susceptibility of this tissue to lipolytic stimuli. Therefore it becomes understandable, that in negative energy balance when lipolysis is upregulated, e.g. in early lactation, these depots are largely mobilised. Further investigations should focus on how this pattern changes in animals suffering from metabolic disturbances like fat mobilisation, hyperketonemia, insulin resistance or inflammatory conditions.

Table 1. Median (1st and 3rd quartile) of mRNA/18s expression in bovine fat tissue from different locations (samples taken at slaughter).

	Subcutaneous	Omental	Renal	Pericardial	Retropertoneal
HSL	0.46 (0.07;803.88)	63.64 (0.61; 829.97)	2.23 (0.76; 188.35)	2.18 (0.89; 16400)	1.37 (0.03; 89.26)
LPL	0.45 (0.08; 6.58)	0.26 (0.09; 3.24)	1.19 (0.09; 3.79)	0.76 (0.09; 1.76)	0.49 (0.13; 5.48)
Adiponectin	0.17 (0.08; 242.8)	0.2 (0.12; 0.61)	0.46 (0.31; 25.5)	3.53 (0.83; 49700)	1.65 (0.31; 260.98)
Leptin	0.0019 (0.0001; 0.13)	0.18 (0.0003;1.3)	0.58 (0.04; 3.85)	0.31 (0.02; 17.84)	0.35 (0.033; 7.18)

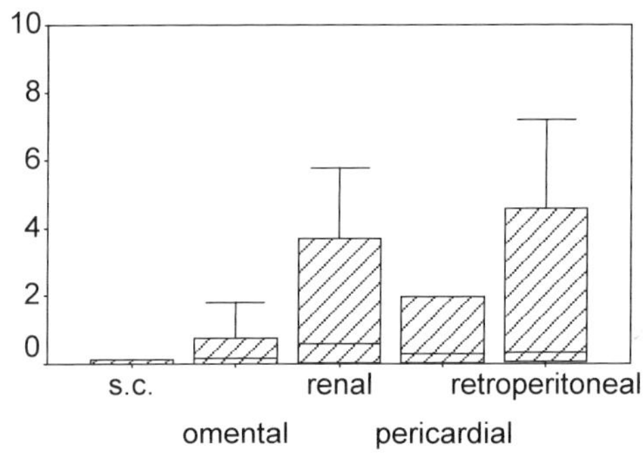

Figure 1. Relative leptin mRNA expression (median; 1st and 3rd quartile, minimum and maximum) (referred to 18s mRNA) in bovine fat tissue of different locations. Significant difference between SC and retroperitoneal fat (P<0.05).

References

Chilliard, Y., M. Bonnet, C. Delavaud, Y. Faulconnier, C. Leroux, J. Dijane and F. Bocquier, 2001. Leptin in ruminants. Gene expression in adipose tissue and mammary gland, and regulation of plasma concentration. Domest. Anim. Endocrinol. 21: 271-295.

Ingvartsten, K.L. and J.B. Andersen, 2000. Symposium: Dry matter intake of lactating dairy cattle. J. Dairy Sci. 83: 1573-1597

Wajchenberg, B.L., 2000. Subcutaneous and visceral adipose tissue: their relation to the metabolic syndrome. Endocr. Rev. 21: 697-738

Exploring the potential for using erythrocyte membranes in the assessment of long-chain polyunsaturated fatty acid status of dairy cows

A.L. Lock[1], C.L. Preseault[1,2] and H.M. Dann[2]
[1]University of Vermont, Burlington, VT, USA; [2]William H. Miner Agricultural Research Institute, Chazy, NY, USA; adam.lock@uvm.edu

Introduction

Noble (1984) reported that ruminants are highly efficient at conserving long-chain polyunsaturated fatty acids (LCPUFA). Under certain situations, however, the modern high producing dairy cow may be deficient in these fatty acids (FA). For example, the amount of C18:2 n-6 excreted in milk (on a daily basis) can exceed the amount of C18:2 n-6 available for absorption (Lock *et al.*, 2006). A primary reason for the interest in assessing, and possibly increasing, the availability of LCPUFA is the potential for n-6 and/or n-3 LCPUFA-supplementation to improve animal production and welfare. Erythrocyte membrane (EM) phospholipid (PL) profile has been used in human studies to assess long-term FA intakes (Poppitt *et al.*, 2005) and can therefore potentially be used as a marker for LCPUFA status of dairy cows. Therefore, the primary objectives were to develop methods and undertake a preliminary examination of the potential for using EM-FA profile in the evaluation of LCPUFA status of dairy cows. For comparison, we also examined milk fat since it also contains a small amount of PL in the milk fat globular membrane.

Material and method

Whole blood was collected from the coccygeal vein of 53 Holstein cows at 5 stages of the lactation cycle: far-off dry (37±2 d pre-calving), close-up dry (10±2 d pre-calving), early lactation (37±4 DIM), mid lactation (104±4 DIM), and late lactation (250±5 DIM). All cows were fed typical diets for their respective physiological state. After isolation and purification of EM (using a modification of Burton *et al.*, 1981) total lipids were extracted using chloroform:methanol. FA methyl esters (FAME) were prepared using a KOH solution and BF_3 in methanol and further purified by thin-layer chromatography (TLC). Milk samples were collected from the early and late lactation groups on the same day that blood sampling occurred. Total lipids were extracted from milk fat cakes using *n*-hexane:isopropanol. TLC was used to separate the PL and triacylglycerol (TAG) fractions from the lipid extract. PL-FAME were prepared by the same method as for EM-FAME with TAG-FAME prepared using sodium methoxide as the methylation reagent. All FAME were analysed by gas-liquid chromatography using a 100 m capillary column. Data were analysed using the Fit Model procedure of JMP.

Results

Stage of lactation affected the concentration (g/100 g FA) of all reported EM-FA (Table 1). Late lactation cows were highest in C18:2 n-6 (31.29; $P<0.01$) whereas far-off dry cows were highest in C18:3 n-3 (0.81; $P<0.01$). C20:4 n-6 and C20:5 n-3 were highest in the close-up group (4.17 and 0.21; $P<0.001$ and $P<0.05$, respectively). Both dry cow groups had higher levels of C22:5 n-3 ($P<0.001$) than lactating cows. No differences ($P>0.05$) were observed in LCPUFA concentrations in either milk fat PL or TAG between early and late lactation cows (Table 2). The concentration of all LCPUFA differed between the PL and TAG fractions of milk fat and was over twofold greater in the PL fraction compared to the TAG fraction. There were no interactions between stage of lactation and source of LCPUFA in milk fat ($P>0.05$). In addition, no relationships were observed between LCPUFA concentration of EM and milk fat PL (data not shown).

Table 1. Selected LCPUFA of EM in cows at defined stages of the lactation cycle.

FA (g/100 g)	Far-off dry (n=10)	Close-up dry (n=6)	Early lactation (n=12)	Mid lactation (n=12)	Late lactation (n=13)	SEM	P
C18:2 n-6	27.03[b]	27.88[ab]	24.80[b]	27.88[b]	31.29[a]	1.73	<0.01
C20:4 n-6	3.86[a]	4.17[a]	2.52[bc]	2.24[c]	3.00[b]	0.37	<0.001
C18:3 n-3	0.81[a]	0.74[ab]	0.58[bc]	0.56[c]	0.57[bc]	0.07	<0.01
C20:5 n-3	0.10[b]	0.21[a]	0.13[b]	0.09[b]	0.11[b]	0.03	<0.05
C22:5 n-3	0.36[a]	0.46[a]	0.09[b]	0.09[b]	0.18[b]	0.06	<0.001

[a,b,c] Means with different superscripts differ significantly at $P<0.05$.

Table 2. Selected LCPUFA of milk fat PL and TAG in cows at defined stages of the lactation cycle.

FA (g/100 g)	Early lactation cows (n=5)		Late lactation cows (n=5)		SEM	P		
	PL	TAG	PL	TAG		Stage	Source	Stage × source
C18:2 n-6	6.15	2.29	4.91	2.13	0.76	0.43	<0.01	0.44
C20:4 n-6	0.64	0.14	0.55	0.16	0.08	0.69	<0.001	0.54
C18:3 n-3	0.29	0.03	0.32	0.03	0.02	0.44	<0.001	0.59
C20:5 n-3	0.06	0.02	0.04	0.02	0.01	0.34	<0.01	0.24
C22:5 n-3	0.28	0.05	0.25	0.05	0.05	0.80	<0.01	0.78

Conclusion

This preliminary evaluation highlights the potential for EM-FA analysis as a useful tool for the assessment of LCPUFA status of dairy cows and indicates possible differences in LCPUFA status at different stages of the lactation cycle. These differences may provide information on dietary status and physiological state of individual animals at different stages of lactation. Further work, however is required in order to separate nutritional and physiological effects on variations in EM-FA composition. For comparative purposes milk fat was examined, in particular the PL fraction; data so far, however, indicate that the PL and TAG fractions of milk fat may not be efficacious in assessing LCPUFA status. Future research should focus on characterising EM-FA changes within individual cows throughout lactation, examine the relationship between FA profiles of EM and other blood lipids and tissues, and also determine the impact of different diets and nutritional supplements aimed at improving the LCPUFA status of dairy cows, particularly during the transition period.

References

Burton, G.W., K.U. Ingold and K.E. Thompson, 1981. An improved procedure for the isolation of ghost membranes from human red blood cells. Lipids 16: 946.

Lock, A.L., K.J. Harvatine, J.K. Drackley and D.E. Bauman, 2006. Concepts in fat and fatty acid digestion in ruminants. Proc. Intermountain Nutr. Conf., pp. 85-100.

Noble, R.C., 1984. Essential fatty acids in the ruminant. In: Fats in Animal Nutrition. J. Wiseman (ed.) Butterworths, London, UK, pp. 185-200.

Poppitt, S.D., P. Kilmartin, P. Butler and G.F, Keogh. 2005. Assessment of erythrocyte phospholipid fatty acid composition as a biomarker for dietary MUFA, PUFA or saturated fatty acid intake in a controlled cross-over intervention trial. Lipids Health. Dis. 4: 30.

Adaptation of hepatic glucose uptake and metabolism in growing lambs fed energy and nitrogen imbalanced diets

C. Loncke, G. Kraft, I. Savary-Auzeloux and I. Ortigues-Marty
INRA, UR 1213, Theix, 63122 Saint Genes-Champanelle, France; isabelle.ortigues@clermont.inra.fr

Introduction

Nutritional allowances to ruminants are defined for both protein and energy supply in the diet, however animal metabolic responses to nitrogen/energy imbalanced diets are difficult to predict. Kraft *et al.* (2009) showed that in lambs fed nitrogen or energy imbalanced diets, adaptations of hepatic amino acid (AA) uptake and metabolism contribute to the nitrogen economy in the whole body. This work aims to quantify in the same lambs if glucose metabolism undergoes similar adaptive mechanisms by measuring the net hepatic fluxes of glucose and of its major precursors.

Material and methods

Six growing lambs (41.5±2.6 kg) were surgically fitted with aorta, portal and hepatic vein catheters and an ultrasonic probe on the portal vein for blood flow measurement. The lambs were fed 3 diets made of 30% hay and 70% of 1 of 3 experimental concentrates, according to a duplicated 3×3 Latin Square design. The control diet (C) was designed to offer a balanced and adequate supply of protein (1.43 g N/d/kg $BW^{0.75}$) and metabolisable energy (ME=853 kJ/d/kg $BW^{0.75}$) for growing lambs according to INRA allowances. Low nitrogen (LN) and energy (LE) diets presented a 23% and 20% deficit in PDI and ME supply respectively compared to the C diet. Blood/plasma samples were collected from the 3 vessels for analysis of blood glucose, propionate (C3) and lactate as well as plasma AA.

Results and discussion

With the C diet (Table 1), the potential contribution of C3, alanine, glutamine and glycine (major glucose precursors) to net hepatic release (NHR) were 80%, 8%, 4% and 2.5% respectively. Glutamate and lactate could not contribute on a net basis to NHR of glucose, which was expected for lactate considering the high concentrate level in the C diet.
As planned, the LE diet was responsible for a reduction in the net portal appearance (NPA) of C3 by 41% relative to the C diet, with no change in the NPA of total AA or glucose (Table 1). Nitrogen balance was reduced [-38.27%, *P*=0.003; G. Kraft, unpublished results 2009]. The net hepatic uptake of C3 was proportional to its NPA and also reduced by 35% (-55.4 mmol C/h; *P*=0.04), while its potential contribution to glucose hepatic release was not changed (78%). The NPA of glucogenic AA tended to be reduced (-20%; *P*=0.08) while their potential contribution was slightly increased (20% vs. 15% in C diet; *P*=0.09). Overall, the hepatic supply in precursors decreased by 39% (-77.9 mmol C/h) compared to the C diet and induced a 33% (-66.2 mmol C/h; *P*=0.048) reduction in NHR of glucose.
With the LN diet, as expected, the NPA of total AA was reduced by 38%, while that of C3 was not significantly modified. Nitrogen balance was unchanged (*P*=0.28; G. Kraft, unpublished results, 2009). Unexpectedly, a positive NPA of glucose was noted, presumably due (Loncke *et al.*, 2009) to slight changes in the dietary undegradable starch content as compared to the C diet (Kraft *et al.*, 2009). Despite the limited drop in carbon supply from glucose precursors to the liver (-15%), NHR of glucose decreased by 41% (*P*=0.018). On a molar basis, this drop (-13.56 mmol/h) was close to the difference in NPA of glucose between the C and LN diets (+18.13 mmol/h). These data were consistent with Piccioli Capelli *et al.* (1997) who observed that duodenal infusion of

glucose induced a decrease in NHR of glucose. Net splanchnic release of glucose remained thus unchanged. The potential contribution of C3 to glucose NHR was high (115%), suggesting that theorically C3 alone would have been sufficient to account for glucose NHR, thereby sparing AA in animals fed a nitrogen deficient diet.

Table 1. Net splanchnic fluxes of glucose and gluconeogenic precursors in growing lambs fed control (C), low protein (LN) and low energy (LE) diets.

	Experimental diets			SEM	P
	C	LN	LE		
Net portal appearance (mmol/h)					
Total amino acids	30.89	18.99	25.38	2.04	0.01
Propionate	59.54	52.44	34.98	4.98	<0.001
Glucose	-11.77	6.36	-9.66	2.80	0.006
Net hepatic release (mmol/h)					
Propionate	-52.80	-47.74	-34.33	4.41	0.002
Glucose	33.11	19.55	22.07	3.55	0.041
Lactate	4.30	-1.43	3.10	1.54	0.29
Glutamate	2.16	3.35	1.30	0.30	<0.001
Glutamine	-1.62	-0.42	-1.84	0.43	0.10
Alanine	-5.32	-3.68	-3.92	0.22	0.001
Glycine	-2.49	-2.80	-2.70	0.33	0.81
Net splanchnic release (mmol/h)					
Glucose	21.34	25.91	12.41	3.59	0.067

Conclusion

This work shows the glucose NHR is strongly regulated in animals fed nitrogen or energy imbalanced diets. It is significantly reduced with the energy deficient diet in coherence with the assumed reduction in whole body anabolism with this treatment (Kraft *et al.*, 2009a). With the nitrogen deficient diet, the potential contribution of C3 for glucose NHR is increased, which spares AA.

References

Kraft, G., D. Gruffat, D. Dardevet, D. Rémond, I. Ortigues-Marty and I. Savary-Auzeloux, 2009. Nitrogen and energy imbalanced diets differentrially affected hepatic protein synthesis and neoglucogenesis in growing lambs, J. Anim. Sci., in press.

Loncke, C., I. Ortigues-Marty, J. Vernet, H. Lapierre, D. Sauvant and P. Nozière, 2009, Empirical prediction of net portal appearance of volatile fatty acids, glucose, and their secondary metabolites (ß-hydroxybutyrate, lactate) from dietary characteristics in ruminants: A meta-analysis approach, J. Anim. Sci. 87: 253-268.

Piccioli Capelli, F.P., C.J. Seal and S. Parker, 1997, Glucose and [^{13}C] leucine metabolism by the portal drained viscera of sheep fed on dried grass with acute intravenous and intraduodenal infusions of glucose, Br. J. Nutr. 78: 931-946.

Milk fatty acid profile from dairy cows fed increasing levels of soybean oil in diets based on tropical forage

F.C.F. Lopes[1], C.G.S. Ribeiro[2], M.T. Ribeiro[1], N.M. Rodriguez[2], H.G.B. Filho[1], R.J.C. Castro[1], P.A.V. Barros[1] and M.A.S. Gama[1]
[1]National Dairy Cattle Research Centre, Embrapa, Juiz de Fora, MG, Brazil; [2]School of Veterinary Medicine, UFMG, Belo Horizonte, MG, Brazil; gama@cnpgl.embrapa.br

Introduction

Due to health concerns, efforts have been done to decrease hypercholesterolemic saturated fatty acids (C12, C14 and C16:0) and increase *cis*-9 C18:1 and *cis*-9 *trans*-11 conjugated linoleic acid (CLA) in milk fat from dairy cows (Dewhurst *et al.*, 2006). Dietary supplementation with plant oils rich in linoleic and α-linolenic acids is a practical way to achieve this goal. However, most of these studies were conducted with temperate forage-based diets, usually conserved as silage or hay. During the conservation process, oxidative losses occur in the PUFA present in phospholipids (Dewhurst *et al.*, 2006). This can partially explain why milk fat from cows fed fresh green forage had higher unsaturated: saturated ratio and *cis*-9 *trans*-11 CLA than cows fed silage-based diets (Elgersma *et al.*, 2006). Therefore, the association of fresh tropical forage with plant oils seems to be an interesting dietary strategy for improving milk fatty acid profile of dairy cows. This study was designed to evaluate the effects of soybean oil levels on fatty acid profile from butter fat of dairy cows fed diets based on Elephant grass (*Pennisetum purpureum*) as tropical forage.

Material and methods

This study was conducted at the National Dairy Cattle Research Centre, Brazil. Twelve multiparous Holstein × Zebu dairy cows (90±25 DIM, initial milk yield and BW of 18.0±4.6 kg/d and 502.6±47.8 kg) were blocked according to milk yield and assigned to the following dietary treatments (DM basis): (1) Control: no soybean oil (SO); (2) T1: diets with 1.5% of SO; (3) T2: diets with 3.0% of SO and 4) T3: diets with 4.5% of SO. The experimental design was a 4×4 Latin square with 15-d treatment periods (10 d for adaptation and 5 d for data collection). Cows were allocated in free-stall barns and individual feed intake was recorded using the Calan headgate system (American Calan Inc., Northwood, NH, USA). Diets were fed once daily as a total mixed ration (TMR) and were composed of chopped Elephant grass and a concentrate mixture (50:50, on a DM basis). The Elephant grass was harvested daily at mid maturity stage and chopped immediately before TMR preparation. The basal concentrate was composed of ground corn, soybean meal, citrus pulp and a mineral-vitamin supplement. Soybean oil was mixed into the concentrate every 10 days throughout the study to prevent oxidative rancidity. Butter was prepared from milk produced by cows fed each dietary treatment at the first day of each collection period. Butter was stored at -20 °C degrees and subsequently analysed for fatty acid profile by gas chromatography using a CPSil-88 column as described by Destaillats *et al.* (2007). Treatment effects were determined by regression analysis using the GLM procedure of SAS® (2000) and declared significant at $P<0.05$.

Results

The concentration of selected FA identified in butter fat from cows fed each dietary treatment is presented in Table 1. Concentration of short and medium chain FA as well as odd chain FA were linearly decreased ($P<0.01$) in response to increasing levels of SO. In contrast, addition of SO to the diet linearly increased ($P<0.01$) the concentration of *trans*-C18:1 isomers in butter fat. The SO levels were also linearly and positively associated with *cis*-9 *trans*-11, *trans*-9 *cis*-11 and *trans*-10 *cis*-12 CLA in milk fat ($P<0.01$).

Table 1. Fatty acid concentration (g/100 g of total FA) in butter fat from dairy cows fed different levels of soybean oil (SO) in diets based on Elephant grass as tropical forage.

Fatty acids (g/100g of total FA)	Soybean oil levels (% DM)				Coefficient of variation (%)	Effect[1]	
	0	1.5	3.0	4.5		L	Q
C4:0 to C10:0	8.99	8.42	6.90	5.94	10.1	**	ns
C12:0 to C16:0	44.8	37.7	33.0	29.6	13.6	**	ns
Odd chain FA (except C19)	3.44	2.99	2.52	2.17	14.7	**	ns
C18:0	7.78	9.12	9.42	9.94	6.36	**	ns
C18:1 *trans-6 to 8*	0.23	0.51	0.66	0.88	15.7	**	ns
C18:1 *trans-9*	0.34	0.52	0.64	0.69	15.3	**	*
C18:1 *trans-10*	0.43	1.04	1.36	1.52	34.9	**	ns
C18:1 *trans-11*	2.00	4.24	6.44	9.35	21.0	**	ns
C18:1 *trans 12*	0.36	0.55	0.72	0.80	13.1	**	ns
C18:1 *trans-13* + *trans-14*	0.50	0.66	0.75	1.18	37.7	**	ns
C18:1 *cis-9* + *trans-15*	20.4	21.8	23.1	22.6	7.61	**	ns
C18:2 *cis-9 cis-12*	2.29	2.4	2.42	2.40	7.23	ns	ns
C18:3 *cis-9 cis-12 cis-15*	0.28	0.29	0.26	0.23	20.0	ns	ns
CLA *cis-9 trans-11*	1.28	2.47	3.74	4.59	18.0	**	ns
CLA *trans-9 cis-11*	0.02	0.04	0.06	0.08	15.1	**	ns
CLA *trans-10 cis-12*	<0.01	0.01	0.03	0.03	49.9	**	ns

[1] Probability of linear (L) or quadratic (Q) effect (*$P<0.05$, **$P<0.01$, ns: not significant).

Conclusion

Inclusion of soybean oil in Elephant grass-based diets positively affects milk fat composition of Holstein × Zebu lactating dairy cows.

Acknowledgement

The authors thank FAPEMIG and CNPQ for financial support as well as Rosimeire A.C. Dornelas and J.M. de Castilho for helping with butter production.

References

Destaillats, F., P.A. Golay, J. Joffre, M. Wispelaere, B. Hug, F. Giuffrida, L. Fauconnot and F. Dionisi, 2007. Comparison of available analytical methods to measure *trans*-octadecenoic acid isomeric profile and content by gas-liquid chromatography in milk fat. J. Chromat. A. 1145: 222-228.

Dewhurst, R.J., K.J. Shingfield, M.R.F. Lee and N.D. Scollan, 2006. Increasing the concentrations of beneficial polyunsaturated fatty acids in milk produced by dairy cows in high-forage systems. Anim. Feed Sci. Technol. 131: 168-206.

Elgersma, A., S. Tamminga and G. Ellen, 2006. Modifying milk composition through forage. Anim. Feed Sci. Technol. 131: 207-225.

SAS, 2000. Statistical analysis system, Release 8.02.2000. SAS Institute Inc. Cary, NC.

Effects of linseed and *Acacia cyanophylla* intake on performance and milk fatty acid composition in Sicilo-Sarde ewes fed oat silage or grazing triticale pasture

O. Maamouri[1,2], N. Atti[1], A. Ferlay[2], K. Kraeim[4], M. Mahouachi[3] and Y. Chilliard[2]
[1]Institut National de la Recherche Agronomique de Tunisie (INRAT), Laboratoire des Productions Animales et Fourragères, Rue Hédi Karray, 2049 Ariana, Tunisia; [2]INRA, UR1213 Herbivores, Site de Theix, 63122 Saint-Genès-Champanelle, France; [3]ESA Kef, Département Productions Animales, Kef, Tunisia; [4]ISA Chott Meriem, Sousse, Tunisia; anne.ferlay@clermont.inra.fr

Introduction

Oilseeds could be incorporated in ruminant diets in order to increase polyunsaturated fatty acids (PUFA) content (notably, omega 3 fatty acids (FA) and CLA) in dairy products and thus to improve their nutritional quality. The changes in PUFA content of dairy products depend on the nature of forage, oilseeds and their interaction (Atti *et al.*, 2006; Chilliard *et al.*, 2007). *Acacia cyanophylla* contain tannins which could partially protect dietary proteins and lipids from rumen metabolism. Thus, the aim of this study was to evaluate the effects of incorporation of linseeds with or without *Acacia cyanophylla* in the diet on milk production and FA composition in ewes fed oat silage or grazing cultivated pasture, which are the common dairy ewe feeding systems in this area.

Material and methods

Fifty-four Sicilo-Sarde ewes from the dairy experimental farm of INRAT were divided into six groups: three groups grazed triticale pasture (G) and the 3 other groups received oat silage (S). For each type of forage, one group received 400 g of concentrate (G and S groups), the second group received 400 g of concentrate and 100 g whole crude linseeds (GL and SL groups), and the third group received 400 g of concentrate, 100 g linseeds and 100 g of acacia foliage (GLA and SLA groups) during 2 mo. Individual milk yield was recorded and samples were analysed for milk fat, protein and urea. Other milk samples were lyophilised until analysis of FA composition. Methyl esters of FA were separated on a 100 m×0.25 mm i.d. fused-silica capillary column (CP-Sil 88). Milk production and composition, and FA composition were analysed by using the MIXED procedure of SAS® (SAS Institute Inc., 2000, Cary, NC, USA) for block designs with repeated measures. The fixed effects were time, diet treatment and their interaction. Differences between diets were performed using the procedure of least squares means (SAS®, 2000).

Results and discussion

Ewes grazing pasture had higher milk (735 vs. 418 ml/d), fat and protein yields than ewes fed silage. Linseed and/or *Acacia cyanophylla* intake did not change milk yield, protein or fat yield. Our results did not agree with those of Woodward *et al.* (1999), reporting an increase in the milk yield from dairy cows fed *Clover corniculatis* containing condensed tannins. The milk urea values were higher for the S groups than for the G groups (31 vs. 21 mg/dl).

The milk 18:3 n-3 fatty acid content was higher for grazing groups than for S ones. High milk 18:3 n-3 content was previously reported in cows (Chilliard *et al.*, 2007) and ewes (Atti *et al.*, 2006) consuming grassland pastures. Parallely, saturated FA decreased with linseeds for both grazing and silage groups as reported by Chilliard *et al.* (2007). Grazing group had a higher CLA content than the silage one, in agreement with Atti *et al.* (2006). Conversely, the *cis*-18:1 content increased with linseeds and acacia for 2 types of forages. Acacia intake did not affect the milk FA profile, except for 18:3 n-3 in the GLA group.

Table 1. Milk yield, composition and FA profile (% total FA) according to experimental diets.

	Diet						SE	Diet effect
	G	GL	GLA	S	SL	SLA		*P*-value[1]
Milk yield, ml	682[a]	746[a]	776[a]	405[b]	390[b]	459[b]	37	0.0001
Fat,%	7.6	8.3	7.8	8.1	8.0	7.5	0.2	0.1678
Fat, g/d	52.1[a]	60[a]	60.8[a]	32.4[b]	32.1[b]	35[b]	3.2	0.0001
Protein,%	5.0	5.0	5.0	5.2	5.4	5.0	0.1	0.0851
Protein, g/d	33.9[a]	36.6[a]	37.9[a]	21.4[b]	21.5[b]	23.3[b]	1.8	0.0001
Urea, mg/dL	21.8[b]	22.2 [b]	18.9[b]	24.7[b]	35.2[a]	31.6[a]	2.8	0.0008
SFA	70.8[a]	64.0[b]	63.8[b]	62.9[b]	56.0[c]	55.8[c]	1.1	<0.0001
MUFA	23.2[d]	28.9[c]	28.9[c]	32.0[b]	37.5[a]	37.5[a]	1.0	<0.0001
cis-18:1	17.9[d]	22.3[c]	22.9[c]	27.7[b]	32.9[a]	32.4[a]	1.1	<0.0001
18:2 n-6	1.43[ab]	1.30[a]	1.39[ab]	1.77[c]	1.45[ab]	1.52[b]	0.05	<0.0001
18:3 n-3	0.65[b]	0.95[c]	1.04[d]	0.44[a]	0.79[c]	0.88[c]	0.04	<0.0001
*c9t*11-CLA	0.61[a]	0.62[a]	0.61[a]	0.35[b]	0.56[a]	0.52[a]	0.04	<0.0001

SFA = saturated fatty acids; MUFA = monounsaturated fatty acids, = sum of *cis* and *trans* isomers.
[a,b,c,d] Means within rows with same superscript letters are not significantly different (P>0.05).
[1] Orthogonal contrasts were conducted: effect of nature of forage was significant for milk, fat and protein yields, milk urea and all milk FA; effect of linseed was significant for all FA.

Conclusion

Grazing pasture increased milk, fat, and protein yields compared to silage intake. Changes in milk fatty acid composition were more dependent on the nature of forage than on whole crude linseed or acacia intake. The pasture-based diets supplemented with linseeds increased ewe milk content of FA (CLA, Omega 3) having positive effects on human health. Acacia intake had no effect on milk yield and composition and probably did not modify rumen biohydrogenation of PUFA.

Acknowledgement

The authors thank workers of experimental farm of INRAT and P. Capitan of INRA Theix.

References

Atti, N., H. Rouissi and M.H. Othmane, 2006. Milk production, milk fatty acid composition and conjugated linoleic acid (CLA) content in dairy ewes raised on feedlot or grazing pasture. Livest. Sci. 104: 121-127.

Chilliard, Y., F. Glasser, A. Ferlay, L. Bernard, J., Rouel and M. Doreau, 2007. Diet, rumen biohydrogenation and nutritional quality of cow and goat milk fat. Eur. J. Lipid Sci. Technol. 109: 828-855.

Woodward, S.L., M.J. Auldist., P.J. Laboyrie and E.B.L. Jansen, 1999. Effect of *L. corniculatus* and condensed tannins on milk yield and milk composition of dairy cows. Proc. N.Z. Soc. Anim. Prod. 59: 152-155.

Effect of supplementation of area specific mineral mixture or common salt on nutrient utilisation and growth in female calves fed wheat straw and concentrates

S.K. Mahanta, A. Kumar, G.H. Pailan and N.C. Verma
Plant Animal Relatioship Division, Indian Grassland and Fodder Research Institute, Jhansi- 284003, India; mahantask@rediffmail.com

Introduction

The low productivity in dairy animals occurs as a result of complex climatic, social and economic problems but under-nutrition is the common cause, affecting production, reproduction and health. Dairy animals in India are mostly reared on cereal crop residues. These crop residues are low in proteins, minerals, vitamins and available energy. Generally, livestock do not receive mineral mixture supplementation on crop residue based diets. However, it is essential to provide macro and micro-minerals in adequate quantities for efficient production and reproduction of dairy animals. Even though macro and micro elements are required in small quantities, they play important roles in various metabolic activities of the animal's body (Garg *et al.*, 2007). When the micro-nutrient contents were analysed in the soil, feed and animal blood samples of Bundelkhand region of India, it was found that crop residue based diets of dairy animals are deficient in calcium, phosphorous, copper and zinc which limits productivity in a major way, therefore, mixing of these mineral sources with common salt will be beneficial to dairy farmers. Accordingly, an area specific mineral mixture (ASMM) comprising dicalcium phosphate, zinc sulphate, copper sulphate and common salt was developed, which is more effective and economical in combating the deficiencies (Singh *et al.*, 2009). Hence the present study was conducted to evaluate ASMM or common salt as mineral supplements in growing female calves fed wheat straw and concentrates.

Material and methods

Fifteen growing crossbred (Jersey × Tharparkar) female calves (95.5±4.35 kg), divided into 3 treatment groups (T_1 to T_3) of 5 animals each, were fed a composite ration of concentrate mixture and wheat straw to meet the nutrient requirements. Calves in the T_1 group received a concentrate mixture (barley:mustard cake in ratio of 65:35) without containing any salt or mineral mixture, while animals in the T_2 group also received the same concentrate mixture plus common salt (at 20 g/ head/d) and the T_3 group received the concentrate mixture plus AASM (at 30 g/head/d). Feeding regime continued for a period of 181 days. After 40 days a digestion trial of 7-day duration was conducted. Blood samples were also collected once through the jugular vein-puncture in the morning (9:00 a.m.) before feeding and watering. Blood samples and feedstuffs were analysed for mineral contents using an Atomic Absorption Spectrometer (Varian AA240). The samples of the digestion trial were analysed for proximate constituents and phosphorus (AOAC, 1990). The data were subjected to statistical analysis following a completely randomised design.

Results

Daily DMI expressed as kg per 100 kg body weight was 2.84, 2.59 and 2.60 kg in animals of T_1, T_2 and T_3 groups, respectively and did not differ significantly amongst the groups (Table 1). Digestibility of DM and OM was similar among the treatment groups, but tended to be higher in the T_3 group. Plasma concentrations of mineral elements like Ca, Cu and Zn in growing calves were within the normal range of variation. However, plasma P concentrations in calves of the T_1

group (4.25 mg/dl) were below the critical level of 4.50 mg/dl. But calves receiving the ASMM supplement (T_3 group) maintained higher (P< 0.05) concentration of minerals. After 181 days of experimental feeding, the female calves in the ASMM supplemented group (T_3) attained higher (P<0.01) average daily gain (399 g) compared to those supplemented with common salt (T_2: 353 g) and non-supplemented (T_1: 320 g) groups.

Table1. Nutrient utilisation, plasma mineral profile and growth performance in calves.

Attributes	Treatment groups			SEM	Diet effect
	T_1	T_2	T_3		P-value
DM intake (unit/d)					
kg	3.10	2.81	2.89	0.29	0.77
kg/ 100 kg body weight	2.84	2.59	2.60	0.08	0.10
g/ kg $W^{0.75}$	91.12	83.23	84.24	2.81	0.13
Digestibility,%					
DM	53.9	55.2	56.2	0.64	0.07
OM	56.4	57.7	58.5	0.65	0.09
Plasma minerals					
Ca, mg/dl	9.30^b	9.90^{ab}	10.35^a	0.26	<0.05
P, mg/dl	4.25^b	4.85^{ab}	5.55^a	0.29	<0.05
Cu, µg/ml	0.67^b	0.76^{ab}	0.92^a	0.06	<0.05
Zn, µg/ml	1.14^b	1.55^{ab}	1.90^a	0.19	<0.05
Growth performance (for 181 days)					
Total weight gain, kg	57.9^c	63.9^b	72.2^a	1.77	<0.01
Average daily gain, g	320^c	353^b	399^a	9.78	<0.01

^{a,b} Means bearing different superscripts in a row differ significantly.

Conclusion

This study revealed that supplementation with micronutrients such as ASMM or common salt improved growth performances and plasma mineral status, without affecting nutrient intake and utilisation much in female calves fed wheat straw and concentrates.

References

AOAC, 1990. Official Methods of Analysis. 15th edn. Association of Official Analytical Chemists, Washington, DC, USA.

Garg, M.R., B.M. Bhandari and P.L. Sherasia, 2007. Assessment of macro and micro-mineral status of milk animals in Jodhpur district of Rajasthan. Indian J. Dairy Sci. 60: 191-198.

Singh, K.K., M.M. Das, S.K. Mahanta, Anil Kumar and A.K. Misra, 2009. Practical aspects of mineral nutrition in livestock. In: Das N., A.K. Misra, S.B. Maity, K.K. Singh and M.M. Das (Eds.), Forage for sustainable livestock production. Satish Serial Publishing House, Delhi, India, p. 404.

Unrefined sunflower oil supplementation selectively influenced the milk fatty acid profile and oxidative status in Simmental cows

T.S. Marenjak[1], *I. Delaš*[2], *N. Poljičak-Milas*[1] *and J. Piršljin*[1]
[1]*Faculty of Veterinary Medicine, University of Zagreb, Heinzelova 55, 10000 Zagreb, Croatia;* [2]*School of Medicine, University of Zagreb, Šalata 3, 10000 Zagreb, Croatia; marenjak@vef.hr*

Introduction

The use of unprotected vegetable oils and whole grains in the dairy cow diet may have a limited effect on the fatty acid composition in ruminants (Chilliard *et al.*, 2001). As a consequence of rumen biohydrogenation processes, unsaturated fatty acids from the diet are rapidly converted to saturated fatty acids by rumen microorganisms (Harfoot and Hazlewood, 1997). However, an increase of polyunsaturated fatty acids (PUFA) in the diet could change the pattern of rumen microbe metabolism, and consequently change the fatty acid synthesis in milk and adipose tissue of ruminants. There are some concerns regarding the feeding of PUFA to animals and possible higher susceptibility to lipid peroxidation if the antioxidant elements are not sufficiently present in the organism. The defects in antioxidant defence may lead to several diseases. Diets high in PUFA may induce fibrosis and oxidative damage of the liver (Brown *et al.*, 2003). Our objective was to identify the possible effects of non-protected sunflower oil on the antioxidative status, milk production and fatty acid composition in Simmental cows fed with non-protected and unrefined sunflower oil.

Material and methods

Ten healthy, multiparous Simmental cows in mid-lactation (78±40 DIM) were assigned for a cross-over feeding trial with one of the diet applied: the WSO diet that contained 26% corn silage, 33% perennial ryegrass haylage, and 41% concentrate in DM, and the SO diet where the concentrate mixture of the basal diet was supplemented with 0.2% of non-protected sunflower oil in DM per cow and day. After one month pre-trial period, cows were randomly selected and divided into two groups, five cows each, and fed one of the selected diets. Fourteen days wash out period between two treatments was applied. The cows were kept indoors, milked and fed twice daily. The milk samples were taken once in a pre-trial period, and the 14th and 21st day after the introduction of sunflower oil in the diet of each animal group. The composite of the regular evening and morning milkings of each cow was stored at -20 °C until analysis for the fatty acid composition according to ISO standards (ISO 14156-IDF 172:2001 and ISO15884-IDF 182:2002) on a GC chromatograph (GC-SRI 8619C, Quadrex Corporation, New Haven). All fatty acids determined in this trial were identified by comparing the retention times with methylated standards (Sigma Aldrich Chemie, GmbH and Supelco, USA and Matreya INC., PA, USA.). The quantification of FAME was done using nonadecanoic acid as an internal standard. The glutathion peroxidase (GPx) activity and concentration of thiobarbituric acid reactive substances (TBARS) were measured in whole blood. The blood was collected 3-4 h after morning feeding from *v. jugularis* in BD vacutainer tubes (LH 85 I.U.) with lithium heparin as the anticoagulant agent. The GPx activity was measured on a SABA 18 (AMS Italy) using commercial kits (RANSEL, RS 506 'Randox', Ireland). The TBARS concentration was determined according to Trotta *et al.* (1982) method on the Helios delta spectrophotometer (Unicam, Cambridge, UK). The data were analysed using the PROC MIXED procedure of SAS® (version 8; SAS Institute Inc. SAS/STAT), with cow, treatment, period, and order as class variables. The statistical model used included treatment, period, and treatment by order interaction. The significance of the difference between means was evaluated by the Tukey-Kramer test. For all data, significance was declared at $P<0.05$.

Results

Three of 30 fatty acids measured were significantly changed under the influence of the SO treatment (Table 1). Among the antioxidant enzymes, blood GPx activity significantly increased, whereas the indicator of the oxidative stress, TBARS concentration significantly decreased in the SO treatment.

Table 1. The milk fatty acids, whole blood GPx activity and TBARS concentration in Simmental cows respective to treatment.

	Treatment[1]		SEM	*P*
	SO	WSO		
Fatty acid, g/100g of fatty acids				
C16:0	25.14	29.99	1.427	0.04
C18:2 *cis*-9, *cis*-12	3.56	2.75	0.240	0.04
C 18:2 *cis*-9, *trans*-11, CLA	1.25	0.80	0.130	0.04
GPx[2], U/ g prot.l	112.98	87.84	0.153	0.05
TBARS, nmol/l	10.19	13.45	0.187	0.05

[1] SO: diet with sunflower oil supplementation, WSO: diet without sunflower oil supplementation.
[2] GPx: glutathion peoroxidase, TBARS: thiobarbituric acid reactive substances.

Conclusion

The results indicate that unrefined sunflower oil in the Simmental cows diet increased the concentration of some beneficial fatty acids, linoleic and conjugated linoleic (CLA) acids in milk. At the same time, unrefined sunflower oil led to lower concentrations of palmitic acid in milk and TBARS in blood with a significant increase of GPx activity that is probably related to the sufficient level of the antioxidant agent that might be provided by the unrefined sunflower oil.

References

Brown, K.E., P.A. Dennery, L.A. Ridnour, C.J. Fimmel, R.D. Kladney, E.M. Brunt and D.R. Spitz, 2003. Effect of iron overload and dietary fat on indices of oxidative stress and hepatic fibrinogenesis in rats. Liver Int. 23: 232-242.

Chilliard, Y., A. Ferlay and M. Doreau, 2001. Effect of different types of forages, animal fat or marine oils in cows' diet on milk fat secretion and composition, especially conjugated linoleic acid (CLA) and polyunsaturatedfa fatty acids. Livest. Prod. Sci.70: 31-48.

Harfoot, C.G. and G.P. Hazlewood, 1997. Lipid metabolism in the rumen. In: Hobson P.N. and C.S. Stewart (Eds.), The Rumen Microbial Ecosystem. Chapman and Hall, New York, NY, USA, pp. 382-426.

Trotta, R.J., S.G. Sullivan and A. Stern, 1982. Lipid peroxidation and hemoglobin degeneration in red blood cells exposed to t-butyl hydroperoxide. Biochem. J. 204: 405-415.

Additive effects of *trans*10,*cis*12-CLA and propionic acid on milk fat content and composition in dairy cows

G. Maxin[1,2], F. Glasser[3], P. Lamberton[1,2] and H. Rulquin[1,2]
[1]INRA, UMR 1080 Production du lait, 35000 Rennes, France; [2]Agrocampus Ouest, UMR 1080 Production du lait, 35000 Rennes, France; [3]INRA, UR 1213 Herbivores, 63122 Saint-Genès-Champanelle, France; gaelle.maxin@rennes.inra.fr

Introduction

Controlling milk fat production is a demand of farmers and dairy product manufacturers. Nutrition is the simplest way to modulate milk fat concentration and yield because its effects are quick and reversible. Some nutrients produced during digestive processes are known to modify milk fat concentration: acetic acid and butyric acid increase milk fat content whereas propionic acid, glucose and *trans*10,*cis*12-CLA decrease milk fat content (Rulquin *et al.*, 2007). Through infusion experiments, individual effects of these nutrients on milk fat synthesis are known and quantified. However, these nutrients are produced simultaneously during diet digestion and very few experiments report their effects on milk fat synthesis when they are supplied together to dairy cows. The objective of this experiment was to study the effects of two of these nutrients, provided alone or together, on milk fat content and composition. The two nutrients chosen were *trans*10,*cis*12-CLA and propionic acid, both known to reduce milk fat secretion, but by a different mechanism.

Material and methods

Four duodenum- and rumen-fistulated Holstein cows (207±65 DIM) were assigned to four treatments in a 4×4 Latin square design with 14 d periods. The four treatments were control (CON), infusion of 1.85 g/d of *trans*10,*cis*12-CLA in the duodenum (CLA), infusion of 500 g/d of propionic acid in the rumen (C3) and infusion of 1.85 g/d of *trans*10,*cis*12-CLA in the duodenum +500 g/d of propionic acid in the rumen (CLA+C3). Solutions were continously infused during 24 h. The *trans*10,*cis*12-CLA was provided by a fatty acid (FA) mixture, Lutalin® (BASF AG, Germany) in which *trans*10,*cis*12-CLA represented 30% of total FA. Cows were milked twice a day and individually fed controlled amounts of a TMR (20% orchardgrass hay, 40% dehydrated corn pellets and 40% concentrate). Milk yield was recorded and fat and protein contents were determined by infrared analysis at each milking. Milk samples from two consecutive milkings were taken and analysed for FA as described by Loor *et al.* (2005). Data were analysed as a 4×4 Latin square using the MIXED procedure of SAS® software (2000). The statistical model included cow (random effect), period, CLA, C3, interaction between CLA and C3 and residual error. Differences between treatment means were considered to be significant when $P \leq 0.05$.

Results and discussion

Feed intake, milk yield and milk protein content were not affected by infusions (Table 1). *Trans*10,*cis*12-CLA infusions, with or without C3, decreased milk fat content and yield by 18%, consistently with previous CLA infusion experiments (review in: Shingfield and Griinari, 2007). C3 infusions did not modify milk fat content and yield. This result is in contrast with several published papers which reported a significant milk fat reduction with similar C3 amounts infused (Rigout *et al.*, 2003; Vanhatalo *et al.*, 2003).
*Trans*10,*cis*12-CLA decreased the percentage of C6:0 to C12:0 and C16:0 (Table 1), and increased the percentage of C18 FA, meaning that the secretion of de novo synthesised FA was more reduced by CLA infusion than that of C18 FA. On the contrary, C3 infusions increased odd-chain FA (Sum

C5 to C19) slightly and decreased C4:0 slightly. The other FA were not affected by C3 infusions. These differences in milk FA composition confirmed that mechanisms involved in milk fat variation are different for *trans*10,*cis*12-CLA and C3. There was no significant interaction between CLA and C3 infusions.

Table 1. Effects of trans10,cis12-CLA and propionic acid infusions on intake, milk yield and composition from dairy cows.

	Treatments[1]				SEM	*P*-value		
	CON	CLA	C3	CLA+C3		CLA	C3	CLA*C3
DMI, kg/d	21.3	21.5	21.6	21.1	1.1	0.66	0.87	0.28
Milk, kg/d	30.1	31.7	30.9	31.0	3.6	0.31	0.95	0.33
Protein,%	3.17	3.16	3.15	3.23	0.12	0.18	0.24	0.11
Fat,%	3.39^{a}	2.76^{b}	3.28^{a}	2.74^{b}	0.32	<0.01	0.58	0.68
Fat yield, kg/d	1.00^{a}	0.85^{b}	0.99^{a}	0.83^{b}	0.09	<0.01	0.59	0.79
Milk fatty acids g/100 g FA								
C4:0	2.76^{a}	2.63^{a}	2.43^{ab}	2.22^{b}	0.15	0.14	<0.05	0.70
Sum C6 to C12	12.0^{a}	10.6^{b}	11.7^{a}	10.6^{b}	0.8	<0.05	0.84	0.77
C14:0	15.0	14.9	14.8	14.9	0.5	0.79	0.50	0.57
C16:0	32.0^{ab}	30.1^{b}	32.6^{a}	30.7^{ab}	1.6	<0.05	0.41	0.94
Odd-chain (Sum C5 to C19)	2.1^{b}	2.1^{b}	2.6^{a}	2.5^{a}	0.08	0.18	<0.01	0.99
Total C18 FA	29.3^{ab}	32.6^{a}	28.5^{b}	31.4^{ab}	1.9	<0.05	0.34	0.82

[1] CON = control, CLA = infusion of *trans*10,*cis*12-CLA, C3 = infusion of propionic acid, CLA+C3 = infusion of *trans*10,*cis*12-CLA + propionic acid.

Conclusion

This study describes the effects of *trans*10,*cis*12-CLA and propionic acid on milk fat content, yield and milk FA composition, and shows that *trans*10,*cis*12-CLA has higher and different effects than C3. With the amounts infused in the present experiment, their effects on milk fat content and composition were additive.

Acknowledgement

The authors thank the experimental farm and laboratory staff. Financial support from Arrivé, Basf NA, CCPA, Centralys, Cybélia, Evialis, Inzo and Primex is acknowledged.

References

Loor, J.J., A. Ferlay, A. Ollier, M. Doreau and Y. Chilliard, 2005. Relationship among *trans* and conjugated fatty acids and bovine milk fat yield due to dietary concentrate and linseed oil. J. Dairy Sci. 88: 726-740.

Rigout, S., C. Hurtaud, S. Lemosquet, A. Bach and H. Rulquin, 2003. Lactational effect of propionic acid and duodenal glucose in cows. J. Dairy Sci. 86: 243-253.

Rulquin, H., C. Hurtaud, S. Lemosquet and J.-L. Peyraud, 2007. Quantification of the effects of energetic nutrients on fat content of cow milk. INRA Prod. Anim. 20: 163-173.

Shingfield, K.J. and J.M. Griinari, 2007. Role of biohydrogenation intermediates in milk fat depression. Eur. J. Lipid Sci. Technol. 109: 799-816.

Vanhatalo, A., T. Varvikko and P. Huhtanen, 2003. Effects of various glucogenic sources on production and metabolic responses of dairy cows fed grass silage-based diets. J. Dairy Sci. 86: 3249-3259.

Selection for muscling in Angus steers increases glycogen and reduces response to adrenaline in muscle

P. McGilchrist[1], P.L. Greenwood[2], D.W. Pethick[1] and G.E. Gardner[1]
[1]Australian Cooperative Research Centre for Beef Genetic Technologies, School of Veterinary & Biomedical Science, Murdoch University, 6150, WA, Australia; [2]Australian Cooperative Research Centre for Beef Genetic Technologies, NSW DPI Beef Industry Centre of Excellence, 2351, Armidale, NSW, Australia; p.mcgilchrist@murdoch.edu.au

Introduction

The quantity of glycogen stored in muscle at slaughter is a major determinant of meat quality and profitability. Muscle glycogen levels below ~0.6% can result in high ultimate pH (>5.7) of beef (Ferguson *et al.*, 2001) leading to dark cutting or dark firm dry (DFD) meat. Meat which is DFD has a darker colour, reduced shelf life, bland flavour and variable tenderness (Ferguson *et al.*, 2001). This condition significantly reduces the value of a carcase since an elevated ultimate pH (>5.7) makes it ineligible for grading by Meat Standards Australia and premiums will not be rewarded. Low muscle glycogen pre-slaughter is caused by low glycogen storage as a result of low metabolisable energy intake (Knee *et al.*, 2004) and/or stress between mustering and slaughter. Stress stimulates the release of endogenous adrenaline, causing the mobilisation and depletion of muscle glycogen stores. When glycogen stores are low, stressors have the greatest influence on the incidence of DFD meat.
The demand for more profitable, high yielding carcasses by the beef industry has increased selection for animals with more muscle and less fat. Animals that have increased muscle hypertrophy generally have more fast-glycolytic type IIX myofibres (Wegner *et al.*, 2000). The adrenaline responsiveness of muscle with more fast-glycolytic fibres is likely to be greater due to their increased glycolytic and glycogenolytic capacity (Wegner *et al.*, 2000), possibly extenuating the problem of DFD meat in high muscling selection line cattle. Therefore, we hypothesise that selection for muscling will decrease muscle glycogen storage and increase the response of muscle to adrenaline.

Material and methods

Muscle glycogen concentration and muscle tissue responsiveness to adrenaline were examined in 10 low muscled and 11 high muscled, 18 mo-old steers from an Angus herd selected for divergence in width, convexity and visual muscle score for 15 yr. Muscle glycogen and lactate concentration was analysed from four muscle biopsies taken from the *M. semimembranosus*, *M. semitendinosus* and *M. longisimus dorsi*, at 14, 90 and 150 d on an *ad-libitum* grain-based diet and at slaughter. The samples were obtained using a purpose built 12V biopsy drill, and were stored at -80 °C until analysed. Glycogen and lactate concentrations were analysed using a linear mixed effects model, with muscling genotype, sampling time and muscle as fixed effects and animal within sire as the random term.
The steers were also habituated in individual pens for 2 weeks, between the 1st and 2nd biopsies, before adrenaline challenges were administered via indwelling jugular catheters at 7 levels (2 per d) ranging from 0.2 to 3.0 μg/kg live weight (LW). Sixteen blood samples were taken between -30 and 130 minutes from adrenaline administration and analysed for plasma lactate concentration. The area under curve (AUC) was determined for the first 20 min post adrenaline administration which represents the muscle response to adrenaline. AUC was analysed using a linear mixed effects model, with muscling genotype as a fixed effect, adrenaline challenge as a covariate, and animal within sire as the random term.

Results

The high muscled steers had 0.095±0.048 g/100g higher basal muscle glycogen than the low muscled steers ($P<0.05$). This equated to ~6.1% more glycogen when averaged over 3 muscles at the 4 sampling time points. At high levels of adrenaline challenge (>2 μg/kg LW) high muscled steers had lower plasma lactate AUC ($P<0.05$; Figure 1). At biopsy 1, before the steers were habituated to human activity, the high muscled cattle also had 0.033±0.013 g/100 g lower basal muscle lactate or ~16% less lactate compared to the low muscled steers ($P<0.05$). This suggests that cattle selected for muscling store more glycogen and mobilise less glycogen in response to high levels of stress, rejecting our initial hypothesis.

One explanation for this response may relate to the density of β_2-adrenoreceptors on the myofibres. Martin *et al.* (1989) showed that rats with more oxidative myofibres (type I) had an increased density of β_2-adrenoreceptors and increased glycogenolysis. Thus selection for muscling in cattle, which reduces the proportion of more oxidative myofibres, may dilute the density of β_2-adrenoreceptors thus reducing the response to adrenaline.

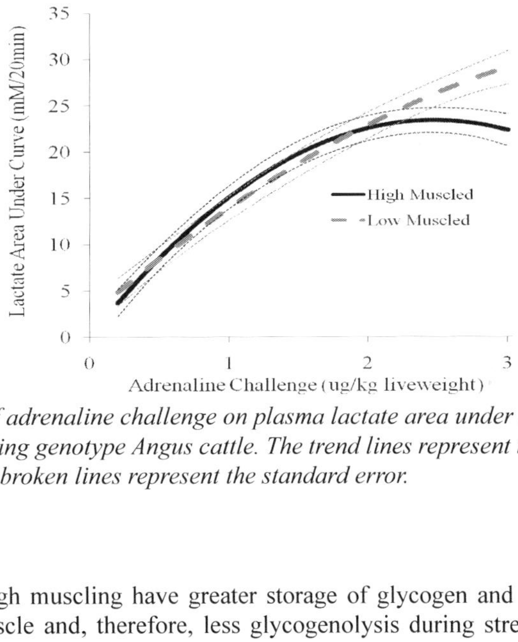

Figure 1. The effect of adrenaline challenge on plasma lactate area under the curve (mM/20min) in high and low muscling genotype Angus cattle. The trend lines represent the least squares mean estimates and the fine broken lines represent the standard error.

Conclusion

Cattle selected for high muscling have greater storage of glycogen and reduced sensitivity to adrenaline in the muscle and, therefore, less glycogenolysis during stress than low muscling counterparts. Hence selection for muscling may decrease the incidence of DFD meat. This response may be driven by lower β_2-adrenoreceptor density in highly muscled cattle.

References

Ferguson D.M., H.L. Bruce, J.M. Thompson, A.F. Egan, D. Perry and W.R. Shorthose, 2001. Factors affecting beef palatability – Farmgate to chilled carcass. Aust. J. Exp. Agric. 41: 879-891.

Knee B.W., L.J. Cummins, P.J. Walker and R. Warner, 2004. Seasonal variation in muscle glycogen in beef steers. Aust. J. Exp. Agric. 44: 729-734.

Martin W., S. Murphree and J. Saffitz, 1989. Beta-adrenergic receptor distribution among muscle fibre types and resistance arterioles of white, red, and intermediate skeletal muscle. Circulation Res. 64: 1096-1105.

Wegner, J., E. Albrecht, I. Fiedler, F. Teuscher, H.J. Papstein and K. Ender, 2000. Growth- and breed-related changes of muscle fibre characteristics in cattle. J. Anim. Sci. 78: 1485-1496.

Selection for muscling in Angus steers increases leanness and adipose tissue response to adrenaline

P. McGilchrist[1], P.L. Greenwood[2], D.W. Pethick[1] and G.E. Gardner[1]
[1]Australian Cooperative Research Centre for Beef Genetic Technologies, School of Veterinary & Biomedical Science, Murdoch University, 6150, WA, Australia; [2]Australian Cooperative Research Centre for Beef Genetic Technologies, NSW DPI Beef Industry Centre of Excellence, Armidale, 2351, NSW, Australia; p.mcgilchrist@murdoch.edu.au

Introduction

Carcass lean meat yield is a key profit driver for beef producers, processors and retailers. It can be improved by selection using retail beef yield estimated breeding values, visual selection, cross-breeding or myostatin gene markers. Higher yielding cattle have proportionately more muscle and less subcutaneous fat (Perry *et al.*, 1993). However, fat reduction in cattle may reduce marbling which accounts for around 10-15% of variance in palatability (Dikeman, 1987) and attracts a premium in some export markets. With continued selection for muscling, it is important that this negative correlation is monitored, as the physiological mechanisms under-pinning reduced adipose tissue and increased beef yield in highly muscled cattle are unclear. One possibility, however, may be via the key regulatory stress hormone adrenaline, which causes mobilisation of adipose triacylglycerol for energy production. Work in obese humans has demonstrated a reduced response to adrenaline in adipose tissue (Jocken and Blaak, 2008), which reduces lipolysis and thus potentiates obesity – the reverse may apply in lean animals. Therefore we hypothesise that selection for muscling will reduce fatness and increase lean meat yield, and increase the adipose tissue response to adrenaline.

Material and methods

Adipose tissue response to adrenaline was examined in 10 low muscled and 11 high muscled, 18 mo-old steers from an Angus herd visually selected for divergence in muscling for over 15 yr (Perry *et al.*, 1993). They were habituated in individual pens for 2 wk on an *ad-libitum* grain-based diet. Adrenaline challenges were administered via indwelling jugular catheters at 7 levels (2/d) ranging from 0.2-3.0 μg/kg live weight. Sixteen blood samples were taken between -30 and 130 minutes relative to adrenaline administration. Plasma was analysed for non-esterified fatty acid (NEFA) concentration and area under curve (AUC) was determined for the first 10 min following adrenaline administration. AUC reflects the adipose response to adrenaline. After completing the challenges, the steers were lot-fed until they met export market specifications, after which they were slaughtered at a commercial abattoir. A carcass butt with rump attached was scanned using computer aided tomography (CT) scanning, with images taken at 10 mm intervals and partitioned into bone, muscle and fat portions. These proportions of total butt weight were determined based on the volume and average CT density of these tissues. CT yield proportions were analysed using a linear mixed effects model with muscling genotype as a fixed effect and sire as the random term. AUC for NEFA was also analysed using a linear mixed effects model with adrenaline challenge and CT scanned muscle percentage as covariates, and animal within sire as the random term.

Results and discussion

The CT scan images showed that the high muscled cattle had 1.71 percent less fat, and 2.18 percent more muscle per kg of butt than the low muscled steers (Table 1). There was no difference in the percentage of bone.

Table 1. Mean ± standard error percentage of fat, muscle and bone of the carcass butt with rump attached for high & low muscling genotype steers determined by CT scanning.

	High muscle	Low muscle	Significance
Fat (%)	14.17±0.47	15.87±0.49	$P<0.05$
Muscle (%)	69.27±0.35	67.08±0.38	$P<0.01$
Bone (%)	16.57±0.33	17.06±0.35	$P=0.32$

As the CT scanned percentage of muscle increased, the NEFA AUC also increased ($P=0.064$) by approximately 50% across the range of muscle yield values (Figure 1). This suggests that high yielding cattle mobilise lipid more readily in response to adrenaline, which supports our initial hypothesis and indicates a catabolic (lipolytic) mechanism contributing to reduced adiposity in these animals.

While the physiological mechanisms contributing to these results are unclear, Jocken and Blaak (2008) found reduced hormone sensitive lipase (HSL) expression coupled with reduced number and function of β_2-adrenoreceptors in adipose tissue in obese humans (Reynisdottir *et al.,* 1994). Gregory *et al.* (1986) also showed that adipose tissue capillary blood flow declines with increasing fatness in sheep, slowing the perfusion of endogenous adrenaline. This may lead to less activation of HSL and β_2-adrenoreceptors, reducing lipolysis.

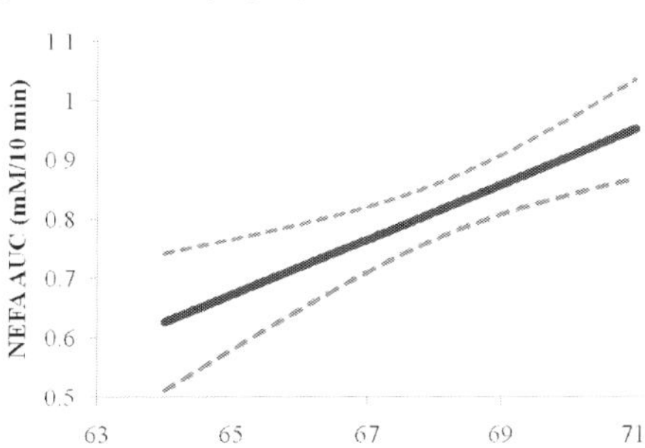

Figure 1. The effect of CT scan muscle as a percentage of the butt-rump on NEFA AUC. The trend line represents the least squares mean estimate with standard error (broken lines).

Conclusion

Animals selected for high muscling have less fat than their low muscling counterparts and are more responsive to adrenaline in adipose, resulting in more lipolysis. This difference in lipolysis may in part explain their reduced fatness, although the specific mechanism underpinning the difference in adrenalin responsiveness and lipolytic rate is unclear.

References

Dikeman, M.E., 1987. Fat reduction in animals and the effects on palatability and consumer acceptance of meat products. Reciprocal Meat Conf. 40: 93-103.

Gregory, N.G., R.J. Christopherson and D. Lister, 1986. Adipose tissue capillary blood flow in relation to fatness in sheep. Res. Vet. Sci. 40: 352-356.

Jocken, J.W.E. and E.E. Blaak, 2008. Catecholamine-induced lipolysis in adipose tissue and skeletal muscle in obesity. Physiol. Behav. 94: 219-230.

Perry, D., A.P. Yeates and W.A. McKiernan, 1993. Muscle score: Its usefulness in describing the potential yield of saleable meat from live steers and their carcasses. Aust. J. Exp. Ag. 33: 275-281.

Reynisdttir, S., H. Wahrenberg, K. Carlstrom, S. Rossner and P. Arner, 1994. Catecholamine resistance in fat cells of women with upper body obesity due to decreased expression of beta 2-adrenoreceptors. Diabetologia 37: 428-435.

Effects of maternal nutritional plane and selenium supply during gestation on neonatal offspring growth and visceral organ mass

A.M. Meyer[1], J.J. Reed[1], T.L. Neville[1], J.B. Taylor[2], D.A. Redmer[1], L.P. Reynolds[1], K.A. Vonnahme[1] and J.S. Caton[1]
[1]Center for Nutrition and Pregnancy, Department of Animal Science, North Dakota State University, Fargo, ND, USA; [2]USDA-ARS, U.S. Sheep Experiment Station, Dubois, ID, USA; joel.caton@ndsu.edu

Introduction

Gestating ruminants often receive improper nutrition during gestation, either due to inadequate nutrient intake while grazing poor quality forages or excessive intake from lush forages or intensive management. Intake of specific nutrients may also be problematic, especially for trace minerals such as selenium (Se) that have regional soil, and thus feedstuff, variability. Previous research has demonstrated that fetal growth and organ development is impaired during intrauterine growth retardation, which can lead to reduced growth and performance later in life (Wu *et al.*, 2006; Caton *et al.*, 2007). Although the early neonatal period is critical for offspring health and development, it is unknown how nutrition of the dam during gestation affects this time period when residual effects upon lactation are negated. Therefore our objective was to determine the effects of both maternal plane of nutrition and dietary Se supply during gestation on birth weight and subsequent early neonatal growth of lambs when placed under similar management postnatally.

Material and methods

Eighty-four Rambouillet ewe lambs (age = 240±17 d, BW = 52.1±6.2 kg), were allocated to a 2 x 3 factorial design. Factors included Se (adequate Se [ASe, 11.5 μg/kg BW] or high Se [HSe, 77.0 μg/kg BW]) initiated at breeding and nutritional plane (60% [RES], 100% [CON], and 140% [HIGH] of requirements) initiated at d 40 of gestation. Ewes were individually fed pelleted diets in a temperature-controlled facility. At parturition, lambs were removed from their dams before suckling, fed artificial colostrum to body weight, and reared on *ad libitum* milk replacer. Lambs were slaughtered on 20.6±0.9 d of age when detailed necropsies were performed. Data were analysed using the GLM procedure of SAS® with effects of Se supply, nutritional plane, and their interaction in the model. Twins were removed (n=6), and sex was included as a covariate if significant. Means were separated using least significant difference.

Results

Maternal nutrient restriction tended (P=0.11) to reduce lamb birth weight (Table 1), which was more pronounced in lambs born to dams fed ASe compared with HSe. Curved crown rump length at birth was also decreased in lambs born to RES ewes. From birth to d 19 of age, gain (g/d) was greater ($P \leq 0.08$) for offspring of CON and HIGH fed ewes compared with RES, resulting in lambs from nutrient restricted dams weighing less than their contemporaries on d 19. Additionally, lamb heart, liver, GI tract, spleen, and perirenal fat masses were decreased ($P \leq 0.06$) by maternal nutrient restriction during gestation. Lambs born to HIGH ewes had greater kidney mass than those born to RES, whereas omental and mesenteric fat mass was less for lambs born to RES and CON fed ewes compared with overnourished ewes. Relative to empty body weight, lambs born to restricted dams had greater ($P \leq 0.08$) lung mass than those born to HIGH and greater kidney mass compared with CON (data not shown). Although there were no main effects of maternal Se on lamb growth or organ masses, relative liver mass of lambs born to restricted ewes was greater when dams were fed ASe.

Table 1. Effect of maternal nutritional plane and Se supply on lamb growth and organ masses.

Item[4]	Selenium[1]		SEM	Nutritional Plane[2]			SEM	*P*-value[3]		
	ASe	HSe		RES	CON	HIGH		Se	Nut	Se x Nut
Lamb wt, kg										
Birth	4.5	4.6	0.1	4.3[a]	4.7[b]	4.6[b]	0.1	0.36	0.08	0.08
d 19	9.9	10.4	0.3	9.2[a]	10.5[b]	10.6[b]	0.4	0.31	0.03	0.59
CCR, cm	50.9	51.1	0.5	49.9[a]	51.4[b]	51.7[b]	0.6	0.71	0.07	0.48
Gain, g/d	287	302	15	261[a]	307[b]	316[b]	18	0.43	0.06	0.21
Organ mass, g										
Heart	66.7	69.7	1.9	63.0[a]	70.7[b]	71.0[b]	2.3	0.25	0.02	0.09
Lung	200	200	9	199	207	194	11	0.97	0.63	0.72
Kidney	64.1	65.7	2.2	60.6[a]	64.6[ab]	69.5[b]	2.7	0.60	0.06	0.11
Liver	301	304	11	278[a]	316[b]	315[b]	13	0.85	0.06	0.17
GI tract	541	553	17	505[a]	562[b]	573[b]	21	0.60	0.04	0.75
Spleen	29.7	30.9	1.1	27.1[a]	32.5[b]	31.2[b]	1.4	0.46	0.02	0.65
PR fat	125	131	9	104[a]	131[b]	149[b]	11	0.62	0.02	0.23
OM fat	140	146	9	123[a]	142[a]	163[b]	12	0.61	0.05	0.55

[1] Ewes fed 11.0 μg/kg BW Se (ASe) or 77.0 μg/kg BW Se (HSe).
[2] Ewes fed 60% (RES), 100% (CON), or 140% (HIGH) of nutrient requirements.
[3] Probability values for effects of selenium (Se), nutritional plane (Nut), and their interaction.
[4] CCR = curved crown rump length at birth, PR fat = perirenal fat, OM fat = omental and mesenteric fat.
[a,b] Within a parameter, means differ ($P<0.10$).

Conclusion

Results indicate that maternal nutrient restriction during gestation has negative consequences upon offspring which occur even in the face of proper postnatal nutrition and may become more evident with advancing neonatal age. Maternal overnutrition and supranutritional Se had fewer effects upon offspring in the current study. Further investigations into the effects of maternal nutrient supply on postnatal responses should allow for identification of optimal management of both gestating and neonatal ruminants.

Acknowledgement

Partially supported by USDA-NRI grants No. 2003-35206-13621 and 2005-35206-15281, from the USDA-CSREES, and by NIH Grant HL 64141.

References

Caton, J.S., K.A. Vonnahme, J.J. Reed, T.L. Neville, C.M. Effertz, C.J. Hammer, J.S. Luther, J.B. Taylor, D.A. Redmer and L.P. Reynolds, 2007. Effects of maternal nutrition on birth weight and postnatal nutrient metabolism in sheep, Vichy, France. EAAP Pub. No. 124: 101-102.

Wu, G., F.W. Bazer, J.M. Wallace and T.E. Spencer, 2006. Intrauterine growth retardation: Implications for the animal sciences. J. Anim. Sci. 84: 2316-2337.

Heifer nutrition during gestation affects expression of IGF-1R, IGF-2 and IGF-2R in omental adipose tissue of their mature off-spring

G.C. Micke[1], T.M. Sullivan[1], S. Lie[2], S. Gentili[2], I.C. McMillen[2] and V.E.A. Perry[1]
[1]School of Veterinary Science, University of Queensland, St Lucia 4072, Australia; [2]Early Origins of Adult Health Research Group, Sansom Institute, School of Pharmacy and Medical Sciences, University of South Australia, SA 5000, Australia; v.perry@uq.edu.au

Introduction

Evidence suggests that prenatal nutrition can result in the permanent programming of genes responsible for adipose tissue formation (Muhlhaulser *et al.*, 2007) as well as affect the sensitivity of adipose tissue to circulating insulin-like growth factors (IGF) (Bispham *et al.*, 2003). In addition to an insulation role, adipose tissue is an endocrine organ associated with the regulation of appetite, growth and reproduction (Hausman *et al.*, 2009). Therefore, the omental fat depot has the ability to affect traits of commercial significance via endocrine mechanisms, despite omental fat not being of direct commercial carcass value itself. As such the aim of this study was to determine the effect of heifer nutrient intake during gestation on the expression of IGF-I, IGF-IR, IGF-II and IGF-IIR in omental adipose tissue of their adult offspring.

Material and methods

The study was a 2 x 2 factorial design. Composite breed beef heifers were allocated to either a high (H/- = 76 MJ metabolisable energy (ME) and 1.4 kg crude protein (CP)) or low (L/- = 62 MJ ME and 0.4 kg CP daily) nutritional treatment at artificial insemination. Half of each nutritional group changed to an opposite nutritional group at the end of the first trimester (-/H = 82 MJ ME and 1.4 kg CP; -/L = 62 MJ ME and 0.4 kg CP daily), resulting in 4 treatment groups: HH (n=16); HL (n=19); LH (n=17); LL (n=19). During the third trimester all heifers were fed the same diets. Progeny were run as 1 group until slaughter at 22 months of age. Two of the progeny were excluded from the trial prior to slaughter for welfare reasons. At slaughter, omental adipose tissue was collected from each animal, snap frozen and stored at -80 °C. The expression of IGF-1, IGF-1R, IGF-2 and IGF-2R were measured using a process of RNA extraction, cDNA synthesis and quantitative real-time PCR, as described by MacLaughlin *et al.* (2007). The abundance of each gene relative to cyclophilin was calculated. Plasma urea (BUN) and non-esterified free fatty acid (NEFA) concentrations 3 weeks pre-slaughter were measured on a Hitachi autoanalyser, using commercially available kits (UREA/BUN, Roche Diagnostic Systems, Germany for urea; NEFA-C, Wako Pure Chemical Industries, Japan for FFA). The effects of treatment group and gender on liveweight, expression of each gene and concentration of BUN and ln(NEFA), were analysed using two-way ANOVA (SPSS for Windows version 11.5). Data are presented as mean ± SEM. Significance was set at P<0.05.

Results

Pre-slaughter liveweight of female LL (576.1±16.0kg) was less than female HL (536.5±14.4kg). Plasma ln(NEFA) was increased in L/- (-1.34±0.04) compared with the H/- (-1.26±0.04) treatment group (P<0.05), and LL (-1.31±0.05) compared to the HL (-1.21±0.05) treatment group (P<0.05) cattle. There was no effect of gender on plasma ln(NEFA), nor treatment group or gender on plasma BUN. There was no effect of treatment group on cyclophilin expression. IGF-1R expression was greater in HH and LH in comparison to LL animals (HH = 0.005±0.0005; HL = 0.0043±0.0004; LH = 0.005±0.0004; LL = 0.0038±0.0002; P<0.05). IGF-2 and IGF-2R expression were subjected to an interaction between treatment group and gender as shown in Figure 1. There was no effect of treatment group, gender or the interaction term on IGF-1 expression.

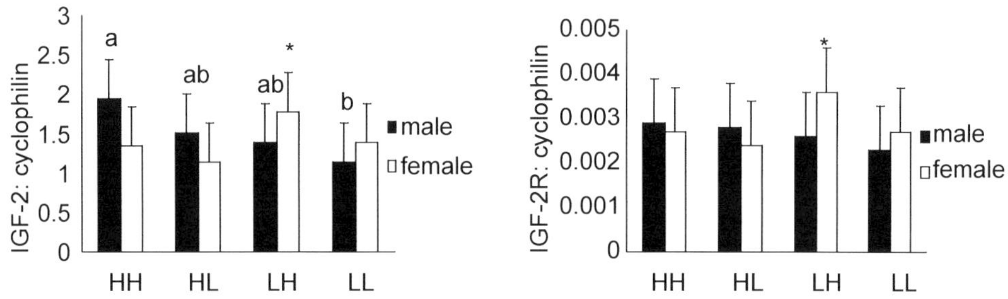

*Figure 1. The relative expression of IGF-2 and IGF-2R mRNA in omental adipose tissue of male and female progeny of heifers fed high (H) or low (L) nutrient diets during early- and mid-gestation. * Denotes a significant effect of maternal nutrition within females and different letters denotes a significant effect within males ($P<0.05$).*

Conclusion

This study shows that the relative expression of IGF-1R, IGF-2 and IGF-2R in mature beef cattle is influenced by their dam's nutrient intake during gestation, with the effects being sex specific. Increased IGF-1R expression in progeny whose dams received high diets during mid-gestation would have resulted in an increased sensitivity to the effects of circulating IGF-1. Both IGF-2 and IGF-2R were decreased in LL progeny however the role of IGF-2 in the bovine postnatal animal is poorly understood and warrants further investigation. Together, these findings suggest that the postnatal endocrine status of cattle may be permanently altered by their *in utero* environment.

Acknowledgement

The authors acknowledge the financial and in kind support of Meat and Livestock Australia, Australian Agricultural Co., Western Australian Cattle Industry Compensation Fund, Ridley AgriProducts and Milne AgriGroup.

References

Bispham, J., G.S. Gopalakrishnan, J. Dandrea, V. Wilson, H. Budge, D.H. Keisler, P.F. Broughton, T. Stephenson and M.E. Symonds, 2003. Maternal endocrine adaptation throughout pregnancy to nutritional manipulation: consequences for maternal plasma leptin and cortisol and the programming of fetal adipose tissue development. Endocrinology 144: 3575-3585.

Hausman, G.J., M.V. Dodson, K. Ajuwon, M. Azain, K.M. Barnes, L.L. Guan, Z. Jiang, S.P. Poulos, R.D. Sainz, S. Smith, M. Spurlock, J. Novakofski, M.E. Fernyhough and W.G. Bergen, 2009. The biology and regulation of preadipocytes and adipocytes in meat animals. J. Anim. Sci. 87: 1218-1246.

MacLaughlin, S.M., S.K. Walker, D.O. Kleeman, J.P. Sibbons, D.N. Tosh, S. Gentili, L. Coulter and I.C. McMillen, 2007. Impact of periconceptional undernutrition on adrenal growth and adrenal insulin-like growth factor and steroidogenic enzyme expression in the sheep fetus during early pregnancy. Endocrinology 148: 1911-1920.

Muhlhausler, B.S., J.A. Duffield and I.C. McMillen, 2007. Increased maternal nutrition increases leptin expression in perirenal and subcutaneous adipose tissue in the postnatal lamb. Endocrinology 148: 6157-6163.

Tissue distribution of the nutrient sensing free fatty acid receptors FFAR2 and FFAR3 mRNA expression in the bovine species

M. Mielenz, A. Hosseini, S. Vorspohl and H. Sauerwein
Institute of Animal Science, Physiology & Hygiene Unit, University of Bonn, D-53115, Bonn, Germany; sauerwein@uni-bonn.de

Introduction

Recently, a family of nutrient sensing free fatty acid receptors has been deorphanised. These receptors are of current interest with regards to inflammation and type 2 diabetes (Stoddart *et al.*, 2008). Two of them, FFAR2 and FFAR3 (formerly GPR43 and GPR41) bind short chain fatty acids (SCFA) which are the most important energy substrates in ruminants. FFAR1 (formerly GPR40) binds medium- and long-chain fatty acids. FFAR2 and 3 are expressed in many tissues of monogastrics. However, the expression of FFAR3 in human adipose tissue (AT) is controversial. For ruminant species, less data are available: In cattle, propionate stimulates leptin mRNA *in vitro* (Soliman *et al.*, 2007), but does not seem to be a physiological regulator of plasma leptin *in vivo* (Bradford *et al.*, 2006). Propionate regulates mRNA expression of a putative FFAR3 in different AT depots of goats (Mielenz *et al.*, 2008). Only recently it was shown that FFAR2 and FFAR3 orthologs are functional receptors in cattle, and because of the higher mRNA abundance of FFAR3 relative to FFAR2, the former was assumed to be a main mediator of SCFA signalling in the mammary gland (Yonezawa *et al.*, 2009). During the transition period, FFAR3 mRNA in subcutaneous AT increases *post partum* (Lemor *et al.*, 2009). Here we aimed at further screening bovine tissues for the presence of FFAR2 and FFAR3 mRNA.

Material and methods

A panel of RNA extracts from 11 different tissues (liver, pancreas, skeletal muscle, mammary gland, spleen, kidney, adrenal gland, heart, lung, subcutaneous AT, visceral AT) from a German Holstein dairy cow, 15 weeks in milk, was used for reverse transcription PCR. After DNase digestion, the RNA was purified using RNeasy Mini Kit. RNA integrity was verified by denaturing gel electrophoresis. Following cDNA synthesis, a qualitative PCR was carried out using 1.5 μl cDNA synthesis product in a 25 μl reaction volume with the following primers and annealing temperatures for FFAR2: forward 5'atgggtttcggcttctacg3', reverse 5'ggtggtctcattctcctttg3', 60 °C; FFAR3: forward 5'acctgatgg-ccctggtg3', reverse 5'ggacgtgagatagatggtgg3', 59 °C; ß-actin: forward 5'ctcttccagccttccttcct3', reverse 5'gggcagtgatctctttctgc3', 60 °C. PCR products were analysed by gel electrophoresis.

Results

The results for FFAR2 and 3 mRNA expression in the tissues tested are shown in Figure 1. Expression of FFAR2 mRNA was detectable in all samples analysed, even in the pancreas for which only poor RNA quality could be obtained. This might explain the lack of FFAR3 mRNA in this tissue but could also be due to physiologically low or absent expression. In skeletal muscle where the signal for FFAR2 mRNA was low, no FFAR3 mRNA was found. High signal intensities for both receptors were observed in the spleen. Unspecific signals were also observed in the mammary gland and spleen.

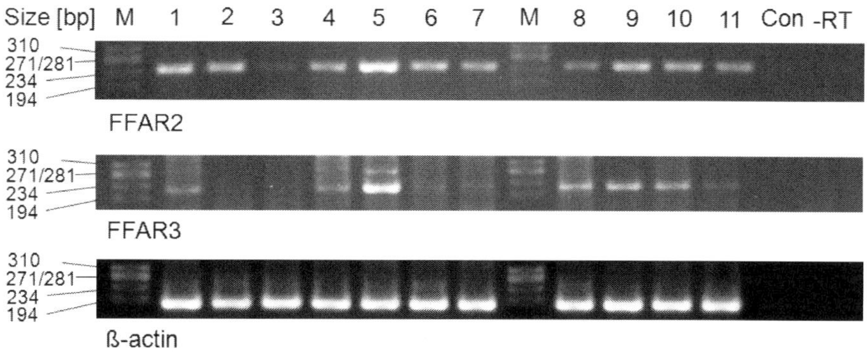

Figure 1. Qualitative mRNA detection of the free fatty acid receptors FFAR2 and FFAR3 in different bovine tissues. 1: liver; 2: pancreas; 3: skeletal muscle (Musculus semitendinosus); 4 mammary gland (parenchyma); 5: spleen; 6: kidney (cortex); 7: adrenal gland (cortex); 8: heart; 9: lung; 10: subcutaneous AT (sternum); 11: visceral AT (kidney); con: control water; –RT: cDNA synthesis control without reverse transcriptase using total RNA from liver; PCR product sizes: FFAR2 = 237 bp, FFAR3 = 215 Bp, ß-actin = 178 bp.

Conclusion

The mRNA of the nutrient sensors FFAR2 and FFAR3 are expressed in different tissues in cattle. With respect to the relevance of SCFA in ruminants, the significance of this observation should be analysed in detail in the future. Our analysis was not quantitative, nevertheless, the signal we obtained for both receptors in the spleen suggests a role for these receptors in inflammatory processes in cattle.

References

Bradford, B.J., M. Oba, R.A. Ehrhardt, Y.R. Boisclair and M.S. Allen, 2006. Propionate is not an important regulator of plasma leptin concentration in dairy cattle. Domest. Anim. Endocrinol. 30: 65-75.

Lemor, A., A. Hosseini, H. Sauerwein and M. Mielenz, 2009. Transition period-related changes in the abundance of the mRNAs of adiponectin and its receptors, of visfatin and of fatty acid binding receptors in adipose tissue of high-yielding dairy cows. Domest. Anim. Endocrinol. doi:10.1016/j.domaniend.2009.01.004.

Mielenz, M., C. Seybold and H. Sauerwein, 2008. Effects of short-term infusion with propionate on the mRNA expression of a putative G-protein coupled receptor 41 (GPR41) in adipose tissue of goats. Livest. Sci. 116: 328-331.

Soliman, M., K. Kimura, M. Ahmed, D. Yamaji, Y. Matsushita, Y. Okamatsu-Ogura, K. Makondo and M. Saito, 2007. Inverse regulation of leptin mRNA expression by short- and long-chain fatty acids in cultured bovine adipocytes. Domest. Anim. Endocrinol. 33: 400-409.

Stoddart, L.A., N.J. Smith and G. Milligan, 2008. International union of pharmacology. Lxxi. Free fatty acid receptors FFA1, -2, and -3: Pharmacology and pathophysiological functions. Pharmacol. Rev. 60: 405-417.

Yonezawa, T., S. Haga, Y. Kobayashi, K. Katoh and Y. Obara, 2009. Short-chain fatty acid signaling pathways in bovine mammary epithelial cells. Regul. Pept. 153: 30-36.

A preliminary milk recording study on restrictedly suckled cows in Burkina Faso

V. Millogo[1], G.A. Ouédraogo[2], K. Svennersten-Sjaunja[1] and S. Agenäs[1]
[1]Department of Animal Nutrition and Management, Swedish University of Agricultural Sciences, P.O. Box 7024, 750 07 Uppsala, Sweden; [2]Département d'élevage, Institut du Développement Rural, Université Polytechnique de Bobo-Dioulasso, BP 1091, Bobo-Dioulasso, Burkina Faso; sigrid.agenas@huv.slu.se

Introduction

Milk recording is one of the important steps for the development of milk production since it is the basis for selective breeding for improved milk yield (Hare *et al.*, 2004). In Burkina Faso, the only available information about milk yield is based on limited survey studies (Millogo *et al.*, 2008). The conditions for milk production in Burkina Faso are challenging and it is important to find suitable animals for milk production in this environment. The local *Bos indicus* cows are well adapted to cope with the local conditions but their production capacity is generally seen as low and a common opinion is that their lactation is very short, only a few months. Therefore cross-breeding with North American or European cows attract interest from farmers and politicians.
The aim of this study was to investigate variation in milk yield between local cows in Burkina Faso, in order to gather information for designing a milk recording system for selective breeding.

Material and methods

Milk recording was carried out once per month from January 2008 to February 2009 on ten dairy farms around Bobo-Dioulasso in the west part of Burkina Faso. On each farm all lactating cows were included in the study, giving a total of 79 cows across the ten farms. Individual yield of saleable milk was determined with a 2,000 ml graduated test tube. Recording was done by the same person on all farms. During the study, twice daily milking was recommended and morning and evening milk yield was determined separately. Average milk yield and standard error of the mean was calculated for milk yield separately for the different breeds identified in the material (Minitab version 15).

Results

The results showed five different dairy breeds (Table 1) and the peak yield appeared between the second and the third month of lactation. Mean values for the different breeds and for the total material is shown in Table 1. The variation between individual cows was high, ranging from the lowest recordings of daily yield of 0.5 l in Zebu cows in lactation months 1 and 7 to Zebu cows with a daily yield of 5.5 l in the 3rd month of lactation. The highest peak yields for the other breeds were 5.8 l for the Gudali, 3.3 l for Azawak and one of the Gir cows had a peak yield of 6.6 l in the second month of lactation. The highest daily yield in a crossbred cow was 5.2 l. All 79 cows were still lactating seven months after calving and some of the cows had longer lactations but data beyond seven months were not included in this material.

Table 1. Average daily milk yield (l) recorded once per month during the first seven months of lactation. Values are shown as mean ± standard error of the mean.

	Zebu cow (N=49)	Cross-bred cows (N=8)	Gudali (N=7)	Azawak (N=3)	Gir cow (N=2)	Overall mean (N=79)
1st month	1.66±0.10	2.81±0.24	2.92±0.49	1.40±0.14	3.00±1.50	2.06±0.11
2nd month	2.08±0.11	3.05±0.21	3.42±0.56	2.01±0.13	4.15±2.45	2.47±0.12
3rd month	3.81±0.46	3.41±0.26	3.81±2.23	2.55±0.21	3.10±1.44	2.71±0.12
4th month	2.22±0.09	3.23±0.71	3.39±2.23	2.76±0.57	3.39±2.23	2.61±0.11
5th month	1.87±0.09	2.86±0.25	2.92±1.64	2.35±0.22	2.90±1.64	2.25±0.11
6th month	1.59±0.08	2.59±0.24	2.81±1.61	2.02±0.16	2.80±1.60	1.96±0.10
7th month	1.32±0.07	2.24±0.23	2.72±1.57	1.71±0.02	2.70±1.57	1.69±0.09

Discussion and conclusion

Milk recording is an important tool to detect individual characteristics of cows, regarding milk yield (Table 1). According to Hare *et al.* (2004), participation in the milk recording programme that provide data for national genetic evaluations of dairy cattle in the United States is voluntary, but the effectiveness of the evaluation system increases with the number of herds that contribute to data. The day-to-day variation in milk yield is a lot higher (18-21%) in restrictedly suckled and hand milked zebu cows than machine milked cows (Millogo *et al.*, 2009). Because of the high day-to-day variation, milk recording data should be collected more frequently in hand-milked populations than in machine milked populations, in order to achieve satisfactory quality of the data. The data from the present study showed that there is a large variation in milk yield in local cows in Burkina Faso, suggesting that it would be possible to improve milk production through selective breeding without cross-breeding with breeds that are less adapted to the conditions in Burkina Faso. However, further data collection is necessary and should include breed, milk yield, milk fat and milk protein. Furthermore, data analysis should target the issue of how frequently milk recording should be carried out on each farm in order to provide good estimates of lactation yield in individual animals.

Ackowledgements

The authors would like to acknowledge the Swedish International Development Authority (SIDA), Department for Research Cooperation (SAREC), for financial support for this research.

References

Hare, E., H.D. Norman and J.R. Wright, 2004. Duration of herd participation in dairy herd improvement milk recording in the United States. J. Dairy Sci. 87: 2743-2747.

Millogo, V., G.A. Ouédraogo, S. Agenäs and K. Svennersten-Sjaunja, 2008. Survey on dairy cattle milk production and milk quality problems in peri-urban areas in Burkina Faso. African J. Agric. Res. 3: 215-224.

Millogo, V., G.A. Ouédraogo, S. Agenäs and K. Svennersten-Sjaunja, 2009. Day-to-day variation in yield, composition and somatic cell count of saleable milk in hand-milked zebu dairy cattle. African J. Agric. Res., in press.

Effect of a myostatin mutation, nutrition and a β-adrenergic agonist (Ractopamine) on carcass and meat quality in lambs

F.E. Milton[1], P.L. Greenwood[2], M.B. McDonagh[3] and V.H. Oddy[1]
[1]Cooperative Research Centre for Sheep Industry Innovation, Australia, University of New England (UNE), Armidale, New South Wales, Australia; [2]Cooperative Research Centre for Sheep Industry Innovation, Australia, NSW DPI Beef Industry Centre of Excellence, UNE, Armidale, New South Wales, Australia; [3]Cooperative Research Centre for Sheep Industry Innovation, Australia, Department of Primary Industries Victoria, Bundoora, Victoria, Australia; fmilton2@une.edu.au

Introduction

To produce a uniformly high yielding and quality lamb meat product of high nutritional value it is important to understand the growth and developmental mechanisms of the animal. A naturally occurring myostatin mutation g+6723G->A is associated with increased muscle and reduced fat content in lambs (Kijas *et al.*, 2007). Higher nutrient intake and metabolic modifiers such as β-adrenergic agonists can also improve growth and carcass characteristics (Koohmaraie *et al.*, 1991). We undertook an experiment to investigate the effects of myostatin (MSTN) genotype and Ractopamine (RAC) at two planes of nutrition and their interactions on animal performance and meat quality of sheep. It is anticipated this will enhance our understanding of the mechanisms involved in the regulation of carcass, meat quality and nutritional attributes.

Material and methods

Eighty second-cross (Border Leicester × Merino ewes mated to White Suffolk or Poll Dorset sires) wether lambs (initial average ± SD live weight (LW) 40.1±3.8 kg) with (n=40) or without (n=40) the g+6723G->A MSTN mutation were selected for study. Lambs with heterozygous G/A genotype were matched with control (G/G) animals from the same flock of the same birth type (single or multiple siblings) and LW range.
The experiment was conducted for 49 d following a 10 d adaptation period. Lambs were stratified by initial LW and genotype, and randomly assigned to individual pens. The experiment was a 2×2×2 factorial with 2 genotypes (MSTN G/A or MSTN G/G), two dietary treatments (low or high, 1.1 × or 1.8 × estimated ME/kg requirements for maintenance), with or without RAC. A pelleted diet comprising 90% DM, 12 MJ estimated ME/kg DM and 17% CP/kg DM was offered daily. The RAC was administered in the feed at a 0.4 mg RAC/kg of initial LW. One lamb from the MSTN G/G, high nutrition and No Rac group was excluded from the analyses because it was a shy feeder. Statistical significance was determined by analysis of variance with each factor (diet, MSTN and RAC) treated as a fixed effect using the R statistical package (http://www.r-project.org/). Reported values for final LW are adjusted using initial LW as a covariate, while all other results are unadjusted.

Results

As anticipated, the final LW, average daily gain (ADG), dry matter intake (DMI), hot carcass weight (HCW) and total soft tissue depth 110 mm from midline in the region of the 12th rib (GR) were significantly (P<0.0001) greater for the lambs offered high nutrition (Table 1). There was no consistent effect of MSTN genotype on reported traits. Inclusion of RAC significantly improved ADG and meat tenderness (Warner Bratzel shear (WBS)) at 1 and 5 days of aging in lambs on high but not on low nutrition. The RAC was associated with an increase (P<0.05) in redness colour of meat and intramuscular fat (IMF%) by 15 and 23%, respectively, in the MSTN G/G, high nutrition group.

Table 1. Growth, feed intake, carcass and meat quality characteristics of wether lambs given high or low diet with or without the g+6723G->A myostatin allele, treated or not treated with the β-agonist Ractopamine.

Variable[4]	High[1]				Low[1]			
	MSTN G/A[2]		MSTN G/G* [2]		MSTN G/A		MSTN G/G*	
	No[3]	RAC[3]	No	RAC	No	RAC	No	RAC
n	10	10	9	10	10	10	10	10
Growth and intake								
Final LW (kg)	53.2^{b}	54.0^{ab}	54.7^{a}	54.3^{ab}	45.2^{c}	42.9^{de}	41.7^{e}	43.5^{d}
ADG (g/d)	286^{a}	307^{b}	273^{a}	306^{b}	59^{b}	87^{b}	65^{b}	68^{b}
DMI (g/d)	1,499.7^{a}	1,499.2^{a}	1,431.6^{b}	1,517.6^{a}	698.4^{b}	702.5^{b}	717.9^{b}	718.4^{b}
Carcass								
HCW (kg)	26.0^{a}	26.8^{a}	27.1^{a}	26.1^{a}	21.9^{b}	20.2^{bc}	19.7^{c}	20.8^{bc}
GR (mm)	13.0^{a}	13.4^{a}	13.5^{a}	13.0^{a}	11.0^{b}	10.1^{bc}	9.9^{c}	10.4^{bc}
WBS1 (kg)	9.4^{b}	5.9^{d}	11.9^{a}	7.3^{cd}	9.1^{b}	10.2^{ab}	9.5^{b}	9.6^{b}
WBS5 (kg)	7.1^{c}	4.8^{e}	9.0^{a}	5.5^{de}	6.6^{cd}	7.7^{abc}	8.7^{ab}	7.4^{bc}
Colour (red)[5]	15.9^{bc}	16.3^{bc}	15.2^{c}	17.9^{a}	16.2^{bc}	15.5^{bc}	16.6^{ab}	15.7^{bc}
IMF (%)	4.8^{a}	4.1^{abc}	3.5^{c}	4.5^{ab}	3.7^{bc}	3.6^{c}	4.4^{abc}	3.9^{bc}

[1] Nutrient intake: High = 1.8; Low = 1.1 estimated maintenance requirement.
[2] Genotype: MSTN G/G* = wild type (control); MSTN G/A = heterozygous causative myostatin mutation allele g+6723G->A.
[3] Ractopamine: No = No; RAC = Ractopamine at 0.4 mg/kg LW included in diet.
[4] n = number of lambs per treatment;. LW = live weight; ADG = average daily gain; DMI = dry matter intake; HCW = hot carcass weight; GR = total soft tissue depth 110 mm from midline in the region of the 12th rib; WBS = Warber Bratzer shear at 1 and 5 days of aging; [5]Colour = Relative redness (a*) measured using Minolta chromometer (Model CR-300); IMF = intramuscular fat.
[a,b,c,d,e] Means within rows with different superscripts are significantly different (P<0.05).

Conclusion

Genotype did not have an effect on lamb performance or carcass attributes at either of the two levels of nutrition. When RAC was included in the diet of the animals there was a reduction in the objective measure of toughness (WBS) and increase in ADG on the high plane of nutrition. This study provides a framework for further exploration of the effects of nutrition, MSTN genotype and Ractopamine treatment on body composition and energetics as measured by CT scanning. It provides an opportunity to further explore the relationship between these treatments and muscle fibre type and nutrient content of meat, and potential mechanisms of action of β-agonists as metabolic modifiers.

Acknowledgement

The authors thank Meat Livestock Australia for assistance in funding this research.

References

Kijas, J.W., R. McCulloch, J.E.H. Edwards, V.H. Oddy, S.H. Lee and J. van der Werf, 2007. Evidence for multiple alleles effecting muscling and fatness at the ovine GDF8 locus. BMC Gen. 8: 1471-2156.

Koohmaraie, M., S.D. Shackelford, N.E. Muggli-Cockett and R.T. Stone, 1991. Effect of the β-adrenergic agonist $L_{644,969}$ on muscle growth, endogenous proteinase activities and postmortem proteslysis in wether lambs. J. Anim. Sci. 69: 4823-4835.

Chromium eases coincident challenges of lactation and heat stress

M. Mirzaee[1], G.R. Ghorbani[1], M. Khorvash[1], H.R. Rahmani[1] and A. Nikkhah[1,2]
[1]Department of Animal Science, Isfahan University of Technology, 84156, Isfahan, Iran; [2]Department of Animal Sciences, College of Agriculture, Zanjan University, Zanjan, Iran; anikkha@yahoo.com

Introduction

An enormous challenge to the dairy industry has been the coincidence of metabolic pressures of early lactation with heat stress. The dramatic rises in milk yield over the last 4 decades have compromised the ability to cope with heat stress, thus depressing production and fertility (Jordan, 2003; West, 1999). Due to the negative impact of genetic selection for productivity on heat tolerance and the inadequate effect of modifying housing systems on eliminating heat stress (Hansen and Arechiga, 1999), effective nutritional strategies remain a promising path to ease metabolic and heat stresses in associative manners. Chromium (Cr) necessity for insulin and immune functions (Mowat, 1997) leads to the hypothesis that Cr supplementation during stressful times of early lactation and heat overload will control body reservs mobilisation and improve liver health. Our objective was to determine the effects of dietary use of a Cr-Met product on DMI, blood metabolites and hormones, and milk production of early lactation cows under heat stress.

Material and methods

Twelve multiparous and 3 primiparous Holstein cows were grouped based on parity and randomly assigned to 3 diets supplemented with 1) 0, 2) 0.05, or 3) 0.10 mg Cr per kg of $BW^{0.75}$. The cows were in 38±6 (mean ± SE) DIM and had 620±20 kg BW at the commencement of the study. Cows were housed in free individual boxes (4×4 m), fed to permit about 7% daily orts, and had unlimited access to fresh water. A same basal total mixed ration (TMR), formulated with the Cornell Net Carbohydrate and Protein System's program, was fed to all groups individually and twice daily at 9.00 h and 16.00 h. Chromium was provided in the form of a Cr-Met supplement (10% Cr and 90% Met, Micro-Plex, Zinpro Inc., Eden Prairie, MN), which was mixed and top dressed with 100 g of ground corn on the morning feed. Cows were cared for under the guidelines of the Iranian Council of Animal Care (1995) and allowed 30 min of outside walk daily before noon's milking. Maximum air temperature (T) and minimum relative humidity (RH) data were obtained from daily reports of Najaf Abad Meteorological Network Station (Najaf Abad, Isfahan, Iran) to calculate the T-H index or THI using the formula: [THI = 0.8 × (maximum T) + (minimum RH/100) × (maximum T – 14.4) + 46.4]; Garcia-Ispierto *et al.*, 2006). The average maximum T, RH and maximum THI during the study were 35.3 °C, 14.76% and 77.7 units, respectively, indicating medium degrees of heat stress. Cows were milked three times daily at 4:00, 12:00, and 20:00 h in the milking parlor. Blood was collected at 11:30 h via the coccygeal vein for 7 wk, and serum was analysed for insulin, glucagon, progesterone, IGF-1, and cortisol using radio immunoassays with commercial kits and an automatic gamma-counter (Biosource, Co., Italy). The serum NEFA and BHBA were measured using commercial kits with a Technician-RA 1000 Auto-analyzer (DRG, Co. Germany). Serum total proteins, albumin, triglycerides, high-density lipoproteins (HDL), very low density lipoproteins (VLDL), urea nitrogen, cholesterol and glucose were using Pars Azmoon Kits (Pars Azmoon, Co. Tehran, Iran). Rectal temperature and respiration rate were measured 4 d a wk during the 7 sampling wk. Respiration rate was counted at 15:00 h for 3 separate min and an average was calculated. Data were analysed as a mixed model for repeated measures with best fit covariance structures.

Results and discussion

Moderate Cr supplementation improved DMI and milk yields of fat, protein and total solids, and reduced serum concentrations of insulin and NEFA without affecting serum glucose, urea, BHBA, glucagon, progesterone, IGF-1, cortisol, globulin, total proteins, high-density and very low-density lipoproteins (Table 1). Cows on the moderate but not the high Cr dose had a mild BW loss, suggesting that Cr effects on nutrient availability involved ingestive, digestive and postabsorptive mechanisms. These data indicate that Cr enhanced insulin efficiency, and improved milk secretion by simultaneous rises in DMI and weight loss.

Conclusion and implications

Dietary inclusion of a Cr-Met supplement for 7 wk in early lactation heat-stressed cows to provide 0.05 mg Cr per kg of $BW^{0.75}$ improved DMI and milk fat, protein and total solid yields. The Cr supply increased respiration rate, caused a mild BW loss, and reduced serum NEFA and insulin. These data suggest a reduced need for insulin to prioritise milk secretion over peripheral nutrient oxidation or deposition. Total tract DM digestibility and blood glucose, urea, BHBA, globulin, total proteins, cholesterol, triglycerides, glucagon, cortisol, and progesterone were unaltered by Cr. Findings suggest that Cr is involved in orchestrating ingestive, digestive, and postabsorptive regulations of nutrient partitioning.

Table 1. Effects of Cr-Met supplement on milk productivity and blood metabolites and hormones.

Item	Trt or mg Cr/Kg $BW^{0.75}$			SEM	*P*-value		
	0	0.05	0.10		Trt	Wk	Trt × Wk
4%FCM, kg/d	35.0^{ab}	37.8^{a}	34.1^{b}	1.1	0.04	0.02	0.97
Milk fat,%	3.46^{ab}	3.69^{a}	3.36^{b}	0.12	0.13	0.0003	0.75
Fat yield, kg/d	1.32^{b}	1.46^{a}	1.27^{b}	0.05	0.03	0.002	0.91
Milk protein,%	2.73^{b}	2.85^{a}	2.82^{a}	0.02	0.003	0.07	0.70
Protein yield, kg/d	1.04^{b}	1.13^{a}	1.06^{ab}	0.03	0.04	0.73	0.97
Milk lactose,%	5.67^{b}	5.79^{a}	5.59^{b}	0.04	0.0002	0.75	0.99
Lactose yield, kg/d	2.16^{ab}	2.29^{a}	2.11^{b}	0.05	0.04	0.82	0.99
Milk total solids,%	12.05^{b}	12.53^{a}	11.98^{b}	0.11	0.001	0.0007	0.89
Total solids yield, kg/d	4.59^{b}	4.96^{a}	4.51^{b}	0.12	0.01	0.28	0.99
DMI, kg/d	21.8^{b}	24.2^{a}	23.7^{a}	0.53	0.004	0.33	1.00
FCM/DMI	1.62^{a}	1.59^{a}	1.45^{b}	0.05	0.03	0.0003	0.94
Respiration rate, per min	55.4^{ab}	58.6^{a}	55.1^{b}	1.2	0.10	0.008	0.99
Serum insulin, µIU/ml	8.9^{a}	8.2^{b}	8.3^{ab}	0.3	0.13	0.96	0.32
Insulin:glucagon	0.123^{a}	0.108^{b}	0.109^{b}	0.005	0.08	0.87	0.34
NEFA, µEq/l	158.9^{a}	144.6^{b}	157.3^{ab}	4.7	0.07	0.11	0.13

References

Garcia-Ispierto, I., F. Lopez-Gatius, P. Santolaria, J.L. Yaniz, C. Nogareda, M. Lopez-Bejar and F. De Rensis, 2006. Relationship between heat stress during the peri-implantation period and early fetal loss in dairy cattle. Theriogen. 65: 799-807.

Hansen P.J. and C.F. Arechiga, 1999. Strategies for managing reproduction in the heat-stressed dairy cow. J. Anim. Sci. 77: 36-50.

Jordan, E.R., 2003. Effects of heat stress on reproduction. J. Dairy Sci. 86:E104-114E.

Mowat, D.N., 1997. Organic Chromium in Animal Nutrition. Chromium Books, Guelph, ON, Canada.

West, J.W., 1999. Nutritional strategies for managing the heat-stressed dairy cows. J. Anim. Sci. 77: 21-29.

The effects of diet on ascorbic acid status of Sudanese camels

H.E. Mohamed, A. Al-Haidary and A.C. Beynen
Department of Animal Production, Faculty of Food and Agricultural Sciences, King Saud University, Riyadh, Saudi Arabia; hmohamedd@ksu.edu.sa

Introduction

A low status of ascorbic acid may enhance the risk for contracting infectious disease. In ruminants, even though they are able to synthesise vitamin C from glucose, the composition of the ration could influence the concentration of plasma ascorbic acid. Kolb *et al.* (1991) showed that values of liver ascorbic acid in bulls and oxes were the highest in December when they were kept inside and fed a stall ration. Thus, there might be an effect of diet on vitamin C metabolism in ruminants. It is possible that camels consuming their habitual diet and kept under practical conditions have a diet-dependent ascorbic acid status. The present experiment was carried out to test this possibility.

Material and methods

This study was conducted on six non-pregnant, non-lactating female Arabian camels (age, 10.3±1.7 years; mean ± SD) with a mean body weight of 450 kg (SD = 24.5). The maximum and minimum temperatures during the study were 31 and 18 °C and relative humidity was between 17 and 61%. The trial had a crossover design with three camels per treatment sequence. Each dietary treatment lasted for three weeks. All camels went through a 14-day pre-experimental period during which they were fed green alfalfa (*Medicago sativa L.*) as the sole source of nutrition. The alfalfa was obtained from a local market and fed in fresh form. Each camel received 5 kg dry matter of alfalfa twice a day at 7 a.m. and 2 p.m. The composition of the habitual diet, based on field observations, was simulated by a mixture consisting of fresh *Acacia mellifera*, *Aristidia funiculata* (hummra) and *Blepharis persica* in a 2:1:1 ratio on fed basis. Each camel received 5 kg dry matter of the mixture two times a day. Water was freely available. During the experiment, the camels were either fed alfalfa or the habitual diet. The macronutrient composition of the feedstuffs was analysed according to the Weende methods. Plasma and leukocyte vitamin C levels were determined according to Behrens and Madere, 1987. The Student paired t-test was used to identify a diet effect. The level of significance was pre-set at $P<0.05$.

Results

Table 1 shows the analysed composition of the alfalfa and the calculated composition of the habitual diet. When the camels consumed the simulated habitual diet, they ingested somewhat more crude fiber and less crude protein than when they ate the alfalfa diet.

Table 1. The micronutrient of the camel's diet.

Parameter	Alfalfa ration	Habitual diet
Dry matter, g/kg diet	956	872
Crude protein, g/kg dm	173	114
Crude fiber, g/kg dm	287	328
Crude fat, g/kg dm	29	29
Ash, g/kg dm	123	102
Nitrogen-free extract, g/kg dm	388	427

Table 2 shows the plasma and leukocyte ascorbic acid concentrations as related to diet. Leukocyte ascorbic acid is the most sensitive index of ascorbic acid status in animals. Diet-related differences in this vitamin single out the possible risk of deficiency in camels on a natural grazing system in Sudan

Table 2. Ascorbic acid concentrations (Means±SD) in plasma and leukocytes of camels (n = 6) fed the two rations.

Ascorbic acid	Alfalfa ration	Habitual diet
Plasma (µg/ml)	39±1.11	53.90±0.97*
Leucocytes (µg/ml)	46.17±3.85	39.85±3.97*

Means on the same row having different superscripts are significantly different at $P<0.05$.

Conclusion

The composition of the diet might affect vitamin C status in camels; and the introduction of green fodder to field dietary regimes may be of help to overcome the possible risk of deficient status. Restricted feeding of lambs lowers ascorbic acid (glucogenic substrate) (Kolb *et al.*, 1993).

References

Behrens, W.A. and R. Madere, 1987. A highly sensitive high-performance liquid chromatography method for the estimation of ascorbic acid and dehydroascorbic acid in tissues, biological fluids and foods. Anal. Biochem. 165: 102-107.

Kolb, E., H. Dittrich, G. Dobeleit, R. Schmalfuss, P. Siebert, E. Stauber and M. Wahren, 1991. Content of beta-carotene, vitamin E and ascorbic acid in blood plasma of female calves, cattle, bulls, castrates and ox throughout the course of the year. Berl. Münch. Tierärztl. Wochenschr. 104: 387-391.

Kolb E, E. Boldt, M. Eltohami, H. Kirbach, M. Ulbrich and M. Wahren, 1993. The content of ascorbic acid in plasma and in 14 tissues of lambs of different ages during normal feeding and feeding restricted from time to time during the suckling period. Berl. Munch. Tierarztl. Wochenschr. 106: 341-347.

Subcutaneous or oral administration of liposome-encapsulated vasoactive intestinal peptide increases dietary intake in small ruminants

G.K. Murdoch[1], R. Soofi-Siawash[2], E. Okine[2], L. Goonewardene[2] and R.J. Christopherson[2]
[1]University of Idaho, Animal and Veterinary Science, 83844-2330, Moscow, ID, USA; [2]University of Alberta, Animal Science, T6G 2A1, Edmonton, AB, Canada; gmurdoch@uidaho.edu

Introduction and background

Enhancement of dietary intake is important during weaning, feedlot introduction, and at the start of lactation. Increasing dietary intake without decreasing digestibility may improve overall growth rate in ruminants. Vasoactive intestinal peptide (VIP) is a neuropeptide associated with parasympathetic nerves of the myenteric plexus which are abundant in the reticulo-omasal orifice of ruminants. However, intrinsic VIP has a short one minute half-life as it is degraded by endogenous peptidases. Electromyographic studies (Okine and Mathison, 1996) found that intra-arterial VIP increases the duration of quiescence of the reticulo-omasal orifice mediating rates of digesta passage from the forestomach. Our lab reported that VIP increased feed intake by 14-22% in sheep (Li *et al.*, 2000a, b). The pharmaceutical industry considers liposome use to be an accepted strategy for protection and delivery of bioactive compounds. Administration of liposome encapsulated VIP was used to examine the effects on feed intake in sheep, initially using a subcutaneous route, as well as the more practical oral delivery which may promote a localised gut rather than a systemic effect.

Material and methods

Liposome encapsulation of VIP (LipoVIP) was achieved using distearoyl-phosphatidylcholine, egg yolk phosphatidylglycerol, cholesterol and PEG linked distearoyl-phosphatidylethanolamine in ethanol solution (5:1:3.5:0.5, 17mmol phospholipids). The solution was evaporated at 45 °C in a rotovap, rehydrated and extruded (100µm Avestin stacked, extruder). VIP was added and was repetitively freeze-thawed and then lyophilized until re-suspension.

Experiment 1: sixteen sheep (4 per treatment), were given daily subcutaneous injection of empty liposomes (control), 2 nmol/kg $BW^{0.75}$ freeVIP, 1 nmol/kg $BW^{0.75}$ LipoVIP or 2 nmol/kg $BW^{0.75}$ lipoVIP. After 3 weeks of treatment the sheep were re-randomised and the experiment was repeated.

Experiment 2: eighteen sheep housed in individual pens, fed a chopped hay diet *ad libitum* and allocated to three treatments (6 sheep per treatment). Oral supplements; Control (empty liposomal lipids), LipoVIP (1 nmol/kg $BW^{0.75}$/d) or LipoVIP (2 nmol/kg $BW^{0.75}$/d), were given in 5 ml of Canola Oil added to 100 g of grain once daily for 14 d, one hour before feeding chopped hay.

Experiment 3: nine sheep fed a chopped hay diet *ad libitum*, received LipoVIP orally (2 nmol LipoVIP/kg $BW^{0.75}$/d) as compared to 9 sheep that received orally an equivalent empty Liposome (control) for 14 days. Oral treatments were given as described in experiment 2. In each experiment, data were analysed by ANOVA using a mixed model.

Results

Subcutaneous VIP administration in each of the treatments increased feed intake roughly 15% ($P<0.005$) as compared to the controls. Over the six days of treatment, mean feed intakes of 0.079, 0.092, 0.093 and 0.089 kg/kg$BW^{0.75}$ per day (SE=0.003) were observed for controls, 2nmol free VIP, 1 nmol LipoVIP and 2 nmol LipoVIP, respectively. Furthermore, subcutaneous 1nmol LipoVIP administration was as effective at increasing feed intake as 2nmol free VIP. Plasma concentrations of VIP ranged from 80 to 160 picoM during the period between 15 and 75 min following injection of 2 nmol free or liposomal VIP. Elevated concentrations (4 to 8-fold higher than basal) persisted longer in animals treated subcutaneously with liposomal VIP than with free VIP (data not shown).

Table 1. Effect of oral VIP administration on daily intake, cumulative intake and digestibility.

	Exp.	Treatment			SE	Effect
		Control	1 nmol LipoVIP	2 nmol LipoVIP		P-value
DM intake kg/kg $BW^{0.75}$	two	0.093	0.095	0.108	0.007	0.06
	three	0.102	-	0.114	0.006	0.15
	combined	0.098	0.095	0.111	0.004	0.001
DM digestibility %	two	59.2	63.2	66.1	3.22	NS
	three	58.6	-	57.8	1.95	NS
	combined	58.9	63.2	61.9	2.59	NS

In experiments 2 and 3 (oral administration) sheep readily consumed the concentrate and the voluntary intake of chopped hay increased during VIP treatment with a similar magnitude as in experiment 1 (subcutaneous administration). This effect was highly significant when the results of the two last experiments were combined for analysis. There was no significant difference in digestibility between the VIP treated (61.9%) and control sheep (58.6%), emphasising that the increase in feed intake was not associated with a reduced dry matter digestibility. Mean plasma VIP concentrations (pico *M*) collected 1, 3 and 5 hours after oral administration were 9.1±3.2 for controls, 13.4±2.5 for LipoVIP, 1 nmol/kg $BW^{0.75}$ and 10.9±2.9 for LipoVIP, 2 nmol/kg $BW^{0.75}$. These mean values were close to the detection limit of the RIA, and were not significantly different.

Conclusion

The oral administration of 2 nmol liposome encapsulated VIP increased the mean daily feed intake without decreasing the measured digestibility indicating some liposomal protection of VIP from degradation in the gut. However, 1 nmol/kg $BW^{0.75}$ of liposomal-VIP was ineffective at increasing the intake above that observed for the lipid control animals. The higher dose may be a requisite to the observed biological activity because of the potential for breakdown and loss of VIP function within the digestive milieu. Lack of increases in plasma VIP after oral administration raises the possibility that enough encapsulated VIP could reach receptor sites in the digestive tissue to influence intake, but not alter circulating levels. Further studies are needed to clarify this point.

Acknowledgement

We thank the University of Alberta, Metabolic Unit staff for animal care and maintenance and the Canadian NSERC Strategic grant (STPGP 258015-02) for financial support.

References

Li, B.T., E.K. Okine and R.J. Christopherson, 2000a. Effect of vasoactive intestinal polypeptide and environmental temperature on forestomach electromyographic activity and gastrointestinal contraction in sheep. Can. J. Anim. Sci. 80: 343-353.

Li, B.T., E.K. Okine and R.J. Christopherson, 2000b. Effect of vasoactive intestinal polypeptide infusion on feed intake, digesta flow and nutrient digestion in sheep with access to feed for limited time. Can. J. Anim. Sci. 80: 105-112.

Okine, E.K. and G.W. Mathison, 1996. Effects of vasoactive intestinal polypeptide on electromyographic activity of the reticulo-omasal orifice in sheep at two levels of intake J. Anim. Sci. 74: 1681-1688.

Hormonal regulation of phosphate homeostasis in goats during transition to rumination

A. Muscher[1], E. Pfeffer[2], G. Breves[1] and K. Huber[1]
[1]Department of Physiology, School of Veterinary Medicine Hannover, Germany; [2]Department of Animal Nutrition, University of Bonn, Germany; alexandra.muscher@tiho-hannover.de

Introduction

Regulation of phosphorus (P) homeostasis in small ruminants is different compared to non-ruminant species. Adaptive responses of hormones like parathyroid hormone (PTH) and calcitriol to feeding variations in dietary P supply are lacking in ruminants (Wan Zahari *et al.*, 1994). Although adaptive differences in plasma PTH concentrations were not observed, the renal Na^+-dependent phosphate (P_i) transport, known as PTH-target in monogastric animals, was decreased in goats fed a high P diet (Huber *et al.*, 2007). Therefore, the hormonal regulation of P homeostasis must be different between ruminants and non-ruminant species. Since ruminants change from pre-ruminant to ruminant status during their growth phase, P homeostasis and its hormonal regulation has to be adapted to this transition period. It is hypothesised that the adaptation of P_i transporters is rather modulated by changes in PTH and calcitriol receptor expression levels than by hormonal concentrations. The aim of the present study was to determine the expression of PTH receptor (PTHR) and calcitriol receptor (VDR) in the jejunum and the kidneys in goats fed a high P diet.

Material and methods

White Saanen goats were divided into two experimental groups at three days of age. Animals of the control group (n = 8) were fed 4 g P/kg dry matter (DM). In the high P group (n = 6), goats received 8 g·P/kg DM in their diet. The P_i concentration was achieved by adding NaH_2PO_4 to the milk. When the goats reached a body weight of 9 kg (about 9 weeks of age), they were exclusively fed a mixture of pelleted concentrates and straw in a 4:1 ratio. The goats were kept on this diet for three weeks. Based on this feeding regime, the goats changed from the non-ruminant to the ruminant stage at the end of the experiment. At 12 weeks of age, the animals were slaughtered. Tissue samples of mid jejunum and kidney cortices were taken for preparation of brush border membrane vesicles (BBMV) and basolateral membranes (BLM). Concentration-dependent uptakes of ^{32}P into renal and jejunal BBMV were measured by rapid filtration technique. Transport capacity (V_{max}) and transporter affinity (K_m) were calculated by nonlinear curve fitting based on the Michaelis-Menten equation. In western blot analyses, the expression of PTHR, VDR and the intestinal Na^+-dependent P_i transporter (NaPi IIb) was analysed in isolated membrane vesicles. The quantification of mRNA of the described renal and jejunal receptors and transporters was performed by northern blot analyses. Plasma PTH and calcitriol concentrations were determined by commercial assays (Immundiagnostik, Bensheim, Germany). Concentrations of P_i and total Ca were determined colorimetrically by standard spectrometric techniques (Kruse-Jarres, 1979; Sarkar and Chauhan, 1967). The Student t-test was performed for data analysis and $P<0.05$ was considered statistically significant.

Results

Plasma P_i concentration of the high P fed goats were 44.5% higher than in the control animals ($P<0.01$) (Table 1) while total Ca levels were not affected ($P=0.15$). Plasma PTH concentrations were significantly elevated ($P<0.01$) whereas calcitriol levels were not changed ($P=0.97$). The expression of jejunal and renal VDR were neither influenced on protein ($P=0.51$; $P=0.41$) nor on mRNA level (renal VDR not quantifiable; *P*=0.11). The renal PTHR however was significantly decreased on both mRNA (*P*<0.01) and protein level (*P*<0.01) under a high P diet. The NaPi IIb expression in the mid jejunum was not influenced on mRNA (*P*=0.45) and protein level (*P*=0.37) either. Concomitantly, the kinetic parameters of the Na^+-dependent P_i transport into jejunal BBMV (V_{max}, *P*=0.31 and K_m, *P*=0.50) were not affected either. The amount of renal Na^+-dependent P_i transporter (NaPi IIa) was significantly reduced on mRNA level (*P*<0.01). On protein level, it could not be assessed due to the lack of any commercial antibody. The V_{max} of the Na^+-dependent P_i transport into renal BBMV was significantly decreased (*P*<0.001) while K_m was not changed (*P*=0.15).

Table 1. Concentrations of inorganic phosphate (P_i), total calcium (Ca), midregional parathyroid hormone (PTH) and calcitriol in plasma of goats fed a high phosphorus (P) diet.

Group	Inorganic P_i, mmol/l	Total Ca, mmol/l	PTH, pmol/l	Calcitriol, pmol/l
Control	2.54±0.56	3.1±0.39	160±47.5	144±56
High P	3.67±0.72**	2.67±0.66	327±104**	146±60

Values are given as means ±SD; 8 animals in the control group and 6 in the high phosphorus group; Significance of differences was *P*<0.01 (**).

Conclusion

The P homeostasis is regulated by variation of hormonal receptor expression in goats during transition to rumination. Therefore, it can be concluded that the adaptive response of renal P_i reabsorption in goats changing from non-ruminant to ruminant stage was comparable to that of non-ruminant species whereas the modulation of the intestinal P_i absorption was similar to the regulation mechanisms in adult ruminants. In pre-ruminant goats, the kidneys are the main excretory pathway for a high dietary P load and the gastrointestinal tract is comparable to that of monogastric animals at that time point. It is assumed that PTH-mediated regulation of renal P_i excretion becomes less important in mature ruminants because high amounts of P_i will be secreted in the saliva, transported into the rumen and eliminated with the faeces.

Acknowledgement

The authors thank Kathrin Hansen, Karin Hustedt and Julia Hattendorf for excellent technical assistance.

References

Huber, K., U. Roesler, A. Holthausen, E. Pfeffer and G. Breves, 2007. Influence of dietary calcium and phosphorus supply on epithelial phosphate transport in preruminant goats. J. Comp. Physiol. [B] 177: 193-203.

Kruse-Jarres, J.D., 1979. Klinische Chemie Klinische Chemie, vol II. Spezielle klinisch-chemische Analytik. Fischer, Stuttgart, Germany.

Sarkar, B.C. and U.P. Chauhan, 1967. A new method for determining micro quantities of calcium in biological materials. Anal. Biochem. 20: 155-166.

Wan Zahari, M., D. Scott, N. Loveridge, W. Buchan and J. Milne, 1994. The effect of high phosphorus intake on calcium and phosphorus retention and bone turnover in growing lambs. Exp. Physiol. 79: 175-181.

Effects of dry matter and energy intake on the concentrations of blood metabolites in dairy cows receiving fresh-cut grass

F.Y. Obese[1,2], *K.L. Macmillan*[2] *and A.R. Egan*[2]
[1]*CSIR- Animal Research Institute, P.O. Box AH 20, Achimota, Ghana;* [2]*School of Agriculture and Food Systems, the University of Melbourne, 3031, Victoria, Australia; fyobese@yahoo.com*

Introduction

Insulin-like growth factor-I (IGF-I), glucose, beta-hydroxybutyrate (BHBA), and non-esterified fatty acids (NEFA) are among the blood constituents which reflect various aspects of metabolic status (Spicer *et al.*, 1990; Mashek *et al.*, 2001). Alterations in their concentrations may be indicative of energy availability or the value of substrates being used. A strong positive relationship has been reported between serum IGF-I and plasma urea nitrogen, and a strong negative relationship between IGF-I and free fatty acids in Holstein-Friesian cows during lactation in cows fed total mixed rations (TMR) (Zulu *et al.*, 2002). There is however, limited data available on the concentrations and relationships between IGF-I and other blood metabolites in Holstein-Friesian cows in the seasonal pasture-based dairy systems in Australia. The objectives of this study were the following: (1) to investigate the effects of dry matter intake (DMI) and metabolisable energy (ME) density on the concentrations of IGF-I, glucose and NEFA (2) to assess the relationships between IGF-I and these blood metabolites in pasture-fed Holstein-Friesian cows.

Material and methods

Thirty-two Holstein-Friesian cows 4 to 5 wk *post partum* were randomly assigned to four dietary groups. Cows in the four groups received daily rations formulated to provide respectively, high (H) or low (L) DMI and ME density (LL: 16.6 kg of DMI and 174 MJ of ME; HL: 17.3 kg of DMI and 181 MJ of ME; LH: 15.4 kg of DMI and 183.1 MJ of ME; HH: 17.9 kg of DMI and 213.3 MJ of ME). The animals were fed on diets comprising fresh cut-pasture, hay and pelleted cereal grain to achieve the two different levels of DMI and ME density. Water was available to the cows *ad libitum*. The study period was for 5 weeks. Plasma from blood samples were collected before morning feeding from each of the 32 cows by coccygeal vessel puncture once weekly in the 5-wk trial. They were assayed for glucose, NEFA and BHBA using a Cobas–Mira auto-analyser (Roche Diagnostica, Basel, Switzerland). IGF-I concentrations were analysed using the DSL-10-2800 Active™ Non-extraction IGF-I ELISA Commercial kit (Diagnostic Systems Laboratories, Inc, Webster, TX, USA). The effects of DM and ME on IGF-I, glucose, NEFA, BHBA concentrations in plasma and interaction between DM and ME at d 14 and d 35 were analysed using GLM procedure in SPSS (2002). Data for d 0, the initial readings obtained before cows had access to their various diets were used as covariates. The model used was the following:
$Y = DM + ME + DM \times ME + \text{Covariate at d } 0$
(Where Y= the variable at day) d 14 and at d 35. The interaction term was included in the model when $P<0.10$. Associations between IGF-I and the other blood metabolites at d14 and d 35 were assessed using linear regression analysis in SPSS (2002). Overall differences between treatment means were declared significant at $P<0.05$.

Results

Cows on high-ME diets had greater plasma concentrations of IGF-I than those on low-ME diets at d 14 (83.7 vs. 45.6±7.7 ng/ml; $P<0.001$) and d 35 (79.3±3.7 vs. 41.4±3.7 ng/ml; $P<0.001$). The level of DMI did not affect ($P>0.05$) IGF-I concentrations at the period. Neither DMI nor ME affected glucose concentrations at d 14, but concentrations were greater in cows on high-ME diets than those on low-ME diets at d 35 (4.14±0.06 vs. 3.72±0.06 mmol/l; $P<0.001$). The NEFA concentrations at d 14 or d 35 were not influenced by DMI or ME density. The NEFA concentrations were low averaging 0.59 and 0.22 mmol/l at d 14 and d 35 respectively. High-ME density significantly reduced mean BHBA concentrations at d 14 (0.50±0.08 vs. 0.68±0.08; $P=0.028$) and d 35 (0.33±0.04 vs. 0.64±0.04 mmol/l; $P<0.001$), but DMI did not affect BHBA concentrations during this period. Plasma IGF-I had a higher R^2 estimate in its model than any other metabolite at d 14 ($R^2=0.528$) and d 35 ($R^2=0.649$), except for BHBA at d 35 ($R^2=0.708$). Plasma IGF-I concentrations were positively correlated with glucose at d 14 (r=0.422; $P<0.05$) and d 35 (r=0.587; $P<0.01$). There were no significant correlations ($P>0.05$) between blood IGF-I and NEFA at any of the periods considered. There was, however, a negative correlation of circulating IGF-I with BHBA concentrations at d 14 (r = -0.411, $P<0.05$) and d 35 (r = -0.588; $P<0.01$).

Conclusion

Nutritional status influenced blood metabolite concentrations with ME density, being more effective than DM intake in affecting changes in plasma IGF-I, glucose and BHBA concentrations, but not NEFA concentrations. The higher R^2 model estimates of plasma IGF-I suggests its higher sensitivity to dietary effects than glucose, BHBA, NEFA or urea. The half life of IGF-I in plasma is prolonged (about 12 h) as a result of its binding with binding proteins. This makes the concentrations of IGF-I stable throughout the day compared to glucose, BHBA, and NEFA concentrations which changes during the day in response to meal and milking. IGF-I may therefore be a sensitive monitor of dietary effects and energy balance during early lactation in pasture-fed dairy cows.

Acknowledgement

The authors thank the University of Melbourne and Primegro Ltd, Australia for financial support.

References

Mashek, D.G., K.L. Ingvartsen, J.B. Andersen, M. Vestergaard and T. Larsen, 2001. Effects of a four-day hyperinsulinemic-euglycemic clamp in early and mid-lactation dairy cows on plasma concentrations of metabolites, hormones, and binding proteins. Dom. Anim. Endocrinol. 21: 169-185.

Spicer, L.J., W.B. Tucker and G.D. Adams, 1990. Insulin-like growth factor-1 in dairy cows: Relationships among energy balance, body condition, ovarian activity, and estrous behaviour. J. Dairy Sci. 73: 929-937.

Zulu, V.C., Y. Sawamukai, K. Nakada, K. Kida and M. Moriyoshi, 2002. Relationship among insulin-like growth factor-I, blood metabolites and *post partum* ovarian function in dairy cows. J. Vet. Med. Sci. 64: 879-885.

Gene expression of adiponectin and its receptors in bovine mammary gland and mammary epithelial cells

Y. Ohtani, T. Yonezawa, A. Hagino and K. Katoh
Laboratory of Animal Physiology, Graduate School of Agricultural Science, Tohoku University, Sendai 981-8555, Japan; yo-ta2@bios.tohoku.ac.jp

Introduction

The mammary gland, which consists of parenchyma and stroma, is drastically changed by pregnancy and lactation. Adipose cells, which exist in stroma, secrete various cytokines (adipocytokine) and regulate the mammary gland and *vice versa*. One of the adipocytokines, adiponectin, is involved in the regulation of energy metabolism, cell proliferation and differentiation in various cells. In particular, its expression is markedly increased at the differentiation of the adipocytes (Scherer *et al.*, 1995). Two subtypes of adiponectin receptors (AdipoR1 and AdipoR2) mediate these functions. In cattle, adiponectin mRNA is expressed in adipose tissues and differentiated adipocytes derived from bovine stromal vascular cells (Feuermann *et al.*, 2006; Mohamed *et al.*, 2006). However, the detailed functions of adiponectin in lactation remain unclear. The aim of the present study, therefore, was to demonstrate the mRNA expression of adiponectin and its receptors during lactational stages, and the effects of lactogenic hormones and growth hormone (GH) on mRNA expression of the receptors in cultured bovine mammary epithelial cells (BMEC).

Material and methods

Four non-pregnant (NP) Holstein heifers and 9 dairy cows (peak lactation (PL: n=3, two months after parturition), late lactation (LL: n=3, eight months after parturition) and dry off (DO: n=3, three years after parturition)) were used. The mammary tissue samples were immediately frozen in liquid nitrogen and stored at −80 °C until analysis. BMEC were established from the mammary gland of one 102-day pregnant Holstein heifer and cloned using the limiting dilution method. BMEC were seeded in 6-well plates and grown in DMEM supplemented with 10% FCS under 5% CO_2 at 37 °C until confluence. After confluence, the cells were starved in serum-free DMEM supplemented with 0.1% BSA. After 24 h serum-deprivation, lactogenic hormones (dexamethasone (Dex), insulin (Ins) and prolactin (Prl)) or GH at various concentrations were added to the medium and incubated for a further 24 h. Total RNA was isolated from mammary tissues and BMEC using TRIzol reagent (Invitrogen). The RNA was treated with RNase-free DNase. Two micrograms of total RNA was reverse-transcribed using a Prime Script RT reagent kit (Takara-bio). Real-time PCR was performed with the SYBR Premix Ex Taq (Takara-Bio) using a DNA engine Opticon 2 real-time PCR detection system (MJ Research). Data were normalised with mRNA of GAPDH or beta-actin, and were calculated by the $2^{-\Delta\Delta Ct}$ method. Data are presented as the means ± S.E.M. Statistical significance was determined using a one-way ANOVA followed by the Duncan multiple range test.

Results

The expression of mRNA for adiponectin and its receptors was detected in all the bovine mammary gland tissues. The expression of adiponectin mRNA at PL and LL was significantly lower than that at NP and DO (Figure 1A). On the contrary, the AdipoR1 mRNA expression at PL was significantly higher than that at DO (Figure 1B), while there was no significant difference in the AdipoR2 mRNA expression during the lactational stages (Figure 1C).

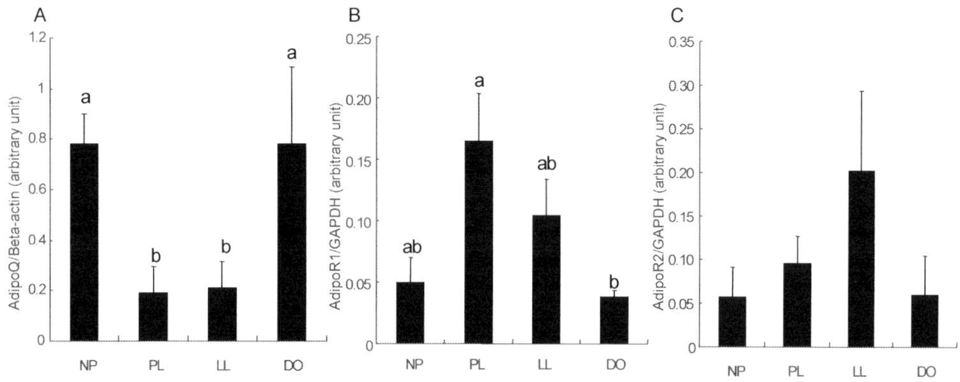

Figure 1. Expression of mRNA for adiponectin (A) and its receptors (AdipoR1 (B) and AdipoR2 (C)) at various stages of Holstein mammary gland tissues. The values labelled with different letters are significantly different ($P<0.05$*).*

In BMEC, adiponectin mRNA was not detected, although AdipoR1 and AdipoR2 mRNA were detected. Treatment with dexamethasone, insulin, prolactin or growth hormone for 24 h increased AdipoR2 mRNA expression, whereas the AdipoR1 expression was not affected by the treatment.

Conclusion

Adiponectin mRNA was expressed in the mammary gland tissues but not in BMEC. The expression of adiponectin mRNA was lowered during lactation in the mammary gland, whereas that of AdipoR1 mRNA was increased. However, the AdipoR2, but not AdipoR1, expression was increased by treatment of lactogenic hormones or GH in BMEC. These data suggest that the mammary epithelial cells express adiponectin receptors, but not adiponectin mRNA. In addition, the expression of AdipoR1 and AdipoR2 may be regulated differently by lactogenic hormones and GH.

References

Feuermann, Y., S.J. Mabjeesh, L. Niv-Spector, D. Levin and A. Shamay, 2006. Prolactin affects leptin action in the bovine mammary gland via the mammary fat pad. J. Endocrinol. 191: 407-413.

Mohamed, S, K. Kazuhiro, A. Mohamed, Y. Daisuke, M. Yukiko, O. Yuko, M. Kennedy and S. Masayuki, 2006. Inverse regulation of leptin mRNA expression by short- and long-chain fatty acids in cultured bovine adipocytes. Domestic Anim. Endocrinol. 33: 400-409.

Scherer, P.E., S. Williams, M. Fogliano, G. Baldini and H.F. Lodish, 1995. A Novel Serum Protein Similar to C1q, Produced Exclusively in Adipocytes. J. Biol. Chem. 270: 26746-26749.

Diet supplementation with different levels of unprotected conjugated linoleic acid (CLA) progressively decreases milk fat content and yield in dairy ewes

D.E. Oliveira[1], M.P. Soares[1], M.A.S. Gama[2], R. Dresch[1], M. Baldin[1] and L.L. Martelo[1]
[1]Santa Catarina State University (UDESC), Chapecó, Santa Catarina, Brazil; [2]National Dairy Cattle Research Centre, Embrapa, Juiz de Fora, Minas Gerais, Brazil; deolivei2@terra.com.br

Introduction

Conjugated linoleic acid (CLA) can be fed to dairy cows in order to increase CLA concentration in milk fat and to decrease energy requirements during the transition period. In dairy ewes, a ruminally protected CLA supplement has shown to increase milk yield and milk CLA content while reducing milk fat content and yield (Lock *et al.*, 2006). Recently, Oliveira *et al.* (2008) observed some adverse effects on milk yield and composition of dairy ewes by feeding a high dose of an unprotected CLA supplement. However, as far as we know no dose-response study using unprotected CLA has been conducted in dairy ewes. This study was designed to evaluate the effects of CLA levels (fed as an unprotected supplement) on milk yield and composition in dairy ewes.

Material and methods

Twenty-nine Lacaune lactating primiparous ewes (30 to 50 d in milk) were assigned to the treatments according to body weight (BW), lambing date and milk production and randomly allotted to the following dietary treatments: (a) Control: 30 g of calcium salts of long chain fatty acids from soybean oil (BW = 50.1±1.9 kg, n=7); (b) CLA10: 20 g of calcium salts of long chain fatty acids from soybean oil plus 10 g of CLA supplement (BW = 50.1±4.8 kg, n=7); (c) CLA20: 10 g of calcium salts of long chain fatty acids from soybean oil plus 20 g of CLA supplement (BW = 50.6±6.6 kg, n=7); (d) CLA30: 30 g of CLA supplement (BW = 54.7±7.0 kg, n=8). The fat supplements were mixed into the concentrate (1.0 kg/d) and fed individually in two equal meals after morning and afternoon milkings. The CLA supplement had about 30% of *cis*-9,*trans*-11 and 30% of *trans*-10,*cis*-12 as methyl esters. All ewes grazed paddocks of a tropical pasture (*Panicum maximum* Jacq. cv. Aruana) as the only source of forage. The experimental period lasted 28 days: 7d for adaptation, 14 d for milk sampling and 7 days for 'wash-out', where all animals received the Control diet. Milk production was recorded daily and milk samples were taken every two days throughout the study. Milk samples were analysed for contents of fat, protein, lactose and somatic cell count (SCC). One ewe from Control, CLA10, CLA30 and two ewes from CLA20 were excluded from the analysis due to health problems (hoof injuries). Data were analysed as repeated measures design using the PROC MIXED procedure of SAS® (2000). The statistical model included treatment, day and treatment-day interaction as sources of variation. Effect of interaction was removed from the model when not significant. Ewe within treatment was considered as a random effect. Differences between treatments were declared significant at $P<0.05$.

Results and discussion

Least squares means for milk yield, composition and SCC in response to dietary treatments are presented in Table 1. Milk yield, protein and lactose content and protein and lactose secretion were unaffected by treatments. Compared to Control, the milk fat content was decreased by 7.1, 16.1 and 28.6% and milk fat yield by 9.5, 18.5 and 27.5% in response to T10, T20 and T30 treatments, respectively. Milk SCC was unchanged by treatments. The highest dose of CLA supplement (30 g/d) in this study was the same used by Oliveira *et al.* (2008) in which, similar milk fat depression was observed. The inhibitory effect of unprotected CLA on milk fat synthesis even at the T20 dose might be due to the hydrolysis of CLA methyl esters releasing methanol in the rumen which could be toxic for the microorganisms thus decreasing the biohydrogenation.

Table 1. Milk yield and composition in dairy ewes fed 0, 10, 20 or 30 g of an unprotected CLA supplement for 21 days.

Variable	Treatments[1]				SE	*P*-value
	Control	T10	T20	T30		
Milk yield, kg/d	1.6	1.6	1.6	1.6	0.07	0.93
Milk fat,%	5.6[a]	5.2[a]	4.7[b]	4.0[c]	0.16	<0.001
Milk protein,%	4.5	4.6	4.7	4.5	0.06	0.33
Milk lactose,%	4.8	4.7	4.6	4.7	0.05	0.21
Fat yield, g/d	89.1[a]	80.6[ab]	72.6[bc]	64.6[c]	4.7	<0.001
Protein yield, g/d	72.1	70.3	72.1	72.6	4.1	0.97
Linear score for SCC	2.7	3.0	3.3	4.0	0.5	0.31

[1] Control = 30 g of calcium salts of long chain fatty acid from soybean oil; T10 = 20 g of calcium salts of long chain fatty acid from soybean oil + 10 g of unprotected CLA supplement; T20 = 10 g of calcium salts of long chain fatty acid from soybean oil + 20 g of unprotected CLA supplement; T30 = 30 g of unprotected CLA.
[a,b,c] Means within rows with the same superscript letters are not significantly different ($P>0.05$).

Conclusion

It can be concluded that milk fat content and yield in dairy ewes were decreased in a dose-dependent manner in response to increasing levels of an unprotected CLA supplement.

References

Lock, A.L., B.M. Teles, J.W. Perfield II, D.E. Bauman and L.A. Sinclair, 2006. A conjugated linoleic supplement containing *trans*-10,*cis*-12 reduces milk fat synthesis in lactating sheep. J. Dairy Sci. 89: 1525-1532.

Oliveira, D.E., M.P. Soares, F.J. Bianchett, R. Fornazier, M.R. Fachinello, M. Girardi, D. Fernandes, D. Soster, M.A.S. Gama, M.G.C.D. Peixoto, S.O. Juchem and L.O. Tedeschi, 2008. Unprotected linoleic conjugated acid (CLA) negatively affects milk production and secretion of milk components in dairy ewes. J. Dairy Sci. 91, E-Suppl. 1: 438.

SAS Institute, 2000. SAS/STAT: guide for personal computers. SAS Institute, Cary, USA.

Blood parameters of sheep fed different levels of detoxificated castor bean waste

L.G.R. Pereira[1], D.R. Menezes[2], R.G. Costa[3], G.G.L. Araújo[1] and M.G. Malheiro[4]
[1]Embrapa Tropical Semi-Arid, Rod. BR 428, km 152 - Zona Rural, Cx. Postal 23, 56302-970, Petrolina, PE, Brazil; [2]Universidade Federal do Vale do São Francisco, Cx. Postal 252, 56304-410, Petrolina, PE, Brazil; [3]Department of Agriculture and Animal Husbandry, Federal University of Paraiba, Bananeiras, Paraíba, Brazil; [4]Universidade do Estado da Bahia, Av. Edgard Chastinet, s/n, 48900-000, Juazeiro, BA, Brazil; luiz.gustavo@cpatsa.embrapa.br

Introduction

Castor beans (*Ricinus communis*) represent a potential candidate for biodiesel production. The cultivation of castor beans has been increasing in some countries, and castor bean waste (CBW) is a co-product generated in oil processing. This extremely alkaline waste is toxic and allergenic and, as such, poses a significant environmental problem. However, the CBW is a potential protein source for ruminant feeding. Ricin present in castor bean waste is an anti-nutritional factor of high toxicity that can be neutralised by alkali substances such as lime (CaO). Studies of detoxification have been performed in laboratory conditions and the most promising approaches need validation in animal assays for large scale application on farms or in industries. Soybean meal is the standard protein concentrate used as reference for comparison with other protein sources. Changes in diet could change the blood metabolic profile, allowing evaluation of nutrient imbalance, especially energy protein ration or metabolic disorders. The present study was aimed at evaluating the influence of partial substitution of soybean meal by detoxificated castor bean waste on blood parameters related to nutritional imbalance, toxicity and metabolic disorders in sheep.

Material and methods

Thirty-two male sheep, castrated, mixed breed, with average weight of 21.7±2.6 kg were used in a randomised design with eight replications. The treatments were standard diet with soybean meal as protein source and three levels of replacement of soybean meal (SBM) by detoxified castor bean waste (CBW) (15, 30 and 45% on dry matter basis). The forage:concentrate ration was 40:60. The components of the diets were Buffel grass hay, corn, soybean meal, urea and detoxificated CBW. The urea was added to diets in order to make them isonitrogenous, and their proportions were 0, 0.3, 0.5 and 0.7% for 0, 15, 30 and 45% of the soybean meal was substituted by detoxified CBW. The CBW detoxification was made according to the recommendations of Anandan *et al.* (2005), using lime (CaO), but the proportion was changed from 40 to 60 g/kg of CBW. The lime was diluted with water at a ratio of 1 kg to 9 l of water. After treatment, the CBW was wrapped in a polyethylene box of 200 l for one night and then dried outside in the sun. Blood was collected before feeding, and three and six hours after feeding. Haemoglobin was measured by the potassium cyanide colorimetric technique. The blood was centrifuged at 1,600 rpm for the withdrawal of serum, which was stored at -20 °C until the time of analysis for metabolic profile using commercial kits. The profile examined included the following metabolites: total protein (biuret method), glucose (the glucose oxidase method) and urea (urease test), glutamic oxaloacetic transaminase (GOT) (automated Frankel Reitmam method), glutamic pyruvic transaminase (GPT) (automated Frankel Reitmam method) and calcium (automated Armenzano III method). SAS® - Statistic Analysis System (SAS, 2003) was used for analysis of variance and regression study. Overall differences between treatment means were declared significant at $P<0.01$.

Results

For all treatments the levels of urea in the serum blood (SUL) were above the recommended level, which is between 24.0 to 50.0 mg/dl suggested as the ideal range for sheep. The SUL decreased linearly ($P<0.01$) with the replacement of SBM by detoxificated CBW (Table 1). For each 15% of inclusion, the SUL was reduced by 7.0 mg/dl. These differences are probably related to the different availability of nitrogen and energy generated by the substitution of SBM by detoxified CBW. The replacement of SBM by CBW decreased linearly GOT values ($P<0.01$), but all values observed were below 280 IU (upper limit considered normal for sheep). The other parameters were not affected ($P>0.01$) by the levels of substitution indicating that the detoxification of castor bean cake with CaO was effective to eliminate the active form of ricin that could have caused changes in metabolic blood profile.

Table 1. Serum urea levels (SUL), hemoglobin, glucose, total proteins, glutamic oxaloacetic transaminase (GOT), glutamic pyruvic transaminase (GPT) and calcium concentration in the blood of sheep fed a combination of soybean meal (SBM) and castor bean waste (CBW)(Means for n = 8).

Parameters	Proportion of CBWas protein source				RE[1]	r^2	P
	0%	15%	30%	45%			
SUL mg/dl[0,3,6]	85.7	66.0	76.2	59.2	Y= 82.19 – 0.46x	0.65	*
Hemoglobin (g/%)[0]	12.2	12.4	12.7	11.3	Y= 12.14	-	ns
Glucose (mg/dl)[0]	68.6	70.6	69.7	67.2	Y= 69.03	-	ns
Total Proteins (g/dl)[0]	7.4	7.4	7.2	7.6	Y= 7.39	-	ns
GOT (UI)[0]	144.9	129.7	117.9	117.1	Y= 141.69 – 0.64x	0.50	*
GPT (UI)[0]	17.0	17.6	16.9	16.3	Y= 16.94	-	ns
Calcium (mg/dl)[0]	9.8	10.4	10.4	10.5	Y= 10.27	-	ns

[1] Regression equations (RE) and coefficient of determination (r^2); * significantly different ($P<0.01$), [ns] not significantly different ($P>0.01$); [0] blood collected before meal; [0,3,6] averages of blood collected before meal and 3 and 6 h after meal.

Conclusion

Changes in blood (SUL and GOT) parameters were probably due to differences in source and availability of nitrogen and energy. The values of hemoglobin, glucose, calcium, GPT and total protein were not changed by the partial substitution of soybean meal by detoxified CBW.

References

Anandan, S., G.K. Anial Kumar, J. Ghosh and K.S. Ramachandra, 2005. Effect of different physical and chemical treatments on detoxification of ricin in castor cake. Anim. Feed Sci. Technol. 120: 159-168.

SAS. Institute, 2003. Statics: user's guide: version 9, 1. SAS Institute, Inc., Cary, NC, USA.

Flax hulls and oil supplementation on the activity of antioxidant enzymes in dairy cows

H.V. Petit[1], C. Côrtes[1], N. Gagnon[1], M.F. Palin[1], S. Tao[2], C. Benchaar[1] and P. Lacasse[1]
[1]Dairy and Swine Research and Development Centre, Agriculture and Agri-Food Canada, P.O. Box 90, Stn Lennoxville, Sherbrooke, QC J1M 1Z3, Canada; [2]China Agricultural University, Beijing, China; helene.petit@agr.gc.ca

Introduction

Flaxseed is an excellent source of n-3 polyunsaturated fatty acids (PUFA) and lignans (Thompson *et al.*, 1991), which are strong antioxidants (Kitts *et al.*, 1999). Rajesha *et al.* (2006) demonstrated in rat models that flax lignans upregulate the expression of hepatic genes encoding for enzymes such as superoxide dismutase (SOD), catalase (CAT), and glutathione peroxidase (GPx) that are involved in defence mechanisms against oxidative stress. Several studies have demonstrated that n-3 PUFA possess anti-inflammatory and antioxidative properties (Fernandes *et al.*, 2008). Both n-3 PUFA and lignans contained in flaxseed could contribute to enhance protection against oxidative stress with various health benefits. The present study was aimed at determining the effects of dietary flax oil and hulls on the activity of antioxidant enzymes (SOD, CAT and GPx) in dairy cows.

Material and methods

Eight multiparous Holstein cows fitted with ruminal cannulae averaging 731 kg of BW and 223 days in milk were assigned to four treatments in a double 4×4 Latin square design. The total mixed diets were a control diet (CO; 33.6% grass silage, 34.3% corn silage, 12.1% corn, 7.8% ground barley, 6.8% soybean meal, 2.9% protein supplement, and 2.5% mineral), CO with 500 g/d of flaxseed oil infused in the abomasum (CO500), a diet with 10% flaxseed hulls in the DM (HU; 31.9% of grass silage, 31.9% corn silage, 6.8% broken corn grain, 7.8% ground barley, 5.8% soybean meal, 10.3% flaxseed hulls, 2.9% protein supplement, and 2.6% mineral) and HU with 500 g/d of flaxseed oil infused in the abomasum (HU500). The amount of oil infused was similar to the one supplied by a diet with 7% flaxseed in the DM. Cows had *ad libitum* access to the diets. Each experimental period lasted 21 d. Blood was collected from the jugular vein on day 20 and biopsy of the mammary gland was performed on day 21. The activity of SOD, CAT and GPx were determined in plasma, erythrocyte, and mammary gland (Cayman Chemical) and total protein was determined with a bicinchoninic acid protein kit (Sigma-Aldrich). Data were analysed using the MIXED procedure of SAS® (2000) according to a replicated 4×4 Latin square design within a 2×2 factorial arrangement of treatments.

Results

There was no interaction between hulls and oil for antioxidant activity in erythrocytes and enzyme activity was similar among treatments (Table 1). The activity of CAT in the mammary gland tended (P=0.07) to increase with oil infusion and the effect tended (P=0.10) to be greater in the absence of hulls as suggested by the interaction P-value. There was no treatment effect on GPx activity in the mammary gland and SOD activity tended (P=0.06) to be higher with flaxseed hull supplementation. Plasma activity of CAT and SOD were similar among diets and GPx activity tended (P=0.05) to decrease with flax oil supplementation.

Table 1. Activity of antioxidant enzymes (CAT, GPx, and SOD)[1].

	Diet[2]				SE	Treatment effect, *P*-value		
	CO	CO500	HU	HU500		Hulls	Oil	Interaction
Erythrocytes								
CAT	8,717	8,650	8,853	8,349	763	0.91	0.71	0.77
GPx	4,564	4,377	4,025	4,584	240	0.50	0.44	0.13
SOD	252	255	244	262	13	0.95	0.40	0.57
Mammary gland								
CAT	1,648	2,147	1,791	1,814	140	0.50	0.07	0.10
GPx	753	869	971	1,098	173	0.18	0.46	0.97
SOD	599	591	656	876	83	0.06	0.23	0.20
Plasma								
CAT	4.1	4.6	6.4	4.3	0.9	0.28	0.38	0.13
GPx	26.4	24.8	31.9	22.3	2.7	0.58	0.05	0.15
SOD	2.7	2.6	3.0	2.5	0.3	0.73	0.25	0.47

[1] nmol/min/ml/mg prot for catalase (CAT) and glutathione peroxydase (GPx) activity and in U/ml/mg prot for superoxide dismutase (SOD) activity.
[2] CO = control diet; CO500 = control diet with 500 g/d of flaxseed oil infused in the abomasum; HU = diet with 10% of flaxseed hulls in the dry matter; HU500 = HU diet with 500 g/d of flaxseed oil infused in the abomasum.

Conclusion

Supplementing dairy cow diets with flax products high in n-3 PUFA and lignans had no effect on the activity of antioxidant enzymes in erythrocytes although n-3 PUFA supplementation tended to decrease plasma GPx activity. In the mammary gland, the activity of CAT and SOD, which are responsible for the removal of free radicals leading to oxidative stress, were modulated by n-3 PUFA and lignans, respectively. The modulation of antioxidant enzymes in the mammary gland by flaxseed components can be helpful for protection against oxidative stress damages occurring in the mammary gland of dairy cows although increased PUFA concentration in membranes may increase peroxidation in tissues. Further analyses in blood parameters related to peroxidation still need to be completed.

Acknowledgement

The authors thank cowshed workers and staff of the laboratory and of the statistics unit.

References

Fernandes, G., M.M. Rahman, A. Bhattacharya, J. Banu, J.X. Kang and G. Halade, 2008. Effect of endogenoux n-3 PUFA on inflammation and oxidative stress. FASEB J. 22:1094.1.

Kitts, D.D., Y.V. Yuan, A.N. Wijewickreme and L.U. Thompson, 1999. Antioxidant activity of the flaxseed lignan secoisolariciresinol diglycoside and its mammalian lignan metabolites enterodiol and enterolactone. Mol. Cell. Biochem. 202: 91-100.

Rajesha, J., N. Kotamballi, C. Murthy, M.K. Kumar, B. Madhusudhan and G.A. Ravishankar, 2006. Antioxidant potentials of flaxseed by *in vivo* model. J. Agric. Food Chem. 54: 3794-3799.

Thompson, L.U., P. Robb, M. Serraino and F. Cheung, 1991. Mammalian lignan production from various foods. Nutr. Cancer 16: 43-52.

Alteration in the activation of NF-κB upon TNF-α and/or IFN-α/γ treatment of C2C12 myotubes

B. Pijet[1], M. Pijet[1], A. Pogorzelska[1], B. Pająk[2] and A. Orzechowski[1,2]
[1]Department of Physiological Sciences, Faculty of Veterinary Medicine, Warsaw University of Life Sciences (SGGW), Nowoursynowska 159, 02-776 Warsaw, Poland; [2]Department of Cell Ultrastructure, Medical Research Center, Polish Academy of Sciences, Pawinskiego 5, 02-106 Warsaw, Poland; arkadiusz_orzechowski@sggw.pl

Introduction

A decrease in skeletal muscle mass is commonly observed in pathological conditions including chronic inflammation and diabetes (Tisdale, 1997; Argiles *et al.*, 2003). In adults, muscle mass and nitrogen balance are generally maintained by anti-catabolic action of insulin. The main symptom of accelerated catabolism in insulin resistant states is reduced muscle mass associated with degradation of muscle proteins (Acharyya *et al.*, 2004, Li *et al.*, 1998). It is believed, that the key mediators of muscle cachexia are tumor necrosis factor alpha (TNF-α) and interferons (IFN-s) produced and secreted by macrophages and other cell types. NF-κB transcription factor seems to play a crucial role in TNF-α-mediated muscle cachexia. It activates ubiquitination of muscle proteins (Li *et al.*, 1998). The aim of present study was to examine the effects of TNF-α and/or IFN on the activity of NF-κB in C2C12 myotubes. Protein expression levels were also determined.

Material and methods

Mouse C2C12 myoblastic cells (ECACC-European Collection of Animal Cell Cultures) were grown at 37 °C, 5% CO_2 in v/v 10% foetal bovine serum FBS/DMEM growth medium supplied with an antibiotic mixture. Cells were seeded at 100 mm diameter Petri dishes for immunoblotting and TransAm assay. After reaching confluence, cells were guided to the transition phase by replacement of growth medium with 2% (v/v) horse serum HS/DMEM differentiation medium alone (control, CTRL) or additionally supplemented sequentially with TNF-α (10 ng/ml) and IFN-α/IFN-γ (10 ng/ml) or insulin (10 nM). The expression of NF-κB p65 protein in nuclei, and IkBα protein in cytosol were determined by western blot method. The transcriptional activity of FoxO1, NF-κB and STAT-1α were measured by TransAm assay. The results were statistically evaluated with the one-way Anova and Tukey multiple range test and compared to the control treatment.

Results

Short-term exposure of C2C12 myotubes to TNF-α (15, 30, 60 min) elevated the level of NF-κB in the nuclear fraction while a marked drop in cytosolic IκB protein was observed at the same time. Interestingly, also chronic treatment (24 h) with TNF-α led to higher protein expression levels of NF-κB in the nuclear fraction. Neither IFN-α nor IFN-γ raised nuclear expression of NF-κB. TNF-α stimulated NF-κB activity in a time-dependent manner although in contrast to protein expression levels, the maximum activity was found at the 60th minute of treatment. One-day pretreatment with INF-α and IFN-γ increased TNF-α-mediated NF-κB activation. Surprisingly, insulin was able to increase TNF-α-, IFN-γ- and IFN-α-mediated NF-κB activation. Insulin effect was also time-dependent. Higher expression levels of nuclear NF-κB during short- and long-term TNF-α treatment prompted us to monitor nuclear levels of NF-κB at additional time-points (1, 2, 6, 12, 24, 36, 48 h). Increased expression of nuclear NF-κB was observed after TNF-α treatment with a maximal response found at 0.5 h and 12/24 h. Further studies are needed to evaluate whether STAT-1α could exert genomic activity after IFN-s addition in the presence of TNF-α.

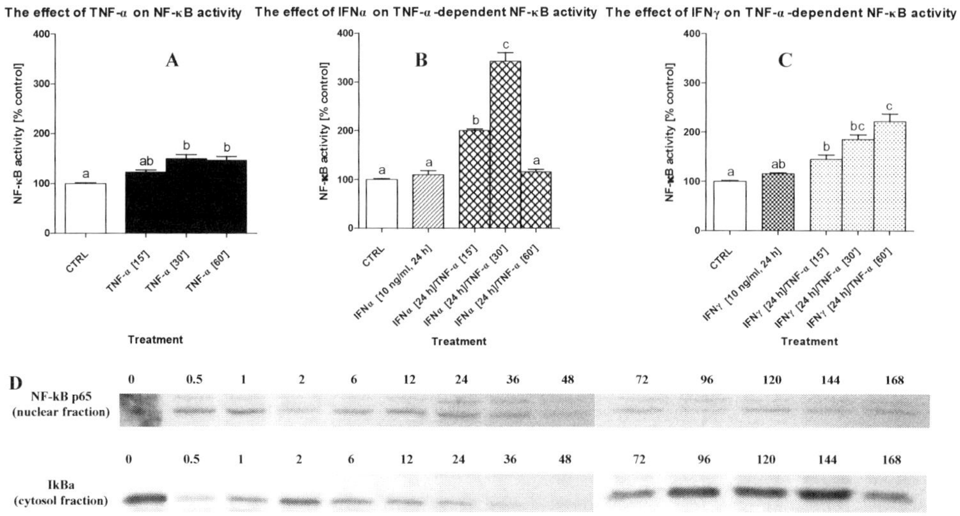

Figure 1. The effects of TNF-α on NF-κB activity (A) and the effect of IFNα (B) and IFNγ (C) on TNF-α-dependent activity of NF-κB in C2C12 myotubes (3rd day of myogenesis). CTRL (control) - cells differentiated for 3 days in 2%HS/DMEM medium. Different lower case letters indicate statistical differences between the means (P<0,05). Biphasic activation of NF-κB induced by TNF-α during seven subsequent days of myogenesis, measured by expression of NF-κB in nuclear fraction and IκBα in cytosolic fraction (D).

Conclusion

TNF-α efficiently activates NF-κB leading to an increased nuclear level of this protein. Simultaneously, lower expression of cytoplasmic IκB is suggestive of a canonical pathway of NF-κB. The biphasic pattern of TNF-α-dependent NF-κB activation suggests that TNF-α activity was extended on the subsequent days of myogenesis. Interferons make C2C12 myotubes more vulnerable to TNF-α whereas insulin opposes this effect.

References

Acharyya, S., K.J. Ladner, L.L. Nelsen, J. Damrauer, P.J. Reiser, S. Swoap and D.C. Guttridge, 2004. Cancer cachexia is regulated by selective targeting of skeletal muscle gene products. J. Clin. Invest. 114: 370-378.

Argiles, J.M, R. Moore-Carrasco, G. Fuster, S. Busquets and F.J. Lopez-Soriano, 2003. Cancer cachexia: the molecular mechanisms. Int. J. Biochem. Cell Biol. 35: 405-409.

Li, Y.-P., R.J. Schwartz, I.D. Waddell, B.R. Holloway and M.B. Reid, 1998. Skeletal muscle myocytes undergo protein loss and reactive oxygen-mediated NF-κB activation in response to tumor necrosis factor α. FASEB J. 12: 871-880.

Tisdale, M.J., 1997. Biology of cachexia. J. Natl. Cancer Inst. 89: 1763-1773.

Leptin impairs expression levels of myogenic regulatory factors (MRF) and potentiates staurosporine effect in C2C12 myotubes

M. Pijet[1], *B. Pijet*[1], *A. Pogorzelska*[1] *and A. Orzechowski*[1,2]
[1]*Department of Physiological Sciences, Faculty of Veterinary Medicine, Warsaw University of Life Sciences (SGGW), Nowoursynowska 159, 02-776 Warsaw, Poland;* [2]*Department of Cell Ultrastructure, Medical Research Center, Polish Academy of Sciences, Pawinskiego 5, 02-106 Warsaw, Poland; arkadiusz_orzechowski@sggw.pl*

Introduction

Leptin is an adipocyte-secreted protein involved in the regulation of body weight. It has been implicated in a variety of physiological functions mainly in food intake and energy expenditure. It acts trough Ob membrane receptors encoded by the *db* gene (Margetic *et al.*, 2002). Interaction of leptin with ObR stimulates the JAK/STAT signalling pathway and leads to phosphorylation and recruitment of STAT3 kinase (Hegyi *et al.*, 2004). It has been recently reported that leptin action could be inhibited at a step downstream of STAT3 phosphorylation and nuclear translocation, and may provide a potential mechanism of leptin resistance in which an increased FoxO1 antagonises STAT3-mediated leptin signalling (Yang *et al.*, 2009). Leptin can also act through certain components of the insulin signalling cascade such as: IRS1/2, Ras/MAPK, PI3-K/PKB (Maroni *et al.*, 2005). In obesity disease, low sensitivity to leptin in skeletal muscles is observed, often linked to insulin-resistance. Some studies suggest that leptin impairs the expression of myogenic regulatory factors (MRF) (Yu *et al.*, 2008). The aim of present study was to verify the hypothesis that leptin affects myogenesis. STAT3 kinase phosphorylation status upon leptin and/or insulin treatment was also determined.

Material and methods

Mouse C2C12 myoblastic cells (ECACC-European Collection of Animal Cell Cultures) were grown at 37 °C, 5% CO_2 in v/v 10% fetal bovine serum FBS/DMEM growth medium supplied with antibiotic mixture. Cells were seeded at 100 mm diameter Petri dishes (for immunoblotting and TransAm Assay) or 96-flat-well plates (for cell viability studies). After reaching confluence, cells were guided to the transition phase by replacement of growth medium with 2% (v/v) horse serum HS/DMEM differentiation medium alone/or additionally supplemented with leptin (50, 100 ng/ml) alone/or with insulin (10 nM) and metabolic inhibitors such as LY294002 (25 uM), PD098059 (100 uM), staurosporine (0.2 uM, STS), *cycloheximide* (5 ug/ml, CHX), actinomycin D (5 ug/ml, AD). Protein expression levels of myogenic regulatory factor (MyoD, myogenin) were determined by western blotting. The transcriptional activity of FoxO1 and STAT3 were measured with the TransAm assay. The results were statistically evaluated with one-way anova and Tukey multiple range test and compared to controls.

Results

Viability studies (MTT assay) revealed that leptin (100 ng/ml) did not affect muscle viability during 5-d differentiation. Neither metabolic inhibitors affected leptin activity. Conversely, STS inhibited metabolic activity of C2C12 myotubes in a time-dependent manner whereas leptin potentiated this effect (Figure 1). Immunoblotting have shown that both PI3-K inhibitor LY294002 and leptin differently reduced protein levels of MyoD. Leptin similarly to LY294002 delayed the rise in the myogenin expression, step limiting transcription factor in terminal differentiation of muscle cells (Figure 2). It is assumed that the observed leptin effects could be associated with antagonistic regulation of the PI3-K/Akt signalling pathway. Further studies are needed in order to assess the role of JAK/STAT and FoxO1 in leptin-mediated retardation of myogenesis.

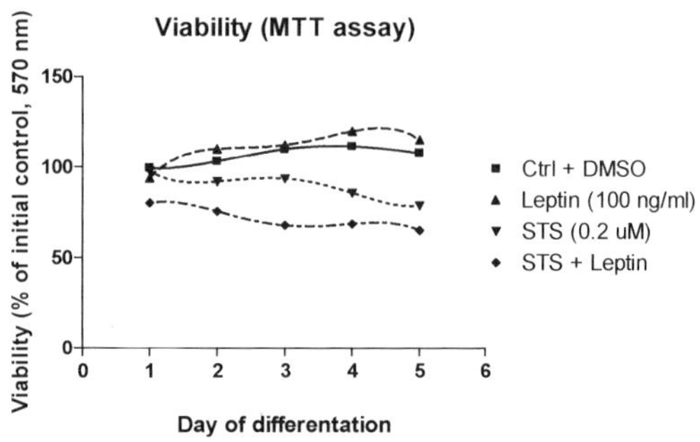

Figure 1. Viability of C2C12 myotubes in the presence of leptin or/and staurosporine. Polynomial fourth order regression curves.

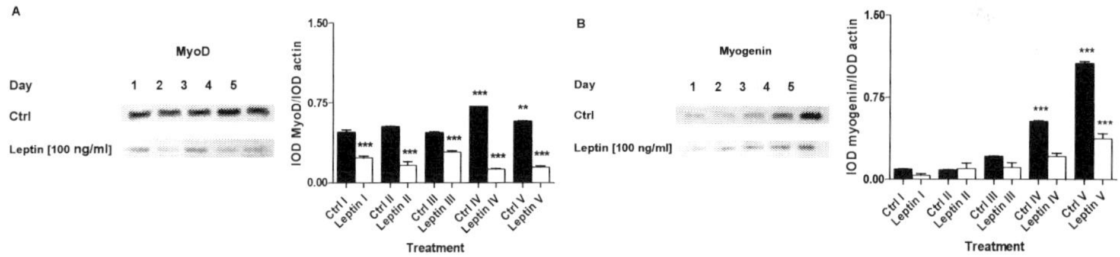

*Figure 2. Effect of leptin (100 ng/ml) on expression levels of MyoD (A) and myogenin (B) proteins in C2C12 myotubes during five subsequent days of myogenesis. Ctrl (control) - cells differentiated during 5 d in 2% HS/DMEM medium. Asterisks indicate means statistically different from control at the respective day of myogenesis (**P<0.01; ***P<0.001).*

Conclusion

Leptin represses myogenesis from mouse C2C12 muscle satellite cells. It also potentiates STS-dependent reduction in cell viability. The molecular mechanisms of these processes are not clear, although, PI3-K/Akt seems to antagonise leptin-mediated effects.

References

Hegyi, K., K. Fülöp, K. Kovács, S. Tóth and A. Falus, 2004. Leptin-induced signal transduction pathways. Cell Biol. Int. 28: 159-69.

Margetic, S., C. Gazzola, G.G. Pegg and Hill R.A., 2002. Leptin: a review of its peripheral actions and interactions. Int. J. Obes. 26: 1407-1433.

Maroni, P., P. Bendinelli and R. Piccoletti, 2005. Intracellular signal transduction pathways induced by leptin in C2C12 cells. Cell Biol. Int. 29: 542-550.

Yang, G., C.-Y. Lim, C. Li, X. Xiao, G.K. Radda, C. Li, X. Cao and W. Han, 2009. FoxO1 inhibits leptin regulation of pro-opiomelanocortin promoter activity by blocking STAT3 interaction with specificity protein 1. J. Biol. Chem. 284: 3719-3727.

Yu, T., G. Luo, L. Zhang, J. Wu, H. Zhang and G. Yang, 2008. Leptin promotes proliferation and inhibits differentiation in porcine skeletal myoblasts. Biosci. Biotechnol. Biochem. 72: 13-21.

Plasma glucagon-like peptide-1 concentration in non-lactating cows during abomasal infusion of linseed oil and in response to glucose and insulin challenges

J.A.A. Pires[1,4], A.E. Relling[2], C.K. Reynolds[3] and R.R. Grummer[1]
[1]Dept. of Dairy Science, University of Wisconsin, 53706, Madison, USA; [2]Dept. of Animal Science, The Ohio State University, 44691, Wooster, USA; [3]School of Agriculture, Policy, and Development, University of Reading, Reading, United Kingdom; [4]Current address: INRA, UR1213 Herbivores, 63122 Saint-Genès-Champanelle, France; rgrummer@wisc.edu

Introduction

In nonruminants, the gastrointestinal hormone glucagon-like peptide-1 (GLP-1) increases glucose-stimulated insulin secretion and has multiple (direct and indirect) effects on feed regulation and metabolism (Holst, 1997). In dairy cows, plasma GLP-1 concentration is influenced by physiological state, dietary fat supplementation and fatty acid profile (Relling and Reynolds, 2007a,b). Unsaturated dietary fat increases GLP-1 to a greater extent than saturated sources (Relling and Reynolds, 2007b). Previous experiments have focused on the effects of abomasal infusion of linseed oil (LO), a fat source rich in C18:3 n-3, on whole-body responses to insulin in fed and feed-restricted non-lactating Holstein cows, as assessed by i.v. glucose tolerance test (GTT) and insulin challenge (IC). Abomasal infusion of LO decreased insulin response during GTT in fed cows, compared to tallow (T) (Pires *et al.*, 2008), and we speculated that this effect could have been mediated by differential secretion of GLP-1. The objective was to determine whether abomasal LO infusion altered plasma concentration of GLP-1 and assess whether it could explain the pattern of insulin response during GTT in the previously mentioned experiments (Pires *et al.*, 2008).

Material and methods

Detailed protocol has been published elsewhere (Pires *et al.*, 2008). In experiment 1, 8 non-lactating, non-gestating, ruminally cannulated Holstein cows were randomly assigned to a sequence of treatments in a cross-over design with 2 periods. Cows were fed at a rate to meet individual maintenance requirements. Treatments were abomasal infusion of LO or T for 5.5 d at a rate of 0.54 g/d per kg of BW (i.e. 369±45 g/d; mean±SD). On d 5 of each period, GTT (0.25g/kg BW of glucose i.v.) and IC (0.1 IU /kg BW of insulin) were performed at 9 AM and 9 PM. In experiment 2, 6 non-lactating, non-gestating, Holstein cows were assigned to treatments in a replicated 3×3 Latin square design. Treatments were abomasal infusion of water, LO or T for 5.5 d, at the same rate as in experiment 1 (i.e. 409±22 g/d; mean ± SD). Feeding was suspended on d 3, leading to 50 and 62 h of fasting prior to GTT and IC, respectively. Plasma GLP-1 concentration was quantified (Benson and Reynolds, 2001) in samples collected daily each period at 7:00 AM, at -15, -5, 10, 30, 60, 90 min relative to glucose infusion for GTT, at -15, -5, 10, 20 min, relative to insulin infusion for IC in both experiments, and also at 45, 60 min of IC in experiment 2. The intra-assay coefficient of variation was <12%. Data were analysed with the MIXED procedure of SAS® with repeated measures, using models previously described (Pires *et al.*, 2008).

Results

In experiment 1 (Table 1), there was no effect of fat source on plasma GLP-1 concentration. In experiment 2 (Table 2), fat infusion increased GLP-1 concentration compared to water, and this effect was independent from fat source. There was a tendency for a time effect during GTT because average GLP-1 concentration at min -15 and -5 was 14.5±2.4 pM and ranged between to 12.4 and 13.0 pM thereafter. There was a significant time effect during IC (basal GLP-1 was 13.0, and 15.6, 14.8, 10.6, 12.2±4.0 pM at 10, 20, 45, and 60 min after insulin injection, respectively).

Table 1. Experiment 1 (fed state). Effect of abomasal infusion of linseed oil (LO) or tallow (T) to non-lactating Holstein cows on plasma GLP-1 concentration (pM).

Samples	Treatment (Trt)		SEM	P		
	LO	T		Time	Trt	Trt × Time
5 d of infusions	31.1	28.9	4.9	0.35	0.76	0.23
GTT on d 5	35.4	34.6	7.0	0.62	0.57	0.33
IC on d 5	35.4	35.3	5.7	0.15	0.99	0.62

Table 2. Experiment 2 (feed restriction). Effect of abomasal infusion of water, linseed oil (LO) or tallow (T) to non-lactating Holstein cows on plasma GLP-1 concentration (pM).

Samples	Treatments (Trt)			SEM	P			
	Water	LO	T		Time	Trt × Time	Water vs. Fat	LO vs. T
5 d of infusions	8.59	14.49	12.74	1.82	0.18	0.33	0.002	0.22
GTT on d 5	9.13	15.70	14.19	2.50	0.06	0.48	0.004	0.38
IC on d 5	9.84	14.17	15.68	4.04	0.04	0.25	0.001	0.36

Discussion and conclusion

The positive effect of fat infusion on plasma GLP-1 concentration found in experiment 2 was in agreement with previous studies reporting a relationship between postruminal fat infusion and plasma GLP-1 concentration (Benson and Reynolds, 2001). Plasma GLP-1 concentrations were lower in experiment 2 than in experiment 1; however, a statistical comparison between these independent experiments cannot be made. Interestingly, there was no significant time effect during 5 d of fat infusions in experiment 2, which suggests that the duration of fasting (it started on d 3) was not sufficient to alter GLP-1 concentration, perhaps because several days of feed withdrawal are required to completely empty rumen contents. Fat source (T vs. LO) did not change plasma GLP-1 concentration in either experiment, therefore, plasma GLP-1 concentration was probably not a factor mediating LO effects on the profile of insulin observed during GTT in experiment 1 (Pires *et al.*, 2008).

References

Benson, J.A. and C.K. Reynolds, 2001. Effects of abomasal infusion of long-chain fatty acids on splanchnic metabolism of pancreatic and gut hormones in lactating dairy cows. J. Dairy Sci. 84: 1488-1500.

Holst, J.J., 1997. Enteroglucagon. Annu. Rev. Physiol. 59: 257-271.

Pires, J.A.A., J.B. Pescara, A.E. Brickner, N. Silva del Rio, A.P. Cunha and R.R. Grummer, 2008. Effects of abomasal infusion of linseed oil on responses to glucose and insulin in Holstein cows. J. Dairy Sci. 91: 1378-1390.

Relling, A.E. and C.K. Reynolds, 2007a. Plasma concentrations of gut peptides in dairy cattle increase after calving. J. Dairy Sci. 90: 325-330.

Relling, A.E. and C.K. Reynolds, 2007b. Feeding rumen-inert fats differing in their degree of saturation decreases intake and increases plasma concentrations of gut peptides in lactating dairy cows. J. Dairy Sci. 90: 1506-1515.

Effects of abomasal infusion of nicotinic acid on responses to glucose and β-agonist challenges in partially feed-restricted lactating cows

J.A.A. Piresa[1,2], L.F. Stumpf[1], I.D. Soutullo[1], J.B. Pescara[1], S.E. Stebulis[1] and R.R. Grummer[1]
[1]Dept. of Dairy Science, University of Wisconsin, Madison, 53706, USA; [2]Current address: INRA, UR1213 Herbivores, F-63122 Saint-Genès-Champanelle, France; rgrummer@wisc.edu

Introduction

The antilipolytic properties of nicotinic acid (NA) have been used to decrease plasma nonesterified fatty acid (NEFA) concentrations in feed-restricted non-lactating Holstein cows (Pires *et al.*, 2007a). An 11 h reduction of plasma NEFA enhanced the response to insulin, implicating elevated plasma NEFA as a causal factor of insulin resistance in the bovine (Pires *et al.*, 2007b). The objectives were to assess the use of NA to chronically (i.e. 74 h) manipulate plasma NEFA concentrations in partially feed-restricted lactating cows, to determine whether the reduction of plasma NEFA altered responses to i.v. glucose tolerance test (GTT), and to assess whether NA would attenuate an acute lipolytic stimuli of a β-agonist challenge (BAC).

Material and methods

Eight lactating dairy cows (244±31 DIM) were assigned to a sequence of two treatments in a cross-over design. Treatments were 74 h of continuous abomasal infusions of water (200 ml/h) or NA (3 mg/h per kg BW) to decrease plasma NEFA concentrations. The rate of NA infusion corresponds to 50% of the amount given as hourly boluses to non-lactating cows (Pires *et al.*, 2007b). Cows were allowed *ad libitum* intake of a TMR (62 Mcal of NEL/kg of DM and 17.5% CP). From 0 to 74 h of each period, cows were feed-restricted to 33% of the intake recorded during prior 5 d (d -7 to -3) to increase plasma NEFA concentration. A GTT (0.25 g/kg of BW of glucose i.v.) was performed at 48 h and a BAC (4 nmol/kg BW of isoproterenol i.v.) at 72 h of each period. Data were analysed with the MIXED procedure of SAS® using models similar to Pires *et al.* (2007b).

Results and discussion

Intake was 24.1, 8.2, 8.0 and 8.0 kg DM/d before restriction, on d 1, 2 and 3, respectively. Nicotinic acid decreased blood NEFA, and increased insulin and glucose (Table 1). Nicotinic acid also led to greater glucose and insulin response areas under the curve (AUC) during GTT and BAC, and NA enhanced NEFA AUC during BAC, despite a lower NEFA peak (Table 2). Milk, fat and protein yields (29.1, 1.2 and 0.93 kg on d -2, respectively) decreased to 17.9 and 11.5, 0.81 and 0.54, 0.56 and 0.39 kg on d 3, for control and NA, respectively ($P<0.05$). Nicotinic acid may have decreased production by inhibiting the supply of NEFA for energy and milk fat synthesis. Milk urea nitrogen was increased by NA on d 2 (12.8 vs. 19.1) and d 3 (11.6 vs. 17.8 mg/dl; $P<0.05$), probably due to a greater reliance on mobilised amino acids. Somatic cell count was increased by NA on d 3 (187 vs. 848 ×1000; $P<0.05$).

Conclusion

Patterns of glucose and insulin concentration observed during 74 h of NA infusion (3 mg/h per kg BW) reflect a state of insulin resistance, which contrast with short-term responses in non-lactating cows (Pires *et al.*, 2007b). Data suggest that long-term infusion of NA affects intermediary metabolism beyond antilipolysis; it did not inhibit acute lipolytic stimuli of BAC (4 nmol/kg BW of isoproterenol).

Table 1. Blood NEFA, glucose and insulin concentrations relative to the initiation of feed restriction and abomasal administration of water or nicotinic acid (NA).

	Hour	Treatment		SEM	P			
		Water	NA		Sequence	Time	Treatment	Treatment × time
NEFA	0	173	160	101	0.25	<0.001	<0.001	0.005
(μEq/l)	24	689^a	393^b					
	48	899^a	330^b					
	72	701^a	320^b					
Glucose	0	65.0^c	60.2^d	2.6	0.26	<0.001	0.51	0.003
(mg/dl)	24	56.9	53.3					
	48	57.5	61.4					
	72	56.6^b	65.2^a					
Insulin	0	9.5	8.1	2.4	0.06	0.03	0.001	<0.001
(μIU/ml)	24	4.4	7.1					
	48	4.4^b	13.4^a					
	72	3.0^b	23.2^a					

[a,b] Within a row: $P<0.05$; [c,d] Within a row: $P<0.10$.

Table 2. Measurements during GTT and BAC, performed respectively at 48 h and 72 h of feed restriction, concomitant to abomasal administration of water or nicotinic acid (NA).

Test	Measurement	Treatment		SEM	P
		Water	NA		
GTT	Glucose (mg/dl)				
(at 48 h)	Peak [1]	145.8	162.5	6.7	0.009
	CR_{60} (%/min) [2]	2.45	2.36	0.33	0.65
	AUC_{180} (×180 min) [3]	5,056	6,562	837	0.03
	Insulin (μIU/ml)				
	AUC_{180} (×180 min) [3]	2,502	6,042	983	0.004
	NEFA (μEq/l				
	AUC_{180} (×180 min) [3]	-62,185	-16,178	14,765	0.02
BAC	Glucose (mg/dl)				
(at 72 h)	AUC_{120} (×120 min) [3]	240	535	169	0.03
	Insulin (μIU/ml)				
	AUC_{120} (×120 min) [3]	222	1283	192	<0.001
	NEFA(μEq/l)				
	Peak [4]	1,461	1,141	107	0.01
	AUC_{120} (×120 min) [2]	22,862	45,521	5,325	0.02

[1] Maximum glucose increase above baseline; [2] Clearance rate during first 60 min of GTT; [3] Area under the curve during 180 min of GTT or 120 min of BAC; [4] Maximum concentration.

References

Pires, J.A.A. and R.R. Grummer, 2007a. The use of nicotinic acid to induce sustained low plasma nonesterified fatty acids in feed-restricted Holstein cows. J. Dairy Sci. 90: 3725-3732.

Pires, J.A.A., J.B. Pescara, and R.R. Grummer, 2007b. Reduction of plasma NEFA concentration by nicotinic acid enhances the response to insulin in feed-restricted Holstein cows. J. Dairy Sci. 90: 4635-4642.

Genomic effects of insulin and insulin signalling inhibitors in evaluation of the mitochondrial contribution to myogenesis

A. Pogorzelska[1], M. Pijet[1], B. Pijet[1] and A. Orzechowski[1,2]
[1]Department of Physiological Sciences, Faculty of Veterinary Medicine, Warsaw University of Life Sciences (SGGW), Nowoursynowska 159, 02-776 Warsaw, Poland; [2]Department of Cell Ultrastructure, Medical Research Center, Polish Academy of Sciences, Pawinskiego 5, 02-106 Warsaw, Poland; arkadiusz_orzechowski@sggw.pl

Introduction

Insulin is a well known anabolic hormone that stimulates glucose transport, protein and glycogen synthesis, inhibits lipolysis, regulates gene transcription and translation, cell growth and differentiation. It was shown that insulin plays important roles in the synthesis of mitochondrial proteins and modulates mitochondrial activity during muscle development (Pawlikowska *et al.*, 2006 and 2007). Interaction of insulin with insulin receptor (IR) leads to activation of at least two main signalling pathways: the mitogen activated protein kinase cascade (MAPK/extracellular regulated kinase ERK) and the phosphatidylinositol-3-kinase (PI3-K/protein kinase Akt). Through activation of PI3-K, insulin stimulates Akt kinase, which in turn targets several substrates including inactivation of glycogen synthase kinase 3β (GSK-3β) and FoxO1 (FKHR) transcription factor. Both GSK-3β and FoxO1 repress cell survival and growth, respectively. They are also believed to inhibit myogenesis. Kamei *et al.* (2004) noted that FoxO1 negatively regulates skeletal muscle mass and type I fiber gene expression and subsequently leads to impaired skeletal muscle function. Akt kinase also stimulates the expression of genes controlling mitochondrial functions (Huang *et al.*, 1999). The aim of the present study was to verify the hypothesis if insulin stimulates mitochondria through GSK-3β and/or FoxO1 suppression. PD98059 or LY294004 metabolic inhibitor was used to inhibit MAPK/ERK and PI3-K/Akt signalling pathway, respectively. Elicited responses were analysed at the genomic (mitochondria controlling genes) and at the signal transduction levels.

Material and methods

The mouse C2C12 myoblastic cell line (ECACC – European Collection of Animal Cell Cultures) was grown at 37 °C, 5% CO_2 in 10% (v/v) foetal bovine serum FBS/DMEM growth medium supplied with an antibiotic mixture. The cells were seeded at 100 mm/20 mm Petri dishes (for immunobloting, real-time PCR, TransAM assay) or 96 flat-well plates (for cell viability studies). After reaching the confluence, cells were guided to the transition phase by replacement of growth medium with 2% (v/v) horse serum HS/DMEM differentiation medium alone (control, CTRL) or additionally supplemented with insulin (10 nM) alone or with a particular inhibitor (50 μM, PD98059), PI3-K (20 μM, LY294002) or GSK-3β (5 mM of LiCl or 10 μM of SB216763). At d 1, 3 and 5 of myogenesis, the following indices were measured: (1) gene expression of *mfn2*, *mtssb*, *mttfa*, *cox-I* by the RT PCR method, (2) viability by the MTT assay, (3) GSK-3β, FoxO1, Akt protein expression levels by the western blot method. The transcriptional activity of FoxO1 was measured with the TransAm assay (Actif Motive). The results were statistically evaluated with one-way ANOVA and Tukey multiple range test and compared to the control treatment.

Results

Viability of muscle cells increased upon insulin treatment alone or with GSK-3β inhibitors SB216763 and LiCl. No additive effects of insulin and GSK-3β inhibitors was found with respect to metabolic activity of muscle cells (Figure 1). Apparently, GSK-3β is the main target of insulin action to control dynamics of mitochondria. Additionally, insulin stimulated the expression of genes (*mfn2, mtssb, mttfa, cox-I*) involved in biogenesis of the mitochondria. The highest expression of selected genes was observed at d 5 of myogenic differentiation. These effects occurred regardless of the inhibition of insulin signalling either with LY294002 or PD98059 whereas GSK-3β inhibitors markedly enhanced gene expression levels (Figure 2).

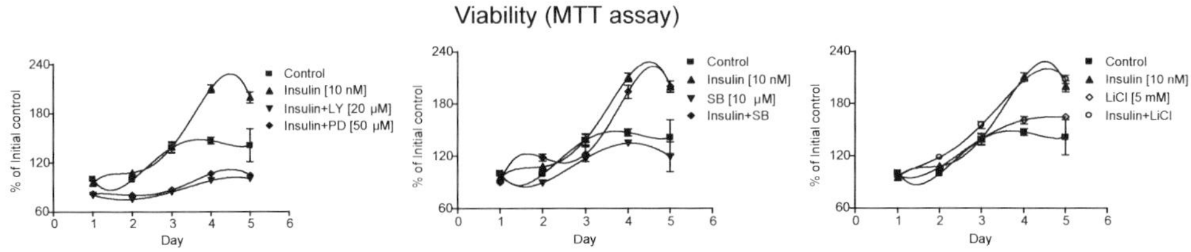

Figure 1. Viability assay of C2C12 cells in the presence of insulin and inhibitors of insulin signalling.

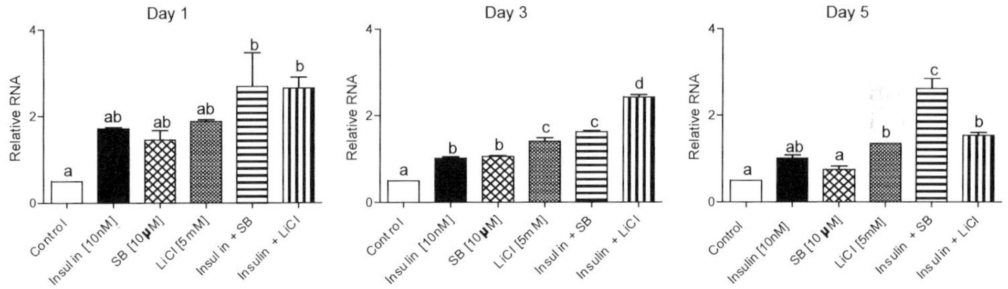

Figure 2. The effects of insulin alone or insulin with GSK-3β inhibitors on mfn2 gene expression.

Conclusion

The results presented indicate that insulin affects myogenesis at the genomic level. It modulates the expression of certain genes whose activity determines mitochondrial function. It seems apparent, that insulin affects cell viability and myogenesis through negative regulation of GSK-3β activity.

References

Huang, X., K.F. Eriksson, A. Vaag, M. Lehtovirta, M. Hansson, E. Laurila, T. Kanninen, B.T. Olesen, I. Kurucz, L. Koranyi and L. Groop, 1999. Insulin-regulated mitochondrial gene expression is associated with glucose flux in human skeletal muscle. Diabetes 48: 1508-1514.

Kamei, Y., S. Miura, M. Suzuki, Y. Kai, J. Mizukami, T. Taniguchi, K. Mochida, T. Hata, J. Matsuda, H. Aburatani, I. Nishino and O. Ezaki, 2004. Skeletal muscle FoxO1 (FKHR) transgenic mice have less skeletal muscle mass, down-regulated type I (Slow twich/red muscle) fiber genes, and impaired glycemic control. J. Biol. Chem. 39: 41114-41123.

Pawlikowska, P., B. Gajkowska, J.-F. Hocquette and A. Orzechowski, 2006. Not only insulin stimulates mitochondriagenesis, but mitochondria are also essential for insulin-mediated myogenesis. Cell Prolif. 39: 127-145.

Pawlikowska, P., B. Gajkowska and A. Orzechowski, 2007. Mitofusin 2 (Mfn2): a key player in insulin-dependent myogenesis *in vitro*. Cell Tissue Res. 327: 571-581.

Use of carcass specific gravity to predict chemical body composition of F1 Boer × Saanen kids

K.T. Resende[1], L. Akinaga[1], I.A.M.A. Teixeira[1], J.M. Pereira Filho[2], T.T. Berchielli[1] and A.C.D. Ferreira
[1]Universidade Estadual de Sao Paulo, UNESP, SP, 14870-000, Jaboticabal, Brazil; [2]Universidade Federal de Campina Grande, UFCG, PB, 58708-110, Paraiba, Brazil; kresende@fcav.unesp.br

Introduction

Knowledge of body composition is of great benefit in nutrition studies in order to understand nutritional requirements and evaluate feeds. Many methods have been reported to assess body composition; each one has advantages and disadvantages that should be pondered before applying for one (Kelly *et al.*, 1968).
According to the literature, the chemical body composition of an animal can be estimated from specific gravity of the dressed carcass (Garret *et al.*, 1959) and by using this technique, body composition assessment would be less expensive and time consuming (Kraybill *et al.*, 1952). In goats, there are few studies evaluating carcass specific gravity to assess body composition. However, this method needs to be better examined in goats, because of their high internal fat deposition (Sahlu *et al.*, 2004). This can affect the accuracy and precision of the method since during carcass specific gravity evaluation internal fat is not considered. Therefore, the aim of this study was to evaluate carcass specific gravity to estimate chemical body composition in Boer × Saanen crossbred kids.

Material and methods

Two trials were carried out in order to reach the objectives of this study. In the first trial the BW of the kids ranged from 5 to 15 kg and in the second trial the BW ranged from 15 to 25 kg. Trial 1: A total of 30 dehorned male Boer x Saanen kids with an initial average BW of 5 kg were used. Six kids were slaughtered at the beginning of the experiment (baseline animals), six kids were slaughtered when the animals reached 10 kg BW (intermediate slaughter animals) and the remainder were randomly allocated to one of three nutritional levels (0, 30, and 60% restriction), and therefore there were 6 kids per nutritional level. The nutritional levels were used in order to accomplish animals with different body composition. The feed intake of the animals in the 0% restriction nutritional level group determined the intake of the animals in the 30% and 60% restriction nutritional level groups. When the animals in the 0% restriction nutritional level group reached 15 kg BW they were slaughtered with the other animals in the other nutritional level groups. Trial 2: A total of 34 dehorned male Boer × Saanen kids with an initial average BW of 15 kg were used. Seven kids represented the baseline animals in this trial, six kids were slaughtered when the animals reached 20 kg BW (intermediate slaughter animals) and the remainder were randomly allocated to one of three nutritional levels, as was done for the first trial, and therefore there were 7 kids per nutritional level. Different nutritional levels were used in order to provide variation in body composition.
Kids were slaughtered and all non-carcass components of the body were weighed and frozen. The carcasses were kept in a cold room for 24 h at 4 °C. After this period the right half carcass was weighed in air and under water. Specific gravity was calculated according to the following equation: carcass specific gravity = carcass weight in the air/(carcass weight in the air - under water carcass weight). The carcasses were then frozen and stored until sampling for chemical analysis. Subsequently non-carcass and carcass were ground and homogenised, and a representative sub sample was freeze-dried. The freeze-dried samples were used to determine DM, fat, ash, CP and GE. The structure of the experiment was assumed in a completely randomized design. Regression equations between body composition and carcass specific gravity were calculated using the REG procedure of SAS® (SAS Inst., Inc., Cary, NC, USA).

Results

The equations to predict body composition that presented better R^2 were the ones that considered carcass specific gravity, age and body fat as independent variables (Table 1). Previous studies reported that carcass specific gravity is a good method to assess body composition when body fat is higher than 20% (Kelly *et al.*, 1968). Since the body fat in the goats used in this study were on average 5% of EBW and 10% for the first and second trial, respectively, the percentage of body fat was added in the equations to improve their precision.

Table 1. Equation describing the relationship between carcass specific gravity and body composition in Boer × Saanen crossbred kids.

	Intercept	Specific gravity	Age	Fat	R^2
Trial 1					
Body ash (%; DM basis)	43.59	-23.94		-0.69	0.91
Body protein (%, DM basis)	93.34	-15.79		-2.02	0.96
Body energy (Kcal/kg, DM basis)	1,083.71	3,227.58		147.80	0.97
Trial 2					
Body dry matter (% EBW)	-34.60	61.21	0.13		0.53
Body water (% EBW)	109.75	-35.10	-0.07		0.34
Body fat (% EBW)	-53.02	56.01	0.17		0.57
Body energy (Kcal/kg, EBW)	-4,463.25	5,764.74	15.04		0.58
Body ash (%; DM basis)	53.19	-36.82	-0.12		0.65
Body energy (Kcal/kg, DM basis)	2,164.72	2,392.51		121.21	0.97
Body fat (%; DM basis)	-144.92	156.37	0.43		0.62

EBW – Empty body weight (kg), Age - (days), Fat = %.

Conclusion

Carcass specific gravity can be used as an indirect method to estimate body composition in goats. In younger goats higher precision in the equations was obtained also using percentage of fat as an independent variable; in older goats beyond carcass specific gravity, age was also used to get a more precise equation.

References

Garrett, W.N., J.H. Meyer and G.P. Lofgreen, 1959. The Comparative Energy Requirements of Sheep and Cattle for Maintenance and Gain. J. Anim. Sci. 18: 528-547.

Kelly, R.F., J.P. Fontenot, P.P. Grahan, W.S. Wilkinson. and C.M. Kincaid, 1968. Estimates of carcass composition of beef cattle fed at different planes of nutrition. J. Anim. Sci. 27: 620-627.

Kraybill, H.F., H.L Bitter. and O.G. Hankins, 1952. Body Composition of Cattle. II. Determination of Fat and Water Content From Measurement of Body Specific Gravity. J. Appl. Physiol. 4: 575-583.

Sahlu, T., A.L. Goetsch, J. Luo, I.V. Nsahlai, J.E. Moore, M.L. Galyean, F.N. Owens, C.L. Ferrell and Z.B. Johnson, 2004. Nutrient requirements of goats: developed equations, other considerations and future research to improve them. Small Rum. Res. 53: 191-219.

Effect of botanical composition of permanent grasslands and feeding practices in three regions of France on liposoluble components in cow milk

A. Reynaud[1], B. Martin[1], A. Ferlay[1], C. Agabriel[2], A. Farruggia[1], J.M. Besle[1], M. Doreau[1] and B. Graulet[1]
[1]INRA, UR1213 Herbivores, F-63122 Saint-Genès Champanelle, France; [2]UR2008.03.102 DGER-USC2005 INRA Elevage et Production des Ruminants, Enita Clermont, F-63370 Lempdes, France; benoit.graulet@clermont.inra.fr

Introduction

The nature of the diet ingested by dairy cows strongly influences the content and composition of the fat-soluble fraction in milk, and thus its nutritional quality. In this concern, a grass-based diet seems to be an effective way to improve the liposoluble components content in milk. However, few data are available on the variability in the milk composition resulting from permanent pasture feeding (Tornambé *et al.*, 2007). So, our aims were to evaluate the variations of liposoluble micronutrients in bulk milk of farms from 3 French regions (1) according to the botanical composition of permanent grasslands in the spring, and (2) to the different feeding practices between the spring and winter in these regions.

Material and methods

The study was conducted on farms in 3 French regions which differ in altitude and likely in the nature of permanent grasslands: Haute-Normandy (HN, 110 m), Isère (IS, 610 m), and Monts-du-Vivarais (MV, 680 m). Eight farms were selected per region. Bulk milk samples were collected between the 10th and 31st May 2007 when cows were fed a diet mainly composed of permanent grassland. The botanical composition of the grazed pasture was determined in each farm by linear surveys (Tornambé *et al.*, 2007). Another sampling period was performed in December in the same farms to assess the effects of feeding practices on milk nutritional quality. Milk production was recorded at sampling and fat content was assayed according to a standard procedure (AOAC, 1997). Spectral index was assessed using a spectrophotometer. Fatty acid (FA) composition was determined by gas chromatography (Ferlay *et al.*, 2008) and carotenoids, vitamins A and E concentrations were determined by UPLC. Data were analysed by ANOVA with region and period as main effects and their interaction.

Results and discussion

The HN grasslands were rich in *Gramineae* and poorly diversified. By contrast, in medium mountains, grasslands were more diversified and dicotyledons reached 35 and 42% in MV and IS, respectively. In the winter, feeding practices varied according to regions: in HN and IS, the diet was based on corn silage, hay or wrapped grass, plus grass silage in IS. In MV, the diet was based on hay or wrapped grass. Milk production and composition were not different between regions and periods. During the spring, the comparison of the concentrations of vitamins and carotenoids, and the colour of milks between regions did not underline any difference. All regions taken together, winter milks were richer in vitamin E but poorer in vitamin A, carotenoids, and less yellow. Feeding cows a grass-based diet even in the winter in MV allowed to limit the differences in carotenoid concentrations between seasons and to maintain the value of spectral index (Nozière *et al.*, 2006). By contrast with the results of Agabriel *et al.* (2007), the vitamin E concentration increased (+60%) in winter milks probably because of dietary vitamin supplementation in this period. Differences in the botanical composition of pastures between regions affected a minor part of the FA composition in spring milks: C4:0 was significantly lower in IS, *c*9,*t*11-C18:2 and *t*11-C18:1 were higher in HN and *c*9,*c*12-C18:2 was lower in HN than in the other regions, in agreement with Tornambé *et al.*, (2007). More notably, in winter milks from HN, C12 to C16:0 FA were lower and monounsaturated FA were higher than in milks from medium mountains. The *c*9,*c*12-C18:2 was lower in MV than in IS and HN in the winter because of the absence of corn silage in the cow diet. In the spring, *c*9,*c*12,*c*15-C18:3 increased (especially in IS) and saturated FA (but not C18:0) decreased in agreement with the literature (Ferlay *et al.*, 2008). Concentrations in *c*9*t*11-C18:2 were higher in the spring than in the winter in all regions and higher in HN than in MV or IS in the spring only.

Table 1. Botanical composition of permanent pastures in the 3 regions (LS Means ± sem).

	HN	IS	MV	sem	P[1]
No. of species (families)	16 (5)	34 (13)	35 (12)	3 (1)	***
% *Gramineae*	81.6	57.7	64.6	4.3	**
% *Fabaceae*	8.9	13.3	13.5	2.7	ns
% *Asteraceae*	5.4	11.3	9.8	2.9	ns
% *Plantaginaceae*	0.1	4.9	4.4	1.9	ns
% *Rosaceae*	-	3.0	0.6	0.7	*

[1] Significance: * $P \leq 0.05$; *** $P \leq 0.001$; ns = not significant.

Table 2. Liposoluble components and colour index of milks (LS Means ± sem).

	HN		IS		MV		sem	P-value[1]		
	Spring	Winter	Spring	Winter	Spring	Winter		Region	Period	Interaction
Spectral index	663[a]	480[b]	605[a]	515[b]	602[a]	601[a]	21	ns	***	***
μg.g⁻¹ of fat										
Vitamin A	9.4[a]	8.7[ab]	9.3[a]	7.2[b]	8.6[ab]	8.1[ab]	0.5	ns	*	ns
Vitamin E	22.5[b]	35.7[a]	22.2[b]	37.5[a]	22.1[b]	36.4[a]	3.1	ns	***	ns
β-carotene	11.4[a]	5.3[b]	10.2[a]	4.8[b]	9.5[a]	6.4[b]	0.6	ns	***	*
g. 100 g^{-1} of total FA										
4:0	2.2[b]	2.7[a]	1.9[c]	2.7[a]	2.6[ab]	2.5[a]	0.1	*	***	***
12:0+14:0+16:0	41.8[c]	47.3[b]	41.3[c]	52.3[a]	43.2[c]	54.1[a]	1.2	**	***	ns
18:0	11.0[a]	9.1[b]	10.7[a]	7.5[c]	10.6[a]	6.9[c]	0.4	**	***	*
*t*11-18:1	2.9[a]	1.2[c]	2.1[b]	0.8[c]	2.0[b]	0.7[c]	0.2	***	***	ns
*c*9,*t*11-18:2	1.2[a]	0.5[c]	0.9[b]	0.4[c]	0.8[b]	0.4[c]	0.1	**	***	ns
*c*9,*c*12-18:2	1.1[b]	1.5[a]	1.6[a]	1.5[a]	1.4[a]	1.2[b]	0.1	***	ns	***
*c*9,*c*12,*c*15-18:3	0.6[a]	0.4[ab]	0.6[a]	0.3[b]	0.5[ab]	0.5[ab]	0.1	ns	***	ns
Saturated FA	61.1[c]	65.6[b]	59.3[c]	69.8[a]	62.4[bc]	70.4[a]	1.2	*	***	ns
Monounsaturated FA	29.0[ab]	26.4[b]	31.1[a]	22.3[c]	28.3[ab]	21.2[c]	1.1	*	***	*
Polyunsaturated FA	4.2[a]	3.2[bc]	4.2[a]	3.0[c]	3.8[ab]	2.9[c]	0.2	ns	***	**

[1] Significance: * $P \leq 0.05$; ** $P \leq 0.01$; *** $P \leq 0.001$; ns = not significant.
[a-c] Statistically different by the SNK test ($P<0.05$).

Conclusion

Among fat-soluble components, some minor milk FA, but with great nutritional interest (t11-C18:1, c9,t11-C18:2), were affected by the botanical composition of pastures. Variations in winter milk composition between regions seemed more dependent nf feeding practices (corn silage vs. conserved grass) in this period.

References

Agabriel, C., A. Cornu, C. Journal, C. Sibra, P. Grolier and B. Martin, 2007. Tanker milk variability according to farm feeding practices: vitamins A and E, carotenoids, color and terpenoids. J. Dairy Sci. 90: 4884-4896.

AOAC, 1997. Official Methods of Analysis, 16th edn. AOAC Int., Gaithersburg, MD, USA.

Ferlay, A., C. Agabriel, C. Sibra, C. Journal, B. Martin and Y. Chilliard, 2008. Tanker milk variability of fatty acids according to farm feeding and husbandry practices in a French semi-mountain area. Dairy Sci. Technol. 88: 193-215.

Nozière P., B. Graulet, A. Lucas, B. Martin, P. Grolier and M. Doreau, 2006. Carotenoids for ruminants: from forages to dairy products. Anim. Feed Sci.Technol. 131: 418-450.

Tornambé, G., A. Ferlay, A. Farruggia, Y. Chilliard, P. Loiseau, J.P. Garel and B. Martin, 2007. Effet de la diversité floristique des pâturages de montagne sur le profil en acides gras et les caractéristiques sensorielles des laits. Renc. Rech. Ruminants 14: 333-336.

Intravenous infusion of a lipid emulsion causes insulin resistance in Merino ewes under hyperinsulinaemic euglycaemic conditions

M.W. Robertson, F.R. Dunshea and B.J. Leury
Department of Agriculture and Food Systems, The University of Melbourne, Parkville, Victoria, 3010 Australia; brianjl@unimelb.edu.au

Introduction

An elevation in blood lipid concentration in humans and rodent species has been shown to cause insulin resistance in several studies (Hawkins *et al.*, 2003; Savage *et al.*, 2007). An increase in blood non-esterified fatty acids (NEFA) in ruminant animals is often associated with periods of negative energy balance, acute stress and/or different physiological states such as pregnancy and lactation. The potential for increased blood NEFA concentration to influence nutrient partitioning through altered insulin sensitivity in ruminants in the non-pregnant, non-lactating state is not well documented. In non-pregnant, non-lactating cows Pires *et al.* (2007) demonstrated some impairment in glucose clearance when cows were infused intravenously with a tallow emulsion but the data were equivocal when similar cows were given an abomasal infusion of linseed oil (Pires *et al.*, 2008). In both these studies insulin sensitivity was estimated by an intravenous glucose tolerance test and insulin challenge. However, no studies have investigated the effect of a lipid infusion on insulin sensitivity in sheep under hyperinsulinaemic euglycaemic clamp (HEC) conditions. Thus, in this study we investigated the effect of a lipid infusion on insulin sensitivity in non-pregnant, non-lactating Merino ewes, to remove the potential confounding associated with pregnancy or lactation, using the HEC technique.

Material and methods

Fifteen non-pregnant, non-lactating Merino ewes (2.5 years old and 46.3±0.8 kg liveweight) were subjected to an overnight fast (13 h) before undergoing an 8-h hyperinsulinaemic (1 mU/kg/min) euglycaemic clamp. The intial 2-h HEC was a baseline clamp and was followed by a 6-h infusion of either saline 0.9% (Control n =9) or Clinoleic 10% lipid emulsion (Lipid n=6) at a rate of 0.01 ml/kg/min. For the lipid infusion only, heparin was added at a rate of 0.2 U per kg/min to increase conversion of infused triglycerides to NEFA, via stimulation of lipoprotein lipase. Blood samples were taken at regular intervals and glucose infusion rate adjusted to maintain glucose concentration within ±10% of pre-clamp blood glucose concentration. Plasma was harvested from blood samples at 15 min intervals and stored at -20 °C before being analysed for glucose, NEFA and insulin. Statistical analysis was conducted using the REML procedure in GENSTAT for all time points and for calculated means of metabolic parameters made over the final hour (1-2 h) of the baseline clamp and over the final hour (5-6h) of the treatment clamp as well as percentage changes within treatments.

Results

Overall there were significant Time × Treatment affects for plasma NEFA (P<0.001), GIR (P<0.001) and insulin sensitivity (P<0.001). After 2.5 h of lipid infusion, NEFA levels had increased significantly (P<0.05) from 100 to 595 μM and concomitantly GIR decreased from 3.1 to 2.35 mg/kg/min and remained around this level for the rest of the lipid infusion. The timing of the fall in GIR (around 2.5 h) following lipid infusion is remarkably similar to human and rodent studies but the magnitude of the reduction (-27%) is less (e.g. -55% in GIR following lipid infusion in humans Boden *et al.*, 1991).

Table 1. Plasma glucose, insulin and NEFA concentrations and glucose infusion rate (GIR) and insulin sensitivity in Merino ewes during a 2 h baseline and 6 h saline or lipid infusion under hyperinsulinaemic euglycaemic clamp (HEC) conditions. Data are the averages calculated over the final hour of the baseline clamp (Baseline) and over the final hour of the treatment clamp (5-6 h).

	Control HEC		Lipid HEC		sed	*P*-value	
	Baseline	5-6 h	Baseline	5-6 h		C vs. L_{Base}	C vs. $L_{5\text{-}6h}$
Glucose mM	3.39	3.09	3.66	3.76	0.25	0.187	0.011
Insulin μU/ml	87.2	87.1	83.1	86.1	6.8	0.422	0.816
NEFA μM	19.1	42.5	99.0	819.8	95.9	0.407	<0.001
GIR (μmol/kg/min)	16.9	22.6	17.2	13.4	1.39	0.830	<0.001
Insulin sensitivity (GIR:insulin)	0.194	0.259	0.207	0.156	0.020	1.000	<0.001

There were no significant differences in any baseline measurement between Control and Lipid treatments (Table 1). There was significantly higher plasma glucose and NEFA concentration and lower GIR and insulin sensitivity at 5-6 h in the Lipid compared with Control treatments (Table 1). When the percentage change in metabolic parameters between Baseline and 5-6 h for both treatments were compared, there was a significant difference in plasma NEFA (+80% and +955% for Control and Lipid HEC, respectively; $P<0.01$), GIR (+34% and -27% for Control and Lipid HEC, respectively; $P<0.001$) and insulin sensitivity (+43% and -29% for Control and Lipid HEC, respectively; $P<0.001$).

Conclusion

Infusion of a lipid emulsion intravenously into non-pregnant, non-lactating Merino ewes under hyperinsulinaemic euglycaemic conditions induces hyperlipidaemia and causes a decrease in GIR necessary to maintain euglycaemia. This presumably occurs as a result of progressive development of insulin resistance but other mechanisms to account for the reduction in GIR maybe involved. Further studies are in progress to investigate the mechanism of this change in insulin sensitivity and whether lipid induced insulin resistance is associated with variation in nutrient partitioning in sheep.

References

Boden G., F. Jadali, J. White, Y. Liang, M. Mozzoli, X. Chen, E. Coleman and C. Smith, 1991. Effect of fat on insulin-stimulated carbohydrate metabolism in normal men. J. Clin. Investig. 88: 960-966.

Hawkins M., N. Barzilai, R. Liu, M. Hu, W. Chen and L. Rossetti, 1997. Role of the glucosamine pathway in fat-induced insulin resistance. J. Clin. Investig. 99: 2173-2182.

Pires J.A.A., A. Souza and R.R. Grummer, 2007. Induction of hyperlipidaemia by intravenous infusion of tallow emulsion causes insulin resistance in Holstein cows. J. Dairy Sci. 90: 2735-2744.

Pires J.A.A., J.B. Pescara, A.E. Brickner, N. Silva Del Rio, A.P. Cunha and R.R. Grummer, 2008. Effects of abomasal infusion of linseed oil on responses to glucose and insulin in holstein cows. J. Dairy Sci. 91: 1378-1390.

Savage D.B., K.B. Petersen, G.I. Shulman, 2007. Disordered lipid metabolism and the pathogenesis of insulin resistance. Physiol. Rev. 87: 507-520.

Effect of feeding solid feed on the hepatic gene expression for the urea cycle and glycogen metabolism in Holstein calves during weaning transition

A.L. Ruiz-Sánchez and M. Oba
Department of Agricultural, Food and Nutritional Science, University of Alberta, Edmonton, T6G 2P5, Canada; masahito.oba@ales.ualberta.ca

Introduction

Dairy calves experience drastic physiological and metabolic changes during the weaning transition (Drackley, 2008). Increasing microbial fermentation from consumption of solid feed is expected to increase organic matter fermentation in the rumen and absorption of ammonia and fermentation acids, and these changes are expected to increase the demand for urea synthesis and gluconeogenesis in the liver. However, hepatic expression of mRNA for arginosuccinate lyase and arginase, which are involved in the urea cycle, decreased with greater consumption of solid feed (Takagi *et al.*, 2008). Further, mRNA expression of hepatic pyruvate carboxylase, a key enzyme for gluconeogenesis, decreased with greater consumption of solid feed (Haga *et al.*, 2008). These observations disagree with expected metabolic adaptation to handle solid feed during the weaning transition, but the calves were fed different diets as they grow in both studies, thus it is not possible to attribute their findings to specific effects of diet or age. Therefore, the objective of the current study was to determine the specific effects of diet on metabolic adaptation of the liver during weaning transition of Holstein calves.

Material and methods

Eight Holstein bull calves at 2 wk of age (44.4±4.0 kg) were blocked by body weight and randomly assigned to either a diet of milk replacer (MR: 22% CP and 17% fat) only or milk replacer plus a commercial calf starter (MR+S: 23% CP). The MR+S calves were fed calf starter *ad libitum*, and MR calves were fed extra milk replacer to provide same metabolisable energy between calves within each block. Dry matter intake was measured daily and body weight was measured twice a week. Both calves in each block were slaughtered three days after a MR+S calf consumed 680 g of starter feed for 3 consecutive days (51.1±6.7 days old). Immediately prior to slaughter, blood samples were collected from a jugular vein to determine plasma glucose concentration. After slaughter, liver samples were taken, snap-frozen and stored at –80 °C until RNA was extracted. The expression of genes encoding for five key enzymes of the urea cycle (carbamoyl phosphate synthetase, ornithine transcarbamoylase, arginosuccinate synthetase, arginosuccinate lyase, and arginase) and four enzymes involved in glycogen metabolism (hexokinase, glycogen synthetase, phosphorylase, glucose-6-phosphatase) were determined using the 7900HT Fast Real-Time PCR System (Applied Biosystems, Foster City, CA, USA). Fold changes in gene expression were calculated for individual calves relative to a calf fed MR. All data were analysed by a randomised block design using initial body weight as the blocking variable.

Results

Average daily gain and plasma glucose concentration on the day of the slaughter were not affected by treatment, averaged at 0.64 kg/d and 6.2 mM, respectively. Contrary to the findings by Takagi *et al.* (2008), the relative mRNA expression of arginosuccinate synthetase was 2.8 times greater for MR+S calves compared to MR animals (Table 1), and it was positively correlated to mRNA expression of carbamoyl phosphate synthetase ($r=0.89$; $P<0.01$), arginosuccinate lyase ($r=0.76$; $P=0.03$), and arginase ($r=0.87$; $P<0.01$). Similarly, the expression of hexokinase mRNA was 1.5 times greater for MR+S calves compared to MR calves.

Table 1. Fold change in mRNA expression of genes encoding for enzymes of the urea cycle and glycogen metabolism in liver tissues of Holstein calves fed milk replacer only or milk replacer plus a commercial calf starter.

	Milk replacer (mean ± SE)	Milk replacer + starter (mean ± SE)	*P*-values
Urea cycle enzymes			
Carbamoyl phosphate synthetase	1.54±0.81	4.07±0.81	0.12
Ornithine transcarbamoylase	0.64±0.24	0.87±0.24	0.56
Arginosuccinate synthetase	1.05±0.39	2.91±0.39	0.04
Aarginosuccinate lyase	1.11±0.79	3.03±0.79	0.18
Arginase	0.67±0.15	1.15±0.15	0.11
Glycogen metabolism enzymes			
Hexokinase	0.84±0.08	1.28±0.08	0.04
Glycogen synthetase	0.89±0.13	0.99±0.13	0.62
Phosphorylase	0.83±0.15	1.17±0.15	0.21
Glucose 6-phosphatase	0.92±0.17	1.12±0.17	0.46

Conclusion

Feeding solid feed increased the expression of the gene encoding for arginosuccinate synthetase and hexokinase in the liver of Holstein dairy calves during weaning transition. Our observation indicates that diet affects metabolic adaptation of the liver during the weaning transition.

Acknowledgement

We acknowledge Anne Laarman for his technical assistance and Zen-Raku-Ren for financial support.

References

Drackley, J.K., 2008. Calf nutrition from birth to breeding. Vet. Clin. Food. Anim. 24: 55-86.

Haga, S., S. Fujimoto, T. Yonezawa, K. Yoshioka, H. Shingu, Y. Kobayashi, T. Takahashi, Y. Otani, K. Katoh and Y. Obara, 2008. Changes in hepatic key enzymes of dairy calves in early weaning production systems. J. Dairy Sci. 91: 3156-3164.

Takagi, M., T. Yonezawa, S. Haga, H. Shingu, Y. Kobayashi, T. Takahashi, Y. Ohtani, Y. Obara and K. Katoh, 2008. Changes of activity and mRNA expression of urea cycle enzymes in the liver of developing Holstein calves. J. Anim. Sci. 86: 1526-1532.

Continuous lactation effects on mammary extraction rates of nutrients in dairy goats

S. Safayi and M.O. Nielsen
Department of Basic Animal and Veterinary Sciences, Faculty of Life Sciences, University of Copenhagen, Groennegaardsvej 7, 1870 Frederiksberg C, Denmark; mon@life.ku.dk

Introduction

Having a dry period appears to be essential for dairy cows to ensure optimal milk production in the following lactation (Madsen *et al.*, 2008). The situation in the dairy goat is less clear, since omission of the dry period (continuous lactation; CL) was reported to have no negative effect on subsequent milk yield in a study by Fowler *et al.* (1991), while Caja *et al.* (2006) reported a 29% decrease in milk yield in the subsequent lactation. We found (Safayi *et al.*, 2009, unpublished data) that in the goat, mammary glands subjected to CL will enter the new lactation with a smaller mammary epithelial cell (MEC) population, but the MEC are more fully differentiated compared to in the normal lactating (NL) glands. If fully differentiated MEC are older and more secretory active, this could in part explain why goats can lactate continuously without major negative impact on milk yield despite a smaller MEC population. The aim of the present study was to determine if CL in the dairy goat will increase the efficiency of nutrient extraction across the mammary gland in the following lactation, which could reflect differences in MEC secretory activity.

Material and methods

Nine dairy goats were used followed over one (5 goats) or two consecutive (4 goats) pregnancy-lactation periods. Goats were previously surgically prepared with exteriorised carotid arteries and milk veins. They were milked manually (at 09:00 h and 15:30 h) and fed twice a day (at 07:30 h and 14:30 h), half the ration being given at each feeding. The experimental design was a randomixed complete block design; the two udder halves in each animal were randomly assigned to two different treatments, thus using the animal as its own control: One udder half was dried-off approximately 9 wk *pre partum* (NL), and the other udder half was milked continuously (CL) until parturition. After morning and afternoon milking, two blood samples were obtained from the exteriorised carotid artery and each of both milk veins, just before the time of drying-off of the NL gland (BDP), within the first two weeks after drying-off (early dry period; EDP), in the mid-dry period (MDP), within the last 2 weeks prior to parturition (late dry period; LDP), and at days 1 (day of parturition, D1), 3, 10, 60, and 180 of lactation. Blood samples were taken for immediate determination of acid-base parameters, and plasma was analysed for glucose, non-esterified fatty acids (NEFA), acetate, beta-hydroxy-butyrate (BOHB), triglycerol (TG), and urea. Mammary extraction (E) rates of metabolites were calculated as mammary arterial-milk vein concentration difference (AVD) divided by arterial concentration. Statistical evaluation of all data was performed using the Proc Mixed procedures in SAS® (2003). Variables in the statistical models included the experimental year, treatment (CL or NL), the sampling time (morning or afternoon), stage at which the sample was taken and its interaction with treatment as fixed effects, and the factors goat within experimental year, and milk vein side within goat at each stage as random effects. *Pre-* and *post partum* values were analysed separately. The PDIFF command in SAS® was used to generate comparisons between treatment means. The level of significance was set at $P<0.05$ and tendencies at $P<0.10$.

Results

There was not any indication of CL affecting the milk yield negatively in the goats in this study. There was no effect of treatment on acetate AVD and E *pre partum*, but for glucose and BHB, they were significantly higher in CL than NL glands. *Post partum* AVD and E for glucose and BHB were higher compared to *pre partum* in both glands, but there was no difference between CL and NL glands during the *post partum* period. NEFA AVD and E were negative *pre partum*, and became more negative across CL than NL glands in the LDP (P=0.01). Both peaked at the same positive level in both glands at parturition, due to a rise in arterial plasma concentration. They remained high in early lactation, but decreased to become negative again in mid-late lactation, thus following overall changes in arterial plasma concentration. AVD and E for TG were higher across CL than NL glands *pre partum*, decreased across the gestation period, and then increased after parturition to the same level in both CL and NL glands. The *post partum* increase was numerically more marked in NL glands so that the AVD and E reached the same level as in CL glands during lactation. AVD for pO2 and pCO2 as well as AVD and E for TO2 and TCO2 were not affected by the treatment, neither *pre-* nor *post partum*. *Pre partum* AVD tended to be higher for pH and lower for BE across CL compared with NL glands.

Conclusion

AVD and/or E differed for several metabolites either significantly or numerically between the CL and NL glands during the late gestation period. This could reflect differences in extraction activity between the CL glands that remained lactating and the NL glands that were dried off *pre partum*. However, there was no significant impact of the *pre partum* milking of the CL glands on efficiency of mammary nutrient extraction in the subsequent lactation. Our previous findings showed that the goat mammary gland subjected to CL had lower cell renewal in the subsequent lactation, but MEC, however, became more fully differentiated (unpublished data). The observation that extractions of nutrients were unaffected by CL do not lend immediate support to the hypothesis that these more fully differentiated MEC could have a more efficient nutrient extraction reflecting a higher metabolic activity in the subsequent lactation. The explanation why milk yield in dairy goats is relatively unaffected in CL (in contrast to dairy cows) thus remains to be established.

References

Caja, G., A.A.K. Salama and X. Such, 2006. Omitting the dry-off period negatively affects colostrum and milk yield in dairy goats. J. Dairy Sci. 89: 4220-4228.

Fowler, P.A., C.H. Knight and M.A. Foster, 1991. Omitting the Dry Period Between Lactations Does Not Reduce Subsequent Milk-Production in Goats. J. Dairy Res. 58: 13-19.

Madsen, T.G., M.O. Nielsen, J.B. Andersen and K.L. Ingvartsen, 2008. Continuous lactation in dairy cows: Effect on milk production and mammary nutrient supply and extraction. J. Dairy Sci. 91: 1791-1801.

Effects of *i.v.* administration of apelin on endocrine in sheep and goats

K. Sato, Y. Kobayashi, T. Takahashi and K. Katoh
Laboratory of Animal Physiology, Graduate School of Agricultural Science, Tohoku University, Sendai 981-8555, Japan; gariyoshikun@bios.tohoku.ac.jp

Introduction

Apelin, a peptide recently isolated from bovine stomach extracts, appears to act as an endogenous ligand for the previously orphaned G-protein-coupled APJ receptor (Tatemoto *et al.*, 1998). Apelin consists of 77 amino acids that are processed to produce several smaller peptide fragments: apelin-36, -17, -13, and -12. Apelin mRNA is expressed in the brain, spinal cord, pituitary, heart, lung, testis, adipose tissue, placenta and mammary gland. Apelin and APJ mRNA have also been demonstrated to exist in the supraoptic and paraventricular nuclei in the hypothalamus. These areas seem to be important for regulating food and water intake (De Mota *et al.*, 2004) because apelin is reported to modulate the plasma AVP and ACTH levels (Charles *et al.*, 2006). We used apelin-13 to investigate whether apelin modulates anterior pituitary and islet hormones in two species of ruminants.

Material and methods

Four crossbred sheep (5 mo old, 40.8±0.76 kg BW, females) and six Saanen goats (4 mo old, 18.9±1.00 kg BW, castrated males) were fed in pens. The animals had *ad libitum* access to water and minerals, and were fed orchardgrass hay once a day at 16:00 h.

In this study, we used [Pyr^1]-Apelin-13 (Peptide Institute, Osaka, Japan) because out of all the apelin peptides, it is the most biologically active (Tatemoto *et al.*, 1998). The C terminal of the peptide is modified pyroglutamic acid.

Each animal was subjected to the administration of vehicle (0.9% NaCl solution) or [Pyr^1]-Apelin-13. One mg of apelin was administered by *i.v.* as a 5 ml solution over 1 min for the sheep, and 0.5 mg for the goats. Saline and apelin were administered in a balanced crossover design with an interval of a week. Venous blood samples, collected through a catheter inserted into the jugular vein with preset intervals, were taken into chilled heparin- or EDTA-containing tubes with aprotinin, and were centrifuged to obtain plasma. The plasma samples were stored at -30 °C before assay for apelin, GH, ACTH, and insulin by radioimmunoassay.

Data were expressed as the mean ± SEM, and $P<0.05$ was considered statistically significant. Significant differences from the baseline, the area under the curve (AUC) and the incremental area (ICA) were analysed by *t*-test.

Results

In sheep, apelin administration did not cause a significant change in the plasma GH concentration, as shown in Table 1, but did cause a significant increase in ICA (Apelin: 524.3±102.6 ng/min/ml and Saline: 169.4±48.1 ng/min/ml, P=0.020). However, apelin administration significantly increased the plasma concentrations of ACTH, insulin, glucose and NEFA ($P<0.05$).

In goats, apelin administration did not induce significant effects on any parameters measured (data not shown).

Table 1. Changes in plasma concentrations of hormones and metabolites in response to apelin administration in sheep.

	Baseline	15 min	30 min	45 min	60 min
Plasma Apelin (ng/ml)					
Control	5.5±0.9	5.1±1.0	3.8±1.5	4.2±1.4	3.8±1.5
Apelin	4.6±0.8	6.4±0.9*	5.1±0.3	4.7±0.2	4.5±0.8
Plasma GH (ng/ml)					
Control	9.36±2.07	5.45±2.51	2.61±0.11	1.49±0.01	1.68±0.06
Apelin	6.77±1.95	10.41±5.48	9.60±3.17	8.58±2.53	2.53±0.52
Plasma ACTH (pg/ml)					
Control	30.1±26.3	18.2±4.3	40.5±2.3	41.4±16.0	44.2±22.8
Apelin	49.8±29.9	97.9±6.3*	140.6±80.7*	51.4±25.5	54.1±23.6
Plasma Insulin(μU/ml)					
Control	0.91±0.01	0.87±0.09	0.87±0.12	0.80±0.07	0.78±0.17
Apelin	1.12±0.07	1.85±0.29*	1.75±0.41*	1.65±0.45	1.33±0.37
Plasma Glucose (mg/dl)					
Control	58.3±3.0	55.0±3.1	52.2±0.4	51.3±0.2*	51.0±0.6*
Apelin	59.7±2.3	80.1±12.3*	72.1±10.4*	72.0±10.0*	67.5±9.4
Plasma NEFA (mEq/l)					
Control	0.29±0.03	0.29±0.04	0.26±0.02	0.28±0.01	0.31±0.04
Apelin	0.30±0.02	0.39±0.05*	0.43±0.06*	0.54±0.05*	0.41±0.01*

Values are means ± SEM (n = 4); * $P<0.05$: significant difference from the baseline (*t*-test).

Conclusion

The present study demonstrates that (1) apelin administration at the dosage used increases the plasma concentrations of ACTH, insulin, glucose and NEFA in sheep, but does not cause any changes in these parameters in goats; (2) apelin may be involved in the regulation of the HPA axis in sheep; and (3) there is a species-difference between sheep and goats in the role apelin plays.

Acknowledgement

This work was supported by the Grant-in-Aid from the Japanese Ministry of Education, Culture, Sports, Science and Technology.

References

Charles, C.J., M.T. Rademaker and A.M. Richards, 2006. Apelin-13 induces a biphasic haemodynamic response and hormonal activation in normal conscious sheep. J. Endocrinol. 189: 701-710.

De Mota, N., A. Reaux-Le Goazigo, S. El Messari, N. Chartrel, D. Roesch, C. Dujardin, C. Kordon, H. Vaudry, F. Moos and C. Llorens-Cortes, 2004. Apelin, a potent diuretic neuropeptide counteracting vasopressin actions through inhibition of vasopressin neuron activity and vasopressin release. Proc. Natl. Acad. Sci. USA 101: 10464-10469.

Tatemoto, K., M. Hosoya, Y. Habata, R. Fuji, T. Kakewaga, M. Zou, Y. Kawamata, S. Fukusumi, S. Hinuma, C. Kitada, T. Kurokawa, H. Onda and M. Fujimoto, 1998. Isolation and characterization of a novel endogenous peptide ligand for the human APJ receptor. Biochem. Biophys. Res. Commun. 251: 471-476.

Trans-10,*cis*-12 conjugated linoleic acid reduces milk fat synthesis and insulin sensitivity in goats during early lactation

Ph. Schmidely[1], S. Hourte[2] and M. Magnin[2]
[1]UMR 791 INRA-AgroParisTech, 75321 Paris Cedex 05, France; [2]BASF Nutrition Animale, 52300, Chateau-Gonthier, France; philippe.schmidely@agroparistech.fr

Introduction

Conjugated linoleic acids (CLA) are positional and geometric isomers of 18:2 fatty acids (FA) with a conjugated double bond, which are produced during ruminal biohydrogenation of polyunsaturated FA. Dietary *trans*-10,*cis*-12 18:2 (10t,12c) inhibits milk fat synthesis during established lactation of cows and ewes, whereas 10t,12c is less effective during the first wks *post partum* (Bauman *et al.*, 2008). In goats (Lock *et al.*, 2008; Shingfield *et al.*, 2009), reduction in milk fat synthesis was dose- related to the supply of 10t,12c but less than in cows and ewes. However, these results were obtained during mid to late lactation and during short duration of 10t,12c supplementation (14 d). Consequently, the primary objective of this trial evaluates temporal response of milk fat content (MFC) and FA profile in dairy goats fed 10t,12c in early lactation. A secondary objective was to test if 10t,12c altered insulin (INS) sensitivity by the use of an euglycemic-hyperinsulinemic clamp (EHC).

Material and methods

Twelve multiparous goats fed a TMR (Hay, dehydrated alfalfa, sugar beet pulp and concentrate) were randomly allocated at 7 d *post partum* (PP) for 73 d to CTL (45 g/d of Ca salt of palm oil) or CLA group (45 g/d of Lutrell, 10% of 9c,11t and 10% of 10t,12c isomer, BASF Nutrition Animale). CLA supplement was top dressed on the TMR on a.m. (15g) and p.m. (30 g) feeding to provide 4.5 g/d 10t,12c. DMI and milk production, and milk composition were recorded daily and weekly respectively. Milk FA profile was determined 10, 15, 25, 55 and 75 d PP. At 78d PP, 3 goats in each group were challenged by an EHC with jugular infusion of increasing doses of INS: 7, 13, and 35 ng/min/kg BW during 90 min for each dose. Glycemia was checked every 10 min and euglycemia was obtained by infusing incremental dosages of glucose (30, 50, and 70 g/l). Steady state conditions for concentration of glucose and insulin (plateau) were obtained during the last 30 min of infusion. DMI, milk production and composition, and milk FA were analysed as a factorial design with repeated measurements (MIXED procedure of SAS®) with time and CLA treatment as fixed effects, and goat as random effect. Plateau insulin concentration and glucose infusion rate were analysed using the GLM procedure with dose of insulin and CLA treatment as fixed effects.

Results

DMI and milk yield were not affected by CLA. MFC was not affected by CLA at 10d PP, but it was progressively reduced by 9.8 g/l at 75 d PP (25% reduction). In the whole trial, an average reduction in MFC was 15% in CLA compared to CTL goats. The proportion of 10t,12c in milk increased from 0.10 to 0.20% of total FA in CLA goats whereas it was below detectable values in the CTL goats. Short chain FA (4 to 8 C) was not affected by treatment (Figure 1), whereas C10:0 to C16:0 in milk were lower in CLA goats. Compared to CTL, C18:0 and 10t-C18:1 proportions were higher in CLA goats, whereas 9c-C18:1 was lower (not shown). During EHC, the INS plateau was higher in CLA than in CTL goats except at the higher infusion rate (Table 1), whereas glucose infusion rates were lower in CLA than in CTL goats.

Table1. Effect of CLA on glucose infusion rate during euglycemic-hyperinsulinemic clamp.

Insulin, ng/min/kg BW	7		13		35	
Treatment	CT	CLA	CTL	CLA	CTL	CLA
INS at plateau, ng/ml	0.71[a]	0.93[b]	1.01[c]	1.27[d]	2.86[e]	2.17[f]
GLU infusion, ng/min/kgBW	0.74[a]	0.45[b]	1.26[c]	0.46[b]	2.36[d]	1.25[c]

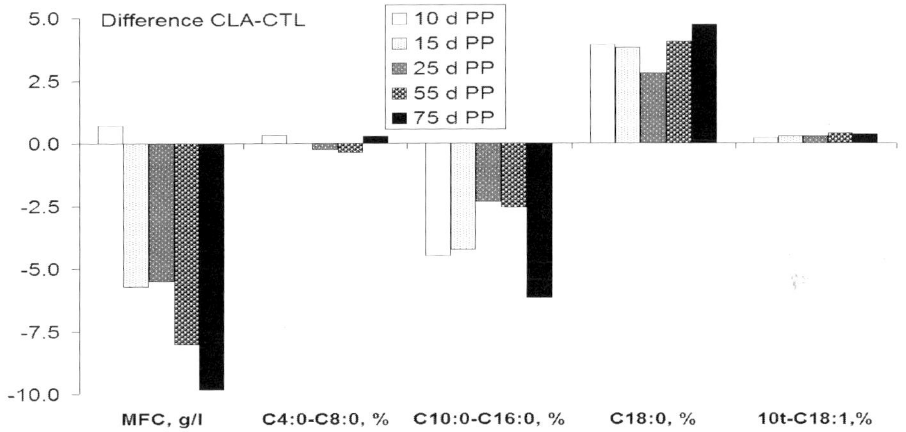

Figure 1. Differences in milk fat content (MFC) and milk FA profile between goats fed CLA vs. CTL diets.

Discussion

In our trial, 4.5g/d 10t,12c-CLA decreased MFC in early lactation (-15% in average) in agreement with data from goats in mid to late lactation: 3 to 6 g/d 10t,12c reduced MFC by 5 to 21% (Lock *et al.*, 2008), whereas 7.5 g/d 10t,12c decreased MFC by 20% (Shingfield *et al.*, 2009) with similar 10t,12c proportions in milk fat in these studies. This indirect comparison was rather different from data of dairy cows where response in early lactation (at least during the first 3 wks PP) was lower than during established lactation. During euglycemic-hyperinsulinemic clamp, lower glucose infusion rate in the CLA goat may reflect a lower insulin sensitivity; this was in contrast to data in cows (Bauman *et al.*, 2008) that suggested insulin resistance is not a major mechanism for lower milk fat content.

References

Bauman, D.E., J.W. Perfield II, K.J. Harvatine and L.H. Baumgard, 2008. Regulation of fat synthesis by Conjugated Linoleic Acid: lactation and the ruminant model. J. Nutr. 138: 403-409.

Lock, A.L., M. Rovai, T.A. Gipson, M.J. de Veth and D.E. Bauman, 2008. A conjugated linoleic acid supplement containing *trans*-10, *cis*-12 conjugated linoleic acid reduces milk fat synthesis in lactating goats. J. Dairy Sci. 91: 3291-3299.

Shingfield, K.J., J. Rouel and Y. Chilliard, 2009. Effect of calcium salts of a mixture of conjugated linoleic acids containing *trans*-10, *cis*-12 in the diet on milk fat synthesis in goats. Br. J. Nutr. 101: 1006-1019.

Effect of two feeding levels on growth, blood metabolites and insulin in postweaning dual purpose cattle

I. Seijas, K. Drescher, L. Pinto-Santini, A. Ruiz-Gaviria, A. Ruiz and N. Martínez
Instituto de Producción Animal, FAGRO, UCV, Maracay, 4579, Venezuela; ilianneseijas@hotmail.com

Introduction

In Latin American tropics cattle growth receives little attention, since the main income of these farms comes from daily sale of milk. The feeding of these animals is based on medium to low quality tropical forages, which limit the availability of critical nutrients for high demanding physiological processes in which strategic supplementation is required (Recabarren, 2005). The effects of feed restriction on growth in crossbred animals (*Bos indicus* × *Bos taurus*) during post weaning have not been studied in depth and the information related to alterations in metabolic and hormonal parameters is preliminary (Raja *et al.*, 1981; Di Marco *et al.*, 2006). There is clear evidence in meat breeds that a combination of hormones can have an effect on animal growth, among which insulin is involved as an anabolic stimulator (Owens *et al.*, 1995). Since dual-purpose systems in the tropics contribute to 50% of the meat for consumers, tissue muscle growth is the highest in economic importance and it is important to assess. Therefore, an experiment was designed to evaluate the effect of two levels of feeding on growth, glucose, cholesterol metabolites and the hormone insulin in dual-purpose post-weaning cattle.

Material and methods

An experiment was carried out using a completely randomised design in an experimental production unit located at 10 ° 17 '5 "N, 64 ° 13' 28" W and 432 m, with castrated crossbred calves (C) (5/8 Holstein – 3/8 Brahman), during 28 wk from 8 months of age (initial body weight of 115.00 ± 19.82 kg). All C had *ad libitum* water, hay (H) of *Cynodon nlemfuensis* (10.39% CP, 81.1% NDF, 36.9% ADF) and a supplement (S)-based on tropical resources (13.3% CP, 34.1% NDF, 9.17% ADF, 1.1% Ca and 1.0% P). C were randomly placed in individual semi-roofed pens of 18 m^2 and divided into two groups: TA (n=5), high feeding level (DM was based on 3.5% of BW, relation H:S was 60:40) and TB (n=5), low feeding level (2.5% and 70:30, respectively). The model included the effect due to T and the residual. Measures of DM intake of H and S (kg/animal/day) were performed daily through the difference between offered and refused food. Body weight (BW, kg), average life weight gain (LWG, kg/animal/day), height (HT, cm), body length (BL, cm) and thoracic perimeter (TP, cm) were performed weekly. Three animals were selected randomly from each treatment in weeks 2 and 28 and a catheter (Intracath ® 1.1 mm thick and 20.3 cm. A long, needle 1.5×5.1 mm) was placed in the jugular vein, in order to measure glucose (GLU) (Henry *et al.*, 1974), cholesterol (COL) (Trinder, 1969) and insulin (INS) (DRG EIA-2935 ®), each 20 minutes of the start of the consumption of feed and during 8 hours. For intake and zoometric variables and given the low n, a Kruskal Wallis test was applied. ANOVA was applied to GLU, COL and INS with evaluation of TA and TB, sampling time and random error. In both cases Statistix 8.0 (2003) was used.

Results

With respect to H and S intake, significant differences ($P<0.05$) were obtained between TA and TB from 1 to 28 wk. The intakes at 28 wk (kg DM/animal/d) were the following: 5.4 H + 3.8 S vs. 3.2 H + 0.88 S for TA and TB, respectively. The BW, HT, BL and TP values remained similar ($P>0.05$) until weeks 12, 21, 21 and 23, respectively, from that wk differences were appreciated ($P<0.01$). At

28 wk, BW were 259 (±36.09) and 199.20 (±22.23) kg (±SD) for TA and TB, respectively, LWG were 0.74±0.07 vs. 0.388±0.06 in TA and TB (P<0.01), respectively. At the second week GLU (mg/dl), CHOL (mg/dl) and INS (μIU/ml) did not show differences between treatments. However, in the 28th wk GLU was 68.39±2.3 and 39.12±2.4 (P<0.01) for TA and TB, respectively; COL reached values of 192.16±5.6 in TA and was different (P<0.01) to TB (157.5±5.6). With regard to INS, there was also a difference (P<0.01) between TA and TB (21.03±0.9; 10.87±1.0) at the 28th wk. TA animals expressed their growth potential by feeding level. BW and the height (HT and BL) of the animal were affected by the permanent availability of GLU and COL. The highest released INS showed that its use led to the formation of tissues characteristic of the phase under study: bone and muscle.

Conclusion

The study shows that the level of feeding affects crossbred animal growth, measured under different parameters, as well as blood profiles, glucose, cholesterol and insulin. Changes of 100% in the INS values found at week 28 shows the impact of the level of replacement of total nutrients on a blood parameter highly influential on animal growth. In future studies, more hormones (GH, IGF1, T3, T4), metabolites (BUN) and enzymes (calpains) need to be quantified, to show any particular effects on those tissues of higher commercial value.

Acknowledgement

This Project was sponsored by the Fondo Nacional de Ciencia, Tecnología e Innovación throughout a Grant G-200500446. The authors thank experimental cowshed workers and staff of the laboratory.

References

Di Marco, O., J. Barcellos and E. Da Costa, 2006. Crescimento de bovinos de corte. Porto Alegre: Departamento de zootecnia da Universidade Federal do Rio Grande do Sul. p. 248.

Henry R., D. Cannon and J. Winkelman, 1974.Clinical Chemistry, Principles and Techniques. 1059 p.

Owens, F., D. Gill, D. Secrist and S. Coleman, 1995. Review of some aspect of growth and development of feedlot cattle. J. Animal. Sci. 73: 3152-3172.

Raja, D. R. Moss, J. Murray, and T. Leng, 1981. Crecimiento y metabolismo de la glucosa en becerros pastoreando pastos tropicales. I. Efecto de la suplementación con maiz y harina de algodón. Producción Animal Tropical. 6: 117-128.

Recabarren, S. 2005. Sensibilidad a la insulina en ovejas prepuberes con alimentación normal y con restricción alimenticia. Arch. Med. Vet. 37: 111-116.

Statistix 8.0, 2003. Analytical Software.

Trinder, P., 1969. Determination of glucose in blood using glucose oxidase with an alternative oxygen acceptor. Ann. Clin. Biochem. 6: 24-25.

Characteristics of galactopoietic and lipolytic effects of exogenous growth hormone-releasing hormone in lactating Japanese Black cows under negative energy balance

H. Shingu[1], S. Kushibiki[1], E. Touno[2], A. Oshibe[2], Y. Ueda[3], M. Shinoda[1] and K. Hodate[4]
[1]National Institute of Livestock and Grassland Science, Tsukuba, 305-0901, Ibaraki, Japan; [2]National Agricultural Research Center for Tohoku Region, Morioka, 020-0198, Iwate, Japan; [3]National Agricultural Research Center for Hokkaido Region, Sapporo, 062-8555, Hokkaido, Japan; [4]Kitasato University, Towada, 034-8628, Aomori, Japan; shinguu@affrc.go.jp

Introduction

In lactating ruminants, growth hormone (GH) exerts galactopoietic and lipolytic effects, resulting in the facilitation of preferential partition of nutrients to the mammary gland for milk production (Bauman and Currie, 1980). In dairy cows under positive energy balance (EB), GH and growth hormone-releasing hormone (GHRH) treatments induce galactopoietic and lipolytic effects. In contrast, in dairy cows under negative EB, the response of milk production to GH and GHRH treatments is negligible. However, there has been evidence to support the speculation that exogenous GH exerts a galactopoietic effect in dairy cows under negative EB (see Carriquiry *et al.*, 2008). Japanese Black cows, or Wagyu (a beef breed), have less plasma GH and nonesterified fatty acids (NEFA) and greater insulin concentrations, compared with Holstein cows (Shingu *et al.*, 2002). Moreover, this beef breed has a smaller proportion of somatotrophs in the adenohypophysis and a greater fat percentage in carcass than dairy breeds (see Shingu *et al.*, 2001), indicating that lactating Japanese Black cows have greater anabolic actions, compared with Holstein cows. However, it is not known whether GHRH would exert the effect in beef cows under negative EB. The present study was aimed at examining the galactopoietic and lipolytic effects of exogenous GHRH in lactating Japanese Black cows under negative EB.

Material and methods

Ten multiparous Japanese Black cows were used from 6 d in milk (DIM), when experimental dietary treatment was started. The types of the diets were total mixed ration formulated for 130% (a high-energy diet: HED) and 80% (a low-energy diet: LED) of ME requirements for maintenance and milk production. Based on daily requirements for maintenance (ME_M) and lactation (ME_L), cow BW and daily milk yield, total daily ME of cows to meet requirements (ME_T) was decided (Japanese Feeding Standard for Beef Cattle, 2000). The theoretical ME to meet the requirement in cows of HED [$ME_{T(HED\ cows)}$] and LED [$ME_{T(LED\ cows)}$] was calculated by multiplying the ME_T by 1.3 and 0.8. When nutritional values of diet residues were defined as ME_R, EB was calculated daily using the equation, $ME_{T(HED\text{-}fed\ cows\ or\ LED\text{-}fed\ cows)} - (ME_T + ME_R)$. From 36 to 56 DIM, all cows received daily subcutaneous injection of 3-mg bovine GHRH. Blood samples were collected from the jugular vein at intervals of 2 or 3 d and milk samples were collected twice daily during periods of pretreatment (29 to 35 DIM) and GHRH treatment. Data were analysed using the GLM procedures (repeated method; SAS® Inst. Inc. Cary, NC). The statistical model included factors of diet, individual cow, period, interaction between diet and period and residual. Differences were considered significant at $P<0.05$.

Results

Average BW of HED- and LED-fed cows was 443 and 456 kg at 6 DIM ($P>0.05$) and 473 and 446 kg at 56 DIM ($P<0.05$), and HED- and LED-fed cows consumed dietary ME equivalent to 128.3% and 78.2% of the ME requirement. These results showed that HED- and LED-fed cows were under positive and negative EB during the experimental period. By administration of GHRH, BW and amounts of intake of diet were unaffected in both HED- and LED-fed cows. GHRH treatment increased ($P<0.01$) milk yield of HED-fed cows, although there was no change in milk yield of LED-fed cows (Table 1). In addition, milk components were unaffected by GHRH treatment in both HED- and LED-fed cows (Table 1). GHRH treatment increased ($P<0.05$) plasma GH concentration in HED- (1.3 to 3.8 ng/ml) and LED-fed cows (1.3 to 2.8 ng/ml), although plasma NEFA concentration was unaffected by administration of GHRH in both HED- (141 vs. 132 μEq/l) and LED-fed cows (250 vs. 239 μEq/l).

Table 1. Milk yield and components in HED- and LED-fed cows by daily GHRH treatment.

	Diet[1]	Period[2]		Pooled SE
		Pre-treatment	GHRH treatment	
Milk yield, kg/d	HED	5.3^{a}	6.2^{b}	0.4
	LED	3.5	3.3	
Milk fat,%	HED	5.0	5.1	0.1
	LED	4.9	5.1	
Milk protein,%	HED	3.8	3.7	0.0
	LED	3.8	3.7	
Lactose,%	HED	4.9	4.8	0.0
	LED	4.8	4.7	

[1] HED: cows offered a high-energy diet to meet 130% of ME requirements (n = 5), LED: cows offered a low-energy diet to meet 80% of ME requirements (n = 5).
[2] For period: pre-treatment, 29 to 35 d in milk; and GHRH treatment, 36 to 56 d in milk.
[a,b] Means within rows with different superscripts are significantly different ($P<0.05$).

Conclusion

This study shows that GHRH is less capable of lipid mobilisation for milk production in the beef breed, irrespective of EB, compared with the dairy breed, suggesting that lactating Japanese Black cows have physiological characteristics of greater anabolic action compared with dairy cows.

References

Bauman, D.E. and W.B. Currie, 1980. Partitioning of nutrients during pregnancy and lactation: A review of mechanisms involving homeostasis and homeorhesis. J. Dairy Sci. 63: 1514-1529.

Carriquiry, M., W.J. Weber and B.A. Crooker, 2008. Administration of bovine somatotropin in early lactation: A meta-analysis of production responses by multiparous Holstein cows. J. Dairy Sci. 91: 2641-2652.

Shingu, H., K. Hodate, S. Kushibiki, Y. Ueda, A. Watanabe, M. Shinoda and M. Matsumoto, 2001. Profiles of growth hormone and insulin secretion, and glucose response to insulin in growing Japanese Black heifers (beef type): Comparison with Holstein heifers (dairy type). Comp. Biochem. Physiol. C 130: 259-270.

Shingu, H., K. Hodate, S. Kushibiki, Y. Ueda, A. Watanabe, M. Shinoda and M. Matsumoto, 2002. Breed differences in growth hormone and insulin secretion between lactating Japanese Black cows (beef type) and Holstein cows (dairy type). Comp. Biochem. Physiol. C 132: 493-504.

Chemerin, highly expressed in adipose tissues, stimulates the glycerol release in bovine differentiated adipocytes *in vitro*

S.H. Song[1], K. Fukui[2], K. Hamano[2], S. Sasaki[2], S.G. Roh[1] and K. Katoh[1]
[1]Laboratory of Animal Physiology, Graduate School of Agricultural Science, Tohoku University, Aoba-ku, Sendai 981-8555, Japan; [2]Faculty of Agriculture, Shinshu University, Nagano-ken 399-4598, Japan; sanggun_roh@bios.tohoku.ac.jp

Introduction

Adipose tissue is now recognised not only as a reservoir of energy-rich molecules, but also as an active endocrine organ that secretes multiple bioactive factors termed adipokines, e.g. leptin, TNF-α and adiponectin. Molecular events comprising the 'adipocyte differentiation program (or adipogenesis)' include the regulation of genes involved in all aspects of adipocyte metabolism. The major transcription factors of adipogenesis, including proliferator-activated receptor-γ2 (PPAR-γ2) and CCAAT/enhancer binding protein α (C/EBPα), play a key role in the complex transcriptional cascade during adipocyte differentiation (Roh *et al.*, 2006). Adipokines have biological functions to regulate the metabolic status through endocrine-, paracrine-, and autocrine-mediated pathways. Recently, we reported that chemerin as a new adipokine is highly expressed in adipose tissues of the mouse and up-regulated during adipocyte differentiation (Roh *et al.*, 2007). The purpose of this study was to investigate (1) the expression patterns of chemerin and its receptor in bovine adipose tissue and differentiated adipocyte and (2) whether chemerin modulates the lipolysis via its own receptor in bovine differentiated adipocytes *in vitro*.

Material and methods

Sixteen different tissues were sampled from 3 female Japanese Black Cattle (69, 82, 188 months of age). All tissues were rapidly separated, immediately frozen in liquid nitrogen, and stored at -80°C until RNA extraction. All experiments were conducted in accordance with the Institute Guide for the Care and Use of Experimental Animals.

For preadipocyte culture and differentiation, subcutaneous adipose tissues were sampled from male Japanese Black Cattle (10 months of age). The preparation and culture of preadipocytes and the induction of adipocyte differentiation were performed as previously described (Hong *et al.*, 2006). Total RNA was extracted from bovine tissues and the confluent preadipocytes and differentiated adipocytes in culture dishes. Semi-quantitative RT-PCR to measure the levels of bovine chemerin, chemerin receptor, PPAR-γ2 and GAPDH mRNA was performed as previously described (Hong *et al.*, 2005).

To investigate the function of chemerin, bovine adipocytes fully differentiated for 12 days were treated with mouse recombinant chemerin for 24 h, and then culture medium was collected for the analysis of glycerol.

Results

The bovine chemerin expression was higher in adipose tissue and the liver, and found at a much lower level in other tissues (data not shown). There was widespread expression of the chemerin receptor gene; its transcript was present in all tissues examined. Oil red staining showed that lipid droplet formations were increased depending on the incubation day of bovine differentiated adipocytes (Figure 1A). Chemerin and its receptor were increased during adipocyte differentiation (Figure 1B). Mouse recombinant chemerin stimulated the glycerol release from bovine differentiated adipocytes in a dose-dependent manner (Figure 2).

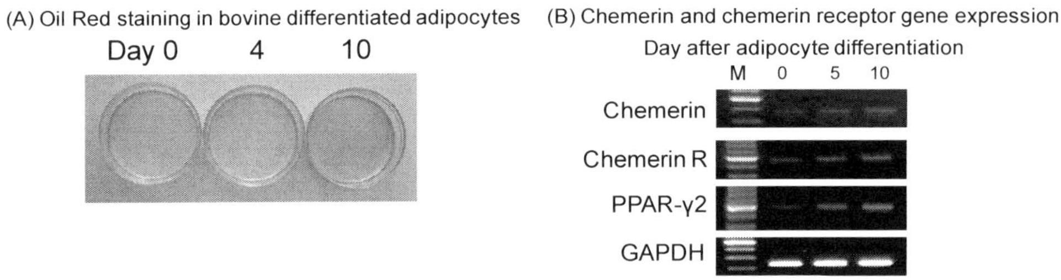

Figure 1. Oil red staining and the expression of bovine chemerin and chemerin receptor mRNA during the adipocyte differentiation of bovine preadipocytes.

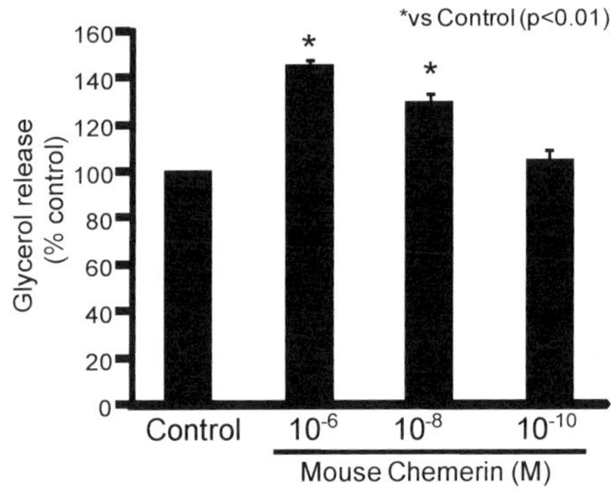

Figure 2. Glycerol release by mouse chemerin in bovine differentiated adipocytes.

Conclusion

This study shows that chemerin and its receptor were expressed in bovine adipocytes and adipose tissues and mouse chemerin stimulated the glycerol release from bovine adipocyte *in vitro*. This suggests that chemerin induces the catabolic process leading to the breakdown of triglycerides stored in fat cells and the release of fatty acids and glycerol by autocrine/paracrine mechanisms. However, further study is in progress to investigate the effect of chemerin in adipocytes.

Acknowledgement

This work was partly supported by Grant-in-Aid for scientific research from the Ministry of Education, Science, Sports and Culture of Japan.

References

Hong, Y.H., Y. Nishimura, D. Hishikawa, H. Tsuzuki, H. Miyahara, C. Gotoh, K.C. Choi, D. Fung, C. Chen, H.G. Lee, K. Katoh, S.G. Roh and S. Sasaki, 2005. Acetate and propionate short chain fatty acids stimulate adipogenesis via GPCR43. Endocrinology 146: 5092-5099.

Hong, Y.H., Y. Ogihara, D. Hishikawa, C. Gotoh, T. Iga, Y. Suzuki, S.H. Song, K. Nakajima, T. Kozakai, S. Sasaki and S.G. Roh, 2006. Molecular cloning and expression analysis of cDNA encoding bovine adipogenin. Anim. Sci. J. 77: 613-619.

Roh, S.G., D. Hishikawa, Y.H. Hong and S. Sasaki, 2006. Control of adipogenesis in ruminants. Anim. Sci. J. 77: 472-477.

Roh, S.G., S.H. Song, K.C. Choi, K. Katoh, V. Wittamer, M. Parmentier and S. Sasaki, 2007. Chemerin - A new adipokine that modulates adipogenesis via its own receptor. Biochem. Biophys. Res. Commun. 362: 1013-1018.

Novel minimal invasive technique for measuring hepatic metabolism quantitatively in dairy cows exemplified by studying hepatic glucose-net production after dexamethasone treatment

A. Starke[1], K. Wussow[1], L. Matthies[1], M. Kusenda[1], R. Busche[1], A. Haudum[1], A. Beineke[2] and J. Rehage[1]
[1]Clinic for Cattle, University of Veterinary Medicine Hannover, Foundation, Bischofsholer Damm 15, 30173 Hannover, Germany; [2]Department for Pathology, University of Veterinary Medicine Hannover, Foundation, Bischofsholer Damm 15, 30173 Hannover, Germany; alexander.starke@tiho.ha

Introduction

A commonly performed method to determine hepatic net production or utilisation of substrates in cattle is to measure the difference in plasma substrate concentrations between afferent (portal vein, an artery) and efferent (a hepatic vein) hepatic vessels in relation to hepatic blood flow. Hepatic blood flow is determined using the indicator dilution method as previously described by Katz and Bergman (1969a,b), whereby para-aminohippuric acid (PAH) is infused into a mesenteric vein and subsequently detected in the portal vein, a hepatic vein and an artery. Measuring plasma substrate concentrations requires frequent sampling in the corresponding abdominal vessels. As a prerequisite, catheters have to be surgically implanted into the portal vein, a hepatic vein, a mesenteric vein and an artery, whereby the animals are exposed to abdominal surgery under general anaesthesia in lateral recumbency. This surgical technique is laborious, demands sophisticated surgical skills, as well as appropriate facilities and equipment. Thus, the objective of this study was to develop a minimally invasive technique under ultrasonographic control for permanent catherisation of the portal vein, a hepatic vein, a mesenteric vein and an artery. The effects of the corticosteroid, dexamethasone, on net hepatic glucose production were studied using the new technique.

Material and methods

Eight clinically healthy non pregnant German Holstein cows (mean ± SD; age 3.6±0.8 years, body weight (BW) 560±59 kg, 382±108 days *post partum*, milk yield 14.4±8.3 kg/d) were included in the study. Cows were housed in single pens on straw bedding, fed a standard diet composed of hay, corn silage, soy bean meal and concentrate three times daily (7 am, 2 and 7 pm) and had free access to water, which was provided *ad libitum*. Indwelling catheters were implanted transcutaneously into the jugular vein and the abdominal aorta six days prior to the start of the experiment. Furthermore, a transcutaneous ultrasound-guided (B-Mode 2-7 MHz convex transducer, 11-23 cm image depth range) catheterisation of the portal and hepatic vein as well as the cranial mesenteric vein was performed. All catheterisations were carried out in standing position after surgical preparation of the abdominal wall and regional infiltration with 10 ml of 2% procaine hydrochloride. Flunixine-meglumine (2.2 mg/kg BW) and enrofloxacine (5 mg/kg BW) were administered intravenously on three consecutive days starting one hour before catheter implantation. To determine hepatic plasma flow rates, the PAH indicator was infused into the mesenteric vein (5 h after application of 2 g of primer (Reynolds *et al.*, 1992), constant infusion rate of 14.4 g/h (Benson and Reynolds, 2001). Blood samples were collected from the portal and jugular vein as well as the abdominal aorta prior to infusion (baseline values) and subsequently at 40 minute intervals starting 50 min after delivering the priming dose. Hence, in total, six simultaneous blood samples for analysis of PAH and Glucose were gathered. The hepatic net production rate of glucose was defined as the concentration difference between afferent and efferent hepatic vessels in relation to hepatic plasma flow (Katz and Bergman, 1969a,b). Hepatic glucose net production was measured the day prior to (day 0) and subsequently on days 1, 2 and 4 after dexamethasone treatment (100 mg/kg BW; intramuscular application). A post-mortem examination was performed in all animals fourteen days after catheter implantation. The results were statistically evaluated using the Wilcoxon Signed Ranks Test (SAS® statistical package; Version 9.1). *P*-values less than 0.05 were considered significant.

Results

Necropsy two weeks post implantation confirmed the correct position of the catheters in all animals. In one cow, a thrombus at the catheter tip in the portal vein prevented blood sampling from day 2 of the study onwards. From the remaining seven animals, blood samples were successfully collected from all vessels until the day of necropsy. In the abdominal cavity a mild circumscipt fibrous perihepatitis at the passage of the catheter into the liver was observed. From day 0 to 4 days after treatment with dexamethasone, mean portal plasma flow increased significantly ($P<0.05$) from 745±231 l/h (day 0; mean+sd) to 1,027±277 l/h (day 4), plasma flow in the hepatic vein from 1,010±330 l/h to 1,512±332 l/h and the arterial fraction of plasma flow from 26%±7 to 32%±7 ($P=0.08$). Glucose concentration increased in all vessels after dexamethasone application ($P<0.001$). However, calculated hepatic net production of glucose (mean+sd; day 0: 1,389±587 g/day; day 1: 1,229±429 g/day; day 2: 1,008±322 g/day; day 4: 1,452±507 g/day) decreased (day 0 vs. 2; $P=0.01$).

Conclusion

The ultrasound-guided percutaneous multiple centesis of hepatic vessels proved to be safe, reliable and simple to conduct. In contrast to the findings of Baird and Heitzman (1971) the results of the net glucose production imply that the increase in plasma glucose concentration following dexamethasone treatment in cows is not due to enhanced hepatic gluconeogenesis. This was in accordance with recent studies in calves (Hammon *et al*, 2003, 2005), in which dexamethasone treatment led to no increase in key enzyme activities of hepatic gluconeogenesis. Similarly to the situation observed in man (Nicod *et al.*, 2003; Dake *et al.*, 2004), bovine hypergylcemia appears to be mainly provoked by dexamethasone induced whole body insulin resistance with reduced peripheral glucose consumption (Kusenda *et al.,* 2009).

References

Baird, G.D. and R.J. Heitzman, 1971. Mode of action of a glucocorticoid on bovine intermediary metabolism. Possible role in controlling hepatic ketogenesis. Biochim. Biophys. Acta 252: 184-198.

Benson, J.A. and C.K. Reynolds, 2001. Effects of abomasal infusion of long-chain fatty acids on splanchnic metabolism of pancreatic and gut hormones in lactating dairy cows. J. Dairy Sci. 84: 1488-500.

Dake, Q., T. Pulinikunnil, A. Ding, G. Sanjoy, A. Ashraf, J.A. Pospisilik, R. Brownsey, R. Wambolt, M. Allard and B. Rodrigues, 2004. Single-dose dexamethasone induces whole-body insulin resistance and alters both cardiac fatty acid and carbohydrate metabolism. Diabetes 53: 1790-1797.

Hammon, H.M., S.N. Sauter, M. Reist, Y. Zbinden, C. Philipona, C. Morel and J.W. Blum, 2003. Dexamethasone and colostrum feeding affect hepatic gluconeogenic enzymes differently in neonatal calves. J. Anim. Sci. 81: 3095-3106.

Hammon, H.M., C. Philipona, Y. Zbinden, J.W. Blum and S. Donkin, 2005. Effects of dexamethasone and growth hormone treatment on hepatic gluconeogenic enzymes in calves. J. Dairy Sci. 88: 2107-2116.

Katz, M.L. and E.N. Bergman, 1969a. Simultaneous measurements of hepatic and portal venous blood flow in the sheep and dog. Am. J. Physiol. 216: 946-952.

Katz, M.L. and E.N. Bergman, 1969b. Hepatic and portal metabolism of glucose, free fatty acids, and ketone bodies in the sheep. Am. J. Physiol. 216: 953-960.

Kusenda, M., A. Starke, M. Kaske, M. Piechota, M. Hoeltershinken and J. Rehage 2009. Insulin resistance after single dose dexamethasone treatment in dairy cows. In: Chilliard, Y., F. Glasser, Y. Faulconnier, F. Bocquier, I. Veissier and M. Doreau (Eds.), Ruminant Physiology. Digestion, metabolism and effects of nutrition on reproduction and welfare. Wageningen Academic Publishers, Wageningen, the Netherlands. p. 562-563.

Nicod, N., V. Giusti, C. Besse and L. Tappy, 2003. Metabolic adaptations to dexamethasone-induced insulin resistance in healthy volunteers. Obesity Res. 11: 625-631.

Reynolds, C.K., H. Lapierre, H.F. Tyrrell, T.H. Elsasser, R.C. Staples, P. Gaudreau and P. Brazeau, 1992. Effects of growth hormone-releasing factor and feed intake on energy metabolism in growing beef steers: net nutrient metabolism by portal-drained viscera and liver. J. Anim. Sci. 70: 752-763.

Effects of ghrelin injection on blood metabolites and hormones of non-lactating and lactating cows

T. Sugino[1], R. Fukumori[1], A. Yokotani[1], F. Itoh[2], H. Shingu[2], N. Moriya[2], Y. Hasegawa[3], M. Kojima[4], K. Kangawa[5], T. Obitsu[1], S. Kushibiki[2] and K. Taniguchi[1]
[1]Graduate School of Biosphere Science, Hiroshima University, Higashi-Hiroshima, 739-8528, Japan; [2]National Institute of Livestock and Grassland Science, Tsukuba, 305-0901, Japan; [3]School of Veterinary Medicine and Animal Sciences, Kitasato University, Towada, 034-8628, Japan; [4]Institute of Life Science, Kurume University, Kurume, 839-0864, Japan; [5]National Cardiovascular Center Research Institute, Osaka, Japan; sugino@hiroshima-u.ac.jp

Introduction

Ghrelin can stimulate food intake and growth hormone (GH) secretion (Kojima *et al.*, 1999, Nakazato *et al.*, 2001), and circulating ghrelin levels have been shown to rise in poor nutritional status of ruminants (Sugino *et al.*, 2004). It is well known that the nutritional and metabolic status of dairy cows dramatically change during the transition period. Thus, we investigated the effects of exogenous ghrelin on the blood metabolites and hormones in non-lactating and lactating cows.

Material and methods

Twelve Holstein dairy cows [six non-lactating (calving number: 2.3±0.6, the days in pregnancy: 193±3.5, BW: 687±18 kg) and six lactating (calving number: 3.3±0.8, the days in milk: 57.2±2.1, daily milk yield: 32.5±0.4 kg, BW: 687±18 kg)] were used. Cows were fed a total mixed ration twice daily (09:00 h and 18:00 h) to meet the nutrient requirements of the Japanese feeding standard. The cows were trained to the assigned meal feeding diet for 10 days at least. At 6 h after feeding (15:00 h), all animals were single injected with synthetic bovine ghrelin (1 μg/kg BW) into the jugular vein using a polyvinyl-chloride catheter. Blood samples (10 ml) were collected into heparinised tubes including aprotinine through the catheter of the other side jugular vein from 14:50 h to 18:00 h. Plasma concentrations of ghrelin, GH and insulin were determined by time-resolved fluoro-immunoassay (TR-FIA, Sugino *et al.*, 2004). Plasma glucose levels were measured by glucose auto analyser.

Results and discussion

Basal ghrelin levels did not differ between non-lactating (1.30±0.23 ng/ml) and lactating cows (1.32±0.22 ng/ml). Basal glucose levels of plasma were lower in lactating compared with non-lactating cows (Table 1). Compared with non-lactating cows, lactating cows showed higher concentrations of basal plasma GH, but less of insulin (Table 1). After the ghrelin injection, plasma ghrelin levels were increased about 25 folds from basal values in both groups. Plasma GH levels were increased by ghrelin injection with no difference between non-lactating and lactating cows. In lactating cows, plasma glucose concentrations were increased by ghrelin injection, and maintained higher plasma glucose levels for 150 min than before ghrelin injection (Figure 1A). The increased glucose levels were followed by the increase of plasma insulin. However, because the increment of insulin did not differ between both groups (Table 1), lower insulin levels in lactating compared with non-lactating cows were kept after ghrelin injection (Figure 1B). It is well known that insulin sensitivity reduces during the lactating period. Thus, less secretion and less sensitivity of insulin in lactating cows might elicit higher plasma glucose levels with ghrelin injection.

Table 1. Plasma GH, insulin and glucose levels in non-lactating and lactating cows.

		Group		SE	*P*-value
		Non-lactating	Lactating		
GH, ng/ml	Basal[1]	10.2	35.9	4.18	<0.05
	Peak[2]	638.0*	648.0*	163.0	0.97
Insulin, ng/ml	Basal[1]	3.99	1.75	0.49	<0.05
	Peak[2]	4.99	2.76*	0.93	<0.05
Glucose, mg/dl	Basal[1]	71.4	58.8	3.24	<0.05
	Peak[2]	74.8	73.2*	2.45	0.65

[1]Basal: the value before ghrelin injection; [2]Peak: the peak value after ghrelin injection; *: $P<0.05$ between basal and peak value.

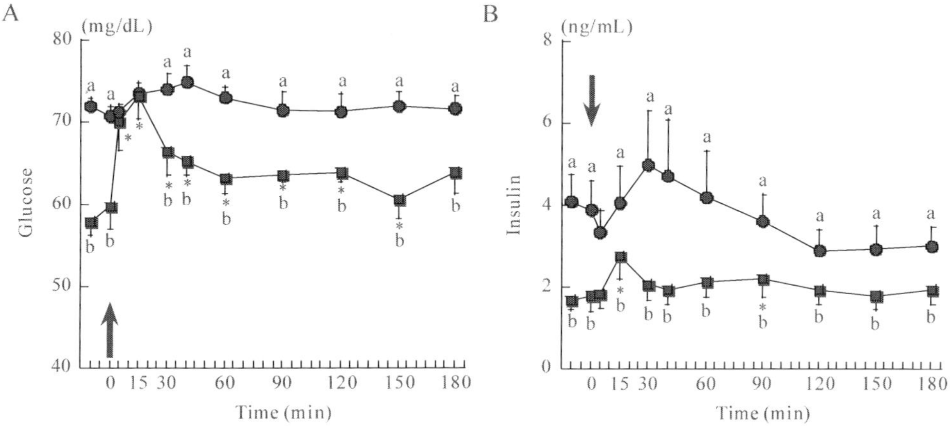

Figure 1. Effects of ghrelin injection on (A) plasma glucose and (B) insulin concentrations in non-lactating (circle) and lactating (square) cows. Each plot shows the average values ± SE of six cows. a, b: P<*0.05, between non-lactating and lactating cows.* * = P<*0.05, between pre- and post-injection of ghrelin. The arrow shows the ghrelin injection time.*

Conclusion

The basal level of ghrelin in blood plasma and the degree of GH secretion stimulated by ghrelin injection are similar between nonlactating and lactating cows, but exogenous ghrelin may increase glucose supply to the mammary gland during lactation.

References

Kojima, M., H. Hosoda, Y. Date, M. Nakazato, H. Matsuo and K. Kangawa, 1999. Ghrelin is a growth-hormone-releasing acylated peptide from stomach. Nature 402: 656-660.

Nakazato, M., N. Murakami, Y. Date, M. Kojima, H. Matsuo, K. Kangawa and S. Matsukura, 2001. A role for ghrelin in the central regulation of feeding. Nature 409: 194-198.

Sugino, T., Y. Hasegawa, Y. Kurose, M. Kojima, K. Kangawa and Y. Terashima, 2004. Effects of ghrelin on food intake and neuroendocrine function in sheep. Anim. Reprod. Sci. 82-83: 183-194.

Cellularity and lipogenic activities in perirenal and intermuscular adipose tissues from Blonde d'Aquitaine, Charolais and Holstein fetuses

H. Taga, M. Bonnet, C. Labonne, I. Cassar-Malek, B. Picard and Y. Chilliard
INRA, UR1213 Herbivores, Site de Theix, F-63122 Saint Genès-Champanelle, France; muriel.bonnet@clermont.inra.fr

Introduction

The adipose tissue mass is critical for the productive efficiency of ruminants, and the anatomical distribution of adipose tissue (AT) is part of carcass and meat qualities. Many investigations have focussed on the post-natal growth of AT depending on rearing factors, due to the economic challenge resulting from the production of ruminants with adequate masses of AT (Bonnet *et al.*, 2007). Recent data obtained in bovines and humans suggest that the growth of foetal brown AT influences the mass of the adult white AT (Gesta *et al.*, 2007; Greenwood and Cafe, 2007). Whether variability in the cellular and molecular processes involved in AT growth exists during foetal life, and contributes to the variability of AT masses in the adult bovine remains to be unravelled. To address this question we measured the cellularity and the lipogenic activities in foetal AT from three bovine breeds and two anatomical sites of the AT, chosen for their differences in adiposity during the post-natal life.

Material and methods

Charolais (n=10), Blonde d'Aquitaine (n=3) and Holstein (n=4) foetuses were generated by artificial insemination of cows and were collected at 260 days of gestation. At slaughter, samples of perirenal and intermuscular AT were frozen in liquid nitrogen and stored at -80 °C. Activities of enzymes involved in *de novo* lipogenesis (fatty acid synthase [FAS], glucose-6-phosphate dehydrogenase [G6PDH], malic enzyme [ME]) and in fatty acid esterification (glycerol-3-phosphate dehydrogenase [G3PDH]) were assayed as described by Chilliard *et al.* (1991). Enzyme activities were normalised by content of soluble proteins assayed in homogenates mainly according to Bradford (1976) using bovine serum albumin as the standard and the Bio-Rad Protein Assay procedure (Bio-Rad, Marnes la Coquette, France). Adipocyte cell and number were assayed from AT stored at 37 °C after slaughter as described by Robelin (1981).

Data were analysed using the MIXED procedure of SAS® software package (version 8.2; SAS Institute, Cary, NC, USA). Fixed effects included breed anatomical site and their interaction. Animal within breed was the random effect. When applicable, a multiple comparison of means was performed using the LSMEANS statement of the MIXED procedure. Differences between breeds and/or anatomical sites were considered to be significant when $P \leq 0.05$.

Results

The diameter of adipocytes and number per gram of lipid were similar among anatomical sites but differed depending on breeds (Table 1). Adipocyte diameter was higher (average of +35%, $P<0.01$) and adipocyte number per gram of lipid was lower (average of -70%, $P<0.01$) in perirenal and intermuscular AT from Holstein than from Blonde d'Aquitaine and Charolais foetuses.

Enzyme activities linked to *de novo* lipogenesis and esterification differed depending on the anatomical site but were only slightly influenced by the breed (Table 1). FAS, EM, G6PDH and G3PDH activities were around 2 fold higher ($P \leq 0.03$) in perirenal AT than in intermuscular AT from Charolais and Blonde d'Aquitaine foetuses, but were similar between anatomical sites in Holstein foetuses. Furthermore, G3PDH activity was higher ($P<0.05$) in intermuscular AT from Holstein than from Blonde d'Aquitaine and Charolais foetuses ($P=0.11$ for breed x anatomical site interaction).

Table 1. Diameter (µm) and number (x10^6/g lipid) of adipocytes, and lipogenic enzyme activities (nmol/min/mg protein) in perirenal (PRAT) and intermuscular (IMAT) adipose tissues from Charolais, Blonde d'Aquitaine and Holstein foetuses at 260 days of gestation.

	Charolais		Blonde d'Aquitaine		Holstein		*P*-values		
	PRAT	IMAT	PRAT	IMAT	PRAT	IMAT	site	breed	int
diameter	41.1 ^{b}	40.9 ^{b}	43.7 ^{b}	42.4 ^{b}	65.2 ^{a}	68.2 ^{a}	0.59	<0.01	0.18
number	24.8 ^{a}	25.8 ^{a}	21.3 ^{a}	22.8 ^{a}	6.4 ^{b}	5.9 ^{b}	0.66	<0.01	0.87
FAS	11.8 ^{b}	5.8^{c}	20.9 ^{a}	9.3 ^{bc}	13.2 ^{ab}	11.7^{abc}	<0.01	0.14	0.23
G6PDH	277^{a}	150 ^{b}	304^{ab}	170^{ab}	229^{ab}	184^{ab}	0.03	0.92	0.65
ME	66.7^{ab}	45.3 ^{c}	74.1^{a}	43.3^{bc}	54.4^{abc}	43.8^{bc}	<0.01	0.69	0.50
G3PDH	1507^{a}	702^{b}	1725^{a}	789^{b}	1538^{a}	1534^{a}	<0.01	0.11	0.11

int= breed × anatomical site interaction.

$^{a, b}$Means within lines with same superscript letters are not significantly different (P>0.05).

Discussion and conclusion

The present results show the higher maturity of AT from Holstein than from Charolais and Blonde d'Aquitaine as well as the higher maturity of perirenal than intermuscular AT at the end of foetal life. Similar differences in the maturity of bovine AT, measured by cellularity and lipogenic enzyme activities, were repeatedly observed in post-natal life (Bonnet *et al.*, 2007). These results thus suggest that the intrauterine growth of brown AT could contribute to the variability of the mass of white AT in growing or adult bovines, by mechanisms that remain to be studied.

Acknowledgement

The authors thank experimental cowshed workers and staffs of the slaughterhouse and the laboratory, especially D. Bany and C. Barboiron.

References

Bonnet, M., Y. Faulconnier, C. Leroux, C. Jurie, I. Cassar-Malek, D. Bauchart, P. Boulesteix, D. Pethick, J.F. Hocquette and Y. Chilliard, 2007. Glucose-6-phosphate dehydrogenase and leptin are related to marbling differences among Limousin and Angus or Japanese x Angus steers. J. Anim. Sci. 85: 2882-2894.

Bradford, M.M., 1976. A rapid and sensitive method for the quantitation of microgram quantities of protein utilizing the principle of protein-dye binding. Anal. Biochem. 72: 248-254.

Chilliard, Y., G. Gagliostro, J. Fléchet, R. Lefaivre and I. Sebastian, 1991. Duodenal rapeseed oil infusion in early and midlactation cows. 5. Milk fatty acid and adipose tissue lipogenic activities. J. Dairy Sci. 74: 1844-1854.

Gesta, S., Y.H. Tseng and C.R. Kahn, 2007. Developmental origin of fat: tracking obesity to its source. Cell. 131: 242-256.

Greenwood, P.L. and L.M. Cafe, 2007. Prenatal and pre-weaning growth and nutrition of cattle: long term consequences for beef production. Animal 1: 1283-1296.

Robelin, J., 1981. Cellularity of bovine adipose tissues - Developmental-changes from 15-percent to 65-percent mature weight. J. Lipid Res. 22: 452-457.

Increasing inclusion of wheat in maize and grass silage-based diets: production responses in dairy cows

M.N. Tahir[1], M. Hetta[1] and C. Swensson[2]
[1]Department of Agricultural Research for Northern Sweden, Swedish University of Agricultural Sciences, 901 83 Umeå, Sweden; [2]Department of Agricultural Buildings, Swedish University of Agricultural Sciences, 230 53 Alnarp, Sweden and Swedish Dairy Association, 223 70 Lund, Sweden; marten.hetta@njv.slu.se

Introduction

Maize silage is becoming a more important feed component in dairy cow diets in Sweden. Good quality maize silage has a high content of starch but low neutral detergent fibre (NDF) digestibility. If the maize silage is used in dairy cow diets with a high proportion of grain, there could be a problem with too high starch concentrations, causing adverse effects on the feed efficiency (FE) (Allen, 2000). The purpose of this experiment was to investigate the response to increasing wheat grain inclusion in dairy cow diets with maize and grass silage on dry matter intake (DMI), milk yield (MY), milk composition (fat and protein) and feed efficiency (FE).

Material and methods

The experiment was conducted with 28 multiparous dairy cows (Swedish Red) with an average initial body weight (BW) of 625±48 kg, milk yield (MY) of 30.6±5 kg ECM and 154±58 d in milk (DIM). The animals were kept in a loose house stable and then allocated to seven balanced Latin squares (4 animals × 4 periods of 21 d) (Morris, 1999) depending on DIM. The first 2 wks in each period served as an adaptation and the rest for data collection. Within each square the animal were offered four diets (Table 1: W8, W16, W24 and W32) with increasing wheat content. The diets were offered *ad libitum* in Roughage Intake Control feeders as total mixed rations (TMR). Feed intake for individual animals was recorded for every visit. The MY and the BW were recorded twice daily in the milking parlour. Milk fat and protein concentrations were measured weekly on the basis of a pooled sample of the two recordings.

Table 1. Mean composition (percent of DM) of dietary treatments and feed components.

	Dietary treatments				Composition of components			
	W8	W16	W24	W32	CP	NDF	Starch	DM
Feed components								
Grass silage	34	31	27	23	13.4	47.7	2.0	24
Maize silage	40	35	31	27	8.9	36.5	35.0	31
Soybean meal	18	18	18	18	52.0	10.0	2.5	85
Wheat pellets	8	16	24	32	13.7	8.6	58.4	85
Composition of diets								
DM	32	34	36	39				
CP	19	19	19	19				
NDF	36	33	30	27				
Starch	19	22	25	28				
ME	11.9	12.1	12.4	12.7				

DM= dry matter (percent of feed). CP=crude protein. NDF=neutral detergent fibre. ME=metabolisable energy (MJ/ kg DM).

Treatment means were estimated by animal within period as an experimental unit. The relationships between the treatment means of percent of wheat in the diets and the response parameters were evaluated with linear regression. Results were considered significant when (P<0.05, β=0) for the regression coefficient (β) using the software MINITAB™ (ver. 15.1).

Results

The inclusion of wheat resulted in the increased DMI and MY in dairy cows (Table 2) (P<0.01). The effects on composition of the milk were minor. The increase in DMI in relation to the percent of wheat in the diet was twice as high as the response in MY resulting in a significant (P<0.01) reduction in lower FE and also a smaller utilisation of the available metabolisable energy in the diets.

Table 2. Means of animal responses to the dietary treatments and linear regression between the proportion of wheat (percent) in the diet and the recorded responses.

Treatment	W8	W16	W24	W32	SE	Intercept	Coef. (β)	P-value (Coef.)	R^2
DMI	18.3	19.9	21.3	23.0	0.43	16.7	0.198	0.00	0.99
MY	28.5	29.7	30.3	30.9	0.96	28	0.091	0.00	0.96
Milk fat%	4.85	4.89	4.73	4.73	0.095	4.93	-0.006	0.19	0.66
Milk protein%	3.73	3.73	3.75	3.79	0.050	3.71	0.002	0.12	0.77
ME/kg ECM[1]	4.74	5.41	6.13	6.95	0.21	4.13	0.089	0.00	0.96
FE	1.61	1.54	1.45	1.38	0.043	1.67	-0.009	0.00	0.99

DMI=dry matter intake (kg DM/cow/day); MY=milk yield (kg ECM/cow/day); ME=metabolisable energy (MJ/kg DM); ECM=energy corrected milk; FE= feed efficiency (kg DMI/kg ECM).
[1] Estimated use of ME per kg of ECM (total ME intake – ME required for maintenance and BW gain)/ MY.

Conclusion

The increasing inclusion of wheat in maize and grass silage-based diets resulted in a smaller increase in MY than expected from the available ME in the diets. The reduction in FE as a result of higher grain content in the diets could depend on the negative effect on rumen efficiency of fibre digestion, as a result of low rumen pH. Further work is in progress to evaluate the interactions between starch and fibre components in dairy cow diets.

Acknowledgement

The authors would like to express their gratitude to the Swedish Farmers Foundation for Agricultural Research for the financial support and the colleagues at the Department of Agricultural Research for Northern Sweden for the collaborative efforts.

References

Allen, M.S., 2000. Effects of diet on short term regulation of food intake by lactating dairy cattle. J. Dairy Sci. 83: 1598-1624.

Morris, T.R., 1999. Experimental design and analysis in animal sciences. CABI Publishing. Oxon. UK, 205 pp.

Intravenous insulin perfusion mimics the meal-dependent rise of renal blood flow in conscious sheep

I. Tebot[1], J.M. Bonnet[2], J.Y. Ayoub[2], C. Paquet[2], S.M. Da Silva[1] and A. Cirio[1]
[1]Area de Fisiología, Facultad de Veterinaria, 11600, Montevideo, Uruguay; [2]Université de Lyon, Ecole Nationale Vétérinaire de Lyon, EA 4173 INSERM ERI 22, 69280 Marcy L'Etoile, France; albertocirio@yahoo.com

Introduction

A circadian rise of renal blood flow (RBF) after meals and independent of blood pressure has been recently reported in caged sheep (Tebot *et al.*, 2009). Plasma insulin concentration increases after meals in sheep and cattle (reviewed by Basset, 1972). The aim of this work was to describe the effect of insulin perfusion on RBF in sheep. Since sympathetically mediated insulin-dependent vasoconstriction in skeletal muscle in humans and rats has been reported (reviewed by Muniyappa *et al.*, 2007), suggesting a role on blood pressure, and variations in renal hemodynamics can be partly dependent on arterial pressure (Miller-Craig *et al.*, 1978), systemic arterial pressure (SAP) was also monitored.

Material and methods

Eight adult Ile de France ewes (54-70 kg BW) in individual pens were fed alfalfa pellets and hay once daily (9:00-11:00 h) with *ad libitum* water. Under general anesthesia, transit-time ultrasonic flow-metering probes (4 mm, R-series, Transonic Systems, Ithaca, NY) were bilaterally implanted around renal arteries for RBF measurement, as previously described (Tebot *et al.*, 2009). Four of the sheep were provided with a telemetry measurement system (Physiotel Transmitter, model TL11M3-D70-CCP, Data Sciences International, St Paul, MN, USA) inserted into one carotid artery for SAP monitoring. Data were transmitted to a processing system (Acqknowledge III for MP150WSW, Biopac Systems Inc., St. Barbara, CA, USA), and mean values of RBF (right + left flows) and SAP were calculated every 10 min. On 3 alternate days, RBF and SAP were continuously recorded between 8:00 h and 18:00 h, and blood samples were taken every 30 min for insulin (RIA kit Insik 5, DiaSorin, Antony, France) and glucose (GOD POD kit, Thermo Fisher Scientific Oy, Vantaa, Finland) analyses. Then, fasted sheep received (a) primed constant rate i.v. perfusions (0.5 ml/min, 10:00-12:00 h) of insulin (Caninsulin, Intervet SA, Beaucouzé, France) at 6 (high dose, HI) or 2 (low dose, LI) mU/kg/min, (b) saline only and (c) no perfusion. Perfusions were one per day on alternate days. Euglycaemia was maintained and hypokalemia was prevented by perfusing a 40 (HI) or 30% (LI) glucose and a 2.5 (HI) or 1.5% (LI) KCl solutions. The perfusion days, RBF and SAP were recorded (8:00 h to 16:00 h) and blood samples were taken every 30 min (9:00 h to 15:00 h). Analysis of variations in RBF, SAP and insulin values was performed by a one-way ANOVA with subsequent Fisher PLSD test (StatView, SAS® Institute, Cary, NC, USA).

Results and discussion

After the onset of meals a long-lasting rise in RBF was observed (Figure 1) without changes in SAP. Insulin level also increased (22.2±1.6 mU/l maximum value, $P<0.05$) remaining significantly high for 7 h. HI and LI perfusions (Figure 2) increased RBF without changes in SAP. As observed after meals, in both perfusions the rise in RBF lasted longer than the insulin increase, and the maximum values (895±59 for HI and 759±53 ml/min for LI) were reached after the onset of the decline in plasma insulin. The magnitude and time-course of the RBF increase under LI perfusion were similar to those found after meals. Owing to the euglycaemic clamp, no changes in plasma glucose during perfusions were observed.

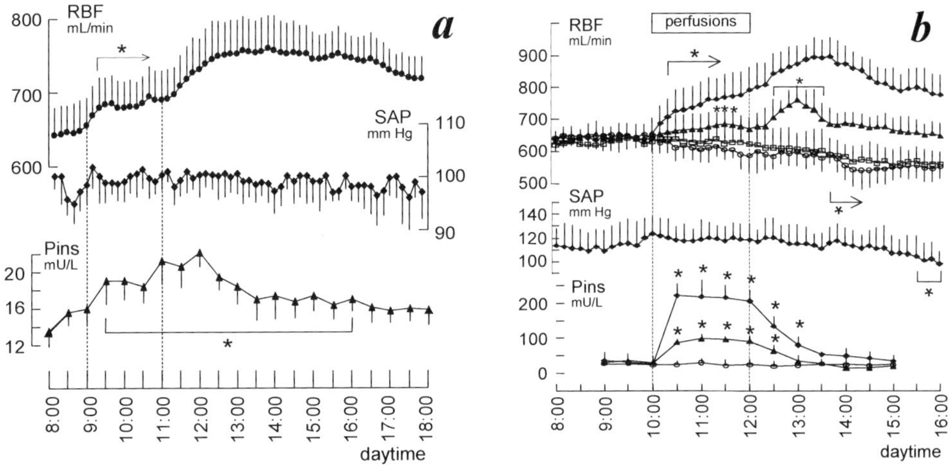

Figure 1. Renal blood flow (RBF), systemic arterial pressure (SAP) and plasma insulin (Pins) during (dotted lines) and after (a) meals and (b) perfusions of high (◆) and low (▲) insulin doses and saline (○). □ = no perfusion. SAP values correspond to high insulin dose. RBF and SAP data are 10 min means ± SD. For RBF and Pins n = 8 sheep (×3 days in (a), × 1 day in (b), for SAP n = 4 sheep ×1 day.
* = P<*0.05 vs. 1 h pre-feeding / pre-perfusion mean values.*

Conclusion

Insulin perfusion in fasted sheep induced a rise of RBF comparable to that associated to meals, independent of blood pressure, and lasting longer than the elevation in plasma insulin. This fact suggests the existence of an insulin-induced mediator of the RBF increase.

Acknowledgement

The authors thank the staff of the Endocrinology Laboratory of the National Veterinary School of Lyon for biochemical analyses. Financial support was provided by the Comité ECOS-*Sud*, action U05S02.

References

Basset, J.M., 1972. Dietary and gastro-intestinal control of hormones regulating carbohydrate metabolism in ruminants. In: McDonald, I.W. and A.I. Warner (Eds.), Digestion and Metabolism in the ruminant. Univ. of New England, Armidale, NSW, Australia, pp. 384-398.

Miller-Craig, M.W., C.N. Bishop and E.B. Raftery, 1978. Circadian variation of blood pressure. Lancet 1: 795-797.

Muniyappa, R., M. Montagnani, K. Kon Koh and M.J. Quon, 2007. Cardiovascular actions of insulin. Endocrine Reviews 28: 463-491.

Tebot, I., J.M. Bonnet, S. Junot, J.Y. Ayoub, C. Paquet and A. Cirio, 2009. Roles of eating, rumination, and arterial pressure in determination of the circadian rhythm of renal blood flow in sheep. J. Anim. Sci. 87: 554-561.

Femur biometry, densitometry and chemical composition of Moxoto goats supplemented with concentrate in a semiarid region

I.A.M.A. Teixeira[1], M.J. Araújo[2], A.N. Medeiros[2], R.G. Costa[2], S.M. Baraldi Artoni[1], C.A.T. Marques[3] and K.T. Resende[1]
[1]Universidade Estadual de Sao Paulo, UNESP, SP, 14870-000, Jaboticabal, Brazil; [2]Universidade Federal da Paraiba, UFPB, PB, 58397-000, Paraiba, Brazil; [3]Universidade Federal do Piaui, UFPI, PI, 64900-000, Bom Jesus, Brazil; izabelle@fcav.unesp.br

Introduction

For decades, much research has been carried out to discover and/or develop simple and accurate biochemical measurements to assess the mineral status of animals, which involved the analysis of soil, water, plant and animal tissue (Underwood and Suttle, 1999). Among these measurements the concentration of minerals in tissues or body fluids are often the best indicators of mineral status of a herd. Although very important, the mineral status of goats has not being completely studied, mainly in semi arid regions of tropical countries. The bone mineral density (BMD) is a tool used to evaluate the quantity of minerals per bone volume, and it is very important for helping to understand the bone mineralisation process. Thus, many studies have been conducted using bone densitometry techniques for the determination of normal values for bone mineral density in different animal species, in order to efficiently assess the bone structure and to diagnose and prevent possible bone lesions. However BMD has not been used for the purpos of evaluating possible nutritional deficiencies. Therefore, this study was carried out to investigate the chemical and biometrical aspects, also the densitometry of femurs from Moxoto goats supplemented with concentrate in the Brazilian semi-arid region, as well as to correlate BMD with body weight, composition of minerals and with the biometric variables of the femur of these animals in order to assess their mineral status.

Material and methods

Thirty-two castrated Moxoto kids (initial BW 15.7±0.8 kg) grazed native pasture from 7:00 am to 4:00 pm. Aafter this period, they were individually housed in a stall, having free access to drinking water, and the amount of supplement (202 g/kg of CP; 3.14 Mcal/kg DM of ME; 8.14 g/kg of Ca; 8.50 g/kg of P; 3.34 g/kg of Mg; 0.27 g/kg of K and 3.05 g/kg of Na), according to the treatment group. The animals were randomly assigned to four supplementation levels (treatment groups: 0, 0.5, 1.0 and 1.5% BW), in order to provide different nutritional levels and consequently different growth rates. The desired supplementation level was maintained by weekly adjustments, depending on BW of the animals. When the animals from the 1.5% BW treatment group reached 25 kg BW, the animals from the other groups were also slaughtered. This procedure was used in order to provide a quantitative restriction and get different nutrient intake (mineral) during the experiment. The animals were stunned through cerebral concussion, immediately before being slaughtered through sectioning of the jugular veins and carotid arteries. After evisceration, the carcasses were weighed and the left femur of each animal was removed for further analysis (chemical, biometrical and densitometry). Samples of concentrate ingredients, extrusa and bone were analysed according to AOAC (1990) and Mahanti and Barnes (1983). In addition, the following evaluation on the femurs, weight, length, bone mineral density (mm Al) and diameter of the diaphysis and the proximal and distal epiphysis were also performed.
A completely randomised design was used, with four treatments, according to the model y = a + bx, which shows the behavior of the dependent variable y (response variable) as a function of the independent variable x (supplementation levels). Statistical analyses were performed using GLM and REG procedures, for the correlation analysis the CORR procedure of SAS® (SAS Inst., Inc., Cary, NC, USA) was used.

Results

Increasing supplementation levels resulted in a better animal performance e.g. BW and EBW gains, and, consequently, in fresh and dry weight of the femur (*P*<0.001). Although dry matter in the femur (%, DM) increased linearly, ash (%), crude protein (%, CP), ether extract (%, EE) and minerals (Ca, P, Mg, Na and K) were not affected by the supplementation levels. The contents of minerals were not affected by treatments, even in the degreased femur evaluation, except for Na, which decreased with increasing supplementation levels (*P*=0.007). The supplementation level did not affect the Ca:P ratio, which on average was 2.7:1.

By assessing the biometry and density of the femur, it was found that the treatments affected these variables positively, except for the thickness of the proximal epiphysis spongy layer (TSL-PE), which was not influenced by the supplementation level, and the thickness of the distal epiphysis spongy layer (TSL- DE), which reduced linearly. Animals fed with increasing concentrate levels had heavier, larger, thicker and denser femurs when compared to unsupplemented animals. Although, the diameter of the proximal epiphysis (Diam-PE) increased with increasing supplementation levels, the thickness of the spongy layer in this region remained unchanged.

It was observed that the BMD values increased linearly since the BW of animals increased in response to the supplementation levels. A highly significant and positive correlation was found between BMD and the contents of ash (g) and major minerals of the femur, indicating that this variable really reflects the concentration of minerals per bone volume.

Table 1 Correlations between bone mineral density (mm Al) and body weight (kg), amount of ash and major minerals (g) in the femur of Moxoto goats.

BMD	Variable						
	BW	ash	Ca	P	Mg	Na	K
BMD - PE	0.82 (<0.001)	0.87 (<0.001)	0.84 (<0.001)	0.83 (<0.001)	0.67 (<0.001)	0.75 (<0.001)	0.72 (<0.001)
BMD - DE	0.50 (0.005)	0.68 (<0.001)	0.66 (<0.001)	0.66 (<0.001)	0.51 (<0.005)	0.59 (<0.001)	0.54 (<0.005)
BMD - D	0.86 (<0.001)	0.86 (<0.001)	0.89 (<0.001)	0.89 (<0.001)	0.70 (<0.001)	0.81 (<0.001)	0.56 (<0.001)

BMD = bone mineral density; PE= proximal epiphyses; DE= distal epiphyses; D=diaphyses; BW=body weight.

Conclusion

The BMD can be used to indicate the mineral status of Moxoto goats as it was correlated with ash, calcium and phosphorus contents. This study also shows that the BMD, associated with the chemical and biometric analysis of the femur helps in assessing the status of minerals, as well as in monitoring the feeding quality of animals. Further studies are necessary to evaluate BMD as an *in vivo* technique to assess mineral status in goats.

References

AOAC, 1990 - Official Methods of Analysis. 15th ed. Assoc. Off. Anal. Chem. Arlington, Va.

Mahanti, H. and R.M. Barnes, 1983. Determination of major and trace elements in bone by ICPEs. Analytica Chemical Acta, 151: 409.

Underwood, E.J. and N.F. Suttle, 1999. The Mineral Nutrition of Livestock, 3rd edn. CABI publishing, NY, USA.

Ad libitum concentrate for dairy cows: performance and calculated energy balance in the 'Kempen System' vs. a conventional Dutch feeding strategy

H. ter Wijlen, H. van Laar and J. Martín-Tereso
Nutreco Ruminant Research Centre, P.O. Box 220, 5830 AE, Boxmeer, the Netherlands; harmen.van.laar@nutreco.com

Introduction

The transition period is unquestionably the most important period in the production cycle of the dairy cow. The difference in energy output in milk and energy intake causes a negative energy balance that affects the risk for diseases like ketosis and fatty liver. Improvement of dry matter intake (DMI) during early lactation, especially from concentrates, may reduce this negative energy balance, if rumen fermentation is maintained. *Ad libitum* compound feed and hay fed separately (Kempen System) has become in the last decade an alternative feeding strategy for small farms in Europe and Asia. This trial studies the effect of an *ad libitum* free choice concentrate feeding strategy *versus* a conventional feeding strategy on animal performance and calculated energy balance, during a full lactation.

Material and methods

Twenty-two animals were blocked according to parity, expected calving date and milk yield (MY) and alternatively joined the control or Kempen diet, starting at an estimated 3 weeks before calving (close-up). In the Kempen close-up period, cows had *ad libitum* access to a grass seed hay/grass silage mix (60/40 wt/wt on DM basis, respectively), supplemented with weekly increasing concentrate (3, 4 and 5 and 4, 5 and 6 kg/d for heifers and multiparous cows respectively). Kempen lactation diet consisted of an *ad libitum* grass seed hay/grass silage mix (65/35 wt/wt on DM basis) and concentrate, the latter being supplied by means of an electronic feeding system without any restrictions in the maximum amount eaten per meal nor in the maximum amount eaten per day. In the control close-up period, cows received *ad libitum* a mixture of corn silage, grass silage, corn cob meal (CCM) and hay (49:40:8:3, DM basis), supplemented with concentrates; 3 kg/d in week 3 and 4 kg/d in week 2 and 1 *pre partum* for all cows. *Post partum*, control animals received an *ad libitum* roughage mix, consisting of corn silage, grass silage, CCM and grass seed hay in a ratio of 47:38:12:3 (DM basis). Directly after parturition 3 kg/d concentrates were supplied, which was gradually increased to 10 and 12 kg/d for respectively heifers and multiparous cows at 28 days in lactation. After this increase, supplementation was individually recalculated per week to meet Centraal Veevoeder Bureau (CVB) energy and protein requirements, based on MY and the forage consumption of the previous week.

After calving, MY was measured daily by automatic milk measurement in the milking parlour. Milk composition was measured weekly in a morning and evening milk sample, both composed milk from 2 milking times. DMI of forage mixes was recorded by automated roughage intake control boxes. Energy balance (EB_c) was calculated subtracting energy output in milk (estimated based on MY and fat and protein content) and energy requirement for maintenance (based on estimated body weights of 575, 625, 675, 725, 750, 775 and 800 kg for 1st, 2nd, 3rd, 4th, 5th, 6th, >6th parity respectively) from energy intake, all based on the equations of the Dutch VEM system (CVB, 2005). Repeated measures ANOVA (PROC MIXED of SAS®, version 9.1; SAS Institute Inc., Carry, NC, USA) was performed for DMI, MY and EB_c, using an autoregressive structure [AR(1)]. Cow was included as the repeated subject and the model included block, week, treatment and week x treatment interaction. Values are presented as least squares means with their SEM.

Results

DMI in the close-up period was not significantly different between treatments (P>0.10; Figure 1). In week 0, the week of calving, the Kempen treatment showed a significant higher DMI (20.3±1.3 kg/d) compared with cows fed the control diet (13.5±1.2 kg/d). In the 3 following weeks, cows fed the Kempen diet also had a higher intake than cows fed the control diet. Furthermore, cows fed the Kempen treatment had a higher DMI in week 12, 15 till 22 and 33.

MY (data already shown in Craninx *et al.*, 2008) showed no significant differences in the first weeks of lactation; from week 17 till week 37 MY was, however, higher for the Kempen treatment, indicating a better lactation persistency.

At calving, body condition score (BCS) did not differ significantly with 3.26 (±0.24) and 3.41 (±0.24) for cows on control and Kempen respectively. *Post partum* BCS did not differ significantly between treatments.

Cows fed the Kempen diet had a significant improved EB_c in the first 3 weeks *post partum* when compared with cows fed the control diet. Already in week 2 *post partum,* cows on the Kempen diet were close to positive EB_c (-0.32±9.43 MJ/d), whereas cows fed the control diet were in positive EB_c in week 6 *post partum* (1.78±9.23 MJ/d). Also from week 15 till week 18 *post partum,* cows on the Kempen diet had a higher EB_c. Although the absolute level of EB_c is based on various estimations and assumptions, the significant difference between treatments in EB_c for the first weeks after lactation suggest a better energy supply for cows on the Kempen diet, although this was not reflected on BCS. In future research this observation needs to be contrasted with the measurement of metabolic parameters of energy balance.

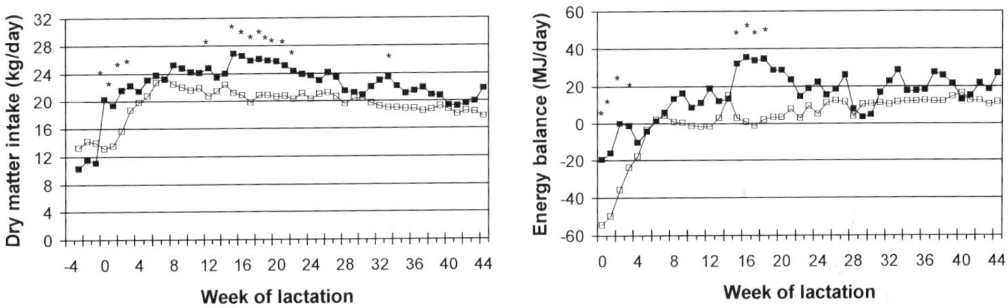

Figure 1. Dry matter intake (left) and calculated energy balance (right) for control (□) and Kempen (■) treatment. Values represent lsmeans per diet per week; * $P<0.05$.

Conclusion

Kempen feeding strategy increases dry matter intake in the first 4 weeks of lactation, leading to a shorter period of negative calculated energy balance. Moreover, Kempen leads to a better milk production persistency without adverse effects on BCS development during lactation.

References

Craninx, M., A. Steen, H. Van Laar, T. Van Nespen, J. Martín-Tereso, B. De Baets and V. Fievez, 2008. Effect of lactation stage on the odd- and branched chain milk fatty acids of dairy cattle under grazing and indoor conditions. J. Dairy Sci. 91: 2622-2677.

CVB (Centraal Veevoederbureau), 2005. Veevoedertabel. Gegevens over chemische samenstelling, verteerbaarheid en voederwaarde van voedermiddelen. CVB, Lelystad, the Netherlands.

Mobilisation of muscle protein and fat tissue in dairy cows around calving investigated by ultrasound measurements

S.G.A. Van der Drift, L. Vernooij and R. Jorritsma
Department of Farm Animal Health, Faculty of Veterinary Medicine, Utrecht University, Marburglaan 2, 3584, CN Utrecht, the Netherlands; s.g.a.vanderdrift@uu.nl

Introduction

The negative energy balance of dairy cows in early lactation results in mobilisation of body reserves to support milk production. Fat tissue is the major energy reserve of the body, but cows also mobilise muscle protein (source of glucogenic amino acids) in especially the first weeks after calving. The mobilisation of body reserves can be monitored by assessing body condition scores (BCS) of cows on a 1 to 5 scale at regular intervals (Edmonson *et al.,* 1989). However, this method is unable to distinguish between fat tissue and muscle mobilisation. To this end, ultrasound measurements of subcutaneous fat tissue and superficial muscles can be applied. The sacral region has been described to be the most suitable site to assess fat thickness (Schröder and Staufenbiel, 2006). The *musculus longissimus* can be used to monitor muscle protein mobilisation (Bruckmaier *et al.,* 1998; Schwager-Suter *et al.,* 2000). In contrast to beef cattle, no extensive information has been published on ultrasound monitoring of dairy cows. The objective of this research was therefore to study the fat tissue and muscle mobilisation of cows in more detail by weekly performed ultrasound measurements.

Material and methods

Research was carried out on dairy farm 'De Tolakker' at the Faculty of Veterinary Medicine, Utrecht University. Ultrasound measurements were performed weekly with a Pie Medical Scanner 100 and linear transducer (freq 5.0 MHz) on 24 dairy cows (mainly HF) from three weeks before until eight weeks after calving from May 2008 to January 2009. Skin spots were brushed and greased with rapeseed oil, but not clipped. At the same time, BCS were assessed by the same, trained person. Backfat thickness was measured in the pelvic region between the *tuber coxae* and *tuber ischadicum*. Longissimus muscle thickness was assessed at the fourth lumbar vertebra. For further analysis, pre-calving back fat thickness and muscle thickness were calculated as the average of the three pre-calving measurements. Total fat and muscle mobilisation were determined as the highest value minus the lowest value of respectively the fat thickness and muscle thickness during the observational period. Box plots were drawn and simple linear regression analysis was performed using SPSS 15.0.

Results

Changes in BCS, fat and muscle thickness during the observational period are displayed in Figure 1 showing considerable variation between cows in the mobilisation of body reserves around calving. Average muscle thickness declined from the start of the observations until approximately week 4 after calving, whereas average backfat thickness was stable before calving but slowly decreased after calving until the end of the observational period (mean values not shown). Average BCS sharply declined between week -1 and wk 1, probably due to physical changes of the cow following the calving process, which was in contrast to the more gradual breakdown of fat tissue and muscle thickness in this period.
Pre-calving backfat and *musculus longissimus* thickness averaged 0.79 cm (SD 0.44 cm) and 3.99 cm (SD 0.61 cm), respectively. Higher pre-calving fat thickness was associated with higher pre-

calving muscle thickness (Pearson correlation coefficient $r = 0.66$, $P<0.001$). After calving, cows mobilised on average 0.44 cm (SD 0.29 cm) of fat tissue and 0.87 cm (SD 0.28 cm) of muscle tissue. Simple linear regression showed that total fat mobilisation was significantly associated with pre-calving fat thickness ($P<0.001$). In contrast, total muscle mobilisation was not at all related with pre-calving muscle thickness ($P=0.966$).

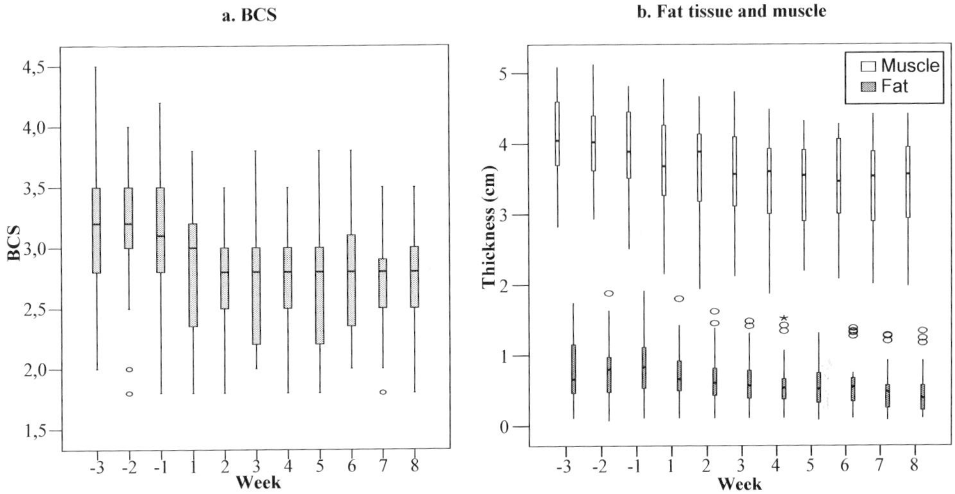

Figure 1. Development of BCS (a) and fat tissue/muscle thickness (b) around calving (n = 24). Boxes represent median and interquartile range; whiskers include all cases except outliers (circles) and extreme cases (asterisks). Calving date is in between week -1 and week 1.

Conclusion

There is a lot of variation in the mobilisation of body reserves of dairy cows around calving. Fat thickness before calving was associated with total fat mobilisation after calving, while muscle mobilisation was not associated with pre-calving muscle thickness. This indicates that fat and protein mobilisation are differently regulated during the negative energy balance.

References

Bruckmaier, R.M., L. Gregoretti, F. Jans, D. Faissler and J.W. Blum, 1998. *Longissimus dorsi* diameter, backfat thickness, body condition scores and skinfold values related to metabolic and endocrine traits in lactating dairy cows fed crystalline fat or free fatty acids. J. Vet. Med. A 45: 397-410.

Edmonson, A.J., I.J. Lean, L.D. Weaver, T. Farver and G. Webster, 1989. A body condition scoring chart for Holstein dairy cows. J. Dairy Sci. 72: 68-78.

Schröder, U.J. and R. Staufenbiel, 2006. Invited review: Methods to determine body fat reserves in the dairy cow with special regard to ultrasonographic measurement of backfat thickness. J. Dairy Sci. 89: 1-14.

Schwager-Suter, R., C. Stricker, D. Erdin and N. Künzi, 2000. Relationship between body condition scores and ultrasound measurements of subcutaneous fat and *m. longissimus dorsi* in dairy cows differing in size and type. Anim. Sci. 71: 465-470.

Expression of genes involved in different metabolic pathways in the liver of metabolically challenged dairy cows during early lactation: a field study

H.A. Van Dorland[1], M. Graber[1], S. Kohler[2], T. Kaufmann[3] and R.M. Bruckmaier[1]
[1]Veterinary Physiology, Vetsuisse Faculty, University of Bern, Bremgartenstrasse 109a, 3001, Bern, Switzerland; [2]Department of Animal Science, Swiss College of Agriculture, Länggasse 85, 3052, Zollikofen, Switzerland; [3]Clinic for Ruminants, Vetsuisse Faculty of the University of Berne, Bremgartenstrasse 109a, 3001, Berne, Switzerland; anette.vandorland@physio.unibe.ch

Introduction

The metabolic and endocrine adaptations to support milk production in early lactation vary between cows (Van Dorland *et al.*, 2009), and may determine whether or not a cow passes successfully (with or without health disorders) through the whole lactation. At present, the underlying metabolic regulation in the liver for an optimal adaptation is still not understood. In the present study, data from a field study were used to investigate hepatic metabolism in week 4 *post partum* (pp) in cows that adapted differently to lactation on the basis of metabolic challenge based on concentration limits of glucose, NEFA, and BHBA (see in the next paragraph).

Material and methods

A field study was carried out in Switzerland with 232 multiparous dairy cows of the breed types Brown Swiss, Simmental × Red Holstein, Red Holstein, and Holstein-Friesian. Blood and liver samples were taken from these cows in week 4 pp (24±2 d pp, +4wk). Blood plasma was assayed for concentrations of metabolites (glucose, NEFA, BHBA) and hormones (insulin, IGF-I). Liver was analysed for mRNA expression levels by real-time qRT-PCR encoding for enzymes of gluconeogenesis (PEPCKm, PC), citric acid cycle (CS), lipid metabolism (ACSL, CPT1A, CPT2, ACADVL), and ketogenesis (HMGCS2, BDH2). To investigate variation in hepatic metabolism underlying adaptation to lactation, two groups of cows were formed based on all three plasma concentrations of glucose, NEFA, and BHBA. In the first group, cows were included with glucose concentrations of <3.0mmol/l, NEFA concentrations of >300 µmol/l, and BHBA concentrations of >1.0 mmol/l, which were referred as being metabolically challenged (GRP-; n=33). The second group included cows with glucose concentrations of >3.0 mmol/l, NEFA concentrations of <300 µmol/l, and BHBA concentrations of <1.0 mmol/l, which were referred as being not metabolically challenged (GRP+; n=40). The limits for metabolite concentrations were chosen based on median concentrations of these metabolites measured in the field study (n=232). Cows that did not belong to these two groups were excluded from this evaluation. Data were analysed with the GLM procedure of SAS® (SAS, 2001), including group as the fixed effect. Prior to data analysis, breed and parity effects were observed to be significant for only 3 of the 16 measured variables, and were therefore excluded from the model. Multiple comparisons between the group means were performed by the Tukey method. Differences between means were considered significant if $P<0.05$.

Results

No significant differences were found between GRP- and GRP + cows for the milk yield at 20 DIM (mean 34±0.7 kg/d), but milk fat content was higher ($P<0.01$) for GRP- than for GRP+ (5.13±0.15 vs. 4.47±0.12, respectively). GRP- compared to GRP+ cows had higher ($P<0.001$) NEFA (528±32 vs. 156±10.0 µmol/l) and BHBA (2.91±0.22 vs. 0.71±0.03 mmol/l) concentrations, and lower ($P<0.001$) glucose (2.54±0.05 vs. 3.49±0.05), insulin (6.80±0.86 vs. 13.8±1.48 µU/l) and IGF-I

(54.2±4.40 vs. 82.9±4.67 ng/ml) concentrations. Table 1 shows the mRNA abundance of metabolic hepatic parameters of dairy cows from GRP- and GRP+.

Table 1. Mean ± SEM of mRNA abundance (delta CT, log_2) of genes involved in different metabolic pathways in the liver of dairy cows adapting differently to lactation.

Parameters		Group[1]		
		GRP- (n=33)	GRP+ (n=40)	*P*-value
Gluconeogenesis related parameters[2]	PEPCKm	11.6±0.12	11.0±0.17	0.01
	bPC	18.6±0.12	17.9±0.12	< 0.001
	G6PC	20.2±0.08	20.2±0.09	0.74
Fatty acid oxidation related parameters[2]	ACSL	17.3±0.10	16.8±0.11	< 0.001
	CPT 1A	15.0±0.11	14.7±0.12	0.12
	CPT 2	16.5±0.07	16.2±0.07	0.02
	ACADVL	18.7±0.07	18.2±0.08	< 0.001
Citric acid cycle related parameter[2]	CS	14.6±0.14	14.4±0.14	0.33
Ketonebody synthesis related parameters[2]	HMGCS2	21.7±0.09	21.6±0.12	0.64
	BDH2	17.8±0.11	17.7±0.14	0.70
Nuclear receptor[2]	PPARα	17.9±0.08	18.2±0.08	0.002

[1] GRP- and GRP+ see Material & methods.

[2] ACADVL, acyl-coenzyme A dehydrogenase very long chain; ACSL, acyl-CoA synthetase long-chain; BDH2, 3-hydroxybutyrate dehydrogenase; CPT 1A or 2, carnitine palmitoyltransferase 1A or 2; CS, citrate synthase; G6PC, Glucose-6-phosphatase; HMGCS2, 3-hydroxy-3-methylglutaryl-coenzyme A synthase 2; PPARα, peroxisome proliferators-activated receptor α; PC, pyruvate carboxylase; PEPCKm, mitochondrial phosphoenolpyruvate carboxykinase.

Conclusion

This study evidenced that metabolically challenged dairy cows (on the basis of concentrations of glucose, BHBA, and NEFA) present differences in metabolic regulation in the liver, compared to metabolically unchallenged dairy cows with a similar milk production in early lactation.

Acknowledgement

This study was supported by Swissgenetics (Zollikofen, Switzerland) and the Federal Office for Agriculture (FOAG, Bern, Switzerland).

References

Van Dorland, H.A., S. Richter, I. Morel, M.G. Doherr, N. Castro and R.M. Bruckmaier, 2009. Variation in hepatic regulation of metabolism during the dry period and in early lacation in dairy cows. J. Dairy Sci., *in press*.

Metabolic and production responses of dairy cows to two levels of rapeseed and soya-bean expeller supplementation on red clover silage based diet

A. Vanhatalo[1], P. Pursiainen[1], M. Tuori[2], M. Rinne[3] and S. Jaakkola[1]
[1]Department of Animal Science, P.O.Box 28, 00014 University of Helsinki, Finland; [2,3]MTT Agrifood Research Finland, 05840 Hyvinkää and 31600 Jokioinen, Finland; aila.vanhatalo@helsinki.fi

Introduction

Increasing dietary crude protein (CP) content of grass silage-cereal based diet with incremental amounts of protein feeds increased silage dry matter (DM) intake and yields of milk and milk protein (Shingfield *et al.*, 2003). However, production responses to rapeseed were higher than to soya-bean meal in this study. Replacing grass silage with red clover silage in the diet has increased non-ammonia N flow into the lower tract, silage DM intake, and milk production in comparison to grass silage based diets (Dewhurst *et al.*, 2003; Vanhatalo *et al.*, 2006). The present study was designed to examine how cows fed exclusively red clover silage as a forage in the diet do respond to increasing amounts of rapeseed and soya-bean expeller as a part of a the fixed amount of cereal-based concentrate.

Material and methods

Four multiparous Finnish Ayrshire cows (11-12 weeks in their 2.-3. lactation) were used as a balanced 4×4 Latin square with 21-d periods. The cows had free access to wilted formic acid preserved red clover silage (270 g DM/kg; metabolisable energy 9.7 MJ/kg DM; CP 188 g/kg DM; neutral detergent fibre (NDF) 342 g/kg DM). Dietary concentrate was given at a level of 10.8 kg/d. Dietary treatments in a 2×2 factorial arrangement consisted of two protein sources (rapeseed expeller and soya-bean expeller) both given at two isonitrogenous levels (low and high). Thus, a portion of the barley:oats (1:1) concentrate in the four dietary treatments was replaced with 1.5 and 3.0 kg/d of rapeseed expeller and 0.95 and 1.9 kg/d of soya-bean expeller, respectively. Feeds, milk and faeces were sampled during the last week, and blood from the tail vein was sampled twice during the last day of each experimental period. Plasma data was pooled over the sampling times, and all data were subjected to the analysis of variance using the mixed procedure of SAS® software. All means are reported as least squares means.

Results and discussion

In contrast to the positive responses obtained with grass silage based diets (Shingfield *et al.*, 2003), the increased protein level in the red clover based diet did not affect silage DM intakes or milk yields (Table 1). However, using rapeseed instead of soya-bean as a protein supplement increased milk protein concentration significantly ($P<0.001$), while milk fat, lactose and urea concentrations were unchanged (data not shown). This resulted in marginal milk protein yield response of 0.31 g per 1 g increase in CP intake with rapeseed expeller on the contrary to non-existent response with soya-bean expeller. Rapeseed diets were associated with higher NDF digestibility ($P<0.05$), and higher plasma methionine ($P<0.05$) and glucose ($P<0.05$) concentrations than soya-bean diets. Organic matter (OM) and N digestibility of the diets, plasma NEFA, BHBA, acetate, essential amino acids (AA), non essential AA or total AA concentrations (data not shown) were not affected ($P>0.10$) by the dietary treatments.

Table 1. Effects of rapeseed and soya-bean expeller on dry matter intake (DMI), milk production, diet digestibility and plasma metabolites in dairy cows fed red clover silage diets.

Protein source	Rapeseed expeller		Soya-bean expeller		SEM	Significance, *P*		
Protein level	Low	High	Low	High		Protein source	Protein level	Source × level
Feed intake, milk production and diet digestibility, kg/d (unless otherwise stated)								
Silage DMI	13.8	12.8	13.3	12.9	0.74	0.68	0.16	0.46
Total DMI	23.2	22.1	22.5	22.4	0.77	0.61	0.20	0.32
CP intake	3.94	4.08	3.81	4.08	0.139	0.43	0.04	0.43
Milk yield	33.1	34.2	33.4	33.6	1.99	0.81	0.46	0.58
Milk protein, g/kg	31.6	31.8	31.0	30.7	0.76	<0.001	0.86	0.14
OM digestibility,%	69.4	69.5	68.7	68.9	0.98	0.53	0.88	0.96
NDF digestibility,%	42.9	45.9	39.8	39.8	2.00	0.05	0.47	0.45
Plasma metabolites, μmol/l (unless otherwise stated)								
Insulin, μU/l	6.24	6.39	5.82	5.38	0.735	0.24	0.81	0.63
Glucose, mmol/l	4.38	4.25	4.04	4.09	0.132	0.05	0.73	0.47
Urea, mmol/l	5.7	6.2	6.2	6.6	0.22	0.08	0.09	0.81
Arginine	92.3	102.3	90.3	83.6	6.05	0.12	0.79	0.19
Histidine	62.3	67.4	59.5	66.1	3.72	0.55	0.12	0.82
Isoleucine	177	189	194	171	12.2	0.97	0.61	0.17
Leucine	195	218	203	197	11.6	0.59	0.51	0.25
Lysine	108	122	113	103	8.2	0.43	0.83	0.20
Methionine	21.6	21.2	19.3	16.1	1.37	0.04	0.24	0.33
Phenylalanine	59.4	64.0	55.7	57.0	2.63	0.08	0.30	0.55
Threonine	119	129	113	112	7.5	0.18	0.58	0.44
Tryptophan	44.8	43.3	42.4	39.1	2.25	0.15	0.28	0.68
Valine	354	391	356	367	23.7	0.61	0.31	0.56

Conclusion

Increased protein supplementation did not elicit expected increases in the silage DM intake or milk yield with red clover silage based diets. The higher milk protein concentrations with rapeseed as compared to soya-bean diets were associated with higher plasma methionine concentrations. This suggests that an inadequate supply of methionine possibly limits further milk production responses on diets, when red clover silage is used as the basal forage.

References

Dewhurst, R.J., W.J. Fisher, J.K.S. Tweed and R.J. Wilkins, 2003. Comparison of grass and legume silages for milk production. 1. Production response with different levels of concentrate. J. Dairy Sci. 86: 2598-2611.

Shingfield, K.J., A. Vanhatalo and P. Huhtanen, 2003. Comparison of heat-treated rapeseed expeller and solvent-extracted soya-bean meal protein supplements for dairy cows given grass silage-based diets. Anim. Sci. 77: 305-317.

Vanhatalo, A., T. Gäddnäs and T. Heikkilä, 2006. Microbial protein synthesis, digestion and lactation responses of cows to grass or grass-red clover silage diet supplemented with barley or oats. Agric. Food Sci. 15: 252-267.

Advances in the understanding of milk cholesterol level regulation

E. Viturro, C. Farke and H.H.D. Meyer
Physiology Weihenstephan, Z I E L, Technical University Munich, Weihenstephaner Berg 3, 85354, Freising, Germany; viturro@wzw.tum.de

Introduction

Milk is not only a source of nutrients in the form of lactose, proteins and lipids, it also serves as an important delivery medium for other crucial molecules, such as vitamins, minerals, bioactive lipids and cholesterol. The milk lipid fraction constitutes a major source of energy and is known to be extremely variable among species, breeds and individuals, depending on factors such as feed (Palmquist *et al.,* 1993) or stage of lactation (Nommsen *et al.,* 1991). Among the lipid fraction, cholesterol is particularly interesting, being the major sterol in whole milk with concentrations ranging from 10-30 mg/dl (0.25-0.77 mM) (Jensen, 2000). This amount represents only 0.5% of the fat fraction, but because of the elevated consumption of milk and dairy products in the human diet, these aliments range second for their contribution to daily cholesterol intake (Royo-Bordonada *et al.*, 2003) and are attracting increasing interest in human health.
Despite of the extensive knowledge available for other species, cholesterol homeostasis has still not been deeply studied in ruminant organisms, a completely necessary knowledge when trying to naturally affect milk cholesterol concentrations.
For that reason, we started a new line of work that points to the comprehension of the molecular mechanisms that regulate cholesterol homeostasis in the dairy cow. In the present overview on the advances made during the last years, we present the data we published concerning the understanding of the following:

- the relation between milk and blood cholesterol levels;
- the variation of milk and blood cholesterol levels during the lactating cycle;
- the mechanisms of gene regulation of cholesterol synthesis in the bovine liver;
- the expression of membrane cholesterol transporters in the bovine organism.

Results

Despite the initial hypothesis that cholesterol might be transferred from blood to milk by a simple passive diffusion, no significant relation was observed between blood and milk cholesterol concentrations in lactating animals (Viturro *et al.*, 2009). Milk was demonstrated to be on a 10-fold lower range (0.2-0.8 mM) with respect to blood (2-6 mM) and no direct relationship between both parameters could be observed. Among individuals with the same blood cholesterol concentration, up to 4-fold differences in milk cholesterol concentrations were shown (and vice versa), pointing to the intervention of an active transport process in the regulation of cholesterol transfer into milk. As candidate genes for this active transport, we studied the expression of the members of the ATP-binding cassette (ABC) family of transporters dedicated to cholesterol transfer in other species. We demonstrated a significant and specific expression of ABCG5 and ABCG8 in the mammary gland (Viturro *et al.*, 2006) and a more ubiquitous expression of ABCA1 (Farke *et al.*, 2006). Moreover, in a recent study (Farke *et al.*, 2008), a significant alteration in the expression levels of ABCA1 between lactation and the dry period was observed.
In parallel, since liver *de-novo* synthesis has a significant influence on cholesterol homeostasis, we studied the expression of liver synthesising enzymes (HMG-CoA reductase, HMG-CoA synthase and Squalene Synthase) during the lactation cycle. A significant and coordinated up-regulation of the expression of HMG-CoA reductase and Squalene Synthase was shown during the first weeks of lactation (E. Viturro, unpublished data, 2009). Since this simultaneous variation might be the

effect of the action of common regulators, we also studied the expression of the SREBP family of gene expression regulators (SREBP1, SREBP2 and SCAP) and found them to be, as expected, intensively over-expressed on the period immediately previous to activation of liver synthesis enzyme expression.

Conclusion

Cholesterol homeostasis in the dairy cow was until now demonstrated to be the result of a precise coordination of very different processes, including membrane transport, liver *de-novo* biosynthesis and gene expression regulation. The results presented open wide fields of investigation for the following years in order to understand milk cholesterol level regulation.

References

Farke C., E. Viturro, H.H. Meyer and C. Albrecht, 2006. Identification of the bovine cholesterol efflux regulatory protein ABCA1 and its expression in various tissues. J. Anim. Sci. 84: 2887-2894.

Farke C., H.H. Meyer, R.M. Bruckmaier and C. Albrecht, 2008. Differential expression of ABC transporters and their regulatory genes during lactation and dry period in bovine mammary tissue. J. Dairy Res. 75: 406-414.

Jensen R.G., 2000. The composition of bovine milk lipids: January 1995 to December 2000. J. Dairy Sci. 85: 295-350.

Nommsen L.A., C.A. Lovelady, M.J. Heinig, B. Lonnerdal and K.G. Dewey, 1991. Determinants of energy, protein, lipid and lactose concentrations in human milk during the first 12 mo of lactation: the DARLING Study. Am. J. Clin. Nutr. 53: 457-465.

Palmquist D.L., A.D. Beaulieu and D.M. Barbano, 1993. Feed and animal factors influencing milk fat composition. J. Dairy Sci. 76: 1753-1771.

Royo-Bordonada M.A., L. Gorgojo, M. Oya, C. Garces, F. Rodriguez-Artalejo, R. Rubio, J.L. del Barrio and J.M. Martin-Moreno, 2003. Food sources of nutrients in the diet of Spanish children: the Four Provinces Study. Br. J. Nutr. 89: 105-114.

Viturro E., C. Farke, H.H. Meyer and C. Albrecht, 2006. Identification, sequence analysis and mRNA tissue distribution of the bovine sterol transporters ABCG5 and ABCG8. J. Dairy Sci. 89: 553-561.

Viturro E., M. Könning, A. Kroemer, G. Schlamberger, S. Wiedemann, M. Kaske and H.H. Meyer, 2009. Cholesterol synthesis in the lactating cow: Induced expression of candidate genes. J.Steroid Biochem. Mol. Biol. In press.

Delayed response of milk fatty acids to micro algae fed in early lactation

B. Vlaeminck[1], M. Hostens[2], G. Opsomer[2] and V. Fievez[1]
[1]Laboratory for Animal Nutrition and Animal Product Quality, Ghent University, Proefhoevestraat 10, 9090, Melle, Belgium; [2]Department of Obstetrics, Reproduction and Herd Health, Ghent University, Salisburylaan 133, 9820, Merelbeke, Belgium; veerle.fievez@ugent.be

Introduction

Rumen by-pass *trans*-10,*cis*-12-18:2 as well as marine products induce milk fat depression (MFD). However, early lactating cows showed to be less responsive to *trans*-10,*cis*-12-18:2 mediated MFD as a three-fold dietary supply was required in the first two to three weeks in lactation (Odens *et al.*, 2007). Hence, a first aim of this study was to assess the effectiveness of marine algae (ALG) to induce MFD during the first twelve weeks of lactation. Furthermore, dietary supply of fish oil induced major changes in milk fatty acid composition that sometimes were transient only. Indeed, *cis*-9,*trans*-11-18:2 in milk fat reached a maximum 5 d after the start of fish oil supplementation in late lactation cows. Concomitant increases in *trans*-10 18:1 at the expense of *trans*-11 18:1 occurred. Hence, a second aim of this study was to assess the persistency of milk fatty acid changes over the first twelve weeks of ALG.

Material and methods

Sixteen cows were randomly assigned to either a control (CON) or micro algae (ALG) supplemented diet, with treatment groups being balanced for parity (3 primiparous and 5 multiparous cows), expected calving date, expected milk production and milk fat content. Cows received a TMR (corn silage/grass silage/soybean meal/sugar beet pulp/corn cob mix/hay/minerals, 532/242/100/45/57/20/4; proportions on DM basis) balanced for net energy and digestible protein with 2 (CON) or 3 (ALG) types of concentrates. The ALG concentrate contained 11% of *Martek DHA gold* (Martek Biosciences Corp., Colombia, MD). Supplementation of the ALG concentrate to the ALG group started three weeks before calving at a rate of 1.8 kg/d and remained 2 kg throughout the entire *post partum* period. Cows were milked by an automated voluntary milking system. Milk parameters were monitored weekly in 24 h milk samples, pooled according to milk yield. Milk samples were stored frozen prior to milk fat extraction. Milk FA were analysed after extraction, methylation and GLC as described by Boeckaert *et al.* (2008). Statistical analysis of milk production and milk fatty acids included the fixed effects of dietary treatment, week of sampling (repeated measures) and their interaction and the random effect of cow assuming an autoregressive order one covariance structure fitted on the basis of Akaike information and Schwarz Bayesian model fit criteria.

Results and discussion

Milk fat decreased with DIM until the fourth week of lactation for both groups, but more dramatically in the ALG group (Figure 1A), remaining 30% lower until the end of ALG supplementation. ALG supplementation dramatically changed milk fat proportions of 18:0, *cis*-9-18:1, *trans*-11-18:1, *trans*-10-18:1, *cis*-9, *trans*-11-18:2, *trans*-9 *cis*-11-18:2 and 22:6n-3 (data not shown). Differences between CON and ALG remained rather constant throughout the whole registration period for 18:0, *cis*-9-18:1 (Figure 1B) and *trans*-11-18:1 (Figure 1C). However, *trans*-10-18:1 (Figure 1C), *cis*-9, *trans*-11-18:2 and *trans*-9,*cis*-11-18:2 (Figure 1D) steadily increased until the sixth week of lactation in the ALG group, but remained stable at a considerably lower (*cis*-9, *trans*-11-18:2) or even negligible (*trans*-10-18:1, *trans*-9, *cis*-11-18:2) level in the CON group. Previously, rumen and milk fatty acids were stable after two/three weeks of ALG supplementation in mid lactating dairy cows (Boeckaert *et al.*, 2008). In the current experiment, supplementation started three weeks prior to calving but temporal variation in several milk fatty acids indicates non steady state conditions due to physiological changes related to lactation stage and/or adaptation at the rumen level.

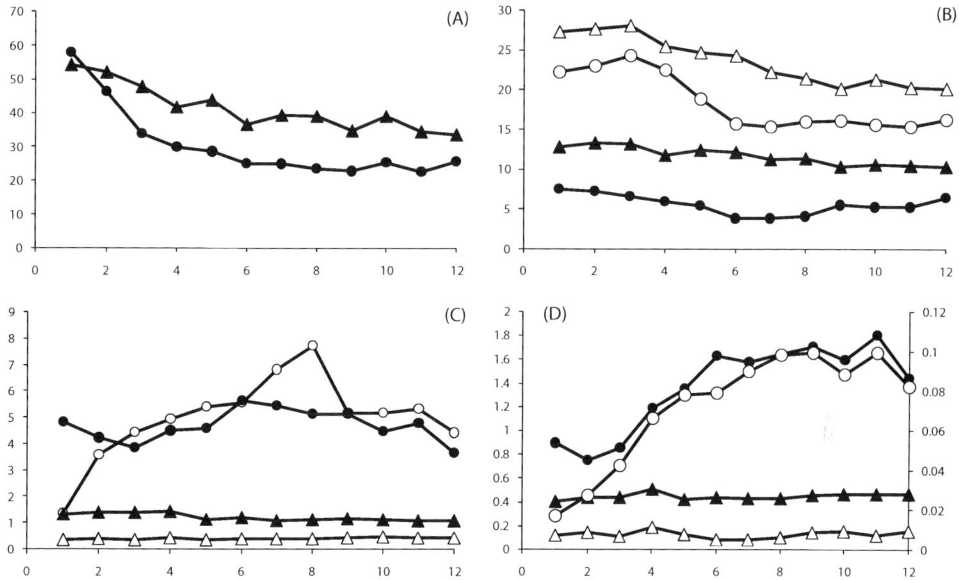

Figure 1. Effect of feeding a control diet (triangles) or algae supplemented diet (circles) to lactating dairy cows from 3 weeks prior to calving until the twelfth week of lactation (X-axis) on milk fat content (g/kg) (A), 18:0 (closed symbols, g/100 g) and cis-9-18:1 (open symbols) (B), trans-11-18:1 (closed symbols) and trans-10-18:1 (open symbols) (C) and cis-9, trans-11-18:2 (closed symbols, primary axis) and trans-9, cis-11-18:2 (open symbols, secondary axis) (D).

Acknowledgement

Research financed by IWT - Institute for the Promotion of Innovation by Science and Technology in Flanders the Fund for Scientific Research-Flanders, Belgium (B. Vlaeminck).

References

Boeckaert, C., B. Vlaeminck, J. Dijkstra, A. Issa-Zacharia, T. Van Nespen, W. Van Straalen and V. Fievez, 2008. Effect of dietary starch or micro algae supplementation on rumen fermentation and milk fatty acid composition. J. Dairy Sci. 91: 4714-4727.

Odens, L.J., R. Burgos,. M. Innocenti, M.J. VanBaale and L.H. Baumgard, 2007. Effects of varying doses of supplemental conjugated linoleic acid on production and energetic variables during the transition period. J. Dairy Sci. 90: 293-305.

Plasma cortisol response to adrenocorticotropin hormone is negatively related to previous wool growth and is greater in twin than single sheep

K.L. Walters[1], F.R. Dunshea[1], A.J. Tilbrook[2] and B.J. Leury[1]
[1]Melbourne School of Land and Environment, The University of Melbourne, Melbourne, VIC, 3010, Australia; [2]Department of Physiology, Monash University, Clayton, VIC, 3800, Australia; fdunshea@unimelb.edu.au

Introduction

Traditional selection programs for Merino sheep are based on outputs such as an increase in fleece weight or reduction of fibre diameter, and take little account of inputs. Thus, much of the information in the literature focusses on factors affecting wool production including, genetics, nutrition and birth type. However the link between sheep metabolism and wool growth efficiency is poorly defined. Previously, a significant negative relationship was shown between an animal's serum cortisol response to exogenous administration of adrenocorticotropin hormone (ACTH) and feed efficiency measured in an unselected line of meat breed rams (Knott *et al.*, 2008). However, it is not known if there is any relationship between cortisol response to ACTH and feed efficiency and/or wool growth in Merino sheep selected for wool growth. The strong relationship between serum cortisol response to exogenous ACTH and feed efficiency suggested that selecting animals based on this approach may be a useful tool for identifying animals which are likely to be more efficient.

Material and methods

Forty-eight Merino wethers (18 mo) were selected based on birth type (Twin vs. Single), extremes of individual wool production (Greasy Fleece Weight (GFW)/kg live weight (LW)) in the previous season (2007), and from six sires involved in a progeny test programme. Sheep were housed indoors and allocated to 4 groups of 12 wethers on the basis of birth type, sire (with each sire represented at least once), and GFW rank. These groups were randomly allocated to 4 pens (4×6 m) with free access to water. Each pen contained an automatic feeder (Lely, Forster Technik, Germany) which enabled individual intake to be set and recorded. All sheep were fed a commercial high quality pellet (12.5 MJ/kg DM ME, 17% CP) with an additional 80 g straw/(sheep d). After acclimation, half of the sheep in each pen were fed *ad libitum* while the other half were restricted to 0.85× maintenance for 49 d after which time the treatments were reversed. Wool growth was measured on a mid side patch over both periods. At the end of both feeding periods an i.m. ACTH (2 μg/kg) challenge was conducted with blood samples being taken immediately before and 45 min after the challenge and assayed for cortisol. Data were analysed using Restricted Maximum Likelihood or regression techniques.

Results

As expected, daily gain and feed intake were significantly greater in wethers fed *ad libitum* compared to those that were restricted (Table 1). There was no effect of GFW on daily gain or absolute feed intake although high GFW ate more per kg of LW, especially when offered feed *ad libitum* as indicated by the interaction, most likely because they were lighter (54.3 vs. 47.1 kg, P<0.001). Wool growth was greater in the high GFW wethers but was not altered by nutrition. Birth status had no effect on daily gain, feed intake or wool growth (data not shown).

Basal plasma cortisol and the magnitude of the cortisol response after ACTH challenge were not influenced (P>0.12) by either feeding level, birth status or GFW status. However, the absolute cortisol concentration after ACTH challenge tended to be higher in twins (71.1 vs. 77.9 nmol/l,

P=0.10) and in high GFW (70.6 vs. 78.4 nmol/l, P=0.06) wethers but was not related to nutrition (71.5 vs. 77.5 nmol/l for *ad libitum* and restricted, P=0.21). The absolute amount of wool obtained in the previous year's shearing was negatively related to the cortisol concentration after ACTH challenge but the intercept was lower for low GFW wethers and twins. The equation describing this relationship was GFW (kg) = 3.66 – 0.0036 post cortisol – 0.387 kg for low GFW – 0.448 kg for twins, P<0.001, R^2=0.47. Feed intake per kg LW was positively related to the cortisol concentration after ACTH challenge in the *ad libitum* fed wethers but not in the restrictively-fed wethers (R^2=0.81 for the multi-regression equation).

Table 1. Effect of selection on previous greasy fleece weight (GFW) and level of nutrition on daily gain, feed intake and wool growth during a 49 d feeding period.[1]

	Low GFW		High GFW		sed	*P*-value	
	Ad libitum	Restricted	*Ad libitum*	Restricted		GFW	Diet
Daily gain, g/d	108	62	103	57	45.9	0.80	0.025
Average intake, g/d	798	322	806	312	43.1	0.97	<0.001
Average intake, g/d × kg[2]	14.5	5.8	16.9	6.7	1.02	0.004	<0.001
Wool growth, g/cm^2	0.044	0.046	0.054	0.049	0.004	0.016	0.90
Wool growth, g/cm^2 × kg	0.80	0.88	1.12	1.06	0.094	<0.001	0.90

[1] There were no interactions except where indicated; [2] Significant interaction (P=0.045).

Discussion and conclusion

These data indicate that there is a negative relationship between previous wool growth and cortisol responsiveness to ACTH but that the response differs between singles and twins. Also, wethers with a greater feed intake have a greater cortisol response to ACTH when allowed to eat *ad libitum* but not when fed restrictively despite there being no effect of feeding level on the cortisol response. Previously, a significant positive relationship was shown between cortisol response to ACTH and residual feed intake efficiency measured in an unselected line of meat breed rams (Knott *et al.*, 2008). These data confirmed that a greater cortisol response to ACTH is related to increased feed intake and lower wool growth although birth state and the potential for wool growth influences this response.

References

Knott, S.A., L.J. Cummins, F.R. Dunshea and B.J. Leury, 2008. Less efficient rams are highly responsive to an adrenocorticotropin hormone (ACTH) induced stress challenge. Domest. Anim. Endocrinol. 34: 261-268.

Effect of reducing dietary crude protein content and supplementing rumen protected lysine on performance of high producing dairy cows during heat stress

X. Wang, H. Zhao, F.C. Wan and Q. Sheng
Institute of Animal Science and Veterinary Medicine, SAAS, Jinan 250100 China; wangxl615@sina.com

Introduction

Heat stress has a dramatic impact on dairy cows where conditions of heat stress can persist for extended periods. Satter *et al.* (2002) proposed that dietary protein could be reduced from 18 to 16% in rations of high producing dairy cows if the rumen protein degradability is carefully balanced to optimise the availability of metabolisable protein. The quality of rumen undegradable protein (RUP) in dairy diets is also receiving a great deal of attention (Chen *et al.*, 1993) since rumen microbes have a steady composition of microbial protein. The increasing dietary lysine from 0.6 to 1.0% of DM increased milk yield in heat stress cows (Huber *et al.*, 1994; Garthwaite *et al.*, 1998). The present study will focus on the effect of dietary protein concentration and supplemental protected lysine on milk production of dairy cows in late lactation exposed to heat stress.

Material and methods

Sixty Holstein cows (232.0±13.51 DIM) averaging over 27 kg of milk/d were randomly assigned into 4 groups and fed a diet containing 18% CP (control) or a 16% CP diet supplemented with either 0, 0.2 or 0.4% rumen-protected lysine (RP-Lysine). The rumen protected product contains 30% Lys (Canadian commercial). Thus, 4 treatment groups, receiving diets of 18% CP and 0.82% total lysine (Control), 16% CP and 0.65% total lysine (treatment 1), 16% CP and 0.85% total lysine (treatment 2) and 16% CP and 1.05% total lysine (treatment 3). Cows were fed the experimental diets for a 2-wk adaptation period, followed by a 10-wk lactation feeding trial. During the study, the minimum and maximum ambient temperature averaged 29.3±0.17 and 38.1±0.31 °C. The minimum and maximum temperature-humidity index (THI) averaged 73.7±0.30 and 83.7±0.21. Milk samples were collected biweekly for determination of milk composition. Rectal temperature and respiratory rate were measured three times per week after PM milking. Meanwhile, the bodyweight and body condition score (BCS) were measured. The blood glucose, NEFA and the blood urea nitrogen (BUN) content were measured biweekly after the morning milking. Data were analysed as a completely randomised design with repeated measures using the Mixed procedure of SAS® (SAS Institute Inc., 1999, Cary, NC, USA).

Results

Table 1 shows the DM intake, milk yield, milk composition and body biological index from the four treatments. DM intake was similar among Control and Treatments ($P>0.05$). Cows fed the control diet had a higher milk yield ($P<0.05$) than treatment 1 and no significance compared with the rest of the treatments. The daily milk yield for treatment 2 or 3 increased 0.70 kg more than treatment 1 though the three treatments had no difference ($P<0.051$).
Cows fed treatment 1 lost more body weight ($P<0.05$) than the other three treatments. This could be due to body protein mobilised to maintain milk production in the face of limiting lysine intake. Cows in the control had the highest respiratory rate and the lowest NEFA or BUN content ($P<0.01$) throughout the study.

Table 1. Milk composition, production and body biological index for the four treatments (n=60).

	Control	Treatment 1	Treatment 2	Treatment 3	SEM	Treatment[1]
DM intake, kg/d	21.40	21.79	21.77	21.75	0.36	$P>0.05$
Milk yield, kg/d	31.55[a]	30.33[b]	31.01[ab]	31.03[ab]	0.67	$P<0.05$
Fat,%	3.10	3.13	3.11	3.20	0.06	NS
Protein,%	3.04	3.03	3.03	3.04	0.03	NS
Body weight change, kg	11.27[a]	-2.71[b]	8.20[a]	9.81[a]	2.30	$P<0.01$
BCS	3.79[a]	3.45[b]	3.61[a]	3.70[a]	0.04	$P<0.05$
Rectal temperature, °C	39.70	39.68	39.48	39.64	0.02	NS
Respiratory rate, /min	70.18[a]	62.68[b]	63.11[b]	64.74[b]	0.69	$P<0.01$
Glucose, mg/dl	66.88	64.94	65.09	65.23	0.96	NS
NEFA, mEq/l	176.12[b]	222.41[a]	225.32[a]	229.10[a]	5.87	$P<0.01$
BUN, mg/dl	11.13[b]	12.48[a]	12.94[a]	12.62[a]	0.30	$P<0.05$

[1] NS = $P>0.05$

[a,b,c] Means the significance of the data in the same row ($P<0.05$, $P<0.01$).

Discussion and conclusion

The trial results were measured under the circumstance that the control diet was designed to CP 18%, NE_L 1.76 Mcal/kg, RUP 39.7% CP and total lysine 0.82% in the late lactation of the trial. Compared to the control, milk yield and composition for treatments were similar even if the dietary CP dropped from 18% to 16% when daily RUP intake including metabolic Lys based on the DM intake was similar. RP-Lysine supplementation might improve milk yield with a low proportion of total dietary Lys (0.62%) and total dietary CP (16%) in the hot season.

References

Chen, K.H., J.T. Huber, C.B. Theurer, D.V. Armstrong, R.C. Wanderley, J.M. Simas, S.C. Chan and J.L. Sullivan, 1993. Effect of protein quality and evaporative cooling on lactational performance of Holstein cows in hot weather. J. Dairy Sci. 76: 819-825.

Garthwaite, B.D., C.G. Schwab and B.K. Sloan, 1998. Amino acid nutrition of the early lactation cow. Proceeding of the Cornell Nutrition Conference, Rochester, NY, pp. 38-50.

Huber, J.T., G. Higginbotham, R.A. Gomez-Alarcon, R.B. Taylor, K.H. Chen, S.C. Chan and Z. Wu, 1994. Heat stress interactions with protein, supplemental fat and fungal cultures. J. Dairy Sci. 77: 2080-2090.

Satter, L.D., T.J. Klopfenstein and G.E. Erickson, 2002. The role of nutrition in reducing nutrient output from ruminants. J. Anim. Sci. 80 (E. Suppl. 2): E143-E156.

Dietary glycerol supplementation to dairy cows: effects on lactation performance and metabolism

A. Werner Omazic[1], J. Bertilsson[1], M. Tråvén[2] and K. Holtenius[1]
Swedish University of Agriculture, [1]Department of Animal Nutrition and Management, Kungsängens Research Centre, 753 23, Uppsala, Sweden; [2]Department of Clinical Sciences, Box 7054, 750 07 Uppsala, Sweden; anna.werner@huv.slu.se

Introduction

Feed intake during the first five weeks of lactation is often insufficient to match the increasing energy demands of lactation. The energy output in the form of milk exceeds energy input in the form of feed, and results in cows with a negative energy balance.
The increasing production of biodiesel in Europe has put its high energetic and glucogenic by-product glycerol forward as an interesting feed for dairy cows, since it is available to farmers at a low cost. Glycerol supplementation to dairy cows has been studied from the time around calving to 3-4 wk after calving when the energy balance has stabilised. The results from such studies have shown both positive and negative effects on feed intake and lactation performance in cows in early lactation (Bodarski *et al.*, 2005 and De Frain *et al.*, 2004). Both raw and refined glycerol have been used in previous studies but there is a lack of comparative studies including both products. The raw glycerol product contains water, ash and methanol in addition to glycerol. The aim of the present study was to evaluate possible differences in the value of raw and refined glycerol as a feed supplement for dairy cows and to establish whether glycerol supplementation can improve energy balance or enhance milk yield in the first four weeks of lactation. The study will also evaluate whether the bad taste and foul smell of raw glycerol have an effect on feed intake.

Material and methods

Thirty dairy cows of the Swedish Red and White Breed in their 1st to 6th lactation were included in the study. The cows were randomly assigned to three different dietary treatments. The treatments consisted of diets with 0.5 kg raw glycerol per day (n=10), 0.5 kg refined glycerol per day (n=10) and the third group was a control group (n=10). The glycerol contained 88.1% glycerol, 9% water, 0.9% ash, and 0.8% methanol (raw glycerol) and 99.5% glycerol (refined), respectively. The trial started on day 4 of lactation and ended on day 32. All cows were fed silage *ad libitum* and concentrate rations were increased from 3 to 10 kg/d for primiparous cows and from 3 to 12 kg/d for multiparous cows over the first 28 d of lactation. Silage and concentrate were fed individually four times per day and feed refusals were collected daily. Glycerol was introduced four days *post partum*. Glycerol was fed together with the concentrate at 09:00 and 17:00. The cows were milked twice a day. Milk yield was registered and samples for milk composition analysis were obtained at evening and morning milkings twice a week. Blood samples were collected once a week. The plasma was analysed for glycerol, glucose, insulin and NEFA. Analysis of variance was performed on all data using the MIXED procedure of SAS® (2003). The statistical model included cow, treatment, age group, and week, with cow as a random variable and treatment, age group and week as fixed variables. Differences were considered significant at $P<0.05$.

Results

Glycerol intake did not affect total dry matter intake during the first four weeks of lactation ($P=0.62$). Average silage intake was 10.5 kg DM/d and average concentrate intake was 7.6 kg/d for all three groups. There was an interaction between silage intake and week after calving ($P<0.001$) and cows fed refined glycerol had significantly lower silage intake during the first week of the experiment than cows fed raw or no glycerol (-3.0 kg DM; P=0.003 and -2.2 kg DM P=0.03, respectively). Feeding refined glycerol increased the milk yield (kg ECM) during the first and second week (P<0.05). Milk composition was not affected by treatment. The concentration of insulin and metabolites did not differ between treatments.

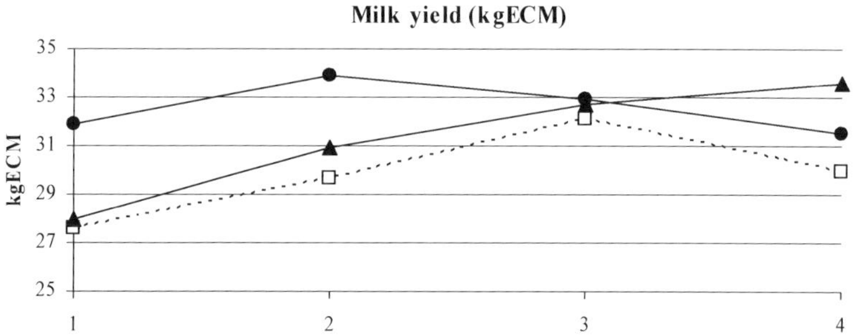

Figure 1. Daily milk yield (kg ECM[1]) for dairy cows fed a silage and concentrate diet with supplementation of raw glycerol (□), refined glycerol (●) or no glycerol (▲). Mean values for weeks 1, 2, 3 and 4 of lactation.

[1] ECM= Energy corrected milk; kg milk × ((383 × fat% + 242 × protein% + 165 × lactose% + 20.7) / 3140)

Discussion and conclusion

Glycerol intake had no effect on total dry matter intake during the first four weeks of lactation. However, intake of refined glycerol decreased silage intake during the first week after calving. Obviously the bad taste and foul smell of raw glycerol did not affect the feed intake. This result creates favourable conditions for using raw glycerol as an energy supplementation since the process to get refined glycerol is more expensive. Intake of refined glycerol increased milk yield during the first and second week, but there was no significant difference in milk composition between treatments. Further work is in progress to establish the effects of raw and refined glycerol supplementation.

Acknowledgement

Support for the research project from AarhusKarlshamn, Sweden, is gratefully acknowledged.

References

Bodarski, R., T. Wertelecki, F. Bommer and S. Gosiewski, 2005. The changes of metabolic status and lactation performance in dairy cows under feeding TMR with glucerol supplement at periparturient period. Electronic J. Polish Agricultural Universities, 8.

De Frain, J.M., A.R. Hippen, K.F. Kalscheur and P.W. Jardon, 2004. Feeding glycerol to transition dairy cows: Effects on blood metabolites and lactation performance. J. Dairy Sci. 87: 4195-4206.

Effect of feeding *Leucaena* hay on thyroid hormones and plasma zinc in dairy goats

J. Wongsanit[1], J.T. Schonewille[2], T. Rukkhamsuk[1], H. Everts[2] and W.H. Hendriks[2]
[1]Faculty of Veterinary Medicine, Kasetsart University, Bangkok, Thailand; [2]Department of Farm Animal Health, Faculty of Veterinary Medicine, Utrecht University, Utrecht, the Netherlands; fvetjrw@ku.ac.th

Introduction

Dairy goat production in Thailand is increasing as a result of increasing goat milk prices and returns. Typically, Pangola hay is used in Thailand as the main source of roughage although its protein content is low. Total protein intake is important to optimise milk production and protein rich feedstuffs either for supplementation or replacement of Pangola hay, are of great interest. *Leucaena leucocephala* is a leguminous tree widely available in Thailand and relatively high in protein as compared to traditional feedstuffs such as Pangola. Currently, there are no data on the effect of the feeding of *Leucaena leucocephala* on milk production in Thai goats under controlled conditions. It is known that *Leucaena leucocephala* contains mimosine which is toxic to non-ruminants and ruminants (Hammond, 1995). The rumen flora can degrade mimosine to 3-hydroxy-4(1H)-pyridone (3, 4 DHP) which is a potent goitrogen. Furthermore, it has been reported that the feeding of *Leucaena leucocephala* is associated with low serum Zn concentration (Megarrity and Jone, 1983). The present study investigated the feeding of *Leucaena leucocephala* on milk yield in goats compared to Pangola hay. In addition, the potential adverse action of mimosine on I and Zn metabolism was concurrently investigated.

Material and methods

Twelve adult, lactating dairy goats with mean lactation number of 2.1 (SE±0.2), weighing 36.4 kg (SE±2.2) were used. The goats were on average 99 (SE±7.0) days in milk and were individually housed in pens during the experiment. The trial had a randomised parallel design with two dietary treatments which was preceded by a 14 d pre-experimental period. The experimental period lasted 7 weeks. All goats were fed a restrictive amount of ration (0.88 kg dry matter/day) based on a hay to commercial concentrate ratio of 50:50 (DM). The chemical composition of the three feed ingredients is shown in Table 1. There were no feed refusals.
During the pre-experimental period, all goats were fed Pangola hay. Thereafter, Pangola hay was replaced by *Leucaena* hay in 6 goats. The rations were offered twice daily at 08:00 and 17:00 h in two equal portions. Feedstuffs were sampled for proximate analysis, Zn and I. Milk yield was recorded daily. Blood samples were taken from each animal at the end of the pre-experimental period and at the third, fifth and seventh week into the study. Serum samples were stored at -20 °C until analysis of serum T_3, T_4, and Zn. Zinc was determined by atomic absorption spectrophotometry (AAS) and I was determined by titration. Serum T_3 and T_4 were determined with a chemiluminescence

Table 1. The chemical composition of experimental feedstuffs.

	CP,% DM	CF,% DM	NDF,% DM	Zn, ppm DM	I, ppm DM
Leucaena hay	21.67	20.25	29.81	36	7
Pangola hay	2.77	31.80	63.63	34	11
Concentrate	16.50	10.08	31.33	143	17

multiwell analyser (CMA). Milk yield, serum T_3, T_4, and Zn concentrations were subjected to repeated measures analyses with dietary treatment as a factor (Wilkinson, 1990). Pre-experimental values (d 14) were used as a co-variable. The level of significance was preset at $P<0.05$.

Results

Zinc intake was similar on both rations, 77 mg/d and 78 mg/d for the Pangola and *Leucaena* hay respectively. Likewise, I intake was 12 and 11 mg/d. There was a significant difference in the mean serum zinc concentration between both groups ($P=0.053$). No significant difference was observed for the serum T_3 ($P=0.917$) and T_4 ($P=0.780$) concentrations between both groups (Table 2). Milk production was not significantly ($P=0.765$) affected by the experimental feedstuffs.

Table 2. Serum concentrations of T_3, T_4, and zinc concentration and milk yield when goats were fed the Pangola hay and Leucaena hay.

Measure	Treatment	Sampling time				Pooled SEM	*P*-value
		0 (d 14)	1 (d 35)	2 (d 49)	3 (d 63)		
Milk, g/d	Pangola	462	212	118	75	100.6	0.765
	Leucaena	642	437	315	198		
T_3, ng/ml	Pangola	0.93	0.48	0.98	1.22	0.139	0.917
	Leucaena	0.91	0.43	1.15	1.12		
T_4, ng/ml	Pangola	51	58	62	60	8.9	0.780
	Leucaena	53	51	60	63		
Zn, ng/ml	Pangola	2,080	2,060	2,070	2,180	88	0.053
	Leucaena	2,110	1,770	1,820	2,260		

Conclusion

The use of *Leucaena* hay instead of Pangola hay did not clearly compromise I and Zn metabolism when the ration contained 50% concentrates rich in Zn and I. The effect of *Leucaena* hay on milk production requires future investigation.

Acknowledgement

The study was supported by Asia-link project (phase III).

References

Hammond, A.C., 1995. *Leucaena* toxicosis and its control in Ruminants. J. Anim. Sci. 73: 1487-1492.

Megarrity, R.G. and R.J. Jones, 1983. Toxicity of *Leucaena leucocephala* in ruminants: the effect of supplemental thyroxine on goats fed on a sole diet of leucaena. Aust. J. Agric. Res. 34: 791-798.

Wilkinson, L., 1990. SYSTAT: The system for Statistics. SYSTAT, Evanston, IL, USA.

Effects of different levels of vitamin A supplementation on antioxidant status of beef cattle with a diet based poor quality silaged corn straw

Z.B. Yang, X.M. Ma, W.R. Yang, F.C. Wan, S.Z. Jiang and T.T. Zhang
Department of Animal Sciences and Technology, Shandong Agricultural University, Taian, Shandong 271018, P.R. China; yangzb@sdau.edu.cn

Introduction

Crop residues play a very important role in beef production in areas where large amounts of corn are grown (Terry *et al.*, 1987). Vitamin A can protect cells from damage caused by unstable substances called free radicals which are believed to contribute to certain chronic diseases and play a role in the degenerative processes seen in aging. Vitamin A is generally supplemented to ruminant diets to insure maximum health and productivity. Unfortunately, considerable supplemental retinol can be destroyed by ruminal microbes. Vitamin A destruction occurs in the rumen with retinol losses up to 80%. The ratio of concentrate in the diet is one factor associated with the extent of rumen destruction (Alosilla *et al.*, 2007). Rode *et al.* (1990) found 80% of vitamin A were lost when cattle were fed 70% concentrate diets, however, losses were only 20% when dietary high-forage diets. In northern China, silaged corn straw was fed beef cattle as exclusive roughage. Because of the poor quality of the silaged corn straw, the growing beef cattle need dietary concentrate over 40% to meet the requirement of the nutrient substance. Meanwhile, little retinol was reserved in silaged corn straw after processing, so the supplementation of vitamin A is especially important. The objective of this study was to investigate the effect of different levels of vitamin A supplementation on antioxidant status of beef cattle with a diet based poor quality silaged corn straw and high-concentrate.

Material and methods

A total of thirty crossbred bulls (Limousin × Ruxi) with an average body weight of 350±10 kg were fed a concentrate diet (80% corn, 16% cottonseed meal, 4% premix; DM basis) without vitamin A supplementation and a free-choice of poor quality silaged corn straw for 10 d. The animals were randomly allotted into five treatments and fed the test diets supplemented with 1,100, 2,200, 3,300, 4,400 and 5,500 IU/kg of vitamin A (retinol palmitate) for 15 d. Blood samples (20.0 ml) were taken from each cattle via jugular vein puncture and collected with a non-heparinised tube, subsequently centrifuged at 3,000× g for 10 min and the serum was collected. Serum samples were analysed for activity of cuprozinc-superoxide dimutase (Cu-Zn SOD), total superoxide dismutase (T-SOD) and glutathione peroxidase (GSH-Px). One unit of SOD was defined as the amount of SOD required to produce 50% inhibition of the rate of nitrite production at 37 °C, and one unit of GSHPx activity was defined as the amount of enzyme per 0.1 ml serum that would catalyse the conversion of 1 μmol/l of reduced glutathione (GSH) to oxidised GSH at 37 °C in five minutes. Total anti-oxidation competence (T-AOC) in serum was determined with the methods of reacting with $Fe_2(SO_4)_3$ and chromogenic agent to form a stable chromophore with a maximal absorbance at 586 nm. The concentration of malondialdehyde (MDA) was analysed with the thiobarbituric acid method, measuring spectrophotometrically malondialdehyde reactive products at 532 nm. All those indices were measured using a commercial kit (Institute of Biological Engineering of Nanjing Jianchen, Nanjing, China). Data were analysed statistically by one-way analysis of variance (ANOVA) with PROC MIXED (SAS®, SAS Institute Inc., 1999, Cary, NC, USA). Values obtained from individual cattle were used as the units for statistical analysis. Differences among 5 treatments were determined by least squares means with PDIFF and adjusted with a Tukey test. In the analyses, significance was considered at $P<0.05$.

Results

The activities of GSH-Px, Cu-Zn SOD and the T-AOC in serum increased with vitamin A supplemented from 1,100 to 4,400 IU/kg in diets of beef cattle ($P<0.05$), and no significant response to increasing levels of dietary vitamin A from 1,100 to 4,400 IU/kg was observed for the activities of T-SOD. However, these four antioxidant indices were decreased by the addition of 5,500 IU/kg of vitamin A ($P<0.05$), and supplementation of dietary vitamin A ranging from 1,100 to 5,500 IU/kg had a quadratic effect on the activities of Cu-Zn SOD ($r=0.97$, $P=0.03$) and T-AOC ($r=0.96$, $P=0.04$) in serum. Meanwhile, there was no significant difference in content of MDA, although it degraded with increasing levels of vitamin A.

Table 1. Activities of GSH-Px, T-SOD, Cu-Zn SOD and T-AOC, content of MDA from beef cattle in response to increasing levels of dietary vitamin A.

	Vitamin A[1], IU/kg					SEM	*P*-value
	1,100	2,200	3,300	4,400	5,500		
GSH-Px, U/ml	15.66[c]	23.49[b]	78.30[a]	85.95[a]	27.15[b]	0.89	<0.05
T-SOD, U/ml	140.66[ab]	142.81[ab]	143.79[ab]	145.43[a]	132.30[b]	3.37	<0.05
Cu-Zn SOD, U/ml	44.18[c]	68.41[b]	93.57[a]	96.35[a]	78.60[b]	3.93	<0.05
MDA, nmol/ml	10.71	10.96	10.81	9.66	9.34	1.4	>0.05
T-AOC, U/ml	6.91[b]	7.92[ab]	9.25[a]	9.13[a]	8.96[a]	0.63	<0.05

[1] Diets supplemented with 1,100, 2,200, 3,300, 4,400 and 5,500 IU/kg of vitamin A (retinol palmitate);
[a,b,c] Means within rows with different superscript letters differ ($P<0.05$).

Discussion and conclusion

This study shows that the activities of GSH-Px, Cu-Zn SOD and T-AOC in serum of beef cattle were all increased as diets supplemented vitamin A from 1,100 to 4,400 IU/kg, but the addition of 5,500 IU/kg of dietary vitamin A decreased these three indices. Large doses of vitamin A can cause birth defects, and may also cause liver abnormalities. This may explain the lack of an effect at 5,500 IU/kg. The content of MDA was not altered by the addition of vitamin A. Vitamin A supplementation may improve antioxidant status of beef cattle with a diet based poor quality silaged corn straw. Further research is recommended to determine the mechanisms of antioxidant status decreases with higher dietary vitamin A in beef cattle.

References

Alosilla C.E. Jr, L.R. McDowell, N.S. Wilkinson, C.R. Staples, W.W. Thatcher, F.G. Martin and M. Blair, 2007. Bioavailability of vitamin A sources for cattle. J. Anim. Sci. 85: 1235-1238.

Rode, L.M., T.A. McAllister and K.J. Cheng. 1990. Microbial degradation of vitamin A in rumen fluid of steers fed concentrate, hay, or straw diets. Can. J. Anim. Sci. 70: 227-233.

Terry, K., R. Larry, F.R. Salvador and L. Mike, 1987. Corn residues in beef production systems. J. Anim. Sci. 65: 1139-1148.

Effect of urea treated *Leucaena leucocephala* leaf meal on growth performence and serum parameters of growing Nanjiang goats

Y.H. Yang[1], Z.S. Wang[1], B. Xue[1], Y.M. Cai[2] and L.Z. Wang[1]
[1]Animal Nutrition Institute, Sichuan Agricultural University, 625014, Ya'an, China; [2]National Institute of Livestock and Grassland Science, 329-2793, Tokyo, Japan; wangzs007@yahoo.com.cn

Introduction

The Nanjiang goat is widely bred in the southwest and northwest of China. Its development is also challenged with acute shortage of protein feedstuff. Therefore, researchers are devoting time to the study on irregular protein feedstuff resources. *Leucaena leucocephala* is an irregular protein source for ruminants containing more than 20% crude protein (Jones, 1979). Its use is limited due to the presence of toxic mimosine, β-[N-(3 hydroxy 4 pyridone)]-α-amino propionic acid (Jones, 1979; Ram *et al.*, 1994). Mimosine is slightly soluble in water, but can be broken down by methanol, ethanol, diluted alkali or acid at room temperature. Information about urea treated *Leucaena leucocephala* leaf meal (LLM) is scanty, therefore, the study was conducted to investigate the influence of substituting parts of the regular protein feed sources with ammoniated LLM on the growth performance and blood parameters in growing Nanjiang goats.

Material and methods

LLM were harvested from Suining city in the Sichuan province of China, and dried in the shade. A portion of the LLM was untreated (ULLM), the other was treated with 4% urea solution (TLLM), diluting 40 g urea in 1l of water and sprayed on 1 kg of dry LLM. The diets contained 50% corn straw silages and 50% concentrate on a DM basis. The concentrate for the control group (CON) was formulated without TLLM or ULLM, whereas the concentrates for the treatment groups were incorporated with 20% TLLM or 20% ULLM, respectively. Chemical analysis showed that the dietary contents of crude protein of CON, TLLM and ULLM were 13.81%, 13.77%, 13.75%, and the calculated DE were 10.44 MJ/kg, 10.47 MJ/kg, 10.47 MJ/kg, respectively. Fifteen 6 month old Nanjiang goats were allocated into three groups based on bodyweight (20.38 kg±1.51 kg). The experimental period lasted 4 weeks, all goats were housed individually and fed twice daily at 8:00 h and 17:00 h. Faecal samples were collected on four consecutive days to determine total tract digestibility. Daily feed intake and fortnightly body weight were recorded for individual goats. Initial and fortnightly blood samples were collected from the jugular vein, serum was separated at 3,000r/min and stored at -20 °C for subsequent analysis. Serum urea nitrogen (UN), total protein (TP), albumin (ALB) concentrations, activities of glutamic oxalacetic transaminase (GOT) and glutamic-pyruvic transaminase (GPT) were analysed with standard assay kits (supplied by Maker Biotechnology Co., Ltd, Chengdu, China). Serum Tri-iodothyronine (T3) and thyroxine (T4) concentrations were analysed with the RIA kit (San V Biotechnology Co., Ltd, Shandong, China). All the data were analysed by one-way ANOVA with treatments as the fixed factor of SPSS 16.0 (SPSS Inc., Chicago, IL) for Windows. The Duncan multiple range test (DMRT) was used to test the significant differences between the means. Significant differences were accepted if $P<0.05$.

Results

The ADG ($P<0.01$) was affected by different treatment groups: +12.48% with TLLM, 18.64% with ULLM compared to CON. The data of DM intake changed slightly in the TLLM group or ULLM group in contrast to CON ($P>0.05$). The F/G showed a descent in the TLLM group ($P=0.13$), but a rise in the ULLM group ($P=0.13$). There was no difference of CP and OM digestibility in the three groups (P>0.05), however, they were numerically higher than in the other two groups. However, in the serum parameters, the level of ALB level of TLLM was significantly higher than the other two groups (P<0.05). The level of serum TP, T3, T4 and the activity of GOT and GPT were not significantly affected by different groups (P>0.05) (Table1).

Table 1. Effect of TLLM or ULLM supplementation on growth performance and serum parameters of the NanJiang goat.

	Group[1]			S.E.	P-value
	CON	TLLM	ULLM		
DM intake, g/d	940.71	917.77	905.05	25.97	0.87
ADG, g/d	104.46^{a}	117.5^{Aa}	82.56^{Bb}	5.18	<0.01
F/G	9.51	7.89	10.99	0.63	0.13
CP digestibility,%	70.60	72.76	68.14	0.96	0.14
OM digestibility,%	70.02	71.15	67.89	2.95	0.22
UN, mmol/ml	2.98	2.97	3.26	0.18	0.75
TP, mg/ml	64.96	64.75	66.18	6.87	0.86
ALB, mg/ml	21.98^{a}	23.78^{Ab}	21.82^{Ba}	0.31	<0.05
GOT, IU/l	120.11	123.92	118.37	2.44	0.65
GPT, IU/l	30.08	34.71	35.62	1.24	0.15
T3, ng/ml	1.68	1.46	1.4	0.16	0.76
T4, ng/ml	45	48.11	48.87	1.43	0.52

[1] TLLM and ULLM = concentrates including 20% *Leucaena leucocephala* leaves meal ammoniated or not; CON = concentrate without TLLM or ULLM.
^{a,b} and ^{A,B} in the same row mean P<0.05 or P<0.01, respectively.

Conclusion

The study demonstrated that urea-treated *Leucaena leucocephala* leaf meal improved growth performance of Nanjiang goats and slightly increased the digestibility of CP and OM, but without apparent effects on the T3 and T4 concentrations.

Acknowledgement

This work is jointly supported by grants from the Chunhui Project by the Ministry of Education of China (No. Z2005-2-61004), and supported by the Educative Department of Sichuan Province.

References

Jones, R.J., 1979. Value of *Leucaena leucocephala* as a feed for ruminants in the tropics. World Anim. Rev. 31: 13-23.

Ram, J.J., P.P. Atreja, R.C. Chopra and A. Chhabra, 1994. Mimosine degradation in calves fed a sole diet of *Leucaena leucocephala* in India. Trop. Anim. Hlth. Prod. 26: 199-206.

Effect of chromium and zinc supplementation on production and blood parameters of lactation Holstein cows under heat stress

S. Zhao, Z.S. Wang, B. Xue, L.Z. Wang and D.W. Wang
Animal Nutrition Institute, Sichuan Agricultural University, Ya'an 625014, China; wangzs007@yahoo.com.cn

Introduction

The climate characteristic of the Sichuan province is humid and hot subtropical in China where the temperature-humidity index (THI) exceeds 72 units from May to October every year, which exerts great challenges on the dairy industry known as heat stress. Some research has shown that heat stress could induce insulin resistance in an animal, decreased SOD enzyme activity and significantly increases the content of MDA (Deng, 2008). Chromium (Cr) and zinc (Zn) are essential trace elements for normal metabolism of carbohydrates, proteins and lipids. Supplementation with organic Cr could significantly increase the feed intake and the milk yield of heat stressed dairy cows (Alsaiady *et al.*, 2004). Zn could reduce the somatic cell count (SCC) and improve milk yield (Wu *et al.*, 2002). It is well documented that Zn has an anti-oxidation function. This experiment was conducted to investigate the effect of Cr and Zn supplementation on performance and blood parameters of Holstein cows under heat stress.

Material and methods

The experiment was conducted from July 21 to September 15, 2008, during the hottest period of the year. Twenty-four Holstein cows (160–185 d *post partum*), with an average initial milk yield of 16.4±1.9 kg/d were divided into 4 groups, according to days in milk, parity and milk yield, with 6 cows in each group. The basal diet contains 55% roughage and 45% concentrate, which were formulated to meet ruminant nutrient requirements recommended by the NRC (2001). Group 1 received no chromium and Zinc. Group 2 received Met-Zn (Zn20%) at the manufacturer's recommended level (5g/head per d). Group 3 received Yeast chromium added at the manufacturer's recommended level (4g/head per d), group 4 received Zinc (5g/head per d) and Yeast chromium (4g/head per d).

Cows were fed 3 times daily at 08:00 h, 14:00 h, 20:00 h. Feed intake was recorded daily throughout the study and milk yield was recorded daily for 3 consecutive days every week. Dry and wet-bulb temperatures were recorded daily from 09:00 h to 21:00 h at 2 h intervals and THI was calculated according to Maust *et al.* (1972). Cows from each group were chosen for bleeding. Blood samples were collected into tubes containing anticoagulant via the jugular vein at 07:00 h on day 1, 28 and 56, and the tubes were kept on ice for 30 min, after which the samples were centrifuged at 3,000× g for 10 min. Plasma was kept at -20 °C for further chemical analysis. Plasma was analysed for glucose, triglyceride (TG), non-esterified fatty acid (NEFA), insulin, cortisol, triiodothyronine (T3), Cu-Zn-SOD, GSH-PX and MDA using standard assay kits.

Statistical analysis was performed using SPSS11.5 (SPSS Inc., Chicago, IL). Data were analysed by one-way ANOVA. LSD multiple comparisons were used to test the differences between treatment, which are denoted by different letter superscripts. Statistical significance was accepted at $P<0.05$.

Results

The calculated THI averaged 79.8 units, which exceeded the upper critical limit (72 units) for dairy cows. Concentrations of plasma TG, NEFA, insulin, cortisol, and T3 were not affected ($P>0.05$) by adding Zn or Cr (Table 1). The addition of Cr and both Zn+Cr increased the cows' plasma glucose concentration and the molar ratio of glucose significantly increased antioxidant enzyme activities. The addition of Cr and Zn alone and when Zn+Cr were given together decreased the MDA content. The three groups (Cr, Zn and Zn+Cr) improved the milk yield by 4%, 6.5% and 8% compared with the control group, respectively but this was not significant.

Table 1. Effect of Cr and Zn supplementation on lactation performance and blood parameters.

Response variables	Least square mean				SEM	P-value
	Control	Zn	Cr	Zn+Cr		
DM intake (kg/d)	18.64[a]	18.96[ab]	19.18[b]	19.17[b]	0.23	0.032
Milk yield (kg/d)	20.34	21.29	21.63	21.92	3.75	0.618
TG (mmol/L)	0.735	0.725	0.717	0.723	0.02	0.183
NEFA (μmol/ml)	0.601	0.589	0.520	0.517	0.07	0.198
Cortisol (ng/ml)	17.94	17.75	17.62	17.80	2.32	0.814
Glucose(mmol/l)	0.321[a]	0.319[a]	0.327[b]	0.336[b]	0.05	0.102
Insulin (μIU/ml)	14.72	13.98	13.72	14.08	0.77	0.099
Glucose/Insulin	0.0208[a]	0.0222[ab]	0.0236[b]	0.0237[b]	0.02	0.015
T3 (ng/ml)	1.64	1.71	1.75	1.74	0.05	0.017
Cu-Zn-SOD (U/ml)	78.4[a]	137.5[c]	118.2[c]	148.8[c]	18.9	<0.001
GSH-PX (U/ml)	107.2[a]	142.8[c]	121.2[bc]	152.1[c]	22.3	0.014
MDA (mmol/ml)	2.61[a]	1.93[c]	2.15[c]	1.96[c]	0.25	0.002

[1] Control =cows receiving no Cr or Zn; Zn, Cr and Zn+Cr = cows receiving 5 g Met-Zn/head per day, 4 g Yeast chromium/head per day and 5 g Met-Zn + 4 g Yeast chromium/head per day respectively.
[a,b,c] In a same row, means with the same superscript letter are not significantly different (P>0.05).

Conclusion

This study shows that both the supplementation of Yeast chromium and met-Zn could improve milk yield by increasing antioxidant enzyme activities and decrease the MDA content in cows under heat stress.

Acknowledgement

This work was supported by the program for public welfare behaviour research fund of the Agricultural Department of China (No. nyhyzx07-036-03).

References

Alsaiady, M.Y., M.A. Alsaiskh, S.I. Almufarrej, T.A.Alshoweimi, H.H. Mogawer and A. Dirrara, 2004. Effect of chelated chromium supplementation on lactation performance and blood parameters of Holstein cows under heat stress. Anim. Feed. Sci. Technol. 117: 223-233.

Deng, F.Q., 2008. Heat stress on lactating dairy cows antioxidant properties and metabolism of trace elements. J. China Anim. Husbandry 44: 61-63.

Maust, L.E., R.E. McDowel and N.W. Hooven, 1972. Effect of summer weather on performance of Holstein cows in the three stage of lactation. J.Dairy. Sci. 55: 1133-1139.

Wu, Z.G. and J. Li, 2002. Effects of dietary levels of zinc-methionine on performance and blood parameters in lactating cows. Chin. J. Vet. Sci. 28: 204-207.

Urinary excretion of volatile fatty acids in sheep sustained by total intragastric infusions

G.-Y. Zhao and Y.-B. Sun
State Key Laboratory of Animal Nutrition, College of Animal Science and Technology, China Agricultural University, Beijing, 100193, P.R. China; zhaogy@cau.edu.cn

Introduction

Dietary carbohydrates can be fermented by rumen microorganisms to produce considerable amounts of volatile fatty acids (VFA) which mainly include acetic acid (C_2), propoinic acid (C_3) and butyric acid (C_4). The VFA absorbed is often believed to be efficiently utilised either for heat production or energy deposition by ruminants. However, Ørskov *et al.* (1991) reported urinary excretion of acetic acid in cattle sustained by intragastric infusions. Other VFA may also be excreted in urine which accounts for an important part of urinary energy loss. The objective of the experiment was to study the urinary excretion of VFA in sheep sustained by total intragastric infusions.

Material and methods

Twelve four-month-old Suffolk × Small tail Han male sheep (live weight 23.17±0.41 kg), fitted with rumen and abomasum fistulas were used in the experiment. The sheep were divided into four groups. Each group had three sheep. The sheep were sustained totally on intragastric infusions which were the same as Sun and Zhao (2009). Casein was used to supply 1.2 times of maintenance (M) nitrogen (N) requirements to all sheep with the assumption that the maintenance N requirement was 350 mg/kg $BW^{0.75}$ per d (MacLeod *et al.*, 1982). The total energy inputs were 495, 540, 585 and 630 MJ energy per d to supply 1.10, 1.20, 1.30 and 1.40 M energy per d for groups I, II, III and IV, respectively, with the assumption that the maintenance energy requirement was 450 KJ/kg $BW^{0.75}$ per d (Ørskov *et al.*, 1979). Four levels of VFA mixture (molar proportions of C2, C3 and C4 =65:25:10), which supplied 333, 378, 423 and 468 KJ energy per d and gave inputs of 0.74, 0.84, 0.94 and 1.04 M energy per d were used as treatments I, II, III and IV, respectively. The VFA mixture was infused into the rumen and casein and other nutrients into the abomasum through fistulas. The infusion continued from 9:00 am to 9:00 pm everyday and lasted for 12 days. The animals had free access to drinking water. During the last 4 days of the experiment, 5% urine was sampled. The urinary VFA was determined by gas chromatography (SP-3420, Beijing Instruments Factory, Beijing). Urinary VFA excretion rate (mol/d) was calculated as the following: Urinary VFA excretion (mol/d)/VFA infused (mol/d)×100. The SAS® 9.0 was used for the analysis of variance between different treatments.

Results

The results indicate that the urinary excretion of C2, C3, C4 and total VFA significantly increased with infusion level ($P<0.05$) which was in agreement with the decrease of urinary pH ($P=0.002$; Table 1). No significant differences were found in urinary excretion rates of C2, C3, C4 and total VFA between different treatments ($P>0.05$), but the urinary excretion rate of C3 was higher than that of C2 and C4 within the same treatments, indicating that the sheep utilised C3 at a lower rate than C2 and C4.

Table 1. Ruminal infusion and urinary excretion of VFA in sheep nourished by intragastric infusions.

Items	Treatments				SE	*P*-value
	I	II	III	IV		
$W^{0.75}$, kg	10.89	10.33	10.33	10.67	0.27	0.434
Urinary pH	9.37^{A}	9.24^{A}	9.01^{A}	8.45^{B}	0.11	0.002
Ruminal VFA infusion, mol/d						
C_2	2.01^{C}	2.17^{C}	2.42^{B}	2.77^{A}	0.06	<0.001
C_3	0.78^{C}	0.83^{C}	0.93^{B}	1.07^{A}	0.02	<0.001
C_4	0.31^{C}	0.33^{C}	0.37^{B}	0.43^{A}	0.01	<0.001
Total VFA[1]	3.09^{C}	3.33^{C}	3.73^{B}	4.26^{A}	0.09	<0.001
Urinary VFA excretion, mol/d						
C_2	0.17^{B}	0.16^{B}	0.33^{A}	0.37^{A}	0.04	0.010
C_3	0.14^{bc}	0.11^{c}	0.23^{ab}	0.26^{a}	0.03	0.021
C_4	0.02	0.01	0.03	0.04	0.01	0.058
Total VFA[1]	0.34^{b}	0.30^{b}	0.59^{a}	0.67^{a}	0.07	0.014
Urinary VFA excretion rate, %						
C_2	8.68	7.71	13.59	13.26	1.81	0.107
C_3	17.71	13.73	24.50	24.17	3.29	0.131
C_4	7.18	4.84	9.43	9.58	1.68	0.230
Total VFA	10.79	8.93	15.90	15.62	2.09	0.109

[1] Total VFA (mol/d) $=C_2$ (mol/d) + C_3 (mol/d) + C_4 (mol/d).

[A,B,C] Means in the same row with capital superscripts differ significantly ($P<0.05$).

[a,b,c] Means in the same row with lowercase superscripts are extremely significantly different ($P<0.01$).

Discussion and conclusion

Considerable amounts of C2, C3 and C4 were excreted in the urine of sheep sustained by total intragastric infusions. Urinary VFA excretion not only loses available energy but may also have a negative influence on the environment. It is necessary to study the mechanism of urinary VFA excretion and ways to reduce urinary VFA loss.

Acknowledgement

We thank National Natural Science Foundation of China for financial support (project No. 30671523).

References

MacLeod, N.A., W. Corrigall, R.A. Stirton and E.R. Ørskov, 1982. Intragastric infusion of nutrients in cattle. Br. J. Nutr. 47: 547-552.

Ørskov, E.R., D.A. Grubb, J.S. Smith, A.J.F. Webster and W. Corrigall, 1979. Efficiency of utilization of volatile fatty acids for maintenance and energy retention in sheep. Br. J. Nutr. 41: 541-551.

Ørskov, E.R., N.A. MacLeod and Y. Nakashima, 1991. Effect of different volatile fatty acids mixtures on energy metabolism in cattle. J. Anim. Sci. 69: 3389-3397.

Sun, Y.B. and G.Y. Zhao, 2009. The relationship between the volatile fatty acids supply and the nitrogen retention in growing sheep nourished by total intragastric infusions. Small Rumin. Res. 81: 8-12.

Study on fasting metabolism in growing water buffaloes (*Bubalus bubalis*) in Guangxi, China

C.X. Zou[1], B.Z.H. Yang[1], X.W. Liang[1], Zh.Sh. Xia[2], K. Liang[1,2], S.J. Wei[1], L.L. Li[1] and Sh.L. Li[1]
[1]Buffalo Research Institute, Chinese Academy of Agricultural Sciences, Nanning 530001, P.R.China;
[2]Guangxi University, Nanning, Guangxi 530005, P.R. China; liangbri@126.com

Introduction

Water buffalo occupies an important position (178 million, FAO, 2007) among domesticated ruminant livestock. In recent years, swamp water buffalo production has become very important in China and other countries such as Thailand, Laos and Cambodia due to the increased demand of water buffalo meat for human consumption. Energy metabolism is the foundation of the energy requirement of animals. There is much literature on diet energy requirements for animals, such as pigs, poultry, goats, sheep and cattle. To our knowledge, there is little available information from experimental studies on the energy metabolism of the growing water buffalo (Tiwari *et al.*, 2000; Khan *et al.*, 1988; Zou *et al.*, 2007). The objective of the present study was to evaluate the net energy requirement for maintenance by measuring fasting heat production in 15 growing water buffalo, using open-circuit respiratory hoods.

Material and methods

Dewormed, healthy, with good body condition, growing female water buffalo (n=15), aged 12, 18, 24 months and body weight (BW) 227.1, 299.0, 358.5 kg, respectively, were used. The water buffalo were divided into three groups (five animals per group) according to their ages (12, 18, 24 months). They were fed 8 kg fresh elephant grass and 4 kg fermented pineapple and 2 kg concentrate per day for 15 days. Then the experimental animals were subjected to fasting and allowed to drink freely for 7 d. From the first day (after fasting for 72 h) of gas collection, the standing time and lying down time of experimental water buffalo were recorded, and the air temperature, air relative humidity and air pressure in the experimental location were also recorded at each collection time. Gas exchanges for the whole animal were collected for 3 d (4 times per day, 06:00 h, 12:00 h, 18:00 h, 24:00 h) which consecutively begun on day 4; open-circuit respiratory hoods were connected, for 5-10 min at each time on each animal. The total volume of the gas in the collection bag was recorded by withdrawing the bag with a hygrometric gas-flow meter connected to a gas pump. During these processes, the gas samples were placed into a 10 ml plastic syringe 3 times. At the same time O_2 and CO_2 concentrations in the gas sample were analysed. Respiratory quotient (R.Q.) was used as an indicator of *in situ* energy metabolism (state) of the animal, and was calculated by dividing the volume of CO_2 produced (L) by the volume of O_2 consumed (L). Fasting B.W. was calculated as the mean of the initial (4th day of starvation) and final (7th day of starvation) B.W.. After converting the respiration gas to standard situation volume, FHP was estimated from the Respiratory Quotient multiplied by the thermal equivalent. Based on the ratio of standing to lying down time, the data were calibrated with the following equations:
HP (standing up)=115% HP (lying down), HP (KJ/24 h)= HP (KJ/min)×1,440 (min).
In the post-absorptive state, the amount of methane production was scant, so if the heat production value of methane production was ignored, there were no effects on the total heat production. Therefore, we referred to Brouwer's (1965) equation to check the reliability of fasting heat production: heat production (KJ) = 16.1753 VO_2 + 5.0208VCO_2 - 5.9873Un.
Statistical analyses of the data obtained in the experiment were performed using Excel 2003 and the data are presented as means. The difference of means was tested by using the Duncan new multiple range test (Steel and Torrie, 1980).

Results

The respiratory quotient (R.Q.) ranged from 0.66 to 0.83 with a mean of 0.73. Each FHP estimated from Brouwer's equation value was similar to FHP estimated from RQ value results (Table 1). The relatonships between FHP and BW was the following: FHP(kJ/d)= 877.864$w^{0.57}$ (n=15, r=0.7075). The metabolic body size of dairy buffalo heifers was $W^{0.57}$. Usually, animal maintenance net energy requirements were 120%~130%×FHP. Therefore, NEm (kJ/.d)= 877.86×125%=1097.33 $W^{0.57}$ or NEm (kJ/d)= 322×125%= 402.50 KJ/kg$w^{0.75\times}$d.

Table 1. Gaseous exchange, urinary N excretion and fasting heat production in growing buffalo after 72 h of fasting (mean±SE).

Month	Head	Fasted BW (kg)	V_{O2} (L/$W^{0.75}$×d)	V_{CO2} (L/$W^{0.75}$×d)	R.Q.	FHP (estimated from R.Q.)	FHP (estimated from Brower`s equation)
12	5	227±14.3	17.94±1.1	12.73±1.263	0.71±0.027	334±21.2	333±21.5
18	5	299±6.9	17.35±2.650	13.57±2.617	0.78±0.044	329±52.4	328±52.2
24	5	359±11.1	16.65±2.091	11.64±1.665	0.70±0.023	306±35.0	305±35.4
	MSE	56.6	1.984	1.973	0.049	37.3	37.7
	P	0.0001	0.6063	0.3109	0.0033	0.4938	0.5046

Conclusion

The relationship between fasting metabolism and body weight exponential was FHP (kJ/d)= 877.864 $W^{0.57}$, n=15, r=0.7075. Metabolic body weight of the growing milk-meat dual-purposed buffalo was $W^{0.57}$ kg.The maintenance net energy requirement of growing buffaloes was the following: NEm(kJ/d)=125%×877.864= 1097.33 $W^{0.57}$ or NEm= 322×125%= 402.50 (KJ/kg$w^{0.75}$×d).

References

Brouwer, E., 1965. Report of sub-committee on constants and factors. Page 441 in Energy Metabolism, EAAP Publ. No.11, Troon, UK.

FAO, 2007. In: FAO Yearbook, Food and agricultural organisation of United Nations. 2007.

Khan, M.Y., L. Kishan, M. Lal and D.C. Joshi, 1988. Energy requirements of Murrah buffalo for maintenance. Pro. World's Buffalo Cong., New Delhi, India (Dec 12-16). pp. 238-243.

Steel, R.D.G. and J.H. Torrie, 1980. Principles and procedures of statistics (2nd edn). McGraw-Hill Book Company: New York, pp. 137-194: 363-372.

Tiwari, C.M., Chandramoni, S.B. Jadhao and M.Y. Khan, 2000. Fasting heat production of growing buffalo calves. Asian Aust. J. Anim. Sci. 13: 307-312.

Zou, C.X., X.W. Liang, B.Z. Yang, K. Liang, J.X. Liu, Z.S. Xia, F. Zhao and S.J. Wei, 2007. Study on energy metabolism of dairy buffalo heifers in Guangxi, China. J. Anim. Feed Sci. 16: 48-52.

Short communications
Nutrition and reproduction

Characterisation of dairy cows carrying 'fertil +/+' or 'fertil -/-' haplotype for one QTL of female fertility located on chromosome 3

S. Coyral-Castel[1,2], C. Ramé[2], C. Fabre-Nys[2], D. Monniaux[2], P. Monget[2], F. Dupont[2], A. Eggen[3], S. Fritz[4], A. Malafosse[4], P. Faverdin[5], C. Disenhaus[5], P. Le Mézec[1] and J. Dupont[2]
[1]Institut de l'élevage, Département Génétique, 149 rue de Bercy, 75595 Paris, France; [2]INRA, UMR Physiologie de la Reproduction et des comportements, 37380 Nouzilly, France; [3]INRA, UMR Génétique animale et biologie intégrative, génétique et génomique bovine, 78350 Jouy-en-Josas, France; [4]UNCEIA, 149 rue de Bercy, 75595 Paris, France; [5]INRA, UMR Production du lait, 35590 St-Gilles, France; jdupont@tours.inra.fr

Introduction

In the last decade, the fertility of the high producing dairy cows (HPDC) has been continuously damaged, especially in Prim'Holstein. Several works realised by geneticists indicate that the decrease of fertility is related, in part, to the intensity of selection on dairy production. Indeed, these HPDC mobilise in an excessive way their energy stores, to the detriment of their fertility. In order to better understand the genetic determinism which underlies this decrease of fertility, a programme of detection of QTL (Quantitative Trait Locus) in cows was undertaken between 1996 and 2000 (Boichard *et al.*, 2003). This program allowed the detection of several QTL involved in the decrease of female fertility, measured as the rate of success after the first artificial insemination (AI1). In the present work, we chose to study animals carrying either the 'fertil +/+' or 'fertil -/-' haplotype for one QTL of female fertility located on chromosome 3 (QTL-F-Fert-BTA3). We choose this QTL because it intervenes in the failures of pregnancy that occurred before 90 days after AI (Guillaume *et al.*, 2007).

Material and methods

Forty-five heifers (n=24 'fertil +/+' and n=21 'fertil -/-'), aged from 8 to 12 months, were bought. On these heifers, we measured growth (body weight, height at the withers, body condition score), cyclicity (progesterone profiles, follicular waves) and AI1 success. After calving, we studied growth, the duration of *post partum* anoestrus, dairy production, milk composition and AI1 success. Moreover, daily food intake and metabolic parameters such as plasma non-esterified fatty acids, glucose and urea were measured. The results are presented as means ± SEM. The T test was used for cyclicity and fertility parameters, body weight before calving and birth weight of the calves, with $P<0.05$ or $P<0.07$ and the Chi^2 test was used for AI1 success with $P<0.05$. Statistical analyses for weight after calving, dairy production, food intake and metabolic parameters are in process.

Results

We found no difference between 'fertil +/+' and 'fertil -/-' heifers, for body weight, height at the withers and body condition score (from 12 months to one month before calving). For cyclicity and fertility, no significant difference was observed for age at puberty, length of the sexual cycle before AI. By echography, we determined the number of follicular waves for 'fertil +/+' and 'fertil -/-' heifers. We observed that 'fertil -/-' heifers with 3 follicular waves were less fertile (i.e. able to lead a gestation up to 90 days) than 'fertil -/-' heifers with 2 follicular waves ($P<0.05$), whereas the number of waves did not affect the fertility of 'fertil +/+' heifers. Then, heifers were inseminated and we studied the non return rate after AI1 at 21 days (NRR21) by plasma progesterone analysis, at 35 days (NRR35) by echography and at 90 days (NRR90) by rectal palpation. The NRR90 was 71% for 'fertil +/+' heifers (n=24) *versus* 55% for 'fertil -/-' heifers (n=20) after AI1.

Fifteen days before calving, there was no significant difference concerning body weight between 'fertil +/+' (672±9.48 kg) and 'fertil -/-' (657.06±9.33 kg) animals ($P>0.05$). We observed that 'fertil -/-' cows lost more weight during the first 35 days after calving than 'fertil +/+' cows (Figure 1). Furthermore, this difference was not explained by variations of the birth weight of the calves (data not shown, $P>0.05$).
During the first 35 days after calving, daily food intake and dairy production seem to be lower for 'fertil -/-' than 'fertil +/+' cows (data not shown). However, milk composition was similar (data not shown). Preliminary results show no difference between 'fertil +/+' and 'fertil -/-' cows for plasma non-esterified fatty acids, glucose and urea. In contrast, the duration of the *post partum* anoestrus is longer ($P<0.07$) for 'fertil -/-' than 'fertil +/+' cows (n=18, 29.89±3.38 *versus* n=23, 23.04±1.84 days, respectively). Finally, we studied AI1 success. The NRR21 was 86.96% for 'fertil +/+' cows (n=23) vs. 61.11% for 'fertil -/-' cows (n=18) and the NRR35 was 69.57% for 'fertil +/+' cows vs. 38.89% for 'fertil -/-' cows. The study of follicular waves, follicular cysts and energy balance (in collaboration with P. Faverdin) are in progress.

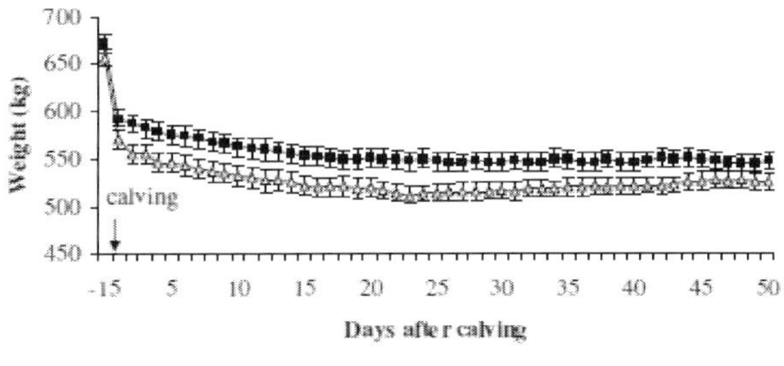

Figure 1. Dairy cows body weight from 15 days before calving to 50 days after calving. ■ 'fertil +/+', Δ 'fertil -/-'.

Discussion and conclusion

The AI1 success was significantly higher for 'fertil +/+' than 'fertil -/-' cows and the same tendency was observed for heifers. This result confirmed the genetic difference. For the moment, this descriptive study shows that 'fertil -/-' cows lose more weight than 'fertil +/+' cows after calving, during 35 days. In this same period, daily food intake and dairy production seem to be lower for 'fertil -/-' than 'fertil +/+' cows. Statistical analyses remain to be determined to affirm if these differences are significant.

References

Boichard, D., C. Grohs, F. Bourgeois, F. Cerqueira, R. Faugeras, A. Neau, R. Rupp, Y. Amigues, M.Y. Boscher and H. Leveziel, 2003. Detection of genes influencing economic traits in three French dairy cow breeds. Genet. Sel. Evol. 35: 77-101.

Guillaume, F., M. Gautier, S. Ben Jemaa, S. Fritz, A. Eggen, D. Boichard and T. Druet, 2007. Refinement of two female fertility QTL using alternative phenotypes in French Holstein dairy cattle. Anim. Genet. 38: 72-74.

The effect of marine algae supplementation in the ration of high yielding dairy cows during transition and its effect on metabolic parameters in the serum and follicular fluid around parturition

M. Hostens[1], V. Fievez[2], B. Vlaeminck[2], S. De Vliegher[1], S. Piepers[1] and G. Opsomer[1]
[1]Department of Obstetrics, Reproduction and Herd Health, Faculty of Veterinary Medicine, Ghent University, Salisburylaan 133, 9820 Merelbeke, Belgium; [2]Laboratory for Animal Nutrition and Animal Product Quality, Faculty of Bioscience Engineering, Ghent University, Proefhoevestraat 10, 9090 Melle, Belgium; Miel.Hostens@UGent.be

Introduction

Milk production has increased with 3,000 kg per lactation over the last two decades in Flanders. Considering a general lactose, fat and protein content of 4.5, 4.0 and 3.3% respectively, the 2009-dairy cow has on average to deal with a daily increase in its production of about 0.73 kg dry matter of milk solids when compared to 1991-dairy cows, leading to a more severe negative energy balance (NEBAL) especially in the beginning of lactation. Many efforts have been made to alleviate the NEBAL of high yielding dairy cows early *post partum* and consequently to improve the bio-energetical markers of the animal's energy status (β-hydroxybutyric acid (BHBA), glucose, insulin, non-esterified fatty acids, aspartate aminotransferase, γ-glutamyl transferase (GGT), cholesterol, albumin) both in the serum as well as in the follicular fluid of high yielding dairy cows. One of the proposed means to diminish the NEBAL *post partum* is the induction of a milk fat depression (MFD) in order to substantially decrease the 'loss' of energy as the production of fat comes with the highest demand for energy (Jensen, 2002). When fed to dairy cows in established lactation, docosahexaenoic acid (DHA) enriched microalgae (Schizochytrium spp.) are able to induce a MFD as described earlier by Boeckaert *et al.* (2008). To our knowledge, there are no experiments in which long core omega-3 fatty acids (LCFA – Ω3) such as DHA were used to induce a MFD in which indicators of the NEBAL in high yielding dairy cows early *post partum* were measured. Our hypothesis was that by feeding DHA, the negative effects of NEBAL on fertility could be diminished by inducing a MFD early *post partum*.

Material and methods

Sixteen healthy primiparous (n=6) and multiparous (n=10) Holstein cows located on the same herd participated in this study. The cows were assigned to 2 different treatment groups and were matched for parity, expected calving date, estimated milk- and milk fat production and genetic origin. All cows received a total mixed ration (corn and grass silage, soybean meal, sugar beet pulp, corn cob mix, hay and minerals; 53.2, 24.2, 10.0, 4.5, 5.7, 2.0% on kg dry matter basis). Cows further received 2 kg of protein rich concentrate and between 6 and 8 kg of balanced dairy concentrate, according to their requirements. In the diet of the ALG group, 2 kg of the latter concentrate was replaced by an iso-energetic concentrate containing marine algae (11% on product basis). Supplementation started 3 wk prior to the expected calving date. Milk parameters were monitored weekly starting on d 5 after parturition and this during 12 consecutive weeks. Metabolic parameters in serum were monitored weekly from the week before parturition until 6 wk after parturition. Metabolic parameters were measured in follicular fluid which was collected by means of transvaginal aspiration from 2 wk until 6 wk after parturition. Only follicles >0.4 and <0.8 cm were punctured in this study and data from atretic follicles were omitted from the study. Data analysis was performed using the PROC MIXED procedure of SAS® (SAS Institute Inc., 2000, Cary, NC, USA). Fixed effects included treatment as the main variable of interest, as well as parity and their interaction. Repeated measurements within the cow were taken into account by adding the cow as the random effect.

Results

The average total roughage intake for all cows was 19.4±1.36 kg on a daily basis. ALG increased milk yield (41.2 vs. 38.2 kg/d, SEM=1.08, *P*=0.057), whereas milk fat yield (1.18 vs. 1.49 kg/d, SEM=0.075, *P*=0.005) and milk fat content (30.6 vs. 41.4 g/kg, SEM=2.26, *P*=0.011) decreased. Protein yield (*P*=0.465) was not affected, whereas a tendency for reduced milk protein content (32.9 vs. 34.7 g/kg, SEM=0.76, *P*=0.094) was observed. Dietary effects on milk production characteristics were independent (*P*>0.05) of week after parturition, although effects were small during the first 2 wk of lactation. Metabolic parameters for NEBAL in serum were not affected by dietary treatment except for the urea level (5.27 vs. 4.72 mmol/L, SEM=0.005, *P*=0.02). A tendency for a dietary treatment effect was observed in the glucose (3.21 vs. 3.47 mmol/L, SEM=0.005, *P*=0.058), BHBA (1.14 *vs* 0.84 mmol/L, SEM=0.003, *P*=0.066) and GGT (48.34 vs. 32.89 IU/L, SEM=0.12, *P*=0.078) levels. Follicular fluid BHBA was significantly increased in the ALG group (1.29 vs. 0.80 mmol/L, SEM=0.016, *P*=0.005). Other indicators for NEBAL in follicular fluid did not differ between groups.

Discussion

Santos *et al.* (2008) stated that improved measures on reproduction have not consistently been observed when feeding fat to dairy cattle probably due to increased milk yield and body weight losses which worsen the NEBAL. While the number of animals used in the present study was small, the results confirm the statement of Santos *et al.* (2008) showing that the supplementation with polyunsaturated fatty acids during transition in high yielding dairy cows may lead to a decrease in milk fat production but an increase in milk production, which may worsen the cows' energy status as measured in the serum as well as in the follicular fluid.

References

Boeckaert, C., B. Vlaeminck, J. Dijkstra, A. Issa-Zacharia, T. Van Nespen, W. Van Straalen, and V. Fievez, 2008. Effect of dietary starch or micro algae supplementation on rumen fermentation and milk fatty acid composition of dairy cows. J. Dairy Sci. 91: 4714-4727.

Jensen, R.G, 2002. The composition of bovine milk lipids: January 1995 to December 2000. J. Dairy Sci. 85: 295-350.

Santos, J.E.P., T.R. Bilby, W.W. Thatcher, C.R. Staples and F.T. Silvestre, 2008. Long chain fatty acids of diet as factors influencing reproduction in cattle. Reprod. Dom. Anim. 43: 23-30.

Proteome and immunoassay analyses elucidate the role of pituitary hormone isoforms and highlight novel signals in response to feed restriction in dairy cows

B. Kuhla[1], D. Albrecht[2], R.M. Bruckmaier[3], T. Viergutz[4] and C.C. Metges[1]
[1]Research Units Nutritional Physiology 'Oskar Kellner' 18196 Dummerstorf, Germany; [2]Institute of Microbiology, Ernst-Moritz-Arndt-University, 17487 Greifswald, Germany; [3]Veterinary Physiology, Vetsuisse Faculty, University of Bern, Bremgartenstrasse. 109a, 3001 Bern, Switzerland; [4]Reproductive Biology, Research Institute for the Biology of Farm Animals 18196 Dummerstorf, Germany; b.kuhla@fbn-dummerstorf.de

Introduction

The hypothalamic-pituitary system of dairy cows regulates a variety of physiological functions highly important for milk production, reproduction, and feed intake. Within the circuits controlling feed intake the pituitary releases peptide hormones into the blood stream to adjust peripheral metabolic processes according to changing environmental and physiological conditions. Under conditions of low feed energy intake, plasma growth hormone concentrations (GH) increases (Block *et al.*, 2003; Vicini *et al.*, 1988) while that of prolactin (PRL), and thyrotropin (TSH) are reduced (Tveit and Larsen, 1983; Vicini *et al.*, 1988) and that of luteinizing hormone (LH) are maintained (McCann *et al.*, 1986) in the bovine. Conversely, enforced secretion of PRL stimulates the mammary gland to produce milk but also amplifies feed intake to meet increased energy demands. However, until recently, posttranslationally generated isoforms or fragments of pituitary hormones have been shown to exert opposing physiological functions as compared to their parent proteins calling for the elucidation of every hormone isoform/fragment in response to feed intake levels. Moreover, data on other pituitary proteins than peptide hormones contributing to the mechanism of feed intake regulation are scarce. Therefore, our first aim was to examine those pituitary isoforms/fragments that are altered after feed restriction, and secondly, to identify additional, so-far unknown pituitary proteins potentially involved in the process for maintaining energy homeostasis. To approach these aims we used global proteome analysis to identify and quantify pituitary peptides and proteins of *ad libitum* and feed deprived dairy cows. To confirm these data, concentrations of pituitary hormones were (at least partially) analysed by Radio-Immuno-Assay (RIA).

Material and methods

Twelve lactating Holstein cows, comparable in age, days in lactation, body weight and milk yield, were fed a total mixed ration (TMR) or straw *ad libitum* for 2.5 days. Average daily NEL intake was reduced to 16% in the straw-fed relative to the TMR group. Animals had free access to water, were milked twice daily, and were killed by exsanguination. Pituitary extracts were subjected to 2-dimensional gel electrophoresis and proteins were stained with colloidal Coomassie, scanned, and processed with computer software for quantification. All gel spots below 40 kDa were punched out, tryptic digested and analysed by MALDI TOF/TOF for identifying proteins by their peptide mass fingerprints using the NCBInr data base.
Pituitary extracts were also analysed for LH by electrochemiluminescence immunoassay (ECLIA). The assay is based on a sandwich technique using ruthenium (II)–labelled monoclonal antibody and purified rabbit polyclonal antibody against the β-subunit of LH.

Results and discussion

To identify pituitary hormones, extracts of the *ad libitum* fed group were pooled, applied on a 2-D gel and 384 spots<40 kDa were analysed by MALDI-TOF measurements. A total of 382 spots were identified. Among them we found 61 different spots containing peptide sequences of pituitary hormones, namely growth hormone (GH; 42), (pre)prolactin (PRL; 15), luteinizing hormone-beta (LH-B; 1), thyrotropin-beta (TSH-B; 1), proopiomelanocortin (POMC; 1) and its cleavage product β-lipotropin (LPH; 1). The comparison of 2D gels of the *ad libitum* and the restricted group using the relative spot volume revealed that each hormone spot differed between feeding groups inconsistently by a factor of 0.8 to 1.2 and each without statistical significance ($P>0.2$; *t*-test). In addition, when the sum of all relative spot volumes for PRL or GH respectively was formed, also no difference between feeding groups was observed. Thus, after 60 h of feed restriction the pool size of these hormone isoforms in the pituitary gland was sustained. Measurement of pituitary LH by ECLIA confirms the results obtained by proteomic analysis (evaluation of pituitary GH and PRL is currently ongoing).

Considering only those protein spots that are differentially expressed between feeding groups by $P<0.05$ (*t*-test), we obtained 8 clearly separated spots representing 6 different proteins. Briefly, glial fibrillary acidic protein (GFAP), purine-rich element binding protein A (PURA), and translation elongation factor Tu (elFTu) were reduced to 70-43% after energy restriction as compared to the control group. By contrast, proline synthetase co-transcribed bacterial homolog, annexin A5, and beta-tubulin were found to be 1.4 to 1.9 times higher expressed in the feed restricted group. Interestingly, besides annexin A5 (which regulates the hormone secretion process) we also identified secretogranin V, a protein of secretory hormone vesicles that tended to be higher expressed to 145% upon energy restriction ($P=0.08$).

Conclusion

our study demonstrates that 60 h dietary energy restriction provokes no significant change of the 61 pituitary hormone isoforms in lactating cows but identified new protein candidates from the pituitary potentially involved in the adaptation process to dietary energy deficiency.

References

Block, S.S., R.P. Rhoads, D.E. Bauman, R.A. Ehrhardt, M.A. McGuire, B.A. Crooker, J.M. Griinari, T.R. Mackle, W.J. Weber, M.E. Van Amburgh and Y.R. Boisclair, 2003. Demonstration of a role for insulin in the regulation of leptin in lactating dairy cows. J. Dairy Sci. 86: 3508-3515.

McCann, J.P. and W. Hansel, 1986. Relationships between insulin and glucose metabolism and pituitary-ovarian functions in fasted heifers. Biol. Reprod. 34: 630-641.

Tveit, B. and F. Larsen, 1983. Suppression and stimulation of TSH and thyroid hormones in bulls during starvation and refeeding. Acta Endocrinol. 103: 223-226.

Vicini, J.L., W.S. Cohick, J.H. Clark, S.N. McCutcheon and D.E. Bauman, 1988. Effects of feed intake and sodium bicarbonate on milk production and concentrations of hormones and metabolites in plasma of cows. J. Dairy Sci. 71: 1232-1238.

Interaction between photoperiod and nutritional status on ovine seasonality

J.B. Menassol[1], D. Chesneau[1], A. Collet[1], B. Malpaux[1] and R.J. Scaramuzzi[1,2]
[1]Physiologie de la Reproduction et des Comportements, UMR 6175 INRA-CNRS-Université François Rabelais de Tours-Haras Nationaux, 37380 Nouzilly, France; [2]Department of Veterinary Basic Sciences, Royal Veterinary College, Herts, United Kingdom; Rex.Scaramuzzi@tours.inra.fr

Introduction

Most mammals, facing variability in environmental conditions, alternate between fertile and non-fertile states throughout the year, a phenomenon known as the seasonality of reproduction. This rhythm is regulated mainly by the annual photoperiod that synchronises an endogenous (circannual) rhythm of reproduction. However other environmental factors such as food availability, social interactions and climate can also affect patterns of seasonal reproduction (Malpaux, 2006). For example nutritional status and photoperiod interact to alter seasonal patterns of reproduction in long-day (Salazar-Ortiz, 2006) and short-day breeders (Zarazaga *et al.*, 2005) but little is known of the underlying mechanisms. Here we report the effect of this interaction on the pattern of seasonal reproduction in the ewe, a short day breeder. Since the 24 h rhythm of melatonin secretion is a key step in the integration of photoperiodic information, we also assessed whether nutrition affected this rhythm, in constant darkness.

Material and methods

Thirty-one intact sexually mature Île-de-France ewes, maintained under artificial light/dark (L/D) cycles mimicking the ambient photoperiod (48N), were fed contrasting diets. The diet consisted of the same feed but the amount differed to achieve body condition scores of 1.7±0.3 and 2.7±0.3 for restricted (R, n=12) and well-fed (WF, n=19) ewes respectively. Ovarian activity was monitored by measuring progesterone by RIA, in blood collected twice weekly by jugular venipuncture. The pattern of melatonin concentrations was determined in 16 animals (8 WF, 8 R) at the summer (L/D: 16/8 h) and the winter (L/D: 8.25/15.75 h) solstices. Melatonin was monitored in constant darkness over 24 h, every 20 min around the theoretical transitions between light and dark and hourly at other times. Samples were collected via jugular catheters in a dim red light, transferred into heparinised tubes. Melatonin was determined by RIA. Data are presented as means ± standard errors and analysed using the t-test or the Mann-Whitney U test where a normal distribution could not be assumed. Differences are regarded as significant when $P<0.05$.

Results

One ewe in the R group had no ovarian activity and by mid-January 2009 two ewes of the WF group were still cycling, they were included in the data. Patterns of ovarian activity are shown in Figure 1. Ewes in the WF group had a longer breeding season than R ewes (134±22 vs. 64±39 d; $P<0.05$). This difference was caused by an earlier start (29 Aug ± 9 d vs. 09 Oct ± 32 d; $P<0.05$) and a delayed end (WF: 09 Jan ± 21 d; R: 10 Dec ± 32 d; $P<0.05$) of the breeding season. The two groups had indistinguishable mid-season dates (WF: 04 Nov ± 12 d; R: 08 Nov ± 25 d). The profiles of melatonin at the summer solstice (Figure 2A), show that the resumption of melatonin secretion occurred earlier in the R group (WF: 18.8±2.1 h; R: 16.8±1.2 h; $P<0.05$). The WF ewes tended to display higher melatonin concentrations than the R ones (WF: 128±95 pg/ml; R: 56±42 pg/ml; $P=0.09$). At the winter solstice (Figure 2B) the levels of melatonin were reduced in R compared to WF ewes (WF: 100±53 pg/ml; R: 43±21 pg/ml; $P<0.05$) but there was no difference in the time of resumption of melatonin secretion.

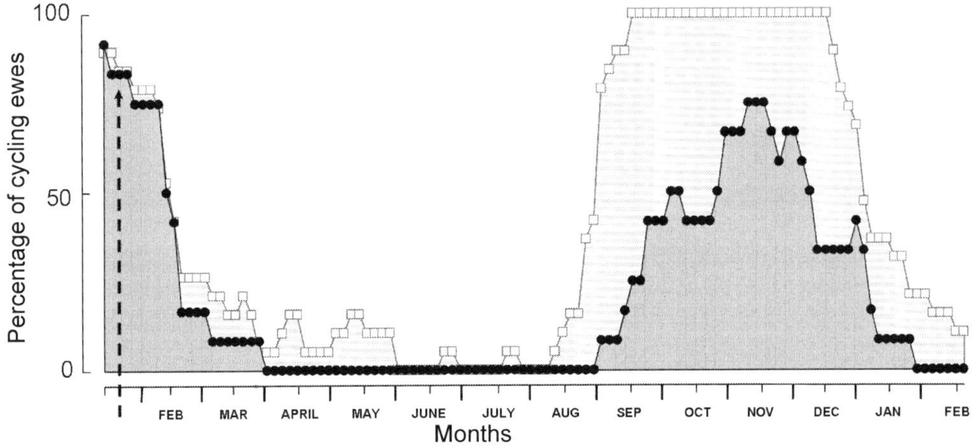

Figure 1. Seasonal patterns of ovarian activity (mid January 2008 to February 2009) in well-fed (□, n=19) and restricted (•, n=12) groups. Dotted line indicates the beginning of the feeding program.

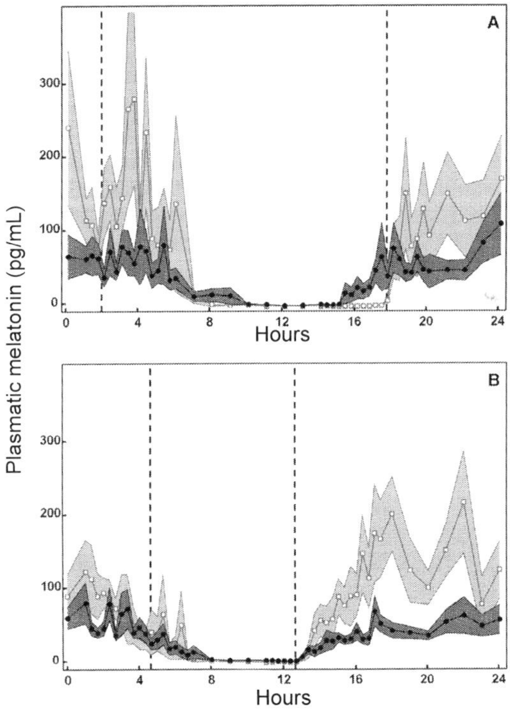

Figure 2. Mean (± SEM) level of melatonin in constant darkness in well fed (□, n=8) and restricted (•, n=8) groups during long (A) and short days (B). Dotted lines show theoretical dawn and dusk.

Discussion and conclusion

Our study establishes that photoperiod and nutritional status interact to modify the rhythm of reproductive activity as has been reported for the mare (Salazar-Ortiz, 2006) and the female goat (Zarazaga *et al.*, 2005). Moreover they show that during long, but not short days, nutritional status affects the timing of circadian melatonin secretion. This change in the circadian timing, as assessed by the melatonin profile, suggesting that nutrition can alter the circadian mechanisms integrating the photoperiodic message and partly explain differences in the annual rhythm of reproduction. We also unexpectedly found that nutritional status affected night time concentrations of melatonin. The origin of this effect is under investigation.

Acknowledgement

R.J. Scaramuzzi is the recipient of an EU Marie Curie Chair of Excellence (MEXC-CT-2006-042499) and J.B. Menassol a PhD grant from the 'Ministère de l'Enseignement Supérieur et de la Recherche'.

References

Malpaux, B., 2006. Seasonal regulation of reproduction in mammals. In: J.D. Neill (Ed.), Knobil and Neill's physiology of reproduction. 3rd edition. Elsevier, the Netherlands, pp. 2231-2281.

Salazar-Ortiz, J., 2006. Bilan des effets du niveau d'alimentation sur la reproduction de la jument. Thèse pour obtenir le grade de Docteur de l'Université de Tours, Tours, France.

Zarazaga, L.A., J.L. Guzman, C. Dominguez, M.C. Pérez and R. Prieto, 2005. Effect of plane nutrition on seasonality of reproduction in Spanish Payoya goats. Anim. Rep. Sci. 87: 253-267.

Influence of nutritional background on neuroendocrine reproductive and appetite responses to central insulin or NPY administration in sheep

D.W. Miller[1], E.J. Bennett[2], J.L. Harrison[1], P.A. Findlay[3] and C.L. Adam[3]
[1]School of Veterinary & Biomedical Sciences, Murdoch University, South St, Murdoch WA 6150, Australia; [2]Scottish Agricultural College, Aberdeen AB21 9YA, United Kingdom; [3]University of Aberdeen, Rowett Institute of Nutrition & Health, Aberdeen AB21 9SB, United Kingdom; d.miller@murdoch.edu.au

Introduction

Short-term nutritional effects on the reproductive axis are strongly influenced by the longer-term nutritional background of the animal (Miller *et al.*, 2007). Nutritional feedback to the hypothalamic region of the brain is also used to make appropriate adjustments to energy intake and expenditure. We have previously demonstrated that luteinising hormone (LH) secretion is stimulated in sheep on an increasing nutritional plane and inhibited in those on a decreasing nutritional plane, even when body weights and conditions scores had converged and were identical between the groups (Miller *et al.*, 2007). This earlier sheep study also indicated a role for peripheral insulin entering the brain and signalling through hypothalamic pathways, notably neuropeptide Y (NPY), to modulate LH secretion (Miller *et al.*, 2007). The present study tested the hypothesis that feed intake and reproductive responses to nutritional feedback depend on an animal's long-term nutritional background. Our approach was to repeatedly examine responses to centrally-administered insulin or NPY in thin sheep as they gained condition and in fat sheep as they lost condition.

Material and methods

Eighteen Suffolk × Greyface castrated, steroid-clamped, adult male sheep, with indwelling cannulae in the third cerebral ventricle, were housed in individual pens. Sixteen weeks prior to the main experiment, nine sheep with average body condition score (BCS) of 1.9 (THIN group) were given sufficient food to maintain body weight and BCS. The remaining animals were given *ad libitum* access to food, reaching BCS of about 3.8 (FAT group) at the start of the main experiment. In wk 1 of the main experiment, sheep in both groups received infusion into the third cerebral ventricle for 8 h/d on two consecutive days, with a control vehicle (artificial cerebrospinal fluid, aCSF) on d 1 and either insulin (0.7 ng/h) or NPY (10 μg/h) on d 2. From wk 1 onwards, the FAT sheep were given a restricted amount of the same diet (700 g/d) so that they were on a decreasing nutritional plane (DNP), and the amount given to the THIN sheep were given *ad libitum* access to the diet so that they were on an increasing nutritional plane (INP). The infusion protocol was then repeated in wk 2, 4, 8 and 12. Once a week, body weight and BCS were measured, and blood and cerebrospinal fluid (CSF) collected. On the days of infusion, blood was taken via temporary jugular catheters every 15 min for 8 h for LH pulse analysis. Voluntary food intake (VFI) was determined by measuring the food refusals of *ad libitum* fed sheep at the end of the 8 h infusion period. Correlation analysis was used to explore relationships between variables using Statview for Windows Version 4.57. Correlations were significant at $P<0.05$.

Results

Correlation analysis was used to explore relationships between variables within the two groups, using data for all individuals across time (Table 1). Significant positive correlations were found between CSF and plasma concentrations of insulin for both INP ($r=0.72$, $P<0.005$) and DNP groups ($r=0.48$, $P<0.05$). In INP but not DNP sheep, plasma insulin correlated positively with BCS, body weight and LH pulse frequency. The proportional LH stimulation by intracerebroventricular (ICV) insulin was negatively correlated with endogenous plasma insulin, and the proportional decrease in food intake during ICV insulin was positively correlated with BCS and body weight.

Table 1. Correlation coefficients for the effects of central insulin or NPY on the response (% compared to control) in LH pulse frequency and voluntary feed intake (VFI) compared to the changing circulating concentrations of plasma insulin, concentration of cerebrospinal fluid (CSF) insulin, body condition score (BCS) and body weight of the sheep over 12 wk on either the increasing nutritional plane (INP) or decreasing nutritional plane (DNP).

	Plasma insulin	CSF insulin	BCS	Body weight
INP				
CSF insulin	0.72[a]			
BCS	0.42[a]	0.39		
Body weight	0.44[a]	0.41[a]	0.78[a]	
LH freq	0.51[a]	0.39	0.49[a]	0.55[a]
LH freq% response to insulin	-0.47[a]	-0.42[a]	-0.51[a]	-0.5[a]
LH freq% response to NPY	0.29	0.21	-0.12	-0.09
VFI% response to insulin	0.29	0.19	0.74[a]	0.82[a]
VFI% response to NPY	0.18	0.12	0.27	0.29
DNP				
CSF insulin	0.48[a]	1		
BCS	-0.37	0.05	1	
Body weight	-0.34	-0.03	0.29	1
LH freq	0.29	-0.12	0.27	0.19
LH freq% response to insulin	-0.21	0.25	0.41[a]	0.38[a]
LH freq% response to NPY	-0.18	0.31	0.35	0.33

[a] Coefficients are significant at $P<0.05$.

Discussion and conclusion

Our results support the hypothesis that appetite and reproductive neuroendocrine responses to acute changes in nutritional feedback depend on the individual's longer-term nutritional background. Central insulin stimulation of the reproductive neuroendocrine axis and inhibition of appetite are dependent on low background adiposity, current nutritional status and insulinemia. Whereas the reproductive neuroendocrine and central appetite axes appear sensitive to positive feedback (insulin/BCS), both appear to be relatively insensitive to negative nutritional feedback (NPY).

References

Miller, D.W., E.J. Bennett, J.L. Harrison, P.A. Findlay and C.L. Adam, 2007. Nutritional influences on reproductive neuroendocrine output: insulin, leptin and orexigenic neuropeptide signaling in the ovine hypothalamus. Endocrinology 148: 5313-5322.

Effect of plane of nutrition on sexual behaviour of Boer and Mubende bucks

S.S. Walusimbi[1,2], J. Ottobre[2], D. Mpairwe[1], M. Day[2], D. Mutetikka[1] and D.K. Ssemambo[3]
[1]Department of Animal Science, Faculty of Agriculture, Makerere University, P.O. Box 7062 Kampala, Uganda; [2]Department of Animal Sciences, The Ohio State University, Plumb Hall, 2027 Coffey Rd., Columbus Ohio, 43210, USA; [3]National Animal Genetic Research Center and Data Bank, Entebbe, P.O. Box 183 Entebbe, Uganda; dmpairwe@agric.mak.ac.ug

Introduction

Goats contribute substantially to lifting the poor out of poverty in Africa. To increase productivity of indigenous breeds, Boer bucks from South Africa were introduced into Uganda. A survey on Boer bucks in Uganda's rangelands reported that they had lower libido compared to indigenous Mubende meat goats (Tumwine, 2005), the breed to which they were being crossbred. Rangelands provide the cheapest source of nutrients for ruminants, but seasonal variation in quantity and quality of feed resources is a challenge. Goats kept under low-input management systems and fed low quality diets require supplementation to optimise energy intake and performance. We hypothesised that supplementing bucks with a balanced concentrate would improve libido. The objective of this study was to examine the effect of increased plane of nutrition on libido and other sexual behaviour parameters of Boer and Mubende bucks.

Material and methods

Twelve Boer and twelve Mubende mature bucks of age 1-3 yrs and weight 30-90 kg were randomly assigned to a low plane (browsing alone) or high plane (browsing plus 1kg daily supplement of 18% CP, 4.23 kcal/kg and minerals) of nutrition in a 2×2 factorial completely randomised design. Libido, scored according to Chenoweth *et al.* (1979), and other sexual behaviour parameters; erection of penis (PE), approach to teaser (AT), genital sniffing intensity (GSI) and Flehmen reaction intensity (FRI) were tested weekly using females in oestrus. Each buck was exposed to a female in oestrus within 15 minute sessions for three days a week for 4 weeks. The time taken by each buck to approach the female from a point of entry into the pen was recorded as the reaction time (RT). Weekly weights were regularly recorded. Analysis of covariance (ANCOVA) for the main effects of plane of nutrition and breed and week was performed using Proc Mixed repeated measures procedures of SAS® (SAS, 2002-2003.ver 9.1.3) with the subject identified as the buck (breed). The compound symmetry covariance structure was selected based on the Schwarz Bayesian criterion. Age and initial weight were used as covariates in assessing the effect of breed and plane of nutrition on libido and sexual behaviour. LS means were separated using the Fisher protected LSD.

Results

Exotic Boer bucks had significantly lower ($P<0.05$) indices of sexual behaviour compared to indigenous Mubende goats (Table 1). Supplementation significantly improved all but one sexual behaviour parameter of Boer bucks, but did not appreciably change any of the sexual behaviour parameters of Mubende bucks. The effect of plane of nutrition on libido and reaction time was dependent on breed. Libido, Flehmen reaction intensity and genital sniffing intensity differed among weeks of testing ($P<0.05$, data not shown). Supplementation improved libido over time in both breeds, but this effect was most dramatic in the Boer bucks (Plane × Breed × Week interaction, $P<0.01$). By week 4, the average libido score in the supplemented Boer bucks was similar to that of the supplemented Mubende bucks (5.17 vs. 5.94, respectively, $P=0.36$). Supplementation also tended to influence changes in body weight ($P=0.07$).

Table 1. Effect of plane of nutrition on sexual behaviour parameters and changes in body weight.

	Mubende		Boer		SEM	*P*-value		
	High	Low	High	Low		Breed	Plane	BreedxPlane
LS	4.68^{a}	4.72^{a}	2.98^{b}	1.50^{c}	0.40	<0.05	<0.1	<0.1
RT (s)	12.07^{a}	34.10^{a}	62.28^{a}	207.96^{c}	33.28	<0.05	<0.1	<0.05
GSI	1.77^{a}	1.24^{b}	1.32^{b}	0.89^{b}	0.17	<0.05	<0.05	NS
FRI	1.91^{a}	1.52^{a}	1.32^{b}	0.79^{c}	0.19	<0.05	<0.05	NS
EP	1.75^{a}	1.58^{a}	1.13^{b}	0.59^{c}	0.18	<0.05	<0.1	NS
AT	1.72^{a}	1.53^{a}	1.38^{b}	1.00^{c}	0.11	<0.05	<0.05	NS
CBW(Kg)	2.80^{a}	-1.00^{a}	3.00^{a}	-0.25^{a}	1.32	NS	<0.1	NS

^{a,b,c} Means in the same row with different superscripts are significantly different ($P<0.05$); NS=Not Significant ($P>0.1$); GSI: Genital Sniffing Intensity (0=not shown, 1=Low (<3 times), 2=Medium (3-5 times), 3=High (>5 times); FRI: Flehmen Reaction Intensity (0=not shown, 1=Low (<3 times), 2=Medium (3-5 times), 3=High (>5 times); EP: Erection of penis (0=absent, 1=in sheath, 2=out of sheath), AT: Approach to teaser (0=no interest, 1=weak, 2=strong); LS: Libido score: buck showed no interest (0), sexual interest shown once (1), sexual interest shown more than once (2), active pursuit with persistent interest (3), one mount – no service (4), two mounts – no service (5), three or more mounts – no service (6), one service – no further interest (7), one service – additional interest (8), two services – no further interest (9), two services – additional interest (10). CBW: change in body weight between 1st and 4th week.

Discussion and conclusion

These data suggest that the reproductive performance of extensively managed, imported Boer bucks is inferior to that of the indigenous Mubende breed. However, improved reproductive performance in imported Boer bucks under extensive goat systems can be achieved with the provision of a higher plane of nutrition. Since the current experiment was limited to four weeks, it is possible that continued supplementation of the Boer bucks for an additional period of time would further improve their reproductive performance or at least would maintain their libido at a level comparable to that of the indigenous Mubende bucks.

Acknowledgement

The authors thank HEPAD/USAID for funding the research and NAGRC for providing the experimental animals. A portion of the salaries and research support were provided by state and federal funds appropriated to the Ohio Agricultural Research and Development Center, The Ohio State University.

References

Chenoweth, P.J., J.S. Brinks and T.M. Nett, 1979. Comparison of three methods of assessing sex drive in yearling beef bulls and relationships with testosterone and LH levels. Theriogenology, 12: 223-233.

SAS, 2002-2003. Statistical Analysis System, Institute Inc, Cary, NC, USA, version 9.1.

Tumwine, A., 2005. Constraints to Boer goat production in Masaka District. A special project report. Veterinary Medicine, Makerere University, pp 33.

17β-oestradiol has dramatic effects on mammary epithelium integrity and loss of lactose in urine in dairy cows in late lactation

S. Agenäs, I. Lundström and K. Holtenius
Swedish University of Agricultural Sciences, Department of Animal Nutrition and Management, Kungsängen research centre, S-753 23 Uppsala, Sweden; sigrid.agenas@huv.slu.se

Introduction

Lactose is only synthesised in the mammary glands and is secreted into the mammary alveoli. Through osmosis lactose attracts water into the alveoli and is thereby closely linked to milk yield. If lactose is lost from the alveoli, milk yield is also affected. The presence of lactose in plasma can only be explained by movements from the lumen of the alveoli via leaky mammary tight junctions into plasma. This happens for example in late lactation, possibly caused by the influence of placental hormones but the effects of placental hormones alone on mammary tight junctions has not been studied. Plasma clearance of lactose is high, Stelwagen *et al.* (1997) has shown a plasma half-life of 44 minutes and that plasma lactose returns to baseline levels 3.1±0.3 h after an increase. This offers an opportunity to study mammary tight junction integrity by determining lactose concentration in urine. The aim of this study was to investigate the effect of exogenous estrogen on mammary tight junctions in cows in late lactation and to establish a correlation between lactose in plasma and urine.

Material and methods

Two studies were performed. One included five non-pregnant cows receiving injections with 18.2 mg 17β-estradiol daily for six days. The dose was based on a dose used by Delbecci *et al.* (2005) who showed decreased milk yield in cows that received 15 mg 17β-estradiol per day. The effect of exogenous 17β-estradiol on milk yield and on lactose in milk, plasma and urine was investigated before, during and after the treatment. The second study included 10 cows and was designed to investigate normal levels of circulating 17β-estradiol, lactose in plasma and lactose in urine around day 200 of the pregnancy and day 300 in lactation. Milk yield and milk composition was registered. Circulating 17β-estradiol and lactose in plasma and urine was determined. The normal values were compared with the values after treatment. In both studies, lactose in plasma and urine was analysed using an enzymatic method (Lactose/D-galactose kit Boehringer Mannheim/R-Biopharm). 17β-estradiol plasma concentrations were determined by a ^{125}I RIA previously validated for bovine plasma (Sirois and Fortune, 1990). The concentration of creatinine was analysed and used as a marker for the urine volume (Valadares *et al.*, 1999). Analysis of variance was performed on data for 17β-estradiol and lactose in plasma, lactose in urine, milk yield and milk composition using PROC MIXED in the SAS® system (SAS inst. Inc. 1996. SAS/Stat Software. Changes and Enhancements through release 9.1).

Results

Milk yield decreased already within the first day of 17β-estradiol injections and was 24% lower during days 8-12 than days 1-7 (P<0.001). Lactose in plasma and urine increased at the end of the injection period. A high correlation between lactose in urine and plasma was found. More than 30% of the total lactose production was lost in urine after oestrogen treatment. Changes in amount of lactose in milk and urine are shown in Figure 1.

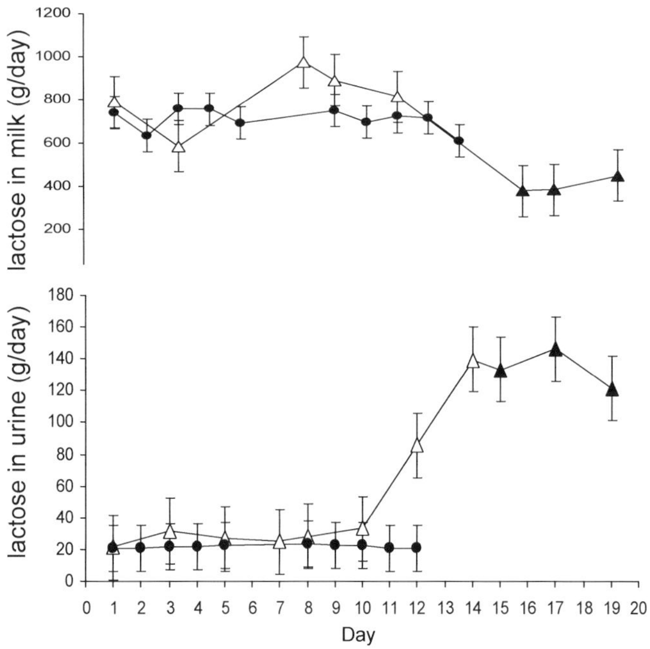

Figure 1. Least squares means ± std dev of lactose in milk and urine. The symbols represent pregnant cows (●) and non-pregnant cows that received exogenous 17β-estradiol injections during day 7-12 of the experiment (Δ). Values that differ significantly (P<0.05) from the control value before treatment are indicated with filled symbols.

Conclusion

The data shows that exogenous 17β-estradiol impairs mammary tight junction integrity and causes a significant loss of lactose in urine. In addition, milk yield decreases. There was a delay between the decrease in milk yield and opening of tight junctions, indicating that other factors are involved.

Ackowledgements

The Swedish Research Council Formas is gratefully acknowledged for financial support.

References

Delbecchi, L., N. Miller, C. Prud'homme, D. Petitclerc, G. F. Wagner and P.Lacasse, 2005. 17 beta-estradiol reduces milk synthesis and increases stanniocalcin gene expression in the mammary gland of lactating cows. Livest. Prod. Sci. 98: 57-66.

Sirois, J. and J.E. Fortune, 1990. Lengthening the bovine estrous cycle with low levels of exogenous progesterone: a model for studying ovarian follicular dominance. Endocrinology, 127: 916-925.

Stelwagen, K., V.C. Farr, H.A. McFadden, C.G. Prosser and S.R. Davis, 1997. Time course of milk accumulation-induced opening of mammary tight junctions, and blood clearance of milk components. Am. J.Physiol. 273: R379-86.

Valadares, R.D., G.A. Broderick, S.D. Valadares Filho and M.K. Clayton, 1999. Effect of replacing alfalfa silage with high moisture corn on ruminal protein synthesis estimated from excretion of total purine derivatives. J. Dairy Sci. 82: 2686-2696.

The effect of protein supplementation on reproductive performance in Moghani ewes maintained on rangeland

M. Bayeriyar[1] *and S. Kargar*[2]
[1]*Department of Animal Sciences, Tabriz University, Tabriz, Iran;* [2]*Department of Animal Sciences, Isfahan University of Technology, Isfahan 84156, Iran; kargar@ag.iut.ac.ir*

Introduction

Reproduction performance of livestock in a harsh environment is determined by four factors: genetic merit, physical environment, nutrition and management. Thus adequate nutrition could improve modest biological types to reach their genetic potential, alleviate the negative effects of a harsh physical environment and minimise the effects of poor management practices (Caturedi *et al.,* 2003). Inversely, poor nutrition will not only reduce performance below genetic potential, but also exacerbate other detrimental environmental effects. Moreover, nutritional factors more than all others, readily lend themselves to manipulations to ensure positive outcomes (Santra *et al.*, 2002). The main objective of this study was to determine the effect of length of the flushing period and the best time to start in order to improve reproductive performances.

Material and methods

A total of 250 adult (3-6 years old) Moghani ewes averaging 42±5.3 kg live weight (LW; mean ± SD) were used under natural day length conditions during the breeding season. Ewes at the first natural estrus, detected by teaser rams, were weighed and assigned randomly to receive a diet that supplied either 60 g (Low: L, n=103) or 120 g (High: H, n=107) crude protein per day per ewe while grazing on rangeland until the next estrus (pre-mating). At the second estrus, the ewes were hand-mated by one of six different fertile Moghani rams and again randomly allocated to either the L or H allowance (post-mating) until day 15 after mating. Hence, according to the post-mating date, there were four dietary protein treatments, LL (n=51); LH (n=52); HL (n=52); and HH (n=55). The numbers of ewes returning to service were determined to calculate non-return rate. Diets were formulated and balanced relative to energy according to NRC (1985) as follows: L (commercial compound feed) or H (50% soybean meal and 50% grass hay) to provide 60 or 120 g of crude protein per ewe per day, respectively. Ewes assigned to the L diet were fed 400 g of compound feed, while those on the H diet received 200 g of soybean meal and 200 g of mixed-grass hay. Hence, the ewes in the L group were fed 0.7 times protein requirement for maintenance, while the ewes in the H group were fed 1.3 times protein requirement as a supplement to grazing on rangeland. The ewes in each group grazed between 09:00 and 16:00 h on rangeland and were housed in separate pens as a group for the rest of the day. During pregnancy, the ewes were maintained on rangeland during the day and supplemented with hay and concentrates. Lambing performances were recorded. In this analysis, animals were used as replicates, the design being unbalanced due to different numbers of ewes on each nutritional treatment (proc GLM; SAS). The effects of nutritional treatment on non-return rate, lambing rate and the number of lambs per ewe lambing (litter size), or per ewe mated was analysed by categorical data modelling (log-linear analysis, SAS). The data for litter size and fecundity were analysed after transformation to avoid heterogeneity of variance (log (x+1)).

Results

As shown in Table 1, there was an interaction (P<0.05) between pre- and post-mating protein supplementation to mating on non-return and lambing rates. Non-return rate, lambing rate tended to be greater (P<0.07 and P<0.06) in ewes fed the LH diet than those fed the HL diets. A similar effect of LH diet was observed on lambing rate compared to the LL ($P<0.06$) diet. The mean litter size and fecundity were greater ($P<0.05$) in ewes fed the LH diet compared to those on the LL diet. The mean litter size of LL ewes was depleted ($P<0.05$) more than any other diets.

Table1. Non-return rate, lambing rate to first estrus, litter size and fecundity in grazing ewes supplemented with low or high protein during pre- and/or post-mating.

	Diet[1]				S.E.M.	Pre-mating	Post-mating	Pre-xpro-
	LL	LH	HL	HH				
Variables								
Non-return rate,%	77.1^{ab}	88.3^{a}	69.6^{b}	77.4^{ab}	3.8	*	*	*
Lambing rate,%	60.0^{b}	81.3^{a}	59.4^{b}	71.0^{ab}	4.5	*	*	*
Litter size	1.09^{a}	1.31^{b}	1.32^{b}	1.40^{b}	0.05	*	-	-
Fecundity	0.66^{a}	1.06^{b}	0.78^{ab}	1.00^{b}	0.07	*	-	-

[1] See text; [a,b] Means within lines with same superscript letters are significantly different ($P<0.05$).

Discussion and conclusion

The results of the present study show that a short-term (2 weeks) increase of high protein supplementation around the time of mating, especially post-mating, improved reproductive performance of grazing ewes in the autumn, when rangeland quality is depleted. The present results apparently contrasted to the results of Torell *et al.* (1974), and ours, because when forage protein content was not limiting Memon *et al.* (1969) found that energy was more important to increase ovulation rate. In fact, the present study shows that dietary protein supplementation may have also improved the overall digestibility, and intake, of the rangeland forages, thus globally improving the ewes' nutritional status.

References

Caturedi, O.H., R. Bhatta, A. Santra, A.S. Mishra and J.S. Mann, 2003. Effect of supplementary feeding of concentration on nutrition on nutrient utilization and production performance of ewes grazing on community rangeland during late gestation and early lactation. Asian Aust. J. Anim. Sci. 16: 983-987.

Memon, G.N., R.J. Antoniewicz, N.J. Benevenga, A.L. Pope and L.E. Casida, 1969. Some effects of differences in dietary energy and protein levels on the ovary and the anterior pituitary gland of the ewe. J. Anim. Sci. 28: 57-62.

National Research Council. 1985. Nutrient requirements of sheep. 6th rev. ed. Natl. Acad. Sci., Washington, DC, USA.

Santra, A., S.A. Karim and O.H. Chaturvedi, 2002. Effect of concentrate supplementation on nutrient intake and performance of lambs of two genotypes grazing a semiarid rangeland. Small Rum. Res. 44: 37-45.

Torell, D.T., I.D. Hume and W.C. Weir, 1974. Factors affecting blood urea nitrogen and its use an index of the nutritional status of sheep. J. Anim. Sci. 39: 435-440.

Delay in muscle development in bovine cloned foetuses

I. Cassar-Malek[1], C. Jurie[1], B. Picard[1], A. Listrat[1], M. Guillomot[2], P. Chavatte-Palmer[2] and Y. Heyman[2]
[1]INRA, URH1213 Herbivores, Site de Theix, F-63122 Saint-Genès-Champanelle, [2]INRA, UMR1198 Biologie du Développement et Reproduction, F-78352, Jouy-en-Josas, France; cassar@clermont.inra.fr

Introduction

Somatic cloning has many potential applications in cattle breeding programs and enables the propagation of animals with desired phenotypic traits such as high quality food products (Heyman, 2005). In some non European countries such as the USA, Argentina, or Japan, cloned cattle and their progeny have entered the food supply chain. Therefore, it is important to foresee the applications of cattle cloning and to evaluate the physiological development of clones and their offspring, especially for the muscle tissue from which meat products derive.

A previous study has shown that cloned cattle exhibit a delay in muscle maturation until the onset of puberty (Jurie *et al.*, 2009), which could be due to a delay in foetal myogenesis. The aim of the present study was to evaluate muscle developmental characteristics in foetal clones compared to animals derived from sexual reproduction. Biochemical and histochemical properties of the *semitendinosus* (ST) muscle were analysed in cloned foetuses and their controls. Two important developmental stages were chosen according to Picard *et al.* (2002): 60 days post-conception [dpc], characterised by intensive cell proliferation and muscular fibre formation and 260 dpc, which corresponds to terminal differentiation of fibres, contractile and on-going metabolic differentiation.

Material and methods

Cloned foetuses were obtained after somatic cell nuclear transfer of skin fibroblasts into enucleated oocytes according to Vignon *et al.* (1998). Their control counterparts of the same breed were generated by artificial insemination. The foetuses were recovered at 60 dpc (n=8/group) and 260 dpc (n=3 clones, n=4 controls) and samples of *ST* muscle were collected.

The contractile properties of muscles were examined by scoring positive cells for myosin heavy chains (MyHC) at 60 dpc, after immuno-histochemical staining (Picard *et al.*, 2002). A combination of electrophoresis/densitometry and western-blot analyses using specific antibodies (Picard *et al.*, 2002) was used at 260 dpc. The composition of the Extracellular Matrix (ECM) and the expression of VEGF growth factor (a cytokine involved in the formation of blood vessels) and its receptors (FLK-1 and Flt-1) were analysed by immuno-histochemistry using specific antibodies. Negative controls were made omitting the primary antibodies or using absorption controls. The metabolic properties of muscles were assessed by enzymology according to Jurie *et al.* (2009).

Results and discussion

Muscle characteristics in cloned foetuses at 60 dpc: At 60 dpc, the MyHC positive cells expressed several types of MyHC and belonged to the first generation of myotubes. The number of myotubes was lower by 29% in the muscle of cloned Charolais foetuses than in control ones (P=0.001). Within muscular fasci the structural organisation was lower in cloned foetuses than in controls. This indicates a delay in myogenesis in cloned foetuses leading to fewer primary myotubes and a lower degree of organisation of the first-generation of myotubes.

Several components of the ECM, e.g. collagens (types I, IV and VI), tenascin X (a glycoprotein) and decorin (a proteoglycan) were investigated by immuno-histochemistry. These molecules interact with each other and are involved in collagen fibrillogenesis and in myogenesis. There was a delay in the spatial deposition of collagen type I, IV and VI in the endomysium of cloned foetuses compared to controls. This is consistent with the delay in contractile differentiation observed in cloned animals.

Conversely, decorin and tenascin X spatial repartition was earlier in the muscle of cloned foetuses vs. controls.

Muscle vascularisation was investigated by revealing the VEGF and its receptors by immunohistochemistry. Only the VEGF and the FLK-1 receptor could be detected in the muscles. No difference between groups was observed in the location nor in the intensity of labellings, suggesting that the delay in myogenesis of cloned foetuses was probably not linked to a defect in muscle vascularisation.

Characteristics of the muscles of cloned foetuses at 260 dpc: The activities of enzymes representative of oxidative and glycolytic metabolisms were compared in the muscles of cloned foetuses vs. their controls. The activities of two glycolytic enzymes, lactate dehydrogenase (EC 1.1.1.27) and phosphofructokinase (EC 2.7.1.11), were lower by 26% (P = 0.015) and 49% (P<0.001), respectively, in cloned foetuses compared to controls. The activities of two oxidative enzymes, isocitrate dehydrogenase (EC 1.1.1.42) and citrate synthase (EC 2.3.3.1) were lower than in controls by 49% (P<0.05) and 33% (P<0.01) respectively, while that of cytochrome-c oxidase (EC 1.9.3.1) was not significantly different in both groups. Thus, muscles of cloned foetuses near term were both less oxidative and less glycolytic than those of controls. This suggests that their metabolic differentiation was delayed leading to reduced energy metabolism. Muscle contractile properties were examined through the study of the MyHC isoforms. Although highly expressed at 60 dpc, foetal MyHC was not detectable in muscle at 260 dpc. Conversely, MyHC adult isoforms were only detectable at 260 dpc. The very low abundance of the MyHC 2x in the muscle of cloned foetuses vs. controls (P<0.05), however, confirms a delay in the appearance of the adult MyHC isoforms, illustrating a delay in contractile differentiation.

Conclusion

Myogenesis is delayed in cloned foetuses as illustrated at 60 dpc by lower numbers and degree of organisation of the first-generation of myotubes, together with the spatial repartition of components of the extracellular matrix. Near term, the muscles of cloned foetuses had not reached the same stage of differentiation/maturation as controls as shown by their lower energy metabolisms and their MyHC pattern. These results are original data showing that disturbances in myogenesis occur early in foetal life in cloned cattle.

Acknowledgement

The authors thank D. LeBourhis, C. Richard, G. Gentès, C. Barboiron, A. Delavaud, D. Chadeyron and C. Nore, for skilled technical assistance, and the staff of the URH experimental slaughter house and the experimental farms at UCEA-Bressonvilliers and Domaine de Gallé-Bourges for animal care. This study was funded by a grant of the GenAnimal call for proposal.

References

Heyman Y., 2005. Nuclear transfer: a new tool for reproductive biotechnology in cattle. Reprod. Nutr. Dev. 45: 353-361.

Jurie C., B. Picard, Y. Heyman, I. Cassar-Malek, P. Chavatte-Palmer, C. Richard and J.-F. Hocquette, 2009. Comparison of cloned and non-cloned Holstein heifers in muscle contractile and metabolic characteristics. Animal 3: 244-250.

Picard B., L. Lefaucheur, C. Berri C and M.J. Duclos, 2002. Muscle fibre ontogenesis in farm animal species. Reprod. Nutr. Dev. 42: 1-17.

Vignon X., P. Chesné, D. LeBourhis, J. Fléchon, Y. Heyman and J.-P. Renard, 1998. Developmental potential of bovine embryos reconstructed from enucleated matured oocytes fused with cultured somatic cells. C. R. Acad. Sci. Paris 321:735-745.

Association between body condition score changes, parity and feeding system and fertility of lactating dairy cows

P. Celi[1,2], *A.R. Rabiee*[2], *T.F. Duffield*[3] *and I.J. Lean*[1,2,3]
[1]*Faculty of Veterinary Science, University of Sydney, Camden, NSW, 2570, Australia;* [2]*SBScibus, P.O. Box 660, Camden, NSW, 2570, Australia;* [3]*Department of Population Medicine, University of Guelph, Ontario, N1G 2W1 Canada; pietroc@camden.usyd.edu.au*

Introduction

The age of dairy cattle influences reproductive performance. Nulliparous cattle are more fertile than lactating older cows and cows of parity greater than five are slower to conceive and require more number of services per pregnancy (Lean *et al.*, 2003). Oestrus synchronisation success may be higher in primiparas than in multiparas (Tenhagen *et al.*, 2003). However, there are contradicting reports on differences in pregnancy rates between primiparous and multiparous cows and it is possible that this apparent discrepancy could be due to differences in nutritional status and body condition score (BCS) (Westwood *et al.*, 2002). This apparent discrepancy may reflect differences in nutritional status and BCS. We hypothesised that mature cows will have greater BCS losses and these will be associated with lower reproductive performance. The objective of the study was to assess associations among BCS, reproductive performance and parity with a focus on mature cattle (6+ yr age) in different feeding systems.

Material and methods

This study was a retrospective study involving data collected between 1995 and 2003 from dairy farms in Australia and Canada. A total of 3515 cows from 57 herds were included in the dataset. To enter the study cows needed a BCS recorded on a scale of 1 (thin) to 5 (fat) (Edmondson *et al.*, 1989) before calving (average 12.0±0.20 (SE) d before) or at calving (d +1 to +3), and between d 33 to 77 (average 58.2±0.40 (SE) d) after calving. The pregnancy data were obtained from three different feeding systems; pasture-based, partial mixed ration (PMR) and North American systems (Total Mixed Ration and Component). Survival time was the number of days from calving to final pregnancy. Parities were grouped into parity 1, 2, 3, 4 and 5 and more than 6 (>6). The BCS losses between first and second body condition scoring, parity, feeding systems and herds were included in the analysis. The outcome of interest was the interval (in days) between calving and final pregnancy. This was assessed using Cox's proportional hazards model. Body condition score difference (BCSD), parity and feeding systems and a random effects term, herd of origin, were used as covariates. Crude and adjusted hazard ratios (HR) derived from the survival analysis reflected the risk of pregnancy after adjusting for the effects of BCSD, parity, feeding systems and herd. The reference group for parity was cows with parity 1, and the reference group for feeding systems was pasture-feeding. The significance of HR obtained from the Cox proportional hazards model was assessed using likelihood ratio.

Results

A total of 2,516 dairy cows from 57 dairy herds had two BCS within the respective intervals defined at study onset. Average BCS of cows around calving, and after calving was 3.24±0.01 (SE) and 2.76±0.01 (SE), respectively. Average parity of cows was 3.0±0.03 (SE) (range from 1 to 12). Number of events (pregnant cows) and censored observations (not known pregnant cows) were 2,122 (60.4%) and 394 (11.2%), respectively. There were 999 cases (28.4%) not used in the final analysis. Cows with parity 6 experienced higher BCS losses. Cows fed with TMR & Component presented lower BCS losses compared to cows fed with PMR and pasture-based system (data not shown). The effect of BCSD on time from calving to pregnancy was significant (*P*=0.004), when herd was not included in the model as a random effects term; however, when herd was included in the model, the effect of BCSD was not significant (*P*=0.258), and was, consequently, removed from the final model. The hazard ratio of calving to pregnancy after adjusting for parity and feeding systems are presented in Table 1.

Table 1. Hazard ratios for the interval from calving to pregnancy with reference groups are parity 1 and pasture fed.

	Hazard ratio ± SE	Significance (*P* value)
Parities (fixed effect)		
Parity 2	1.150±0.056	0.004
Parity 3	1.093±0.056	0.083
Parities 4 and 5	1.031±0.054	0.563
Parities ≥6	0.864±0.057	0.027
Feeding system (fixed effect)		
Partial mixed ration (PMR)	1.130±0.285	0.628
North American (TMR & Component)	0.563±0.067	<0.001
	Estimated effect ± SE	
Herd (random effect)	0.167±0.036	

Cows with parity 6 and greater are less likely to be pregnant than cows with parity 1, 2, 3 or 4 and 5 (Table 1). This effect was independent of BCSD. The pregnancy risk for cows fed in the North American systems (TMR & Component) is significantly less than those from the pasture-based and PMR feeding systems (*P*<0.001, Table 1).

Discussion and conclusion

The results of this study showed that BCS losses after calving are not associated with the interval from calving to pregnancy, when data were adjusted for the random effect of herd. The longer interval from calving to pregnancy for cows > parity 6 compared to younger cows was not influenced by BCSD. Interval from calving to pregnancy for cows from pasture based systems is shorter than those from North American systems, but not significantly different to cows fed PMR diets.

Acknowledgement

We thank C. Westwood, M. Curtis and N. Moss for providing their data.

References

Edmondson, A.J., I.J. Lean, L.D. Weaver, T. Farver and G. Webster, 1989. A body condition scoring chart for Holstein dairy cows. J. Dairy Sci. 72: 68-78.

Lean, I.J., J.A. Porter, A.R. Rabiee, W.F. Morgan, W. Tranter, N. Moss and R.J. Rheinberger, 2003. Comparison of effects of GnRH and prostaglandin in combination and prostaglandin on conception rates and time to conception in dairy cows. Aust. Vet. J. 81: 488-493.

Tenhagen, B.A., M. Wittke, M. Drillich and W. Heuwieser, 2003. Time of ovulation and conception rate in primiparous and multiparous cows after synchronization of ovulation with GnRH and PGF2alpha. Reprod. Domest. Anim. 38: 451-454.

Westwood, C.T., I.J. Lean and J.K. Garvin, 2002. Factors influencing fertility of Holstein dairy cows: a multivariate description. J. Dairy Sci. 85: 3225-3237.

Leptin and NEFA concentrations in yearling Jezersko-Solchava ewes during puberty and in the first reproductive season

V. Cestnik[1], M. Kosec[1], Z. Jenko[2] and N. Čebulj-Kadunc[1]
[1]University of Ljubljana, Veterinary Faculty, Gerbičeva 60; SI-1000 Ljubljana, Slovenia; [2]Gubčeva 7, SI-6250 Ilirska Bistrica, Slovenia; nina.cebulj@vf.uni-lj.si

Introduction

In the northern hemisphere, the majority of sheep breeds enter puberty in the first autumn after their birth, at the ages of 6 to 12 months. The first breeding season usually ends in early spring (Rhodes and Nathanielsz, 1990). Concentration of leptin, which is mostly synthesised in white adipose tissue, depends on body mass and fat supplies and serves as an indicator of body energy balance. Leptin affects the central nervous system or endocrine glands and causes hormonal changes, stimulating lipolysis and inhibiting lipogenesis. It also stimulates the reproductive axis by stimulating GnRH secretion, playing a role in the onset of puberty and seasonal reproductive activity (Chilliard *et al.*, 2005; Zieba *et al.*, 2005). Triglyceride metabolism is also reflected by NEFA (non-esterified fatty acids) blood concentrations, rising after triglyceride degradation, which inhibits leptin synthesis (Marie *et al.*, 2001). The aim of our work was to evaluate the fluctuations in leptin and NEFA concentration in connection to puberty onset and seasonal oestrus in yearling ewes.

Material and methods

The study was performed in a flock of 10 yearling ewes of autochthonous Jezersko-Solchava breed (JS) over a period of 12 months, starting with July. The ewes were held on pasture from May to October and stabled for the rest of the year, fed with hay. The concentrates and vitamin-mineral mixture were added daily during the whole study; drinking water was provided *ad libitum.* The animals were weighed monthly, the weight gain was in the limits expected for the breed. Blood for leptin and NEFA determination was sampled once, for progesterone twice a month. Sampling was performed in the morning, within 2 h after the morning ration. Leptin, NEFA and progesterone serum concentrations were measured by commercial kits (Multi species leptin RIA kit, LINCO, USA; NEFA C-ACS-ACOD, Waco Chemicals, Germany and Ovucheck EIA, Biovet, Canada). Ovarian status was evaluated considering progesterone concentrations twice monthly with 10 d intervals. Progesterone levels above 6.4 nmol/L were considered as indicators of luteal phase, below 3.4 nmol/l as the indicator of oestrus during the breeding season and of anoestrus during seasonal inactivity of gonads. Statistical calculations were performed with the SPSS computer programme and the results are presented as mean ± standard error of the mean $\overline{X}$ ± SE).

Results

Leptin concentrations increased from June till November ($\overline{X}$ = 3.38 ng/ml Human Equivalent (HE)), decreasing to a minimum in January ($\overline{X}$ = 1.77 ng/ml HE; P<0.001) and rising again in February. In this period, the number of ewes exhibiting oestrous cycles rose from 20% in September to 70% in November and December. The percentage of animals after that declined from January to March, when only 10% of ewes were still cyclic. In April all the animals entered anoestrus. The lowest NEFA concentrations were measured in September ($\overline{X}$ = 0.14 mmol/l) and the highest in March ($\overline{X}$ = 0.60 mmol/l). No correlation was established between the leptin and NEFA concentrations (R=-0.055; P>0.05), but the polynomial curves demonstrate a mirror trend.

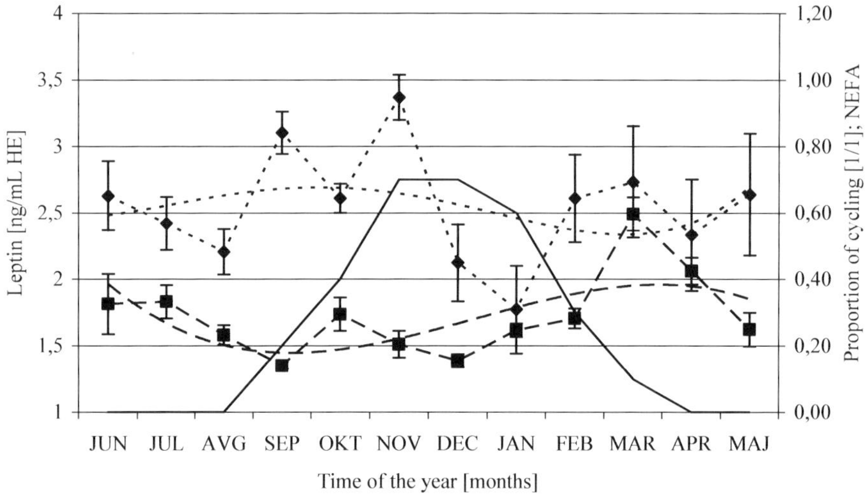

Figure 1. Leptin and NEFA concentrations (mean values and polynomial curves) and proportions of cycling Jezersko-Solchava ewes throughout the year (Legend: — proportion of cycling ewes; - ♦ - leptin; - - leptin (polynomial); –■– NEFA; – – NEFA (polynomial).

Discussion and conclusion

Puberty onset in JS ewes can be expected from September on, reaching a peak in November and December. The ewes entered seasonal anoestrus in April. Transition to puberty, accompanied by high leptin values, and to seasonal anoestrus after the fall of leptin in January, demonstrates the role of leptin in puberty onset and regulation of seasonal reproductive activity. The correlation between leptin and NEFA values was not established, but the mirror appearance of polynomial trend curves demonstrates an increase of NEFA and decrease of leptin concentrations during the spring period. The changes could be attributed to intensified mobilisation of fat depots, since an opposite situation was observed in the summer months, characterised by a positive energy balance.

References

Chilliard, Y., C. Delavaud and M. Bonnet, 2005. Leptin expression in ruminants: nutritional and physiological regulations in relation with energy metabolism. Dom. Anim. Endocrinol. 29: 3-22.

Marie, M., P.A. Findlay, L. Thomas and C.L. Adam, 2001. Daily patterns of plasma leptin in sheep: effects of photoperiod and food intake. J. Endocrinol. 170: 277-286.

Rhodes, L. and P.W. Nathanielsz, 1990. Myometrical activity and plasma progesterone and oxytocin concentrations in cycling and early-pregnant ewes. Biol. Reprod. 41: 834-841.

Zieba, D.A., M. Amstalden and G.L. Williams, 2005. Regulatory roles of leptin in reproduction and metabolism. A comparative review. Dom. Anim. Endocrinol. 29: 166-185.

Possible implications of feeding soybean meal on fertility and milk production of high yielding dairy cows in the early *post partum* period: preliminary results

S. Cools, L. Vanhaecke and G. Opsomer
Faculty of Veterinary Medicine, Salisburylaan 133, 9820 Merelbeke, Belgium; s.cools@ugent.be

Introduction

Although isoflavones are components ubiquitously present in the ration of ruminants, there are vegetable feed materials with a relatively high (e.g. soya beans) vs. low (e.g. rape) content of these molecules (Chapin *et al.*, 1996). The group of isoflavones contains a plethora of molecules of which some are known to influence cows' physiology especially in relation with fertility. In 2005, Zdunczyk *et al.* (2005) showed herds, in which cows had a significantly higher isoflavone plasma concentration, were dealing with a significantly higher amount of cows suffering from silent heats. These cows were fed a traditional totally mixed ration (TMR) containing grass silage, maize silage, soya beans and concentrates. Besides effects during the follicular phase, there are also studies describing effects of isoflavones on the bovine luteal gland. Piotrowska *et al.* (2006) mentioned that cows which were fed 2.5 kg of soya beans daily, showed a lowered progesterone concentration both in plasma and in the luteal tissue. Consequently, feeding soya beans could impair cows' fertility. Yet, in Belgium soybean meal (after oil extraction) instead of whole soya beans is often fed as an important protein source to dairy cows. Since the oil extraction does not eliminate the isoflavones out of the soybean meal, their negative effects may still be present in the meal. The purpose of our study was to examine whether feeding soybean meal can influence the luteal functionality in high yielding dairy cows.

Material and methods

Thirty Holstein dairy heifers (n=13)/cows (n=17) (average 305-d milk yield of 9,350 kg (SD=1,919 kg), 4.11% fat (SD=0.61%), 3.41% protein (SD=0.24%)) were randomly divided into 2 groups: group 1 contained 9 heifers and 11 multiparous cows, in group 2 the ratio was respectively 4 to 6. Both groups were fed 2 different protein rich concentrates: soybean meal (n=20; group 1) and an alternative, based on rape meal (n=10, group 2), known to be respectively relatively high vs. low in isoflavone content. The ratio of energy vs. protein content was the same for both products, but each kg of soybean meal was equivalent to 1.5 kg of the alternative. Both groups had *ad libitum* access to the same roughage ration which was based on a partial TMR (maize and grass silage, sugar beet pulp, corn cob mix, hay and minerals; 53.2, 24.2, 10.0, 4.5, 5.7, 2.0% on kg DM basis). Surplus individual concentrate gift depended on milk production. Feeding treatments were applied from 2 wk *pre partum*. To monitor the plasma concentration of the isoflavones, blood samples were taken out of the tail vein weekly in the *post partum* period starting at calving till the fourth oestrus. Plasma concentrations of 8 isoflavone molecules were analysed by means of Liquid Chromatography Mass Spectrometry (LC-MS-MS) after extraction with acetone and incubation with glucuronidase. The results are expressed semiquantitatively as 'area ratios'. This is the ratio between the area under the curve (AUC) generated by the endogenous amount of the molecule in the blood plasma and the AUC generated by a standardised amount of the same molecule. Besides, production parameters as daily milk production and body condition score (assessed 2 times a week) were analysed. Luteal function was monitored by measuring maximal luteal area via rectal ultrasonography at d 9 and 16 of the first 3 consecutive oestrus cycles, combined with monitoring of the blood serum progesterone concentration. Statistical analysis was performed by means of SPSS 15 (for windows, SPSS Inc.,

Chicago, IL, USA). Both, the independent samples T-test and the analysis of variance (ANOVA) were performed. Overall differences between means were declared significant at $P<0.05$. Trends towards significance were considered at $P<0.10$.

Results

The average daily consumption of soybean meal and rapeseed meal was respectively 1.71 kg (SD=0.05 kg) and 3.01 kg (SD=1.12 kg). In Table 1, an overview is given of the analyses of the isoflavone plasma concentration, relatively expressed by means of the mean area ratio. Analysis was performed with the univariate ANOVA.

Table 1. Mean plasma concentration (area ratio) of the 4 most important isoflavone-molecules for the soy and rape group.

	Soy/rape	Mean area ratio	SD	Soy vs. Rape effect, *P*-value
Daidzein	Soy (n=20)	2.20	0.63	$P<0.01$
	Rape (n=10)	1.85	0.68	
Glycitein	Soy	2.22	0.54	$P<0.01$
	Rape	1.87	0.56	
Genistein	Soy	0.97	0.35	$P<0.01$
	Rape	0.85	0.33	
Coumestrol	Soy	0.99	0.59	$0.01<P<0.04$
	Rape	0.84	0.43	

The mean daily milk production during the first 60 d *post partum* was 33.64 kg (SD=12.77) and 35.04 kg (SD=12.97) respectively for the soy and rape group. Based on the univariate analysis of variance, this difference was statistically significant ($P<0.01$). The distribution of the cows according to their BCS was identical at the start of the study among both groups. During the first 60 d after calving, there was an average decrease in BCS of 0.74, but no significant difference was noticed between the soy and the rape group during this period ($P>0.05$), based on an independent samples T-test of the differences.
There was no statistical difference in the maximal size of the corpus luteum measured on d 9 and 16 of the first 3 cycles after calving between the rape and soybean meal fed cows. Also parity and cycle number were not significantly associated with the size of the luteal surface measured on d 9 and 16 of the cycle.

Discussion and conclusion

This study shows that, in cows fed soybean meal, the plasma concentration of some isoflavone molecules (e.g. daidzein, glycitein, genistein) was significantly increased, while the plasma concentration of, e.g. coumestrol was significantly lowered in comparison with cows which were supplemented with rape as the main protein source. Although some of these molecules were hypothesised to have a deleterious effect on luteogenesis, we could not demonstrate, based on these preliminary results, a difference in the size of the corpus luteum during the first 3 cycles after calving.

References

Chapin, R.E., J.T. Stevens, C.L. Hughes, W.R. Kelce, R.A. Hess and G.P. Daston, 1996. Endocrine modulation of reproduction. Fund. Appl. Toxicol. 29: 1-17.

Piotrowska, K.K., I. Woclawek-Potocka, M.M. Bah, M.K. Piskula, W. Pilawski, A. Bober and D.J. Skarzynski, 2006. Phytoestrogens and their metabolites inhibit the sensitivity of the bovine corpus luteum to luteotropic factors. J. Reprod. Dev. 52: 33-41.

Zdunczyk, S., M. Piskula, T. Janowski, W. Baranski and M. Ras, 2005. Concentrations of isoflavones in blood plasma of dairy cows with different incidence of silent heat. B. Vet. I. Pulawy 49: 189-191.

MAP kinases ERK1/2, but not AMP-activated protein kinase, are involved in the effects of unsaturated fatty acids on goat granulosa cells steroidogenesis *in vitro*

S. Coyral-Castel, C. Ramé, A. Fatet and J. Dupont
INRA, Unité de Physiologie de la Reproduction et des Comportements, UMR 6175, 37380 Nouzilly, France; jdupont@tours.inra.fr

Introduction

Recent studies have shown that unsaturated fatty acids (UFA) could be beneficial for ruminant reproduction (Wathes *et al.*, 2007), especially on ovarian functions (Dupont *et al.*, 2008). There are saturated and unsaturated fatty acids. Unsaturated fatty acids include mono-unsaturated fatty acids (MUFA) such as oleic acid (OA) and polyunsaturated fatty acids (PUFA). In this latter group, fatty acids are classified according to the position of the first double bond from the methyl end. So, we can separate n-3 PUFA and n-6 PUFA. Alpha-linolenic acid (ALA) and linoleic acid (LA) are precursors of n-3 PUFA and n-6 PUFA, respectively. The effects of UFA on steroidogenesis and the related mechanisms are still unclear. In peripheral tissues, fatty acids can activate the Adenosine-5'-Monophosphate-activated Protein Kinase (AMPK, a key regulator of energy balance) and the Mitogen Activated Protein Kinases Extracellular-Regulated Kinases 1/2 (MAPK ERK1/2) signalling pathways (Clark *et al.*, 2004; Yonezawa *et al.*, 2008). These kinases are known to regulate progesterone secretion in rodents (Tosca *et al.*, 2005). Thus, they could be a link between UFA and steroid production. The aim of this *in vitro* study was to better understand mechanisms which underline the role of UFA on ovarian function.

Material and methods

Small follicles (<4 mm) of 60 goat ovaries from local slaughterhouse were dissected and granulosa cells were cultured in a McCoy medium. Granulosa cells were incubated in a serum-free medium in the presence or absence of OA (10^{-5} M), ALA (10^{-4} M) or LA (10^{-5} M) with or without Insulin-like Growth Factor-1 (IGF-1, 5×10^{-8} M) or Follicle Stimulating Hormone (FSH, 10^{-8} M). Progesterone (P4) and oestradiol (E2) concentrations in the culture medium were measured by RIA after 48 h. By the end of 24 h, cell proliferation was determined by ^{3}H-thymidine incorporation. Phosphorylation of AMPKα (phosphoAMPK/AMPKα), MAPK ERK1/2 (phosphoERK/ERKtotal) and the protein level of steroidogenic enzymes (StAR, P450scc, 3βHSD, P450aromatase) were determined by western blot. Each experiment was repeated 3 times and cells for treatment were selected randomly. The results are expressed as mean ± SEM. Bars with different letters indicate significant differences with ANOVA at P<0.05.

Results

After 48 h of culture, OA (10^{-5} M) and LA (10^{-5} M) (P<0.05, Figure 1A), but not ALA (10^{-4} M, data not shown), increase P4 and E2 (data not shown) secretion in the basal state and in the presence of IGF-1 (5×10^{-8} M) or FSH (10^{-8} M).
We did not find any significant effects of these UFA on AMPKα phosphorylation at short-term (from 0 to 60 min) but we observed (P<0.05, Figure 1B) a rapid increase, a few minutes of stimulation, of Thr202 and Tyr204 phosphorylation of MAPK ERK1/2.
Moreover, we observed, by ^{3}H-thymidine incorporation, that OA and LA decrease IGF-1-induced cell proliferation, without altering cell viability. However, OA and LA do not affect the protein level

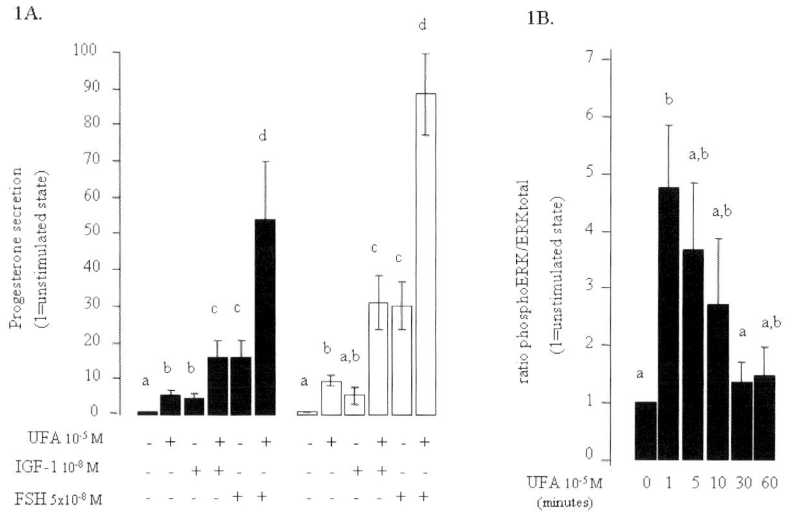

*Figure 1. In vitro effect of OA ■ (10^{-5} M) or LA □ (10^{-5} M) on **(A)** progesterone secretion by granulosa cells, in the presence or absence of IGF-1 (10^{-8} M) or FSH (5×10^{-8} M) and **(B)** ERK phosphorylation, from 0 to 60 min of OA ■ or LA □ (10^{-5} M) stimulation;* $P<0.05$.

of P450scc, 3β-HSD, P450aromatase and StAR (a cholesterol carrier) in the presence or absence of IGF-1 or FSH (data not shown).

Discussion and conclusion

Our results show that *in vitro* supplementation of OA and LA, but not ALA, improves steroid production in goat granulosa cells. However, OA and LA do not affect the protein level of steroidogenic enzymes. We can hypothesise that these UFA modify the activity of these enzymes. The *in vivo* effects of UFA are known. However, the molecular mechanisms of these UFA are still unclear. This study shows direct effects of UFA on steroid production. The MAPK ERK1/2, but not AMPK signalling pathway is involved in the control of steroidogenesis by UFA in these cells.

References

Clark, H., D. Carling and D. Saggerson, 2004. Covalent activation of heart AMP-activated protein kinase in response to physiological concentrations of long-chain fatty acids. Eur. J. Biochem. 271: 2215-2224.

Dupont, J., P. Froment, C. Ramé, P. Pierre, S. Coyral-Castel and C. Chabrolle, 2008. Role of the fatty acids in ovarian functions: involvement of Peroxisome Proliferator-Activated Receptors (PPAR) and adipokines. Gynecol. Obstet. Fertil. 36: 1230-1238.

Tosca, L., P. Froment, P. Solnais, P. Ferre, F. Foufelle and J. Dupont, 2005. Adenosine 5'-Monophosphate-activated Protein Kinase regulates progesterone secretion in rat granulosa cells. Endocrinology 146: 4500-4513.

Wathes, D.C., D.R.E. Abayasekara and R.J. Aitken, 2007. Polyunsaturated fatty acids in male and female reproduction. Biol. Reprod. 77: 190-201.

Yonezawa, T., S. Haga, Y. Kobayashi, K. Katoh and Y. Obara, 2008. Unsaturated fatty acids promote proliferation via ERK1/2 and Akt pathway in bovine mammary epithelial cells. Biochem. Biophys. Res. Commun. 367: 729-735.

Post-natal consequences of a maternal nutritional restriction in the periconceptional period in sheep: effects on male lambs

N. Debus[1], P. Chavatte-Palmer[2], G. Viudes[1], V. Berthelot[3], S. Camous[2] and P. Hassoun[1]
[1]INRA, UMR868 Elevage des Ruminants en Régions Chaudes, 34060 Montpellier Cedex 1, France; [2]INRA, UMR1198 Biologie du Développement et Reproduction, 78350 Jouy en Josas, France; [3]INRA, UMR Physiologie de la nutrition et alimentation, AgroParisTech, 75231 Paris Cedex 05, France; debus@supagro.inra.fr

Introduction

Epidemiological studies in humans have demonstrated that the incidence of metabolic diseases in adults such as obesity, hypertension, insulin resistance and the metabolic syndrome is markedly increased when maternal nutrition is altered at critical periods of foetal development. Periconceptional nutritional environment and/or embryo *in vitro* culture conditions have been shown in sheep to affect offspring outcomes in terms of weight gain, hypothalamo-pituitary-adrenal axis and many other parameters. Most of these data, however, are limited to observations in the late foetal or neonatal period, despite their relevance for the breeding industry (Wu *et al.*, 2006).
The aim of this project was to study post-natal effects of a maternal nutritional restriction from 15 days prior to breeding up to 30 days post-breeding in ewes. Pregnancy parameters and post-natal development of male offspring are presented here.

Material and methods

One hundred and sixteen ewes of Merinos d'Arles breed were used and allocated to one of two groups: undernourished (R50) (n=64) or control (R100) (n=52). They were synchronised in the autumn using intra-vaginal pessaries. Females were fed with half (R50) or 100% (R100) of their nutritional needs (both energy and protein requirements) from day 15 prior to breeding to 30 days of pregnancy. Thereafter, both groups were fed to meet their nutritional requirements. Pregnancy check was performed by PSP60 (60 kD Pregnancy Serum Protein) assay in maternal circulation at 31-42 days and thereafter by ultrasound around 120 days. Maternal weight was monitored throughout pregnancy. Lambs were weighed at birth and thereafter every month until three months of age. Body condition score was also determined monthly all over the experiment. Blood samples were collected monthly to monitor plasma leptin concentrations. Carcass weight, perirenal adipose tissue, kidney, testicle and adrenal weight were recorded at slaughter. Data were analysed by ANOVA or the Student t test. The results are expressed as means ± SD.

Results

R50 ewes lost weight during the time of undernutrition, resulting in a 15% difference in weight between groups at the end of the undernutrition period, and were significantly lighter than R100 throughout gestation ($P<0.0001$). Lambing results are summarised in Table 1. Based on PSP60 assays, 10 ewes were not pregnant at 40 days (88.4% and 92.7% fertility in R50 and R100 groups, respectively). There was no statistical difference between groups for pregnancy rates, prolificity, sex ratio and birth weight of lambs, but pregnancy was significantly longer in R50 (Table 1, $P<0.01$). The growth rate of male lambs was not statistically different between groups. They were slaughtered at a mean live weight of 38.2±2.6 kg, mean age of 136±21 days. Fasted plasma leptin concentrations increased gradually in both groups until 4 months of age. Concentrations were significantly lower in R50 lambs at birth (6.15±0.7 vs. 7.42±1.6 ng/ml, $P<0.001$) and tended to be higher in R50 at 4 months of age (10.61±2.1 vs. 9.79±2.4 ng/ml, $P=0.09$). There was no significant difference between groups for carcass, adrenal, testicle and kidney weight. In contrast, perirenal fat was significantly heavier in R50 lambs (9.96±2.7 vs. 8.48±2.0 g, $P<0.05$) and the carcass weight/live weight ratio was increased (45.4±1.7 vs. 44.4±1.6%, $P<0.05$) compared to R100 lambs.

Table 1. Lambing results.

Group	R100	R50
Number of lambings	41	47
Singletons	26	26
Doubles	13	20
Triple	2	1
Gestation length (days)	149±2.3	151±2.0*
Lambs born	58	69
Stillbirth	7	10
% mortality	12.07%	14.19%
Live lambs	51	59
Females	29	25
Males	22	34
Prolificity	141.5%	146.8%

* Indicates a significant difference between groups, $P<0.05$.

Discussion and conclusion

In contrast to previous studies where periconceptional undernutrition induced premature parturition (Kumarasamy *et al.*, 2005), gestational length was increased in the treated ewes. These opposite results may be due to the use of a different breed of sheep and/or to the nutriments used. Despite similar birth weights, growth rates, weights and ages at slaughter between the 2 groups, perirenal adipose tissue development was decreased at birth and increased at 4 months in R50 lambs, indicating, together with plasma leptin concentrations, a faster growth of adipose tissue and a potential increase in meat fat. Food consumption was monitored in female lambs and was not statistically different between groups. Females are currently being monitored for long term investigations. Indeed, more effects may be detected in adulthood, as reported after vitamin B12 depletion in the periconceptional period (Sinclair *et al.*, 2007).

References

Kumarasamy, V., M.D. Mitchell, F.H. Bloomfield, M.H. Oliver, M.E. Campbell, J.R.G. Challis and J.E. Harding, 2005. Effects of periconceptional undernutrition on the initiation of parturition in sheep. Am. J. Physiol. 288: R67-R72.

Sinclair, K.D., C. Allegrucci, R. Singh, D.S. Gardner, S. Sebastian, J. Bispham, A. Thurston, J.F. Huntley, W.D. Rees, C.A. Maloney, R.G. Lea, J. Craigon, T.G. McEvoy and L.E. Young, 2007. DNA methylation, insulin resistance, and blood pressure in offspring determined by maternal periconceptional B vitamin and methionine status. Proc. Natl. Acad. Sci. 104: 19351-19356.

Wu, G., F.W. Bazer, J.M. Wallace and T.E. Spencer, 2006. Intrauterine growth retardation: Implications for the animal sciences. J. Anim. Sci. 84: 2316-2337.

The infusion of glucose reduces circulating oestradiol and the level of aromatase in granulosa cells of ewes in the luteal phase of the oestrous cycle

C. Gallet[1], *J. Dupont*[1], *D. Monniaux*[1], *B.K. Campbell*[2] *and R.J. Scaramuzzi*[1,3]
[1]*UMR Physiologie de la Reproduction et des Comportements, INRA, 37380 Nouzilly, France;* [2]*Division of Obstetrics and Gynaecology, Queen's Medical Centre, University of Nottingham, United Kingdom;* [3]*Department of Veterinary Basic Sciences, Royal Veterinary College, Herts, United Kingdom; rex.scaramuzzi@tours.inra.fr*

Introduction

Short-term nutritional supplementation of ewes in the late luteal phase of the oestrous cycle stimulates folliculogenesis (Somchit *et al.*, 2007), an effect mimicked by glucose (Downing *et al.*, 1995). The mechanism of this effect involves direct actions of nutrition mediated by the intrafollicular insulin glucose system (Scaramuzzi *et al.*, 2006). However, little is known about insulin signalling or its effects on FSH signalling in granulosa cells. In this experiment, we studied the effect of glucose on the phosphorylation of Akt and AMPK, two kinases involved in insulin signalling and the level of aromatase in granulosa and of oestradiol in jugular blood.

Material and methods

The experiment used 20 Ile-de-France ewes fed a maintenance diet of straw during the experiment. Oestrus was synchronised using progestagen sponges and 8 d after oestrus the ewes were cannulated bilaterally (jugular vein). One cannula was used for blood collection and the other for infusion. The next day one group of ewes was infused with saline (n=7) and the other with glucose (n=5; 10 mM/h) for 72 h. At the end of the infusion the ovaries were collected under anaesthesia. All follicles >2 mm in diameter were dissected within an hour and their granulosa cells were recovered. An aliquot of granulosa cells was examined histologically to identify non-atretic follicles. The granulosa cells were lysed and analysed by western blotting to determine the level of aromatase and the ratio of phosphorylated to total Akt and AMPK. Jugular venous blood was collected regularly during the experiment and the plasma was analysed for glucose (colourimetry), insulin (ELISA), FSH and oestradiol (RIA). Data were analysed by t-test or mixed model ANOVA.

Results

The concentration of glucose rose significantly within 3 h of the start of infusion and remained significantly elevated until 27 h returning to pre-infusion levels by 48 h (Figure 1). The concentration of insulin rose significantly within 3 h and remained significantly elevated for 48 h (Figure 1). The infusion of glucose significantly increased the number of follicles greater than 2 mm in diameter (Table 1) but had no effect on their diameter. The jugular venous concentration of FSH was not significantly affected by treatment (not shown), however, oestradiol was significantly decreased by 27 h and then remained significantly reduced (Figure 1) until the end of the infusion. The infusion of glucose significantly reduced aromatase and the ratios of phosphorylated to total Akt and AMPK in granulosa cells (Table 1).

Discussion

These data show that a short-term infusion of glucose rapidly stimulated follicular growth but inhibited aromatase and an insulin pathway in granulosa cells. The reduced levels of phosphorylated Akt and AMPK in response to a 72 h infusion of glucose suggest that the PI3K (phosphatidylinositol 3-kinase) pathway was inhibited in granulosa perhaps because of desensitisation associated with the continued infusion of glucose. These data also suggest that there may be functional cross-talk between FSH and insulin signalling in granulosa cells.

Table 1. The effect of glucose infused for 72 h during the late luteal phase of the oestrous cycle on the mean ± sem number of follicles and the granulosa cell levels of aromatase and the level of Akt and AMPK phosphorylation.

	Diameter, mm	Folls >2mm	Aromatase	P to T ratio; Akt	P to T ratio; AMPK
Control	4.00±0.26	16.7±2.60	1.53±0.24	0.78±0.096	1.61±0.06
Glucose	3.57±0.33	36.4±6.49[a]	0.90±0.13[a]	0.46±0.089[a]	1.08±0.08[a]

[a] Indicates a significant difference between treatments (P<0.05).

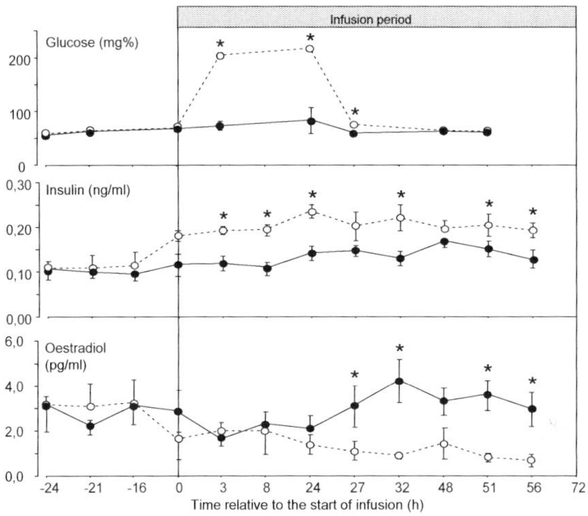

*Figure 1. The effect of glucose infused for 72 h during the late luteal phase on plasma concentrations of glucose, insulin and oestradiol. Solid line = control, dashed line = glucose. An asterisk indicates a significant difference (*P*<0.05) between treatments within times.*

Acknowledgement

R.J. Scaramuzzi is the recipient of an EU Marie Curie Chair of Excellence (MEXC-CT-2006-042499).

References

Downing, J.A., J. Joss and R.J. Scaramuzzi, 1995. Ovulation rate and the concentrations of gonadotrophins and metabolic hormones in ewes infused with glucose during the late luteal phase of the oestrous cycle. J. Endocrinol. 147: 403-410.

Scaramuzzi, R.J., B.K. Campbell, J.A. Downing, N.R. Kendall, M. Khalid, M. Muñoz-Gutiérrez and A. Somchit, 2006. A review of the effects of supplementary nutrition in the ewe, on the concentrations of reproductive and metabolic hormones and the mechanisms that regulate folliculogenesis and ovulation rate. Nutr. Reprod. Develop. 46: 339-354.

Somchit, A., M. Khalid, B.K. Campbell and R.J. Scaramuzzi, 2007. The effects of short-term nutritional supplementation with lupins (*Lupinus luteus*) on the number of ovarian follicles and the concentrations of follicular hormones and glucose in ewes during the luteal phase of the oestrous cycle. Theriogenology. 68: 1037-1046.

Factors decreasing pregnancy rate after embryo transfer in lactating dairy cows

H. Kadokawa[1], Y. Kimura[2], N. Tameoka[3], M. Uchiza[3] and M. Yonai[4]
[1]Faculty of Agriculture, Yamaguchi University, Yoshida 1677-1, Yamaguchi, 753-8515, Japan; [2]Niikappu, National Livestock Breeding Centre, Shizunai-Misono, Hokkaido 056-0141, Japan; [3]Iwate, National Livestock Breeding Centre, Anaguchi, Morioka, Iwate 020-0123, Japan; [4]National Agricultural Research Centre for Tohoku Region, Morioka, Iwate 020-0198, Japan; hiroya@yamaguchi-u.ac.jp

Introduction

Pregnancy rate (PR) is one of most important factors determining the reproductive performance of dairy herds, but the mechanisms for reducing PR remain to be determined. Nutritional condition, reproductive diseases, and mastitis are likely to decrease conception rate (CR) after artificial insemination (AI) (Chebel *et al.*, 2004; Schrick *et al.*, 2001). However, the quality of embryos can not be evaluated in AI. In contrast, the factors that affect PR after embryo transfer (ET) in cattle, where only females without diseases and genital organ problems are utilised as recipients of normal embryos developing for 7 days after insemination are less characterised. Therefore, this study was conducted among dairy cow recipients to survey the factors affecting PR after ET.

Material and methods

In this retrospective survey, all mid-lactation Holstein recipients (n=396) were housed in the National Livestock Breeding Center. Visual observation, rectal palpation, and ultrasonography were utilised to confirm the absence of diseases and abnormalities of genital organs. Single, 7-day-old embryos collected from non-lactating Holstein were transferred to the recipients. The transferred embryos were either fresh (n=192) or frozen-thawed (n=204), and 82 embryos were biopsied for sexing. The morphological qualities of all transferred embryos examined microscopically based on the scale of the International Embryo Transfer Society were grade 1 (excellent n=368); grade 2 (good n=110) and grade 3 (fair n=28). Recipients were examined and given a body condition score (BCS) based on a five-point scale on the day of ET (D 0), and then their BCS was categorised into three groups: 'lean' (BCS of 2.5 or less), 'normal' (BCS 2.75 to 3.25), or 'fat' (BCS of 3.5). Mastitis was identified at each milking based on detected abnormalities of milk or any quarter with the aid of a modified Californian mastitis test. Cows were categorised as 'mastitis-positive' if mastitis was observed from D -30 to D -11, or D 1 to D 30. These periods were set because of the significant effects of mastitis before and after the day of AI (Schrick *et al.*, 2001), and because of no significant effect of mastitis on pregnancy loss from gestation after day 38 in lactating dairy cows (López-Gatius *et al.*, 2002). Embryos were not transferred if mastitis was observed from D -10 to D 0. The affected cows were treated by intramammary infusion of commercially available antibiotics. Pregnancy was confirmed two months after ET. Multivariate logistic regression analysis was utilised to analyse the effects on PR of various factors in order to obtain the odds ratio and 95% confidence interval. The chi-square test was utilised to evaluate the relationship between PR and the onset timing of mastitis utilising data from only ET of non-biopsied embryos.

Results

The number of pregnant cows was 133, thus the overall PR was 33.6%. The number of 'mastitis-positive' cows was 63 and the remaining 333 cows were 'mastitis-negative'. The numbers of 'lean', 'normal', and 'fat' cows were 6, 377, and 13, respectively. The effects of all three embryo categories were significant ($P<0.05$) (Table 1). Parity and body condition had no significant effect on PR. Mastitis-positive cows had only 0.45 times lower PR than mastitis-negative cows ($P<0.05$). Mastitis occurred both before and after ET suppressed PR significantly ($P<0.05$) in fresh ET (negative: 47.1%, positive from day -30 to -11: 0%, positive from day 1 to 30:0%) and in frozen-thawed embryo ET (negative: 36.2%, positive from day -30 to -11:18.2%, positive from day 1 to 30:10.5%).

Table 1. Odds ratios of variables in the multivariate logistic regression model for pregnancy rate after embryo transfer in lactating dairy cows.

Factor	Class	Odds ratio	95% Confidence interval	P-value
Grade of embryo	Grade 1	1	-	-
	Grade 2	0.66	0.38 to 1.14	0.137
	Grade 3	0.13	0.04 to 0.49	0.003
Freshness	Fresh	1	-	-
	Frozen-thawed	0.52	0.30 to 0.90	0.020
Biopsy for sexing	INTACT	1	-	-
	Biopsied	0.47	0.24 to 0.92	0.027
Parity	Primiparous	1	-	-
	Multiparous	0.84	0.53 to 1.33	0.444
Mastitis	Negative	1	-	-
	Positive	0.45	0.20 to 0.97	0.043
Body condition	Lean	1.08	0.18 to 6.42	0.933
	Normal	1	-	-
	Fat	2.58	0.80 to 8.3	0.111

Discussion and conclusion

Apart from the embryo categories, only mastitis occurring 30 days before and after ET, not even Day -10 to Day 0, was the significant maternal factor suppressing the PR after ET in primiparous recipients lacking both disease and abnormalities of genital organs. Mastitis can suppress PR in ET as well as CR in AI, suggesting the importance of avoiding this common disease for both milk production and reproduction in the dairy industry.

References

Chebel, R.C., J.E.P. Santos, J.P. Reynolds, R.L.A. Cerri, S.O. Juchem and M. Overton, 2004. Factors affecting conception rate after artificial insemination and pregnancy loss in lactating dairy cows. Anim. Reprod. Sci. 84: 239-255.

López-Gatius, F., P. Santolaria, J. Yániz, J. Rutllant and M. López-Béjar, 2002. Factors affecting pregnancy loss from gestation Day 38 to 90 in lactating dairy cows from a single herd. Theriogenology 57: 1251-1261.

Schrick, F.N., M.E. Hockett, A.M. Saxton, M.J. Lewis, H.H. Dowlen and S.P. Oliver, 2001. Influence of subclinical mastitis during early lactation on reproductive parameters. J. Dairy Sci. 84: 1407-1412.

Periconception nutrition: effects on gestation length, lamb survival, body and organ growth

D.O. Kleemann[1], J.M. Kelly[1], S.R. Rudiger[1], J.L. Morrison[2], I.C. McMillen[2], S. Zhang[2], S.M. MacLaughlin[2], S. Hiendleder[3], D.H. Smith[1], R.J. Grimson[1], K.S. Jaensch[1], F.D. Brien[1], K.J. Lennon[3] and S.K. Walker[1]
[1]South Australian Research and Development Institute, Holland Road, Rosedale SA 5350 Australia; [2]Sansom Institute, University of South Australia, Adelaide SA 5000 Australia; [3]University of Adelaide, Roseworthy SA 5371 Australia; kleemann.dave@saugov.sa.gov.au

Introduction

Perturbation of the oocyte/embryo's early environment can have downstream consequences for the fetus and lamb (Walker *et al.,* 1992). Maternal undernutrition during the periconception period may alter the timing and activation of the fetal hypothalamo-pituitary-adrenal axis in late pregnancy and possibly timing of parturition and organ maturation (MacLaughlin *et al.*, 2007). In addition, successful transition of the fetus to an extra-uterine environment is reliant on thermogenic metabolic adaptation involving brown adipose tissue and thyroid hormones (Symonds, 1995). Whether programming of the fetal hypothalamo-pituitary-thyroid axis is compromised by the nutritional environment during the periconception period is unknown. In this report we examine the effects of periconception nutrition on gestational length, lamb growth and survival.

Material and methods

Three nutritional treatments (0.7 maintenance (M), 1.0M and 1.5M) were each imposed on 155 mixed-age Merino ewes starting 17 days before and ending 6 days after the time of artificial insemination (day 0). Ewes were then grazed as one flock and fed to maintain a body condition score of 3.0-3.5 during gestation. Animals were weighed at the beginning and end of nutritional treatment and on day 126 of gestation. Litter size was determined via ultrasound on day 75. Lambing paddocks were inspected twice daily to record ewe and lamb identity, gestation length, litter size, birthweight, sex and mortality. A subset of lambs (n=21) from each nutritional treatment was sacrificed at 5 days of age to measure body and organ weights. GLM and CATMOD procedures in SAS® (2002) were used to examine main effects (nutrition, litter size, sex, age of ewe, nutritional replicate) and first-order interactions for continuous and categorical data, respectively.

Results

Mean ewe liveweights for the 0.7M, 1.0M and 1.5M nutritional groups on -d17 were 62.3±0.6, 62.3±0.6 and 62.1±0.6, respectively, which dropped to 55.1±0.5, 56.1±0.5 and 59.4±0.5 kg by d6. Corresponding values on day 126 of gestation were 76.0±0.9, 77.1±0.9 and 77.9±0.8 kg. Periconception nutrition significantly affected neck thymus (F-test, $P<0.01$) and ovary ($P=0.05$) weights, which were both increased by the 1.5M treatment (Table 1). Gestation length, birthweight, body and other organ weights on day 5, and lamb survival were not affected by periconception nutrition ($P>0.05$). In addition, organ weights relative to body weight were not influenced by nutritional treatment ($P>0.05$), whereas values for adrenals, brain, lungs and CRL were higher and chest thymus lower for twins compared with singletons ($P<0.05$; data not shown). Nutrition × litter size interactions with P values of <0.1 were detected for birthweight, CRL and pituitary weight (Table 1).

Table 1. LSMeans (±SE) for gestation length (GL), birthweight(BW d0), survival (Surv d5), crown rump length (CRL) and body (BW d5) and organ weights (d5) in response to periconception nutrition and litter size. Number of lambs born is given in parenthesis.

	Periconception nutrition/litter size					
	0.7M		1.0M		1.5M	
	Singleton	Twin	Singleton	Twin	Singleton	Twin
GL(d)	149.8±0.3	148.8±0.4	149.4±0.3	149.3±0.4	149.5±0.2	148.7±0.4
BW d0 (kg)	5.74±0.14	4.65±0.15	5.51±0.13	5.03±0.15	5.80±0.12	4.78±0.13
Survival (%) d5	95.9 (49)	90.7 (54)	93.5 (62)	79.6 (54)	88.2 (68)	82.8 (64)
BW d5 (kg)	7.48±0.30	6.56±0.30	7.33±0.37	6.61±0.27	7.94±0.33	6.37±0.25
CRL (cm)	62.2±1.5	63.5±1.4	65.4±1.7	62.2±1.3	67.7±1.6	62.5±1.2
Adrenal (mg)	532±31	529±31	505±38	535±28	561±35	471±26
Thyroid (mg)	553±44	499±39	569±51	568±35	675±43	516±33
Perirenal fat (g)	20.3±1.9	18.1±1.9	24.6±2.3	18.7±1.7	25.4±2.1	19.5±1.6
Brain (g)	69.1±1.6	68.1±1.6	67.4±1.9	68.0±1.4	70.4±1.8	69.7±1.3
Neck thymus (g)	15.3±1.9	12.1±1.9	16.8±2.3	14.0±1.7	22.4±2.1	17.8±1.6
Lungs (g)	147±7	144±7	146±9	135±7	157±8	129±6
Heart (g)	55.9±2.7	48.9±2.7	54.9±3.2	49.3±2.4	58.5±3.0	45.7±2.2
Liver (g)	233±13	198±13	209±16	185±12	239±15	192±11
Kidneys (g)	22.6±1.6	21.6±1.6	23.5±1.9	20.3±1.4	24.3±1.7	19.9±1.4
Pituitary (mg)	192±12	189±12	176±15	176±11	217±13	165±10
Ovaries (mg)	161±41	134±42	160±46	125±33	288±57	203±38

Discussion and conclusion

Periconception nutrition that varied from 0.7M to 1.5M for a period of 23 days and starting 17d before the time of artificial insemination altered maternal weight but did not influence gestation length, lamb survival, birth weight, body and most organ weights at 5 days post parturition. Interestingly, ovary and thymus mass were increased by high nutrition. Relative weights for adrenals, lungs and brain were higher in twins compared with singletons. Further communications will report if parameters involved in control of the hypothalamo-pituitary adrenal/thyroid axis were changed by maternal nutrition imposed during oocyte and embryo development.

References

MacLaughlin, S.M., S.K. Walker, D.O. Kleemann, J.P. Sibbons, D.N. Tosh, S. Gentili, C.L. Coulter and I.C. McMillen, 2007. Impact of periconceptional undernutrition on adrenal growth and adrenal IGF and steriodogenic enzyme expression in the sheep fetus during early pregnancy. Endocrinol. 148: 1911-1920.

Symonds, M.E., 1995. Pregnancy, parturition and neonatal development: interactions between nutrition and thyroid hormones. Proc. Nutr. Soc. 54: 329-343.

Walker, S.K., T.M. Heard and R.F. Seamark, 1992. *In vitro* culture of sheep embryos without co-culture: successes and perspectives. Theriogenology 37: 111-126.

Regulatory changes of chemokines in the bovine corpus luteum during the oestrous cycle

H. Kliem, M. Djurkovic, B. Berisha, H.H.D. Meyer and D. Schams
Physiology Weihenstephan, Technical University Munich, Weihenstephaner Berg 3, 85354 Freising, Germany; heike.kliem@wzw.tum.de

Introduction

Immune cells are known to influence the maintenance and regression of the bovine corpus luteum (CL) (Townson *et al.*, 2002). They accumulate within the CL around the time of luteolysis, but are also present throughout the whole oestrous cycle. Especially T-cells and macrophages invade the CL (Pate, 1995) attracted by a chemotactic signal. The family of chemokines is a group of proteins inducing chemotaxis and activation of different leukocytes. It has already been shown by Townson *et al.* (2002) that production of the monocyte chemoattractant protein-1 (MCP-1), which belongs to the CC-chemokines, is increased in the bovine CL during the oestrous cycle. It could be possible that other chemokines are also involved in the attraction of immune cells. Thus we focussed on the gene expression of the chemokines RANTES (regulated upon activation normal T cell expressed and secreted), its receptors C-chemokine-receptor (CCR)-1, CCR-3 and further on Lymphotactin (Lptn) and its X-chemokine-receptor (XCR)-1 in the bovine CL during the oestrous cycle. RANTES attracts eosinophiles and monocytes in the ovary unlike Lptn, which acts more on T-cells (Hedrick and Zlotnik, 1998; Townson and Liptak, 2003).

Material and methods

The bovine CL (n = 6-7 / group) were collected at the local slaughterhouse within 10–20 min of slaughter. The stage of the oestrous cycle was determined as previously described (Berisha *et al.*, 2000). CL were assigned to the following stages: day 1–2, 3–4, 5–7, 8–12, 13–16, >18 (after regression) of oestrous cycle and of early and late gravidity (<4 and >4 months). Total RNA from the CL was extracted with peqGOLD TriFast (PeqLab, Erlangen, Germany) according to the manufacturer's instructions. The mRNA expression was analysed by real-time PCR. The changes in mRNA expression of the examined factors were assayed by normalisation to the ubiquitin (UBQ) internal control (ΔCP). The statistical significance of differences in mRNA expressions of the examined factors was assessed by one-way anova followed by the Holm Sidak as a multiple comparison test. Data which failed the normality or equal variance test were tested by one-way anova on ranks followed by the Kruskal–Wallis test (Sigma Stat 3.0). All results are shown as 40-ΔCP ± SEM (40=highest cycle number). The subtraction was made for graphic reasons. Differences were considered significant if $P<0.05$.

Results

The mRNA expression of RANTES increased significantly at days 5-7 with a further rise from days 8-12 till day >18. This level was also reached during pregnancy. Both receptors CCR-1 and CCR-3 were not significantly regulated (data not shown). Lptn revealed an opposite expression pattern with a significant increase from day 1-2 till day 8-12. At day 13-18 a significant down-regulation and a further increase at day >18 was seen. The lowest expression levels were measured during gravidity, which were significant to the levels during the oestrous cycle. The receptor XCR-1 was significantly increased at day 13-18 till day >8 (Figure 1).

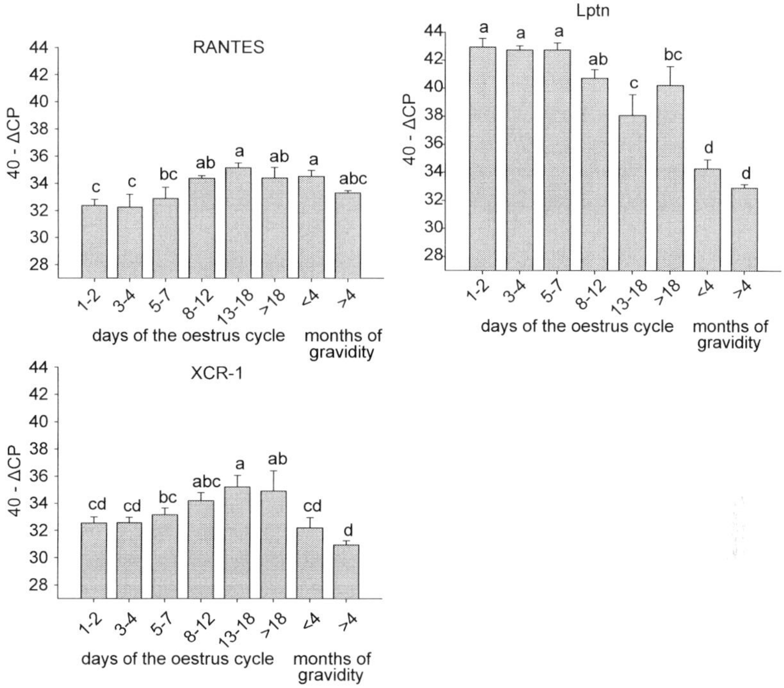

Figure 1. mRNA expression of RANTES, Lptn and XCR-1 shown as 40-ΔCP ± SEM. Different superscript letters indicate significant differences (P<0.05).

Discussion and c onclusion

The increase of RANTES towards the time of luteolysis at about day 18 could indicate a major role in the attraction of T-cells and macrophages, which are known to increase during regression of the CL (Pate, 1995). The same expression level was found during gravidity. At this phase immune cells could be necessary to remove death luteal or endothelial cell within the CL. Lptn showed an opposite expression pattern than RANTES. It is secreted by activated T-cells and attracts further lymphocytes. It seems to be the first signal for immune cells to invade the developing CL. Its receptor XCR-1 is only expressed by lymphocytes and neutrophils indicating an increase of these cells towards luteolysis. We were able to show with these results that the invasion of immune cells into the CL is controlled by several interacting chemokines.

References

Berisha, B., D. Schams, M. Kosmann, W. Amselgruber and R. Einspanier, 2000. Expression and tissue concentration of vascular endothelial growth factor, its receptors, and localization in the bovine corpus luteum during estrous cycle and pregnancy. Biol. Reprod. 63: 1106-1114.

Hedrick, J.A. and A. Zlotnik, 1998. Lymphotactin. Clin. Immunol. Immunopathol. 87: 218-222.

Pate, J.L., 1995. Involvement of immune cells in regulation of ovarian function., J. Reprod. Fertil. Suppl. 49: 365-377.

Townson, D.H. and A.R. Liptak, 2003. Chemokines in the corpus luteum: implications of leukocyte chemotaxis. Reprod. Biol. Endocrinol. 1: 94-100.

Townson, D.H., C.L. O'Connor and J.K. Pru, 2002. Expression of monocyte chemoattractant protein-1 and distribution of immune cell populations in the bovine corpus luteum throughout the estrous cycle. Biol. Reprod. 66: 361-366.

Consequences of maternal feeding restriction during goat's pregnancy on kid morphology and weight at birth

B. Laporte[1], P. Chavatte-Palmer[2], S. Roussel-Huchette[1], J. Perault[2] and C. Duvaux-Ponter[1]
[1]UMR INRA-AgroParisTech Physiologie de la Nutrition et Alimentation, 16 rue Claude Bernard, 75231 Paris cedex 05, France; [2]UMR INRA-ENVA, Domaine de Vilvert, 78350 Jouy-en-Josas, France; berengere.laporte@agroparistech.fr

Introduction

Dietary restriction during pregnancy can lead to various consequences on offspring growth and development (Wu *et al.*, 2006). The aim of this experiment was to study the effects of nutritional restriction in pregnant goats on the morphology of their kids at birth.

Material and methods

The Control group was fed *ad libitum* a Total Mixed Ration (TMR) which covered requirements (INRA, 1988). Thirty Underfed (UN) and thirty Control (C) pregnant Alpine or Saanen goats were used. The underfed goats received from 50% (from d 90 to d 120) to 80% (from d 121 to d 150) of the amount of the TMR given to Control goats during the last third of pregnancy.
A total of 28 male and 36 female kids were studied, all from twin or triplet litters. They were weighed and a flexible tape was used to measure the following parameters: abdominal circumference, thoracic girth, height at the withers and crown rump length. Body Mass Index (BMI: weight/[crown-rump length]^2) was calculated. All the variables were analysed using the PROC MIXED procedure of SAS® (SAS Institute Inc., Cary, NC, USA). The model included litter size (twins or triplets), breed, sex, feeding treatment of dam (UN or C), interaction between feeding treatment and the other parameters. A random dam effect was used in the statistical model.

Results and discussion

UN males tended to be lighter at birth than C males (4.2±0.15 kg vs. 4.7±0.16 kg, P=0.07) but there was no difference between treatments for females (3.8±0.15 kg vs. 3.9±0.14 kg). BMI of UN kids tended to be lower than BMI of C kids (16.7±0.47 vs. 18.1±0.38, P= 0.06). There was an interaction between treatment and breed: BMI was lower for Saanen UN than for Saanen C kids. Abdominal circumference was lower (P=0.01) in UN kids compared to C kids regardless of sex whereas there was no difference for the other measurements.
Feed restriction during the last third of pregnancy can cause changes in the body composition of the offspring in sheep (Robinson *et al.*, 1999) and may reflect changes in developmental physiology and metabolism as shown by Husted *et al.* (2007) in lambs which were affected by late gestational maternal underfeeding. Moreover, maternal underfeeding has been shown to have differential effects according to offspring sex on metabolic parameters (Owens *et al.*, 2007). These preliminary results show a difference in abdominal circumference which may indicate either lower intra-abdominal adipose tissue reserves or lower development of digestive organs in UN compared to C kids.

Conclusion

Nutritional restriction during the last third of pregnancy reduced birth weight and the size of some morphologic components at birth. These changes in morphology may reflect changes in developmental physiology and metabolism. Further investigations are needed on UN kid body tissue composition and on the permanency of these effects in adulthood.

References

INRA, 1988. Alimentation des Bovins, Ovins, Caprins. INRA, Paris, 471 pp.

Husted, S.M., M.O. Nielsen, M.P. Tygesen, A. Kiani, D. Blache and K.L. Ingvartsen, 2007. Programming of intermediate metabolism in young lambs affected by late gestational maternal undernourishment. Am. J. Physiol. Endocrinol. Metab. 293: E548-E557.

Owens, J.A., P. Thavaneswaran, M.J. De Blasio, I.C. McMillen, J.S. Robinson and K.L. Gatford, 2007. Sex-specific effects of placental restriction on components of the metabolic syndrome in young adult sheep. Am. J. Physiol. Endocrinol. Metab. 292: E1879-E1889.

Robinson, J.J., K.D. Sinclair and T.G. McEvoy, 1999. Nutritional effets on foetal growth. Anim. Sci. 68: 315-331.

Wu, G., F.W. Bazer, J.M. Wallace and T.E. Spencer, 2006. Board-invited review: intrauterine growth retardation: implications for the animal sciences. J. Anim. Sci. 84: 2316-2337.

Maternal efficiency in beef cattle is not compromised by selection for leanness or feed efficiency

M. Laurence[1,2], A. Barnes[1,2], E. Taylor[1,2], D.W. Pethick[1,2], F. Jones[2,3], J. Speijers[3] and J. Accioly[2,3]
[1]Murdoch University, Western Australia; [2]Beef CRC for Genetic Technologies; [3]Department of Agriculture and Food Western Australia; m.laurence@murdoch.edu.au

Introduction

Beef cattle producers are concerned that selecting for carcass traits such as leanness, or for increased feed efficiency, might be deleterious to maternal efficiency and limit the use of genetic improvement technologies such as Estimated Breeding Values (EBV). We define maternal efficiency by production parameters such as days to calving, birth weight, growth rate, weaning weight, as well as efficiency measures such as total intake and kg weaned per megajoule of metabolisable energy consumed per cow calf unit – kg weaned/MJ ME intake. Selecting for leanness in cattle is of economic benefit due to the relationship to higher yielding carcasses (Nkrumah *et al.*, 2004). Net Feed Intake (NFI) is a trait used to measure feed efficiency in beef cattle, and is calculated as the actual amount of feed eaten by an individual animal less the expected amount of feed consumed based on the animal's growth rate and body weight (Koch *et al.*, 1963). Low NFI (high efficiency) is economically desirable due to the potential to reduce feed costs and increase stocking rates. Both traits affect the body condition of dams and this is closely linked to maternal traits in cattle (Morrison *et al.*, 1999; Roche *et al.*, 2000; Meikle *et al.*, 2004). This experiment aims to quantify the impact on the breeder herd of selection for leanness or feed efficiency over three breeding cycles. The impact of level of nutrition was also assessed.

Material and methods

Two hundred BREEDPLAN registered, stud Angus heifers, selected for a divergence in either fatness (Fat or Lean) or feed efficiency (High NFI/ Low NFI), were subjected to either a high or low plane of nutrition, on an extensive grazing system at Vasse Research Centre, Busselton, Western Australia. The experiment followed three breeding cycles. The hypothesis tested was that animals selected for leanness, or superior feed efficiency, would maintain maternal efficiency in good, but not in poor nutritional environments, and the economic benefits of selecting for these animals would be lost when energy intake was restricted. Nutritional treatments were controlled by set stocking and providing the low nutrition treatments with restricted food on offer. Low nutrition animals were fed to 90% maintenance requirements and high nutrition animals were fed to 120% maintenance requirements. Body weight and condition score was monitored fortnightly and a minimum 20% difference between nutritional treatments was maintained throughout the experiment. Estimated pasture intake was calculated by visually assessing food on offer (FOO) before and after animals grazed an area. Weekly calibration cuts were done to verify the visual assessments. Pasture cuts were done fortnightly for assessment of pasture quality. Linear mixed REML models were fitted to the data using Genstat 11 (2008, VSN International, Hertfordshire, UK) to look for the main effects of nutrition and genotype and any interaction between the two. Appropriate covariates and random effects were fitted to each analysis.

Results

The experiment is now in its third year (second calving). In 2007 and 2008 neither level of nutrition, nor the genotype of the animal, had an effect on days to calving (time from mating start date to calf birth date) or the birth weight of calves. In both years the calves of dams on low nutrition grew significantly more slowly than the calves of dams on high nutrition ($P<0.05$). The genotype of the cow had no effect on the growth rate of the calf. In 2007 and 2008 the calves of dams on low nutrition weighed less at weaning than those calves from dams on high nutrition ($P<0.05$).

The genotype of the cow had no effect on calf weaning weight. Measures of group intake showed that Fat and Low NFI (superior efficiency) animals ate less pasture than Lean or High NFI (feed inefficient) in 2007 ($P<0.001$). In 2007 across all genotypes, animals on low nutrition weaned significantly heavier calves per MJ ME intake than animals on high nutrition ($P=0.03$). Low NFI animals weaned heavier calves per MJ ME intake than High NFI animals ($P=0.03$). Main effect means are illustrated in Figure 1a and b. These results are not yet available for the 2008 season.

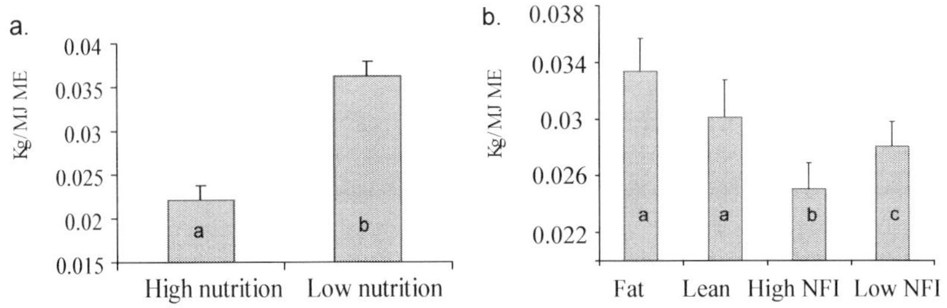

Figure 1. Main effects of nutrition (a; n=60/ nutritional treatment) and genotype (b; n=30/ genotype) on kg beef weaned/ MJ ME intake in 2007. Means with different letters differ significantly (P<0.05).

Discussion and conclusion

Animals selected for leanness, despite their greater total intake, weaned as much calf per MJ ME consumed as Fat animals. Low NFI animals not only had a lower total feed intake, but weaned heavier calves per MJ ME intake, than High NFI animals. Where level of nutrition has an effect, such as on growth rate and weaning of calves, and in kg weaned/MJ ME intake, the effect is across all genotypes. Animals selected for a divergence in leanness weaned heavier calves per MJ ME intake than animals divergently selected for NFI ($P<0.05$). This is a consequence of five generations of focussed selection for a divergence in feed efficiency in the NFI lines without any focus on improving growth rates or body composition. These early results suggest that the economic benefits of selection for leanness or feed efficiency persist in a restricted energy environment, without compromising maternal efficiency, thus far disproving our hypothesis. Note must be made that this result is from the first parity in an experiment that will include three parities.

Acknowledgement

The authors sincerely thank the beef cattle staff at Vasse Research Centre, Busselton, WA, USA.

References

Koch, R.M., L.A. Swiger, D. Chambers and K.E. Gregory, 1963. Efficiency of feed use in beef cattle. J. Anim. Sci. 22: 486-494.

Meikle, A., M. Kulcsar, Y. Chilliard, H. Febel, C. Delavaud, D. Cavestany and P. Chilibroste, 2004. Effects of parity and body condition at parturition on endocrine and reproductive parameters of the cow. Reproduction 127: 727-737.

Morrison, D.G., J.C. Spitzer and J.L. Perkins, 1999. Influence of *prepartum* body condition score change on reproduction in multiparous beef cows calving in moderate body condition. J. Anim. Sci. 77: 1048-1054.

Nkrumah, J.D., J.A. Basarab, M.A. Price, E.K. Okine, A. Ammoura, S. Guercio, C. Hansen, C. Li, B. Benkel, B. Murdoch and S.S. Moore, 2004. Different measures of energetic efficiency and their phenotypic relationships with growth, feed intake, and ultrasound and carcass merit in hybrid cattle. J. Anim. Sci. 82: 2451-2459.

Roche, J.F., D. Mackey and M.D. Diskin, 2000. Reproductive management of *postpartum* cows. Anim. Reprod. Sci. 60-61: 703-712.

Effect of feeding strategies during the winter on fertility of dairy heifers first calving at 3 years of age

Y. Le Cozler[1], J.R. Peccatte[2] and L. Delaby[3]
[1]INRA, UMR1080 Dairy Production, 35000 Rennes, France; [2]AGROCAMPUS OUEST, UMR1080 Dairy Production, 35000 Rennes, France; [3]IUE 326, INRA, Le Pin au Haras Borculo, 61310 Exmes, France; yannick.lecozler@agrocampus-ouest.fr

Introduction

The seasonal feeding system is based on optimisation of fresh grass intake. Such a system is aimed at reducing feeding costs and at limiting labour (Hoch *et al.*, 2003). To reduce even more such the costs, feeding levels of heifers are generally low when housed (i.e. during the winter). During the following pasture season, compensatory growth allows animals to compensate for, at least partially, the previously limited daily gains. In such systems, breeding is also seasonal and it is not unusual that first calving occurs at 36 months of age. However, very little information is available on the effect of feed restriction during the winter season on the subsequent performances. A long-term experiment was performed from 1988 until 2002 at the INRA experimental station of Pin au Haras (Normandy). The first results showed that feeding levels around insemination have a limited impact on fertility when animals were normally fed previously (Peccatte *et al.*, 2006), but when applied during two successive winters, feeding restriction negatively affected fertility at first insemination (Y. Le Cozler *et al.*, unpublished data). The results from the present paper are based on preliminary results of the last trial, which aimed at analysing heifer fertility at first calving at 36 months of age and restrictedly fed only during the first winter following birth.

Material and methods

The experiment was undertaken at the INRA experimental farm at Le Pin au Haras, in Normandy, France. It comprised a total of 78 heifers born during three successive calving winter seasons (December to March 1995, 1996 and 1997), of Normandy (n=34) and Holstein (n=44) breeds. During the first year of the experiment, all animals from 0 to 6 months of age were housed in a free-stall barn along with heifers of similar age, and received a similar treatment. From April to December, they were turned out to pasture and rotationally grazed on a perennial ryegrass sward. Heifers entered into the experiment when housed and fed indoor at the end of summer 1. They were allotted according to their origin, birth date and body weight (BW). Experimental treatment consisted of three different feeding levels, in order to achieve either 600 (H), 400 (M) or 200 (L) g/d growth weight rate during the winter. Heifers got access to a diet based on corn silage, straw, rapeseed meal, urea and minerals. The difference between H, M and L treatments was due to the total amount of feed offered and the proportion of ingredients. Thereafter, all animals had the same treatment. Heifers were inseminated after oestrus synchronisation during winter 2. Individual BW was measured at birth and on average, every 3 wk thereafter and was corrected according to age (extrapolation). During the winter, daily feed allowance and refusals were recorded on a pen basis. Analyses were performed with CHISQ option of the FREQ and LOGISTIC procedures of SAS® (SAS Institute Inc., 1996, Cary, NC, USA).

Results and discussion

Neither BW at birth nor BW at the beginning of the experiment (around 330 d of age) differed between treatments: 41 (±5) and 286 (±25) kg, on average, respectively. Average daily weight gain (ADG) during winter 1, from 330 to 450 d of age on average, was 618 (±148), 441 (±137) and

268 (±159) g/d respectively for H, M and L heifers (Table 1). During the compensatory growth phase (pasture season, from 450 to 570 d of age), ADG significantly differed between treatments: 502 (±256), 586 (±225) and 620 (±238) g/d for H, M and L heifers, respectively. At breeding, BW varied according to growth rate during the winter: 536, 529 and 518 kg for H, M and L heifers, respectively ($P<0.05$), but as a result of oestrous synchronisation, age did not differ (798 d on average). Neither overall fertility nor fertility at first AI was affected by treatment (80 and 49%, respectively), but H heifers tended to have reduced performance at first AI. As previously noted (Y. Le Cozler *et al.*, unpublished data), fertility at first AI was reduced for Holstein heifers in comparison to Normandy heifers.

Table 1. Body weight (BW) at the beginning and end of the experimental period, at 1st artificial insemination (AI), and reproductive performances of Holstein and Normandy heifers according to average daily gain from 330 to 450 d of age.[1]

	Average daily gain (ADG) from 330 to 450 d of age, g/d			Breed	
	618 (= H)	441 (= M)	268 (= L)	Holstein	Normande
N	26	26	26	44	34
BW start, kg	286 (25)	286 (21)	285 (26)	290 (26)	280 (19)
BW end, kg	360[a] (25)	341[b] (32)	315[c] (35)	346[a] (35)	328[b] (34)
Insemination					
Age, d	800 (38)	801 (20)	797 (21)	800 (21)	798 (19)
BW, kg	536 (38)	529 (47)	518 (49)	533 (47)	522 (42)
Successful AI,%					
1st AI[2]	38	65	46	41[a]	62[b]
Overall	74	88	77	72	82
AI/gestating heifer	2.5 (1.6)	1.9 (1.3)	2.0 (1.3)	2.3 (1.4)	1.9 (1.4)

[1] Within a row and treatment (ADG or breed), values with the same superscript letters do not differ at $P<0.05$.

[2] H heifers tended to have lower performance at 1st AI than L and M heifers.

Discussion and conclusion

In a dairy system where first calving occurs at 36 months of age, reducing growth performances during two successive winters had a negative effect on fertility when average daily gain was lower than 600 g/d during two successive winters (Y. Le Cozler *et al.*, unpublished data). The present study showed no effect of feed restriction during the first winter on fertility and similarly, the results from Peccatte *et al.* (2006) showed no effect of feed restriction during winter 2. These results showed that it is possible to reduce rearing cost by a winter feed restriction when feed is not available. Such a restriction may be applied at a high level during the winter, but only during one winter. Long-term effects (lactation, reproduction, culling, etc.) of such procedures should be soon published.

References

Hoch, T., C. Begon, I. Cassar-Malek, B. Picard and I. Savary-Auzeloux, 2003. Mécanismes et conséquences de la croissance compensatrice chez les ruminants. INRA Prod. Anim. 16: 49-59.

Peccatte, J.R., G. Michel, L. Delaby and J.L. Troccon, 2006. Influence du profil de croissance autour de la période d'insémination sur la fertilité des génisses laitières de race Holstein et Normande conduites en vêlage 3 ans. Renc. Rech. Ruminants 13: 286.

Expression of adipokines in bovine ovaries: effect of human recombinant adiponectin and resistin on ovarian cells *in vitro*

V. Maillard, S. Uzbekova, F. Guignot, C. Ramé, C. Perreau and J. Dupont
INRA, Unité de Physiologie de la Reproduction et des Comportements, UMR 6175, 37380 Nouzilly, France; jdupont@tours.inra.fr

Introduction

Adiponectin and resistin are hormones produced by adipocytes (adipokines) with important roles in lipid metabolism and glucose homeostasis (Mitchell *et al.*, 2005). Whereas the resistin receptor is still unidentified, adiponectin is known to exert its action by binding to two seven transmembrane specific receptors, AdipoR1 and AdipoR2 (Yamauchi *et al.*, 2003). These two adipokines have recently been shown to be involved in the control of different reproductive organs, such as the placenta, pituitary, hypothalamus and ovaries in humans, pigs and rodents (Campos *et al.*, 2008; Mitchell *et al.*, 2005). Few data are available about the involvement of adiponectin and even lesser of resistin in bovine reproduction. Thus, the objectives of the present study were to investigate in bovine ovarian cells: (1) the expression of adiponectin receptors and resistin and (2) the effects of adiponectin or resistin on oocyte maturation, early embryo development and granulosa cell (GC) proliferation and steroidogenesis.

Material and methods

Bovine cumulus-oocyte complexes (COC) and GC were retrieved from 3-7 mm follicles. *In vitro* maturation of COC was performed during 22 h in basic or enriched TCM199 medium supplemented or not with human recombinant adiponectin (10 μg/ml). Nuclear maturation of oocytes was determined by chromatin DAPI-staining. Cleavage and blastocyst rates were assessed 48 h and 8 d after *in vitro* fertilisation of COC, respectively. Bovine GC were cultured in McCoy 5A medium with 5% serum. After 18 h of serum starvation, cells were incubated in the presence or absence of human recombinant resistin (667 ng/ml) and/or IGF1 (10^{-8} M) and insulin (10^{-8} M) in serum-free medium for the appropriate time. Cell proliferation (after 24 h) and progesterone secretion in the culture medium (after 48 h) were measured by [3H] thymidine incorporation and RIA protocols, respectively. Expression of adiponectin receptors, resistin and phosphorylation of MAPK ERK1/2 and p38, AMPKα and AKT1/2/3 were assessed by western blotting. Results are representative of at least three independent cultures. Statistical analyses were carried out using one factorial ANOVA and Fisher PLSD tests or a Chi^2 test, with significant differences for $P<0.05$.

Results

The presence of AdipoR1/R2 proteins was shown in GC and theca cells from small (≤6 mm) and large follicles (>7 mm), the corpus luteum, oocytes and cumulus cells. Resistin protein was detected in the ovaries, small and large follicles, and in the corpus luteum (Figure 1A).
Nuclear maturation, cleavage and blastocyst rates were, however, unchanged between the groups of COC matured in the presence or absence of adiponectin.
In GC, resistin treatment did not significantly modify either cell proliferation or basal and IGF1 or insulin-stimulated progesterone secretion. However, we observed a significant increase of MAPK ERK1/2 phosphorylation after 1 min of resistin stimulation (Figure 1B). The other signalling pathways studied (MAPK p38, AMPK and AKT) were not affected.

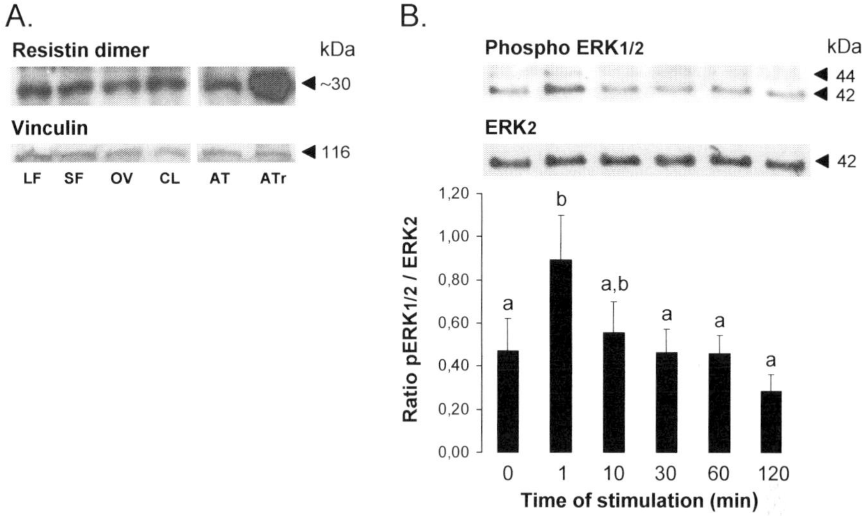

Figure 1. (A) Detection of the resistin dimer by immunoblotting in large (LF) and small (SF) follicles, corpus luteum (CL), ovaries (OV) and adipose tissue (AT) of dairy cattle. Female rat adipose tissue (ATr) was used as a positive control. Vinculin protein was used as a loading control. (B) Effect of recombinant resistin (667 ng/ml) on ERK1/2 phosphorylation in bovine granulosa cells after 1 to 120 min of stimulation. The results are expressed as mean ± SE of the pERK1/2 /ERK2 ratio. Bars with different letters are significantly different ($P<0.05$).

Conclusion

In this study the detection of adiponectin receptors and resistin in bovine ovarian cells suggests a possible role of adipokines in bovine reproduction. Our results show that resistin could act on GC via the MAPK ERK1/2 pathway without involvement in cell proliferation and progesterone secretion. In contrast, the meiotic maturation and early embryo development do not seem to be affected by adiponectin.

The role of resistin on oestradiol secretion in GC, oocyte maturation and embryo development still has to be studied. Adiponectin effects on ovarian cells will be investigated in the presence of various growth factors and other adipokines.

References

Campos, D.B., M.F. Palin, V. Bordignon and B.D. Murphy, 2008. The 'beneficial' adipokines in reproduction and fertility. Int. J. Obes. (Lond.) 32: 223-231.

Mitchell, M., D.T. Armstrong, R.L. Robker and R.J. Norman, 2005. Adipokines: implications for female fertility and obesity. Reproduction 130: 583-597.

Yamauchi, T., J. Kamon, Y. Ito, A. Tsuchida, T. Yokomizo, S. Kita, T. Sugiyama, M. Miyagishi, K. Hara, M. Tsunoda, K. Murakami, T. Ohteki, S. Uchida, S. Takekawa, H. Waki, N.H. Tsuno, Y. Shibata, Y. Terauchi, P. Froguel, K. Tobe, S. Koyasu, K. Taira, T. Kitamura, T. Shimizu, R. Nagai, T. Kadowaki. 2003. Cloning of adiponectin receptors that mediate antidiabetic metabolic effects. Nature 423: 762-769.

Effects of prostaglandin $F_2\alpha$ ($PGF_2\alpha$) intrauterine injection on oestrus synchronisation in Bali cattle *(Bos sondaicus)*

A. Malik[1], Sudarmaji[1], H. Wahid[2], Y. Rosnina[2] and M. Afdal[3]
[1]Department of Animal Science, Faculty of Agriculture, Islamic Kalimantan University, Banjarmasin, Indonesia; [2]Department of Veterinary Clinical Studies, Faculty of Veterinary Medicine, Universiti Putra Malaysia, Serdang, Selangor Darul Ehsan, Malaysia; [3]Department of Animal Nutrition, Faculty of Animal Husbandry, Jambi University, Indonesia; sidol_99@yahoo.com

Introduction

The Bali cattle (*Bos sondaicus*), domesticated from *Bos banteng* in Java, have been reported to have a higher reproductive performance than other indigenous Indonesian cows (Malik *et al.*, 2007). To improve the efficiency of the Bali cattle breeding programme, oestrus synchronisation method can be applied. In a previous report (Gaines *et al.*, 1993), an increase in mean index calving from oestrus synchronised cows was the most significant benefit, gained from the synchronisation programme. $PGF_2\alpha$ causes regression of the corpus luteum (CL) in cows from day 5 until day 17 of the oestrus cycle, and can be used to synchronise any female's oestrus cycle from day 7 onwards (Wright and Malmo, 1992). Several different strategies exist to synchronise cows with $PGF_2\alpha$, the strategy used in this study is called the 2-shot prostaglandin protocol, with different intramuscular (IM) and intrauterine IU injections. Therefore, the present study was aimed at determining the effects of PGF2α intrauterine injection on the number of cows exhibiting oestrus and pregnancy rate in Bali cattle.

Material and methods

Fourty Bali cows, with average body weight of 327 kg, and an average age of 5 years were divided into two groups of twenty cows each. In Group 1, first $PGF_2\alpha$, was injected intramuscularly with 25 mg Dinoprost® (Glandins, Tad Pharmazeutisches werk Gmbh, West Germany). In Group 2, 5 mg dosage intrauterine injections were made with a modified AI gun. Second injections of prostaglandin were given 11 days after the first injection. Oestrus was detected and recorded after the first injection without insemination. After the second injection, timed AI was used. The cows were inseminated 72 h after the second injection using frozen Bali semen obtained from Balai Inseminasi Buatan, Singosari, Malang, Indonesia. Pregnancy diagnosis was by palpation per rectum 3 months after AI. Statistical analysis of the number of cows exhibiting oestrus and pregnancy rate were done using the SAS® procedure (SAS institute, 1997).

Results and discussion

Even though the data showed that the IU injection (70%) was numerically higher than IM (60%) on the number of cows that showed oestrus, it was not statistically significant ($P>0.05$; Table 1). This result was in agreement with Louis *et al.* (1974) who found that $PGF_2\alpha$ intrauterine injection in Holstein cows required a lower amount of prostaglandin to induce luteolysis.
It is suggested that not all cows show oestrus at the same time. This may be due to the different stages of follicular development at the first injection. However, after the second injection, all cows showed oestrus in both IM and IU groups. The IM injection of $PGF_2\alpha$ requires a higher dosage because this hormone will be first absorbed into the circulation before reaching its target organ, which is the ovary. However, intrauterine injection of $PGF_2\alpha$ requires a lower dose, because the uterus is close to the target organ. Furthermore, the intrauterine route was effective in inducing increased CL blood flow, regardless of whether the treatment induced luteolysis.

Table 1. Induction of oestrus by administration of $PGF_2\alpha$ in Bali cattle.

Injection	Population, cows	Dosage of Dinoprost®, ml	Oestrus response,%		Pregnancy rate,%
			1st injection	2nd injection	
IM	20	25	60	100	65
IU	20	5	70	100	60
			NS		NS

NS: non significant.

Acknowledgement

Special thanks to Balitbangda Banjar Region, South Kalimantan, Indonesia for funding this study.

References

Gaines, J.D., J. Galland, D. Schaefer, D. Nusbaum and D. Peschel, 1993. The economic effects of estrus synchronization in beef heifers on average weaning weight of calves. Theriogenology 39: 69-75.

Louis, T.M., H.D. Hafs and D.A. Morrow, 1974. Intrauterine administration of prostaglandin $F_2\alpha$ in cows: progesterone, estrogens, LH, oestrus and ovulation. J. Anim. Sci. 38:347-353.

Malik, A., Sudarmaji and A. Gunawan, 2007. The effect of prostaglandin injection on estrus percentage and pregnancy rate of the Bali and PO cows in South Kalimanytan (in Indonesian). J. Majalah Ilmiah Peternakan 10: #1. Retrieved from http://ejournal.unud.ac.id/?module=daftarvolume&idf=12&idj=31

Wright, P.J. and J. Malmo, 1992. Applied pharmacology and therapeutics II: Pharmacologic manipulation of fertility. Vet. Clin. North Am.: Food Anim. Prac. 8: 57-89.

Use of metabolic profiles in transition cows and cows with low conception rates on a small-scale dairy farm

T.S. Marenjak, Ž. Ipša, N. Poljičak-Milas, J. Piršljin and B. Beer Ljubić
Faculty of Veterinary Medicine University of Zagreb, Heinzelova 55, 10000 Zagreb, Croatia; marenjak@vef.hr

Introduction

Identification of blood parameters that are the most relevant for spotting nutritional and metabolic disorders might be a critical tool for an early diagnosis of post-partal diseases. Analytes that are commonly used for the clinicopathologic evaluations are not usually the best tools for the evaluation of nutritional status (Herdt, 2000). Also, it has been suggested that early lactation cows should be excluded from the analyses due to the state of the metabolic and hormonal imbalance in that period, and questionable representation of the results. However, the blood analysis of that group of cows may be very valuable (Kida, 2002). In small dairy herds it is sometimes difficult to select an equal number of cows per group and to establish the standard values for the specific population. The objective of our study was to identify the blood parameters that could provide relevant data for spotting animals at risk by implying blood testing at the transition and after repeat breeding.

Material and methods

Eleven Holstein and seven Simmental cows in transition (summer–autumn, 2007), three to five years old with average body weights (BW) of 500-650 kg and yearly milk production of 6,700 kg/cow/day were included in the study. The animals were kept indoors, fed three times daily, and after the parturition were milked two times daily (Westfalia Separator AG, GmbH). The blood was collected three times during the transition period (10 days before the expected parturition – period 1) 7 and 15 days *post partum* (p.p.) (period 2 and period 3), and after the repeat breeding. The blood was always taken three hours after the morning feeding from the *v. subcutanea abdominis* in BD vacutainer tubes (SST, Plymouth, UK), transported to the laboratory at +4 °C where plasma was separated after centrifugation at 3,000 rpm for 15 min. Serum was stored at -20 °C until analysis. The body condition scoring (BCS) was performed on the five scale bases divided by 0.25 units, 1 for very thin and 5 for obese (NRC, 2001). The albumin, blood urea, glucose, total cholesterol, total bilirubin, minerals calcium (Ca), phosphorus (P), magnesium (Mg), enzymes gamma-glutamyl-transferase (GGT), aspartate-aminotranspherase (AST), alcalic phosphatase (AP) and lactate dehydrogenase (LDH) were determined on an Olympus AU 600 automatic analyser (Olympus Diagnostica GmbH, Hamburg) and beta-hydroxybutirate (BHB) was determined with a Technicon RA-100 (Technicon, Tarrytown, New York, USA). The HDL and LDL cholesterol (HDLc and LDLc), and non-esterified fatty acids (NEFA) were determined on the SABA 18 automatic analyser (AMS, Italy) using the Randox Ltd commercial kit (Antrim, Unated Kingdom) and enzyme activity of superoxide dismutase (SOD) using the RANSOD commercial kit (Randox, Ireland). The thiobarbituric acid reactive substance (TBARS) concentrations were determined with a Helios delta spectrophotometer (Unicam, Cambridge, UK). The data were analysed by ANOVA repeated measure of STATISTICA 7.1 (Statsoft, USA, 2004) applying the Sheffe test for the evaluation of the difference between means. The significance was declared at $P<0.05$ if not stated otherwise.

Results

Only blood parameters that had been significantly changed are presented in Tables 1 and 2.

Table 1. The blood parameters (mean±SEM) and body condition score (BCS) significantly changed in cows at transition and cows after the first service.

Periods	1 (10 d.a.p)	2 (7 d.p.p)	3 (15 d.p.p.)	4 (AI 80 d.p.p.)
Blood parameters				
Glucose mmol/l	3.47±0.09^{a}	2.70±0.17^{b**}	2.35±0.16^{b***}	2.45±0.19
Ca mmol/l	2.35±0.05^{a**}	2.06±0.03^{b}	2.05±0.05^{b}	2.23±0.08
Mg mmol/l	0.85±0.69	0.70±0.45^{a*}	0.90±0.01^{b}	0.93±0.07
Cholesterol mmol/l	2.17±0.06^{a*}	1.83±0.26^{ac**}	2.95±0.19^{b}	5.03±0.68^{d***}
HDLc mmol/l	0.49±0.02^{c}	0.44±0.06^{c}	0.67±0.09^{c}	0.95±0.09^{d***}
LDLc mmol/l	1.98±0.05^{a*}	1.68±0.24^{ab**}	2.70±0.17^{c}	4.55±0.66^{d***}
BCS	3.79±0.11^{a**}	3.49±0.07	3.17±0.15^{b}	/

Values with different superscript ^{ab} were significantly different; $^{*}P<0.05$; $^{**}P<0.01$; $^{***}P<0.001$; d.a.p.: days *ante partum*; d.p.p.: days *post partum*; AI-artificial insemination.

Table 2. The blood parameters (mean ± SEM) significantly changed in cows after the repeat breeding.

Periods	1 (≤150 d.p.p)	2 (150- 200 d.p.p.)	3 (≥ 200 d.p.p.)
Blood parameters			
Bilirubin µmol/l	3.40±0.42	3.46±0.27^{a}	3.22±0.10^{b}
LDH U/l	1084.18±145.10^{a}	1106.86±130.51	1236.30±180.44^{b}
AP U/l	34.25±12.91^{a}	31.83±6.46^{a}	52.25±20.80^{b}

^{a,b} Values with different superscript were significantly different, $P<0.05$.

Discussion and conclusion

Seven blood parameters and BCS had markedly changed over the transition period and three after the repeated breeding. The suboptimal glucose, total cholesterol and albumin level and significant decrease in BCS are indicative of the energy deficiency that could have been consequences of suboptimal transition. The significantly lower magnesium and calcium concentrations might be indicative of lower food intake, with inappropriate Ca:P ratio over the transition that might cause the sublinical post-partal diseases. Variability within the physiology range of total bilirubin, LDH and AP within the repeated breeding may be indicative of reproductive disorders and the potential causes.

References

Herdt, T.H., 2000. Variability characteristics and test selection in herd-level nutritional and metabolic profile testing. Vet. Clin. North. Am. Food. Anim. Pract. 16: 387-403.

Kida, K., 2002. Use of every ten-day criteria for metabolic profile test after calving and dry off in dairy herds. J. Vet. Med. Sci. 64: 1003-1010.

NRC, 2001. Nutrient requirements of dairy cattle. 7th revised ed. National Academic Press, Washington, D.C. pp. 408.

Long term *in vitro* quantitative evaluation of spermatozoid concentrations from Iranian Lori rams: a new model for aging investigation

S. Mohammadzadeh[1], A. Mohammadzadeh[2], S.M. Moosavi[1], A. Chegeni[3] and A. Kiani[1]
[1]Animal Science Group, Lorestan University, Khorramabad, Iran; [2]Medical college of Gilan University, Rasht, Iran; [3]Lorestan Research Center of Jehad-Keshavarzi, Khormababad, Iran; mohammadzade@lu.ac.ir

Introduction

There are different ways to preserve spermatozoa; keeping them in liquid, under freezing conditions and inside the animal's body (*in situ*). Preserving sperm for a long time *in situ* has several advantages. It does not need microbial evaluation or a defreezing stage therefore avoiding the associated damages. Possibilities of preserving mouse sperm (Fuller and Whittinham, 1998; Doronin *et al.*, 2002) as well as human (Zavos, 1980), ram (Maxwell and Salamon, 1993) and rabbit sperm (Gulyas, 1966, Roca *et al.*, 2000) have been previously reported. However, a decrease in the number of viable spermatozoid at 4 °C during 3-4 d has been reported. The aim of this study was to determine the changes in the number of spermatozoid extracted from the ram's epididymis preserved at 4 °C in a 3% citrate solution.

Material and method

Four adult young rams (Iranian sheep, Lori), 10 months of age, were selected randomly. Rams were slaughtered and their testes were cut off from the spermatic cord. The deferent ductus were separated after dissecting the fat tissues and vessels. Ductus contents were extracted then placed in 1 ml plastic vials in which 3% citrate solution was added. Stalk samples (5.7×10^6 million/ml) were prepared using a hoemocytometer. Samples were kept at 4 °C throughout the experiment and the number of spermatozoa was determined every 10 days for 5 months. Data were statistically analysed using the General Linear procedure (GLM) SAS® v. 8.2.

Results

Figure 1 shows the number of spermatozoid (million/ml) during a 5 month preserving period. No significant changes ($P<0.05$) were observed in the number of spermatozoids up to 35 d of the preserving period (5.5×10^6). However, from day 35 until 105 d the number of spermatozoid slowly decreased ($P<0.05$) reaching 4.8×10^6 up to105 d. After then, the number of sperms decreased more drastically ($P<0.05$) and reached 3×10^6.

Discussion and conclusion

Related to storage duration, there are two explanations for the aging process of spermatozoid: first, structural changes in the cell such as abnormal morphological spermatozoid and cell membrane integrity, second, significant elimination of spermatozoid number. Related to the second explanation, Doronin *et al.* (2004) reported that viability of normal sperm stored at 6 °C did not change for a long period (i.e.18 d) of the experiment (*in situ*). If we accept that a significant elimination in the number of cells is related to aging, the result of this study shows that preserving ram sperm *in vitro* at 4 °C in a 3% citrate sodium solution for a period of 100 days is feasible. Spermatozoids could be alive without any significant decrease in number. The decrease in the number of spermatozoid after one month (5.5×10^6) is fairly slow and this condition continues up to 3 months of the preserving period 5×10^6. Therefore it is concluded that extracted ram spermatozoid from the epididymis may be preserved unfrozen at 4 °C for at least three months. However, the fecundity of these preserved sperm cells still has to be confirmed.

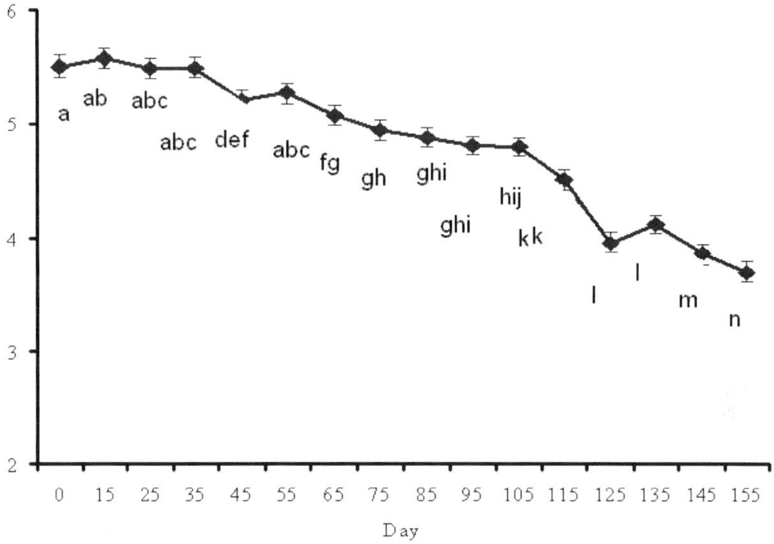

Figure 1. Change of spermatozoa number (million/ml) from the ductus deferens of rams stored in a 3% citrate solution at 4 °C (mean + SEM).

References

Doronin, U.K., S. Mohammadzadeh and G.U. Makcydov, 2002. Lifespan curves of activated oocyte *Misgurnus fossilis* L. on different temperatures. Biological mechanism of aging. The 5th International Symposium, Ukraine, Kharkev, pp: 35-36.

Fuller, S.J. and D.G. Whittinham, 1998. Effect of cooling mouse spermatozoa at 4 °C on fertilization and embryonic development. J. Reprod. Fert. 108: 139-145.

Gulyas, B.J., 1966. *In vitro* maintenance of rabbit sperm at room temperature. Nature 211: 987.

Maxwell, W.M.C. and S. Salamon, 1993. Liquid storage of ram semen. Reprod. Fert. Dev. 5: 613-638.

Roca, J., S. Martinez, J.M. Vazquez, X. Lucas, I. Parrilla and E.A. Martinez, 2000. Viability and fertility on rabbit spermatozoa diluted in tris–buffer extenders and stored at 15 °C. Anim. Reprod. Sci. 64: 103-112.

Zavos, P.M., J.C. Goodpasture, L.J.D. Zanaveld and M.R. Cohen, 1980. Motility and enzyme activity of human spermatozoa stored for 24 h at +5 and –196 °C. Fert. Steril. 34: 607.

Effects of different fat types on concentration of oestradiol and progesterone in the blood of ewes

A. Moharrery
Animal Science Department, Agricultural College, Shahrekord University, Iran; alimoh@mailcity.com

Introduction

Few studies have investigated the feasibility of supplementing fat to meat-type ruminants during lactation. Increased incidence of metabolic disorders and poor reproductive performance can occur from inadequate energy intake during early lactation (Grummer and Carroll, 1991). To deal with this problem, one approach would be to substitute fat as an energy dense nutrient for grain. Therefore inclusion of high fat in the diet may benefit oocytes not only during growth and development in vivo, but also during the short period of oocyte maturation (Fouladi-Nashta *et al.*, 2007). Many different types of supplemental fat have been fed to ruminants under experimental conditions. Many of the whole oil seeds and unprocessed vegetable oils contain a large proportion of long chain, polyunsaturated fatty acids (PUFA) such as linoleic acid (C18:2). Tallow can vary greatly in the ratio of saturated to unsaturated fatty acids and in the proportion of linoleic acid (range of 2 to 8.5%). The aim of the study was to investigate how dietary fat affects the concentration of progesterone and oestradiol (as important reproductive hormones) in the blood of ewes.

Material and methods

Ten nursing Lory-Bakhtiary ewes (Iranian native breed) with mean body weight 49.4 ± 4.0 Kg and parturient in same day were used in this experiment starting four weeks after their parturition. The ewes were individually housed in metabolic cages and their lambs had access to the corresponding ewe freely for suckling. During the experiment, the ewes were fed a constant basal diet as control (C) and 7% additional fat including sunflower oil (O), saponified sunflower oil (SO), tallow (T) and saponified tallow (ST). Saponification used alkaline calcium. The fat content diets contained 2.60 Mcal ME/Kg DM and 14.8% crude protein (DM basis) and were offered to the ewes as a total mixed ration (TMR). All diets were iso-nitrogenous and ewes had *ad libitum* access to the diets. Diets were fed in a balanced changeover arrangement and were offered to two ewes in each treatment over five fourteen-day periods. Milk yield was determined by hand milking on 3 consecutive d two times per day and during these days lambs had no access to their mothers. Blood samples (*Vena jugularis externa)* were collected at the final days of each period. Oestradiol and progesterone were measured in the blood serum by the enzyme-linked immunosorbent assay (ELISA) method. Data on oestradiol and progesterone were analysed as a 5×5 Latin square using the MIXED procedure of SAS (2003). The statistical model included ewe, period, treatment and residual error. Fixed effects included period and treatment. Ewe was the random effect. Overall differences between treatment means were declared significant at $P<0.05$. The correlation among oestradiol and progesterone was determined and correlation coefficients were tested using a *t*-test (SAS®, 2003).

Results and discussion

The mean values for ewes maintained on treatments of C, O, SO, T, and ST for oestradiol and progesterone are shown in Table 1. Feed intake was significantly lower when the ewes were fed the SO diet ($P<0.05$) compared to the other diets. Milk yield (391 g/d at mean; $P=0.34$) were similar among treatments. A significant difference was found between fat supplemented diets and control for progesterone concentration in blood serum ($P<0.05$) but, oestradiol concentration was not affected when using fat supplemented diets ($P>0.05$).

Not only were concentrations of cholesterol increased in the blood (Table 1), but also follicular fluid concentrations increase during fat supplementation to the ration of ewes (Thatcher *et al.*, 1994). The fat supplemented groups had significantly higher production of progesterone (P=0.09) compared with the control group. The implication from these studies is that additional circulating cholesterol stimulated by feeding supplemental fat increases the synthesis of progesterone by follicular and luteal cells but has no effect on secretion of oestradiol. No significant correlations were observed between levels of progesterone and oestradiol in the blood serum of lactating ewes (r=0.25; P=0.13).

Table 1. Oestradiol and progesterone level in the blood serum of ewes on different diets.

	Diet[1]					SE	Ration effect
	C	O	SO	T	ST		*P*-value
DM intake, kg/d	1.75 ^{a}	1.86 ^{a}	1.48^{b}	1.76 ^{a}	1.90 ^{a}	0.12	0.04
Milk, g/d	391	354	420	376	416	20.3	0.34
Cholesterol, mg/dl	192 ^{b}	216 ^{a}	201 ^{ab}	198 ^{ab}	195 ^{ab}	18.3	0.12
Oestradiol, pg/ml	4.03	4.03	3.99	4.02	3.95	0.24	0.95
Progesterone, ng/ml	1.48 ^{b}	2.85 ^{a}	2.75 ^{a}	2.39 ^{ab}	2.37 ^{ab}	1.00	0.09

[1] C = control ration without additional fat; O = basal diet plus sunflower oil; SO = basal diet plus saponified sunflower; T = basal diet plus tallow; ST = basal diet plus saponified tallow.
[a,b] Means within rows with same superscript letters are not significantly different (P>0.05).

Conclusion

This study shows that the effect of dietary sunflower oil (both saponified and unsaponified) was greater than tallow for changing the progesterone level in the blood. In addition, dietary fat did not change the serum oestradiol concentration during the lactating period. Whether this stimulation in progesterone level for oil supplemented diets is complementary or antagonistic to reproductive events after the restoration of estrous cycles in the *post partum* period warrants further detailed investigation.

References

Fouladi-Nashta, A.A., C.G. Gutierrez, J.G. Gong, P.C. Garnsworthy and R. Webb, 2007. Impact of dietary fatty acids on oocyte quality and development in lactating dairy cows. Biol. Reprod. 77: 9-12.

Grummer, R.R. and D.J. Carroll, 1991. Effects of dietary fat on metabolic disorders and reproductive performance of dairy cattle. J. Anim. Sci. 69: 3838.

Thatcher, W.W., C.R. Staples, G. Danet-Desnoyers, B. Oldick and E.-P. Schmitt, 1994. Embryo health and mortality in sheep and cattle. J. Anim. Sci. 72 (Suppl. 3): 16.

Expression of P450-aromatase in the corpus luteum of small ruminants

J.A. Mondragón[1], C. Miranda[1], R. Ocadiz-Delgado[2], J. García-Mena[2], P. Gariglio[2] and M.C. Romano[1]
[1]Departamento de Fisiología, Biofísica y Neurociencias; [2]Departamento de Genética y Biología Molecular, Cinvestav, Apdo Postal 14-740, 07310, México, D.F., México; mromano@fisio.cinvestav.mx

Introduction

In the sheep, the placenta rapidly becomes a steroidogenic organ that replaces the corpus luteum (CL) for the synthesis of progesterone and also estrogens. On the contrary, the goat placenta does not produce enough progesterone to maintain gestation and therefore pregnancy depends on the presence of CL after mid-gestation, the ovariectomy interrupts gestation (Al Gubory *et al.*, 1999). The immunolocalization of aromatase has been shown in the ovaries of non-pregnant Japanese Shiba goats (Weng *et al.*, 2005).
The expression of P450-Aro mRNA has not been investigated in the goat CL, or in the ovine. Although the expression of aromatase had been shown in the sheep ovary by RT-PCR *in vitro* studies, the distribution of the expression and the involved cell populations has not yet been studied. On the basis of this information, the present research was designed to characterise further the expression of P450-Aro mRNA in small ruminant CL.

Material and methods

Animals and tissue collection: Ovaries from five non pregnant Criollo goats and six Pelibuey sheep were obtained at a local slaughterhouse. The CL were collected and cut into pieces, washed in sterile PBS and transferred to 4% paraformaldehyde (pH 7.4). Some specimens were fixed in 10% formaldehyde for standard histology. After fixation the tissues were dehydrated through a series of ethanol solutions (30-100%) followed by ethanol:xylol (50:50) and 100% xylol treatment, and finally tissues were paraffin-embedded by standard procedures. Sections of each specimen were cut to 5 micrometer thickness, mounted on silane coated slides and processed for *in situ* RT-PCR. Three goat corpora lutea per each period of pregnancy considered here were immediately frozen in liquid nitrogen for *in vitro* RT-PCR studies. Foetal age estimation was done measuring crown-rump length as reported by Sivachelvan *et al.* (1996).
In vitro expression of P450-Aro: The tissues were processed according to the procedure described in Mondragón *et al.* (2007). The presence of mRNA for aromatase and actin were examined by reverse transcription-polymerase chain reaction (RT-PCR). Total RNA was extracted using the Trizol reagent (Invitrogen, USA). Specific primers for goat aromatase gene (forward: 5'- GGC ATC ATA TTT AAC AAT CCA GCA-3'; reverse: 5'-CAG ACA TGG-TGT CTG GCG CTG CGA TCA- 3') and for actin gene (expression control, forward: 5'- CCA AGG CCA ACC GCG AGA AGA TGA C-3′; reverse: 5′- AGG GTA CAT GGT GGT GCC GCC AGA C- 3') were used. The PCR products were separated on 2% agarose gels and visualised by ethidium bromide staining.
In situ RT-PCR: The tissues were processed according to the procedure developed by Nuovo (1996) and described in Mondragón *et al.* (2007). The reaction was sealed using AmpliCover discs and clips (Perkin Elmer, USA). In the negative controls, the reverse transcription reaction was omitted. For detection of *in situ* PCR products an indirect immunolabelling method using a primary Anti-Digoxigenin antibody (Fab fragments; Roche, USA) conjugated to alkaline phosphatase was chosen to detect the PCR product. All photomicrographs were obtained using a Hyper HAD Color Video Camera (Model SSC-DC30; Sony Corporation, Japan).

Results

In vitro RT-PCR: The expression of P450-Aro mRNA was determined by *in vitro* RT-PCR in CL from pregnant goat ovaries at different stages of pregnancy. The representative gel (not shown) demonstrated the P450-Aro mRNA expression in the first and third months of gestation (38.6 and 41% respectively, related to actin expression).
In situ RT-PCR: In Figure 1 a strong *in situ* RT-PCR signal was observed in sheep CL. Many positive cells could be observed in a panoramic view of the microphotography (100×). Higher magnification of the same zone showed that there were at least two different sizes of luteal positive cells, both of them expressed the Aro mRNA (200× and 400×). The abundant P450-Aro mRNA signal in the cytoplasm of luteal cells (arrows), having a negative nuclei is evident at 400× (arrow heads). Similar results were found in the goat corpora lutea (not shown).

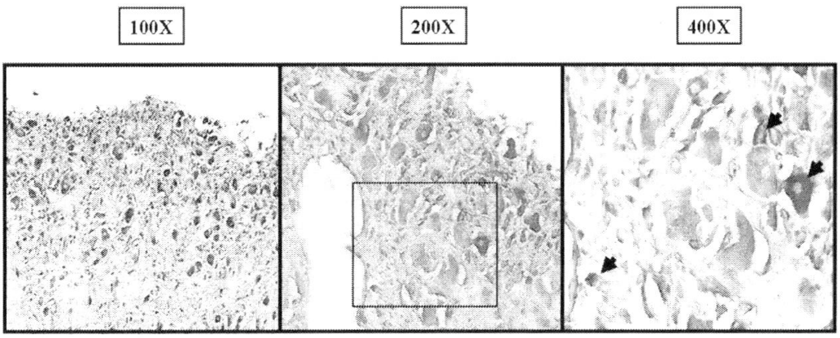

Figure 1. In situ *RT-PCR showing the expression of P450-aromatase in sheep CL.*

Discussion and conclusion

The present results show for the first time the P450-Aro mRNA expression in the CL of the pregnant goat. The expression of P450-Aro mRNA in CL of pregnant goats found in the present paper strongly suggests that this tissue has the capacity to synthesise estrogens throughout pregnancy as has been demonstrated for other mammals. The presence of positive *in situ* RT-PCR signals for Aro mRNA in the ovine CL strongly suggests that the goat and ovine luteal tissue has the capacity to produce estrogens during pregnancy. Additional experiments are going on to detect aromatase by immunocytochemistry and the functionality of the sheep and goat CL.

Acknowledgement

Partially supported by CONACyT grant 60774.

References

Al Gubory, K.H., A. Solari and B. Mirman, 1999. Effects of luteectomy on the maintenance of pregnancy, circulating progesterone concentrations and lambing performance in sheep. Reprod. Fertil. Dev. 11: 317-322.

Mondragón, J.A., R. Ocadiz-Delgado, C. Miranda, J. Valencia, A.M. Rosales, P. Gariglio and M.C. Romano, 2007. Expression of P450-aromatase in goat placenta throughout pregnancy. Theriogenology 68: 646-653.

Nuovo, G., 1996. The foundation of successful *in situ* PCR. Front. Biosci. 1: 4-15.

Sivachelvan, M.N., M. Ghali Ali and G.A. Chibuzo, 1996. Foetal age estimation in sheep and goats. Small Ruminant Res. 19: 69-76.

Weng, Q., M.S. Medan, L. Ren, G. Watanabe, T. Tsubota and K. Taya, 2005. Immunolocalization of steroidogenic enzymes in the corpus luteum and placenta of the Japanese Shiba goat. J. Reprod. Dev. 51(2): 247-252.

Correlation between quantitative three dimensional Doppler parameters and real blood flow within the utero-placental unit: evaluation in a pregnant sheep experimental model

O. Morel[1,2,3], F. Pachy[1,3], V. Tsatsaris[3,4,5], M. Bonneau[6], P. Laigre[1] and P. Chavatte-Palmer[1,3]
[1]INRA, UMR 1198 Developmental biology and Reproduction, Jouy-en-Josas, France; [2]Lariboisière Hospital, APHP, Department of Obstetric and Gynaecology, Paris, France; [3]PremUp Foundation, Paris, France; [4]Port Royal Maternity, APHP, Paris, France; [5]INSERM U767, Paris Descartes University, Paris, France; [6]INRA, CRII, Jouy-en-Josas, France; olivier.morel@lrb.aphp.fr; olivier.morel17@wanadoo.fr

Introduction

Intra-uterine growth retardation (IUGR) in animals and humans is associated with increased neonatal morbidity and mortality and long term consequences in adults such as obesity and associated metabolic disorders in humans and decreased meat quality in animals (Sibai *et al.*, 2005, Wu *et al.*, 2006). IUGR may be due to maternal undernutrition but is often related to insufficient placental transfers despite normal maternal nutrition. Blood flow to and from the placenta is one of the essential elements of transplacental transfers to the fœtus. So far, however, this parameter has remained impossible to evaluate without invasive surgery.

Very recent progress in three-dimensional (3D) ultrasound techniques have enabled the quantification of the power Doppler signal within an organ of interest. This method has been proven to be highly reproducible when applied on the utero-placental unit, and might be of critical interest for the understanding and screening of pre-eclampsia and IUGR in humans as well as for studies on placental function and IUGR in animals (Mercé *et al.*, 2004, Zalud *et al.*, 2007). Indeed, in humans, a significant decrease in the quantified 3D Doppler parameters of the placenta in cases of pre-eclampsia or IUGR was observed in preliminary reports (personal communication). The relationship between measured Doppler parameters and real flow has been confirmed *in vitro* using blood-mimicking fluids (Raine-Fenning *et al.*, 2008). *In vivo* evaluation remains necessary, however, to correlate Doppler parameters with true blood perfusion.

The aim of this study was to evaluate the correlation between 3D Doppler parameters and blood flow within the utero-placental unit in pregnant sheep.

Material and methods

Nine pregnant sheep of the Pre-Alpes breed, carrying singletons, were used at 100 to 140 days gestation. The pregnant horn was exposed under general anaesthesia by laparotomy. A flow quantitative sensor (Transonic, Emka®) allowing real time assessment of the blood flow and a controllable vascular occlusion system were then placed around the common uterine artery, while all the other uterine arterial supplies were ligated. Several occlusion levels were applied, and the blood flow perfusing the horn was permanently measured by the Transonic sensor. 3D Doppler acquisitions of placentomes were concomitantly realised with the ultrasound probe placed directly in contact with the larger curvature of the pregnant horn (VOLUSON E8, 3.5MHz 3D probe, General Electric Medical System, Zipf, Austria). Acquisition parameters were standardised. Each placentome was rebuilt 3-dimensionally using the VOCAL™ software (GE Medical systems, Zipf, Austria): the limits were outlined after rotating the volume each 15°. Three quantitative 3D Doppler parameters were automatically generated for each placentome, i.e. vascularisation index (VI), defined as the percentage of power Doppler information; flow index (FI), defined as the mean intensity of the power Doppler information; vascularisation flow index (VFI), defined as

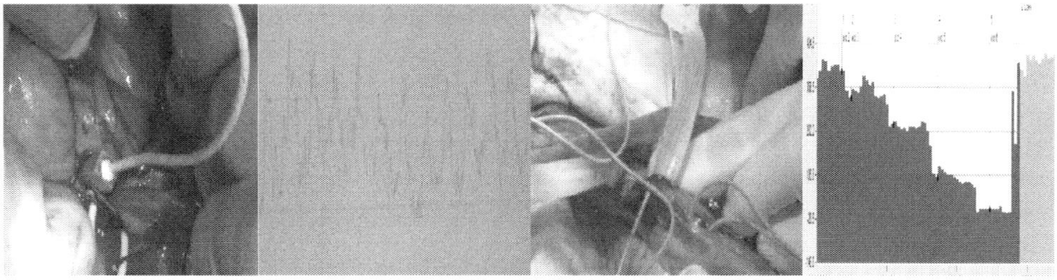

Figure 1. Surgical view of the systems placed around the common uterine artery. (Left panel) flow quantitative sensor (Transonic, Emka®); (mid-left & right panel) real time assessment of the blood flow; (mid-right panel) controllable vascular occlusion.

a combination of the latter two. Acquisitions with artefacts due to foetal sheep movements were excluded from the analysis. The correlation between real vascular flow within the uterine artery as measured with the flow sensor and 3D Doppler parameters was evaluated. Intra and inter observer repeatability were evaluated for each parameter.

Results

Thirty-five acquisitions were analysed. All the 3D Doppler parameters were significantly correlated with the real vascular flow measured concomitantly by the flow sensor. A higher correlation was observed for VI and VFI ($r = +0.86$ [0.74-0.93] and +0.82 [0.67-0.90] respectively; $P<0.0001$) than for FI ($r = +0.64$ [0.39-0.80]; $P<0.0001$). Repeatability was high with intra-class correlation coefficients of at least O.84 for intra-observer comparison and 0.65 for inter observer.

Discussion and conclusion

To our knowledge, this is the first experimental study showing a significant correlation between real blood flux and quantitative three-dimensional Doppler parameters measured within the utero-placental unit. These results confirm the potential great interest of the use of 3D Doppler ultrasound for the assessment of placental vascular insufficiencies in clinical cases and in a research set-up, to evaluate placental blood flow in sheep models of intra-uterine growth retardation. They now have to be evaluated in experimental and clinical evaluations of chronicle placental insufficiencies.

References

Merce, L.T., M.J. Barco and S. Bau, 2004. Reproducibility of the study of placental vascularization by three-dimensional power Doppler. J. Perinat. Med. 32: 228-233.

Raine-Fenning, N.J., N.M. Nordin, K.V. Ramnarine, B.K. Campbell, J.S. Clewes, A. Perkins and I.R. Johnson, 2008. Determining the relationship between three-dimensional power Doppler data and true blood flow characteristics: an *in-vitro* flow phantom experiment Ultrasound. Obstet. Gynecol. 32: 540-550.

Sibai, B., G. Dekker and M. Kupferminc, 2005. Pre-eclampsia. Lancet 365: 785-799.

Wu, G., F.W. Bazer, J.M. Wallace and T.E. Spencer, 2006. Board-invited review: Intra uterine growth retardation: Implications for the animal sciences. J. Anim. Sci. 84: 2316-2337.

Zalud, I. and S. Shaha, 2007. Evaluation of the utero-placental circulation by three-dimensional Doppler ultrasound in the second trimester of normal pregnancy. J. Matern. Fetal Neonatal Med. 20: 299-305.

Productive and reproductive performance of grazing dual purpose cows with or without access to *Leucaena leucocephala* in the tropics

I. Peniche-González, C. Aguilar-Pérez, J. Ku-Vera, A. Ayala-Burgos and Z. González-López
Facultad de Medicina Veterinaria y Zootecnia, Universidad Autónoma de Yucatán, México; caperez@uady.mx

Introduction

Dual purpose (DP) systems contribute to most of the milk and beef produced in the tropics. Low grass quality and availability during the dry season are the main constraints for production. Milk yield and reproductive parameters in DP cows have been successfully improved using cereal-based supplements (Aguilar-Pérez *et al.*, 2009). *Leucaena leucocephala*, a tropical legume widely available in Central America and the Caribbean, has a high nutritional value for ruminants (Shelton and Dalzell, 2007) and its use could help to reduce the grain dependency for supplementing cows. This work was carried out to assess the effect of browsing plots of *Leucaena* (L), in addition to the grazing of *Cynodon nlemfuensis* (Stargrass, SG), on milk yield and composition, bodyweight (BW), body condition score (BCS) and reproductive performance of crossbred DP cows. It was hypothesised that L could substitute 50% of the concentrate without reduction of cow's performance.

Material and methods

The experiment was conducted in Yucatan, Mexico (climate AW_0, tropical sub humid with summer rains) from March to December 2008. Twenty-four multiparous *B. Taurus* × *B. indicus* cows were used in a 2×2 factorial trial from calving to 98 d *post partum* (pp). Treatment factors were diet: with L (L, n=12) vs. without L (NL, n=12) and calving season: dry (DSC, n=11) vs. rainy (RSC, n=13). BW and BCS (scale 1-9) were registered at calving and every 14 d afterwards. BW at calving was 491±71 kg for L and 503±62 kg for NL. BCS at calving was 5.3±1.5 and 4.9±1.0 for L and NL, respectively. Cows were milked mechanically at 6:00 and 15:00 h using the calf to stimulate; 'milk let-down'. Milk yield was measured every 14 d assisted by oxytocin (40 IU/cow) and without calf presence. A cereal-based concentrate (CP 17.7±0.4%, ME 13.0±0.1 MJ/kg DM) was offered during milking (0.9% vs. 0.5% of calving BW for NL and L, respectively). L group accessed to L plots (CP 24.4±2.0%, ME 11.1±0.7 MJ/kg DM) from 8:00 to 12:00 h, whereas NL grazed exclusively SG paddocks (CP 8.1±1.4%, ME 8.4±0.7 MJ/kg DM). Stocking rate in L was 2 and 2.8 cows/ha during the dry and rainy seasons, respectively. After afternoon milking all cows grazed SG as one herd from 17:00 to 5:00 h. Stocking rate in SG was kept at 3.2 cows/ha. Paddocks of SG and L were irrigated during the dry season. Grass availability was measured every 15 d using a 0.25 m^2 metal square (Cox, 1980). A fertile bull was permanently kept with cows. Ovaries of each cow were examined by transrectal ultrasonography every week, beginning on d 14 pp. The number and diameter of follicles were recorded and classified as small (3-5 mm), medium (6-9 mm) or large (>9 mm). Pregnancy was diagnosed by ultrasonography while ovaries were checked. Milk yield was analysed by ANOVA for repeated measurements using the MIXED procedure of SAS® (SAS Institute Inc., Cary, NC, USA); milk yield from d 8 was used as a covariance; milk composition, BW, BCS and follicle numbers at specific time points were analysed by GLM, which included fixed effects of diet, calving season and their interactions. Effects of diet and calving season on pregnancy rates at 98 d were analysed by Fisher exact tests. Effect of diet on days open, was analysed by Kaplan-Meier survival analysis using the LIFETEST procedure of SAS® (SAS Institute Inc., Cary, NC, USA) with a Log-Rank test to indicate significance.

Results

Milk yield and composition were not affected by diet. Days open were not different (P=0.944) between L (80.5±5.2 d) and NL (87.1±5.3 d). Losses of BW and BCS were not different between diets and did not affect reproductive parameters (Table 1). Cows calved in the rainy season lost less BCS and had higher milk fat and protein concentrations probably due to higher forage availability. There was a trend (P=0.08) for a better pregnancy rate using L in the dry season probably due to its greater intake at this time of the year as a consequence of smaller grass availability.

Table 1. Productive and reproductive parameters of grazing dual purpose cows calved during the dry or rainy season with or without access to plots of Leucaena leucocephala.

	DSC		RSC		SEM	*P*-value		
	NL (n=5)	L (n=6)	NL (n=7)	L (n=6)		Diet	Season	D×S
BW change, kg	-34.2	-22.6	-26.3	-44	5.32	0.78	0.54	0.19
BCS change	-2	-2.3	-0.9	-0.7	0.22	0.95	<0.01	0.62
Milk yield, kg/d	11.8	11.9	12.5	12.1	0.32	0.84	0.61	0.79
Milk fat,%	3	3	3.8	3.8	0.07	0.88	<0.01	0.82
Milk protein,%	2.8	2.7	3.1	3.1	0.05	0.53	<0.01	0.82
Milk lactose,%	5.9	6.3	4.7	4.6	0.09	0.33	<0.01	0.18
Pregnancy at 98 d, number and (%[1])	2/5 (20)	5/6 (50)	3/7 (30)	0/6 (0)	-	0.66	0.99	0.08
No. follicles								
3-5 mm	11.3	9.3	3.1	1.8	0.62	0.12	<0.01	0.69
6-9 mm	1.5	2.1	1.5	1.4	0.30	0.35	0.28	0.30
>9 mm	0.7	0.9	1.0	1.2	0.15	0.48	0.31	0.96

[1] Regarding total (10) pregnant cows; NL = without access to *Leucaena*, L = with access to *Leucaena*; Means were declared significant at P<0.05.

Conclusion

A 50% reduction in the amount of cereal-based concentrates in the ration is possible by allowing cows to browse plots of *Leucaena leucocephala,* without negative effects on milk and reproductive parameters in DP systems based on irrigated pastures in the tropics. The results have to be further warranted due to the limited number of cows employed.

Acknowledgement

The authors thank PROMEP, Ministry of Education of Mexico, for supporting the study.

References

Aguilar-Pérez, C., J. Ku-Vera., F. Centurión-Castro and P.C. Garnsworthy, 2009. Energy balance, milk production and reproduction in grazing crossbred cows in the tropics with and without cereal supplementation. Livest. Sci. (in press).

Cox, G.W., 1980. Laboratory manual of general ecology. San Diego State University, San Diego, USA, 237 pp.

Shelton, M. and S. Dalzell, 2007. Production, economic and environmental benefits of leucaena pasture. Trop. Grass. 41: 174-190.

Dietary protein during gestation affects fetal growth and circulating indicators of placental function

V.E.A. Perry, G.C. Micke and T.M. Sullivan
School of Veterinary Science, University of Queensland, Goondiwindi, QLD, Australia 4390; v.perry@uq.edu.au

Introduction

There has been a major experimental focus on determining the impact of low protein diets on experimental laboratory animals but few on large ruminants of agricultural importance. Protein is the most deficient nutrient in the Australian Rangelands. Dietary protein alters placental growth which in turn influences fetal growth. The aim of this experiment was to extrapolate from the endocrine and metabolic profiles involved in this complex system, the extent to which protein influenced the feto placental unit, and by what mechanism this was achieved. Throughout gestation, maternal circulating levels of IGF-I, IGF-II, leptin, bovine placental lactogen (bPL), bovine pregnancy associated glycoprotein (bPAG), estrone sulphate (ES) and progesterone (P4) were sequentially determined and related to measures of fetal growth, both during pregnancy (via ultrasonagraphy), and at birth.

Material and methods

One-hundred twenty *Bos indicus* cross heifers were inseminated with semen from the same bull on a single day and allotted to four treatment groups. These were fed sorghum and cotton seed meal diets that were high (13.4 CP/kgDM) or low (4.7 CP/kgDM) protein and contained 7.7 MJME/kgDM during the first and second trimesters. During the third trimester all heifers received a ration of 11.2 CP/kgDM and 7.6 MJME/kg. A total of 71 heifers calved (Low/Low n=19, Low/High n=17, High/Low n=18, High/High n=17). Foetuses were monitored throughout gestation using transrectal real-time ultrasonagraphy on a total of eight occasions between days 39 and 235 of gestation yielding 568 measurement events (Micke *et al.* 2009). Blood was collected 14 d prior to A.I. and monthly during gestation. ES, leptin, IGF-I, IGF- II and P4 concentrations were assayed at days of gestation (gd) 82, 124, 236, 271 and calving, bPAG was assayed at gd 28, 82,179, 271, bPL at gd 124, 179, 236, 271 and calving, as described by Sullivan *et al.* (2009). At calving, measures of the calf were taken prior to suckling. Measures recorded were calf birth weight (BW), crown rump length (CRL), and abdominal circumference (AC) at the level of the umbilical cord. Cranial measures recorded were biparietal diameter (BPD) and crown nose length (CNL). Limb measurements were taken on both fore- and hind-limbs (Micke *et al.* 2009). Statistical analyses were carried out using Stata SE Version 9.2 (Stata Corporation, College Station, TX). GLM were developed using nutritional treatment and calf gender as fixed effects and placental and metabolic hormones as random effects to test the effect of concentrations of maternal hormones IGF-I, IGF-II, leptin, P4, ES, bPL and bPAG, nutritional treatment and calf gender on foetal measures at days 39, 95, 123, 179, 234 of gestation and at calving. Correlations were calculated between variables and where significant (P<0.05), the predictor that was least correlated with the outcome was excluded from the model.

Results

Both circulating maternal hormones and metabolic hormones were influenced by dietary treatment (P<0.05): in the first trimester low protein increased ES and bPAG but reduced bPL and P4 whilst in the second trimester low protein increased bPL in HL treatment (Figure 1A) and P4; low protein decreased IGF-I on gd 28, 82 and 179, IGF-II on gd 82, gd 271(Figure 1B) and leptin at calving. Foetal growth was influenced by dietary treatment from gd 39 (P<0.05) until birth. Metabolic

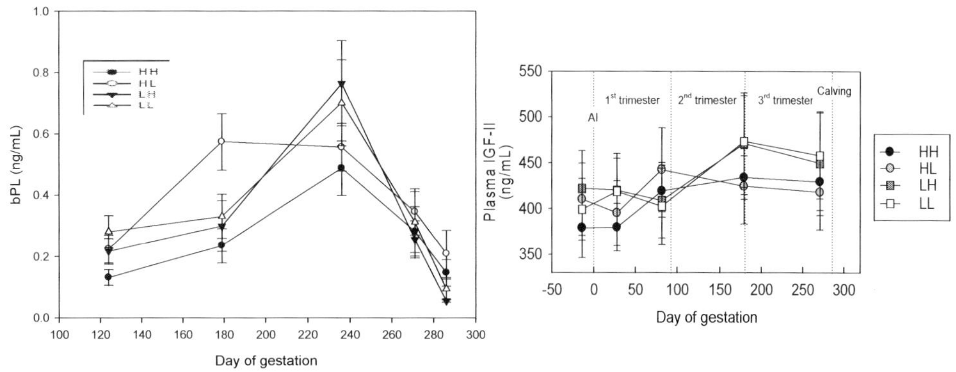

Figure 1. Plasma measurements per treatment groups (HH, HL, LH, LL) during gestation (SEM=error bars). (A): bovine placental lactogen concentrations (bPL), (B): IGF –II.

hormones, leptin, IGF-I and IGF-II, were not associated with measures of circulating ES, P4, bPAG or bPL.

IGF-II was significantly associated (P<0.05) with measures of foetal growth (CRL, BPD, and limb measures) throughout the first trimester but ceased to be associated from gd 123. bPL was significantly associated with foetal measures of growth at gd 123 (P=0.06), 234 (P=0.07) and 271 (P=0.01). ES and leptin at gd 271 were significantly associated with foetal measures at birth (P =0.004 and 0.06 respectively).

Dietary protein influenced measures of foetal growth in first trimester; by gd 39 low protein was associated (P=0.05) with reduced CRL and reduced umbilical cord (UC) measures at gd 95. At gd 123 low protein was associated (P=0.02) with reduced CNL and UC measures. At birth low protein in the second trimester was significantly associated with reduced birth weight (P=0.005).

Discussion and conclusion

Maternal dietary protein significantly affected foetal growth and circulating indicators of placental function from gd 39 until birth. At birth, foetal weight was reduced by low protein in the second trimester. Maternal IGF II may be a useful indicator of healthy foetal growth in the first trimester as it was consistently associated with many measures of foetal growth. Estimates of foetal size using transrectal ultrasound were less accurate and less numerous in the second trimester which may have yielded less favourable associations with measures of placental function. The only measure associated with foetal growth at gd 123, gd 234 and gd 271 was bPL, with ES and leptin only being associated with calf measures at birth when taken at gd 271. Leptin, IGF I and II were not associated with the measures of placental function as anticipated.

Acknowledgement

We are indebted to AACo, MLA, WACICF for funding and J.A. Green, G.B. Martin, C.R. Wallace, K.L. Gatford and J.A. Owens for the completion of hormonal assays.

References

Micke, G.C., T.M. Sullivan, R.J. Soares Magalhaes, P.J. Rolls, S.T. Norman and V.E.A. Perry. 2009. Heifer nutrition during early- and mid-pregnancy alters fetal growth trajectory and birth weight. Anim. Reprod. Sci., in press.

Sullivan, T.M., G.C. Micke, R.S. Magalhaes, G.B. Martin, C.R. Wallace, J.A. Green and V.E.A. Perry, 2009. Dietary protein during gestation affects circulating indicators of placental function and fetal development in heifers. Placenta 30: 348-354.

Blood chemistry modifications and the appearance of pregnancy toxaemia in nutritionally restricted dairy goats

A.A. Ponter[1], B. Laporte[2], J. Promp[2], C. Ficheux[1], J. Tessier[2], J. Perault[1], S. Roussel-Huchette[2], P. Chavatte-Palmer[3] and C. Duvaux-Ponter[2]

[1]UMR INRA-ENVA Biologie du Développement et Reproduction, Ecole Nationale Vétérinaire d'Alfort, 7 avenue du Général-de-Gaulle, 94704 Maisons-Alfort Cedex, France; [2]UMR INRA-AgroParisTech Physiologie de la Nutrition et Alimentation, 16 rue Claude Bernard, 75231 Paris Cedex 05, France; [3]UMR INRA-ENVA 1198 Biologie du Développement et Reproduction, Domaine de Vilvert, 78350 Jouy-en-Josas, France; aponter@vet-alfort.fr

Introduction

Pregnancy toxaemia is a major problem for goat and sheep farmers (Mavrogianni and Brozos, 2008). It is partly due to the very high nutritional requirements in late gestation for multiple foetus carrying animals. The objective of the present experiment was to study the effect of restricted feeding on the metabolism of goats during the last third of gestation, to try to identify the risk factors and to follow them as animals develop pregnancy toxaemia.

Material and methods

A total of 60 Alpine and Saanen dairy goats were synchronised prior to mating so that parturition occurred over a 12 d period. Starting from -8 wk before parturition the goats were allocated to one of two dietary treatments, control (C) or restricted (R) according to breed, age, live weight (LW) and body condition score (BCS). The C group was fed *ad libitum* a TMR which covered requirements (INRA, 1988) for the last third of gestation (C, n=30) and the R group (R, n=30) was given the same TMR but the quantity corresponded to 50% of the amount eaten by the C group between -8 and -5 wk, 60% at -4 wk, 70% at -3 wk and 80% from -2 wk to parturition. The goats of the R group were given free access to straw between meals. LW, BCS and average feed intake were measured and blood samples collected weekly. Plasma was analysed for glucose, non-esterified fatty acids (NEFA), β-hydroxybutyrate (BHB) and urea. Litter weight was measured at birth.
The LW, BCS and metabolite data were analysed using the PROC MIXED procedure of SAS® (SAS Institute Inc., Cary, NC, USA) for repeated measures including a random female effect. The effects of time, treatment and their interaction were tested. A principal component analysis (PCA) was performed on the data obtained from the R group of goats. The variables analysed were the following: LW, BCS, litter weight, plasma glucose, urea, NEFA and BHB and average feed intake.

Results and discussion

Both LW and BCS were affected by dietary treatment (C>R, $P<0.001$). Plasma glucose concentrations decreased in both groups during the experimental period ($P<0.001$), while NEFA, BHB and urea concentrations increased ($P<0.001$). Glucose was higher in C than R while the opposite was observed for NEFA, BHB and urea ($P<0.01$). Four goats in the R group developed pregnancy toxaemia between -4 and -3 wk before parturition. They were isolated and given the appropriate treatment. All goats were studied at -7, -5, -4 and -3 wk before parturition by PCA (Figure 1). The first two principal components (PC 1 and 2) accounted for 34% and 23% of the variation in the dataset. PC 1 was mainly explained by high NEFA (r^2=0.83) and BHB (r^2=0.81) and low plasma glucose (r^2=-0.79) and PC 2 by high LW (r^2=0.85) and intake (r^2=0.84). The goats which showed clinical signs of pregnancy toxaemia (goats: 4026, 6110, 5123 and 5008) had a similar evolutionary pattern.

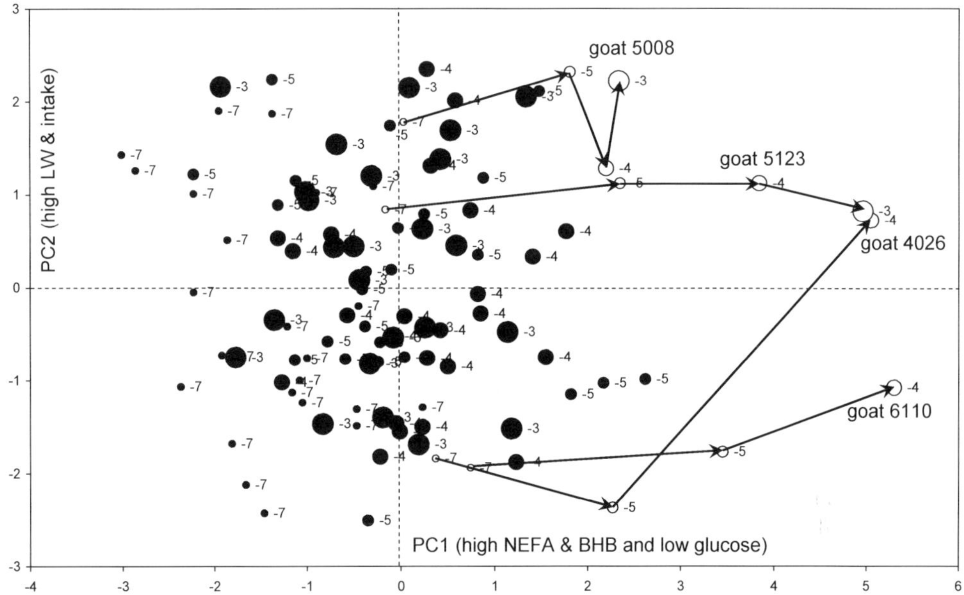

Figure 1. Principal component analysis of parameters measured during the period of dietary restriction at -7, -5, -4 and -3 wk before parturition (n=30). The size of the circle increases as the goat approaches parturition. The dietary restricted goats received the same TMR as controls but the quantity corresponded initially to 50% of the amount eaten by the control group increasing up to 80% from -2 wk to parturition. The four goats which became toxaemic are represented by their number and an empty circle. The progression in time is indicated by arrows. The small numbers represent the week before parturition.

Conclusion

Although the number of goats which became toxaemic was low in the present experiment, it may be possible to predict which animals are at risk from this disease. As early as -7 wk before parturition the future toxaemic animals were positioned on or to the right of the zero of PC 1 on the PCA plot: low glucose and elevated NEFA and BHB and they continued to evolve to the right of the axis as the disease developed.

Acknowledgement

Financial support was provided by DIFAGRI PICOT SA.

References

INRA, 1988. Alimentation des Bovins, Ovins, Caprins. INRA, Paris, 471 pp.

Mavrogianni, V.S. and C. Brozos, 2008. Reflections on the causes and the diagnosis of peri-parturient losses of ewes. Small Rum. Res. 76: 77-82.

Genetic strain x diet interactions on physiological parameters associated with milk production, energy partitioning, and reproduction

J.R. Roche[1], C.R. Burke[1], J.K. Kay[1], C.V.C. Phyn[1], S. Meier[1] and M.C. Lucy[2]
[1]DairyNZ, Hamilton, New Zealand; [2]University of Missouri, Columbia, MO, USA; john.roche@dairynz.co.nz

Introduction

Until recently, breeding objectives focused mainly on milk production traits (Lucy *et al.*, 2009). However, interactions between genetic strain (G) and diet/feeding system (D) have been observed for milk production, feed efficiency, live weight (LW), body condition score (BCS), and fertility (Roche *et al.*, 2006). Considering this G×D interaction, the most appropriate strain of cow will depend on the system of dairying under in use. As cows are managed in a diverse range of environments worldwide, it has become increasingly important to understand the relationship between a cow genetic strain and environment, with the primary environmental variable being diet. The historical, singular selection for milk production has resulted in greater milk yield, but at the expense of body tissue mobilisation (Roche *et al.*, 2006). This increased nutrient partitioning to milk production has been associated with lower reproductive performance (Chagas *et al.*, 2007) and reduced profitability in grazing systems. It is, therefore, important to understand the physiological differences underpinning this genetic effect on energy partitioning, and whether nutrition can alter metabolism to reduce body tissue contribution, thereby improving reproduction.
The somatotropic axis, consisting of growth hormone (GH), GH receptors (GHR), insulin-like growth factor (IGF)-I, and IGF-binding proteins, is involved in the regulation of nutrient partitioning in early-lactation dairy cows (Etherton and Bauman, 1998). The axis is uncoupled in early lactation, resulting in reduced expression of GHR in the liver and resultant low concentrations of IGF-1, and elevated concentrations of GH in blood. Lucy *et al.* (2009) reported a significant G x week interaction in the length of uncoupling, with North American (NA) Holstein-Friesian (HF) cows presenting with an uncoupled axis 12 weeks in milk, while axis re-coupling had occurred between weeks 4 and 8 in New Zealand (NZ) HF. Lucy *et al.* (2009) also alluded to a G x D interaction in the re-coupling of the somatotropic axis, with additional feed increasing blood IGF-1 concentrations in NZHF, but not NAHF. However, their experimental design did not allow a quantification of diet on mRNA expression of the genes involved in the somatotropic axis. The objective of the present study was to determine if there was an interaction in the effect of feeding on either the degree of uncoupling or the timing of re-coupling of the somatotropic axis NA and NZ HF strains.

Material and methods

Sixty HF cows of two divergent genetic strains (NZ and NA; n=30) were offered a generous allowance of fresh pasture and allocated 0, 3, or 6 kg DM/day of concentrates from calving to 12 weeks in milk. Milk yield and composition, LW, and BCS (1 to 10 scale) were recorded weekly. Blood was sampled by coccygeal venipuncture prior to AM milking on one day each week, the plasma aspirated, and analysed for GH and IGF-1 concentration using double-antibody radioimmunoassay. Liver tissue was collected from all cows during weeks 1, 4, 8, and 12 of lactation by percutaneous punch biopsy, snap frozen in liquid nitrogen, and stored at -80 °C awaiting RNA extraction. Real-time reverse transcription-PCR was performed on a Corbett Rotorgene 3000 (Corbett Life Science, Sydney, Australia). Primers, dual-labelled fluorescent probes, and standards were synthesised for cattle GHR-1a, GHR-total (GHR-tot), and IGF-1. For the analysis of mRNA expression, data from weeks 4, 8 and 12 were compared relative to week 1 post-calving, and analysed using a mixed model that included week as a repeated effect within cow, and G, D, and

G x D as fixed effects. The same model was applied to milk production variables, BCS, LW, and blood hormone concentrations.

Results and discussion

Compared with NZ cows, NA cows were heavier ($P<0.05$; 522 and 551 kg, respectively) and produced more milk ($P<0.05$; 30.1 and 33.2 kg/day, respectively) with less fat ($P<0.01$; 4.37 and 3.97%, respectively) and protein ($P<0.01$; 3.49 and 3.38%, respectively). They were also thinner ($P<0.01$; BCS = 4.5 and 4.0, respectively). These data are consistent with previous studies (Roche *et al.* 2006; Lucy *et al.*, 2009) and reflect differences in genetic selection priorities in the different farming systems. Concentrate feeding increased milk yield ($P<0.001$; 1.1 kg/kg concentrate DM) and protein ($P<0.01$; 0.03%/kg concentrate DM). This was also consistent with the published effect of concentrates on milk production (Roche *et al.*, 2006). There was no interaction between strain and diet for milk production.

Plasma GH concentration was greater ($P<0.01$; 3.5 and 2.6 ng/ml, respectively) and IGF-1 concentration less in NA cows ($P<0.01$; 9.3 and 11.4 ng/ml, respectively). Supplementing cows with concentrates reduced GH ($P<0.01$) and increased IGF-1 concentration ($P<0.05$), but there was no effect of level of concentrate offered (3.5, 2.8, and 2.7 ng GH/ml and 9.0, 11.1, and 10.9 ng IGF-1/ml for 0, 3, and 6 kg DM/day, respectively).

Blood data mimicked mRNA expression results. On average across the treatment period, NA cows had reduced hepatic expression of GHR-tot, GHR-1A, and IGF-1 mRNA ($P<0.1$), consistent with an uncoupled somatotropic axis. However, there was a tendency for an interaction between strain and week ($P<0.1$, 0.05 and 0.15, for GHR-tot, GHR-1A, and IGF-1 mRNA, respectively), with little change in mRNA expression in NA cows with successive measurements post-calving, but increased expression in NZ cows by wk 12 *post partum*. These results were similar to those presented by Lucy *et al.* (2009). Concentrate supplementation increased GHR-1A and IGF-1 mRNA expression, but the effect was much less pronounced in NA cows (G×D; $P<0.1$ for IGF-1 mRNA). These results support those of Lucy *et al.* (2009), that additional feed increases plasma IGF-1 concentration in NZ, but not NA, cows.

In conclusion, the somatotropic axis was uncoupled for longer in NA cows. Diet failed to alter the hepatic expression of IGF-1 mRNA in NA cows, but supplementing NZ cows with concentrates re-coupled the somatotropic axis by week 4 post-calving. These data were consistent with the BCS data reported by Roche *et al.* (2006) in these strains and the recent results of Lucy *et al.* (2009) in similar strains of HF cows.

Acknowledgement

The authors gratefully acknowledge the support of NZ dairy farmers through DairyNZ.

References

Chagas, L.M., J.J. Bass, D. Blache, C.R. Burke, J.K. Kay, D.R. Lindsay, M.C. Lucy, G.B. Martin, S. Meier, F.M. Rhodes, J.R. Roche, W.W. Thatcher and R. Webb, 2007. Invited review: New perspectives on the roles of nutrition and metabolic priorities in the subfertility of high-producing dairy cows. J. Dairy Sci. 90: 4022-4032.

Etherton, T.D. and D.E. Bauman, 1998. Biology of somatotropin in growth and lactation of domestic animals. Physiol. Rev. 78: 745-761.

Lucy, M.C., G.A. Verkerk, B.E. Whyte, K.A. Macdonald, L. Burton, R.T. Cursons, J.R. Roche and C.W. Holmes, 2009. Somatotropic axis components and nutrient partitioning in genetically diverse dairy cows managed under different feed allowances in a pasture system. J. Dairy Sci. 92: 526-539.

Roche, J.R., D.P. Berry and E.S. Kolver, 2006. Holstein-Friesian strain and feed effects on milk production, body weight, and body condition score profiles in grazing dairy cows. J. Dairy Sci. 89: 3532-3543.

Change in serum blood components as affected by breeding period and dietary protected protein in ewes

G.M.A. Solouma[1], A.K.I. Abd El Moty[2], A.Y. Kassab[1], A.A. Abdel-Ghani[2] and E.B. Soliman[2]
[1]Department of Animal Production, Faculty of Agriculture, Sohag University, Egypt; [2]Department of Animal Production, Faculty of Agriculture, Minia University, Egypt; gsolouma@hotmail.com

Introduction

In animal production, sheep play an important role especially for meat and wool production. Composition of blood during breeding period can be considered essential for understanding animal reproductive performances. Many investigators reported that serum levels of most blood components significantly differed during the breeding period (Brzostowski *et al.*, 1996; Abu El-Ella, 2006). On the contrary, blood components are affected by feeding with protected protein in the diet (Pailan and Kaura, 1996; El-Reweny, 2006). The aim of this investigation was to study the changes of blood components as affected by both breeding period and by methods (heat treated or sodium hydroxide treatment) of protein protection.

Material and methods

A total number of 36 Sohagi ewes, about 3-4 yr old and 40.60±1.60 kg body weight, were used in this experiment during the breeding period (42 d). Ewes were divided randomly into three treatments as follows: control (C), heat treated (T1) and sodium hydroxide treated (T2) protein fraction. Ewes were offered their requirements through concentrate diet (80%) and wheat straw (20%) according to NRC (1985). Canola meal, which represented 25% of the concentrate feed mixture, was untreated in the control group (C). The canola fraction was subjected to 135-145 °C in a forced air oven for 4 h according to Stern *et al.* (1985) in treatment T1. Canola meal, in treatment T2, was sprayed by a solution of sodium hydroxide at the rate of 3% DM of canola meal according to Mir *et al.* (1984). Ewes' blood samples that were collected during the breeding period were performed at the following rhythm from oestrus (E): 8 d before mating (-8E), at oestrus (E: mating day), 8 d after mating (+8E) and 34 d after mating (+34E) for each dietary treatment to study the effect of protected protein and breeding period on some blood concentration of the following components, i.e. Triiodothyronine (T3), Thyroxin (T4) and progesterone (P4), total protein (Tp), albumin (AL), glucose (Gu), creatinine (Cr) and urea-N (Ur). The results were statistically analysed using the GLM of SAS® (SAS Institute Inc., 1998, Cary, NC, USA) for complete randomised design. Significant differences among treatment means within the experiment were analysed using the Duncan test (Duncan, 1955).

Results

The results in Table 1 revealed that the highest levels of thyroid hormone concentrations were observed at oestrus. Serum T3 concentrations in different treatments (regardless of breeding period) were higher ($P<0.01$) only in T1. While T4 concentrations were slightly increased in T1 and T2 treatments. As expected, progesterone values varied ($P<0.01$) according to breeding stage; P4 values at -8E, +8E and +34E were significantly ($P<0.01$) higher than at E. Regardless of breeding period, P4 concentrations were significantly ($P<0.01$) different; the highest ($P<0.01$) values of P4 were observed in T1. the values of total protein and its fractions revealed no significant differences across breeding period and, also, among different treatments. The present results indicate that a highest mean level of serum Gu coincided with the day of oestrus and remained high until +8E. Higher Gu concentrations were obtained with T1 when compared with the other treatments. Throughout the breeding period, the differences in urea-N and Cr concentrations were not significant while Urea-N and Cr concentrations decreased, as a result of protected protein supplementation, in both T1 and T2.

Table 1. Effect of breeding period and protected protein supply on blood components.

	Breeding period					Feeding treatments			
	-8E	E	+8E	+34E	± SE	C	T1	T2	± SE
T3, ng/dl	60.3^{D}	152.7^{A}	144.9^{B}	113.0^{C}	0.96**	116.2^{B}	119.8^{A}	117.2^{B}	0.89**
T4, µg/dl	2.70^{D}	7.13^{A}	5.57^{B}	5.03^{CB}	0.15**	4.96^{B}	5.28^{A}	5.08^{B}	0.10**
P4, ng/ml	2.33^{C}	0.32^{D}	3.34^{B}	6.53^{A}	0.04**	2.99^{B}	3.23^{A}	3.18^{A}	0.05**
Tp, g/dl	6.19	6.27	6.14	6.16	0.05 N.S	6.19	6.19	6.18	0.03 N.S
AL, g/dl	2.99	2.99	3.00	2.98	0.06 N.S	2.98	2.99	3.02	0.04 N.S
GL, g/dl	3.20	3.25	3.14	3.18	0.05 N.S	3.22	3.20	3.16	0.05 N.S
Gu, mg/dl	70.04^{B}	71.69^{A}	71.00^{A}	70.72^{B}	0.22**	69.69^{C}	71.79^{A}	71.23^{B}	0.18**
Ur, mg/dl	14.60	14.75	14.70	14.72	0.11 NS	15.46^{A}	14.44^{B}	14.32^{B}	0.12**
Cr, mg/dl	1.13	1.13	1.10	1.11	0.02 NS	1.18^{A}	1.08^{B}	1.09^{B}	0.01**

^{A,B,C,D} Means with same letters in the same line in each parameter are significantly different at * $P<0.05$, ** $P<0.01$, NS = not significant; C = control, T1 = heat treatment, T2 = sodium hydroxide treatment.

Conclusion

From the present results it can be concluded that during the breeding period, feeding on protected protein in the diet led to positive and generally significant effects on serum blood components such as thyroid and progesterone hormones and some of the blood metabolites such as glucose, Urea-N and creatinine.

References

Abu El-Ella, A.A., 2006. Response of Barki ewes to treatment with gonadotrophin hormones and energy supplementation (flushing). J. Sheep Goat Desert Anim. Sci. 1: 73-88.

Brzostowski, H., S. Milewski, A. Wasilewska and Z. Tanski, 1996. The influence of the reproductive cycle on levels of some metabolism indices in ewes. Arch. Vet. Pol. 35, 52-59.

Duncan, D.B., 1955. Multiple range and Multiple F- test. Biometrics 11: 1-42.

El-Reweny, A.M.S., 2006. Effect of protected protein on production and reproduction performance in sheep. Ph. D. Thesis, Tanta University, Kafr El-Shiekh, Egypt.

Mir, Z., G.K. MacLeod, J.G. Buchanan-Smith, D.G. Gryeve and W.L. Grovum, 1984. Methods for protecting soybean and canola proteins from degradation in the rumen. Can. J. Anim. Sci. 64: 853-865.

NRC, 1985. Nutrient requirements of sheep, 6th edition. National Research Council, Washington, DC, USA.

Pailan, C.H. and H. Kaura, 1996. Influence of dietary protein content and digestibility on milk yield and blood constituents in lactating goats. Small Rumin. Res. 20: 47-51.

Stern, M.D., K.A. Stantos and L.D. Satter, 1985. Protein degradation in rumen and amino acids absorption in small intestine of lactating dairy cattle fed heat treated whole soybeans. J. Dairy Sci. 68: 45-56.

Short-term nutritional supplementation with lupin grain increases total IRS-2 and IRS-4 and decreases aromatase in ovine granulosa cells

A. Somchit[1,2], B.K. Campbell[3], M. Khalid[1] and R.J. Scaramuzzi[1,4]
[1]Departments of Veterinary Basic and Veterinary Clinical Sciences, Royal Veterinary College, Herts, United Kingdom; [3]Division of Obstetrics and Gynaecology, Queen's Medical Centre, University of Nottingham, United Kingdom; [4]UMR Physiologie de la Reproduction et des Comportements, INRA, 37380 Nouzilly, France; rex.scaramuzzi@tours.inra.fr

Introduction

Folliculogenesis in the ewe is stimulated by short-term nutritional supplementation during the late luteal phase of the oestrous cycle (Somchit *et al.*, 2007). The mechanism of this effect probably involves direct follicular actions of nutrition mediated by the intrafollicular insulin glucose system (Scaramuzzi *et al.*, 2006). However, little is known about insulin signalling pathways in granulosa cells. This experiment studied the effect of short-term nutritional supplementation with lupin grain on the levels of insulin receptor substrate -2 and -4 (IRS-2, IRS-4) and aromatase in granulosa cells, oestradiol in follicular fluid and oestradiol and FSH in jugular venous plasma.

Material and methods

Welsh Mountain ewes fed a maintenance diet of straw, were used. Oestrus was synchronised using progestagen sponges (12 d) and 8 d after oestrus one group of ewes (n=11) were fed a supplement of 500 g of lupin grain a day for 5 d. A second unsupplemented group (n=13) were controls. At the end of the feeding period luteolysis was induced with 125 μg of PGF, in all ewes, and the ovaries collected 40 h later in the follicular phase of the oestrous cycle just before the expected start of the LH surge. All follicles >1 mm in diameter were quickly dissected and the granulosa cells and follicular fluid recovered. The granulosa cells were lysed and the protein fraction analysed by western blotting to determine the level of aromatase and total IRS-2 and IRS-4 and oestradiol was measured in follicular fluid. Jugular venous blood was collected regularly (Figure 1) and the plasma analysed for glucose (colourimetry) and LH, insulin, FSH and oestradiol (RIA). Follicles >2 mm in diameter were classified as oestrogenic (follicular fluid oestradiol >150 ng/ml) or non-oestrogenic (follicular fluid oestradiol <150 ng/ml). Data were analysed by mixed model ANOVA using SPSS (SPSS Inc., Chicago, IL, USA).

Results

The concentration of insulin (control; 0.37±0.05 vs. lupins; 0.53±0.03 ng/ml) and glucose (control; 3.10±0.05 vs. lupins; 3.29±0.06 mM/L) rose significantly at 2 and 3.5 d respectively, after the start of supplementation and remained significantly elevated for the duration of supplementation. Supplementation significantly increased the number of follicles greater than 1 mm in diameter (Table 1).

Table 1. The effect of supplementation with lupin grain for 5 d during the late luteal phase of the oestrous cycle on the mean ± sem number of follicles, the concentration of oestradiol in follicular fluid from oestrogenic follicles and the granulosa cell levels of aromatase, IRS-2 and IRS-4 in oestrogenic follicles. An asterisk indicates a significant difference (P<0.05) between treatments.

	Follicles >1 mm	Oestradiol (ng/ml)	Aromatase	IRS-2	IRS-4
Straw	23.2±1.93	2.26±0.54	0.772±0.135	0.311±0.084	0.372±0.052
Lupins	31.3±2.30*	1.22±0.30*	0.530±0.101*	0.754±0.084*	0.651±0.083*

The jugular venous concentration of FSH was not significantly affected by supplementation (control; 1.37±0.10 vs. lupins; 1.29±0.08 ng/ml) but both the follicular fluid level of oestradiol in oestrogenic follicles (Table 1) and the jugular venous concentration (Figure 1) of oestradiol were significantly decreased by supplementation. Supplementation significantly increased the levels of total IRS-2 and IRS-4 and significantly reduced the level of aromatase (Table 1) in oestrogenic follicles. Non-oestrogenic follicles had very low aromatase that was not affected by supplementation but their levels of both IRS were significantly elevated by supplementation.

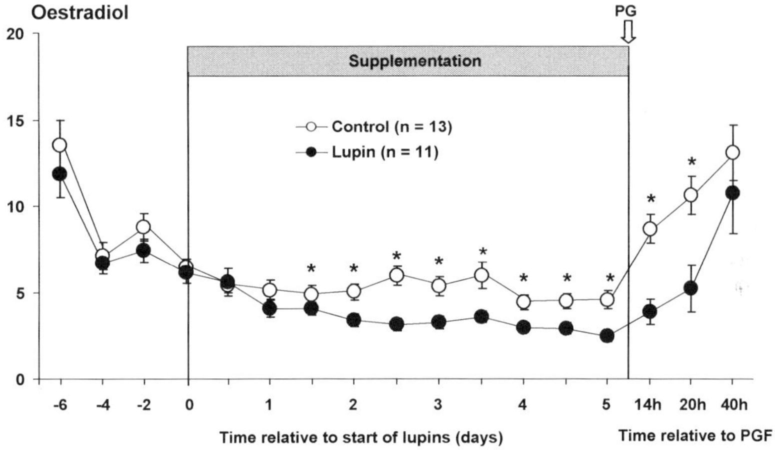

Figure 1. The effect of short-term supplementation with lupin grain on oestradiol. An asterisk indicates a significant difference between treatments within times at P<*0.05.*

Discussion

These data suggest that short-term supplementation with lupin grain has a direct inhibitory action on granulosa cells reducing aromatase levels and oestradiol secretion in the absence of a change in circulating FSH. This effect appears to be mediated by the intracellular insulin system in granulosa cells and suggests that the high energy content of lupin grain stimulated insulin secretion and this effect was associated with increased IRS-2 and IRS-4 in granulosa cells. The changes in IRS-2 and IRS-4 were in turn associated with reduced FSH-stimulated aromatase activity and reduced oestradiol secretion.

Acknowledgement

The research was supported by the BBSRC (Grant BB/0018420/1). R.J. Scaramuzzi is the recipient of an EU Marie Curie Chair of Excellence (MEXC-CT-2006-042499).

References

Scaramuzzi, R.J., B.K. Campbell, J.A. Downing, N.R. Kendall, M. Khalid, M. Muñoz-Gutiérrez and A. Somchit, 2006. A review of the effects of supplementary nutrition in the ewe, on the concentrations of reproductive and metabolic hormones and the mechanisms that regulate folliculogenesis and ovulation rate. Nutr. Reprod. Develop. 46: 339-354.

Somchit, A., M. Khalid, B.K. Campbell and R.J. Scaramuzzi, 2007. The effects of short-term nutritional supplementation with lupins (*Lupinus luteus*) on the number of ovarian follicles and the concentrations of follicular hormones and glucose in ewes during the luteal phase of the oestrous cycle. Theriogenology 68: 1037-1046.

Dairy heifer growth and time to mating weight when fed elephant grass as sole feed: A simulation model

F. Tibayungwa, J.Y.T. Mugisha and M. Nabasirye
Makerere University, P.O. Box 7062, Kampala, Uganda; ftibayungwa@agric.mak.ac.ug

Introduction

In Uganda there is concern to improve nutrition and incomes among resource-poor households through dairy production. However, due to small land acreage of about 0.5 ha per household, the common practice is stall-feeding one to two dairy cattle with elephant grass (*Pennisetum purpureum*) as the sole feed. This study was aimed at using a simulation model to explore the effect of forage crude protein (CP) and metabolisable energy (ME) on heifer growth and time to recommended mating weight of 300 kg under the Ugandan smallholder dairy system.

Material and methods

A hypothetical Holstein heifer weighing 70 kg liveweight (W) was used in this model. Forage characteristics used are given in Table 1.

Table 1. Nutrient composition and digestibility (g/kgDM), Energy (MJ/kgDM) and CP degradation fractions (g/kg CP) of elephant grass[1].

DM	DOMD	CP	GE	ADIN	a	b	c	Reference
180		53	16		213	672	0.04	Muia *et al.* (2001)
	524	115.4		1.3	211	541	0.03	Kabi *et al.* (2005)
155	571	118						Kariuki *et al.* (1998)

[1] DOMD =Digestible organic matter; ADIN = acid detergent insoluble nitrogen; GE = gross energy; *a* = water soluble fraction; *b* = potentially degradable nitrogen other than water soluble fraction; *c* = degradation rate per hour of the *b* fraction.

Daily gain (DG) is dependent on ME for growth and metabolisable protein (MP) for growth and is calculated based on AFRC(1993) Equations (1), (2) and (3):

$$MP_f = C6(168.07–0.16869W + 0.0001633W^2)\times(1.12–0.1223\Delta W)\times1.695\Delta W \qquad (1)$$

MP_f is MP requirement W gain (g/d), $C6$ is a correction factor (0.9), ΔW is changed in W.

$$\Delta W = E_f /(C4\times C2(4.1+0.033W–0.000009W^2)+0.1475E_f) \qquad (2)$$

E_f is Net Energy retained (MJ/d), $C2$ is a correction factor (1.2) for mature body size of heifers, $C4$ is the correction factor for ME for heifers (1.1).

$$ME = 0.0157\times DOMD \qquad (3)$$

All other parameters are calculated by the model. DG is determined by whichever of MP and ME is limiting. The model was coded in VENSIM® 5, based on differential equations with a 1-day time step.

Results

Increased CP level in forage led to higher growth rates and reduced time to target mating weight of 300 kg. CP levels of 60, 80 and 100 g/kgDM correspond to 988, 628, 461 days to target weight. However, when CP exceeded 110 g/kgDM, growth was limited by ME. Figure 1B shows the output where CP is limiting growth. Higher ME level led to higher DG and reduction in days to attain 300 kg. In Figure 1A, the higher DG from 0 to 150 days is due to lower maintenance requirements at smaller W (Kertz *e. al.*,1998).

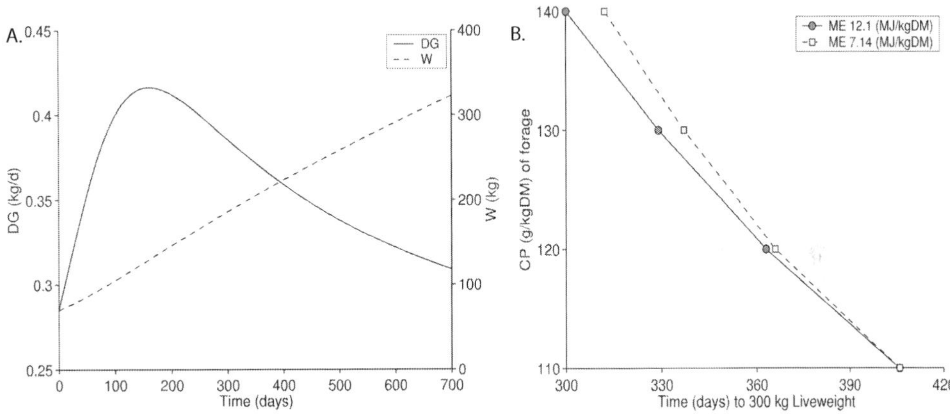

Figure 1. A. DG and W at CP 80 g/kgDM and ME 7.14 MJ/kgDM; B.Effect of CP on time to target W of 300 kg at two levels of ME

Conclusion

This study indicates that when elephant grass is the sole feed, increased CP intake leads to increased growth rate and shorter time to attain the target of 300 kg W, but ME is limiting if CP exceeds 110 g/kgDM. Lower levels of CP severely delayed attainment of target W. The strategy would be to feed elephant grass of high CP up to 110 g/kgDM if fed as the sole feed.

References

AFRC, 1993. Energy and Protein Requirements of Ruminants. Agricultural and Food Research Council. An advisory manual prepared by AFRC technical committee on response to nutrients. CAB International, Wallingford, UK, 159 pp.

Kertz, A.F., B.A. Barton and L.F. Reutzel, 1998. Relative efficiencies of wither height and bodyweight increase from birth until first calving in Holstein cattle. J. Dairy Sci. 8: 1479-1482.

Muia, J.M.K., S. Tamminga, P.N. Mbugua and J.N. Kariuki, 2001. Rumen degradation and estimation of microbial protein yield and intestinal digestion of napier grass (*Pennisetum purpureum*) and various concentrates. Anim. Feed Sci. Technol., 93: 177-192.

Kabi, F., F.B. Bareeba, O. Havrevoll and I.D.T. Mpofu, 2005. Evaluation of protein degradation characteristics and metabolisable protein of elephant grass (*Pennisetum purpureum*) and locally available protein supplements. Livest. Prod. Sci. 95: 143-153.

Kariuki, J.N., C.K. Gachuiri, G.K. Gitau, S. Tamminga, J. van Bruchem, J.M.K. Muia and K.R.G. Irungu, 1998. Effect of feeding napier grass, lucerne and sweet potato vines as sole diets to dairy heifers on nutrient intake, weight gain and rumen degradation. Livest. Prod. Sci. 55: 13-20.

Effect of supplemental n-3 fatty acid source on semen quality in Iranian Holstein bulls

A. Towhidi[1], A. Khoshvaght[1], A. Zare Shahneh[1] and M. Nourozi[2]
[1]Department of Animal Science, Faculty of Agronomy and Animal Science, University of Tehran, Karaj, Iran; [2]Research Centre for Agriculture and Natural Resources, Mashhad, Iran; atowhidi@ut.ac.ir

Introduction

Lipid composition of semen is unique in its content of long chain polyunsaturated fatty acids (LC-PUFA). These LC-PUFA are essential components of all cell membranes and also give rise to many bioactive molecules, such as eicosanoids (Sardesi, 1992). In most mammals, spermatozoa, like brain and retina, have considerable amounts of n-3 LC-PUFA, mainly docosahexaenoic acid (C22:6n-3), playing an essential role in development and function (Neuringer *et al.*, 1988) and regulation of cellular movement, lipid metabolism, and fusion capacity of sperm (Stubbs and Smith, 1984). Differences in phospholipid polyunsaturated fatty acid composition may affect the flexibility and compressibility of cellular membranes (Neuringer *et al.*, 1988). There is some considerable evidence that the lipid composition of the sperm membrane is a major determinant of motility, cold sensitivity and overall viability (Hammerstedt, 1993). Decreased sperm quality with ageing in the bull is related to decreasing DHA content in sperm fatty acids (Kelso *et al.,* 1997). Feeding n-3 fatty acids to rams (Towhidi *et al.*, 2008) and goats (Dolatpanah *et al.*, 2008) improved semen quality. But, there is not any report on the effects of feeding n-3 fatty acids on sperm quality in bulls.

Material and methods

Twelve Iranian Holstein bulls were divided into two groups and fed either a control diet or a fish oil-supplemented diet (2% in dry matter intake) as n-3 fatty acid sources. Both of the diets were isocaloric and isonitrogenous and formulated according to NRC (2001). Semen samples were collected twice weekly from June 10, 2008 to September 11, 2008 by artificial vagina. Semen characteristics were evaluated every two weeks. The collected semen was diluted 1:100 with 0.9% NaCl solution. Sperm concentration was determined with an IMV photometer. Ejaculate volume was measured with graduated tubes. Motility and progressive forward motility were evaluated by phase contrast microscopy at ×200 and ×400 magnifications, respectively. Eosin-Nigrosin staining used to evaluate sperm abnormalities (head and tail abnormalities) and viability (dye materials leak dead sperm cell membrane) at ×400 magnification. Data were analysed using Proc MIXED in a completely randomised design in SAS® (2001).

Results

Feeding fish oil to bulls significantly influenced semen quality. Mean volume, concentration, number of spermatozoa per ejaculate, percentage of sperm motility and progressive motility in the treated group were higher ($P < 0.05$) than in the control. Percentage of abnormality in the treated group significantly ($P<0.05$) decreased compared to the control group (Table 1).

Table 1. Mean (±SEM) semen parameters in fish oil and control group.

Semen parameters	Treated group	Control group	*P*-value
Semen volume, ml	10.91±0.41	9.17±0.41	0.05
Motility,%	78.44±1.45	73.14±1.45	0.05
Progressive motility,%	74.44±0.01	71.10±0.01	0.0001
Concentration, 10^9 cells/ml	1.32±0.05	1.03±0.05	0.003
Total sperm, 10^9 cells	14.65±0.52	9.69±0.52	0.0001
Abnormality	9.25±0.67	14.44±0.67	0.0001
Viability	43.62±0.49	42.26±0.49	0.09

Discussion and conclusion

The present study is the first report of the effect of dietary fish oil on sperm characteristics in Holstein bulls. Recently, we indicated that feeding n-3 fatty acids increased ram (Towhidi *et al.*, 2008) and goat (Dolatpanah *et al.*, 2008) semen quality. The result of the current study was in agreement with the previous studies. Supplement fish oil seems be efficient for improving semen quality in bulls.

References

Dolatpanah, M.B., A. Towhidi, A. Farshad, A. Rashidi and A. Rezayazdi, 2008. Effect of dietary fish oil on semen quality of goats. Asian. Aust. J. Anim. Sci. 21: 29-34.

Hammerstedt, R.H., 1993. Maintenance of bioenergetic balance in sperm and prevention of lipid peroxidation: a review of the effect on design of storage preservation system. Reprod. Fertil. Develop. 5: 657-690.

Kelso, K.A., A. Redpath, N.H.C. Noble and B.K. Speake, 1997. Lipid and antioxidant changes in spermatozoa and seminal plasma throughout the reproductive period of bulls. J. Reprod. Fertil. 110: 53-59.

Neuringer, M., G.J. Anderson and W.E. Connor, 1988. The essentiality of n-3 fatty acids for the development and function of the retina and brain. Annu. Rev. Nutr. 5: 517-541.

NRC, 2001.7th rev. ed. Natl. Acad. Press, Washington, DC, USA.

Sardesi, V.M., 1992. Biochemical and nutritional aspects of eicosanoids. J. Nutr. Biochem. 3: 562-579.

Stubbs, C.D. and A.D. Smith, 1984. The modification of mammalian polyunsaturated fatty acid composition in relation to membrane fluidity and function. Biochim. Biophys. Acta. 779: 89-137.

Towhidi, A., F. Samadian, K. Reza-Yazdi and F. Ghaziani, 2008. Feeding n-3 fatty acid source improves semen quality by increase n3/n6 fatty acids ratio in sheep. Proc. 10th WCAP (World Conference on Animal Production), 23-28 November 2008, Cape Town, South Africa, (in press).

The effect of short-term treatment of ewes with either intravenous glucose or a supplement of soya and maize during the luteal phase on the number of follicles and the AMPK signalling pathway in granulosa and theca cells

N. Zouaidi[1,2], G. Khaldi[2], J. Dupont[1] and R.J. Scaramuzzi[1,3]
[1]INRA, UMR Physiologie de la Reproduction et des Comportements, 37380 Nouzilly, France; [2]INAT, Tunis-Mahrajene, 1082, Tunisie; [3]Department of Veterinary Sciences, Royal Veterinary College, Herts, United Kingdom; nesrine.zouaidi@tours.inra.fr

Introduction

Folliculogenesis in the ewe is stimulated by short-term nutritional supplementation during the late luteal phase of the oestrous cycle (Somchit *et al.*, 2007). The mechanism of this effect probably involves direct follicular actions of nutrition through the insulin glucose system (Scaramuzzi *et al.*, 2006) and possibly 5'AMP-activated kinase (AMPK), in the follicle (Dupont *et al.*, 2008). This experiment investigated the effect of two short-term nutritional treatments on the number of follicles and the levels of AMPK in granulosa and theca cells.

Material and methods

Two experiments were carried out with ewes fed a maintenance diet of straw. Oestrus was synchronised using progestagen sponges. Experiment 1 used 36 Ile-de-France ewes; 8 d after oestrus one group (n=17) was infused with glucose at 10 mM/h for 72 h. A second saline-infused group (n=19) was controls. Experiment 2 used 16 Barbarine ewes; one group (n=8) was fed a supplement of 500 g of maize (400 g) and soya (100 g) a day for 5 d. A second unsupplemented group (n=8) was controls. At the end of the treatment period ovaries were collected (luteal phase) or luteolysis was induced with 125 μg of PGF and the ovaries were collected 30 h later (follicular phase). All follicles >1 mm in diameter were quickly dissected and the granulosa and theca cells were recovered. The cells were lysed and the protein fraction was analysed by western blotting to determine the phosphorylation level of AMPK. In a preliminary study, the different subunits of AMPK in granulosa and theca cells, from Ile-de-France ewes, obtained from a local slaughterhouse, were characterised by RT-PCR and western blot. Data were analysed by the t-test or the Mann Whitney U test within experiments.

Results

The nutritional treatments had no effect on body weight. Both treatments had no effect on the number of corpora lutea (CL) but both increased the total number of follicles (Table 1). Glucose increased the number of small follicles while the dietary supplement increased the number of small and medium sized follicles (Table 1). None of the treatments increased the number of large follicles. Granulosa and theca cell extracts of follicles collected from untreated Ile-de-France ewes from a local slaughterhouse contained the protein and its mRNA for the three (α, β and γ) subunits of AMPK (Figures 1A and 1B). All known isoforms of the subunits were detected by RT-PCR (Figure 1B) and the protein was detected by western blot for the α1, β1/2 and γ1, 2 and 3 subunits (Figure 1).

Discussion and conclusion

These data show that both forms of short-term nutritional supplementation increased the total number of follicles. The stimulatory effect of glucose was restricted to small follicles, while the soya-maize diet increased the number of small and medium follicles. This difference may reflect different treatment periods used in the two experiments. These data report for the first time the presence of the different subunits of AMPK in granulosa and theca cells from the ewe and suggest that intrafollicular AMPK may have a role in the mechanism of nutritional influences on folliculogenesis in the ewe. Further investigations will examine this hypothesis.

Table 1. The effect of glucose infused for 72 h at 10 mM/h (experiment 1) or supplementation with soya and maize for 5 d (experiment 2) during the late luteal phase of the oestrus cycle on the average (± sem) number of corpora lutea and follicles. Within an experiment, an asterisk indicates a significant difference (P<0.05) between treatments.

Experiment	Treatment	CLs	Follicles < 2 mm	Follicles 2-4 mm	Follicles >4 mm	Follicles total
one	Glucose n=17	2.76±0.45	26.6±2.07	9.41±1.2	1.94±0.32	38.64±2.96
	Saline n=19	2.26±0.34	14.5±0.97*	8.36±0.79	1.94±0.31	25.15±1.49*
two	Soya & Maize n=8	2.50±0.82	31.1±1.55	9.87±0.95	1.37±0.37	42.62±4.38
	Straw n=8	1.25±0.36	16.2±1.57*	3.87±1.61*	0.50±0.26	20.62±1.13*

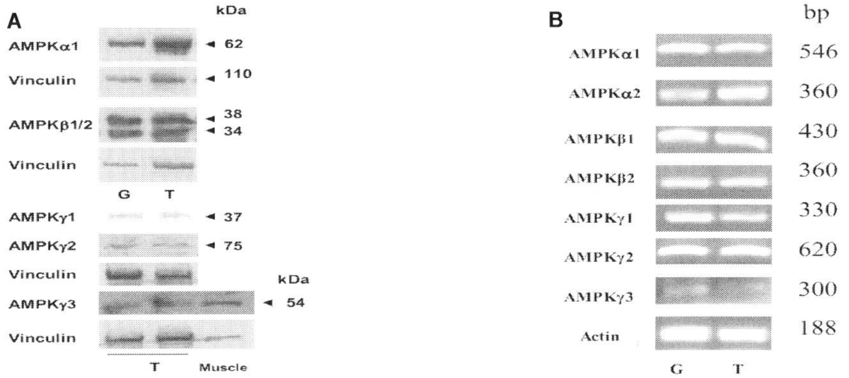

Figure 1. Detection of the AMPK (A) and RT-PCR analysis of the mRNA for AMPK subunits (B) in ovine granulosa (G) and theca (T) cells.

Acknowledgement

The authors thank Dr. D. Monniaux, J.-B. Menassol, S. Le Clanche and Z. Boussada. R.J. Scaramuzzi is the recipient of an EU Marie Curie Chair of Excellence (MEXC-CT-2006-042499).

References

Dupont, J., C. Chabrolle, C. Ramé, L. Tosca and S. Coyral-Castel, 2008. Role of the peroxisome proliferator-activated receptors, adenosine monophosphate-activated kinase, and adiponectin in the ovary. PPAR Res. 2008: 176275.

Scaramuzzi, R.J., B.K. Campbell, J.A. Downing, N.R. Kendall, M. Khalid, M. Muñoz-Gutiérrez and A. Somchit, 2006. A review of the effects of supplementary nutrition in the ewe, on the concentrations of reproductive and metabolic hormones and the mechanisms that regulate folliculogenesis and ovulation rate. Reprod. Nutr. Develop. 46: 339-354.

Somchit, A., M. Khalid, B.K. Campbell and R.J. Scaramuzzi, 2007. The effects of short-term nutritional supplementation with lupins (*Lupinus luteus*) on the number of ovarian follicles and the concentrations of follicular hormones and glucose in ewes during the luteal phase of the oestrous cycle. Theriogenology 68: 1037-1046.

Short communications

Nutrition and welfare

How repeated acidosis challenges affect sheep's behaviour and reactivity?

L. Commun[1,2], M.M. Mialon[1], C. Martin[1], M. Silberberg[1] and I. Veissier[1]
[1]INRA, UR1213 Herbivores, Theix, F-63122 Saint Genès Champanelle, France; [2]ENV de Lyon, Unité Gestion des Elevages, F-69280 Marcy l'Etoile, France; loic.commun@clermont.inra.fr

Introduction

A minimum threshold of 10% long fibre roughage on a DM basis for beef cattle was proposed to avoid pathological conditions and poor welfare (EU-SCAHAW, 2001). With high-energy diets without enough fibre, subacute ruminal acidosis (SARA) can appear. This pathology, frequent in cattle with a high production level, has a big economic impact (Stone, 1999). Several studies have shown SARA consequences like intake depression, low fibre digestion inducing a decrease in production and health consequences (Plaizier *et al.*, 2008). The aim of this study was to describe the response in sheep behaviour and reactivity to SARA and whether this response changed with repeated acidosis challenges.

Material and methods

Twelve ruminally cannulated wethers were exposed to 3 successive 28-d experimental periods each composed of a 23-d recovery period with a low energy diet (LD) and of a 5-d acidosis experiment with a high energy diet (HD). Pelleted ground wheat and chopped hay were offered separately in the two compartments of the manger according to a ratio concentrate:forage of 20:80 for the LD and 60:40 for the HD. The acidosis experiments consisted in offering 90% of *ad libitum* intake in order to induce a fast intake of the diet and limit refusals. The LD diet was offered in 2 distributions with 2/3 of DM at 8:00 h and 1/3 at 16:00 h. Refusals were weighed at 8:00 h. In the HD diet, hay was distributed in 3 meals: 20%, 30% and 50% at 8:00 h, 10:00 h and 16:00 h, whereas all wheat was distributed at 10:00 h and refusals of wheat were removed at 16:00 h. Refusals were weighed at 8:00, 10:00, 12:00, 14:00, 16:00 h during these acidosis experiments. Ruminal pH was recorded continuously. The day before and the 4th day of each experiment, the time budget was estimated from 24-h video recordings. Animals were also subjected to 4 test situations: 1 involving social meeting, 2 involving suddenness and 1 involving nociception. The social test consisted in a 5-min meeting with another sheep. The 2 sudden stimuli were the opening of a familiar black umbrella and a sudden horn noise and these were given when sheep were eating; the first reaction and the time to resume eating were noted. Nociception was measured using a CO_2 laser applied until a muscular twitch was observed (Veissier *et al.*, 2000). Each test was performed once the week before each experiment and then once during each experiment. Six sheep received a daily supplementation of a live yeast product and a placebo. Data were analysed using the Proc Mixed of SAS®. The fixed effects were the diet (HD or LD), the number of the experimental period (1 to 3) and their interaction, and yeast (yeast *vs* placebo). Sheep was included as a random factor.

Results and discussion

Yeast had no effect on behaviour traits so no results concerning this factor are presented here.
Ruminal pH: The time spent with a ruminal pH under 5.6 was significantly higher during acidosis experiments than during recovery periods (754 vs. 303 min, $P<0.001$).
Feeding behaviour: During the recovery periods, sheep ate all the distributed feeds, whereas refusals were observed during experiments. During the experiments, distributed wheat and hay were totally ingested during the first 2 days but wheat intake decreased progressively the days after. This decrease in consumption of wheat was less important during experiments 2 and 3.

General behaviour and reactivity: During the experiments (Table 1), sheep spent significantly more time active (i.e. standing awake, $P<0.001$) and eating ($P<0.05$) than during recovery periods, mainly due to the increase in number of feedings. During the social test, the number of interactions tended to be higher during experiments ($P<0.1$) with an increase in agonistic behaviour like number of knocks ($P<0.05$), whereas the percentage of threats was lower ($P<0.01$). In the 2 suddenness tests, no significant difference was observed between experiment and recovery periods. The response time to the CO_2 laser was significantly longer during the experiments (9.3 vs. 6.1 s., $P<0.001$) and this was more marked for experiment 3.

Table 1. Effect of diet on daily behaviour and on responses to reactivity tests.

	Diet		Experiment			SE	Significance		
	HD	LD	1	2	3		Diet	Period	Diet×Period
Behaviours (min/d)									
Lying	747	832	677^{b}	852^{a}	712^{b}	55	**	***	
Standing awake	307	236	388^{a}	242^{b}	292^{b}	29	***	***	†
Reactivity tests									
Suddenness (sec)									
Umbrella test	1.9	1.7	1.5	2.3	1.9	0.4			
horn test	6.2	4.5	8.5	7.6	2.4	1.4			*
Social meeting									
Nb interactions	45.7	39.6	51.5	41.2	44.3	5.3	†		†
Nb agonist. behav.	39.4	33.1	46.9	32.7	38.4	4.9	†		†
% of threat	10.7	17.3	16.6	11.3	7.6	2.0	**		
% of knock	72.3	66.1	75.4	66.2	75.5	3.9	†		
Nociception (sec)									
Laser test	9.3	6.1	9.5^{b}	6.9^{c}	11.5^{a}	0.7	***	**	

[a,b,c] Values with no common superscript letter differ significantly ($P<0.05$); †$P<0.1$; *$P<0.05$; **$P<0.01$; ***$P<0.001$.

To conclude, when sheep underwent acidosis experiments, they were more agitated, more aggressive towards other sheep, and seemed less sensitive to pain. The lower sensitivity to pain when experiments are repeated could participate to better withstand acidosis explaining partly no decrease in consumption of wheat.

Acknowledgement

The authors thank experimental staff and Lallemand Animal Nutrition for providing yeasts.

References

EU-SCAHAW, 2001. The welfare of cattle kept for beef production. Retrieved April 25, 2001, from http://ec.europa.eu.int/food/fs/scah/out54_en.pdf

Plaizier, J., D. Krause, G. Gozho and B. Mcbride, 2008. SARA in Dairy Cows: the Physiological Causes, Incidence and Consequences. Vet. J. 176: 21-31.

Stone, W. C., 1999. The effect of subclinical acidosis on milk components. Cornell Nutrition Conference for Feed Manufacturers. Cornell University, Ithaca, NY, pp. 40-46.

Veissier, I., J. Rushen, D. Collwell and A.M. de Passillé, 2000. A laser-based method for measuring thermal nociception of cattle. Appl. Anim. Behav. Sci. 66: 289-304.

Relationships between feed intake variability and rumen pH in mid-lactating goats fed an acidogenic diet

S. Giger-Reverdin, C. Duvaux-Ponter and D. Sauvant
UMR 791, INRA AgroParisTech, PNA, 16 rue Claude Bernard, 75005 Paris, France; sylvie.giger-reverdin@agroparistech.fr

Introduction

High producing ruminants need to be fed with diets of high nutritive value to meet their requirements. However, these diets might induce subacute acidosis which can be deleterious for the animals and their production. Moreover, although the occurrence of acidosis varies between animals fed the same diet (Nocek, 1997), it has seldom been related to feeding behaviour. The aim of this paper was to relate the variability in intake behaviour of dairy goats to their rumen pH when fed an acidogenic diet.

Material and methods

Ten cannulated stall-housed dairy goats in mid-lactation (two months after kidding at the beginning of the trial) received twice a day a total mixed ration with 35% hay, 15% pressed sugar beet pulp and 50% concentrate on a dry matter basis. The concentrate part was a mixture of barley (15%), wheat (12.5%), oats (12.5%), soyabean meal (15%), sugar beet pulp (15.5%), soyabean hulls (13%), corn gluten feed (13%), molasses (2%) and vitamin and mineral mixture (1.5%).
Intake was recorded every two minutes by weighing devices fitted under the feed trough. Rumen pH was measured directly in the rumen before the morning feeding (T0) and 3 hours (T3) after. For each afternoon feeding, the quantity of diet eaten 90 min after feeding divided by the total feed intake for this feeding was calculated (fractional intake, P90). Dry matter intake was also recorded every day and expressed on a liveweight basis. Intake was recorded during five days and pH during four days.
The goat effect was analysed by using a repeated analysis of variance (proc mixed procedure of SAS®) as days were considered as repetitions.

Results and discussion

The chemical composition of the diet was on a dry matter basis: enzymatic starch (9.7%), crude protein (13.2%), NDF (42.0%), ADF (21.5%) and ADL (1.7%).
Individual dry matter intake varied from 31.7 g/kg LW to 57.1, with a mean value of 44.0±5.85. Around half of the intake occurred during the first 90 min after feeding (0.473±0.186%). pH mean value at T0 was 6.37 and varied from 5.64 to 6.93. At T3, the range of variation was from 5.64 to 6.42, with a mean value of 6.06.
The goat effect was statistically significant, and the estimated mean values of the main parameters for each goat are given in Table 1.
There was a close negative relationship between the estimated values of rumen pH at T3 and P90 (Figure 1), but none between rumen pH at T3 and dry matter intake. The range of mean pH values was quite important as it varied from 6.34 to 5.82 which means from a normal situation to subacute acidosis (Owens *et al.*, 1998).
The variability between animals for fractional intake was quite high and this fractional intake had a major influence on rumen pH. These data might explain part of the variability observed in the occurrence of acidosis within a herd.

Table 1. Estimated mean values of intake parameters (DMI and fraction of the meal eaten 90 min after feeding, P90) and of rumen pH values before the feeding (T0) or 3 h (T3) after.

Goat	DMI (g/kg LW)	P90	pHT0	pHT3
1	34.5	0.171	6.45	6.31
2	41.7	0.578	6.28	5.82
3	45.2	0.526	6.31	5.89
4	47.1	0.638	6.58	5.91
5	38.7	0.530	6.05	6.04
6	48.9	0.330	6.17	6.34
7	53.1	0.564	6.59	5.95
8	49.9	0.305	6.08	6.12
9	38.6	0.661	6.56	6.11
10	42.2	0.423	6.61	6.17

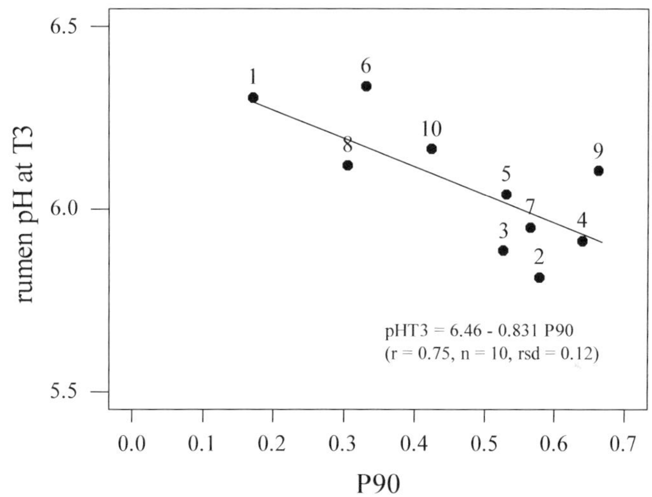

Figure 1. Relationship between rumen pH 3h after feeding and the fraction of the meal eaten 90 min after feeding in dairy goats.

Conclusion

Feeding behaviour needs to be taken into account in rumen studies, especially when dealing with acidogenic diets. Indeed, acidosis occurrence might be explained by variability in feed intake. It could be of practical interest to select animals with a low rate of intake when fed very high concentrate diets in order to prevent them from experiencing acidosis.

References

Nocek, J.E., 1997. Bovine acidosis: implications on laminitis. J. Dairy Sci. 80: 1005-1028.

Owens, F.N., D.S. Secrist, W.J. Hill and D.R. Gill, 1998. Acidosis in cattle: a review. J. Anim. Sci. 76: 275-286.

Sheep avoid eating saltbushes with high sulphur concentrations

H.C. Norman, D.K. Revell and D.G. Masters
CSIRO Livestock Industries and CRC Future Farm Industries, Private Bag 5, Wembley WA 6913, Australia; hayley.norman@csiro.au

Introduction

Old man saltbush (*Atriplex nummularia* Lindl.) is an Australian native shrub that is used internationally in saline and arid grazing systems. A number of chemical constraints limit productivity of ruminants grazing saltbush. These include low to moderate organic matter digestibility (OMD), high salt accumulation in the leaves (up to 35% of DM), high sulphur (S) concentrations and production of secondary compounds such as nitrates and oxalates (Masters *et al.*, 2001). Sheep grazing diverse stands of old man saltbush have demonstrated preference for some individual shrubs while barely grazing others (Norman *et al.*, 2004). In 2005, a project was initiated to select cultivars of old man saltbush with higher nutritive value and biomass production. The aim of this experiment was to investigate factors that influence sheep to eat some shrubs in preference to others. We hypothesised that organic matter digestibility, soluble salt concentrations and S will all influence relative preference.

Material and methods

The experiment was conducted in Yealering, Western Australia (32.60°S, 117.62°E). The 8 ha stand of saltbush consisted of approximately 7,200 shrubs, planted on a mildly saline gravelly loam soil (3.2 dSm^{-1} ECe in top 50 cm) with a highly saline water table (15. 8 dSm^{-1}) 1.0 m below the soil surface. Ten month old Merino wethers grazed the shrubs in autumn 2004 at a stocking intensity of 8 sheep/ha. The sheep were removed when the majority of shrubs had been completely defoliated. During the initial stages of grazing the 50 most preferred shrubs were tagged. At the end of grazing the least preferred shrubs were also tagged. Where a preferred and ungrazed shrub were adjacent (within 5 m of each other), leaves from the pair of shrubs (16 pairs in total) were sampled. The pairs of shrubs were monitored again in autumn 2005, when grazing was repeated with a similar class of sheep at the same stocking intensity. Again, the sheep demonstrated the same preferences and the leaves from shrubs were sampled.

The representative samples were dried in an oven at 65 °C for 48 h and ground with a Cyclotech mill with a 1 mm screen. Near infrared spectroscopy (Coleman and Henry, 2002), calibrated with a minimum of 200 saltbush samples, was used to predict OMD, total ash, soluble ash, nitrogen (N) and S. The laboratory methods, including *in vivo* calibration of *in vitro* prediction of OMD were provided by H.C. Norman *et al.* (unpublished results, 2009). For each year, the means of preferred and ungrazed plants were compared by ANOVA, with pair included as a block. The effect of pair was not significant in any of the analyses.

Results and discussion

Sulphur concentration in the leaves was associated with relative preference in both 2004 ($P<0.01$) and 2005 ($P<0.1$). In both years, the sheep selected saltbushes with lower S. In 2004, the mean estimated OMD of leaves of the preferred shrubs was significantly lower ($P<0.05$) than the mean of the ungrazed shrubs. The hypothesis, that organic matter digestibility, soluble salt concentrations and S will all influence relative preference, was therefore partially supported by the data.

The levels of S in the old man saltbush leaves exceeds recommended levels, diets containing 2.5 to 3.5 g/kg DM of S are deemed to be high for sheep (Standing Committee on Agriculture, 1990).

Table 1. Mean nutritive value traits of leaves from preferred and ungrazed old man saltbush shrubs, sampled after grazing with sheep in autumn 2004 and autumn 2005.

Trait	2004 grazing period					2005 grazing period				
	Preferred		Ungrazed		*P*-value	Preferred		Ungrazed		*P*-value
	mean	se	mean	se		mean	se	mean	se	
OMD	68.09	0.356	69.45	0.411	0.012	67.91	0.276	67.94	0.429	n.s. [1]
S	0.47	0.007	0.50	0.004	0.002	0.48	0.006	0.49	0.008	0.084
N	2.26	0.111	2.27	0.077	n.s.	2.18	0.099	2.30	0.112	n.s.
N:S ratio	4.77	0.209	4.55	0.156	n.s.	4.50	0.170	4.65	0.186	n.s.
Soluble ash	23.96	0.545	24.30	0.570	n.s.	25.05	0.484	25.53	0.583	n.s.
Total_ash	28.17	0.578	27.82	0.650	n.s.	28.72	0.603	28.95	0.687	n.s.

[1] n.s. = not significant, $P>0.1$.

Sulphur is used with N for the production of ruminal microbial protein. An optimal N: S ratio for sheep is 12.5: 1 and if the ratio is lower, excess S is converted to sulphide (Standing Committee on Agriculture, 1990). If saltbush is the major component of the diet (as was the case in this study), the 5 g/kg concentration that was measured in this experiment, together with the N:S ratio of 4.5:1, could lead to reduced voluntary feed intake, impaired rumen motility, damage to the central nervous system and reduced copper absorption (Bird, 1972; Underwood and Suttle, 1999). Sheep may select against shrubs with high S concentrations because they can associate negative post-ingestive consequences with malodorous compounds, given that many S-containing compounds are volatile. The salt levels in the leaves were also excessive but were not associated with preference between individual shrubs

References

Bird, P.R., 1972. Sulphur metabolism and excretion studies in ruminants. 10. Sulphide toxicity in sheep. Aust. J. Biol. Sci. 25: 1087-1098.

Coleman, S.W. and D.A. Henry, 2002. Nutritive value of herbage. In: Freer, M., Dove, H. (Eds.), Sheep Nutrition, Melbourne, Australia, CSIRO Publishing. pp. 1-26.

Masters, D.G., H.C. Norman and R.A. Dynes, 2001. Opportunities and limitations for animal production from saline land. Asian-Australas. J. Anim. Sci. 14: 199-211.

Norman, H.C., C. Freind, D.G. Masters, A.J. Rintoul, R.A. Dynes and I.H. Williams, 2004. Variation within and between two saltbush species in plant composition and subsequent selection by sheep. Aust. J. Agric. Res. 55: 999-1007.

Standing Committee on Agriculture, 1990. Feeding Standards for Australian Livestock. Melbourne, Australia: CSIRO Publications.

Underwood, E.J. and N.F. Suttle, 1999. The Mineral Nutrition of Livestock. 3rd edition. New York, USA: CAB International, Wallingford, UK.

Influence of diet-induced sub-acute ruminal acidosis on the oxidative status of plasma in dairy cows

N. Wullepit[1,2]*, W. Fokkink*[3]*, V. Fievez*[1]*, J.R. Newbold*[3]*, D. Fremaut*[2] *and S. De Smet*[1]
[1]*LANUPRO, Ghent University, Proefhoevestraat 10, Melle, Belgium;* [2]*Faculty of Biosciences and Landscape Architecture, University college Ghent, Voskenslaan 270, 9000 Ghent, Belgium;* [3]*Provimi Research and Technology Centre, Lenneke Marelaan 2, 1932 Sint-Stevens-Woluwe, Belgium; nicolas.wullepit@hogent.be*

Introduction

In typical diets of high yielding dairy cattle, the proportion of energetic dense feed components is often maintained at the highest possible level. Due to high amounts of readily fermentable carbohydrates, low amounts of dietary roughage and a non adapted ruminal environment, cows can develop subacute ruminal acidosis (SARA). SARA may be involved in the aetiology of several diseases. It can damage the ruminal and intestinal wall, cause diarrhoea, and hence retard nutrient absorption (Owens *et al.*, 1998). It is also associated with inflammations of different organs and tissues in dairy cows, such as liver abscesses (Krause and Oetzel, 2006). The above, together with a shift in rumen bacteria and a possible defaunation (Gazi *et al.*, 2007), could cause an imbalance between the production of reactive oxygen metabolites (ROM) and the neutralising capacity of oxidant mechanisms (Lykkesfeldt and Svendsen, 2007). The aim of this study was to investigate the possible effects of SARA on some parameters related to the oxidative status in plasma of dairy cows.

Material and methods

Twelve multiparous fistulated Holstein dairy cows were used in a complete randomised block design. Cows were paired according to milk yield, parity and days in milk (DIM), and allocated, within pair, at random to two groups. The control group (A) underwent a SARA induction experiment (average initial DIM and milk yield of 222±45 d and 43±4 kg/d). The treatment group (B) underwent the same SARA induction experiment but received feed additives (0.33 kg/d) aimed at preventing SARA (average initial DIM and milk yield of 227±47 d and 38±7 kg/d). The additive consisted mainly of sodium bicarbonate, yeast and synthetic vitamin E. The diet included 50% forage (67% maize and 33% grass silage) and 50% concentrates on a DM basis. Concentrates were a 'normal pH' concentrate (sugar beet pulp basis) and a 'low pH' concentrate (ground wheat basis). From three weeks before the start of the experiment all cows were fed the normal pH concentrate. During the SARA induction experiment, the 'normal pH' concentrate was, on a weekly basis, replaced by the 'low pH' concentrate. The normal pH/low pH concentrate proportions for weeks 1 to 6 were 4:0; 3:1; 2:2; 1:3; 0:4; 0:4, respectively. The amount of concentrate was adapted daily to the consumed amount of forage at a fixed rate per pair, and was divided into 2 meals (7:00 h and 17:00 h) with a limit of 5 kg per meal. Amounts of concentrate in excess of these 10 kg were supplied at noon (12:00 h), except for week 6 where all the concentrate was offered in 2 meals. Any refused concentrate was removed, weighed and inserted directly into the rumen 1 h after offering. In week 6 the experiment was ended due to clinical acidotic signs of the 7 cows.
Blood samples were collected into evacuated tubes at 14:00 h one day before and one day after every dietary change. Samples were centrifuged and plasma was stored at -20 °C pending analysis. The following parameters related to oxidative status were measured in plasma: ferric reducing ability of plasma (FRAP), α-tocopherol level, glutathione peroxidase activity (GSH-Px), superoxide dismutase activity (SOD), thiobarbituric acid reactive substances (TBARS) and conjugated dienes (CD). Statistical analysis was performed using the Linear Mixed Models procedure of SPSS for windows version 12.0. The mixed model included cow as the random effect; induction week

and treatment as the fixed effects and the 2-way interaction term. Week 0 (before the start of the treatment) was included as the covariate. Comparison of means was performed using Bonferroni as the *post-hoc* test.

Results and discussion

The mean plasma α-tocopherol level was significantly higher ($P<0.01$) in the treatment group supplemented with vitamin E (14.3±0.7 μg/ml) compared to the control group (7.3±0.7 μg/ml). No other treatment effects were observed. The oxidative status parameters α-tocopherol, GSH-Px, CD and TBARS were significantly affected by the SARA induction experiment (effect of week, Table 1), whereas SOD levels were not affected and the total antioxidant capacity of plasma, expressed as FRAP-values, showed a trend. There were no significant interaction terms. The declining levels of plasma α-tocopherol and FRAP, and the diminishing resistance against plasma lipoprotein oxidation, expressed at CD-values, during the experiment could be interpreted as indices of increased oxidative imbalance. Also the higher GSH-Px activity and increased lipid peroxidation (higher TBARS values) at induction wk 5 reflect an altered oxidative status.

Table 1. Effect of SARA induction on oxidative status markers in dairy cow plasma.

		Induction week[1]							Week
	Mean	1	2	3	4	5	6	SE	*P*-value
FRAP, μmol Fe^{2+} per liter	403	418	414	409	399	391	386	13.0	0.064
α-tocopherol, μg/ml	11.0	11.7^{ab}	12.8^{a}	12.0^{ab}	10.7^{bc}	9.66^{c}	9.24^{c}	0.553	<0.01
L	0.084	0.081^{ab}	0.080^{a}	0.084^{ab}	0.084^{ab}	0.089^{b}	0.085^{ab}	0.002	0.031
SOD, U/ml	3.17	3.07	3.15	3.13	3.25	3.43	3.01	0.281	0.650
CD-ET_{50}[2], sec	4667	4965^{a}	4887^{a}	4760^{ab}	4683^{ab}	4447^{ab}	4261^{b}	150	<0.01
TBARS, nmol/ml	3.22	2.61^{a}	3.44^{ab}	2.50^{a}	2.54^{a}	4.52^{b}	3.73^{ab}	0.461	<0.01

[1] The normal pH/low pH concentrate proportions for weeks 1 to 6 were 4:0; 3:1; 2:2; 1:3; 0:4; 0:4, respectively.

[2] ET_{50} = estimated time to reach 50% of maximal lipoprotein oxidation.

[a-c] Means within rows with different superscript letters are significantly different ($P<0.05$).

Conclusion

This study shows that dietary induction of SARA causes an unfavourable shift in several plasma oxidative parameters. This indicates the involvement of oxidative stress in the development or the outcome of SARA. Further work is in progress to evaluate the effect of SARA on some metabolic parameters and to seek possible links.

References

Gazi, M.R., M. Yokota, Y. Tanaka, S. Kanda and H. Itabashi, 2007. Effects of protozoa on the antioxidant activity in the ruminal fluid and blood plasma of cattle. Anim. Sci. J. 78: 34-40.

Krause, K.M. and G.R. Oetzel, 2006. Understanding and preventing subacute ruminal acidosis in dairy herds: A review. Anim. Feed Sci. Technol. 126: 215-236.

Lykkesfeldt, J. and O. Svendsen, 2007. Oxidants and antioxidants in disease: Oxidative stress in farm animals. Vet. J. 173: 502-511.

Owens, F.N., D.S. Secrist, W.J. Hill and D.R. Gill, 1998. Acidosis in cattle: A review. J. Anim. Sci. 76: 275-286.

Milk yield and quality of two genetic groups of dairy cattle under two cooling strategies during severe summer conditions

L. Avendaño-Reyes[1], F.D. Alvarez-Valenzuela[1], A. Correa-Calderón[1], J.A. Hernández-Rivera[1], R. Pérez-Velázquez[1], P.H. Robinson[2] and J.S. Fadel[2]
[1]Instituto de Ciencias Agrícolas, Universidad Autónoma de Baja California, Mexicali, Baja California, México; [2]Department of Animal Science, University of California, Davis, CA, USA; lar62@hotmail.com

Introduction

For many years, the superiority of the Holstein breed for milk production has limited the use of crossbreeding in dairy cattle. However, in warm environments, Holstein cows decrease their productivity. Intensification of dairy cattle and increasing concern for animal care and well-being have extended interest from dairy producers on better understanding of the basic principles of environmental stress and its impact on milk yield. Furthermore, consumers are questioning the methods of milk production, focusing not only on food safety aspects, but also on the well-being of the cows within these systems. During the summer, the inability of a Holstein cow to dissipate heat diminishes feed consumption, production, reproduction, and immune function (Fuquay, 1981). While all European breeds of dairy cattle are vulnerable to such heat stress, Jersey cattle is becoming more popular in many regions of the United States and Mexico because during summer months, their milk production is not as depressed as Holstein cattle (Bryant *et al.*, 2007; Keister *et al.*, 2002). Moreover, VanRaden and Sanders (2003) reported that fat yield from Jersey×Holstein crossbreds were slightly higher than purebred Holstein cows, and considering cheese yield pricing, profits from this cross exceeded that of Holstein for mating at breed basis. Environmental modifications such as spray and fans have been used with different success to ameliorate unfavorable effects of heat stress (West, 2003). Implementing strategies is more practical for intensive livestock systems. The objective of the present study was to assess two cooling strategies under severe heat stress conditions on animal welfare based on milk yield and milk components of pure Holstein and Jersey×Holstein cows under severe heat stress conditions.

Material and methods

Fifteen pure Holstein (H) and fifteen Jersey×Holstein (J×H) cows were assigned to two treatment groups (group one, 7 H and 8 J×H; group two, 8H and 7 J×H) and were subjected to one of two cooling strategies (CS) based on time of cooling (A, 3 h cooling; B, 4 h cooling) on daily basis in a commercial dairy herd located in the Mexicali valley, Baja California, Mexico, which is part of the Sonoran Desert. Groups of cows moved to each CS during four periods of 21 d from middle June to the end of August. During the last 7 d of each period, data on milk production, fat, protein and solid non fat percentages were averaged over the three days of measurements. All response variables from the experiment were analysed using the MIXED procedure of SAS® (2000). The statistical model included cow, period, genetic group, CS, interaction CS×genetic group, and residual error. Fixed effects included period and treatments. Cow was the random effect. Overall differences between treatment means were declared significant at $P<0.05$. Trends towards significance were considered at $P<0.10$.

Results

Summer temperatures in the zone of study varied from 19 to 49 °C, and relative humidity was on average 43%. Maximum and minimum THI were 91 and 69 units respectively. Interaction between genetic group and CS was not significant (P>0.05). Cows subjected to 4-h cooling produced almost 1 kg of more milk than cows under a 3-h cooling (21.86 vs. 20.74 kg of milk; P<0.01), meanwhile pure H cows (22.45 kg) tended to produce higher (P=0.09) milk yield than J×H (20.42 kg) cows (Table 1). Cooling strategy did not affect (P>0.05) fat, protein or non fat solids; however, J×H cows produced more fat (P<0.01) and protein (P=0.02) percent than H cows. Four hours cooling resulted in significantly higher milk energy output than 3 h cooling (16.34 vs. 15.37 Mcal respectively; P<0.01).

Table 1. Effect of genetic group and cooling strategies on production variables of dairy cows during the summer.

	Cooling strategy[1]		SE	*P*-value	Genetic group[2]		SE	*P*-value
	A	B			H	H×J		
Milk, kg	20.74	21.86	0.61	<0.01	22.45	20.42	0.76	0.091
Fat, %	3.76	3.80	0.09	0.53	3.50	4.05	0.12	<0.01
Protein, %	3.56	3.57	0.02	0.44	3.53	3.60	0.02	<0.05
Solids-non-fat,%	9.40	9.46	0.05	0.21	9.37	9.50	0.06	0.120
Milk energy output, Mcal	15.37	16.34	0.41	<0.01	16.21	15.82	0.56	0.615

[1] A = 3-h cooling; B = 4-h cooling; [2]H = Holstein; H×J = Holstein×Jersey.

Conclusion

This study shows that milk yield and milk energy output of dairy cows increased in response to increasing time of cooling under severe heat stress conditions. Also, J×H cows enhanced fat and protein in milk compared to H cows. Based on these production parameters, increasing cooling time improved dairy welfare of cows under hot environmental temperatures. More information about physiological measures is needed to provide more support to these results.

References

Bryant, J.R., N. Lopez-Villalobos, J.E. Pryce, C.W. Holmes, D.L. Johnson and D.J. Garrick, 2007. Effect of environment on the expression of breed and heterosis effects for production traits. J. Dairy Sci. 90: 1548-1553.

Fuquay, J.W., 1981. Heat stress as it affects animal production. J. Anim. Sci. 52: 164-174.

Keister, Z.O., K.D. Moss, H.M. Zhang, T. Teegerstrom, R.A. Edling, R.J. Collier and R.L. Ax, 2002. Physiological responses in thermal stressed Jersey cows subjected to different management strategies. J. Dairy Sci. 85: 3217-3224.

SAS Institute, 2000. User's Guide, version 8. SAS Institute Inc., Cary, NC, USA.

VanRaden, P.M. and A.H. Sanders, 2003. Economic merit of crossbred and purebred US dairy cattle. J. Dairy Sci. 86: 1036-1044.

West, J.W., 2003. Effects of heat stress on production in dairy cattle. J. Dairy Sci. 86: 2131-2144.

The impact of long term grain feeding on the core body temperature of cattle

D.T. Beatty, G.E. Gardner and R. Jacob
Murdoch University, South Street, Murdoch, Western Australia, 6150; dbeatty@mla.com.au

Introduction

Australia produces beef cattle for export markets which require meat to have high marbling and intramuscular fat contents. In order to produce such meat qualities cattle are fed high energy grain based diets for extended periods of time. Compared to cattle on pasture based roughage diets, cattle on high concentrate rations may be exposed to increased heat loads due to increased heat production from digestion and metabolism of the high energy diet. The impact of prolonged feeding of high energy diets (up to 300 d) on the core body temperature (T_{core}) of cattle is unknown. It has been shown that cattle fed high energy diets are more susceptible to heat stress (Gaughan *et al.*, 1997) and stresses such as transport and handling cause an increase in the T_{core} of cattle and that may impact on meat quality (Ferguson *et al.*, 2001). If T_{core} is increased due to long term grain feeding, then there may be potential animal welfare implications associated with heat stress in these cattle during any stressful event which is likely to raise T_{core}. This paper measured the T_{core} of cattle over extended periods of time whilst animals were on either high energy grain based diets or pasture diets.

Material and methods

Eighteen 13 mo old *Bos taurus* Angus steers (369±14 kg; mean±sem) were randomly assigned to one of three treatment groups. Group 1 steers (n=6) were fed a grain based feedlot ration (12.6 MJ/kg and 13.2% protein) for 300 d in a commercial feedlot. Group 2 steers (n=6) were fed on pasture (average 9 MJ/kg and 12% protein) for 150 d at a nearby farm and then transferred to the feedlot grain based ration for a further 150 d. Group 3 steers (n=6) were fed on the same pasture for 300 d. The proximity of the farm and feedlot was such that climatic conditions were similar for all groups. To measure the T_{core} of steers, temperature telemeters (iButton, DS1922L, Maximum Dallas, California, USA) were surgically implanted into the peritoneum of each steer following the methodology described by Beatty *et al.* (2006). Temperature telemeters logged temperature every hour for the duration of the experiment.

For all groups, the monthly mean T_{core} was calculated for each steer. A linear mixed model was used to fit a repeated measurements model to the temperature data (GenStat, VSN International Ltd, Hemel Hempstead, UK, version 10.1). The models can be described in a GenStat formula where time is the repeated-measurement factor using the observed levels 1-11 mo. Month was in the model as a factor with one of 11 levels and diet was a factor with 3 levels (grain 300 d, grass 150 d + grain 150 d and grass 300 d).

Results

There was a significant effect of diet (P<0.01) but no effect of month (P>0.05) on T_{core}. Figure 1 shows that group 1 steers (grain 300 d) had a higher mean monthly T_{core} compared to group 2 (grass 150 d + grain 150 d) and group 3 steers (grass 300 d). The mean monthly T_{core} for group 2 and group 3 steers were similar for the first 150 d of the experiment. When the group 2 steers were introduced to the grain based diet their mean monthly T_{core} was consistently higher than group 3 steers.

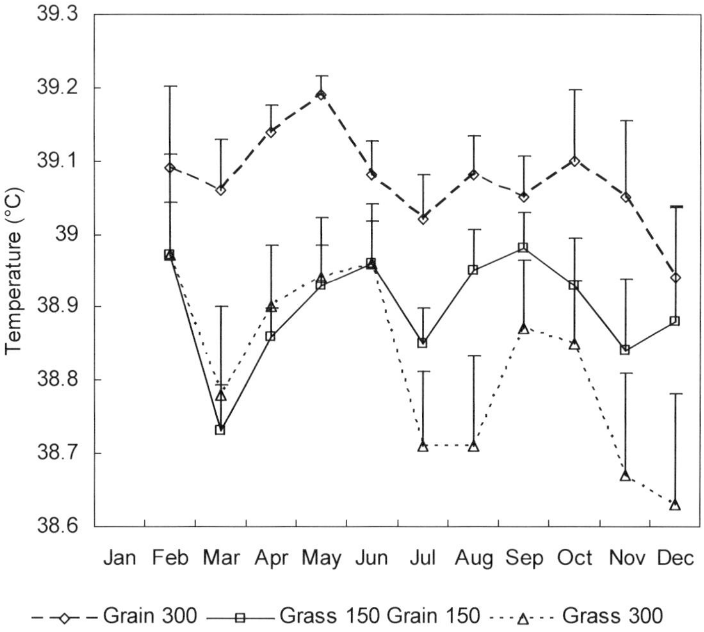

Figure 1. Monthly core body temperature (mean + sem) of group 1 (grain 300 d, n=6) group 2 (grass 150 d + grain 150 d, n=6) and group 3 (grass 300 d, n=6) steers over the duration of the experiment.

Discussion and conclusion

The feeding of high energy grain based diets significantly influences the basal T_{core} of cattle. Interesting, it would appear that it does not require long term grain feeding to influence T_{core} as changes in mean monthly T_{core} were evident after the first month in this experiment. This may have animal welfare implications during hot environmental conditions as well as an impact on meat quality due to heat toughening. Further research is required to assess the impacts that added stresses such as transport, handling and slaughter may have on T_{core} and animal welfare.

Acknowledgement

The authors would like to thank Meat and Livestock Australia and Beef CRC for funding this project and Victoria Surridge and Paul OMeehan for use of farm and feedlot facilities.

References

Beatty, D.T., A. Barnes, E. Taylor, D. Pethick, M. McCarthy and S.K. Maloney, 2006. Physiological responses of *Bos taurus* and *Bos indicus* cattle to prolonged continuous heat and humidity. J. Anim. Sci. 84: 972-985.

Ferguson, D.D., H.L. Bruce, J.M. Thompson, A.F. Egan, D. Perry and W.R. Shorthose, 2001. Factors affecting beef palatability: farm gate to chilled carcass. Aust. J. Exp. Agric. 41: 879-891.

Gaughan, J.B., T.L. Mader, S.M. Holt, M.J. Jovey and K.J. Rowan, 1997. Heat tolerance of Boran and Tuli crossbred steers. J. Anim. Sci. 77: 2398-2405.

Stress physiology in cattle is modified by temperament and hormonal growth promotant

L.M. Cafe[1], D.M. Ferguson[2], D.L. Robinson[1] and P.L. Greenwood[1]
Australian Cooperative Research Centre for Beef Genetic Technologies; [1]NSW Department of Primary Industries, Beef Industry Centre, UNE, Armidale, NSW 2351, Australia; [2]CSIRO Livestock Industries, FD McMaster Laboratories, Armidale, NSW 2350, Australia; linda.cafe@dpi.nsw.gov.au

Introduction

In cattle, the behavioural expression of fear in response to stressful events is commonly referred to as temperament. Temperament is important for the safety of stockpersons and for the welfare of cattle during handling. Temperament can also affect growth, carcass and beef quality characteristics (King *et al.*, 2006, Cafe *et al.*, 2008).
Primary physiological responses to stress in animals are the activation of the sympatho-adrenal-medullary (SAM) and hypothalamo-pituitary-adrenal (HPA) axes. Exogenous stimulation or challenge of the HPA axis with adrenocorticotropin (ACTH) is used to assess HPA axis reactivity in cattle (Verkerk *et al.*, 1994).
Although it is likely that temperament is related to stress-susceptibility (Curley *et al.*, 2008), and that temperament modifies performance via this mechanism (Ferguson *et al.*, 2006), this assumption requires further scientific testing. In this paper we report findings on the physiological response of cattle differing in temperament, and also on hormonal growth promotant (HGP) status, to routine handling followed by an ACTH challenge.

Material and methods

Brahman steers (n=81) were sourced at weaning from central Queensland, backgrounded at Glen Innes Research Station, New South Wales and finished at *Tullimba* feedlot, New South Wales. They were selected for their tenderness genotype as having either 0 or 2 copies of the favourable alleles of two calpain-system gene markers (Greenwood *et al.*, 2009). At feedlot entry (live weight±SD, 334±40 kg), half the steers were implanted with an HGP (revalor-H®) composed of 200 mg trenbolone acetate and 20 mg 17β-oestradiol.
Flight speed (exit speed from crush, m/s) of the cattle was measured approximately monthly during backgrounding and weekly at the feedlot. The feedlot flight speed was used to stratify the cattle into 3 temperament categories based on the mean and SD of their feedlot flight speed measurements: (1) Quiet, 1 SD below the mean; (2) Flighty, 1 SD above the mean; (3) Average, all other cattle.
HPA axis responsiveness was assessed using an ACTH challenge near the end of the feedlot period. The concentration of cortisol and metabolites (glucose, lactate and non-esterified fatty acids (NEFA)) in plasma were measured immediately prior to and 60 min after intramuscular injection of 2.5 μg/kg live weight of synthetic ACTH (Synacthen®).
Data were analysed using linear mixed models in the Genstat software package (VSN International Ltd, Hemel Hempstead, UK). In the final model, the fixed effects were calpain-system markers, HGP treatment, and temperament category. Property of origin, backgrounding replicate, feedlot replicate, and day and timing of ACTH challenge were fitted as random.

Results

Pre-challenge cortisol concentration was higher with increasing flight speed, but there was no difference in the cortisol response to ACTH between temperament categories (Table 1). Glucose

and NEFA increased with ACTH challenge whilst lactate decreased. Glucose, lactate and NEFA were significantly higher in the flighty cattle both pre- and post-ACTH challenge.
The HGP treatment reduced cortisol concentration both pre-challenge (30 vs. 52, sed 4.2 nmol/l, $P<0.001$) and post-challenge (124 vs. 172, sed 5.2 nmol/L, $P<0.001$), but had no effect on metabolites at either time point. There was no association between any of the physiological measures and the calpain-system gene markers (data not presented).

Table 1. Feedlot flight speed (FFS, m/s), and plasma cortisol (nmol/l) and metabolite (mmol/l concentrations before and 60 min after an ACTH challenge in Brahman steers categorised into temperament groups based on FFS.

Category	n	FFS	Pre-challenge				Post-challenge			
			Cortisol	Glucose	Lactate	NEFA	Cortisol	Glucose	Lactate	NEFA
Quiet	8	1.2[a]	23[a]	5.9[a]	1.4[a]	0.12[a]	150	6.2[a]	0.8[a]	0.14[a]
Average	59	1.9[b]	42[b]	6.1[a]	1.5[a]	0.15[ab]	140	6.5[a]	1.2[a]	0.20[ab]
Flighty	14	3.3[c]	58[c]	6.9[b]	2.7[b]	0.18[b]	154	7.6[b]	2.0[b]	0.24[b]
sed		0.14	7.4	0.27	0.35	0.024	9.1	0.33	0.44	0.033
P-values		<0.001	<0.001	<0.001	<0.001	0.092	n.s.	<0.001	0.019	0.036

Within columns, means with different superscripts differ significantly.

Conclusion

The results show an increased physiological response to handling in cattle with poorer temperament, suggesting greater susceptibility of the flightier animals to the stress associated with handling. The cortisol (HPA axis) response to ACTH did not diffffer with temperament. Metabolites differed with temperament before and after an ACTH challenge suggesting that these responses were not driven by the HPA axis alone. Treatment with HGP reduced HPA axis responsiveness to handling and to the ACTH challenge.

References

Cafe, L.M., D.M. Ferguson, D.L. Robinson and P.L. Greenwood, 2008. Temperament of young Brahman cattle assessed during backgrounding persists and is related to performance. Anim. Prod. Aust. 27: 28.

Curley, K.O. Jr, D.A. Neuendorff, A.W. Lewis, J.J. Cleere, T.H. Welsh Jr and R.D. Randel, 2008. Functional characteristics of the bovine hypothalamic-pituitary-adrenal axis vary with temperament. Horm. Behav. 53: 20-27.

Ferguson D.M., D. Johnston, H.M. Burrow and A. Reverter, 2006. Relationships between temperament, feedlot performance and beef quality. In: Australian Beef - the Leader. Beef CRC, Armidale, March 2006, pp. 161-165.

Greenwood, P.L., L.M. Cafe, D.W. Pethick, D.L. Robinson and J.M. Thompson, 2009. Production and processing studies on calpain-system gene markers in cattle. Ruminant physiology: digestion, metabolism, and effects of nutrition on reproduction and welfare. In: Chilliard, Y., F. Glasser, Y. Faulconnier, F. Bocquier, I. Veissier and M. Doreau (Eds.), Ruminant Physiology. Digestion, metabolism and effects of nutrition on reproduction and welfare. Wageningen Academic Publishers, Wageningen, the Netherlands. p. 518-519.

King, D.A., C.E. Schuehle Pfeiffer, R.D. Randel, T.H. Welsh Jr, R.A. Oliphint, B.E. Baird, K.O. Curley Jr, R.C. Vann, D.S. Hale and J.W. Savell, 2006. Influence of animal temperament and stress responsiveness on the carcass quality and beef tenderness of feedlot cattle. Meat Sci. 74: 546-556.

Verkerk, G.A., K.L. MacMillan and L.M. McLeay, 1994. Adrenal cortex response to adrenocorticotropic hormone in dairy cattle. Domest. Anim. Endocrinol. 11: 115-123.

The effects of Yerba Mate (*Ilex paraguarensis*) supplementation on the productive performance of lambs

P. Celi and H.W. Raadsma
The University of Sydney, 425 Werombi Road Camden, 2570, NSW, Australia; pietroc@camden.usyd.edu.au

Introduction

The animal feed industry is under increasing consumer pressure to reduce the use of antibiotics as feed additives. This is a natural consequence of the increasing demand of safe products for human consumption. The use of herbs as additives in livestock nutrition, as an alternative to other chemical compounds, is becoming a new goal in livestock production (Makkar *et al.*, 2007). It is known that Yerba Mate (YM) tea (*Ilex paraguarensis*) exerts antioxidant activity due to its content of several compounds such as polyphenols (Heck and De Mejia, 2007). The use of alternative feedstuff like YM, in ruminant nutrition represents a novel management tool that is green, clean, ethical (Martin and Kadokawa, 2006) and extremely easy to use. However, the effect of YM supplementation in ruminants has not yet been studied and thus, the role and the activity of natural antioxidants not commonly present in the diets of ruminants warrant investigation. The hypothesis of this study was that the supplementation of YM tea in lambs' diet will improve their productive performances. We proposed to test this hypothesis by supplementing lambs' diet with YM, and to monitor their oxidative and metabolic status, productive performances and immune response.

Material and methods

The study was conducted at the animal handling facilities of the Camden campus, Faculty of Veterinary Science. We used 40 Merino × Awassi × Dorset lambs homogeneous for age (10 weeks) and live weight (LW; 21.8±0.6 kg). Lambs were fed a pelleted concentrate diet (ME 12.5 MJ/kg DM and 16% crude protein content; Control Diet) *ad libitum* through automatic feeders that automatically recorded feed intake (VFI). Lambs were gradually acclimatised to the pelleted diet over a two weeks period and then were randomly divided into two groups: one group (n=20) was fed the control diet and one group (n=20) was fed the control diet with 2.5% YM (*Ilex paraguarensis*). Lambs received the dietary treatments for a total of 8 weeks and were monitored weekly for LW. Blood samples (8 ml) were taken from all lambs, at Week 0 (before the start of the administration of the dietary treatments) and on Weeks 2, 4, 6, and 8. Blood samples were centrifuged and plasma was analysed for total antioxidant capacity, advanced oxidation protein products (AOPP), non-esterified fatty acids (NEFA), triglycerides and cholesterol. Wool growth was measured according to the midside patch technique. Wool was shorn from a 10×10 cm area at the beginning of the trial (Week 0) and discarded, while the wool clipped on week 8 was saved in a sealable clip-on plastic bag. The sample was weighed and then washed to determine the clean fleece weight. The clean sample weights were then converted to an individual average daily growth rate. Cell-mediated immune status was evaluated by means of skin tests which were performed on all animals at the beginning (week 0) and at the end (week 8) of the experimental period. During each test, 1 mg of phytohaemagglutinin (PHA) dissolved in 1 ml of sterile saline solution was injected intra-dermally on the upper side of each shoulder. The average increase in skinfold thickness of each animal was calculated from the two measurements made with callipers. Changes in LW, VFI and plasma concentration of metabolites were analysed by means of ANOVA for repeated measures procedure. Analysis included between-subjects main effect of diet, within-subjects main effect of time of sampling and interaction time of sampling × diet. The effects were considered to be significant at $P<0.05$; differences between means were tested using least significant difference.

Results

Overall, YM lambs ate significantly higher (P<0.001) levels of pellet than the control lambs, however, LW and growth rate were not significantly different between the two groups. Wool growth rates were affected by the diet treatments (P=0.003), with the YM lambs having significantly higher growth rates when compared to the control group (0.09±0.005 *vs* 0.068±0.005 g/day). There was a significant effect of diet on plasma AOPP concentrations (P<0.05), with YM lambs presenting higher AOPP levels than the control lambs (3.3±0.03 mmol/l *vs* 3.1±0.03 mmol/l). Overall, the YM lambs showed significantly lower plasma NEFA (P<0.001) and triglycerides (P<0.001) concentrations than the control lambs, (1.92±0.06 mmol/l, 2.29±0.06 mmol/L; P<0.001) and (0.277±0.001 mmol/l, 0.287±0.001 mmol/L, P<0.001) respectively. No significant effect of diet was noted for PHA test, plasma cholesterol and total antioxidant capacity levels.

Discussion and conclusion

Yerba Mate could be recommended as a natural novel feed supplement with the potential for improving feed intake and wool growth in lambs. Several compounds have been identified in YM such as tannins (Heck and Mejia, 2007). The inclusion of moderate levels of tannin in the diet can increase VFI (Wang *et al.*, 1996). Tannins have also been shown to increase protein rumen undegradable protein and making feed protein available post-ruminally for production purposes (Makkar *et al.*, 2007). This might explain the greater wool production in YM supplemented lambs. Further work is required to investigate the mechanisms by which YM supplementation improves feed intake and wool production. These studies can lead to the development of new feeding strategies that can exploit the beneficial effects of YM.

Acknowledgement

This work was supported by the Faculty of Veterinary Science (University of Sydney) and by the Australian Wool Education Trust (AWET).

References

Heck, C.I. and E.G. de Mejia, 2007. Yerba Mate Tea (*Ilex paraguariensis*): A Comprehensive Review on Chemistry, Health Implications, and Technological Considerations. J. Food Sci. 72: R138-R151.

Makkar, H.P.S., G. Francis and K. Becker, 2007. Bioactivity of phytochemicals in some lesser-known plants and their effects and potential applications in livestock and aquaculture production systems. Animal 1: 1371-1391.

Martin, G.B. and H. Kadokawa, 2006. 'Clean, green and ethical' animal production. case study: reproductive efficiency in small ruminants. J. Reprod. Develop. 52: 145-152.

Wang, Y., G.B. Douglas, G.C. Waghorn, T.N. Barry, A.G. Foote and R.W. Purchas, 1996. Effect of condensed tannins upon the performance of lambs grazing *Lotus corniculatus* and lucerne (*Medicago sativa*). J. Agric. Sci. 126: 87-98.

Response of white blood cell stress-related gene expression to heat stress in lactating dairy cattle

K. DiGiacomo[1], F.R. Dunshea[1], B.J. Leury[1], L.H. Baumgard[2] and R.P. Rhoads[2]
[1]University of Melbourne, Parkville, Australia; [2]University of Arizona, Tucson, AZ, USA; rhoadsr@email.arizona.edu

Introduction

Agricultural animals, particularly lactating dairy cattle, frequently experience heat stress (HS) which negatively effects animal health and production. Annually, milk yield per cow continues to steadily increase and this is accompanied with a parallel rise in metabolic heat production further increasing the cow's susceptibility to HS. The negative effects of HS on the global dairy economy are staggering and the estimated annual economic loss due to HS in the USA alone exceeds $ 900 million (St-Pierre *et al.*, 2003). Historically, the majority of HS research has focused on management (i.e. housing, cooling systems, etc.) and nutritional strategies to reduce its negative effects. However, the effects of HS on genetic and cellular responses have not been extensively examined.

During periods of environmental and physiological stress specific intracellular pathways are activated to illicit functional adaptations which serve to protect the cell from damage and/or death. Such protective mechanisms involve two transcription factors, heat shock factor-1 (HSF-1) and hypoxia-inducible factor-1 (HIF-1). It is well-established that HSF-1 is the main regulator of heat shock genes while HIF-1 participates in the regulation of hundreds of genes involved in cell survival, metabolism, proliferation and angiogenesis. The positive effects of HSF and HIF-1 on cell survival stem, in part, from their ability to induce a highly conserved family of proteins, termed Heat shock proteins (Hsp), which are expressed in nearly all organisms and cell types. The Hsp assist in the re-folding or degradation of damaged proteins while protecting proteins from denaturation and further stress-induced damage. Studies from other species indicate HS acclimation is associated with an increased ability to induce Hsp (Kresfelder *et al.,* 2006) and in agricultural species, such characteristics may lead to enhanced well-being and productivity. We chose to examine the effect of hyperthermia on white blood cells (WBC) because animals are more susceptible to immune challenges during periods of HS and harvesting WBC is a minimally invasive means of monitoring cellular response during HS. In an initial effort to generate support for the above hypothesis, we sought to characterise expression of stress-related genes in the WBC population during HS in lactating dairy cattle.

Material and methods

Four lactating multiparous Holstein cows (milk yield ~40 kg/d; 2.75 BCS and days in milk ~ 115) were housed in the University of Arizona's environmental chambers. Cows were exposed to 7 d of thermoneutral (TN) conditions (18 °C, 20% humidity, 12/12 h light and dark cycle), then 7 d of cyclical HS (29 °C to 40 °C, 20% humidity, 12/12 h light and dark cycle). Cows were fed a TMR diet consisting primarily of alfalfa hay and steam-flaked corn. Body temperature indices measured at 06:00, 13:00 and 16:00 h daily and included rectal temperature (Tr) and respiration rate (RR). Feed intake and milk yield was measured daily. Blood samples from the coccygeal vein were obtained at 13:00 h during the TN period, on d 2 (24 h post-HS initiation) and d 7 of HS and WBC were immediately harvested for isolation of total RNA using the RiboPure™-Blood Kit (Ambion Inc. Austin, TX, USA). The RNA content of each sample was calculated based on absorbance at 260 nm, quality evaluated by calculating the ratio of absorbance at 260 nm and 280 nm, followed by Experion chip electrophoresis analysis (BioRad laboratories inc, Hercules, CA, USA). The SuperScript First Strand Synthesis System for RT-PCR (Invitrogen) was used to reverse transcribe

250 ng of RNA to cDNA. Real-time PCR using Sybr Green I dye was used to evaluate expression of Hsp72, Hsp90, HSF-1, HIF-1α, S15 and RPS9 genes. Statistical analysis was performed using a general linear model with Minitab. Data are reported as LSmeans and considered significant if $P<0.05$.

Results and discussion

During HS, Tr increased ($P<0.05$) by >2.0 °C and RR doubled ($P<0.05$) compared to the TN period. By d 2 of HS feed intake decreased ($P<0.05$) by 27% compared to TN values, but milk yield did not decline (40.1 kg/d, SD±5.3 kg/d). On d 7 of HS, feed intake was decreased by 36% and milk yield decreased ($P<0.05$) by 20% compared to TN values. The body temperature indices and decreased nutrient intake and milk synthesis are indicative of severe heat stress. Abundance of HSF-1 and HIF-1α mRNA remained similar between TN and throughout the HS period. At the onset of HS Hsp72 mRNA abundance did not differ from the TN period but increased (3-fold; $P<0.05$) following 7 d of HS. Hsp90 exhibited a similar numerical trend by remaining alike at TN and 24 h after the initiation of HS while increasing 2-fold after 7 d of HS (although tending towards significance; $P=0.13$).

HSF-1 and HIF-1 interact with promoter regions to regulate Hsp gene expression (Huang *et al.*, 2009; Kresfelder *et al.*, 2006). Discordant changes in gene expression between HSF, HIF and the Hsp may indicate that regulation of HSF and HIF activity in bovine WBC during HS occurs via post-translational rather than transcriptional mechanisms. Indeed, in the non-stressed state HSF-1 is bound to either Hsp70 or 90 and is inactive. The onset of HS draws away the Hsp's leaving HSF-1 free to translocate from the cytoplasm to the nucleus where they initiate gene transcription (Kresfelder *et al.*, 2006). Activity of HIF-1 is also controlled predominately in a post-translational manner (Hirota and Semenza, 2006). We do not understand the basis for the temporal pattern in Hsp gene expression but the simplest explanation may relate to an increase in Tr (core) after 7 d of HS and the presence of a greater heat load.

Conclusion

During chronic HS, abundance of WBC Hsp72 mRNA is increased significantly in lactating dairy cows despite invariant expression of HSF and HIF-1α genes. Further studies are warranted to determine if differences between WBC Hsp gene expression corresponds to differences in health and production.

References

Hirota, K. and G.L. Semenza, 2006. Regulation of angiogenesis by hypoxia-induced factor 1. Crit. Rev. Oncology/Hematology 59: 15-26.

Huang, W.J., L.M. Xia, F. Zhu, B. Huang, C. Zhou, H.F. Zhu, B. Wang, B. Chen, P. Lei, G.X. Shen and D.A. Tian, 2009. Transcriptional upregulation of HSP70-2 by HIF-1 in cancer cells in response to hypoxia. Int. J. Cancer 124: 298-305.

Kresfelder, T.L., N. Claassen and M.J. Cronje, 2006. Hsp70 induction and *hsp70* gene polymorphisms as indicators of acclimatization under hyperthermic conditions. J.Thermal Biol. 31: 406-415.

St-Pierre, N.R., B. Cobanov and G. Schnitkey, 2003. Economic losses from heat stress by US livestock industries. J. Dairy Sci. 86: E52-E77.

Feeding selenomethionine improves viability in Iranian Holstein suckling calves

M. Ebrahimi, A. Towhidi and A. Nikkhah
Department of Animal Science, Faculty of Agronomy and Animal Science, University of Tehran, P.O. Box#3158777871-4111, Karaj, Iran; mrzh_ebrah@yahoo.com

Introduction

Nutritional degenerative myopathy (white muscle disease) and poor growth rate of calves are common manifestations of Se deficiency (Underwood, 1977). As an integral part of the enzyme glutathione peroxidase (GSH-Px), Se prevents oxidative damage to body tissues (Hoekstra, 1974). Gunter *et al.* (2003) reported that the activity of glutathione peroxidase (GSH-Px) in the erythrocytes was significantly greater in the supplemented groups of cows with selenium than in the control group. Backall and Scholz (1981) reported that Holstein and Angus cows with adequate whole-blood selenium status (90 to 105 ng/ml) had whole-blood GSH-Px-1 activity between 27 to 35 EU/g of hemoglobin, whereas cows with marginal status had whole-blood GSH-Px-1 activity of 11.5 EU/g of hemoglobin. It was also shown that there are increased immunoglobulin concentrations (IgG) in the plasma of pregnant cows supplemented with selenium and also in their calves (Awadeh *et al.*, 1998). However, the mechanism of selenium effect on immunoglobulin concentrations is not clear. Accordingly, the objectives of this study were to determine the effect of supplemented milk with Se-Met on glutathione peroxidase and immunoglobulin concentrations (IgG), and therefore viability in Holstein suckling calves.

Material and methods

On February 4, 2006, ten Iranian Holstein suckling male calves with approximately one month of age were selected and randomly assigned to two groups (n=5). Calves had free choice starter with the following: (1) unsupplemented milk, (2) supplemented milk with 0.3 mg Se/kg (dry matter intake of milk) as Se-Met for two months during the winter. Milk intakes were set in a fixed amount at 10% of BW and were adjusted based on BW weekly. Calves had *ad libitum* access to water. Starter was formulated to meet the animal's requirements for energy, protein, vitamins, and minerals (except Se) for calves (NRC, 2001). On the 60th day of the trial, blood samples were also collected, and heparinised whole blood samples were stored at -80 °C for determining glutathione peroxidase activity. Two unheparinised blood samples were also collected on the 50th and 60th day after the treatment from all calves. Blood samples were centrifuged (2000 g for 15 min at 4 °C) and sera were separated and stored at -20 °C to measure IgG concentrations. Selenium-dependent glutathione peroxidase (GSH-PX-1) activity (as enzyme unit per milligram of hemoglobin) in blood was assayed with the method of Paglia and Valentine (1967). Concentrations of IgG in the serum of calves were measured by radial immunodiffusion (VMRD Inc, Pullman, WA). The effect of treatment on glutathione peroxidase activity in whole blood and serum IgG concentration were analysed using Proc GLM in a completely randomised design of SAS® (1990).

Results

Glutathione peroxidase activity of calves was affected by Se-Met (P<0.01) and it was 110% greater in Se-Met group than in the control group (Table 1). The results also showed that calves in the Se-Met group were in an adequate whole-blood glutathione peroxidase, while calves in the control group were in marginal whole blood glutathione peroxidase activity. Serum concentration of IgG was increased by Se-Met supplementation (P<0.05; Table 1).

Table 1. The effect of Se-Met on different blood parameters.

	Control	Se-Met	SEM	*P*-value
GSH-Px, EU/g Hb	19.39	40.7	3.96	0.01
IgG, mg/ml	32.21	44.37	3.65	0.04

Conclusion

The results indicate that Se-Met supplementation to the milk of suckling calves with a marginal selenium status increased whole blood glutathione peroxidase activity and the synthesis of IgG, thus, supplemental Se-Met apparently improves the viability of suckling calves.

References

Awadeh, F.T., R.L. Kincaid and K.A. Johnson, 1998. Effect of level and source of dietary selenium on concentrations of thyroid hormones and immunoglobulins in beef cows and calves. J. Anim. Sci. 76: 1204-1215.

Backall, K.A. and R.W. Scholz, 1981. Reference values for a field test to estimate inadequate glutathione peroxidase activity and selenium status in the blood of cattle. Am. J. Vet. Res. 40: 733-738.

Gunter S.A., P.A. Beck and J.M. Phillips, 2003. Effects of supplementary selenium source on the performance and blood measurements in beef cows and their calves. J. Anim. Sci. 81: 856–864.

Hoekstra, W.G., 1974. Biochemical role of Selenium. In: Hoekstra W.G., J.W. Suttie, H.E. Ganther and W. Mertz (Eds), Trace element metabolism in animals, Vol. II. University Park Press, Baltimore, MD, USA, pp. 61-77.

NRC, 2001. Nutrient requirements of the young calf, 7th rev. ed. Natl Acad. Press, Washington, DC, USA, pp. 214-233.

Paglia, D.E and W.N. Valentine, 1967. Studies on the quantitative and qualitative characterization of erythrocyte glutathione peroxidase. J. Lab. Clin. Med. 70: 158-169.

SAS, 1990. User's Guide, Statistics, Version 6. SAS Institute, Inc., Cary, NC, USA.

Underwood, E.J., 1977. Trace elements in human and animal nutrition, 4th ed. Academic Press, Inc., New York, NY, 545 pp.

Effects of age on transportation and preslaughter stress responsiveness in Moroccan dromedary camels

M. El Khasmi[1], F. Riad[1], A. Safwate[1], H. El Tahri[1], M. Farh[1], N. El Abbadi[2], M. Bengoumi[3], V. Coxam[4] and B. Faye[5]

[1]Equipe Hormones et Métabolisme, Université Hassan II-Mohammedia, Faculté des Sciences Ben M'Sik, Casablanca, Maroc; [2]Centre National de l'Énergie des Sciences et des Techniques Nucléaires, Maâmoura, Maroc; [3]Institut Agronomique et Vétérinaire Hassan II, BP 6202, Rabat, Maroc; [4]INRA, Clermont-Theix, France; [5]CIRAD-EMVT, Montpellier, France; elkhasmi_mohammed@hotmail.com

Introduction

The dromedary camel is a good source of meat especially in areas where the climate adversely affects the performance of other animals (Kadim *et al.*, 2006). However, camel meat comes mostly from old females and males that are primarily kept for milk, racing, and transportation rather than for meat production (Kurtu, 2004). In addition, transportation and preslaughter stress induce high circulating levels of cortisol in goats (Kannan *et al.*, 2000) and beef calves (King *et al.*, 1991), which result in changes of muscle metabolism. The aim of this study was to compare the transportation and preslaughter stress responsiveness of young camels with those of old ones slaughtered at comparable conditions.

Material and methods

Twenty-eight male Moroccan camels (n=28) were classified into four groups: young control group (YC) and young stressed group (YS) (2 to 3 yr of age, average weight of 194 kg) and old control group (OC) and old stressed group (OS) (8 to 9 yr of age, average weight of 430 kg). All the animals were clinically healthy, feed deprived overnight, and slaughtered after 2 h road transportation (70 km) thus having been held in a lairage for 1 h, according to traditional procedures. Animals were slaughtered at the Tit-Mellil Municipality slaughterhouse. Blood samples were taken at 06:00 h into heparinised tubes. After hematocrit measurement, plasma was separated by centrifugation at $750\times g$ for 15 min, pipetted into different aliquots then stored at -40 °C until analysis. In plasma, calcium, sodium, and potassium concentrations were measured using an atomic absorption spectrophotometer. Phosphorus level was measured by colorimetry. Plasma hormone (cortisol, thyroxine and 25-hydroxyvitamin D) concentrations were analysed by the radioimmunoassay method at the National Center of Science and Nuclear Technical Energy, Maamoura. The data were expressed in SI units and analysed by repeated measurements ANOVA. All values were expressed as mean and standard error (SE), and $P<0.05$ was seen as statistically significant.

Results

We found no significant difference in plasma calcium, phosphorus (mg/l), sodium and potassium (mM) and hematocrit (%) between young and old control camels (101±2.6 vs. 100±2.5; 60±1.7 vs. 60±1.6; 168±6 vs. 170±6 and 7.0±0.5 vs. 6.6±0.6 and 29±0.83 vs. 30±0.60, respectively) nor between young and old stressed animals (100±2.7 vs. 98±2.4; 61±1.6 vs. 60±1.5; 164±5 vs. 169±6 and 6.8±0.6 vs. 6.8±0.6 and 30±0.41 vs. 31±0.77, respectively).
Plasma levels of 25-hydroxyvitamin D showed no significant differences between all groups (Table 1), however, plasma thyroxine levels in stressed groups were significantly higher than those of controls ($P<0.05$) (Table 1). In old and young stressed camels, the mean value of cortisol was significantly higher ($P<0.01$) than the corresponding mean from control animals, but in old stressed camels, this value was significantly greater ($P<0.05$) than those of the young stressed ones (Table 1).

Table 1. Effects of transportation and preslaughter stress on plasma concentrations of cortisol (nM), thyroxine (nM) and 25-hydroxyvitamin D [25(OH)D] (ng/ml) in young and old control camels (YC and OC respectively) and young and old stressed groups (YS and OS respectively).

Animal groups	YC	OC	YS	OS
Cortisol	60.41±5.3	58.73±6.1	180.1±36.6	278.74±50.4*
Thyroxine	165.56±10.4	184.25±11.32	221.41±23.4	230.7±29.16*
25(OH)D	374.28±63.41	340.46±56.28	355.16±60.27	326.31±51.27

* $P<0.05$: comparison between YS and OS.

Discussion and conclusion

In our animals, transportation and preslaughter stress increased the plasma levels of cortisol without any variation of plasma thyroxine and 25-hydroxyvitamin D levels. However, the magnitude of cortisol response was higher in old camels than in young ones. Thus, considerable attention must be given to decrease transportation and preslaughter stress for old camel welfare. The mechanisms involved that can explicit whether the old camels are more stressed than the young ones remain to be determined.

References

Kadim, I.T., O. Mahgoub, W. Al-Marzooqi, S. Al-Zadgali, K. Annamali and M.H. Mansour, 2006. Effects of age on composition and quality of muscle *longissimus thoracis* of the Omani Arabian camel (*Camelus dromedaries*). Meat Sci. 73: 619-625.

Kannan, G., T.H. Terrill, B. Kouakou, O.S. Gazal, S. Gelaye, E.A. Amoah and S. Samake, 2000. Transportation of goats: Effects on physiological stress responses and live weight loss. J. Anim. Sci. 78: 1450-1457.

King, B.D., R.D.H. Cohen, C.L. Guenther and E.D. Janzen, 1991. The effect of age and method of castration on plasma cortisol in beef calves. Can. J. Anim. Sci. 71: 257-263.

Kurtu, M.Y., 2004. An assessment of the productivity for meat and carcass yield of camel (*Camelus dromedarious*) and the consumption of camel meat in the Eastern region of Ethiopia. Trop. Anim. Health Prod. 36: 65-76.

Influence of two drying off methods on udder health in Holstein cows given short dry periods

M.H. Ghafari, G.R. Ghorbani, H.R. Rahmani, M. Yari, A.H. Ghafari, A. Akbariyan and M. Mirzaee
Department of Animal Science, Isfahan University of Technology, Isfahan 84156 Iran;
moterza.h.g@gmail.com

Introduction

The dry period is a crucial stage of the lactation cycle with respect to udder health. During the dry period, remodelling of mammary tissue takes place, which allows the reinitiation of lactation to be at a maximal level after the next calving. It has been proposed that the process of mammary involution is complete by day 25 of the dry period in dairy cows (Gulay, 2005). Shortening the dry period and feeding one high energy diet increase DMI, eliminate unnecessary stress, and decrease the incidence of metabolic disorders during the transition period (Rastani *et al.*, 2005). Shorter (<30 d) dry periods had no negative influences on DMI, BW or apparent health problems during the *post partum* period (Gulay *et al.*, 2003). Methods of drying off may have an effect on udder protective factors, such as bovine lactoferrin, which has bacteriostatic and bactericidal properties attributed to its ability to chelate iron or to bind to the bacterial surface and damage the outer membrane of Gram-negative bacteria (Machnicki, 1991). The aim of the present study was to assess the effects of two drying off methods on udder health in Holstein cows given a 30 d dry period.

Material and methods

Eighteen multiparous dairy cows were dried off approximately 4 weeks prior to the expected calving time. Treatments were arranged in a completely randomised design with two treatments and nine replicates. The two drying off methods, abrupt (AD) or intermittent (ID) cessation of milking, were used in this study. For intermittent milk cessation treatment, cows were dried off in the following manner. The afternoon milkings were terminated from 7 to 3 d prior to drying off, and both morning and afternoon milkings were ceased from 3 to 0 d prior to drying off. Dry period lengths (mean±s.e) were 30±3 d. Days in milk for drying off methods AD and ID were 304.4±9 and 305.2±9, respectively. Milk samples were collected from front and rear quarters from 7 to 3 d prior to drying off, at drying off, 2 weeks and 1 week prior to calving, at calving and 7 d post-calving. Milk yield on the drying off day for methods AD and ID were 6.84 and 5.81, respectively. Udder secretion samples were stored in a freezer (-18 °Ċ) prior to lactoferrin quantitation. Additonal samples were stored at 4 °Ċ for somatic cell count (SCC) and total bacteria count (TBC). The SCC in milk samples were determined electronically using a Fossomatic cell counter and TBC in milk samples were determined using a Bacterioscan (Photo Denmark). The concentration of lactoferrin in the samples was measured by ELISA using a commercial kit (at Isfahan University of Technology, Iran). Data were analysed using the mixed procedure for repeated measures of SAS® version 8.2.

Results

The results indicate that there was a significant effect of drying off methods on lactoferrin concentration; it was higher in dairy cows dried off by method ID in samples collected 2 weeks and 1 week prior to calving time ($P<0.01$). Total bacteria count was not different between the two drying off methods. Milk total bacteria counts in rear or front quarters was 5.60 and 5.38, respectively ($\log_{10}$ TBC per ml, s.e.±0.085). Milk total bacteria counts were higher in rear quarters compared to front quarters ($P<0.05$). Milk somatic cell counts were higher in cows dried off by abrupt milk cessation ($P<0.05$).

Table 1. The effect of abrupt (AD) and intermittent (ID) drying off methods on lactoferrin (LF), somatic cell count (SCC) and total bacteria count (TBC) in udder secretions.

Sampling time	LF (mg/ml)		SEM	$\log_{10}$ SCC (ml)		SEM	$\log_{10}$ TBC(ml)		SEM
	AD	ID		AD	ID		AD	ID	
On 7 d prior to drying off	0.03	0.05	0.137	5.46	5.24	0.120	5.35	5.22	0.120
On 3 d prior to drying off	0.04	0.12	0.136	5.71^{b}	5.05^{a}	0.122	5.50	5.52	0.118
On drying off	0.05	0.15	0.135	5.66^{b}	5.28^{a}	0.121	5.57	5.57	0.119
1 wk prior to calving	4.04^{b}	5.06^{a}	0.138	-	-		-	-	
2 wk prior to calving	2.32^{b}	3.15^{a}	0.135	-	-		-	-	
Calving	1.51	1.06	0.136	6.19	6.27	0.123	5.83	5.89	0.122
On 7 d post-calving	0.05	0.23	0.134	5.51	5.19	0.120	5.21	5.23	0.120

[a,b] Means in row with different superscript letters are significantly (P<0.05) different.

Discussion and conclusion

These results show that drying off methods influence the concentration of lactoferrin and somatic cell count in udder secretions of Holstein cows given short dry periods. The known antibacterial action of lactoferrin against mastitis pathogens suggests that intermittent milk cessation treatment is preferred for cows given short dry periods.

Acknowledgement

The authors gratefully acknowledge funding from Isfahan University of Technology.

References

Gulay, M.S., 2005. Altering the lactation cycle: is a 60-day dry period too long? Turk. J. Vet. Anim. Sci. 29: 197-205.

Gulay, M.S., M.J. Hayen, K.C. Bachman, T. Belloso, M. Liboni and H.H. Head, 2003. Milk production and feed intake of Holstein cows given short (30-d) or normal (60-d) dry periods. J. Dairy Sci. 86: 2030-2038.

Machnicki, M., 1991. Biological properties of lactoferrin. Folia Biol. 37: 65-76.

Rastani, R.R., R.R. Grummer, S.J. Bertics, A.Gümen, M.C. Wiltbank, D.G. Mashek and M.C. Schwab, 2005. Reducing dry period length to simplify feeding transition cows: milk production, energy balance, and metabolic profiles. J. Dairy Sci. 88: 1004-1014.

Pre-slaughter stress and lipoperoxidation: protective effect of vitamin E and plant extracts rich in polyphenols given to finishing cattle

M. Gobert[1], C. Bourguet[1], C. Terlouw[1], V. Deiss[1], O. Berdeaux[2], B. Comte[3], D. Gruffat[1], D. Bauchart[1] and D. Durand[1]

[1]INRA, UR 1213 Herbivores, Site de Theix, 63122 St-Genès-Champanelle, France; [2]INRA, Unité Mixte de Recherche Flavic, Site de Dijon, 21000 Dijon, France; [3]INRA, Unité de Nutrition Humaine, Site de Theix, 63122 St-Genès-Champanelle, France; denis.durand@clermont.inra.fr

Introduction

Stress of animals can induce oxidative stress (Chirase *et al.*, 2004) and, consequently, increases production of free radicals. During rearing, animal stress may be detrimental for reproduction and growing performances and, at slaughter, for the quality of meat products due to lipid and protein peroxidations (Aurousseau, 2002). Supplemention of ruminant diets with oil seeds rich in n-3 polyunsaturated fatty acids (PUFA) represents a rapid and efficient way to improve the nutritional quality of products. However, PUFA are susceptible to peroxidation. Our previous studies on dairy cows and on stressed sheep fed on a n-3 PUFA-rich diet showed that the simultaneous oral administration of lipophilic vitamin E (vit E) and hydrophilic plant extracts rich in polyphenols (PERP) reduced plasma lipoperoxidation (Gobert *et al.*, 2008; Durand *et al.*, unpublished data). The present study was aimed at evaluating the protective effects of vit E and PERP against plasma lipoperoxidation in finishing cows given a n-3 PUFA-rich diet and submitted to emotional and physical pre-slaughter stress.

Material and methods

Thirty-two Normandy culled cows were given a straw (30%) and concentrate (70%)-based diet supplemented with lipids (40 g oil/kg diet DM) provided by extruded linseeds (C group, n=16), or the same lipids associated with vit E (155 IU/kg) and PERP (7 g/kg diet DM) (EP group, n=16) during their 100 d finishing period. Eight cows from each feeding group were slaughtered under minimized stress (stress-) conditions (5 min transport from the cattle shed to the abattoir accompanied by a conspecific in the lorry). Eight cows from each feeding group were submitted to additional (stress+) physical exercise (15 min transport, 28 min walking in an unknown environment accompanied by 2 purposely noisy stockpersons, and again 15 min transport) and to psychological stress (non accompanied by a conspecific). Blood samples were collected immediately after stunning (before bleeding) by venepuncture. Stress levels were evaluated by plasma levels of cortisol, glucose and non-esterified fatty acids (NEFA). Plasma α-tocopherol (vit E) was measured by HPLC-UV spectrophotometry and lipoperoxidation intensity by plasma malondialdehyde (MDA) as described by Gobert *et al.* (2008). Plasma free-hydroxynonenal (4-HNE) and free-hydroxyhexenal (4-HHE), specific markers of n-6 and n-3 PUFA peroxidations respectively, were determined by GC-MS. All data were submitted to ANOVA analysis using the SAS® software (SAS Institute Inc., 2000, Cary, NC, USA).

Results and discussion

Plasma cortisol level (Table 1) was 1.8-fold higher in the stress+ cows (*P*=0.007; average 106.8 ng/ml) than in the stress- cows. This level increase was lower compared to those observed in sheep driven with the aid of shepherd dogs (×6.7, Durand *et al.*, unpublished data). Plasma glucose and NEFA (Table 1) were significantly higher in the stress+ cows (+8 and +81% respectively, *P*=0.01), probably due to increased energy needs to sustain physical exercise.

Table 1. Effect of diets (C vs. *EP) and pre-slaughter treatments (Stress-* vs. *Stress+) on stress, lipoperoxidation biomarkers and antioxidant status in plasma collected just after slaughter.*

Diets	C		EP		SEM	*P*-values		
Treatments	Stress-	Stress+	Stress-	Stress+		Antioxidant	Stress	Antioxidant ×Stress
Cortisol, ng/ml	72.11	105.26	47.50	108.37	17.31	ns	0.007	ns
Glucose, mg/dl	72.88	82.21	73.24	75.99	2.47	ns	0.01	ns
NEFA, mmol/l	0.14	0.25	0.13	0.24	0.04	ns	0.008	ns
MDA, µg/ml	0.06	0.07	0.06	0.05	0.02	ns	ns	ns
HNE, ng/ml	3.50	2.58	1.95	2.16	0.57	0.07	ns	ns
HHE, ng/ml	0.45	0.42	0.38	0.34	0.06	ns	ns	ns
Vit E, µg/ml	3.17	2.91	7.75	8.77	0.73	<0.0001	ns	ns

Intensity of plasma lipoperoxidation (Table 1) was not increased in the stress+ group in contrast to sheep submitted to larger stressing conditions (Durand *et al.*, unpublished data). Moreover, plasma HNE concentrations (Table 1) tended to decrease in plasma of cows given dietary vit E and PERP (P=0.07). Thus the addition of the two antioxidant sources had a protective action against lipoperoxidation, as reported in plasma of dairy cows (Gobert *et al.*, 2008) and sheep (Durand *et al.*, unpublished data). The protective action was observed in all cows, whether submitted to minimised or added pre-slaughter stress. Plasma vit E (Table 1) was not decreased by pre-slaughter stress and showed relatively high concentration in EP diet (up to 8.3 µg/ml). When vit E exerts its antioxidant activity, it is transformed and consequently, plasma levels decreased. The lack of stress conditions on both plasma peroxidised products and vit E levels was thus coherent.

Conclusion

Although physical exercise during pre-slaughter stress increased the use of energy reserves in cows, this did not favour plasma lipoperoxidation in contrast to earlier observations on plasma of sheep. Independently of the pre-slaughter stress conditions used in our study, the addition of vit E and PERP to the diet tends to protect plasma lipids against peroxidation in cows as reported earlier in sheep.

References

Aurousseau, B., 2002. Oxygen radicals in farm animals. Physiological effects and consequences on animal products. Inra Prod. Anim. 1: 67-82.

Chirase, N.K., W. Greene and C.W. Purdy, 2004. Effect of transport stress on respiratory disease, serum antioxidant status, and serum concentrations of lipid peroxidation biomarkers in beef cattle. Am. J. Vet. Res. 65: 860-864.

Gobert, M., B. Martin, A. Ferlay, Y. Chilliard, B. Graulet, P. Pradel, D. Bauchart and D. Durand, 2008. Plant extracts rich in polyphenols and vitamin E protect cows fed an n-3 PUFA-rich diet against lipoperoxidation. Proc. Nutr. Soc. 67 (OCE5): E165.

The effects of feeding *Chromolaena odorata* to goat dams during pregnancy on the acceptance of this feedstuff by their offspring

P.V. Hai[1,2], J.T. Schonewille[1], D.V. Tien[2], H. Everts[1] and W.H. Hendriks[1]
[1]Department of Farm Animal Health, Faculty of Veterinary Medicine, Utrecht University, Utrecht, the Netherlands; [2]Faculty of Animal Science, Hue University, Vietnam; v.h.phan@students.uu.nl

Introduction

Goats are found in many parts of Vietnam and are an important source of income for many small scale farmers. The feeding system is based on the ingestion of grasses and leaves under free ranging conditions. Obviously, protein intake is amongst others, important to optimise meat production and as such protein rich alternative feedstuffs are of great importance. *Chromolaena odorata* (Siam weed, Table 1) is classified as the most abundant weed plant in the tropical humid areas. This weed is widely available but free ranging goats do not consume *C. odorata* in large quantities, probably because of its strong smell. Overcoming this aversion against *C. odorata* is therefore of great interest since it would increase production efficiency and increase the income of small scale farmers. It is generally accepted that offspring mimic the feeding behaviour of their parents. There are indications that maternal influence on feeding behaviour can already start *in utero*. Indeed, it has been shown in pregnant sheep that various metabolites, including those with chemosensory properties, originating from the ingested feed could be detected in foetal plasma (Nolte *et al.*, 1992; Simitzis *et al.*, 2008). Consequently, it could be suggested that the foetuses are primed *in utero* for the consumption of post-weaning feedstuff intake. The current experiment investigated the acceptance of *C. odorata* by offspring from goat dams fed this feedstuff.

Table 1. Chemical composition of fresh leaf of Chromolaena odorata.

	Chromolaena odorata
Dry matter,%	18.17
Crude protein,% DM	20.63
Ether extract,% DM	3.05
Ash,% DM	6.27
Crude fiber,% DM	12.90

Material and methods

Ten female goats, with an initial mean body weight of 29±3.8 kg and aged between 12-14 mo, were successfully synchronised. From noon to about 18:00 h, all goats were allowed to range freely in the forest with barbed wire fence (~1 ha, TT.Hue province, Vietnam) and special care was taken that there was no *C. odorata* available during the free range period. From 18:00 h to noon the next day, goats were individually housed in pens (1.2×2 m) and had free access to a mineral block and fresh water. Five goats (treatment group) were offered 50 g of dried young branches and leaves of *C. odorata* for 30 min each day (at 10:00 h) for the last 3 mo of pregnancy until a week before parturition. Mean intake of *C. odorata* during pregnancy was 8.05±3.02 g/d. After weaning (3 mo after birth), one kid from each goat dam (control and treatment) was selected and was provided each day with 50 g of dried *C. odorata* leaf over a 4 wk period. After 15 min feed refusals were weighed and feed intake was calculated. During the experiment, kids were individually housed in pens. Kids were fed *ad libitum* hay and 350 to 500 g commercial concentrates each day and all feeds were withdrawn 9 h before intake of *C. odorata* was measured. Water and mineral blocks were available *ad libitum*. Feeding activities of the individually housed goat kids were monitored with a camera system (webcamXP5 v5.3.2). Intake of dried *C. odorata* leaves was analysed using SPSS 17.0 (SPSS Inc., Chicago, IL, USA) to compare the mean values per week (ANOVA).

Results and discussion

Goat kids in the treatment group had a significant higher intake of *C. odorata* than those of the control group in wk 3 and 4. *C. odorata* intake of goat kids in the treatment group increased each week, whereas there was a gradual decline in *C. odorata* intake in the control group.
Feeding behaviour characteristics such as feed intake frequency, time spent licking and smelling indicated a stronger interest and a better adaptation of goat kids in the experimental group towards *C. odorata* in comparison to the controls.

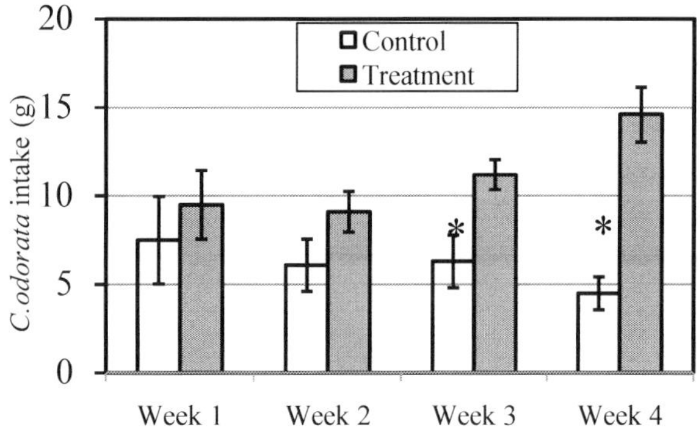

Figure 1. C. odorata *intake during 15 min (mean ± SD) in 3-mo old goat kids born to control goats or to goats fed* C. odorata *for the last 3 mo of pregnancy. * shown the significant differences between groups within a week.*

Conclusion

Goat kids from mothers fed *C. odorata* during the pregnancy show more interest toward this feedstuff than kids from mothers which did not ingest this plant. These results indicate that the foetus *in utero* can be primed by their mother's diet.

Acknowledgement

The study was supported by Asia-link project (phase III).

References

Nolte, D.L., F.D. Provenza, R. Callan and K.E. Panter, 1992. Garlic in the ovine fetal environment. Physiol. Behav. 52: 1091-1093.

Simitzis, P.E., S.G. Deligeorgis, J.A. Bizelis and K. Fegeros, 2008. Feeding preferences in lambs influenced by prenatal flavour exposure. Physiol. Behav. 93: 529-536.

Effects of the methionine analogue isopropyl ester of 2-hydroxy-4-methylthio-butanoic acid (HMBi) on blood parameters of cows under heat-stressed conditions

Z. Han[1], G. Zhou[1], Z. Jin[1], Y. Chen[1], Y. Wang[1], E. Devillard[2] and H. Peng[3]
[1]Nanjing Agricultural University, Jiangsu Province, China; [2]Adisseo France, 03600 Commentry, France; [3]Adisseo Life Science, Shanghai, China; haihong.peng@adisseo.com

Introduction

High temperatures can lead to heat stress in dairy cows, causing significant production losses. In response to heat-stress, metabolic rate of animals and thyroid function are reduced. Indeed, secretions of thyroid hormones T3, T4 and cortisol are typically decreased in the summer (Wilks *et al.*, 1990). The increase in secretion of heat-shock proteins (HSP), such as HSP70, could improve the resistance to heat-stress of animals (Collier *et al.*, 2008). Decreasing blood levels of creatine lactic acid dehydrogenase (LDH), alkaline phosphatase (ALP), superoxide dismutase (SOD), glutathione peroxidase (GSH-Px) and red cell K^+ or increasing phosphokinase (CPK) levels have also been proposed as indicators of heat-stress (Verlecar *et al.*, 2007). When balancing dairy cow diets by limiting amino acids (AA) such as methionine, nitrogen utilisation and energy status of the cows are improved, leading to an increase of milk production and milk components (St-Pierre and Sylvester, 2005), an improvement of reproduction (Thiaucourt, 1996) and a reduction of metabolic disorders (Bauchart *et al.*, 1998). However, there is no information available on how an AA balanced diet, reached by supplementing methionine, would affect response to heat-stress. The aim of this study was to investigate the effects of including methionine, provided as its analogue isopropyl ester (HMBi), on heat-stress markers measured in cows under heat-stressed conditions.

Material and methods

The trial was conducted during the hottest season in Nanjing (Jiangsu province, South China) over a 9 week-period (June-August 2008). The temperatures increased during the trial period from 26.8 to 34 to 36 °C, respectively on weeks 0, 4 week 8 of the trial. The humidity ranged from 60-75%. Thirtysix Holstein cows were allocated into 3 groups based on the age, number of lactations, yield and day in milk. The cows were fed a diet containing maize silage, alfalfa hay, grass hay, concentrates and mineral pellet. After a 1-wk covariate on the same diet, cows were fed test diets for 9 weeks. The test diets (control, treatment 1 and treatment 2) contained 0, 13 or 30 g of MetaSmart™ Dry (containing 57% HMBi) (Adisseo, China). Treatment 1 provided the animals with an AA balanced ration with 6.54% metabolisable lysine and 2.20% metabolisable methionine in metabolisable protein (MP) based on CPM model, whereas treatment 2 provided excessive methionine (6.52% of metabolisable Lysine and 2.41% of metabolisable methionine). Rectal temperatures of the animals were recorded at the beginning of the trial and once a week after commencing the trial. Blood samples were collected via the tail vein on wk0, wk4 and wk8 of the trail. Blood parameters (ALP, LDH and CPK) were measured with an auto-analyzer. K+, SOD, GSH-Px were analysed by spectrophotometry at 420 nm. The HSP70 Kit was used for detecting HSP70 in plasma and the RIA method was used for detecting T3, T4 and cortisol in plasma. All data were analysed by Anova using the statistical software SPSS16.

Results and discussion

At the beginning of the trial (wk 0), there was no difference between the control and the two treatments, regarding all blood parameters measured ($P>0.05$). The average rectal temperatures of the control, treatment 1 and treatment 2 cows were 39.37, 39.20 and 39.22 °C respectively over the trial period whereas the normal temperature range is from 37.5 to 39 °C. The serum contents of ALP, GSH-Px, HSP70, LDH, SOD, K^+, T3, T4 and cortisol in the control animals were lower on wk4 and wk8, than on wk0, as expected in response to heat stress. During the overall trial period, there was no change in all these blood parameters when animals were supplemented with either 13 or 30 g of MetaSmart™ Dry. Serum CPK content in control animals was significantly increased ($P<0.001$) in response to heat-stress conditions, whereas it was unchanged in animals supplemented with MetaSmart™ Dry ($P>0.05$). The release of CPK in the blood could be related to cell damage occurring with heat-stress, and would be lowered when dietary AA are balanced. There was no difference in responses for any of the parameters studied between the two levels of MetaSmart™ Dry, indicating, that once the ration was balanced for metabolisable methionine and lysine, there was no additional benefit on heat stress parameters to increasing methionine levels further.

Table 1. Blood parameters of animals under heat-stressed conditions supplemented with 0, 13 or 30 g of MetaSmart™ Dryon wk8 of the trial.

Treatments (MetaSmart g)	Control 0	Treatment1 13	Treatment2 30	*P*-value
APL (U/l)	54.08^a	60.75^b	60.58^b	<0.001
CPK (U/l)	170.17^a	112.98^b	111.89^b	<0.001
GSH-Px (U/l)	139.43^a	148.95^b	148.51^b	<0.001
LDH (U/l)	892.00^a	1,157.80^b	1,159.42^b	<0.001
SOD (U/l)	136.99^a	153.38^b	153.91^b	<0.001
HSP70 (ng/ml)	17.53^a	25.47^b	25.52^b	<0.001
K^+ (mg/l	593.60 ^{a}	691.17^b	642.81^b	<0.05
T3 (ng/ml)	1.89^a	2.64^b	2.66^b	<0.001
T4 (ng/ml)	90.62^a	118.02^b	117.73^b	<0.001
Cortisol (ng/ml)	4.31^a	6.40^b	6.41^b	<0.001

^{a,b} Numbers in the same row with different superscripts are significantly different.

Conclusion

Cows challenged to temperatures up to 36 °C showed typical blood heat stress markers, when the ration was not balanced in AA. Providing the limiting methionine led to favourable levels of blood heat stress markers, suggesting that balancing rations for AA could help to counteract heat stress in cows. Further studies are required to investigate the beneficial effects of feeding dairy cows with AA balanced rations under heat-stressed conditions and how this could impact on dairy production and cow reproduction.

References

Bauchart, D., D. Durand, D. Gruffat and Y. Chilliard, 1998. Mechanism of liver steatosis in early lactation cows-effects of hepatoprotector agents. Proceedings of the Cornell Nutrition Conference, pp. 27-37.

Collier, R.J., J.L. Collier, R.P. Rhoads and L.H. Baumgard, 2008. Genes involved in the bovine heat stress responses. J. Dairy Sci. 91: 445-454.

St-Pierre, N.R. and J.T. Sylvester, 2005. Effects of 2-hydroxy-4-(Methylthio) Butanoic (HMB) and Its isopropyl Ester on Milk Production and Composition by Holstein Cows. J. Dairy Sci. 88: 2487-2497.

Thiaucourt, L., 1996. L'opportunité de la méthionine protégée en production laitière. Bulletin des GTV 2B, p. 45.

Verlecar, X.N., K.B. Jena and G.B.N. Chainy, 2007. Biochemical markers of oxidative stress in Perna viridis exposed to mercury and temperature. Chemico-Biolog. Interact. 167: 219-226.

Wilks, D.L., C.E. Coppock, J.K. Lanham, K.N. Brooks, C.C. Baker and W.L. Bryson, 1990. Responses of lactating holstein cows to chilled drinking water in high ambient temperatures. J. Dairy Sci. 73: 1091-1099.

Effect of level of endophyte-infected perennial ryegrass intake on plasma prolactin and some physiological parameters in Merino ewes

M.L.E. Henry[1], *S. Kemp*[2], *I.J. Clarke*[3], *F.R. Dunshea*[1] *and B.J. Leury*[1]
[1]*Department of Agriculture and Food Systems, University of Melbourne, VIC, Australia;* [2]*PastureWise, Caroline Springs, VIC, Australia;* [3]*Department of Physiology, Monash University, Clayton, VIC, Australia; brianjl@unimelb.edu.au*

Introduction

Perennial ryegrass toxicity (PRGT) can affect livestock grazing perennial ryegrass (PRG) infected with the endophyte *Neotyphodium lolli*. This endophyte produces two major classes of toxins harmful to livestock, ergot alkaloids (e.g. ergovaline) and indole-diterpenes (e.g. lolitrem B) which can produce tremors and hyperthermia (e.g. Spiers *et al.*, 2005). In southeast Australia, outbreaks of PRGT have caused serious production losses (Cawdell-Smith *et al.*, 2007) but there is little known about the sub clinical effects of toxin ingestion on sheep production. Most research has focussed on grazing studies in which toxin intake is not known (e.g. Fletcher *et al.*, 1999) and/or on clinical cases of PRGT. Thus, we investigated the effect of three different levels of endophyte infected PRG seed (analysed for ergovaline and lolitrem B), under controlled thermoneutral conditions, on plasma prolactin and some physiological parameters, all of which have been shown to be affected by ergot alkaloid ingestion.

Material and methods

Eighteen Merino ewes (16 mo; 48±0.9 kg) were housed indoors at 22 °C, to reduce confounding associated with exposure to varying environmental conditions. Sheep were allocated to three dietary alkaloid levels (Level) (n=6 per treatment): Control, roughage plus barley; Low, Control plus PR seed with ergovaline and lolitrem B at 50 and 22 μg/kg LW, respectively; and High, Control plus PR seed with ergovaline and lolitrem B at 100 and 44 μg/kg LW, for a period of three weeks following an adaption period of two weeks on the Control diet for all sheep. These treatments were imposed to investigate the Low and High effects of primarily ergovaline; Lolitrem B concentration was below the threshold for inducing tremors. At 9 am each day all sheep were offered the three diets at 1.5 maintenance. At the end of adaptation (week 0) and the three week experimental period (week 3), jugular blood samples, rectal temperature, respiration rate and heart rate were taken every 4 h (commencing at 8 am) for 24 h. Plasma was harvested and stored at -20 °C before prolactin analysis using an ovine specific RIA. Statistical analysis was performed using REML in GenStat with the model including Level, Week and Time.

Results

All sheep consumed the majority of feed offered and all of the infected PRG seed. There were significant Level x Week interactions for all parameters (Table 1). There were also significant (P<0.001) diurnal affects for all physiological measurements which corresponded to the expected effects of once daily feed intake. Plasma prolactin concentration was significantly (P<0.001) reduced by alkaloid ingestion and this was most pronouned for the High treatment. Rectal temperature was significantly elevated by alkaloid ingestion but the magnitude of this increase was similar for the Low and High treatments (-0.41, +0.79 and +0.82% for Control, Low and High, respectively). Respiration rate was significantly elevated in a dose dependant manner (+19, +0.47 and +115% for Control, Low and High, respectively). There was a small significant increase in heart rate with ingestion of alkaloid but the response was variable.

Table 1. Effect of level of endophyte-infected perennial ryegrass on plasma prolactin[1], rectal temperature (RTemp, °C), respiration rate (Resp Rate, breaths/min) and heart rate (HR, beats min^{-1}) in Merino ewes.

	Week 0			Week 3			sed[3]	*P*-value[2]
	Control	Low	High	Control	Low	High		
Log (Pro+1)	2.104	1.913	2.063	1.97	0.946	0.092	0.162	<0.001
	(126)	(80.8)	(115)	(92.3)	(7.8)	(0.24)		
RTemp	39.27	39.09	39.18	39.11	39.42	39.50	0.11	<0.001
Resp. rate	64	74	64	78	109	138	9	<0.001
HR	77	67	73	86	81	92	4	<0.05

[1] Data were transformed for analyses with back-transformed values in parentheses (ng/ml).
[2] *P*-value for interaction between level and week.
[3] Average standard error of the difference for interaction between level and week.

Discussion and conclusion

Short term feeding of endophyte alkaloids in PRG seed under thermoneutral conditions decreased plasma prolactin and increased respiration rate and rectal temperature. This is consistent with the effects of ergovaline and not Lolitrem B, as the intake of the latter was relatively low in this study. Ergovaline, a dopamine agonist, reduces plasma prolactin by inhibiting prolactin secretion (e.g. Nihsen *et al.*,, 2004) and increases core temperature possibly through enhanced vasoconstriction. Elevated respiration rate is likely to occur as a thermoregulatory response to increased rectal temperature. An important observation was the pronounced decrease in plasma prolactin and increase in respiration rate, under this sampling regime, with increasing ergovaline intake, demonstrating in Merino ewes that this effect is dose dependant. Thus, plasma prolactin and respiration rate may be sensitive parameters to assess the quantity of alkaloid ingestion. Further experiments will investigate longer term feeding of alkaloids, and different alkaloid ratios, as well as the interaction with hotter environmental conditions common in southeast Australia.

References

Cawdell-Smith, A.J., W.J.K. Bakau, C.J. Scrivener, K.F.M. Reed, B.J. Blaney and W.L. Bryden, 2007. Ergot toxins and animal disease in Australia. In: Panter, K.E., Wierenga, T.L. and Pfister, J.A. (eds.) Poisonous plants: global research and solutions. CABI, Wallingford; UK. pp. 519-524.

Fletcher, L.R., B.L. Sutherland and C.G. Fletcher, 1999. The impact of endophyte on the health and productivity of sheep grazing ryegrass-based pastures. In: Grassland Research and Practice Series - New Zealand Grassland Association, pp. 11-17.

Nihsen, M.E., E.L. Piper, C.P. West, R.J. Crawford Jr., T.M. Denard, Z.B. Johnson, C.A. Roberts, D.A. Spiers and C.F. Rosenkrans Jr., 2004. Growth rate and physiology of steers grazing tall fescue inoculated with novel endophytes. J. Anim. Sci. 82: 878-883.

Spiers, D.E., P.A. Eichen and G.E. Rottinghaus, 2005. A model of fescue toxicosis: responses of rats to intake of endophyte-infected tall fescue. J. Anim. Sci. 83: 1423-1434.

Effect of physical processing of diet on eating and ruminating behaviors of dairy cows in early lactation

A. Hosseinkhani[1], H. Daghigh Kia[1] and S.A.R. Vakili[2]
[1]Department of Animal Science, Faculty of Agriculture, University of Tabriz, Tabriz, Iran; [2]Department of Animal Science, Faculty of Agriculture, Ferdowsi University of Mashhad, Mashhad, Iran; hosseinkhani18@gmail.com

Introduction

One of the main goals of dairy producers is to promote dry matter intake (DMI) to support milk production without any increased ruminal acidosis risk. High producing dairy cows typically spend about 4 to 6 h per day for feeding, and this period of time spent for feeding is divided into 9 to 14 feeding sessions, or meals, over the course of the day (Botheras, 2007). The objective of this experiment was to understand whether the particle size and dry matter content of the diet affects DMI, eating and ruminating behaviour and rumen health of dairy cows.

Material and methods

This experiment was conducted at the Ferdowsi University of Mashhad Dairy Research Farm. Eight multiparous Holstein cows were allocated to the treatments in a changeover design with periods of 21 days in early lactation period. The average day in milk (DIM) and milk production of the cows were 28±12 d (mean ± SD) and 43±3.5 kg/d, respectively. Cows were housed individually in a tie-stall and fed a TMR (Total Mixed Ration) diet *ad libitum* twice daily and had free access to drinking water through one automatic water bowl for each cow. The balanced diets (NRC, 2001) had the same chemical composition. Diet main ingredients were as follows (g/kg): Lucerne hay (200), maize silage (150), barley grain (310), cottonseed (90), soybean meal (120), safflower meal (60), wheat bran (30) and protected fat (20). Two particle sizes of Alfalfa hay (5 and 20 mm theoretical cutting size) and two levels of TMR (dry matter (without and with water addition up to 50% of DM) were applied in the treatments as experimental factors. Water was sprinkled to the diet during diet preparation every day. The amount of feed offered was adjusted daily to obtain around 10% orts (as fed basis). All animals were milked three times a day (7, 13 and 23 h) in a herringbone-milking parlor. Eating and ruminating behavioural activities were observed manually by a team of observers and recorded for 24 h (5 min intervals) during 14-15th days of each experimental period. Rumen fluid was obtained by the Rumenocentesis method and its pH was determined using a digital pH-meter (Metrohm, 744). The data were analysed using the mixed model procedure of SAS® (1999).

Results

Experimental treatments had no effect on DMI (Table 1). Experimental treatments had no effect on eating behaviour of cows as well, but affected some ruminating behaviours. Particle size reduction of hay decreased the total daily time spent for rumination (min/d). Both Lucerne hay particle size reduction and water addition to the diets resulted in significant changes in meal number of rumination. Water addition to the diet decreased ruminating meal duration (P=0.06) and use of short hay in the diet increased ruminating meal size (P=0.07). Reduction of hay particle size had no effect on rumen pH, however, water addition resulted in a significant reduction in rumen pH (Table 1). Daily pattern of chewing behaviours of the experimental cows showed that water addition resulted in a 2h delay in reaching the peak of ruminating time (Figure 1).

Table 1. DMI, eating and ruminating behaviours and ruminal pH of experimental cows.

Behaviour		Long hay		Short hay		SEM	Effect[1]		
		Dry	Wet	Dry	Wet		PS	DM	PS×DM
	DMI (kg/d)	23.64	23.80	23.48	22.34	0.50	0.61	0.53	0.22
Eating									
	Time (min/d)	350.6	343.8	346.3	321.4	12.36	0.31	0.23	0.37
	Rate (g/min)	68.2	69.8	69.4	70.8	6.07	0.59	0.64	0.47
	Meal number	11.25	10.50	11.50	11	0.72	0.41	0.31	0.58
	Meal duration (min/meal)	31.7	32.1	30.3	29.8	2.14	0.12	0.20	0.31
	Meal size (kg)	2.12	2.21	2.09	2.10	0.09	0.65	0.96	0.44
Ruminating									
	Time (min/d)	465.5	463.3	401.3	413.8	22.9	0.02	0.92	0.85
	Meal number	11	12.75	10.25	11	0.90	<0.01	<0.01	<0.01
	Meal duration (min/meal)	41.99	36.54	40.02	39.59	2.8	0.69	0.06	0.87
	Meal size (kg)	2.15	1.88	2.38	2.26	0.57	0.07	0.21	0.97
Rumen pH		6.22	5.93	6.18	5.96	0.08	0.61	0.02	0.35

[1] Probability of the main effect of hay Particle Size (PS), TMR Dry Mater (DM) and the interaction of Particle Size and Dry Mater (PS×DM).

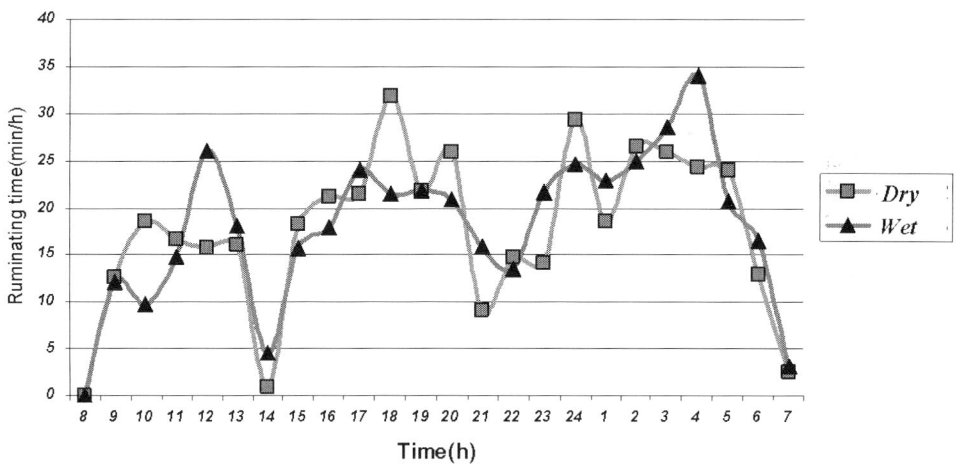

Figure 1. Hourly average ruminating time for cows fed dry or wet diets.

Conclusion

Although water addition to TMR is one of the several techniques used to minimise diet selection of dairy cows (Stone, 2004), it may change some eating behaviors of cows resulting in rumen pH reduction. The delay to reach the peak of rumination time because of water addition to the TMR can be problematic for diets with high and rapid fermentation rates such as barley based diets. This may be the reason for lower pH in the wet diets in the present study.

References

Botheras, N.A., 2007. The feeding behavior of dairy cows: considerations to improve cow welfare and productivity. Tri-State Dairy Nutrition Conference.

SAS Institute Inc., 1999. SAS Users Guide. SAS Institute Inc.,Cary, NC.

Stone, W.C., 2004. Nutritional approaches to minimize subacute ruminal acidosis and laminitis in dairy cattle. J. Dairy Sci. 87: 13-26.

The validity of glucometer produced for humans in farm animals

Ö. Kaynar[1] *and A. Hayirli*[2]
Departments of Biochemistry and Animal Nutrition & Nutritional Disorders, Atatürk University, Erzurum 25700, Turkey; ahayirli_2000@yahoo.com

Introduction

Glucose is an immediate energy source of the most of mammalian and avian cells to function properly. After absorption of carbohydrates from the GI tract, blood glucose level rises. Under stress due to various factors (i.e. starvation, fear and pain), glycogenolysis and gluconeogenesis lead blood glucose level to rise as well. Blood glucose is controlled strictly by actions of many hormones and its fate depends on energy balance and production (Goff, 2004). Thus, blood glucose level is a critical indicator of well being, metabolic status and carbohydrate and lipid-related metabolic disorders including ketosis, hepatic lipidosis, pregnancy toxaemia and diabetes mellitus. Spectrophotometry is widely used in laboratories for determination of blood glucose level. A glucometer is, however, a rapid, easy and inexpensive device available for monitoring blood sugar for humans. Such an instrument would also help veterinarians monitoring the well-being of animals in the fields, and consequently allow them to take measures and actions for prevention and treatment of metabolic disorders when needed. But, first, the reliability and accuracy of the glucometer should be attained. This study was therefore conducted to determine the validity of a glucometer in veterinary medicine.

Material and methods

Blood samples of dogs (n=49), cows (n=38) and quails (n=40) were drained into vacutainers containing sodium fluoride-potassium oxalate. A drop of blood samples was immediately added to the test strip of a commercially available glucometer (Precision Xtra®, Abbott Laboratories, Abbott Park, IL) and the strip was inserted into the glucometer until the result appeared (5 s). Also, after centrifugation at 3000× g for 15 min at 20 °C, aliquots were subjected to spectrophotometric assay (Raabo and Terkildsen, 1960). Glucose data generated from dogs via glucometer and spectrophotometer were utilised for 'the development model' and those from quails and cows were used for 'the validation model' using the PROC REG and MEAN procedures (SAS®, 2002). The regression lines were tested if slope and intercept were equal to 1 and 0, respectively after plotting predicted vs. actual spectrophotometer data and if both slope and intercept were equal to 0 after plotting actual spectrophotometer data vs. bias (predicted – actual) (Hayirli *et al.*, 2003).

Results and discussion

Dog blood glucose data measured with a glucometer were employed for the model development because humans and dogs have similar values. Compared with dogs, cows were hypoglycaemic, whereas quails were hyperglycaemic (Kahn and Line, 2005), suggesting that the quail and cow are ideal subjects to test the validity of a glucometer (Table 1). Regressing glucometer data (x) on spectrophotometer data (y) in dogs resulted best in the following linear equation: $y = 0.92x + 6.21$, R^2=0.88, P<0.0001 for slope and P<0.05 for intercept. The prediction model (x) under-predicted the actual spectrophotometer data (y) in both quails by 4.76% [$y = 1.15x - 22.73$, R^2=0.80, P<0.03 for slope and P<0.12 for intercept] and cows by 2.57% [$y = 0.88x + 9.59$, R^2=0.59, P<0.05 for slope and P<0.11 for intercept] (Figures 1A and 1B). The results suggest that a glucometer produced for human medicine under-predicts blood glucose level of quails and cows. Due to narrow physiological glucose range and prediction error, precaution is necessary to diagnose sub-clinical metabolic disorders (e.g. ketosis) in ruminants when a glucometer is used.

Table 1. Blood glucose level (mg/dl) measured with a glucometer and spectrophotometer and the validity of a glucometer in quails and cows.

		Descriptive statistics		
		Mean	SD	Range
Dog	Glucometer value	88.21	9.79	68-122
	Spectrophotometer value	87.53	9.62	64-119
Quail	Glucometer value	229.91	20.46	183-269
	Spectrophotometer value	229.06	24.33	182-283
	Predicted value	218.15	18.86	174-254
	Bias	-10.91	11.25	-38.38-8.43
Cow	Glucometer value	66.14	10.34	40-88
	Spectrophotometer value	68.96	11.01	44-90
	Predicted value	67.19	9.53	43.08-87.33
	Bias	-1.77	7.17	-14.16-18.04

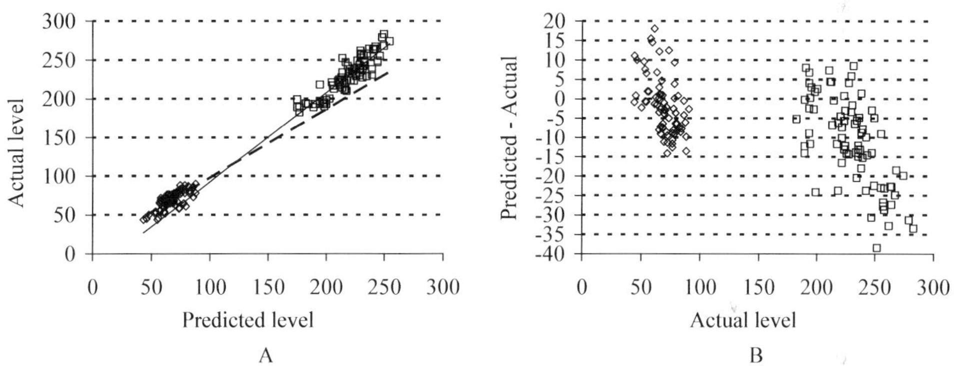

Figure 1. Plots of actual vs. predicted blood glucose levels (mg/dl) (A) and actual blood glucose levels vs. bias (B) in quails (□, straight line) and cows (◊, broken line).

References

Goff, J.P., 2004. Disorders of carbohydrate and fat metabolism. In: Reece W.O. (Ed), Duke's physiology of domestic animals. Cornell University Press, Ithaca, NY, pp. 553-561.

Hayirli, A., R.R. Grummer, E.V. Nordheim and P.M. Crump, 2003. Models for predicting dry matter intake of Holsteins during the prefresh transition period. J. Dairy Sci. 86: 1771-1779.

Kahn, C.M. and S. Line, 2005. The Merck veterinary manual. 9th ed. Merial Ltd., Whitehouse Station, NJ, USA.

Raabo, E. and T.C. Terkildsen, 1960. On the enzymatic determination of blood glucose. Scand. J. Clin. Lab. Invest. 12: 402-407.

SAS, 2002. User's Guide. Statistics, Version 9th. Statistical Analysis System. SAS Inst., Inc., Cary, NC, USA.

Effect of chromium supplementation on production and blood parameters of early-lactation Holstein cows under heat stress

A.Q. Lai, Z.S. Wang, B. Xue, L.Z. Wang and D.Y. Peng
Animal Nutrition Institute, Sichuan Agricultural University, Ya'an 625014, China; wangzs007@yahoo.com.cn

Introduction

Sichuan, one of the southwest provinces in China, is characterised as a humid and hot subtropical climate. Such climatic conditions exert great challenges on the dairy industry known as heat stress. Chromium (Cr) is an essential trace element for the maintenance of normal metabolism of carbohydrates, proteins and lipids. The necessity of Cr supplementation is highly related to stress which may deplete body reserve of Cr (Anderson, 1994). It has reported that supplementation of organic Cr remarkably improved lactation performance and physiological status of stressed dairy cows (Alsaiady *et al.*, 2004). During heat stress, dairy cows in early lactation are subject to greater physical and metabolic stress, which could be alleviated by Cr supplementation. However, evidence for the benefit of Cr supplementation with heat-stress and early-lactation cows is extremely limited. The objective of this study was to investigate the effect of chromium picolinate (CrP) supplementation on production and blood parameters of early-lactation Holstein cows under heat stress.

Material and methods

The experiment was conducted from July 21 to September 23, 2008, the hottest period of the year. Twenty-four lactating Holstein cows, weighing 593±28 kg, were divided into 4 groups, according to days in milk, parity and milk yield, with 6 cows in each group. Cows were from second and forth lactation, and were from 15 to 24 d *post partum*, with an average initial milk yield of 21.4±2.4 kg/d. Cows received a basal diet containing 56% roughage and 44%concentrate, which was formulated to meet nutrient requirements recommended by the NRC (2001). Cows in the 4 groups correspondingly received the following 4 CrP supplementation dosages: 0, 0.03, 0.06 and 0.12 mg /kg of $BW^{0.75}$. Cows had *ad libitum* access to the diets and water. Cows were fed three times daily at 08.00, 14:00 and 20:00 h, to allow a proportionate excess of 10%. For each feeding, cows were fed on the mixture of corn silage, wafering alfalfa hay, wafering oat hay, beet pellet, brewer grain and concentrate, about 40 min later, cows were fed on whip grass and alfalfa hay. Cr was added at 08:00 and 20:00 h via concentrate. Feed intake was recorded daily throughout the study and milk yield was recorded daily for 3 consecutive days every week. Dry-bulb and wet-bulb temperatures were recorded daily from 09:00 to 21:00 h at 2-h intervals and temperature–humidity index (THI) was calculated according to the formulae described by Maust *et al.* (1972).

Five cows from each group were randomly chosen for bleeding. Blood samples were collected into tubes containing anticoagulant via jugular vein at 07:00 h on d 1, 28 and 56, and the tubes were kept on ice for 30 min, after that the samples were centrifuged at 3000× g for 10 min. Plasma was kept at -20 °C for further chemical analysis. Plasma was analysed for glucose, NEFA, insulin, cortisol, T4, T3 and IGF-I using standard assay kits.

Treatment was the experimental unit for feed intake, milk yield and blood parameters. Data were analysed by single factorial variance analysis using the GLM procedure of SPSS 12.0 software (SPSS Inc., Chicago, IL, USA). Overall differences between treatment means were declared significant at $P<0.05$. Trends towards significance were considered at $P<0.10$.

Results

The calculated THI averaged 79.61 units, which exceeded the upper critical limit (72 units) for dairy cows (Table 1). Cr supplementation increased feed intake and milk yield ($P<0.001$ and $P=0.013$, respectively). Plasma glucose concentration, molar ratio of glucose to insulin and the conversion rate of T4 to T3 were significantly increased for M and H treatment cows than C treatment cows ($P=0.019$, $P=0.013$ and $P=0.015$, respectively). Cr supplementation tended to increase IGF-1 and decrease insulin concentrations ($P=0.057$ and $P=0.079$, respectively). The results of feed intake, glucose, glucose/insulin, T3, as well as NEFA indicated that the benefits of Cr supplementation at 0.06 mg of Cr/kg of $BW^{0.75}$ were superior or equal to that of the control and the other two dosages.

Table 1. Effect of Cr supplementation on production and blood parameters.

	Treatment least squares mean[1]				SEM	Diet effect
	C	L	M	H		*P*-value
Feed intake, kg/d	17.64^{b}	17.96^{a}	18.18^{a}	18.17^{a}	0.14	<0.001
Milk yield, kg/d	24.34^{b}	25.29^{ab}	25.63^{a}	25.52^{a}	0.43	0.013
Glucose, mmol/l	3.01^{b}	3.10^{ab}	3.24^{a}	3.22a	0.08	0.019
Insulin, μIU/ml	14.47	13.92	13.39	13.36	0.67	0.079
Glucose/insulin	0.207^{b}	0.231^{ab}	0.251^{a}	0.254^{a}	0.02	0.013
T3, ng/ml	1.61^{b}	1.70^{ab}	1.75^{a}	1.74^{a}	0.05	0.015
T4, ng/ml	66.53	66.24	65.31	65.49	2.54	0.932
IGF-1, ng/ml	74.32	82.83	92.44	93.35	7.77	0.057
Cortisol, ng/ml	19.28	18.80	18.81	18.73	1.32	0.978
NEFA, mmol/l	0.599	0.569	0.537	0.526	0.05	0.249

[1] C = cow not receiving Cr; L, M and H = cows receiving 0.03, 0.06 and 0.12 mg of Cr as CrP/kg of $BW^{0.75}$, respectively.

[a,b] Means within rows with same superscript letters are not significantly different ($P>0.05$).

Discussion and conclusion

This study indicates that Cr supplementation, particularly at 0.06mg of Cr/kg of $BW^{0.75}$, increased feed intake and milk yield, and also had an effect on certain blood parameters. Cr supplementation partially alleviated heat stress in dairy cows in early lactation. Cr supplementation could promote the welfare of early-lactation Holstein cows under heat stress.

Acknowledgement

This work was supported by the programme for public welfare behaviour research fund of the Agricultural Department of China (No. nyhyzx07-036-03).

References

Alsaiady, M.Y., M.A. Alsaiskh, S.I. Almufarrej, T.A. Alshoweimi, H.H. Mogawer and A. Dirrara, 2004. Effect of chelated chromium supplementation on lactation performance and blood parameters of Holstein cows under heat stress. Anim. Feed Sci. Technol. 117: 223-233.

Anderson, R.A., 1994. Stress effects on chromium nutrition of human and farm animals. Proceedings of the Alltech's 10th Annual Symposium. Nottingham University Press, Nottingham, UK, 267 pp.

Maust, L.E., R.E. McDowel and N.W. Hooven, 1972. Effect of summer weather on performance of Holstein cows in the three stage of lactation. J. Dairy. Sci. 55: 1133-1139.

Effect of cassava (*Manihot esculenta*) foliage on nutrition, parasite infection and growth of lambs

C. Marie-Magdeleine[1], M. Mahieu[1], L. Philiber[1], P. Despois[2] and H. Archimède[1]
[1]INRA UR143 Unité de Recherches Zootechniques, Centre INRA-Antilles-Guyane, Domaine Duclos, 97170 Petit Bourg, Guadeloupe (French West Indies), [2]INRA UE503, Unité Expérimentale en Production et Santé Animale, Centre INRA-Antilles-Guyane, Domaine Duclos, 97170 Petit Bourg, Guadeloupe (French West Indies); harry.archimede@antilles.inra.fr

Introduction

Cassava foliage is rich in condensed tannins and proteins (Phengvichith and Ledin, 2007). Moreover some studies (Dung *et al.*, 2005; Nguyen *et al.*, 2005; Seng *et al.*, 2007) on goats indicate a protective action of cassava foliage against nematodes. The first objective of this study was thus to test the effect of cassava foliage against *H. contortus* in lambs and to explain the observed action. The second objective was to evaluate cassava foliage as a source of protein.

Material and methods

A completely randomised design for a 78-day feeding trial was performed to test the nutritional value and the anthelmintic properties of 12-month-old cassava foliage. Thirty 6-month-old Martinik lambs (body weight: 20.3±1.6 kg) were divided into three groups of ten animals and placed in individual pens. The lambs were fitted with faecal bags in order to study the digestibility of the diet. The diets consisted of 45-day-old *Dichanthium* spp. hay with a supplementation regime according to the experimental treatment as follows (as fed basis): (1) hay + lucerne pellets (450 g/lamb/day) + ground wilted cassava tuber (450 g/lamb/day); (2) hay + wilted cassava foliage (650 g/lamb/day) + ground wilted cassava tuber (450 g/lamb/day); (3) hay + wilted cassava foliage (650 g/lamb/day) + ground wilted cassava tuber (450 g/lamb/day) + polyethylene glycol (40 g/lamb/day). The polyethylene glycol (PEG) was dissolved in water and given *per os* to the lambs. At the end of a 14-day period of adaptation to the diets, all lambs were artificially infected with L3 larvae of *H. contortus*. Intake and total tract digestibility of the diets were monitored throughout the trial. Growth of lambs, faecal egg count, packed cell volume (PCV), and abomasal adult worm count (FEC) were also monitored.

Results

The main results are presented in Table 1. No significant difference was found between the kinetics of the eosinophils and PCV of the three groups of lambs. The mean FEC tended to be higher for lucerne compared with the two other groups (P=0.09). FEC was also lower in animals fed cassava leaves compared to cassava leaves + PEG but the differences were not significant. The total dry matter intake was significantly higher in the group fed the diet containing lucerne than in the other groups, whereas there was no significant difference between groups fed wilted cassava foliage (WCF). The intake of CP was not significantly different among treatments. The intake of NDF and ADF was significantly higher in the control group fed lucerne than in the groups supplemented with WCF and WCF-PEG. Conversely, the intake of ADL was significantly higher with the WCF diets. There were no significant differences in the digestibility of DM and OM between the three diets. The digestibility of CP was significantly higher for the diet containing lucerne than for the WCF diets without PEG. When PEG was added to the WCF, the total tract digestibility of CP was higher than the diet with lucerne.

Table 1. Effect of diets fed to lamb on faecal egg count, intake, daily weight gain and feed conversion ratio (Least squares means and standard error of mean).

	Lucerne	Cassava	Cassava + PEG	SEM	*P*-values
Initial LW (kg)	19.8[a]	20.7[a]	20.5[a]	0.9	0.27
Final LW (kg)	32.7[a]	30.2[a]	30.8[a]	1.1	0.11
Live weight gain LWG (g/day)	163.5[a]	120.8[b]	134.8 [a]	7.7	0.001
FCR (kg feed/kg LWG)	5.7[a]	5.6 [a]	6.8[a]	3.5	0.55
Log mean faecal egg count	8.3[a]	7.6[b]	7.9[b]	0.3	0.04
Total worm in abomasum	939.0[a]	460.0[a]	573.0[a]	236	0.16
DMI/$LW^{0.75}$	89.9[a]	78.4[b]	76.6[b]	0.9	0.05
Digestible DMI/$LW^{0.75}$	57.6[a]	51.0[b]	49.5[b]	0.7	0.12
Digestible Crude Protein/ $LW^{0.75}$	4.7[a]	4.6[a]	6.0[b]	0.1	0.0001

[a,b,c,d] Means that within a row, values with different letters differ significantly ($P<0.05$).

Discussion and conclusion

The anthelmintic properties of wilted cassava leaves, which are linked to tannins, were demonstrated: worm colonisation in the *abomasum* and *H. contortus* egg development were lower in animals fed cassava foliage compared with lucerne, and FEC was lower with cassava leaves compared to PEG treatment. Due to the high fill value (ADL), the consumption of wilted cassava leaves provides limited total energy intake in lambs and has a depressive action on growth compared to the lucerne diet. An alternative strategy would be to use younger cassava foliage to optimise the anthelmintic properties and nutritional value.

References

Dung, N.T., N.T. Mui and I. Ledin, 2005. Effect of replacing a commercial concentrate with cassava hay (*Manihot esculenta* Crantz) on the performance of growing goats. Anim. Feed Sci. Technol. 119: 271-281.

Nguyen, T.M., D. Van Binh and E.R. Orskov, 2005. Effect of foliages containing condensed tannins and on gastrointestinal parasites. Anim. Feed Sci. Technol. 121: 77-87.

Phengvichith, V. and I. Ledin, 2007. Effect of a diet high in energy and protein on growth, carcass characteristics and parasite resistance in goats. Trop. Anim. Health Prod. 39: 59-70.

Seng, S., P.J. Waller, I. Ledin and J. Höglund, 2007. The effects of short-term feeding of fresh cassava (*Manihot esculenta*) leaves on gastrointestinal nematode parasite infections in goats in Cambodia. Trop. Biomed. 24: 47-54.

The effects of high levels of rumen degradable protein on rumen fermentation and rumen histamine concentrations in dairy cows

R. Pilachai[1], J.T. Schonewille[1], A. Chaiyotwittayakun[2], S. Aiumlamai[2], C. Wachirapakorn[3], H. Everts[1] and W.H. Hendriks[1]
[1]Faculty of Veterinary Medicine, Utrecht University, Utrecht, the Netherlands; [2]Faculty of Veterinary Medicine, Khon Kaen University, Thailand; [3]Faculty of Agriculture, Khon Kaen University, Thailand; rittichai_pilachai@yahoo.com

Introduction

In the north-east of Thailand, dairy cows are typically fed high concentrate rations (75-80% of total dry matter), which is associated with an average milk yield of 3500 kg/cow/305 days. The concentrates typically contain 50-70% of cassava chips and rice straw is the main source of roughage for dairy cattle. Consequently, the rations are low in crude protein (CP) which may be a major constraint for milk production. However, supplementation of these rations with CP might lead to excessive rates of fermentation thereby enhancing the risk of subacute rumen acidosis (SARA) (Owen *et al.*, 1998). Secondly, it is well known that a chronic low rumen pH is related to high histamine levels and associated with laminitis in dairy cows (Nocek, 1997). However, the exact relationship between the feeding of high levels of CP on rumen pH and laminitis are not yet described under the unique Thai feeding conditions. We hypothesise that the supplementation of CP results in rumen acidosis which may lead to laminitis. In the present study, crude protein was supplemented in the form of formaldehyde treated- and untreated soybean meal. It was anticipated that the fermentation rate of the untreated soybean meal is much higher than formaldehyde treated soybean meal and offers the possibility to distinguish between the effect of protein and rate of fermentation.

Material and methods

Six fistulated non-pregnant non-lactating Holstein x American Brahman cows were randomly assigned to 3 dietary treatments in a replicated 3 x 3 Latin square design. Each experimental period included 14 days for adaptation to the diet and 7 days for sampling. The ingredient composition of the experimental concentrates is shown in Table 1. The crude protein contents in the rations were 112, 259, and 266 g/kg DM, for the control diet (Low-CP), high rumen undegradable protein (High-RUP), and high rumen degradable protein (High-RDP) diets, respectively. The rations were offered in 2 equal portions at 08:00 and 17:00 h. The cows were fed 7.7 kg DM of the concentrates and 2.7 kg DM of rice straw. Rumen fluid samples were taken at 0, 1, 2, 3, 4, 5, 7 and 9 h after feeding on day 18 and day 21 and pH was measured immediately after sampling. A sub-sample was analysed for VFA using gas chromatography and for ammonia by the macro Kjeldahl method. Rumen histamine concentration was analysed using the HPLC technique of Eerola *et al.* (1993). Blood samples were taken from the jugular vein for measurement of plasma urea nitrogen (PUN). Throughout the experiment all cows were clinically examined for laminitis daily. All data were analysed using the GLM procedure (SAS®). Differences among treatments were considered significant at $P<0.05$.

Results

High levels of RDP significantly increased the ammonia, total VFA ($P<0.001$) and histamine concentrations ($P=0.002$) in the rumen. However, rumen pH was not significantly ($P=0.226$) affected by the supplementation of RDP (Table 2). Plasma urea nitrogen concentration was higher ($P<0.001$) in high-RDP and high-RUP as compared with cows fed the control diet. During the last experimental period, two cows, one cow fed high-RDP and the other fed high-RUP rations, showed signs of laminitis.

Table 1. Ingredient composition of the experimental concentrates.

Ingredient composition,% of DM	Experimental concentrates[1]		
	Low-CP	High CP	
		High-RUP	High-RDP
Constant components[2]	26.3	26.3	26.3
Cassava chips	54.9	18.9	18.8
Soybean meal	18.8	-	54.9
Formaldehyde treated soybean meal	-	54.8	-

[1] Low-CP = low crude protein; High-RDP = high rumen degradable protein; High-RUP = high rumen undegradable protein; [2] The constant components consisted (% of DM): 17.9 rice bran, 4.5 molasses, 1.7 premix, 1.1 salt, and 1.1 calcium phosphate.

Table2. Effects of experimental concentrates on selected indices of rumen fermentation, rumen histamine concentrations and plasma urea nitrogen (PUN) in dairy cows.

Item	Experimental concentrates[1]			SEM	*P*-value
	Low-CP	High CP			
		High-RUP	High-RDP		
Ammonia[2], m*M*	8.6[a]	14.2[a]	29.7[b]	2.88	<0.001
Total VFA[2], m*M*	110[a]	113[a]	128[b]	2.6	<0.001
pH[2]	6.55	6.38	6.29	0.093	0.226
Rumen Histamine[3], m*M*	0.45[a]	0.37[a]	1.01[b]	0.091	0.002
PUN[3], m*M*	1.47[a]	4.38[b]	4.38[b]	0.042	<0.001

[a,b,c] Values in the same row with different superscripts differ significantly ($P<0.05$).
[1] Low-CP = control diet; High-RDP = high rumen degradable protein; High-RUP = high rumen undegradable protein.
[2] Means of 5 samples taken at 1, 3, 5, 7 and 9 h after feeding.
[3] Mean of 2 samples taken at 2 and 4 h after feeding.

Discussion and conclusion

This study indicates that the supplementation of high rumen degradable soybean meal increased rumen fermentation (more VFA). However, the effects of soybean meal supplementation on pH and occurrence of laminitis were not clearly observed, probably due to the relatively low level of DM intake and to the small number of cows.

Acknowledgement

This study was supported by Asia-link project (phase III).

References

Eerola, S., R. Hinkkanen, E. Lindfors and T. Hirvi, 1993. Liquid chromatographic determination of biogenic amines in dry sausages. J. AOAC Int. 76: 575-577.

Nocek, J.E, 1997. Bovine acidosis: Implications on laminitis. J. Dairy Sci. 80: 1005-1028.

Owens, F.N., D.S. Secrist, W.J. Hill and D.R. Gill, 1998. Acidosis in cattle: A review. J. Anim. Sci. 76: 275-286.

Effect of vitamin E supplementation on SCC in periparturient dairy goats

L. Pinotti, V. Dell'Orto and A. Baldi
Department of Veterinary Sciences and Technology for Food Safety, Veterinary Faculty, University of Milan, Via Celoria 10, 20133, Milano, Italy; luciano.pinotti@unimi.it

Introduction

Deficiencies in vitamin E or selenium, have been associated with high somatic cell counts (SCC) in individual cows and in bulk milk (Baldi, 2000), and also with increased incidence and severity of intramammary infections (IMI) and mastitis (Smith *et al.*, 1997). In dairy goats however some differences exist. The milk SCC for uninfected goats is higher than milk SCC for uninfected cows and sheep. On average, milk SCC from goat mammary glands free from intra-mammary infection ranges from 270 to 2000×10^3/ml while it ranges from 659 to 4213×10^3/ml for infected glands (Paape *et al.*, 2001). Furthermore the possible positive effects of vitamin E on SCC in goats is not well known. In the light of this, the objective of the present study was to determine the effect of vitamin E supplementation on SCC in periparturient dairy goats.

Material and methods

Forty pregnant multiparous Saanen goats of uniform weight (mean 63.5±3 kg) at the experimental farm of the University of Milan were selected for the 63-day experiment. Four weeks before expected parturition, the goats were assigned randomly to one of the two experimental groups: control (CRT), no vitamin E supplementation; vitamin E (VITE), supplemented with 200 IU/day vitamin E (DL-α-tocopheryl acetate) in a rumen-protected form. The quantity of vitamin E used was based on an experiment with dairy cows (Baldi *et al.*, 2000) and scaled down using metabolic body weight at the beginning of the experiment. The treatment was administered to each animal individually before the morning feed to ensure complete consumption, starting 4 weeks prior to expected kidding and continuing for 5 weeks after parturition. The goats were fed basal diet formulated to provide 2.00 and 2.40 Mcal of ME/per kg dry matter (DM), 11.50% and 14.30% of crude protein per kg DM, for pre-kidding and lactation phases, respectively. Both pre-kidding and lactation diets contained 25-35 IU/kg of DM of naturally occurring vitamin E. During the experiment, blood samples were collected from about four weeks (33±3 days) before kidding to one week before kidding, and on days 0, 7, 14, 21 and 28 post-kidding. Plasma was analysed for alpha-tocopherol (Baldi *et al.*, 2000). From week 1 to week 5 of lactation, milk yield and composition were measured weekly. Milk samples of each animal were treated with a preservative (sodium azide) and stored at 5 °C pending analysis for milk fat, milk protein (Milkoscan, Foss Technology, Denmark), and SCC (Fossmatic Somatic Cell Counter, Foss Technology, Denmark). Logarithmic transformations of the SCC in milk were used in the statistical analysis. Data were analysed using the MIXED procedure of Statistical Analysis System® Institute (SAS, 1999). The REPEATED statement was used for variables measured over time (milk yield, and milk components). The first measurement of milk yield and composition were used as covariates. The random error term used for all mixed models was goat within treatment group. In addition to the main effects of treatment, the level of SCC [high SCC $>500\times10^3$/ml (HSCC) and low SCC $<500\times10^3$/ml (LSCC)] at first milking were included in the model. The interactions between treatment and SCC level within the final models were tested. Differences with P values <0.05 were considered significant; P values ≤0.15 were considered to indicate a trend.

Results and discussion

Vitamin E administration resulted in plasma α-tocopherol levels that were 2 times higher than in CRT goats throughout the study period (2.8±0.12 vs. 1.4±0.09 mg/l), except at kidding when the difference reached 2.7±0.10 vs. 0.9±0.08 mg/l. Milk yield and composition did not differ throughout the experiment (yield 3033 vs. 3100 g/day; Fat, 3.89 vs. 3.90%; Protein, 3.73 vs. 3.79%, in CRT and VITE respectively). Mean milk SCC tended (P=0.10) to be lower in VITE group (5.65 $\log_{10}$/ml) than in the CRT group (5.71 $\log_{10}$/ml). It should be noted that a significant time × treatment effect was observed on the 3rd and 4th week of lactation (Figure 1), when SCC decreased in the VITE compared to the CRTgroup.

When considering SCC levels, vitamin E tended (P<0.14) to reduce SCC in HSCC animals (6.08 vs. 5.96 $\log_{10}$/ml in HSCC-CRT and HSCC-VITE respectively), but not in LSCC (5.35 vs. 5.35 $\log_{10}$/ml in LSCC-CRT and LSCC-VITE respectively). It is noteworthy that again a significant (P<0.05) time × treatment effect was observed on weeks 3, 4 and 5 of lactation, when SCC were reduced in HSCC-VITE compared to HSCC-CRT goats.

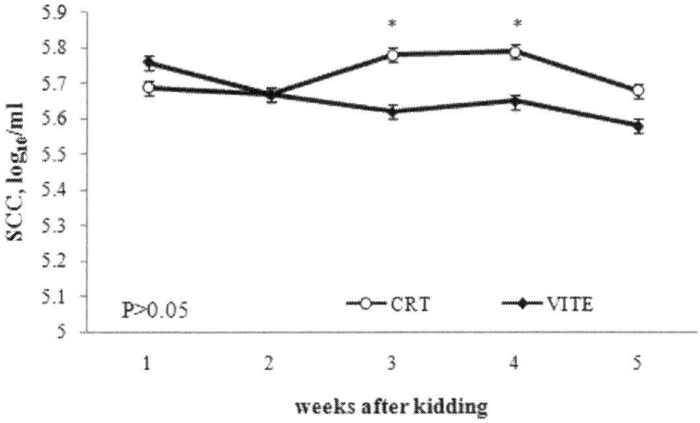

Figure 1. Milk SCC in CRT and VITE groups throughout the experiment.

Conclusion

Although a larger number of goats would be required to demonstrate significant differences among the treatments, our results are in agreement with those reported by others in dairy cows who found that a dietary vitamin E supplement may reduce SCC.

References

Baldi, A., G. Savoini, L. Pinotti, E. Monfardini, F. Cheli and V. Dell'Orto, 2000. Effect of vitamin E and different energy sources on vitamin E status, milk quality and reproduction in transition cows. J. Vet. Med. A 47: 599-608.

Paape, M.J., B. Poutrel, A. Contreras, J.C. Marco and A.V. Capuco, 2001. Milk somatic cells and lactation in small ruminants. J. Dairy Sci. 84: E237-244E.

Smith, K.L., J.S. Hogan and W.P. Weiss, 1997. Dietary vitamin E and selenium affect mastitis and milk quality. J. Anim. Sci. 75: 1659-1665.

SAS, 1999: SAS/STAT® User's Guide, Version 6.03, 6th Edition. SAS Inst., Inc., Cary, NC, USA.

Serum constituents and thyroid hormones in sheep fed *Kochia scoparia* hay

A. Riasi[1] *and M. Danesh Mesgaran*[2]
[1]*Department of Animal Science, Faculty of Agriculture, University of Birjand, Iran;* [2]*Department of Animal Science, Faculty of Agriculture, Ferdowsi University of Mashhad, Iran; riasi2008@gmail.com*

Introduction

Kochia scoparia is a halophytic plant that typically grows in salty lands and seashores. During the last two decades, there has been increased interest in planting *Kochia* spp. and other halophytes in arid and semi-arid regions of Iran to improve animal production particularly when higher quality feeds are too expensive or not available in sufficient amounts (Rezvani Moghadam and Koocheki, 2003). The general nutritional characteristics of *Kochia scoparia* have been well defined and reported (Riasi *et al.*, 2008). However, the physiological effects of this plant need greater attention for effective use as a forage resource (Rankins *et al.*, 1991). The aim of study was to compare the effects of *Kochia scoparia* and lucerne hay on blood serum constituents and thyroid hormones of Baloochi ewes.

Material and methods

Ten Balouchi ewes (50±3 kg BW) were transferred to metabolism cages and randomly allocated to two dietary treatments (100% *Kochia* hay or 100% lucerne hay) for 45 days. *Kochia* and lucerne were harvested at mid bloom stage. The animals had *ad libitum* access to feed and water. During the last 3 days of experiment, blood samples were taken just before morning feeding and analysed for Ca, Mg (AOAC, 2000), bilirubin, GGT, GPT, GOT, T_3, and T_4 (using standard laboratory kits). The 3 days of data were averaged for each sheep before statistical analysis. During the final day of the experiment, blood samples were taken at different times (0, 2, 4, 6 h) after morning feeding and analysed for glucose and blood urea nitrogen (BUN). All data from experiment were analysed as a completely random design using the MIXED procedure of SAS® (1998). The differences between treatment means were declared significant at $P<0.05$. Trend towards significance were considered at $P<0.10$.

Results

The serum glucose ($P<0.01$) and BUN ($P<0.05$) levels were significantly lower at the different times after feeding *Kochia* compared with lucerne hay (Table 1). Linear relationship for the sampling time tended to be significant ($P<0.09$) for glucose and was not significant ($P>0.05$) for BUN. The quadratic relationships were significant ($P<0.05$) for both glucose and BUN. No differences were found between *Kochia scoparia* and lucerne hay in regards to serum Mg, glucose, bilirubin, GGT and T_3 hormone. However, there were some differences for the other serum constituents. Compared with lucerne hay, feeding *Kochia* significantly ($P<0.05$) elevated the activity of GOT (98.25±2.03 vs. 89.33±2.11 U/L), and GPT (39.17±1.23 vs. 30.56±1.71), and the concentration of T_4 hormone (16.72±0.24 vs. 10.52±0.46 μg/dl). Feeding *Kochia* significantly ($P<0.05$) decreased the serum concentration of Ca (6.3±0.12 vs. 9.4±0.28 mg/dl).

Table 1. Glucose and BUN (mg/dl) in blood serum at different times after feeding Kochia scoparia *or Lucerne hay to 10 ewes.*

		Forages		Treat		Linear		Quadratic	
		Kochia	Lucerne	*P*-value	SE	*P*-value	SE	*P*-value	SE
Glucose									
	0 h	40.4	49.3	<0.01	1.3	<0.09	0.6	<0.05	0.3
	2 h	42.3	51.3						
	4 h	42.1	55.1						
	6 h	41.2	52.4						
BUN									
	0 h	16.2	19.2	<0.05	0.45	>0.05	0.5	<0.05	0.4
	2 h	17.4	19.9						
	4 h	17.3	20.3						
	6 h	19.0	18.1						

Discussion and conclusion

This study showed that *Kochia scoparia* affects some serum constituents and thyroid hormones of Iranian Baloochi ewes. Increasing the cellular enzymes (GOT and GPT) could be due to the mild toxicosis of *Kochia scoparia* that might have been unrecognised without blood clinical profiles. It appears that the energy, CP and Ca content of *Kochia scoparia* are not high enough to cover ruminant requirements. It is concluded that using this plant can be recommended for short-term feeding of sheep or only as a maintenance feed. There is a need to investigate the anti-nutritional factors of *Kochia scoparia* and the regimens for preventing its toxicity.

Acknowledgement

This work was conducted with the financial support of Ferdowsi University of Mashhad, Iran,

References

AOAC, 2000. Official Methods of Analysis, 17th ed. Official Methods of Analysis of AOAC International, Gaithersburg, MD, USA.

Rankins Jr, D.L., G.S. Smith and D.M. Hallford, 1991. Altered metabolic hormones, impaired nitrogen retention, and hepatotoxicosis in lambs fed *Kochia scoparia* hay. J. Anim. Sci. 69: 2932-2940.

Rezvani Moghadam, P. and A. Koocheki, 2003. A comprehensive survey of halophytes in Khorasan province of Iran. In: Lieth H. and M. Mochtchenko (Eds.), Cash crop halophytes: recent studies. Kluwer Academic Publishers, pp. 189-195.

Riasi, A., M. Danesh Mesgaran, M.D. Stern and M.J. Riuz Moreno, 2008. Chemical composition, *in situ* ruminal degradability and post-ruminal disappearance of dry matter and crude protein from the halophytic plants *Kochia scoparia*, *Atriplex dimorphostegia*, *Suaeda arcuata* and *Gamanthus gamacarpus*. Anim. Feed Sci. Technol. 141: 209-219.

The potential of pomegranate peels to decrease the incidence of oxidative-stress related diseases in cattle

A. Shabtay, H. Eitam, A. Orlov and A. Brosh
Institute of Animal Science, Department of Ruminant Science & Genetics, Newe Ya'ar Research Center, Agricultural Research Organization, P.O. Box 1021 Ramat Yishay 30095, Israel; shabtay@volcani.agri.gov.il

Introduction

Pomegranate peel (PG) is a nutritive-rich by-product whose amounts are extensively growing due to exponential increase in production of pomegranate juice and 'ready to eat' arils. PG is a rich source for antioxidants and thus may serve in the prevention of oxidative stress related cattle diseases. Bovine respiratory disease complex (BRD), the main leading cause of morbidity and mortality among young calves, is an example of such a disease. BRD is triggered by stress events such as transportation and weaning and has been associated with the decrease in antioxidative capacity of the body and the generation of free radicals. Supplementing feedlot calves with antioxidants such as vitamin E (tocopherols) decreases the incidence of BRD. Our objective was to test the effect of dietary supplementation with fresh PG on bull calves' weight gain, vitamin E concentration and antioxidant capacity in their plasma.

Material and methods

Freshly produced PG were subjected to various storage conditions, namely, 4 °C refrigeration, open air shade, oven-drying at 60 °C and ensiling. Residual levels of antioxidant compounds (polyphenols, flavonoids, condensed tannins, hydrolysable tannins, punicalagin, tocopherols and minerals) were analysed. Twelve Holstein-Friesian bull calves participated in this study. The calves were divided randomly to two groups of six individuals, each housed in a different pen. The average body weights (BW) of the calves were 378±46 kg and 385±33 kg for the control and experimental calves, respectively. The average ages of the calves were 334±21 days and 321±15 days for the control and experimental calves, respectively. The control group was served a fattening ration *ad libitum* (ME= 2.75Mcal/kg DM, CP=13.5%, DM=71%). The experimental group was served the same fattening ration, with the addition of *ad libitum* PG given separately from the fattening ration. The experiment lasted eight weeks. Intake of PG was measured once daily during the whole experimental period, whereas the intake of the fattening ration was measured in two week intervals. Fattening ration intake was adjusted to twenty-four hours, to enable the calculation of average daily intake. At day 0 and in two week intervals the calves were weighed. Total antioxidant capacity of the plasma was measured by chemiluminscence. Plasma tocopherols were measured using HPLC equipped with a fluorescence detector.

Results and discussion

Dietary supplementation with fresh PG promoted significant increases in feed intake (58.9±2.7 vs. 66.2±3 kg DM/d for control and PG groups, respectively; $P<0.001$), plasma antioxidant capacity (Figure 1) and α-tocopherol concentration (Table 1), with positive tendency towards increased weight gain of bull calves (1.26±0.07 vs. 1.46±0.13 kg/d for control and PG groups, respectively; $P=0.053$).

It was demonstrated recently (Gorelik *et al.*, 2005) that polyphenols protect various antioxidants from oxidation in a physiological digestive environment. Accordingly, the PG-derived polyphenols, in our study, may have positively affected the efficacy of α-tocopherol absorption from the diet.

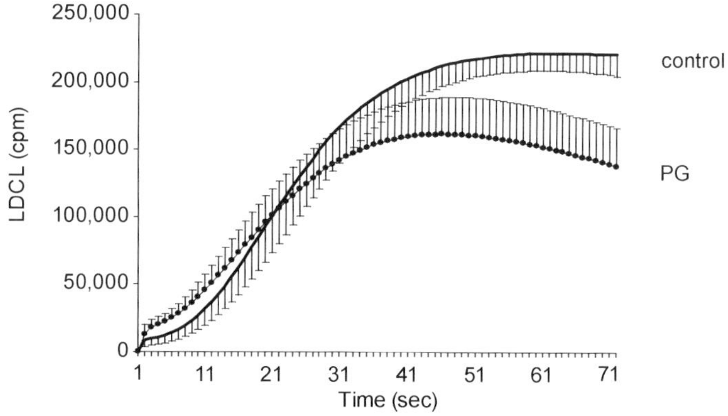

Figure 1. PG supplementation for 8 weeks improved the antioxidant capacity (lower cpm) of the plasma in bull calves.

Table 1. Plasma α-tocopherol concentration in control and PG-consuming calves along the feeding experiment. Data are given as means±SD. Different letters indicate significant differences (P<0.05) within the PG-consuming group.

Weeks in experiment	α-tocopherol (μg/ml)	
	Control	PG
0	1.75±0.19	1.72±0.29[a]
2	1.69±0.11	2.25±0.32[ab]
4	1.76±0.11	2.71±0.41[bc]
8	1.82±0.15	2.92±0.36

The results reported herein reflect the potential of PG to enhance the calves' feed intake and weight gain, in spite of their high tannin content. They also reflect its capability to increase antioxidant capacity of the plasma and absorption of vitamin E from the diet. Feedlot calves are exposed to various stresses, such as transportation, weaning and regrouping, all with deleterious effects on nutrition and the redox state of the body, and are known to promote oxidative stress related diseases. The current study gains insight into the potential of PG to protect calves from oxidative stress related diseases, thus improving their welfare.

References

Gorelik, S., T. Lapidot, I. Shaham, R. Granit, M. Ligumsky, R. Kohen and J. Kanner, 2005. J. Agric. Food Chem. 53: 3397-3402.

Improved water TDS can improve dairy cattle performance under heat stress

M. Shapasand[1], A.R. Alizadeh[1], M. Yosefi[1] and J. Amini[2]
[1]Islamic Azad University, Saveh Branch, Saveh, Iran; [2]Zanjan Agricultural Jahad Organization, Zanjan, Iran; majid_shapasand@yahoo.com

Introduction

A lactating dairy cow has one of the largest requirements for water of any animal. This is because 56 to 81% of its body weight is water, and it needs to commute the major loss of water through milk production (milk contains 87% water) each day. Under heat stress, water intake could significantly increase by 120-200%. This increased water intake helps dissipate heat through the lungs (respiration) and by sweating. Total dissolved solids (TDS) and salinity are physiochemical properties of water used to assess water quality (NRC, 2001). Research to determine the effects of TDS on the performance of lactating dairy cattle has produced varying results on water intake, feed intake, and milk production. However, high TDS (H.TDS) water is more likely to decrease milk production during the summer (Jaster *et al.*, 1978; Challis *et al.*, 1987; Solomon *et al.*, 1995). The TDS guidelines suggest water containing <5,000 ppm TDS may be fed to lactating cattle, but water containing >7,000 ppm is unacceptable for all cattle (NRC, 2001). The study objective was to determine the effects of consumption of water with different levels of TDS on feed intake, milk production and physiological responses (rectal temperature and respiration rate) in early lactating dairy cows under heat stress.

Material and methods

Ten multiparous Holstein cows in early lactation, average milk yield of 37 kg and 96±34 DIM, from an industrial dairy herd in the Central province of Iran were randomly assigned to two treatments consisting of water containing different levels of TDS: Treatment 1: 937 ppm (L.TDS) and Treatment 2: 3437 ppm (H.TDS). The experiment lasted for 7 weeks (2 weeks for adaptation and 5 weeks for data collection) during July-August 2008. Water samples (L.TDS and H.TDS) were taken 3 times during the experiment directly from the pipeline and sent for analysis. All cows were fed a similar diet approved by NRC (2001), and had access to water *ad libitum*. The diet contained (% of DM): 27.4% corn silage, 24.7% alfalfa hay, 3.8% beet sugar pulp, 16.6% barley grain, 5.7% corn grain, 2.3% cottonseed meal, 10.6% soybean meal, 9.3% wheat bran, 0.5% fatty acids, 0.26% salt, 0.63% calcium carbonate, 0.42% sodium bicarbonate, 0.63% vitamin-mineral premix. The analysis of the diet included: 17.5% CP, 33.47% NDF, 0.8% Ca, 0.4% P and 1.62 NEL (Mcal/kg of DM). Milk samples from each cow were collected weekly and were analysed for fat, protein, lactose and SNF contents by Lacto Star analyser.

All data from the experiment were analysed as a completely randomised design using the MIXED procedure of SAS® (2003). The statistical model included cow, treatment and residual error. Fixed effects included treatment. Cow was the random effect.

Data of environmental temperature and humidity were collected daily by a meteorological station near the dairy farm. Temperature-Humidity Index (THI) was calculated as a physiological measure of animal comfort used to characterise or quantify thermo neutral zones, and has been proposed as a tool for estimating the heat stress suffered by lactating dairy cows (Bohamanova *et al.*, 2007). A THI value above 72 units can indicate heat stress in cows (Garcia-Ispierto *et al.*, 2006). Rectal temperature and respiration rate were recorded in the afternoon as indicators of heat stress.

Results

Average THI was 81.97. This value of THI implies heat stress conditions during this study. According to Table 1, improved water quality and using L.TDS water dramatically increased milk production in cows ($P<0.05$). DMI and milk fat% had an inclination to be significant ($P<0.1$). Changing water TDS did not significantly affect water consumption, milk composition, rectal temperature and respiration rate.

Table 1. Daily milk production, water consumption, DMI, milk composition and physiological parameters of cows drinking L.TDS or H.TDS water for 5 sampling weeks.

	L.TDS	H.TDS	SE	Treatment effect *P*-value
Milk, kg/d	30.14[a]	27.94[b]	1.84	0.0005
Water consumption, l/d	118.21	120.99	3.94	NS
DMI, kg/d	21.07	20.71	0.31	<0.1 NS
Rectal temperature, °C	39.16	39.28	0.05	NS
Respiration rate/min	86.96	85.78	1.23	NS
Milk fat,%	3.85	4.24	0.22	<0.1 NS
Milk protein,%	2.85	2.9	0.05	NS
Milk lactose,%	4.61	4.7	0.07	NS
Solids not fat (SNF),%	8.1	8.23	0.14	NS

[a,b] Means in row with different superscripts differ ($P<0.05$); NS = non significant.

Discussion and conclusion

Improving water quality under heat stress conditions, by supplying L.TDS water instead of H.TDS water, can increase milk production. Because all the conditions (diet, THI and management) for experimental cows were the same, it appears that decreased water TDS can dramatically improve milk yield by improving cow comfort and welfare under heat stress. However, further studies are needed to determine correlations among water quality and other parameters.

References

Bohamanova, J., I. Misztal and J.B. Cole, 2007. Temperature-Humidity Indices as indicators of milk production losses due to heat stress. J. Dairy Sci. 90: 1947-1956.

Challis, D.J., J.S. Zeinstra and M.J. Anderson, 1987. Some effects of water quality on the performance of high yielding cows in an arid climate. Vet. Rec. 120: 12-15.

Garcia-Ispierto, I., F. Lopez-Gatius, P. Santolaria, J. Yaniz, C. Nogareda, M. Lopez-Bejar and F. De Rensis, 2006. Relationship between heat stress during the peri-implantation period and early fetal loss in dairy cattle. Theriogenology 65: 799-807.

Jaster, E.H., J.D. Schuh and T.N. Wegner, 1978. Physiological effects of saline drinking water on high producing dairy cows. J. Dairy Sci. 61: 66-71.

Solomon, R., J. Miron, D. Ben-Ghedalia and Z. Zomberg, 1995. Performance of high producing dairy cows offered drinking water of high and low salinity in the Arava desert. J. Dairy Sci. 78: 620-624.

Adrenal response to ACTH challenge in early lactating dairy cows characterised by different inflammatory conditions

E. Trevisi, A. Minuti, R. Lombardelli and G. Bertoni
Institute of Zootechnics, Faculty of Agriculture, U.C.S.C., 29100, Piacenza, Italy; erminio.trevisi@unicatt.it

Introduction

Chronically stressed cows seem to have a hyper-reactive adrenal cortex (Broom, 1988), so the response to an acute stressor would be greater in maladapted animals. This justifies the use of ACTH challenge as a method to detect chronic stress condition in dairy cows. Nevertheless not all researchers agree with this; in fact the cortisol level after ACTH challenge could be affected by several factors and, within them, the circulating transcortin level (Greenspan and Forsham, 1986) and inflammatory conditions (Priftis *et al.*, 2007) might have some relevance. To clarify these latter aspects we investigated the cortisol changes of lactating cows, submitted to low dose ACTH challenge in early stages of lactation, and its relationship with inflammatory conditions evaluated by liver response.

Material and methods

The trial was carried out in a commercial farm on 23 Holstein Friesian multiparous dairy cows. After calving, each subject was monitored daily for health status and milk yield. At the 7th and 14th d in milk (DIM), individual blood samples (Li-heparin tubes) were taken from the jugular vein, before feed distribution, to assess metabolic and inflammatory status according to Bertoni *et al.* (2008). At the 34th ($\pm$4) d, about 3 h after the morning meal, cows were captured using any care to avoid fright, immediately bled and i.v. injected with a solution containing 20 µg of a synthetic analogue of ACTH (Novartis Pharma AG, Stein, CH). Blood samples were also taken 30 and 60 min after ACTH injection. Plasma cortisol was measured by the RIA method (Coat-A-Count; DPC, Los Angeles, CA, USA) at 0, 30 and 60 min after ACTH challenge; moreover samples taken before treatment (0 min) were still analysed for metabolic and inflammatory status. The integrated responses of cortisol over 60 min were evaluated as area under the curve (AUC). The cows were retrospectively grouped into quartiles based on the liver activity index (LAI), that included albumin, cholesterol (as an index of lipoproteins) and vitamin A (as an index of retinol binding protein) levels in the first month of lactation, as described by Bertoni *et al.* (2008). For the statistical evaluation, data were subjected to ANOVA using REPEATED statement in the MIXED procedure of SAS® (SAS Institute Inc., Cary, NC, USA, release 9.1) including group (lower, intermediate lower, intermediate upper, upper quartiles: LO; IN-LO; IN-UP, UP respectively), DIM and group by DIM interaction as the fixed effects, and cow within group as the random variable.

Results and discussion

According to the retrospective ranking, in the 1st mo of lactation the cows of the lowest LAI index (LO) showed the lowest level of plasma cholesterol, albumin and vitamin A (Table 1) in comparison to the other three quartiles (mainly vs. IN-UP and UP; $P<0.05$). Moreover, LO quartile also showed the highest level of plasma haptoglobin at the 7th DIM (Table 1), suggesting that cows suffered more severe inflammatory conditions immediately after calving and consequently impaired liver synthesis (e.g. reduction of the negative acute phase proteins).
LO quartile *vs* others also showed a higher level of NEFA (1.19 vs. 0.75 mmol/l at the 7th DIM in LO) and BHB (Table 1) in the week after calving, demonstrating a higher lipomobilisation. The

latter could be due to a lower DMI, as suggested by lower plasma urea in this group (3.4 vs. 4.4 mmol/l in LO and UP respectively in the first month of lactation $P<0.05$). LO vs. other quartiles finally showed the lowest milk yield (e.g. 33.8 vs. 37.9 kg/d at 34th DIM; $P<0.1$) and the highest health problem frequency in the first month of lactation (75% *vs* 36% of cows). Basal cortisol did not show differences between quartiles (6.91±3.2 ng/ml) but it was slightly higher than usual. Otherwise the adrenal response after ACTH challenge was lower in LO, when measured before peak (30th min; Table 1) or evaluated as AUC (1477 vs. 1925 ng/100 ml×60 min in LO vs. others respectively $P<0.01$). Therefore, the adrenal response occurring in early lactating cows challenged with a low dose of ACTH appears inversely related to the severity of inflammatory conditions occurring after calving. A possible explanation of this could be found in the reduced level of the binding protein of cortisol (transcortin) in the plasma, since the free cortisol remains around 4-8% of the total. Transcortin is synthesised in the liver and its level is reduced in severe liver disease (Greenspan and Forsham, 1986) and – in our view – during an acute phase reaction. For this reason, the changes of transcortin during inflammatory phenomena could be very similar to those of negative acute phase proteins, as lipoproteins or retinol binding protein. Therefore, the lower rise of blood cortisol after ACTH challenge in more inflamed cows (LO quartile) was probably a consequence of the lower availability of transcortin rather than a reduced reactivity of the adrenal gland.

Table 1. Average values of some plasma parameters in cows grouped into quartiles of LAI index.

		Units	LO	IN-LO	IN-UP	UP	SEM
ACTH-Cortisol	(30 min)	ng/ml	33.8	43.2*	44.7*	45.6*	2.120
Albumin	(7th -34th DIM)	g/l	33.5	35.5*	35.6*	37.7*	0.757
Vitamin A	(7th -34th DIM)	µg/100 ml	30.7	38.6	48.8*	54.2*	3.471
Haptoglobin	(7th DIM)	g/Ll	0.94	0.82*	0.44*	0.56*	0.082
BHB	(7th DIM)	mmol/l	1.25	0.54*	0.52*	0.47*	0.086

* $P<0.05$ vs. LO quartile; BHB = ß-hydroxybutyrate.

Conclusion,

Early lactating cows affected by marked inflammation seem to show a lower rise of cortisol during the low dose ACTH challenge, than expected. Thus, during the transition period, the evaluation of chronic stress conditions by ACTH challenge would be interpreted in the light of inflammatory status. However a wide study in field conditions appears necessary to confirm the above relationship.

Acknowledgement

The research was partially supported by Associazione Italiana Allevatori.

References

Bertoni, G., E. Trevisi, X. Han and M. Bionaz, 2008. Effects of inflammatory conditions on liver activity in puerperium period and consequences for performance in dairy cows. J. Dairy Sci. 91: 3300-3310.

Broom, D.M., 1988. The scientific assessment of animal welfare. Appl. Anim. Behav. Sci. 20: 5-19.

Greenspan, F.S. and P.H. Forsham, 1986. Basic & clinical endocrinology. Lange Medical Publications, Los Altos, CA, USA.

Priftis, KN., A. Papadimitriou, P. Nicolaidou and G.P. Chrousos, 2007. The hypothalamic–pituitary–adrenal axis in asthmatic children. Trends Endocrinol. Metabol. 19: 32-38.

Performance and physiological responses of Holstein calves undergoing heat stress to supplementation with Chromium – Methionine

M. Yari, G.R. Ghorbani, M. Alikhani, H.R. Rahmani, M. Khorvash, M. Mirzaee, F. Hashem Zade and M. Ghafari
Department of Animal Science, Isfahan University of Technology, Isfahan 84156 Iran; khorvashm@yahoo.com

Introduction

Chromium (Cr) in the trivalent state is an active component of Glucose Tolerance Factor and thus is essential for the maintenance of normal metabolism of carbohydrates and lipids. In functional ruminants, positive production responses to supplemental Cr appear to depend on the presence of a stressor, but in dairy calves before weaning, a small positive effect of Cr on performance was reported (Depew *et al.*, 1998). This study was conducted to delineate the effects of supplemental Cr as Chromium-Methionine (Cr-Met) on performance and physiological responses in Holstein dairy calves from birth through 3 wk after weaning.

Material and methods

Twenty-four Holstein calves (41.5±1.9 kg) born at the Lavark Research Station (Isfahan University of Technology, Isfahan, Iran) between May and July 2007 were used to study performance and physiological responses (PeRs and PhRs) to a conventional whole milk and starter feeding regime supplemented with Cr-Met. Calf birth date and sex were balanced between treatments. Beginning on d 4 and until weaning, each calf received 2 l whole milk at 14:00 and 20:30 h, and had *ad libitum* access to a starter (CP=21.6% and ME=2.9 mcal/kg DM) and fresh water. Treatments were a supplementation of 0, 0.02 and 0.04 mg Cr as Cr-Met/kg of $BW^{0.75}$ for calves in control (C), low (L) and high (H) groups. Cr-Met was added to the milk before weaning and to the starter thereafter. Calves were weighed when 4 d old then once a week. Once a week the amount of Cr-Met was calculated based on $BW^{0.75}$. Calves were weaned when they ate 1 kg starter for 6 consecutive days. After weaning, calves had *ad libitum* access to the starter. During wk 2 and 3 after weaning calves had *ad libitum* access to chopped alfalfa hay (CP, 16%; ADF, 37.5%). Climatic data were obtained from a meteorological station located 8 km from the station. The maximum Temperature – Humidity Index (THI) was calculated (Garcia-Ispierto *et al.*, 2006). Rectal Temperature (RT) and Rate of Respiration (RR) were taken at 15:00 h on 3 d each week (Coleman *et al.*, 1995). Data were analysed by a mixed model for repeated measures in SAS® (SAS Institute Inc., Cary, NC, USA). Effects were declared significant at $P<0.05$. Trends were considered at $P<0.10$.

Results

The maximum daily THI and temperature throughout the experiment were 77 and 34±0.3 °C. Before weaning PeR were not affected by treatments ($P>0.1$). After weaning and throughout the experiment calves receiving Cr-Met had lower average daily DM intake ($P<0.1$) and it decreased quadratically ($P<0.05$) with increasing Cr-Met, but daily weight gain and feed efficiency were not affected ($P>0.1$) (Table 1). Before weaning RT and RR were not affected by treatments ($P>0.1$), but they were affected by the age ($P<0.05$) and they decreased as calves aged. There was an interaction between treatment and week for RR ($P=0.05$) but not for RT. During wk 5 and 6 after birth, L calves had a lower RR than the C calves ($P=0.01$ and $P=0.1$ respectively) but there was no difference between L calves and H calves ($P>0.1$). The RT was not affected by the post weaning Cr-Met supplementation ($P=0.32$), but there was an interaction between treatment and week ($P=0.05$): the RT decreased as calves aged (P=0.05). During wk 1 after weaning L calves had lower RT than C calves (38.97±0.15 vs. 39.34±0.15 °C, P=0.02) but the difference between the groups L and H was not affected (P>0.1).

Table 1. Effects of Cr-Met supplementation on PeR (performance responses) in dairy calves.

Variable[1]	Treatments			SEM	P-value		
	C	L	H		C vs. Cr	L	Q
Weaning age, d	48.60	50.70	48.30	2.90	0.74	0.69	0.25
BW, kg							
First	41.8	41.60	41.10	1.90			
Wean	65.10	65.40	64.50	2.40	0.95	0.99	0.74
Final	89.20	87.10	87.40	3.80	0.56	0.78	0.70
ADG, kg							
First to wean	0.52	0.49	0.54	0.04	0.71	0.79	0.25
Wean to final	1.10	1.04	1.10	0.11	0.49	0.61	0.39
First to final	0.72	0.65	0.71	0.05	0.32	0.77	0.15
ADDMI[2]							
First to wean	0.97	0.96	1.00	0.05	0.97	0.71	0.60
Wean to final	2.44	2.20	2.35	0.13	0.07	0.35	0.04
First to final	1.45	1.32	1.44	0.07	0.08	0.57	0.02
Gain:Feed							
First to wean	0.54	0.51	0.54	0.04	0.74	0.94	0.43
Wean to final	0.46	0.48	0.47	0.05	0.54	0.36	0.70
First to final	0.50	0.50	0.50	0.03	0.58	0.28	0.43

[1] First d 4 after birth; Wean = weaning day; Final = d 21 after weaning.
[2] ADDMI = average daily dry matter intake.

Discussion and conclusion

The calves underwent moderate heat stress (THI=77). The RR and RT of dairy calves decreased as they aged. The reduced RR during wk 5 and 6 after birth indicates beneficial Cr-Met effects on heat stress in dairy calves. Abrupt weaning, which can be considered as an acute stress for dairy calves, might increase the Cr requirement of calves. After weaning the calves that did not receive Cr-Met consumed more food than the calves receiving Cr-Met; this may help them to overcome the stress situation.

Acknowledgement

The authors gratefully acknowledge the Sana Dam-e Pars Company.

References

Coleman, D.A., B.R. Moss and T.A. McCaskey, 1996. Supplemental shade for dairy calves reared in commercial calf hutches in a southern climate. J. Dairy. Sci. 79: 2038-2043.

Depew, C.L., L.D. Bunting, J.M. Fernandez and D.L. Tompson, 1998. Performance and metabolic response of young dairy calves fed diets supplemented with chromium tripicolinate. J. Dairy Sci. 81: 2916-2923.

Garcia-Ispierto, I., F. Lopez-Gatius, P. Santolaria, J.L. Yaniz, C. Nogareda, M. Lopez-Bejar and F. De Rensis, 2006. Relationship between heat stress during the peri-implantation period and early fetal loss in dairy cattle. Theriogenology 65: 799-807.

Forage intake enhances omasal epithelium growth associated with accelerated epithelial cell cycle progression and increased cyclin D1 in weanling goats

H. Zhao, J. Lu and Z. Shen
Laboratory of Animal Physiology and Biochemistry, College of Veterinary Medicine, Nanjing Agricultural University, China; zmshen@njau.edu.cn

Introduction

The omasal epithelium is important in the absorption of water, electrolytes (Harrison, 1971; Engelhardt and Hauffe, 1975) and short chain fatty acids (Ali *et al.*, 2006). It also plays a significant physiological role in ruminants: supplying energy substances after absorption and removing potent buffer substance from the ingesta prior to entering the abomasum. It is well documented that dietary forage intake significantly influences omasum development and growth in young and adult ruminant animals (Fluharty *et al.*, 1999; McLeod *et al.*, 2000). Sheep receiving a high forage:concentrate ratio diet have heavier omasum organ weight than sheep consuming low forage:concentrate ratio diets (McLeod *et al.*, 2000). Shen *et al.* (2004) reported that a high concentrate (energy-rich) diet stimulates rumen epithelial proliferation and papillae growth in young goats, which are associated with the IGF system. However, it remains unclear whether the forage or concentrate intake influences omasal epithelial cell proliferation and growth. This paper therefore studied the effects of diet on omasal epithelial cell cycle progression and the mRNA expression of cyclin D1, a regulation protein of mammalian cells required for cell cycle progression and targetting of proliferative signals in G_1-phase (Balin *et al.*, 1993), and cyclin dependent kinase 4 (CDK4) in the omasal epithelium of weanling goats receiving diets of forage only or forage supplemented with concentrate.

Material and methods

Eighteen crossbred weanling goats (70-d, BW 14.27±0.69 kg/d) were randomly assigned to two dietary groups fed peanut straw *ad libitum* (PS group, n=9, ME intake 0.6 MJ/kg$^{0.75}$/d, nitrogen intake 1.0 g/kg$^{0.75}$/d) or PS supplemented with 400 g/d of concentrate (PSC400 group, n=9, ME intake 1.00 MJ/kg$^{0.75}$/d, nitrogen intake 2.4 g/kg$^{0.75}$/d) for 42 d. The forage: concentrate ratio was 100:0 in the PS and 60:40 in the PS400 groups, respectively. The intake of peanut straw and concentrate were measured daily over the feeding trial period.

At slaughter, the body weight, and the weight of full and empty omasum organ mass were determined (n=9). The fresh omasal epithelium was taken for detection of the cell cycle by flow cytometry (n=9) and for determination of the effects of dietary forage: concentrate ratio on epithelium cyclin D1 and CDK4 mRNA expression (n=9). The mRNA expression was measured by reverse transcription-polymerase chain reaction (RT-PCR). The significance was evaluated using the Student *t* test. $P<0.05$ was considered to be significant. All of the statistical analyses were performed with SPSS 12.0.

Results and discussion

The average daily gain was about 5 times higher in PSC400 than in the PS group ($P<0.05$). The full weight of omasum organ mass in PS was greater than that in PSC400 ($P<0.05$). The empty weight showed no difference between groups, although, when expressed as a percentage of empty omasal weight: empty body weight, the ratio was enhanced more in PS than in PSC400 ($P<0.05$), indicating the promoting effects of forage diet on omasal organ growth.

The forage diet exerted effects on cell cycle progression of the omasal epithelium. In the PC group, the G_0/G_1-phase (restriction point, control when the cells enter S-phase) cell number was lower ($P<0.05$), while the S-phase (the phase of DNA and protein synthesis, $P<0.05$) and G_2/M-phase (cells enter mitosis, $P<0.05$) cell number was higher than in PSC400, indicating that the forage diet stimulating the omasal epithelial cell proliferation accelerated its cell cycle progression.
Cyclin D1 is known as the cell cycle regulating protein. CDK4 is one of the cyclin-dependent kinases. The cyclin-D1-CDK4 complex is involved in regulating the progression of the cell cycle through the G_1/S checkpoint. In this study, the cyclin D1 and CDK4 encoding genes were detected in omasal epithelial cells. The abundance of cyclin D1 mRNA in PS increased by 19% ($P<0.05$) compared with that in PSC400. In agreement with the alteration of D1 mRNA, the CDK4 mRNA abundance in PS was increased by 14% ($P<0.05$) too.
Furthermore, a negative correlation coefficient between the mRNA abundance of cyclin D1 and G_0/G_1-phase cell number (r=-0.59, $P<0.05$), and positive correlation coefficients between mRNA abundance of cyclin D1 and S-phase (r=0.4, $P=0.1$) and G_2/M-phase cell number (r=0.54, $P<0.05$) were observed.

Conclusion

The forage diet promotes the omasal epithelium growth by stimulating cell proliferation, which was asociated with accelerated epithelial cell cycle progression and increased expression of cyclin D1 and CDK 4 mRNA in the omasal epithelium of weanling goats.

Acknowledgement

The authors thank the Chinese National Nature Science Foundation for financial support (No. 30471252).

References

Ali, O., Z. Shen, U. Tietjen and H. Martens, 2006.Transport of acetate and sodium in sheep omasum: mutual, but asymmetric interactions. J. Comp. Physiol [B]. 176: 477-487.

Balin, V., J. Lukas, M.J. Marcote, M. Pagano and G. Draetta, 1993. Cyclin D1 is a nuclear protein required for cell cycle progression in G1. Genes. Dev. 7: 812-21.

Fluharty, F.L. and K.E. McClure, 1999. Effects of dietary energy intake and protein concentration on performance and visceral organ mass in Lambs. J. Anim. Sci. 75: 604-610.

McLeod, K.R. and R.L. Baldwin, 2000. Effects of diet forage: concentrate ratio and metabolizable energy intake on visceral organ growth and *in vitro* oxidative capacity of gut tissues in sheep. J. Anim Sci. 78: 760-770.

Shen, Z., H. Syfert, B. Loehrke, F. Schneider, R. Zitnan, A. Chudy, S. Kuhla, H.M. Hammon, J.W. Blum, H. Martens, H. Hagemeister and J. Voigt, 2004. An energy-rich diet caused rumen papillae proliferation associated with more igf type 1 receptors and increased plasma igf-1 concentrations in young goats. J. Nutr. 134: 11-17.

Author index

C

D

E

F

G

H

L

M

N

O

S

T

Y

Z